ISBN 978-1-332-98948-5
PIBN 10447071

1 MONTH OF FREE READING

at

www.ForgottenBooks.com

By purchasing this book you are eligible for one month membership to ForgottenBooks.com, giving you unlimited access to our entire collection of over 1,000,000 titles via our web site and mobile apps.

To claim your free month visit:

www.forgottenbooks.com/free447071

* Offer is valid for 45 days from date of purchase. Terms and conditions apply.

T H E   F R A M I N G H A M   S T U D Y

An Epidemiological Investigation of Cardiovascular Disease

Section 34:   Some Risk Factors Related to the Annual Incidence
              of Cardiovascular Disease and Death Using Pooled
              Repeated Biennial Measurements: Framingham
              Heart Study, 30-Year Followup

U.S. DEPARTMENT OF HEALTH AND HUMAN SERVICES
Public Health Service
National Institutes of Health

Some Risk Factors Related to the Annual Incidence of
Cardiovasclar Disease and Death Using Pooled
Repeated Biennal Measurements: Framingham Heart Study,
30-Year follow-Up

Since · 1948 the Framingham Heart Study has followed a sample
of adults who in 1948 were residents of the town of Framingham,
Massachusetts for development of cardiovascular disease.   A
standardized   clinic   examination   was   repeated   at two-year
intervals.· By this means a wide variety of information has been
collected which includes the characteristics of people in this
study   cohort   both   before   and   after the development of
cardiovascular disease, as well as a standardized clinical
evaluation of changes in their cardiovascular status.   In
addition, information about death and the development and course
of disease has been obtained from a number of sources outside
the clinic.

The design and execution of the Study were described in
Sections 1 and 2 of this series of monographs. A complete
listing of prior monographs in this series is included at the
end of this section. For some characteristics, mean values are
given for the first seven examinations in Sections 3 and 4, and
bivariate correlations are given for the second examination in
Section 5.   This material is updated in Section 29.   The
incidence of various events is given in Sections 6, 7, and 8.
Criteria for events are given in Section 8, and a description of
the characteristics in Section 3.   These ·are repeated and
updated in Sections 26, 30 and in this report.

·The major concern of the Study has been the evaluation of
the relationship of potential risk factors determined in healthy.
individuals to the subsequent development of disease. This is
done in a systematic fashion in Sections 9 through 22, using the
14 year followup. The same general material updated to include
16 years of followup is presented in Sections 26 and 27.
Section 30 reports similar results for 18 years of followup.
The present report is an update for 30 years of followup.
Evaluation ·of survival following cardiovascular events has been
examined in Sections 25 and 32.

Like earlier monographs in this series, · this section· is _ .— - _..
exclusively a display of data which ·provides detailed
information for those who wish to interpret and use them for
program planning, teaching, reviews and for comparison with
their own data.   The additional followup provides a more
substantial body of data in the geriatric age groups, for women,
and for the less common cardiovascular endpoints.

METHODOLOGY

One method of examining the relationship between potential
risk factors and the development of disease is to measure the
risk factors at a single moment in time and to follow

individuals in time to observe the incidence of disease. For example, in this study we could utilize the measurement of risk factors at exam 1 and report incidence over 30 years of followup according to different levels of these exam 1 measurements to ascertain whether a risk factor is associated with the subsequent development of disease. In this case, information obtained in exams 2 through 15 would be ignored. Since individuals change over time, this interim information which could influence the outcome would be lost.

The methodology in this section utilizes <u>all</u> measurements determined on Exams 1 through 15 for those risk factors recorded virtually every two years and relates the risk factors to the occurrence of an event within two years after the exam. Other characteristics measured on only a few examinations are not included in this monograph. We refer to this approach of employing the biennial observations as the <u>cross-sectional pooling</u> method.

This method evaluates each two-year interval as a new short-term followup study. After being characterized at entry into the study, persons are characterized anew at each following biennial examination. Hence, a person who attended twelve of the fifteen examinations during the 30-year followup contributes the information of twelve persons who enter the study at the beginning of a two-year cycle with the risk factors measured at the twelve examinations. More accurately, this person contributes the information of twelve person exams. To implement this approach, an observation is generated for each examination. The information obtained on the 15 two-year intervals is then pooled to obtain a file from which two-year predictions can be examined.

This method is to be distinguished from the long-term perspective described earlier in which observations only from exam 1 are employed to examine the development of disease over 30 years of followup. The cross-sectional pooling method as implemented in this section considers only the next two years of followup, given an individual's current age, sex, and risk factor status. The inherent assumption is that only the current risk factor status of an individual is needed to predict the risk of disease in the next two years.

DESCRIPTION OF TABLES

The tables in this section of the monograph series, Section 34, present data for 23 events and 19 potential risk factors. The tables are numerically sequenced according to the combination of an event and a risk factor with the numbers indexed as follows:

Numbering Scheme for Event-Risk Factor Combinations
Framingham Study Monograph 30-year Followup

EVENTS

1. Coronary Heart Disease ⊆ CHD
2. Coronary Heart Disease other than Angina Pectoris
3. Myocardial Infarction
4. Myocardial Infarction unrecognized
5. Myocardial Infarction recognized
6. Coronary Insufficiency
7. Angina Pectoris
8. Angina Pectoris uncomplicated
9. Sudden Coronary Death among persons free of CHD
10. Sudden Coronary Death among all persons
11. Coronary Heart Disease Death among persons free of CHD
12. Coronary Heart Disease Death among all persons
13. Stroke and Transient Ischemic Attack
14. Atherothrombotic Brain Infarction
15. Transient Ischemic Attack
16. Stroke Death among persons free of Stroke and TIA
17. Stroke Death among all persons
18. Intermittent Claudication
19. Congestive Heart Failure
20. Cardiovascular Disease
21. Cardiovascular Disease Death among persons free of CVD
22. Cardiovascular Disease Death among all persons
23. Death among all persons

RISK FACTORS

1. Systolic Blood Pressure, First examiner
2. Diastolic Blood Pressure, First examiner
3A. Hypertension with antihypertensive treatment
3B. Hypertension ignoring treatment
4. Serum Cholesterol
5. Hematocrit
6. Blood glucose
7. Diabetes mellitus
8. Glucose in Urine
9. Glucose intolerance
10. Metropolitan Relative Weight
11. Vital Capacity
12. Heart Rate
13. Cigarettes smoked per day
14. Albumin in Urine
15. Heart enlargement by x-ray
16. Left Ventricular Hypertrophy
17. Intraventricular conduction defect
18. Nonspecific T-wave or ST-segment abnormality by ECG

-3-

The numbering scheme on the tables indicates the number of the event first and then the number of the risk factor. For example, the table for Myocardial Infarction and Cigarettes smoked per day is numbered as 3-13.

Each table displays for each sex descriptive information on the relationship between the risk factor and the risk of the specified subsequent event in a two-year period. Logistic regression coefficients, for the risk factor, including univariate, bivariate with age at exam, and multivariate analyses, are displayed at the bottom of the table. The descriptive portion of the table displays annual rates for 6 age groups: 35-44, 45-54, 55-64, 65-74, 75-84, 85-94. Age-adjusted annual rates computed by the direct method are also given for age groups 35-64 and 65-94. Age-specific logistic regressions, using the above specified 10-year age groups, and regressions for age groups 35-64 and 65-94 are also based upon the two-year cross-sectional pooling method and indicate the risk of the event in the next two years of followup among persons free of the event at the beginning the two-year interval. However, the rates are expressed as the average annual rate per 1000.

The header of each table indicates the sex, event, risk factor, and the population at risk for each table. Each risk factor-event combination is presented on one page with the top half for males and the bottom half for females.

Population at Risk

These tables report results only for persons free of cardiovascular disease at exam 1, so that any person with coronary heart disease, cerebrovascular disease, intermittent claudication, or congestive heart failure is excluded from consideration. In addition, for non-fatal events the population at risk consists generally of persons free of the event under consideration at the beginning of a two-year interval. For coronary heart disease (CHD) events (tables 1-8), persons must be free of CHD. For cerebrovascular events (tables 13-15), persons must be free of stroke and transient ischemic attack (TIA). For intermittent claudication (IC) in table 18, congestive heart failure (CHF) in table 19, and cardiovascular disease (CVD) in table 20, persons must be free of these respective events. For fatal events, two tables are displayed, one for persons free of the disease at each exam interval and one for all persons free of cardiovascular disease at exam 1. For overall death, only the latter table is provided. For CHD death events in tables 9 and 11 persons must be free of coronary heart disease while in tables 10 and 12 all persons free of cardiovascular disease at exam 1 are at risk including those with interim CHD events.

Risk Factor Description

The range of each risk factor, from the lowest to the highest value observed in the fifteen exams, is displayed in each table. The tables for hematocrit and vital capacity-height

-4-

index are the only tables which have different ranges for men and women.

Each individual is characterized by his or her value at an exam. If that value is unknown, the most recent, known value at a previous exam is used. An exception to this rule is diabetes mellitus (risk factor 7). Once a person is diagnosed as being diabetic, that person retains that diagnosis on all subsequent exams.

Results Reported

A table takes into account risk factors on all of the first fifteen examinations and the incidence of the specified event in the fifteen biennial intervals of the 30-year followup and consolidates this by the cross-sectional pooling method into an average annual incidence rate by age, sex, and level of the risk factor. The statistics shown are:

Person Exams

On each of the first fifteen examinations, persons who are free of cardiovascular disease at exam 1 and currently free of the event under consideration are considered at risk of developing the event in the next two-year interval. The person-exams column reports the number of all such exams from the persons at risk over the 30-year period. An individual who entered the study at exam 1 may contribute up to 15 observations, one for each exam at which he or she is at risk. Also, individuals can contribute to more than one age group over the 30 years of followup as long as that person is at risk of developing disease. Hence, the tables of this report do not yield the counts of people taking the exams. See Section 9 of the monograph series to obtain counts of persons distributed by various risk factors for the first seven exams.

of Events

This column contains the number of subjects in the population at risk at an examination who develop a primary event in the two-year interval between that examination and the next. The numbers developing the event in a two-year interval are summed over the first fifteen examinations. The entries in this column are suppressed for an age group when fewer than five events occur in that age group.

Annual Rate

The annual rate is reported as the number of events per 1000 person years. It is obtained by dividing the number of events by twice the person exams to adjust for the two-year followup. This column is suppressed when fewer than five events occur in an age group.

Crude Annual Rate

In the middle of each page on the right, summary results are displayed for age groups 35-64 and 65-94. First, a column for

the sum of the events in the age group is given by level of the
risk factor. Next, the crude annual rate per 1000 person
yearsis provided. It is computed by dividing the number of
events by twice the population at risk (i.e. the total number of
person exams). The latter can be obtained by summing the
age-specific persons exams columns for the respective ten-year
age groups.

Age-adjusted Annual Rate

The age-adjusted rates per 1000 person years are computed by
the direct method (Fleiss, 1981) using the ten-year age group
information provided. The standardizing age distribution
employed for this computation is the total number of persons in
each ten-year age group. Hence, each age-specific rate is
weighted by the total number of persons in the age group, and
the sum of the weighted products is then divided by the total
number of persons in the standardizing age distribution. For
example, the age-adjusted annual rate of CHD (Table 1-1) for
persons 35-64 with systolic blood pressure in the range 74-119
is computed as

$$\frac{4472*(7/(2*1223)) + 7724*(21/(2*1769)) + 8049*(31/(2*1440))}{4472 + 7724 + 8049}$$

Logistic regression coefficients of the risk factor

In the bottom portion of the table, logistic regression
coefficients for the risk factor, computed by the maximum
likelihood method, are reported. First, age-specific
coefficients are given, then univariate, bivariate with age, and
multivariate coefficients are reported for the 35-64 and 65-94
age groups.

The age-specific logistic regression coefficients are
reported for each ten-year age group. The logistic model
employed for these age-specific computations is

Pr(event in the next two years) =
$p = 1/(1 + exp(-a - bX))$

where a and b are the logistic parameters to be estimated and X
is the value of the risk factor under consideration for a person
in the population at risk. See Schlesselman (1982) for further
explanation of the logistic model. According to the
cross-sectional pooling approach, a person may contribute
several observations to each computation as long as they remain
in the age group. The coefficient indicates the logarithm of
the change in the odds of developing the event in the next two
years for each unit increase in the risk factor among persons
free of the event. To obtain the risk of developing disease in
the next two years, the logistic model must be employed to
estimate p.

Iterative maximum likelihood estimation procedures available
in the "Logist" procedure of Supplemental Library of the
Statistical Analysis System (SAS) were employed to obtain

estimates of the coefficients. This procedure utilizes zeros as initial values for the iterative procedure. When the procedure would not converge with these initial estimates, we employed starting values obtained from linear regression models.

When fewer than 5 events occur in an age group, the regression results are suppressed. In other situations where the regression coefficients are inestimable or nearly so, such as the case where all individuals who develop disease have the same value of the risk factor, dashes are displayed.

For continuous variables, the actual value of the risk factor is employed. The classes for categorical variables are indicated in the top portion of the table, and their coding is explained in the description of the risk factors in this report.

The univariate results report the logistic coefficients for the combined 35-64 and 65-94 age groups. Only the risk factor under consideration is included in the model in these computations.

The bivariate coefficients are computed from logistic models with age and the risk factor under consideration. These results are analogous to the age-adjusted rate computations, provided that the relationship between the risk factor and the event follows a logistic model with the $\log(p/(1-p))$ changing in a linear fashion as the risk factor increases. The table displays only the coefficient for the risk factor under consideration and not the age coefficient.

The multivariate coefficients indicate the partial or "independent" effect of the risk factor after taking into account the risk factors known to be associated with cardiovascular disease: age, systolic blood pressure, serum cholesterol, cigarettes smoked per day, glucose intolerance, and electrocardiographic left ventricular hypertrophy. The table reports only the coefficient for the risk factor under consideration from the multiple logistic model:
$$p = 1/(1 + \exp(-b_0 - b_i X_i))$$
where X_i represents the risk factor under consideration and the other risk factors listed above. Generally, there are seven variables in this model. In cases where the risk factor under consideration is one of the known risk factors, there are six variables. Further, when the risk factor is highly correlated with one of the known risk factors, the model contains six variables. For example, when the risk factor is diastolic blood pressure or hypertension (tables 2, 3A and 3B), systolic blood pressure is dropped. For blood glucose, diabetes mellitus, and glucose in the urine (tables 6-8), glucose intolerance is dropped.

Standard Error of the Logistic Coefficient

The standard error of the logistic regression coefficients obtained from the second derivative of the information matrix and computed by the Logist procedure in SAS are reported in this column.

P-value

The p-value is obtained by computing the ratio of the logistic coefficient to its standard error and comparing it to the standard normal distribution.

Standardized Coefficient

The standardized logistic regression coefficient is computed as the product of the maximum likelihood coefficient obtained from the Logist procedure in SAS and the standard deviation of the risk factor in the population at risk.
Standardized $b_i = b_i*SD(X_i)$ See Truett, Cornfield, Kannel (1967) and Schlesselman (1982) for a reference each standardized coefficient measures the change in the log odds or log $(p/(1-p))$ of risk for each change of one standard deviation of the risk variable and is often used as a measure of the importance of changes in that variable relative to changes in other risk variables.   There is considerable controversy regarding the use of standardized coefficients (Greenland, Schlesselman, and Criqui (1986)) and caution is advised.

Standardized Standard Error

The standard error of the standardized logistic regression coefficient is computed by equating the ratio of the standardized coefficient to its standard error with the respective unstandardized ratio so that the two ratios have the same p-value. Hence, it is computed as
SE (standardized b_i) = (SE $(b_i)/b_i$) * standardized b_i
Note that this is equivalent to assuming that the variance in the risk factor is estimated with high accuracy.

CRITERIA FOR EVENTS

1. Coronary Heart Disease

Subjects were diagnosed as having developed coronary heart disease (CHD) if upon review of the case a panel of three investigators agreed on one of the following definite manifestations of CHD: myocardial infarction, coronary insufficiency, angina pectoris, sudden death from CHD, non-sudden death from CHD. Persons with pre-existing CHD at Exam 1 were excluded from the population at risk of developing CHD. Pre-existing CHD at Exam 1 was identified by any one of the following diagnoses at Exam 1: definite angina pectoris, definite history of myocardial infarction, definite myocardial infarction by electrocardiogram, doubtful myocardial infarction by electrocardiogram, definite coronary insufficiency by electrocardiogram and history.·

The various manifestations of CHD are these:

Angina pectoris

Brief recurrent chest discomfort of up to 15 minutes duration, precipitated by exertion or emotion and relieved by rest or by nitroglycerine, was regarded as angina pectoris (AP) if two physicians interviewing the subject agreed that this condition was definitely present. This diagnosis was based solely on evaluation of subjective manifestations. Abnormality of the resting or exercise electrocardiogram was not required for this diagnosis.

Angina pectoris uncomplicated

If angina pectoris developed between two successive examinations but the person remained free of any other manifestation of coronary heart disease, the diagnosis was designated as angina pectoris uncomplicated.

Coronary heart disease other than angina pectoris (Coronary Attack)

Because an interest is often expressed in the incidence of "heart attacks," this category is established·to represent the occurrence of myocardial infarction, coronary insufficiency, or death from coronary heart disease.

Myocardial infarction

Recent or acute myocardial infarction (MI) was designated when there were serial changes in the electrocardiograms indicating the evolution of an infarction, including: S-T segment elevation in the electrocardiographic tracing associated with terminal inversion of T waves and the loss of initial QRS potentials (that is, development of "pathologic" Q waves of 0.04 second duration or greater), followed by serial changes indicating reversion towards normal. An old or remote myocardial infarction was considered to be present when the electrocardiogram showed a stable pattern including a pathologic Q wave of 0.04 second or greater or loss of initial QRS potential (R wave) in those leads in which this would not be expected to occur. Also, an interim unrecognized MI was indicated when changes from a previous tracing showed development of loss of R-wave potential or appearance of pathologic Q waves not otherwise explained, in persons in whom neither the patient nor his physician considered the possibility of MI. If the patient was asymptomatic for chest pain or upper abdominal pain during the interval at which the unrecognized MI occurred, the event was classified as silent, unrecognized. More weight was given to this finding if a T-wave abnormality was also associated with Q-wave abnormality.

Beginning in 1956, a hospital report for a subject showing a rise in the serum glutamic oxalacetic transaminase to a level of at least 60 units along with a history of prolonged ischemic chest pain was accepted as evidence of myocardial infarction. Subsequently, in 1962, pathologic elevation of another enzyme was included: lactic dehydrogenase greater than 500 units. At a later date, with a change in laboratory techniques, SGOT was abnormal at 50 and LDH at 200. CPK greater than 200 units or CPK-MB band positive was included if these tests were done.

An autopsy report showing an acute, new, or recent infarction of the myocardium was accepted as evidence of an incident myocardial infarction. Because it is not possible to date an old infarction found on autopsy such evidence was not used in the clinical diagnosis of a new event, unless there was an interim clinical event suspected of being an infarction.

Coronary insufficiency

The coronary insufficiency syndrome was designated when a history of prolonged ischemic chest pain was accompanied by transient ischemic S-T segment and T-wave abnormality in the electrocardiographic tracing but not accompanied by development of Q-wave abnormality or by serum enzyme changes characteristic of myocardial necrosis. Virtually all these subjects were hospitalized for suspected MI.

Coronary heart disease death

Death from coronary heart disease was diagnosed as either sudden or nonsudden.

Nonsudden death from CHD

If the terminal episode lasted longer than one hour, if the available information implied that the cause of death was probably CHD, and if no other cause could be ascribed, this was called nonsudden death from CHD. In making this diagnosis, the review panel used prior clinical information as well as information concerning the final illness.

Sudden death from coronary heart disease

If a subject, apparently well, was observed to have died within a few minutes (operationally documented as under one hour) from onset of symptoms and if the cause of death could not

-10-

reasonably be attributed on the basis of the full clinical information and the information concerning death to some potentially lethal disease other than coronary heart disease, this was called sudden death and was attributed to coronary heart disease.

Stroke

The diagnosis of overt vascular disease of the brain was based on the occurrence of a stroke. Minimal criteria for a stroke consisted of abrupt onset of a localizing neurologic deficit (such as hemiparesis, aphasia, homonymous hemianopia). For stroke due to intracranial hemorrhage, a change in the state of consciousness, headache, and signs of meningeal irritation in association with a bloody spinal fluid under increased pressure with or without other localizing neurological deficits. On recent examinations, CAT scan information has been used to confirm hemorrhage. A diagnosis of embolus to the brain was made if a source for embolus (that is, atrial fibrillation, rheumatic heart disease with mitral stenosis, recent myocardial infarction, bacterial endocarditis) was present, the clinical course consistent (that is, rapid onset and clearing, slightly bloody spinal fluid, a more localized deficit), or the occurrence of associated peripheral emboli elsewhere noted. A consultant neurologist and the clinical staff of the study reviewed hospital and clinic protocols. Starting with Exam 8 neurologists have examined patients suspected of stroke in the hospital.

Atherothrombotic brain infarction

Specifically, thrombotic brain infarction was defined as the sudden onset of a localizing neurologic deficit (for example, aphasia, homonymous hemianopia, a central type of facial weakness, hemiparesis) documented by a physician, lasting longer than 24 hours, in the absence of: 1) known source of embolism (atrial fibrillation, rheumatic heart disease with mitral stenosis, myocardial infarction within preceding six months, bacterial endocarditis), 2) intracranial hemorrhage (intracerebral, subarachnoid), 3) known hypercoagulable states (for example, erythremia) 4) other disease processes causing focal brain deficits (brain tumor, subdural hematoma, hypoglycemia). All stroke data were evaluated by a neurologist. On recent examinations CAT-scan information has been used to exclude intracranial hemmorrhage.

Transient Ischemic Attack

A transient ischemic attack was designated when a history of a focal neurologic deficit was documented which lasted less than 24 hours in duration.

Stroke Death

Death attributed to stroke was designated when a documented focal neurologic deficit of greater than 24 hours duration preceeded death and was responsible for the fatality.

-11-

Intermittent claudication   *Blood flow issues*

Minimum criteria for the subjective diagnosis of intermittent claudication consisted of a cramping discomfort in the calf clearly provoked by walking some distance with the pain appearing sooner when walking quickly or uphill and being relieved within a few minutes by rest. Interviews for symptoms of claudication were conducted by the physician using structured forms for uniformity of the assessment. In addition, a second physician confirmed all cases suspected of claudication at time of examination.

Congestive heart failure

A definite diagnosis of congestive heart failure required that a minimum of two major or one major and two minor criteria present concurrently. The presence of other conditions capable of producing the symptoms and signs was considered in evaluating the findings.

Major criteria:

1) Paroxysmal nocturnal dyspnea.
2) Distended neck veins (in other than the supine position).
3) Rales.
4) Increasing heart size by X-ray
5) Acute pulmonary edema described in hospital record.
6) Ventricular S(3) gallop.
7) Increased venous pressure (greater than 16 cm H_2O from right atrium).
8) Circulation time (greater than 24 seconds, arm to tongue).
9) Hepatojugular reflux.
10) Pulmonary edema, visceral congestion, cardiomegaly shown on autopsy.

Minor criteria:

1) Bilateral ankle edema.
2) Night cough.
3) Dyspnea on ordinary exertion.
4) Hepatomegaly.
5) Pleural effusion.
6) Decrease in vital capacity by one-third from maximum record.
7) Tachycardia (120 beats per minute or more).

Arbitrary major or minor criterion:

Weight loss (ten pounds or more in five days) while on therapy for congestive heart failure.

4 = enlarged liver

-12-

5 = lung fluid

3 = no air when breathing

Cardiovascular disease

Cardiovascular disease is considered to have developed if there is a definite manifestation of coronary heart disease, intermittent claudication, congestive heart failure, stroke or transient ischemic attack in the absence of a previous manifestation of any of these diseases or rheumatic heart disease. Criteria for all these events have been given. A person having more than one cardiovascular manifestation within the followup period is counted as a case only at the time of the first event.

Death from cardiovascular disease

This cause of death was designated when any disease of the heart or blood vessels was considered responsible.

Death

The fact of death was supported by a death certificate. Additional information was obtained from records supplied by hospital, attending physician, pathologist, medical examiner, or family. A panel of staff physicians reviewed all evidence to arrive at the cause of death.

3 - Rales = sounds in lungs

-13-

DESCRIPTION OF CHARACTERISTICS

1. __Blood pressure, first examiner, systolic (mm Hg)__

2. __Blood pressure, first examiner, diastolic (mm Hg)__

Systolic and diastolic readings in the left arm of the subject were taken with a mercurial sphygmomanometer and a 14-cm cuff long enough to fit the most obese arm. The recommendations of the American and British Heart Associations were followed for reading the pressure (see Standardization of Blood Pressure Readings, American Heart Journal, July 1939, 18:95). Palpatory method was used to check auscultatory systolic readings. Measurement was expressed in millimeters of mercury on the scale of the manometer. Although the original protocol did not specify the accuracy of the measurement to be employed most readings were made to the nearest even number. After Exam 5 this became the standard practice.

At the beginning of the Study two readings were taken on each subject: 1) admission blood pressure taken by the nurse and 2) final blood pressure taken by the examining physician. Beginning in April 1950 three readings were taken on each subject: 1) admission blood pressure taken by the nurse, 2) blood pressure taken by the first examining physician at the start of his interview, and 3) another blood pressure taken at the end of the examination by a second physician after drawing blood. All of these pressures were taken on the left arm with the subject seated and the arm at heart level. In coding blood pressure at Exam 1 for subjects through Record Number 2938, the nurse's blood pressure is used as the reading by the first examining physician.

Minor changes in procedure have occurred in some earlier exams. A second blood pressure reading by the same examining physician has been substituted for a reading by a second examiner when a second observer was not present.

3. __Hypertension__

At examination, a subject had two blood pressure readings taken by the examining physician(s). If both readings were "abnormal" the subject had definite hypertension; if both readings were "normal" the subject had normotension; with any other combination of readings the subject had borderline hypertension.

A blood pressure reading was "abnormal" if either the systolic or the diastolic component was "abnormal". The reading was "normal" if both systolic and diastolic parts were "normal". A systolic pressure was called "normal" when under 140 mm Hg, "abnormal" when 160 or greater. A diastolic pressure was called "normal" when under 90 mm Hg, "abnormal" when 95 or greater. Thus, a person was a definite hypertensive if both of the following conditions held:

-14-

1. the first systolic pressure was 160 or greater <u>or</u> the first diastolic pressure was 95 or greater.
<u>and</u>
2. the second systolic pressure was 160. or greater <u>or</u> the second diastolic pressure was 95 or greater.

When on rare occasion a subject had only one blood pressure reading taken by the examining physician, hypertensive status was determined on the basis of this one reading.

For purposes of analysis in this monograph, hypertension was evaluated in two fashions: (1) one utilizing only blood pressure readings as described above and (2) one in which individuals not treated with anti-hypertensive medication were classified on the basis of blood pressure while persons treated with anti-hypertensive medication were considered as definite hypertensives.

4. Serum cholesterol (mg/100 ml)

At the beginning of the study, serum cholesterol concentrations were determined by the colorimetric method of Sperry (Schoenheimer and Sperry: J. Biol. Chem. 106, 745, 1934. Sperry: Am. J. Clin. Path. & Tech. Supplement 2, 91, 1938). On December 12, 1952 after the start of Exam 2, the method of Abell-Kendall was adopted (Abell, L.L., Levy, B.B., Brodie, B.B., and Kendall, F.E., J. Biol. Chem. 195, 357-366, 1952).

5. Hematocrit

Blood was collected in a balanced oxalate tube and spun at 5000 rpm for 20 minutes. Hematocrit was read in the tube against a special scale constructed for this purpose. This method proved more reliable than the standard capillary method. On exams 1 to 3, hemoglobin was measured instead of hematocrit. These measurements were converted into hematocrit by dividing by 3.129.

6. Blood glucose (mg/100 ml)

The amount of glucose present in a casual specimen of the subject's whole blood (collected in a solution of potassium oxalate and sodium flouride) was determined using the method of Nelson (Nelson, B., J. Biol.Chem. 153, 375, 1944; Somogyi, M., J. Biol. Chem. 160, 61, 1945; Somogyi, M., J. Biol. Chem. 160, 69, 1945). Blood glucose was measured on Exams 1 to 4, 6, 8, 9, and 10, 12-15 (not on Exams 5, 7 and 11).

7. Diabetes mellitus

A subject was diagnosed as having diabetes if he or she was under treatment by a private physician for diabetes or the subject had a record of an abnormal glucose tolerance test or had on at least two exams a casual blood glucose determination of 150 mg/100 ml or more.

Treatment by private physician meant that the subject was taking insulin or oral hypoglycemic agents. In addition, if the subject was taking oral agents, the diagnosis of diabetes was made only if records from the Framingham Study, private physician, or hospital showed several elevated blood glucose determinations before treatment started. For this purpose an elevated blood glucose meant a value of 150 mg/100 ml or higher.

The study did not give subjects glucose tolerance tests; therefore the record of such tests had to be obtained from a private physician or hospital. The standard glucose tolerance test was taken as the ingestion of 100 grams of glucose after a 12-hour fast. The test was considered abnormal if the blood glucose concentration was 160 mg/100 ml or more at one hour after challenge, 140 or more at two hours, and was still higher at three hours than when the test began.

8. Glucose in urine, definite or trace

Procedure for testing the presence of glucose in a subject's freshly voided urine was as follows:

To five drops of urine in test tube add 10 drops of water and one Clinitest tablet. Allow to stand 15 seconds after boiling has ceased and then read promptly. Compare color to chart provided with tablets. Read as: negative, 1+, 2+, 3+, or 4+.

In coding the result of the color comparison, definite urine glucose is 2+, 3+, or 4+; trace urine glucose is 1+.

After exam 11, the Ames Combustix (Labstix) method was employed.

9. Glucose intolerance

A subject was said to demonstrate glucose intolerance at examination if any one of the following conditions obtained:

a)   A diagnosis of diabetes mellitus at that examination or any preceding examination;

b)   A determination of glucose in the urine sample, either- definite or trace by Clinitest or Combustix, at that examination;

c)   A measurement of the amount of glucose present in a casual specimen of whole blood equal to 120 milligrams or more per 100 milliliters at that examination.

10. Metropolitan relative weight (percent)

This relative weight at exam was computed for a subject by forming the ratio of his body weight at that exam to the desirable weight for his particular sex-height group according to standards set by the Metropolitan Life Insurance Company.

The ratio is expressed as a whole number in percent. The reference weight, for a given sex-height group, is the midpoint of the range for medium frame shown in the table of desirable weights on page 12 of "Four Steps to Weight Control" (1969), distributed by the Metropolitan Life Insurance Company. By linear extrapolation, reference weights were assigned to those sex-height groups not covered by the Metropolitan table, women 55 inches and men 60 inches tall. Since the desirable weights are shown for persons with their clothes and shoes on, adjustments are made to apply the reference weight to the subjects in the Framingham Study, who were weighed without clothes and measured for height without shoes: five pounds are subtracted from the weight for men, four pounds for women; one inch was subtracted from the height for men, two inches for women. The resulting set of reference weights is:

Height (inches)	Weight (pounds) Men	Women	Height (inches)	Weight (pounds) Men	Women
55		94	65	131	128
56		97	66	135	132
57		100	67	140	136
58		103	68	144	140
59		106	69	148	

Height (inches)	Weight (pounds) Men	Women	Height (inches)	Weight (pounds) Men	Women
60	116	109	70	152	
61	119	112	71	157	
62	122 ·	116	72	161	
63	125	120	73	166	
64	128	124	74	170	

11. Vital capacity-height index (ml/inch)

A nurse or technician instructed the subject to take the deepest breath possible and exhale to the fullest extent into the tube of a water-sealed spirometer (Collins Vitalometer). Three trials were made for each subject; the highest reading was recorded. This measurement of vital capacity was read to the next lower 0.1 liter on the scale of the spirometer. The vital capacity in milliliters was divided by the subject's height in inches to get the vital capacity-height index.

12. Heart rate (per minute)

The heart rate is the ventricular rate as determined by the physician from the electrocardiogram made with the subject in the recumbent position. Electrocardiograms were made at each exam and the ventricular rate was coded for data processing for all exams except 2 and 3.

13. Cigarettes smoked (number per day)

Histories of cigarette smoking were obtained on every exam except Exam 6. However, information from the first three exams was coded to represent usage at Exam 1, leaving Exams 2 and 3

-17-

uncoded.   On the initial examinations there was no fixed rule
for deciding whether a person was an ex-cigarette smoker, but
persons who had only recently stopped were always considered
still smoking.   On later examinations anyone smoking cigarettes
within a year was considered a current smoker. For current
smoker, the number of cigarettes usually smoked each day was
recorded.   Additional details are given in Sections 3 and 26 of
the monograph series.

14. Albumin in urine, definite

Procedure for testing for the presence of albumin in a
subject's freshly voided urine was as follows:

To 10 cc of urine in test tube add 2 cc reagent (Quantitest
reagents: beta naphthalene sulphonic acid and glacial acetic
acid).   Allow to stand. Result is read in milligrams of albumin
per 100 milliliters by comparing turbidity with standards in
lighted rack. In coding the result of the turbidity comparison,
definite urine albumin is 20 mg/100 ml or higher; trace urine
albumin is 10 mg/ 100 ml.

This method was employed for examinations 1 to 8, though the
results were not coded on Exam 4. Since Exam 9, the Combustix
(Labstix) method has been utilized to determine urine albumin.

15. Heart enlargement by x-ray

Roentgenograms were taken at each exam with a Westinghouse
Autoflex · 300-milliampere X-ray· unit and with a Machlett
superdynamax   tube.   The subject was positioned erect in
max-respiration phase with full anterior chest against the film
cassette.·   The X-ray tube was placed at level of seventh
thoracic vertebra, a distance of two meters from anode to film.
The film size was 14 x 17 inches. Beginning in Exam 4 the
machine was actuated by a device attached to the subject through
electrodes applied to the left and right axillary margin at
level of the seventh rib, to take the X-ray at the moment of
full   diastole.   (Peak of R wave plus 0.008 second to
termination of exposure insured the shadow of the cardiac
silhouette in full systole.)   The transthoracic Lead I
electrocardiographic tracing showing the point of X-ray exposure
was   printed at the top of the exposed X-ray film before _ _ _ _
development. ·

All films were interpreted by a roentgenology consultant who
did not have access to clinical data on the subject before
viewing his films. Beginning in early 1952 one consultant, Dr.
Lloyd E. Hawes, viewed all films taken at the Framingham Heart
Clinic and made the X-ray interpretations.

The diagnosis of left ventricular hypertrophy signified that
this chamber was enlarged as shown by an extension of the left
apex of the heart towards the left, an elevation of the apex and
a squaring of the apex. The transverse diameter of the heart
was usually more than half the transverse width of the chest.

-18-

However, in a subject with a large chest the square prominent shape of the left apex of the heart may have signified left ventricular hypertrophy even with transverse diameter of the heart less than half the width of the chest. The diagnosis of left ventricular hypertrophy could be made with normal measurement of the transverse diameter of the heart.

The diagnosis of generalized cardiac enlargement was made when the heart shadow was rounded and expanded both to the right and the left, giving an appearance not of left ventricular hypertrophy alone but of either a dilatation of all chambers or a dilatation of the atria and right ventricle as in mitral heart disease.

A diagnosis of generalized cardiac enlargement and left ventricular hypertrophy was made if the heart was rounded overall and enlarged, and the apex projected further out and was square and prominent.

A subject with heart enlargement had either generalized cardiac enlargement or left ventricular hypertrophy or both.

16. Left ventricular hypertrophy by electrocardiogram

Electrocardiograms were taken at each exam with a Sanborn Visocardiette. The subject was in the recumbent position. Thirteen leads were used, three standard, three augmented unipolar limb leads, and seven precordial leads of Wilson. See description of technique, pages 31 to 44, Katz, Electrocardiography, Lee and Febiger, Philadelphia, 1947.

Left ventricular hypertrophy consisted of a tracing exhibiting a slight prolongation of "ventricular activation time" (at least 0.05 second on the left) associated with R-wave potentials of at least 20 mm in standard leads, at least 11 mm in augmented unipolar leads, at least 25 mm in any deflection in precordial leads or at least 35 mm combining any V (1), V (2) S-wave deflection with R-wave deflections from V (5) or V (6). This had to be accompanied by depressed S-T segments or flattened to inverted T waves reflecting potentials from left precordial leads.

A category of possible left ventricular hypertrophy was designated when tracings exhibited characteristics similar to those above except in less striking degree or when not all were present. Increased R-wave potential without associated S-T and T-wave abnormality was included in this category.

17. Intraventricular conduction defect with QRS interval more than 0.11 second

Complete right intraventricular heart block was diagnosed when QRS duration exceeded 0.11 second and an R-R' wave was noted in any one or more right precordial leads, with a delayed peak of R' wave (equal to 0.03 second or greater) after onset of QRS complex, accompanied by broad S waves in left precordial leads and a corresponding broad S wave in Lead I. Complete <u>left</u> intraventricular heart block was designated when the QRS interval exceeded 0.11 second and tall slurred R waves were observed along with absent Q waves in any of the left precordial leads and a reciprocal deep S wave in the right precordial leads. The peak of the notched R' wave occurred at greater than 0.06 second after the onset of the QRS complex in one or more left precordial leads. Complete <u>indeterminate</u> intraventricular heart block was designated when the QRS interval was prolonged beyond 0.11 second and the characteristic changes indicative of either right or left intraventricular heart block were absent, but the lengthened QRS interval was not obtained at the expense of a shortened P-R interval (as in the Wolff-Parkinson-White syndrome). No hemiblock or bifascicular block designations were distinguished.

18. <u>Definite Nonspecific T-wave or ST-segment abnormality by electrocardiogram</u>

Nonspecific abnormality was designated in the absence of a prominent R wave when there was ST-segment depression (exceeding one millimeter below the base line) and/or primary T-wave inversion or flattening. This diagnosis was not made if a more specific explanation could be made for these changes (such as intraventricular block, myocardial infarction, or left ventricular hypertrophy) or if the changes occurred in those leads where such variation is acceptable.

Glucose Intolerance:

    1= glucose intolerant
    0= no glucose intolerance

Left Ventricular Hypertrophy:
    1=no
    2=borderline
    3=definite

Hypertension ignoring antihypertensive treatment

    1=none
    2=borderline
    3=definite

Hypertension with antihypertensive treatment:

    1=none
    2=borderline (not treated)
    3=definite (not treated)
    4=treated

In 'the logistic regressions treated persons are coded as 3, definite hypertensives.

Diabetes:
    1=diabetes previously diagnosed
    0=nondiabetic

Glucose in Urine:

    0=negative
    1=positive, trace or doubtful, light, medium, or dark

Albuminim Urine:

    0=negative or trace
    1=positive

Heart Enlargement by X-ray:

    1=none
    2=possible GCE or definite LVH without definite GCE
    3=definite GCE

Nonspecific T-wave or ST-segment abnormality:

    0=none or doubtful
    1=definite on either T-wave or ST-segment

tricular conduction defect:

    1=no IV Block, either complete or incomplete on left
      or right
    2=incomplete on left,right, or indeterminate and
      not complete on any
    · 3=complete on left or right or indeterminate

Sections of the Monograph Series

Section 1. Introduction and general background
Tavia Gordon and William B. Kannel

Section 2. Follow-up to the eighth examination
Tavia Gordon

Section 3. Some characteristics of the Framingham cohort on
the first 7 examinations by broad age groups
Christine Cole, Esther Allen, Manning
Feinleib

Notes: Dewey Shurtleff, William P. Castelli,
Lloyd E. Hawes, and Patricia M. McNamara

Section 4. Some characteristics of the Framingham cohort on
the first 7 examinations by 5-year age groups
Dewey Shurtleff

Section 5. Bivariate correlations among some
characteristics of the Framingham cohort at Exam
2
Dewey Shurtleff

Section 6. Smoothed average annual incidence rates by age
at exam preceding event, and sex
Manning Feinleib

Section 7. Average two-year incidence rates by age at exam
preceding event and sex, with total number of
events and total population at risk
Manning Feinleib

Section 8. Two-year incidence by exam interval by age at
Exam 1 and sex
Manning Feibleib

Criteria: William B. Kannel, William P.
Castelli, and Patricia M. McNamara

Section 9. Population alive with specified events by sex,
age and level of characteristic at exam
Dewey Shurtleff

Section 10. Incidence of Coronary Heart Disease by sex, age
and level of characteristic at exam for 22
characteristics
Dewey Shurtleff

Section 11. Incidence of Myocardial Infarction by sex, age
and level of characteristic at exam for 22
characteristics
Dewey Shurtleff

References

1. Fleiss, JL (1981). _Statistical Methods for Rates and Proportions_ New York: John Wiley and Sons, 237 - 255.

2. Harrell, FE Jr (1983). The Logist procedure, _SUGI: Supplemental Library & User's Guide._ Cary, No. Carolina: SAS Institute Inc, 181-202.

3. Truett J, Cornfield J, Kannel W (1967). A multivariate analysis of the risk of coronary heart disease in Framingham. _J. Chron Dis_ 20: 511-24.

4. Schlesselman JJ (1982). _Case-control studies: Design, Conduct, Analysis_ New York: Oxford University Press, 227-269.

5. Greenland S, Schlesselman JJ, Criqui MA (1986). The fallacy of employing standardized regression coefficients and correlations as measures of effect. _Am J. Epidemiology_ 123: 203-208.

6. Standardization of blood pressure readings, _Am Heart J_ 18: 95, July 1939.

7. Sperry (1938) _Am J Clin Path and Tech. Supplement 2:_ 91.

8. Abell LL, Levy BB, Brodie BB, Kendall FE (1952). _J Biol Chem_ 195: 357-66.

9. Nelson B (1944). _J Biol Chem_ 153: 375.

10. Somogyi M (1945). _J Biol Chem_ 160: 61.

11. Somogyi M (1945). _J Biol Chem_ 160: 69

12. Katz (1947). _Electrocardiography._ Philadelphia: Lee and Febiger, 31-44.

TABLE 1-1   DATE=02OCT85
INUAL INCIDENCE RATE FOR EVENT PER 1000 PERSONS AT RISK AT EXAM
) LEVEL OF RISK FACTOR AT EXAM: FRAMINGHAM STUDY 30-YEAR FOLLOWUP

MALE
IT: CORONARY HEART DISEASE
: FACTOR: SYSTOLIC BLOOD PRESSURE, FIRST EXAMINER (MM HG)
IONS AT RISK: PERSONS FREE OF CORONARY HEART DISEASE

EXAM		45-54 AT EXAM			55-64 AT EXAM			65-74 AT EXAM		
: ANNUAL	ANNUAL'	PERSON	# OF	ANNUAL	PERSON	# OF	ANNUAL	PERSON	# OF	ANNUAL
ITS RATE	RATE	EXAMS	EVENTS	RATE	EXAMS	EVENTS	RATE	EXAMS	EVENTS	RATE
:	3	1769	21	6	1440	31	11	548	10	9
:	5	3417	62	9	3166	95	15	1341	47	18
:	7	1732	50	14	2090	92	22	1194	63	26
:	10	567	26	23	916	54	29	552	34	31
:	0	239	8	17	437	35	40	278	26	47
:	-	0	0	-	0	0	-	0	0	-
:	5	7724	167	11	8049	307	19	3913	180	23

EXAM		85-94 AT EXAM			35-64 COMBINED			65-94 COMBINED		
: ANNUAL	ANNUAL	PERSON	# OF	ANNUAL	# OF		AGE-	# OF		AGE-
ITS RATE	RATE	EXAMS	EVENTS	RATE	EVENTS	CRUDE	ADJUSTED	EVENTS	CRUDE	ADJUSTED
:	21	7	0	0	59	7	7	15	11	11
	22	15	0	0	177	10	11	62	18	19
:	32	11	0	0	153	16	16	84	28	27
	36	4	0	0	84	25	23	47	33	34
	60	2	0	0	43	30	22	36	49	49
:	-	0	0	-	0	-	-	0	-	-
:	30	39	0	0	516	13	13	244	25	25

LOGISTIC REGRESSIONS

!EFFICIENT :ISK FACTOR	SE COEFFICIENT	P-VALUE	STANDARDIZED COEFFICIENT	STD SE
1.016	0.008	0.039	0.275	0.133
1.017	0.003	0.001	0.334	0.067
1.017	0.002	0.001	0.381	0.050
1.018	0.003	0.001	0.404	0.066
1.016	0.006	0.004	0.354	0.124
1.020'	0.002	0.001	0.411	0.036
1.017	0.003	0.001	0.399	0.058
1.017	0.002	0.001	0.338	0.037
1.018	0.003	0.001	0.481	0.058
1.014	0.002	0.001	0.286	0.041
1.015	0.003	0.001	0.341	0.062

FEMALE
IT: CORONARY HEART DISEASE
. FACTOR: SYSTOLIC BLOOD PRESSURE, FIRST EXAMINER (MM HG)
ONS AT RISK: PERSONS FREE OF CORONARY HEART DISEASE

EXAM		45-54 AT EXAM			55-64 AT EXAM			65-74 AT EXAM		
ANNUAL	ANNUAL	PERSON	# OF	ANNUAL	PERSON	# OF	ANNUAL	PERSON	# OF	ANNUAL
TS RATE	RATE	EXAMS	EVENTS	RATE	EXAMS	EVENTS	RATE	EXAMS	EVENTS	RATE
	0	2714	11	2	1979	19	5	647	10	8
	1	3982	22	3	4056	52	6	1747	38	11
	1	2053	23	6	3025	64	11	2013	59	15
	0	768	8	5	1560	51	16	1155	24	10
	10	442	8	9	893	40	22	686	43	31
	-	0	0	-	0	0	-	0	0	-
	1	9959	72	4	11513	226	10	6248	174	14

EXAM		85-94 AT EXAM			35-64 COMBINED			65-94 COMBINED		
ANNUAL	ANNUAL	PERSON	# OF	ANNUAL	# OF		AGE-	# OF		AGE-
TS RATE	RATE	EXAMS	EVENTS	RATE	EVENTS	CRUDE	ADJUSTED	EVENTS	CRUDE	ADJUSTED
	15	8	0	0	30	2	3	14	9	10
	18	38	3	39	78	4	4	58	13	13
	15	29	4	69	89	8	7	82	15	16
	31	31	1	16	59	12	9	53	16	15
	31	17	2	59	49	18	15	62	32	31
	-	0	0	-	0	-	-	0	-	-
	22	123	10	41	305	6	6	269	16	16

LOGISTIC REGRESSIONS

!FFICIENT :SK FACTOR	SE COEFFICIENT	P-VALUE	STANDARDIZED COEFFICIENT	STD SE
033	0.012	0.007	0.572	0.213
017	0.004	0.001	0.385	0.087
018	0.002	0.001	0.451	0.054
013	0.003	0.001	0.324	0.070
009	0.004	0.024	0.233	0.105
004	0.013	0.784	0.089	0.326
022	0.002	0.001	0.530	0.041
012	0.002	0.001	0.302	0.057
018	0.002	0.001	0.418	0.045
012	0.002	0.001	0.286	0.057
015	0.002	0.001	0.548	0.050
008	0.002	0.001	0.199	0.060

TABLE 1-2   DATE=02OCT85
AVERAGE ANNUAL INCIDENCE RATE FOR EVENT PER 1000 PERSONS AT RISK AT EXAM
BY AGE AND LEVEL OF RISK FACTOR AT EXAM: FRAMINGHAM STUDY 30-YEAR FOLLOWUP

SEX: MALE
EVENT: CORONARY HEART DISEASE
RISK FACTOR: DIASTOLIC BLOOD PRESSURE, FIRST EXAMINER (MM HG)
PERSONS AT RISK: PERSONS FREE OF CORONARY HEART DISEASE

RISK FACTOR	35-44 AT EXAM			45-54 AT EXAM			55-64 AT EXAM			65-74 AT EXAM		
	PERSON EXAMS	# OF EVENTS	ANNUAL RATE	PERSON EXAMS	# OF EVENTS	ANNUAL RATE	PERSON EXAMS	# OF EVENTS	ANNUAL RATE	PERSON EXAMS	# OF EVENTS	ANNUAL RATE
20-74	1012	2	1	1478	23	8	1847	54	15	1225	45	18
75-84	1604	14	4	2766	46	8	2845	81	14	1323	56	21
85-94	1187	11	5	2062	49	12	1964	92	23	893	38	21
95-104	499	12	12	1008	30	15	982	49	25	343	30	44
105-160	170	3	9	410	19	23	411	31	38	129	11	43
UNKNOWN	0	0	-	0	0	-	0	0	-	0	0	-
TOTAL	4472	42	5	7724	167	11	8049	307	19	3913	180	23

RISK FACTOR	75-84 AT EXAM			85-94 AT EXAM			35-64 COMBINED			65-94 COMBINED		
	PERSON EXAMS	# OF EVENTS	ANNUAL RATE	PERSON EXAMS	# OF EVENTS	ANNUAL RATE	# OF EVENTS	CRUDE	AGE-ADJUSTED	# OF EVENTS	CRUDE	AGE-ADJUSTED
20-74	422	20	24	22			79	9	9	67	20	19
75-84	331	21	32	12			141	10	10	77	23	23
85-94	204	13	32	5			152	15	15	52	24	24
95-104	44	4	45	0			91	18	18	34	44	44
105-160	16	3	94	0			53	27	26	14	48	53
UNKNOWN	0	0	-	0			0	-		0	-	
TOTAL	1017	61	30	39			516	13	13	244	25	25

LOGISTIC REGRESSIONS

AGE AT EXAM		COEFFICIENT OF RISK FACTOR	SE COEFFICIENT	P-VALUE	STANDARDIZED COEFFICIENT	STD SE
35-44		0.045	0.012	0.001	0.512	0.133
45-54		0.026	0.006	0.001	0.300	0.072
55-64		0.025	0.004	0.001	0.304	0.053
65-74		0.024	0.006	0.001	0.290	0.070
75-84		0.020	0.011	0.066	0.239	0.130
85-94						
35-64	UNIVARIATE	0.028	0.003	0.001	0.325	0.041
65-94	UNIVARIATE	0.021	0.005	0.001	0.258	0.062
35-64	BIVARIATE	0.027	0.003	0.001	0.321	0.040
65-94	BIVARIATE	0.024	0.005	0.001	0.296	0.063
35-64	MULTIVARIATE	0.024	0.004	0.001	0.286	0.043
65-94	MULTIVARIATE	0.020	0.005	0.001	0.248	0.064

SEX: FEMALE
EVENT: CORONARY HEART DISEASE
RISK FACTOR: DIASTOLIC BLOOD PRESSURE, FIRST EXAMINER (MM HG)
PERSONS AT RISK: PERSONS FREE OF CORONARY HEART DISEASE

RISK FACTOR	35-44 AT EXAM			45-54 AT EXAM			55-64 AT EXAM			65-74 AT EXAM		
	PERSON EXAMS	# OF EVENTS	ANNUAL RATE	PERSON EXAMS	# OF EVENTS	ANNUAL RATE	PERSON EXAMS	# OF EVENTS	ANNUAL RATE	PERSON EXAMS	# OF EVENTS	ANNUAL RATE
20-74	2164	2	0	2639	15	3	2969	41	7	1831	50	14
75-84	2030	1	0	3541	16	2	2833	41	8	2135	52	12
85-94	987	1	1	2361	25	5	2840	64	11	1439	43	15
95-104	324	3	5	926	7	4	1272	33	13	600	13	11
105-160	111	0	0	492	9	9	579	27	23	243	16	33
UNKNOWN	0	0	-	0	0	-	0	0	-	0	0	-
TOTAL	5616	7	1	9959	72	4	11513	226	10	6248	174	14

RISK FACTOR	75-84 AT EXAM			85-94 AT EXAM			35-64 COMBINED			65-94 COMBINED		
	PERSON EXAMS	# OF EVENTS	ANNUAL RATE	PERSON EXAMS	# OF EVENTS	ANNUAL RATE	# OF EVENTS	CRUDE	AGE-ADJUSTED	# OF EVENTS	CRUDE	AGE-ADJUSTED
20-74	724	28	19	53	4	38	58	4	4	82	16	16
75-84	596	29	24	43	4	47	78	7	7	85	15	15
85-94	443	21	24	20	2	50	90	7	7	66	47	18
95-104	146	5	17	7	0	0	43	9	8	18	12	12
105-160	42	2	24	0	0	-	36	15	13	18	32	30
UNKNOWN	0	0	-	0	0	-	0	-		0	-	
TOTAL	1951	85	22	123	10	41	305	6	6	269	16	16

LOGISTIC REGRESSIONS

AGE AT EXAM		COEFFICIENT OF RISK FACTOR	SE COEFFICIENT	P-VALUE	STANDARDIZED COEFFICIENT	STD SE
35-44		0.037	0.027	0.175	0.412	0.303
45-54		0.027	0.008	0.001	0.329	0.101
55-64		0.030	0.005	0.001	0.370	0.059
65-74		0.011	0.006	0.084	0.129	0.074
75-84		-0.001	0.009	0.932	-0.009	0.111
85-94		0.005	0.028	0.869	0.055	0.335
35-64	UNIVARIATE	0.034	0.004	0.001	0.414	0.048
65-94	UNIVARIATE	0.005	0.005	0.338	0.059	0.061
35-64	BIVARIATE	0.030	0.004	0.001	0.363	0.050
65-94	BIVARIATE	0.007	0.005	0.180	0.082	0.061
35-64	MULTIVARIATE	0.024	0.004	0.001	0.293	0.054
65-94	MULTIVARIATE	0.003	0.005	0.545	0.037	0.061

TABLE 1-3A   DATE=02OCT85

AVERAGE ANNUAL INCIDENCE RATE FOR EVENT PER 1000 PERSONS AT RISK AT EXAM
BY AGE AND LEVEL OF RISK FACTOR AT EXAM: FRAMINGHAM STUDY 30-YEAR FOLLOWUP

SEX: MALE
EVENT: CORONARY HEART DISEASE
RISK FACTOR: HYPERTENSION WITH ANTIHYPERTENSIVE TREATMENT
PERSONS AT RISK: PERSONS FREE OF CORONARY HEART DISEASE

35-44 AT EXAM			45-54 AT EXAM			55-64 AT EXAM			65-74 AT EXAM		
PERSON EXAMS	# OF EVENTS	ANNUAL RATE	PERSON EXAMS	# OF EVENTS	ANNUAL RATE	PERSON EXAMS	# OF EVENTS	ANNUAL RATE	PERSON EXAMS	# OF EVENTS	ANNUAL RATE
2651	19	4	4122	59	7	3773	90	12	1512	34	11
1301	11	4	2205	52	12	2301	110	24	1255	64	25
500	11	11	1139	42	18	1268	68	27	552	40	36
20	1	25	258	14	27	707	39	28	594	42	35
0	0	-	0	0	-	0	0	-	0	0	-
4472	42	5	7724	167	11	8049	307	19	3913	180	23

75-84 AT EXAM			85-94 AT EXAM			35-64 COMBINED			65-94 COMBINED		
PERSON EXAMS	# OF EVENTS	ANNUAL RATE	PERSON EXAMS	# OF EVENTS	ANNUAL RATE	# OF EVENTS	CRUDE	AGE-ADJUSTED	# OF EVENTS	CRUDE	AGE-ADJUSTED
570	14	19	18			168	8	8	48	13	13
333	19	29	11			173	15	15	84	26	26
125	10	40	3			121	21	20	51	38	38
189	18	48	7			54	27	27	61	39	38
0	0	-	0			0	-		0	-	
1017	61	30	39			516	13	13	244	25	25

LOGISTIC REGRESSIONS

	COEFFICIENT OF RISK FACTOR	SE COEFFICIENT	P-VALUE	STANDARDIZED COEFFICIENT	STD SE
	0.559	0.197	0.004	0.388	0.137
	0.528	0.095	0.001	0.406	0.073
	0.420	0.070	0.001	0.342	0.057
	0.571	0.097	0.001	0.467	0.079
	0.459	0.166	0.006	0.376	0.136
VARIATE	0.534	0.054	0.001	0.415	0.042
VARIATE	0.555	0.083	0.001	0.454	0.068
ARIATE	0.451	0.054	0.001	0.351	0.042
ARIATE	0.553	0.083	0.001	0.453	0.068
TIVARIATE	0.399	0.057	0.001	0.311	0.044
TIVARIATE	0.494	0.085	0.001	0.405	0.070

SEX: FEMALE
EVENT: CORONARY HEART DISEASE
RISK FACTOR: HYPERTENSION WITH ANTIHYPERTENSIVE TREATMENT
PERSONS AT RISK: PERSONS FREE OF CORONARY HEART DISEASE

35-44 AT EXAM			45-54 AT EXAM			55-64 AT EXAM			65-74 AT EXAM		
PERSON EXAMS	# OF EVENTS	ANNUAL RATE	PERSON EXAMS	# OF EVENTS	ANNUAL RATE	PERSON EXAMS	# OF EVENTS	ANNUAL RATE	PERSON EXAMS	# OF EVENTS	ANNUAL RATE
4250	3	0	5622	24	2	4863	45	5	1751	38	9
1001	2	1	2579	25	5	3290	71	11	1911	43	11
316	2	3	1164	11	5	1835	62	17	1036	29	16
49	0	0	592	12	10	1525	48	16	1550	72	23
0	9	-	0	0	-	0	0	-	0	0	-
5616	7	1	9959	72	4	11513	226	10	6248	174	14

75-84 AT EXAM			85-94 AT EXAM			35-64 COMBINED			65-94 COMBINED		
PERSON EXAMS	# OF EVENTS	ANNUAL RATE	PERSON EXAMS	# OF EVENTS	ANNUAL RATE	# OF EVENTS	CRUDE	AGE-ADJUSTED	# OF EVENTS	CRUDE	AGE-ADJUSTED
395	12	15	31	2	32	72	2	3	44	10	11
585	22	19	28	2	36	98	7	7	67	13	13
375	23	31	24	1	21	75	11	10	53	18	18
596	28	23	40	5	63	60	14	10	105	24	24
0	0	-	0			0	-		0	-	
1951	85	22	123	10	41	305	6	6	269	16	16

LOGISTIC REGRESSIONS

TABLE 1-3B   DATE=02OCT85
AVERAGE ANNUAL INCIDENCE RATE FOR EVENT PER 1000 PERSONS AT RISK AT EXAM
BY AGE AND LEVEL OF RISK FACTOR AT EXAM: FRAMINGHAM STUDY 30-YEAR FOLLOWUP

SEX: MALE
EVENT: CORONARY HEART DISEASE
RISK FACTOR: HYPERTENSION IGNORING TREATMENT
PERSONS AT RISK: PERSONS FREE OF CORONARY HEART DISEASE

RISK FACTOR	35-44 AT EXAM			45-54 AT EXAM			55-64 AT EXAM			65-74 AT EXAM		
	PERSON EXAMS	# OF EVENTS	ANNUAL RATE	PERSON EXAMS	# OF EVENTS	ANNUAL RATE	PERSON EXAMS	# OF EVENTS	ANNUAL RATE	PERSON EXAMS	# OF EVENTS	ANNUAL RATE
NO	2653	19	4	4154	60	7	3909	96	12	1631	40	12
MILD	1310	11	4	2284	58	13	2583	124	24	1499	83	28
DEFINITE	509	12	12	1286	49	19	1557	87	28	783	57	36
UNKNOWN	0	0	—	0	0	—	0	0	—	0	0	—
TOTAL	4472	42	5	7724	167	11	8049	307	19	3913	180	23

RISK FACTOR	75-84 AT EXAM			85-94 AT EXAM			55-64 COMBINED			65-94 COMBINED		
	PERSON EXAMS	# OF EVENTS	ANNUAL RATE	PERSON EXAMS	# OF EVENTS	ANNUAL RATE	# OF EVENTS	CRUDE	AGE-ADJUSTED	# OF EVENTS	CRUDE	AGE-ADJUSTED
NO	412	18	22.	21	2		175	8	8	58	14	14
MILD	418	23	28	15			193	16	15	108	28	28
DEFINITE	187	20	53	3			148	22	21	78	40	41
UNKNOWN	0	0	—	0			0	—		0	—	
TOTAL	1017	61	30	39			516	13	13	244	25	25

LOGISTIC REGRESSIONS

AGE AT EXAM		COEFFICIENT OF RISK FACTOR	SE COEFFICIENT	P-VALUE	STANDARDIZED COEFFICIENT	STD SE
35-44		0.569	0.197	0.004	0.393	0.136
45-54		0.500	0.096	0.001	0.376	0.072
55-64		0.434	0.072	0.001	0.334	0.056
65-74		0.552	0.099	0.001	0.417	0.075
75-84		0.484	0.176	0.006	0.356	0.129
85-94						
35-64	UNIVARIATE	0.517	0.055	0.001	0.388	0.041
65-94	UNIVARIATE	0.544	0.086	0.001	0.408	0.064
35-64	BIVARIATE	0.458	0.056	0.001	0.343	0.042
65-94	BIVARIATE	0.549	0.086	0.001	0.412	0.065
35-64	MULTIVARIATE	0.393	0.059	0.001	0.295	0.044
65-94	MULTIVARIATE	0.476	0.089	0.001	0.357	0.067

SEX: FEMALE
EVENT: CORONARY HEART DISEASE
RISK FACTOR: HYPERTENSION IGNORING TREATMENT
PERSONS AT RISK: PERSONS FREE OF CORONARY HEART DISEASE

RISK FACTOR	35-44 AT EXAM			45-54 AT EXAM			55-64 AT EXAM			65-74 AT EXAM		
	PERSON EXAMS	# OF EVENTS	ANNUAL RATE	PERSON EXAMS	# OF EVENTS	ANNUAL RATE	PERSON EXAMS	# OF EVENTS	ANNUAL RATE	PERSON EXAMS	# OF EVENTS	ANNUAL RATE
NO	4272	3	0	5804	27	2	5245	56	5	2035	37	9
MILD	1016	2	1	2763	28	5	3860	90	12	2549	73	14
DEFINITE	328	2	3	1392	17	6	2408	80	17	1664	64	19
UNKNOWN	0	0	—	0	0	—	0	0	—	0	0	—
TOTAL	5616	7	1	9959	72	4	11513	226	10	6248	174	14

RISK FACTOR	75-84 AT EXAM			85-94 AT EXAM			55-64 COMBINED			65-94 COMBINED		
	PERSON EXAMS	# OF EVENTS	ANNUAL RATE	PERSON EXAMS	# OF EVENTS	ANNUAL RATE	# OF EVENTS	CRUDE	AGE-ADJUSTED	# OF EVENTS	CRUDE	AGE-ADJUSTED
NO	507	15	15.	37	2	27	86	3	5	54	10	11
MILD	827	33	20	45	5	56	120	8	7	111	16	16
DEFINITE	617	37	30	41	3	37	99	12	10	104	22	22
UNKNOWN	0	0	—	0			0	—		0	—	
TOTAL	1951	85	22	123	10	41	305	6	6	269	16	16

LOGISTIC REGRESSIONS

AGE AT EXAM		COEFFICIENT OF RISK FACTOR	SE COEFFICIENT	P-VALUE	STANDARDIZED COEFFICIENT	STD SE
35-44		1.080	0.458	0.018	0.617	0.261
45-54		0.510	0.146	0.001	0.370	0.106
55-64		0.566	0.084	0.001	0.440	0.065
65-74		0.376	0.101	0.001	0.289	0.078
75-84		0.380	0.152	0.012	0.288	0.115
85-94		0.116	0.417	0.780	0.093	0.333
35-64	UNIVARIATE	0.729	0.070	0.001	0.539	0.052
65-94	UNIVARIATE	0.385	0.082	0.001	0.295	0.063
35-64	BIVARIATE	0.543	0.073	0.001	0.402	0.054
65-94	BIVARIATE	0.366	0.083	0.001	0.281	0.063
35-64	MULTIVARIATE	0.440	0.077	0.001	0.326	0.057
65-94	MULTIVARIATE	0.269	0.084	0.001	0.206	0.065

TABLE 1-4   DATE=02OCT85
AVERAGE ANNUAL INCIDENCE RATE FOR EVENT PER 1000 PERSONS AT RISK AT EXAM
BY AGE AND LEVEL OF RISK FACTOR AT EXAM: FRAMINGHAM STUDY 30-YEAR FOLLOWUP

SEX: MALE
EVENT: CORONARY HEART DISEASE
RISK FACTOR: SERUM CHOLESTEROL
PERSONS AT RISK: PERSONS FREE OF CORONARY HEART DISEASE

RISK FACTOR	35-44 AT EXAM PERSON EXAMS	# OF EVENTS	ANNUAL RATE	45-54 AT EXAM PERSON EXAMS	# OF EVENTS	ANNUAL RATE	55-64 AT EXAM PERSON EXAMS	# OF EVENTS	ANNUAL RATE	65-74 AT EXAM PERSON EXAMS	# OF EVENTS	ANNUAL RATE
84-204	1253	4	2	1767	20	6	2255	57	13	1234	51	21
205-234	1247	7	3	2306	49	11	2405	94	20	1171	52	22
235-264	872	11	6	1745	38	11	1889	80	21	873	41	23
265-294	475	9	9	1001	23	11	918	41	22	428	21	25
295-1124	291	7	12	557	31	28	454	28	31	191	14	37
UNKNOWN	334	4	6	348	6	9	128	7	27	16	1	31
TOTAL	4472	42	5	7724	167	11	8049	307	19	3913	180	23

RISK FACTOR	75-84 AT EXAM PERSON EXAMS	# OF EVENTS	ANNUAL RATE	85-94 AT EXAM PERSON EXAMS	# OF EVENTS	ANNUAL RATE	35-64 COMBINED # OF EVENTS	CRUDE	AGE-ADJUSTED	65-94 COMBINED # OF EVENTS	CRUDE	AGE-ADJUSTED
84-204	390	21	27	17			81	8	8	73	22	22
205-234	332	22	33	11			150	13	13	75	25	24
235-264	177	13	37	5			129	14	14	55	26	26
265-294	96	3	16	6			73	15	15	24	23	23
295-1124	22	2	45	0			66	25	26	16	38	38
UNKNOWN	0	0	-	0			17	10	15	1	31	24
TOTAL	1017	61	30	39			516	13	13	244	25	25

LOGISTIC REGRESSIONS

AGE AT EXAM		COEFFICIENT OF RISK FACTOR	SE COEFFICIENT	P-VALUE	STANDARDIZED COEFFICIENT	STD SE
35-44		0.011	0.003	0.001	0.502	0.113
45-54		0.009	0.002	0.001	0.383	0.064
55-64		0.005	0.001	0.001	0.220	0.056
65-74		0.003	0.002	0.126	0.115	0.075
75-84		0.003	0.003	0.451	0.098	0.130
85-94						
35-64	UNIVARIATE	0.007	0.001	0.001	0.298	0.039
65-94	UNIVARIATE	0.003	0.002	0.117	0.102	0.065
35-64	BIVARIATE	0.008	0.001	0.001	0.322	0.040
65-94	BIVARIATE	0.003	0.002	0.064	0.121	0.065
35-64	MULTIVARIATE	0.007	0.001	0.001	0.290	0.041
65-94	MULTIVARIATE	0.003	0.002	0.068	0.119	0.065

SEX: FEMALE
EVENT: CORONARY HEART DISEASE
RISK FACTOR: SERUM CHOLESTEROL
PERSONS AT RISK: PERSONS FREE OF CORONARY HEART DISEASE

RISK FACTOR	35-44 AT EXAM PERSON EXAMS	# OF EVENTS	ANNUAL RATE	45-54 AT EXAM PERSON EXAMS	# OF EVENTS	ANNUAL RATE	55-64 AT EXAM PERSON EXAMS	# OF EVENTS	ANNUAL RATE	65-74 AT EXAM PERSON EXAMS	# OF EVENTS	ANNUAL RATE
84-204	2350	0	0	2094	10	2	1625	22	7	789	14	9
205-234	1417	2	1	2558	18	4	2615	42	8	1401	30	11
235-264	806	1	1	2259	11	2	2938	43	7	1703	44	13
265-294	372	2	3	1473	12	4	2240	55	12	1287	37	14
295-1124	171	2	6	1101	13	6	1866	58	16	1031	48	23
UNKNOWN	500	0	0	474	8	8	229	6	13	37	1	14
TOTAL	5616	7	1	9959	72	4	11513	226	10	6248	174	14

RISK FACTOR	75-84 AT EXAM PERSON EXAMS	# OF EVENTS	ANNUAL RATE	85-94 AT EXAM PERSON EXAMS	# OF EVENTS	ANNUAL RATE	35-64 COMBINED # OF EVENTS	CRUDE	AGE-ADJUSTED	65-94 COMBINED # OF EVENTS	CRUDE	AGE-ADJUSTED
84-204	361	12	17	46	3	33	32	3	4	29	12	11
205-234	486	23	24	28	3	54	62	5	4	56	15	15
235-264	542	29	27	35	2	29	55	5	4	75	16	17
265-294	350	15	21	9	2	111	69	8	7	54	16	17
295-1124	205	6	15	5	0	0	73	12	10	54	22	21
UNKNOWN	7	0	0	0	0	-	14	6	8	1	11	11
TOTAL	1951	85	22	123	10	41	305	6	6	269	16	16

LOGISTIC REGRESSIONS

AGE AT EXAM		COEFFICIENT OF RISK FACTOR	SE COEFFICIENT	P-VALUE	STANDARDIZED COEFFICIENT	STD SE
35-44		0.023	0.005	0.001	0.908	0.180
45-54		0.005	0.002	0.011	0.213	0.083
55-64		0.007	0.001	0.001	0.321	0.062
65-74		0.007	0.002	0.001	0.320	0.073
75-84		-0.000	0.003	0.949	-0.007	0.111
85-94		0.002	0.007	0.745	0.107	0.327
35-64	UNIVARIATE	0.009	0.001	0.001	0.423	0.050
65-94	UNIVARIATE	0.004	0.001	0.003	0.179	0.060
35-64	BIVARIATE	0.007	0.001	0.001	0.307	0.049
65-94	BIVARIATE	0.005	0.001	0.001	0.213	0.061
35-64	MULTIVARIATE	0.006	0.001	0.001	0.263	0.049
65-94	MULTIVARIATE	0.005	0.001	0.001	0.222	0.062

TABLE 1-5

AVERAGE ANNUAL INCIDENCE RATE FOR EVENT PER 1000 PERSONS AT RISK AT EXAM
BY AGE AND LEVEL OF RISK FACTOR AT EXAM: FRAMINGHAM STUDY 30-YEAR FOLLOWUP

SEX: MALE
EVENT: CORONARY HEART DISEASE
RISK FACTOR: HEMATOCRIT
PERSONS AT RISK: PERSONS FREE OF CORONARY HEART DISEASE

RISK FACTOR	35-44 AT EXAM			45-54 AT EXAM			55-64 AT EXAM			65-74 AT EXAM		
	PERSON EXAMS	# OF EVENTS	ANNUAL RATE	PERSON EXAMS	# OF EVENTS	ANNUAL RATE	PERSON EXAMS	# OF EVENTS	ANNUAL RATE	PERSON EXAMS	# OF EVENTS	ANNUAL RATE
25-42	623	7	6	942	11	6	1018	40	20	450	17	19
43-44	546	0	0	967	20	10	985	36	18	493	21	21
45-46	983	3	2	1854	39	11	1955	70	18	963	43	22
47-48	775	4	3	1510	31	10	1599	55	17	861	38	22
49-70	1175	23	10	2066	58	14	2338	99	21	1127	60	27
UNKNOWN	370	5	7	385	8	10	156	7	22	19	1	26
TOTAL	4472	42	5	7724	167	11	8049	307	19	3913	180	23

RISK FACTOR	75-84 AT EXAM			85-94 AT EXAM			35-64 COMBINED		AGE-ADJUSTED	65-94 COMBINED		AGE-ADJUSTED
	PERSON EXAMS	# OF EVENTS	ANNUAL RATE	PERSON EXAMS	# OF EVENTS	ANNUAL RATE	# OF EVENTS	CRUDE	AGE-ADJUSTED	# OF EVENTS	CRUDE	AGE-ADJUSTED
25-42	214	15	35	16			58	11	12	33	24	22
43-44	119	10	42	5			56	11	11	31	25	25
45-46	240	14	29	7			112	12	12	58	24	24
47-48	197	10	25	7			90	12	11	48	23	22
49-70	247	12	24	4			180	16	16	73	26	27
UNKNOWN	0	0	-	0			20	11	14	1	26	20
TOTAL	1017	61	30	39			516	13	13	244	25	25

LOGISTIC REGRESSIONS

AGE AT EXAM		COEFFICIENT OF RISK FACTOR	SE COEFFICIENT	P-VALUE	STANDARDIZED COEFFICIENT	STD SE
35-44		0.139	0.045	0.002	0.510	0.166
45-54		0.055	0.022	0.013	0.198	0.080
55-64		0.020	0.016	0.210	0.074	0.059
65-74		0.038	0.021	0.067	0.142	0.077
75-84		-0.023	0.031	0.456	-0.097	-0.131
85-94						
35-64	UNIVARIATE	0.040	0.012	0.001	0.149	0.046
65-94	UNIVARIATE	0.017	0.017	0.332	0.064	0.066
35-64	BIVARIATE	0.039	0.012	0.002	0.142	0.046
65-94	BIVARIATE	0.023	0.017	0.176	0.090	0.067
35-64	MULTIVARIATE	0.015	0.013	0.229	0.056	0.046
65-94	MULTIVARIATE	0.014	0.018	0.423	0.055	0.068

SEX: FEMALE
EVENT: CORONARY HEART DISEASE
RISK FACTOR: HEMATOCRIT
PERSONS AT RISK: PERSONS FREE OF CORONARY HEART DISEASE

RISK FACTOR	35-44 AT EXAM			45-54 AT EXAM			55-64 AT EXAM			65-74 AT EXAM		
	PERSON EXAMS	# OF EVENTS	ANNUAL RATE	PERSON EXAMS	# OF EVENTS	ANNUAL RATE	PERSON EXAMS	# OF EVENTS	ANNUAL RATE	PERSON EXAMS	# OF EVENTS	ANNUAL RATE
21-39	1697	3	1	1843	8	2	1553	27	9	646	20	15
40-41	967	0	0	1741	11	3	2061	36	9	1069	21	10
42-43	1043	2	1	1955	13	3	2260	32	7	1256	30	12
44-45	856	2	1	2194	16	4	2847	57	10	1609	50	16
46-65	487	0	0	1681	16	5	2521	67	13	1626	52	16
UNKNOWN	566	0	0	545	8	7	271	7	13	42	1	12
TOTAL	5616	7	1	9959	72	4	11513	226	10	6248	174	14

RTSK FACTOR	75-84 AT EXAM			85-94 AT EXAM			35-64 COMBINED		AGE-ADJUSTED	65-94 COMBINED		AGE-ADJUSTED
	PERSON EXAMS	# OF EVENTS	ANNUAL RATE	PERSON EXAMS	# OF EVENTS	ANNUAL RATE	# OF EVENTS	CRUDE	AGE-ADJUSTED	# OF EVENTS	CRUDE	AGE-ADJUSTED
21-39	297	14	24	24	1	63	38	4	5	37	19	18
40-41	373	20	27	31	2	32	47	5	5	43	13	13
42-43	353	8	11	19	3	79	47	4	4	41	13	13
44-45	486	20	21	26	1	19	75	6	6	71	17	17
46-65	436	23	26	23	1	22	83	9	7	76	18	18
UNKNOWN	6	0	-	0	0	-	15	5	8	1	10	9
TOTAL	1951	85	22	123	10	41	305	6	6	269	16	16

LOGISTIC REGRESSIONS

AGE AT EXAM		COEFFICIENT OF RISK FACTOR	SE COEFFICIENT	P-VALUE	STANDARDIZED COEFFICIENT	STD SE
35-44		-0.054	0.101	0.592	-0.195	0.364
45-54		0.075	0.036	0.037	0.268	0.128
55-64		0.051	0.020	0.009	0.178	0.069
65-74		0.039	0.022	0.080	0.137	0.078
75-84		-0.005	0.028	0.852	-0.021	0.111
85-94		-0.116	0.086	0.178	-0.461	0.342
35-64	UNIVARIATE	0.084	0.017	0.001	0.301	0.060
65-94	UNIVARIATE	0.011	0.017	0.524	0.040	0.062
35-64	BIVARIATE	0.049	0.017	0.004	0.176	0.062
65-94	BIVARIATE	0.016	0.017	0.367	0.056	0.062
35-64	MULTIVARIATE	0.021	0.017	0.218	0.076	0.061
65-94	MULTIVARIATE	0.012	0.017	0.501	0.042	0.062

AVERAGE ANNUAL INCIDENCE RATE FOR EVENT PER 1000 PERSONS AT RISK AT EXAM
BY AGE AND LEVEL OF RISK FACTOR AT EXAM: FRAMINGHAM STUDY 30-YEAR FOLLOWUP

SEX: MALE
EVENT: CORONARY HEART DISEASE
RISK FACTOR: BLOOD GLUCOSE (MG/DL)
PERSONS AT RISK: PERSONS FREE OF CORONARY HEART DISEASE

RISK FACTOR	35-44 AT EXAM			45-54 AT EXAM			55-64 AT EXAM			65-74 AT EXAM		
	PERSON EXAMS	# OF EVENTS	ANNUAL RATE	PERSON EXAMS	# OF EVENTS	ANNUAL RATE	PERSON EXAMS	# OF EVENTS	ANNUAL RATE	PERSON EXAMS	# OF EVENTS	ANNUAL RATE
29-69	992	9	5	1490	33	11	1087	39	18	421	11	13
70-89	2501	24	5	4172	71	9	4056	158	19	1761	76	22
90-109	626	5	4	1348	35	13	1866	71	19	1046	50	24
110-129	105	1	5	305	12	20	537	20	19	335	19	28
130-524	48	0	0	193	10	26	408	16	20	334	23	34
UNKNOWN	200	3	8	216	6	14	95	3	16	16	1	31
TOTAL	4472	42	5	7724	167	11	8049	307	19	3913	180	23

RISK FACTOR	75-84 AT EXAM			85-94 AT EXAM			35-64 COMBINED			65-94 COMBINED		
	PERSON EXAMS	# OF EVENTS	ANNUAL RATE	PERSON EXAMS	# OF EVENTS	ANNUAL RATE	# OF EVENTS	CRUDE	AGE-ADJUSTED	# OF EVENTS	CRUDE	AGE-ADJUSTED
29-69	62	4	32	1			81	11	12	15	15	17
70-89	377	26	34	11			253	12	12	102	24	24
90-109	328	19	29	14			111	14	13	70	25	25
110-129	125	5	20	4			33	17	16	24	26	26
130-524	125	7	28	9			26	20	18	32	34	33
UNKNOWN	0	0	-	0			12	12	13	1	31	24
TOTAL	1017	61	30	39			516	13	13	244	25	25

LOGISTIC REGRESSIONS

AGE AT EXAM		COEFFICIENT OF RISK FACTOR	SE COEFFICIENT	P-VALUE	STANDARDIZED COEFFICIENT	STD SE
35-44		-0.007	0.011	0.536	-0.122	0.197
45-54		0.006	0.002	0.005	0.144	0.052
55-64		0.000	0.002	0.917	0.006	0.058
65-74		0.005	0.002	0.001	0.186	0.054
75-84		0.002	0.004	0.660	0.055	0.124
85-94						
35-64	UNIVARIATE	0.004	0.001	0.002	0.105	0.034
65-94	UNIVARIATE	0.005	0.001	0.001	0.173	0.049
35-64	BIVARIATE	0.002	0.002	0.259	0.043	0.038
65-94	BIVARIATE	0.005	0.001	0.001	0.166	0.050
35-64	MULTIVARIATE	0.001	0.002	0.341	0.036	0.038
65-94	MULTIVARIATE	0.004	0.002	0.005	0.144	0.051

SEX: FEMALE
EVENT: CORONARY HEART DISEASE
RISK FACTOR: BLOOD GLUCOSE (MG/DL)
PERSONS AT RISK: PERSONS FREE OF CORONARY HEART DISEASE

RISK FACTOR	35-44 AT EXAM			45-54 AT EXAM			55-64 AT EXAM			65-74 AT EXAM		
	PERSON EXAMS	# OF EVENTS	ANNUAL RATE	PERSON EXAMS	# OF EVENTS	ANNUAL RATE	PERSON EXAMS	# OF EVENTS	ANNUAL RATE	PERSON EXAMS	# OF EVENTS	ANNUAL RATE
29-69	1066	2	1	1797	8	2	1369	32	12	521	16	15
70-89	3412	4	1	5744	39	3	6098	98	8	3102	77	12
90-109	672	1	1	1625	6	2	2712	47	9	1732	50	14
110-129	96	0	0	501	3	5	730	21	14	501	10	10
130-524	39	0	0	159	9	28	438	21	24	360	20	28
UNKNOWN	331	0	0	333	7	11	166	7	21	32	1	16
TOTAL	5616	7	1	9959	72	4	11513	226	10	6248	174	14

RISK FACTOR	75-84 AT EXAM			85-94 AT EXAM			35-64 COMBINED			65-94 COMBINED		
	PERSON EXAMS	# OF EVENTS	ANNUAL RATE	PERSON EXAMS	# OF EVENTS	ANNUAL RATE	# OF EVENTS	CRUDE	AGE-ADJUSTED	# OF EVENTS	CRUDE	AGE-ADJUSTED
29-69	105	7	33	3	0	0	42	5	6	23	18	19
70-89	823	28	17	44	3	34	141	5	5	108	14	13
90-109	640	21	16	51	4	39	54	5	5	75	15	15
110-129	219	14	32	13	1	58	24	11	8	25	17	16
130-524	163	15	46	12	2	83	30	24	20	37	35	33
UNKNOWN	1	0	0	0	0	-	14	8	13	1	15	12
TOTAL	1951	85	22	123	10	41	305	6	6	269	16	16

LOGISTIC REGRESSIONS

AGE AT EXAM		COEFFICIENT OF RISK FACTOR	SE COEFFICIENT	P-VALUE	STANDARDIZED COEFFICIENT	STD SE
35-44		-0.016	0.033	0.635	-0.237	0.499
45-54		0.015	0.002	0.001	0.303	0.047
55-64		0.008	0.002	0.001	0.209	0.040
65-74		0.006	0.002	0.001	0.180	0.053
75-84		0.008	0.002	0.001	0.273	0.067
85-94		0.002	0.008	0.812	0.071	0.297
35-64	UNIVARIATE	0.012	0.001	0.001	0.259	0.027
65-94	UNIVARIATE	0.007	0.001	0.001	0.222	0.039
35-64	BIVARIATE	0.009	0.001	0.001	0.206	0.029
65-94	BIVARIATE	0.007	0.001	0.001	0.209	0.040
35-64	MULTIVARIATE	0.008	0.001	0.001	0.169	0.030
65-94	MULTIVARIATE	0.006	0.001	0.001	0.190	0.041

TABLE 1-7   DATE=08OCT85
AVERAGE ANNUAL INCIDENCE RATE FOR EVENT PER 1000 PERSONS AT RISK AT EXAM
BY AGE AND LEVEL OF RISK FACTOR AT EXAM: FRAMINGHAM STUDY 30-YEAR FOLLOWUP

SEX: MALE
EVENT: CORONARY HEART DISEASE
RISK FACTOR: DIABETES MELLITUS
PERSONS AT RISK: PERSONS FREE OF CORONARY HEART DISEASE

RISK FACTOR	35-44 AT EXAM			45-54 AT EXAM			55-64 AT EXAM			65-74 AT EXAM		
	PERSON EXAMS	# OF EVENTS	ANNUAL RATE	PERSON EXAMS	# OF EVENTS	ANNUAL RATE	PERSON EXAMS	# OF EVENTS	ANNUAL RATE	PERSON EXAMS	# OF EVENTS	ANNUAL RATE
NO	4424	42	5	7486	157	10	7549	277	18	3504	151	22
YES	48	0	0	238	10	21	500	30	30	409	29	35
UNKNOWN	0	0	-	0	0	-	0	0	-	0	0	-
TOTAL	4472	42	5	7724	167	11	8049	307	19	3913	180	23

RISK FACTOR	75-84 AT EXAM			85-94 AT EXAM			35-64 COMBINED			65-94 COMBINED		
	PERSON EXAMS	# OF EVENTS	ANNUAL RATE	PERSON EXAMS	# OF EVENTS	ANNUAL RATE	# OF EVENTS	CRUDE	AGE- ADJUSTED	# OF EVENTS	CRUDE	AGE- ADJUSTED
NO	900	55	31	34			476	12	12	208	23	24
YES	117	6	26	5			40	25	20	36	34	34
UNKNOWN	0	0	-	0			0	-	-	0	-	
TOTAL	1017	61	30	39			516	13	13	244	25	25

LOGISTIC REGRESSIONS

AGE AT EXAM		COEFFICIENT OF RISK FACTOR	SE COEFFICIENT	P-VALUE	STANDARDIZED COEFFICIENT	STD SE
35-44		-	-	-	-	-
45-54		0.717	0.333	0.031	0.124	0.058
55-64		0.516	0.198	0.009	0.125	0.048
65-74		0.527	0.210	0.012	0.161	0.064
75-84		-0.186	0.442	0.674	-0.059	0.141
85-94						
35-64	UNIVARIATE	0.760	0.169	0.001	0.147	0.033
65-94	UNIVARIATE	0.391	0.187	0.036	0.121	0.058
35-64	BIVARIATE	0.494	0.171	0.004	0.095	0.033
65-94	BIVARIATE	0.376	0.187	0.044	0.116	0.058
35-64	MULTIVARIATE	0.441	0.175	0.012	0.085	0.034
65-94	MULTIVARIATE	0.349	0.190	0.067	0.108	0.059

SEX: FEMALE
EVENT: CORONARY HEART DISEASE
RISK FACTOR: DIABETES MELLITUS
PERSONS AT RISK: PERSONS FREE OF CORONARY HEART DISEASE

RISK FACTOR	35-44 AT EXAM			45-54 AT EXAM			55-64 AT EXAM			65-74 AT EXAM		
	PERSON EXAMS	# OF EVENTS	ANNUAL RATE	PERSON EXAMS	# OF EVENTS	ANNUAL RATE	PERSON EXAMS	# OF EVENTS	ANNUAL RATE	PERSON EXAMS	# OF EVENTS	ANNUAL RATE
NO	5573	6	1	9760	64	3	10985	203	9	5780	148	13
YES	43	1	12	199	8	20	528	23	22	468	26	28
UNKNOWN	0	0	-	0	0	-	0	0	-	0	0	-
TOTAL	5616	7	1	9959	72	4	11513	226	10	6248	174	14

RISK FACTOR	75-84 AT EXAM			85-94 AT EXAM			35-64 COMBINED			65-94 COMBINED		
	PERSON EXAMS	# OF EVENTS	ANNUAL RATE	PERSON EXAMS	# OF EVENTS	ANNUAL RATE	# OF EVENTS	CRUDE	AGE- ADJUSTED	# OF EVENTS	CRUDE	AGE- ADJUSTED
NO	1738	67	19	106	8	38	273	5	5	223	15	15
YES	213	18	42	17	2	59	32	21	19	46	33	32
UNKNOWN	0	0	-	0	0	-	0	-	-	0	-	
TOTAL	1951	85	22	123	10	41	305	6	6	269	16	16

LOGISTIC REGRESSIONS

AGE AT EXAM		COEFFICIENT OF RISK FACTOR	SE COEFFICIENT	P-VALUE	STANDARDIZED COEFFICIENT	STD SE
35-44		3.095	1.091	0.005	0.270	0.095
45-54		1.848	0.382	0.001	0.259	0.053
55-64		0.883	0.225	0.001	0.185	0.047
65-74		0.806	0.218	0.001	0.212	0.057
75-84		0.834	0.276	0.003	0.260	0.086
85-94		0.491	0.838	0.558	0.170	0.290
35-64	UNIVARIATE	1.420	0.191	0.001	0.236	0.032
65-94	UNIVARIATE	0.851	0.167	0.001	0.236	0.046
35-64	BIVARIATE	1.058	0.193	0.001	0.176	0.032
65-94	BIVARIATE	0.803	0.168	0.001	0.223	0.047
35-64	MULTIVARIATE	0.824	0.198	0.001	0.137	0.033
65-94	MULTIVARIATE	0.668	0.173	0.001	0.185	0.048

TABLE 1-8   DATE=08OCT85

AVERAGE ANNUAL INCIDENCE RATE FOR EVENT PER 1000 PERSONS AT RISK AT EXAM
BY AGE AND LEVEL OF RISK FACTOR AT EXAM: FRAMINGHAM STUDY 30-YEAR FOLLOWUP

SEX: MALE
EVENT: CORONARY HEART DISEASE
RISK FACTOR: GLUCOSE IN URINE, DEFINITE OR TRACE
PERSONS AT RISK: PERSONS FREE OF CORONARY HEART DISEASE

RISK FACTOR	35-44 AT EXAM PERSON EXAMS	# OF EVENTS	ANNUAL RATE	45-54 AT EXAM PERSON EXAMS	# OF EVENTS	ANNUAL RATE	55-64 AT EXAM PERSON EXAMS	# OF EVENTS	ANNUAL RATE	65-74 AT EXAM PERSON EXAMS	# OF EVENTS	ANNUAL RATE
NO	4380	42	5	7535	158	10	7815	290	19	3767	162	22
YES	60	0	0	168	9	27	210	16	38	140	18	64
UNKNOWN	32	0	0	21	0	0	24	1	21	6	0	0
TOTAL	4472	42	5	7724	167	11	8049	307	19	3913	180	23

RISK FACTOR	75-84 AT EXAM PERSON EXAMS	# OF EVENTS	ANNUAL RATE	85-94 AT EXAM PERSON EXAMS	# OF EVENTS	ANNUAL RATE	35-64 COMBINED # OF EVENTS	CRUDE	AGE-ADJUSTED	65-94 COMBINED # OF EVENTS	CRUDE	AGE-ADJUSTED
NO	986	57	29	38			490	12	12	221	23	23
YES	28	4	71	1			25	29	25	23	68	69
UNKNOWN	3	0	0	0			1	6	8	0	0	0
TOTAL	1017	61	30	39			516	13	13	244	25	25

LOGISTIC REGRESSIONS

AGE AT EXAM	COEFFICIENT OF RISK FACTOR	SE COEFFICIENT	P-VALUE	STANDARDIZED COEFFICIENT	STD SE
35-44	-	-	-	-	-
45-54	0.972	0.352	0.006	0.142	0.051
55-64	0.761	0.267	0.004	0.121	0.043
65-74	1.189	0.265	0.001	0.221	0.049
75-84	0.999	0.557	0.073	0.164	0.091
85-94	-	-	-	-	-
35-64 UNIVARIATE	0.866	0.211	0.001	0.126	0.031
65-94 UNIVARIATE	1.181	0.235	0.001	0.214	0.043
35-64 BIVARIATE	0.750	0.212	0.001	0.109	0.031
65-94 BIVARIATE	1.204	0.235	0.001	0.218	0.043
35-64 MULTIVARIATE	0.633	0.224	0.005	0.092	0.033
65-94 MULTIVARIATE	1.098	0.242	0.001	0.199	0.044

SEX: FEMALE
EVENT: CORONARY HEART DISEASE
RISK FACTOR: GLUCOSE IN URINE, DEFINITE OR TRACE
PERSONS AT RISK: PERSONS FREE OF CORONARY HEART DISEASE

RISK FACTOR	35-44 AT EXAM PERSON EXAMS	# OF EVENTS	ANNUAL RATE	45-54 AT EXAM PERSON EXAMS	# OF EVENTS	ANNUAL RATE	55-64 AT EXAM PERSON EXAMS	# OF EVENTS	ANNUAL RATE	65-74 AT EXAM PERSON EXAMS	# OF EVENTS	ANNUAL RATE
NO	5537	6	1	9841	68	3	11346	218	10	6151	165	13
YES	41	1	12	94	3	16	144	8	28	107	9	42
UNKNOWN	38	0	0	24	1	21	23	0	0	10	0	0
TOTAL	5616	7	1	9959	72	4	11513	226	10	6248	174	14

RISK FACTOR	75-84 AT EXAM PERSON EXAMS	# OF EVENTS	ANNUAL RATE	85-94 AT EXAM PERSON EXAMS	# OF EVENTS	ANNUAL RATE	35-64 COMBINED # OF EVENTS	CRUDE	AGE-ADJUSTED	65-94 COMBINED # OF EVENTS	CRUDE	AGE-ADJUSTED
NO	1919	80	21	120	9	38	292	5	6	254	16	15
YES	23	5	109	3	1	167	12	22	-20	15	56	60
UNKNOWN	9	0	0	0	0	-	1	6	6	0	0	0
TOTAL	1951	85	22	123	10	41	305	6	6	269	16	16

LOGISTIC REGRESSIONS

AGE AT EXAM	COEFFICIENT OF RISK FACTOR	SE COEFFICIENT	P-VALUE	STANDARDIZED COEFFICIENT	STD SE
35-44	3.137	1.092	0.004	0.268	0.093
45-54	1.556	0.599	0.009	0.151	0.058
55-64	1.100	0.370	0.003	0.122	0.041
65-74	1.200	0.357	0.001	0.156	0.046
75-84	1.854	0.518	0.001	0.201	0.056
85-94	1.819	1.273	0.153	0.282	0.197
35-64 UNIVARIATE	1.403	0.301	0.001	0.142	0.030
65-94 UNIVARIATE	1.377	0.281	0.001	0.173	0.035
35-64 BIVARIATE	1.290	0.304	0.001	0.130	0.031
65-94 BIVARIATE	1.401	0.282	0.001	0.176	0.035
35-64 MULTIVARIATE	0.973	0.313	0.002	0.098	0.032
65-94 MULTIVARIATE	1.190	0.294	0.001	0.149	0.037

TABLE 1-9   DATE=01MAR86

AVERAGE ANNUAL INCIDENCE RATE FOR EVENT PER 1000 PERSONS AT RISK AT EXAM
BY AGE AND LEVEL OF RISK FACTOR AT EXAM: FRAMINGHAM STUDY 30-YEAR FOLLOWUP

SEX: MALE
EVENT: CORONARY HEART DISEASE
RISK FACTOR: GLUCOSE INTOLERANCE
PERSONS AT RISK: PERSONS FREE OF CORONARY HEART DISEASE

RISK FACTOR	35-44 AT EXAM			45-54 AT EXAM			55-64 AT EXAM			65-74 AT EXAM		
	PERSON EXAMS	# OF EVENTS	ANNUAL RATE	PERSON EXAMS	# OF EVENTS	ANNUAL RATE	PERSON EXAMS	# OF EVENTS	ANNUAL RATE	PERSON EXAMS	# OF EVENTS	ANNUAL RATE
NO	4335	41	5	7221	144	10	7145	264	18	3228	135	21
YES	137	1	4	503	23	23	904	43	24	685	45	33
UNKNOWN	0	0	-	0	0	-	0	0	-	0	0	-
TOTAL	4472	42	5	7724	167	11	8049	307	19	3913	180	23

RISK FACTOR	75-84 AT EXAM			85-94 AT EXAM			35-64 COMBINED			65-94 COMBINED		
	PERSON EXAMS	# OF EVENTS	ANNUAL RATE	PERSON EXAMS	# OF EVENTS	ANNUAL RATE	# OF EVENTS	CRUDE	AGE-ADJUSTED	# OF EVENTS	CRUDE	AGE-ADJUSTED
NO	602	51	32	27			449	12	12	187	23	23
YES	215	10	23	12			67	22	19	57	31	31
UNKNOWN	0	0	-	0				-	-	0	-	-
TOTAL	1017	61	30	39			516	13	13	244	25	25

LOGISTIC REGRESSIONS

AGE AT EXAM		COEFFICIENT OF RISK FACTOR	SE COEFFICIENT	P-VALUE	STANDARDIZED COEFFICIENT	STD SE
35-44		-0.261	1.016	0.797	-0.045	0.175
45-54		0.857	0.229	0.001	0.211	0.057
55-64		0.264	0.168	0.117	0.083	0.053
65-74		0.477	0.178	0.007	0.181	0.067
75-84		-0.331	0.355	0.351	-0.135	0.145
85-94						
35-64	UNIVARIATE	0.612	0.134	0.001	0.162	0.035
65-94	UNIVARIATE	0.322	0.156	0.039	0.125	0.060
35-64	BIVARIATE	0.390	0.135	0.004	0.104	0.036
65-94	BIVARIATE	0.294	0.156	0.060	0.114	0.061
35-64	MULTIVARIATE	0.335	0.139	0.016	0.089	0.037
65-94	MULTIVARIATE	0.254	0.158	0.109	0.098	0.061

SEX: FEMALE
EVENT: CORONARY HEART DISEASE
RISK FACTOR: GLUCOSE INTOLERANCE
PERSONS AT RISK: PERSONS FREE OF CORONARY HEART DISEASE

RISK FACTOR	35-44 AT EXAM			45-54 AT EXAM			55-64 AT EXAM			65-74 AT EXAM		
	PERSON EXAMS	# OF EVENTS	ANNUAL RATE	PERSON EXAMS	# OF EVENTS	ANNUAL RATE	PERSON EXAMS	# OF EVENTS	ANNUAL RATE	PERSON EXAMS	# OF EVENTS	ANNUAL RATE
NO	5522	6	1	9588	62	3	10503	190	9	5398	138	13
YES	94	1	5	371	10	13	1010	36	18	850	36	21
UNKNOWN	0	0	-	0	0	-	0	0	-	0	0	-
TOTAL	5616	7	1	9959	72	4	11513	226	10	6248	174	14

RISK FACTOR	75-84 AT EXAM			85-94 AT EXAM			35-64 COMBINED			65-94 COMBINED		
	PERSON EXAMS	# OF EVENTS	ANNUAL RATE	PERSON EXAMS	# OF EVENTS	ANNUAL RATE	# OF EVENTS	CRUDE	AGE-ADJUSTED	# OF EVENTS	CRUDE	AGE-ADJUSTED
NO	1581	57	48	100	6	30	258	5	5	201	14	14
YES	370	28	38	23	4	87	47	16	14	68	27	26
UNKNOWN	0	0	-	0	0	-	0	-	-	0	-	-
TOTAL	1951	85	22	123	10	41	305	6	6	269	16	16

LOGISTIC REGRESSIONS

AGE AT EXAM		COEFFICIENT OF RISK FACTOR	SE COEFFICIENT	P-VALUE	STANDARDIZED COEFFICIENT	STD SE
35-44		2.291	1.085	0.035	0.294	0.139
45-54		1.448	0.345	0.001	0.274	0.065
55-64		0.696	0.185	0.001	0.197	0.052
65-74		0.522	0.191	0.006	0.179	0.065
75-84		0.783	0.238	0.001	0.307	0.093
85-94		1.193	0.693	0.085	0.467	0.271
35-64	UNIVARIATE	1.174	0.161	0.001	0.266	0.037
65-94	UNIVARIATE	0.683	0.144	0.001	0.244	0.051
35-64	BIVARIATE	0.809	0.163	0.001	0.184	0.037
65-94	BIVARIATE	0.640	0.145	0.001	0.228	0.052
35-64	MULTIVARIATE	0.713	0.166	0.001	0.162	0.038
65-94	MULTIVARIATE	0.570	0.148	0.001	0.203	0.053

TABLE 1-10   DATE=02OCT85
AVERAGE ANNUAL INCIDENCE RATE FOR EVENT PER 1000 PERSONS AT RISK AT EXAM
BY AGE AND LEVEL OF RISK FACTOR AT EXAM: FRAMINGHAM STUDY 30-YEAR FOLLOWUP

SEX: MALE
EVENT: CORONARY HEART DISEASE
RISK FACTOR: METROPOLITAN RELATIVE WEIGHT (PERCENT)
PERSONS AT RISK: PERSONS FREE OF CORONARY HEART DISEASE

RISK FACTOR	35-44 AT EXAM			45-54 AT EXAM			55-64 AT EXAM			65-74 AT EXAM		
	PERSON EXAMS	# OF EVENTS	ANNUAL RATE	PERSON EXAMS	# OF EVENTS	ANNUAL RATE	PERSON EXAMS	# OF EVENTS	ANNUAL RATE	PERSON EXAMS	# OF EVENTS	ANNUAL RATE
54-104	723	2	1	1131	11	5	1155	43	19	686	22	16
105-114	965	8	4	1540	34	11	1655	50	15	851	30	18
115-124	1122	8	4	2037	43	11	2175	77	18	1069	56	26
125-134	915	12	7	1645	40	12	1674	64	19	766	39	25
135-272	747	12	8	1370	39	14	1389	73	26	541	33	30
UNKNOWN	0	0	-	1	0	-	1	0	0	0	0	-
TOTAL	4472	42	5	7724	167	11	8049	307	19	3913	180	23

RISK FACTOR	75-84 AT EXAM			85-94 AT EXAM			35-64 COMBINED			65-94 COMBINED		
	PERSON EXAMS	# OF EVENTS	ANNUAL RATE	PERSON EXAMS	# OF EVENTS	ANNUAL RATE	# OF EVENTS	CRUDE	AGE-ADJUSTED	# OF EVENTS	CRUDE	AGE-ADJUSTED
54-104	224	15	33	18			56	9	10	37	20	19
105-114	252	15	30	9			92	11	11	47	21	21
115-124	296	12	20	12			128	12	12	69	25	25
125-134	155	11	35	0			116	14	14	50	27	27
135-272	90	8	44	0			124	18	17	41	32	33
UNKNOWN	0	0	-	0			0	0	0	0	0	-
TOTAL	1017	61	30	39			516	13	13	244	25	25

LOGISTIC REGRESSIONS

AGE AT EXAM		COEFFICIENT OF RISK FACTOR	SE COEFFICIENT	P-VALUE	STANDARDIZED COEFFICIENT	STD SE
35-44		0.023	0.009	0.008	0.377	0.143
45-54		0.013	0.005	0.004	0.213	0.074
55-64		0.009	0.004	0.012	0.142	0.057
65-74		0.013	0.005	0.008	0.197	0.074
75-84		0.004	0.009	0.623	0.064	0.130
85-94						
35-64	UNIVARIATE	0.012	0.003	0.001	0.190	0.043
65-94	UNIVARIATE	0.010	0.004	0.016	0.154	0.064
35-64	BIVARIATE	0.012	0.003	0.001	0.190	0.043
65-94	BIVARIATE	0.011	0.004	0.006	0.177	0.065
35-64	MULTIVARIATE	0.009	0.003	0.001	0.148	0.045
65-94	MULTIVARIATE	0.007	0.004	0.125	0.102	0.067

SEX: FEMALE
EVENT: CORONARY HEART DISEASE
RISK FACTOR: METROPOLITAN RELATIVE WEIGHT (PERCENT)
PERSONS AT RISK: PERSONS FREE OF CORONARY HEART DISEASE

RISK FACTOR	35-44 AT EXAM			45-54 AT EXAM			55-64 AT EXAM			65-74 AT EXAM		
	PERSON EXAMS	# OF EVENTS	ANNUAL RATE	PERSON EXAMS	# OF EVENTS	ANNUAL RATE	PERSON EXAMS	# OF EVENTS	ANNUAL RATE	PERSON EXAMS	# OF EVENTS	ANNUAL RATE
54-104	1694	1	0	2095	15	4	2045	32	8	1216	33	14
105-114	1430	2	1	2331	10	2	2409	32	7	1145	31	14
115-124	1058	2	1	2107	14	3	2479	43	9	1395	29	10
125-134	566	0	0	1372	7	3	1743	35	10	982	32	16
135-272	860	2	1	2054	26	6	2837	84	15	1510	49	16
UNKNOWN	8	0	0	0	0	-	0	0	-	0	0	-
TOTAL	5616	7	1	9959	72	4	11513	226	10	6248	174	14

RISK FACTOR	75-84 AT EXAM			85-94 AT EXAM			35-64 COMBINED			65-94 COMBINED		
	PERSON EXAMS	# OF EVENTS	ANNUAL RATE	PERSON EXAMS	# OF EVENTS	ANNUAL RATE	# OF EVENTS	CRUDE	AGE-ADJUSTED	# OF EVENTS	CRUDE	AGE-ADJUSTED
54-104	517	17	16	41	4	49	48	4	5	54	15	15
105-114	375	9	12	28	0	0	44	4	4	40	13	13
115-124	399	27	34	21	3	71	59	5	5	59	16	17
125-134	266	15	28	16	1	31	42	4	5	48	19	19
135-272	394	17	22	17	2	59	112	10	9	68	18	18
UNKNOWN	0	0	-	0	0	-	0	0	0	0	0	-
TOTAL	1951	85	22	123	10	41	305	6	6	269	16	16

LOGISTIC REGRESSIONS

AGE AT EXAM		COEFFICIENT OF RISK FACTOR	SE COEFFICIENT	P-VALUE	STANDARDIZED COEFFICIENT	STD SE
35-44		0.019	0.014	0.167	0.380	0.275
45-54		0.015	0.004	0.001	0.325	0.091
55-64		0.011	0.003	0.001	0.244	0.058
65-74		0.005	0.003	0.105	0.117	0.072
75-84		0.007	0.005	0.164	0.147	0.106
85-94		0.006	0.017	0.703	0.125	0.328
35-64	UNIVARIATE	0.015	0.002	0.001	0.319	0.046
65-94	UNIVARIATE	0.005	0.003	0.091	0.100	0.059
35-64	BIVARIATE	0.012	0.002	0.001	0.264	0.048
65-94	BIVARIATE	0.006	0.003	0.036	0.124	0.059
35-64	MULTIVARIATE	0.006	0.002	0.012	0.137	0.054
65-94	MULTIVARIATE	0.003	0.003	0.253	0.070	0.061

TABLE 1-11　　　　　　　DATE=02OCT85
AVERAGE ANNUAL INCIDENCE RATE FOR EVENT PER 1000 PERSONS AT RISK AT EXAM
BY AGE AND LEVEL OF RISK FACTOR AT EXAM: FRAMINGHAM STUDY 30-YEAR FOLLOWUP

SEX: MALE
EVENT: CORONARY HEART DISEASE
RISK FACTOR: VITAL CAPACITY - HEIGHT INDEX (ML/INCH)
PERSONS AT RISK: PERSONS FREE OF CORONARY HEART DISEASE

RISK FACTOR	35-44 AT EXAM			45-54 AT EXAM			55-64 AT EXAM			65-74 AT EXA	
	PERSON EXAMS	# OF EVENTS	ANNUAL RATE	PERSON EXAMS	# OF EVENTS	ANNUAL RATE	PERSON EXAMS	# OF EVENTS	ANNUAL RATE	PERSON EXAMS	# OF EVENTS
8-44	269	2	4	1062	33	16	2045	93	23	1752	85
45-49	411	3	3	1125	29	13	1480	48	16	749	44
50-54	747	1	1	1658	42	13	1835	76	21	712	26
55-59	1022	11	5	1739	28	8	1416	56	20	442	14
60-85	2010	24	6	2135	35	8	1255	34	14	246	10
UNKNOWN	13	1	38	5	0	0	18	0	0	12	1
TOTAL	4472	42	5	7724	167	11	8049	307	19	3913	180

RISK FACTOR	75-84 AT EXAM			85-94 AT EXAM			35-64 COMBINED			65-94 COMB	
	PERSON EXAMS	# OF EVENTS	ANNUAL RATE	PERSON EXAMS	# OF EVENTS	ANNUAL RATE	# OF EVENTS	CRUDE	AGE-ADJUSTED	# OF EVENTS	CRUDE
8-44	649	46	35	24			128	19	16	133	27
45-49	172	4	12	6			80	13	12	48	26
50-54	97	5	26	7			119	14	14	32	20
55-59	65	6	46	2			95	11	12	20	20
60-85	26	0	0	0			93	9	10	10	18
UNKNOWN	8	0	0	0			1	14	8	1	25
TOTAL	1017	61	30	39			516	13	13	244	25

LOGISTIC REGRESSIONS

AGE AT EXAM		COEFFICIENT OF RISK FACTOR	SE COEFFICIENT	P-VALUE	STANDARDIZED COEFFICIENT	STD SE
35-44		0.011	0.018	0.523	0.101	0.159
45-54		-0.026	0.008	0.002	-0.235	0.076
55-64		-0.011	0.006	0.062	-0.107	0.058
65-74		-0.013	0.008	0.114	-0.120	0.076
75-84		-0.021	0.014	0.135	-0.199	0.133
85-94						
35-64	UNIVARIATE	-0.027	0.004	0.001	-0.261	0.043
65-94	UNIVARIATE	-0.012	0.007	0.014	-0.160	0.065
35-64	BIVARIATE	-0.012	0.005	0.016	-0.112	0.046
65-94	BIVARIATE	-0.013	0.007	0.059	-0.127	0.067
35-64	MULTIVARIATE	-0.003	0.005	0.587	-0.027	0.049
65-94	MULTIVARIATE	-0.010	0.007	0.158	-0.097	0.069

SEX: FEMALE
EVENT: CORONARY HEART DISEASE
RISK FACTOR: VITAL CAPACITY - HEIGHT INDEX (ML/INCH)
PERSONS AT RISK: PERSONS FREE OF CORONARY HEART DISEASE

RISK FACTOR	35-44 AT EXAM			45-54 AT EXAM			55-64 AT EXAM			65-74 AT EXAM	
	PERSON EXAMS	# OF EVENTS	ANNUAL RATE	PERSON EXAMS	# OF EVENTS	ANNUAL RATE	PERSON EXAMS	# OF EVENTS	ANNUAL RATE	PERSON EXAMS	# OF EVENTS R
4-29	122	2	8	534	7	7	1461	49	17	1747	79
30-34	289	1	2	1217	10	4	2418	54	11	1715	39
35-39	769	1	1	2077	22	5	3048	57	9	1529	24
40-44	1427	1	0	2804	19	3	2706	51	9	854	22
45-92	2988	2	0	3318	14	2	1859	15	4	397	10
UNKNOWN	21	0	0	17	0	0	21	0	0	6	0
TOTAL	5616	7	1	9959	72	4	11513	226	10	6248	174

RISK FACTOR	75-84 AT EXAM			85-94 AT EXAM			35-64 COMBINED			65-94 COMBI	
	PERSON EXAMS	# OF EVENTS	ANNUAL RATE	PERSON EXAMS	# OF EVENTS	ANNUAL RATE	# OF EVENTS	CRUDE	AGE-ADJUSTED	# OF EVENTS	CRUDE
4-29	930	41	22	80	8	50	58	14	-11	128	23
30-34	569	30	26	29	2	34	65	8	7	71	15
35-39	314	13	21	13	0	0	80	7	6	37	10
40-44	111	1	5	1	0	0	71	5	5	23	12
45-92	25	0	0	0	0	-	31	2	2	10	12
UNKNOWN	2	0	0	0	0	-	0	0	0	0	0
TOTAL	1951	85	22	123	10	41	305	6	6	269	16

LOGISTIC REGRESSIONS

AGE AT EXAM		COEFFICIENT OF RISK FACTOR	SE COEFFICIENT	P-VALUE	STANDARDIZED COEFFICIENT	STD SE
35-44		-0.106	0.044	0.016	-0.775	0.322
45-54		-0.048	0.015	0.002	-0.357	0.114
55-64		-0.043	0.009	0.001	-0.323	0.065
65-74		-0.036	0.010	0.001	-0.263	0.075
75-84		-0.020	0.016	0.198	-0.141	0.109
85-94		-0.090	0.045	0.047	-0.606	0.304
35-64	UNIVARIATE	-0.068	0.007	0.001	-0.542	0.055
65-94	UNIVARIATE	-0.039	0.008	0.001	-0.293	0.061
35-64	BIVARIATE	-0.042	0.008	0.001	-0.331	0.060
65-94	BIVARIATE	-0.034	0.008	0.001	-0.253	0.064
35-64	MULTIVARIATE	-0.029	0.008	0.001	-0.231	0.065
65-94	MULTIVARIATE	-0.029	0.009	0.001	-0.218	0.066

TABLE 1-12   DATE=02OCT85
AVERAGE ANNUAL INCIDENCE RATE FOR EVENT PER 1000 PERSONS AT RISK AT EXAM
BY AGE AND LEVEL OF RISK FACTOR AT EXAM: FRAMINGHAM STUDY 30-YEAR FOLLOWUP

SEX: MALE
EVENT: CORONARY HEART DISEASE
RISK FACTOR: HEART RATE (PER MINUTE)
PERSONS AT RISK: PERSONS FREE OF CORONARY HEART DISEASE

RISK FACTOR	35-44 AT EXAM			45-54 AT EXAM			55-64 AT EXAM			65-74 AT EXAM		
	PERSON EXAMS	# OF EVENTS	ANNUAL RATE	PERSON EXAMS	# OF EVENTS	ANNUAL RATE	PERSON EXAMS	# OF EVENTS	ANNUAL RATE	PERSON EXAMS	# OF EVENTS	ANNUAL RATE
30-67	1284	6	2	1949	37	9	2309	69	15	1183	50	21
68-75	1492	11	4	2490	44	9	2490	100	20	1159	37	16
76-83	722	8	6	1394	34	12	1374	54	20	624	34	27
84-91	579	14	12	1097	29	13	1064	40	19	487	30	31
92-220	395	3	4	794	23	14	812	44	27	460	29	32
UNKNOWN	0	0	-	0	0	-	0	0	-	0	0	-
TOTAL	4472	42	5	7724	167	11	8049	307	19	3913	180	23

RISK FACTOR	75-84 AT EXAM			85-94 AT EXAM			35-64 COMBINED			65-94 COMBINED		
	PERSON EXAMS	# OF EVENTS	ANNUAL RATE	PERSON EXAMS	# OF EVENTS	ANNUAL RATE	# OF EVENTS	CRUDE	AGE-ADJUSTED	# OF EVENTS	CRUDE	AGE-ADJUSTED
30-67	301	17	28	11			112	10	10	67	22	22
58-75	296	19	32	12			155	12	12	57	19	19
76-83	192	11	29	10			96	14	14	45	27	27
84-91	123	8	33	5			85	15	15	40	33	33
92-220	105	6	29	1			70	17	17	35	31	31
UNKNOWN	0	0	-	0			0	-	-	0	-	-
TOTAL	1017	61	30	39			516	13	13	244	25	25

LOGISTIC REGRESSIONS

AGE AT EXAM		COEFFICIENT OF RISK FACTOR	SE COEFFICIENT	P-VALUE	STANDARDIZED COEFFICIENT	STD SE
35-44		0.028	0.011	0.010	0.354	0.137
45-54		0.014	0.006	0.019	0.176	0.075
55-64		0.012	0.004	0.004	0.160	0.055
65-74		0.015	0.005	0.005	0.199	0.071
75-84		0.002	0.010	0.823	0.029	0.131
85-94						
35-64	UNIVARIATE	0.014	0.003	0.001	0.183	0.042
65-94	UNIVARIATE	0.012	0.005	0.009	0.162	0.062
35-64	BIVARIATE	0.014	0.003	0.001	0.181	0.042
65-94	BIVARIATE	0.012	0.005	0.009	0.162	0.062
35-64	MULTIVARIATE	0.004	0.004	0.275	0.050	0.045
65-94	MULTIVARIATE	0.010	0.005	0.036	0.134	0.064

SEX: FEMALE
EVENT: CORONARY HEART DISEASE
RISK FACTOR: HEART RATE (PER MINUTE)
PERSONS AT RISK: PERSONS FREE OF CORONARY HEART DISEASE

RISK FACTOR	35-44 AT EXAM			45-54 AT EXAM			55-64 AT EXAM			65-74 AT EXAM		
	PERSON EXAMS	# OF EVENTS	ANNUAL RATE	PERSON EXAMS	# OF EVENTS	ANNUAL RATE	PERSON EXAMS	# OF EVENTS	ANNUAL RATE	PERSON EXAMS	# OF EVENTS	ANNUAL RATE
30-67	1092	0	0	1897	17	4	2305	36	8	1103	32	15
68-75	1803	2	1	3244	23	4	3468	64	9	1668	53	16
76-83	1051	3	1	1911	10	3	2351	45	10	1358	30	11
84-91	939	2	1	1604	13	4	1780	41	12	1089	28	13
92-220	731	0	0	1303	9	3	1609	40	12	1030	31	15
UNKNOWN	0	0	-	0	0	-	0	0	-	0	0	-
TOTAL	5616	7	1	9959	72	4	11513	226	10	6248	174	14

RISK FACTOR	75-84 AT EXAM			85-94 AT EXAM			35-64 COMBINED			65-94 COMBINED		
	PERSON EXAMS	# OF EVENTS	ANNUAL RATE	PERSON EXAMS	# OF EVENTS	ANNUAL RATE	# OF EVENTS	CRUDE	AGE-ADJUSTED	# OF EVENTS	CRUDE	AGE-ADJUSTED
30-67	368	17	24	26	4	77	53	5	5	53	18	18
68-75	462	17	18	32	0	0	89	5	6	70	16	16
76-83	428	21	25	24	0	0	58	5	6	51	14	14
84-91	330	15	23	23	4	87	56	6	7	47	16	16
92-220	371	15	20	18	2	56	49	7	6	48	17	17
UNKNOWN	0	0	-	0	0	-	0	-	-	0	-	-
TOTAL	1959	85	22	123	10	41	305	6	6	269	16	16

LOGISTIC REGRESSIONS

AGE AT EXAM		COEFFICIENT OF RISK FACTOR	SE COEFFICIENT	P-VALUE	STANDARDIZED COEFFICIENT	STD SE
35-44		0.014	0.027	0.592	0.180	0.335
45-54		-0.003	0.010	0.773	-0.034	0.120
55-64		0.009	0.005	0.071	0.115	0.064
65-74		-0.002	0.006	0.722	-0.028	0.078
75-84		-0.005	0.008	0.518	-0.074	0.114
85-94		0.005	0.024	0.832	0.070	0.328
35-64	UNIVARIATE	0.007	0.004	0.121	0.086	0.056
65-94	UNIVARIATE	-0.003	0.005	0.566	-0.036	0.063
35-64	BIVARIATE	0.006	0.004	0.142	0.081	0.055
65-94	BIVARIATE	-0.003	0.005	0.518	-0.040	0.063
35-64	MULTIVARIATE	-0.002	0.005	0.631	-0.029	0.059
65-94	MULTIVARIATE	-0.007	0.005	0.143	-0.093	0.064

TABLE 1-13   DATE=04OCT85

AVERAGE ANNUAL INCIDENCE RATE FOR EVENT PER 1000 PERSONS AT RISK AT EXAM
BY AGE AND LEVEL OF RISK FACTOR AT EXAM: FRAMINGHAM STUDY 30-YEAR FOLLOWUP

SEX: MALE
EVENT: CORONARY HEART DISEASE
RISK FACTOR: CIGARETTES SMOKED (NUMBER PER DAY)
PERSONS AT RISK: PERSONS FREE OF CORONARY HEART DISEASE

RISK FACTOR	35-44 AT EXAM			45-54 AT EXAM			55-64 AT EXAM			65-74 AT EXAM		
	PERSON EXAMS	# OF EVENTS	ANNUAL RATE	PERSON EXAMS	# OF EVENTS	ANNUAL RATE	PERSON EXAMS	# OF EVENTS	ANNUAL RATE	PERSON EXAMS	# OF EVENTS	ANNUAL RATE
NONE	1383	6	2	3068	47	8	4295	138	16	2611	120	23
1-10	433	4	5	716	12	8	769	21	14	374	15	20
11-20	1485	15	5	2051	55	13	1682	81	24	526	25	24
21-40	987	12	6	1602	46	14	1119	57	25	369	17	23
41-90	158	5	16	252	5	10	162	10	31	27	2	37
UNKNOWN	26	0	0	35	2	29	22	0	0	6	1	83
TOTAL	4472	42	5	7724	167	11	8049	307	19	3913	180	23

RISK FACTOR	75-84 AT EXAM			85-94 AT EXAM			35-64 COMBINED			65-94 COMBINED		
	PERSON EXAMS	# OF EVENTS	ANNUAL RATE	PERSON EXAMS	# OF EVENTS	ANNUAL RATE	# OF EVENTS	CRUDE	AGE-ADJUSTED	# OF EVENTS	CRUDE	AGE-ADJUSTED
NONE	797	53	53	27			191	11	10	175	25	25
1-10	92	5	27	4			37	10	10	20	21	21
11-20	77	1	6	5			151	14	16	27	22	21
21-40	46	2	22	1			115	16	17	19	23	23
41-90	5	0	0	0			20	17	20	2	31	29
UNKNOWN	0	0	-	0			2	12	11	1	83	65
TOTAL	1017	61	30	39			516	13	13	244	25	25

LOGISTIC REGRESSIONS

AGE AT EXAM		COEFFICIENT OF RISK FACTOR	SE COEFFICIENT	P-VALUE	STANDARDIZED COEFFICIENT	STD SE
35-44		0.029	0.010	0.005	0.407	0.144
45-54		0.015	0.005	0.002	0.228	0.073
55-64		0.016	0.004	0.001	0.218	0.052
65-74		0.002	0.006	0.719	0.027	0.075
75-84		-0.040	0.023	0.084	-0.361	0.209
85-94						
35-64	UNIVARIATE	0.012	0.003	0.001	0.171	0.042
65-94	UNIVARIATE	-0.004	0.006	0.552	-0.040	0.068
35-64	BIVARIATE	0.018	0.003	0.001	0.259	0.041
65-94	BIVARIATE	-0.001	0.006	0.808	-0.017	0.069
35-64	MULTIVARIATE	0.020	0.003	0.001	0.290	0.042
65-94	MULTIVARIATE	0.001	0.006	0.902	0.008	0.069

SEX: FEMALE
EVENT: CORONARY HEART DISEASE
RISK FACTOR: CIGARETTES SMOKED (NUMBER PER DAY)
PERSONS AT RISK: PERSONS FREE OF CORONARY HEART DISEASE

RISK FACTOR	35-44 AT EXAM			45-54 AT EXAM			55-64 AT EXAM			65-74 AT EXAM		
	PERSON EXAMS	# OF EVENTS	ANNUAL RATE	PERSON EXAMS	# OF EVENTS	ANNUAL RATE	PERSON EXAMS	# OF EVENTS	ANNUAL RATE	PERSON EXAMS	# OF EVENTS	ANNUAL RATE
NONE	2582	2	0	5528	41	4	7595	146	10	4863	136	14
1-10	1314	1	0	1726	11	3	1566	31	10	626	14	11
11-20	1371	3	1	2009	14	3	1696	35	10	568	22	19
21-40	298	1	2	622	5	4	602	12	10	172	1	3
41-90	22	0	0	37	1	14	16	1	63	0	0	-
UNKNOWN	29	0	0	37	0	0	38	0	0	19	1	26
TOTAL	5616	7	1	9959	72	4	11513	226	10	6248	174	14

RISK FACTOR	75-84 AT EXAM			85-94 AT EXAM			35-64 COMBINED			65-94 COMBINED		
	PERSON EXAMS	# OF EVENTS	ANNUAL RATE	PERSON EXAMS	# OF EVENTS	ANNUAL RATE	# OF EVENTS	CRUDE	AGE-ADJUSTED	# OF EVENTS	CRUDE	AGE-ADJUSTED
NONE	1710	75	22	117	10	43	189	6	6	221	17	16
1-10	137	7	26	4	0	0	43	5	5	21	14	14
11-20	64	3	18	2	0	0	52	5	5	25	19	14
21-40	16	0	0	0	0	-	18	6	6	1	3	2
41-90	8	0	-	0	0	-	3	20	32	0	-	-
UNKNOWN	4	0	0	0	0	-	0	0	0	7	22	20
TOTAL	1951	85	22	123	10	41	305	6	6	269	16	16

LOGISTIC REGRESSIONS

AGE AT EXAM		COEFFICIENT OF RISK FACTOR	SE COEFFICIENT	P-VALUE	STANDARDIZED COEFFICIENT	STD SE
35-44		0.057	0.029	0.046	0.563	0.282
45-54		0.006	0.011	0.588	0.061	0.113
55-64		0.006	0.007	0.364	0.058	0.064
65-74		-0.002	0.010	0.841	-0.016	0.079
75-84		-0.018	0.026	0.490	-0.091	0.131
85-94		-				
35-64	UNIVARIATE	0.001	0.006	0.877	0.009	0.057
65-94	UNIVARIATE	-0.009	0.010	0.332	-0.066	0.068
35-64	BIVARIATE	0.009	0.006	0.091	0.094	0.056
65-94	BIVARIATE	-0.005	0.010	0.614	-0.034	0.068
35-64	MULTIVARIATE	0.014	0.006	0.013	0.139	0.056
65-94	MULTIVARIATE	0.001	0.010	0.953	0.004	0.068

TABLE 1-14   DATE=19OCT85

AVERAGE ANNUAL INCIDENCE RATE FOR EVENT PER 1000 PERSONS AT RISK AT EXAM
BY AGE AND LEVEL OF RISK FACTOR AT EXAM: FRAMINGHAM STUDY 30-YEAR FOLLOWUP

SEX: MALE
EVENT: CORONARY HEART DISEASE
RISK FACTOR: ALBUMIN IN URINE, DEFINITE
PERSONS AT RISK: PERSONS FREE OF CORONARY HEART DISEASE

35-44 AT EXAM			45-54 AT EXAM			55-64 AT EXAM			65-74 AT EXAM		
PERSON EXAMS	# OF EVENTS	ANNUAL RATE	PERSON EXAMS	# OF EVENTS	ANNUAL RATE	PERSON EXAMS	# OF EVENTS	ANNUAL RATE	PERSON EXAMS	# OF EVENTS	ANNUAL RATE
4366	41	5	7599	161	11	7838	299	19	3792	171	23
70	1	7	99	6	30	184	7	19	115	9	39
36	0	0	26	0	0	27	1	19	6	0	0
4472	42	5	7724	167	11	8049	307	19	3913	180	23

75-84 AT EXAM			85-94 AT EXAM			35-64 COMBINED			65-94 COMBINED		
PERSON EXAMS	# OF EVENTS	ANNUAL RATE	PERSON EXAMS	# OF EVENTS	ANNUAL RATE	# OF EVENTS	CRUDE	AGE-ADJUSTED	# OF EVENTS	CRUDE	AGE-ADJUSTED
989	60	30	32	7		501	13	13	234	24	25
25	1	20				14	20	21	10	34	35
3	0	0	0	39		1	6	8	0	0	0
1017	61	30				516	13	13	244	25	25

LOGISTIC REGRESSIONS

	COEFFICIENT OF RISK FACTOR	SE COEFFICIENT	P-VALUE	STANDARDIZED COEFFICIENT	STD SE
	0.424	1.019	0.677	0.053	0.127
	1.092	0.429	0.011	0.123	0.048
	-0.003	0.390	0.994	-0.000	0.058
	0.587	0.356	0.099	0.099	0.060
	-0.438	1.029	0.670	-0.068	0.160
NIVARIATE	0.464	0.276	0.093	0.061	0.036
NIVARIATE	0.357	0.334	0.286	0.060	0.057
IVARIATE	0.334	0.278	0.230	0.044	0.036
IVARIATE	0.338	0.335	0.313	0.057	0.057
ULTIVARIATE	-0.268	0.306	0.381	-0.035	0.040
ULTIVARIATE	0.005	0.344	0.988	0.001	0.058

SEX: FEMALE
EVENT: CORONARY HEART DISEASE
RISK FACTOR: ALBUMIN IN URINE, DEFINITE
PERSONS AT RISK: PERSONS FREE OF CORONARY HEART DISEASE

35-44 AT EXAM			45-54 AT EXAM			55-64 AT EXAM			65-74 AT EXAM		
PERSON EXAMS	# OF EVENTS	ANNUAL RATE	PERSON EXAMS	# OF EVENTS	ANNUAL RATE	PERSON EXAMS	# OF EVENTS	ANNUAL RATE	PERSON EXAMS	# OF EVENTS	ANNUAL RATE
5353	6	1	9656	68	4	11263	219	10	6118	167	14
225	1	2	278	3	5	225	7	16	120	7	29
38	0	0	25	1	20	25	0	0	10	0	0
5616	7	1	9959	72	4	11513	226	10	6248	174	14

75-84 AT EXAM			85-94 AT EXAM			35-64 COMBINED			65-94 COMBINED		
PERSON EXAMS	# OF EVENTS	ANNUAL RATE	PERSON EXAMS	# OF EVENTS	ANNUAL RATE	# OF EVENTS	CRUDE	AGE-ADJUSTED	# OF EVENTS	CRUDE	AGE-ADJUSTED
1904	85	22	120	9	38	293	6	6	261	16	16
38	0	0	3	1	167	11	8	9	8	25	24
9	0	0	0	0	-	1	6	7	0	0	0
1931	85	22	123	10	41	305	6	6	269	16	16

LOGISTIC REGRESSIONS

COEFFICIENT OF RISK FACTOR	SE COEFFICIENT	P-VALUE	STANDARDIZED COEFFICIENT	STD SE

TABLE 1-15   DATE=04OCT85
AVERAGE ANNUAL INCIDENCE RATE FOR EVENT PER 1000 PERSONS AT RISK AT EXAM
BY AGE AND LEVEL OF RISK FACTOR AT EXAM: FRAMINGHAM STUDY 30-YEAR FOLLOWUP

SEX: MALE
EVENT: CORONARY HEART DISEASE
RISK FACTOR: HEART ENLARGEMENT BY X-RAY
PERSONS AT RISK: PERSONS FREE OF CORONARY HEART DISEASE

RISK FACTOR	35-44 AT EXAM			45-54 AT EXAM			55-64 AT EXAM			65-74 AT EXAM		
	PERSON EXAMS	# OF EVENTS	ANNUAL RATE	PERSON EXAMS	# OF EVENTS	ANNUAL RATE	PERSON EXAMS	# OF EVENTS	ANNUAL RATE	PERSON EXAMS	# OF EVENTS	ANNUAL RATE
NO	4063	37	5	6383	125	10	5967	205	17	2584	103	20
MILD	185	4	11	650	19	15	962	42	22	578	21	18
DEFINITE	219	1	2	687	22	16	1119	62	28	751	56	37
UNKNOWN	5	0	0	4	1	125	1	0	0	0	0	-
TOTAL	4472	42	5	7724	167	11	8049	307	19	3913	180	23

RISK FACTOR	75-84 AT EXAM			85-94 AT EXAM			35-64 COMBINED			65-94 COMBINED		
	PERSON EXAMS	# OF EVENTS	ANNUAL RATE	PERSON EXAMS	# OF EVENTS	ANNUAL RATE	# OF EVENTS	CRUDE	AGE-ADJUSTED	# OF EVENTS	CRUDE	AGE-ADJUSTED
NO	589	26	22	29			365	11	12	130	20	20
MILD	178	11	31	5			65	18	17	33	22	21
DEFINITE	250	24	48	5			85	21	18	81	40	40
UNKNOWN	0	0	-	0			1	50	48	0	0	-
TOTAL	1017	61	30	39			516	13	13	244	25	25

LOGISTIC REGRESSIONS

AGE AT EXAM		COEFFICIENT OF RISK FACTOR	SE COEFFICIENT	P-VALUE	STANDARDIZED COEFFICIENT	STD SE
35-44		0.016	0.328	0.962	0.007	0.153
45-54		0.273	0.110	0.013	0.167	0.067
55-64		0.255	0.072	0.001	0.184	0.052
65-74		0.310	0.088	0.001	0.247	0.070
75-84		0.416	0.148	0.005	0.351	0.125
85-94						
35-64	UNIVARIATE	0.347	0.058	0.001	0.221	0.037
65-94	UNIVARIATE	0.351	0.074	0.001	0.284	0.060
35-64	BIVARIATE	0.237	0.059	0.001	0.151	0.038
65-94	BIVARIATE	0.337	0.075	0.001	0.272	0.060
35-64	MULTIVARIATE	0.175	0.064	0.006	0.111	0.041
65-94	MULTIVARIATE	0.217	0.080	0.007	0.175	0.064

SEX: FEMALE
EVENT: CORONARY HEART DISEASE
RISK FACTOR: HEART ENLARGEMENT BY X-RAY
PERSONS AT RISK: PERSONS FREE OF CORONARY HEART DISEASE

RISK FACTOR	35-44 AT EXAM			45-54 AT EXAM			55-64 AT EXAM			65-74 AT EXAM		
	PERSON EXAMS	# OF EVENTS	ANNUAL RATE	PERSON EXAMS	# OF EVENTS	ANNUAL RATE	PERSON EXAMS	# OF EVENTS	ANNUAL RATE	PERSON EXAMS	# OF EVENTS	ANNUAL RATE
NO	5164	5	0	8017	47	5	7522	109	7	3191	64	10
MILD	199	0	0	943	8	4	1769	44	12	1150	26	11
DEFINITE	245	2	0	993	17	9	2218	73	16	1907	84	22
UNKNOWN	8	0	0	6	0	0	4	0	0	0	0	-
TOTAL	5616	7	1	9959	72	4	11513	226	10	6248	174	14

RISK FACTOR	75-84 AT EXAM			85-94 AT EXAM			35-64 COMBINED			65-94 COMBINED		
	PERSON EXAMS	# OF EVENTS	ANNUAL RATE	PERSON EXAMS	# OF EVENTS	ANNUAL RATE	# OF EVENTS	CRUDE	AGE-ADJUSTED	# OF EVENTS	CRUDE	AGE-ADJUSTED
NO	702	15	11	32	3	47	161	4	5	82	18	11
MILD	373	20	27	30	3	50	52	9	7	49	16	15
DEFINITE	876	50	29	61	4	33	92	13	11	138	24	24
UNKNOWN	0	0	-	0	0	-	0	0	0	0	0	-
TOTAL	1951	85	22	123	10	41	305	6	6	269	16	16

LOGISTIC REGRESSIONS

AGE AT EXAM		COEFFICIENT OF RISK FACTOR	SE COEFFICIENT	P-VALUE	STANDARDIZED COEFFICIENT	STD SE
35-44		1.027	0.439	0.019	0.454	0.194
45-54		0.529	0.142	0.001	0.338	0.091
55-64		0.425	0.075	0.001	0.338	0.060
65-74		0.412	0.086	0.001	0.362	0.076
75-84		0.453	0.136	0.001	0.405	0.122
85-94		-0.205	0.385	0.594	-0.173	0.324
35-64	UNIVARIATE	0.635	0.064	0.001	0.443	0.045
65-94	UNIVARIATE	0.436	0.071	0.001	0.389	0.063
35-64	BIVARIATE	0.429	0.067	0.001	0.299	0.047
65-94	BIVARIATE	0.406	0.072	0.001	0.363	0.064
35-64	MULTIVARIATE	0.312	0.073	0.001	0.218	0.051
65-94	MULTIVARIATE	0.323	0.075	0.001	0.288	0.067

TABLE 1-16   DATE=01NOV85
AVERAGE ANNUAL INCIDENCE RATE FOR EVENT PER 1000 PERSONS AT RISK AT EXAM
BY AGE AND LEVEL OF RISK FACTOR AT EXAM: FRAMINGHAM STUDY 30-YEAR FOLLOWUP

SEX: MALE
EVENT: CORONARY HEART DISEASE
RISK FACTOR: LEFT VENTRICULAR HYPERTROPHY BY ELECTROCARDIOGRAM
PERSONS AT RISK: PERSONS FREE OF CORONARY HEART DISEASE

RISK FACTOR	35-44 AT EXAM PERSON EXAMS	# OF EVENTS	ANNUAL RATE	45-54 AT EXAM PERSON EXAMS	# OF EVENTS	ANNUAL RATE	55-64 AT EXAM PERSON EXAMS	# OF EVENTS	ANNUAL RATE	65-74 AT EXAM PERSON EXAMS	# OF EVENTS	ANNUAL RATE
NO	4366	42	5	7497	154	10	7693	283	18	3703	153	21
MILD	78	0	0	132	6	23	210	8	19	90	10	56
DEFINITE	28	0	0	95	7	37	146	16	55	120	17	71
UNKNOWN	0	0		0	0	-	0	0	-	0	0	-
TOTAL	4472	42	5	7724	167	11	8049	307	19	3913	180	23

RISK FACTOR	75-84 AT EXAM PERSON EXAMS	# OF EVENTS	ANNUAL RATE	85-94 AT EXAM PERSON EXAMS	# OF EVENTS	ANNUAL RATE	35-64 COMBINED # OF EVENTS	CRUDE	AGE-ADJUSTED	65-94 COMBINED # OF EVENTS	CRUDE	AGE-ADJUSTED
NO	949	56	30	34			479	12	12	212	23	23
MILD	37	1	14	2			14	17	16	11	43	47
DEFINITE	31	4	65	3			23	43	36	21	68	69
UNKNOWN	0	0	-	0			0	-		0	-	
TOTAL	1017	61	30	39			516	13	13	244	25	25

LOGISTIC REGRESSIONS

AGE AT EXAM		COEFFICIENT OF RISK FACTOR	SE COEFFICIENT	P-VALUE	STANDARDIZED COEFFICIENT	STD SE
35-44		-	-		-	-
45-54		0.544	0.188	0.003	0.152	0.051
55-64		0.246	0.140	0.080	0.084	0.048
65-74		0.697	0.143	0.001	0.238	0.049
75-84		-0.032	0.332	0.923	-0.013	0.135
85-94						
35-64	UNIVARIATE	0.605	0.107	0.001	0.162	0.029
65-94	UNIVARIATE	0.610	0.116	0.001	0.230	0.044
35-64	BIVARIATE	0.515	0.107	0.001	0.138	0.029
65-94	BIVARIATE	0.603	0.117	0.001	0.227	0.044
35-64	MULTIVARIATE	0.298	0.117	0.011	0.080	0.031
65-94	MULTIVARIATE	0.377	0.127	0.003	0.142	0.048

SEX: FEMALE
EVENT: CORONARY HEART DISEASE
RISK FACTOR: LEFT VENTRICULAR HYPERTROPHY BY ELECTROCARDIOGRAM
PERSONS AT RISK: PERSONS FREE OF CORONARY HEART DISEASE

RISK FACTOR	35-44 AT EXAM PERSON EXAMS	# OF EVENTS	ANNUAL RATE	45-54 AT EXAM PERSON EXAMS	# OF EVENTS	ANNUAL RATE	55-64 AT EXAM PERSON EXAMS	# OF EVENTS	ANNUAL RATE	65-74 AT EXAM PERSON EXAMS	# OF EVENTS	ANNUAL RATE
NO	5563	6	1	9798	69	4	11137	208	9	5966	147	12
MILD	38	0	0	107	1	5	226	7	15	134	9	34
DEFINITE	15	1	33	54	2	19	150	11	37	148	18	61
UNKNOWN	0	0	-	0	0	-	0	0	-	0	0	-
TOTAL	5616	7	1	9959	72	4	11513	226	10	6248	174	14

RISK FACTOR	75-84 AT EXAM PERSON EXAMS	# OF EVENTS	ANNUAL RATE	85-94 AT EXAM PERSON EXAMS	# OF EVENTS	ANNUAL RATE	35-64 COMBINED # OF EVENTS	CRUDE	AGE-ADJUSTED	65-94 COMBINED # OF EVENTS	CRUDE	AGE-ADJUSTED
NO	1833	76	21	111	9	41	283	5	5	232	15	15
MILD	49	4	41	6	0	0	8	11	8	13	34	35
DEFINITE	69	5	36	6	1	83	14	32	29	24	54	55
UNKNOWN	0	0	-	0	0	-	0	-		0	-	
TOTAL	1951	85	22	123	10	41	305	6	6	269	16	16

LOGISTIC REGRESSIONS

AGE AT EXAM		COEFFICIENT OF RISK FACTOR	SE COEFFICIENT	P-VALUE	STANDARDIZED COEFFICIENT	STD SE
35-44		1.189	0.750	0.113	0.204	0.129
45-54		0.410	0.376	0.275	0.089	0.082
55-64		0.451	0.156	0.004	0.135	0.047
65-74		0.752	0.139	0.001	0.244	0.045
75-84		0.408	0.235	0.082	0.146	0.084
85-94		-0.269	0.838	0.748	-0.128	0.397
35-64	UNIVARIATE	0.894	0.135	0.001	0.190	0.029
65-94	UNIVARIATE	0.713	0.107	0.001	0.251	0.038
35-64	BIVARIATE	0.710	0.137	0.001	0.151	0.029
65-94	BIVARIATE	0.689	0.107	0.001	0.243	0.038
35-44	MULTIVARIATE	0.331	0.156	0.034	0.070	0.033
65-94	MULTIVARIATE	0.586	0.113	0.001	0.206	0.040

TABLE 1-17   DATE=23NOV85

AVERAGE ANNUAL INCIDENCE RATE FOR EVENT PER 1000 PERSONS AT RISK AT EXAM BY AGE AND LEVEL OF RISK FACTOR AT EXAM: FRAMINGHAM STUDY 30-YEAR FOLLOWUP

SEX: MALE
EVENT: CORONARY HEART DISEASE
RISK FACTOR: INTRAVENTRICULAR CONDUCTION DEFECT (WITH QRS INTERVAL MORE THAN 0.11 SECOND)
PERSONS AT RISK: PERSONS FREE OF CORONARY HEART DISEASE

RISK FACTOR	35-44 AT EXAM PERSON EXAMS	# OF EVENTS	ANNUAL RATE	45-54 AT EXAM PERSON EXAMS	# OF EVENTS	ANNUAL RATE	55-64 AT EXAM PERSON EXAMS	# OF EVENTS	ANNUAL RATE	65-74 AT EXAM PERSON EXAMS	# OF EVENTS	ANNUAL RATE
NO	4161	39	5	7213	152	11	7386	272	18	3566	156	22
INCOMPLETE	275	3	5	446	12	13	456	24	26	183	14	38
COMPLETE	36	0	0	65	3	23	207	11	27	164	10	30
UNKNOWN	0	0	-	0	0	-	0	0	-	0	0	-
TOTAL	4472	42	5	7724	167	11	8049	307	19	3913	180	23

RISK FACTOR	75-84 AT EXAM PERSON EXAMS	# OF EVENTS	ANNUAL RATE	85-94 AT EXAM PERSON EXAMS	# OF EVENTS	ANNUAL RATE	35-64 COMBINED # OF EVENTS	CRUDE	AGE-ADJUSTED	65-94 COMBINED # OF EVENTS	CRUDE	AGE-ADJUSTED
NO	918	55	30	31	2	-	463	12	12	213	24	24
INCOMPLETE	21	1	24	2			39	17	17	15	34	35
COMPLETE	78	5	32	6			14	23	19	16	32	31
UNKNOWN	0	0	-	0			0	-		0	-	
TOTAL	1017	61	30	39			516	13	13	244	25	25

LOGISTIC REGRESSIONS

AGE AT EXAM	COEFFICIENT OF RISK FACTOR	SE COEFFICIENT	P-VALUE	STANDARDIZED COEFFICIENT	STD SE
35-44	-0.074	0.547	0.892	-0.022	0.162
45-54	0.325	0.221	0.141	0.093	0.065
55-64	0.248	0.129	0.054	0.096	0.050
65-74	0.259	0.145	0.074	0.115	0.064
75-84	0.022	0.239	0.926	0.012	0.130
85-94					
35-64 UNIVARIATE	0.311	0.110	0.005	0.104	0.037
65-94 UNIVARIATE	0.211	0.121	0.082	0.099	0.057
35-64 BIVARIATE	0.227	0.109	0.036	0.076	0.036
65-94 BIVARIATE	0.188	0.121	0.121	0.088	0.057
35-64 MULTIVARIATE	0.191	0.115	0.096	0.064	0.038
65-94 MULTIVARIATE	0.217	0.123	0.078	0.102	0.058

SEX: FEMALE
EVENT: CORONARY HEART DISEASE
RISK FACTOR: INTRAVENTRICULAR CONDUCTION DEFECT (WITH QRS INTERVAL MORE THAN 0.11 SECOND)
PERSONS AT RISK: PERSONS FREE OF CORONARY HEART DISEASE

RISK FACTOR	35-44 AT EXAM PERSON EXAMS	# OF EVENTS	ANNUAL RATE	45-54 AT EXAM PERSON EXAMS	# OF EVENTS	ANNUAL RATE	55-64 AT EXAM PERSON EXAMS	# OF EVENTS	ANNUAL RATE	65-74 AT EXAM PERSON EXAMS	# OF EVENTS	ANNUAL RATE
NO	5529	5	0	9771	69	4	11183	213	10	6027	167	14
INCOMPLETE	75	2	13	144	1	3	187	3	8	115	2	9
COMPLETE	12	0	0	44	2	23	143	10	35	106	5	24
UNKNOWN	0	0	-	0	0	-	0	0	-	0	0	-
TOTAL	5616	7	1	9959	72	4	11513	226	10	6248	174	14

RISK FACTOR	75-84 AT EXAM PERSON EXAMS	# OF EVENTS	ANNUAL RATE	85-94 AT EXAM PERSON EXAMS	# OF EVENTS	ANNUAL RATE	35-64 COMBINED # OF EVENTS	CRUDE	AGE-ADJUSTED	65-94 COMBINED # OF EVENTS	CRUDE	AGE-ADJUSTED
NO	1847	80	22	117	9	38	287	5	5	256	16	16
INCOMPLETE	36	2	28	1	1	500	6	7	7	5	16	20
COMPLETE	68	3	22	5	0	0	12	30	23	8	22	23
UNKNOWN	0	0	-	0	0	-	0	-		0	-	
TOTAL	1951	85	22	123	10	41	305	6	6	269	16	16

LOGISTIC REGRESSIONS

AGE AT EXAM	COEFFICIENT OF RISK FACTOR	SE COEFFICIENT	P-VALUE	STANDARDIZED COEFFICIENT	STD SE
35-44	2.049	0.608	0.001	0.301	0.089
45-54	0.775	0.371	0.037	0.138	0.066
55-64	0.586	0.168	0.001	0.149	0.043
65-74	0.177	0.231	0.443	0.051	0.067
75-84	0.040	0.277	0.885	0.016	0.108
85-94	0.067	0.779	0.931	0.027	0.315
35-64 UNIVARIATE	0.801	0.149	0.001	0.167	0.031
65-94 UNIVARIATE	0.153	0.173	0.379	0.048	0.055
35-64 BIVARIATE	0.626	0.149	0.001	0.131	0.031
65-94 BIVARIATE	0.120	0.173	0.488	0.038	0.055
35-64 MULTIVARIATE	0.554	0.161	0.001	0.116	0.034
65-94 MULTIVARIATE	0.108	0.175	0.538	0.034	0.056

TABLE 1-18   DATE=02OCT85

AVERAGE ANNUAL INCIDENCE RATE FOR EVENT PER 1000 PERSONS AT RISK AT EXAM
BY AGE AND LEVEL OF RISK FACTOR AT EXAM: FRAMINGHAM STUDY 30-YEAR FOLLOWUP

SEX: MALE
EVENT: CORONARY HEART DISEASE
RISK FACTOR: DEFINITE NONSPECIFIC T-WAVE OR ST-SEGMENT ABNORMALITY BY ELECTROCARDIOGRAM
PERSONS AT RISK: PERSONS FREE OF CORONARY HEART DISEASE

RISK FACTOR	35-44 AT EXAM PERSON EXAMS	# OF EVENTS	ANNUAL RATE	45-54 AT EXAM PERSON EXAMS	# OF EVENTS	ANNUAL RATE	55-64 AT EXAM PERSON EXAMS	# OF EVENTS	ANNUAL RATE	65-74 AT EXAM PERSON EXAMS	# OF EVENTS	ANNUAL RATE
NO	4362	37	4	7352	152	10	7462	273	18	3472	147	21
YES	110	5	23	372	15	20	587	34	29	441	33	37
UNKNOWN	0	0	-	0	0	-	0	0	-	0	0	-
TOTAL	4472	42	5	7724	167	11	8049	307	19	3913	180	23

RISK FACTOR	75-84 AT EXAM PERSON EXAMS	# OF EVENTS	ANNUAL RATE	85-94 AT EXAM PERSON EXAMS	# OF EVENTS	ANNUAL RATE	35-64 COMBINED # OF EVENTS	CRUDE	AGE-ADJUSTED	65-94 COMBINED # OF EVENTS	CRUDE	AGE-ADJUSTED
NO	830	52	31	28			462	12	12	199	23	23
YES	187	9	24	11			54	25	24	45	35	35
UNKNOWN	0	0	-	0			0	-	-	0	-	-
TOTAL	1017	61	30	39			516	13	13	244	25	25

LOGISTIC REGRESSIONS

AGE AT EXAM		COEFFICIENT OF RISK FACTOR	SE COEFFICIENT	P-VALUE	STANDARDIZED COEFFICIENT	STD SE
35-44		1.717	0.487	0.001	0.266	0.075
45-54		0.688	0.276	0.013	0.147	0.059
55-64		0.482	0.187	0.010	0.125	0.049
65-74		0.604	0.200	0.002	0.191	0.063
75-84		-0.279	0.370	0.451	-0.108	0.144
85-94						
35-64	UNIVARIATE	0.768	0.147	0.001	0.172	0.033
65-94	UNIVARIATE	0.453	0.171	0.008	0.152	0.057
35-64	BIVARIATE	0.580	0.149	0.001	0.130	0.033
65-94	BIVARIATE	0.408	0.172	0.018	0.137	0.058
35-64	MULTIVARIATE	0.242	0.156	0.121	0.054	0.035
65-94	MULTIVARIATE	0.318	0.174	0.068	0.106	0.058

SEX: FEMALE
EVENT: CORONARY HEART DISEASE
RISK FACTOR: DEFINITE NONSPECIFIC T-WAVE OR ST-SEGMENT ABNORMALITY BY ELECTROCARDIOGRAM
PERSONS AT RISK: PERSONS FREE OF CORONARY HEART DISEASE

RISK FACTOR	35-44 AT EXAM PERSON EXAMS	# OF EVENTS	ANNUAL RATE	45-54 AT EXAM PERSON EXAMS	# OF EVENTS	ANNUAL RATE	55-64 AT EXAM PERSON EXAMS	# OF EVENTS	ANNUAL RATE	65-74 AT EXAM PERSON EXAMS	# OF EVENTS	ANNUAL RATE
NO	5517	7	1	9543	66	3	10647	194	9	5528	136	12
YES	99	0	0	416	6	7	866	32	18	720	38	26
UNKNOWN	0	0	-	0	0	-	0	0	-	0	0	-
TOTAL	5616	7	1	9959	72	4	11513	226	10	6248	174	14

RISK FACTOR	75-84 AT EXAM PERSON EXAMS	# OF EVENTS	ANNUAL RATE	85-94 AT EXAM PERSON EXAMS	# OF EVENTS	ANNUAL RATE	35-64 COMBINED # OF EVENTS	CRUDE	AGE-ADJUSTED	65-94 COMBINED # OF EVENTS	CRUDE	AGE-ADJUSTED
NO	1632	64	20	97	8	41	267	5	5	208	14	14
YES	319	21	33	26	2	38	38	14	10	61	29	28
UNKNOWN	0	0	-	0	0	-	0	-	-	0	-	-
TOTAL	1951	85	22	123	10	41	305	6	6	269	16	16

LOGISTIC REGRESSIONS

AGE AT EXAM		COEFFICIENT OF RISK FACTOR	SE COEFFICIENT	P-VALUE	STANDARDIZED COEFFICIENT	STD SE
35-44		-	-	-	-	-
45-54		0.743	0.429	0.084	0.149	0.086
55-64		0.726	0.194	0.001	0.192	0.051
65-74		0.793	0.188	0.001	0.253	0.060
75-84		0.546	0.259	0.035	0.202	0.096
85-94		-0.076	0.823	0.927	-0.031	0.338
35-64	UNIVARIATE	0.992	0.176	0.001	0.218	0.039
65-94	UNIVARIATE	0.722	0.149	0.001	0.241	0.050
35-64	BIVARIATE	0.671	0.177	0.001	0.148	0.039
65-94	BIVARIATE	0.678	0.150	0.001	0.226	0.050
35-44	MULTIVARIATE	0.497	0.182	0.006	0.109	0.040
65-94	MULTIVARIATE	0.620	0.152	0.001	0.207	0.051

TABLE 2-1   DATE=02APR85
AVERAGE ANNUAL INCIDENCE RATE FOR EVENT PER 1000 PERSONS AT RISK AT EXAM
BY AGE AND LEVEL OF RISK FACTOR AT EXAM: FRAMINGHAM STUDY 30-YEAR FOLLOWUP

SEX: MALE
EVENT: CORONARY HEART DISEASE OTHER THAN ANGINA PECTORIS
RISK FACTOR: SYSTOLIC BLOOD PRESSURE. FIRST EXAMINER (MM HG)
PERSONS AT RISK: PERSONS FREE OF CORONARY HEART DISEASE

RISK FACTOR	35-44 AT EXAM PERSON EXAMS	# OF EVENTS	ANNUAL RATE	45-54 AT EXAM PERSON EXAMS	# OF EVENTS	ANNUAL RATE	55-64 AT EXAM PERSON EXAMS	# OF EVENTS	ANNUAL RATE	65-74 AT EXAM PERSON EXAMS	# OF EVENTS	ANNUAL RATE
74-119	1223	6	2	1769	15	4	1440	18	6	548	5	5
120-139	2165	17	4	3417	44	6	3166	60	9	1341	35	13
140-159	841	7	4	1732	37	11	2090	62	15	1194	48	20
160-179	191	4	10	567	20	18	916	34	19	552	27	24
180-300	52	0	0	239	5	10	437	24	27	278	23	41
UNKNOWN	0	0	-	0	0	-	0	0	-	0	0	-
TOTAL	4472	34	4	7724	121	8	8049	198	12	3913	138	18

RISK FACTOR	75-84 AT EXAM PERSON EXAMS	# OF EVENTS	ANNUAL RATE	85-94 AT EXAM PERSON EXAMS	# OF EVENTS	ANNUAL RATE	35-64 COMBINED # OF EVENTS	CRUDE	AGE-ADJUSTED	65-94 COMBINED # OF EVENTS	CRUDE	AGE-ADJUSTED
74-119	120	5	21	7			39	4	4	10	7	8
120-139	345	12	17	15			121	7	7	47	14	14
140-159	315	17	27	11			106	11	11	66	22	22
160-179	153	10	33	4			58	17	17	39	28	28
180-300	84	10	60	2			29	20	15	33	45	45
UNKNOWN	0	0	-	0			0	-		0	-	
TOTAL	1017	54	27	39			353	9	9	195	20	20

LOGISTIC REGRESSIONS

AGE AT EXAM		COEFFICIENT OF RISK FACTOR	SE COEFFICIENT	P-VALUE	STANDARDIZED COEFFICIENT	STD SE
35-44		0.016	0.009	0.068	0.270	0.148
45-54		0.017	0.004	0.001	0.330	0.078
55-64		0.018	0.003	0.001	0.397	0.060
65-74		0.020	0.003	0.001	0.460	0.073
75-84		0.018	0.006	0.002	0.400	0.130
85-94						
35-64	UNIVARIATE	0.020	0.002	0.001	0.407	0.043
65-94	UNIVARIATE	0.020	0.003	0.001	0.450	0.063
35-64	BIVARIATE	0.017	0.002	0.001	0.342	0.044
65-94	BIVARIATE	0.020	0.003	0.001	0.456	0.063
35-64	MULTIVARIATE	0.015	0.002	0.001	0.298	0.048
65-94	MULTIVARIATE	0.016	0.003	0.001	0.375	0.067

SEX: FEMALE
EVENT: CORONARY HEART DISEASE OTHER THAN ANGINA PECTORIS
RISK FACTOR: SYSTOLIC BLOOD PRESSURE. FIRST EXAMINER (MM HG)
PERSONS AT RISK: PERSONS FREE OF CORONARY HEART DISEASE

RISK FACTOR	35-44 AT EXAM PERSON EXAMS	# OF EVENTS	ANNUAL RATE	45-54 AT EXAM PERSON EXAMS	# OF EVENTS	ANNUAL RATE	55-64 AT EXAM PERSON EXAMS	# OF EVENTS	ANNUAL RATE	65-74 AT EXAM PERSON EXAMS	# OF EVENTS	ANNUAL RATE
74-119	2493			2714	4	1	1979	8	2	647	7	5
120-139	2244			3982	12	2	4056	23	3	1747	19	5
140-159	673			2053	13	3	3025	32	5	2013	33	8
160-179	156			768	2	1	1560	22	7	1155	15	6
180-300	50			442	1	1	893	23	13	686	28	20
UNKNOWN	0			0	0	-	0	0	-	0	0	-
TOTAL	5616			9959	32	2	11513	108	5	6248	102	8

RISK FACTOR	75-84 AT EXAM PERSON EXAMS	# OF EVENTS	ANNUAL RATE	85-94 AT EXAM PERSON EXAMS	# OF EVENTS	ANNUAL RATE	35-64 COMBINED # OF EVENTS	CRUDE	AGE-ADJUSTED	65-94 COMBINED # OF EVENTS	CRUDE	AGE-ADJUSTED
74-119	133	3	11	8	0	0	12	1	1	10	6	6
120-139	463	10	11	38	3	39	36	2	2	32	7	7
140-159	626	10	8	29	4	69	45	4	3	47	9	9
160-179	451	19	21	31	1	16	24	5	5	35	11	10
180-300	276	13	24	17	1	29	25	9	8	42	21	21
UNKNOWN	0	0	-	0	0	-	0			0	-	
TOTAL	1951	55	14	123	9	37	142	3	3	166	10	10

LOGISTIC REGRESSIONS

AGE AT EXAM		COEFFICIENT OF RISK FACTOR	SE COEFFICIENT	P-VALUE	STANDARDIZED COEFFICIENT	STD SE
35-44						
45-54		0.011	0.006	0.094	0.246	0.147
55-64		0.019	0.003	0.001	0.457	0.076
65-74		0.015	0.004	0.001	0.374	0.089
75-84		0.011	0.005	0.053	0.268	0.126
85-94		-0.003	0.014	0.856	-0.064	0.352
35-64	UNIVARIATE	0.022	0.002	0.001	0.514	0.059
65-94	UNIVARIATE	0.014	0.003	0.001	0.337	0.071
35-64	BIVARIATE	0.017	0.003	0.001	0.390	0.065
65-94	BIVARIATE	0.013	0.003	0.001	0.312	0.071
35-64	MULTIVARIATE	0.015	0.003	0.001	0.360	0.071
65-94	MULTIVARIATE	0.009	0.003	0.003	0.224	0.074

RISK FACTOR	75-84 AT EXAM PERSON EXAMS	# OF EVENTS	ANNUAL RATE	85-94 AT EXAM PERSON EXAMS	# OF EVENTS	ANNUAL RATE	35-64 COMBINED # OF EVENTS	CRUDE	AGE ADJUSTED	65-94 COMBINED # OF EVENTS	CRUDE	AGE ADJUSTED
20-74	422	17	20	22			51	6	6	55	16	16
75-84	331	18	27	12			101	7	7	57	17	17
85-94	204	12	29	5			108	10	11	42	19	19
95-104	44	4	45	0			57	11	12	28	36	37
105-160	16	3	94	0			36	18	17	13	45	50
UNKNOWN	0	0	-	0			0	-		0	-	
TOTAL	1017	54	27	39			353	9	9	195	20	20

LOGISTIC REGRESSIONS

AGE AT EXAM		COEFFICIENT OF RISK FACTOR	SE COEFFICIENT	P-VALUE	STANDARDIZED COEFFICIENT	STD SE
35-44		0.039	0.013	0.004	0.440	0.151
45-54		0.026	0.007	0.001	0.308	0.084
55-64		0.025	0.005	0.001	0.298	0.066
65-74		0.025	0.007	0.001	0.307	0.079
75-84		0.024	0.012	0.036	0.287	0.137
85-94						
35-64	UNIVARIATE	0.027	0.004	0.001	0.320	0.049
65-94	UNIVARIATE	0.022	0.006	0.001	0.270	0.068
35-64	BIVARIATE	0.027	0.004	0.001	0.315	0.048
65-94	BIVARIATE	0.026	0.006	0.001	0.329	0.069
35-64	MULTIVARIATE	0.024	0.004	0.001	0.288	0.051
65-94	MULTIVARIATE	0.021	0.006	0.001	0.257	0.070

SEX: FEMALE
EVENT: CORONARY HEART DISEASE OTHER THAN ANGINA PECTORIS
RISK FACTOR: DIASTOLIC BLOOD PRESSURE, FIRST EXAMINER (MM HG)
PERSONS AT RISK: PERSONS FREE OF CORONARY HEART DISEASE

RISK FACTOR	35-44 AT EXAM PERSON EXAMS	# OF EVENTS	ANNUAL RATE	45-54 AT EXAM PERSON EXAMS	# OF EVENTS	ANNUAL RATE	55-64 AT EXAM PERSON EXAMS	# OF EVENTS	ANNUAL RATE	65-74 AT EXAM PERSON EXAMS	# OF EVENTS	ANNUAL RATE
20-74	2164			2639	6	1	2969	17	3	1831	30	8
75-84	2030			3541	7	1	3453	28	4	2135	30	7
85-94	987			2361	14	3	2840	31	5	1439	23	8
95-104	324			926	1	1	1272	20	8	600	9	8
105-160	111			492	4	4	979	12	10	243	10	21
UNKNOWN	0			0	0	-	0	0		0	0	-
TOTAL	5616			9959	32	2	11513	108	5	6248	102	8

RISK FACTOR	75-84 AT EXAM PERSON EXAMS	# OF EVENTS	ANNUAL RATE	85-94 AT EXAM PERSON EXAMS	# OF EVENTS	ANNUAL RATE	35-64 COMBINED # OF EVENTS	CRUDE	AGE ADJUSTED	65-94 COMBINED # OF EVENTS	CRUDE	AGE ADJUSTED
20-74	724	21	15	53	4	38	24	2	2	55	11	10
75-84	596	17	14	43	3	35	35	2	2	50	9	9
85-94	443	11	12	20	2	50	45	3	3	36	9	10
95-104	146	4	14	7	0	0	22	4	4	13	9	9
105-160	42	2	24	0	0	-	16	7	6	12	21	21
UNKNOWN	0	0	-	0	0	-	0			0	-	
TOTAL	1951	55	14	123	9	37	142	3	3	166	10	10

LOGISTIC REGRESSIONS

AGE AT EXAM		COEFFICIENT OF RISK FACTOR	SE COEFFICIENT	P-VALUE	STANDARDIZED COEFFICIENT	STD SE
35-44						
45-54		0.023	0.013	0.065	0.286	0.155
55-64		0.031	0.007	0.001	0.383	0.084
65-74		0.013	0.008	0.105	0.156	0.096
75-84		-0.007	0.012	0.521	-0.089	0.139
85-94		0.002	0.029	0.955	0.019	0.349
35-64	UNIVARIATE	0.034	0.006	0.001	0.410	0.070
65-94	UNIVARIATE	0.003	0.006	0.644	0.036	0.078
35-64	BIVARIATE	0.029	0.006	0.001	0.356	0.073
65-94	BIVARIATE	0.006	0.006	0.349	0.073	0.078
35-64	MULTIVARIATE	0.026	0.006	0.001	0.322	0.077
65-94	MULTIVARIATE	0.002	0.006	0.713	0.029	0.077

TABLE 2-3A   DATE=31AUG85
AVERAGE ANNUAL INCIDENCE RATE FOR EVENT PER 1000 PERSONS AT RISK AT EXAM
BY AGE AND LEVEL OF RISK FACTOR AT EXAM: FRAMINGHAM STUDY 30-YEAR FOLLOWUP

SEX: MALE
EVENT: CORONARY HEART DISEASE OTHER THAN ANGINA PECTORIS
RISK FACTOR: HYPERTENSION WITH ANTIHYPERTENSIVE TREATMENT
PERSONS AT RISK: PERSONS FREE OF CORONARY HEART DISEASE

RISK FACTOR	35-44 AT EXAM PERSON EXAMS	# OF EVENTS	ANNUAL RATE	45-54 AT EXAM PERSON EXAMS	# OF EVENTS	ANNUAL RATE	55-64 AT EXAM PERSON EXAMS	# OF EVENTS	ANNUAL RATE	65-74 AT EXAM PERSON EXAMS	# OF EVENTS	ANNUAL RATE
NO	2651	15	3	4122	42	5	3782	56	7	1516	21	7
BORDER	1301	9	3	2205	36	8	2299	74	16	1252	49	20
DEFINITE	500	9	9	1139	32	14	1266	43	17	552	32	29
TREATED	20	1	25	258	11	21	702	25	18	593	36	30
UNKNOWN	0	0	-	0	0	-	0	0	-	0	0	-
TOTAL	4472	34	4	7724	121	8	8049	198	12	3913	138	18

RISK FACTOR	75-84 AT EXAM PERSON EXAMS	# OF EVENTS	ANNUAL RATE	85-94 AT EXAM PERSON EXAMS	# OF EVENTS	ANNUAL RATE	35-64 COMBINED # OF EVENTS	CRUDE	AGE-ADJUSTED	65-94 COMBINED # OF EVENTS	CRUDE	AGE-ADJUSTED
NO	573	12	16	18			113	5	5	33	9	9
BORDER	333	15	23	11			119	10	10	65	20	21
DEFINITE	123	9	37	3			84	14	14	42	31	32
TREATED	188	18	48	7			37	19	21	55	35	34
UNKNOWN	0	0	-	0			0	-	-	0	-	-
TOTAL	1017	54	27	39			353	9	9	195	20	20

LOGISTIC REGRESSIONS

AGE AT EXAM		COEFFICIENT OF RISK FACTOR	SE COEFFICIENT	P-VALUE	STANDARDIZED COEFFICIENT	STD SE
35-44		0.590	0.218	0.007	0.409	0.151
45-54		0.563	0.111	0.001	0.433	0.085
55-64		0.424	0.086	0.001	0.345	0.070
65-74		0.692	0.113	0.001	0.567	0.093
75-84		0.547	0.179	0.002	0.447	0.146
85-94						
35-64	UNIVARIATE	0.544	0.065	0.001	0.423	0.050
65-94	UNIVARIATE	0.665	0.095	0.001	0.544	0.078
35-64	BIVARIATE	0.470	0.065	0.001	0.366	0.051
65-94	BIVARIATE	0.663	0.095	0.001	0.543	0.078
35-64	MULTIVARIATE	0.425	0.068	0.001	0.331	0.053
65-94	MULTIVARIATE	0.591	0.097	0.001	0.484	0.080

SEX: FEMALE
EVENT: CORONARY HEART DISEASE OTHER THAN ANGINA PECTORIS
RISK FACTOR: HYPERTENSION WITH ANTIHYPERTENSIVE TREATMENT
PERSONS AT RISK: PERSONS FREE OF CORONARY HEART DISEASE

RISK FACTOR	35-44 AT EXAM PERSON EXAMS	# OF EVENTS	ANNUAL RATE	45-54 AT EXAM PERSON EXAMS	# OF EVENTS	ANNUAL RATE	55-64 AT EXAM PERSON EXAMS	# OF EVENTS	ANNUAL RATE	65-74 AT EXAM PERSON EXAMS	# OF EVENTS	ANNUAL RATE
NO	4250			5622	13	1	4888	19	2	1767	19	5
BORDER	1001			2579	13	3	3282	30	5	1903	24	6
DEFINITE	316			1166	1	0	1835	33	9	1035	16	8
TREATED	49			592	5	4	1508	26	9	1543	43	14
UNKNOWN	0			0	0	-	0	0	-	0	0	-
TOTAL	5616			9959	32	2	11513	108	5	6248	102	8

RISK FACTOR	75-84 AT EXAM PERSON EXAMS	# OF EVENTS	ANNUAL RATE	85-94 AT EXAM PERSON EXAMS	# OF EVENTS	ANNUAL RATE	35-64 COMBINED # OF EVENTS	CRUDE	AGE-ADJUSTED	65-94 COMBINED # OF EVENTS	CRUDE	AGE-ADJUSTED
NO	402	7	9	31	2	32	53	1	1	28	6	6
BORDER	575	12	10	28	2	36	43	3	3	38	8	7
DEFINITE	371	17	23	21	1	24	35	5	5	34	12	12
TREATED	603	19	16	43	4	47	31	7	5	66	15	15
UNKNOWN	0	0	-	0	0	-	0	-	-	0	-	-
TOTAL	1951	55	14	123	9	37	142	3	3	166	10	10

LOGISTIC REGRESSIONS

AGE AT EXAM		COEFFICIENT OF RISK FACTOR	SE COEFFICIENT	P-VALUE	STANDARDIZED COEFFICIENT	STD SE
35-44						
45-54		0.268	0.216	0.215	0.206	0.166
55-64		0.747	0.125	0.001	0.624	0.104
65-74		0.421	0.132	0.001	0.347	0.109
75-84		0.437	0.198	0.027	0.344	0.156
85-94		0.102	0.423	0.809	0.086	0.355
35-64	UNIVARIATE	0.830	0.102	0.001	0.658	0.081
65-94	UNIVARIATE	0.437	0.106	0.001	0.357	0.087
35-64	BIVARIATE	0.599	0.105	0.001	0.475	0.083
65-94	BIVARIATE	0.397	0.106	0.001	0.325	0.087
35-64	MULTIVARIATE	0.560	0.109	0.001	0.444	0.086
65-94	MULTIVARIATE	0.294	0.108	0.006	0.241	0.088

TABLE 2-3B   DATE=04DEC85
AVERAGE ANNUAL INCIDENCE RATE FOR EVENT PER 1000 PERSONS AT RISK AT EXAM
BY AGE AND LEVEL OF RISK FACTOR AT EXAM: FRAMINGHAM STUDY 30-YEAR FOLLOWUP

SEX: MALE
EVENT: CORONARY HEART DISEASE OTHER THAN ANGINA PECTORIS
RISK FACTOR: HYPERTENSION IGNORING TREATMENT
PERSONS AT RISK: PERSONS FREE OF CORONARY HEART DISEASE

RISK FACTOR	35-44 AT EXAM			45-54 AT EXAM			55-64 AT EXAM			65-74 AT EXAM		
	PERSON EXAMS	# OF EVENTS	ANNUAL RATE	PERSON EXAMS	# OF EVENTS	ANNUAL RATE	PERSON EXAMS	# OF EVENTS	ANNUAL RATE	PERSON EXAMS	# OF EVENTS	ANNUAL RATE
NO	2653	15	3	4154	43	5	3909	60	8	1631	27	8
MILD	1310	9	3	2284	40	9	2583	82	16	1499	65	22
DEFINITE	509	10	10	1286	38	15	1557	56	18	783	46	29
UNKNOWN	0	0	-	0	0	-	0	0	-	0	0	-
TOTAL	4472	34	4	7724	121	8	8049	198	12	3913	138	18

RISK FACTOR	75-84 AT EXAM			85-94 AT EXAM			35-64 COMBINED			65-94 COMBINED		
	PERSON EXAMS	# OF EVENTS	ANNUAL RATE	PERSON EXAMS	# OF EVENTS	ANNUAL RATE	# OF EVENTS	CRUDE	AGE-ADJUSTED	# OF EVENTS	CRUDE	AGE-ADJUSTED
NO	412	16	19	21			118	6	6	43	10	10
MILD	418	19	23	15			131	11	10	86	22	22
DEFINITE	187	19	51	3			104	16	15	66	34	35
UNKNOWN	0	0	-	0			0	-		0	-	
TOTAL	1017	54	27	39			353	9	9	195	20	20

LOGISTIC REGRESSIONS

AGE AT EXAM		COEFFICIENT OF RISK FACTOR	SE COEFFICIENT	P-VALUE	STANDARDIZED COEFFICIENT	STD SE
35-44		0.599	0.218	0.006	0.414	0.151
45-54		0.534	0.112	0.001	0.402	0.085
55-64		0.441	0.089	0.001	0.340	0.069
65-74		0.622	0.114	0.001	0.470	0.086
75-84		0.523	0.187	0.005	0.384	0.137
85-94						
35-64	UNIVARIATE	0.532	0.066	0.001	0.399	0.050
65-94	UNIVARIATE	0.602	0.096	0.001	0.452	0.072
35-64	BIVARIATE	0.478	0.067	0.001	0.359	0.050
65-94	BIVARIATE	0.611	0.096	0.001	0.458	0.072
35-64	MULTIVARIATE	0.422	0.070	0.001	0.316	0.053
65-94	MULTIVARIATE	0.516	0.100	0.001	0.387	0.075

SEX: FEMALE
EVENT: CORONARY HEART DISEASE OTHER THAN ANGINA PECTORIS
RISK FACTOR: HYPERTENSION IGNORING TREATMENT
PERSONS AT RISK: PERSONS FREE OF CORONARY HEART DISEASE

RISK FACTOR	35-44 AT EXAM			45-54 AT EXAM			55-64 AT EXAM			65-74 AT EXAM		
	PERSON EXAMS	# OF EVENTS	ANNUAL RATE	PERSON EXAMS	# OF EVENTS	ANNUAL RATE	PERSON EXAMS	# OF EVENTS	ANNUAL RATE	PERSON EXAMS	# OF EVENTS	ANNUAL RATE
NO	4272			5804	13	1	5245	26	2	2035	21	5
MILD	1016			2763	15	3	3860	42	5	2549	42	8
DEFINITE	328			1392	4	1	2408	40	8	1664	39	12
UNKNOWN	0			0	0	-	0	0	-	0	0	-
TOTAL	5616			9959	32	2	11513	108	5	6248	102	8

RISK FACTOR	75-84 AT EXAM			85-94 AT EXAM			35-64 COMBINED			65-94 COMBINED		
	PERSON EXAMS	# OF EVENTS	ANNUAL RATE	PERSON EXAMS	# OF EVENTS	ANNUAL RATE	# OF EVENTS	CRUDE	AGE-ADJUSTED	# OF EVENTS	CRUDE	AGE-ADJUSTED
NO	507	9	9	37	2	27	40	1	1	32	6	6
MILD	827	18	11	45	5	56	57	4	3	65	10	10
DEFINITE	617	28	23	41	2	24	45	5	4	69	15	14
UNKNOWN	0	0	-	0	0	-	0	-		0	-	
TOTAL	1951	55	14	123	9	37	142	3	3	166	10	10

LOGISTIC REGRESSIONS

AGE AT EXAM		COEFFICIENT OF RISK FACTOR	SE COEFFICIENT	P-VALUE	STANDARDIZED COEFFICIENT	STD SE
35-44						
45-54		0.284	0.226	0.209	0.206	0.164
55-64		0.597	0.121	0.001	0.464	0.094
65-74		0.410	0.132	0.002	0.314	0.101
75-84		0.546	0.193	0.005	0.414	0.146
85-94		-0.055	0.435	0.899	-0.044	0.347
35-64	UNIVARIATE	0.713	0.103	0.001	0.527	0.076
65-94	UNIVARIATE	0.447	0.105	0.001	0.343	0.081
35-64	BIVARIATE	0.504	0.106	0.001	0.373	0.078
65-94	BIVARIATE	0.418	0.106	0.001	0.321	0.081
35-64	MULTIVARIATE	0.452	0.110	0.001	0.335	0.082
65-94	MULTIVARIATE	0.318	0.108	0.003	0.244	0.083

AVERAGE ANNUAL INCIDENCE RATE FOR EVENT PER 1000 PERSONS AT RISK AT EXAM
BY AGE AND LEVEL OF RISK FACTOR AT EXAM: FRAMINGHAM STUDY 30-YEAR FOLLOWUP

SEX: MALE
EVENT: CORONARY HEART DISEASE OTHER THAN ANGINA PECTORIS
RISK FACTOR: SERUM CHOLESTEROL
PERSONS AT RISK: PERSONS FREE OF CORONARY HEART DISEASE

RISK FACTOR	35-44 AT EXAM			45-54 AT EXAM			55-64 AT EXAM			65-74 AT EXAM		
	PERSON EXAMS	# OF EVENTS	ANNUAL RATE	PERSON EXAMS	# OF EVENTS	ANNUAL RATE	PERSON EXAMS	# OF EVENTS	ANNUAL RATE	PERSON EXAMS	# OF EVENTS	ANNUAL RATE
84-204	1253	3	1	1767	19	5	2255	37	8	1234	44	18
205-234	1247	6	2	2306	32	7	2405	66	14	1171	39	17
235-264	872	10	6	1745	24	7	1889	50	13	873	30	17
265-294	475	7	7	1001	20	10	918	23	13	428	16	19
295-1124	291	6	10	557	23	21	454	18	20	191	8	21
UNKNOWN	334	2	3	348	3	4	128	4	16	16	1	31
TOTAL	4472	34	4	7724	121	8	8049	198	12	3913	138	18

RTSK FACTOR	75-84 AT EXAM			85-94 AT EXAM			35-64 COMBINED			65-94 COMBINED		
	PERSON EXAMS	# OF EVENTS	ANNUAL RATE	PERSON EXAMS	# OF EVENTS	ANNUAL RATE	# OF EVENTS	CRUDE	AGE-ADJUSTED	# OF EVENTS	CRUDE	AGE-ADJUSTED
84-204	590	19	24	17			59	6	5	64	20	19
205-234	332	19	29	11			104	9	9	59	19	20
235-264	177	11	31	5			84	9	9	42	20	21
265-294	96	3	16	6			50	10	11	19	18	18
295-1124	22	2	45	0			47	18	18	10	23	26
UNKNOWN	0	0	-	0			9	6	9	1	31	24
TOTAL	1017	54	27	39			353	9	9	195	20	20

LOGISTIC REGRESSIONS

AGE AT EXAM		COEFFICIENT OF RISK FACTOR	SE COEFFICIENT	P-VALUE	STANDARDIZED COEFFICIENT	STD SE
35-44		0.011	0.003	0.001	0.509	0.121
45-54		0.009	0.002	0.001	0.384	0.073
55-64		0.004	0.002	0.008	0.182	0.069
65-74		0.001	0.002	0.791	0.023	0.087
75-84		0.003	0.004	0.391	0.118	0.138
85-94						
35-64	UNIVARIATE	0.007	0.001	0.001	0.295	0.047
65-94	UNIVARIATE	0.001	0.002	0.639	0.034	0.073
35-64	BIVARIATE	0.007	0.001	0.001	0.315	0.048
65-94	BIVARIATE	0.001	0.002	0.419	0.059	0.073
35-64	MULTIVARIATE	0.007	0.001	0.001	0.281	0.048
65-94	MULTIVARIATE	0.001	0.002	0.426	0.058	0.073

SEX: FEMALE
EVENT: CORONARY HEART DISEASE OTHER THAN ANGINA PECTORIS
RISK FACTOR: SERUM CHOLESTEROL
PERSONS AT RISK: PERSONS FREE OF CORONARY HEART DISEASE

RISK FACTOR	35-44 AT EXAM			45-54 AT EXAM			55-64 AT EXAM			65-74 AT EXAM		
	PERSON EXAMS	# OF EVENTS	ANNUAL RATE	PERSON EXAMS	# OF EVENTS	ANNUAL RATE	PERSON EXAMS	# OF EVENTS	ANNUAL RATE	PERSON EXAMS	# OF EVENTS	ANNUAL RATE
84-204	2350			2094	2	0	1625	11	3	789	5	3
205-234	1417			2558	7	1	2615	19	4	1401	19	7
235-264	806			2259	4	1	2938	18	3	1703	27	8
265-294	372			1473	6	2	2240	33	7	1287	22	9
295-1124	171			1101	10	5	1866	27	7	1031	28	14
UNKNOWN	500			474	3	3	229	0	0	37	1	14
TOTAL	5616			9959	32	2	11513	108	5	6248	102	8

RTSK FACTOR	75-84 AT EXAM			85-94 AT EXAM			35-64 COMBINED			65-94 COMBINED		
	PERSON EXAMS	# OF EVENTS	ANNUAL RATE	PERSON EXAMS	# OF EVENTS	ANNUAL RATE	# OF EVENTS	CRUDE	AGE-ADJUSTED	# OF EVENTS	CRUDE	AGE-ADJUSTED
84-204	361	9	12	46	3	33	13	1	1	17	7	8
205-234	486	15	15	28	2	36	26	2	2	56	9	9
235-264	542	17	16	35	2	29	23	3	3	46	10	10
265-294	350	9	13	9	2	111	39	5	4	33	10	11
295-1124	205	5	12	5	0	0	38	6	5	33	13	13
UNKNOWN	7	0	0	0	0	-	3	1	1	1	11	11
TOTAL	1951	55	14	123	9	37	142	3	3	166	10	10

LOGISTIC REGRESSIONS

AGE AT EXAM		COEFFICIENT OF RISK FACTOR	SE COEFFICIENT	P-VALUE	STANDARDIZED COEFFICIENT	STD SE
35-44						
45-54		0.007	0.002	0.001	0.332	0.091
55-64		0.007	0.002	0.001	0.340	0.086
65-74		0.008	0.002	0.001	0.355	0.094
75-84		0.000	0.003	0.958	0.007	0.137
85-94		0.002	0.008	0.812	0.082	0.344
35-64	UNIVARIATE	0.010	0.001	0.001	0.448	0.069
65-94	UNIVARIATE	0.004	0.002	0.019	0.180	0.076
35-64	BIVARIATE	0.007	0.001	0.001	0.349	0.063
65-94	BIVARIATE	0.005	0.002	0.002	0.233	0.077
35-64	MULTIVARIATE	0.007	0.001	0.001	0.311	0.063
65-94	MULTIVARIATE	0.005	0.002	0.002	0.242	0.078

TABLE 2-5   DATE=04DEC85
AVERAGE ANNUAL INCIDENCE RATE FOR EVENT PER 1000 PERSONS AT RISK AT EXAM
BY AGE AND LEVEL OF RISK FACTOR AT EXAM: FRAMINGHAM STUDY 30-YEAR FOLLOWUP

SEX: MALE
EVENT: CORONARY HEART DISEASE OTHER THAN ANGINA PECTORIS
RISK FACTOR: HEMATOCRIT
PERSONS AT RISK: PERSONS FREE OF CORONARY HEART DISEASE

RISK FACTOR	35-44 AT EXAM			45-54 AT EXAM			55-64 AT EXAM			65-74 AT EXAM		
	PERSON EXAMS	# OF EVENTS	ANNUAL RATE	PERSON EXAMS	# OF EVENTS	ANNUAL RATE	PERSON EXAMS	# OF EVENTS	ANNUAL RATE	PERSON EXAMS	# OF EVENTS	ANNUAL RATE
25-42	623	7	6	942	8	4	1018	26	13	450	13	14
43-44	546	0	0	967	14	7	985	20	10	493	16	16
45-46	983	3	2	1854	29	8	1955	47	12	963	31	16
47-48	775	4	3	1510	27	9	1599	34	11	861	25	15
49-70	1175	18	8	2066	39	9	2338	67	14	1127	52	23
UNKNOWN	370	2	5	385	4	5	156	4	13	19	1	26
TOTAL	4472	34	4	7724	121	8	8049	198	12	3915	138	18

RISK FACTOR	75-84 AT EXAM			85-94 AT EXAM			35-64 COMBINED			65-94 COMBINED		
	PERSON EXAMS	# OF EVENTS	ANNUAL RATE	PERSON EXAMS	# OF EVENTS	ANNUAL RATE	# OF EVENTS	CRUDE	AGE-ADJUSTED	# OF EVENTS	CRUDE	AGE-ADJUSTED
25-42	214	14	33	16			41	8	7	28	21	18
43-44	119	9	38	5			34	7	7	25	20	21
45-46	240	10	21	7			79	8	8	42	17	17
47-48	197	10	25	7			65	8	8	35	16	17
49-70	247	11	22	4			124	11	11	64	23	24
UNKNOWN	0	0	-	0			10	5	8	1	26	21
TOTAL	1017	54	27	39			353	9	9	195	20	20

LOGISTIC REGRESSIONS

AGE AT EXAM		COEFFICIENT OF RISK FACTOR	SE COEFFICIENT	P-VALUE	STANDARDIZED COEFFICIENT	STD SE
35-44		0.116	0.049	0.017	0.426	0.179
45-54		0.053	0.026	0.043	0.188	0.095
55-64		0.027	0.019	0.160	0.103	0.073
65-74		0.053	0.024	0.027	0.195	0.088
75-84		-0.021	0.033	0.532	-0.086	0.139
85-94						
35-64	UNIVARIATE	0.045	0.013	0.003	0.164	0.055
65-94	UNIVARIATE	0.023	0.019	0.231	0.089	0.074
35-64	BIVARIATE	0.043	0.015	0.004	0.158	0.055
65-94	BIVARIATE	0.032	0.019	0.096	0.123	0.074
35-64	MULTIVARIATE	0.017	0.015	0.264	0.062	0.055
65-94	MULTIVARIATE	0.024	0.020	0.228	0.091	0.076

SEX: FEMALE
EVENT: CORONARY HEART DISEASE OTHER THAN ANGINA PECTORIS
RISK FACTOR: HEMATOCRIT
PERSONS AT RISK: PERSONS FREE OF CORONARY HEART DISEASE

RISK FACTOR	35-44 AT EXAM			45-54 AT EXAM			55-64 AT EXAM			65-74 AT EXAM		
	PERSON EXAMS	# OF EVENTS	ANNUAL RATE	PERSON EXAMS	# OF EVENTS	ANNUAL RATE	PERSON EXAMS	# OF EVENTS	ANNUAL RATE	PERSON EXAMS	# OF EVENTS	ANNUAL RATE
21-39	1697			1843	3	1	1553	14	5	646	12	9
40-41	967			1741	2	1	2061	18	4	1069	12	6
42-43	1043			1955	6	2	2260	18	4	1256	16	6
44-45	856			2194	8	2	2847	23	4	1609	28	9
46-65	487			1681	10	3	2521	34	7	1626	33	10
UNKNOWN	566			545	3	3	271	1	2	42	1	12
TOTAL	5616			9959	32	2	11513	108	5	6248	102	8

RISK FACTOR	75-84 AT EXAM			85-94 AT EXAM			35-64 COMBINED			65-94 COMBINED		
	PERSON EXAMS	# OF EVENTS	ANNUAL RATE	PERSON EXAMS	# OF EVENTS	ANNUAL RATE	# OF EVENTS	CRUDE	AGE-ADJUSTED	# OF EVENTS	CRUDE	AGE-ADJUSTED
21-39	297	10	17	24	3	63	18	2	2	25	13	12
40-41	373	13	17	31	2	32	20	2	2	27	9	9
42-43	353	6	8	19	3	79	24	2	2	25	8	9
44-45	486	9	9	26	1	19	32	3	3	38	9	9
46-65	436	17	19	23	0	0	44	5	4	50	12	12
UNKNOWN	6	0	0	0	0	-	4	1	2	1	10	9
TOTAL	1951	55	14	123	9	37	142	3	3	166	10	10

LOGISTIC REGRESSIONS

AGE AT EXAM		COEFFICIENT OF RISK FACTOR	SE COEFFICIENT	P-VALUE	STANDARDIZED COEFFICIENT	STD SE
35-44						
45-54		0.135	0.053	0.011	0.480	0.190
55-64		0.056	0.028	0.049	0.193	0.098
65-74		0.049	0.029	0.093	0.171	0.102
75-84		-0.009	0.035	0.806	-0.033	0.136
85-94		-0.158	0.092	0.085	-0.631	0.366
35-64	UNIVARIATE	0.102	0.024	0.001	0.367	0.087
65-94	UNIVARIATE	0.007	0.022	0.748	0.025	0.079
35-64	BIVARIATE	0.065	0.025	0.009	0.233	0.089
65-94	BIVARIATE	0.014	0.022	0.517	0.051	0.078
35-64	MULTIVARIATE	0.018	0.024	0.452	0.066	0.088
65-94	MULTIVARIATE	0.010	0.022	0.656	0.035	0.078

51

TABLE 2-6   DATE=04DEC85
AVERAGE ANNUAL INCIDENCE RATE FOR EVENT PER 1000 PERSONS AT RISK AT EXAM
BY AGE AND LEVEL OF RISK FACTOR AT EXAM: FRAMINGHAM STUDY 30-YEAR FOLLOWUP

SEX: MALE
EVENT: CORONARY HEART DISEASE OTHER THAN ANGINA PECTORIS
RISK FACTOR: BLOOD GLUCOSE (MG/DL)
PERSONS AT RISK: PERSONS FREE OF CORONARY HEART DISEASE

RISK FACTOR	35-44 AT EXAM			45-54 AT EXAM			55-64 AT EXAM			65-74 AT EXAM		
	PERSON EXAMS	# OF EVENTS	ANNUAL RATE	PERSON EXAMS	# OF EVENTS	ANNUAL RATE	PERSON EXAMS	# OF EVENTS	ANNUAL RATE	PERSON EXAMS	# OF EVENTS	ANNUAL RATE
29-69	992	7	4	1490	24	8	1087	23	11	421	7	8
70-89	2501	19	4	4172	50	6	4056	103	13	1761	61	17
90-109	626	5	4	1348	24	9	1866	46	12	1046	36	17
110-129	105	1	5	305	11	18	537	14	13	335	15	22
130-524	48	0	0	193	9	23	408	9	11	334	18	27
UNKNOWN	200	2	5	216	3	7	95	3	16	16	1	31
TOTAL	4472	34	4	7724	121	8	8049	198	12	3913	138	18

RISK FACTOR	75-84 AT EXAM			85-94 AT EXAM			35-64 COMBINED			65-94 COMBINED		
	PERSON EXAMS	# OF EVENTS	ANNUAL RATE	PERSON EXAMS	# OF EVENTS	ANNUAL RATE	# OF EVENTS	CRUDE	AGE-ADJUSTED	# OF EVENTS	CRUDE	AGE-ADJUSTED
29-69	62	4	32	1			54	8	8	11	11	13
70-89	377	21	28	11			172	8	8	82	19	19
90-109	328	18	27	14			75	10	9	55	20	19
110-129	125	5	20	4			26	14	13	20	22	22
130-524	125	6	24	9			18	14	13	26	28	27
UNKNOWN	0	0	-	0			8	8	10	1	31	25
TOTAL	1017	54	27	39			353	9	9	195	20	20

LOGISTIC REGRESSIONS

AGE AT EXAM		COEFFICIENT OF RISK FACTOR	SE COEFFICIENT	P-VALUE	STANDARDIZED COEFFICIENT	STD SE
35-44		-0.001	0.011	0.931	-0.016	0.185
45-54		0.008	0.002	0.001	0.185	0.052
55-64		-0.000	0.003	0.979	-0.002	0.073
65-74		0.006	0.002	0.001	0.216	0.057
75-84		0.000	0.004	0.993	0.001	0.140
85-94						
35-64	UNIVARIATE	0.005	0.002	0.002	0.121	0.039
65-94	UNIVARIATE	0.006	0.002	0.001	0.192	0.052
35-64	BIVARIATE	0.003	0.002	0.100	0.071	0.043
65-94	BIVARIATE	0.005	0.002	0.001	0.183	0.053
35-64	MULTIVARIATE	0.003	0.002	0.141	0.063	0.043
65-94	MULTIVARIATE	0.005	0.002	0.004	0.157	0.055

SEX: FEMALE
EVENT: CORONARY HEART DISEASE OTHER THAN ANGINA PECTORIS
RISK FACTOR: BLOOD GLUCOSE (MG/DL)
PERSONS AT RISK: PERSONS FREE OF CORONARY HEART DISEASE

RISK FACTOR	35-44 AT EXAM			45-54 AT EXAM			55-64 AT EXAM			65-74 AT EXAM		
	PERSON EXAMS	# OF EVENTS	ANNUAL RATE	PERSON EXAMS	# OF EVENTS	ANNUAL RATE	PERSON EXAMS	# OF EVENTS	ANNUAL RATE	PERSON EXAMS	# OF EVENTS	ANNUAL RATE
29-69	1066			1797	5	1	1369	17	6	521	9	9
70-89	3412			5744	15	1	6098	46	4	3102	44	7
90-109	672			1625	2	1	2712	23	4	1732	31	9
110-129	96			301	1	2	730	11	8	501	3	3
130-524	39			159	6	19	438	10	11	360	14	19
UNKNOWN	331			333	3	5	166	1	3	32	1	16
TOTAL	5616			9959	32	1	11513	108	5	6248	102	8

RISK FACTOR	75-84 AT EXAM			85-94 AT EXAM			35-64 COMBINED			65-94 COMBINED		
	PERSON EXAMS	# OF EVENTS	ANNUAL RATE	PERSON EXAMS	# OF EVENTS	ANNUAL RATE	# OF EVENTS	CRUDE	AGE-ADJUSTED	# OF EVENTS	CRUDE	AGE-ADJUSTED
29-69	105	5	24	3	0	0	22	3	3	14	11	12
70-89	823	20	12	44	3	34	63	2	2	67	8	9
90-109	640	10	8	51	3	29	25	2	4	44	9	9
110-129	219	10	23	13	1	38	12	5	4	14	10	8
130-524	163	10	31	12	2	83	16	13	12	26	24	23
UNKNOWN	1	0	0	0	0	-	4	2	3	1	15	12
TOTAL	1951	55	14	123	9	37	142	3	3	166	10	10

LOGISTIC REGRESSIONS

AGE AT EXAM		COEFFICIENT OF RISK FACTOR	SE COEFFICIENT	P-VALUE	STANDARDIZED COEFFICIENT	STD SE
35-44						
45-54		0.015	0.003	0.001	0.290	0.069
55-64		0.009	0.002	0.001	0.234	0.050
65-74		0.008	0.002	0.001	0.239	0.059
75-84		0.009	0.002	0.001	0.294	0.074
85-94		0.002	0.008	0.808	0.075	0.308
35-64	UNIVARIATE	0.012	0.002	0.001	0.269	0.035
65-94	UNIVARIATE	0.009	0.001	0.001	0.265	0.044
35-64	BIVARIATE	0.010	0.002	0.001	0.214	0.039
65-94	BIVARIATE	0.008	0.002	0.001	0.246	0.045
35-64	MULTIVARIATE	0.008	0.002	0.001	0.175	0.039
65-94	MULTIVARIATE	0.008	0.002	0.001	0.229	0.046

AVERAGE ANNUAL INCIDENCE RATE FOR EVENT PER 1000 PERSONS AT RISK AT EXAM
BY AGE AND LEVEL OF RISK FACTOR AT EXAM: FRAMINGHAM STUDY 30-YEAR FOLLOWUP

SEX: MALE
EVENT: CORONARY HEART DISEASE OTHER THAN ANGINA PECTORIS
RISK FACTOR: DIABETES MELLITUS
PERSONS AT RISK: PERSONS FREE OF CORONARY HEART DISEASE

RISK FACTOR	35-44 AT EXAM PERSON EXAMS	# OF EVENTS	ANNUAL RATE	45-54 AT EXAM PERSON EXAMS	# OF EVENTS	ANNUAL RATE	55-64 AT EXAM PERSON EXAMS	# OF EVENTS	ANNUAL RATE	65-74 AT EXAM PERSON EXAMS	# OF EVENTS	ANNUAL RATE
NO	4424	34	4	7486	114	8	7549	182	12	3504	113	16
YES	48	0	0	238	7	15	500	16	16	409	25	31
UNKNOWN	0	0	-				0	0	-	0	0	-
TOTAL	4472	34	4	7724	121	8	8049	198	12	3913	138	18

RISK FACTOR	75-84 AT EXAM PERSON EXAMS	# OF EVENTS	ANNUAL RATE	85-94 AT EXAM PERSON EXAMS	# OF EVENTS	ANNUAL RATE	35-64 COMBINED # OF EVENTS	CRUDE	AGE-ADJUSTED	65-94 COMBINED # OF EVENTS	CRUDE	AGE-ADJUSTE
NO	900	49	27	34			330	8	9	164	18	19
YES	117	5	21	5			23	15	12	31	29	29
UNKNOWN	0	0	-	0			0	-		0	-	
TOTAL	1017	54	27	39			353	9	9	195	20	20

LOGISTIC REGRESSIONS

AGE AT EXAM		COEFFICIENT OF RISK FACTOR	SE COEFFICIENT	P-VALUE	STANDARDIZED COEFFICIENT	STD SE
35-44		-	-	-	-	-
45-54		0.673	0.395	0.089	0.116	0.068
55-64		0.291	0.265	0.272	0.070	0.064
65-74		0.670	0.227	0.003	0.205	0.070
75-84		-0.254	0.480	0.596	-0.081	0.153
85-94						
35-64	UNIVARIATE	0.558	0.219	0.011	0.108	0.042
65-94	UNIVARIATE	0.480	0.201	0.017	0.148	0.062
35-64	BIVARIATE	0.315	0.221	0.154	0.061	0.043
65-94	BIVARIATE	0.460	0.202	0.023	0.142	0.062
35-64	MULTIVARIATE	0.259	0.225	0.250	0.050	0.044
65-94	MULTIVARIATE	0.428	0.206	0.038	0.132	0.064

SEX: FEMALE
EVENT: CORONARY HEART DISEASE OTHER THAN ANGINA PECTORIS
RISK FACTOR: DIABETES MELLITUS
PERSONS AT RISK: PERSONS FREE OF CORONARY HEART DISEASE

RISK FACTOR	35-44 AT EXAM PERSON EXAMS	# OF EVENTS	ANNUAL RATE	45-54 AT EXAM PERSON EXAMS	# OF EVENTS	ANNUAL RATE	55-64 AT EXAM PERSON EXAMS	# OF EVENTS	ANNUAL RATE	65-74 AT EXAM PERSON EXAMS	# OF EVENTS	ANNUAL RATE
NO	5573			9760	27	1	10985	94	4	5780	84	7
YES	43			199	5	13	528	14	13	468	18	19
UNKNOWN	0			0	0	-	0	0	-	0	0	-
TOTAL	5616			9959	32	2	11513	108	5	6248	102	8

RISK FACTOR	75-84 AT EXAM PERSON EXAMS	# OF EVENTS	ANNUAL RATE	85-94 AT EXAM PERSON EXAMS	# OF EVENTS	ANNUAL RATE	35-64 COMBINED # OF EVENTS	CRUDE	AGE-ADJUSTED	65-94 COMBINED # OF EVENTS	CRUDE	AGE-ADJUSTE
NO	1738	42	12	106	7	33	122	2	2	133	9	9
YES	213	13	31	17	2	59	20	13	13	33	24	22
UNKNOWN	0	0	-	0	9	37	0	-		0	-	
TOTAL	1951	55	14	123	9	37	142	3	3	166	10	10

LOGISTIC REGRESSIONS

AGE AT EXAM		COEFFICIENT OF RISK FACTOR	SE COEFFICIENT	P-VALUE	STANDARDIZED COEFFICIENT	STD SE
35-44						
45-54		2.229	0.492	0.001	0.312	0.069
55-64		1.149	0.290	0.001	0.240	0.061
65-74		0.998	0.264	0.001	0.263	0.070
75-84		0.965	0.326	0.003	0.301	0.102
85-94		0.634	0.848	0.455	0.220	0.294
35-64	UNIVARIATE	1.745	0.244	0.001	0.290	0.041
65-94	UNIVARIATE	1.028	0.199	0.001	0.285	0.055
35-64	BIVARIATE	1.353	0.247	0.001	0.225	0.041
65-94	BIVARIATE	0.954	0.200	0.001	0.264	0.055
35-64	MULTIVARIATE	1.082	0.253	0.001	0.180	0.042
65-94	MULTIVARIATE	0.817	0.207	0.001	0.227	0.057

TABLE 2-5   DATE=04DEC85
AVERAGE ANNUAL INCIDENCE RATE FOR EVENT PER 1000 PERSONS AT RISK AT EXAM
BY AGE AND LEVEL OF RISK FACTOR AT EXAM: FRAMINGHAM STUDY 30-YEAR FOLLOWUP

SEX: MALE
EVENT: CORONARY HEART DISEASE OTHER THAN ANGINA PECTORIS
RISK FACTOR: GLUCOSE IN URINE, DEFINITE OR TRACE
PERSONS AT RISK: PERSONS FREE OF CORONARY HEART DISEASE

RISK FACTOR	35-44 AT EXAM			45-54 AT EXAM			55-64 AT EXAM			65-74 AT EXAM		
	PERSON EXAMS	# OF EVENTS	ANNUAL RATE	PERSON EXAMS	# OF EVENTS	ANNUAL RATE	PERSON EXAMS	# OF EVENTS	ANNUAL RATE	PERSON EXAMS	# OF EVENTS	ANNUAL RATE
NO	4380	34	4	7535	113	7	7815	183	12	3767	122	16
YES	60	0	0	168	8	24	210	14	33	140	16	57
UNKNOWN	32	0	0	21	0	0	24	1	21	6	0	0
TOTAL	4472	34	4	7724	121	8	8049	198	12	3913	138	18

RISK FACTOR	75-84 AT EXAM			85-94 AT EXAM			35-64 COMBINED			65-94 COMBINED		
	PERSON EXAMS	# OF EVENTS	ANNUAL RATE	PERSON EXAMS	# OF EVENTS	ANNUAL RATE	# OF EVENTS	CRUDE	AGE-ADJUSTED	# OF EVENTS	CRUDE	AGE-ADJUSTED
NO	986	51	26	38			330	8	8	175	18	18
YES	28	3	54	1			22	25	22	20	59	60
UNKNOWN	3	0	0	0			1	6	8	0	0	0
TOTAL	1017	54	27	39			353	9	9	195	20	20

LOGISTIC REGRESSIONS

AGE AT EXAM	COEFFICIENT OF RISK FACTOR	SE COEFFICIENT	P-VALUE	STANDARDIZED COEFFICIENT	STD SE
35-44	-	-	-	-	-
45-54	1.189	0.374	0.001	0.174	0.055
55-64	1.092	0.287	0.001	0.174	0.046
65-74	1.349	0.281	0.001	0.251	0.052
75-84	0.788	0.628	0.209	0.129	0.103
85-94					
35-64 UNIVARIATE	1.134	0.226	0.001	0.165	0.033
65-94 UNIVARIATE	1.264	0.250	0.001	0.229	0.045
35-64 BIVARIATE	1.031	0.227	0.001	0.150	0.033
65-94 BIVARIATE	1.296	0.251	0.001	0.235	0.046
35-64 MULTIVARIATE	0.969	0.236	0.001	0.141	0.034
65-94 MULTIVARIATE	1.180	0.259	0.001	0.214	0.047

SEX: FEMALE
EVENT: CORONARY HEART DISEASE OTHER THAN ANGINA PECTORIS
RISK FACTOR: GLUCOSE IN URINE, DEFINITE OR TRACE
PERSONS AT RISK: PERSONS FREE OF CORONARY HEART DISEASE

RISK FACTOR	35-44 AT EXAM			45-54 AT EXAM			55-64 AT EXAM			65-74 AT EXAM		
	PERSON EXAMS	# OF EVENTS	ANNUAL RATE	PERSON EXAMS	# OF EVENTS	ANNUAL RATE	PERSON EXAMS	# OF EVENTS	ANNUAL RATE	PERSON EXAMS	# OF EVENTS	ANNUAL RATE
NO	5537			9841	30	2	11346	102	4	6131	96	8
YES	41			94	2	11	144	6	21	107	6	28
UNKNOWN	38			24	0	0	23	0	0	10	0	0
TOTAL	5616			9959	32	2	11513	108	5	6248	102	8

RISK FACTOR	75-84 AT EXAM			85-94 AT EXAM			35-64 COMBINED			65-94 COMBINED		
	PERSON EXAMS	# OF EVENTS	ANNUAL RATE	PERSON EXAMS	# OF EVENTS	ANNUAL RATE	# OF EVENTS	CRUDE	AGE-ADJUSTED	# OF EVENTS	CRUDE	AGE-ADJUSTED
NO	1919	52	14	120	8	33	133	2	2	156	10	10
YES	23	3	65	3	1	167	9	16	15	10	38	39
UNKNOWN	9	0	0	0			0	0	0	0	0	0
TOTAL	1951	55	14	123	9	37	142	3	3	166	10	10

LOGISTIC REGRESSIONS

AGE AT EXAM	COEFFICIENT OF RISK FACTOR	SE COEFFICIENT	P-VALUE	STANDARDIZED COEFFICIENT	STD SE
35-44					
45-54	1.961	0.738	0.008	0.190	0.071
55-64	1.567	0.429	0.001	0.174	0.048
65-74	1.318	0.433	0.002	0.171	0.056
75-84	1.684	0.635	0.008	0.182	0.069
85-94	1.946	1.278	0.128	0.301	0.198
35-64 UNIVARIATE	1.897	0.350	0.001	0.192	0.035
65-94 UNIVARIATE	1.429	0.339	0.001	0.179	0.043
35-64 BIVARIATE	1.773	0.353	0.001	0.179	0.036
65-94 BIVARIATE	1.467	0.340	0.001	0.184	0.043
35-64 MULTIVARIATE	1.408	0.364	0.001	0.142	0.037
65-94 MULTIVARIATE	1.246	0.353	0.001	0.156	0.044

TABLE 2-9   DATE=01MAR86

AVERAGE ANNUAL INCIDENCE RATE FOR EVENT PER 1000 PERSONS AT RISK AT EXAM
BY AGE AND LEVEL OF RISK FACTOR AT EXAM: FRAMINGHAM STUDY 30-YEAR FOLLOWUP

SEX: MALE
EVENT: CORONARY HEART DISEASE OTHER THAN ANGINA PECTORIS
RISK FACTOR: GLUCOSE INTOLERANCE
PERSONS AT RISK: PERSONS FREE OF CORONARY HEART DISEASE

RISK FACTOR	35-44 AT EXAM PERSON EXAMS	# OF EVENTS	ANNUAL RATE	45-54 AT EXAM PERSON EXAMS	# OF EVENTS	ANNUAL RATE	55-64 AT EXAM PERSON EXAMS	# OF EVENTS	ANNUAL RATE	65-74 AT EXAM PERSON EXAMS	# OF EVENTS	ANNUAL RATE
NO	4335	33	4	7221	102	7	7145	173	12	3228	102	16
YES	137	1	4	503	19	19	904	25	14	685	36	26
UNKNOWN	0	0	.	0	0	-	0	0	-	0	0	-
TOTAL	4472	34	4	7724	121	8	8049	198	12	3913	138	18

RISK FACTOR	75-84 AT EXAM PERSON EXAMS	# OF EVENTS	ANNUAL RATE	85-94 AT EXAM PERSON EXAMS	# OF EVENTS	ANNUAL RATE	35-64 COMBINED # OF EVENTS	CRUDE	AGE-ADJUSTED	65-94 COMBINED # OF EVENTS	CRUDE	AGE-ADJUSTED
NO	802	45	28	27			308	8	8	148	18	18
YES	215	9	21	12			45	15	14	47	26	26
UNKNOWN	0	0	-	0			0	-		0	-	
TOTAL	1017	54	27	39			353	9	9	195	20	20

LOGISTIC REGRESSIONS

AGE AT EXAM		COEFFICIENT OF RISK FACTOR	SE COEFFICIENT	P-VALUE	STANDARDIZED COEFFICIENT	STD SE
35-44		-0.042	1.019	0.967	-0.007	0.176
45-54		1.008	0.254	0.001	0.249	0.063
55-64		0.136	0.217	0.529	0.043	0.069
65-74		0.531	0.199	0.008	0.202	0.075
75-84		-0.308	0.374	0.410	-0.126	0.153
85-94						
35-64	UNIVARIATE	0.584	0.162	0.001	0.155	0.043
65-94	UNIVARIATE	0.361	0.172	0.035	0.140	0.066
35-64	BIVARIATE	0.383	0.163	0.019	0.102	0.043
65-94	BIVARIATE	0.324	0.172	0.060	0.125	0.067
35-64	MULTIVARIATE	0.342	0.168	0.042	0.091	0.045
65-94	MULTIVARIATE	-0.272	0.175	0.119	0.105	0.068

SEX: FEMALE
EVENT: CORONARY HEART DISEASE OTHER THAN ANGINA PECTORIS
RISK FACTOR: GLUCOSE INTOLERANCE
PERSONS AT RISK: PERSONS FREE OF CORONARY HEART DISEASE

RISK FACTOR	35-44 AT EXAM PERSON EXAMS	# OF EVENTS	ANNUAL RATE	45-54 AT EXAM PERSON EXAMS	# OF EVENTS	ANNUAL RATE	55-64 AT EXAM PERSON EXAMS	# OF EVENTS	ANNUAL RATE	65-74 AT EXAM PERSON EXAMS	# OF EVENTS	ANNUAL RATE
NO	5522			9588	25	1	10503	89	4	5398	79	7
YES	94			371	7	9	1010	19	9	850	23	14
UNKNOWN	0			0	0	-	0	0	-	0	0	-
TOTAL	5616			9959	32	2	11513	108	5	6248	102	8

RISK FACTOR	75-84 AT EXAM PERSON EXAMS	# OF EVENTS	ANNUAL RATE	85-94 AT EXAM PERSON EXAMS	# OF EVENTS	ANNUAL RATE	35-64 COMBINED # OF EVENTS	CRUDE	AGE-ADJUSTED	65-94 COMBINED # OF EVENTS	CRUDE	AGE-ADJUSTED
NO	1581	38	12	100	5	25	115	2	2	122	9	9
YES	370	17	23	23	4	87	27	9	9	44	18	17
UNKNOWN	0	0	-	0	0	-	0	-		0	-	
TOTAL	1951	55	14	123	9	37	142	3	3	166	10	10

LOGISTIC REGRESSIONS

AGE AT EXAM	COEFFICIENT OF RISK FACTOR	SE COEFFICIENT	P-VALUE	STANDARDIZED COEFFICIENT	STD SE
35-44					
45-54	1.996	0.431	0.001	0.378	0.082
55-64	0.808	0.255	0.002	0.229	0.072
65-74	0.627	0.240	0.009	0.215	0.082
75-84	0.671	0.298	0.024	0.263	0.117

SEX: MALE
EVENT: CORONARY HEART DISEASE OTHER THAN ANGINA PECTORIS
RISK FACTOR: METROPOLITAN RELATIVE WEIGHT (PERCENT)
PERSONS AT RISK: PERSONS FREE OF CORONARY HEART DISEASE

RISK FACTOR	35-44 AT EXAM			45-54 AT EXAM			55-64 AT EXAM			65-74 AT EXAM		
	PERSON EXAMS	# OF EVENTS	ANNUAL RATE	PERSON EXAMS	# OF EVENTS	ANNUAL RATE	PERSON EXAMS	# OF EVENTS	ANNUAL RATE	PERSON EXAMS	# OF EVENTS	ANNUAL RATE
54-104	723	2	1	1131	9	4	1155	32	14	686	17	12
105-114	965	7	4	1540	25	8	1655	32	10	851	24	14
115-124	1122	7	3	2037	33	8	2175	49	11	1069	42	20
125-134	915	10	5	1645	27	8	1674	39	12	766	27	18
135-272	747	8	5	1370	27	10	1389	46	17	541	28	26
UNKNOWN	0	0	-	1	0	0	1	0	0	0	0	-
TOTAL	4472	34	4	7724	121	8	8049	198	12	3913	138	18

RISK FACTOR	75-84 AT EXAM			85-94 AT EXAM			35-64 COMBINED			65-94 COMBINED		
	PERSON EXAMS	# OF EVENTS	ANNUAL RATE	PERSON EXAMS	# OF EVENTS	ANNUAL RATE	# OF EVENTS	CRUDE	AGE-ADJUSTED	# OF EVENTS	CRUDE	AGE-ADJUSTED
54-104	224	13	29	18			43	7	7	30	16	16
105-114	252	13	26	9			64	8	8	39	18	17
115-124	296	11	19	12			89	8	8	54	20	20
125-134	155	9	29	0			76	9	9	36	20	20
135-272	90	8	44	0			81	12	12	36	29	29
UNKNOWN	0	0	-	0			0	0	0	0	-	
TOTAL	1017	54	27	39			353	9	9	195	20	20

LOGISTIC REGRESSIONS

AGE AT EXAM		COEFFICIENT OF RISK FACTOR	SE COEFFICIENT	P-VALUE	STANDARDIZED COEFFICIENT	STD SE
35-44		0.016	0.010	0.116	0.256	0.163
45-54		0.010	0.005	0.078	0.156	0.088
55-64		0.005	0.004	0.258	0.080	0.071
65-74		0.013	0.005	0.013	0.208	0.084
75-84		0.006	0.009	0.491	0.095	0.137
85-94						
35-64	UNIVARIATE	0.008	0.003	0.014	0.129	0.052
65-94	UNIVARIATE	0.010	0.005	0.026	0.159	0.071
35-64	BIVARIATE	0.008	0.003	0.016	0.127	0.053
65-94	BIVARIATE	0.012	0.005	0.008	0.191	0.072
35-64	MULTIVARIATE	0.005	0.003	0.138	0.081	0.055
65-94	MULTIVARIATE	0.007	0.005	0.141	0.109	0.074

SEX: FEMALE
EVENT: CORONARY HEART DISEASE OTHER THAN ANGINA PECTORIS
RISK FACTOR: METROPOLITAN RELATIVE WEIGHT (PERCENT)
PERSONS AT RISK: PERSONS FREE OF CORONARY HEART DISEASE

RISK FACTOR	35-44 AT EXAM			45-54 AT EXAM			55-64 AT EXAM			65-74 AT EXAM		
	PERSON EXAMS	# OF EVENTS	ANNUAL RATE	PERSON EXAMS	# OF EVENTS	ANNUAL RATE	PERSON EXAMS	# OF EVENTS	ANNUAL RATE	PERSON EXAMS	# OF EVENTS	ANNUAL RATE
54-104	1694			2095	5	1	2045	16	4	1216	19	8
105-114	1430			2331	7	2	2409	18	4	1145	19	8
115-124	1058			2107	4	1	2479	24	5	1395	16	6
125-134	566			1372	3	1	1743	14	4	982	17	9
135-272	860			2054	13	3	2837	36	6	1510	31	10
UNKNOWN	8			0	0	-	0	0	-	0	0	-
TOTAL	5616			9959	32	2	11513	108	5	6248	102	8

RISK FACTOR	75-84 AT EXAM			85-94 AT EXAM			35-64 COMBINED			65-94 COMBINED		
	PERSON EXAMS	# OF EVENTS	ANNUAL RATE	PERSON EXAMS	# OF EVENTS	ANNUAL RATE	# OF EVENTS	CRUDE	AGE-ADJUSTED	# OF EVENTS	CRUDE	AGE-ADJUSTED
54-104	517	10	10	41	4	49	22	2	2	33	9	9
105-114	375	7	9	28	0	0	25	2	2	26	8	8
115-124	399	18	23	21	3	71	28	2	2	37	10	11
125-134	266	10	19	16	1	31	17	2	2	28	11	11
135-272	394	10	13	17	1	29	50	4	4	42	11	11
UNKNOWN	0	0	-	0	0	-	0	0	0	0	-	
TOTAL	1951	55	14	123	9	37	142	3	3	166	10	10

LOGISTIC REGRESSIONS

AGE AT EXAM		COEFFICIENT OF RISK FACTOR	SE COEFFICIENT	P-VALUE	STANDARDIZED COEFFICIENT	STD SE
35-44						
45-54		0.017	0.006	0.004	0.370	0.138
55-64		0.007	0.004	0.066	0.161	0.087
65-74		0.008	0.004	0.045	0.182	0.090
75-84		0.007	0.006	0.259	0.147	0.130
85-94		-0.007	0.018	0.718	-0.128	0.354
35-64	UNIVARIATE	0.013	0.003	0.001	0.276	0.069
65-94	UNIVARIATE	0.006	0.003	0.096	0.123	0.074
35-64	BIVARIATE	0.010	0.003	0.003	0.212	0.072
65-94	BIVARIATE	0.007	0.003	0.029	0.161	0.074
35-64	MULTIVARIATE	0.006	0.004	0.095	0.131	0.078
65-94	MULTIVARIATE	0.005	0.003	0.163	0.106	0.076

TABLE 2-11   DATE=04DEC85
AVERAGE ANNUAL INCIDENCE RATE FOR EVENT PER 1000 PERSONS AT RISK AT EXAM
BY AGE AND LEVEL OF RISK FACTOR AT EXAM: FRAMINGHAM STUDY 30-YEAR FOLLOWUP

SEX: MALE
EVENT: CORONARY HEART DISEASE OTHER THAN ANGINA PECTORIS
RISK FACTOR: VITAL CAPACITY - HEIGHT INDEX (ML/INCH)
PERSONS AT RISK: PERSONS FREE OF CORONARY HEART DISEASE

RISK FACTOR	35-44 AT EXAM PERSON EXAMS	# OF EVENTS	ANNUAL RATE	45-54 AT EXAM PERSON EXAMS	# OF EVENTS	ANNUAL RATE	55-64 AT EXAM PERSON EXAMS	# OF EVENTS	ANNUAL RATE	65-74 AT EXAM PERSON EXAMS	# OF EVENTS	ANNUAL RATE
8-44	269	2	4	1062	23	11	2045	64	16	1752	64	18
45-49	411	2	2	1125	21	9	1480	32	11	749	36	24
50-54	747	0	2	1658	32	10	1835	45	12	712	21	15
55-59	1022	9	4	1739	21	6	1416	31	11	442	8	9
60-85	2010	21	5	2135	24	6	1255	26	10	246	8	16
UNKNOWN	13	0		5	0		18	0	0	12	1	42
TOTAL	4472	34	4	7724	121	8	8049	198	12	3913	138	18

RISK FACTOR	75-84 AT EXAM PERSON EXAMS	# OF EVENTS	ANNUAL RATE	85-94 AT EXAM PERSON EXAMS	# OF EVENTS	ANNUAL RATE	35-64 COMBINED # OF EVENTS	CRUDE	AGE-ADJUSTED	65-94 COMBINED # OF EVENTS	CRUDE	AGE-ADJUSTED
8-44	649	42	32	24			89	13	14	108	22	21
45-49	172	2	6	6			55	9	8	38	20	20
50-54	97	5	26	7			77	9	9	27	17	17
55-59	65	5	38	2			61	7	8	13	13	15
60-85	26	0	0	0			71	7	7	8	15	13
UNKNOWN	8	0	0	0			0	0	0	1	25	33
TOTAL	1017	54	27	39			353	9	9	195	20	20

LOGISTIC REGRESSIONS

AGE AT EXAM		COEFFICIENT OF RISK FACTOR	SE COEFFICIENT	P-VALUE	STANDARDIZED COEFFICIENT	STD SE
35-44		0.014	0.020	0.487	0.121	0.175
45-54		-0.024	0.010	0.016	-0.215	0.089
55-64		-0.015	0.008	0.042	-0.144	0.071
65-74		-0.015	0.009	0.110	-0.137	0.086
75-84		-0.027	0.015	0.070	-0.255	0.141
85-94						
35-64	UNIVARIATE	-0.026	0.005	0.001	-0.251	0.052
65-94	UNIVARIATE	-0.021	0.008	0.005	-0.202	0.073
35-64	BIVARIATE	-0.012	0.006	0.033	-0.119	0.056
65-94	BIVARIATE	-0.016	0.008	0.035	-0.157	0.075
35-64	MULTIVARIATE	-0.002	0.006	0.739	-0.020	0.059
55-94	MULTIVARIATE	-0.012	0.008	0.122	-0.119	0.077

SEX: FEMALE
EVENT: CORONARY HEART DISEASE OTHER THAN ANGINA PECTORIS
RISK FACTOR: VITAL CAPACITY - HEIGHT INDEX (ML/INCH)
PERSONS AT RISK: PERSONS FREE OF CORONARY HEART DISEASE

RISK FACTOR	35-44 AT EXAM PERSON EXAMS	# OF EVENTS	ANNUAL RATE	45-54 AT EXAM PERSON EXAMS	# OF EVENTS	ANNUAL RATE	55-64 AT EXAM PERSON EXAMS	# OF EVENTS	ANNUAL RATE	65-74 AT EXAM PERSON EXAMS	# OF EVENTS	ANNUAL RATE
4-29	122			534	5	3	1461	24	8	1747	45	13
30-34	289			1217	5	2	2418	24	5	1715	24	7
35-39	769			2077	4	1	3048	25	4	1529	14	5
40-44	1427			2804	10	2	2706	29	5	654	13	8
45-92	2988			3310	8	1	1859	6	2	397	6	8
UNKNOWN	21			17	0	0	21	0	0	6	0	0
TOTAL	5616			9959	32	2	11513	108	5	6248	102	8

RTSK FACTOR	75-84 AT EXAM PERSON EXAMS	# OF EVENTS	ANNUAL RATE	85-94 AT EXAM PERSON EXAMS	# OF EVENTS	ANNUAL RATE	35-64 COMBINED # OF EVENTS	CRUDE	AGE-ADJUSTED	65-94 COMBINED # OF EVENTS	CRUDE	AGE-ADJUSTED
4-29	930	28	15	80	7	44	27	6	5	80	15	14
30-34	569	21	18	29	2	34	29	4	3	47	10	10
35-39	314	6	10	13	0	0	32	3	2	20	5	6
40-44	111	0	0	1	0	0	39	3	1	13	7	6
45-92	25	0	0	0	0	-	15	1	1	6	7	6
UNKNOWN	2	0	0	0	0	0	0	0	0	0	0	0
TOTAL	1951	55	14	123	9	37	142	3	3	166	10	10

LOGISTIC REGRESSIONS

AGE AT EXAM		COEFFICIENT OF RISK FACTOR	SE COEFFICIENT	P-VALUE	STANDARDIZED COEFFICIENT	STD SE
35-44						
45-54		-0.031	0.023	0.192	-0.226	0.173
55-64		-0.041	0.012	0.001	-0.311	0.094
65-74		-0.032	0.013	0.016	-0.237	0.098
75-84		-0.039	0.019	0.038	-0.275	0.133
85-94		-0.069	0.047	0.147	-0.461	0.318
35-64	UNIVARIATE	-0.064	0.010	0.001	-0.505	0.080
65-94	UNIVARIATE	-0.045	0.010	0.001	-0.335	0.076
35-64	BIVARIATE	-0.032	0.011	0.004	-0.258	0.088
65-94	BIVARIATE	-0.035	0.011	0.001	-0.265	0.080
35-64	MULTIVARIATE	-0.017	0.012	0.153	-0.135	0.094
65-94	MULTIVARIATE	-0.031	0.011	0.005	-0.234	0.084

TABLE 2-12   DATE=04DEC85
AVERAGE ANNUAL INCIDENCE RATE FOR EVENT PER 1000 PERSONS AT RISK AT EXAM
BY AGE AND LEVEL OF RISK FACTOR AT EXAM: FRAMINGHAM STUDY 30-YEAR FOLLOWUP

SEX: MALE
EVENT: CORONARY HEART DISEASE OTHER THAN ANGINA PECTORIS
RISK FACTOR: HEART RATE (PER MINUTE)
PERSONS AT RISK: PERSONS FREE OF CORONARY HEART DISEASE

RISK FACTOR	35-44 AT EXAM			45-54 AT EXAM			55-64 AT EXAM			65-74 AT EXAM		
	PERSON EXAMS	# OF EVENTS	ANNUAL RATE	PERSON EXAMS	# OF EVENTS	ANNUAL RATE	PERSON EXAMS	# OF EVENTS	ANNUAL RATE	PERSON EXAMS	# OF EVENTS	ANNUAL RATE
30-67	1284	6	2	1949	23	6	2309	41	9	1183	36	15
68-75	1492	6	2	2490	30	6	2490	65	13	1159	27	12
76-83	722	7	5	1394	21	8	1374	33	12	624	27	22
84-91	579	12	10	1097	25	11	1064	26	12	487	23	24
92-220	395	3	4	794	22	14	812	33	20	460	25	27
UNKNOWN	0	0	-	0	0	-	0	0	-	0	0	-
TOTAL	4472	34	4	7724	121	8	8049	198	12	3913	138	18

RISK FACTOR	75-84 AT EXAM			85-94 AT EXAM			35-64 COMBINED			65-94 COMBINED		
	PERSON EXAMS	# OF EVENTS	ANNUAL RATE	PERSON EXAMS	# OF EVENTS	ANNUAL RATE	# OF EVENTS	CRUDE	AGE-ADJUSTED	# OF EVENTS	CRUDE	AGE-ADJUSTED
30-67	301	17	28	11			70	6	6	53	18	18
68-75	296	16	27	12			101	8	8	44	15	15
76-83	192	8	21	10			61	9	9	35	21	21
84-91	123	7	28	5			63	11	11	32	26	26
92-220	105	6	29	1			58	14	14	31	27	27
UNKNOWN	0	0	-	0			0	-	-	0	-	
TOTAL	1017	54	27	39			353	9	9	195	20	20

LOGISTIC REGRESSIONS

AGE AT EXAM		COEFFICIENT OF RISK FACTOR	SE COEFFICIENT	P-VALUE	STANDARDIZED COEFFICIENT	STD SE
35-44		0.031	0.012	0.009	0.390	0.150
45-54		0.024	0.007	0.001	0.300	0.085
55-64		0.017	0.005	0.001	0.215	0.067
65-74		0.018	0.006	0.002	0.250	0.079
75-84		0.001	0.010	0.888	0.020	0.139
85-94						
35-64	UNIVARIATE	0.021	0.004	0.001	0.263	0.050
65-94	UNIVARIATE	0.014	0.005	0.005	0.193	0.068
35-64	BIVARIATE	0.021	0.004	0.001	0.260	0.049
65-94	BIVARIATE	0.014	0.005	0.005	0.194	0.068
35-64	MULTIVARIATE	0.009	0.004	0.029	0.116	0.053
65-94	MULTIVARIATE	0.012	0.005	0.017	0.168	0.070

SEX: FEMALE
EVENT: CORONARY HEART DISEASE OTHER THAN ANGINA PECTORIS
RISK FACTOR: HEART RATE (PER MINUTE)
PERSONS AT RISK: PERSONS FREE OF CORONARY HEART DISEASE

RISK FACTOR	35-44 AT EXAM			45-54 AT EXAM			55-64 AT EXAM			65-74 AT EXAM		
	PERSON EXAMS	# OF EVENTS	ANNUAL RATE	PERSON EXAMS	# OF EVENTS	ANNUAL RATE	PERSON EXAMS	# OF EVENTS	ANNUAL RATE	PERSON EXAMS	# OF EVENTS	ANNUAL RATE
30-67	1092			1897	8	2	2305	11	2	1103	21	10
68-75	1803			3244	12	2	3468	29	4	1668	28	8
76-83	1051			1911	4	1	2351	22	5	1358	14	5
84-91	939			1604	4	1	1780	24	7	1089	20	9
92-220	731			1303	4	2	1609	22	7	1030	19	9
UNKNOWN	0			0	0	-	0	0	-	0	0	-
TOTAL	5616			9959	32	2	11513	108	5	6248	102	8

RISK FACTOR	75-84 AT EXAM			85-94 AT EXAM			35-64 COMBINED			65-94 COMBINED		
	PERSON EXAMS	# OF EVENTS	ANNUAL RATE	PERSON EXAMS	# OF EVENTS	ANNUAL RATE	# OF EVENTS	CRUDE	AGE-ADJUSTED	# OF EVENTS	CRUDE	AGE-ADJUSTED
30-67	360	12	17	26	4	77	19	2	2	37	12	12
68-75	462	9	10	32	0	0	41	2	2	37	9	9
76-83	428	15	18	24	0	0	27	3	3	29	8	8
84-91	330	7	11	23	4	87	29	3	3	31	11	11
92-220	371	12	16	18	1	28	26	4	5	32	11	11
UNKNOWN	0	0	-	0	0	-	0			0	-	
TOTAL	1951	55	14	123	9	37	142	3	3	166	10	10

LOGISTIC REGRESSIONS

AGE AT EXAM		COEFFICIENT OF RISK FACTOR	SE COEFFICIENT	P-VALUE	STANDARDIZED COEFFICIENT	STD SE
35-44						
45-54		-0.015	0.015	0.339	-0.180	0.188
55-64		0.020	0.007	0.003	0.251	0.085
65-74		0.001	0.008	0.941	0.007	0.100
75-84		-0.004	0.010	0.692	-0.055	0.140
85-94		-0.008	0.026	0.751	-0.112	0.354
35-64	UNIVARIATE	0.014	0.006	0.025	0.174	0.078
65-94	UNIVARIATE	-0.001	0.006	0.824	-0.018	0.079
35-64	BIVARIATE	0.013	0.006	0.031	0.165	0.077
65-94	BIVARIATE	-0.002	0.006	0.762	-0.024	0.079
35-64	MULTIVARIATE	0.002	0.007	0.777	0.024	0.083
65-94	MULTIVARIATE	-0.006	0.006	0.286	-0.085	0.080

RISK FACTOR	PERSON EXAMS	# OF EVENTS	ANNUAL RATE	PERSON EXAMS	# OF EVENTS	ANNUAL RATE	# OF EVENTS	CRUDE	AGE-ADJUSTED	# OF EVENTS	CRUDE	AGE-ADJUSTED
11-20	1485	13	4	2051	43	10	1682	60	18	526	21	20
21-40	987	10	5	1602	35	11	1119	39	17	369	13	18
41-90	158	3	9	252	4	8	162	8	25	27	2	37
UNKNOWN	26	0	0	55	2	29	22	0	0	6	1	83
TOTAL	4472	34	4	7724	121	8	8049	198	12	3913	138	18

RISK FACTOR	75-84 AT EXAM			85-94 AT EXAM			35-64 COMBINED			6-94 COMBINED		
	PERSON EXAMS	# OF EVENTS	ANNUAL RATE	PERSON EXAMS	# OF EVENTS	ANNUAL RATE	# OF EVENTS	CRUDE	AGE-ADJUSTED	# OF EVENTS	CRUDE	AGE-ADJUSTED
NONE	797	46	29	27			109	6	5	136	20	20
1-10	92	5	27	6			27	7	7	18	19	19
11-20	77	1	6	5			116	11	12	23	19	18
21-40	46	2	22	1			84	11	12	15	18	19
41-90	5	0	-	0			15	13	15	2	31	29
UNKNOWN	0	0	-	0			2	12	11	1	83	65
TOTAL	1017	54	27	39			353	9	9	195	20	20

LOGISTIC REGRESSIONS

AGE AT EXAM		COEFFICIENT OF RISK FACTOR	SE COEFFICIENT	P-VALUE	STANDARDIZED COEFFICIENT	STD SE
35-44		0.028	0.011	0.012	0.400	0.159
45-54		0.020	0.006	0.001	0.294	0.084
55-64		0.022	0.004	0.001	0.309	0.061
65-74		0.006	0.007	0.400	0.069	0.082
75-84		-0.033	0.023	0.146	-0.299	0.205
85-94						
35-64	UNIVARIATE	0.018	0.003	0.001	0.261	0.049
65-94	UNIVARIATE	-0.001	0.007	0.859	-0.013	0.074
35-64	BIVARIATE	0.023	0.003	0.001	0.337	0.048
65-94	BIVARIATE	0.002	0.007	0.790	0.020	0.074
35-64	MULTIVARIATE	0.025	0.003	0.001	0.366	0.048
65-94	MULTIVARIATE	0.005	0.007	0.464	0.055	0.075

SEX: FEMALE
EVENT: CORONARY HEART DISEASE OTHER THAN ANGINA PECTORIS
RISK FACTOR: CIGARETTES SMOKED (NUMBER PER DAY)
PERSONS AT RISK: PERSONS FREE OF CORONARY HEART DISEASE

RISK FACTOR	35-44 AT EXAM			45-54 AT EXAM			55-64 AT EXAM			65-74 AT EXAM		
	PERSON EXAMS	# OF EVENTS	ANNUAL RATE	PERSON EXAMS	# OF EVENTS	ANNUAL RATE	PERSON EXAMS	# OF EVENTS	ANNUAL RATE	PERSON EXAMS	# OF EVENTS	ANNUAL RATE
NONE	2582			5528	14	1	7595	55	4	4863	79	8
1-10	1314			1726	6	2	1566	20	6	626	9	7
11-20	1371			2009	8	2	1696	22	6	568	13	11
21-40	298			622	3	2	602	9	7	172	0	0
41-90	22			37	1	14	16	2	63	0	0	-
UNKNOWN	29			37	0	0	38	0	0	19	1	26
TOTAL	5616			9959	32	2	11513	108	3	6248	102	8

RISK FACTOR	75-84 AT EXAM			85-94 AT EXAM			35-64 COMBINED			65-94 COMBINED		
	PERSON EXAMS	# OF EVENTS	ANNUAL RATE	PERSON EXAMS	# OF EVENTS	ANNUAL RATE	# OF EVENTS	CRUDE	AGE-ADJUSTED	# OF EVENTS	CRUDE	AGE-ADJUSTED
NONE	1710	48	14	117	9	38	69	2	2	136	10	10
1-10	137	4	15	2	0	0	26	3	3	13	8	9
11-20	84	3	18	0	0	-	31	3	3	16	12	12
21-40	16	0	0	0	0	-	13	4	4	0	0	0
41-90	0	0	-	0	0	-	3	20	32	0	-	-
UNKNOWN	4	0	0	0	0	-	0	0	0	1	22	20
TOTAL	1951	55	14	123	9	37	142	3	3	166	10	10

LOGISTIC REGRESSIONS

AGE AT EXAM		COEFFICIENT OF RISK FACTOR	SE COEFFICIENT	P-VALUE	STANDARDIZED COEFFICIENT	STD SE
35-44						
45-54		0.027	0.014	0.062	0.274	0.147
55-64		0.029	0.008	0.001	0.272	0.077
65-74		-0.003	0.014	0.838	-0.021	0.103
75-84		-0.003	0.028	0.915	-0.015	0.141
85-94		-	-	-	-	-
35-64	UNIVARIATE	0.024	0.007	0.001	0.242	0.071
65-94	UNIVARIATE	-0.009	0.012	0.439	-0.067	0.086
35-64	BIVARIATE	0.032	0.007	0.001	0.322	0.069
65-94	BIVARIATE	-0.002	0.012	0.851	-0.016	0.087
35-64	MULTIVARIATE	0.036	0.007	0.001	0.354	0.069
65-94	MULTIVARIATE	0.004	0.012	0.754	0.027	0.086

TABLE 2-14   DATE=04DEC85
AVERAGE ANNUAL INCIDENCE RATE FOR EVENT PER 1000 PERSONS AT RISK AT EXAM
BY AGE AND LEVEL OF RISK FACTOR AT EXAM: FRAMINGHAM STUDY 30-YEAR FOLLOWUP

SEX: MALE
EVENT: CORONARY HEART DISEASE OTHER THAN ANGINA PECTORIS
RISK FACTOR: ALBUMIN IN URINE, DEFINITE
PERSONS AT RISK: PERSONS FREE OF CORONARY HEART DISEASE

RISK FACTOR	35-44 AT EXAM PERSON EXAMS	# OF EVENTS	ANNUAL RATE	45-54 AT EXAM PERSON EXAMS	# OF EVENTS	ANNUAL RATE	55-64 AT EXAM PERSON EXAMS	# OF EVENTS	ANNUAL RATE	65-74 AT EXAM PERSON EXAMS	# OF EVENTS	ANNUAL RATE
NO	4366	34	4	7599	116	8	7838	191	12	3792	130	17
YES	70	0	0	99	5	25	184	6	16	115	8	35
UNKNOWN	36	0	0	26	0	0	27	1	19	6	0	0
TOTAL	4472	34	4	7724	121	8	8049	198	12	3913	138	18

RISK FACTOR	75-84 AT EXAM PERSON EXAMS	# OF EVENTS	ANNUAL RATE	85-94 AT EXAM PERSON EXAMS	# OF EVENTS	ANNUAL RATE	35-64 COMBINED # OF EVENTS	CRUDE	AGE-ADJUSTED	65-94 COMBINED # OF EVENTS	CRUDE	AGE-ADJUSTED
NO	989	53	27	32			341	9	9	186	19	19
YES	25	1	20	7			11	16	16	9	31	31
UNKNOWN	3	0	0	0			1	6	7	0	0	0
TOTAL	1017	54	27	39			353	9	9	195	20	20

LOGISTIC REGRESSIONS

AGE AT EXAM		COEFFICIENT OF RISK FACTOR	SE COEFFICIENT	P-VALUE	STANDARDIZED COEFFICIENT	STD SE
35-44		-	-		-	-
45-54		1.233	0.468	0.008	0.139	0.053
55-64		0.300	0.421	0.477	0.045	0.063
65-74		0.745	0.377	0.048	0.126	0.064
75-84		-0.307	1.030	0.766	-0.048	0.160
85-94						
35-64	UNIVARIATE	0.607	0.311	0.051	0.080	0.041
65-94	UNIVARIATE	0.484	0.352	0.169	0.082	0.060
35-64	BIVARIATE	0.492	0.312	0.115	0.064	0.041
65-94	BIVARIATE	0.456	0.353	0.196	0.077	0.060
35-64	MULTIVARIATE	-0.060	0.335	0.857	-0.008	0.044
65-94	MULTIVARIATE	0.070	0.365	0.847	0.012	0.062

SEX: FEMALE
EVENT: CORONARY HEART DISEASE OTHER THAN ANGINA PECTORIS
RISK FACTOR: ALBUMIN IN URINE, DEFINITE
PERSONS AT RISK: PERSONS FREE OF CORONARY HEART DISEASE

RISK FACTOR	35-44 AT EXAM PERSON EXAMS	# OF EVENTS	ANNUAL RATE	45-54 AT EXAM PERSON EXAMS	# OF EVENTS	ANNUAL RATE	55-64 AT EXAM PERSON EXAMS	# OF EVENTS	ANNUAL RATE	65-74 AT EXAM PERSON EXAMS	# OF EVENTS	ANNUAL RATE
NO	5353			9656	32	2	11263	104	5	6118	97	8
YES	225			278	0	0	225	4	9	120	5	21
UNKNOWN	38			25	0	0	25	0	0	10	0	0
TOTAL	5616			9959	32	2	11513	108	5	6248	102	8

RISK FACTOR	75-84 AT EXAM PERSON EXAMS	# OF EVENTS	ANNUAL RATE	85-94 AT EXAM PERSON EXAMS	# OF EVENTS	ANNUAL RATE	35-64 COMBINED # OF EVENTS	CRUDE	AGE-ADJUSTED	65-94 COMBINED # OF EVENTS	CRUDE	AGE-ADJUSTED
NO	1904	55	14	120	8	33	138	3	3	160	10	10
YES	38	0	0	3	1	167	4	3	4	6	19	18
UNKNOWN	9	0	0	0	0	0	0	0	0	0	0	0
TOTAL	1951	55	14	123	9	37	142	3	3	166	10	10

LOGISTIC REGRESSIONS

AGE AT EXAM		COEFFICIENT OF RISK FACTOR	SE COEFFICIENT	P-VALUE	STANDARDIZED COEFFICIENT	STD SE
35-44						
45-54		-	-		-	-
55-64		0.664	0.514	0.197	0.092	0.071
65-74		0.993	0.468	0.034	0.136	0.064
75-84		-	-		-	-
85-94		1.946	1.278	0.128	0.301	0.198
35-64	UNIVARIATE	0.045	0.509	0.929	0.007	0.082
65-94	UNIVARIATE	0.658	0.424	0.120	0.091	0.058
35-64	BIVARIATE	0.293	0.510	0.566	0.048	0.083
65-94	BIVARIATE	0.651	0.425	0.125	0.090	0.059
35-64	MULTIVARIATE	-0.230	0.551	0.676	-0.037	0.089
65-94	MULTIVARIATE	0.308	0.439	0.483	0.042	0.061

· 60

TABLE 2-15   DATE=04DEC85
AVERAGE ANNUAL INCIDENCE RATE FOR EVENT PER 1000 PERSONS AT RISK AT EXAM
BY AGE AND LEVEL OF RISK FACTOR AT EXAM: FRAMINGHAM STUDY 30-YEAR FOLLOWUP

SEX: MALE
EVENT: CORONARY HEART DISEASE OTHER THAN ANGINA PECTORIS
RISK FACTOR: HEART ENLARGEMENT BY X-RAY
PERSONS AT RISK: PERSONS FREE OF CORONARY HEART DISEASE

RISK FACTOR	35-44 AT EXAM			45-54 AT EXAM			55-64 AT EXAM			65-74 AT EXAM		
	PERSON EXAMS	# OF EVENTS	ANNUAL RATE	PERSON EXAMS	# OF EVENTS	ANNUAL RATE	PERSON EXAMS	# OF EVENTS	ANNUAL RATE	PERSON EXAMS	# OF EVENTS	ANNUAL RATE
NO	4063	30	4	6383	94	7	5967	130	11	2584	77	15
MILD	185	3	8	650	11	8	962	25	13	578	16	14
DEFINITE	219	1	2	687	16	12	1119	43	19	751	45	30
UNKNOWN	5	0	0	4	0	0	1	0	0	0	0	-
TOTAL	4472	34	4	7724	121	8	8049	198	12	3913	138	18

RISK FACTOR	75-84 AT EXAM			85-94 AT EXAM			35-64 COMBINED			65-94 COMBINED		
	PERSON EXAMS	# OF EVENTS	ANNUAL RATE	PERSON EXAMS	# OF EVENTS	ANNUAL RATE	# OF EVENTS	CRUDE	AGE-ADJUSTED	# OF EVENTS	CRUDE	AGE-ADJUSTED
NO	589	20	17	29			254	8	8	98	15	15
MILD	178	11	31	5			39	11	10	28	18	18
DEFINITE	250	23	46	5			60	15	13	69	34	34
UNKNOWN	0	0	-	0			0	0	0	0	-	
TOTAL	1017	54	27	39			353	9	9	195	20	20

LOGISTIC REGRESSIONS

AGE AT EXAM		COEFFICIENT OF RISK FACTOR	SE COEFFICIENT	P-VALUE	STANDARDIZED COEFFICIENT	STD SE
35-44		0.034	0.359	0.924	0.016	0.168
45-54		0.222	0.132	0.092	0.135	0.080
55-64		0.283	0.088	0.001	0.203	0.064
65-74		0.346	0.099	0.001	0.275	0.073
75-84		0.529	0.156	0.001	0.447	0.132
85-94						
35-64	UNIVARIATE	0.333	0.070	0.001	0.212	0.045
65-94	UNIVARIATE	0.416	0.082	0.001	0.336	0.066
35-64	BIVARIATE	0.234	0.072	0.001	0.149	0.046
65-94	BIVARIATE	0.396	0.082	0.001	0.320	0.066
35-64	MULTIVARIATE	0.164	0.077	0.033	0.104	0.049
65-94	MULTIVARIATE	0.253	0.088	0.004	0.204	0.071

SEX: FEMALE
EVENT: CORONARY HEART DISEASE OTHER THAN ANGINA PECTORIS
RISK FACTOR: HEART ENLARGEMENT BY X-RAY
PERSONS AT RISK: PERSONS FREE OF CORONARY HEART DISEASE

RISK FACTOR	35-44 AT EXAM			45-54 AT EXAM			55-64 AT EXAM			65-74 AT EXAM		
	PERSON EXAMS	# OF EVENTS	ANNUAL RATE	PERSON EXAMS	# OF EVENTS	ANNUAL RATE	PERSON EXAMS	# OF EVENTS	ANNUAL RATE	PERSON EXAMS	# OF EVENTS	ANNUAL RATE
NO	5164			8017	20	1	7522	51	3	3191	39	6
MILD	199			943	5	3	1769	23	7	1150	15	7
DEFINITE	245			993	7	4	2218	34	8	1907	48	13
UNKNOWN	8			6	0	0	4	0	0	0	0	-
TOTAL	5616			9959	32	2	11513	108	5	6248	102	8

RISK FACTOR	75-84 AT EXAM			85-94 AT EXAM			35-64 COMBINED			65-94 COMBINED		
	PERSON EXAMS	# OF EVENTS	ANNUAL RATE	PERSON EXAMS	# OF EVENTS	ANNUAL RATE	# OF EVENTS	CRUDE	AGE-ADJUSTED	# OF EVENTS	CRUDE	AGE-ADJUSTED
NO	702	4	6	32	3	47	72	2	2	50	6	7
MILD	373	14	19	30	3	50	28	5	4	32	10	10
DEFINITE	876	33	19	61	3	25	42	6	5	84	15	14
UNKNOWN	0	0	-	0	0	-	0	0	0	0	-	
TOTAL	1951	55	14	123	9	37	142	3	3	166	10	10

LOGISTIC REGRESSIONS

AGE AT EXAM		COEFFICIENT OF RISK FACTOR	SE COEFFICIENT	P-VALUE	STANDARDIZED COEFFICIENT	STD SE
35-44						
45-54		0.541	0.211	0.010	0.345	0.135
55-64		0.422	0.108	0.001	0.336	0.086
65-74		0.373	0.111	0.001	0.328	0.098
75-84		0.514	0.172	0.003	0.460	0.154
85-94		-0.350	0.403	0.385	-0.294	0.338
35-64	UNIVARIATE	0.44	0.094	0.001	0.452	0.065
65-94	UNIVARIATE	0.626	0.089	0.001	0.380	0.080
35-64	BIVARIATE	0.419	0.097	0.001	0.293	0.068
65-94	BIVARIATE	0.373	0.091	0.001	0.333	0.081
35-64	MULTIVARIATE	.337	0.104	0.001	0.235	0.072
65-94	MULTIVARIATE	.281	0.095	0.003	0.251	0.085

TABLE 2-16   DATE=02APR85

AVERAGE ANNUAL INCIDENCE RATE FOR EVENT PER 1000 PERSONS AT RISK AT EXAM
BY AGE AND LEVEL OF RISK FACTOR AT EXAM: FRAMINGHAM STUDY 30-YEAR FOLLOWUP

SEX: MALE
EVENT: CORONARY HEART DISEASE OTHER THAN ANGINA PECTORIS
RISK FACTOR: LEFT VENTRICULAR HYPERTROPHY BY ELECTROCARDIOGRAM
PERSONS AT RISK: PERSONS FREE OF CORONARY HEART DISEASE

RISK FACTOR	35-44 AT EXAM PERSON EXAMS	# OF EVENTS	ANNUAL RATE	45-54 AT EXAM PERSON EXAMS	# OF EVENTS	ANNUAL RATE	55-64 AT EXAM PERSON EXAMS	# OF EVENTS	ANNUAL RATE	65-74 AT EXAM PERSON EXAMS	# OF EVENTS	ANNUAL RATE
NO	4366	34	4	7497	112	7	7693	180	12	3703	112	15
BORDER	78	0	0	132	5	19	210	7	17	90	9	50
DEFINITE	28	0	0	95	4	21	146	11	38	120	17	71
UNKNOWN	0	0	-	0	0	-	0	0	-	0	0	-
TOTAL	4472	34	4	7724	121	8	8049	198	12	3913	138	18

RISK FACTOR	75-84 AT EXAM PERSON EXAMS	# OF EVENTS	ANNUAL RATE	85-94 AT EXAM PERSON EXAMS	# OF EVENTS	ANNUAL RATE	35-64 COMBINED # OF EVENTS	CRUDE	AGE-ADJUSTED	65-94 COMBINED # OF EVENTS	CRUDE	AGE-ADJUSTED
NO	949	49	26	34			326	8	8	164	17	17
BORDER	37	1	14	2			12	14	14	10	39	42
DEFINITE	31	4	65	3			15	28	23	21	68	69
UNKNOWN	0	0	-	0			0	-	-	0	-	-
TOTAL	1017	54	27	39			353	9	9	195	20	20

LOGISTIC REGRESSIONS

AGE AT EXAM		COEFFICIENT OF RISK FACTOR	SE COEFFICIENT	P-VALUE	STANDARDIZED COEFFICIENT	STD SE
35-44		-				
45-54		0.613	0.223	0.006	0.156	0.057
55-64		0.575	0.154	0.001	0.177	0.047
65-74		0.871	0.132	0.001	0.324	0.049
75-84		0.377	0.282	0.182	0.145	0.109
85-94						
35-64	UNIVARIATE	0.612	0.126	0.001	0.164	0.034
65-94	UNIVARIATE	0.745	0.119	0.001	0.281	0.045
35-64	BIVARIATE	0.529	0.127	0.001	0.142	0.034
65-94	BIVARIATE	0.738	0.119	0.001	0.278	0.045
35-64	MULTIVARIATE	0.300	0.138	0.029	0.080	0.037
65-94	MULTIVARIATE	0.496	0.130	0.001	0.187	0.049

SEX: FEMALE
EVENT: CORONARY HEART DISEASE OTHER THAN ANGINA PECTORIS
RISK FACTOR: LEFT VENTRICULAR HYPERTROPHY BY ELECTROCARDIOGRAM
PERSONS AT RISK: PERSONS FREE OF CORONARY HEART DISEASE

RISK FACTOR	35-44 AT EXAM PERSON EXAMS	# OF EVENTS	ANNUAL RATE	45-54 AT EXAM PERSON EXAMS	# OF EVENTS	ANNUAL RATE	55-64 AT EXAM PERSON EXAMS	# OF EVENTS	ANNUAL RATE	65-74 AT EXAM PERSON EXAMS	# OF EVENTS	ANNUAL RATE
NO	5563			9798	32	2	11137	101	5	5966	86	7
BORDER	38			107	0	0	226	3	7	134	7	26
DEFINITE	15			54	0	0	150	4	13	148	9	30
UNKNOWN	0			0	0	-	0	0	-	0	0	-
TOTAL	5616			9959	32	2	11513	108	5	6248	102	8

RISK FACTOR	75-84 AT EXAM PERSON EXAMS	# OF EVENTS	ANNUAL RATE	85-94 AT EXAM PERSON EXAMS	# OF EVENTS	ANNUAL RATE	35-64 COMBINED # OF EVENTS	CRUDE	AGE-ADJUSTED	65-94 COMBINED # OF EVENTS	CRUDE	AGE-ADJUSTED
NO	1833	48	13	114	8	36	134	3	3	142	9	9
BORDER	49	3	31	6	0	0	3	4	3	10	26	27
DEFINITE	69	4	29	6	1	83	5	11	12	14	31	31
UNKNOWN	0	0	-	0	0	-	0	-	-	0	-	-
TOTAL	1951	55	14	123	9	37	142	3	3	166	10	10

LOGISTIC REGRESSIONS

AGE AT EXAM		COEFFICIENT OF RISK FACTOR	SE COEFFICIENT	P-VALUE	STANDARDIZED COEFFICIENT	STD SE
35-44		-				
45-54		-				
55-64		0.521	0.242	0.031	0.138	0.064
65-74		0.810	0.164	0.001	0.270	0.055
75-84		0.469	0.245	0.055	0.186	0.097
85-94		0.303	0.612	0.620	0.144	0.290
35-64	UNIVARIATE	0.720	0.220	0.001	0.153	0.047
65-94	UNIVARIATE	0.702	0.133	0.001	0.247	0.047
35-64	BIVARIATE	0.512	0.224	0.022	0.109	0.048
65-94	BIVARIATE	0.664	0.134	0.001	0.234	0.047
35-64	MULTIVARIATE	0.148	0.243	0.542	0.031	0.052
65-94	MULTIVARIATE	0.547	0.141	0.001	0.193	0.050

TABLE 2-17   DATE=04DEC85
AVERAGE ANNUAL INCIDENCE RATE FOR EVENT PER 1000 PERSONS AT RISK AT EXAM
BY AGE AND LEVEL OF RISK FACTOR AT EXAM: FRAMINGHAM STUDY 30-YEAR FOLLOWUP

SEX: MALE
EVENT: CORONARY HEART DISEASE OTHER THAN ANGINA PECTORIS
RISK FACTOR: INTRAVENTRICULAR CONDUCTION DEFECT (WITH QRS INTERVAL MORE THAN 0.11 SECOND)
PERSONS AT RISK: PERSONS FREE OF CORONARY HEART DISEASE

RISK FACTOR	35-44 AT EXAM			45-54 AT EXAM			55-64 AT EXAM			65-74 AT EXAM		
	PERSON EXAMS	# OF EVENTS	ANNUAL RATE	PERSON EXAMS	# OF EVENTS	ANNUAL RATE	PERSON EXAMS	# OF EVENTS	ANNUAL RATE	PERSON EXAMS	# OF EVENTS	ANNUAL RATE
NO	4161	52	4	7213	111	8	7386	173	12	3566	126	18
INCOMPLETE	275	2	4	446	7	8	456	18	20	183	7	19
COMPLETE	36	0	0	65	3	23	207	7	17	164	5	15
UNKNOWN	0	0	-	0	0	-	0	0	-	0	0	-
TOTAL	4472	34	4	7724	121	8	8049	198	12	3913	138	18

RISK FACTOR	75-84 AT EXAM			85-94 AT EXAM			35-64 COMBINED			65-94 COMBINED		
	PERSON EXAMS	# OF EVENTS	ANNUAL RATE	PERSON EXAMS	# OF EVENTS	ANNUAL RATE	# OF EVENTS	CRUDE	AGE-ADJUSTED	# OF EVENTS	CRUDE	AGE-ADJUSTED
NO	918	49	27	31			316	8	8	177	20	20
INCOMPLETE	21	1	24	2			27	11	12	8	19	20
COMPLETE	78	4	26	6			10	16	16	10	20	18
UNKNOWN	0	0	-	0			0	-	-	0	-	-
TOTAL	1017	54	27	39			353	9	9	195	20	20

LOGISTIC REGRESSIONS

AGE AT EXAM	COEFFICIENT OF RISK FACTOR	SE COEFFICIENT	P-VALUE	STANDARDIZED COEFFICIENT	STD SE
35-44	-0.249	0.673	0.712	-0.074	0.199
45-54	0.320	0.259	0.217	0.094	0.076
55-64	0.301	0.153	0.049	0.116	0.059
65-74	-0.040	0.201	0.842	-0.018	0.089
75-84	-0.027	0.262	0.919	-0.015	0.143
85-94					
35-64 UNIVARIATE	0.328	0.130	0.012	0.109	0.043
65-94 UNIVARIATE	0.011	0.154	0.944	0.005	0.073
35-64 BIVARIATE	0.252	0.129	0.050	0.084	0.043
65-94 BIVARIATE	-0.020	0.154	0.896	-0.009	0.072
35-64 MULTIVARIATE	0.233	0.136	0.086	0.078	0.045
65-94 MULTIVARIATE	0.004	0.156	0.982	0.002	0.074

SEX: FEMALE
EVENT: CORONARY HEART DISEASE OTHER THAN ANGINA PECTORIS
RISK FACTOR: INTRAVENTRICULAR CONDUCTION DEFECT (WITH QRS INTERVAL MORE THAN 0.11 SECOND)
PERSONS AT RISK: PERSONS FREE OF CORONARY HEART DISEASE

RISK FACTOR	35-44 AT EXAM			45-54 AT EXAM			55-64 AT EXAM			65-74 AT EXAM		
	PERSON EXAMS	# OF EVENTS	ANNUAL RATE	PERSON EXAMS	# OF EVENTS	ANNUAL RATE	PERSON EXAMS	# OF EVENTS	ANNUAL RATE	PERSON EXAMS	# OF EVENTS	ANNUAL RATE
NO	5529			9771	32	2	11183	102	5	6027	98	8
INCOMPLETE	75			144	0	0	187	2	5	115	1	4
COMPLETE	12			44	0	0	143	4	14	106	3	14
UNKNOWN	0	0	-	0	0	-	0	0	-	0	0	-
TOTAL	5616			9959	32	2	11513	108	5	6248	102	8

RISK FACTOR	75-84 AT EXAM			85-94 AT EXAM			35-64 COMBINED			65-94 COMBINED		
	PERSON EXAMS	# OF EVENTS	ANNUAL RATE	PERSON EXAMS	# OF EVENTS	ANNUAL RATE	# OF EVENTS	CRUDE	AGE-ADJUSTED	# OF EVENTS	CRUDE	AGE-ADJUSTED
NO	1847	52	14	117	9	38	136	3	3	159	10	10
INCOMPLETE	36	1	14	1	0	0	2	2	2	2	7	7
COMPLETE	68	2	15	5	0	0	4	10	6	5	14	14
UNKNOWN	0	0	-	0	0	-	0	-	-	0	-	-
TOTAL	1951	55	14	123	9	37	142	3	3	166	10	10

LOGISTIC REGRESSIONS

AGE AT EXAM	COEFFICIENT OF RISK FACTOR	SE COEFFICIENT	P-VALUE	STANDARDIZED COEFFICIENT	STD SE
35-44					
45-54					
55-64	0.516	0.251	0.040	0.131	0.064
65-74	0.172	0.301	0.567	0.050	0.087
75-84	0.019	0.348	0.957	0.007	0.135
85-94					
35-64 UNIVARIATE	0.574	0.255	0.024	0.120	0.053
65-94 UNIVARIATE	0.103	0.227	0.652	0.033	0.072
35-64 BIVARIATE	0.383	0.253	0.130	0.080	0.053
65-94 BIVARIATE	0.053	0.227	0.816	0.017	0.072
35-64 MULTIVARIATE	0.427	0.255	0.095	0.089	0.053
65-94 MULTIVARIATE	0.051	0.229	0.823	0.016	0.073

TABLE 2-18   DATE=04DEC85
AVERAGE ANNUAL INCIDENCE RATE FOR EVENT PER 1000 PERSONS AT RISK AT EXAM
BY AGE AND LEVEL OF RISK FACTOR AT EXAM: FRAMINGHAM STUDY 30-YEAR FOLLOWUP

SEX: MALE
EVENT: CORONARY HEART DISEASE OTHER THAN ANGINA PECTORIS
RISK FACTOR: DEFINITE NONSPECIFIC T-WAVE OR ST-SEGMENT ABNORMALITY BY ELECTROCARDIOGRAM
PERSONS AT RISK: PERSONS FREE OF CORONARY HEART DISEASE

RISK FACTOR	35-44 AT EXAM			45-54 AT EXAM			55-64 AT EXAM			65-74 AT EXAM		
	PERSON EXAMS	# OF EVENTS	ANNUAL RATE	PERSON EXAMS	# OF EVENTS	ANNUAL RATE	PERSON EXAMS	# OF EVENTS	ANNUAL RATE	PERSON EXAMS	# OF EVENTS	ANNUAL RATE
NO	4362	29	3	7352	113	8	7462	179	12	3472	110	16
YES	110	5	23	372	8	11	587	19	16	441	28	32
UNKNOWN	0	0	-	0	0	-	0	0	-	0	0	-
TOTAL	4472	34	4	7724	121	8	8049	198	12	3913	138	18

RISK FACTOR	75-84 AT EXAM			85-94 AT EXAM			35-64 COMBINED			65-94 COMBINED		
	PERSON EXAMS	# OF EVENTS	ANNUAL RATE	PERSON EXAMS	# OF EVENTS	ANNUAL RATE	# OF EVENTS	CRUDE	AGE-ADJUSTED	# OF EVENTS	CRUDE	AGE-ADJUSTED
NO	830	45	27	28			321	8	8	155	18	18
YES	187	9	24	11			32	15	16	40	31	31
UNKNOWN	0	0	-	0			0	-	-	0	-	-
TOTAL	1017	54	27	39			353	9	9	195	20	20

LOGISTIC REGRESSIONS

AGE AT EXAM		COEFFICIENT OF RISK FACTOR	SE COEFFICIENT	P-VALUE	STANDARDIZED COEFFICIENT	STD SE
35-44		1.962	0.494	0.001	0.304	0.077
45-54		0.342	0.370	0.355	0.073	0.079
55-64		0.308	0.245	0.209	0.080	0.064
65-74		0.729	0.218	0.001	0.230	0.069
75-84		-0.126	0.374	0.737	-0.049	0.145
85-94						
35-64	UNIVARIATE	0.595	0.188	0.002	0.133	0.042
65-94	UNIVARIATE	0.587	0.183	0.001	0.197	0.061
35-64	BIVARIATE	0.422	0.189	0.026	0.094	0.042
65-94	BIVARIATE	0.527	0.184	0.004	0.177	0.062
35-64	MULTIVARIATE	0.037	0.199	0.853	0.008	0.044
65-94	MULTIVARIATE	0.445	0.187	0.017	0.149	0.062

SEX: FEMALE
EVENT: CORONARY HEART DISEASE OTHER THAN ANGINA PECTORIS
RISK FACTOR: DEFINITE NONSPECIFIC T-WAVE OR ST-SEGMENT ABNORMALITY BY ELECTROCARDIOGRAM
PERSONS AT RISK: PERSONS FREE OF CORONARY HEART DISEASE

RISK FACTOR	35-44 AT EXAM			45-54 AT EXAM			55-64 AT EXAM			65-74 AT EXAM		
	PERSON EXAMS	# OF EVENTS	ANNUAL RATE	PERSON EXAMS	# OF EVENTS	ANNUAL RATE	PERSON EXAMS	# OF EVENTS	ANNUAL RATE	PERSON EXAMS	# OF EVENTS	ANNUAL RATE
NO	5517			9543	28	1	10647	90	4	5528	79	7
YES	99			416	4	5	866	18	10	720	23	16
UNKNOWN	0			0	0	-	0	0	-	0	0	-
TOTAL	5616			9959	32	2	11513	108	5	6248	102	8

RISK FACTOR	75-84 AT EXAM			85-94 AT EXAM			35-64 COMBINED			65-94 COMBINED		
	PERSON EXAMS	# OF EVENTS	ANNUAL RATE	PERSON EXAMS	# OF EVENTS	ANNUAL RATE	# OF EVENTS	CRUDE	AGE-ADJUSTED	# OF EVENTS	CRUDE	AGE-ADJUSTED
NO	1632	42	13	97	7	36	120	2	2	128	9	9
YES	319	13	20	26	2	38	22	8	6	38	18	17
UNKNOWN	0	0	-	0	0	37	0	-	-	0	-	-
TOTAL	1951	55	14	123	9	37	142	3	3	166	10	10

LOGISTIC REGRESSIONS

AGE AT EXAM		COEFFICIENT OF RISK FACTOR	SE COEFFICIENT	P-VALUE	STANDARDIZED COEFFICIENT	STD SE
35-44						
45-54		1.194	0.537	0.026	0.239	0.107
55-64		0.912	0.261	0.001	0.241	0.069
65-74		0.822	0.240	0.001	0.263	0.077
75-84		0.475	0.323	0.142	0.176	0.120
85-94		0.069	0.834	0.934	0.028	0.342
35-64	UNIVARIATE	1.239	0.234	0.001	0.273	0.051
65-94	UNIVARIATE	0.723	0.188	0.001	0.242	0.063
35-64	BIVARIATE	0.890	0.236	0.001	0.196	0.052
65-94	BIVARIATE	0.652	0.189	0.001	0.218	0.063
35-64	MULTIVARIATE	0.642	0.244	0.009	0.141	0.054
65-94	MULTIVARIATE	0.592	0.190	0.002	0.198	0.064

TABLE 3-1   DATE=30MAR85

AVERAGE ANNUAL INCIDENCE RATE FOR EVENT PER 1000 PERSONS AT RISK AT EXAM
BY AGE AND LEVEL OF RISK FACTOR AT EXAM: FRAMINGHAM STUDY 30-YEAR FOLLOWUP

SEX: MALE
EVENT: MYOCARDIAL INFARCTION
RISK FACTOR: SYSTOLIC BLOOD PRESSURE, FIRST EXAMINER (MM HG)
PERSONS AT RISK: PERSONS FREE OF CORONARY HEART DISEASE

RISK FACTOR	35-44 AT EXAM			45-54 AT EXAM			55-64 AT EXAM			65-74 AT EXAM		
	PERSON EXAMS	# OF EVENTS	ANNUAL RATE	PERSON EXAMS	# OF EVENTS	ANNUAL RATE	PERSON EXAMS	# OF EVENTS	ANNUAL RATE	PERSON EXAMS	# OF EVENTS	ANNUAL RATE
74-119	1223	3	1	1769	13	4	1440	14	5	548	2	2
120-139	2165	12	3	3417	26	4	3166	42	7	1341	27	10
140-159	841	5	3	1732	28	8	2090	48	11	1194	34	14
160-179	191	3	8	567	13	11	916	24	13	552	19	17
180-300	52	0	0	239	1	2	437	14	16	278	15	27
UNKNOWN	0	0	-	0	0	-	0	0	-	0	0	-
TOTAL	4472	23	3	7724	81	5	8049	142	9	3913	97	12

RISK FACTOR	75-84 AT EXAM			85-94 AT EXAM			35-64 COMBINED			65-94 COMBINED		
	PERSON EXAMS	# OF EVENTS	ANNUAL RATE	PERSON EXAMS	# OF EVENTS	ANNUAL RATE	# OF EVENTS	CRUDE	AGE-ADJUSTED	# OF EVENTS	CRUDE	AGE-ADJUSTED
74-119	120	2	8	7			30	3	4	4	3	3
120-139	345	9	13	15			80	5	5	36	11	11
140-159	315	15	24	11			81	9	8	50	16	16
160-179	153	5	16	4			40	12	11	24	17	17
180-300	84	6	36	2			15	10	7	21	29	29
UNKNOWN	0	0	-	0			0	-	-	0	-	-
TOTAL	1017	37	18	39			246	6	6	135	14	13

LOGISTIC REGRESSIONS

AGE AT EXAM		COEFFICIENT OF RISK FACTOR	SE COEFFICIENT	P-VALUE	STANDARDIZED COEFFICIENT	STD SE
35-44		0.021	0.010	0.036	0.353	0.168
45-54		0.012	0.005	0.018	0.236	0.099
55-64		0.016	0.003	0.001	0.347	0.072
65-74		0.019	0.004	0.001	0.433	0.086
75-84		0.016	0.007	0.020	0.362	0.155
85-94						
35-64	UNIVARIATE	0.018	0.003	0.001	0.362	0.053
65-94	UNIVARIATE	0.018	0.003	0.001	0.419	0.075
35-64	BIVARIATE	0.014	0.003	0.001	0.290	0.054
65-94	BIVARIATE	0.018	0.003	0.001	0.421	0.075
35-64	MULTIVARIATE	0.013	0.003	0.001	0.256	0.059
65-94	MULTIVARIATE	0.015	0.004	0.001	0.342	0.080

SEX: FEMALE
EVENT: MYOCARDIAL INFARCTION
RISK FACTOR: SYSTOLIC BLOOD PRESSURE, FIRST EXAMINER (MM HG)
PERSONS AT RISK: PERSONS FREE OF CORONARY HEART DISEASE

RISK FACTOR	35-44 AT EXAM			45-54 AT EXAM			55-64 AT EXAM			65-74 AT EXAM		
	PERSON EXAMS	# OF EVENTS	ANNUAL RATE	PERSON EXAMS	# OF EVENTS	ANNUAL RATE	PERSON EXAMS	# OF EVENTS	ANNUAL RATE	PERSON EXAMS	# OF EVENTS	ANNUAL RATE
74-119	2493			2714	1	0	1979	4	1	647	5	4
120-139	2244			3982	10	1	4056	13	2	1747	11	3
140-159	673			2053	9	2	3025	16	3	2013	24	6
160-179	156			768	1	1	1560	16	5	1155	9	4
180-300	50			442	0	0	893	15	8	686	20	15
UNKNOWN	0			0	0	-	0	0	-	0	0	-
TOTAL	5616			9959	21	1	11513	64	3	6248	69	6

RISK FACTOR	75-84 AT EXAM			85-94 AT EXAM			35-64 COMBINED			65-94 COMBINED		
	PERSON EXAMS	# OF EVENTS	ANNUAL RATE	PERSON EXAMS	# OF EVENTS	ANNUAL RATE	# OF EVENTS	CRUDE	AGE-ADJUSTED	# OF EVENTS	CRUDE	AGE-ADJUSTED
74-119	133	3	11	8			5	0	0	8	5	6
120-139	465	4	6	38	2	26	24	1	1	19	4	4
140-159	626	6	5	29	3	52	25	2	2	33	6	6
160-179	451	9	10	31	0	0	17	3	2	18	5	5
180-300	276	6	16	17	1	29	16	6	5	30	15	15
UNKNOWN	0	0	-	0	0	-	0	-	-	0	-	-
TOTAL	1951	33	8	123	6	24	87	2	2	108	6	7

LOGISTIC REGRESSIONS

AGE AT EXAM		COEFFICIENT OF RISK FACTOR	SE COEFFICIENT	P-VALUE	STANDARDIZED COEFFICIENT	STD SE
35-44						
45-54		0.011	0.008	0.156	0.255	0.180
55-64		0.021	0.004	0.001	0.518	0.093
65-74		0.017	0.004	0.001	0.410	0.106
75-84		0.011	0.006	0.083	0.277	0.160
85-94		-0.003	0.017	0.862	-0.074	0.427
35-64	UNIVARIATE	0.023	0.003	0.001	0.547	0.072
65-94	UNIVARIATE	0.015	0.004	0.001	0.361	0.086
35-64	BIVARIATE	0.019	0.003	0.001	0.441	0.080
65-94	BIVARIATE	0.014	0.004	0.001	0.339	0.087
35-64	MULTIVARIATE	0.018	0.004	0.001	0.416	0.088
65-94	MULTIVARIATE	0.011	0.004	0.003	0.267	0.090

TABLE 3-2   DATE=30MAR85
AVERAGE ANNUAL INCIDENCE RATE FOR EVENT PER 1000 PERSONS AT RISK AT EXAM
BY AGE AND LEVEL OF RISK FACTOR AT EXAM: FRAMINGHAM STUDY 30-YEAR FOLLOWUP

SEX: MALE
EVENT: MYOCARDIAL INFARCTION
RISK FACTOR: DIASTOLIC BLOOD PRESSURE, FIRST EXAMINER (MM HG)
PERSONS AT RISK: PERSONS FREE OF CORONARY HEART DISEASE

RISK FACTOR	35-44 AT EXAM			45-54 AT EXAM			55-64 AT EXAM			65-74 AT EXAM		
	PERSON EXAMS	# OF EVENTS	ANNUAL RATE	PERSON EXAMS	# OF EVENTS	ANNUAL RATE	PERSON EXAMS	# OF EVENTS	ANNUAL RATE	PERSON EXAMS	# OF EVENTS	ANNUAL RATE
20-74	1012	1	0	1478	14	5	1847	23	6	1225	29	12
75-84	1604	6	2	2766	22	4	2845	43	8	1323	30	11
85-94	1187	7	3	2062	23	6	1964	46	12	893	20	11
95-104	499	8	8	1008	14	7	982	19	10	343	12	17
105-160	170	1	3	410	8	10	411	11	13	129	6	23
UNKNOWN	0	0	-	0	0	-	0	0	-	0	0	-
TOTAL	4472	23	3	7724	81	5	8049	142	9	3913	97	12

RISK FACTOR	75-84 AT EXAM			85-94 AT EXAM			35-64 COMBINED			65-94 COMBINED		
	PERSON EXAMS	# OF EVENTS	ANNUAL RATE	PERSON EXAMS	# OF EVENTS	ANNUAL RATE	# OF EVENTS	CRUDE	AGE-ADJUSTED	# OF EVENTS	CRUDE	AGE-ADJUSTED
20-74	422	12	14	22	2		38	4	4	42	13	12
75-84	331	12	18	12			71	5	5	42	13	12
85-94	204	9	22	5			76	7	8	29	13	13
95-104	44	2	23	0			41	8	8	14	18	18
105-160	16	2	63	0			20	10	10	8	28	31
UNKNOWN	0	0	-	0			0	-	-	0	-	
TOTAL	1017	37	18	39			246	6	6	135	14	13

LOGISTIC REGRESSIONS

AGE AT EXAM		COEFFICIENT OF RISK FACTOR	SE COEFFICIENT	P-VALUE	STANDARDIZED COEFFICIENT	STD SE
35-44		0.050	0.015	0.001	0.565	0.175
45-54		0.018	0.009	0.042	0.214	0.105
55-64		0.020	0.006	0.002	0.244	0.078
65-74		0.015	0.008	0.057	0.185	0.097
75-84		0.025	0.014	0.069	0.296	0.163
85-94						
35-64	UNIVARIATE	.023	0.005	0.001	0.271	0.059
65-94	UNIVARIATE	0.015	0.007	0.027	0.184	0.083
35-64	BIVARIATE	0.025	0.005	0.001	0.267	0.058
65-94	BIVARIATE	0.017	0.007	0.012	0.212	0.084
35-64	MULTIVARIATE	0.022	0.005	0.001	0.257	0.061
65-94	MULTIVARIATE	0.013	0.007	0.074	0.153	0.086

SEX: FEMALE
EVENT: MYOCARDIAL INFARCTION
RISK FACTOR: DIASTOLIC BLOOD PRESSURE, FIRST EXAMINER (MM HG)
PERSONS AT RISK: PERSONS FREE OF CORONARY HEART DISEASE

RISK FACTOR	35-44 AT EXAM			45-54 AT EXAM			55-64 AT EXAM			65-74 AT EXAM		
	PERSON EXAMS	# OF EVENTS	ANNUAL RATE	PERSON EXAMS	# OF EVENTS	ANNUAL RATE	PERSON EXAMS	# OF EVENTS	ANNUAL RATE	PERSON EXAMS	# OF EVENTS	ANNUAL RATE
20-74	2164			2639	3	1	2969	10	2	1831	20	5
75-84	2030			3541	3	0	3853	14	2	2135	20	5
85-94	987			2361	13	3	2840	17	3	1439	16	6
95-104	324			926	0	0	1272	14	6	600	5	4
105-160	111			492	2	2	579	9	8	243	8	16
UNKNOWN	0			0	0	-	0	0	-	0	0	-
TOTAL	5616			9959	21	1	11513	64	3	6248	69	6

RISK FACTOR	75-84 AT EXAM			85-94 AT EXAM			35-64 COMBINED			65-94 COMBINED		
	PERSON EXAMS	# OF EVENTS	ANNUAL RATE	PERSON EXAMS	# OF EVENTS	ANNUAL RATE	# OF EVENTS	CRUDE	AGE-ADJUSTED	# OF EVENTS	CRUDE	AGE-ADJUSTED
20-74	724	11	8	53	2	19	14	1	1	33	6	6
75-84	596	12	10	43	3	35	17	1	1	35	6	7
85-94	443	6	7	20	1	25	30	2	2	23	6	7
95-104	146	2	7	7	0	0	15	3	3	7	5	5
105-160	42	2	24	0	0	-	11	5	4	10	18	18
UNKNOWN	0	0	-	0			0	-	-	0	-	
TOTAL	1951	33	8	123	6	24	87	2	2	108	6	7

LOGISTIC REGRESSIONS

AGE AT EXAM		COEFFICIENT OF RISK FACTOR	SE COEFFICIENT	P-VALUE	STANDARDIZED COEFFICIENT	STD SE
35-44						
45-54		0.025	0.015	0.108	0.304	0.189
55-64		0.039	0.008	0.001	0.479	0.105
65-74		0.016	0.009	0.092	0.193	0.115
75-84		-0.003	0.015	0.816	-0.041	0.177
85-94		0.004	0.035	0.900	0.055	0.425
35-64	UNIVARIATE	0.039	0.007	0.001	0.472	0.086
65-94	UNIVARIATE	0.007	0.008	0.362	0.087	0.095
35-64	BIVARIATE	0.035	0.007	0.001	0.425	0.091
65-94	BIVARIATE	0.010	0.008	0.207	0.120	0.095
35-64	MULTIVARIATE	0.032	0.008	0.001	0.395	0.096
65-94	MULTIVARIATE	0.006	0.008	0.414	0.077	0.095

TABLE 3-3A   DATE=23NOV85

AVERAGE ANNUAL INCIDENCE RATE FOR EVENT PER 1000 PERSONS AT RISK AT EXAM
BY AGE AND LEVEL OF RISK FACTOR AT EXAM: FRAMINGHAM STUDY 30-YEAR FOLLOWUP

SEX: MALE
EVENT: MYOCARDIAL INFARCTION
RISK FACTOR: HYPERTENSION WITH ANTIHYPERTENSIVE TREATMENT
PERSONS AT RISK: PERSONS FREE OF CORONARY HEART DISEASE

RISK FACTOR	35-44 AT EXAM PERSON EXAMS	# OF EVENTS	ANNUAL RATE	45-54 AT EXAM PERSON EXAMS	# OF EVENTS	ANNUAL RATE	55-64 AT EXAM PERSON EXAMS	# OF EVENTS	ANNUAL RATE	65-74 AT EXAM PERSON EXAMS	# OF EVENTS	ANNUAL RATE
NO	2651	9	2	4122	29	4	3773	43	6	1512	15	5
MILD	1301	6	2	2205	26	6	2301	57	12	1255	38	15
DEFINITE	500	7	7	1139	18	8	1268	30	12	552	21	19
TREATED	20	1	25	258	8	16	707	12	8	594	23	19
UNKNOWN	0	0	-	0	0	-	0	0	-	0	0	-
TOTAL	4472	23	3	7724	81	5	8049	142	9	3913	97	12

RISK FACTOR	75-84 AT EXAM PERSON EXAMS	# OF EVENTS	ANNUAL RATE	85-94 AT EXAM PERSON EXAMS	# OF EVENTS	ANNUAL RATE	35-64 COMBINED # OF EVENTS	CRUDE	AGE-ADJUSTED	65-94 COMBINED # OF EVENTS	CRUDE	AGE-ADJUSTED
NO	370	7	9	18			81	4	4	22	6	6
MILD	333	12	18	11			89	8	8	51	16	16
DEFINITE	125	6	24	3			55	9	9	27	20	20
TREATED	189	12	32	7			21	11	15	35	22	22
UNKNOWN	0	0	-	0			0	-	-	0	-	-
TOTAL	1017	37	18	39			246	6	6	135	14	14

LOGISTIC REGRESSIONS

AGE AT EXAM		COEFFICIENT OF RISK FACTOR	SE COEFFICIENT	P-VALUE	STANDARDIZED COEFFICIENT	STD SE
35-44		0.752	0.261	0.004	0.522	0.181
45-54		0.493	0.135	0.001	0.379	0.104
55-64		0.320	0.101	0.002	0.260	0.082
65-74		0.612	0.132	0.001	0.501	0.108
75-84		0.561	0.215	0.009	0.459	0.176
85-94						
35-64	UNIVARIATE	0.475	0.077	0.001	0.370	0.060
65-94	UNIVARIATE	0.598	0.112	0.001	0.490	0.092
35-64	BIVARIATE	0.395	0.078	0.001	0.508	0.061
65-94	BIVARIATE	0.596	0.112	0.001	0.488	0.092
35-64	MULTIVARIATE	0.361	0.082	0.001	0.281	0.063
65-94	MULTIVARIATE	0.520	0.115	0.001	0.426	0.094

SEX: FEMALE
EVENT: MYOCARDIAL INFARCTION
RISK FACTOR: HYPERTENSION WITH ANTIHYPERTENSIVE TREATMENT
PERSONS AT RISK: PERSONS FREE OF CORONARY HEART DISEASE

RISK FACTOR	35-44 AT EXAM PERSON EXAMS	# OF EVENTS	ANNUAL RATE	45-54 AT EXAM PERSON EXAMS	# OF EVENTS	ANNUAL RATE	55-64 AT EXAM PERSON EXAMS	# OF EVENTS	ANNUAL RATE	65-74 AT EXAM PERSON EXAMS	# OF EVENTS	ANNUAL RATE
NO	4250			5622	8	1	4863	8	1	1751	11	3
MILD	1001			2579	10	2	3290	13	2	1911	19	5
DEFINITE	316			1166	0	0	1835	23	6	1036	10	5
TREATED	49			592	3	3	1525	20	7	1550	29	9
UNKNOWN	0			0	0	-	0	0	-	0	0	-
TOTAL	5616			9959	21	1	11513	64	3	6248	69	6

RISK FACTOR	75-84 AT EXAM PERSON EXAMS	# OF EVENTS	ANNUAL RATE	85-94 AT EXAM PERSON EXAMS	# OF EVENTS	ANNUAL RATE	35-64 COMBINED # OF EVENTS	CRUDE	AGE-ADJUSTED	65-94 COMBINED # OF EVENTS	CRUDE	AGE-ADJUSTED
NO	395	5	5	31	1	16	17	1	1	17	4	4
MILD	585	9	8	28	2	16	23	2	2	30	6	6
DEFINITE	375	10	13	24	0	0	24	4	3	20	7	7
TREATED	596	9	8	40	3	38	23	5	4	41	9	9
UNKNOWN	0	0	-	0			0	-	-	0	-	-
TOTAL	1951	33	8	123	6	24	87	2	2	108	6	6

LOGISTIC REGRESSIONS

AGE AT EXAM		COEFFICIENT OF RISK FACTOR	SE COEFFICIENT	P-VALUE	STANDARDIZED COEFFICIENT	STD SE
35-44						
45-54		0.239	0.268	0.373	0.183	0.206
55-64		1.068	0.182	0.001	0.892	0.152
65-74		0.438	0.161	0.007	0.360	0.133
75-84		0.226	0.238	0.342	0.177	0.186
85-94		0.100	0.512	0.845	0.084	0.430
35-64	UNIVARIATE	0.998	0.136	0.001	0.792	0.108
65-94	UNIVARIATE	0.382	0.129	0.003	0.312	0.105
35-64	BIVARIATE	0.793	0.140	0.001	0.629	0.111
65-94	BIVARIATE	0.347	0.130	0.008	0.283	0.106
35-64	MULTIVARIATE	0.773	0.146	0.001	0.613	0.116
65-94	MULTIVARIATE	0.255	0.132	0.053	0.208	0.108

TABLE 3-3B — DATE=23OCT85

AVERAGE ANNUAL INCIDENCE RATE FOR EVENT PER 1000 PERSONS AT RISK AT EXAM
BY AGE AND LEVEL OF RISK FACTOR AT EXAM: FRAMINGHAM STUDY 30-YEAR FOLLOWUP

SEX: MALE
EVENT: MYOCARDIAL INFARCTION
RISK FACTOR: HYPERTENSION IGNORING TREATMENT
PERSONS AT RISK: PERSONS FREE OF CORONARY HEART DISEASE

RISK FACTOR	35-44 AT EXAM			45-54 AT EXAM			55-64 AT EXAM			65-74 AT EXAM		
	PERSON EXAMS	# OF EVENTS	ANNUAL RATE	PERSON EXAMS	# OF EVENTS	ANNUAL RATE	PERSON EXAMS	# OF EVENTS	ANNUAL RATE	PERSON EXAMS	# OF EVENTS	ANNUAL RATE
NO	2653	9	2	4154	30	4	3909	45	6	1631	19	6
MILD	1310	6	2	2284	28	6	2583	60	12	1499	48	16
DEFINITE	509	8	8	1286	23	9	1557	37	12	783	30	19
UNKNOWN	0	0	-	0	0	-	0	0	-	0	0	-
TOTAL	4472	23	3	7724	81	5	8049	142	9	3913	97	12

RISK FACTOR	75-84 AT EXAM			85-94 AT EXAM			35-64 COMBINED			65-94 COMBINED		
	PERSON EXAMS	# OF EVENTS	ANNUAL RATE	PERSON EXAMS	# OF EVENTS	ANNUAL RATE	# OF EVENTS	CRUDE	AGE-ADJUSTED	# OF EVENTS	CRUDE	AGE-ADJUSTED
NO	412	10	12	21	1	5	84	4	4	29	7	7
MILD	418	16	19	15	1		94	8	8	65	17	17
DEFINITE	187	11	29	3			68	10	10	41	21	21
UNKNOWN	0	0	-	0			0	-	-	0	-	-
TOTAL	1017	37	18	39			246	6	6	135	14	13

LOGISTIC REGRESSIONS

AGE AT EXAM		COEFFICIENT OF RISK FACTOR	SE COEFFICIENT	P-VALUE	STANDARDIZED COEFFICIENT	STD. SE
35-44		0.763	0.262	0.004	0.527	0.181
45-54		0.462	0.137	0.001	0.348	0.103
55-64		0.382	0.105	0.001	0.294	0.081
65-74		0.572	0.134	0.001	0.432	0.101
75-84		0.461	0.223	0.039	0.339	0.164
85-94						
35-64	UNIVARIATE	0.489	0.079	0.001	0.367	0.059
65-94	UNIVARIATE	0.542	0.114	0.001	0.407	0.086
35-64	BIVARIATE	0.431	0.080	0.001	0.324	0.060
65-94	BIVARIATE	0.546	0.114	0.001	0.410	0.086
35-64	MULTIVARIATE	0.390	0.084	0.001	0.292	0.063
65-94	MULTIVARIATE	0.453	0.119	0.001	0.340	0.089

SEX: FEMALE
EVENT: MYOCARDIAL INFARCTION
RISK FACTOR: HYPERTENSION IGNORING TREATMENT
PERSONS AT RISK: PERSONS FREE OF CORONARY HEART DISEASE

RISK FACTOR	35-44 AT EXAM			45-54 AT EXAM			55-64 AT EXAM			65-74 AT EXAM		
	PERSON EXAMS	# OF EVENTS	ANNUAL RATE	PERSON EXAMS	# OF EVENTS	ANNUAL RATE	PERSON EXAMS	# OF EVENTS	ANNUAL RATE	PERSON EXAMS	# OF EVENTS	ANNUAL RATE
NO	4272			5804	8	1	5245	14	1	2035	12	3
MILD	1016			2763	12	2	3860	21	3	2549	32	6
DEFINITE	328			1392	1	0	2408	29	6	1664	25	8
UNKNOWN	0			0	0	-	0	0	-	0	0	-
TOTAL	5616			9959	21	1	11513	64	3	6248	69	6

RISK FACTOR	75-84 AT EXAM			85-94 AT EXAM			35-64 COMBINED			65-94 COMBINED		
	PERSON EXAMS	# OF EVENTS	ANNUAL RATE	PERSON EXAMS	# OF EVENTS	ANNUAL RATE	# OF EVENTS	CRUDE	AGE-ADJUSTED	# OF EVENTS	CRUDE	AGE-ADJUSTED
NO	507	6	4	37	1	14	23	1	1	19	4	4
MILD	827	10	6	45	1	44	33	2	2	46	7	7
DEFINITE	617	17	14	41	1	12	31	4	3	43	9	9
UNKNOWN	0	0	-	0	0	-	0	-	-	0	-	-
TOTAL	1951	33	8	123	6	24	87	2	2	108	6	7

LOGISTIC REGRESSIONS

AGE AT EXAM		COEFFICIENT OF RISK FACTOR	SE COEFFICIENT	P-VALUE	STANDARDIZED COEFFICIENT	STD. SE
35-44						
45-54		0.197	0.285	0.489	0.143	0.296
55-64		0.764	0.161	0.001	0.593	0.125
65-74		0.429	0.161	0.008	0.329	0.123
75-84		0.316	0.247	0.037	0.390	0.187
85-94		-0.054	0.526	0.918	-0.043	0.420
35-64	UNIVARIATE	0.796	0.131	0.001	0.589	0.097
65-94	UNIVARIATE	0.443	0.130	0.001	0.340	0.100
35-64	BIVARIATE	0.606	0.135	0.001	0.448	0.100
65-94	BIVARIATE	0.417	0.130	0.001	0.319	0.100
35-64	MULTIVARIATE	0.567	0.142	0.001	0.419	0.105
65-94	MULTIVARIATE	0.334	0.133	0.012	0.256	0.102

TABLE 3-4   DATE=30MAR85
AVERAGE ANNUAL INCIDENCE RATE FOR EVENT PER 1000 PERSONS AT RISK AT EXAM
BY AGE AND LEVEL OF RISK FACTOR AT EXAM: FRAMINGHAM STUDY 30-YEAR FOLLOWUP

SEX: MALE
EVENT: MYOCARDIAL INFARCTION
RISK FACTOR: SERUM CHOLESTEROL
PERSONS AT RISK: PERSONS FREE OF CORONARY HEART DISEASE

RISK FACTOR	35-44 AT EXAM			45-54 AT EXAM			55-64 AT EXAM			65-74 AT EXAM		
	PERSON EXAMS	# OF EVENTS	ANNUAL RATE	PERSON EXAMS	# OF EVENTS	ANNUAL RATE	PERSON EXAMS	# OF EVENTS	ANNUAL RATE	PERSON EXAMS	# OF EVENTS	ANNUAL RATE
84-204	1253	3	1	1767	13	4	2255	31	7	1234	29	12
205-234	1247	4	2	2306	24	5	2405	46	10	1171	28	12
235-264	872	4	2	1745	14	4	1889	33	9	873	21	12
265-294	475	5	5	1001	14	7	918	16	9	428	12	14
295-1124	291	5	9	557	14	13	454	12	13	191	6	16
UNKNOWN	334	2	3	348	2	3	128	4	16	16	1	31
TOTAL	4472	23	3	.7724	81	5	8049	142	9	3913	97	12

RISK FACTOR	75-84 AT EXAM			85-94 AT EXAM			35-64 COMBINED			65-94 COMBINED		
	PERSON EXAMS	# OF EVENTS	ANNUAL RATE	PERSON EXAMS	# OF EVENTS	ANNUAL RATE	# OF EVENTS	CRUDE	AGE-ADJUSTED	# OF EVENTS	CRUDE	AGE-ADJUSTED
84-204	390	15	19	17			47	4	5	45	14	14
205-234	332	12	18	11			74	6	6	40	13	13
235-264	177	8	23	5			51	6	6	29	14	14
265-294	96	1	5	6			35	7	7	13	12	12
295-1124	22	1	23	0			31	12	12	7	16	17
UNKNOWN	0	0	-	0			8	5	8	1	31	24
TOTAL	1017	37	18	39			246	6	6	135	14	13

LOGISTIC REGRESSIONS

AGE AT EXAM		COEFFICIENT OF RISK FACTOR	SE COEFFICIENT	P-VALUE	STANDARDIZED COEFFICIENT	STD SE
35-44		0.012	0.003	0.001	0.521	0.144
45-54		0.007	0.002	0.002	0.283	0.093
55-64		0.004	0.002	0.077	0.146	0.082
65-74		0.002	0.003	0.439	0.079	0.102
75-84		0.000	0.004	0.975	0.005	0.167
85-94						
35-64	UNIVARIATE	0.006	0.001	0.001	0.236	0.058
65-94	UNIVARIATE	0.001	0.002	0.630	0.042	0.087
35-64	BIVARIATE	0.006	0.001	0.001	0.256	0.059
65-94	BIVARIATE	0.001	0.002	0.518	0.056	0.087
35-64	MULTIVARIATE	0.005	0.001	0.001	0.225	0.059
65-94	MULTIVARIATE	0.001	0.002	0.506	0.058	0.088

SEX: FEMALE
EVENT: MYOCARDIAL INFARCTION
RISK FACTOR: SERUM CHOLESTEROL
PERSONS AT RISK: PERSONS FREE OF CORONARY HEART DISEASE

RISK FACTOR	35-44 AT EXAM			45-54 AT EXAM			55-64 AT EXAM			65-74 AT EXAM		
	PERSON EXAMS	# OF EVENTS	ANNUAL RATE	PERSON EXAMS	# OF EVENTS	ANNUAL RATE	PERSON EXAMS	# OF EVENTS	ANNUAL RATE	PERSON EXAMS	# OF EVENTS	ANNUAL RATE
84-204	2350			2094	2	0	1625	8	2	789	3	2
205-234	1417			2558	5	1	2615	13	2	1401	5	5
235-264	806			2259	2	0	2938	11	2	1703	18	5
265-294	372			1471	3	1	2240	16	4	1287	15	6
295-1124	171			1101	6	3	1866	16	4	1031	19	9
UNKNOWN	500			474	3	3	229	0	0	37	0	0
TOTAL	5616			9959	21	1	11513	64	3	6248	69	6

RISK FACTOR	75-84 AT EXAM			85-94 AT EXAM			35-64 COMBINED			65-94 COMBINED		
	PERSON EXAMS	# OF EVENTS	ANNUAL RATE	PERSON EXAMS	# OF EVENTS	ANNUAL RATE	# OF EVENTS	CRUDE	AGE-ADJUSTED	# OF EVENTS	CRUDE	AGE-ADJUSTED
84-204	361	7	10	46	1	11	10	1	1	11	5	4
205-234	486	8	8	28	2	36	18	1	1	24	6	6
235-264	542	8	7	35	2	29	14	1	1	28	6	6
265-294	350	6	9	9	1	56	19	2	2	22	7	7
295-1124	205	4	10	5	0	0	23	4	3	23	9	9
UNKNOWN	7	1	-	0	0	-	3	1	1	0	0	0
TOTAL	1951	33	8	123	6	24	87	2	2	108	6	7

LOGISTIC REGRESSIONS

AGE AT EXAM		COEFFICIENT OF RISK FACTOR	SE COEFFICIENT	P-VALUE	STANDARDIZED COEFFICIENT	STD SE
35-44						
45-54		0.007	0.002	0.001	0.315	0.096
55-64		0.006	0.003	0.015	0.278	0.115
65-74		0.008	0.003	0.002	0.351	0.113
75-84		0.001	0.004	0.866	0.029	0.175
85-94		0.005	0.009	0.551	0.243	0.408
35-64	UNIVARIATE	0.008	0.002	0.001	0.379	0.077
65-94	UNIVARIATE	0.005	0.002	0.027	0.207	0.093
35-64	BIVARIATE	0.007	0.002	0.001	0.316	0.075
65-94	BIVARIATE	0.006	0.002	0.007	0.254	0.094
35-64	MULTIVARIATE	0.006	0.002	0.001	0.276	0.077
65-94	MULTIVARIATE	0.006	0.002	0.007	0.255	0.095

TABLE 3-5   DATE=30MAR85

AVERAGE ANNUAL INCIDENCE RATE FOR EVENT pER 1000 PERSONS AT RISK AT EXAM
BY AGE AND LEVEL OF RISK FACTOR AT EXAM, FRAMINGHAM STUDY 30-YEAR FOLLOWUP

SEX: MALE
EVENT: MYOCARDIAL INFARCTION
RISK FACTOR: HEMATOCRIT
PERSONS AT RISK: PERSONS FREE OF CORONARY HEART DISEASE

RISK FACTOR	35-44 AT EXAM PERSON EXAMS	# OF EVENTS	ANNUAL RATE	45-54 AT EXAM PERSON EXAMS	# OF EVENTS	ANNUAL RATE	55-64 AT EXAM PERSON EXAMS	# OF EVENTS	ANNUAL RATE	65-74 AT EXAM PERSON EXAMS	# OF EVENTS	ANNUAL RATE
25-42	623	5	4	942	4	2	1018	17	8	450	10	11
43-44	546	0	0	967	10	5	885	13	7	493	12	12
45-46	983	2	1	1854	19	5	1853	32	8	963	21	11
47-48	773	2	1	1510	20	7	1599	29	9	861	18	10
49-70	1175	12	5	2066	25	6	2338	67	10	1127	35	16
UNKNOWN	370	3	3	385	3	4	156	4	13			
TOTAL	4471	23	3	7724	81	5	8049	142	9	3913	97	12

RISK FACTOR	75-84 AT EXAM PERSON EXAMS	# OF EVENTS	ANNUAL RATE	85-94 AT EXAM PERSON EXAMS	# OF EVENTS	ANNUAL RATE	35-64 COMBINED # OF EVENTS	CRUDE	AGE ADJUSTED	65-94 COMBINED # OF EVENTS	CRUDE	AGE ADJUSTED
25-42	214	9	21	16			26	5	5	19	14	15
43-44	119	7	29	5			23	5	5	19	15	15
45-46	240	6	13	7			53	7	6	28	12	12
47-48	197	8	20	7			51			26	12	12
49-70	247	7	14	4			84	8	7	42	15	15
UNKNOWN	0	0	-	0			9	5	7	1	26	20
TOTAL	1017	37	18	39			246	6	6	135	14	13

LOGISTIC REGRESSIONS

AGE AT EXAM	COEFFICIENT OF RISK FACTOR	SE COEFFICIENT	P-VALUE	STANDARDIZED COEFFICIENT	STD SE
35-44	0.115	0.060	0.055	0.422	0.220
45-54	0.056	0.032	0.080	0.199	0.114
55-64	0.040	0.023	0.078	0.153	0.087
65-74	0.047	0.028	0.100	0.172	0.105
75-84	-0.046	0.039	0.243	-0.191	0.164
85-94					
35-64 UNIVARIATE	0.053	0.018	0.003	0.195	0.066
65-94 UNIVARIATE	0.012	0.023	0.608	0.045	0.088
35-64 BIVARIATE	0.051	0.018	0.004	0.188	0.066
65-94 BIVARIATE	0.017	0.023	0.460	0.065	0.088
35-64 MULTIVARIATE	0.027	0.018	0.130	0.100	0.066
65-94 MULTIVARIATE	0.012	0.023	0.608	0.046	0.090

SEX: FEMALE
EVENT: MYOCARDIAL INFARCTION
RISK FACTOR: HEMATOCRIT
PERSONS AT RISK: PERSONS FREE OF CORONARY HEART DISEASE

RISK FACTOR	35-44 AT EXAM PERSON EXAMS	# OF EVENTS	ANNUAL RATE	45-54 AT EXAM PERSON EXAMS	# OF EVENTS	ANNUAL RATE	55-64 AT EXAM PERSON EXAMS	# OF EVENTS	ANNUAL RATE	65-74 AT EXAM PERSON EXAMS	# OF EVENTS	ANNUAL RATE
21-39	1697			1843	1	0	1553	8	3	646	7	5
40-41	967			1741	0	0	2361	9	2	1069	10	5
42-43	1043			1955	5	1	2260	8	2	1256	13	5
44-45	856			2194	7	2	2847	14	2	1609	16	5
46-65	487			1681	5	1	2521	24	5	1626	23	7
UNKNOWN	566			545	3	3	271	1	2	42	0	0
TOTAL	5616			9959	21	1	11513	64	3	6243	69	6

RISK FACTOR	75-84 AT EXAM PERSON EXAMS	# OF EVENTS	ANNUAL RATE	85-94 AT EXAM PERSON EXAMS	# OF EVENTS	ANNUAL RATE	35-64 COMBINED # OF EVENTS	CRUDE	AGE ADJUSTED	65-94 COMBINED # OF EVENTS	CRUDE	AGE ADJUSTED
21-39	297	6	10	24	2	42	10	1	1	15	8	7
40-41	373	9	12	31	2	32	9	1	1	18	7	5
42-43	353	3	4	19	2	53	13	1	1	23	6	5
44-45	486	7	7	26	0	0	22	1	2	31	7	7
46-65	436	8	9	23	0	-	29	3	2			
UNKNOWN	6	0	0	0	0	-	4	1	2	0	0	0
TOTAL	1951	33	8	123	6	24	87	2	2	108	6	7

LOGISTIC REGRESSIONS

AGE AT EXAM	COEFFICIENT OF RISK FACTOR	SE COEFFICIENT	P-VALUE	STANDARDIZED COEFFICIENT	STD SE
35-44	0.177	0.066	0.008	0.630	0.237
45-54	0.090	0.036	0.013	0.314	0.127
55-64	0.061	0.035	0.083	0.213	0.123
65-74	-0.043	0.044	0.318	-0.169	0.169
75-84	-0.198	0.112	0.078	-0.789	0.447
85-94					
35-64 UNIVARIATE	0.134	0.031	0.001	0.482	0.112
65-94 UNIVARIATE	0.001	0.027	0.969	0.004	0.097
35-64 BIVARIATE	0.100	0.032	0.002	0.360	0.114
65-94 BIVARIATE	0.007	0.027	0.782	0.027	0.096
35-64 MULTIVARIATE	0.040	0.031	0.196	0.144	0.112
65-94 MULTIVARIATE	0.002	0.027	0.951	0.006	0.096

TABLE 3-6   DATE=30MAR85

AVERAGE ANNUAL INCIDENCE RATE FOR EVENT PER 1000 PERSONS AT RISK AT EXAM
BY AGE AND LEVEL OF RISK FACTOR AT EXAM: FRAMINGHAM STUDY 30-YEAR FOLLOWUP

SEX: MALE
EVENT: MYOCARDIAL INFARCTION
RISK FACTOR: BLOOD GLUCOSE (MG/DL)
PERSONS AT RISK: PERSONS FREE OF CORONARY HEART DISEASE

35-44 AT EXAM PERSON EXAMS	# OF EVENTS	ANNUAL RATE	45-54 AT EXAM PERSON EXAMS	# OF EVENTS	ANNUAL RATE	55-64 AT EXAM PERSON EXAMS	# OF EVENTS	ANNUAL RATE	65-74 AT EXAM PERSON EXAMS	# OF EVENTS	ANNUAL RATE
992	5	3	1490	15	5	1087	18	8	421	4	5
2501	15	3	4172	35	4	4056	67	8	1761	45	13
626	1	1	1348	18	7	1866	36	10	1046	23	11
105	0	0	305	7	11	537	11	10	335	10	15
48	0	0	193	4	10	408	7	9	334	14	21
200	2	5	216	2	5	95	3	16	16	1	31
4472	23	3	7724	81	5	8049	142	9	3913	97	12

75-84 AT EXAM PERSON EXAMS	# OF EVENTS	ANNUAL RATE	85-94 AT EXAM PERSON EXAMS	# OF EVENTS	ANNUAL RATE	35-64 COMBINED # OF EVENTS	CRUDE	AGE-ADJUSTED	65-94 COMBINED # OF EVENTS	CRUDE	AGE-ADJUSTED
62	4	32	1			38	5	5	8	8	10
577	15	20	11			117	5	5	60	14	14
328	12	18	14			55	7	7	35	13	12
125	3	12	4			18	10	8	13	14	14
125	3	12	9			11	8	7	18	19	19
0	0	-	0			7	7	9	1	31	24
1017	37	18	39			246	6	6	135	14	13

LOGISTIC REGRESSIONS

	COEFFICIENT OF RISK FACTOR	SE COEFFICIENT	P-VALUE	STANDARDIZED COEFFICIENT	STD SE
	-0.012	0.017	0.464	-0.210	0.287
	0.006	0.003	0.030	0.150	0.069
	0.001	0.003	0.620	0.039	0.079
	0.007	0.002	0.001	0.241	0.063
	-0.006	0.006	0.324	-0.206	0.209
VARIATE	0.005	0.002	0.014	0.116	0.047
VARIATE	0.006	0.002	0.001	0.194	0.060
ARIATE	0.002	0.002	0.254	0.060	0.053
ARIATE	0.006	0.002	0.002	0.189	0.061
TIVARIATE	0.002	0.002	0.274	0.058	0.053
TIVARIATE	0.005	0.002	0.013	0.156	0.063

SEX: FEMALE
EVENT: MYOCARDIAL INFARCTION
RISK FACTOR: BLOOD GLUCOSE (MG/DL)
PERSONS AT RISK: PERSONS FREE OF CORONARY HEART DISEASE

35-44 AT EXAM PERSON EXAMS	# OF EVENTS	ANNUAL RATE	45-54 AT EXAM PERSON EXAMS	# OF EVENTS	ANNUAL RATE	55-64 AT EXAM PERSON EXAMS	# OF EVENTS	ANNUAL RATE	65-74 AT EXAM PERSON EXAMS	# OF EVENTS	ANNUAL RATE
1066			1797	3	1	1369	9	3	521	6	6
3412			5744	9	1	6098	22	2	3102	28	5
672			1625	1	0	2712	18	3	1732	24	7
96			501	1	2	730	7	5	501	2	2
39			159	4	13	438	7	8	360	9	13
331			333	3	5	166	1	3	32	0	0
5616			9959	21	1	11513	64	3	6248	69	6

75-84 AT EXAM PERSON EXAMS	# OF EVENTS	ANNUAL RATE	85-94 AT EXAM PERSON EXAMS	# OF EVENTS	ANNUAL RATE	35-64 COMBINED # OF EVENTS	CRUDE	AGE-ADJUSTED	65-94 COMBINED # OF EVENTS	CRUDE	AGE-ADJUSTED
105	2	10	3	0	0	12	1	2	8	6	7
823	13	8	44	2	23	33	1	1	43	5	6
640	7	5	51	1	10	19	2	1	32	7	7
219	5	11	13	1	38	8	4	3	8	5	5
163	6	18	12	2	83	11	9	8	17	16	15
1	0	-	0	0	-	4	2	3	0	0	0
1951	33	8	123	6	24	87	2	2	108	6	7

LOGISTIC REGRESSIONS

COEFFICIENT OF RISK FACTOR	SE COEFFICIENT	P-VALUE	STANDARDIZED COEFFICIENT	STD SE
0.013	0.005	0.013	0.249	0.100
0.011	0.002	0.001	0.266	0.058
0.008	0.003	0.002	0.223	0.073
0.006	0.003	0.071	0.199	0.110

TABLE 3-7 . DATE=23OCT85
AVERAGE ANNUAL INCIDENCE RATE FOR EVENT PER 1000 PERSONS AT RISK AT EXAM
BY AGE AND LEVEL OF RISK FACTOR AT EXAM: FRAMINGHAM STUDY 30-YEAR FOLLOWUP

SEX: MALE
EVENT: MYOCARDIAL INFARCTION
RISK FACTOR: DIABETES MELLITUS
PERSONS AT RISK: PERSONS FREE OF CORONARY HEART DISEASE

RISK FACTOR	35-44 AT EXAM PERSON EXAMS	# OF EVENTS	ANNUAL RATE	45-54 AT EXAM PERSON EXAMS	# OF EVENTS	ANNUAL RATE	55-64 AT EXAM PERSON EXAMS	# OF EVENTS	ANNUAL RATE	65-74 AT EXAM PERSON EXAMS	# OF EVENTS	ANNUAL RATE
NO	4424	23	3	7486	77	5	7549	128	8	3504	79	11
YES	48	0	0	238	4	8	500	14	14	409	18	22
UNKNOWN	0	0	-	0	0	-	0	0	-	0	0	-
TOTAL	4472	23	3	7724	81	5	8049	142	9	3913	97	12

RISK FACTOR	75-84 AT EXAM PERSON EXAMS	# OF EVENTS	ANNUAL RATE	85-94 AT EXAM PERSON EXAMS	# OF EVENTS	ANNUAL RATE	35-64 COMBINED # OF EVENTS	CRUDE	AGE-ADJUSTED	65-94 COMBINED # OF EVENTS	CRUDE	AGE-ADJUSTED
NO	900	35	19	34			228	6	6	114	13	13
YES	117	2	9	5			18	11	9	21	20	20
UNKNOWN	0	0	-	0			0	-		0		
TOTAL	1017	37	18	39			246	6	6	135	14	13

LOGISTIC REGRESSIONS

AGE AT EXAM	COEFFICIENT OF RISK FACTOR	SE COEFFICIENT	P-VALUE	STANDARDIZED COEFFICIENT	STD SE
35-44	-				
45-54	0.498	0.517	0.336	0.086	0.089
55-64	0.513	0.285	0.072	0.124	0.069
65-74	0.691	0.267	0.010	0.211	0.082
75-84	-0.844	0.734	0.250	-0.270	0.234
85-94	-				
35-64 UNIVARIATE	0.682	0.248	0.006	0.132	0.048
65-94 UNIVARIATE	0.446	0.242	0.065	0.138	0.075
35-64 BIVARIATE	0.425	0.250	0.089	0.082	0.048
65-94 BIVARIATE	0.434	0.242	0.073	0.134	0.075
35-64 MULTIVARIATE	0.413	0.254	0.105	0.080	0.049
65-94 MULTIVARIATE	0.374	0.246	0.129	0.115	0.076

SEX: FEMALE
EVENT: MYOCARDIAL INFARCTION
RISK FACTOR: DIABETES MELLITUS
PERSONS AT RISK: PERSONS FREE OF CORONARY HEART DISEASE

RISK FACTOR	35-44 AT EXAM PERSON EXAMS	# OF EVENTS	ANNUAL RATE	45-54 AT EXAM PERSON EXAMS	# OF EVENTS	ANNUAL RATE	55-64 AT EXAM PERSON EXAMS	# OF EVENTS	ANNUAL RATE	65-74 AT EXAM PERSON EXAMS	# OF EVENTS	ANNUAL RATE
NO	5573			9760	18	1	10985	54	2	5780	54	5
YES	43			199	3	8	528	10	9	468	15	16
UNKNOWN	0			0	0	-	0	0	-	0	0	-
TOTAL	5616			9959	21	1	11513	64	3	6248	69	6

RISK FACTOR	75-84 AT EXAM PERSON EXAMS	# OF EVENTS	ANNUAL RATE	85-94 AT EXAM PERSON EXAMS	# OF EVENTS	ANNUAL RATE	35-64 COMBINED # OF EVENTS	CRUDE	AGE-ADJUSTED	65-94 COMBINED # OF EVENTS	CRUDE	AGE-ADJUSTED
NO	1738	27	8	106	4	19	73	1	1	85	6	6
YES	213	6	14	17	2	59	14	9	9	23	16	16
UNKNOWN	0	0	-	0	0	24	0	-		0	-	
TOTAL	1951	33	8	123	6	24	87	2	2	108	6	7

LOGISTIC REGRESSIONS

AGE AT EXAM	COEFFICIENT OF RISK FACTOR	SE COEFFICIENT	P-VALUE	STANDARDIZED COEFFICIENT	STD SE
35-44					
45-54	2.114	0.628	0.001	0.296	0.088
55-64	1.363	0.347	0.001	0.285	0.073
65-74	1.256	0.296	0.001	0.331	0.078
75-84	0.608	0.457	0.184	0.190	0.143
85-94	1.224	0.909	0.178	0.424	0.315
35-64 UNIVARIATE	1.896	0.294	0.001	0.315	0.049
65-94 UNIVARIATE	1.106	0.238	0.001	0.307	0.066
35-64 BIVARIATE	1.528	0.297	0.001	0.254	0.049
65-94 BIVARIATE	1.040	0.240	0.001	0.288	0.066
35-64 MULTIVARIATE	1.240	0.305	0.001	0.206	0.051
65-94 MULTIVARIATE	0.908	0.247	0.001	0.252	0.068

TABLE 5-8   DATE=23OCT85
AVERAGE ANNUAL INCIDENCE RATE FOR EVENT PER 1000 PERSONS AT RISK AT EXAM
BY AGE AND LEVEL OF RISK FACTOR AT EXAM: FRAMINGHAM STUDY 30-YEAR FOLLOWUP

SEX: MALE
EVENT: MYOCARDIAL INFARCTION
RISK FACTOR: GLUCOSE IN URINE, DEFINITE OR TRACE
PERSONS AT RISK: PERSONS FREE OF CORONARY HEART DISEASE

35-44 AT EXAM			45-54 AT EXAM			55-64 AT EXAM			65-74 AT EXAM		
PERSON EXAMS	# OF EVENTS	ANNUAL RATE	PERSON EXAMS	# OF EVENTS	ANNUAL RATE	PERSON EXAMS	# OF EVENTS	ANNUAL RATE	PERSON EXAMS	# OF EVENTS	ANNUAL RATE
4380	23	3	7555	76	5	7815	129	8	3767	84	11
60	0	0	168	5	15	210	12	29	140	13	46
32	0	0	21	0	0	24	1	21	6	0	0
4472	23	3	7724	81	5	8049	142	9	3913	97	12

75-84 AT EXAM			85-94 AT EXAM			35-64 COMBINED			65-94 COMBINED		
PERSON EXAMS	# OF EVENTS	ANNUAL RATE	PERSON EXAMS	# OF EVENTS	ANNUAL RATE	# OF EVENTS	CRUDE	AGE-ADJUSTED	# OF EVENTS	CRUDE	AGE-ADJUSTED
986	36	18	38			228	6	6	120	13	12
28	1	18	1			17	19	17	15	44	44
3	0	0	0			1	6	8	0	0	0
1017	37	18	39			246	6	6	135	14	13

LOGISTIC REGRESSIONS

	COEFFICIENT OF RISK FACTOR	SE COEFFICIENT	P-VALUE	STANDARDIZED COEFFICIENT	STD SE
	1.102	0.468	0.019	0.161	0.068
	1.284	0.310	0.001	0.205	0.050
	1.501	0.311	0.001	0.279	0.058
	-0.023	1.032	0.982	-0.004	0.169
/ARIATE	1.240	0.256	0.001	0.181	0.037
/ARIATE	1.333	0.286	0.001	0.242	0.052
\RIATE	1.129	0.257	0.001	0.165	0.038
\RIATE	1.349	0.286	0.001	0.245	0.052
[IVARIATE	1.088	0.268	0.001	0.159	0.039
[IVARIATE	1.221	0.294	0.001	0.221	0.053

SEX: FEMALE
EVENT: MYOCARDIAL INFARCTION
RISK FACTOR: GLUCOSE IN URINE, DEFINITE OR TRACE
PERSONS AT RISK: PERSONS FREE OF CORONARY HEART DISEASE

35-44 AT EXAM			45-54 AT EXAM			55-64 AT EXAM			65-74 AT EXAM		
PERSON EXAMS	# OF EVENTS	ANNUAL RATE	PERSON EXAMS	# OF EVENTS	ANNUAL RATE	PERSON EXAMS	# OF EVENTS	ANNUAL RATE	PERSON EXAMS	# OF EVENTS	ANNUAL RATE
5537			9841	20	1	11346	60	3	6131	66	5
41			94	1	5	144	4	14	107	3	14
38			24	0	0	23	0	0	10	0	0
5616			9959	21	1	11513	64	3	6248	69	6

75-84 AT EXAM			85-94 AT EXAM			35-64 COMBINED			65-94 COMBINED		
PERSON EXAMS	# OF EVENTS	ANNUAL RATE	PERSON EXAMS	# OF EVENTS	ANNUAL RATE	# OF EVENTS	CRUDE	AGE-ADJUSTED	# OF EVENTS	CRUDE	AGE-ADJUSTED
1919	32	8	120	5	21	81	2	2	103	6	6
23	1	22	3	1	167	6	11	10	3	19	18
9	0	0	0	0	-	0	0	0	0	0	0
1951	33	8	123	6	24	87	2	2	108	6	7

LOGISTIC REGRESSIONS

	COEFFICIENT OF RISK FACTOR	SE COEFFICIENT	P-VALUE	STANDARDIZED COEFFICIENT	STD SE
	1.664	1.030	0.106	0.161	0.100
	1.682	0.523	0.001	0.187	0.058
	0.975	0.599	0.103	0.127	0.078
	0.986	1.038	0.342	0.107	0.112
	2.442	1.307	0.062	0.378	0.202
VARIATE	1.978	0.427	0.001	0.200	0.043
VARIATE	1.118	0.467	0.017	0.140	0.059
ARIATE	1.856	0.450	0.001	0.188	0.043
ARIATE	1.148	0.467	0.014	0.144	0.059
TIVARIATE	1.440	0.443	0.001	0.146	0.045
TIVARIATE	0.918	0.480	0.056	0.115	0.060

TABLE 3-9   DATE=01MAR86

AVERAGE ANNUAL INCIDENCE RATE FOR EVENT PER 1000 PERSONS AT RISK AT EXAM
BY AGE AND LEVEL OF RISK FACTOR AT EXAM: FRAMINGHAM STUDY 30-YEAR FOLLOWUP

SEX: MALE
EVENT: MYOCARDIAL INFARCTION
RISK FACTOR: GLUCOSE INTOLERANCE
PERSONS AT RISK: PERSONS FREE OF CORONARY HEART DISEASE

RISK FACTOR	35-44 AT EXAM			45-54 AT EXAM			55-64 AT EXAM			65-74 AT EXAM		
	PERSON EXAMS	# OF EVENTS	ANNUAL RATE	PERSON EXAMS	# OF EVENTS	ANNUAL RATE	PERSON EXAMS	# OF EVENTS	ANNUAL RATE	PERSON EXAMS	# OF EVENTS	ANNUAL RATE
NO	4335	23	5	7221	68	5	7145	121	8	3228	71	11
YES	137	0	0	503	13	13	904	21	12	685	26	19
UNKNOWN	0			0			0			0		
TOTAL	4472	23	5	7724	81	5	8049	142	9	3913	97	12

RISK FACTOR	75-84 AT EXAM			85-94 AT EXAM			35-64 COMBINED			65-94 COMBINED		
	PERSON EXAMS	# OF EVENTS	ANNUAL RATE	PERSON EXAMS	# OF EVENTS	ANNUAL RATE	# OF EVENTS	CRUDE	AGE-ADJUSTED	# OF EVENTS	CRUDE	AGE-ADJUSTED
NO	802	33	21	27			212	6	6	104	13	13
YES	215	4	9	12			34	11	10	31	17	17
UNKNOWN	0			0			0			0		
TOTAL	1017	37	18	39			246	6	6	135	14	14

LOGISTIC REGRESSIONS

AGE AT EXAM	COEFFICIENT OF RISK FACTOR	SE COEFFICIENT	P-VALUE	STANDARDIZED COEFFICIENT	STD SE
35-44	-	-	-	-	-
45-54	1.026	0.306	0.001	0.253	0.076
55-64	0.322	0.239	0.177	0.102	0.075
65-74	0.562	0.253	0.016	0.214	0.089
75-84	-0.817	0.535	0.127	-0.334	0.219
85-94					
35-64 UNIVARIATE	0.675	0.187	0.001	0.179	0.050
65-94 UNIVARIATE	0.291	0.208	0.162	0.113	0.081
35-64 BIVARIATE	0.463	0.189	0.014	0.123	0.050
65-94 BIVARIATE	0.269	0.209	0.197	0.104	0.081
35-64 MULTIVARIATE	0.443	0.194	0.022	0.118	0.051
65-94 MULTIVARIATE	0.206	0.211	0.329	0.080	0.082

SEX: FEMALE
EVENT: MYOCARDIAL INFARCTION
RISK FACTOR: GLUCOSE INTOLERANCE
PERSONS AT RISK: PERSONS FREE OF CORONARY HEART DISEASE

RISK FACTOR	35-44 AT EXAM			45-54 AT EXAM			55-64 AT EXAM			65-74 AT EXAM		
	PERSON EXAMS	# OF EVENTS	ANNUAL RATE	PERSON EXAMS	# OF EVENTS	ANNUAL RATE	PERSON EXAMS	# OF EVENTS	ANNUAL RATE	PERSON EXAMS	# OF EVENTS	ANNUAL RATE
NO	5522			9588	16	1	10503	50	2	5398	52	5
YES	94			371	5	7	1010	14	7	850	17	10
UNKNOWN	0			0			0			0		
TOTAL	5616			9959	21	1	11513	64	3	6248	69	6

RISK FACTOR	75-84 AT EXAM			85-94 AT EXAM			35-64 COMBINED			65-94 COMBINED		
	PERSON EXAMS	# OF EVENTS	ANNUAL RATE	PERSON EXAMS	# OF EVENTS	ANNUAL RATE	# OF EVENTS	CRUDE	AGE-ADJUSTED	# OF EVENTS	CRUDE	AGE-ADJUSTED
NO	1581	23	7	100	2	10	67	1	1	77	5	5
YES	370	10	14	23	4	87	20	7	7	31	12	12
UNKNOWN	0			0			0			0		
TOTAL	1951	33	8	123	6	24	87	2	2	108	6	6

LOGISTIC REGRESSIONS

AGE AT EXAM	COEFFICIENT OF RISK FACTOR	SE COEFFICIENT	P-VALUE	STANDARDIZED COEFFICIENT	STD SE
35-44					
45-54	2.101	0.515	0.001	0.398	0.098
55-64	1.078	0.304	0.001	0.305	0.086
65-74	0.741	0.282	0.009	0.254	0.097
75-84	0.632	0.383	0.099	0.248	0.150
85-94	2.334	0.902	0.010	0.914	0.353
35-64 UNIVARIATE	1.657	0.256	0.001	0.376	0.058
65-94 UNIVARIATE	0.844	0.215	0.001	0.301	0.077
35-64 BIVARIATE	1.290	0.260	0.001	0.293	0.059
65-94 BIVARIATE	0.783	0.216	0.001	0.279	0.077
35-64 MULTIVARIATE	1.154	0.264	0.001	0.262	0.060
65-94 MULTIVARIATE	0.717	0.220	0.001	0.255	0.078

TABLE 3-10   DATE=30MAR85

AVERAGE ANNUAL INCIDENCE RATE FOR EVENT PER 1000 PERSONS AT RISK AT EXAM
BY AGE AND LEVEL OF RISK FACTOR AT EXAM: FRAMINGHAM STUDY 30-YEAR FOLLOWUP

SEX: MALE
EVENT: MYOCARDIAL INFARCTION
RISK FACTOR: METROPOLITAN RELATIVE WEIGHT (PERCENT)
PERSONS AT RISK: PERSONS FREE OF CORONARY HEART DISEASE

RISK FACTOR	35-44 AT EXAM			45-54 AT EXAM			55-64 AT EXAM			65-74 AT EXAM		
	PERSON EXAMS	# OF EVENTS	ANNUAL RATE	PERSON EXAMS	# OF EVENTS	ANNUAL RATE	PERSON EXAMS	# OF EVENTS	ANNUAL RATE	PERSON EXAMS	# OF EVENTS	ANNUAL RATE
54-104	723	1	1	1131	7	3	1155	23	10	686	9	7
105-114	965	7	4	1540	16	5	1655	22	7	851	21	12
115-124	1122	6	3	2037	22	5	2175	37	9	1069	29	14
125-134	915	7	4	1645	18	5	1674	32	10	766	18	12
135-272	747	2	1	1370	18	7	1389	28	10	541	20	18
UNKNOWN	0	0	-	1	0	-	1	0	0	0	0	-
TOTAL	4472	23	3	7724	81	5	8049	142	9	3913	97	12

RISK FACTOR	75-84 AT EXAM			85-94 AT EXAM			35-64 COMBINED			65-94 COMBINED		
	PERSON EXAMS	# OF EVENTS	ANNUAL RATE	PERSON EXAMS	# OF EVENTS	ANNUAL RATE	# OF EVENTS	CRUDE	AGE-ADJUSTED	# OF EVENTS	CRUDE	AGE-ADJUSTED
54-104	224	9	20	18			31	5	5	18	10	10
105-114	252	7	14	9			45	5	6	29	13	13
115-124	296	9	15	12			65	6	6	38	14	14
125-134	155	9	29	0			57	7	7	27	15	15
135-272	90	3	17	0			48	7	7	23	18	18
UNKNOWN	0	0	-	0			0	0	0	0	-	-
TOTAL	1017	37	18	59			246	6	6	135	14	13

LOGISTIC REGRESSIONS

AGE AT EXAM		COEFFICIENT OF RISK FACTOR	SE COEFFICIENT	P-VALUE	STANDARDIZED COEFFICIENT	STD SE
35-44		-0.002	0.013	0.889	-0.029	0.210
45-54		0.008	0.007	0.219	0.133	0.108
55-64		0.002	0.005	0.670	0.036	0.084
65-74		0.013	0.006	0.038	0.206	0.099
75-84		0.011	0.011	0.334	0.156	0.162
85-94						
35-64	UNIVARIATE	0.004	0.004	0.288	0.067	0.063
65-94	UNIVARIATE	0.011	0.005	0.037	0.177	0.085
35-64	BIVARIATE	0.004	0.004	0.306	0.065	0.064
65-94	BIVARIATE	0.013	0.005	0.022	0.195	0.085
35-64	MULTIVARIATE	0.002	0.004	0.663	0.029	0.066
65-94	MULTIVARIATE	0.007	0.006	0.233	0.106	0.089

SEX: FEMALE
EVENT: MYOCARDIAL INFARCTION
RISK FACTOR: METROPOLITAN RELATIVE WEIGHT (PERCENT)
PERSONS AT RISK: PERSONS FREE OF CORONARY HEART DISEASE

RISK FACTOR	35-44 AT EXAM			45-54 AT EXAM			55-64 AT EXAM			65-74 AT EXAM		
	PERSON EXAMS	# OF EVENTS	ANNUAL RATE	PERSON EXAMS	# OF EVENTS	ANNUAL RATE	PERSON EXAMS	# OF EVENTS	ANNUAL RATE	PERSON EXAMS	# OF EVENTS	ANNUAL RATE
54-104	1694			2095	4	1	2045	14	3	1216	13	5
105-114	1430			2331	4	1	2409	8	2	1145	15	7
115-124	1058			2107	4	1	2479	12	2	1595	9	3
125-134	566			1372	1	0	1743	11	3	982	9	5
135-272	860			2054	8	2	2837	19	3	1510	23	8
UNKNOWN	8			0	0	-	0	0	-	0	0	-
TOTAL	5616			9959	21	1	11513	64	3	6248	69	6

RISK FACTOR	75-84 AT EXAM			85-94 AT EXAM			35-64 COMBINED			65-94 COMBINED		
	PERSON EXAMS	# OF EVENTS	ANNUAL RATE	PERSON EXAMS	# OF EVENTS	ANNUAL RATE	# OF EVENTS	CRUDE	AGE-ADJUSTED	# OF EVENTS	CRUDE	AGE-ADJUSTED
54-104	517	7	7	41	2	24	19	2	2	22	6	6
105-114	375	3	4	28	0	0	12	1	1	18	6	6
115-124	399	13	16	21	3	71	16	1	1	25	7	7
125-134	266	3	6	16	1	31	12	2	2	13	5	6
135-272	394	7	9	17	0	0	28	2	2	30	8	8
UNKNOWN	0	0	-	0	0	-	0	0	0	0	-	-
TOTAL	1951	33	8	123	6	24	87	2	2	108	6	7

LOGISTIC REGRESSIONS

AGE AT EXAM		COEFFICIENT OF RISK FACTOR	SE COEFFICIENT	P-VALUE	STANDARDIZED COEFFICIENT	STD SE
35-44						
45-54		0.011	0.008	0.187	0.239	0.181
55-64		0.006	0.005	0.260	0.130	0.115
65-74		0.011	0.005	0.028	0.231	0.106
75-84		0.007	0.008	0.363	0.151	0.166
85-94		-0.003	0.022	0.895	-0.056	0.423
35-64	UNIVARIATE	0.010	0.004	0.015	0.222	0.091
65-94	UNIVARIATE	0.008	0.004	0.060	0.167	0.089
35-64	BIVARIATE	0.007	0.004	0.104	0.156	0.096
65-94	BIVARIATE	0.009	0.004	0.024	0.201	0.089
35-64	MULTIVARIATE	0.002	0.005	0.669	0.044	0.102
65-94	MULTIVARIATE	0.006	0.004	0.136	0.135	0.091

TABLE 3-11   DATE=30MAR85
AVERAGE ANNUAL INCIDENCE RATE FOR EVENT PER 1000 PERSONS AT RISK AT EXAM
BY AGE AND LEVEL OF RISK FACTOR AT EXAM: FRAMINGHAM STUDY 30-YEAR FOLLOWUP

SEX: MALE
EVENT: MYOCARDIAL INFARCTION
RISK FACTOR: VITAL CAPACITY - HEIGHT INDEX (ML/INCH)
PERSONS AT RISK: PERSONS FREE OF CORONARY HEART DISEASE

RISK FACTOR	35-44 AT EXAM			45-54 AT EXAM			55-64 AT EXAM			65-74 A	
	PERSON EXAMS	# OF EVENTS	ANNUAL RATE	PERSON EXAMS	# OF EVENTS	ANNUAL RATE	PERSON EXAMS	# OF EVENTS	ANNUAL RATE	PERSON EXAMS	# EV
8-44	269	2	4	1062	18	8	2045	48	12	1752	44
45-49	411	1	1	1125	13	6	1480	20	7	749	26
50-54	747	0	0	1658	18	5	1835	31	8	712	13
55-59	1022	4	2	1739	16	5	1416	27	10	442	6
60-85	2010	16	4	2135	16	4	1255	16	6	246	7
UNKNOWN	13	0	0	5	0	0	18	0	0	12	1
TOTAL	4472	23	3	7724	81	5	8049	142	9	3913	97

RISK FACTOR	75-84 AT EXAM			85-94 AT EXAM			35-64 COMBINED			65-94 (	
	PERSON EXAMS	# OF EVENTS	ANNUAL RATE	PERSON EXAMS	# OF EVENTS	ANNUAL RATE	# OF EVENTS	CRUDE	AGE-ADJUSTED	# OF EVENTS	CI
8-44	649	29	22	24			68	10	9	74	15
45-49	172	2	6	6			34	6	5	28	15
50-54	97	4	21	7			49	6	5	17	10
55-59	65	2	15	2			47	6	6	8	8
60-85	26	0	0	0			48	4	5	7	13
UNKNOWN	8	0	0	0			0	0	0	1	25
TOTAL	1017	37	18	39			246	6	6	135	14

LOGISTIC REGRESSIONS

AGE AT EXAM		COEFFICIENT OF RISK FACTOR	SE COEFFICIENT	P-VALUE	STANDARDIZED COEFFICIENT	STD SE
35-44		0.019	0.024	0.433	0.167	0.213
45-54		-0.025	0.012	0.037	-0.227	0.109
55-64		-0.018	0.009	0.044	-0.167	0.083
65-74		-0.010	0.011	0.336	-0.099	0.103
75-84		-0.028	0.018	0.125	-0.259	0.169
85-94						
35-64	UNIVARIATE	-0.028	0.006	0.001	-0.270	0.062
65-94	UNIVARIATE	-0.017	0.009	0.054	-0.167	0.087
35-64	BIVARIATE	-0.014	0.007	0.049	-0.131	0.066
65-94	BIVARIATE	-0.015	0.009	0.110	-0.143	0.089
35-64	MULTIVARIATE	-0.004	0.007	0.621	-0.034	0.070
65-94	MULTIVARIATE	-0.012	0.010	0.212	-0.115	0.092

SEX: FEMALE
EVENT: MYOCARDIAL INFARCTION
RISK FACTOR: VITAL CAPACITY - HEIGHT INDEX (ML/INCH)
PERSONS AT RISK: PERSONS FREE OF CORONARY HEART DISEASE

RISK FACTOR	35-44 AT EXAM			45-54 AT EXAM			55-64 AT EXAM			65-74 A]	
	PERSON EXAMS	# OF EVENTS	ANNUAL RATE	PERSON EXAMS	# OF EVENTS	ANNUAL RATE	PERSON EXAMS	# OF EVENTS	ANNUAL RATE	PERSON EXAMS	EVE
4-29	122			534	2	2	1461	13	4	1747	31
30-34	289			1217	3	1	2418	17	4	1715	18
35-39	769			2077	3	1	3048	11	2	1529	9
40-44	1427			2804	7	1	2706	19	4	854	6
45-92	2988			3310	6	1	1859	4	1	392	5
UNKNOWN	21			17	0	0	21	0	0	6	0
TOTAL	5616			9959	21	1	11513	64	3	6248	69

RISK FACTOR	75-84 AT EXAM			85-94 AT EXAM			35-64 COMBINED			65-94 C	
	PERSON EXAMS	# OF EVENTS	ANNUAL RATE	PERSON EXAMS	# OF EVENTS	ANNUAL RATE	# OF EVENTS	CRUDE	AGE-ADJUSTED	# OF EVENTS	CR
4-29	930	20	11	80	4	25	15	4	2	55	10
30-34	569	10	9	29	2	34	20	3	2	30	6
35-39	314	3	5	13	0	0	15	1	1	12	5
40-44	111	0	0	1	0	0	26	2	2	6	6
45-92	25	0	0	0	0	-	11	1	1	5	6
UNKNOWN	2	0	-	0	0	-	0	0	0	0	0
TOTAL	1951	33	8	123	6	24	87	2	2	108	6

LOGISTIC REGRESSIONS

AGE AT EXAM		COEFFICIENT OF RISK FACTOR	SE COEFFICIENT	P-VALUE	STANDARDIZED COEFFICIENT	STD SE
35-44						
45-54		-0.021	0.029	0.477	-0.153	0.215
55-64		-0.039	0.016	0.014	-0.297	0.121
65-74		-0.039	0.016	0.015	-0.287	0.118
75-84		-0.053	0.024	0.026	-0.371	0.167
85-94		-0.062	0.056	0.272	-0.417	0.379
35-64	UNIVARIATE	-0.058	0.013	0.001	-0.463	0.102
65-94	UNIVARIATE	-0.050	0.012	0.001	-0.380	0.094
35-64	BIVARIATE	-0.028	0.014	0.046	-0.226	0.113
65-94	BIVARIATE	-0.043	0.013	0.001	-0.324	0.098
35-64	MULTIVARIATE	-0.008	0.015	0.609	-0.063	0.122
65-94	MULTIVARIATE	-0.039	0.014	0.005	-0.291	0.103

TABLE 3-12   DATE=30MAR85

AVERAGE ANNUAL INCIDENCE RATE FOR EVENT PER 1000 PERSONS AT RISK AT EXAM
BY AGE AND LEVEL OF RISK FACTOR AT EXAM: FRAMINGHAM STUDY 30-YEAR FOLLOWUP

SEX: MALE
EVENT: MYOCARDIAL INFARCTION
RISK FACTOR: HEART RATE (PER MINUTE)
PERSONS AT RISK: PERSONS FREE OF CORONARY HEART DISEASE

RISK FACTOR	35-44 AT EXAM PERSON EXAMS	# OF EVENTS	ANNUAL RATE	45-54 AT EXAM PERSON EXAMS	# OF EVENTS	ANNUAL RATE	55-64 AT EXAM PERSON EXAMS	# OF EVENTS	ANNUAL RATE	65-74 AT EXAM PERSON EXAMS	# OF EVENTS	ANNUAL RATE
30-67	1284	5	2	1949	17	4	2309	27	6	1183	29	12
68-75	1492	6	2	2490	21	4	2490	53	11	1159	19	8
76-83	722	5	3	1394	14	5	1374	25	9	624	22	18
84-91	579	4	3	1097	16	7	1064	20	9	487	12	12
92-220	395	3	4	794	13	8	812	17	10	460	15	16
UNKNOWN	0	0	-	0	0	-	0	0	-	0	0	-
TOTAL	4472	23	3	7724	81	5	8049	142	9	3913	97	12

RISK FACTOR	75-84 AT EXAM PERSON EXAMS	# OF EVENTS	ANNUAL RATE	85-94 AT EXAM PERSON EXAMS	# OF EVENTS	ANNUAL RATE	35-64 COMBINED # OF EVENTS	CRUDE	AGE-ADJUSTED	65-94 COMBINED # OF EVENTS	CRUDE	AGE-ADJUSTED
30-67	301	12	20	11	-	-	49	4	4	41	14	14
68-75	296	12	20	12	-	-	80	6	6	32	11	11
76-83	192	5	13	10	-	-	44	6	6	27	16	17
84-91	123	5	20	5	-	-	40	7	7	17	14	14
92-220	105	3	14	1	-	-	33	8	8	18	16	15
UNKNOWN	0	0	-	0	-	-	0	-	-	0	-	-
TOTAL	1017	37	18	39	-	-	246	6	6	135	14	13

LOGISTIC REGRESSIONS

AGE AT EXAM		COEFFICIENT OF RISK FACTOR	SE COEFFICIENT	P-VALUE	STANDARDIZED COEFFICIENT	STD SE
35-44		0.026	0.015	0.080	0.326	0.187
45-54		0.019	0.008	0.023	0.213	0.105
55-64		0.010	0.006	0.098	0.134	0.081
65-74		0.012	0.007	0.101	0.159	0.097
75-84		-0.005	0.013	0.676	-0.072	0.172
85-94						
35-64	UNIVARIATE	0.015	0.005	0.002	0.186	0.061
65-94	UNIVARIATE	0.007	0.006	0.238	0.099	0.084
35-64	BIVARIATE	0.015	0.005	0.002	0.184	0.060
65-94	BIVARIATE	0.007	0.006	0.239	0.099	0.084
35-64	MULTIVARIATE	0.003	0.005	0.527	0.041	0.065
65-94	MULTIVARIATE	0.006	0.006	0.340	0.082	0.086

SEX: FEMALE
EVENT: MYOCARDIAL INFARCTION
RISK FACTOR: HEART RATE (PER MINUTE)
PERSONS AT RISK: PERSONS FREE OF CORONARY HEART DISEASE

RISK FACTOR	35-44 AT EXAM PERSON EXAMS	# OF EVENTS	ANNUAL RATE	45-54 AT EXAM PERSON EXAMS	# OF EVENTS	ANNUAL RATE	55-64 AT EXAM PERSON EXAMS	# OF EVENTS	ANNUAL RATE	65-74 AT EXAM PERSON EXAMS	# OF EVENTS	ANNUAL RATE
30-67	1092			1897	6	2	2305	3	1	1103	15	7
68-75	1803			3244	6	1	3468	17	2	1668	19	6
76-83	1051			1911	4	1	2351	16	3	1358	10	4
84-91	939			1604	2	1	1780	12	3	1089	16	7
92-220	731			1303	3	1	1609	16	5	1030	9	4
UNKNOWN	0			0	0	-	0	0	-	0	0	-
TOTAL	5616			9959	21	1	11513	64	3	6248	69	6

RISK FACTOR	75-84 AT EXAM PERSON EXAMS	# OF EVENTS	ANNUAL RATE	85-94 AT EXAM PERSON EXAMS	# OF EVENTS	ANNUAL RATE	35-64 COMBINED # OF EVENTS	CRUDE	AGE-ADJUSTED	65-94 COMBINED # OF EVENTS	CRUDE	AGE-ADJUSTED
30-67	360	7	10	26	2	38	9	1	1	24	8	8
68-75	462	4	4	32	0	0	23	1	1	23	5	5
76-83	428	12	14	24	0	0	21	2	2	22	6	6
84-91	330	4	6	23	3	65	15	2	2	23	8	8
92-220	371	6	8	18	1	28	19	3	2	16	6	5
UNKNOWN	0	0	-	0	0	-	0	-	-	0	-	-
TOTAL	1951	33	8	123	6	24	87	2	2	108	6	7

LOGISTIC REGRESSIONS

AGE AT EXAM		COEFFICIENT OF RISK FACTOR	SE COEFFICIENT	P-VALUE	STANDARDIZED COEFFICIENT	STD SE
35-44						
45-54		-0.014	0.019	0.459	-0.171	0.231
55-64		0.027	0.008	0.001	0.348	0.105
65-74		-0.006	0.009	0.533	-0.078	0.124
75-84		-0.002	0.013	0.862	-0.031	0.178
85-94		0.012	0.030	0.695	0.161	0.411
35-64	UNIVARIATE	0.020	0.007	0.007	0.251	0.094
65-94	UNIVARIATE	-0.004	0.007	0.627	-0.048	0.099
35-64	BIVARIATE	0.019	0.007	0.010	0.242	0.093
65-94	BIVARIATE	-0.004	0.007	0.586	-0.053	0.098
35-64	MULTIVARIATE	0.006	0.008	0.448	0.079	0.104
65-94	MULTIVARIATE	-0.008	0.007	0.262	-0.111	0.099

- 77

TABLE 3-13   DATE=01APR85

AVERAGE ANNUAL INCIDENCE RATE FOR EVENT PER 1000 PERSONS AT RISK AT EXAM
BY AGE AND LEVEL OF RISK FACTOR AT EXAM: FRAMINGHAM STUDY 30-YEAR FOLLOWUP

SEX: MALE
EVENT: MYOCARDIAL INFARCTION
RISK FACTOR: CIGARETTES SMOKED (NUMBER PER DAY)
PERSONS AT RISK: PERSONS FREE OF CORONARY HEART DISEASE

RISK FACTOR	35-44 AT EXAM			45-54 AT EXAM			55-64 AT EXAM			65-74 AT EXAM		
	PERSON EXAMS	# OF EVENTS	ANNUAL RATE	PERSON EXAMS	# OF EVENTS	ANNUAL RATE	PERSON EXAMS	# OF EVENTS	ANNUAL RATE	PERSON EXAMS	# OF EVENTS	ANNUAL RATE
NONE	1383	3	1	3068	13	2	4295	56	7	2611	69	13
1-10	433	3	3	716	7	5	769	8	5	374	7	9
11-20	1485	11	4	2051	33	8	1682	45	13	526	13	12
21-40	987	5	3	1602	25	8	1119	30	13	369	6	8
41-90	158	1	3	232	2	4	162	3	9	27	1	19
UNKNOWN	26	0	0	35	1	14	22	0	0	6	1	83
TOTAL	4472	23	3	7724	81	5	8049	142	9	3913	97	12

RISK FACTOR	75-84 AT EXAM			85-94 AT EXAM			35-64 COMBINED			65-94 COMBINED		
	PERSON EXAMS	# OF EVENTS	ANNUAL RATE	PERSON EXAMS	# OF EVENTS	ANNUAL RATE	# OF EVENTS	CRUDE	AGE-ADJUSTED	# OF EVENTS	CRUDE	AGE-ADJUSTED
NONE	797	31	19	27			72	4	4	101	15	14
1-10	92	3	16	6			18	5	5	10	11	10
11-20	77	1	6	5			89	9	9	14	12	11
21-40	46	2	22	1			60	8	9	8	10	11
41-90	5	0	0	0			6	5	6	1	16	15
UNKNOWN	0	0	-	0			1	6	5	1	83	65
TOTAL	1017	37	18	39			246	6	6	135	14	13

LOGISTIC REGRESSIONS

AGE AT EXAM		COEFFICIENT OF RISK FACTOR	SE COEFFICIENT	P-VALUE	STANDARDIZED COEFFICIENT	STD SE
35-44		0.014	0.014	0.308	0.204	0.200
45-54		0.024	0.007	0.001	0.364	0.099
55-64		0.021	0.005	0.001	0.291	0.072
65-74		-0.007	0.010	0.475	-0.079	0.110
75-84		-0.018	0.023	0.428	-0.165	0.208
85-94						
35-64	UNIVARIATE	0.017	0.004	0.001	0.254	0.058
65-94	UNIVARIATE	-0.011	0.009	0.225	-0.119	0.098
35-64	BIVARIATE	0.023	0.004	0.001	0.335	0.057
65-94	BIVARIATE	-0.009	0.009	0.305	-0.101	0.098
35-64	MULTIVARIATE	0.025	0.004	0.001	0.370	0.057
65-94	MULTIVARIATE	-0.006	0.009	0.484	-0.069	0.099

SEX: FEMALE
EVENT: MYOCARDIAL INFARCTION
RISK FACTOR: CIGARETTES SMOKED (NUMBER PER DAY)
PERSONS AT RISK: PERSONS FREE OF CORONARY HEART DISEASE

RISK FACTOR	35-44 AT EXAM			45-54 AT EXAM			55-64 AT EXAM			65-74 AT EXAM		
	PERSON EXAMS	# OF EVENTS	ANNUAL RATE	PERSON EXAMS	# OF EVENTS	ANNUAL RATE	PERSON EXAMS	# OF EVENTS	ANNUAL RATE	PERSON EXAMS	# OF EVENTS	ANNUAL RATE
NONE	2582			5528	10	1	7595	30	2	4863	53	5
1-10	1314			1726	2	1	1566	11	4	626	7	6
11-20	1371			2009	6	1	1696	15	4	568	9	8
21-40	298			622	3	2	602	6	5	172	0	0
41-90	22			37	0	0	16	2	63	0	0	-
UNKNOWN	29			37	0	0	38	0	0	19	0	-
TOTAL	5616			9959	21	1	11513	64	3	6248	69	6

RISK FACTOR	75-84 AT EXAM			85-94 AT EXAM			35-64 COMBINED			65-94 COMBINED		
	PERSON EXAMS	# OF EVENTS	ANNUAL RATE	PERSON EXAMS	# OF EVENTS	ANNUAL RATE	# OF EVENTS	CRUDE	AGE-ADJUSTED	# OF EVENTS	CRUDE	AGE-ADJUSTED
NONE	1710	28	8	117	6	26	40	1	1	87	7	6
1-10	137	3	11	4	0	0	15	1	2	10	7	7
11-20	84	2	12	2	0	0	22	2	2	11	8	9
21-40	16	0	0	0	0	-	10	3	3	0	0	0
41-90	0	0	-	0	0	-	2	13	27	0	-	-
UNKNOWN	4	0	0	0	0	-	0	0	0	0	0	0
TOTAL	1951	33	8	123	6	24	87	2	2	108	6	7

LOGISTIC REGRESSIONS

AGE AT EXAM		COEFFICIENT OF RISK FACTOR	SE COEFFICIENT	P-VALUE	STANDARDIZED COEFFICIENT	STD SE
35-44						
45-54		0.026	0.018	0.143	0.267	0.182
55-64		0.041	0.010	0.001	0.388	0.092
65-74		-0.003	0.016	0.865	-0.021	0.124
75-84		0.009	0.032	0.787	0.044	0.161
85-94		-	-	-	-	-
35-64	UNIVARIATE	0.035	0.008	0.001	0.348	0.084
65-94	UNIVARIATE	-0.006	0.015	0.672	-0.044	0.103
35-64	BIVARIATE	0.042	0.008	0.001	0.415	0.082
65-94	BIVARIATE	0.000	0.015	0.991	0.001	0.103
35-64	MULTIVARIATE	0.046	0.008	0.001	0.454	0.083
65-94	MULTIVARIATE	0.007	0.015	0.649	0.047	0.103

AVERAGE ANNUAL INCIDENCE RATE FOR EVENT PER 1000 PERSONS AT RISK AT EXAM
BY AGE AND LEVEL OF RISK FACTOR AT EXAM: FRAMINGHAM STUDY 30-YEAR FOLLOWUP

SEX: MALE
EVENT: MYOCARDIAL INFARCTION
RISK FACTOR: ALBUMIN IN URINE, DEFINITE
PERSONS AT RISK: PERSONS FREE OF CORONARY HEART DISEASE

RISK FACTOR	35-44 AT EXAM PERSON EXAMS	# OF EVENTS	ANNUAL RATE	45-54 AT EXAM PERSON EXAMS	# OF EVENTS	ANNUAL RATE	55-64 AT EXAM PERSON EXAMS	# OF EVENTS	ANNUAL RATE	65-74 AT EXAM PERSON EXAMS	# OF EVENTS	ANNUAL RATE
NO	4366	23	3	7599	78	5	7838	137	9	3792	92	12
YES	70	0	0	99	3	15	184	4	11	115	5	22
UNKNOWN	36	0	0	26	0	0	27	1	19	6	0	0
TOTAL	4472	23	3	7724	81	5	8049	142	9	3913	97	12

RISK FACTOR	75-84 AT EXAM PERSON EXAMS	# OF EVENTS	ANNUAL RATE	85-94 AT EXAM PERSON EXAMS	# OF EVENTS	ANNUAL RATE	35-64 COMBINED # OF EVENTS	CRUDE	AGE-ADJUSTED	65-94 COMBINED # OF EVENTS	CRUDE	AGE-ADJUSTED
NO	989	36	18	32			238	6	6	129	13	13
YES	25	1	20	7			7	10	10	6	20	21
UNKNOWN	3	0	0	0			1	6	8	0	0	0
TOTAL	1017	37	18	39			246	6	6	135	14	13

LOGISTIC REGRESSIONS

AGE AT EXAM		COEFFICIENT OF RISK FACTOR	SE COEFFICIENT	P-VALUE	STANDARDIZED COEFFICIENT	STD SE
35-44		-	-	-	-	-
45-54		1.103	0.597	0.065	0.124	0.067
55-64		0.222	0.513	0.664	0.033	0.077
65-74		0.603	0.469	0.199	0.102	0.079
75-84		0.098	1.035	0.924	0.015	0.161
85-94						
35-64	UNIVARIATE	0.509	0.387	0.189	0.067	0.051
65-94	UNIVARIATE	0.435	0.426	0.307	0.074	0.072
35-64	BIVARIATE	0.384	0.388	0.323	0.050	0.051
65-94	BIVARIATE	0.421	0.427	0.324	0.071	0.072
35-64	MULTIVARIATE	-0.150	0.425	0.723	-0.020	0.056
65-94	MULTIVARIATE	0.042	0.439	0.923	0.007	0.074

SEX: FEMALE
EVENT: MYOCARDIAL INFARCTION
RISK FACTOR: ALBUMIN IN URINE, DEFINITE
PERSONS AT RISK: PERSONS FREE OF CORONARY HEART DISEASE

RISK FACTOR	35-44 AT EXAM PERSON EXAMS	# OF EVENTS	ANNUAL RATE	45-54 AT EXAM PERSON EXAMS	# OF EVENTS	ANNUAL RATE	55-64 AT EXAM PERSON EXAMS	# OF EVENTS	ANNUAL RATE	65-74 AT EXAM PERSON EXAMS	# OF EVENTS	ANNUAL RATE
NO	5353			9656	21	1	11263	62	3	6118	65	5
YES	225			278	0	0	225	2	4	120	4	17
UNKNOWN	38			25	0	0	25	0	0	10	0	0
TOTAL	5616			9959	21	1	11513	64	3	6248	69	6

RISK FACTOR	75-84 AT EXAM PERSON EXAMS	# OF EVENTS	ANNUAL RATE	85-94 AT EXAM PERSON EXAMS	# OF EVENTS	ANNUAL RATE	35-64 COMBINED # OF EVENTS	CRUDE	AGE-ADJUSTED	65-94 COMBINED # OF EVENTS	CRUDE	AGE-ADJUSTED
NO	1904	33	9	120	5	21	85	2	2	103	6	6
YES	38	0	0	3	1	167	2	1	2	5	16	15
UNKNOWN	9	0	0	0	0	-	0	0	0	0	0	0
TOTAL	1951	33	8	123	6	24	87	2	2	108	6	7

LOGISTIC REGRESSIONS

AGE AT EXAM		COEFFICIENT OF RISK FACTOR	SE COEFFICIENT	P-VALUE	STANDARDIZED COEFFICIENT	STD SE
35-44						
45-54		-	-	-	-	-
55-64		0.483	0.722	0.504	0.067	0.100
65-74		1.167	0.524	0.026	0.160	0.072
75-84		-	-	-	-	-
85-94		2.442	1.307	0.062	0.378	0.202
35-64	UNIVARIATE	-0.164	0.716	0.819	-0.027	0.116
65-94	UNIVARIATE	0.917	0.465	0.049	0.126	0.064
35-64	BIVARIATE	0.068	0.718	0.924	0.011	0.116
65-94	BIVARIATE	0.911	0.466	0.051	0.126	0.064
35-64	MULTIVARIATE	-0.525	0.766	0.493	-0.085	0.124
65-94	MULTIVARIATE	0.583	0.481	0.225	0.080	0.066

TABLE 3-15 · DATE=23OCT85
AVERAGE ANNUAL INCIDENCE RATE FOR EVENT PER 1000 PERSONS AT RISK AT EXAM
BY AGE AND LEVEL OF RISK FACTOR AT EXAM: FRAMINGHAM STUDY 30-YEAR FOLLOWUP

SEX: MALE
EVENT: MYOCARDIAL INFARCTION
RISK FACTOR: HEART ENLARGEMENT BY X-RAY
PERSONS AT RISK: PERSONS FREE OF CORONARY HEART DISEASE

| RISK FACTOR | 35-44 AT EXAM | | | 45-54 AT EXAM | | | 55-64 AT EXAM | | | 6 |
	PERSON EXAMS	# OF EVENTS	ANNUAL RATE	PERSON EXAMS	# OF EVENTS	ANNUAL RATE	PERSON EXAMS	# OF EVENTS	ANNUAL RATE	PERSO EXAM!
NO	4063	20	2	6383	67	5	5967	92	8	258
MILD	185	2	5	650	6	5	962	20	10	57
DEFINITE	219	1	2	687	8	6	1119	30	13	7!
UNKNOWN	5	0	0	4	0	0	1	0	0	
TOTAL	4472	23	3	7724	81	5	8049	142	9	391

| RISK FACTOR | 75-84 AT EXAM | | | 85-94 AT EXAM | | | 35-64 COMBINED | | | # C |
	PERSON EXAMS	# OF EVENTS	ANNUAL RATE	PERSON EXAMS	# OF EVENTS	ANNUAL RATE	# OF EVENTS	CRUDE	AGE-ADJUSTED	EVE
NO	589	15	13	29			179	5	6	7
MILD	178	5	14	5			28	8	7	
DEFINITE	250	17	34	5			39	10	8	4
UNKNOWN	0	0	-	0			0	0	0	
TOTAL	1017	37	18	39			246	6	6	13

LOGISTIC REGRESSIONS

AGE AT EXAM		COEFFICIENT OF RISK FACTOR	SE COEFFICIENT	P-VALUE	STANDARDIZED COEFFICIENT
35-44		0.142	0.402	0.724	0.066
45-54		0.025	0.180	0.889	0.015
55-64		0.284	0.104	0.006	0.205
65-74		0.335	0.117	0.004	0.266
75-84		0.516	0.187	0.006	0.436
85-94					
35-64	UNIVARIATE	0.297	0.085	0.001	0.189
65-94	UNIVARIATE	0.392	0.098	0.001	0.316
35-64	BIVARIATE	0.190	0.087	0.029	0.121
65-94	BIVARIATE	0.381	0.098	0.001	0.308
35-64	MULTIVARIATE +	0.134	0.093	0.150	0.085
65-94	MULTIVARIATE	0.232	0.105	0.027	0.187

SEX: FEMALE
EVENT: MYOCARDIAL INFARCTION
RISK FACTOR: HEART ENLARGEMENT BY X-RAY
PERSONS AT RISK: PERSONS FREE OF CORONARY HEART DISEASE

| RISK FACTOR | 35-44 AT EXAM | | | 45-54 AT EXAM | | | 55-64 AT EXAM | | | 6! |
	PERSON EXAMS	# OF EVENTS	ANNUAL RATE	PERSON EXAMS	# OF EVENTS	ANNUAL RATE	PERSON EXAMS	# OF EVENTS	ANNUAL RATE	PERSO! EXAMS
NO	5164			8017	14	1	7522	31	2	319!
MILD	199			943	3	2	1769	11	3	115(
DEFINITE	245			993	4	2	2218	22	5	190!
UNKNOWN	8			6	0	0	4	0	0	(
TOTAL	5616			9959	21	1	11513	64	3	624!

| RISK FACTOR | 75-84 AT EXAM | | | 85-94 AT EXAM | | | 35-64 COMBINED | | | # OF |
	PERSON EXAMS	# OF EVENTS	ANNUAL RATE	PERSON EXAMS	# OF EVENTS	ANNUAL RATE	# OF EVENTS	CRUDE	AGE-ADJUSTED	EVE!
NO	702	7	5	32	2	31	46	1	1	3!
MILD	373	11	15	30	2	33	14	2	2	2!
DEFINITE	876	15	9	61	2	16	27	4	3	4!
UNKNOWN	0	0	-	0	0	-	0	0	0	(
TOTAL	1951	33	8	123	6	24	87	2	2	10!

LOGISTIC REGRESSIONS

AGE AT EXAM		COEFFICIENT OF RISK FACTOR	SE COEFFICIENT	P-VALUE	STANDARDIZED COEFFICIENT
35-44					
45-54		0.437	0.271	0.107	0.279
55-64		0.441	0.140	0.002	0.351
65-74		0.315	0.135	0.019	0.277
75-84		0.199	0.202	0.323	0.178
85-94		-0.340	0.486	0.484	-0.286
35-64	UNIVARIATE	0.637	0.119	0.001	0.445
65-94	UNIVARIATE	0.291	0.108	0.007	0.260
35-64	BIVARIATE	0.424	0.124	0.001	0.295
65-94	BIVARIATE	0.239	0.110	0.030	0.213
35-64	MULTIVARIATE	0.314	0.134	0.019	0.219
65-94	MULTIVARIATE	0.149	0.115	0.195	0.133

4366	23	3	7497	77	5	7693	130	8	3703	79	11
78	0	0	132	3	11	210	5	12	90	6	33
28	0	0	95	1	5	146	7	24	120	12	50
0	0	-	0	0	-	0	0	-	0	0	-
4472	23	3	7724	81	5	8049	142	9	3913	97	12

75-84 AT EXAM			85-94 AT EXAM			35-64 COMBINED			65-94 COMBINED		
PERSON EXAMS	# OF EVENTS	ANNUAL RATE	PERSON EXAMS	# OF EVENTS	ANNUAL RATE	# OF EVENTS	CRUDE	AGE-ADJUSTED	# OF EVENTS	CRUDE	AGE-ADJUSTED
949	34	18'	34	2		230	6	6	114	12	12
37	1	14	2			8	10	9	7	27	29
31	2	32	3			8	15	11	14	45	46
0	0	-	0			0	-	-	0	-	-
1017	37	18	39			246	6	6	135	14	13

LOGISTIC REGRESSIONS

	COEFFICIENT OF RISK FACTOR	SE COEFFICIENT	P-VALUE	STANDARDIZED COEFFICIENT	STD SE
	0.254	0.358	0.479	0.064	0.091
	0.508	0.187	0.007	0.156	0.058
	0.845	0.154	0.001	0.315	0.057
	0.223	0.370	0.546	0.086	0.143
VARIATE	0.476	0.165	0.004	0.127	0.044
VARIATE	0.708	0.141	0.001	0.267	0.053
ARIATE	0.386	0.166	0.020	0.103	0.044
ARIATE	0.703.	0.141	0.001	0.265	0.053
TIVARIATE	0.184'	0.178	0.301	0.049	0.048
TIVARIATE	0.459	0.155	0.003	0.173	0.058

SEX: FEMALE
EVENT: MYOCARDIAL INFARCTION
RISK FACTOR: LEFT VENTRICULAR HYPERTROPHY BY ELECTROCARDIOGRAM
PERSONS AT RISK: PERSONS FREE OF CORONARY HEART DISEASE

35-44 AT EXAM			45-54 AT EXAM			55-64 AT EXAM			65-74 AT EXAM		
PERSON EXAMS	# OF EVENTS	ANNUAL RATE	PERSON EXAMS	# OF EVENTS	ANNUAL RATE	PERSON EXAMS	# OF EVENTS	ANNUAL RATE	PERSON EXAMS	# OF EVENTS	ANNUAL RATE
5563	38	15	9798	21	1	11137	60	3	5966	59	5
			107	0	0	226	1	2	134	4	15
			54	0	0	150	5	10	148	6	20
0			0	0	-	0	0	-	0	0	-
5616			9959			11513	64	3	6248	69	6

75-84 AT EXAM			85-94 AT EXAM			35-64 COMBINED			65-94 COMBINED		
PERSON EXAMS	# OF EVENTS	ANNUAL RATE	PERSON EXAMS	# OF EVENTS	ANNUAL RATE	# OF EVENTS	CRUDE	AGE-ADJUSTED	# OF EVENTS	CRUDE	AGE-ADJUSTED
1833	31	8'	111	5	23	82	2	2	95	6	6
49	0	0	4	0	0	1	1	1	4	11	11
69	2	14	6	1	83	4	9	11	9	20	20
0	0	-	0	0	-	0	-	-	0	-	-
1951	33	8	123	6	24	87	2	2	108	6	7

LOGISTIC REGRESSIONS

	COEFFICIENT OF RISK FACTOR	SE COEFFICIENT	P-VALUE	STANDARDIZED COEFFICIENT	STD SE
	0.564	0.302	0.062	0.149	0.080
	0.764	0.202	0.001	0.255	0.068
	0.146	0.394	0.712	0.058	0.156
	0.596	0.625	0.340	0.282	0.296
ARIATE	0.794	0.264	0.003	0.169	0.056
ARIATE	0.616	0.171	0.001	0.217	0.060

TABLE 3-17   DATE=23NOV85
AVERAGE ANNUAL INCIDENCE RATE FOR EVENT PER 1000 PERSONS AT RISK AT EXAM
BY AGE AND LEVEL OF RISK FACTOR AT EXAM: FRAMINGHAM STUDY 30-YEAR FOLLOWUP

SEX: MALE
EVENT: MYOCARDIAL INFARCTION
RISK FACTOR: INTRAVENTRICULAR CONDUCTION DEFECT (WITH QRS INTERVAL MORE THAN 0.11 SECOND)
PERSONS AT RISK: PERSONS FREE OF CORONARY HEART DISEASE

RISK FACTOR	35-44 AT EXAM			45-54 AT EXAM			55-64 AT EXAM			65-74 AT EXAM		
	PERSON EXAMS	# OF EVENTS	ANNUAL RATE	PERSON EXAMS	# OF EVENTS	ANNUAL RATE	PERSON EXAMS	# OF EVENTS	ANNUAL RATE	PERSON EXAMS	# OF EVENTS	ANNUAL RATE
NO	4161	22	3	7213	75	5	7384	126	9	3566	89	12
INCOMPLETE	275	1	2	446	3	6	454	14	15	183	3	14
COMPLETE	36	0	0	65	1	8	207	2	5	164	3	9
UNKNOWN	0	0	-	0	0	-	0	0	-	0	0	-
TOTAL	4472	23	3	7724	81	5	8049	142	9	3913	97	12

RISK FACTOR	75-84 AT EXAM			85-94 AT EXAM			35-64 COMBINED			65-94 COMBINED		
	PERSON EXAMS	# OF EVENTS	ANNUAL RATE	PERSON EXAMS	# OF EVENTS	ANNUAL RATE	# OF EVENTS	CRUDE	AGE-ADJUSTED	# OF EVENTS	CRUDE	AGE-ADJUSTED
NO	918	33	18	31			223	6	6	123	14	14
INCOMPLETE	21	0	0	2			20	3	3	3	12	11
COMPLETE	78	4	26	6			3	3	3	7	14	12
UNKNOWN	0	0	-	0			0	-	-	0	-	-
TOTAL	1017	37	18	39			246	6	6	135	14	14

LOGISTIC REGRESSIONS

AGE AT EXAM		COEFFICIENT OF RISK FACTOR	SE COEFFICIENT	P-VALUE	STANDARDIZED COEFFICIENT	STD SE
35-44		-0.524	0.959	0.585	-0.155	0.284
45-54		0.129	0.353	0.715	0.038	0.103
55-64		0.118	0.203	0.560	0.045	0.078
65-74		-0.096	0.248	0.700	-0.043	0.110
75-84		0.134	0.279	0.635	0.073	0.152
85-94						
35-64	UNIVARIATE	0.142	0.175	0.418	0.047	0.258
65-94	UNIVARIATE	-0.002	0.186	0.989	-0.001	0.087
35-64	BIVARIATE	0.066	0.173	0.702	0.022	0.058
65-94	BIVARIATE	-0.020	0.186	0.915	-0.009	0.087
35-64	MULTIVARIATE	-0.000	0.185	0.998	-0.000	0.062
65-94	MULTIVARIATE	-0.003	0.189	0.988	-0.001	0.089

SEX: FEMALE
EVENT: MYOCARDIAL INFARCTION
RISK FACTOR: INTRAVENTRICULAR CONDUCTION DEFECT (WITH QRS INTERVAL MORE THAN 0.11 SECOND)
PERSONS AT RISK: PERSONS FREE OF CORONARY HEART DISEASE

RISK FACTOR	35-44 AT EXAM			45-54 AT EXAM			55-64 AT EXAM			65-74 AT EXAM		
	PERSON EXAMS	# OF EVENTS	ANNUAL RATE	PERSON EXAMS	# OF EVENTS	ANNUAL RATE	PERSON EXAMS	# OF EVENTS	ANNUAL RATE	PERSON EXAMS	# OF EVENTS	ANNUAL RATE
NO	5573			9771	21	1	11183	62	5	6027	67	6
INCOMPLETE				144	0	0	187	1	3	115	1	4
COMPLETE	12			44	0	0	143	1	3	106	1	5
UNKNOWN	0			0	0	-	0	0	-	0	0	-
TOTAL	5616			9959	21	1	11513	64	5	6248	69	6

RISK FACTOR	75-84 AT EXAM			85-94 AT EXAM			35-64 COMBINED			65-94 COMBINED		
	PERSON EXAMS	# OF EVENTS	ANNUAL RATE	PERSON EXAMS	# OF EVENTS	ANNUAL RATE	# OF EVENTS	CRUDE	AGE-ADJUSTED	# OF EVENTS	CRUDE	AGE-ADJUSTED
NO	1847	32	9	117	6	26	85	2	2	105	7	7
INCOMPLETE	36	1	14	1	0	0	1	1	1	2	7	7
COMPLETE	68	0	0	5	0	0	1	3	1	3	3	4
UNKNOWN	0	0	-	0	0	-	0	-	-	0	-	-
TOTAL	1951	33	8	123	6	24	87	2	2	108	6	6

LOGISTIC REGRESSIONS

AGE AT EXAM		COEFFICIENT OF RISK FACTOR	SE COEFFICIENT	P-VALUE	STANDARDIZED COEFFICIENT	STD SE
35-44		-	-	-	-	-
45-54		-	-	-	-	-
55-64		0.084	0.461	0.855	0.021	0.117
65-74		-0.118	0.462	0.798	-0.034	0.133
75-84		-0.643	0.784	0.412	-0.249	0.304
85-94						
35-64	UNIVARIATE	0.102	0.475	0.830	0.021	0.099
65-94	UNIVARIATE	-0.314	0.397	0.429	-0.100	0.126
35-64	BIVARIATE	-0.069	0.469	0.883	-0.014	0.098
65-94	BIVARIATE	-0.356	0.395	0.368	-0.113	0.125
35-64	MULTIVARIATE	0.011	0.469	0.982	0.002	0.098
65-94	MULTIVARIATE	-0.363	0.399	0.363	-0.115	0.127

TABLE 3-18   DATE=23OCT85
AVERAGE ANNUAL INCIDENCE RATE FOR EVENT PER 1000 PERSONS AT RISK AT EXAM
BY AGE AND LEVEL OF RISK FACTOR AT EXAM: FRAMINGHAM STUDY 30-YEAR FOLLOWUP

SEX: MALE
EVENT: MYOCARDIAL INFARCTION
RISK FACTOR: DEFINITE NONSPECIFIC T-WAVE OR ST-SEGMENT ABNORMALITY BY ELECTROCARDIOGRAM
PERSONS AT RISK: PERSONS FREE OF CORONARY HEART DISEASE

RISK FACTOR	35-44 AT EXAM			45-54 AT EXAM			55-64 AT EXAM			65-74 AT EXAM		
	PERSON EXAMS	# OF EVENTS	ANNUAL RATE	PERSON EXAMS	# OF EVENTS	ANNUAL RATE	PERSON EXAMS	# OF EVENTS	ANNUAL RATE	PERSON EXAMS	# OF EVENTS	ANNUAL RATE
NO	4362	19	2	7352	78	5	7462	129	9	3472	83	12
YES	110	4	18	372	3	4	587	13	11	441	14	16
UNKNOWN	0	0	-	0	0	-	0	0	-	0	0	-
TOTAL	4472	23	5	7724	81	5	8049	142	9	3913	97	12

RISK FACTOR	75-84 AT EXAM			85-94 AT EXAM			35-64 COMBINED			65-94 COMBINED		
	PERSON EXAMS	# OF EVENTS	ANNUAL RATE	PERSON EXAMS	# OF EVENTS	ANNUAL RATE	# OF EVENTS	CRUDE	AGE-ADJUSTED	# OF EVENTS	CRUDE	AGE-ADJUSTED
NO	830	30	18	28		11	226	6	6	113	13	13
YES	187	7	19	11			20	9	10	22	17	17
UNKNOWN	0	0	-	0			0			0		
TOTAL	1017	37	18	39			246	6	6	135	14	13

LOGISTIC REGRESSIONS

AGE AT EXAM		COEFFICIENT OF RISK FACTOR	SE COEFFICIENT	P-VALUE	STANDARDIZED COEFFICIENT	STD SE
35-44		2.155	0.559	0.001	0.334	0.087
45-54		-0.277	0.591	0.639	-0.059	0.126
55-64		0.255	0.294	0.390	0.066	0.077
65-74		0.292	0.293	0.320	0.092	0.093
75-84		0.036	0.428	0.932	0.014	0.166
85-94						
35-64	UNIVARIATE	0.469	0.235	0.046	0.105	0.053
65-94	UNIVARIATE	0.286	0.237	0.228	0.096	0.079
35-64	BIVARIATE	0.284	0.237	0.230	0.064	0.053
65-94	BIVARIATE	0.250	0.238	0.294	0.084	0.080
35-64	MULTIVARIATE	-0.106	0.252	0.675	-0.024	0.056
65-94	MULTIVARIATE	0.171	0.241	0.477	0.057	0.081

SEX: FEMALE
EVENT: MYOCARDIAL INFARCTION
RISK FACTOR: DEFINITE NONSPECIFIC T-WAVE OR ST-SEGMENT ABNORMALITY BY ELECTROCARDIOGRAM
PERSONS AT RISK: PERSONS FREE OF CORONARY HEART DISEASE

RISK FACTOR	35-44 AT EXAM			45-54 AT EXAM			55-64 AT EXAM			65-74 AT EXAM		
	PERSON EXAMS	# OF EVENTS	ANNUAL RATE	PERSON EXAMS	# OF EVENTS	ANNUAL RATE	PERSON EXAMS	# OF EVENTS	ANNUAL RATE	PERSON EXAMS	# OF EVENTS	ANNUAL RATE
NO	5517			9543	19	1	10647	54	3	5528	56	5
YES	99			416	2	2	866	10	6	720	13	9
UNKNOWN	0			0	0	-	0	0	-	0	0	-
TOTAL	5616			9959	21	1	11513	64	3	6248	69	6

RISK FACTOR	75-84 AT EXAM			85-94 AT EXAM			35-64 COMBINED			65-94 COMBINED		
	PERSON EXAMS	# OF EVENTS	ANNUAL RATE	PERSON EXAMS	# OF EVENTS	ANNUAL RATE	# OF EVENTS	CRUDE	AGE-ADJUSTED	# OF EVENTS	CRUDE	AGE-ADJUSTED
NO	1632	25	8	97	4	21	75	1	2	85	6	6
YES	319	8	13	26	2	38	12	4	3	23	11	10
UNKNOWN	0	0	-	0	0	-	0			0		
TOTAL	1951	33	8	123	6	24	87	2	2	108	6	7

LOGISTIC REGRESSIONS

AGE AT EXAM		COEFFICIENT OF RISK FACTOR	SE COEFFICIENT	P-VALUE	STANDARDIZED COEFFICIENT	STD SE
35-44						
45-54		0.884	0.745	0.235	0.177	0.149
55-64		0.829	0.346	0.017	0.219	0.091
65-74		0.586	0.310	0.059	0.187	0.099
75-84		0.503	0.411	0.221	0.186	0.152
85-94		0.661	0.896	0.460	0.271	0.367
35-64	UNIVARIATE	1.097	0.312	0.001	0.241	0.069
65-94	UNIVARIATE	0.622	0.237	0.009	0.208	0.079
35-64	BIVARIATE	0.766	0.314	0.015	0.168	0.069
65-94	BIVARIATE	0.557	0.239	0.020	0.186	0.080
35-64	MULTIVARIATE	0.629	0.332	0.196	0.094	0.073
65-94	MULTIVARIATE	0.473	0.240	0.048	0.158	0.080

AVERAGE ANNUAL INCIDENCE RATE FOR EVENT PER 1000 PERSONS AT RISK AT EXAM
BY AGE AND LEVEL OF RISK FACTOR AT EXAM: FRAMINGHAM STUDY 30-YEAR FOLLOWUP

SEX: MALE
EVENT: UNRECOGNIZED MYOCARDIAL INFARCTION
RISK FACTOR: SYSTOLIC BLOOD PRESSURE, FIRST EXAMINER (MM HG)
PERSONS AT RISK: PERSONS FREE OF CORONARY HEART DISEASE

RISK FACTOR	35-44 AT EXAM PERSON EXAMS	# OF EVENTS	ANNUAL RATE	45-54 AT EXAM PERSON EXAMS	# OF EVENTS	ANNUAL RATE	55-64 AT EXAM PERSON EXAMS	# OF EVENTS	ANNUAL RATE	65-74 AT EXAM PERSON EXAMS	# OF EVENTS	ANNUAL RATE
74-119	1223	1	0	1769	1	0	1440	5	1	548	0	0
120-139	2165	2	0	3417	3	0	3166	7	1	1341	8	3
140-159	841	2	1	1732	8	2	2090	17	4	1194	13	5
160-179	191	0	0	567	2	2	916	5	3	552	10	9
180-300	52	0	0	239	1	2	437	7	8	278	6	11
UNKNOWN	0	0	-	0	0	-	0	0	-	0	0	-
TOTAL	4472	5	1	7724	15	1	8049	39	2	3913	37	5

RISK FACTOR	75-84 AT EXAM PERSON EXAMS	# OF EVENTS	ANNUAL RATE	85-94 AT EXAM PERSON EXAMS	# OF EVENTS	ANNUAL RATE	35-64 COMBINED # OF EVENTS	CRUDE	AGE- ADJUSTED	65-94 COMBINED # OF EVENTS	CRUDE	AGE- ADJUSTED
74-119	120	2	8	7			5	1	0	2	1	2
120-139	345	4	6	15			12	1	0	12	4	4
140-159	315	7	11	11			27	3	3	20	7	6
160-179	153	2	7	4			7	2	2	12	8	9
180-300	84	2	12	2			8	5	4	8	11	11
UNKNOWN	0	0	-	0			0	-		0	-	
TOTAL	1017	17	8	39			59	1	1	54	5	6

LOGISTIC REGRESSIONS

AGE AT EXAM		COEFFICIENT OF RISK FACTOR	SE COEFFICIENT	P-VALUE	STANDARDIZED COEFFICIENT	STD SE
35-44		0.014	0.023	0.541	0.238	0.390
45-54		0.026	0.010	0.010	0.504	0.197
55-64		0.025	0.005	0.001	0.562	0.120
65-74		0.024	0.006	0.001	0.555	0.128
75-84		0.007	0.011	0.513	0.154	0.235
85-94						
35-64	UNIVARIATE	0.028	0.005	0.001	0.573	0.092
65-94	UNIVARIATE	0.020	0.005	0.001	0.458	0.113
35-64	BIVARIATE	0.024	0.005	0.001	0.491	0.095
65-94	BIVARIATE	0.020	0.005	0.001	0.467	0.115
35-64	MULTIVARIATE	0.021	0.005	0.001	0.420	0.105
65-94	MULTIVARIATE	0.014	0.005	0.009	0.320	0.122

SEX: FEMALE
EVENT: UNRECOGNIZED MYOCARDIAL INFARCTION
RISK FACTOR: SYSTOLIC BLOOD PRESSURE, FIRST EXAMINER (MM HG)
PERSONS AT RISK: PERSONS FREE OF CORONARY HEART DISEASE

RISK FACTOR	35-44 AT EXAM PERSON EXAMS	# OF EVENTS	ANNUAL RATE	45-54 AT EXAM PERSON EXAMS	# OF EVENTS	ANNUAL RATE	55-64 AT EXAM PERSON EXAMS	# OF EVENTS	ANNUAL RATE	65-74 AT EXAM PERSON EXAMS	# OF EVENTS	ANNUAL RATE
74-119	2493			2714	0	0	1979	2	1	647	2	2
120-139	2244			3982	5	1	4056	2	0	1747	2	1
140-159	673			2053	2	0	3025	1	1	2013	7	2
160-179	156			768	1	1	1560	8	5	1155	5	1
180-300	50			442	0	0	893	7	4	686	6	4
UNKNOWN	0			0	0	-	0	0	-	0	0	-
TOTAL	5616			9959	8	0	11513	27	1	6248	22	2

RISK FACTOR	75-84 AT EXAM PERSON EXAMS	# OF EVENTS	ANNUAL RATE	85-94 AT EXAM PERSON EXAMS	# OF EVENTS	ANNUAL RATE	35-64 COMBINED # OF EVENTS	CRUDE	AGE- ADJUSTED	65-94 COMBINED # OF EVENTS	CRUDE	AGE- ADJUSTED
74-119	133	2	8	8			2	0	0	4	3	3
120-139	465	2	2	38			7	0	0	7	2	1
140-159	626	1	1	29			10	1	0	10	2	2
160-179	451	4	4	31			9	2	2	7	2	2
180-300	276	4	7	17			7	3	2	10	5	5
UNKNOWN	0	0	-	0			0	-		0	-	
TOTAL	1951	13	3	123			35	1	0	38	2	2

LOGISTIC REGRESSIONS

AGE AT EXAM		COEFFICIENT OF RISK FACTOR	SE COEFFICIENT	P-VALUE	STANDARDIZED COEFFICIENT	STD SE
35-44						
45-54		0.009	0.013	0.522	0.196	0.307
55-64		0.024	0.005	0.001	0.592	0.134
65-74		0.014	0.008	0.088	0.327	0.192
75-84		0.011	0.010	0.287	0.269	0.253
85-94						
35-64	UNIVARIATE	0.025	0.005	0.001	0.592	0.107
65-94	UNIVARIATE	0.012	0.006	0.056	0.282	0.148
35-64	BIVARIATE	0.020	0.005	0.001	0.481	0.120
65-94	BIVARIATE	0.010	0.006	0.100	0.247	0.150
35-64	MULTIVARIATE	0.022	0.005	0.001	0.519	0.128
65-94	MULTIVARIATE	0.008	0.006	0.225	0.188	0.155

TABLE 4-2　　　　　　　　　DATE=02AUG85

AVERAGE ANNUAL INCIDENCE RATE FOR EVENT PER 1000 PERSONS AT RISK AT EXAM
BY AGE AND LEVEL OF RISK FACTOR AT EXAM: FRAMINGHAM STUDY 30-YEAR FOLLOWUP

SEX: MALE
EVENT: UNRECOGNIZED MYOCARDIAL INFARCTION
RISK FACTOR: DIASTOLIC BLOOD PRESSURE, FIRST EXAMINER (MM HG)
PERSONS AT RISK: PERSONS FREE OF CORONARY HEART DISEASE

RISK FACTOR	35-44 AT EXAM			45-54 AT EXAM			55-64 AT EXAM			65-74 AT EXAM		
	PERSON EXAMS	# OF EVENTS	ANNUAL RATE	PERSON EXAMS	# OF EVENTS	ANNUAL RATE	PERSON EXAMS	# OF EVENTS	ANNUAL RATE	PERSON EXAMS	# OF EVENTS	ANNUAL RATE
20-74	1012	0	0	1478	1	0	1847	5	1	1225	14	6
75-84	1604	2	1	2766	3	1	2845	11	2	1323	9	3
85-94	1187	0	0	2062	6	1	1964	12	3	893	5	3
95-104	499	3	3	1008	3	1	982	5	3	343	5	7
105-160	170	0	0	410	2	2	411	6	7	129	4	16
UNKNOWN	0	0	-	0	0	-	0	0	-	0	0	-
TOTAL	4472	5	1	7724	15	1	8049	39	2	3913	37	5

RISK FACTOR	75-84 AT EXAM			85-94 AT EXAM			35-64 COMBINED			65-94 COMBINED		
	PERSON EXAMS	# OF EVENTS	ANNUAL RATE	PERSON EXAMS	# OF EVENTS	ANNUAL RATE	# OF EVENTS	CRUDE	AGE-ADJUSTED	# OF EVENTS	CRUDE	AGE-ADJUSTED
20-74	422	6	7	22			6	1	0	20	6	6
75-84	331	5	8	12			16	1	1	14	4	4
85-94	204	3	7	5			18	2	2	8	4	4
95-104	44	2	23	0			11	2	2	7	9	10
105-160	16	1	31	0			8	4	4	5	17	19
UNKNOWN	0	0	-	0			0			0	-	
TOTAL	1017	17	8	39			59	1	1	54	5	6

LOGISTIC REGRESSIONS

AGE AT EXAM		COEFFICIENT OF RISK FACTOR	SE COEFFICIENT	P-VALUE	STANDARDIZED COEFFICIENT	STD SE
35-44		0.054	0.032	0.096	0.604	0.363
45-54		0.046	0.018	0.012	0.541	0.215
55-64		0.053	0.012	0.004	0.401	0.140
65-74		0.016	0.013	0.204	0.197	0.155
75-84		0.030	0.020	0.128	0.357	0.234
85-94						
35-64	UNIVARIATE	0.039	0.009	0.001	0.459	0.112
65-94	UNIVARIATE	0.018	0.011	0.093	0.216	0.129
35-64	BIVARIATE	0.038	0.009	0.001	0.448	0.110
65-94	BIVARIATE	0.022	0.011	0.037	0.272	0.130
35-64	MULTIVARIATE	0.034	0.010	0.001	0.405	0.115
65-94	MULTIVARIATE	0.014	0.011	0.175	0.175	0.129

SEX: FEMALE
EVENT: UNRECOGNIZED MYOCARDIAL INFARCTION
RISK FACTOR: DIASTOLIC BLOOD PRESSURE, FIRST EXAMINER (MM HG)
PERSONS AT RISK: PERSONS FREE OF CORONARY HEART DISEASE

RISK FACTOR	35-44 AT EXAM			45-54 AT EXAM			55-64 AT EXAM			65-74 AT EXAM		
	PERSON EXAMS	# OF EVENTS	ANNUAL RATE	PERSON EXAMS	# OF EVENTS	ANNUAL RATE	PERSON EXAMS	# OF EVENTS	ANNUAL RATE	PERSON EXAMS	# OF EVENTS	ANNUAL RATE
20-74	2164			2639	1	0	2969	3	1	1831	5	1
75-84	2030			3541	1	0	3853	5	1	2135	8	2
85-94	987			2361	5	1	2840	7	1	1439	4	1
95-104	324			926	0	0	1272	8	3	600	2	2
105-160	111			492	1	1	579	4	3	243	3	6
UNKNOWN	0			0	0	-	0	0	-	0	0	-
TOTAL	5616			9959	8	0	11513	27	1	6248	22	2

RISK FACTOR	75-84 AT EXAM			85-94 AT EXAM			35-64 COMBINED			65-94 COMBINED		
	PERSON EXAMS	# OF EVENTS	ANNUAL RATE	PERSON EXAMS	# OF EVENTS	ANNUAL RATE	# OF EVENTS	CRUDE	AGE-ADJUSTED	# OF EVENTS	CRUDE	AGE-ADJUSTED
20-74	724	4	3	53			4	0	0	10	2	2
75-84	596	3	3	43			6	0	0	12	2	2
85-94	445	3	3	20			12	1	1	8	2	2
95-104	146	1	3	7			8	2	1	3	2	2
105-160	42	2	24	0			5	2	2	5	9	10
UNKNOWN	0	0	-	0			0	-		0	-	
TOTAL	1951	13	3	123			35	1	0	38	2	2

LOGISTIC REGRESSIONS

AGE AT EXAM		COEFFICIENT OF RISK FACTOR	SE COEFFICIENT	P-VALUE	STANDARDIZED COEFFICIENT	STD SE
35-44						
45-54		0.027	0.024	0.266	0.334	0.301
55-64		0.050	0.012	0.001	0.619	0.150
65-74		0.020	0.016	0.224	0.242	0.199
75-84		0.033	0.021	0.128	0.390	0.256
85-94						
35-64	UNIVARIATE	0.049	0.010	0.001	0.596	0.126
65-94	UNIVARIATE	0.019	0.013	0.126	0.234	0.153
35-64	BIVARIATE	0.046	0.011	0.001	0.562	0.134
65-94	BIVARIATE	0.024	0.013	0.062	0.287	0.154
35-64	MULTIVARIATE	0.049	0.011	0.001	0.598	0.140
65-94	MULTIVARIATE	0.021	0.012	0.080	0.261	0.149

TABLE 4-3A   DATE=05AUG85

AVERAGE ANNUAL INCIDENCE RATE FOR EVENT PER 1000 PERSONS AT RISK AT EXAM
BY AGE AND LEVEL OF RISK FACTOR AT EXAM: FRAMINGHAM STUDY 30-YEAR FOLLOWUP

SEX: MALE
EVENT: UNRECOGNIZED MYOCARDIAL INFARCTION
RISK FACTOR: HYPERTENSION WITH ANTIHYPERTENSIVE TREATMENT
PERSONS AT RISK: PERSONS FREE OF CORONARY HEART DISEASE

RISK FACTOR	35-44 AT EXAM			45-54 AT EXAM			55-64 AT EXAM			65-74 AT EXAM		
	PERSON EXAMS	# OF EVENTS	ANNUAL RATE	PERSON EXAMS	# OF EVENTS	ANNUAL RATE	PERSON EXAMS	# OF EVENTS	ANNUAL RATE	PERSON EXAMS	# OF EVENTS	ANNUAL RATE
NO	2651	2	0	4122	2	0	3782	7	1	1516	4	1
BORDER	1301	0	0	2205	5	1	2299	17	4	1252	16	6
DEFINITE	500	5	5	1139	4	2	1266	9	4	552	9	8
TREATED	20	0	0	258	4	8	702	6	4	593	8	7
UNKNOWN	0	0	-	0	0	-	0	0	-	0	0	-
TOTAL	4472	5	1	7724	15	1	8049	39	2	3913	37	5

RISK FACTOR	75-84 AT EXAM			85-94 AT EXAM			35-64 COMBINED			65-94 COMBINED		
	PERSON EXAMS	# OF EVENTS	ANNUAL RATE	PERSON EXAMS	# OF EVENTS	ANNUAL RATE	# OF EVENTS	CRUDE	AGE-ADJUSTED	# OF EVENTS	CRUDE	AGE-ADJUSTED
NO	373	3	4	18			11	1	0	7	2	2
BORDER	333	6	9	11			22	2	2	22	7	7
DEFINITE	123	1	4	3			16	3	3	10	7	7
TREATED	188	7	19	7			10	5	5	15	10	9
UNKNOWN	0	0	-	0			0	-		0	-	
TOTAL	1017	17	8	39			59	1	1	54	5	6

LOGISTIC REGRESSIONS

AGE AT EXAM		COEFFICIENT OF RISK FACTOR	SE COEFFICIENT	P-VALUE	STANDARDIZED COEFFICIENT	STD SE
35-44		1.132	0.571	0.047	0.786	0.397
45-54		1.167	0.345	0.001	0.896	0.265
55-64		0.626	0.196	0.001	0.510	0.160
65-74		0.692	0.216	0.001	0.567	0.177
75-84		0.553	0.314	0.078	0.453	0.257
85-94						
35-64	UNIVARIATE	0.569	0.148	0.001	0.504	0.131
65-94	UNIVARIATE	0.444	0.162	0.006	0.398	0.145
35-64	BIVARIATE	0.536	0.150	0.001	0.475	0.132
65-94	BIVARIATE	0.444	0.161	0.006	0.398	0.145
35-64	MULTIVARIATE	0.484	0.154	0.002	0.429	0.137
65-94	MULTIVARIATE	0.454	0.169	0.007	0.407	0.152

SEX: FEMALE
EVENT: UNRECOGNIZED MYOCARDIAL INFARCTION
RISK FACTOR: HYPERTENSION WITH ANTIHYPERTENSIVE TREATMENT
PERSONS AT RISK: PERSONS FREE OF CORONARY HEART DISEASE

RISK FACTOR	35-44 AT EXAM			45-54 AT EXAM			55-64 AT EXAM			65-74 AT EXAM		
	PERSON EXAMS	# OF EVENTS	ANNUAL RATE	PERSON EXAMS	# OF EVENTS	ANNUAL RATE	PERSON EXAMS	# OF EVENTS	ANNUAL RATE	PERSON EXAMS	# OF EVENTS	ANNUAL RATE
NO	4250			5622	4	0	4888	1	0	1767	3	1
BORDER	1001			2579	3	1	3282	5	0	1903	9	2
DEFINITE	316			1166	0	0	1835	12	5	1035	3	1
TREATED	49			592	1	1	1508	11	4	1543	7	2
UNKNOWN	0			0	0	-	0	0	-	0	0	-
TOTAL	5616			9959	8	0	11513	27	1	6248	22	2

RISK FACTOR	75-84 AT EXAM			85-94 AT EXAM			35-64 COMBINED			65-94 COMBINED		
	PERSON EXAMS	# OF EVENTS	ANNUAL RATE	PERSON EXAMS	# OF EVENTS	ANNUAL RATE	# OF EVENTS	CRUDE	AGE-ADJUSTED	# OF EVENTS	CRUDE	AGE-ADJUSTED
NO	402	1	1	31			5	0	0	4	1	1
BORDER	575	3	3	28			6	0	0	13	3	2
DEFINITE	371	6	8	21			12	2	1	9	3	3
TREATED	603	5	2	43			12	3	2	12	3	2
UNKNOWN	0	0	-	0			0	-		0	-	
TOTAL	1951	15	3	123			35	1	0	38	2	2

LOGISTIC REGRESSIONS

AGE AT EXAM		COEFFICIENT OF RISK FACTOR	SE COEFFICIENT	P-VALUE	STANDARDIZED COEFFICIENT	STD SE
35-44						
45-54		0.022	0.457	0.962	0.017	0.351
55-64		1.879	0.421	0.001	1.569	0.352
65-74		0.291	0.273	0.286	0.240	0.225
75-84		0.630	0.439	0.151	0.496	0.346
85-94						
35-64	UNIVARIATE	0.485	0.186	0.009	0.425	0.163
65-94	UNIVARIATE	0.317	0.203	0.119	0.267	0.171
35-64	BIVARIATE	0.325	0.192	0.091	0.285	0.168
65-94	BIVARIATE	0.298	0.205	0.147	0.251	0.173
35-64	MULTIVARIATE	0.389	0.198	0.050	0.340	0.174
65-94	MULTIVARIATE	0.319	0.210	0.129	0.269	0.177

TABLE 4-3B   DATE=06DEC85
AVERAGE ANNUAL INCIDENCE RATE FOR EVENT PER 1000 PERSONS AT RISK AT EXAM
BY AGE AND LEVEL OF RISK FACTOR AT EXAM: FRAMINGHAM STUDY 30-YEAR FOLLOWUP

SEX: MALE
EVENT: UNRECOGNIZED MYOCARDIAL INFARCTION
RISK FACTOR: HYPERTENSION IGNORING TREATMENT
PERSONS AT RISK: PERSONS FREE OF CORONARY HEART DISEASE

RISK FACTOR	35-44 AT EXAM			45-54 AT EXAM			55-64 AT EXAM			65-74 AT EXAM		
	PERSON EXAMS	# OF EVENTS	ANNUAL RATE	PERSON EXAMS	# OF EVENTS	ANNUAL RATE	PERSON EXAMS	# OF EVENTS	ANNUAL RATE	PERSON EXAMS	# OF EVENTS	ANNUAL RATE
NO	2653	2	0	4154	2	0	3909	7	1	1631	5	2
MILD	1310	0	0	2284	6	1	2583	19	4	1499	19	6
DEFINITE	509	3	3	1286	7	3	1557	13	4	783	13	8
UNKNOWN	0	0	-	0	0	-	0	0	-	0	0	-
TOTAL	4472	5	1	7724	15	1	8049	39	2	3913	37	5

RISK FACTOR	75-84 AT EXAM			85-94 AT EXAM			35-64 COMBINED			65-94 COMBINED		
	PERSON EXAMS	# OF EVENTS	ANNUAL RATE	PERSON EXAMS	# OF EVENTS	ANNUAL RATE	# OF EVENTS	CRUDE	AGE-ADJUSTED	# OF EVENTS	CRUDE	AGE-ADJUSTED
NO	412	5	6	21			11	1	1	10	2	2
MILD	418	8	10	15			25	2	2	27	7	7
DEFINITE	187	4	11	3			23	3	3	17	9	9
UNKNOWN	0	0	-	0			0	-		0	-	
TOTAL	1017	17	8	39			59	1	1	54	5	5

LOGISTIC REGRESSIONS

AGE AT EXAM		COEFFICIENT OF RISK FACTOR	SE COEFFICIENT	P-VALUE	STANDARDIZED COEFFICIENT	STD SE
35-44		1.145	0.573	0.046	0.791	0.396
45-54		1.110	0.338	0.001	0.835	0.254
55-64		0.701	0.201	0.001	0.540	0.155
65-74		0.741	0.219	0.001	0.559	0.165
75-84		0.297	0.325	0.361	0.218	0.239
85-94						
35-64	UNIVARIATE	0.898	0.164	0.001	0.673	0.123
65-94	UNIVARIATE	0.603	0.179	0.001	0.452	0.135
35-64	BIVARIATE	0.825	0.165	0.001	0.619	0.124
65-94	BIVARIATE	0.613	0.180	0.001	0.460	0.135
35-64	MULTIVARIATE	0.726	0.172	0.001	0.544	0.129
65-94	MULTIVARIATE	0.455	0.187	0.015	0.342	0.140

SEX: FEMALE
EVENT: UNRECOGNIZED MYOCARDIAL INFARCTION
RISK FACTOR: HYPERTENSION IGNORING TREATMENT
PERSONS AT RISK: PERSONS FREE OF CORONARY HEART DISEASE

RISK FACTOR	35-44 AT EXAM			45-54 AT EXAM			55-64 AT EXAM			65-74 AT EXAM		
	PERSON EXAMS	# OF EVENTS	ANNUAL RATE	PERSON EXAMS	# OF EVENTS	ANNUAL RATE	PERSON EXAMS	# OF EVENTS	ANNUAL RATE	PERSON EXAMS	# OF EVENTS	ANNUAL RATE
NO	4272			5804	4	0	5245	3	0	2035	3	1
MILD	1016			2763	3	1	3860	8	1	2549	11	2
DEFINITE	328			1392	1	0	2408	16	3	1664	8	2
UNKNOWN	0			0	0	-	0	0	-	0	0	-
TOTAL	5616			9959	8	0	11513	27	1	6248	22	2

RISK FACTOR	75-84 AT EXAM			85-94 AT EXAM			35-64 COMBINED			65-94 COMBINED		
	PERSON EXAMS	# OF EVENTS	ANNUAL RATE	PERSON EXAMS	# OF EVENTS	ANNUAL RATE	# OF EVENTS	CRUDE	AGE-ADJUSTED	# OF EVENTS	CRUDE	AGE-ADJUSTED
NO	507	2	2	37			7	0	0	5	1	1
MILD	827	3	2	45			11	1	1	17	2	3
DEFINITE	617	8	6	41			17	2	2	16	3	3
UNKNOWN	0	0	-	0			0	-		0	-	
TOTAL	1951	13	3	123			35	1	1	38	2	2

LOGISTIC REGRESSIONS

AGE AT EXAM		COEFFICIENT OF RISK FACTOR	SE COEFFICIENT	P-VALUE	STANDARDIZED COEFFICIENT	STD SE
35-44						
45-54		0.124	0.470	0.791	0.090	0.341
55-64		1.212	0.278	0.001	0.942	0.216
65-74		0.495	0.286	0.084	0.380	0.220
75-84		0.782	0.418	0.062	0.592	0.317
85-94						
35-64	UNIVARIATE	1.094	0.216	0.001	0.810	0.160
65-94	UNIVARIATE	0.563	0.222	0.011	0.432	0.171
35-64	BIVARIATE	0.890	0.222	0.001	0.658	0.164
65-94	BIVARIATE	0.524	0.224	0.019	0.402	0.171
35-64	MULTIVARIATE	0.979	0.232	0.001	0.725	0.172
65-94	MULTIVARIATE	0.467	0.226	0.039	0.358	0.174

TABLE 4-4   DATE=05AUG85
AVERAGE ANNUAL INCIDENCE RATE FOR EVENT PER 1000 PERSONS AT RISK AT EXAM
BY AGE AND LEVEL OF RISK FACTOR AT EXAM: FRAMINGHAM STUDY 30-YEAR FOLLOWUP

SEX: MALE
EVENT: UNRECOGNIZED MYOCARDIAL INFARCTION
RISK FACTOR: SERUM CHOLESTEROL
PERSONS AT RISK: PERSONS FREE OF CORONARY HEART DISEASE

RISK FACTOR	35-44 AT EXAM			45-54 AT EXAM			55-64 AT EXAM			65-74 AT EXAM		
	PERSON EXAMS	# OF EVENTS	ANNUAL RATE	PERSON EXAMS	# OF EVENTS	ANNUAL RATE	PERSON EXAMS	# OF EVENTS	ANNUAL RATE	PERSON EXAMS	# OF EVENTS	ANNUAL RATE
84-204	1253	0	0	1767	1	0	2255	10	2	1234	14	6
205-234	1247	2	1	2306	4	1	2405	14	3	1171	11	5
235-264	872	2	1	1745	2	1	1889	9	2	873	4	2
265-294	475	0	0	1001	4	2	918	2	1	428	7	8
295-1124	291	1	2	557	4	4	454	4	4	191	1	3
UNKNOWN	334	0	0	348	0	0	128	0	0	16	0	0
TOTAL	4472	5	1	7724	15	1	8049	39	2	3913	37	5

RISK FACTOR	75-84 AT EXAM			85-94 AT EXAM			35-64 COMBINED			65-94 COMBINED		
	PERSON EXAMS	# OF EVENTS	ANNUAL RATE	PERSON EXAMS	# OF EVENTS	ANNUAL RATE	# OF EVENTS	CRUDE	AGE-ADJUSTED	# OF EVENTS	CRUDE	AGE-ADJUSTED
84-204	390	6	8	17			11	1	1	20	6	6
205-234	332	6	9	11			20	2	2	17	6	6
235-264	177	4	11	5			13	1	1	8	4	4
265-294	96	0	0	6			6	1	1	7	7	6
295-1124	22	1	23	0			9	3	4	2	5	7
UNKNOWN	0	0	-	0			0	0	0	0	0	0
TOTAL	1017	17	8	39			59	1	1	54	5	6

LOGISTIC REGRESSIONS

AGE AT EXAM		COEFFICIENT OF RISK FACTOR	SE COEFFICIENT	P-VALUE	STANDARDIZED COEFFICIENT	STD SE
35-44		0.012	0.006	0.049	0.541	0.275
45-54		0.011	0.004	0.002	0.462	0.152
55-64		0.001	0.004	0.712	0.058	0.158
65-74		-0.002	0.004	0.569	-0.095	0.168
75-84		0.002	0.006	0.787	0.065	0.242
85-94						
35-64	UNIVARIATE	0.006	0.003	0.026	0.250	0.115
65-94	UNIVARIATE	-0.002	0.003	0.636	-0.065	0.138
35-64	BIVARIATE	0.007	0.003	0.016	0.278	0.115
65-94	BIVARIATE	-0.001	0.003	0.792	-0.036	0.139
35-64	MULTIVARIATE	0.006	0.003	0.041	0.239	0.117
65-94	MULTIVARIATE	-0.001	0.003	0.829	-0.030	0.138

SEX: FEMALE
EVENT: UNRECOGNIZED MYOCARDIAL INFARCTION
RISK FACTOR: SERUM CHOLESTEROL
PERSONS AT RISK: PERSONS FREE OF CORONARY HEART DISEASE

RISK FACTOR	35-44 AT EXAM			45-54 AT EXAM			55-64 AT EXAM			65-74 AT EXAM		
	PERSON EXAMS	# OF EVENTS	ANNUAL RATE	PERSON EXAMS	# OF EVENTS	ANNUAL RATE	PERSON EXAMS	# OF EVENTS	ANNUAL RATE	PERSON EXAMS	# OF EVENTS	ANNUAL RATE
84-204	2350			2094	2	0	1625	4	1	789	1	1
205-234	1417			2558	2	0	2615	7	1	1401	8	3
235-264	806			2259	0	0	2938	3	1	1703	8	2
265-294	372			1473	2	1	2240	8	2	1287	4	2
295-1124	171			1101	1	0	1866	5	1	1031	1	0
UNKNOWN	500			474	1	1	229	0	0	37	0	0
TOTAL	5616			9959	8	0	11513	27	1	6248	22	2

RISK FACTOR	75-84 AT EXAM			85-94 AT EXAM			35-64 COMBINED			65-94 COMBINED		
	PERSON EXAMS	# OF EVENTS	ANNUAL RATE	PERSON EXAMS	# OF EVENTS	ANNUAL RATE	# OF EVENTS	CRUDE	AGE-ADJUSTED	# OF EVENTS	CRUDE	AGE-ADJUSTED
84-204	361	4	6	46			6	0	0	5	2	2
205-234	486	3	3	28			9	1	0	12	3	3
235-264	542	2	2	35			3	0	0	12	3	2
265-294	350	4	4	9			10	1	1	7	2	2
295-1124	205	1	2	5			6	1	0	2	1	0
UNKNOWN	7	0	0	0			1	0	0	0	0	0
TOTAL	1951	13	3	123			35	1	0	38	2	2

LOGISTIC REGRESSIONS

AGE AT EXAM		COEFFICIENT OF RISK FACTOR	SE COEFFICIENT	P-VALUE	STANDARDIZED COEFFICIENT	STD SE
35-44						
45-54		-0.001	0.008	0.905	-0.046	0.388
55-64		0.001	0.004	0.877	0.030	0.191
65-74		-0.005	0.005	0.360	-0.202	0.221
75-84		-0.004	0.007	0.587	-0.154	0.284
85-94						
35-64	UNIVARIATE	0.004	0.003	0.187	0.184	0.139
65-94	UNIVARIATE	-0.004	0.004	0.236	-0.198	0.167
35-64	BIVARIATE	-0.000	0.004	0.978	-0.005	0.176
65-94	BIVARIATE	-0.003	0.004	0.443	-0.129	0.168

TABLE 4-5   DATE=06DEC85

AVERAGE ANNUAL INCIDENCE RATE FOR EVENT PER 1000 PERSONS AT RISK AT EXAM
BY AGE AND LEVEL OF RISK FACTOR AT EXAM: FRAMINGHAM STUDY 30-YEAR FOLLOWUP

SEX: MALE
EVENT: UNRECOGNIZED MYOCARDIAL INFARCTION
RISK FACTOR: HEMATOCRIT
PERSONS AT RISK: PERSONS FREE OF CORONARY HEART DISEASE

RISK FACTOR	35-44 AT EXAM			45-54 AT EXAM			55-64 AT EXAM			65-74 AT EXAM		
	PERSON EXAMS	# OF EVENTS	ANNUAL RATE	PERSON EXAMS	# OF EVENTS	ANNUAL RATE	PERSON EXAMS	# OF EVENTS	ANNUAL RATE	PERSON EXAMS	# OF EVENTS	ANNUAL RATE
25-42	623	1	1	942	1	1	1018	4	2	450	4	4
43-44	546	0	0	967	2	1	985	6	3	493	3	3
45-46	983	1	1	1854	4	1	1953	10	3	963	12	6
47-48	775	1	1	1510	2	1	1599	3	1	861	7	4
49-70	1175	2	1	2066	6	1	2338	16	3	1127	11	5
UNKNOWN	370	0	0	385	0	0	156	0	0	19	0	0
TOTAL	4472	5	1	7724	15	1	8049	39	2	3913	37	5

RISK FACTOR	75-84 AT EXAM			85-94 AT EXAM			35-64 COMBINED			65-94 COMBINED		
	PERSON EXAMS	# OF EVENTS	ANNUAL RATE	PERSON EXAMS	# OF EVENTS	ANNUAL RATE	# OF EVENTS	CRUDE	AGE-ADJUSTED	# OF EVENTS	CRUDE	AGE-ADJUSTED
25-42	214	6	14	16			6	1	1	10	7	6
43-44	119	4	17	5			8	2	2	7	6	3
45-46	240	0	0	7			15	2	2	12	5	5
47-48	197	5	13	7			6	1	1	12	6	6
49-70	247	2	4	4			24	2	2	13	5	5
UNKNOWN	0	0	-	0			0	0	0	0	0	0
TOTAL	1017	17	8	39			59	1	1	54	5	5

LOGISTIC REGRESSIONS

AGE AT EXAM		COEFFICIENT OF RISK FACTOR	SE COEFFICIENT	P-VALUE	STANDARDIZED COEFFICIENT	STD SE
35-44		0.075	0.123	0.544	0.275	0.453
45-54		0.130	0.071	0.067	0.468	0.254
55-64		0.064	0.043	0.136	0.243	0.163
65-74		0.035	0.045	0.440	0.129	0.167
75-84		-0.085	0.056	0.127	-0.355	0.232
85-94						
35-64	UNIVARIATE	0.083	0.036	0.019	0.306	0.131
65-94	UNIVARIATE	-0.016	0.035	0.645	-0.062	0.136
35-64	BIVARIATE	0.080	0.035	0.023	0.295	0.130
65-94	BIVARIATE	-0.006	0.035	0.872	-0.022	0.136
35-64	MULTIVARIATE	0.050	0.035	0.153	0.186	0.130
65-94	MULTIVARIATE	-0.012	0.036	0.736	-0.047	0.138

SEX: FEMALE
EVENT: UNRECOGNIZED MYOCARDIAL INFARCTION
RISK FACTOR: HEMATOCRIT
PERSONS AT RISK: PERSONS FREE OF CORONARY HEART DISEASE

RISK FACTOR	35-44 AT EXAM			45-54 AT EXAM			55-64 AT EXAM			65-74 AT EXAM		
	PERSON EXAMS	# OF EVENTS	ANNUAL RATE	PERSON EXAMS	# OF EVENTS	ANNUAL RATE	PERSON EXAMS	# OF EVENTS	ANNUAL RATE	PERSON EXAMS	# OF EVENTS	ANNUAL RATE
21-39	1697			1843	1	0	1553	2	1	646	1	1
40-41	967			1741	0	0	2061	3	1	1069	3	1
42-43	1043			1955	2	1	2260	2	0	1256	8	3
44-45	856			2194	3	1	2847	6	1	1609	3	1
46-65	487			1681	1	0	2521	13	3	1626	7	2
UNKNOWN	566			545	1	1	271	1	2	42	0	0
TOTAL	5616			9959	8	0	11513	27	1	6248	22	2

RISK FACTOR	75-84 AT EXAM			85-94 AT EXAM			35-64 COMBINED			65-94 COMBINED		
	PERSON EXAMS	# OF EVENTS	ANNUAL RATE	PERSON EXAMS	# OF EVENTS	ANNUAL RATE	# OF EVENTS	CRUDE	AGE-ADJUSTED	# OF EVENTS	CRUDE	AGE-ADJUSTED
21-39	297	1	2	24			3	0	0	3	2	1
40-41	373	6	8	31			3	0	0	10	3	3
42-43	353	2	3	19			4	0	0	11	3	3
44-45	486	2	2	26			9	1	1	5	1	1
46-65	436	2	2	23			14	1	1	7	2	2
UNKNOWN	6	0	0	0			2	1	1	0	0	0
TOTAL	1951	13	3	123			35	1	1	38	2	2

LOGISTIC REGRESSIONS

AGE AT EXAM		COEFFICIENT OF RISK FACTOR	SE COEFFICIENT	P-VALUE	STANDARDIZED COEFFICIENT	STD SE
35-44						
45-54		0.086	0.109	0.429	0.307	0.387
55-64		0.157	0.056	0.005	0.545	0.193
65-74		0.057	0.062	0.355	0.200	0.217
75-84		-0.063	0.067	0.348	-0.244	0.260
85-94						
35-64	UNIVARIATE	0.172	0.049	0.001	0.618	0.175
65-94	UNIVARIATE	-0.020	0.045	0.651	-0.073	0.161
35-64	BIVARIATE	0.134	0.050	0.007	0.483	0.180
65-94	BIVARIATE	-0.010	0.044	0.818	-0.037	0.159
35-64	MULTIVARIATE	0.098	0.050	0.051	0.351	0.180
65-94	MULTIVARIATE	-0.001	0.043	0.974	-0.005	0.156

TABLE 4-6   DATE=06DEC85
AVERAGE ANNUAL INCIDENCE RATE FOR EVENT PER 1000 PERSONS AT RISK AT EXAM
BY AGE AND LEVEL OF RISK FACTOR AT EXAM: FRAMINGHAM STUDY 30-YEAR FOLLOWUP

SEX: MALE
EVENT: UNRECOGNIZED MYOCARDIAL INFARCTION
RISK FACTOR: BLOOD GLUCOSE (MG/DL)
PERSONS AT RISK: PERSONS FREE OF CORONARY HEART DISEASE

RISK FACTOR	35-44 AT EXAM			45-54 AT EXAM			55-64 AT EXAM			65-74 AT EXAM		
	PERSON EXAMS	# OF EVENTS	ANNUAL RATE	PERSON EXAMS	# OF EVENTS	ANNUAL RATE	PERSON EXAMS	# OF EVENTS	ANNUAL RATE	PERSON EXAMS	# OF EVENTS	ANNUAL RATE
29-69	992	0	0	1490	4	1	1087	4	2	421	2	2
70-89	2501	5	1	4172	5	1	4056	17	2	1761	18	5
90-109	626	0	0	1348	2	1	1866	11	3	1046	9	4
110-129	105	0	0	305	2	3	537	3	3	335	1	1
130-524	48	0	0	193	2	5	408	4	5	334	7	10
UNKNOWN	200	0	0	216	0	0	95	0	0	16	0	0
TOTAL	4472	5	1	7724	15	1	8049	39	2	3913	37	5

RISK FACTOR	75-84 AT EXAM			85-94 AT EXAM			35-64 COMBINED			65-94 COMBINED		
	PERSON EXAMS	# OF EVENTS	ANNUAL RATE	PERSON EXAMS	# OF EVENTS	ANNUAL RATE	# OF EVENTS	CRUDE	AGE-ADJUSTED	# OF EVENTS	CRUDE	AGE-ADJUSTED
29-69	62	2	16	1			8	1	1	4	4	5
70-89	377	7	9	11			27	1	1	25	6	6
90-109	328	5	8	14			13	2	1	14	5	5
110-129	125	1	4	4			5	3	2	2	2	2
130-524	125	2	8	9			6	5	4	9	10	10
UNKNOWN	0	0	-	0			0	0	0	0	0	0
TOTAL	1017	17	8	39			59	1	1	54	5	5

LOGISTIC REGRESSIONS

AGE AT EXAM	COEFFICIENT OF RISK FACTOR	SE COEFFICIENT	P-VALUE	STANDARDIZED COEFFICIENT	STD SE
35-44	-0.011	0.034	0.742	-0.190	0.579
45-54	0.011	0.004	0.004	0.262	0.091
55-64	0.005	0.004	0.183	0.148	0.111
65-74	0.006	0.003	0.037	0.211	0.101
75-84	-0.002	0.008	0.848	-0.050	0.260
85-94					
35-64	UNIVARIATE	0.009	0.003	0.001	0.209	0.064
65-94	UNIVARIATE	0.005	0.003	0.075	0.171	0.096
35-64	BIVARIATE	0.007	0.003	0.019	0.168	0.072
65-94	BIVARIATE	0.005	0.003	0.109	0.161	0.100
35-64	MULTIVARIATE	0.006	0.003	0.029	0.159	0.073
65-94	MULTIVARIATE	0.004	0.003	0.217	0.126	0.102

SEX: FEMALE
EVENT: UNRECOGNIZED MYOCARDIAL INFARCTION
RISK FACTOR: BLOOD GLUCOSE (MG/DL)
PERSONS AT RISK: PERSONS FREE OF CORONARY HEART DISEASE

RISK FACTOR	35-44 AT EXAM			45-54 AT EXAM			55-64 AT EXAM			65-74 AT EXAM		
	PERSON EXAMS	# OF EVENTS	ANNUAL RATE	PERSON EXAMS	# OF EVENTS	ANNUAL RATE	PERSON EXAMS	# OF EVENTS	ANNUAL RATE	PERSON EXAMS	# OF EVENTS	ANNUAL RATE
29-69	1066			1797	0	0	1369	3	1	521	1	1
70-89	3412			5744	4	0	6098	9	1	3102	11	2
90-109	672			1625	1	0	2712	9	2	1732	8	2
110-129	96			301	1	2	730	4	3	501	0	0
130-524	39			159	1	3	438	1	1	360	2	3
UNKNOWN	331			333	1	2	166	1	3	32	0	0
TOTAL	5616			9959	8	8	11513	27	1	6248	22	2

RISK FACTOR	75-84 AT EXAM			85-94 AT EXAM			35-64 COMBINED			65-94 COMBINED		
	PERSON EXAMS	# OF EVENTS	ANNUAL RATE	PERSON EXAMS	# OF EVENTS	ANNUAL RATE	# OF EVENTS	CRUDE	AGE-ADJUSTED	# OF EVENTS	CRUDE	AGE-ADJUSTED
29-69	105	1	5	3			3	0	0	2	2	2
70-89	823	3	2	44			13	0	0	17	2	2
90-109	640	3	2	51			10	1	1	11	2	2
110-129	219	2	5	13			5	2	2	3	2	2
130-524	163	3	9	12			2	2	2	5	4	4
UNKNOWN	1	0	0	0			2	1	2	0	0	0
TOTAL	1951	13	3	123			35	1	1	58	2	2

LOGISTIC REGRESSIONS

AGE AT EXAM	COEFFICIENT OF RISK FACTOR	SE COEFFICIENT	P-VALUE	STANDARDIZED COEFFICIENT	STD SE
35-44					
45-54	0.013	0.008	0.094	0.256	0.153
55-64	0.010	0.004	0.008	0.247	0.092
65-74	-0.001	0.008	0.937	-0.017	0.221
75-84	0.008	0.004	0.071	0.259	0.143
85-94					
35-64	UNIVARIATE	0.012	0.003	0.001	0.269	0.067
65-94	UNIVARIATE	0.005	0.004	0.143	0.162	0.110
35-64	BIVARIATE	0.010	0.003	0.005	0.210	0.074
65-94	BIVARIATE	0.004	0.004	0.269	0.128	0.116
35-64	MULTIVARIATE	0.008	0.003	0.024	0.167	0.074
65-94	MULTIVARIATE	0.003	0.004	0.491	0.084	0.122

TABLE 4-7   DATE=06DEC85

AVERAGE ANNUAL INCIDENCE RATE FOR EVENT PER 1000 PERSONS AT RISK AT EXAM
BY AGE AND LEVEL OF RISK FACTOR AT EXAM: FRAMINGHAM STUDY 30-YEAR FOLLOWUP

SEX: MALE
EVENT: UNRECOGNIZED MYOCARDIAL INFARCTION
RISK FACTOR: DIABETES MELLITUS
PERSONS AT RISK: PERSONS FREE OF CORONARY HEART DISEASE

RISK FACTOR	35-44 AT EXAM PERSON EXAMS	# OF EVENTS	ANNUAL RATE	45-54 AT EXAM PERSON EXAMS	# OF EVENTS	ANNUAL RATE	55-64 AT EXAM PERSON EXAMS	# OF EVENTS	ANNUAL RATE	65-74 AT EXAM PERSON EXAMS	# OF EVENTS	ANNUAL RATE
NO	4424	5	1	7486	13	1	7549	32	2	3504	30	4
YES	48	0	0	238	2	4	500	7	7	409	7	9
UNKNOWN	0	0	-	0	0	-	0	0	-	0	0	-
TOTAL	4472	5	1	7724	15	1	8049	39	2	3913	37	5

RISK FACTOR	75-84 AT EXAM PERSON EXAMS	# OF EVENTS	ANNUAL RATE	85-94 AT EXAM PERSON EXAMS	# OF EVENTS	ANNUAL RATE	35-64 COMBINED # OF EVENTS	CRUDE	AGE-ADJUSTED	65-94 COMBINED # OF EVENTS	CRUDE	AGE-ADJUSTED
NO	900	15	8	34		5	50	1	1	45	5	5
YES	117	2	9	5			9	6	4	9	8	8
UNKNOWN	0	0	-	0			0	-	-	0	-	-
TOTAL	1017	17	8	39			59	1	1	54	5	5

LOGISTIC REGRESSIONS

AGE AT EXAM		COEFFICIENT OF RISK FACTOR	SE COEFFICIENT	P-VALUE	STANDARDIZED COEFFICIENT	STD SE
35-44		-	-		-	-
45-54		1.583	0.762	0.038	0.274	0.132
55-64		1.205	0.420	0.004	0.291	0.101
65-74		0.701	0.423	0.097	0.215	0.129
75-84		0.026	0.759	0.973	0.008	0.242
85-94						
35-64	UNIVARIATE	1.503	0.364	0.001	0.290	0.070
65-94	UNIVARIATE	0.521	0.368	0.157	0.161	0.114
35-64	BIVARIATE	1.190	0.369	0.001	0.230	0.071
65-94	BIVARIATE	0.497	0.368	0.177	0.154	0.114
35-64	MULTIVARIATE	1.058	0.380	0.005	0.204	0.073
65-94	MULTIVARIATE	0.460	0.374	0.219	0.142	0.116

SEX: FEMALE
EVENT: UNRECOGNIZED MYOCARDIAL INFARCTION
RISK FACTOR: DIABETES MELLITUS
PERSONS AT RISK: PERSONS FREE OF CORONARY HEART DISEASE

RISK FACTOR	35-44 AT EXAM PERSON EXAMS	# OF EVENTS	ANNUAL RATE	45-54 AT EXAM PERSON EXAMS	# OF EVENTS	ANNUAL RATE	55-64 AT EXAM PERSON EXAMS	# OF EVENTS	ANNUAL RATE	65-74 AT EXAM PERSON EXAMS	# OF EVENTS	ANNUAL RATE
NO	5573			9760	8	0	10985	25	1	5780	19	2
YES	43			199	0	0	528	2	2	468	3	3
UNKNOWN	0			0	0	-	0	0	-	0	0	-
TOTAL	5616			9959	8	0	11513	27	1	6248	22	2

RISK FACTOR	75-84 AT EXAM PERSON EXAMS	# OF EVENTS	ANNUAL RATE	85-94 AT EXAM PERSON EXAMS	# OF EVENTS	ANNUAL RATE	35-64 COMBINED # OF EVENTS	CRUDE	AGE-ADJUSTED	65-94 COMBINED # OF EVENTS	CRUDE	AGE-ADJUSTED
NO	1738	10	3	106			33	1	1	32	2	2
YES	213	3	7	17			2	1	1	6	4	4
UNKNOWN	0	0	-	0			0	-	-	0	-	-
TOTAL	1951	13	3	123			35	1	1	38	2	2

LOGISTIC REGRESSIONS

AGE AT EXAM		COEFFICIENT OF RISK FACTOR	SE COEFFICIENT	P-VALUE	STANDARDIZED COEFFICIENT	STD SE
35-44		-	-		-	-
45-54		-	-		-	-
55-64		0.511	0.736	0.488	0.107	0.154
65-74		0.671	0.623	0.282	0.177	0.164
75-84		0.904	0.662	0.172	0.282	0.207
85-94						
35-64	UNIVARIATE	0.730	0.729	0.317	0.121	0.121
65-94	UNIVARIATE	0.721	0.447	0.106	0.200	0.124
35-64	BIVARIATE	0.295	0.731	0.687	0.049	0.122
65-94	BIVARIATE	0.614	0.449	0.171	0.170	0.124
35-64	MULTIVARIATE	0.004	0.737	0.996	0.001	0.122
65-94	MULTIVARIATE	0.413	0.463	0.373	0.115	0.128

AVERAGE ANNUAL INCIDENCE RATE FOR EVENT PER 1000 PERSONS AT RISK AT EXAM
BY AGE AND LEVEL OF RISK FACTOR AT EXAM: FRAMINGHAM STUDY 30-YEAR FOLLOWUP

SEX: MALE
EVENT: UNRECOGNIZED MYOCARDIAL INFARCTION
RISK FACTOR: GLUCOSE IN URINE, DEFINITE OR TRACE
PERSONS AT RISK: PERSONS FREE OF CORONARY HEART DISEASE

RISK FACTOR	35-44 AT EXAM			45-54 AT EXAM			55-64 AT EXAM			65-74 AT EXAM		
	PERSON EXAMS	# OF EVENTS	ANNUAL RATE	PERSON EXAMS	# OF EVENTS	ANNUAL RATE	PERSON EXAMS	# OF EVENTS	ANNUAL RATE	PERSON EXAMS	# OF EVENTS	ANNUAL RATE
NO	4380	5	1	7535	14	1	7815	36	2	3767	32	4
YES	60	0	0	168	1	3	210	3	7	140	5	18
UNKNOWN	32	0	0	21	0	0	24	0	0	6	0	0
TOTAL	4472	5	1	7724	15	1	8049	39	2	3913	37	5

RISK FACTOR	75-84 AT EXAM			85-94 AT EXAM			35-64 COMBINED			65-94 COMBINED		
	PERSON EXAMS	# OF EVENTS	ANNUAL RATE	PERSON EXAMS	# OF EVENTS	ANNUAL RATE	# OF EVENTS	CRUDE	AGE-ADJUSTED	# OF EVENTS	CRUDE	AGE-ADJUSTED
NO	986	17	9	38			55	1	1	49	5	5
YES	28	0	0	1			4	5	4	5	15	14
UNKNOWN	5	0	0	0			0	0	0	0	0	0
TOTAL	1017	17	8	39			59	1	1	54	5	5

LOGISTIC REGRESSIONS

AGE AT EXAM		COEFFICIENT OF RISK FACTOR	SE COEFFICIENT	P-VALUE	STANDARDIZED COEFFICIENT	STD SE
35-44		-	-	-	-	-
45-54		1.168	1.038	0.260	0.171	0.152
55-64		1.142	0.605	0.059	0.182	0.097
65-74		1.464	0.489	0.003	0.272	0.091
75-84		-	-			
85-94		-				
35-64	UNIVARIATE	1.193	0.520	0.022	0.174	0.076
65-94	UNIVARIATE	1.082	0.476	0.023	0.196	0.086
35-64	BIVARIATE	1.046	0.521	0.045	0.152	0.076
65-94	BIVARIATE	1.118	0.477	0.019	0.203	0.087
35-64	MULTIVARIATE	0.965	0.529	0.068	0.141	0.077
65-94	MULTIVARIATE	0.958	0.488	0.050	0.174	0.089

SEX: FEMALE
EVENT: UNRECOGNIZED MYOCARDIAL INFARCTION
RISK FACTOR: GLUCOSE IN URINE, DEFINITE OR TRACE
PERSONS AT RISK: PERSONS FREE OF CORONARY HEART DISEASE

RISK FACTOR	35-44 AT EXAM			45-54 AT EXAM			55-64 AT EXAM			65-74 AT EXAM		
	PERSON EXAMS	# OF EVENTS	ANNUAL RATE	PERSON EXAMS	# OF EVENTS	ANNUAL RATE	PERSON EXAMS	# OF EVENTS	ANNUAL RATE	PERSON EXAMS	# OF EVENTS	ANNUAL RATE
NO	5537			9841	7	0	11346	26	1	6131	22	2
YES	41			94	1	5	144	1	3	107	0	0
UNKNOWN	38			24	0	0	23	0	0	10	0	0
TOTAL	5616			9959	8	0	11513	27	1	6248	22	2

RISK FACTOR	75-84 AT EXAM			85-94 AT EXAM			35-64 COMBINED			65-94 COMBINED		
	PERSON EXAMS	# OF EVENTS	ANNUAL RATE	PERSON EXAMS	# OF EVENTS	ANNUAL RATE	# OF EVENTS	CRUDE	AGE-ADJUSTED	# OF EVENTS	CRUDE	AGE-ADJUSTED
NO	1919	13	3	120			33	1	1	38	2	2
YES	23	0	0	3			2	4	3	0	0	0
UNKNOWN	9	0	0	0			0	0	0	0	0	0
TOTAL	1951	13	3	123			35	1	1	38	2	2

LOGISTIC REGRESSIONS

AGE AT EXAM		COEFFICIENT OF RISK FACTOR	SE COEFFICIENT	P-VALUE	STANDARDIZED COEFFICIENT	STD SE
35-44						
45-54		1.168	1.038	0.260	0.113	0.100
55-64		1.113	1.023	0.276	0.124	0.114
65-74		-	-	-	-	-
75-84		-	-	-	-	-
85-94		-	-	-	-	-
35-64	UNIVARIATE	1.765	0.731	0.016	0.178	0.074
65-94	UNIVARIATE	-	-	-	-	-
35-64	BIVARIATE	1.620	0.753	0.027	0.164	0.074
65-94	BIVARIATE	-	-	-	-	-
35-64	MULTIVARIATE	1.242	0.746	0.096	0.126	0.075
65-94	MULTIVARIATE	-				

AVERAGE ANNUAL INCIDENCE RATE FOR EVENT PER 1000 PERSONS AT RISK AT EXAM
BY AGE AND LEVEL OF RISK FACTOR AT EXAM: FRAMINGHAM STUDY 30-YEAR FOLLOWUP

SEX: MALE
EVENT: UNRECOGNIZED MYOCARDIAL INFARCTION
RISK FACTOR: GLUCOSE INTOLERANCE
PERSONS AT RISK: PERSONS FREE OF CORONARY HEART DISEASE

RISK FACTOR	35-44 AT EXAM			45-54 AT EXAM			55-64 AT EXAM			65-74 AT EXAM		
	PERSON EXAMS	# OF EVENTS	ANNUAL RATE	PERSON EXAMS	# OF EVENTS	ANNUAL RATE	PERSON EXAMS	# OF EVENTS	ANNUAL RATE	PERSON EXAMS	# OF EVENTS	ANNUAL RATE
NO	4335	5	1	7221	11	1	7145	30	2	3228	28	4
YES	137	0	0	503	4	4	904	9	5	685	9	7
UNKNOWN	0	0	-	0	0	-	0	0	-	0	0	-
TOTAL	4472	5	1	7724	15	1	8049	39	2	3913	37	5

RISK FACTOR	75-84 AT EXAM			85-94 AT EXAM			35-64 COMBINED			65-94 COMBINED		
	PERSON EXAMS	# OF EVENTS	ANNUAL RATE	PERSON EXAMS	# OF EVENTS	ANNUAL RATE	# OF EVENTS	CRUDE	AGE-ADJUSTED	# OF EVENTS	CRUDE	AGE-ADJUSTED
NO	802	14	9	27			46	1	1	42	5	5
YES	215	3	7	12			13	4	3	12	7	7
UNKNOWN	0	0	-	0			0	-	-	0	-	
TOTAL	1017	17	8	39			59	1	1	54	5	5

LOGISTIC REGRESSIONS

AGE AT EXAM		COEFFICIENT OF RISK FACTOR	SE COEFFICIENT	P-VALUE	STANDARDIZED COEFFICIENT	STD SE
35-44		-	-	-	-	-
45-54		1.659	0.586	0.005	0.409	0.145
55-64		0.869	0.382	0.023	0.274	0.121
65-74		0.420	0.386	0.276	0.160	0.147
75-84		-0.228	0.641	0.723	-0.093	0.262
85-94						
35-64	UNIVARIATE	1.237	0.315	0.001	0.328	0.084
65-94	UNIVARIATE	0.243	0.329	0.461	0.094	0.128
35-64	BIVARIATE	0.975	0.319	0.002	0.259	0.085
65-94	BIVARIATE	0.196	0.330	0.553	0.076	0.128
35-64	MULTIVARIATE	0.906	0.323	0.005	0.241	0.086
65-94	MULTIVARIATE	0.121	0.333	0.717	0.047	0.129

SEX: FEMALE
EVENT: UNRECOGNIZED MYOCARDIAL INFARCTION
RISK FACTOR: GLUCOSE INTOLERANCE
PERSONS AT RISK: PERSONS FREE OF CORONARY HEART DISEASE

RISK FACTOR	35-44 AT EXAM			45-54 AT EXAM			55-64 AT EXAM			65-74 AT EXAM		
	PERSON EXAMS	# OF EVENTS	ANNUAL RATE	PERSON EXAMS	# OF EVENTS	ANNUAL RATE	PERSON EXAMS	# OF EVENTS	ANNUAL RATE	PERSON EXAMS	# OF EVENTS	ANNUAL RATE
NO	5522			9588	6	0	10503	23	1	5398	18	2
YES	94			371	2	3	1010	4	2	850	4	2
UNKNOWN	0			0	0	-	0	0	-	0	0	-
TOTAL	5616			9959	8	0	11513	27	1	6248	22	2

RISK FACTOR	75-84 AT EXAM			85-94 AT EXAM			35-64 COMBINED			65-94 COMBINED		
	PERSON EXAMS	# OF EVENTS	ANNUAL RATE	PERSON EXAMS	# OF EVENTS	ANNUAL RATE	# OF EVENTS	CRUDE	AGE-ADJUSTED	# OF EVENTS	CRUDE	AGE-ADJUSTED
NO	1581	8	3	100			29	1	1	28	2	2
YES	370	5	7	23			6	2	2	10	4	4
UNKNOWN	0	0	-	0			0	-	-	0	-	
TOTAL	1951	13	3	123			35	1	1	38	2	2

LOGISTIC REGRESSIONS

AGE AT EXAM		COEFFICIENT OF RISK FACTOR	SE COEFFICIENT	P-VALUE	STANDARDIZED COEFFICIENT	STD SE
35-44						
45-54		2.158	0.818	0.008	0.409	0.155
55-64		0.594	0.543	0.274	0.168	0.154
65-74		0.346	0.554	0.532	0.119	0.190
75-84		0.991	0.573	0.084	0.389	0.225
85-94						
35-64	UNIVARIATE	1.282	0.449	0.004	0.291	0.102
65-94	UNIVARIATE	0.714	0.370	0.053	0.255	0.132
35-64	BIVARIATE	0.852	0.453	0.060	0.193	0.103
65-94	BIVARIATE	0.618	0.372	0.096	0.220	0.133
35-64	MULTIVARIATE	0.705	0.457	0.123	0.160	0.104
65-94	MULTIVARIATE	0.484	0.381	0.204	0.172	0.136

AVERAGE ANNUAL INCIDENCE RATE FOR EVENT PER 1000 PERSONS AT RISK AT EXAM
BY AGE AND LEVEL OF RISK FACTOR AT EXAM: FRAMINGHAM STUDY 30-YEAR FOLLOWUP

SEX: MALE
EVENT: UNRECOGNIZED MYOCARDIAL INFARCTION
RISK FACTOR: METROPOLITAN RELATIVE WEIGHT (PERCENT)
PERSONS AT RISK: PERSONS FREE OF CORONARY HEART DISEASE

RISK FACTOR	35-44 AT EXAM PERSON EXAMS	# OF EVENTS	ANNUAL RATE	45-54 AT EXAM PERSON EXAMS	# OF EVENTS	ANNUAL RATE	55-64 AT EXAM PERSON EXAMS	# OF EVENTS	ANNUAL RATE	65-74 AT EXAM PERSON EXAMS	# OF EVENTS	ANNUAL RATE
54-104	723	1	1	1131	1	0	1155	9	4	686	5	4
105-114	965	2	1	1540	2	1	1655	3	1	851	9	5
115-124	1122	0	0	2037	3	1	2175	11	3	1069	8	4
125-134	915	2	1	1645	3	1	1674	10	3	766	7	5
135-272	747	0	0	1370	6	2	1389	6	2	541	8	7
UNKNOWN	0	0	-	1	0	0	1	0	0	0	0	-
TOTAL	4472	5	1	7724	15	1	8049	39	2	3913	37	5

RISK FACTOR	75-84 AT EXAM PERSON EXAMS	# OF EVENTS	ANNUAL RATE	85-94 AT EXAM PERSON EXAMS	# OF EVENTS	ANNUAL RATE	35-64 COMBINED # OF EVENTS	CRUDE	AGE-ADJUSTED	65-94 COMBINED # OF EVENTS	CRUDE	AGE-ADJUSTED
54-104	224	5	11	18			11	2	2	10	5	5
105-114	252	4	8	9			7	1	1	13	6	6
115-124	296	3	5	12			14	1	1	11	4	4
125-134	155	2	6	0			15	2	2	9	5	5
135-272	90	3	17	0			12	2	2	11	9	9
UNKNOWN	0	0	-	0			0	0	0	0	-	
TOTAL	1017	17	8	39			59	1	1	54	5	5

LOGISTIC REGRESSIONS

AGE AT EXAM		COEFFICIENT OF RISK FACTOR	SE COEFFICIENT	P-VALUE	STANDARDIZED COEFFICIENT	STD SE
35-44		-0.021	0.029	0.469	-0.345	0.477
45-54		0.024	0.014	0.089	0.390	0.229
55-64		-0.003	0.010	0.763	-0.049	0.162
65-74		0.010	0.010	0.332	0.156	0.161
75-84		0.019	0.015	0.212	0.281	0.225
85-94						
35-64	UNIVARIATE	0.003	0.008	0.670	0.055	0.129
65-94	UNIVARIATE	0.012	0.009	0.175	0.179	0.132
35-64	BIVARIATE	0.003	0.008	0.679	0.054	0.130
65-94	BIVARIATE	0.014	0.009	0.101	0.217	0.133
35-64	MULTIVARIATE	-0.002	0.008	0.783	-0.037	0.132
65-94	MULTIVARIATE	0.009	0.009	0.281	0.147	0.136

SEX: FEMALE
EVENT: UNRECOGNIZED MYOCARDIAL INFARCTION
RISK FACTOR: METROPOLITAN RELATIVE WEIGHT (PERCENT)
PERSONS AT RISK: PERSONS FREE OF CORONARY HEART DISEASE

RISK FACTOR	35-44 AT EXAM PERSON EXAMS	# OF EVENTS	ANNUAL RATE	45-54 AT EXAM PERSON EXAMS	# OF EVENTS	ANNUAL RATE	55-64 AT EXAM PERSON EXAMS	# OF EVENTS	ANNUAL RATE	65-74 AT EXAM PERSON EXAMS	# OF EVENTS	ANNUAL RATE
54-104	1694			2095	2	0	2045	8	2	1216	3	1
105-114	1430			2331	3	1	2409	1	0	1145	6	3
115-124	1058			2107	1	0	2679	5	1	1395	4	1
125-134	566			1372	1	0	1743	3	1	982	2	1
135-272	860			2054	1	0	2837	10	2	1510	7	2
UNKNOWN	8			0	0	-	0	0	-	0	0	-
TOTAL	5616			9959	8	0	11513	27	1	6248	22	2

RISK FACTOR	75-84 AT EXAM PERSON EXAMS	# OF EVENTS	ANNUAL RATE	85-94 AT EXAM PERSON EXAMS	# OF EVENTS	ANNUAL RATE	35-64 COMBINED # OF EVENTS	CRUDE	AGE-ADJUSTED	65-94 COMBINED # OF EVENTS	CRUDE	AGE-ADJUSTED
54-104	517	3	3	41			10	1	1	7	2	2
105-114	375	2	3	28			4	0	0	8	3	2
115-124	399	2	3	21			6	1	0	7	2	2
125-134	266	2	4	16			4	1	0	5	2	2
135-272	394	4	5	17			11	1	1	11	3	3
UNKNOWN	0	0	-	0			0	0	0	0	-	
TOTAL	1951	13	3	123			35	1	1	38	2	2

LOGISTIC REGRESSIONS

AGE AT EXAM		COEFFICIENT OF RISK FACTOR	SE COEFFICIENT	P-VALUE	STANDARDIZED COEFFICIENT	STD SE
35-44						
45-54		-0.017	0.020	0.392	-0.372	0.435
55-64		0.009	0.008	0.252	0.194	0.169
65-74		0.015	0.008	0.058	0.327	0.173
75-84		0.008	0.012	0.523	0.167	0.261
85-94						
35-64	UNIVARIATE	0.008	0.007	0.256	0.170	0.150
65-94	UNIVARIATE	0.009	0.007	0.156	0.206	0.145
35-64	BIVARIATE	0.004	0.007	0.564	0.091	0.157
65-94	BIVARIATE	0.012	0.007	0.075	0.258	0.145
35-64	MULTIVARIATE	-0.001	0.008	0.854	-0.030	0.164
65-94	MULTIVARIATE	0.009	0.007	0.186	0.193	0.146

TABLE 4-11   DATE=06DEC85
AVERAGE ANNUAL INCIDENCE RATE FOR EVENT PER 1000 PERSONS AT RISK AT EXAM
BY AGE AND LEVEL OF RISK FACTOR AT EXAM: FRAMINGHAM STUDY 30-YEAR FOLLOWUP

SEX: MALE
EVENT: UNRECOGNIZED MYOCARDIAL INFARCTION
RISK FACTOR: VITAL CAPACITY - HEIGHT INDEX (ML/INCH)
PERSONS AT RISK: PERSONS FREE OF CORONARY HEART DISEASE

RISK FACTOR	35-44 AT EXAM			45-54 AT EXAM			55-64 AT EXAM			65-74 AT EXAM		
	PERSON EXAMS	# OF EVENTS	ANNUAL RATE	PERSON EXAMS	# OF EVENTS	ANNUAL RATE	PERSON EXAMS	# OF EVENTS	ANNUAL RATE	PERSON EXAMS	# OF EVENTS	ANNUAL RATE
8-44	289	0	0	1062	6	3	2045	13	3	1752	17	5
45-49	411	0	0	1125	1	0	1480	5	2	749	8	5
50-54	747	0	0	1658	3	1	1835	13	4	712	6	4
55-59	1022	0	0	1739	3	1	1416	6	2	442	3	3
60-85	2010	5	1	2135	2	0	1255	2	1	246	3	6
UNKNOWN	13	0	0		0	0	18	0	0	12	0	0
TOTAL	4472	5	1	7724	15	1	8049	39	2	3913	37	5

RISK FACTOR	75-84 AT EXAM			85-94 AT EXAM			35-64 COMBINED			65-94 COMBINED		
	PERSON EXAMS	# OF EVENTS	ANNUAL RATE	PERSON EXAMS	# OF EVENTS	ANNUAL RATE	# OF EVENTS	CRUDE	AGE-ADJUSTED	# OF EVENTS	CRUDE	AGE-ADJUSTED
8-44	649	14	11	24			19	3	2	31	6	6
45-49	172	0	0	6			6	1	1	8	4	4
50-54	97	3	15	7			16	2	2	9	6	6
55-59	65	0	0	2			9	1	1	3	3	3
60-85	26	0	0	0			9	1	1	3	6	5
UNKNOWN	8	0	0	0			0	0	0	0	0	0
TOTAL	1017	17	8	39			59	1	1	54	5	5

LOGISTIC REGRESSIONS

AGE AT EXAM	COEFFICIENT OF RISK FACTOR	SE COEFFICIENT	P-VALUE	STANDARDIZED COEFFICIENT	STD SE
35-44	0.058	0.053	0.273	0.515	0.470
45-54	-0.047	0.027	0.080	-0.423	0.245
55-64	-0.025	0.017	0.126	-0.238	0.155
65-74	-0.016	0.017	0.349	-0.153	0.163
75-84	-0.014	0.026	0.588	-0.133	0.245
85-94					
35-64 UNIVARIATE	-0.038	0.013	0.003	-0.370	0.124
65-94 UNIVARIATE	-0.020	0.014	0.163	-0.189	0.135
35-64 BIVARIATE	-0.020	0.014	0.142	-0.196	0.133
65-94 BIVARIATE	-0.014	0.015	0.336	-0.134	0.139
35-64 MULTIVARIATE	-0.002	0.014	0.866	-0.024	0.140
65-94 MULTIVARIATE	-0.007	0.015	0.635	-0.068	0.145

SEX: FEMALE
EVENT: UNRECOGNIZED MYOCARDIAL INFARCTION
RISK FACTOR: VITAL CAPACITY - HEIGHT INDEX (ML/INCH)
PERSONS AT RISK: PERSONS FREE OF CORONARY HEART DISEASE

RISK FACTOR	35-44 AT EXAM			45-54 AT EXAM			55-64 AT EXAM			65-74 AT EXAM		
	PERSON EXAMS	# OF EVENTS	ANNUAL RATE	PERSON EXAMS	# OF EVENTS	ANNUAL RATE	PERSON EXAMS	# OF EVENTS	ANNUAL RATE	PERSON EXAMS	# OF EVENTS	ANNUAL RATE
4-29	122			534			1461	6	2	1747	12	3
30-34	289			1217	2	1	2418	9	2	1715	4	1
35-39	769			2077	1	0	3048	3	0	1529	3	1
40-44	1427			2804	1	0	2706	7	1	854	1	1
45-92	2988			3310	3	0	1859	2	1	397	2	3
UNKNOWN	21			17	0	0	21	0	0	6	0	0
TOTAL	5616			9959	8	0	11513	27	1	6248	22	2

RISK FACTOR	75-84 AT EXAM			85-94 AT EXAM			35-64 COMBINED			65-94 COMBINED		
	PERSON EXAMS	# OF EVENTS	ANNUAL RATE	PERSON EXAMS	# OF EVENTS	ANNUAL RATE	# OF EVENTS	CRUDE	AGE-ADJUSTED	# OF EVENTS	CRUDE	AGE-ADJUSTED
4-29	930	7	4	80			7	2	1	21	6	4
30-34	569	4	4	29			11	1	1	9	2	2
35-39	314	2	3	13			4	0	0	5	1	1
40-44	111	0	0	1			8	1	1	1	1	0
45-92	25	0	0	0			5	0	0	2	2	2
UNKNOWN	2	0	0	0			0	0	0	0	0	0
TOTAL	1951	13	3	123			35	1	1	38	2	2

LOGISTIC REGRESSIONS

AGE AT EXAM	COEFFICIENT OF RISK FACTOR	SE COEFFICIENT	P-VALUE	STANDARDIZED COEFFICIENT	STD SE
35-44					
45-54	-0.027	0.047	0.569	-0.198	0.347
55-64	-0.059	0.024	0.014	-0.446	0.182
65-74	-0.057	0.028	0.039	-0.422	0.205
75-84	-0.042	0.038	0.274	-0.293	0.267
85-94					
35-64 UNIVARIATE	-0.076	0.020	0.001	-0.606	0.158
65-94 UNIVARIATE	-0.067	0.021	0.001	-0.501	0.154
35-64 BIVARIATE	-0.045	0.022	0.041	-0.357	0.175
65-94 BIVARIATE	-0.055	0.022	0.011	-0.415	0.163
35-64 MULTIVARIATE	-0.033	0.023	0.159	-0.262	0.186
65-94 MULTIVARIATE	-0.051	0.022	0.024	-0.381	0.169

TABLE 4-12   DATE=06DEC85

AVERAGE ANNUAL INCIDENCE RATE FOR EVENT PER 1000 PERSONS AT RISK AT EXAM
BY AGE AND LEVEL OF RISK FACTOR AT EXAM: FRAMINGHAM STUDY 30-YEAR FOLLOWUP

SEX: MALE
EVENT: UNRECOGNIZED MYOCARDIAL INFARCTION
RISK FACTOR: HEART RATE (PER MINUTE)
PERSONS AT RISK: PERSONS FREE OF CORONARY HEART DISEASE

RISK FACTOR	35-44 AT EXAM			45-54 AT EXAM			55-64 AT EXAM			65-74 AT EXAM		
	PERSON EXAMS	# OF EVENTS	ANNUAL RATE	PERSON EXAMS	# OF EVENTS	ANNUAL RATE	PERSON EXAMS	# OF EVENTS	ANNUAL RATE	PERSON EXAMS	# OF EVENTS	ANNUAL RATE
30-67	1284	1	0	1949	1	0	2309	4	1	1183	11	5
68-75	1492	3	1	2490	5	1	2490	17	3	1159	9	4
76-83	722	0	0	1394	2	1	1374	6	2	624	4	3
84-91	579	0	0	1097	4	2	1064	6	3	487	7	7
92-220	395	1	1	794	3	2	812	6	4	460	6	7
UNKNOWN	0	0	-	0	0	-	0	0	-	0	0	-
TOTAL	4472	5	1	7724	15	1	8049	39	2	3913	37	5

RISK FACTOR	75-84 AT EXAM			85-94 AT EXAM			35-64 COMBINED			65-94 COMBINED		
	PERSON EXAMS	# OF EVENTS	ANNUAL RATE	PERSON EXAMS	# OF EVENTS	ANNUAL RATE	# OF EVENTS	CRUDE	AGE-ADJUSTED	# OF EVENTS	CRUDE	AGE-ADJUSTED
30-67	301	7	12	11			6	1	1	18	6	6
68-75	296	5	8	12			25	2	2	14	5	5
76-83	192	1	3	10			8	1	1	5	3	3
84-91	123	3	12	5			10	2	2	10	8	8
92-220	105	1	5	1			10	2	2	7	6	6
UNKNOWN	0	0	-	0			0	-	-	0	-	-
TOTAL	1017	17	8	39			59	1	1	54	5	5

LOGISTIC REGRESSIONS

AGE AT EXAM		COEFFICIENT OF RISK FACTOR	SE COEFFICIENT	P-VALUE	STANDARDIZED COEFFICIENT	STD SE
35-44		0.026	0.032	0.415	0.324	0.398
45-54		0.041	0.018	0.022	0.512	0.223
55-64		0.024	0.011	0.030	0.312	0.143
65-74		0.017	0.011	0.136	0.226	0.151
75-84		-0.021	0.020	0.292	-0.283	0.268
85-94						
35-64	UNIVARIATE	0.029	0.009	0.002	0.364	0.115
65-94	UNIVARIATE	0.006	0.010	0.515	0.086	0.133
35-64	BIVARIATE	0.028	0.009	0.002	0.357	0.114
65-94	BIVARIATE	0.006	0.010	0.516	0.086	0.133
35-64	MULTIVARIATE	0.012	0.010	0.218	0.151	0.122
65-94	MULTIVARIATE	0.005	0.010	0.650	0.061	0.134

SEX: FEMALE
EVENT: UNRECOGNIZED MYOCARDIAL INFARCTION
RISK FACTOR: HEART RATE (PER MINUTE)
PERSONS AT RISK: PERSONS FREE OF CORONARY HEART DISEASE

RISK FACTOR	35-44 AT EXAM			45-54 AT EXAM			55-64 AT EXAM			65-74 AT EXAM		
	PERSON EXAMS	# OF EVENTS	ANNUAL RATE	PERSON EXAMS	# OF EVENTS	ANNUAL RATE	PERSON EXAMS	# OF EVENTS	ANNUAL RATE	PERSON EXAMS	# OF EVENTS	ANNUAL RATE
30-67	1092			1897	3	1	2305	2	0	1103	8	4
68-75	1803			3244	2	0	3468	6	1	1668	7	2
76-83	1051			1911	2	1	2351	7	1	1358	2	1
84-91	939			1604	1	0	1780	4	1	1089	3	1
92-220	731			1303	0	0	1609	8	2	1030	2	1
UNKNOWN	0			0	0	-	0	0	-	0	0	-
TOTAL	5616			9959	8	0	11513	27	1	6248	22	2

RISK FACTOR	75-84 AT EXAM			85-94 AT EXAM			35-64 COMBINED			65-94 COMBINED		
	PERSON EXAMS	# OF EVENTS	ANNUAL RATE	PERSON EXAMS	# OF EVENTS	ANNUAL RATE	# OF EVENTS	CRUDE	AGE-ADJUSTED	# OF EVENTS	CRUDE	AGE-ADJUSTED
30-67	360	3	4	26			5	0	0	12	4	4
68-75	462	1	1	32			8	0	0	8	2	2
76-83	428	5	6	24			9	1	1	7	2	2
84-91	330	1	2	23			5	1	1	6	2	2
92-220	371	3	4	18			8	1	1	5	2	2
UNKNOWN	0	0	-	0			0	-	-	0	-	-
TOTAL	1951	13	3	123			35	1	1	38	2	2

LOGISTIC REGRESSIONS

AGE AT EXAM		COEFFICIENT OF RISK FACTOR	SE COEFFICIENT	P-VALUE	STANDARDIZED COEFFICIENT	STD SE
35-44						
45-54		-0.032	0.032	0.323	-0.393	0.398
55-64		0.031	0.012	0.007	0.401	0.150
65-74		-0.044	0.019	0.022	-0.575	0.250
75-84		0.010	0.019	0.590	0.139	0.259
85-94						
35-64	UNIVARIATE	0.022	0.011	0.055	0.277	0.144
65-94	UNIVARIATE	-0.017	0.013	0.203	-0.222	0.175
35-64	BIVARIATE	0.021	0.011	0.065	0.265	0.144
65-94	BIVARIATE	-0.017	0.013	0.190	-0.226	0.173
35-64	MULTIVARIATE	0.011	0.012	0.362	0.143	0.157
65-94	MULTIVARIATE	-0.018	0.013	0.156	-0.245	0.173

TABLE 4-13   DATE=05AUG85
AVERAGE ANNUAL INCIDENCE RATE FOR EVENT PER 1000 PERSONS AT RISK AT EXAM
BY AGE AND LEVEL OF RISK FACTOR AT EXAM: FRAMINGHAM STUDY 30-YEAR FOLLOWUP

SEX: MALE
EVENT: UNRECOGNIZED MYOCARDIAL INFARCTION
RISK FACTOR: CIGARETTES SMOKED (NUMBER PER DAY)
PERSONS AT RISK: PERSONS FREE OF CORONARY HEART DISEASE

RISK FACTOR	35-44 AT EXAM PERSON EXAMS	# OF EVENTS	ANNUAL RATE	45-54 AT EXAM PERSON EXAMS	# OF EVENTS	ANNUAL RATE	55-64 AT EXAM PERSON EXAMS	# OF EVENTS	ANNUAL RATE	65-74 AT EXAM PERSON EXAMS	# OF EVENTS	ANNUAL RATE
NONE	1383	0	0	3068	3	0	4295	12	1	2611	25	5
1-10	433	0	0	716	1	1	769	3	2	374	3	4
11-20	1485	2	1	2051	5	1	1682	15	4	526	4	4
21-40	987	2	1	1602	5	2	1119	8	4	369	4	5
41-90	158	1	3	252	1	2	162	1	3	27	1	19
UNKNOWN	26	0	0	35	0	0	22	0	0	6	0	0
TOTAL	4472	5	1	7724	15	1	8049	39	2	3913	37	5

RISK FACTOR	75-84 AT EXAM PERSON EXAMS	# OF EVENTS	ANNUAL RATE	85-94 AT EXAM PERSON EXAMS	# OF EVENTS	ANNUAL RATE	35-64 COMBINED # OF EVENTS	CRUDE	AGE-ADJUSTED	65-94 COMBINED # OF EVENTS	CRUDE	AGE-ADJUSTED
NONE	797	14	9	27			15	1	0	39	6	6
1-10	92	1	5	6			4	1	1	4	4	4
11-20	77	1	6	5			22	2	2	5	4	4
21-40	46	1	11	1			15	2	3	5	6	6
41-90	5	0	0	0			3	3	3	1	16	15
UNKNOWN	0	0	-	0			0	0	0	0	0	0
TOTAL	1017	17	8	39			59	1	1	54	5	6

LOGISTIC REGRESSIONS

AGE AT EXAM		COEFFICIENT OF RISK FACTOR	SE COEFFICIENT	P-VALUE	STANDARDIZED COEFFICIENT	STD SE
35-44		0.068	0.028	0.015	0.961	0.396
45-54		0.038	0.014	0.008	0.566	0.213
55-64		0.024	0.010	0.015	0.331	0.133
65-74		0.008	0.013	0.518	0.099	0.152
75-84		-0.011	0.051	0.716	-0.102	0.282
85-94						
35-64	UNIVARIATE	0.026	0.008	0.001	0.577	0.113
65-94	UNIVARIATE	0.002	0.012	0.847	0.026	0.134
35-64	BIVARIATE	0.032	0.007	0.001	0.471	0.109
65-94	BIVARIATE	0.006	0.012	0.628	0.065	0.134
35-64	MULTIVARIATE	0.035	0.007	0.001	0.510	0.108
65-94	MULTIVARIATE	0.011	0.012	0.374	0.120	0.135

SEX: FEMALE
EVENT: UNRECOGNIZED MYOCARDIAL INFARCTION
RISK FACTOR: CIGARETTES SMOKED (NUMBER PER DAY)
PERSONS AT RISK: PERSONS FREE OF CORONARY HEART DISEASE

RISK FACTOR	35-44 AT EXAM PERSON EXAMS	# OF EVENTS	ANNUAL RATE	45-54 AT EXAM PERSON EXAMS	# OF EVENTS	ANNUAL RATE	55-64 AT EXAM PERSON EXAMS	# OF EVENTS	ANNUAL RATE	65-74 AT EXAM PERSON EXAMS	# OF EVENTS	ANNUAL RATE
NONE	2582			5528	3	0	7595	15	1	4863	20	2
1-10	1314			1726	2	1	1566	4	1	626	1	1
11-20	1371			2009	3	1	1696	5	1	568	1	1
21-40	298			622	0	0	602	2	2	172	0	0
41-90	22			37	0	0	16	1	31	0	0	-
UNKNOWN	29			37	0	0	38	0	0	19	0	0
TOTAL	5616			9959	8	0	11513	27	1	6248	22	2

RISK FACTOR	75-84 AT EXAM PERSON EXAMS	# OF EVENTS	ANNUAL RATE	85-94 AT EXAM PERSON EXAMS	# OF EVENTS	ANNUAL RATE	35-64 COMBINED # OF EVENTS	CRUDE	AGE-ADJUSTED	65-94 COMBINED # OF EVENTS	CRUDE	AGE-ADJUSTED
NONE	1710	13	4	117			18	1	0	36	3	3
1-10	137	0	0	4			6	1	1	1	1	1
11-20	84	0	0	2			8	1	1	1	1	1
21-40	16	0	0	0			2	1	1	0	0	0
41-90	0	0	-	0			1	7	13	0	-	-
UNKNOWN	4	0	0	0			0	0	0	0	0	0
TOTAL	1951	13	3	123			35	1	0	38	2	2

LOGISTIC REGRESSIONS

AGE AT EXAM		COEFFICIENT OF RISK FACTOR	SE COEFFICIENT	P-VALUE	STANDARDIZED COEFFICIENT	STD SE
35-44						
45-54		0.019	0.030	0.520	0.198	0.308
55-64		0.035	0.015	0.024	0.330	0.146
65-74		-0.063	0.052	0.224	-0.480	0.394
75-84		-	-	-	-	-
85-94						
35-64	UNIVARIATE	0.026	0.014	0.070	0.255	0.141
65-94	UNIVARIATE	-0.092	0.055	0.095	-0.649	0.389
35-64	BIVARIATE	0.034	0.014	0.012	0.339	0.136
65-94	BIVARIATE	-0.081	0.055	0.137	-0.574	0.386
35-64	MULTIVARIATE	0.034	0.014	0.012	0.342	0.137
65-94	MULTIVARIATE	-0.078	0.054	0.154	-0.549	0.385

TABLE 4-13   DATE=05AUG85
AVERAGE ANNUAL INCIDENCE RATE FOR EVENT PER 1000 PERSONS AT RISK AT EXAM
BY AGE AND LEVEL OF RISK FACTOR AT EXAM: FRAMINGHAM STUDY 30-YEAR FOLLOWUP

SEX: MALE
EVENT: UNRECOGNIZED MYOCARDIAL INFARCTION
RISK FACTOR: CIGARETTES SMOKED (NUMBER PER DAY)
PERSONS AT RISK: PERSONS FREE OF CORONARY HEART DISEASE

RISK FACTOR	35-44 AT EXAM			45-54 AT EXAM			55-64 AT EXAM			65-74 AT EXAM		
	PERSON EXAMS	# OF EVENTS	ANNUAL RATE	PERSON EXAMS	# OF EVENTS	ANNUAL RATE	PERSON EXAMS	# OF EVENTS	ANNUAL RATE	PERSON EXAMS	# OF EVENTS	ANNUAL RATE
NONE	1383	0	0	3068	3	0	4295	12	1	2611	25	5
1-10	433	0	0	716	1	1	769	3	2	374	3	4
11-20	1485	2	1	2051	5	1	1682	15	4	526	4	4
21-40	987	2	1	1602	5	2	1119	8	4	369	4	5
41-90	158	1	3	252	1	2	162	1	3	27	.1	19
UNKNOWN	26	0	0	35	0	0	22	0	0	6	0	0
TOTAL	4472	5	1	7724	15	1	8049	39	2	3913	37	5

RISK FACTOR	75-84 AT EXAM			85-94 AT EXAM			35-64 COMBINED			65-94 COMBINED		
	PERSON EXAMS	# OF EVENTS	ANNUAL RATE	PERSON EXAMS	# OF EVENTS	ANNUAL RATE	# OF EVENTS	CRUDE	AGE-ADJUSTED	# OF EVENTS	CRUDE	AGE-ADJUSTED
NONE	797	14	9	27			15	1	0	39	6	6
1-10	92	1	5	6			4	1	1	4	4	4
11-20	77	1	6	5			22	2	2	5	4	6
21-40	46	1	11	1			15	3	3	4	6	6
41-90	5	0	0	0			3	3	3	1	16	15
UNKNOWN	0	0	-	0			0	0	0	0	0	0
TOTAL	1017	17	8	39			59	1	1	54	5	6

LOGISTIC REGRESSIONS

AGE AT EXAM		COEFFICIENT OF RISK FACTOR	SE COEFFICIENT	P-VALUE	STANDARDIZED COEFFICIENT	STD SE
35-44		0.068	0.028	0.015	0.961	0.396
45-54		0.038	0.014	0.008	0.566	0.213
55-64		0.024	0.010	0.013	0.331	0.133
65-74		0.008	0.013	0.518	0.099	0.152
75-84		-0.011	0.031	0.716	-0.102	0.282
85-94						
35-64	UNIVARIATE	0.026	0.008	0.001	0.377	0.113
65-94	UNIVARIATE	0.002	0.012	0.847	0.026	0.134
35-64	BIVARIATE	0.032	0.007	0.001	0.471	0.109
65-94	BIVARIATE	0.006	0.012	0.628	0.065	0.134
35-64	MULTIVARIATE	0.035	0.007	0.001	0.510	0.108
65-94	MULTIVARIATE	0.011	0.012	0.374	0.120	0.135

SEX: FEMALE
EVENT: UNRECOGNIZED MYOCARDIAL INFARCTION
RISK FACTOR: CIGARETTES SMOKED (NUMBER PER DAY)
PERSONS AT RISK: PERSONS FREE OF CORONARY HEART DISEASE

RISK FACTOR	35-44 AT EXAM			45-54 AT EXAM			55-64 AT EXAM			65-74 AT EXAM		
	PERSON EXAMS	# OF EVENTS	ANNUAL RATE	PERSON EXAMS	# OF EVENTS	ANNUAL RATE	PERSON EXAMS	# OF EVENTS	ANNUAL RATE	PERSON EXAMS	# OF EVENTS	ANNUAL RATE
NONE	2582			5528	3	0	7595	15	1	4863	20	2
1-10	1314			1726	2	1	1566	4	1	626	1	1
11-20	1371			2009	3	1	1696	5	1	568	1	1
21-40	298			622	0	0	602	2	2	172	0	0
41-90	22			37	0	0	16	1	31	0	0	-
UNKNOWN	29			37	0	0	38	0	0	19	0	0
TOTAL	5616			9959	8	0	11513	27	1	6248	22	2

RISK FACTOR	75-84 AT EXAM			85-94 AT EXAM			35-64 COMBINED			65-94 COMBINED		
	PERSON EXAMS	# OF EVENTS	ANNUAL RATE	PERSON EXAMS	# OF EVENTS	ANNUAL RATE	# OF EVENTS	CRUDE	AGE-ADJUSTED	# OF EVENTS	CRUDE	AGE-ADJUSTED
NONE	1710	13	4	117			18	1	1	36	3	3
1-10	137	0	0	4			6	1	1	1	1	1
11-20	84	0	0	2			8	1	1	0	0	4
21-40	16	0	0	0			2	1	13	0	0	0
41-90	0	0	-	0			1	7	13	0	0	-
UNKNOWN	4	0	0	0			0	0	0	0	0	0
TOTAL	1951	13	3	123			35	1	0	38	2	2

LOGISTIC REGRESSIONS

AGE AT EXAM		COEFFICIENT OF RISK FACTOR	SE COEFFICIENT	P-VALUE	STANDARDIZED COEFFICIENT	STD SE
35-44						
45-54		0.019	0.030	0.520	0.198	0.308
55-64		0.035	0.015	0.024	0.330	0.146
65-74		-0.063	0.052	0.224	-0.480	0.394
75-84		-	-		-	-
85-94						
35-64	UNIVARIATE	0.026	0.014	0.070	0.255	0.141
65-94	UNIVARIATE	-0.092	0.055	0.095	-0.649	0.389
35-64	BIVARIATE	0.034	0.014	0.012	0.339	0.136
65-94	BIVARIATE	-0.081	0.055	0.137	-0.574	0.386
35-64	MULTIVARIATE	0.034	0.014	0.012	0.342	0.137
65-94	MULTIVARIATE	-0.078	0.054	0.154	-0.549	0.385

AVERAGE ANNUAL INCIDENCE RATE FOR EVENT PER 1000 PERSONS AT RISK AT EXAM
BY AGE AND LEVEL OF RISK FACTOR AT EXAM: FRAMINGHAM STUDY 30-YEAR FOLLOWUP

SEX: MALE
EVENT: UNRECOGNIZED MYOCARDIAL INFARCTION
RISK FACTOR: HEART ENLARGEMENT BY X-RAY
PERSONS AT RISK: PERSONS FREE OF CORONARY HEART DISEASE

RISK FACTOR	35-44 AT EXAM			45-54 AT EXAM			55-64 AT EXAM			65-74 AT EXAM		
	PERSON EXAMS	# OF EVENTS	ANNUAL RATE	PERSON EXAMS	# OF EVENTS	ANNUAL RATE	PERSON EXAMS	# OF EVENTS	ANNUAL RATE	PERSON EXAMS	# OF EVENTS	ANNUAL RATE
NO	4063	5	1	6383	12	1	5967	20	2	2584	17	3
MILD	185	0	0	650	1	1	962	6	3	578	4	3
DEFINITE	219	0	0	687	2	1	1119	13	6	751	16	11
UNKNOWN	5	0	0	4	0	0	1	0	0	0	0	-
TOTAL	4472	5	1	7724	15	1	8049	39	2	3913	37	5

RISK FACTOR	75-84 AT EXAM			85-94 AT EXAM			35-64 COMBINED			65-94 COMBINED		
	PERSON EXAMS	# OF EVENTS	ANNUAL RATE	PERSON EXAMS	# OF EVENTS	ANNUAL RATE	# OF EVENTS	CRUDE	AGE-ADJUSTED	# OF EVENTS	CRUDE	AGE-ADJUSTED
NO	589	8	7	29			37	1	1	25	4	4
MILD	178	1	3	5			7	2	2	5	3	3
DEFINITE	250	8	16	5			15	4	3	24	12	12
UNKNOWN	0	0	-	0			0	0	0	0	-	
TOTAL	1017	17	8	39			59	1	1	54	5	5

LOGISTIC REGRESSIONS

AGE AT EXAM		COEFFICIENT OF RISK FACTOR	SE COEFFICIENT	P-VALUE	STANDARDIZED COEFFICIENT	STD SE
35-44		-	-	-	-	-
45-54		0.172	0.381	0.652	0.105	0.232
55-64		0.626	0.178	0.001	0.450	0.128
65-74		0.592	0.182	0.001	0.471	0.145
75-84		0.437	0.272	0.108	0.369	0.230
85-94						
35-64	UNIVARIATE	0.594	0.152	0.001	0.379	0.097
65-94	UNIVARIATE	0.564	0.151	0.001	0.455	0.122
35-64	BIVARIATE	0.467	0.155	0.003	0.298	0.099
65-94	BIVARIATE	0.541	0.152	0.001	0.437	0.122
35-64	MULTIVARIATE	0.324	0.167	0.052	0.207	0.106
65-94	MULTIVARIATE	0.372	0.163	0.023	0.300	0.132

SEX: FEMALE
EVENT: UNRECOGNIZED MYOCARDIAL INFARCTION
RISK FACTOR: HEART ENLARGEMENT BY X-RAY
PERSONS AT RISK: PERSONS FREE OF CORONARY HEART DISEASE

RISK FACTOR	35-44 AT EXAM			45-54 AT EXAM			55-64 AT EXAM			65-74 AT EXAM		
	PERSON EXAMS	# OF EVENTS	ANNUAL RATE	PERSON EXAMS	# OF EVENTS	ANNUAL RATE	PERSON EXAMS	# OF EVENTS	ANNUAL RATE	PERSON EXAMS	# OF EVENTS	ANNUAL RATE
NO	5164			8017	6	0	7522	12	1	3191	11	2
MILD	199			943	1	1	1769	3	1	1150	3	1
DEFINITE	245			993	1	1	2218	12	3	1907	8	2
UNKNOWN	8			6	0	0	4	0	0	0	0	-
TOTAL	5616			9959	8	0	11513	27	1	6248	22	2

RISK FACTOR	75-84 AT EXAM			85-94 AT EXAM			35-64 COMBINED			65-94 COMBINED		
	PERSON EXAMS	# OF EVENTS	ANNUAL RATE	PERSON EXAMS	# OF EVENTS	ANNUAL RATE	# OF EVENTS	CRUDE	AGE-ADJUSTED	# OF EVENTS	CRUDE	AGE-ADJUSTED
NO	702	2	1	32			18	0	0	14	2	2
MILD	373	4	5	30			4	1	1	7	2	2
DEFINITE	876	7	4	61			13	2	1	17	3	3
UNKNOWN	0	0	-	0			0	0	0	0	-	
TOTAL	1951	13	3	123			35	1	1	38	2	2

LOGISTIC REGRESSIONS

AGE AT EXAM		COEFFICIENT OF RISK FACTOR	SE COEFFICIENT	P-VALUE	STANDARDIZED COEFFICIENT	STD SE
35-44						
45-54		0.178	0.499	0.721	0.114	0.318
55-64		0.612	0.213	0.004	0.488	0.169
65-74		0.088	0.240	0.712	0.078	0.211
75-84		0.396	0.338	0.241	0.385	0.302
85-94						
35-64	UNIVARIATE	0.728	0.185	0.001	0.508	0.129
65-94	UNIVARIATE	0.260	0.181	0.151	0.232	0.162
35-64	BIVARIATE	0.485	0.192	0.011	0.338	0.134
65-94	BIVARIATE	0.176	0.185	0.341	0.157	0.165
35-64	MULTIVARIATE	0.404	0.204	0.048	0.282	0.143
65-94	MULTIVARIATE	0.053	0.193	0.786	0.047	0.172

TABLE 4-16   DATE=05AUG85

AVERAGE ANNUAL INCIDENCE RATE FOR EVENT PER 1000 PERSONS AT RISK AT EXAM
BY AGE AND LEVEL OF RISK FACTOR AT EXAM: FRAMINGHAM STUDY 30-YEAR FOLLOWUP

SEX: MALE
EVENT: UNRECOGNIZED MYOCARDIAL INFARCTION
RISK FACTOR: LEFT VENTRICULAR HYPERTROPHY BY ELECTROCARDIOGRAM
PERSONS AT RISK: PERSONS FREE OF CORONARY HEART DISEASE

RISK FACTOR	35-44 AT EXAM PERSON EXAMS	# OF EVENTS	ANNUAL RATE	45-54 AT EXAM PERSON EXAMS	# OF EVENTS	ANNUAL RATE	55-64 AT EXAM PERSON EXAMS	# OF EVENTS	ANNUAL RATE	65-74 AT EXAM PERSON EXAMS	# OF EVENTS	ANNUAL RATE
NO	4366	5	1	7497	14	1	7693	32	2	3703	27	4
BORDER	78	0	0	132	0	0	210	4	10	90	4	22
DEFINITE	28	0	0	95	1	5	146	3	10	120	6	25
UNKNOWN	0	0	-	0	0	-	0	0	-	0	0	-
TOTAL	4472	5	1	7724	15	1	8049	39	2	3913	37	5

RISK FACTOR	75-84 AT EXAM PERSON EXAMS	# OF EVENTS	ANNUAL RATE	85-94 AT EXAM PERSON EXAMS	# OF EVENTS	ANNUAL RATE	35-64 COMBINED # OF EVENTS	CRUDE	AGE-ADJUSTED	65-94 COMBINED # OF EVENTS	CRUDE	AGE-ADJUSTED
NO	949	14	7	34			51	1	1	41	4	5
BORDER	37	1	14	2			4	5	4	5	19	20
DEFINITE	31	2	32	3			4	7	6	8	26	26
UNKNOWN	0	0	-	0			0	-	-	0	-	-
TOTAL	1017	17	8	39			59	1	1	54	5	6

LOGISTIC REGRESSIONS

AGE AT EXAM		COEFFICIENT OF RISK FACTOR	SE COEFFICIENT	P-VALUE	STANDARDIZED COEFFICIENT	STD SE
35-44		-	-	-	-	-
45-54		0.732	0.558	0.190	0.186	0.142
55-64		0.925	0.259	0.001	0.285	0.080
65-74		1.050	0.213	0.001	0.391	0.079
75-84		0.748	0.378	0.048	0.289	0.146
85-94						
35-64	UNIVARIATE	0.940	0.233	0.001	0.252	0.062
65-94	UNIVARIATE	0.963	0.184	0.001	0.363	0.070
35-64	BIVARIATE	0.829	0.234	0.001	0.222	0.063
65-94	BIVARIATE	0.954	0.185	0.001	0.360	0.070
35-64	MULTIVARIATE	0.435	0.261	0.095	0.116	0.070
65-94	MULTIVARIATE	0.755	0.208	0.001	0.285	0.078

SEX: FEMALE
EVENT: UNRECOGNIZED MYOCARDIAL INFARCTION
RISK FACTOR: LEFT VENTRICULAR HYPERTROPHY BY ELECTROCARDIOGRAM
PERSONS AT RISK: PERSONS FREE OF CORONARY HEART DISEASE

RISK FACTOR	35-44 AT EXAM PERSON EXAMS	# OF EVENTS	ANNUAL RATE	45-54 AT EXAM PERSON EXAMS	# OF EVENTS	ANNUAL RATE	55-64 AT EXAM PERSON EXAMS	# OF EVENTS	ANNUAL RATE	65-74 AT EXAM PERSON EXAMS	# OF EVENTS	ANNUAL RATE
NO	5563			9798	8	0	11137	25	1	5966	19	2
BORDER	38			107	0	0	226	1	2	134	1	4
DEFINITE	15			54	0	0	150	1	3	148	2	7
UNKNOWN	0			0	0	-	0	0	-	0	0	-
TOTAL	5616			9959	8	0	11513	27	1	6248	22	2

RISK FACTOR	75-84 AT EXAM PERSON EXAMS	# OF EVENTS	ANNUAL RATE	85-94 AT EXAM PERSON EXAMS	# OF EVENTS	ANNUAL RATE	35-64 COMBINED # OF EVENTS	CRUDE	AGE-ADJUSTED	65-94 COMBINED # OF EVENTS	CRUDE	AGE-ADJUSTED
NO	1833	11	3	111			33	1	0	33	2	2
BORDER	49	0	0	6			1	1	0	1	3	3
DEFINITE	69	2	14	6			1	2	1	4	9	9
UNKNOWN	0	0	-	0			0	-	-	0	-	-
TOTAL	1951	13	3	123			35	1	0	38	2	2

LOGISTIC REGRESSIONS

AGE AT EXAM		COEFFICIENT OF RISK FACTOR	SE COEFFICIENT	P-VALUE	STANDARDIZED COEFFICIENT	STD SE
35-44		-	-	-	-	-
45-54		-	-	-	-	-
55-64		0.571	0.459	0.214	0.151	0.121
65-74		0.740	0.356	0.038	0.247	0.119
75-84		0.731	0.407	0.072	0.290	0.161
85-94						
35-64	UNIVARIATE	0.674	0.456	0.139	0.143	0.097
65-94	UNIVARIATE	0.703	0.266	0.008	0.248	0.094
35-64	BIVARIATE	0.448	0.463	0.333	0.095	0.098
65-94	BIVARIATE	0.649	0.268	0.015	0.229	0.094
35-64	MULTIVARIATE	-0.069	0.498	0.889	-0.015	0.106
65-94	MULTIVARIATE	0.496	0.283	0.080	0.175	0.100

AVERAGE ANNUAL INCIDENCE RATE FOR EVENT PER 1000 PERSONS AT RISK AT EXAM
BY AGE AND LEVEL OF RISK FACTOR AT EXAM: FRAMINGHAM STUDY 30-YEAR FOLLOWUP

SEX: MALE
EVENT: UNRECOGNIZED MYOCARDIAL INFARCTION
RISK FACTOR: INTRAVENTRICULAR CONDUCTION DEFECT (WITH QRS INTERVAL MORE THAN 0.11 SECOND)
PERSONS AT RISK: PERSONS FREE OF CORONARY HEART DISEASE

RISK FACTOR	35-44 AT EXAM PERSON EXAMS	# OF EVENTS	ANNUAL RATE	45-54 AT EXAM PERSON EXAMS	# OF EVENTS	ANNUAL RATE	55-64 AT EXAM PERSON EXAMS	# OF EVENTS	ANNUAL RATE	65-74 AT EXAM PERSON EXAMS	# OF EVENTS	ANNUAL RATE
NO	4161	5	1	7213	15	1	7386	34	2	3566	34	5
INCOMPLETE	275	0	0	446	0	0	456	4	4	183	2	5
COMPLETE	36	0	0	65	0	0	207	1	2	164	1	3
UNKNOWN	0	0	-	0	0	-	0	0	-	0	0	-
TOTAL	4472	5	1	7724	15	1	8049	39	2	3913	37	5

RISK FACTOR	75-84 AT EXAM PERSON EXAMS	# OF EVENTS	ANNUAL RATE	85-94 AT EXAM PERSON EXAMS	# OF EVENTS	ANNUAL RATE	35-64 COMBINED # OF EVENTS	CRUDE	AGE-ADJUSTED	65-94 COMBINED # OF EVENTS	CRUDE	AGE-ADJUSTED
NO	918	15	8	31			54	1	1	49	5	5
INCOMPLETE	21	0	0	2			4	2	2	2	5	4
COMPLETE	78	2	13	6			1	2	1	3	6	5
UNKNOWN	0	0	-	0			0	-		0	-	
TOTAL	1017	17	8	39			59	1	1	54	5	5

LOGISTIC REGRESSIONS

AGE AT EXAM	COEFFICIENT OF RISK FACTOR	SE COEFFICIENT	P-VALUE	STANDARDIZED COEFFICIENT	STD SE
35-44	-	-	-	-	-
45-54	-	-	-	-	-
55-64	0.258	0.348	0.459	0.099	0.134
65-74	-0.126	0.409	0.753	-0.056	0.182
75-84	0.182	0.392	0.642	0.099	0.214
85-94					
35-64 UNIVARIATE	0.110	0.363	0.762	0.037	0.121
65-94 UNIVARIATE	0.031	0.284	0.914	0.014	0.134
35-64 BIVARIATE	0.010	0.355	0.976	0.004	0.118
65-94 BIVARIATE	-0.007	0.283	0.980	-0.003	0.133
35-64 MULTIVARIATE	-0.050	0.361	0.890	-0.017	0.120
65-94 MULTIVARIATE	0.018	0.289	0.949	0.009	0.136

SEX: FEMALE
EVENT: UNRECOGNIZED MYOCARDIAL INFARCTION
RISK FACTOR: INTRAVENTRICULAR CONDUCTION DEFECT (WITH QRS INTERVAL MORE THAN 0.11 SECOND)
PERSONS AT RISK: PERSONS FREE OF CORONARY HEART DISEASE

RISK FACTOR	35-44 AT EXAM PERSON EXAMS	# OF EVENTS	ANNUAL RATE	45-54 AT EXAM PERSON EXAMS	# OF EVENTS	ANNUAL RATE	55-64 AT EXAM PERSON EXAMS	# OF EVENTS	ANNUAL RATE	65-74 AT EXAM PERSON EXAMS	# OF EVENTS	ANNUAL RATE
NO	5529			9771	8	0	11183	27	1	6027	22	2
INCOMPLETE	75			144	0	0	187	0	0	115	0	0
COMPLETE	12			44	0	0	143	0	0	106	0	0
UNKNOWN	0			0	0	-	0	0	-	0	0	-
TOTAL	5616			9959	8	0	11513	27	1	6248	22	2

RISK FACTOR	75-84 AT EXAM PERSON EXAMS	# OF EVENTS	ANNUAL RATE	85-94 AT EXAM PERSON EXAMS	# OF EVENTS	ANNUAL RATE	35-64 COMBINED # OF EVENTS	CRUDE	AGE-ADJUSTED	65-94 COMBINED # OF EVENTS	CRUDE	AGE-ADJUSTED
NO	1847	12	3	117			35	1	1	37	2	2
INCOMPLETE	36	1	14	1			0	0	0	0	3	3
COMPLETE	68	0	0	5			0	0	0	0	-	0
UNKNOWN	0	0	-	0			0	-		0	-	
TOTAL	1951	13	3	123			35	1	1	38	2	2 -

LOGISTIC REGRESSIONS

AGE AT EXAM	COEFFICIENT OF RISK FACTOR	SE COEFFICIENT	P-VALUE	STANDARDIZED COEFFICIENT	STD SE
35-44					
45-54					
55-64					
65-74	-	-	-	-	-
75-84	-0.081	0.769	0.917	-0.031	0.298
85-94					
35-64 UNIVARIATE	-	-	-	-	-
65-94 UNIVARIATE	-0.529	0.803	0.510	-0.168	0.255
35-64 BIVARIATE	-	-	-	-	-
65-94 BIVARIATE	-0.587	0.795	0.460	-0.186	0.252
35-64 MULTIVARIATE	-	-	-	-	-
65-94 MULTIVARIATE	-0.630	0.798	0.430	-0.200	0.253

TABLE 4-18   DATE=06DEC85
AVERAGE ANNUAL INCIDENCE RATE FOR EVENT PER 1000 PERSONS AT RISK AT EXAM
BY AGE AND LEVEL OF RISK FACTOR AT EXAM: FRAMINGHAM STUDY 30-YEAR FOLLOWUP

SEX: MALE
EVENT: UNRECOGNIZED MYOCARDIAL INFARCTION
RISK FACTOR: DEFINITE NONSPECIFIC T-WAVE OR ST-SEGMENT ABNORMALITY BY ELECTROCARDIOGRAM
PERSONS AT RISK: PERSONS FREE OF CORONARY HEART DISEASE

RISK FACTOR	35-44 AT EXAM PERSON EXAMS	# OF EVENTS	ANNUAL RATE	45-54 AT EXAM PERSON EXAMS	# OF EVENTS	ANNUAL RATE	55-64 AT EXAM PERSON EXAMS	# OF EVENTS	ANNUAL RATE	65-74 AT EXAM PERSON EXAMS	# OF EVENTS	ANNUAL RATE
NO	4362	5	1	7352	14	2	7462	34	2	3472	29	4
YES	110	0	0	372	1	1	587	5	4	441	8	9
UNKNOWN	0	0	-	0	0	-	0	0	-	0	0	-
TOTAL	4472	5	1	7724	15	2	8049	39	2	3913	37	5

RISK FACTOR	75-84 AT EXAM PERSON EXAMS	# OF EVENTS	ANNUAL RATE	85-94 AT EXAM PERSON EXAMS	# OF EVENTS	ANNUAL RATE	35-64 COMBINED # OF EVENTS	CRUDE	AGE-ADJUSTED	65-94 COMBINED # OF EVENTS	CRUDE	AGE-ADJUSTED
NO	830	14	8	28			53	1	1	43	5	5
YES	187	3	8	11			6	3	2	11	9	9
UNKNOWN	0	0	-	0			0	-	-	0	-	-
TOTAL	1017	17	8	39			59	1	1	54	5	5

LOGISTIC REGRESSIONS

AGE AT EXAM		COEFFICIENT OF RISK FACTOR	SE COEFFICIENT	P-VALUE	STANDARDIZED COEFFICIENT	STD SE
35-44		-	-	-	-	-
45-54		0.346	1.036	0.739	0.074	0.222
55-64		0.630	0.481	0.190	0.164	0.125
65-74		0.786	0.403	0.051	0.248	0.127
75-84		-0.051	0.641	0.937	-0.020	0.249
85-94						
35-64	UNIVARIATE	0.711	0.432	0.100	0.159	0.097
65-94	UNIVARIATE	0.557	0.341	0.102	0.187	0.114
35-64	BIVARIATE	0.476	0.434	0.273	0.106	0.097
65-94	BIVARIATE	0.483	0.343	0.159	0.162	0.115
35-64	MULTIVARIATE	0.016	0.440	0.972	0.003	0.098
65-94	MULTIVARIATE	0.436	0.347	0.208	0.146	0.116

SEX: FEMALE
EVENT: UNRECOGNIZED MYOCARDIAL INFARCTION
RISK FACTOR: DEFINITE NONSPECIFIC T-WAVE OR ST-SEGMENT ABNORMALITY BY ELECTROCARDIOGRAM
PERSONS AT RISK: PERSONS FREE OF CORONARY HEART DISEASE

RISK FACTOR	35-44 AT EXAM PERSON EXAMS	# OF EVENTS	ANNUAL RATE	45-54 AT EXAM PERSON EXAMS	# OF EVENTS	ANNUAL RATE	55-64 AT EXAM PERSON EXAMS	# OF EVENTS	ANNUAL RATE	65-74 AT EXAM PERSON EXAMS	# OF EVENTS	ANNUAL RATE
NO	5517			9543	8	0	10647	24	1	5528	20	2
YES	99			416	0	0	866	3	2	720	2	1
UNKNOWN	0			0	0	-	0	0	-	0	0	-
TOTAL	5616			9959	8	0	11513	27	1	6248	22	2

RISK FACTOR	75-84 AT EXAM PERSON EXAMS	# OF EVENTS	ANNUAL RATE	85-94 AT EXAM PERSON EXAMS	# OF EVENTS	ANNUAL RATE	35-64 COMBINED # OF EVENTS	CRUDE	AGE-ADJUSTED	65-94 COMBINED # OF EVENTS	CRUDE	AGE-ADJUSTED
NO	1632	8	2	97			32	1	1	30	2	2
YES	319	5	8	26			3	1	1	8	4	3
UNKNOWN	0	0	-	0			0	-	-	0	-	-
TOTAL	1951	13	3	123			35	1	1	38	2	2

LOGISTIC REGRESSIONS

AGE AT EXAM		COEFFICIENT OF RISK FACTOR	SE COEFFICIENT	P-VALUE	STANDARDIZED COEFFICIENT	STD SE
35-44		-				
45-54		-				
55-64		0.431	0.613	0.482	0.114	0.162
65-74		-0.265	0.743	0.721	-0.085	0.237
75-84		1.173	0.573	0.041	0.434	0.212
85-94						
35-64	UNIVARIATE	0.558	0.604	0.356	0.123	0.133
65-94	UNIVARIATE	0.601	0.399	0.132	0.201	0.133
35-64	BIVARIATE	0.174	0.607	0.774	0.038	0.133
65-94	BIVARIATE	0.501	0.401	0.212	0.167	0.134
35-64	MULTIVARIATE	0.001	0.612	0.998	0.000	0.135
65-94	MULTIVARIATE	0.490	0.402	0.223	0.164	0.134

TABLE 5-4   DATE=07AUG85

AVERAGE ANNUAL INCIDENCE RATE FOR EVENT PER 1000 PERSONS AT RISK AT EXAM
BY AGE AND LEVEL OF RISK FACTOR AT EXAM: FRAMINGHAM STUDY 30-YEAR FOLLOWUP

SEX: MALE
EVENT: RECOGNIZED MYOCARDIAL INFARCTION
RISK FACTOR: SYSTOLIC BLOOD PRESSURE, FIRST EXAMINER (MM HG)
PERSONS AT RISK: PERSONS FREE OF CORONARY HEART DISEASE

RISK FACTOR	35-44 AT EXAM			45-54 AT EXAM			55-64 AT EXAM			65-74 AT EXAM		
	PERSON EXAMS	# OF EVENTS	ANNUAL RATE	PERSON EXAMS	# OF EVENTS	ANNUAL RATE	PERSON EXAMS	# OF EVENTS	ANNUAL RATE	PERSON EXAMS	# OF EVENTS	ANNUAL RATE
74-119	1223	2	0	1749	12	3	1440	11	4	548	2	2
120-139	2165	10	2	3417	23	3	3166	35	6	1341	19	7
140-159	841	3	2	1732	20	6	2090	31	7	1194	21	9
160-179	191	3	8	567	11	10	916	19	10	552	9	8
180-300	52	0	0	239	0	0	437	7	8	278	9	16
UNKNOWN	0	0	-	0	0	-	0	0	-	0	0	-
TOTAL	4472	18	2	7724	66	4	8049	103	6	3913	60	8

RISK FACTOR	75-84 AT EXAM			85-94 AT EXAM			35-64 COMBINED			65-94 COMBINED		
	PERSON EXAMS	# OF EVENTS	ANNUAL RATE	PERSON EXAMS	# OF EVENTS	ANNUAL RATE	# OF EVENTS	CRUDE	AGE-ADJUSTED	# OF EVENTS	CRUDE	AGE-ADJUSTED
74-119	120	0	0		7		25	3	3	2	1	2
120-139	345	5	7		15		68	4	4	24	7	7
140-159	315	8	13		11		54	6	6	30	10	10
160-179	153	3	10		4		33	10	10	12	8	8
180-300	84	4	24		2		7	5	3	13	18	18
UNKNOWN	0	0	-		0		0	-	-	0	-	-
TOTAL	1017	20	10		39		187	5	4	81	8	8

LOGISTIC REGRESSIONS

AGE AT EXAM	COEFFICIENT OF RISK FACTOR	SE COEFFICIENT	P-VALUE	STANDARDIZED COEFFICIENT	STD SE
35-44	0.023	0.011	0.041	0.380	0.186
45-54	0.008	0.006	0.162	0.160	0.114
55-64	0.011	0.004	0.006	0.242	0.088
65-74	0.015	0.005	0.003	0.354	0.112
75-84	0.023	0.009	0.011	0.513	0.203
85-94					
35-64 UNIVARIATE	0.014	0.003	0.001	0.276	0.063
65-94 UNIVARIATE	0.016	0.004	0.001	0.376	0.096
35-64 BIVARIATE	0.010	0.003	0.001	0.208	0.065
65-94 BIVARIATE	0.017	0.004	0.001	0.377	0.096
35-64 MULTIVARIATE	0.009	0.003	0.006	0.190	0.070
65-94 MULTIVARIATE	0.015	0.005	0.001	0.348	0.103

SEX: FEMALE
EVENT: RECOGNIZED MYOCARDIAL INFARCTION
RISK FACTOR: SYSTOLIC BLOOD PRESSURE, FIRST EXAMINER (MM HG)
PERSONS AT RISK: PERSONS FREE OF CORONARY HEART DISEASE

RISK FACTOR	35-44 AT EXAM			45-54 AT EXAM			55-64 AT EXAM			65-74 AT EXAM		
	PERSON EXAMS	# OF EVENTS	ANNUAL RATE	PERSON EXAMS	# OF EVENTS	ANNUAL RATE	PERSON EXAMS	# OF EVENTS	ANNUAL RATE	PERSON EXAMS	# OF EVENTS	ANNUAL RATE
74-119	2493			2714	1	0	1979	2	1	647	3	2
120-139	2244			3982	5	1	4056	11	1	1747	7	2
140-159	673			2053	7	2	3025	8	1	2013	17	
160-179	156			768	0	0	1560	8	3	1155	6	3
180-300	50			442	0	0	893	8	4	686	14	10
UNKNOWN	0			0	0	-	0	0	-	0	0	-
TOTAL	5616			9959	13	1	11513	37	2	6248	47	4

RISK FACTOR	75-84 AT EXAM			85-94 AT EXAM			35-64 COMBINED			65-94 COMBINED		
	PERSON EXAMS	# OF EVENTS	ANNUAL RATE	PERSON EXAMS	# OF EVENTS	ANNUAL RATE	# OF EVENTS	CRUDE	AGE-ADJUSTED	# OF EVENTS	CRUDE	AGE-ADJUSTED
74-119	133	1	4		8		3	0	0	3	3	2
120-139	465	4	4		38		17	1	1	12	3	2
140-159	626	5	4		29		15	1	1	23	4	4
160-179	451	5	4		31		8	2	1	11	3	4
180-300	276	5	9		17		9	3	4	20	10	10
UNKNOWN	0	0	-		0		0	-	-	0	-	-
TOTAL	1951	20	5		123		52	1	1	70	4	4

LOGISTIC REGRESSIONS

AGE AT EXAM	COEFFICIENT OF RISK FACTOR	SE COEFFICIENT	P-VALUE	STANDARDIZED COEFFICIENT	STD SE
35-44					
45-54					
55-64	0.013	0.010	0.193	0.288	0.221
65-74	0.018	0.005	0.001	0.450	0.127
75-84	0.018	0.005	0.001	0.443	0.127
85-94	0.011	0.008	0.173	0.278	0.204
35-64 UNIVARIATE	0.022	0.004	0.001	0.509	0.095
65-94 UNIVARIATE	0.016	0.004	0.001	0.397	0.105
35-64 BIVARIATE	0.017	0.004	0.001	0.409	0.105
65-94 BIVARIATE	0.016	0.004	0.001	0.383	0.106
35-64 MULTIVARIATE	0.015	0.005	0.003	0.351	0.118
65-94 MULTIVARIATE	0.013	0.005	0.006	0.306	0.111

TABLE 5-2   DATE=06AUG85

AVERAGE ANNUAL INCIDENCE RATE FOR EVENT PER 1000 PERSONS AT RISK AT EXAM
BY AGE AND LEVEL OF RISK FACTOR AT EXAM: FRAMINGHAM STUDY 30-YEAR FOLLOWUP

SEX: MALE
EVENT: RECOGNIZED MYOCARDIAL INFARCTION
RISK FACTOR: DIASTOLIC BLOOD PRESSURE, FIRST EXAMINER (MM HG)
PERSONS AT RISK: PERSONS FREE OF CORONARY HEART DISEASE

RISK FACTOR	35-44 AT EXAM PERSON EXAMS	# OF EVENTS	ANNUAL RATE	45-54 AT EXAM PERSON EXAMS	# OF EVENTS	ANNUAL RATE	55-64 AT EXAM PERSON EXAMS	# OF EVENTS	ANNUAL RATE	65-74 AT EXAM PERSON EXAMS	# OF EVENTS	ANNUAL RATE
20-74	1012	1	0	1478	13	4	1847	18	5	1225	15	6
75-84	1604	4	1	2766	19	3	2845	32	6	1323	21	8
85-94	1187	7	3	2062	17	4	1964	34	9	893	15	8
95-104	499	5	5	1008	11	5	982	14	7	343	7	10
105-160	170	1	3	410	6	7	411	5	6	129	2	8
UNKNOWN	0	0	-	0	0	-	0	0	-	0	0	-
TOTAL	4472	18	2	7724	66	4	8049	103	6	3913	60	8

RISK FACTOR	75-84 AT EXAM PERSON EXAMS	# OF EVENTS	ANNUAL RATE	85-94 AT EXAM PERSON EXAMS	# OF EVENTS	ANNUAL RATE	35-64 COMBINED # OF EVENTS	CRUDE	AGE-ADJUSTED	65-94 COMBINED # OF EVENTS	CRUDE	AGE-ADJUSTED
20-74	422	6	7	22			32	4	4	22	7	6
75-84	331	7	11	12			55	4	4	28	8	9
85-94	204	6	15	5			58	6	6	21	10	9
95-104	44	0	0	0			30	6	6	7	9	8
105-160	16	1	31	0			12	6	6	3	10	13
UNKNOWN	0	0	-	0			0	-	-	0	-	-
TOTAL	1017	20	10	39			187	5	4	81	8	8

LOGISTIC REGRESSIONS

AGE AT EXAM		COEFFICIENT OF RISK FACTOR	SE COEFFICIENT	P-VALUE	STANDARDIZED COEFFICIENT	STD SE
35-44		0.049	0.018	0.005	0.551	0.198
45-54		0.011	0.010	0.290	0.127	0.120
55-64		0.015	0.008	0.061	0.176	0.094
65-74		0.014	0.010	0.159	0.173	0.123
75-84		0.020	0.019	0.295	0.231	0.220
85-94						
35-64	UNIVARIATE	0.017	0.006	0.003	0.204	0.069
65-94	UNIVARIATE	0.013	0.009	0.144	0.157	0.107
35-64	BIVARIATE	0.017	0.006	0.003	0.201	0.068
65-94	BIVARIATE	0.014	0.009	0.126	0.166	0.109
35-64	MULTIVARIATE	0.017	0.006	0.005	0.202	0.072
65-94	MULTIVARIATE	0.011	0.009	0.232	0.135	0.112

SEX: FEMALE
EVENT: RECOGNIZED MYOCARDIAL INFARCTION
RISK FACTOR: DIASTOLIC BLOOD PRESSURE, FIRST EXAMINER (MM HG)
PERSONS AT RISK: PERSONS FREE OF CORONARY HEART DISEASE

RISK FACTOR	35-44 AT EXAM PERSON EXAMS	# OF EVENTS	ANNUAL RATE	45-54 AT EXAM PERSON EXAMS	# OF EVENTS	ANNUAL RATE	55-64 AT EXAM PERSON EXAMS	# OF EVENTS	ANNUAL RATE	65-74 AT EXAM PERSON EXAMS	# OF EVENTS	ANNUAL RATE
20-74	2164			2639	2	0	2969	7	1	1851	15	4
75-84	2030			3541	2	0	3853	9	1	2135	12	3
85-94	987			2361	8	2	2848	10	2	1439	12	4
95-104	324			926	0	0	1272	6	2	600	3	3
105-160	111			492	1	1	579	5	4	243	5	10
UNKNOWN	0			0	0	-	0	0	-	0	0	-
TOTAL	5616			9959	13	1	11513	37	2	6248	47	4

RISK FACTOR	75-84 AT EXAM PERSON EXAMS	# OF EVENTS	ANNUAL RATE	85-94 AT EXAM PERSON EXAMS	# OF EVENTS	ANNUAL RATE	35-64 COMBINED # OF EVENTS	CRUDE	AGE-ADJUSTED	65-94 COMBINED # OF EVENTS	CRUDE	AGE-ADJUSTED
20-74	724	7	5	53			10	1	0	23	4	4
75-84	596	9	8	43			11	1	0	23	4	4
85-94	443	3	3	20			18	1	2	15	4	4
95-104	146	1	3	7			7	1	1	5	3	3
105-160	42	0	0	0			7	3	2	5	9	8
UNKNOWN	0	0	-	0			0	-	-	0	-	-
TOTAL	1951	20	5	123			52	1	1	70	4	4

LOGISTIC REGRESSIONS

AGE AT EXAM		COEFFICIENT OF RISK FACTOR	SE COEFFICIENT	P-VALUE	STANDARDIZED COEFFICIENT	STD SE
35-44						
45-54		0.023	0.020	0.242	0.284	0.243
55-64		0.029	0.012	0.012	0.361	0.144
65-74		0.014	0.011	0.229	0.168	0.140
75-84		-0.030	0.020	0.127	-0.364	0.239
85-94						
35-64	UNIVARIATE	0.031	0.010	0.001	0.375	0.118
65-94	UNIVARIATE	0.000	0.010	0.999	0.000	0.120
35-64	BIVARIATE	0.026	0.010	0.008	0.322	0.122
65-94	BIVARIATE	0.002	0.010	0.846	0.023	0.120
35-64	MULTIVARIATE	0.020	0.011	0.056	0.249	0.130
65-94	MULTIVARIATE	-0.002	0.010	0.810	-0.029	0.120

TABLE 5-3A DATE=03SEP85

AVERAGE ANNUAL INCIDENCE RATE FOR EVENT PER 1000 PERSONS AT RISK AT EXAM
BY AGE AND LEVEL OF RISK FACTOR AT EXAM: FRAMINGHAM STUDY 30-YEAR FOLLOWUP

SEX: MALE
EVENT: RECOGNIZED MYOCARDIAL INFARCTION
RISK FACTOR: HYPERTENSION WITH ANTIHYPERTENSIVE TREATMENT
PERSONS AT RISK: PERSONS FREE OF CORONARY HEART DISEASE

35-44 AT EXAM			45-54 AT EXAM			55-64 AT EXAM			65-74 AT EXAM		
PERSON EXAMS	# OF EVENTS	ANNUAL RATE	PERSON EXAMS	# OF EVENTS	ANNUAL RATE	PERSON EXAMS	# OF EVENTS	ANNUAL RATE	PERSON EXAMS	# OF EVENTS	ANNUAL RATE
2651	7	1	4122	27	3	3782	36	5	1516	11	4
1301	6	2	2205	21	5	2299	40	9	1252	22	9
500	4	4	1139	14	6	1266	19	8	552	12	11
20	1	25	258	4	6	702	8	6	593	15	13
0	0	-	0	0	-	0	0	-	0	0	-
4472	18	2	7724	66	4	8049	103	6	3913	60	8

75-84 AT EXAM			85-94 AT EXAM			35-64 COMBINED			65-94 COMBINED		
PERSON EXAMS	# OF EVENTS	ANNUAL RATE	PERSON EXAMS	# OF EVENTS	ANNUAL RATE	# OF EVENTS	CRUDE	AGE-ADJUSTED	# OF EVENTS	CRUDE	AGE-ADJUST
373	4	5	18	4	-	70	3	3	15	4	4
533	6	9	11	-	-	67	6	6	29	9	9
123	4	16	3	-	-	57	6	6	16	12	12
188	6	16	7	-	-	13	7	11	21	13	14
0	0	-	0			0	-	-	0	-	-
1017	20	10	39			187	5	4	81	8	8

LOGISTIC REGRESSIONS

	COEFFICIENT OF RISK FACTOR	SE COEFFICIENT	P-VALUE	STANDARDIZED COEFFICIENT	STD SE
	0.644	0.297	0.030	0.447	0.206
	0.344	0.151	0.023	0.264	0.116
	0.205	0.119	0.085	0.167	0.097
	0.555	0.165	0.001	0.453	0.135
	0.565	0.291	0.052	0.462	0.238
RIATE	0.347	0.089	0.001	0.270	0.069
RIATE	0.554	0.142	0.001	0.454	0.117
IATE	0.275	0.090	0.002	0.214	0.070
IATE	0.553	0.142	0.001	0.453	0.117
VARIATE	0.258	0.094	0.006	0.200	0.073
VARIATE	0.511	0.146	0.001	0.419	0.128

SEX: FEMALE
EVENT: RECOGNIZED MYOCARDIAL INFARCTION
RISK FACTOR: HYPERTENSION WITH ANTIHYPERTENSIVE TREATMENT
PERSONS AT RISK: PERSONS FREE OF CORONARY HEART DISEASE

35-44 AT EXAM			45-54 AT EXAM			55-64 AT EXAM			65-74 AT EXAM		
PERSON EXAMS	# OF EVENTS	ANNUAL RATE	PERSON EXAMS	# OF EVENTS	ANNUAL RATE	PERSON EXAMS	# OF EVENTS	ANNUAL RATE	PERSON EXAMS	# OF EVENTS	ANNUAL RATE
4250			5622	4	0	4888	7	1	1767	8	2
1001			2579	7	1	3282	10	2	1903	10	3
316			1166	0	0	1835	11	3	1035	7	3
49			592	2	2	1508	9	3	1543	22	7
0			0	0	-	0	0	-	0	0	-
5616			9959	13	1	11513	37	2	6248	47	4

75-84 AT EXAM			85-94 AT EXAM			35-64 COMBINED			65-94 COMBINED		
PERSON EXAMS	# OF EVENTS	ANNUAL RATE	PERSON EXAMS	# OF EVENTS	ANNUAL RATE	# OF EVENTS	CRUDE	AGE-ADJUSTED	# OF EVENTS	CRUDE	AGE-ADJUSTED
402	4	5	31			12	0	0	13	-	3
575	5	4	28			17	1	1	16	3	3
371	4	5	21			12	2	2	11	4	3
603	7	6	43			11	3	2	10	7	7
0	0	-	0			0	-	-	0	-	-
1951	20	5	123			52	1	1	70	4	4

LOGISTIC REGRESSIONS

	COEFFICIENT OF RISK FACTOR	SE COEFFICIENT	P-VALUE	STANDARDIZED COEFFICIENT	STD SE
	0.363	0.334	0.277	0.279	0.257
	0.711	0.210	0.001	0.594	0.176
	0.516	0.199	0.010	0.425	0.164
	0.095	0.292	0.746	0.075	0.230
ATE	0.796	0.167	0.001	0.631	0.133
ATE	0.371	0.159	0.020	0.304	0.130
TE	0.607	0.173	0.001	0.481	0.137
TE	0.346	0.160	0.030	0.283	0.131
RIATE	0.522	0.181	0.004	0.414	0.144
RIATE	0.250	0.163	0.125	0.204	0.133

TABLE 5-3B   DATE=10DEC85
AVERAGE ANNUAL INCIDENCE RATE FOR EVENT PER 1000 PERSONS AT RISK AT EXAM
BY AGE AND LEVEL OF RISK FACTOR AT EXAM: FRAMINGHAM STUDY 30-YEAR FOLLOWUP

SEX: MALE
EVENT: RECOGNIZED MYOCARDIAL INFARCTION
RISK FACTOR: HYPERTENSION IGNORING TREATMENT
PERSONS AT RISK: PERSONS FREE OF CORONARY HEART DISEASE

RISK FACTOR	35-44 AT EXAM PERSON EXAMS	# OF EVENTS	ANNUAL RATE	45-54 AT EXAM PERSON EXAMS	# OF EVENTS	ANNUAL RATE	55-64 AT EXAM PERSON EXAMS	# OF EVENTS	ANNUAL RATE	65-74 AT EXAM PERSON EXAMS	# OF EVENTS	ANNUAL RATE
NO	2653	7	1	4154	28	3	3909	38	5	1631	14	4
MILD	1510	6	2	2284	22	5	2583	41	8	1499	29	10
DEFINITE	509	5	5	1286	16	6	1557	24	8	783	17	11
UNKNOWN	0	0	-	0	0	-	0	0	-	0	0	-
TOTAL	4472	18	2	7724	66	4	8049	103	6	3913	60	8

RISK FACTOR	75-84 AT EXAM PERSON EXAMS	# OF EVENTS	ANNUAL RATE	85-94 AT EXAM PERSON EXAMS	# OF EVENTS	ANNUAL RATE	35-64 COMBINED # OF EVENTS	CRUDE	AGE-ADJUSTED	65-94 COMBINED # OF EVENTS	CRUDE	AGE-ADJUSTED
NO	412	5	6	21			73	3	4	19	5	5
MILD	418	8	10	15			69	6	5	38	10	10
DEFINITE	187	7	19	3			45	7	7	24	12	12
UNKNOWN	0	0	-	0			0	-	-	0	-	-
TOTAL	1017	20	10	39			187	5	5	81	8	8

LOGISTIC REGRESSIONS

AGE AT EXAM	COEFFICIENT OF RISK FACTOR	SE COEFFICIENT	P-VALUE	STANDARDIZED COEFFICIENT	STD SE
35-44	0.654	0.298	0.028	0.452	0.206
45-54	0.314	0.154	0.042	0.236	0.116
55-64	0.255	0.124	0.039	0.197	0.095
65-74	0.458	0.168	0.007	0.346	0.127
75-84	0.584	0.301	0.053	0.429	0.222
85-94					
35-64 UNIVARIATE	0.356	0.092	0.001	0.267	0.069
65-94 UNIVARIATE	0.489	0.146	0.001	0.367	0.109
35-64 BIVARIATE	0.302	0.092	0.001	0.227	0.069
65-94 BIVARIATE	0.489	0.146	0.001	0.367	0.109
35-64 MULTIVARIATE	0.279	0.097	0.004	0.209	0.073
65-94 MULTIVARIATE	0.447	0.152	0.003	0.335	0.114

SEX: FEMALE
EVENT: RECOGNIZED MYOCARDIAL INFARCTION
RISK FACTOR: HYPERTENSION IGNORING TREATMENT
PERSONS AT RISK: PERSONS FREE OF CORONARY HEART DISEASE

RISK FACTOR	35-44 AT EXAM PERSON EXAMS	# OF EVENTS	ANNUAL RATE	45-54 AT EXAM PERSON EXAMS	# OF EVENTS	ANNUAL RATE	55-64 AT EXAM PERSON EXAMS	# OF EVENTS	ANNUAL RATE	65-74 AT EXAM PERSON EXAMS	# OF EVENTS	ANNUAL RATE
NO	4272			5804	4	0	5245	11	1	2035	9	2
MILD	1016			2763	9	2	3860	13	2	2549	21	4
DEFINITE	328			1392	0	0	2408	13	3	1664	17	5
UNKNOWN	0			0	0	-	0	0	-	0	0	-
TOTAL	5616			9959	13	1	11513	37	2	6248	47	4

RISK FACTOR	75-84 AT EXAM PERSON EXAMS	# OF EVENTS	ANNUAL RATE	85-94 AT EXAM PERSON EXAMS	# OF EVENTS	ANNUAL RATE	35-64 COMBINED # OF EVENTS	CRUDE	AGE-ADJUSTED	65-94 COMBINED # OF EVENTS	CRUDE	AGE-ADJUSTED
NO	507	4	4	37			16	1	1	14	3	3
MILD	827	7	4	45			22	1	1	29	4	4
DEFINITE	617	9	7	41			14	2	1	27	6	6
UNKNOWN	0	0	-	0			0	-	-	0	-	-
TOTAL	1951	20	5	123			52	1	1	70	4	4

LOGISTIC REGRESSIONS

AGE AT EXAM	COEFFICIENT OF RISK FACTOR	SE COEFFICIENT	P-VALUE	STANDARDIZED COEFFICIENT	STD SE
35-44					
45-54	0.240	0.358	0.503	0.174	0.260
55-64	0.474	0.205	0.021	0.369	0.159
65-74	0.395	0.194	0.041	0.303	0.149
75-84	0.351	0.307	0.254	0.265	0.233
85-94					
35-64 UNIVARIATE	0.600	0.169	0.001	0.444	0.125
65-94 UNIVARIATE	0.575	0.160	0.019	0.288	0.122
35-64 BIVARIATE	0.420	0.174	0.016	0.311	0.129
65-94 BIVARIATE	0.356	0.160	0.026	0.273	0.123
35-64 MULTIVARIATE	0.309	0.185	0.096	0.228	0.137
65-94 MULTIVARIATE	0.264	0.164	0.107	0.202	0.126

TABLE 5-4　　　　　　　DATE=07AUG85
AVERAGE ANNUAL INCIDENCE RATE FOR EVENT PER 1000 PERSONS AT RISK AT EXAM
BY AGE AND LEVEL OF RISK FACTOR AT EXAM: FRAMINGHAM STUDY 30-YEAR FOLLOWUP

SEX: MALE
EVENT: RECOGNIZED MYOCARDIAL INFARCTION
RISK FACTOR: SERUM CHOLESTEROL
PERSONS AT RISK: PERSONS FREE OF CORONARY HEART DISEASE

RISK FACTOR	35-44 AT EXAM			45-54 AT EXAM			55-64 AT EXAM			65- PERSON EXAMS
	PERSON EXAMS	# OF EVENTS	ANNUAL RATE	PERSON EXAMS	# OF EVENTS	ANNUAL RATE	PERSON EXAMS	# OF EVENTS	ANNUAL RATE	
84-204	1253			1767	12	3	2255	21	5	1234
205-234	1247	3	1	2306	20	4	2405	32	7	1171
235-264	872	2	1	1745	12	3	1889	24	6	873
265-294	475	5	5	1001	10	5	918	14	8	428
295-1124	291	4	7	557	10	9	454	8	9	191
UNKNOWN	334	2	3	348	2	3	128	4	16	16
TOTAL	4472	18	2	7724	66	4	8049	103	6	3913

RISK FACTOR	75-84 AT EXAM			85-94 AT EXAM			35-64 COMBINED			
	PERSON EXAMS	# OF EVENTS	ANNUAL RATE	PERSON EXAMS	# OF EVENTS	ANNUAL RATE	# OF EVENTS	CRUDE	AGE- ADJUSTED	# OF EVE
84-204	390	9	12	17			36	3	3	25
205-234	332	6	9	11			54	5	5	23
235-264	177	4	11	5			38	4	4	21
265-294	96	1	5	6			29	6	6	6
295-1124	22	0	0	0			22	8	9	5
UNKNOWN	0	0	-	0			8	5	8	1
TOTAL	1017	20	10	39			187	5	4	81

LOGISTIC REGRESSIONS

AGE AT EXAM		COEFFICIENT OF RISK FACTOR	SE COEFFICIENT	P-VALUE	STANDARDIZED COEFFICIENT
35-44		0.011	0.004	0.002	0.508
45-54		0.005	0.003	0.048	0.216
55-64		0.004	0.002	0.064	0.178
65-74		0.005	0.003	0.152	0.183
75-84		-0.001	0.006	0.835	-0.047
85-94					
35-64	UNIVARIATE	0.005	0.002	0.001	0.229
65-94	UNIVARIATE	0.003	0.003	0.313	0.112
35-64	BIVARIATE	0.006	0.002	0.001	0.246
65-94	BIVARIATE	0.003	0.003	0.295	0.117
35-64	MULTIVARIATE	0.005	0.002	0.001	0.222
65-94	MULTIVARIATE	0.003	0.003	0.300	0.116

SEX: FEMALE
EVENT: RECOGNIZED MYOCARDIAL INFARCTION
RISK FACTOR: SERUM CHOLESTEROL
PERSONS AT RISK: PERSONS FREE OF CORONARY HEART DISEASE

RISK FACTOR	35-44 AT EXAM			45-54 AT EXAM			55-64 AT EXAM			65- PERSON EXAMS
	PERSON EXAMS	# OF EVENTS	ANNUAL RATE	PERSON EXAMS	# OF EVENTS	ANNUAL RATE	PERSON EXAMS	# OF EVENTS	ANNUAL RATE	
84-204	2350			2094	0	0	1625	4	1	789
205-234	1417			2558	3	1	2615	6	1	1401
235-264	806			2259	2	0	2938	8	1	1703
265-294	372			1473	1	0	2240	8	2	1287
295-1124	171			1101	5	2	1866	11	3	1031
UNKNOWN	500			474	2	2	229	0	0	37
TOTAL	5616			9959	13	1	11513	37	2	6248

RISK FACTOR	75-84 AT EXAM			85-94 AT EXAM			35-64 COMBINED			
	PERSON EXAMS	# OF EVENTS	ANNUAL RATE	PERSON EXAMS	# OF EVENTS	ANNUAL RATE	# OF EVENTS	CRUDE	AGE- ADJUSTED	# OF EVEN
84-204	361	3	4	46			4	0	0	6
205-234	486	5	5	28			9	1	1	12
235-264	542	6	6	35			11	1	1	16
265-294	350	3	4	9			9	1	1	15
295-1124	205	3	7	5			17	3	3	21
UNKNOWN	7	0	0	0			2	1	1	0
TOTAL	1951	20	5	123			52	1	1	70

LOGISTIC REGRESSIONS

AGE AT EXAM		COEFFICIENT OF RISK FACTOR	SE COEFFICIENT	P-VALUE	STANDARDIZED COEFFICIENT	
35-44						
45-54		0.008	0.002	0.001	0.366	
55-64		0.009	0.003	0.002	0.432	
65-74		0.013	0.003	0.001	0.570	
75-84		0.003	0.005	0.513	0.144	
85-94						
35-64	UNIVARIATE	0.010	0.002	0.001	0.455	C
65-94	UNIVARIATE	0.009	0.003	0.001	0.403	C
35-64	BIVARIATE	0.009	0.002	0.001	0.407	0
65-94	BIVARIATE	0.010	0.003	0.001	0.439	0
35-64	MULTIVARIATE	0.008	0.002	0.001	0.363	0
65-94	MULTIVARIATE	0.010	0.003	0.001	0.437	0

TABLE 5-5   DATE=10DEC85

AVERAGE ANNUAL INCIDENCE RATE FOR EVENT PER 1000 PERSONS AT RISK AT EXAM
BY AGE AND LEVEL OF RISK FACTOR AT EXAM: FRAMINGHAM STUDY 30-YEAR FOLLOWUP

SEX: MALE
EVENT: RECOGNIZED MYOCARDIAL INFARCTION
RISK FACTOR: HEMATOCRIT
PERSONS AT RISK: PERSONS FREE OF CORONARY HEART DISEASE

RISK FACTOR	35-44 AT EXAM PERSON EXAMS	# OF EVENTS	ANNUAL RATE	45-54 AT EXAM PERSON EXAMS	# OF EVENTS	ANNUAL RATE	55-64 AT EXAM PERSON EXAMS	# OF EVENTS	ANNUAL RATE	65-74 AT EXAM PERSON EXAMS	# OF EVENTS	ANNUAL RATE
25-42	623	4	3	942	3	2	1018	13	6	450	6	7
43-44	546	0	0	967	8	4	985	7	4	493	9	9
45-46	983	1	1	1854	15	4	1953	22	6	963	9	5
47-48	775	1	1	1510	18	6	1599	26	8	861	11	6
49-70	1175	10	4	2066	19	5	2338	31	7	1127	24	11
UNKNOWN	370	2	3	385	3	4	156	4	13	19	1	26
TOTAL	4472	18	2	7724	66	4	8049	103	6	3913	60	8

RISK FACTOR	75-84 AT EXAM PERSON EXAMS	# OF EVENTS	ANNUAL RATE	85-94 AT EXAM PERSON EXAMS	# OF EVENTS	ANNUAL RATE	35-64 COMBINED # OF EVENTS	CRUDE	AGE-ADJUSTED	65-94 COMBINED # OF EVENTS	CRUDE	AGE-ADJUSTED
25-42	214	3	7	16			20	4	4	9	7	7
43-44	119	3	13	5			15	3	3	12	10	10
45-46	240	6	13	7			38	4	4	16	7	7
47-48	197	3	8	7			45	4	6	14	7	7
49-70	247	5	10	4			60	5	5	29	11	10
UNKNOWN	0	0	-	0			9	5	7	1	26	21
TOTAL	1017	20	10	39			187	5	5	81	8	8

LOGISTIC REGRESSIONS

AGE AT EXAM	COEFFICIENT OF RISK FACTOR	SE COEFFICIENT	P-VALUE	STANDARDIZED COEFFICIENT	STD SE
35-44	0.127	0.069	0.063	0.467	0.252
45-54	0.037	0.035	0.291	0.134	0.127
55-64	0.031	0.027	0.256	0.116	0.102
65-74	0.053	0.036	0.141	0.196	0.133
75-84	-0.009	0.054	0.869	-0.037	0.225
85-94					
35-64 UNIVARIATE	0.042	0.021	0.039	0.156	0.076
65-94 UNIVARIATE	0.031	0.030	0.298	0.119	0.114
35-64 BIVARIATE	0.041	0.020	0.047	0.150	0.075
65-94 BIVARIATE	0.033	0.030	0.274	0.126	0.115
35-64 MULTIVARIATE	0.019	0.021	0.350	0.072	0.077
65-94 MULTIVARIATE	0.029	0.030	0.339	0.112	0.117

SEX: FEMALE
EVENT: RECOGNIZED MYOCARDIAL INFARCTION
RISK FACTOR: HEMATOCRIT
PERSONS AT RISK: PERSONS FREE OF CORONARY HEART DISEASE

RISK FACTOR	35-44 AT EXAM PERSON EXAMS	# OF EVENTS	ANNUAL RATE	45-54 AT EXAM PERSON EXAMS	# OF EVENTS	ANNUAL RATE	55-64 AT EXAM PERSON EXAMS	# OF EVENTS	ANNUAL RATE	65-74 AT EXAM PERSON EXAMS	# OF EVENTS	ANNUAL RATE
21-39	1697			1843	0	0	1553	6	2	646	6	5
40-41	967			1741	0	0	2061	6	1	1069	7	3
42-43	1043			1955	3	1	2260	6	1	1256	5	2
44-45	856			2194	4	1	2847	8	1	1609	13	4
46-65	487			1681	4	1	2521	11	2	1626	16	5
UNKNOWN	566			545	2	2	271	0	0	42	0	0
TOTAL	5616			9959	13	1	11513	37	2	6248	47	4

RISK FACTOR	75-84 AT EXAM PERSON EXAMS	# OF EVENTS	ANNUAL RATE	85-94 AT EXAM PERSON EXAMS	# OF EVENTS	ANNUAL RATE	35-64 COMBINED # OF EVENTS	CRUDE	AGE-ADJUSTED	65-94 COMBINED # OF EVENTS	CRUDE	AGE-ADJUSTED
21-39	297	5	8	24			7	1	1	12	6	6
40-41	373	3	4	31			9	1	1	11	4	4
42-43	353	1	1	19			9	1	1	7	2	2
44-45	486	5	5	26			13	1	1	18	4	4
46-65	436	6	7	23			15	2	1	22	5	5
UNKNOWN	6	0	0	0			2	1	1	0	0	0
TOTAL	1951	20	5	123			52	1	1	70	4	4

LOGISTIC REGRESSIONS

AGE AT EXAM	COEFFICIENT OF RISK FACTOR	SE COEFFICIENT	P-VALUE	STANDARDIZED COEFFICIENT	STD SE
35-44					
45-54	0.231	0.082	0.005	0.823	0.294
55-64	0.042	0.048	0.380	0.146	0.166
65-74	0.062	0.043	0.143	0.218	0.149
75-84	-0.030	0.057	0.600	-0.115	0.220
85-94					
35-64 UNIVARIATE	0.108	0.040	0.007	0.390	0.144
65-94 UNIVARIATE	0.013	0.033	0.702	0.046	0.121
35-64 BIVARIATE	0.077	0.041	0.061	0.276	0.147
65-94 BIVARIATE	0.017	0.033	0.606	0.062	0.120
35-64 MULTIVARIATE	0.010	0.039	0.793	0.037	0.141
65-94 MULTIVARIATE	0.004	0.033	0.900	0.015	0.120

SEX: MALE
EVENT: RECOGNIZED MYOCARDIAL INFARCTION
RISK FACTOR: BLOOD GLUCOSE (MG/DL)
PERSONS AT RISK: PERSONS FREE OF CORONARY HEART DISEASE

35-44 AT EXAM			45-54 AT EXAM			55-64 AT EXAM			65-74 AT EXAM		
PERSON EXAMS	# OF EVENTS	ANNUAL RATE	PERSON EXAMS	# OF EVENTS	ANNUAL RATE	PERSON EXAMS	# OF EVENTS	ANNUAL RATE	PERSON EXAMS	# OF EVENTS	ANNUAL RATE
992	5	3	1490	11	4	1087	14	6	421	2	2
2501	10	2	4172	30	4	4056	50	6	1761	27	8
626	1	1	1348	16	6	1866	25	7	1046	14	7
105	0	0	305	5	8	537	8	7	335	9	13
48	0	0	193	2	5	408	3	4	334	7	10
200	2	5	216	2	5	95	3	16	16	1	31
4472	18	2	7724	66	4	8049	103	6	3913	60	8

75-84 AT EXAM			85-94 AT EXAM			35-64 COMBINED			65-94 COMBINED		
PERSON EXAMS	# OF EVENTS	ANNUAL RATE	PERSON EXAMS	# OF EVENTS	ANNUAL RATE	# OF EVENTS	CRUDE	AGE-ADJUSTED	# OF EVENTS	CRUDE	AGE-ADJUSTED
62	2	16	1			30	4	5	4	4	5
377	8	11	11			90	4	4	35	8	8
328	7	11	14			42	5	5	21	8	7
125	2	8	4			13	7	6	11	12	12
125	1	4	9			5	4	3	9	10	10
0	0	-	0			7	7	9	1	31	25
1017	20	10	39			187	5	5	81	8	8

LOGISTIC REGRESSIONS

	COEFFICIENT OF RISK FACTOR	SE COEFFICIENT	P-VALUE	STANDARDIZED COEFFICIENT	STD SE
	-0.013	0.019	0.512	-0.216	0.329
	0.004	0.004	0.376	0.086	0.097
	-0.001	0.004	0.800	-0.027	0.106
	0.007	0.002	0.001	0.247	0.076
	-0.012	0.010	0.246	-0.380	0.328
IATE	0.002	0.003	0.359	0.059	0.064
IATE	0.006	0.002	0.007	0.200	0.075
ITE	-0.000	0.003	0.942	-0.005	0.072
ITE	0.006	0.002	0.008	0.200	0.075
IRIATE	-0.000	0.003	0.969	-0.003	0.072
IRIATE	0.005	0.002	0.028	0.168	0.076

SEX: FEMALE
EVENT: RECOGNIZED MYOCARDIAL INFARCTION
RISK FACTOR: BLOOD GLUCOSE (MG/DL)
PERSONS AT RISK: PERSONS FREE OF CORONARY HEART DISEASE

35-44 AT EXAM			45-54 AT EXAM			55-64 AT EXAM			65-74 AT EXAM		
PERSON EXAMS	# OF EVENTS	ANNUAL RATE	PERSON EXAMS	# OF EVENTS	ANNUAL RATE	PERSON EXAMS	# OF EVENTS	ANNUAL RATE	PERSON EXAMS	# OF EVENTS	ANNUAL RATE
1066			1797	3	1	1369	6	2	521	5	5
3412			5744	5	0	6098	13	1	3102	17	3
672			1625	0	0	2712	9	2	1732	16	5
96			301	0	0	730	3	2	501	2	2
39			159	3	9	438	6	7	360	7	10
331			333	2	3	166	0	0	32	0	0
5616			9959	13	1	11513	37	2	6248	47	4

75-84 AT EXAM			85-94 AT EXAM			35-64 COMBINED			65-94 COMBINED		
PERSON EXAMS	# OF EVENTS	ANNUAL RATE	PERSON EXAMS	# OF EVENTS	ANNUAL RATE	# OF EVENTS	CRUDE	AGE-ADJUSTED	# OF EVENTS	CRUDE	AGE-ADJUSTED
105	1	5	3			9	1	1	6	5	5
823	8	5	44			20	1	1	26	3	3
640	5	4	51			9	1	1	21	4	4
219	3	7	13			3	1	1	5	3	3
163	3	9	12			9	7	6	12	11	11
1	0	0	0			2	1	1	0	0	0
1951	20	5	123			52	1	1	70	4	4

LOGISTIC REGRESSIONS

	COEFFICIENT OF RISK FACTOR	SE COEFFICIENT	P-VALUE	STANDARDIZED COEFFICIENT	STD SE
	0.012	0.007	0.064	0.242	0.130
	0.011	0.003	0.001	0.273	0.072
	0.010	0.003	0.001	0.278	0.076
	0.004	0.005	0.405	0.136	0.163
ITE	0.013	0.002	0.001	0.276	0.054
ITE	0.008	0.002	0.001	0.246	0.066
TE	0.011	0.003	0.001	0.231	0.058
TE	0.008	0.002	0.001	0.235	0.067
RIATE	0.008	0.003	0.003	0.176	0.060
RIATE	0.008	0.002	0.001	0.228	0.069

TABLE 5-7   DATE=07AUG85

AVERAGE ANNUAL INCIDENCE RATE FOR EVENT PER 1000 PERSONS AT RISK AT EXAM
BY AGE AND LEVEL OF RISK FACTOR AT EXAM: FRAMINGHAM STUDY 30-YEAR FOLLOWUP

SEX: MALE
EVENT: RECOGNIZED MYOCARDIAL INFARCTION
RISK FACTOR: DIABETES MELLITUS
PERSONS AT RISK: PERSONS FREE OF CORONARY HEART DISEASE

RISK FACTOR	35-44 AT EXAM			45-54 AT EXAM			55-64 AT EXAM			65-74 AT EXAM		
	PERSON EXAMS	# OF EVENTS	ANNUAL RATE	PERSON EXAMS	# OF EVENTS	ANNUAL RATE	PERSON EXAMS	# OF EVENTS	ANNUAL RATE	PERSON EXAMS	# OF EVENTS	ANNUAL RATE
NO	4424	18	2	7486	64	4	7549	96	6	3504	49	7
YES	48	0	0	238	2	4	500	7	7	409	11	13
UNKNOWN	0	0		0	0		0	0		0	0	
TOTAL	4472	18	2	7724	66	4	8049	103	6	3913	60	8

RISK FACTOR	75-84 AT EXAM			85-94 AT EXAM			35-64 COMBINED			65-94 COMBINED		
	PERSON EXAMS	# OF EVENTS	ANNUAL RATE	PERSON EXAMS	# OF EVENTS	ANNUAL RATE	# OF EVENTS	CRUDE	AGE-ADJUSTED	# OF EVENTS	CRUDE	AGE-ADJUSTED
NO	900	20	11	34			178	5	4	69	8	8
YES	117	0	0	5			9	6	4	12	11	11
UNKNOWN	0	0	-	0			0	-	-	0	-	
TOTAL	1017	20	10	39			187	5	4	81	8	8

LOGISTIC REGRESSIONS

AGE AT EXAM		COEFFICIENT OF RISK FACTOR	SE COEFFICIENT	P-VALUE	STANDARDIZED COEFFICIENT	STD SE
35-44		-	-	-	-	-
45-54		-0.017	0.721	0.981	-0.003	0.125
55-64		0.097	0.394	0.805	0.024	0.095
65-74		0.467	0.338	0.048	0.204	0.103
75-84		-	-	-	-	-
85-94		-	-	-	-	-
35-64	UNIVARIATE	0.227	0.344	0.509	0.044	0.066
65-94	UNIVARIATE	0.381	0.316	0.228	0.118	0.098
35-64	BIVARIATE	-0.011	0.346	0.974	-0.002	0.067
65-94	BIVARIATE	0.379	0.316	0.231	0.117	0.098
35-64	MULTIVARIATE	-0.376	0.474	0.427	-0.073	0.092
65-94	MULTIVARIATE	0.235	0.520	0.652	0.073	0.161

SEX: FEMALE
EVENT: RECOGNIZED MYOCARDIAL INFARCTION
RISK FACTOR: DIABETES MELLITUS
PERSONS AT RISK: PERSONS FREE OF CORONARY HEART DISEASE

RISK FACTOR	35-44 AT EXAM			45-54 AT EXAM			55-64 AT EXAM			65-74 AT EXAM		
	PERSON EXAMS	# OF EVENTS	ANNUAL RATE	PERSON EXAMS	# OF EVENTS	ANNUAL RATE	PERSON EXAMS	# OF EVENTS	ANNUAL RATE	PERSON EXAMS	# OF EVENTS	ANNUAL RATE
NO	5573			9760	10	1	10985	29	1	5780	35	3
YES	43			199	3	8	528	8	8	468	12	13
UNKNOWN	0		-	0	0	-	0	0	-	0	0	-
TOTAL	5616			9959	13	1	11513	37	2	6248	47	4

RISK FACTOR	75-84 AT EXAM			85-94 AT EXAM			35-64 COMBINED			65-94 COMBINED		
	PERSON EXAMS	# OF EVENTS	ANNUAL RATE	PERSON EXAMS	# OF EVENTS	ANNUAL RATE	# OF EVENTS	CRUDE	AGE-ADJUSTED	# OF EVENTS	CRUDE	AGE-ADJUSTED
NO	1738	17	5	186	17		40	1	1	53	3	3
YES	213	3	7	17			12	8	9	17	12	12
UNKNOWN	0	0	-	0			0	-	-	0	-	-
TOTAL	1951	20	5	123			52	1	1	70	4	4

LOGISTIC REGRESSIONS

AGE AT EXAM		COEFFICIENT OF RISK FACTOR	SE COEFFICIENT	P-VALUE	STANDARDIZED COEFFICIENT	STD SE
35-44		-	-	-	-	-
45-54		2.703	0.662	0.001	0.378	0.093
55-64		1.740	0.402	0.001	0.368	0.084
65-74		1.463	0.338	0.001	0.385	0.089
75-84		0.369	0.630	0.558	0.115	0.197
85-94						
35-64	UNIVARIATE	2.342	0.331	0.001	0.389	0.055
65-94	UNIVARIATE	1.271	0.282	0.001	0.352	0.078
35-64	BIVARIATE	2.021	0.336	0.001	0.336	0.056
65-94	BIVARIATE	1.227	0.283	0.001	0.340	0.079
35-64	MULTIVARIATE	1.450	0.784	0.064	0.241	0.130
65-94	MULTIVARIATE	1.289	0.578	0.026	0.357	0.160

TABLE 5-8   DATE=10DEC85
AVERAGE ANNUAL INCIDENCE RATE FOR EVENT PER 1000 PERSONS AT RISK AT EXAM
BY AGE AND LEVEL OF RISK FACTOR AT EXAM: FRAMINGHAM STUDY 30-YEAR FOLLOWUP

SEX: MALE
EVENT: RECOGNIZED MYOCARDIAL INFARCTION
RISK FACTOR: GLUCOSE IN URINE, DEFINITE OR TRACE
PERSONS AT RISK: PERSONS FREE OF CORONARY HEART DISEASE

35-44 AT EXAM			45-54 AT EXAM			55-64 AT EXAM			65-74 AT EXAM		
PERSON EXAMS	# OF EVENTS	ANNUAL RATE	PERSON EXAMS	# OF EVENTS	ANNUAL RATE	PERSON EXAMS	# OF EVENTS	ANNUAL RATE	PERSON EXAMS	# OF EVENTS	ANNUAL RATE
4380	18	2	7535	62	4	7815	93	6	3767	52	7
60	0	0	168	4	12	210	9	21	140	8	29
32	0	0	21	0	0	24	1	21	6	0	0
4472	18	2	7724	66	4	8049	103	6	3913	60	8

75-84 AT EXAM			85-94 AT EXAM			35-64 COMBINED			65-94 COMBINED		
PERSON EXAMS	# OF EVENTS	ANNUAL RATE	PERSON EXAMS	# OF EVENTS	ANNUAL RATE	# OF EVENTS	CRUDE	AGE-ADJUSTED	# OF EVENTS	CRUDE	AGE-ADJUSTED
986	19	10	38			173	4	4	71	7	7
28	1	18	1			13	15	13	10	30	30
3	0	0	0			1	6	8	0	0	0
1017	20	10	39			187	5	5	81	8	8

LOGISTIC REGRESSIONS

	COEFFICIENT OF RISK FACTOR	SE COEFFICIENT	P-VALUE	STANDARDIZED COEFFICIENT	STD SE
	1.078	0.522	0.039	0.158	0.076
	1.313	0.356	0.001	0.210	0.057
	1.466	0.390	0.001	0.272	0.072
	0.634	1.044	0.544	0.104	0.171
ARIATE	1.241	0.292	0.001	0.181	0.043
ARIATE	1.431	0.347	0.001	0.260	0.063
RIATE	1.140	0.293	0.001	0.166	0.043
RIATE	1.435	0.347	0.001	0.260	0.063
IVARIATE	1.115	0.307	0.001	0.163	0.045
IVARIATE	1.330	0.354	0.001	0.241	0.064

SEX: FEMALE
EVENT: RECOGNIZED MYOCARDIAL INFARCTION
RISK FACTOR: GLUCOSE IN URINE, DEFINITE OR TRACE
PERSONS AT RISK: PERSONS FREE OF CORONARY HEART DISEASE

35-44 AT EXAM			45-54 AT EXAM			55-64 AT EXAM			65-74 AT EXAM		
PERSON EXAMS	# OF EVENTS	ANNUAL RATE	PERSON EXAMS	# OF EVENTS	ANNUAL RATE	PERSON EXAMS	# OF EVENTS	ANNUAL RATE	PERSON EXAMS	# OF EVENTS	ANNUAL RATE
5537			9841	13	1	11346	34	1	6131	44	4
41			94	0	0	144	3	10	107	3	14
38			24	0	0	23	0	0	10	0	0
5616			9959	13	1	11513	37	2	6248	47	4

75-84 AT EXAM			85-94 AT EXAM			35-64 COMBINED			65-94 COMBINED		
PERSON EXAMS	# OF EVENTS	ANNUAL RATE	PERSON EXAMS	# OF EVENTS	ANNUAL RATE	# OF EVENTS	CRUDE	AGE-ADJUSTED	# OF EVENTS	CRUDE	AGE-ADJUSTED
1919	19	5	120			48	1	1	65	4	4
23	1	22	3			4	7	7	5	19	18
9	0	0	0			0	0	0	0	0	0
1951	20	5	123			52	1	1	70	4	4

LOGISTIC REGRESSIONS

	COEFFICIENT OF RISK FACTOR	SE COEFFICIENT	P-VALUE	STANDARDIZED COEFFICIENT	STD SE
	1.957	0.608	0.001	0.218	0.068
	1.384	0.605	0.022	0.180	0.079
	1.514	1.048	0.149	0.164	0.113
ARIATE	2.090	0.524	0.001	0.211	0.053
ARIATE	1.583	0.473	0.001	0.199	0.059
RIATE	1.977	0.526	0.001	0.200	0.053
RIATE	1.604	0.473	0.001	0.201	0.059
IVARIATE	1.540	0.544	0.005	0.156	0.055
IVARIATE	1.403	0.489	0.004	0.176	0.061

111

TABLE 5-9   DATE=01MAR86
AVERAGE ANNUAL INCIDENCE RATE FOR EVENT PER 1000 PERSONS AT RISK AT EXAM
BY AGE AND LEVEL OF RISK FACTOR AT EXAM: FRAMINGHAM STUDY 30-YEAR FOLLOWUP

SEX: MALE
EVENT: RECOGNIZED MYOCARDIAL INFARCTION
RISK FACTOR: GLUCOSE INTOLERANCE
PERSONS AT RISK: PERSONS FREE OF CORONARY HEART DISEASE

RISK FACTOR	35-44 AT EXAM			45-54 AT EXAM			55-64 AT EXAM			65-74 AT EXAM		
	PERSON EXAMS	# OF EVENTS	ANNUAL RATE	PERSON EXAMS	# OF EVENTS	ANNUAL RATE	PERSON EXAMS	# OF EVENTS	ANNUAL RATE	PERSON EXAMS	# OF EVENTS	ANNUAL RATE
NO	4335	18	2	7221	57	4	7145	91	6	3228	43	7
YES	137	0	0	503	9	9	904	12	7	685	17	12
UNKNOWN	0	0	-	0	0	-	0	0	-	0	0	-
TOTAL	4472	18	2	7724	66	4	8049	103	6	3913	60	8

RISK FACTOR	75-84 AT EXAM			85-94 AT EXAM			35-64 COMBINED			65-94 COMBINED		
	PERSON EXAMS	# OF EVENTS	ANNUAL RATE	PERSON EXAMS	# OF EVENTS	ANNUAL RATE	# OF EVENTS	CRUDE	AGE-ADJUSTED	# OF EVENTS	CRUDE	AGE-ADJUSTED
NO	802	19	12	27			166	4	4	62	8	8
YES	215	1	2	12			21	7	6	19	10	11
UNKNOWN	0	0	-	0			0			0		
TOTAL	1017	20	10	39			187	5	5	81	8	8

LOGISTIC REGRESSIONS

AGE AT EXAM		COEFFICIENT OF RISK FACTOR	SE COEFFICIENT	P-VALUE	STANDARDIZED COEFFICIENT	STD SE
35-44		-	-	-	-	-
45-54		0.828	0.362	0.022	0.204	0.089
55-64		0.042	0.309	0.892	0.013	0.098
65-74		0.634	0.290	0.029	0.241	0.110
75-84		-1.647	1.029	0.109	-0.673	0.420
85-94						
35-64	UNIVARIATE	0.432	0.233	0.064	0.115	0.062
65-94	UNIVARIATE	0.316	0.265	0.233	0.122	0.103
35-64	BIVARIATE	0.236	0.235	0.315	0.063	0.062
65-94	BIVARIATE	0.312	0.266	0.240	0.121	0.103
35-64	MULTIVARIATE	0.229	0.243	0.346	0.061	0.064
65-94	MULTIVARIATE	0.265	0.268	0.323	0.102	0.104

SEX: FEMALE
EVENT: RECOGNIZED MYOCARDIAL INFARCTION
RISK FACTOR: GLUCOSE INTOLERANCE
PERSONS AT RISK: PERSONS FREE OF CORONARY HEART DISEASE

RISK FACTOR	35-44 AT EXAM			45-54 AT EXAM			55-64 AT EXAM			65-74 AT EXAM		
	PERSON EXAMS	# OF EVENTS	ANNUAL RATE	PERSON EXAMS	# OF EVENTS	ANNUAL RATE	PERSON EXAMS	# OF EVENTS	ANNUAL RATE	PERSON EXAMS	# OF EVENTS	ANNUAL RATE
NO	5522			9588	10	1	10503	27	1	5398	34	3
YES	94			371	3	4	1010	10	5	850	13	8
UNKNOWN	0			0	0	-	0	0	-	0	0	-
TOTAL	5616			9959	13	1	11513	37	2	6248	47	4

RISK FACTOR	75-84 AT EXAM			85-94 AT EXAM			35-64 COMBINED			65-94 COMBINED		
	PERSON EXAMS	# OF EVENTS	ANNUAL RATE	PERSON EXAMS	# OF EVENTS	ANNUAL RATE	# OF EVENTS	CRUDE	AGE-ADJUSTED	# OF EVENTS	CRUDE	AGE-ADJUSTED
NO	1581	15	5	100			38	1	1	49	3	3
YES	370	5	7	23			14	5	5	21	8	8
UNKNOWN	0	0	-	0			0	-	-	0	-	-
TOTAL	1951	20	5	123			52	1	1	70	4	4

LOGISTIC REGRESSIONS

AGE AT EXAM		COEFFICIENT OF RISK FACTOR	SE COEFFICIENT	P-VALUE	STANDARDIZED COEFFICIENT	STD SE
35-44						
45-54		2.055	0.660	0.002	0.389	0.125
55-64		1.356	0.372	0.001	0.384	0.105
65-74		0.896	0.328	0.006	0.307	0.113
75-84		0.358	0.520	0.491	0.140	0.204
85-94						
35-64	UNIVARIATE	1.864	0.314	0.001	0.423	0.071
65-94	UNIVARIATE	0.902	0.263	0.001	0.322	0.094
35-64	BIVARIATE	1.542	0.319	0.001	0.350	0.072
65-94	BIVARIATE	0.861	0.264	0.001	0.307	0.094
35-64	MULTIVARIATE	1.374	0.329	0.001	0.312	0.075
65-94	MULTIVARIATE	0.819	0.268	0.002	0.292	0.096

TABLE 5-10   DATE=10DEC85

AVERAGE ANNUAL INCIDENCE RATE FOR EVENT PER 1000 PERSONS AT RISK AT EXAM
BY AGE AND LEVEL OF RISK FACTOR AT EXAM: FRAMINGHAM STUDY 30-YEAR FOLLOWUP

SEX: MALE
EVENT: RECOGNIZED MYOCARDIAL INFARCTION
RISK FACTOR: METROPOLITAN RELATIVE WEIGHT (PERCENT)
PERSONS AT RISK: PERSONS FREE OF CORONARY HEART DISEASE

RISK FACTOR	35-44 AT EXAM			45-54 AT EXAM			55-64 AT EXAM			65-74 AT EXAM		
	PERSON EXAMS	# OF EVENTS	ANNUAL RATE	PERSON EXAMS	# OF EVENTS	ANNUAL RATE	PERSON EXAMS	# OF EVENTS	ANNUAL RATE	PERSON EXAMS	# OF EVENTS	ANNUAL RATE
54-104	723	0	0	1131	6	3	1155	14	6	686	4	3
105-114	965	5	3	1540	14	5	1655	19	6	851	12	7
115-124	1122	6	3	2037	19	5	2175	26	6	1069	21	10
125-134	915	5	3	1645	15	5	1674	22	7	766	11	7
135-272	747	2	1	1370	12	4	1389	22	8	541	12	11
UNKNOWN	0	0	-	1	0	0	1	0	0	0	0	-
TOTAL	4472	18	2	7724	66	4	8049	103	6	3913	60	8

RISK FACTOR	75-84 AT EXAM			85-94 AT EXAM			35-64 COMBINED			65-94 COMBINED		
	PERSON EXAMS	# OF EVENTS	ANNUAL RATE	PERSON EXAMS	# OF EVENTS	ANNUAL RATE	# OF EVENTS	CRUDE	AGE-ADJUSTED	# OF EVENTS	CRUDE	AGE-ADJUSTED
54-104	224	4	9	18			20	3	3	8	4	4
105-114	252	3	6	9			38	5	5	16	7	7
115-124	296	6	10	12			51	5	5	27	10	10
125-134	155	7	23	0			42	5	5	18	10	10
135-272	90	0	0	0			36	5	5	12	10	9
UNKNOWN	0	0	-	0			0	0	0	0	-	
TOTAL	1017	20	10	39			187	5	5	81	8	8

LOGISTIC REGRESSIONS

AGE AT EXAM	COEFFICIENT OF RISK FACTOR	SE COEFFICIENT	P-VALUE	STANDARDIZED COEFFICIENT	STD SE
35-44	0.003	0.014	0.823	0.052	0.234
45-54	0.004	0.008	0.577	0.068	0.122
55-64	0.004	0.006	0.493	0.067	0.098
65-74	0.015	0.008	0.062	0.232	0.125
75-84	0.002	0.015	0.877	0.035	0.224
85-94					
35-64 UNIVARIATE	0.004	0.004	0.329	0.071	0.072
65-94 UNIVARIATE	0.011	0.007	0.116	0.171	0.108
35-64 BIVARIATE	0.004	0.005	0.347	0.068	0.073
65-94 BIVARIATE	0.011	0.007	0.106	0.176	0.109
35-64 MULTIVARIATE	0.003	0.005	0.499	0.051	0.076
65-94 MULTIVARIATE	0.005	0.007	0.523	0.073	0.115

SEX: FEMALE
EVENT: RECOGNIZED MYOCARDIAL INFARCTION
RISK FACTOR: METROPOLITAN RELATIVE WEIGHT (PERCENT)
PERSONS AT RISK: PERSONS FREE OF CORONARY HEART DISEASE

RISK FACTOR	35-44 AT EXAM			45-54 AT EXAM			55-64 AT EXAM			65-74 AT EXAM		
	PERSON EXAMS	# OF EVENTS	ANNUAL RATE	PERSON EXAMS	# OF EVENTS	ANNUAL RATE	PERSON EXAMS	# OF EVENTS	ANNUAL RATE	PERSON EXAMS	# OF EVENTS	ANNUAL RATE
54-104	1694			2095	2	0	2045	6	1	1216	10	4
105-114	1430			2331	1	0	2409	7	1	1145	9	4
115-124	1058			2107	3	1	2479	7	1	1395	5	2
125-134	566			1572	0	0	1743	8	2	982	7	4
135-272	860			2054	7	2	2837	9	2	1510	16	5
UNKNOWN	8			0	0	-	0	0	-	0	0	-
TOTAL	5616			9959	13	1	11513	37	2	6248	47	4

RISK FACTOR	75-84 AT EXAM			85-94 AT EXAM			35-64 COMBINED			65-94 COMBINED		
	PERSON EXAMS	# OF EVENTS	ANNUAL RATE	PERSON EXAMS	# OF EVENTS	ANNUAL RATE	# OF EVENTS	CRUDE	AGE-ADJUSTED	# OF EVENTS	CRUDE	AGE-ADJUSTED
54-104	517	4	4	41			9	1	1	15	4	4
105-114	375	1	1	28			8	1	1	10	3	3
115-124	399	11	14	21			10	1	1	18	5	5
125-134	266	1	2	16			8	1	1	4	3	3
135-272	394	3	4	17			17	1	1	19	5	5
UNKNOWN	0	0	-	0			0	0	0	0	-	
TOTAL	1951	20	5	123			52	1	1	70	4	4

LOGISTIC REGRESSIONS

AGE AT EXAM	COEFFICIENT OF RISK FACTOR	SE COEFFICIENT	P-VALUE	STANDARDIZED COEFFICIENT	STD SE
35-44					
45-54	0.021	0.009	0.019	0.441	0.188
55-64	0.004	0.007	0.615	0.079	0.157
65-74	0.008	0.006	0.175	0.179	0.132
75-84	0.007	0.010	0.516	0.139	0.213
85-94					
35-64 UNIVARIATE	0.012	0.005	0.027	0.254	0.115
65-94 UNIVARIATE	0.007	0.005	0.200	0.143	0.111
35-64 BIVARIATE	0.009	0.006	0.102	0.196	0.120
65-94 BIVARIATE	0.008	0.005	0.135	0.167	0.111
35-64 MULTIVARIATE	0.004	0.006	0.504	0.087	0.130
65-94 MULTIVARIATE	0.005	0.005	0.377	0.101	0.114

TABLE 5-11   DATE=10DEC85
AVERAGE ANNUAL INCIDENCE RATE FOR EVENT PER 1000 PERSONS AT RISK AT EXAM
BY AGE AND LEVEL OF RISK FACTOR AT EXAM, FRAMINGHAM STUDY 30-YEAR FOLLOWUP

SEX: MALE
EVENT: RECOGNIZED MYOCARDIAL INFARCTION
RISK FACTOR: VITAL CAPACITY - HEIGHT INDEX (ML/INCH)
PERSONS AT RISK: PERSONS FREE OF CORONARY HEART DISEASE

RISK FACTOR	35-44 AT EXAM PERSON EXAMS	# OF EVENTS	ANNUAL RATE	45-54 AT EXAM PERSON EXAMS	# OF EVENTS	ANNUAL RATE	55-64 AT EXAM PERSON EXAMS	# OF EVENTS	ANNUAL RATE	65-74 AT EXAM PERSON EXAMS	# OF EVENTS	ANNUAL RATE
8-44	269	2	4	1062	12	6	2045	35	9	1752	27	8
45-49	411	1	1	1125	12	5	1480	15	5	749	18	12
50-54	747	0	0	1658	15	5	1835	18	5	712	7	5
55-59	1022	2	2	1739	13	4	1416	21	7	442	3	5
60-85	2010	11	3	2135	14	3	1255	14	6	246	4	8
UNKNOWN	13	0	0	5	0	0	18	0	0	12	1	42
TOTAL	4472	18	2	7724	66	4	8049	103	6	3913	60	8

RISK FACTOR	75-84 AT EXAM PERSON EXAMS	# OF EVENTS	ANNUAL RATE	85-94 AT EXAM PERSON EXAMS	# OF EVENTS	ANNUAL RATE	35-64 COMBINED # OF EVENTS	CRUDE	AGE-ADJUSTED	65-94 COMBINED # OF EVENTS	CRUDE	AGE-ADJUSTED
8-44	649	15	12	24			49	7	6	43	9	9
45-49	172	2	6	6			28	5	4	20	11	11
50-54	97	1	5	2			33	5	5	8	5	5
55-59	65	2	15	0			38	5	5	5	5	6
60-85	26	0	0	0			39	4	4	4	7	6
UNKNOWN	8	0	0	0			0	0	0	1	25	33
TOTAL	1017	20	10	39			187	5	5	81	8	8

LOGISTIC REGRESSIONS

AGE AT EXAM		COEFFICIENT OF RISK FACTOR	SE COEFFICIENT	P-VALUE	STANDARDIZED COEFFICIENT	STD SE
35-44		0.008	0.027	0.758	0.074	0.238
45-54		-0.020	0.013	0.141	-0.178	0.121
55-64		-0.015	0.010	0.157	-0.138	0.097
65-74		-0.007	0.014	0.633	-0.062	0.130
75-84		-0.038	0.024	0.116	-0.357	0.227
85-94						
35-64	UNIVARIATE	-0.024	0.007	0.001	-0.235	0.071
65-94	UNIVARIATE	-0.016	0.012	0.183	-0.149	0.112
35-64	BIVARIATE	-0.011	0.008	0.154	-0.109	0.076
65-94	BIVARIATE	-0.015	0.012	0.203	-0.146	0.115
35-64	MULTIVARIATE	-0.004	0.008	0.640	-0.037	0.080
65-94	MULTIVARIATE	-0.015	0.012	0.227	-0.143	0.118

SEX: FEMALE
EVENT: RECOGNIZED MYOCARDIAL INFARCTION
RISK FACTOR: VITAL CAPACITY - HEIGHT INDEX (ML/INCH)
PERSONS AT RISK: PERSONS FREE OF CORONARY HEART DISEASE

RISK FACTOR	35-44 AT EXAM PERSON EXAMS	# OF EVENTS	ANNUAL RATE	45-54 AT EXAM PERSON EXAMS	# OF EVENTS	ANNUAL RATE	55-64 AT EXAM PERSON EXAMS	# OF EVENTS	ANNUAL RATE	65-74 AT EXAM PERSON EXAMS	# OF EVENTS	ANNUAL RATE
4-29	122			534	1	1	1461	7	2	1747	19	5
30-34	289			1217	1	0	2418	8	2	1715	14	4
35-39	769			2077	2	0	3048	8	1	1529	6	2
40-44	1427			2804	6	1	2706	12	2	854	5	3
45-92	2988			3310	3	0	1859	2	1	397	3	4
UNKNOWN	21			17	0	0	21	0	0	6	0	0
TOTAL	5616			9959	13	1	11513	37	2	6248	47	4

RISK FACTOR	75-84 AT EXAM PERSON EXAMS	# OF EVENTS	ANNUAL RATE	85-94 AT EXAM PERSON EXAMS	# OF EVENTS	ANNUAL RATE	35-64 COMBINED # OF EVENTS	CRUDE	AGE-ADJUSTED	65-94 COMBINED # OF EVENTS	CRUDE	AGE-ADJUSTED
4-29	930	13	7	80			8	2	1	34	6	6
30-34	569	6	5	29			9	1	1	21	5	5
35-39	314	1	2	13			11	1	1	7	2	2
40-44	111	0	0	1			18	1	1	5	3	2
45-92	25	0	0	0			6	0	0	3	4	3
UNKNOWN	2	0	0	0			0	0	0	0	0	0
TOTAL	1951	20	5	123			52	1	1	70	4	4

LOGISTIC REGRESSIONS

AGE AT EXAM		COEFFICIENT OF RISK FACTOR	SE COEFFICIENT	P-VALUE	STANDARDIZED COEFFICIENT	STD SE
35-44						
45-54		-0.017	0.037	0.648	-0.125	0.274
55-64		-0.024	0.022	0.259	-0.183	0.162
65-74		-0.030	0.019	0.127	-0.219	0.144
75-84		-0.059	0.030	0.050	-0.416	0.212
85-94						
35-64	UNIVARIATE	-0.045	0.017	0.007	-0.361	0.134
65-94	UNIVARIATE	-0.041	0.016	0.008	-0.308	0.117
35-64	BIVARIATE	-0.017	0.019	0.364	-0.134	0.148
65-94	BIVARIATE	-0.036	0.016	0.027	-0.270	0.122
35-64	MULTIVARIATE	0.012	0.020	0.566	0.092	0.160
65-94	MULTIVARIATE	-0.031	0.017	0.074	-0.231	0.129

TABLE 5-12   DATE=10DEC85
AVERAGE ANNUAL INCIDENCE RATE FOR EVENT PER 1000 PERSONS AT RISK AT EXAM
BY AGE AND LEVEL OF RISK FACTOR AT EXAM: FRAMINGHAM STUDY 30-YEAR FOLLOWUP

SEX: MALE
EVENT: RECOGNIZED MYOCARDIAL INFARCTION
RISK FACTOR: HEART RATE (PER MINUTE)
PERSONS AT RISK: PERSONS FREE OF CORONARY HEART DISEASE

RISK FACTOR	35-44 AT EXAM PERSON EXAMS	# OF EVENTS	ANNUAL RATE	45-54 AT EXAM PERSON EXAMS	# OF EVENTS	ANNUAL RATE	55-64 AT EXAM PERSON EXAMS	# OF EVENTS	ANNUAL RATE	65-74 AT EXAM PERSON EXAMS	# OF EVENTS	ANNUAL RATE
30-67	1284	4	2	1949	16	4	2309	23	5	1183	18	8
68-75	1492	3	1	2490	16	3	2490	36	7	1159	10	4
76-83	722	5	3	1394	12	4	1374	19	7	624	18	14
84-91	579	4	3	1097	12	5	1064	14	7	487	5	5
92-220	395	2	3	794	10	6	812	11	7	460	9	10
UNKNOWN	0	0	-	0	0	-	0	0	-	0	0	-
TOTAL	4472	18	2	7724	66	4	8049	103	6	3913	60	8

RISK FACTOR	75-84 AT EXAM PERSON EXAMS	# OF EVENTS	ANNUAL RATE	85-94 AT EXAM PERSON EXAMS	# OF EVENTS	ANNUAL RATE	35-64 COMBINED # OF EVENTS	CRUDE	AGE-ADJUSTED	65-94 COMBINED # OF EVENTS	CRUDE	AGE-ADJUSTED
30-67	301	5	8	11			43	4	4	23	8	8
68-75	296	7	12	12			55	4	4	18	6	6
76-83	192	4	10	10			36	5	5	22	13	13
84-91	123	2	8	5			30	5	5	7	6	6
92-220	105	2	10	1			23	6	6	11	10	10
UNKNOWN	0	0	-	0			0	-	-	0	-	-
TOTAL	1017	20	10	39			187	5	5	81	8	8

LOGISTIC REGRESSIONS

AGE AT EXAM		COEFFICIENT OF RISK FACTOR	SE COEFFICIENT	P-VALUE	STANDARDIZED COEFFICIENT	STD SE
35-44		0.026	0.017	0.122	0.326	0.211
45-54		0.013	0.009	0.175	0.161	0.118
55-64		0.005	0.008	0.549	0.058	0.097
65-74		0.008	0.009	0.366	0.113	0.125
75-84		0.007	0.016	0.683	0.089	0.219
85-94						
35-64	UNIVARIATE	0.010	0.006	0.081	0.124	0.071
65-94	UNIVARIATE	0.008	0.008	0.326	0.106	0.108
35-64	BIVARIATE	0.010	0.006	0.082	0.122	0.070
65-94	BIVARIATE	0.008	0.008	0.326	0.106	0.108
35-64	MULTIVARIATE	0.000	0.006	0.938	0.006	0.076
65-94	MULTIVARIATE	0.007	0.008	0.395	0.093	0.110

SEX: FEMALE
EVENT: RECOGNIZED MYOCARDIAL INFARCTION
RISK FACTOR: HEART RATE (PER MINUTE)
PERSONS AT RISK: PERSONS FREE OF CORONARY HEART DISEASE

RISK FACTOR	35-44 AT EXAM PERSON EXAMS	# OF EVENTS	ANNUAL RATE	45-54 AT EXAM PERSON EXAMS	# OF EVENTS	ANNUAL RATE	55-64 AT EXAM PERSON EXAMS	# OF EVENTS	ANNUAL RATE	65-74 AT EXAM PERSON EXAMS	# OF EVENTS	ANNUAL RATE
30-67	1092			1897	3	1	2305	1	0	1103	7	3
68-75	1803			3244	4	1	3468	11	2	1668	12	4
76-83	1051			1911	2	1	2351	9	2	1358	8	3
84-91	939			1604	1	0	1780	8	2	1089	13	6
92-220	731			1303	3	1	1609	8	2	1030	7	3
UNKNOWN	0			0	0	-	0	0	-	0	0	-
TOTAL	5616			9959	13	1	11513	37	2	6248	47	4

RISK FACTOR	75-84 AT EXAM PERSON EXAMS	# OF EVENTS	ANNUAL RATE	85-94 AT EXAM PERSON EXAMS	# OF EVENTS	ANNUAL RATE	35-64 COMBINED # OF EVENTS	CRUDE	AGE-ADJUSTED	65-94 COMBINED # OF EVENTS	CRUDE	AGE-ADJUSTED
30-67	360	4	6	26			4	0	0	12	4	4
68-75	462	3	3	32			15	1	1	15	3	3
76-83	428	7	5	24			12	1	1	15	4	4
84-91	330	3	5	23			10	1	1	17	6	6
92-220	371	3	4	18			11	2	1	11	4	4
UNKNOWN	0	0	-	0			0	-	-	0	-	-
TOTAL	1951	20	5	123			52	1	1	70	4	4

LOGISTIC REGRESSIONS

AGE AT EXAM		COEFFICIENT OF RISK FACTOR	SE COEFFICIENT	P-VALUE	STANDARDIZED COEFFICIENT	STD SE
35-44						
45-54		-0.004	0.023	0.867	-0.047	0.282
55-64		0.023	0.011	0.030	0.302	0.139
65-74		0.009	0.011	0.419	0.113	0.140
75-84		-0.011	0.017	0.511	-0.156	0.237
85-94						
35-64	UNIVARIATE	0.018	0.010	0.059	0.232	0.123
65-94	UNIVARIATE	0.003	0.009	0.739	0.039	0.118
35-64	BIVARIATE	0.018	0.010	0.067	0.224	0.122
65-94	BIVARIATE	0.003	0.009	0.768	0.035	0.118
35-64	MULTIVARIATE	0.003	0.011	0.783	0.038	0.138
65-94	MULTIVARIATE	-0.003	0.009	0.709	-0.045	0.119

TABLE 5-13   DATE=07AUG85

AVERAGE ANNUAL INCIDENCE RATE FOR EVENT PER 1000 PERSONS AT RISK AT EXAM
BY AGE AND LEVEL OF RISK FACTOR AT EXAM: FRAMINGHAM STUDY 30-YEAR FOLLOWUP

SEX: MALE
EVENT: RECOGNIZED MYOCARDIAL INFARCTION
RISK FACTOR: CIGARETTES SMOKED (NUMBER PER DAY)
PERSONS AT RISK: PERSONS FREE OF CORONARY HEART DISEASE

RISK FACTOR	35-44 AT EXAM			45-54 AT EXAM			55-64 AT EXAM			65-74 AT EXAM		
	PERSON EXAMS	# OF EVENTS	ANNUAL RATE	PERSON EXAMS	# OF EVENTS	ANNUAL RATE	PERSON EXAMS	# OF EVENTS	ANNUAL RATE	PERSON EXAMS	# OF EVENTS	ANNUAL RATE
NONE	1383	3	1	3068	10	2	4295	44	5	2611	44	8
1-10	433	3	3	716	6	4	769	5	3	374	4	5
11-20	1485	9	3	2951	28	7	1682	30	9	526	9	9
21-40	987	3	2	1602	20	6	1119	22	10	369	2	3
41-90	158	0	0	252	1	2	162	2	6	27	0	0
UNKNOWN	26	0	0	35	1	14	22	0	0	6	1	83
TOTAL	4472	18	2	7724	66	4	8049	103	6	3913	60	8

RISK FACTOR	75-84 AT EXAM			85-94 AT EXAM			35-64 COMBINED			65-94 COMBINED		
	PERSON EXAMS	# OF EVENTS	ANNUAL RATE	PERSON EXAMS	# OF EVENTS	ANNUAL RATE	# OF EVENTS	CRUDE	AGE ADJUSTED	# OF EVENTS	CRUDE	AGE ADJUSTED
NONE	797	17	11	27			57	3	3	62	9	9
1-10	92	2	11	6			14	4	3	6	6	6
11-20	77	0	0	5			67	6	7	9	7	7
21-40	46	1	11	1			45	3	7	3	4	5
41-90	5	0	0	0			3	3	3	0	0	0
UNKNOWN	0	0	-	0			1	6	5	1	83	65
TOTAL	1017	20	10	39			187	5	4	81	8	8

LOGISTIC REGRESSIONS

AGE AT EXAM		COEFFICIENT OF RISK FACTOR	SE COEFFICIENT	P-VALUE	STANDARDIZED COEFFICIENT	STD SE
35-44		-0.003	0.017	0.857	-0.043	0.239
45-54		0.021	0.007	0.006	0.310	0.112
55-64		0.020	0.006	0.001	0.271	0.085
65-74		-0.019	0.014	0.159	-0.225	0.160
75-84		-0.024	0.034	0.466	-0.223	0.305
85-94						
35-64	UNIVARIATE	0.014	0.005	0.002	0.211	0.068
65-94	UNIVARIATE	-0.022	0.013	0.088	-0.242	0.142
35-64	BIVARIATE	0.020	0.005	0.001	0.284	0.067
65-94	BIVARIATE	-0.022	0.013	0.092	-0.241	0.143
35-64	MULTIVARIATE	0.022	0.005	0.001	0.317	0.067
65-94	MULTIVARIATE	-0.020	0.013	0.119	-0.223	0.143

SEX: FEMALE
EVENT: RECOGNIZED MYOCARDIAL INFARCTION
RISK FACTOR: CIGARETTES SMOKED (NUMBER PER DAY)
PERSONS AT RISK: PERSONS FREE OF CORONARY HEART DISEASE

RISK FACTOR	35-44 AT EXAM			45-54 AT EXAM			55-64 AT EXAM			65-74 AT EXAM		
	PERSON EXAMS	# OF EVENTS	ANNUAL RATE	PERSON EXAMS	# OF EVENTS	ANNUAL RATE	PERSON EXAMS	# OF EVENTS	ANNUAL RATE	PERSON EXAMS	# OF EVENTS	ANNUAL RATE
NONE	2582			5528	7	1	7595	15	1	4863	33	3
1-10	1314			1726	9	0	1566	7	2	626	6	5
11-20	1371			2009	3	1	1696	10	3	568	8	7
21-40	298			622	3	2	602	4	3	172	0	0
41-90	22			37	0	0	16	1	51	0	0	-
UNKNOWN	29			37	0	0	38	0	0	19	0	0
TOTAL	5616			9959	13	1	11513	37	2	6248	47	4

RISK FACTOR	75-84 AT EXAM			85-94 AT EXAM			35-64 COMBINED			65-94 COMBINED		
	PERSON EXAMS	# OF EVENTS	ANNUAL RATE	PERSON EXAMS	# OF EVENTS	ANNUAL RATE	# OF EVENTS	CRUDE	AGE ADJUSTED	# OF EVENTS	CRUDE	AGE ADJUSTED
NONE	1710	15	4	117			22	1	1	51	4	3
1-10	157	3	11	4			7	1	1	9	6	6
11-20	84	2	12	2			14	1	2	10	8	8
21-40	16	0	0	0			8	3	2	0	-	0
41-90	0	0	-	0			1	7	13	0	-	-
UNKNOWN	4	0	0	0			0	0	0	0	0	0
TOTAL	1951	20	5	123			52	1	1	70	4	4

LOGISTIC REGRESSIONS

AGE AT EXAM		COEFFICIENT OF RISK FACTOR	SE COEFFICIENT	P-VALUE	STANDARDIZED COEFFICIENT	STD SE
35-44						
45-54		0.050	0.022	0.176	0.306	0.226
55-64		0.045	0.012	0.001	0.424	0.117
65-74		0.012	0.017	0.501	0.088	0.131
75-84		0.036	0.031	0.239	0.183	0.155
85-94						
35-64	UNIVARIATE	0.041	0.011	0.001	0.404	0.105
65-94	UNIVARIATE	0.012	0.015	0.405	0.088	0.106
35-64	BIVARIATE	0.046	0.010	0.001	0.459	0.103
65-94	BIVARIATE	0.017	0.015	0.256	0.121	0.107
35-64	MULTIVARIATE	0.053	0.010	0.001	0.522	0.104
65-94	MULTIVARIATE	0.025	0.015	0.098	0.176	0.107

AVERAGE ANNUAL INCIDENCE RATE FOR EVENT PER 1000 PERSONS AT RISK AT EXAM
BY AGE AND LEVEL OF RISK FACTOR AT EXAM: FRAMINGHAM STUDY 30-YEAR FOLLOWUP

SEX: MALE
EVENT: RECOGNIZED MYOCARDIAL INFARCTION
RISK FACTOR: ALBUMIN IN URINE, DEFINITE
PERSONS AT RISK: PERSONS FREE OF CORONARY HEART DISEASE

RISK FACTOR	35-44 AT EXAM PERSON EXAMS	# OF EVENTS	ANNUAL RATE	45-54 AT EXAM PERSON EXAMS	# OF EVENTS	ANNUAL RATE	55-64 AT EXAM PERSON EXAMS	# OF EVENTS	ANNUAL RATE	65-74 AT EXAM PERSON EXAMS	# OF EVENTS	ANNU RAT
0	4366	18	2	7599	64	4	7838	99	6	3792	58	8
ES	70	0	0	99	2	10	184	3	8	115	2	9
UNKNOWN	36	0	0	26	0	0	27	1	19	6	0	0
TOTAL	4472	18	2	7724	66	4	8049	103	6	3913	60	8

RISK FACTOR	75-84 AT EXAM PERSON EXAMS	# OF EVENTS	ANNUAL RATE	85-94 AT EXAM PERSON EXAMS	# OF EVENTS	ANNUAL RATE	35-64 COMBINED # OF EVENTS	CRUDE	AGE- ADJUSTED	65-94 COMBINED # OF EVENTS	CRUDE	A
0	989	19	10	32	7		181	5	5	78	8	
ES	25	1	20	7			5	7	7	3	10	
UNKNOWN	3	0	0	0			1	6	7	0	0	
TOTAL	1017	20	10	39			187	5	5	81	8	

LOGISTIC REGRESSIONS

AGE AT EXAM		COEFFICIENT OF RISK FACTOR	SE COEFFICIENT	P-VALUE	STANDARDIZED COEFFICIENT	STD SE
35-44		-	-	-	-	-
35-54		0.887	0.725	0.221	0.100	0.082
35-64		0.259	0.661	0.693	0.039	0.088
35-74		0.131	0.726	0.857	0.022	0.123
35-84		0.755	1.047	0.471	0.117	0.162
35-94						
35-64	UNIVARIATE	0.443	0.457	0.332	0.058	0.060
35-94	UNIVARIATE	0.235	0.594	0.693	0.040	0.101
35-64	BIVARIATE	0.331	0.457	0.469	0.043	0.060
35-94	BIVARIATE	0.232	0.594	0.696	0.039	0.101
35-64	MULTIVARIATE	-0.140 *	0.515	0.786	-0.018	0.068
35-94	MULTIVARIATE	-0.071	0.605	0.907	-0.012	0.103

SEX: FEMALE
EVENT: RECOGNIZED MYOCARDIAL INFARCTION
RISK FACTOR: ALBUMIN IN URINE, DEFINITE
PERSONS AT RISK: PERSONS FREE OF CORONARY HEART DISEASE

RISK FACTOR	35-44 AT EXAM PERSON EXAMS	# OF EVENTS	ANNUAL RATE	45-54 AT EXAM PERSON EXAMS	# OF EVENTS	ANNUAL RATE	55-64 AT EXAM PERSON EXAMS	# OF EVENTS	ANNUAL RATE	65-74 AT EXAM PERSON EXAMS	# OF EVENTS	ANNU RATE
S	5353			9656	13	1	11263	36	2	6118	44	4
	225			278	0	0	225	1	2	120	3	13
UNKNOWN	38			25	0	0	25	0	0	10	0	0
TOTAL	5616			9959	13	1	11513	37	2	6248	47	4

RISK FACTOR	75-84 AT EXAM PERSON EXAMS	# OF EVENTS	ANNUAL RATE	85-94 AT EXAM PERSON EXAMS	# OF EVENTS	ANNUAL RATE	35-64 COMBINED # OF EVENTS	CRUDE	AGE- ADJUSTED	65-94 COMBINED # OF EVENTS	CRUDE	AD
S	1904	20	5	120	3		51	1	1	66	4	
	38	0	0				1	1	1	4	12	
UNKNOWN	9	0	0	0			0	0	0	0	0	
TOTAL	1951	20	5	123			52	1	1	70	4	

LOGISTIC REGRESSIONS

AGE AT EXAM		COEFFICIENT OF RISK FACTOR	SE COEFFICIENT	P-VALUE	STANDARDIZED COEFFICIENT	STD SE
35-44		-	-	-	-	-
35-54		-	-	-	-	-
35-64		0.331	1.016	0.745	0.046	0.141
35-74		1.264	0.604	0.036	0.174	0.083
35-84		-	-	-	-	-
35-94		-	-	-	-	-
35-64	UNIVARIATE	-0.346	1.010	0.732	-0.056	0.164
35-94	UNIVARIATE	1.137	0.521	0.029	0.157	0.072
35-64	BIVARIATE	-0.137	1.011	0.892	-0.022	0.164
35-94	BIVARIATE	1.134	0.522	0.030	0.156	0.072
35-64	MULTIVARIATE	-1.527	1.377	0.268	-0.247	0.225

TABLE 5-15   DATE=10DEC85
AVERAGE ANNUAL INCIDENCE RATE FOR EVENT PER 1000 PERSONS AT RISK AT EXAM
BY AGE AND LEVEL OF RISK FACTOR AT EXAM: FRAMINGHAM STUDY 30-YEAR FOLLOWUP

SEX: MALE
EVENT: RECOGNIZED MYOCARDIAL INFARCTION
RISK FACTOR: HEART ENLARGEMENT BY X-RAY
PERSONS AT RISK: PERSONS FREE OF CORONARY HEART DISEASE

RISK FACTOR	35-44 AT EXAM			45-54 AT EXAM			55-64 AT EXAM			65-74 AT EXAM		
	PERSON EXAMS	# OF EVENTS	ANNUAL RATE	PERSON EXAMS	# OF EVENTS	ANNUAL RATE	PERSON EXAMS	# OF EVENTS	ANNUAL RATE	PERSON EXAMS	# OF EVENTS	ANNUAL RATE
NO	4063	15	2	6383	55	4	5967	72	6	2584	38	7
MILD	185	2	5	650	5	4	962	14	7	578	6	5
DEFINITE	219	1	2	687	6	4	1119	17	8	751	16	11
UNKNOWN	5	0	0	4	0	0	1	0	0	0	0	-
TOTAL	4472	18	2	7724	66	4	8049	103	6	3913	60	8

RISK FACTOR	75-84 AT EXAM			85-94 AT EXAM			35-64 COMBINED			65-94 COMBINED		
	PERSON EXAMS	# OF EVENTS	ANNUAL RATE	PERSON EXAMS	# OF EVENTS	ANNUAL RATE	# OF EVENTS	CRUDE	AGE-ADJUSTED	# OF EVENTS	CRUDE	AGE-ADJUSTED
NO	589	7	6	29			142	4	4	46	7	7
MILD	178	4	11	5			21	6	6	10	7	6
DEFINITE	250	9	18	5			24	6	5	25	12	12
UNKNOWN	0	0	-	0			0	0	0	0	-	-
TOTAL	1017	20	10	39			187	5	5	81	8	8

LOGISTIC REGRESSIONS

AGE AT EXAM		COEFFICIENT OF RISK FACTOR	SE COEFFICIENT	P-VALUE	STANDARDIZED COEFFICIENT	STD SE
35-44		0.301	0.404	0.456	0.141	0.189
45-54		-0.013	0.204	0.951	-0.008	0.125
55-64		0.126	0.130	0.332	0.090	0.093
65-74		0.154	0.155	0.319	0.123	0.123
75-84		0.565	0.253	0.025	0.477	0.214
85-94						
35-64	UNIVARIATE	0.179	0.104	0.085	0.114	0.066
65-94	UNIVARIATE	0.261	0.128	0.042	0.211	0.104
35-64	BIVARIATE	0.078	0.106	0.461	0.050	0.067
65-94	BIVARIATE	0.260	0.129	0.043	0.210	0.104
35-64	MULTIVARIATE	0.050	0.113	0.659	0.032	0.072
65-94	MULTIVARIATE	0.128	0.137	0.350	0.103	0.111

SEX: FEMALE
EVENT: RECOGNIZED MYOCARDIAL INFARCTION
RISK FACTOR: HEART ENLARGEMENT BY X-RAY
PERSONS AT RISK: PERSONS FREE OF CORONARY HEART DISEASE

RISK FACTOR	35-44 AT EXAM			45-54 AT EXAM			55-64 AT EXAM			65-74 AT EXAM		
	PERSON EXAMS	# OF EVENTS	ANNUAL RATE	PERSON EXAMS	# OF EVENTS	ANNUAL RATE	PERSON EXAMS	# OF EVENTS	ANNUAL RATE	PERSON EXAMS	# OF EVENTS	ANNUAL RATE
NO	5164			8017	8	0	7522	19	2	3191	17	3
MILD	199			943	2	1	1769	8	2	1150	7	3
DEFINITE	245			993	3	2	2218	10	2	1907	23	6
UNKNOWN	8			6	0	0	4	0	0	0	0	-
TOTAL	5616			9959	13	1	11513	37	2	6248	47	4

RISK FACTOR	75-84 AT EXAM			85-94 AT EXAM			35-64 COMBINED			65-94 COMBINED		
	PERSON EXAMS	# OF EVENTS	ANNUAL RATE	PERSON EXAMS	# OF EVENTS	ANNUAL RATE	# OF EVENTS	CRUDE	AGE-ADJUSTED	# OF EVENTS	CRUDE	AGE-ADJUSTED
NO	702	5	4	32			28	1	1	23	3	3
MILD	373	7	9	30			10	2	1	16	5	5
DEFINITE	876	8	5	61			14	2	2	31	5	6
UNKNOWN	0	0	-	0			0	0	0	0	-	-
TOTAL	1951	20	5	123			52	1	1	70	4	4

LOGISTIC REGRESSIONS

AGE AT EXAM		COEFFICIENT OF RISK FACTOR	SE COEFFICIENT	P-VALUE	STANDARDIZED COEFFICIENT	STD SE
35-44						
45-54		0.570	0.327	0.081	0.364	0.209
55-64		0.309	0.188	0.100	0.246	0.150
65-74		0.419	0.164	0.011	0.369	0.144
75-84		0.077	0.253	0.760	0.069	0.227
85-94						
35-64	UNIVARIATE	0.572	0.157	0.001	0.399	0.109
65-94	UNIVARIATE	0.305	0.134	0.023	0.272	0.120
35-64	BIVARIATE	0.378	0.162	0.020	0.264	0.113
65-94	BIVARIATE	0.271	0.137	0.047	0.242	0.122
35-64	MULTIVARIATE	0.214	0.179	0.231	0.149	0.125
65-94	MULTIVARIATE	0.202	0.143	0.158	0.180	0.127

TABLE 5-16   DATE=07AUG85

AVERAGE ANNUAL INCIDENCE RATE FOR EVENT.PER 1000 PERSONS AT RISK AT EXAM BY AGE AND LEVEL OF RISK FACTOR AT EXAM: FRAMINGHAM STUDY 30-YEAR FOLLOWUP

SEX: MALE
EVENT: RECOGNIZED MYOCARDIAL INFARCTION
RISK FACTOR: LEFT VENTRICULAR HYPERTROPHY BY ELECTROCARDIOGRAM
PERSONS AT RISK: PERSONS FREE OF CORONARY HEART DISEASE

OR	35-44 AT EXAM			45-54 AT EXAM			55-64 AT EXAM			65-74 AT EXAM		
	PERSON EXAMS	# OF EVENTS	ANNUAL RATE	PERSON EXAMS	# OF EVENTS	ANNUAL RATE	PERSON EXAMS	# OF EVENTS	ANNUAL RATE	PERSON EXAMS	# OF EVENTS	ANNUAL RATE
ER	4366	18	2	7497	63	4	7693	98	6	3703	52	7
NITE	78	0	0	132	3	11	210	1	2	90	2	11
	28	0	0	95	0	0	146	4	14	120	6	25
DHN	0	0	-	0	0	-	0	0	-	0	0	-
L	4472	18	2	7724	66	4	8049	103	6	3913	60	8

OR	75-84 AT EXAM			85-94 AT EXAM			35-64 COMBINED			65-94 COMBINED		
	PERSON EXAMS	# OF EVENTS	ANNUAL RATE	PERSON EXAMS	# OF EVENTS	ANNUAL RATE	# OF EVENTS	CRUDE	AGE-ADJUSTED	# OF EVENTS	CRUDE	AGE-ADJUSTED
ER	949	20	11	34	2		179	5	4	73	8	8
NITE	37	0	0	2			4	5	5	2	8	9
	31	0	0	3			4	7	6	6	19	20
DHN	0	0	-	0			0	-		0	-	
L	1017	20	10	39			187	5	4	81	8	8

LOGISTIC REGRESSIONS

AT	COEFFICIENT OF RISK FACTOR	SE COEFFICIENT	P-VALUE	STANDARDIZED COEFFICIENT	STD SE
	-	-	-	-	-
	0.056	0.465	0.904	0.014	0.118
	0.223	0.271	0.409	0.069	0.083
	0.619	0.216	0.003	0.238	0.080
	-	-	-	-	-
UNIVARIATE	0.201	0.234	0.389	0.054	0.063
UNIVARIATE	0.427	0.214	0.046	0.161	0.081
BIVARIATE	0.118	0.235	0.616	0.031	0.063
BIVARIATE	0.426	0.214	0.047	0.161	0.081
MULTIVARIATE	-0.011	0.246	0.964	-0.003	0.066
MULTIVARIATE	0.152	0.231	0.510	0.057	0.087

SEX: FEMALE
EVENT: RECOGNIZED MYOCARDIAL INFARCTION
RISK FACTOR: LEFT VENTRICULAR HYPERTROPHY BY ELECTROCARDIOGRAM
PERSONS AT RISK: PERSONS FREE OF CORONARY HEART DISEASE

R	35-44 AT EXAM			45-54 AT EXAM			55-64 AT EXAM			65-74 AT EXAM		
	PERSON EXAMS	# OF EVENTS	ANNUAL RATE	PERSON EXAMS	# OF EVENTS	ANNUAL RATE	PERSON EXAMS	# OF EVENTS	ANNUAL RATE	PERSON EXAMS	# OF EVENTS	ANNUAL RATE
R	5563	58		9798	13	1	11137	35	2	5966	40	3
ITE	15			107	0	0	226	0	0	134	3	11
				54	0	0	150	2	7	148	4	14
IHN	0			0			0	0	-	0	0	-
	5616			9959	13	1	11513	37	2	6248	47	4

R	75-84 AT EXAM			85-94 AT EXAM			35-64 COMBINED			65-94 COMBINED		
	PERSON EXAMS	# OF EVENTS	ANNUAL RATE	PERSON EXAMS	# OF EVENTS	ANNUAL RATE	# OF EVENTS	CRUDE	AGE-ADJUSTED	# OF EVENTS	CRUDE	AGE-ADJUSTED
R	1833	20	5	111			49	1	1	62	4	4
ITE	49	0	0	6			0	0	0	3	8	8
	69	0	0	6			3	7	10	5	11	12
WN	0	0	-	0			0			0	-	
	1951	20	5	123			52	1	1	70	4	4

LOGISTIC REGRESSIONS

	COEFFICIENT OF RISK FACTOR	SE COEFFICIENT	P-VALUE	STANDARDIZED COEFFICIENT	STD SE
	-	-	-	-	-
	0.554	0.398	0.164	0.146	0.105
	0.765	0.242	0.002	0.255	0.081
	-	-	-	-	-
UNIVARIATE	0.860	0.322	0.008	0.183	0.068
UNIVARIATE	0.553	0.220	0.012	0.195	0.078
BIVARIATE	0.684	0.327	0.036	0.145	0.069
BIVARIATE	0.527	0.221	0.017	0.186	0.078
MULTIVARIATE	0.301	0.358	0.399	0.064	0.076
MULTIVARIATE	0.374	0.235	0.112	0.132	0.083

TABLE 5-17   DATE=10DEC85
AVERAGE ANNUAL INCIDENCE RATE FOR EVENT PER 1000 PERSONS AT RISK AT EXAM
BY AGE AND LEVEL OF RISK FACTOR AT EXAM: FRAMINGHAM STUDY 30-YEAR FOLLOWUP

SEX: MALE
EVENT: RECOGNIZED MYOCARDIAL INFARCTION
RISK FACTOR: INTRAVENTRICULAR CONDUCTION DEFECT (WITH QRS INTERVAL MORE THAN 0.11 SECOND)
PERSONS AT RISK: PERSONS FREE OF CORONARY HEART DISEASE

RISK FACTOR	35-44 AT EXAM PERSON EXAMS	# OF EVENTS	ANNUAL RATE	45-54 AT EXAM PERSON EXAMS	# OF EVENTS	ANNUAL RATE	55-64 AT EXAM PERSON EXAMS	# OF EVENTS	ANNUAL RATE	65-74 AT EXAM PERSON EXAMS	# OF EVENTS	ANNUAL RATE
NO	4161	17	2	7213	60	4	7386	92	6	3566	55	8
INCOMPLETE	275	1	2	446	5	6	456	10	11	183	3	8
COMPLETE	36	0	0	65	1	8	207	1	2	164	2	6
UNKNOWN	0	0	-	0	0	-	0	0	-	0	0	-
TOTAL	4472	18	2	7724	66	4	8049	103	6	3913	60	8

RISK FACTOR	75-84 AT EXAM PERSON EXAMS	# OF EVENTS	ANNUAL RATE	85-94 AT EXAM PERSON EXAMS	# OF EVENTS	ANNUAL RATE	35-64 COMBINED # OF EVENTS	CRUDE	AGE-ADJUSTED	65-94 COMBINED # OF EVENTS	CRUDE	AGE-ADJUSTED
NO	918	18	10	31			169	5	5	74	8	8
INCOMPLETE	21	0	0	2			16	7	7	3	7	6
COMPLETE	78	2	13	6			2	3	4	4	8	7
UNKNOWN	0	0	-	0			0	-	-	0	-	-
TOTAL	1017	20	10	39			187	5	5	81	8	8

LOGISTIC REGRESSIONS

AGE AT EXAM	COEFFICIENT OF RISK FACTOR	SE COEFFICIENT	P-VALUE	STANDARDIZED COEFFICIENT	STD SE
35-44	-0.309	0.952	0.753	-0.089	0.282
45-54	0.306	0.551	0.583	0.090	0.103
55-64	0.055	0.248	0.823	0.021	0.095
65-74	-0.076	0.310	0.807	-0.034	0.138
75-84	0.084	0.389	0.830	0.046	0.212
85-94					
35-64 UNIVARIATE	0.151	0.199	0.449	0.050	0.067
65-94 UNIVARIATE	-0.026	0.243	0.916	-0.012	0.114
35-64 BIVARIATE	0.083	0.197	0.674	0.028	0.066
65-94 BIVARIATE	-0.029	0.243	0.904	-0.014	0.114
35-64 MULTIVARIATE	0.018	0.215	0.932	0.006	0.072
65-94 MULTIVARIATE	-0.027	0.247	0.913	-0.013	0.116

SEX: FEMALE
EVENT: RECOGNIZED MYOCARDIAL INFARCTION
RISK FACTOR: INTRAVENTRICULAR CONDUCTION DEFECT (WITH QRS INTERVAL MORE THAN 0.11 SECOND)
PERSONS AT RISK: PERSONS FREE OF CORONARY HEART DISEASE

RISK FACTOR	35-44 AT EXAM PERSON EXAMS	# OF EVENTS	ANNUAL RATE	45-54 AT EXAM PERSON EXAMS	# OF EVENTS	ANNUAL RATE	55-64 AT EXAM PERSON EXAMS	# OF EVENTS	ANNUAL RATE	65-74 AT EXAM PERSON EXAMS	# OF EVENTS	ANNUAL RATE
NO	5529			9771	13	1	11183	35	2	6027	45	4
INCOMPLETE	75			144	0	0	187	1	3	115	1	4
COMPLETE	12			44	0	0	143	1	3	106	1	5
UNKNOWN	0			0	0	-	8	0	-	0	0	-
TOTAL	5616			9959	13	1	11513	37	2	6248	47	4

RISK FACTOR	75-84 AT EXAM PERSON EXAMS	# OF EVENTS	ANNUAL RATE	85-94 AT EXAM PERSON EXAMS	# OF EVENTS	ANNUAL RATE	35-64 COMBINED # OF EVENTS	CRUDE	AGE-ADJUSTED	65-94 COMBINED # OF EVENTS	CRUDE	AGE-ADJUSTED
NO	1847	20	5	117			50	1	1	68	4	4
INCOMPLETE	36	0	0	1			1	1	1	1	3	3
COMPLETE	68	0	0	5			1	3	1	1	3	4
UNKNOWN	0	0	-	0			0	-	-	0	-	-
TOTAL	1951	20	5	123			52	1	1	70	4	4

LOGISTIC REGRESSIONS

AGE AT EXAM	COEFFICIENT OF RISK FACTOR	SE COEFFICIENT	P-VALUE	STANDARDIZED COEFFICIENT	STD SE
35-44					
45-54					
55-64	0.428	0.455	0.347	0.109	0.115
65-74	0.125	0.457	0.785	0.036	0.132
75-84					
85-94					
35-64 UNIVARIATE	0.442	0.463	0.342	0.092	0.097
65-94 UNIVARIATE	-0.221	0.456	0.627	-0.070	0.145
35-64 BIVARIATE	0.280	0.461	0.543	0.058	0.096
65-94 BIVARIATE	-0.252	0.455	0.579	-0.080	0.144
35-64 MULTIVARIATE	0.392	0.464	0.399	0.082	0.097
65-94 MULTIVARIATE	-0.239	0.460	0.603	-0.076	0.146

TABLE 5-18   DATE=12DEC85
AVERAGE ANNUAL INCIDENCE RATE FOR EVENT PER 1000 PERSONS AT RISK AT EXAM
BY AGE AND LEVEL OF RISK FACTOR AT EXAM: FRAMINGHAM STUDY 30-YEAR FOLLOWUP

SEX: MALE
EVENT: RECOGNIZED MYOCARDIAL INFARCTION
RISK FACTOR: DEFINITE NONSPECIFIC T-WAVE OR ST-SEGMENT ABNORMALITY BY ELECTROCARDIOGRAM
PERSONS AT RISK: PERSONS FREE OF CORONARY HEART DISEASE

RISK FACTOR	35-44 AT EXAM PERSON EXAMS	# OF EVENTS	ANNUAL RATE	45-54 AT EXAM PERSON EXAMS	# OF EVENTS	ANNUAL RATE	55-64 AT EXAM PERSON EXAMS	# OF EVENTS	ANNUAL RATE	65-74 AT EXAM PERSON EXAMS	# OF EVENTS	ANNUAL RATE
NO	4362	14	2	7552	64	4	7462	95	6	3472	54	8
YES	110	4	18	372	2	3	587	8	7	441	6	7
UNKNOWN	0	0	-	0	0	-	0	0	-	0	0	-
TOTAL	4472	18	2	7724	66	4	8049	103	6	3913	60	8

RISK FACTOR	75-84 AT EXAM PERSON EXAMS	# OF EVENTS	ANNUAL RATE	85-94 AT EXAM PERSON EXAMS	# OF EVENTS	ANNUAL RATE	35-64 COMBINED # OF EVENTS	CRUDE	AGE-ADJUSTED	65-94 COMBINED # OF EVENTS	CRUDE	AGE-ADJUSTED
NO	830	16	10	28			173	5	5	70	8	8
YES	187	4	11	11			14	7	8	11	9	8
UNKNOWN	0	0	-	0			0	-		0	-	
TOTAL	1017	20	10	39			187	5	5	81	8	8

LOGISTIC REGRESSIONS

AGE AT EXAM	COEFFICIENT OF RISK FACTOR	SE COEFFICIENT	P-VALUE	STANDARDIZED COEFFICIENT	STD SE
35-44	2.461	0.575	0.001	0.381	0.089
45-54	-0.485	0.720	0.500	-0.104	0.154
55-64	0.069	0.371	0.852	0.018	0.096
65-74	-0.136	0.433	0.754	-0.043	0.137
75-84	0.106	0.565	0.851	0.041	0.219
85-94					
35-64 UNIVARIATE	0.377	0.280	0.178	0.084	0.063
65-94 UNIVARIATE	0.064	0.327	0.845	0.021	0.110
35-64 BIVARIATE	0.209	0.281	0.458	0.047	0.063
65-94 BIVARIATE	0.056	0.329	0.864	0.019	0.110
35-64 MULTIVARIATE	-0.146	0.306	0.633	-0.033	0.068
65-94 MULTIVARIATE	-0.039	0.331	0.907	-0.013	0.111

SEX: FEMALE
EVENT: RECOGNIZED MYOCARDIAL INFARCTION
RISK FACTOR: DEFINITE NONSPECIFIC T-WAVE OR ST-SEGMENT ABNORMALITY BY ELECTROCARDIOGRAM
PERSONS AT RISK: PERSONS FREE OF CORONARY HEART DISEASE

RISK FACTOR	35-44 AT EXAM PERSON EXAMS	# OF EVENTS	ANNUAL RATE	45-54 AT EXAM PERSON EXAMS	# OF EVENTS	ANNUAL RATE	55-64 AT EXAM PERSON EXAMS	# OF EVENTS	ANNUAL RATE	65-74 AT EXAM PERSON EXAMS	# OF EVENTS	ANNUAL RATE
NO	5517			9543	11	1	10647	30	1	5528	36	3
YES	99			416	2	2	866	7	4	720	11	8
UNKNOWN	0			0	0	-	0	0	-	0	0	-
TOTAL	5616			9959	13	1	11513	37	2	6248	47	4

RISK FACTOR	75-84 AT EXAM PERSON EXAMS	# OF EVENTS	ANNUAL RATE	85-94 AT EXAM PERSON EXAMS	# OF EVENTS	ANNUAL RATE	35-64 COMBINED # OF EVENTS	CRUDE	AGE-ADJUSTED	65-94 COMBINED # OF EVENTS	CRUDE	AGE-ADJUSTED
NO	1632	17	5	97			43	1	1	55	4	4
YES	319	3	5	26			9	3	3	15	7	7
UNKNOWN	0			0			0	-		0	-	
TOTAL	1951	20	5	123			52	1	1	70	4	4

LOGISTIC REGRESSIONS

AGE AT EXAM	COEFFICIENT OF RISK FACTOR	SE COEFFICIENT	P-VALUE	STANDARDIZED COEFFICIENT	STD SE
35-44					
45-54	1.432	0.770	0.063	0.286	0.154
55-64	1.059	0.421	0.012	0.279	0.111
65-74	0.862	0.347	0.013	0.275	0.111
75-84	-0.103	0.629	0.870	-0.038	0.233
85-94					
35-64 UNIVARIATE	1.365	0.368	0.001	0.300	0.081
65-94 UNIVARIATE	0.626	0.293	0.033	0.209	0.098
35-64 BIVARIATE	1.070	0.371	0.004	0.235	0.082
65-94 BIVARIATE	0.581	0.295	0.049	0.194	0.098
35-64 MULTIVARIATE	0.672	0.396	0.090	0.148	0.087
65-94 MULTIVARIATE	0.467	0.296	0.115	0.156	0.099

TABLE 6-1   DATE=12DEC85

AVERAGE ANNUAL INCIDENCE RATE FOR EVENT PER 1000 PERSONS AT RISK AT EXAM
BY AGE AND LEVEL OF RISK FACTOR AT EXAM: FRAMINGHAM STUDY 30-YEAR FOLLOWUP

SEX: MALE
EVENT: CORONARY INSUFFICIENCY
RISK FACTOR: SYSTOLIC BLOOD PRESSURE, FIRST EXAMINER (MM HG)
PERSONS AT RISK: PERSONS FREE OF CORONARY HEART DISEASE

RISK FACTOR	35-44 AT EXAM			45-54 AT EXAM			55-64 AT EXAM			65-74 AT EXAM		
	PERSON EXAMS	# OF EVENTS	ANNUAL RATE	PERSON EXAMS	# OF EVENTS	ANNUAL RATE	PERSON EXAMS	# OF EVENTS	ANNUAL RATE	PERSON EXAMS	# OF EVENTS	ANNUAL RATE
74-119	1223	1	0	1769	1	0	1440	2	1	548	1	1
120-139	2165	3	1	3417	8	1	3166	4	1	1341	2	1
140-159	841	3	2	1732	6	2	2090	8	2	1194	6	3
160-179	191	1	3	567	3	3	916	4	2	552	3	3
180-300	52	0	0	239	1	2	437	4	5	278	1	2
UNKNOWN	0	0	-	0	0	-	0	0	-	0	0	-
TOTAL	4472	8	1	7724	19	1	8049	22	1	3913	13	2

RISK FACTOR	75-84 AT EXAM			85-94 AT EXAM			35-64 COMBINED			65-94 COMBINED		
	PERSON EXAMS	# OF EVENTS	ANNUAL RATE	PERSON EXAMS	# OF EVENTS	ANNUAL RATE	# OF EVENTS	CRUDE	AGE-ADJUSTED	# OF EVENTS	CRUDE	AGE-ADJUSTED
74-119	120			7			4	0	0	3	1	1
120-139	345			15			15	1	1	3	1	1
140-159	315			11			17	2	2	6	2	2
160-179	155			4			8	2	2	4	4	3
180-300	84			2			5	3	3	1	1	1
UNKNOWN	0			0			0	-	-	0	-	-
TOTAL	1017			39			49	1	1	16	2	2

LOGISTIC REGRESSIONS

AGE AT EXAM		COEFFICIENT OF RISK FACTOR	SE COEFFICIENT	P-VALUE	STANDARDIZED COEFFICIENT	STD SE
35-44		0.027	0.015	0.085	0.448	0.260
45-54		0.018	0.010	0.067	0.350	0.191
55-64		0.019	0.008	0.016	0.414	0.172
65-74		0.015	0.010	0.154	0.338	0.237
75-84						
85-94						
35-64	UNIVARIATE	0.020	0.006	0.001	0.404	0.113
65-94	UNIVARIATE	0.017	0.009	0.073	0.380	0.212
35-64	BIVARIATE	0.019	0.006	0.001	0.388	0.116
65-94	BIVARIATE	0.017	0.009	0.073	0.380	0.211
35-64	MULTIVARIATE	0.016	0.006	0.008	0.331	0.125
65-94	MULTIVARIATE	0.016	0.010	0.115	0.360	0.229

SEX: FEMALE
EVENT: CORONARY INSUFFICIENCY
RISK FACTOR: SYSTOLIC BLOOD PRESSURE, FIRST EXAMINER (MM HG)
PERSONS AT RISK: PERSONS FREE OF CORONARY HEART DISEASE

RISK FACTOR	35-44 AT EXAM			45-54 AT EXAM			55-64 AT EXAM			65-74 AT EXAM		
	PERSON EXAMS	# OF EVENTS	ANNUAL RATE	PERSON EXAMS	# OF EVENTS	ANNUAL RATE	PERSON EXAMS	# OF EVENTS	ANNUAL RATE	PERSON EXAMS	# OF EVENTS	ANNUAL RATE
74-119	2493			2714	2	0	1979	1	0	647	1	1
120-139	2244			3982	1	0	4056	8	1	1747	3	1
140-159	673			2053	2	0	3025	6	1	2013	6	1
160-179	156			768	1	1	1560	3	1	1155	2	1
180-300	50			442	0	0	893	6	3	686	3	2
UNKNOWN	0			0	0	-	0	0	-	0	0	-
TOTAL	5616			9959	6	0	11513	24	1	6248	15	1

RISK FACTOR	75-84 AT EXAM			85-94 AT EXAM			35-64 COMBINED			65-94 COMBINED		
	PERSON EXAMS	# OF EVENTS	ANNUAL RATE	PERSON EXAMS	# OF EVENTS	ANNUAL RATE	# OF EVENTS	CRUDE	AGE-ADJUSTED	# OF EVENTS	CRUDE	AGE-ADJUSTED
74-119	133	0	0	8			3	0	0	1	1	1
120-139	445	4	1	38			9	0	0	4	1	1
140-159	626	1	1	29			8	1	1	8	1	2
160-179	431	2	2	31			4	1	1	4	1	1
180-300	276	1	2	17			6	2	2	4	2	2
UNKNOWN	0	0	-	0			0	-	-	0	-	-
TOTAL	1951	5	1	123			30	1	1	21	1	1

LOGISTIC REGRESSIONS

AGE AT EXAM		COEFFICIENT OF RISK FACTOR	SE COEFFICIENT	P-VALUE	STANDARDIZED COEFFICIENT	STD SE
35-44						
45-54		-0.001	0.018	0.958	-0.022	0.414
55-64		0.017	0.007	0.008	0.425	0.160
65-74		0.008	0.010	0.413	0.198	0.242
75-84		0.012	0.016	0.445	0.306	0.401
85-94						
35-64	UNIVARIATE	0.020	0.005	0.001	0.469	0.129
65-94	UNIVARIATE	0.009	0.008	0.279	0.220	0.203
35-64	BIVARIATE	0.014	0.006	0.025	0.327	0.146
65-94	BIVARIATE	0.009	0.008	0.294	0.214	0.204
35-64	MULTIVARIATE	0.011	0.007	0.117	0.248	0.158
65-94	MULTIVARIATE	0.007	0.009	0.446	0.161	0.211

TABLE 6-2   DATE=03MAY85

AVERAGE ANNUAL INCIDENCE RATE FOR EVENT PER 1000 PERSONS AT RISK AT EXAM
BY AGE AND LEVEL OF RISK FACTOR AT EXAM: FRAMINGHAM STUDY 30-YEAR FOLLOWUP

SEX: MALE
EVENT: CORONARY INSUFFICIENCY
RISK FACTOR: DIASTOLIC BLOOD PRESSURE, FIRST EXAMINER (MM HG)
PERSONS AT RISK: PERSONS FREE OF CORONARY HEART DISEASE

35-44 AT EXAM			45-54 AT EXAM			55-64 AT EXAM			65-74 AT EXAM		
PERSON EXAMS	# OF EVENTS	ANNUAL RATE	PERSON EXAMS	# OF EVENTS	ANNUAL RATE	PERSON EXAMS	# OF EVENTS	ANNUAL RATE	PERSON EXAMS	# OF EVENTS	ANNUAL RATE
1012	1	0	1478	0	0	1847	3	1	1225	4	2
1604	2	1	2766	5	1	2845	9	2	1325	4	2
1187	2	1	2062	8	2	1964	3	1	893	2	1
499	3	3	1008	4	2	982	2	1	343	3	4
170	0	0	410	2	2	411	5	6	129	0	0
0	0	-	0	0	-	0	0	-	0	0	-
4472	8	1	7724	19	1	8049	22	1	3913	13	2

75-84 AT EXAM			85-94 AT EXAM			35-64 COMBINED			65-94 COMBINED		
PERSON EXAMS	# OF EVENTS	ANNUAL RATE	PERSON EXAMS	# OF EVENTS	ANNUAL RATE	# OF EVENTS	CRUDE	AGE-ADJUSTED	# OF EVENTS	CRUDE	AGE-ADJUSTED
422			22			4	0	0	5	1	2
331			12			16	1	1	6	2	2
204			5			13	1	1	2	1	1
44			0			9	2	2	3	4	3
16			0			7	4	3	0	0	0
0			0			0	-	-	0	-	-
1017			39			49	1	1	16	2	2

LOGISTIC REGRESSIONS

	COEFFICIENT OF RISK FACTOR	SE COEFFICIENT	P-VALUE	STANDARDIZED COEFFICIENT	STD SE
	0.043	0.027	0.106	0.488	0.302
	0.033	0.017	0.055	0.391	0.204
	0.033	0.015	0.033	0.398	0.186
	0.003	0.023	0.878	0.042	0.274
VARIATE	0.035	0.011	0.001	0.413	0.125
VARIATE	0.004	0.020	0.830	0.053	0.247
ARIATE	0.035	0.011	0.001	0.408	0.124
ARIATE	0.004	0.021	0.852	0.047	0.251
TIVARIATE	0.031	0.011	0.005	0.367	0.131
TIVARIATE	0.003	0.021	0.873	0.041	0.258

SEX: FEMALE
EVENT: CORONARY INSUFFICIENCY
RISK FACTOR: DIASTOLIC BLOOD PRESSURE, FIRST EXAMINER (MM HG)
PERSONS AT RISK: PERSONS FREE OF CORONARY HEART DISEASE

35-44 AT EXAM			45-54 AT EXAM			55-64 AT EXAM			65-74 AT EXAM		
PERSON EXAMS	# OF EVENTS	ANNUAL RATE	PERSON EXAMS	# OF EVENTS	ANNUAL RATE	PERSON EXAMS	# OF EVENTS	ANNUAL RATE	PERSON EXAMS	# OF EVENTS	ANNUAL RATE
2164			2639	2	0	2969	3	1	1831	6	2
2030			3541	1	0	3853	10	1	2135	3	1
987			2361	2	0	2840	6	1	1439	4	1
324			926	0	0	1272	5	2	600	1	1
111			492	1	1	579	2	2	243	1	2
0			0	0	-	0	0	-	0	0	-
5616			9959	6	0	11513	24	1	6248	15	1

75-84 AT EXAM			85-94 AT EXAM			35-64 COMBINED			65-94 COMBINED		
PERSON EXAMS	# OF EVENTS	ANNUAL RATE	PERSON EXAMS	# OF EVENTS	ANNUAL RATE	# OF EVENTS	CRUDE	AGE-ADJUSTED	# OF EVENTS	CRUDE	AGE-ADJUSTED
724	4	3	53			5	1	0	10	2	2
596	0	0	43			11	1	0	3	1	1
443	1	1	20			8	1	0	6	2	1
146	0	0	7			3	1	0	1	1	1
42	0	0	0			3	1	1	1	2	2
0	0	-	0			0	-	-	0	-	-
1951	5	1	123			30	1	0	21	1	1

LOGISTIC REGRESSIONS

	COEFFICIENT OF RISK FACTOR	SE COEFFICIENT	P-VALUE	STANDARDIZED COEFFICIENT	STD SE
	0.014	0.031	0.650	0.172	0.379
	0.019	0.015	0.222	0.230	0.189
	-0.001	0.021	0.959	-0.013	0.259
	-0.101	0.045	0.024	-1.207	0.536
VARIATE	0.024	0.013	0.073	0.290	0.162
VARIATE	-0.018	0.019	0.346	-0.215	0.228
ARIATE	0.018	0.014	0.182	0.224	0.168
ARIATE	-0.017	0.019	0.363	-0.208	0.229
TIVARIATE	0.012	0.014	0.382	0.151	0.172
TIVARIATE	-0.022	0.019	0.250	-0.265	0.230

TABLE 6-3A   DATE=03MAY85
AVERAGE ANNUAL INCIDENCE RATE FOR EVENT PER 1000 PERSONS AT RISK AT EXAM
BY AGE AND LEVEL OF RISK FACTOR AT EXAM: FRAMINGHAM STUDY 30-YEAR FOLLOWUP

SEX: MALE
EVENT: CORONARY INSUFFICIENCY
RISK FACTOR: HYPERTENSION WITH ANTIHYPERTENSIVE TREATMENT
PERSONS AT RISK: PERSONS FREE OF CORONARY HEART DISEASE

RISK FACTOR	35-44 AT EXAM			45-54 AT EXAM			55-64 AT EXAM			65-74 AT EXAM		
	PERSON EXAMS	# OF EVENTS	ANNUAL RATE	PERSON EXAMS	# OF EVENTS	ANNUAL RATE	PERSON EXAMS	# OF EVENTS	ANNUAL RATE	PERSON EXAMS	# OF EVENTS	ANNUAL RATE
NO	2651	3	1	4122	7	1	3782	5	1	1516	2	1
BORDER	1301	3	1	2205	3	1	2299	7	2	1252	7	3
DEFINITE	500	2	2	1139	7	3	1266	7	3	552	3	3
TREATED	20	0	0	258	2	4	702	3	2	593	1	1
UNKNOWN	0	0	-	0	0	-	0	0	-	0	0	-
TOTAL	4472	8	1	7724	19	1	8049	22	1	3913	13	2

RISK FACTOR	75-84 AT EXAM			85-94 AT EXAM			35-64 COMBINED			65-94 COMBINED		
	PERSON EXAMS	# OF EVENTS	ANNUAL RATE	PERSON EXAMS	# OF EVENTS	ANNUAL RATE	# OF EVENTS	CRUDE	AGE-ADJUSTED	# OF EVENTS	CRUDE	AGE-ADJUSTED
NO	373			18			15	1	1	3	1	1
BORDER	333			11			13	1	1	7	2	2
DEFINITE	123			3			16	3	3	5	4	4
TREATED	188			7			5	3	2	1	1	1
UNKNOWN	0			0			0	-		0	-	
TOTAL	1017			39			49	1	1	16	2	2

LOGISTIC REGRESSIONS

AGE AT EXAM		COEFFICIENT OF RISK FACTOR	SE COEFFICIENT	P-VALUE	STANDARDIZED COEFFICIENT	STD SE
35-44		0.620	0.445	0.164	0.431	0.309
45-54		0.697	0.279	0.013	0.536	0.215
55-64		0.658	0.262	0.012	0.536	0.214
65-74		0.371	0.342	0.279	0.304	0.280
75-84						
85-94						
35-64	UNIVARIATE	0.326	0.157	0.038	0.288	0.139
65-94	UNIVARIATE	0.270	0.285	0.343	0.242	0.255
35-64	BIVARIATE	0.316	0.158	0.046	0.280	0.140
65-94	BIVARIATE	0.271	0.285	0.342	0.243	0.255
35-64	MULTIVARIATE	0.248	0.162	0.125	0.220	0.143
65-94	MULTIVARIATE	0.246	0.288	0.393	0.221	0.258

SEX: FEMALE
EVENT: CORONARY INKUFFICIENCY
RISK FACTOR: HYPERTENSION WITH ANTIHYPERTENSIVE TREATMENT
PERSONS AT RISK: PERSONS FREE OF CORONARY HEART DISEASE

RISK FACTOR	35-44 AT EXAM			45-54 AT EXAM			55-64 AT EXAM			65-74 AT EXAM		
	PERSON EXAMS	# OF EVENTS	ANNUAL RATE	PERSON EXAMS	# OF EVENTS	ANNUAL RATE	PERSON EXAMS	# OF EVENTS	ANNUAL RATE	PERSON EXAMS	# OF EVENTS	ANNUAL RATE
NO	4250			5622	3	0	4888	6	1	1767	4	1
BORDER	1001			2579	2	0	3282	8	1	1903	4	1
DEFINITE	316			1166	0	0	1835	6	2	1035	2	1
TREATED	49			592	1	1	1508	4	1	1543	5	2
UNKNOWN	0			0	0	-	0	0	-	0	0	-
TOTAL	5616			9959	6	0	11513	24	1	6248	15	1

RISK FACTOR	75-84 AT EXAM			85-94 AT EXAM			35-64 COMBINED			65-94 COMBINED		
	PERSON EXAMS	# OF EVENTS	ANNUAL RATE	PERSON EXAMS	# OF EVENTS	ANNUAL RATE	# OF EVENTS	CRUDE	AGE-ADJUSTED	# OF EVENTS	CRUDE	AGE-ADJUSTED
NO	402	1	1	31			9	0	0	5	1	1
BORDER	575	0	0	28			10	1	0	4	1	1
DEFINITE	371	2	3	21			6	1	1	4	1	1
TREATED	603	2	2	43			5	1	1	8	2	2
UNKNOWN	0	0	-	0			0	-		0	-	
TOTAL	1951	5	1	123			30	1	0	21	1	1

LOGISTIC REGRESSIONS

AGE AT EXAM		COEFFICIENT OF RISK FACTOR	SE COEFFICIENT	P-VALUE	STANDARDIZED COEFFICIENT	STD SE
35-44						
45-54		0.090	0.518	0.862	0.069	0.398
55-64		0.428	0.246	0.082	0.357	0.206
65-74		0.105	0.318	0.741	0.086	0.262
75-84		0.590	0.695	0.396	0.465	0.547
85-94						
35-64	UNIVARIATE	0.428	0.200	0.032	0.375	0.175
65-94	UNIVARIATE	0.059	0.261	0.821	0.050	0.220
35-64	BIVARIATE	0.266	0.207	0.198	0.233	0.181
65-94	BIVARIATE	0.054	0.261	0.835	0.046	0.220
35-64	MULTIVARIATE	0.231	0.214	0.280	0.202	0.187
65-94	MULTIVARIATE	0.030	0.266	0.910	0.025	0.225

TABLE 6-3B DATE=03DEC85
AVERAGE ANNUAL INCIDENCE RATE FOR EVENT PER 1000 PERSONS AT RISK AT EXAM
BY AGE AND LEVEL OF RISK FACTOR AT EXAM: FRAMINGHAM STUDY 30-YEAR FOLLOWUP

SEX: MALE
EVENT: CORONARY INSUFFICIENCY
RISK FACTOR: HYPERTENSION IGNORING TREATMENT
PERSONS AT RISK: PERSONS FREE OF CORONARY HEART DISEASE

RISK FACTOR	35-44 AT EXAM			45-54 AT EXAM			55-64 AT EXAM			65-74 AT EXAM		
	PERSON EXAMS	# OF EVENTS	ANNUAL RATE	PERSON EXAMS	# OF EVENTS	ANNUAL RATE	PERSON EXAMS	# OF EVENTS	ANNUAL RATE	PERSON EXAMS	# OF EVENTS	ANNUAL RATE
NO	2653	3	1	4154	7	1	3909	5	1	1631	3	1
MILD	1310	3	1	2284	5	1	2583	9	2	1499	7	2
DEFINITE	509	2	2	1286	7	3	1557	8	3	783	3	2
UNKNOWN	0	0	-	0	0	-	0	0	-	0	0	-
TOTAL	4472	8	1	7724	19	1	8049	22	1	3913	13	2

RISK FACTOR	75-84 AT EXAM			85-94 AT EXAM			35-64 COMBINED			65-94 COMBINED		
	PERSON EXAMS	# OF EVENTS	ANNUAL RATE	PERSON EXAMS	# OF EVENTS	ANNUAL RATE	# OF EVENTS	CRUDE	AGE-ADJUSTED	# OF EVENTS	CRUDE	AGE-ADJUSTED
NO	412			21			15	1	1	4	1	1
MILD	418			15			17	1	1	7	2	2
DEFINITE	187			3			17	3	2	5	3	3
UNKNOWN	0			0			0	-	-	0	-	
TOTAL	1017			39			49	1	1	16	2	2

LOGISTIC REGRESSIONS

AGE AT EXAM		COEFFICIENT OF RISK FACTOR	SE COEFFICIENT	P-VALUE	STANDARDIZED COEFFICIENT	STD SE
35-44		0.630	0.447	0.158	0.435	0.309
45-54		0.588	0.281	0.036	0.443	0.211
55-64		0.672	0.266	0.012	0.518	0.205
65-74		0.368	0.358	0.304	0.278	0.271
75-84						
85-94						
35-64	UNIVARIATE	0.645	0.176	0.001	0.484	0.132
65-94	UNIVARIATE	0.483	0.326	0.138	0.362	0.244
35-64	BIVARIATE	0.627	0.177	0.001	0.471	0.133
65-94	BIVARIATE	0.482	0.325	0.138	0.362	0.244
35-64	MULTIVARIATE	0.553	0.183	0.003	0.415	0.137
65-94	MULTIVARIATE	0.445	0.336	0.185	0.334	0.252

SEX: FEMALE
EVENT: CORONARY INSUFFICIENCY
RISK FACTOR: HYPERTENSION IGNORING TREATMENT
PERSONS AT RISK: PERSONS FREE OF CORONARY HEART DISEASE

RISK FACTOR	35-44 AT EXAM			45-54 AT EXAM			55-64 AT EXAM			65-74 AT EXAM		
	PERSON EXAMS	# OF EVENTS	ANNUAL RATE	PERSON EXAMS	# OF EVENTS	ANNUAL RATE	PERSON EXAMS	# OF EVENTS	ANNUAL RATE	PERSON EXAMS	# OF EVENTS	ANNUAL RATE
NO	4272			5804	3	0	5245	7	1	2035	4	1
MILD	1016			2763	2	0	3860	10	1	2549	6	1
DEFINITE	328			1392	1	0	2408	7	1	1664	5	2
UNKNOWN	0			0	0	-	0	0	-	0	0	-
TOTAL	5616			9959	6	0	11513	24	1	6248	15	1

RISK FACTOR	75-84 AT EXAM			85-94 AT EXAM			35-64 COMBINED			65-94 COMBINED		
	PERSON EXAMS	# OF EVENTS	ANNUAL RATE	PERSON EXAMS	# OF EVENTS	ANNUAL RATE	# OF EVENTS	CRUDE	AGE-ADJUSTED	# OF EVENTS	CRUDE	AGE-ADJUSTED
NO	507	1	1	37			10	0	0	5	1	1
MILD	827	1	1	45			12	1	1	8	1	1
DEFINITE	617	3	2	41			8	1	1	8	2	2
UNKNOWN	0	0	-	0			0	-	-	0	-	
TOTAL	1951	5	1	123			30	1	1	21	1	1

LOGISTIC REGRESSIONS

AGE AT EXAM		COEFFICIENT OF RISK FACTOR	SE COEFFICIENT	P-VALUE	STANDARDIZED COEFFICIENT	STD SE
35-44						
45-54		0.196	0.532	0.712	0.143	0.386
55-64		0.390	0.254	0.124	0.303	0.197
65-74		0.214	0.337	0.525	0.164	0.259
75-84		0.644	0.650	0.321	0.488	0.492
85-94						
35-64	UNIVARIATE	0.558	0.223	0.012	0.413	0.165
65-94	UNIVARIATE	0.298	0.288	0.301	0.228	0.221
35-64	BIVARIATE	0.325	0.229	0.157	0.240	0.170
65-94	BIVARIATE	0.291	0.289	0.313	0.223	0.222
35-64	MULTIVARIATE	0.216	0.239	0.366	0.160	0.177
65-94	MULTIVARIATE	0.227	0.294	0.440	0.174	0.225

TABLE 6-4   DATE=03MAY85

AVERAGE ANNUAL INCIDENCE RATE FOR EVENT PER 1000 PERSONS AT RISK AT EXAM
BY AGE AND LEVEL OF RISK FACTOR AT EXAM: FRAMINGHAM STUDY 30-YEAR FOLLOWUP

SEX: MALE
EVENT: CORONARY INSUFFICIENCY
RISK FACTOR: SERUM CHOLESTEROL
PERSONS AT RISK: PERSONS FREE OF CORONARY HEART DISEASE

RISK FACTOR	35-44 AT EXAM			45-54 AT EXAM			55-64 AT EXAM			65-74 AT EXAM		
	PERSON EXAMS	# OF EVENTS	ANNUAL RATE	PERSON EXAMS	# OF EVENTS	ANNUAL RATE	PERSON EXAMS	# OF EVENTS	ANNUAL RATE	PERSON EXAMS	# OF EVENTS	ANNUAL RATE
84-204	1253	0	0	1767	2	1	2255	6	1	1234	3	1
205-234	1247	2	1	2306	2	0	2405	4	1	1171	5	2
235-264	872	4	2	1745	3	1	1889	7	2	873	3	2
265-294	475	2	2	1001	5	2	918	2	1	428	2	2
295-1124	291	0	0	557	5	4	454	3	3	191	0	0
UNKNOWN	334	0	0	348	0	0	128	0	0	16	0	0
TOTAL	4472	8	1	7724	19	1	8049	22	1	3913	13	2

RISK FACTOR	75-84 AT EXAM			85-94 AT EXAM			35-64 COMBINED			65-94 COMBINED		
	PERSON EXAMS	# OF EVENTS	ANNUAL RATE	PERSON EXAMS	# OF EVENTS	ANNUAL RATE	# OF EVENTS	CRUDE	AGE-ADJUSTED	# OF EVENTS	CRUDE	AGE-ADJUSTED
84-204	390			17			8	1	1	3	1	1
205-234	332			11			8	1	1	7	2	2
235-264	177			5			16	2	2	4	2	2
265-294	96			6			9	2	2	2	2	2
295-1124	22			0			8	3	3	0	0	0
UNKNOWN	0			0			0	0	0	0	0	0
TOTAL	1017			39			49	1	1	16	2	2

LOGISTIC REGRESSIONS

AGE AT EXAM		COEFFICIENT OF RISK FACTOR	SE COEFFICIENT	P-VALUE	STANDARDIZED COEFFICIENT	STD SE
35-44		0.009	0.006	0.152	0.380	0.266
45-54		0.013	0.003	0.001	0.565	0.125
55-64		0.004	0.005	0.414	0.166	0.203
65-74		0.002	0.007	0.764	0.082	0.274
75-84						
85-94						
35-64	UNIVARIATE	0.010	0.002	0.001	0.421	0.100
65-94	UNIVARIATE	0.004	0.006	0.559	0.143	0.245
35-64	BIVARIATE	0.010	0.002	0.001	0.428	0.101
65-94	BIVARIATE	0.003	0.006	0.570	0.140	0.246
35-64	MULTIVARIATE	0.009	0.002	0.001	0.389	0.102
65-94	MULTIVARIATE	0.004	0.006	0.495	0.161	0.236

SEX: FEMALE
EVENT: CORONARY INSUFFICIENCY
RISK FACTOR: SERUM CHOLESTEROL
PERSONS AT RISK: PERSONS FREE OF CORONARY HEART DISEASE

RISK FACTOR	35-44 AT EXAM			45-54 AT EXAM			55-64 AT EXAM			65-74 AT EXAM		
	PERSON EXAMS	# OF EVENTS	ANNUAL RATE	PERSON EXAMS	# OF EVENTS	ANNUAL RATE	PERSON EXAMS	# OF EVENTS	ANNUAL RATE	PERSON EXAMS	# OF EVENTS	ANNUAL RATE
84-204	2350			2094	0	0	1625	2	1	789	1	1
205-234	1417			2558	1	0	2615	4	1	1401	2	1
235-264	806			2259	1	0	2938	4	1	1703	4	1
265-294	372			1473	2	1	2240	10	2	1287	4	2
295-1124	171			1101	2	1	1866	4	1	1031	4	2
UNKNOWN	500			474	0	0	229	0	0	37	0	0
TOTAL	5616			9959	6	0	11513	24	1	6248	15	1

RISK FACTOR	75-84 AT EXAM			85-94 AT EXAM			35-64 COMBINED			65-94 COMBINED		
	PERSON EXAMS	# OF EVENTS	ANNUAL RATE	PERSON EXAMS	# OF EVENTS	ANNUAL RATE	# OF EVENTS	CRUDE	AGE-ADJUSTED	# OF EVENTS	CRUDE	AGE-ADJUSTED
84-204	361	0	0	46			2	0	0	1	0	1
205-234	486	2	2	28			5	0	0	4	1	1
235-264	542	2	2	35			5	0	0	6	1	1
265-294	350	1	1	9			12	1	1	5	2	2
295-1124	205	0	0	5			6	1	1	4	2	2
UNKNOWN	7	0	0	0			0	0	0	0	0	0
TOTAL	1951	5	1	123			30	1	0	21	1	1

LOGISTIC REGRESSIONS

AGE AT EXAM		COEFFICIENT OF RISK FACTOR	SE COEFFICIENT	P-VALUE	STANDARDIZED COEFFICIENT	STD SE
35-44						
45-54		0.008	0.002	0.001	0.378	0.105
55-64		0.006	0.004	0.179	0.253	0.188
65-74		0.009	0.005	0.117	0.375	0.239
75-84		0.000	0.010	0.977	0.013	0.447
85-94						
35-64	UNIVARIATE	0.008	0.002	0.001	0.375	0.090
65-94	UNIVARIATE	0.007	0.005	0.114	0.324	0.205
35-64	BIVARIATE	0.008	0.002	0.001	0.363	0.098
65-94	BIVARIATE	0.008	0.005	0.099	0.340	0.206
35-64	MULTIVARIATE	0.007	0.002	0.001	0.340	0.101
65-94	MULTIVARIATE	0.007	0.005	0.116	0.330	0.210

TABLE 6-5   DATE=03DEC85
AVERAGE ANNUAL INCIDENCE RATE FOR EVENT PER 1000 PERSONS AT RISK AT EXAM
BY AGE AND LEVEL OF RISK FACTOR AT EXAM: FRAMINGHAM STUDY 30-YEAR FOLLOWUP

SEX: MALE
EVENT: CORONARY INSUFFICIENCY
RISK FACTOR: HEMATOCRIT
PERSONS AT RISK: PERSONS FREE OF CORONARY HEART DISEASE

RISK FACTOR	35-44 AT EXAM PERSON EXAMS	# OF EVENTS	ANNUAL RATE	45-54 AT EXAM PERSON EXAMS	# OF EVENTS	ANNUAL RATE	55-64 AT EXAM PERSON EXAMS	# OF EVENTS	ANNUAL RATE	65-74 AT EXAM PERSON EXAMS	# OF EVENTS	ANNUAL RATE
25-42	623	1	1	942	2	1	1018	6	3	450	0	0
43-44	546	0	0	967	1	1	985	3	2	493	4	4
45-46	983	1	1	1854	4	1	1953	8	2	963	4	2
47-48	775	2	1	1510	5	2	1599	0	0	861	2	1
49-70	1175	4	2	2066	7	2	2338	5	1	1127	3	1
UNKNOWN	370	0	0	385	0	0	156	0	0	19	0	0
TOTAL	4472	8	1	7724	19	1	8049	22	1	3913	13	2

RISK FACTOR	75-84 AT EXAM PERSON EXAMS	# OF EVENTS	ANNUAL RATE	85-94 AT EXAM PERSON EXAMS	# OF EVENTS	ANNUAL RATE	35-64 COMBINED # OF EVENTS	CRUDE	AGE-ADJUSTED	65-94 COMBINED # OF EVENTS	CRUDE	AGE-ADJUSTED
25-42	214			16			9	2	2	1	1	0
43-44	119			5			4	1	1	4	3	3
45-46	240			7			13	1	1	5	2	2
47-48	197			7			7	1	1	2	1	1
49-70	247			4			16	1	1	4	1	1
UNKNOWN	0			0			0	0	0	0	0	0
TOTAL	1017			39			49	1	1	16	2	2

LOGISTIC REGRESSIONS

AGE AT EXAM		COEFFICIENT OF RISK FACTOR	SE COEFFICIENT	P-VALUE	STANDARDIZED COEFFICIENT	STD SE
35-44		0.124	0.097	0.201	0.454	0.355
45-54		0.069	0.064	0.281	0.247	0.229
55-64		-0.100	0.051	0.048	-0.380	0.193
65-74		-0.018	0.075	0.810	-0.066	0.275
75-84						
85-94						
35-64	UNIVARIATE	-0.007	0.039	0.864	-0.024	0.143
65-94	UNIVARIATE	0.014	0.066	0.827	0.055	0.252
35-64	BIVARIATE	-0.007	0.039	0.855	-0.026	0.142
65-94	BIVARIATE	0.013	0.067	0.846	0.049	0.255
35-64	MULTIVARIATE	-0.035	0.039	0.367	-0.129	0.143
65-94	MULTIVARIATE	-0.006	0.068	0.931	-0.022	0.259

SEX: FEMALE
EVENT: CORONARY INSUFFICIENCY
RISK FACTOR: HEMATOCRIT
PERSONS AT RISK: PERSONS FREE OF CORONARY HEART DISEASE

RISK FACTOR	35-44 AT EXAM PERSON EXAMS	# OF EVENTS	ANNUAL RATE	45-54 AT EXAM PERSON EXAMS	# OF EVENTS	ANNUAL RATE	55-64 AT EXAM PERSON EXAMS	# OF EVENTS	ANNUAL RATE	65-74 AT EXAM PERSON EXAMS	# OF EVENTS	ANNUAL RATE
21-39	1697			1843	1	0	1553	2	1	646	2	2
40-41	967			1741	1	0	2061	6	1	1069	2	1
42-43	1043			1955	2	1	2260	7	2	1256	0	0
44-45	856			2194	0	0	2847	4	1	1609	6	2
46-65	487			1681	2	1	2521	5	1	1626	5	2
UNKNOWN	566			545	0	0	271	0	0	42	0	0
TOTAL	5616			9959	6	0	11513	24	1	6248	15	1

RISK FACTOR	75-84 AT EXAM PERSON EXAMS	# OF EVENTS	ANNUAL RATE	85-94 AT EXAM PERSON EXAMS	# OF EVENTS	ANNUAL RATE	35-64 COMBINED # OF EVENTS	CRUDE	AGE-ADJUSTED	65-94 COMBINED # OF EVENTS	CRUDE	AGE-ADJUSTED
21-39	297	1	2	24			3	1	0	3	2	2
40-41	373	1	1	31			7	1	1	3	1	1
42-43	353	1	1	19			9	1	1	1	0	0
44-45	486	0	0	26			4	0	0	7	2	2
46-65	436	2	2	23			7	1	1	7	2	2
UNKNOWN	6	0	0	0			0	0	0	0	0	0
TOTAL	1951	5	1	123			30	1	1	21	2	1

LOGISTIC REGRESSIONS

AGE AT EXAM		COEFFICIENT OF RISK FACTOR	SE COEFFICIENT	P-VALUE	STANDARDIZED COEFFICIENT	STD SE
35-44						
45-54		0.035	0.116	0.765	0.124	0.415
55-64		-0.004	0.059	0.951	-0.013	0.204
65-74		0.043	0.075	0.563	0.151	0.262
75-84		-0.023	0.113	0.839	-0.090	0.440
85-94						
35-64	UNIVARIATE	0.040	0.052	0.437	0.144	0.185
65-94	UNIVARIATE	0.025	0.061	0.677	0.092	0.221
35-64	BIVARIATE	-0.001	0.052	0.991	-0.002	0.188
65-94	BIVARIATE	0.027	0.061	0.657	0.098	0.220
35-64	MULTIVARIATE	-0.033	0.051	0.516	-0.120	0.185
65-94	MULTIVARIATE	0.020	0.062	0.754	0.070	0.224

127

TABLE 6-6   DATE=03DEC85
AVERAGE ANNUAL INCIDENCE RATE FOR EVENT PER 1000 PERSONS AT RISK AT EXAM
BY AGE AND LEVEL OF RISK FACTOR AT EXAM: FRAMINGHAM STUDY 30-YEAR FOLLOWUP

SEX: MALE
EVENT: CORONARY INSUFFICIENCY
RISK FACTOR: BLOOD GLUCOSE (MG/DL)
PERSONS AT RISK: PERSONS FREE OF CORONARY HEART DISEASE

RISK FACTOR	35-44 AT EXAM			45-54 AT EXAM			55-64 AT EXAM			65-74 AT EXAM		
	PERSON EXAMS	# OF EVENTS	ANNUAL RATE	PERSON EXAMS	# OF EVENTS	ANNUAL RATE	PERSON EXAMS	# OF EVENTS	ANNUAL RATE	PERSON EXAMS	# OF EVENTS	ANNUAL RATE
29-69	992	2	1	1490	4	1	1087	1	0	421	0	0
70-89	2501	2	0	4172	7	1	4056	16	2	1761	5	1
90-109	626	3	2	1348	2	1	1866	3	1	1046	5	2
110-129	105	1	5	305	4	7	537	1	1	335	2	3
130-524	48	0	0	193	2	5	408	1	1	334	1	1
UNKNOWN	200	0	0	216	0	0	95	0	0	16	0	0
TOTAL	4472	8	1	7724	19	1	8049	22	1	3913	13	2

RISK FACTOR	75-84 AT EXAM			85-94 AT EXAM			35-64 COMBINED			65-94 COMBINED		
	PERSON EXAMS	# OF EVENTS	ANNUAL RATE	PERSON EXAMS	# OF EVENTS	ANNUAL RATE	# OF EVENTS	CRUDE	AGE-ADJUSTED	# OF EVENTS	CRUDE	AGE-ADJUSTED
29-69	62			1			7	1	1	0	0	0
70-89	377			11			25	1	1	5	1	1
90-109	328			14			8	1	1	7	2	3
110-129	125			4			6	3	4	2	2	2
130-524	125			9			3	2	2	2	2	2
UNKNOWN	0			0			0	0	0	0	0	0
TOTAL	1017			39			49	1	1	16	-2	2

LOGISTIC REGRESSIONS

AGE AT EXAM	COEFFICIENT OF RISK FACTOR	SE COEFFICIENT	P-VALUE	STANDARDIZED COEFFICIENT	STD SE
35-44	0.010	0.008	0.211	0.165	0.132
45-54	0.010	0.004	0.019	0.225	0.096
55-64	0.000	0.008	0.965	0.009	0.209
65-74	0.006	0.005	0.294	0.187	0.178
75-84					
85-94					
35-64 UNIVARIATE	0.007	0.003	0.046	0.166	0.083
65-94 UNIVARIATE	0.007	0.004	0.078	0.249	0.141
35-64 BIVARIATE	0.006	0.004	0.075	0.155	0.087
65-94 BIVARIATE	0.007	0.004	0.074	0.250	0.140
35-64 MULTIVARIATE	0.005	0.004	0.122	0.133	0.086
65-94 MULTIVARIATE	0.007	0.004	0.099	0.236	0.143

SEX: FEMALE
EVENT: CORONARY INSUFFICIENCY
RISK FACTOR: BLOOD GLUCOSE (MG/DL)
PERSONS AT RISK: PERSONS FREE OF CORONARY HEART DISEASE

RISK FACTOR	35-44 AT EXAM			45-54 AT EXAM			55-64 AT EXAM			65-74 AT EXAM		
	PERSON EXAMS	# OF EVENTS	ANNUAL RATE	PERSON EXAMS	# OF EVENTS	ANNUAL RATE	PERSON EXAMS	# OF EVENTS	ANNUAL RATE	PERSON EXAMS	# OF EVENTS	ANNUAL RATE
29-69	1066			1797	1	0	1369	3	1	521	1	1
70-89	3412			5744	3	0	6098	13	1	3102	11	2
90-109	672			1625	1	0	2712	2	0	1732	3	1
110-129	96			501	0	0	730	4	3	501	0	0
130-524	39			159	1	3	438	2	2	360	0	0
UNKNOWN	331			333	0	0	166	0	0	32	0	0
TOTAL	5616			9959	6	0	11513	24	1	6248	15	1

RISK FACTOR	75-84 AT EXAM			85-94 AT EXAM			35-64 COMBINED			65-94 COMBINED		
	PERSON EXAMS	# OF EVENTS	ANNUAL RATE	PERSON EXAMS	# OF EVENTS	ANNUAL RATE	# OF EVENTS	CRUDE	AGE-ADJUSTED	# OF EVENTS	CRUDE	AGE-ADJUSTED
29-69	105	2	10	3			4	0	1	3	2	3
70-89	823	1	1	44			16	1	1	12	2	1
90-109	640	0	0	51			3	0	0	4	1	1
110-129	219	1	2	15			4	2	2	1	1	1
130-524	163	1	3	12			3	2	2	1	1	1
UNKNOWN	1	0	0	0			0	0	0	0	0	0
TOTAL	1951	5	1	123			30	1	1	21	1	1

LOGISTIC REGRESSIONS

AGE AT EXAM	COEFFICIENT OF RISK FACTOR	SE COEFFICIENT	P-VALUE	STANDARDIZED COEFFICIENT	STD SE
35-44					
45-54	0.010	0.011	0.353	0.196	0.212
55-64	0.011	0.004	0.003	0.265	0.098
65-74	-0.028	0.018	0.120	-0.786	0.506
75-84	0.013	0.004	0.002	0.439	0.142
85-94					
35-64 UNIVARIATE	0.013	0.003	0.001	0.274	0.069
65-94 UNIVARIATE	0.005	0.005	0.310	0.153	0.151
35-64 BIVARIATE	0.010	0.003	0.003	0.221	0.075
65-94 BIVARIATE	0.005	0.005	0.330	0.149	0.153
35-64 MULTIVARIATE	0.008	0.004	0.022	0.177	0.077
65-94 MULTIVARIATE	0.005	0.005	0.346	0.145	0.153

TABLE 6-7   DATE=03DEC85

AVERAGE ANNUAL INCIDENCE RATE FOR EVENT PER 1000 PERSONS AT RISK AT EXAM
BY AGE AND LEVEL OF RISK FACTOR AT EXAM: FRAMINGHAM STUDY 30-YEAR FOLLOWUP

SEX: MALE
EVENT: CORONARY INSUFFICIENCY
RISK FACTOR: DIABETES MELLITUS
PERSONS AT RISK: PERSONS FREE OF CORONARY HEART DISEASE

RISK FACTOR	35-44 AT EXAM			45-54 AT EXAM			55-64 AT EXAM			65-74 AT EXAM		
	PERSON EXAMS	# OF EVENTS	ANNUAL RATE	PERSON EXAMS	# OF EVENTS	ANNUAL RATE	PERSON EXAMS	# OF EVENTS	ANNUAL RATE	PERSON EXAMS	# OF EVENTS	ANNUAL RATE
NO	4424	8	1	7486	18	1	7549	20	1	3504	9	1
YES	48	0	0	238	1	2	500	2	2	409	4	5
UNKNOWN	0	0	-	0	0	-	0	0	-	0	0	-
TOTAL	4472	8	1	7724	19	1	8049	22	1	3913	13	2

RISK FACTOR	75-84 AT EXAM			85-94 AT EXAM			35-64 COMBINED			65-94 COMBINED		
	PERSON EXAMS	# OF EVENTS	ANNUAL RATE	PERSON EXAMS	# OF EVENTS	ANNUAL RATE	# OF EVENTS	CRUDE	AGE-ADJUSTED	# OF EVENTS	CRUDE	AGE-ADJUSTED
NO	900			34			46	1	1	10	1	1
YES	117			5			3	2	2	6	6	6
UNKNOWN	0			0			0	-	-	0	-	
TOTAL	1017			39			49	1	1	16	2	2

LOGISTIC REGRESSIONS

AGE AT EXAM		COEFFICIENT OF RISK FACTOR	SE COEFFICIENT	P-VALUE	STANDARDIZED COEFFICIENT	STD SE
35-44		-	-	-	-	-
45-54		0.560	1.030	0.587	0.097	0.178
55-64		0.413	0.743	0.578	0.100	0.179
65-74		1.622	0.551	0.003	0.496	0.169
75-84						
85-94						
35-64	UNIVARIATE	0.481	0.597	0.421	0.093	0.115
65-94	UNIVARIATE	1.621	0.518	0.002	0.501	0.160
35-64	BIVARIATE	0.383	0.602	0.525	0.074	0.116
65-94	BIVARIATE	1.629	0.519	0.002	0.503	0.160
35-64	MULTIVARIATE	0.298	0.611	0.626	0.057	0.118
65-94	MULTIVARIATE	1.649	0.532	0.002	0.509	0.164

SEX: FEMALE
EVENT: CORONARY INSUFFICIENCY
RISK FACTOR: DIABETES MELLITUS
PERSONS AT RISK: PERSONS FREE OF CORONARY HEART DISEASE

RISK FACTOR	35-44 AT EXAM			45-54 AT EXAM			55-64 AT EXAM			65-74 AT EXAM		
	PERSON EXAMS	# OF EVENTS	ANNUAL RATE	PERSON EXAMS	# OF EVENTS	ANNUAL RATE	PERSON EXAMS	# OF EVENTS	ANNUAL RATE	PERSON EXAMS	# OF EVENTS	ANNUAL RATE
NO	5573			9760	5	0	10985	22	1	5780	15	1
YES	43			199	1	3	528	2	2	468	0	0
UNKNOWN	0			0	0	-	0	0	-	0	0	-
TOTAL	5616			9959	6	0	11513	24	1	6248	15	1

RISK FACTOR	75-84 AT EXAM			85-94 AT EXAM			35-64 COMBINED			65-94 COMBINED		
	PERSON EXAMS	# OF EVENTS	ANNUAL RATE	PERSON EXAMS	# OF EVENTS	ANNUAL RATE	# OF EVENTS	CRUDE	AGE-ADJUSTED	# OF EVENTS	CRUDE	AGE-ADJUSTED
NO	1738	3	1	106			27	1	1	19	1	1
YES	213	2	5	17			3	2	2	2	1	1
UNKNOWN	0	0	-	0			0	-	-	0	-	
TOTAL	1951	5	1	123			30	1	1	21	1	1

LOGISTIC REGRESSIONS

AGE AT EXAM		COEFFICIENT OF RISK FACTOR	SE COEFFICIENT	P-VALUE	STANDARDIZED COEFFICIENT	STD SE
35-44						
45-54		2.288	1.098	0.037	0.320	0.154
55-64		0.639	0.740	0.388	0.134	0.155
65-74		-	-		-	-
75-84		1.701	0.916	0.063	0.531	0.286
85-94						
35-64	UNIVARIATE	1.337	0.610	0.028	0.222	0.101
65-94	UNIVARIATE	0.140	0.744	0.851	0.039	0.206
35-64	BIVARIATE	0.913	0.613	0.136	0.152	0.102
65-94	BIVARIATE	0.120	0.747	0.873	0.033	0.207
35-64	MULTIVARIATE	0.563	0.640	0.379	0.094	0.106
65-94	MULTIVARIATE	0.041	0.756	0.957	0.011	0.210

TABLE 6-8   DATE=03DEC85
AVERAGE ANNUAL INCIDENCE RATE FOR EVENT PER 1000 PERSONS AT RISK AT EXAM
BY AGE AND LEVEL OF RISK FACTOR AT EXAM: FRAMINGHAM STUDY 30-YEAR FOLLOWUP

SEX: MALE
EVENT: CORONARY INSUFFICIENCY
RISK FACTOR: GLUCOSE IN URINE, DEFINITE OR TRACE
PERSONS AT RISK: PERSONS FREE OF CORONARY HEART DISEASE

35-44 AT EXAM			45-54 AT EXAM			55-64 AT EXAM			65-74 AT EXAM		
PERSON EXAMS	# OF EVENTS	ANNUAL RATE	PERSON EXAMS	# OF EVENTS	ANNUAL RATE	PERSON EXAMS	# OF EVENTS	ANNUAL RATE	PERSON EXAMS	# OF EVENTS	ANNUAL RATE
7580	8	1	7535	18	1	7815	20	1	3767	12	2
60	0	0	168	1	3	210	2	5	140	1	4
32	0	0	21	0	0	24	0	0	6	0	0
472	8	1	7724	19	1	8049	22	1	3913	13	2

75-84 AT EXAM			85-94 AT EXAM			35-64 COMBINED			65-94 COMBINED		
PERSON EXAMS	# OF EVENTS	ANNUAL RATE	PERSON EXAMS	# OF EVENTS	ANNUAL RATE	# OF EVENTS	CRUDE	AGE-ADJUSTED	# OF EVENTS	CRUDE	AGE-ADJUSTED
486			38			46	1	1	14	1	1
28			1			3	3	3	2	6	6
3			0			0	0	0	0	0	0
117			39			49	1	1	16	2	2

LOGISTIC REGRESSIONS

	COEFFICIENT OF RISK FACTOR	SE COEFFICIENT	P-VALUE	STANDARDIZED COEFFICIENT	STD SE
	0.917	1.030	0.374	0.134	0.151
	1.321	0.745	0.076	0.211	0.119
	0.811	1.044	0.437	0.131	0.194
TE	1.082	0.598	0.070	0.158	0.087
TE	1.408	0.760	0.064	0.255	0.138
E	1.039	0.599	0.083	0.151	0.087
E	1.404	0.760	0.065	0.255	0.138
IATE	0.937	0.604	0.121	0.137	0.088
IATE	1.296	0.771	0.093	0.235	0.140

SEX: FEMALE
EVENT: CORONARY INSUFFICIENCY
RISK FACTOR: GLUCOSE IN URINE, DEFINITE OR TRACE
PERSONS AT RISK: PERSONS FREE OF CORONARY HEART DISEASE

35-44 AT EXAM			45-54 AT EXAM			55-64 AT EXAM			65-74 AT EXAM		
PERSON EXAMS	# OF EVENTS	ANNUAL RATE	PERSON EXAMS	# OF EVENTS	ANNUAL RATE	PERSON EXAMS	# OF EVENTS	ANNUAL RATE	PERSON EXAMS	# OF EVENTS	ANNUAL RATE
9837			9841	6	0	11346	23	1	6131	15	1
41			94	0	0	144	1	3	107	0	0
38			24	0	0	23	0	0	10	0	0
16			9959	6	0	11513	24	1	6248	15	1

75-84 AT EXAM			85-94 AT EXAM			35-64 COMBINED			65-94 COMBINED		
PERSON EXAMS	# OF EVENTS	ANNUAL RATE	PERSON EXAMS	# OF EVENTS	ANNUAL RATE	# OF EVENTS	CRUDE	AGE-ADJUSTED	# OF EVENTS	CRUDE	AGE-ADJUSTED
19	4	1	120			29	1	1	20	1	1
23	1	22	3			1	2	1	1	4	5
9	0	0	0			0	0	0	0	0	0
51	5	1	123			30	1	1	21	1	1

LOGISTIC REGRESSIONS

	COEFFICIENT OF RISK FACTOR	SE COEFFICIENT	P-VALUE	STANDARDIZED COEFFICIENT	STD SE
	1.236	1.025	0.228	0.138	0.114
	3.080	1.138	0.007	0.333	0.123
TE	1.197	1.019	0.240	0.121	0.103
TE	1.127	1.028	0.273	0.142	0.129
E	1.051	1.020	0.303	0.106	0.103
E	1.135	1.029	0.270	0.142	0.129
IATE	0.637	1.038	0.540	0.064	0.105
IATE	1.027	1.044	0.327	0.129	0.132

AVERAGE ANNUAL INCIDENCE RATE FOR EVENT PER 1000 PERSONS AT RISK AT EXAM
BY AGE AND LEVEL OF RISK FACTOR AT EXAM: FRAMINGHAM STUDY 30-YEAR FOLLOWUP

SEX: MALE
EVENT: CORONARY INSUFFICIENCY
RISK FACTOR: GLUCOSE INTOLERANCE
PERSONS AT RISK: PERSONS FREE OF CORONARY HEART DISEASE

RISK FACTOR	35-44 AT EXAM			45-54 AT EXAM			55-64 AT EXAM			65-74 AT EXAM		
	PERSON EXAMS	# OF EVENTS	ANNUAL RATE	PERSON EXAMS	# OF EVENTS	ANNUAL RATE	PERSON EXAMS	# OF EVENTS	ANNUAL RATE	PERSON EXAMS	# OF EVENTS	ANNUAL RATE
NO	4335	7	1	7221	16	1	7145	19	1	3228	9	1
YES	137	1	4	503	3	3	904	3	2	685	4	3
UNKNOWN	0	0	-	0	0	-	0	0	-	0	0	-
TOTAL	4472	8	1	7724	19	1	8049	22	1	3913	13	2

RISK FACTOR	75-84 AT EXAM			85-94 AT EXAM			35-64 COMBINED			65-94 COMBINED		
	PERSON EXAMS	# OF EVENTS	ANNUAL RATE	PERSON EXAMS	# OF EVENTS	ANNUAL RATE	# OF EVENTS	CRUDE	AGE-ADJUSTED	# OF EVENTS	CRUDE	AGE-ADJUSTED
NO	802			27			42	1	1	10	1	1
YES	215			12			7	2	3	6	3	3
UNKNOWN	0			0			0	-	-	0	-	-
TOTAL	1017			39			49	1	1	16	2	2

LOGISTIC REGRESSIONS

AGE AT EXAM		COEFFICIENT OF RISK FACTOR	SE COEFFICIENT	P-VALUE	STANDARDIZED COEFFICIENT	STD SE
35-44		1.514	1.073	0.158	0.261	0.185
45-54		0.994	0.631	0.115	0.245	0.156
55-64		0.222	0.622	0.721	0.070	0.196
65-74		0.742	0.602	0.218	0.282	0.229
75-84						
85-94						
35-64	UNIVARIATE	0.705	0.409	0.085	0.187	0.109
65-94	UNIVARIATE	0.986	0.518	0.057	0.382	0.200
35-64	BIVARIATE	0.632	0.415	0.128	0.168	0.110
65-94	BIVARIATE	1.000	0.520	0.054	0.387	0.201
35-64	MULTIVARIATE	0.568	0.418	0.174	0.151	0.111
65-94	MULTIVARIATE	0.975	0.523	0.062	0.378	0.202

SEX: FEMALE
EVENT: CORONARY INSUFFICIENCY
RISK FACTOR: GLUCOSE INTOLERANCE
PERSONS AT RISK: PERSONS FREE OF CORONARY HEART DISEASE

RISK FACTOR	35-44 AT EXAM			45-54 AT EXAM			55-64 AT EXAM			65-74 AT EXAM		
	PERSON EXAMS	# OF EVENTS	ANNUAL RATE	PERSON EXAMS	# OF EVENTS	ANNUAL RATE	PERSON EXAMS	# OF EVENTS	ANNUAL RATE	PERSON EXAMS	# OF EVENTS	ANNUAL RATE
NO	5522			9588	5	0	10503	21	1	5398	15	1
YES	94			371	1	1	1010	3	1	850	0	0
UNKNOWN	0			0	0	-	0	0	-	0	0	-
TOTAL	5616			9959	6	0	11513	24	1	6248	15	1

RISK FACTOR	75-84 AT EXAM			85-94 AT EXAM			35-64 COMBINED			65-94 COMBINED		
	PERSON EXAMS	# OF EVENTS	ANNUAL RATE	PERSON EXAMS	# OF EVENTS	ANNUAL RATE	# OF EVENTS	CRUDE	AGE-ADJUSTED	# OF EVENTS	CRUDE	AGE-ADJUSTED
NO	1581	3	1	100			26	1	1	19	1	1
YES	370	2	3	23			4	1	1	2	1	1
UNKNOWN	0	0	-	0			0	-	-	0	-	-
TOTAL	1951	5	1	123			30	1	1	21	1	1

LOGISTIC REGRESSIONS

AGE AT EXAM		COEFFICIENT OF RISK FACTOR	SE COEFFICIENT	P-VALUE	STANDARDIZED COEFFICIENT	STD SE
35-44						
45-54		1.645	1.097	0.134	0.312	0.208
55-64		0.397	0.618	0.521	0.112	0.175
65-74		-			-	
75-84		1.050	0.915	0.251	0.412	0.359
85-94						
35-64	UNIVARIATE	0.984	0.538	0.067	0.223	0.122
65-94	UNIVARIATE	-0.513	0.744	0.491	-0.183	0.265
35-64	BIVARIATE	0.556	0.541	0.305	0.126	0.123
65-94	BIVARIATE	-0.535	0.746	0.473	-0.191	0.266
35-64	MULTIVARIATE	0.357	0.555	0.520	0.081	0.126
65-94	MULTIVARIATE	-0.568	0.750	0.449	-0.203	0.267

TABLE 6-10   DATE=03DEC85
AVERAGE ANNUAL INCIDENCE RATE FOR EVENT PER 1000 PERSONS AT RISK AT EXAM
BY AGE AND LEVEL OF RISK FACTOR AT EXAM: FRAMINGHAM STUDY 30-YEAR FOLLOWUP

SEX: MALE
EVENT: CORONARY INSUFFICIENCY
RISK FACTOR: METROPOLITAN RELATIVE WEIGHT (PERCENT)
PERSONS AT RISK: PERSONS FREE OF CORONARY HEART DISEASE

RISK FACTOR	35-44 AT EXAM			45-54 AT EXAM			55-64 AT EXAM			65-74 AT EXAM		
	PERSON EXAMS	# OF EVENTS	ANNUAL RATE	PERSON EXAMS	# OF EVENTS	ANNUAL RATE	PERSON EXAMS	# OF EVENTS	ANNUAL RATE	PERSON EXAMS	# OF EVENTS	ANNUAL RATE
54-104	723	0	0	1131	0	0	1155	3	1	686	2	1
105-114	965	1	1	1540	5	2	1655	5	2	851	2	1
115-124	1122	0	0	2037	5	1	2175	6	1	1069	2	1
125-134	915	1	1	1645	6	2	1674	3	1	766	4	3
135-272	747	6	4	1370	3	1	1389	5	2	541	3	3
UNKNOWN	0	0	-	1	0	0	1	0	0	0	0	-
TOTAL	4472	8	1	7724	19	1	8049	22	1	3913	13	2

RISK FACTOR	75-84 AT EXAM			85-94 AT EXAM			35-64 COMBINED			65-94 COMBINED		
	PERSON EXAMS	# OF EVENTS	ANNUAL RATE	PERSON EXAMS	# OF EVENTS	ANNUAL RATE	# OF EVENTS	CRUDE	AGE-ADJUSTED	# OF EVENTS	CRUDE	AGE-ADJUSTED
54-104	224			18			3	0	1	2	1	1
105-114	252			9			11	1	1	4	2	2
115-124	296			12			11	1	1	2	1	1
125-134	155			0			10	1	1	2	2	2
135-272	90			0			14	2	2	4	3	3
UNKNOWN	0			0			0	0	0	0	-	
TOTAL	1017			39			49	1	1	16	2	2

LOGISTIC REGRESSIONS

AGE AT EXAM		COEFFICIENT OF RISK FACTOR	SE COEFFICIENT	P-VALUE	STANDARDIZED COEFFICIENT	STD SE
35-44		0.055	0.018	0.002	0.887	0.286
45-54		0.015	0.013	0.267	0.239	0.216
55-64		-0.002	0.013	0.895	-0.028	0.215
65-74		0.025	0.016	0.115	0.396	0.251
75-84						
85-94						
35-64	UNIVARIATE	0.016	0.008	0.058	0.255	0.134
65-94	UNIVARIATE	0.024	0.015	0.098	0.375	0.226
35-64	BIVARIATE	0.016	0.008	0.059	0.254	0.134
65-94	BIVARIATE	0.024	0.015	0.100	0.375	0.228
35-64	MULTIVARIATE	0.011	0.009	0.200	0.179	0.139
65-94	MULTIVARIATE	0.019	0.015	0.195	0.301	0.232

SEX: FEMALE
EVENT: CORONARY INSUFFICIENCY
RISK FACTOR: METROPOLITAN RELATIVE WEIGHT (PERCENT)
PERSONS AT RISK: PERSONS FREE OF CORONARY HEART DISEASE

RISK FACTOR	35-44 AT EXAM			45-54 AT EXAM			55-64 AT EXAM			65-74 AT EXAM		
	PERSON EXAMS	# OF EVENTS	ANNUAL RATE	PERSON EXAMS	# OF EVENTS	ANNUAL RATE	PERSON EXAMS	# OF EVENTS	ANNUAL RATE	PERSON EXAMS	# OF EVENTS	ANNUAL RATE
54-104	1694			2095	0	0	2045	1	0	1216	2	1
105-114	1430			2331	3	1	2409	7	1	1145	2	1
115-124	1058			2107	0	0	2479	3	1	1395	5	2
125-134	566			1372	0	0	1743	2	1	982	2	1
135-272	860			2054	3	1	2837	11	2	1510	4	1
UNKNOWN	8			0	0	-	0	0	-	0	0	-
TOTAL	5616			9959	6	0	11513	24	1	6248	15	1

RISK FACTOR	75-84 AT EXAM			85-94 AT EXAM			35-64 COMBINED			65-94 COMBINED		
	PERSON EXAMS	# OF EVENTS	ANNUAL RATE	PERSON EXAMS	# OF EVENTS	ANNUAL RATE	# OF EVENTS	CRUDE	AGE-ADJUSTED	# OF EVENTS	CRUDE	AGE-ADJUSTED
54-104	517	0	0	41			1	0	0	2	1	1
105-114	375	1	1	28			10	1	1	3	1	1
115-124	399	0	0	21			3	0	0	5	1	1
125-134	266	2	4	16			2	0	0	4	2	2
135-272	394	2	3	17			14	1	1	7	2	2
UNKNOWN	0	0	-	0			0	0	0	0	-	
TOTAL	1951	5	1	123			30	1	1	21	1	1

LOGISTIC REGRESSIONS

AGE AT EXAM		COEFFICIENT OF RISK FACTOR	SE COEFFICIENT	P-VALUE	STANDARDIZED COEFFICIENT	STD SE
35-44						
45-54		0.029	0.011	0.006	0.623	0.226
55-64		0.013	0.007	0.070	0.300	0.165
65-74		0.006	0.011	0.550	0.142	0.238
75-84		0.027	0.017	0.105	0.570	0.351
85-94						
35-64	UNIVARIATE	0.020	0.006	0.001	0.435	0.128
65-94	UNIVARIATE	0.013	0.008	0.116	0.289	0.184
35-64	BIVARIATE	0.018	0.006	0.004	0.385	0.135
65-94	BIVARIATE	0.014	0.008	0.105	0.300	0.185
35-64	MULTIVARIATE	0.017	0.007	0.015	0.361	0.149
65-94	MULTIVARIATE	0.013	0.009	0.149	0.281	0.195

TABLE 6-11   DATE=03DEC85

AVERAGE ANNUAL INCIDENCE RATE FOR EVENT PER 1000 PERSONS AT RISK AT EXAM
BY AGE AND LEVEL OF RISK FACTOR AT EXAM: FRAMINGHAM STUDY 30-YEAR FOLLOWUP

SEX: MALE
EVENT: CORONARY INSUFFICIENCY
RISK FACTOR: VITAL CAPACITY - HEIGHT INDEX (ML/INCH)
PERSONS AT RISK: PERSONS FREE OF CORONARY HEART DISEASE

RISK FACTOR	35-44 AT EXAM			45-54 AT EXAM			55-64 AT EXAM			65-74 AT EXAM		
	PERSON EXAMS	# OF EVENTS	ANNUAL RATE	PERSON EXAMS	# OF EVENTS	ANNUAL RATE	PERSON EXAMS	# OF EVENTS	ANNUAL RATE	PERSON EXAMS	# OF EVENTS	ANNUAL RATE
8-44	269	0	0	1062	2	1	2045	3	1	1752	7	2
45-49	411	0	0	1125	5	2	1480	7	2	749	3	2
50-54	747	0	0	1658	8	2	1835	6	2	712	2	1
55-59	1022	4	2	1739	2	1	1416	2	1	442	0	0
60-85	2010	4	1	2135	2	0	1255	4	2	246	1	2
UNKNOWN	13	0	0	5	0	0	18	0	0	12	0	0
TOTAL	4472	8	1	7724	19	1	8049	22	1	3913	13	2

RISK FACTOR	75-84 AT EXAM			85-94 AT EXAM			35-64 COMBINED			65-94 COMBINED		
	PERSON EXAMS	# OF EVENTS	ANNUAL RATE	PERSON EXAMS	# OF EVENTS	ANNUAL RATE	# OF EVENTS	CRUDE	AGE-ADJUSTED	# OF EVENTS	CRUDE	AGE-ADJUSTED
8-44	649			24			5	1	1	10	2	2
45-49	172			6			12	2	2	3	1	1
50-54	97			7			14	2	2	2	1	1
55-59	65			2			8	1	1	0	0	0
60-85	26			0			10	1	1	1	2	2
UNKNOWN	8			0			0	0	0	0	0	0
TOTAL	1017			39			49	1	1	16	2	2

LOGISTIC REGRESSIONS

AGE AT EXAM		COEFFICIENT OF RISK FACTOR	SE COEFFICIENT	P-VALUE	STANDARDIZED COEFFICIENT	STD SE
35-44		0.021	0.041	0.604	0.187	0.362
45-54		-0.046	0.024	0.052	-0.420	0.216
55-64		0.014	0.023	0.540	0.133	0.216
65-74		-0.037	0.028	0.189	-0.352	0.268
75-84						
85-94						
35-64	UNIVARIATE	-0.013	0.015	0.391	-0.121	0.141
65-94	UNIVARIATE	-0.057	0.025	0.023	-0.549	0.241
35-64	BIVARIATE	-0.007	0.016	0.641	-0.070	0.150
65-94	BIVARIATE	-0.061	0.025	0.017	-0.582	0.244
35-64	MULTIVARIATE	0.003	0.016	0.832	0.033	0.154
65-94	MULTIVARIATE	-0.060	0.027	0.025	-0.577	0.257

SEX: FEMALE
EVENT: CORONARY INSUFFICIENCY
RISK FACTOR: VITAL CAPACITY - HEIGHT INDEX (ML/INCH)
PERSONS AT RISK: PERSONS FREE OF CORONARY HEART DISEASE

RISK FACTOR	35-44 AT EXAM			45-54 AT EXAM			55-64 AT EXAM			65-74 AT EXAM		
	PERSON EXAMS	# OF EVENTS	ANNUAL RATE	PERSON EXAMS	# OF EVENTS	ANNUAL RATE	PERSON EXAMS	# OF EVENTS	ANNUAL RATE	PERSON EXAMS	# OF EVENTS	ANNUAL RATE
4-29	122			534	0	0	1461	4	1	1747	4	1
30-34	289			1217	2	1	2418	3	1	1715	5	1
35-39	769			2077	2	0	3048	10	2	1529	3	1
40-44	1427			2804	1	0	2706	6	1	854	3	2
45-92	2988			3310	1	0	1859	1	0	397	1	1
UNKNOWN	21			17	0	0	21	0	0	6	0	0
TOTAL	5616			9959	6	0	11513	24	1	6248	15	1

RISK FACTOR	75-84 AT EXAM			85-94 AT EXAM			35-64 COMBINED			65-94 COMBINED		
	PERSON EXAMS	# OF EVENTS	ANNUAL RATE	PERSON EXAMS	# OF EVENTS	ANNUAL RATE	# OF EVENTS	CRUDE	AGE-ADJUSTED	# OF EVENTS	CRUDE	AGE-ADJUSTED
4-29	930	3	2	80			4	1	1	8	1	1
30-34	569	2	2	29			5	1	1	6	1	1
35-39	314	0	0	13			12	1	1	3	1	1
40-44	111	0	0	1			7	1	1	3	2	1
45-92	25	0	0	0			2	0	0	1	1	1
UNKNOWN	2	0	0	0			0	0	0	0	0	0
TOTAL	1951	5	1	123			30	1	1	21	1	1

LOGISTIC REGRESSIONS

AGE AT EXAM		COEFFICIENT OF RISK FACTOR	SE COEFFICIENT	P-VALUE	STANDARDIZED COEFFICIENT	STD SE
35-44						
45-54		-0.055	0.053	0.295	-0.408	0.390
55-64		-0.021	0.027	0.440	-0.156	0.201
65-74		0.034	0.035	0.340	0.251	0.262
75-84		-0.062	0.059	0.293	-0.440	0.418
85-94						
35-64	UNIVARIATE	-0.057	0.022	0.009	-0.453	0.174
65-94	UNIVARIATE	0.001	0.029	0.973	0.007	0.219
35-64	BIVARIATE	-0.022	0.024	0.369	-0.173	0.193
65-94	BIVARIATE	0.005	0.030	0.880	0.034	0.229
35-64	MULTIVARIATE	-0.008	0.026	0.755	-0.064	0.203
65-94	MULTIVARIATE	0.006	0.031	0.847	0.045	0.236

AVERAGE ANNUAL INCIDENCE RATE FOR EVENT PER 1000 PERSONS AT RISK AT EXAM
BY AGE AND LEVEL OF RISK FACTOR AT EXAM: FRAMINGHAM STUDY 30-YEAR FOLLOWUP

SEX: MALE
EVENT: CORONARY INSUFFICIENCY
RISK FACTOR: HEART RATE (PER MINUTE)
PERSONS AT RISK: PERSONS FREE OF CORONARY HEART DISEASE

RISK FACTOR	35-44 AT EXAM			45-54 AT EXAM			55-64 AT EXAM			65-74 AT EXAM		
	PERSON EXAMS	# OF EVENTS	ANNUAL RATE	PERSON EXAMS	# OF EVENTS	ANNUAL RATE	PERSON EXAMS	# OF EVENTS	ANNUAL RATE	PERSON EXAMS	# OF EVENTS	ANNUAL RATE
30-67	1284	1	0	1949	4	1	2309	11	2	1183	3	1
68-75	1492	0	0	2490	4	1	2490	2	0	1159	3	1
76-83	722	1	1	1394	3	1	1374	2	1	624	1	1
84-91	579	6	5	1097	5	2	1064	2	1	487	3	3
92-220	395	0	0	794	3	2	812	5	3	460	3	3
UNKNOWN	0	0	-	0	0	-	0	0	-	0	0	-
TOTAL	4472	8	1	7724	19	1	8049	22	1	3913	13	2

RISK FACTOR	75-84 AT EXAM			85-94 AT EXAM			35-64 COMBINED			65-94 COMBINED		
	PERSON EXAMS	# OF EVENTS	ANNUAL RATE	PERSON EXAMS	# OF EVENTS	ANNUAL RATE	# OF EVENTS	CRUDE	AGE-ADJUSTED	# OF EVENTS	CRUDE	AGE-ADJUSTED
30-67	301			11			16	1	1	4	1	1
68-75	296			12			6	0	0	3	1	1
76-83	192			10			6	1	1	2	1	1
84-91	123			5			13	2	2	3	2	2
92-220	105			1			8	2	2	4	4	4
UNKNOWN	0			0			0	-	-	0	-	-
TOTAL	1017			39			49	1	1	16	2	2

LOGISTIC REGRESSIONS

AGE AT EXAM		COEFFICIENT OF RISK FACTOR	SE COEFFICIENT	P-VALUE	STANDARDIZED COEFFICIENT	STD SE
35-44		0.040	0.023	0.088	0.494	0.290
45-54		0.020	0.017	0.252	0.246	0.215
55-64		0.001	0.016	0.953	0.013	0.213
65-74		0.023	0.018	0.206	0.310	0.245
75-84						
85-94						
35-64	UNIVARIATE	0.015	0.011	0.153	0.192	0.135
65-94	UNIVARIATE	0.022	0.016	0.185	0.294	0.222
35-64	BIVARIATE	0.015	0.011	0.156	0.190	0.134
65-94	BIVARIATE	0.022	0.016	0+185	0.295	0.222
35-64	MULTIVARIATE	-0.000	0.011	0.995	-0.001	0.145
65-94	MULTIVARIATE	0.014	0.017	0.405	0.193	0.232

SEX: FEMALE
EVENT: CORONARY INSUFFICIENCY
RISK FACTOR: HEART RATE (PER MINUTE)
PERSONS AT RISK: PERSONS FREE OF CORONARY HEART DISEASE

RISK FACTOR	35-44 AT EXAM			45-54 AT EXAM			55-64 AT EXAM			65-74 AT EXAM		
	PERSON EXAMS	# OF EVENTS	ANNUAL RATE	PERSON EXAMS	# OF EVENTS	ANNUAL RATE	PERSON EXAMS	# OF EVENTS	ANNUAL RATE	PERSON EXAMS	# OF EVENTS	ANNUAL RATE
30-67	1092			1897	2	1	2305	5	1	1103	4	2
68-75	1803			3244	2	0	3468	4	1	1668	3	1
76-83	1051			1911	1	0	2351	4	1	1358	1	0
84-91	939			1604	1	0	1780	6	2	1089	3	1
92-220	731			1303	0	0	1609	3	1	1030	4	2
UNKNOWN	0			0	0	-	0	0	-	0	0	-
TOTAL	5616			9959	6	0	11513	24	1	6248	15	1

RISK FACTOR	75-84 AT EXAM			85-94 AT EXAM			35-64 COMBINED			65-94 COMBINED		
	PERSON EXAMS	# OF EVENTS	ANNUAL RATE	PERSON EXAMS	# OF EVENTS	ANNUAL RATE	# OF EVENTS	CRUDE	AGE-ADJUSTED	# OF EVENTS	CRUDE	AGE-ADJUSTED
30-67	560	2	3	26			7	1	1	7	2	2
68-75	462	0	0	32			8	0	0	3	1	1
76-83	428	2	3	24			5	0	0	3	1	1
84-91	330	1	2	23			7	1	1	4	1	1
92-220	371	0	0	18			3	0	0	4	1	1
UNKNOWN	0			0			0	-	-	0	-	-
TOTAL	1951	5	1	123			30	1	1	21	1	1

LOGISTIC REGRESSIONS

AGE AT EXAM		COEFFICIENT OF RISK FACTOR	SE COEFFICIENT	P-VALUE	STANDARDIZED COEFFICIENT	STD SE
35-44						
45-54		-0.033	0.037	0.374	-0.411	0.462
55-64		0.001	0.016	0.930	0.018	0.203
65-74		0.001	0.020	0.958	0.014	0.257
75-84		-0.058	0.038	0.130	-0.805	0.532
85-94						
35-64	UNIVARIATE	-0.004	0.015	0.767	-0.055	0.187
65-94	UNIVARIATE	-0.016	0.018	0.350	-0.219	0.234
35-64	BIVARIATE	-0.005	0.015	0.751	-0.058	0.184
65-94	BIVARIATE	-0.017	0.018	0.346	-0.220	0.234
35-64	MULTIVARIATE	-0.015	0.015	0.319	-0.193	0.194
65-94	MULTIVARIATE	-0.017	0.018	0.330	-0.229	0.235

0	0	0	0	0	0	0
39	49	1	1	16	2	2

LOGISTIC REGRESSIONS

SE COEFFICIENT	P-VALUE	STANDARDIZED COEFFICIENT	STD SE
0.022	0.013	0.781	0.315
0.015	0.748	0.072	0.223
0.014	0.286	0.202	0.189
0.021	0.577	0.138	0.248
0.009	0.082	0.227	0.130
0.021	0.766	0.070	0.235
0.009	0.047	0.259	0.130
0.021	0.785	0.065	0.237
0.009	0.042	0.265	0.130
0.021	0.694	0.093	0.237

INSUFFICIENCY
GARETTES SMOKED (NUMBER PER DAY)
: PERSONS FREE OF CORONARY HEART DISEASE

45-54 AT EXAM			55-64 AT EXAM			65-74 AT EXAM		
PERSON EXAMS	# OF EVENTS	ANNUAL RATE	PERSON EXAMS	# OF EVENTS	ANNUAL RATE	PERSON EXAMS	# OF EVENTS	ANNUAL RATE
5528	3	0	7595	11	1	4863	11	1
1726	2	1	1566	5	2	626	2	2
2009	1	0	1696	5	1	568	2	2
622	0	0	602	3	2	172	0	
37	0	0	16	0	0	0	0	-
37	0	0	38	0	0	19	0	0
9959	6	0	11513	24	1	6248	15	1

85-94 AT EXAM			35-64 COMBINED			65-94 COMBINED		
PERSON EXAMS	# OF EVENTS	ANNUAL RATE	# OF EVENTS	CRUDE	AGE- ADJUSTED	# OF EVENTS	CRUDE	AGE- ADJUSTED
117	14	0	0	1	1			
4	7	1	1	2	2	2		
2	6	1	0	2	2	0		
0	3	1	1	0	0	-		
0	0	0	0	0	-			
0	0	0	0	0	0	0		
123	30	1	0	21	1	1		

LOGISTIC REGRESSIONS

SE COEFFICIENT	P-VALUE	STANDARDIZED COEFFICIENT	STD SE
0.041	0.931	-0.036	0.419
0.018	0.179	0.226	0.168
0.033	0.951	0.016	0.253
-	-	-	-
0.017	0.488	0.116	0.167
0.054	0.802	-0.060	0.238
0.016	0.195	0.209	0.161
0.034	0.838	-0.049	0.240
0.016	0.143	0.236	0.161
0.034	0.948	-0.016	0.239

SEX: MALE
EVENT: CORONARY INSUFFICIENCY
RISK FACTOR: ALBUMIN IN URINE, DEFINITE
PERSONS AT RISK: PERSONS FREE OF CORONARY HEART DISEASE

35-44 AT EXAM			45-54 AT EXAM			55-64 AT EXAM			65-74 AT EXAM		
PERSON EXAMS	# OF EVENTS	ANNUAL RATE	PERSON EXAMS	# OF EVENTS	ANNUAL RATE	PERSON EXAMS	# OF EVENTS	ANNUAL RATE	PERSON EXAMS	# OF EVENTS	ANNUAL RATE
4366	8	1	7599	18	1	7838	22	1	3792	12	2
70	0	0	99	1	5	184	0	0	115	1	4
36	0	0	26	0	0	27	0	0	6	0	0
4472	8	1	7724	19	1	8049	22	1	3913	13	2

75-84 AT EXAM			85-94 AT EXAM			35-64 COMBINED			65-94 COMBINED		
PERSON EXAMS	# OF EVENTS	ANNUAL RATE	PERSON EXAMS	# OF EVENTS	ANNUAL RATE	# OF EVENTS	CRUDE	AGE-ADJUSTED	# OF EVENTS	CRUDE	AGE-ADJUSTED
989	.		32	7		48	1	1	15	2	2
25						1	2		1	3	3
3			0	0		0	0	0	0	0	0
1017			39			49	1	1	16	2	2

LOGISTIC REGRESSIONS

	COEFFICIENT OF RISK FACTOR	SE COEFFICIENT	P-VALUE	STANDARDIZED COEFFICIENT	STD SE
	-	-	-	-	-
	1.458	1.032	0.158	0.164	0.116
	1.016	1.045	0.331	0.172	0.177
IATE	0.156	1.012	0.877	0.021	0.133
IATE	0.784	1.036	0.449	0.133	0.176
ATE	0.111	1.012	0.912	0.015	0.133
ATE	0.788	1.036	0.447	0.134	0.176
ARIATE	-0.426	1.024	0.678	-0.056	0.134
ARIATE	0.359	1.056	0.734	0.061	0.179

SEX: FEMALE
EVENT: CORONARY INSUFFICIENCY
RISK FACTOR: ALBUMIN IN URINE, DEFINITE
PERSONS AT RISK: PERSONS FREE OF CORONARY HEART DISEASE

35-44 AT EXAM			45-54 AT EXAM			55-64 AT EXAM			65-74 AT EXAM		
PERSON EXAMS	# OF EVENTS	ANNUAL RATE	PERSON EXAMS	# OF EVENTS	ANNUAL RATE	PERSON EXAMS	# OF EVENTS	ANNUAL RATE	PERSON EXAMS	# OF EVENTS	ANNUAL RATE
5353			9656	6	0	11263	24	1	6118	15	1
225			278	0	0	225	0	0	120	0	0
38			25	0	0	25	0	0	10	0	0
5616			9959	6	0	11513	24	1	6248	15	1

75-84 AT EXAM			85-94 AT EXAM			35-64 COMBINED			65-94 COMBINED		
PERSON EXAMS	# OF EVENTS	ANNUAL RATE	PERSON EXAMS	# OF EVENTS	ANNUAL RATE	# OF EVENTS	CRUDE	AGE-ADJUSTED	# OF EVENTS	CRUDE	AGE-ADJUSTED
1904	5	1	120	3		30	1	1	21	1	1
38	0	0				0	0	0	0	0	0
9	0	0				0	0	0	0	0	0
1951	5	1	123			30	1	1	21	1	1

LOGISTIC REGRESSIONS

	COEFFICIENT OF RISK FACTOR	SE COEFFICIENT	P-VALUE	STANDARDIZED COEFFICIENT	STD SE
	-	-	-	-	-
IATE		-	-	-	
IATE					
ATE					
ATE		-		-	
ARIATE	-		-		
ARIATE					

29		38	1	1	7	1	1	
5		5	1	2	5	3	3	
5		6	1	1	4	2	2	
0		0	0	0	0	-	-	
39		49	1	1	16	2	2	

LOGISTIC REGRESSIONS

COEFFICIENT OF RISK FACTOR	SE COEFFICIENT	P-VALUE	STANDARDIZED COEFFICIENT	STD SE
-0.070	0.800	0.930	-0.033	0.374
0.245	0.325	0.450	0.149	0.198
0.022	0.293	0.940	0.016	0.211
0.234	0.322	0.468	0.186	.
0.132	0.207	0.524	0.084	0.132
0.350	0.281	0.213	0.283	0.227
0.090	0.211	0.669	0.057	0.134
0.358	0.282	0.205	0.289	0.228
0.002*	0.221	0.993	0.001	0.141
0.296	0.299	0.393	0.207	0.242

SEX: FEMALE
EVENT: CORONARY INSUFFICIENCY
RISK FACTOR: HEART ENLARGEMENT BY X-RAY
PERSONS AT RISK: PERSONS FREE OF CORONARY HEART DISEASE

AT EXAM # OF EVENTS	ANNUAL RATE	45-54 AT EXAM PERSON EXAMS	# OF EVENTS	ANNUAL RATE	55-64 AT EXAM PERSON EXAMS	# OF EVENTS	ANNUAL RATE	65-74 AT EXAM PERSON EXAMS	# OF EVENTS	ANNUAL RATE
		8017	4	0	7522	14	1	3191	6	1
		943	1	1	1749	5	1	1150	1	0
		993	1	1	2218	5	1	1907	8	2
		6	0	0	4	0	0	0	0	-
		9959	6	0	11513	24	1	6248	15	1

AT EXAM # OF EVENTS	ANNUAL RATE	85-94 AT EXAM PERSON EXAMS	# OF EVENTS	ANNUAL RATE	55-64 COMBINED # OF EVENTS	CRUDE	AGE- ADJUSTED	65-94 COMBINED # OF EVENTS	CRUDE	AGE- ADJUSTED
1	9	32			18	0	0	7	1	1
1	1	30			6	1	1	3	1	1
3	2	61			6	1	1	11	2	2
0	-	0			0	0	0	0	-	-
5	1	123			30	1	1	21	1	1

LOGISTIC REGRESSIONS

COEFFICIENT OF RISK FACTOR	SE COEFFICIENT	P-VALUE	STANDARDIZED COEFFICIENT	STD SE
0.399	0.515	0.438	0.255	0.329
0.129	0.245	0.598	0.103	0.195
0.424	0.290	0.144	0.373	0.255
0.418	0.548	0.446	0.374	0.490
0.397	0.217	0.068	0.277	0.152
0.400	0.247	0.106	0.357	0.221
0.136	0.224	0.544	0.095	0.156
0.396	0.251	0.115	0.354	0.224
0.017*	0.238	0.944	0.012	0.166
0.368	0.261	0.159	0.328	0.233

TABLE 6-16   DATE=03MAY85
AVERAGE ANNUAL INCIDENCE RATE FOR EVENT PER 1000 PERSONS AT RISK AT EXAM
BY AGE AND LEVEL OF RISK FACTOR AT EXAM: FRAMINGHAM STUDY 30-YEAR FOLLOWUP

SEX: MALE
EVENT: CORONARY INSUFFICIENCY
RISK FACTOR: LEFT VENTRICULAR HYPERTROPHY BY ELECTROCARDIOGRAM
PERSONS AT RISK: PERSONS FREE OF CORONARY HEART DISEASE

RISK FACTOR	35-44 AT EXAM			45-54 AT EXAM			55-64 AT EXAM			65-74 AT EXAM		
	PERSON EXAMS	# OF EVENTS	ANNUAL RATE	PERSON EXAMS	# OF EVENTS	ANNUAL RATE	PERSON EXAMS	# OF EVENTS	ANNUAL RATE	PERSON EXAMS	# OF EVENTS	ANNUAL RATE
NO	4366	8	1	7497	18	1	7693	19	1	3703	13	2
BORDER	78	0	0	132	1	4	210	2	5	90	0	0
DEFINITE	28	0	0	95	0	0	146	1	3	120	0	0
UNKNOWN	0	0	-	0	0	-	0	0	-	0	0	-
TOTAL	4472	8	1	7724	19	1	8049	22	1	3913	13	2

RISK FACTOR	75-84 AT EXAM			85-94 AT EXAM			35-64 COMBINED			65-94 COMBINED		
	PERSON EXAMS	# OF EVENTS	ANNUAL RATE	PERSON EXAMS	# OF EVENTS	ANNUAL RATE	# OF EVENTS	CRUDE	AGE-ADJUSTED	# OF EVENTS	CRUDE	AGE-ADJUSTED
NO	949			34			45	1	1	15	2	2
BORDER	37			2			3	4	4	0	0	0
DEFINITE	31			3			1	2	1	1	3	3
UNKNOWN	0			0			0	-	-	0	-	-
TOTAL	1017			39			49	1	1	16	2	2

LOGISTIC REGRESSIONS

AGE AT EXAM		COEFFICIENT OF RISK FACTOR	SE COEFFICIENT	P-VALUE	STANDARDIZED COEFFICIENT	STD SE
35-44		-	-	-	-	-
45-54		0.150	0.799	0.851	0.038	0.203
55-64		0.694	0.403	0.085	0.214	0.124
65-74		-	-	-	-	-
75-84						
85-94						
35-64	UNIVARIATE	0.498	0.357	0.164	0.133	0.096
65-94	UNIVARIATE	0.218	0.556	0.696	0.082	0.210
35-64	BIVARIATE	0.464	0.358	0.195	0.124	0.096
65-94	BIVARIATE	0.220	0.556	0.692	0.083	0.210
35-64	MULTIVARIATE	0.162	0.387	0.676	0.043	0.104
65-94	MULTIVARIATE	-0.096	0.586	0.869	-0.036	0.221

SEX: FEMALE
EVENT: CORONARY INSUFFICIENCY
RISK FACTOR: LEFT VENTRICULAR HYPERTROPHY BY ELECTROCARDIOGRAM
PERSONS AT RISK: PERSONS FREE OF CORONARY HEART DISEASE

RISK FACTOR	35-44 AT EXAM			45-54 AT EXAM			55-64 AT EXAM			65-74 AT EXAM		
	PERSON EXAMS	# OF EVENTS	ANNUAL RATE	PERSON EXAMS	# OF EVENTS	ANNUAL RATE	PERSON EXAMS	# OF EVENTS	ANNUAL RATE	PERSON EXAMS	# OF EVENTS	ANNUAL RATE
NO	5563			9798	6	0	11137	21	1	5966	13	1
BORDER	38			107	0	0	226	2	4	134	2	7
DEFINITE	15			54	0	0	150	1	3	148	0	0
UNKNOWN	0			0	0	-	0	0	-	0	0	-
TOTAL	5616			9959	6	0	11513	24	1	6248	15	1

RISK FACTOR	75-84 AT EXAM			85-94 AT EXAM			35-64 COMBINED			65-94 COMBINED		
	PERSON EXAMS	# OF EVENTS	ANNUAL RATE	PERSON EXAMS	# OF EVENTS	ANNUAL RATE	# OF EVENTS	CRUDE	AGE-ADJUSTED	# OF EVENTS	CRUDE	AGE-ADJUSTED
NO	1833	4	1	111	6		27	1	0	18	1	1
BORDER	49	1	10	6			2	3	2	3	8	8
DEFINITE	69	0	0	6			1	2	1	0	0	0
UNKNOWN	0	0	-	0			0	-	-	0	-	-
TOTAL	1951	5	1	123			30	1	0	21	1	1

LOGISTIC REGRESSIONS

AGE AT EXAM		COEFFICIENT OF RISK FACTOR	SE COEFFICIENT	P-VALUE	STANDARDIZED COEFFICIENT	STD SE
35-44						
45-54		-	-	-	-	-
55-64		0.827	0.399	0.038	0.218	0.105
65-74		0.407	0.554	0.463	0.136	0.185
75-84		0.451	0.789	0.567	0.179	0.313
85-94						
35-64	UNIVARIATE	0.948	0.394	0.016	0.201	0.084
65-94	UNIVARIATE	0.385	0.453	0.395	0.136	0.160
35-64	BIVARIATE	0.734	0.400	0.067	0.156	0.085
65-94	BIVARIATE	0.376	0.454	0.408	0.132	0.160
35-64	MULTIVARIATE	0.495	0.440	0.260	0.105	0.093
65-94	MULTIVARIATE	0.366	0.470	0.436	0.129	0.165

TABLE 6-17   DATE=03DEC85

AVERAGE ANNUAL INCIDENCE RATE FOR EVENT PER 1000 PERSONS AT RISK AT EXAM
BY AGE AND LEVEL OF RISK FACTOR AT EXAM: FRAMINGHAM STUDY 30-YEAR FOLLOWUP

SEX: MALE
EVENT: CORONARY INSUFFICIENCY
RISK FACTOR: INTRAVENTRICULAR CONDUCTION DEFECT (WITH QRS INTERVAL MORE THAN 0.11 SECOND)
PERSONS AT RISK: PERSONS FREE OF CORONARY HEART DISEASE

RISK FACTOR	35-44 AT EXAM PERSON EXAMS	# OF EVENTS	ANNUAL RATE	45-54 AT EXAM PERSON EXAMS	# OF EVENTS	ANNUAL RATE	55-64 AT EXAM PERSON EXAMS	# OF EVENTS	ANNUAL RATE	65-74 AT EXAM PERSON EXAMS	# OF EVENTS	ANNUAL RATE
NO	4161	7	1	7213	17	1	7386	19	1	3566	12	2
INCOMPLETE	275	1	2	446	0	0	456	5	3	183	1	3
COMPLETE	36	0	0	65	2	15	207	0	0	164	0	0
UNKNOWN	0	0	-	0	0	-	0	0	-	0	0	-
TOTAL	4472	8	1	7724	19	1	8049	22	1	3913	13	2

RISK FACTOR	75-84 AT EXAM PERSON EXAMS	# OF EVENTS	ANNUAL RATE	85-94 AT EXAM PERSON EXAMS	# OF EVENTS	ANNUAL RATE	35-64 COMBINED # OF EVENTS	CRUDE	AGE-ADJUSTED	65-94 COMBINED # OF EVENTS	CRUDE	AGE-ADJUSTED
NO	918			31			43	1	1	15	2	2
INCOMPLETE	21			2			4	2	2	1	2	2
COMPLETE	78			6			2	3	6	0	0	0
UNKNOWN	0			0			0	-	-	0	-	-
TOTAL	1017			39			49	1	1	16	2	2

LOGISTIC REGRESSIONS

AGE AT EXAM		COEFFICIENT OF RISK FACTOR	SE COEFFICIENT	P-VALUE	STANDARDIZED COEFFICIENT	STD SE
35-44		0.417	0.930	0.654	0.124	0.275
45-54		0.881	0.458	0.054	0.258	0.134
55-64		0.169	0.492	0.730	0.065	0.189
65-74		-0.354	0.823	0.667	-0.157	0.366
75-84						
85-94						
35-64	UNIVARIATE	0.480	0.312	0.123	0.160	0.104
65-94	UNIVARIATE	-0.527	0.810	0.516	-0.248	0.381
35-64	BIVARIATE	0.450	0.311	0.147	0.150	0.104
65-94	BIVARIATE	-0.523	0.812	0.519	-0.246	0.382
35-64	MULTIVARIATE	0.463	0.313	0.140	0.154	0.105
65-94	MULTIVARIATE	-0.563	0.812	0.488	-0.265	0.382

SEX: FEMALE
EVENT: CORONARY INSUFFICIENCY
RISK FACTOR: INTRAVENTRICULAR CONDUCTION DEFECT (WITH QRS INTERVAL MORE THAN 0.11 SECOND)
PERSONS AT RISK: PERSONS FREE OF CORONARY HEART DISEASE

RISK FACTOR	35-44 AT EXAM PERSON EXAMS	# OF EVENTS	ANNUAL RATE	45-54 AT EXAM PERSON EXAMS	# OF EVENTS	ANNUAL RATE	55-64 AT EXAM PERSON EXAMS	# OF EVENTS	ANNUAL RATE	65-74 AT EXAM PERSON EXAMS	# OF EVENTS	ANNUAL RATE
NO	5529			9771	6	0	11183	22	1	6027	14	1
INCOMPLETE	75			144	0	0	187	0	0	115	0	0
COMPLETE	12			44	0	0	143	2	7	106	1	5
UNKNOWN	0			0	0	-	0	0	-	0	0	-
TOTAL	5616			9959	6	0	11513	24	1	6248	15	1

RISK FACTOR	75-84 AT EXAM PERSON EXAMS	# OF EVENTS	ANNUAL RATE	85-94 AT EXAM PERSON EXAMS	# OF EVENTS	ANNUAL RATE	35-64 COMBINED # OF EVENTS	CRUDE	AGE-ADJUSTED	65-94 COMBINED # OF EVENTS	CRUDE	AGE-ADJUSTED
NO	1847	5	1	117			28	1	1	20	1	1
INCOMPLETE	36	0	0	1			0	0	0	0	0	0
COMPLETE	68	0	0	5			2	5	3	1	3	4
UNKNOWN	0	0	-	0			0	-	-	0	-	-
TOTAL	1951	5	1	123			30	1	1	21	1	1

LOGISTIC REGRESSIONS

AGE AT EXAM		COEFFICIENT OF RISK FACTOR	SE COEFFICIENT	P-VALUE	STANDARDIZED COEFFICIENT	STD SE
35-44		-	-	-	-	-
45-54		-	-	-	-	-
55-64		0.874	0.395	0.027	0.221	0.100
65-74		0.579	0.555	0.297	0.167	0.160
75-84		-	-	-	-	-
85-94		-	-	-	-	-
35-64	UNIVARIATE	0.976	0.398	0.014	0.204	0.083
65-94	UNIVARIATE	0.268	0.550	0.626	0.085	0.174
35-64	BIVARIATE	0.766	0.396	0.053	0.160	0.083
65-94	BIVARIATE	0.256	0.551	0.642	0.081	0.175
35-64	MULTIVARIATE	0.776	0.401	0.053	0.162	0.084
65-94	MULTIVARIATE	0.255	0.556	0.649	0.080	0.176

TABLE 6-18   DATE=04DEC85
AVERAGE ANNUAL INCIDENCE RATE FOR EVENT PER 1000 PERSONS AT RISK AT EXAM
BY AGE AND LEVEL OF RISK FACTOR AT EXAM: FRAMINGHAM STUDY 30-YEAR FOLLOWUP

SEX: MALE
EVENT: CORONARY INSUFFICIENCY
RISK FACTOR: DEFINITE NONSPECIFIC T-WAVE OR ST-SEGMENT ABNORMALITY BY ELECTROCARDIOGRAM
PERSONS AT RISK: PERSONS FREE OF CORONARY HEART DISEASE

RISK FACTOR	35-44 AT EXAM			45-54 AT EXAM			55-64 AT EXAM			65-74 AT EXAM		
	PERSON EXAMS	# OF EVENTS	ANNUAL RATE	PERSON EXAMS	# OF EVENTS	ANNUAL RATE	PERSON EXAMS	# OF EVENTS	ANNUAL RATE	PERSON EXAMS	# OF EVENTS	ANNUAL RATE
NO	4362	6	1	7352	17	1	7462	19	1	3472	7	1
YES	110	2	9	372	2	3	587	3	3	441	6	7
UNKNOWN	0	0	-	0	0	-	0	0	-	0	0	-
TOTAL	4472	8	1	7724	19	1	8049	22	1	3913	13	2

RISK FACTOR	75-84 AT EXAM			85-94 AT EXAM			35-64 COMBINED			65-94 COMBINED		
	PERSON EXAMS	# OF EVENTS	ANNUAL RATE	PERSON EXAMS	# OF EVENTS	ANNUAL RATE	# OF EVENTS	CRUDE	AGE-ADJUSTED	# OF EVENTS	CRUDE	AGE-ADJUSTED
NO	830			28			42	1	1	10	1	1
YES	187			11			7	3	4	6	5	5
UNKNOWN	0			0			0	-	-	0	-	-
TOTAL	1017			39			49	1	1	16	2	2

LOGISTIC REGRESSIONS

AGE AT EXAM		COEFFICIENT OF RISK FACTOR	SE COEFFICIENT	P-VALUE	STANDARDIZED COEFFICIENT	STD SE
35-44		2.599	0.822	0.002	0.403	0.127
45-54		0.847	0.749	0.258	0.181	0.160
55-64		0.699	0.623	0.261	0.182	0.162
65-74		1.921	0.559	0.001	0.608	0.177
75-84						
85-94						
35-64	UNIVARIATE	1.100	0.409	0.007	0.246	0.092
65-94	UNIVARIATE	1.410	0.518	0.007	0.472	0.173
35-64	BIVARIATE	1.040	0.413	0.012	0.233	0.092
65-94	BIVARIATE	1.442	0.522	0.006	0.483	0.175
35-64	MULTIVARIATE	0.674	0.424	0.112	0.151	0.095
65-94	MULTIVARIATE	1.259	0.528	0.017	0.422	0.177

SEX: FEMALE
EVENT: CORONARY INSUFFICIENCY
RISK FACTOR: DEFINITE NONSPECIFIC T-WAVE OR ST-SEGMENT ABNORMALITY BY ELECTROCARDIOGRAM
PERSONS AT RISK: PERSONS FREE OF CORONARY HEART DISEASE

RISK FACTOR	35-44 AT EXAM			45-54 AT EXAM			55-64 AT EXAM			65-74 AT EXAM		
	PERSON EXAMS	# OF EVENTS	ANNUAL RATE	PERSON EXAMS	# OF EVENTS	ANNUAL RATE	PERSON EXAMS	# OF EVENTS	ANNUAL RATE	PERSON EXAMS	# OF EVENTS	ANNUAL RATE
NO	5517			9543	6	0	10647	20	1	5528	9	1
YES	99			416	0	0	866	4	2	720	6	4
UNKNOWN	0			0	0	-	0	0	-	0	0	-
TOTAL	5616			9959	6	0	11513	24	1	6248	15	1

RISK FACTOR	75-84 AT EXAM			85-94 AT EXAM			35-64 COMBINED			65-94 COMBINED		
	PERSON EXAMS	# OF EVENTS	ANNUAL RATE	PERSON EXAMS	# OF EVENTS	ANNUAL RATE	# OF EVENTS	CRUDE	AGE-ADJUSTED	# OF EVENTS	CRUDE	AGE-ADJUSTED
NO	1632	3	1	97			26	1	1	13	1	1
YES	319	2	3	26			4	1	1	8	4	4
UNKNOWN	0	0	-	0			0	-	-	0	-	-
TOTAL	1951	5	1	123			30	1	1	21	1	1

LOGISTIC REGRESSIONS

AGE AT EXAM		COEFFICIENT OF RISK FACTOR	SE COEFFICIENT	P-VALUE	STANDARDIZED COEFFICIENT	STD SE
35-44		-	-	-	-	-
45-54		-	-	-	-	-
55-64		0.902	0.549	0.100	0.238	0.145
65-74		1.640	0.529	0.002	0.524	0.169
75-84		1.231	0.915	0.178	0.456	0.338
85-94						
35-64	UNIVARIATE	1.054	0.538	0.050	0.232	0.118
65-94	UNIVARIATE	1.439	0.451	0.001	0.481	0.151
35-64	BIVARIATE	0.680	0.541	0.208	0.150	0.119
65-94	BIVARIATE	1.434	0.453	0.002	0.479	0.151
35-64	MULTIVARIATE	0.506	0.544	0.353	0.111	0.120
65-94	MULTIVARIATE	1.411	0.455	0.002	0.471	0.152

SEX: MALE
EVENT: ANGINA PECTORIS
RISK FACTOR: SYSTOLIC BLOOD PRESSURE, FIRST EXAMINER (MM HG)
PERSONS AT RISK: PERSONS FREE OF CORONARY HEART DISEASE

RISK FACTOR	35-44 AT EXAM			45-54 AT EXAM			55-64 AT EXAM			65-74 AT EXAM		
	PERSON EXAMS	# OF EVENTS	ANNUAL RATE	PERSON EXAMS	# OF EVENTS	ANNUAL RATE	PERSON EXAMS	# OF EVENTS	ANNUAL RATE	PERSON EXAMS	# OF EVENTS	ANNUAL RATE
74-119	1223	3	1	1769	11	3	1440	17	6	548	5	5
120-139	2165	9	2	3417	29	4	3166	46	7	1341	19	7
140-159	841	7	4	1732	26	8	2090	41	10	1194	22	9
160-179	191	2	5	567	11	10	916	29	16	552	8	7
180-300	52	0	0	239	3	6	437	16	18	278	8	14
UNKNOWN	0	0	-	0	0	-	0	0	-	0	0	-
TOTAL	4472	21	2	7724	80	5	8049	149	9	3913	62	8

RISK FACTOR	75-84 AT EXAM			85-94 AT EXAM			35-64 COMBINED			65-94 COMBINED		
	PERSON EXAMS	# OF EVENTS	ANNUAL RATE	PERSON EXAMS	# OF EVENTS	ANNUAL RATE	# OF EVENTS	CRUDE	AGE-ADJUSTED	# OF EVENTS	CRUDE	AGE-ADJUSTED
74-119	120	0	0	7			31	3	4	5	4	4
120-139	345	4	6	15			84	5	5	23	7	7
140-159	315	4	6	11			74	8	8	27	9	9
160-179	153	1	3	4			42	13	11	10	7	7
180-300	84	0	0	2			19	13	9	8	11	11
UNKNOWN	0	0	-	0			0	-	-	0	-	-
TOTAL	1017	9	4	39			250	6	6	73	7	7

LOGISTIC REGRESSIONS

AGE AT EXAM		COEFFICIENT OF RISK FACTOR	SE COEFFICIENT	P-VALUE	STANDARDIZED COEFFICIENT	STD SE
35-44		0.023	0.010	0.029	0.378	0.173
45-54		0.015	0.005	0.004	0.284	0.097
55-64		0.016	0.003	0.001	0.345	0.070
65-74		0.011	0.005	0.024	0.258	0.114
75-84		-0.006	0.016	0.699	-0.134	0.348
85-94						
35-64	UNIVARIATE	0.019	0.003	0.001	0.381	0.052
65-94	UNIVARIATE	0.010	0.005	0.036	0.226	0.108
35-64	BIVARIATE	0.015	0.003	0.001	0.309	0.053
65-94	BIVARIATE	0.010	0.005	0.035	0.226	0.107
35-64	MULTIVARIATE	0.013	0.003	0.001	0.269	0.058
65-94	MULTIVARIATE	0.012	0.005	0.020	0.264	0.114

SEX: FEMALE
EVENT: ANGINA PECTORIS
RISK FACTOR: SYSTOLIC BLOOD PRESSURE, FIRST EXAMINER (MM HG)
PERSONS AT RISK: PERSONS FREE OF CORONARY HEART DISEASE

RISK FACTOR	35-44 AT EXAM			45-54 AT EXAM			55-64 AT EXAM			65-74 AT EXAM		
	PERSON EXAMS	# OF EVENTS	ANNUAL RATE	PERSON EXAMS	# OF EVENTS	ANNUAL RATE	PERSON EXAMS	# OF EVENTS	ANNUAL RATE	PERSON EXAMS	# OF EVENTS	ANNUAL RATE
74-119	2493	0	0	2714	9	2	1979	13	3	647	7	5
120-139	2244	3	1	3982	12	2	4056	35	4	1747	22	6
140-159	673	2	1	2053	12	3	3025	39	6	2013	33	8
160-179	156	0	0	768	7	5	1560	32	10	1155	11	5
180-300	50	0	0	442	7	8	893	21	12	686	18	13
UNKNOWN	0	0	-	0	0	-	0	0	-	0	0	-
TOTAL	5616	5	0	9959	47	2	11513	140	6	6248	91	7

RISK FACTOR	75-84 AT EXAM			85-94 AT EXAM			35-64 COMBINED			65-94 COMBINED		
	PERSON EXAMS	# OF EVENTS	ANNUAL RATE	PERSON EXAMS	# OF EVENTS	ANNUAL RATE	# OF EVENTS	CRUDE	AGE-ADJUSTED	# OF EVENTS	CRUDE	AGE-ADJUSTED
74-119	133	1	4	8			22	2	2	8	5	5
120-139	465	7	8	58			50	2	3	29	6	6
140-159	626	9	7	29			53	5	4	42	8	8
160-179	451	10	11	31			39	8	6	21	6	4
180-300	276	4	7	17			28	10	8	23	12	12
UNKNOWN	0	0	-	0			0	-	-	0	-	-
TOTAL	1951	31	8	123			192	4	3	123	7	7

LOGISTIC REGRESSIONS

AGE AT EXAM		COEFFICIENT OF RISK FACTOR	SE COEFFICIENT	P-VALUE	STANDARDIZED COEFFICIENT	STD SE
35-44		0.025	0.018	0.158	0.429	0.304
45-54		0.019	0.004	0.001	0.443	0.101
55-64		0.017	0.003	0.001	0.420	0.068
65-74		0.007	0.004	0.084	0.173	0.100
75-84		0.007	0.007	0.302	0.176	0.171
85-94						
35-64	UNIVARIATE	0.022	0.002	0.001	0.516	0.051
65-94	UNIVARIATE	0.008	0.003	0.031	0.184	0.086
35-64	BIVARIATE	0.018	0.002	0.001	0.416	0.056
65-94	BIVARIATE	0.008	0.004	0.026	0.191	0.086
35-64	MULTIVARIATE	0.013	0.003	0.001	0.312	0.064
65-94	MULTIVARIATE	0.005	0.004	0.191	0.116	0.089

TABLE 7-2   DATE=05APR85
AVERAGE ANNUAL INCIDENCE RATE FOR EVENT PER 1000 PERSONS AT RISK AT EXAM
BY AGE AND LEVEL OF RISK FACTOR AT EXAM: FRAMINGHAM STUDY 30-YEAR FOLLOWUP

SEX: MALE
EVENT: ANGINA PECTORIS
RISK FACTOR: DIASTOLIC BLOOD PRESSURE, FIRST EXAMINER (MM HG)
PERSONS AT RISK: PERSONS FREE OF CORONARY HEART DISEASE

RISK FACTOR	35-44 AT EXAM			45-54 AT EXAM			55-64 AT EXAM			65-74 AT EXAM		
	PERSON EXAMS	# OF EVENTS	ANNUAL RATE	PERSON EXAMS	# OF EVENTS	ANNUAL RATE	PERSON EXAMS	# OF EVENTS	ANNUAL RATE	PERSON EXAMS	# OF EVENTS	ANNUAL RATE
20-74	1012	2	1	1478	15	5	1847	27	7	1225	13	5
75-84	1604	4	1	2766	19	3	2845	35	6	1323	25	9
85-94	1187	7	3	2062	23	6	1964	42	11	893	14	8
95-104	499	6	6	1008	15	7	982	30	15	343	8	12
105-160	170	2	6	410	8	10	411	15	18	129	2	8
UNKNOWN	0	0	-	0	0	-	0	0	-	0	0	-
TOTAL	4472	21	2	7724	80	5	8049	149	9	3913	62	8

RISK FACTOR	75-84 AT EXAM			85-94 AT EXAM			35-64 COMBINED			65-94 COMBINED		
	PERSON EXAMS	# OF EVENTS	ANNUAL RATE	PERSON EXAMS	# OF EVENTS	ANNUAL RATE	# OF EVENTS	CRUDE	AGE-ADJUSTED	# OF EVENTS	CRUDE	AGE-ADJUSTED
20-74	422	4	5	22			44	5	5	19	6	6
75-84	331	4	6	12			58	4	4	29	9	8
85-94	204	1	2	5			72	7	7	15	7	7
95-104	44	0	0	0			51	10	10	8	10	9
105-160	16	0	0	0			25	13	12	2	7	6
UNKNOWN	0	0	-	0			0	-	-	0	-	-
TOTAL	1017	9	4	39			250	6	6	73	7	7

LOGISTIC REGRESSIONS

AGE AT EXAM		COEFFICIENT OF RISK FACTOR	SE COEFFICIENT	P-VALUE	STANDARDIZED COEFFICIENT	STD SE
35-44		0.050	0.016	0.002	0.568	0.182
45-54		0.019	0.009	0.039	0.218	0.106
55-64		0.026	0.006	0.001	0.315	0.075
65-74		0.017	0.010	0.084	0.207	0.120
75-84		-0.016	0.029	0.573	-0.190	0.340
85-94						
35-64	UNIVARIATE	0.027	0.005	0.001	0.313	0.058
65-94	UNIVARIATE	0.013	0.009	0.176	0.153	0.113
35-64	BIVARIATE	0.026	0.005	0.001	0.308	0.057
65-94	BIVARIATE	0.012	0.009	0.203	0.146	0.115
35-64	MULTIVARIATE	0.024	0.005	0.001	0.283	0.061
65-94	MULTIVARIATE	0.014	0.010	0.155	0.171	0.121

SEX: FEMALE
EVENT: ANGINA PECTORIS
RISK FACTOR: DIASTOLIC BLOOD PRESSURE, FIRST EXAMINER (MM HG)
PERSONS AT RISK: PERSONS FREE OF CORONARY HEART DISEASE

RISK FACTOR	35-44 AT EXAM			45-54 AT EXAM			55-64 AT EXAM			65-74 AT EXAM		
	PERSON EXAMS	# OF EVENTS	ANNUAL RATE	PERSON EXAMS	# OF EVENTS	ANNUAL RATE	PERSON EXAMS	# OF EVENTS	ANNUAL RATE	PERSON EXAMS	# OF EVENTS	ANNUAL RATE
20-74	2164	1	0	2639	12	2	2969	27	5	1831	30	8
75-84	2030	1	0	3561	9	1	3853	40	5	2135	27	6
85-94	987	1	1	2361	13	3	2840	40	7	1439	22	8
95-104	324	2	3	926	6	3	1272	17	7	600	5	4
105-160	111	0	0	492	7	7	579	16	14	243	7	14
UNKNOWN	0	0	-	0	0	-	0	0	-	0	0	-
TOTAL	5616	5	0	9959	47	2	11513	140	6	6248	91	7

RISK FACTOR	75-84 AT EXAM			85-94 AT EXAM			35-64 COMBINED			65-94 COMBINED		
	PERSON EXAMS	# OF EVENTS	ANNUAL RATE	PERSON EXAMS	# OF EVENTS	ANNUAL RATE	# OF EVENTS	CRUDE	AGE-ADJUSTED	# OF EVENTS	CRUDE	AGE-ADJUSTED
20-74	724	7	5	53			40	3	3	37	7	7
75-84	596	12	10	43			50	3	2	40	7	7
85-94	445	11	12	28			54	4	4	33	9	9
95-104	146	1	3	7			25	5	5	6	4	4
105-160	42	0	0	6			23	10	9	7	12	11
UNKNOWN	0	0	-	0			0	-	-	0	-	-
TOTAL	1951	31	8	123			192	4	3	123	7	7

LOGISTIC REGRESSIONS

AGE AT EXAM		COEFFICIENT OF RISK FACTOR	SE COEFFICIENT	P-VALUE	STANDARDIZED COEFFICIENT	STD SE
35-44		0.052	0.028	0.067	0.574	0.315
45-54		0.027	0.010	0.006	0.339	0.124
55-64		0.027	0.006	0.001	0.331	0.076
65-74		-0.000	0.009	0.956	-0.006	0.106
75-84		0.012	0.015	0.402	0.148	0.176
85-94						
35-64	UNIVARIATE	0.032	0.005	0.001	0.394	0.061
65-94	UNIVARIATE	0.003	0.007	0.692	0.036	0.090
35-64	BIVARIATE	0.028	0.005	0.001	0.344	0.063
65-94	BIVARIATE	0.003	0.007	0.733	0.031	0.091
35-64	MULTIVARIATE	0.019	0.006	0.001	0.235	0.069
65-94	MULTIVARIATE	-0.001	0.007	0.881	-0.013	0.090

TABLE 7-3A   DATE=31AUG85

AVERAGE ANNUAL INCIDENCE RATE FOR EVENT PER 1000 PERSONS AT RISK AT EXAM
BY AGE AND LEVEL OF RISK FACTOR AT EXAM: FRAMINGHAM STUDY 30-YEAR FOLLOWUP

SEX: MALE
EVENT: ANGINA PECTORIS
RISK FACTOR: HYPERTENSION WITH ANTIHYPERTENSIVE TREATMENT
PERSONS AT RISK: PERSONS FREE OF CORONARY HEART DISEASE

RISK FACTOR	35-44 AT EXAM PERSON EXAMS	# OF EVENTS	ANNUAL RATE	45-54 AT EXAM PERSON EXAMS	# OF EVENTS	ANNUAL RATE	55-64 AT EXAM PERSON EXAMS	# OF EVENTS	ANNUAL RATE	65-74 AT EXAM PERSON EXAMS	# OF EVENTS	ANNUAL RATE
NO	2651	9	2	4122	30	4	3782	44	6	1516	16	5
BORDER	1301	7	3	2205	28	6	2299	53	12	1252	23	9
DEFINITE	500	4	4	1159	16	7	1266	31	12	552	12	11
TREATED	20	1	25	258	6	12	702	21	15	593	11	9
UNKNOWN	0	0	-	0	0	-	0	0	-	0	0	-
TOTAL	4472	21	2	7724	80	5	8049	149	9	3913	62	8

RISK FACTOR	75-84 AT EXAM PERSON EXAMS	# OF EVENTS	ANNUAL RATE	85-94 AT EXAM PERSON EXAMS	# OF EVENTS	ANNUAL RATE	35-64 COMBINED # OF EVENTS	CRUDE	AGE-ADJUSTED	65-94 COMBINED # OF EVENTS	CRUDE	AGE-ADJUSTED
NO	373	5	4	18			83	4	4	19	5	5
BORDER	333	5	8	11			88	8	8	29	9	9
DEFINITE	123	0	0	3			51	9	8	12	9	9
TREATED	188	1	3	7			28	14	16	13	8	8
UNKNOWN	0	0	-	0			0			0	-	
TOTAL	1017	9	4	39			250	6	6	73	7	7

LOGISTIC REGRESSIONS

AGE AT EXAM	COEFFICIENT OF RISK FACTOR	SE COEFFICIENT	P-VALUE	STANDARDIZED COEFFICIENT	STD SE
35-44	0.517	0.279	0.063	0.359	0.194
45-54	0.399	0.137	0.004	0.307	0.105
55-64	0.411	0.099	0.001	0.335	0.081
65-74	0.313	0.157	0.046	0.256	0.128
75-84	-0.248	0.420	0.554	-0.203	0.344
85-94					
35-64 UNIVARIATE	0.483	0.077	0.001	0.376	0.060
65-94 UNIVARIATE	0.257	0.144	0.074	0.211	0.118
35-64 BIVARIATE	0.402	0.077	0.001	0.312	0.060
65-94 BIVARIATE	0.259	0.144	0.072	0.212	0.118
35-64 MULTIVARIATE	0.362	0.081	0.001	0.281	0.063
65-94 MULTIVARIATE	0.270	0.146	0.065	0.221	0.120

SEX: FEMALE
EVENT: ANGINA PECTORIS
RISK FACTOR: HYPERTENSION WITH ANTIHYPERTENSIVE TREATMENT
PERSONS AT RISK: PERSONS FREE OF CORONARY HEART DISEASE

RISK FACTOR	35-44 AT EXAM PERSON EXAMS	# OF EVENTS	ANNUAL RATE	45-54 AT EXAM PERSON EXAMS	# OF EVENTS	ANNUAL RATE	55-64 AT EXAM PERSON EXAMS	# OF EVENTS	ANNUAL RATE	65-74 AT EXAM PERSON EXAMS	# OF EVENTS	ANNUAL RATE
NO	4250	2	0	5622	14	1	4888	30	3	1767	17	5
BORDER	1001	2	1	2579	15	3	3282	46	7	1903	23	6
DEFINITE	316	1	2	1166	10	4	1835	34	9	1035	15	7
TREATED	49	0	0	592	8	7	1508	30	10	1543	36	12
UNKNOWN	0	0	-	0	0	-	0	0	-	0	0	-
TOTAL	5616	5	0	9959	47	2	11515	140	6	6248	91	6

RISK FACTOR	75-84 AT EXAM PERSON EXAMS	# OF EVENTS	ANNUAL RATE	85-94 AT EXAM PERSON EXAMS	# OF EVENTS	ANNUAL RATE	35-64 COMBINED # OF EVENTS	CRUDE	AGE-ADJUSTED	65-94 COMBINED # OF EVENTS	CRUDE	AGE-ADJUSTED
NO	402	6	7	31			46	2	2	23	5	5
BORDER	575	5	8	28			63	5	4	32	6	6
DEFINITE	371	6	8	21			45	6	6	21	7	7
TREATED	603	10	8	45			38	9	7	47	11	11
UNKNOWN	0	0	-	0			0	-		0	-	
TOTAL	1951	31	8	123			192	4	3	123	7	7

LOGISTIC REGRESSIONS

AGE AT EXAM	COEFFICIENT OF RISK FACTOR	SE COEFFICIENT	P-VALUE	STANDARDIZED COEFFICIENT	STD SE
35-44	0.944	0.542	0.082	0.553	0.318
45-54	0.706	0.175	0.001	0.542	0.134
55-64	0.546	0.104	0.001	0.456	0.087
65-74	0.388	0.138	0.005	0.319	0.114
75-84	0.049	0.233	0.834	0.038	0.183
85-94					
35-64 UNIVARIATE	0.773	0.087	0.001	0.613	0.069
65-94 UNIVARIATE	0.313	0.118	0.008	0.257	0.097
35-64 BIVARIATE	0.587	0.090	0.001	0.466	0.071
65-94 BIVARIATE	0.324	0.119	0.006	0.265	0.097
35-64 MULTIVARIATE	0.469	0.096	0.001	0.372	0.076
65-94 MULTIVARIATE	0.249	0.121	0.039	0.203	0.099

TABLE 7-3B   DATE=27NOV85
AVERAGE ANNUAL INCIDENCE RATE FOR EVENT PER 1000 PERSONS AT RISK AT EXAM
BY AGE AND LEVEL OF RISK FACTOR AT EXAM: FRAMINGHAM STUDY 30-YEAR FOLLOWUP

SEX: MALE
EVENT: ANGINA PECTORIS
RISK FACTOR: HYPERTENSION IGNORING TREATMENT
PERSONS AT RISK: PERSONS FREE OF CORONARY HEART DISEASE

RISK FACTOR	35-44 AT EXAM			45-54 AT EXAM			55-64 AT EXAM			65-74 AT EXAM		
	PERSON EXAMS	# OF EVENTS	ANNUAL RATE	PERSON EXAMS	# OF EVENTS	ANNUAL RATE	PERSON EXAMS	# OF EVENTS	ANNUAL RATE	PERSON EXAMS	# OF EVENTS	ANNUAL RATE
NO	2653	9	2	4154	30	4	3909	46	6	1631	17	5
MILD	1310	7	3	2284	32	7	2583	61	12	1499	29	10
DEFINITE	509	5	5	1286	18	7	1557	42	13	783	16	10
UNKNOWN	0	0	-	0	0	-	0	0	-	0	0	-
TOTAL	4472	21	2	7724	80	5	8049	149	9	3913	62	8

RISK FACTOR	75-84 AT EXAM			85-94 AT EXAM			35-64 COMBINED			65-94 COMBINED		
	PERSON EXAMS	# OF EVENTS	ANNUAL RATE	PERSON EXAMS	# OF EVENTS	ANNUAL RATE	# OF EVENTS	CRUDE	AGE-ADJUSTED	# OF EVENTS	CRUDE	AGE-ADJUSTED
NO	412	3	4	21			85	4	4	20	5	5
MILD	418	3	6	15			100	8	8	36	9	9
DEFINITE	187	1	3	3			65	10	9	17	9	9
UNKNOWN	0	0	-	0			0	-		0	-	
TOTAL	1017	9	4	39			250	6	6	73	7	7

LOGISTIC REGRESSIONS

AGE AT EXAM		COEFFICIENT OF RISK FACTOR	SE COEFFICIENT	P-VALUE	STANDARDIZED COEFFICIENT	STD SE
35-44		0.527	0.280	0.060	0.364	0.193
45-54		0.364	0.139	0.009	0.274	0.109
55-64		0.428	0.102	0.001	0.330	0.079
65-74		0.346	0.165	0.036	0.261	0.125
75-84		-0.002	0.456	0.997	-0.001	0.335
85-94						
35-64	UNIVARIATE	0.466	0.079	0.001	0.350	0.059
65-94	UNIVARIATE	0.312	0.153	0.042	0.234	0.115
35-64	BIVARIATE	0.407	0.079	0.001	0.305	0.059
65-94	BIVARIATE	0.311	0.153	0.042	0.233	0.115
35-64	MULTIVARIATE	0.356	0.084	0.001	0.267	0.063
65-94	MULTIVARIATE	0.342	0.157	0.029	0.256	0.117

SEX: FEMALE
EVENT: ANGINA PECTORIS
RISK FACTOR: HYPERTENSION IGNORING TREATMENT
PERSONS AT RISK: PERSONS FREE OF CORONARY HEART DISEASE

RISK FACTOR	35-44 AT EXAM			45-54 AT EXAM			55-64 AT EXAM			65-74 AT EXAM		
	PERSON EXAMS	# OF EVENTS	ANNUAL RATE	PERSON EXAMS	# OF EVENTS	ANNUAL RATE	PERSON EXAMS	# OF EVENTS	ANNUAL RATE	PERSON EXAMS	# OF EVENTS	ANNUAL RATE
NO	4272	2	0	5804	17	1	5245	38	4	2035	23	6
MILD	1016	2	1	2763	14	3	3860	57	7	2549	40	8
DEFINITE	328	1	2	1392	14	5	2408	45	9	1664	28	8
UNKNOWN	0	0	-	0	0	-	0	0	-	0	0	-
TOTAL	5616	5	0	9959	47	2	11513	140	6	6248	91	7

RISK FACTOR	75-84 AT EXAM			85-94 AT EXAM			35-64 COMBINED			65-94 COMBINED		
	PERSON EXAMS	# OF EVENTS	ANNUAL RATE	PERSON EXAMS	# OF EVENTS	ANNUAL RATE	# OF EVENTS	CRUDE	AGE-ADJUSTED	# OF EVENTS	CRUDE	AGE-ADJUSTED
NO	507	6	6	57			57	2	2	29	6	6
MILD	827	15	9	45			75	5	4	55	8	8
DEFINITE	617	10	8	41			60	7	6	39	8	8
UNKNOWN	0	0	-	0			0	-		0	-	
TOTAL	1951	31	8	123			192	4	4	123	7	7

LOGISTIC REGRESSIONS

AGE AT EXAM		COEFFICIENT OF RISK FACTOR	SE COEFFICIENT	P-VALUE	STANDARDIZED COEFFICIENT	STD SE
35-44		0.995	0.548	0.069	0.568	0.313
45-54		0.623	0.179	0.001	0.452	0.130
55-64		0.473	0.106	0.001	0.367	0.082
65-74		0.197	0.138	0.153	0.151	0.106
75-84		0.130	0.241	0.590	0.098	0.182
85-94						
35-64	UNIVARIATE	0.685	0.088	0.001	0.507	0.065
65-94	UNIVARIATE	0.194	0.119	0.103	0.149	0.091
35-64	BIVARIATE	0.514	0.091	0.001	0.380	0.067
65-94	BIVARIATE	0.200	0.119	0.093	0.154	0.091
35-64	MULTIVARIATE	0.362	0.098	0.001	0.268	0.073
65-94	MULTIVARIATE	0.119	0.121	0.328	0.091	0.093

AVERAGE ANNUAL INCIDENCE RATE FOR EVENT PER 1000 PERSONS AT RISK AT EXAM
BY AGE AND LEVEL OF RISK FACTOR AT EXAM: FRAMINGHAM STUDY 30-YEAR FOLLOWUP

SEX: MALE
EVENT: ANGINA PECTORIS
RISK FACTOR: SERUM CHOLESTEROL
PERSONS AT RISK: PERSONS FREE OF CORONARY HEART DISEASE

RISK FACTOR	35-44 AT EXAM			45-54 AT EXAM			55-64 AT EXAM			65-74 AT EXAM		
	PERSON EXAMS	# OF EVENTS	ANNUAL RATE	PERSON EXAMS	# OF EVENTS	ANNUAL RATE	PERSON EXAMS	# OF EVENTS	ANNUAL RATE	PERSON EXAMS	# OF EVENTS	ANNUAL RATE
84-204	1253	4	2	1767	6	2	2255	25	6	1234	13	5
205-234	1247	3	1	2306	25	5	2405	45	9	1171	19	8
235-264	872	4	2	1745	24	7	1889	38	10	873	16	9
265-294	475	5	5	1001	9	4	918	23	13	428	6	7
295-1124	291	2	3	557	13	12	454	12	13	191	8	21
UNKNOWN	334	3	4	348	3	4	128	6	23	16	0	0
TOTAL	4472	21	2	7724	80	5	8049	149	9	3913	62	8

RISK FACTOR	75-84 AT EXAM			85-94 AT EXAM			35-64 COMBINED			65-94 COMBINED		
	PERSON EXAMS	# OF EVENTS	ANNUAL RATE	PERSON EXAMS	# OF EVENTS	ANNUAL RATE	# OF EVENTS	CRUDE	AGE-ADJUSTED	# OF EVENTS	CRUDE	AGE-ADJUSTED
84-204	390	3	4	17			35	3	4	17	5	5
205-234	332	4	6	11			73	7	6	24	8	8
235-264	177	2	6	5			66	7	7	18	9	8
265-294	96	0	0	6			37	8	8	6	6	6
295-1124	22	0	0	0			27	10	10	8	19	17
UNKNOWN	0	0	-	0			12	7	12	0	0	0
TOTAL	1017	9	4	39			250	6	6	73	7	7

LOGISTIC REGRESSIONS

AGE AT EXAM		COEFFICIENT OF RISK FACTOR	SE COEFFICIENT	P-VALUE	STANDARDIZED COEFFICIENT	STD SE
35-44		0.007	0.004	0.096	0.317	0.190
45-54		0.008	0.002	0.001	0.331	0.090
55-64		0.006	0.002	0.003	0.236	0.078
65-74		0.008	0.003	0.006	0.336	0.122
75-84		-0.004	0.009	0.616	-0.172	0.343
85-94						
35-64	UNIVARIATE	0.006	0.001	0.001	0.267	0.056
65-94	UNIVARIATE	0.007	0.003	0.013	0.282	0.113
35-64	BIVARIATE	0.007	0.001	0.001	0.289	0.058
65-94	BIVARIATE	0.007	0.003	0.014	0.278	0.114
35-64	MULTIVARIATE	0.006	0.001	0.001	0.258	0.058
65-94	MULTIVARIATE	0.007	0.003	0.013	0.281	0.113

SEX: FEMALE
EVENT: ANGINA PECTORIS
RISK FACTOR: SERUM CHOLESTEROL
PERSONS AT RISK: PERSONS FREE OF CORONARY HEART DISEASE

RISK FACTOR	35-44 AT EXAM			45-54 AT EXAM			55-64 AT EXAM			65-74 AT EXAM		
	PERSON EXAMS	# OF EVENTS	ANNUAL RATE	PERSON EXAMS	# OF EVENTS	ANNUAL RATE	PERSON EXAMS	# OF EVENTS	ANNUAL RATE	PERSON EXAMS	# OF EVENTS	ANNUAL RATE
84-204	2350	0	0	2094	8	2	1625	12	4	789	9	6
205-234	1417	2	1	2558	13	3	2615	26	5	1401	15	5
235-264	806	0	0	2259	8	2	2958	31	5	1703	23	7
265-294	372	2	3	1475	6	2	2240	28	6	1287	20	8
295-1124	171	1	3	1101	5	2	1866	37	10	1031	24	12
UNKNOWN	500	0	0	474	7	7	229	6	13	37	0	0
TOTAL	5616	5	0	9959	47	2	11513	140	6	6248	91	7

RISK FACTOR	75-84 AT EXAM			85-94 AT EXAM			35-64 COMBINED			65-94 COMBINED		
	PERSON EXAMS	# OF EVENTS	ANNUAL RATE	PERSON EXAMS	# OF EVENTS	ANNUAL RATE	# OF EVENTS	CRUDE	AGE-ADJUSTED	# OF EVENTS	CRUDE	AGE-ADJUSTED
84-204	361	3	4	46			20	2	2	12	5	5
205-234	486	8	8	28			41	3	3	24	6	6
235-264	542	12	11	35			39	3	3	35	8	8
265-294	350	7	10	9			36	4	4	27	8	8
295-1124	205	1	2	5			43	7	6	25	10	9
UNKNOWN	7	0	0	0			13	5	8	0	0	0
TOTAL	1951	31	8	123			192	4	3	123	7	7

LOGISTIC REGRESSIONS

AGE AT EXAM		COEFFICIENT OF RISK FACTOR	SE COEFFICIENT	P-VALUE	STANDARDIZED COEFFICIENT	STD SE
35-44		0.023	0.005	0.001	0.921	0.196
45-54		0.003	0.003	0.386	0.117	0.135
55-64		0.007	0.002	0.001	0.325	0.078
65-74		0.006	0.002	0.011	0.256	0.101
75-84		-0.000	0.004	0.980	-0.004	0.181
85-94						
35-64	UNIVARIATE	0.008	0.001	0.001	0.379	0.061
65-94	UNIVARIATE	0.004	0.002	0.027	0.194	0.088
35-64	BIVARIATE	0.006	0.001	0.001	0.288	0.059
65-94	BIVARIATE	0.004	0.002	0.032	0.190	0.089
35-64	MULTIVARIATE	0.005	0.001	0.001	0.247	0.059
65-94	MULTIVARIATE	0.004	0.002	0.030	0.195	0.090

TABLE 7-5   DATE=27NOV85
AVERAGE ANNUAL INCIDENCE RATE FOR EVENT PER 1000 PERSONS AT RISK AT EXAM
BY AGE AND LEVEL OF RISK FACTOR AT EXAM: FRAMINGHAM STUDY 30-YEAR FOLLOWUP

SEX: MALE
EVENT: ANGINA PECTORIS
RISK FACTOR: HEMATOCRIT
PERSONS AT RISK: PERSONS FREE OF CORONARY HEART DISEASE

RISK FACTOR	35-44 AT EXAM			45-54 AT EXAM			55-64 AT EXAM			65-74 AT EXAM		
	PERSON EXAMS	# OF EVENTS	ANNUAL RATE	PERSON EXAMS	# OF EVENTS	ANNUAL RATE	PERSON EXAMS	# OF EVENTS	ANNUAL RATE	PERSON EXAMS	# OF EVENTS	ANNUAL RATE
25-42	623	3	2	942	5	3	1018	22	11	450	5	6
43-44	546	0	0	967	9	5	985	17	9	493	8	8
45-46	983	2	1	1854	14	4	1953	31	8	963	16	8
47-48	775	0	0	1510	14	5	1599	28	9-	861	16	9
49-70	1175	12	5	2066	33	8	2338	45	10	1127	17	8
UNKNOWN	370	4	5	385	5	6	156	6	19	19	0	0
TOTAL	4472	21	2	7724	80	5	8049	149	9	3913	62	8

RISK FACTOR	75-84 AT EXAM			85-94 AT EXAM			35-64 COMBINED			65-94 COMBINED		
	PERSON EXAMS	# OF EVENTS	ANNUAL RATE	PERSON EXAMS	# OF EVENTS	ANNUAL RATE	# OF EVENTS	CRUDE	AGE-ADJUSTED	# OF EVENTS	CRUDE	AGE-ADJUSTED
25-42	214	1	2	16			30	6	6	7	5	5
43-44	119	1	4	5			26	5	5	9	7	7
45-46	240	5	10	7			47	5	5	22	9	9
47-48	197	1	3	7			42	5	5	17	8	8
49-70	247	1	2	4			90	8	8	18	7	6
UNKNOWN	0	0		0			15	3	11	0	0	0
TOTAL	1017	9	4	39			250	6	6	73	7	7

LOGISTIC REGRESSIONS

AGE AT EXAM	COEFFICIENT OF RISK FACTOR	SE COEFFICIENT	P-VALUE	STANDARDIZED COEFFICIENT	STD SE
35-44	0.121	0.067	0.070	0.444	0.244
45-54	0.077	0.032	0.018	0.275	0.116
55-64	0.006	0.022	0.778	0.024	0.085
65-74	0.017	0.035	0.619	0.064	0.129
75-84	-0.028	0.079	0.727	-0.115	0.330
85-94					
35-64 UNIVARIATE	0.037	0.018	0.040	0.136	0.066
65-94 UNIVARIATE	0.011	0.031	0.734	0.040	0.119
35-64 BIVARIATE	0.035	0.018	0.051	0.128	0.066
65-94 BIVARIATE	0.008	0.031	0.793	0.032	0.120
35-64 MULTIVARIATE	0.015	0.018	0.399	0.056	0.067
65-94 MULTIVARIATE	-0.005	0.032	0.876	-0.019	0.124

SEX: FEMALE
EVENT: ANGINA PECTORIS
RISK FACTOR: HEMATOCRIT
PERSONS AT RISK: PERSONS FREE OF CORONARY HEART DISEASE

RISK FACTOR	35-44 AT EXAM			45-54 AT EXAM			55-64 AT EXAM			65-74 AT EXAM		
	PERSON EXAMS	# OF EVENTS	ANNUAL RATE	PERSON EXAMS	# OF EVENTS	ANNUAL RATE	PERSON EXAMS	# OF EVENTS	ANNUAL RATE	PERSON EXAMS	# OF EVENTS	ANNUAL RATE
21-39	1697	2	1	1843	5	1	1553	14	5	646	14	11
40-41	967	0	0	1741	10	3	2061	21	5	1069	11	5
42-43	1043	2	1	1955	9	2	2260	19	4	1256	16	6
44-45	856	1	1	2194	9	2	2847	41	7	1609	27	8
46-65	487	0	0	1681	7	2	2521	39	8	1626	23	7
UNKNOWN	566	0	0	545	7	6	271	6	11	42	0	0
TOTAL	5616	5	0	9959	47	2	11513	140	6	6248	91	7

RISK FACTOR	75-84 AT EXAM			85-94 AT EXAM			35-64 COMBINED			65-94 COMBINED		
	PERSON EXAMS	# OF EVENTS	ANNUAL RATE	PERSON EXAMS	# OF EVENTS	ANNUAL RATE	# OF EVENTS	CRUDE	AGE-ADJUSTED	# OF EVENTS	CRUDE	AGE-ADJUSTED
21-39	297	4	7	24			21	2	3	18	9	10
40-41	373	7	9	31			31	3	3	18	6	6
42-43	353	2	3	19			30	3	3	18	6	5
44-45	486	11	11	26			51	4	4	38	9	9
46-65	436	7	8	23			46	5	4	31	7	8
UNKNOWN	6	0	0	0			13	5	7	0	0	0
TOTAL	1951	31	8	123			192	4	4	123	7	7

LOGISTIC REGRESSIONS

AGE AT EXAM	COEFFICIENT OF RISK FACTOR	SE COEFFICIENT	P-VALUE	STANDARDIZED COEFFICIENT	STD SE
35-44	-0.065	0.118	0.586	-0.232	0.426
45-54	0.033	0.045	0.468	0.117	0.161
55-64	0.054	0.025	0.033	0.186	0.087
65-74	-0.002	0.030	0.947	-0.007	0.106
75-84	0.012	0.047	0.799	0.046	0.182
85-94					
35-64 UNIVARIATE	0.074	0.021	0.001	0.268	0.076
65-94 UNIVARIATE	0.004	0.025	0.865	0.016	0.091
35-64 BIVARIATE	0.042	0.022	0.054	0.151	0.078
65-94 BIVARIATE	0.003	0.025	0.899	0.012	0.091
35-64 MULTIVARIATE	0.026	0.022	0.233	0.093	0.078
65-94 MULTIVARIATE	0.000	0.025	0.996	0.000	0.091

TABLE 7-7 DATE=27NOV85
AVERAGE ANNUAL INCIDENCE RATE FOR EVENT PER 1000 PERSONS AT RISK AT EXAM
BY AGE AND LEVEL OF RISK FACTOR AT EXAM: FRAMINGHAM STUDY 30-YEAR FOLLOWUP

SEX: MALE
EVENT: ANGINA PECTORIS
RISK FACTOR: DIABETES MELLITUS
PERSONS AT RISK: PERSONS FREE OF CORONARY HEART DISEASE

RISK FACTOR	35-44 AT EXAM PERSON EXAMS	# OF EVENTS	ANNUAL RATE	45-54 AT EXAM PERSON EXAMS	# OF EVENTS	ANNUAL RATE	55-64 AT EXAM PERSON EXAMS	# OF EVENTS	ANNUAL RATE	65-74 AT EXAM PERSON EXAMS	# OF EVENTS	ANNUAL RATE
	4424	21	2	7486	76	5	7549	135	9	3504	54	8
	48	0	0	238	4	8	500	14	14	409	8	10
UNKNOWN TOTAL	4472	21	2	7724	80	5	8049	149	9	3913	62	8

RISK FACTOR	75-84 AT EXAM PERSON EXAMS	# OF EVENTS	ANNUAL RATE	85-94 AT EXAM PERSON EXAMS	# OF EVENTS	ANNUAL RATE	35-64 COMBINED # OF EVENTS	CRUDE	AGE-ADJUSTED	65-94 COMBINED # OF EVENTS	CRUDE	AGE-ADJUSTED
	900	7	4	34			232	6	6	62	7	7
	117	2	9	5			18	11	9	11	10	10
UNKNOWN TOTAL	1017	9	4	39			250	6	6	73	7	7

LOGISTIC REGRESSIONS

AGE AT EXAM	COEFFICIENT OF RISK FACTOR	SE COEFFICIENT	P-VALUE	STANDARDIZED COEFFICIENT	STD SE
44	-	-	-	-	-
54	0.511	0.517	0.323	0.088	0.089
64	0.459	0.285	0.107	0.111	0.069
74	0.243	0.382	0.526	0.074	0.117
84	0.797	0.808	0.324	0.254	0.258
94					
64 UNIVARIATE	0.664	0.247	0.007	0.128	0.048
94 UNIVARIATE	0.401	0.330	0.225	0.124	0.102
64 BIVARIATE	0.402	0.249	0.107	0.078	0.048
94 BIVARIATE	0.408	0.331	0.217	0.126	0.102
64 MULTIVARIATE	0.357	0.254	0.160	0.069	0.049
94 MULTIVARIATE	0.403	0.335	0.230	0.125	0.104

SEX: FEMALE
EVENT: ANGINA PECTORIS
RISK FACTOR: DIABETES MELLITUS
PERSONS AT RISK: PERSONS FREE OF CORONARY HEART DISEASE

RISK FACTOR	35-44 AT EXAM PERSON EXAMS	# OF EVENTS	ANNUAL RATE	45-54 AT EXAM PERSON EXAMS	# OF EVENTS	ANNUAL RATE	55-64 AT EXAM PERSON EXAMS	# OF EVENTS	ANNUAL RATE	65-74 AT EXAM PERSON EXAMS	# OF EVENTS	ANNUAL RATE
	5573	5	0	9760	43	2	10985	127	6	5780	82	7
	43	0	0	199	4	10	528	13	12	468	9	10
UNKNOWN TOTAL	5616	5	0	9959	47	2	11513	140	6	6248	91	7

RISK FACTOR	75-84 AT EXAM PERSON EXAMS	# OF EVENTS	ANNUAL RATE	85-94 AT EXAM PERSON EXAMS	# OF EVENTS	ANNUAL RATE	35-64 COMBINED # OF EVENTS	CRUDE	AGE-ADJUSTED	65-94 COMBINED # OF EVENTS	CRUDE	AGE-ADJUSTED
	1738	26	7	106			175	3	3	109	7	7
	213	5	12	17			17	11	9	14	10	10
UNKNOWN TOTAL	1951	31	8	123			192	4	4	123	7	2

LOGISTIC REGRESSIONS

AGE AT EXAM	COEFFICIENT OF RISK FACTOR	SE COEFFICIENT	P-VALUE	STANDARDIZED COEFFICIENT	STD SE
35-44					
45-54	1.534	0.528	0.004	0.215	0.074
55-64	0.769	0.295	0.009	0.161	0.062
65-74	0.309	0.354	0.383	0.081	0.093
75-84	0.459	0.494	0.352	0.143	0.154
85-94					
35-64 UNIVARIATE	1.216	0.257	0.001	0.202	0.043

TABLE 7-8   DATE=27NOV85

AVERAGE ANNUAL INCIDENCE RATE FOR EVENT PER 1000 PERSONS AT RISK AT EXAM
BY AGE AND LEVEL OF RISK FACTOR AT EXAM, FRAMINGHAM STUDY 30-YEAR FOLLOWUP

SEX: MALE
EVENT: ANGINA PECTORIS
RISK FACTOR: GLUCOSE IN URINE, DEFINITE OR TRACE
PERSONS AT RISK: PERSONS FREE OF CORONARY HEART DISEASE

RISK FACTOR	35-44 AT EXAM			45-54 AT EXAM			55-64 AT EXAM			65-74 AT EXAM		
	PERSON EXAMS	# OF EVENTS	ANNUAL RATE	PERSON EXAMS	# OF EVENTS	ANNUAL RATE	PERSON EXAMS	# OF EVENTS	ANNUAL RATE	PERSON EXAMS	# OF EVENTS	ANNUAL RATE
NO	4380	21	2	7535	76	5	7815	145	9	3767	57	8
YES	60	0	0	168	4	12	210	4	10	140	5	18
UNKNOWN	32	0	0	21	0	0	24	0	0	6	0	0
TOTAL	4472	21	2	7724	80	5	8049	149	9	3913	62	8

RISK FACTOR	75-84 AT EXAM			85-94 AT EXAM			35-64 COMBINED			65-94 COMBINED		
	PERSON EXAMS	# OF EVENTS	ANNUAL RATE	PERSON EXAMS	# OF EVENTS	ANNUAL RATE	# OF EVENTS	CRUDE	AGE-ADJUSTED	# OF EVENTS	CRUDE	AGE-ADJUSTED
NO	986	7	4	38			242	6	6	65	7	7
YES	28	2	36	1			8	9	8	8	24	25
UNKNOWN	3	0	0	0			0	0	0	0	0	0
TOTAL	1017	9	4	39			250	6	6	73	7	7

LOGISTIC REGRESSIONS

AGE AT EXAM		COEFFICIENT OF RISK FACTOR	SE COEFFICIENT	P-VALUE	STANDARDIZED COEFFICIENT	STD SE
35-44		-	-	-	-	-
45-54		0.873	0.519	0.093	0.128	0.076
55-64		0.027	0.512	0.958	0.004	0.082
65-74		0.880	0.475	0.064	0.164	0.088
75-84		2.376	0.826	0.004	0.389	0.135
85-94						
35-64	UNIVARIATE	0.404	0.363	0.265	0.059	0.053
65-94	UNIVARIATE	1.284	0.383	0.001	0.233	0.070
35-64	BIVARIATE	0.285	0.364	0.432	0.042	0.053
65-94	BIVARIATE	1.279	0.383	0.001	0.232	0.070
35-64	MULTIVARIATE	-0.041	0.419	0.922	-0.006	0.061
65-94	MULTIVARIATE	1.281	0.388	0.001	0.232	0.070

SEX: FEMALE
EVENT: ANGINA PECTORIS
RISK FACTOR: GLUCOSE IN URINE, DEFINITE OR TRACE
PERSONS AT RISK: PERSONS FREE OF CORONARY HEART DISEASE

RISK FACTOR	35-44 AT EXAM			45-54 AT EXAM			55-64 AT EXAM			65-74 AT EXAM		
	PERSON EXAMS	# OF EVENTS	ANNUAL RATE	PERSON EXAMS	# OF EVENTS	ANNUAL RATE	PERSON EXAMS	# OF EVENTS	ANNUAL RATE	PERSON EXAMS	# OF EVENTS	ANNUAL RATE
NO	5537	5	0	9841	45	2	11346	138	6	6151	88	7
YES	41	0	0	94	1	5	144	2	7	107	3	14
UNKNOWN	38	0	0	24	1	21	23	0	0	0	0	0
TOTAL	5616	5	0	9959	47	2	11513	140	6	6248	91	7

RISK FACTOR	75-84 AT EXAM			85-94 AT EXAM			35-64 COMBINED			65-94 COMBINED		
	PERSON EXAMS	# OF EVENTS	ANNUAL RATE	PERSON EXAMS	# OF EVENTS	ANNUAL RATE	# OF EVENTS	CRUDE	AGE-ADJUSTED	# OF EVENTS	CRUDE	AGE-ADJUSTED
NO	1919	29	8	120	3		188	4	4	118	7	7
YES	23	2	43	3			3	5	4	5	19	21
UNKNOWN	9	0	0	0			1	6	8	0	0	0
TOTAL	1951	31	8	123			192	4	4	123	7	7

LOGISTIC REGRESSIONS

AGE AT EXAM		COEFFICIENT OF RISK FACTOR	SE COEFFICIENT	P-VALUE	STANDARDIZED COEFFICIENT	STD SE
35-44		-	-	-	-	-
45-54		0.850	1.016	0.403	0.082	0.098
55-64		0.134	0.717	0.851	0.015	0.080
65-74		0.644	0.595	0.251	0.089	0.077
75-84		1.826	0.763	0.017	0.198	0.083
85-94						
35-64	UNIVARIATE	0.428	0.585	0.464	0.043	0.059
65-94	UNIVARIATE	0.980	0.465	0.035	0.123	0.058
35-64	BIVARIATE	0.311	0.586	0.596	0.031	0.059
65-94	BIVARIATE	0.976	0.465	0.036	0.123	0.058
35-64	MULTIVARIATE	0.002	0.595	0.997	0.000	0.060
65-94	MULTIVARIATE	0.791	0.476	0.096	0.099	0.060

TABLE 7-9   DATE=01MAR86
AVERAGE ANNUAL INCIDENCE RATE FOR EVENT PER 1000 PERSONS AT RISK AT EXAM
BY AGE AND LEVEL OF RISK FACTOR AT EXAM: FRAMINGHAM STUDY 30-YEAR FOLLOWUP

SEX: MALE
EVENT: ANGINA PECTORIS
RISK FACTOR: GLUCOSE INTOLERANCE
PERSONS AT RISK: PERSONS FREE OF CORONARY HEART DISEASE

RISK FACTOR	35-44 AT EXAM PERSON EXAMS	# OF EVENTS	ANNUAL RATE	45-54 AT EXAM PERSON EXAMS	# OF EVENTS	ANNUAL RATE	55-64 AT EXAM PERSON EXAMS	# OF EVENTS	ANNUAL RATE	6! PERSON EXAMS
NO	4335	20	2	7221	70	5	7145	128	9	322!
YES	137	1	4	503	10	10	904	21	12	68!
UNKNOWN	0	0	-	0	0	-	0	0	-	
TOTAL	4472	21	2	7724	80	5	8049	149	9	391!

RISK FACTOR	75-84 AT EXAM PERSON EXAMS	# OF EVENTS	ANNUAL RATE	85-94 AT EXAM PERSON EXAMS	# OF EVENTS	ANNUAL RATE	35-64 COMBINED # OF EVENTS	CRUDE	AGE-ADJUSTED	# OF EVE!
NO	802	7	4	27	12		218	6	6	5!
YES	215	2 *	5	12			32	10	9	1!
UNKNOWN	0	0	-	0			0	-	-	!
TOTAL	1017	9	4	39			250	6	6	7!

LOGISTIC REGRESSIONS

AGE AT EXAM		COEFFICIENT OF RISK FACTOR	SE COEFFICIENT	P-VALUE	STANDARDIZED COEFFICIENT
35-44		0.461	1.028	0.654	0.080
45-54		0.729	0.341	0.033	0.180
55-64		0.265	0.238	0.265	0.084
65-74		0.416	0.300	0.165	0.158
75-84		0.064	0.805	0.936	0.026
85-94					
35-64	UNIVARIATE	0.585	0.191	0.002	0.155
65-94	UNIVARIATE	0.382	0.274	0.163	0.148
35-64	BIVARIATE	0.367	0.193	0.057	0.097
65-94	BIVARIATE	0.396	0.275	0.150	0.153
35-64	MULTIVARIATE	0.281	0.201	0.163	0.074
65-94	MULTIVARIATE	0.412	0.277	0.137	0.159

SEX: FEMALE
EVENT: ANGINA PECTORIS
RISK FACTOR: GLUCOSE INTOLERANCE
PERSONS AT RISK: PERSONS FREE OF CORONARY HEART DISEASE

RISK FACTOR	35-44 AT EXAM PERSON EXAMS	# OF EVENTS	ANNUAL RATE	45-54 AT EXAM PERSON EXAMS	# OF EVENTS	ANNUAL RATE	55-64 AT EXAM PERSON EXAMS	# OF EVENTS	ANNUAL RATE	6! PERSON EXAMS
NO	5522	5	0	9588	43	2	10503	119	6	539!
YES	94	0	0	371	4	5	1010	21	10	85!
UNKNOWN	0	0	-	0	0	-	0	0	-	
TOTAL	5616	5	0	9959	47	2	11513	140	6	624!

RISK FACTOR	75-84 AT EXAM PERSON EXAMS	# OF EVENTS	ANNUAL RATE	85-94 AT EXAM PERSON EXAMS	# OF EVENTS	ANNUAL RATE	35-64 COMBINED # OF EVENTS	CRUDE	AGE-ADJUSTED	# OF EVE!
NO	1581	20	6	100	23		167	3	3	9!
YES	370	11	15	23			25	8	6	2!
UNKNOWN	0	0	-	0			0	-	-	
TOTAL	1951	31	8	123			192	4	4	12!

LOGISTIC REGRESSIONS

AGE AT EXAM		COEFFICIENT OF RISK FACTOR	SE COEFFICIENT	P-VALUE	STANDARDIZED COEFFICIENT
35-44		-	-	-	-
45-54		0.884	0.525	0.093	0.167
55-64		0.617	0.239	0.010	0.174
65-74		0.146	0.293	0.618	0.050
75-84		0.872	0.380	0.022	0.342
85-94					
35-64	UNIVARIATE	0.966	0.216	0.001	0.219
45-94	UNIVARIATE	0.380	0.226	0.093	0.135

SEX: MALE
EVENT: ANGINA PECTORIS
RISK FACTOR: METROPOLITAN RELATIVE WEIGHT (PERCENT)
PERSONS AT RISK: PERSONS FREE OF CORONARY HEART DISEASE

35-44 AT EXAM			45-54 AT EXAM			55-64 AT EXAM			65-74 AT EXAM		
PERSON EXAMS	# OF EVENTS	ANNUAL RATE	PERSON EXAMS	# OF EVENTS	ANNUAL RATE	PERSON EXAMS	# OF EVENTS	ANNUAL RATE	PERSON EXAMS	# OF EVENTS	ANNUAL RATE
723	0	0	1131	5	2	1155	16	7	686	5	4
965	4	2	1540	14	5	1655	26	8	851	11	6
1122	4	2	2037	21	5	2175	39	9	1069	18	8
915	6	3	1645	20	6	1674	33	10	766	18	12
747	7	3	1370	20	7	1389	35	13	541	10	9
0	0	-	1	0	0	0	0	0	0	0	-
4472	21	2	7724	80	5	8049	149	9	3913	62	8

75-84 AT EXAM			85-94 AT EXAM			35-64 COMBINED			65-94 COMBINED		
PERSON EXAMS	# OF EVENTS	ANNUAL RATE	PERSON EXAMS	# OF EVENTS	ANNUAL RATE	# OF EVENTS	CRUDE	AGE-ADJUSTED	# OF EVENTS	CRUDE	AGE-ADJUSTED
224	2	4	18			21	3	4	7	4	4
252	2	4	9			44	5	5	14	6	6
296	1	2	12			64	6	6	20	7	7
155	3	10	0			59	7	7	21	11	11
90	1	6	8			62	9	9	11	9	8
0	0	-	0			0	0	0	0	-	
1017	9	4	39			250	6	6	73	7	7

LOGISTIC REGRESSIONS

	COEFFICIENT OF RISK FACTOR	SE COEFFICIENT	P-VALUE	STANDARDIZED COEFFICIENT	STD SE
	0.032	0.012	0.008	0.515	0.194
	0.018	0.006	0.004	0.297	0.104
	0.013	0.005	0.011	0.200	0.079
	0.015	0.008	0.048	0.242	0.122
	0.011	0.022	0.599	0.168	0.320
NIVARIATE	0.016	0.004	0.001	0.264	0.060
NIVARIATE	0.015	0.007	0.036	0.235	0.112
IVARIATE	0.016	0.004	0.001	0.264	0.060
IVARIATE	0.015	0.007	0.041	0.231	0.113
ULTIVARIATE	0.015	0.004	0.001	0.237	0.063
ULTIVARIATE	0.011	0.008	0.127	0.179	0.117

SEX: FEMALE
EVENT: ANGINA PECTORIS
RISK FACTOR: METROPOLITAN RELATIVE WEIGHT (PERCENT)
PERSONS AT RISK: PERSONS FREE OF CORONARY HEART DISEASE

35-44 AT EXAM			45-54 AT EXAM			55-64 AT EXAM			65-74 AT EXAM		
PERSON EXAMS	# OF EVENTS	ANNUAL RATE	PERSON EXAMS	# OF EVENTS	ANNUAL RATE	PERSON EXAMS	# OF EVENTS	ANNUAL RATE	PERSON EXAMS	# OF EVENTS	ANNUAL RATE
1694	0	0	2095	10	2	2045	18	4	1216	16	7
1430	2	1	2331	5	1	2409	17	4	1145	19	8
1058	2	1	2107	12	3	2479	23	5	1395	16	6
566	0	0	1372	4	1	1743	24	7	982	19	10
860	1	1	2054	16	4	2837	58	10	1510	21	7
8	0	0	0	0	-	0	0	-	0	0	-
5616	5	0	9959	47	2	11513	140	6	6248	91	7

75-84 AT EXAM			85-94 AT EXAM			35-64 COMBINED			65-94 COMBINED		
PERSON EXAMS	# OF EVENTS	ANNUAL RATE	PERSON EXAMS	# OF EVENTS	ANNUAL RATE	# OF EVENTS	CRUDE	AGE-ADJUSTED	# OF EVENTS	CRUDE	AGE-ADJUSTED
517	7	7	41			28	2	3	23	6	7
375	3	3	28			24	2	3	21	7	7
399	9	11	21			37	3	3	25	7	7
266	5	9	16			28	4	3	24	9	9
394	8	10	17			75	7	6	30	8	8
0	0	-	0			0	0	0	0	-	
1951	31	8	123			192	4	4	123	7	7

LOGISTIC REGRESSIONS

	COEFFICIENT OF RISK FACTOR	SE COEFFICIENT	P-VALUE	STANDARDIZED COEFFICIENT	STD SE
	0.020	0.016	0.201	0.406	0.317
	0.013	0.005	0.020	0.273	0.118
	0.014	0.003	0.001	0.317	0.069
	0.001	0.005	0.889	0.015	0.105
	0.009	0.008	0.282	0.182	0.169
IVARIATE	0.016	0.003	0.001	0.355	0.055
IVARIATE	0.003	0.004	0.416	0.071	0.088
VARIATE	0.014	0.003	0.001	0.307	0.058
VARIATE	0.003	0.004	0.449	0.067	0.088
LTIVARIATE	0.007	0.003	0.020	0.158	0.068
LTIVARIATE	0.001	0.004	0.799	0.023	0.091

TABLE 7-11   DATE=27NOV85
AVERAGE ANNUAL INCIDENCE RATE FOR EVENT PER 1000 PERSONS AT RISK AT EXAM
BY AGE AND LEVEL OF RISK FACTOR AT EXAM: FRAMINGHAM STUDY 30-YEAR FOLLOWUP

SEX: MALE
EVENT: ANGINA PECTORIS
RISK FACTOR: VITAL CAPACITY - HEIGHT INDEX (ML/INCH)
PERSONS AT RISK: PERSONS FREE OF CORONARY HEART DISEASE

RISK FACTOR	35-44 AT EXAM			45-54 AT EXAM			55-64 AT EXAM			65-74 AT EXAM		
	PERSON EXAMS	# OF EVENTS	ANNUAL RATE	PERSON EXAMS	# OF EVENTS	ANNUAL RATE	PERSON EXAMS	# OF EVENTS	ANNUAL RATE	PERSON EXAMS	# OF EVENTS	ANNUAL RATE
8-44	269	2	4	1062	18	8	2045	38	9	1752	29	8
45-49	411	2	2	1125	13	6	1480	23	8	749	14	9
50-54	747	1	1	1658	20	6	1835	39	11	712	8	6
55-59	1022	8	4	1739	13	4	1416	34	12	442	8	9
60-85	2010	7	2	2135	16	4	1255	15	6	246	3	6
UNKNOWN	13	1	38	5	0	0	18	0	0	12	0	0
TOTAL	4472	21	2	7724	80	5	8049	149	9	3913	62	8

RISK FACTOR	75-84 AT EXAM			85-94 AT EXAM			35-64 COMBINED			65-94 COMBINED		
	PERSON EXAMS	# OF EVENTS	ANNUAL RATE	PERSON EXAMS	# OF EVENTS	ANNUAL RATE	# OF EVENTS	CRUDE	AGE-ADJUSTED	# OF EVENTS	CRUDE	AGE-ADJUSTED
8-44	649	6	5	24			58	9	8	37	8	9
45-49	172	2	6	6			38	6	7	16	9	9
50-54	97	0	0	7			60	7	7	8	5	4
55-59	65	1	8	2			55	7	7	9	9	9
60-85	26	0	0	0			38	4	4	3	6	5
UNKNOWN	8	0	0	0			1	14	8	0	0	0
TOTAL	1017	9	4	39			250	6	6	73	7	7

LOGISTIC REGRESSIONS

AGE AT EXAM		COEFFICIENT OF RISK FACTOR	SE COEFFICIENT	P-VALUE	STANDARDIZED COEFFICIENT	STD SE
35-44		-0.022	0.024	0.361	-0.198	0.217
45-54		-0.033	0.012	0.006	-0.296	0.108
55-64		0.001	0.009	0.926	0.008	0.083
65-74		-0.004	0.014	0.795	-0.033	0.128
75-84		0.008	0.036	0.818	0.077	0.334
85-94						
35-64	UNIVARIATE	-0.025	0.006	0.001	-0.237	0.062
65-94	UNIVARIATE	-0.001	0.012	0.951	-0.007	0.118
35-64	BIVARIATE	-0.009	0.007	0.190	-0.087	0.066
65-94	BIVARIATE	-0.002	0.013	0.851	-0.023	0.121
35-64	MULTIVARIATE	-0.003	0.007	0.644	-0.032	0.070
65-94	MULTIVARIATE	-0.002	0.013	0.881	-0.018	0.133

SEX: FEMALE
EVENT: ANGINA PECTORIS
RISK FACTOR: VITAL CAPACITY - HEIGHT INDEX (ML/INCH)
PERSONS AT RISK: PERSONS FREE OF CORONARY HEART DISEASE

RISK FACTOR	35-44 AT EXAM			45-54 AT EXAM			55-64 AT EXAM			65-74 AT EXAM		
	PERSON EXAMS	# OF EVENTS	ANNUAL RATE	PERSON EXAMS	# OF EVENTS	ANNUAL RATE	PERSON EXAMS	# OF EVENTS	ANNUAL RATE	PERSON EXAMS	# OF EVENTS	ANNUAL RATE
4-29	122	2	8	534	5	5	1461	27	9	1747	40	11
30-34	289	1	2	1217	7	3	2418	34	7	1715	20	6
35-39	769	0	0	2077	18	4	3048	37	6	1529	14	5
40-44	1427	1	0	2804	10	2	2706	32	6	854	12	7
45-92	2988	1	0	3310	7	1	1859	10	3	397	5	6
UNKNOWN	21	0	0	17	0	0	21	0	0	6	0	0
TOTAL	5616	5	0	9959	47	2	11513	140	6	6248	91	7

RISK FACTOR	75-84 AT EXAM			85-94 AT EXAM			35-64 COMBINED			65-94 COMBINED		
	PERSON EXAMS	# OF EVENTS	ANNUAL RATE	PERSON EXAMS	# OF EVENTS	ANNUAL RATE	# OF EVENTS	CRUDE	AGE-ADJUSTED	# OF EVENTS	CRUDE	AGE-ADJUSTED
4-29	930	14	8	80			34	8	7	55	10	10
30-34	569	9	8	29			42	5	4	29	6	6
35-39	314	7	11	13			55	4	4	21	6	6
40-44	111	1	5	1			43	3	3	13	7	6
45-92	25	0	0	0			18	1	2	5	6	5
UNKNOWN	2	0	0	0			0	0	0	0	0	0
TOTAL	1951	31	8	123			192	4	4	123	7	7

LOGISTIC REGRESSIONS

AGE AT EXAM		COEFFICIENT OF RISK FACTOR	SE COEFFICIENT	P-VALUE	STANDARDIZED COEFFICIENT	STD SE
35-44		-0.134	0.049	0.006	-0.980	0.359
45-54		-0.065	0.019	0.001	-0.478	0.138
55-64		-0.037	0.011	0.001	-0.279	0.083
65-74		-0.033	0.014	0.018	-0.244	0.104
75-84		0.011	0.026	0.672	0.077	0.182
85-94						
35-64	UNIVARIATE	-0.068	0.009	0.001	-0.541	0.069
65-94	UNIVARIATE	-0.024	0.012	0.047	-0.178	0.090
35-64	BIVARIATE	-0.045	0.009	0.001	-0.355	0.075
65-94	BIVARIATE	-0.028	0.012	0.024	-0.209	0.093
35-64	MULTIVARIATE	-0.032	0.010	0.002	-0.256	0.081
65-94	MULTIVARIATE	-0.024	0.013	0.059	-0.181	0.096

TABLE 7-12   DATE=27NOV85
AVERAGE ANNUAL INCIDENCE RATE FOR EVENT PER 1000 PERSONS AT RISK AT EXAM
BY AGE AND LEVEL OF RISK FACTOR AT EXAM: FRAMINGHAM STUDY 30-YEAR FOLLOWUP

SEX: MALE
EVENT: ANGINA PECTORIS
RISK FACTOR: HEART RATE (PER MINUTE)
PERSONS AT RISK: PERSONS FREE OF CORONARY HEART DISEASE

RISK FACTOR	35-44 AT EXAM			45-54 AT EXAM			55-64 AT EXAM			65-74 AT EXAM		
	PERSON EXAMS	# OF EVENTS	ANNUAL RATE	PERSON EXAMS	# OF EVENTS	ANNUAL RATE	PERSON EXAMS	# OF EVENTS	ANNUAL RATE	PERSON EXAMS	# OF EVENTS	ANNUAL RATE
30-67	1284	2	1	1949	19	5	2309	39	8	1183	20	8
68-75	1492	7	2	2490	25	5	2490	46	9	1139	14	6
76-83	722	4	3	1394	17	6	1374	30	11	624	10	8
84-91	579	6	5	1097	13	6	1064	19	9	487	12	12
92-220	395	2	3	794	6	4	812	15	9	460	6	7
UNKNOWN	0	0	-	0	0	-	0	0	-	0	0	-
TOTAL	4472	21	2	7724	80	5	8049	149	9	3913	62	8

RISK FACTOR	75-84 AT EXAM			85-94 AT EXAM			35-64 COMBINED			65-94 COMBINED		
	PERSON EXAMS	# OF EVENTS	ANNUAL RATE	PERSON EXAMS	# OF EVENTS	ANNUAL RATE	# OF EVENTS	CRUDE	AGE-ADJUSTED	# OF EVENTS	CRUDE	AGE-ADJUSTED
30-67	301	1	2	11			60	5	5	21	7	7
68-75	296	3	5	12			78	6	6	18	6	6
76-83	192	3	8	10			51	7	7	13	8	8
84-91	123	1	4	5			38	7	7	14	11	11
92-220	105	1	5	1			23	6	6	7	6	6
UNKNOWN	0	0	-	0			0	-	-	0	-	-
TOTAL	1017	9	4	39			250	6	6	73	7	7

LOGISTIC REGRESSIONS

AGE AT EXAM		COEFFICIENT OF RISK FACTOR	SE COEFFICIENT	P-VALUE	STANDARDIZED COEFFICIENT	STD SE
35-44		0.030	0.015	0.053	0.371	0.191
45-54		0.003	0.009	0.766	0.033	0.111
55-64		0.005	0.006	0.456	0.060	0.081
65-74		0.005	0.009	0.622	0.062	0.125
75-84		0.030	0.021	0.156	0.402	0.283
85-94						
35-64	UNIVARIATE	0.006	0.005	0.191	0.081	0.062
65-94	UNIVARIATE	0.009	0.008	0.292	0.119	0.113
35-64	BIVARIATE	0.006	0.005	0.190	0.081	0.062
65-94	BIVARIATE	0.009	0.008	0.291	0.119	0.113
35-64	MULTIVARIATE	-0.001	0.005	0.844	-0.013	0.066
65-94	MULTIVARIATE	0.005	0.009	0.560	0.068	0.117

SEX: FEMALE
EVENT: ANGINA PECTORIS
RISK FACTOR: HEART RATE (PER MINUTE)
PERSONS AT RISK: PERSONS FREE OF CORONARY HEART DISEASE

RISK FACTOR	35-44 AT EXAM			45-54 AT EXAM			55-64 AT EXAM			65-74 AT EXAM		
	PERSON EXAMS	# OF EVENTS	ANNUAL RATE	PERSON EXAMS	# OF EVENTS	ANNUAL RATE	PERSON EXAMS	# OF EVENTS	ANNUAL RATE	PERSON EXAMS	# OF EVENTS	ANNUAL RATE
30-67	1092	0	0	1897	10	3	2305	28	6	1103	14	6
68-75	1803	2	1	3244	14	2	3468	40	6	1668	31	9
76-83	1051	2	1	1911	7	2	2351	27	6	1358	18	7
84-91	939	1	1	1604	9	3	1780	23	6	1089	13	6
92-220	731	0	0	1303	5	2	1609	22	7	1030	15	7
UNKNOWN	0	0	-	0	0	-	0	0	-	0	0	-
TOTAL	5616	5	0	9959	47	2	11513	140	6	6248	91	7

RISK FACTOR	75-84 AT EXAM			85-94 AT EXAM			35-64 COMBINED			65-94 COMBINED		
	PERSON EXAMS	# OF EVENTS	ANNUAL RATE	PERSON EXAMS	# OF EVENTS	ANNUAL RATE	# OF EVENTS	CRUDE	AGE-ADJUSTED	# OF EVENTS	CRUDE	AGE-ADJUSTED
30-67	360	5	7	26			38	4	4	19	6	6
68-75	462	8	9	32			58	3	3	39	9	9
76-83	428	7	8	24			36	3	3	25	7	7
84-91	330	8	12	23			33	4	4	21	7	7
92-220	371	3	4	18			27	4	4	19	7	7
UNKNOWN	0	0	-	0			0	-	-	0	-	-
TOTAL	1951	31	8	123			192	4	4	123	7	7

LOGISTIC REGRESSIONS

AGE AT EXAM		COEFFICIENT OF RISK FACTOR	SE COEFFICIENT	P-VALUE	STANDARDIZED COEFFICIENT	STD SE
35-44		0.010	0.033	0.763	0.126	0.416
45-54		-0.002	0.012	0.897	-0.019	0.147
55-64		0.000	0.007	0.963	0.004	0.085
65-74		-0.005	0.008	0.547	-0.065	0.108
75-84		-0.007	0.013	0.595	-0.100	0.188
85-94						
35-64	UNIVARIATE	0.000	0.006	0.960	0.004	0.072
65-94	UNIVARIATE	-0.004	0.007	0.529	-0.058	0.093
35-64	BIVARIATE	0.000	0.006	0.997	0.000	0.072
65-94	BIVARIATE	-0.004	0.007	0.538	-0.057	0.093
35-64	MULTIVARIATE	-0.007	0.006	0.270	-0.085	0.077
65-94	MULTIVARIATE	-0.006	0.007	0.391	-0.080	0.093

TABLE 7-13   DATE=05APR85
AVERAGE ANNUAL INCIDENCE RATE FOR EVENT PER 1000 PERSONS AT RISK AT EXAM
BY AGE AND LEVEL OF RISK FACTOR AT EXAM: FRAMINGHAM STUDY 30-YEAR FOLLOWUP

SEX: MALE
EVENT: ANGINA PECTORIS
RISK FACTOR: CIGARETTES SMOKED (NUMBER PER DAY)
PERSONS AT RISK: PERSONS FREE OF CORONARY HEART DISEASE

RISK FACTOR	35-44 AT EXAM			45-54 AT EXAM			55-64 AT EXAM			65-74 AT EXAM		
	PERSON EXAMS	# OF EVENTS	ANNUAL RATE	PERSON EXAMS	# OF EVENTS	ANNUAL RATE	PERSON EXAMS	# OF EVENTS	ANNUAL RATE	PERSON EXAMS	# OF EVENTS	ANNUAL RATE
NONE	1383	5	1	3068	29	5	4295	72	8	2611	47	9
1-10	433	3	3	716	6	4	769	12	8	374	4	5
11-20	1485	9	3	2051	25	6	1682	32	10	526	5	5
21-40	987	3	2	1602	18	6	1119	28	13	369	6	8
41-90	158	3	9	252	2	4	162	5	15	27	0	0
UNKNOWN	26	0	0	35	0	0	22	0	0	6	0	0
TOTAL	4472	21	2	7724	80	5	8049	149	9	3913	62	8

RISK FACTOR	75-84 AT EXAM			85-94 AT EXAM			35-64 COMBINED			65-94 COMBINED		
	PERSON EXAMS	# OF EVENTS	ANNUAL RATE	PERSON EXAMS	# OF EVENTS	ANNUAL RATE	# OF EVENTS	CRUDE	AGE-ADJUSTED	# OF EVENTS	CRUDE	AGE-ADJUSTED
NONE	797	9	6	27			104	6	5	57	8	8
1-10	92	0	0	6			21	5	5	4	4	4
11-20	77	0	0	5			66	6	7	6	5	5
21-40	46	0	0	1			49	7	8	6	7	6
41-90	5	0	0	0			10	9	9	0	0	0
UNKNOWN	0	0	-	0			0	0	0	0	0	0
TOTAL	1017	9	4	39			250	6	6	73	7	7

LOGISTIC REGRESSIONS

AGE AT EXAM		COEFFICIENT OF RISK FACTOR	SE COEFFICIENT	P-VALUE	STANDARDIZED COEFFICIENT	STD SE
35-44		0.021	0.015	0.143	0.301	0.205
45-54		0.004	0.007	0.556	0.065	0.110
55-64		0.012	0.005	0.050	0.163	0.075
65-74		-0.012	0.012	0.318	-0.145	0.145
75-84		-				
85-94						
35-64	UNIVARIATE	0.005	0.004	0.221	0.075	0.062
65-94	UNIVARIATE	-0.013	0.012	0.275	-0.148	0.136
35-64	BIVARIATE	0.011	0.004	0.008	0.162	0.061
65-94	BIVARIATE	-0.014	0.012	0.242	-0.160	0.137
35-64	MULTIVARIATE	0.014	0.004	0.001	0.200	0.062
65-94	MULTIVARIATE	-0.014	0.012	0.237	-0.161	0.136

SEX: FEMALE
EVENT: ANGINA PECTORIS
RISK FACTOR: CIGARETTES SMOKED (NUMBER PER DAY)
PERSONS AT RISK: PERSONS FREE OF CORONARY HEART DISEASE

RISK FACTOR	35-44 AT EXAM			45-54 AT EXAM			55-64 AT EXAM			65-74 AT EXAM		
	PERSON EXAMS	# OF EVENTS	ANNUAL RATE	PERSON EXAMS	# OF EVENTS	ANNUAL RATE	PERSON EXAMS	# OF EVENTS	ANNUAL RATE	PERSON EXAMS	# OF EVENTS	ANNUAL RATE
NONE	2582	2	0	5528	33	3	7595	100	7	4863	72	7
1-10	1314	1	0	1726	5	1	1566	14	4	626	7	6
11-20	1371	2	1	2009	7	2	1696	20	6	568	11	10
21-40	298	0	0	622	2	2	602	6	5	172	1	3
41-90	22	0	0	37	0	0	16	0	0	0	0	-
UNKNOWN	29	0	0	37	0	0	38	0	0	19	0	0
TOTAL	5616	5	0	9959	47	2	11513	140	6	6248	91	7

RISK FACTOR	75-84 AT EXAM			85-94 AT EXAM			35-64 COMBINED			65-94 COMBINED		
	PERSON EXAMS	# OF EVENTS	ANNUAL RATE	PERSON EXAMS	# OF EVENTS	ANNUAL RATE	# OF EVENTS	CRUDE	AGE-ADJUSTED	# OF EVENTS	CRUDE	AGE-ADJUSTED
NONE	1710	28	8	117			135	4	4	101	8	7
1-10	137	3	11	4			20	2	2	10	7	7
11-20	84	0	0	2			29	5	3	11	8	8
21-40	16	0	0	0			8	3	3	1	3	2
41-90	0	0	-	0			0	0	0	0		
UNKNOWN	4	0	0	0			0	0	0	0	0	0
TOTAL	1951	31	8	123			192	4	3	123	7	7

LOGISTIC REGRESSIONS

AGE AT EXAM		COEFFICIENT OF RISK FACTOR	SE COEFFICIENT	P-VALUE	STANDARDIZED COEFFICIENT	STD SE
35-44		0.017	0.041	0.680	0.167	0.406
45-54		-0.022	0.017	0.196	-0.223	0.173
55-64		-0.009	0.010	0.354	-0.085	0.092
65-74		-0.004	0.014	0.788	-0.030	0.110
75-84		-0.066	0.066	0.320	-0.335	0.337
85-94						
35-64	UNIVARIATE	-0.019	0.008	0.024	-0.189	0.084
65-94	UNIVARIATE	-0.009	0.014	0.518	-0.064	0.099
35-64	BIVARIATE	-0.010	0.008	0.216	-0.101	0.081
65-94	BIVARIATE	-0.010	0.014	0.468	-0.073	0.100
35-64	MULTIVARIATE	-0.004	0.008	0.607	-0.042	0.082
65-94	MULTIVARIATE	-0.006	0.014	0.652	-0.045	0.100

· 153

TABLE 7-14   DATE=27NOV85
AVERAGE ANNUAL INCIDENCE RATE FOR EVENT PER 1000 PERSONS AT RISK AT EXAM
BY AGE AND LEVEL OF RISK FACTOR AT EXAM: FRAMINGHAM STUDY 30-YEAR FOLLOWUP

SEX: MALE·
EVENT: ANGINA PECTORIS
RISK FACTOR: ALBUMIN IN URINE, DEFINITE
PERSONS AT RISK: PERSONS FREE OF CORONARY HEART DISEASE

RISK FACTOR	35-44 AT EXAM PERSON EXAMS	# OF EVENTS	ANNUAL RATE	45-54 AT EXAM PERSON EXAMS	# OF EVENTS	ANNUAL RATE	55-64 AT EXAM PERSON EXAMS	# OF EVENTS	ANNUAL RATE	65-74 AT EXAM PERSON EXAMS	# OF EVENTS	ANNUAL RATE
NO	4366	20	2	7599	78	5	7838	148	9	3792	60	8
YES	70	1	7	99	2	10	184	1	3	115	2	9
UNKNOWN	36·	0	0	26	0	0	27	0	0	6·	0	0
TOTAL	4472	21	2	7724	80	5	8049	149	9	3913·	62	8

RISK FACTOR	75-84 AT EXAM PERSON EXAMS	# OF EVENTS	ANNUAL RATE	85-94 AT EXAM PERSON EXAMS	# OF EVENTS	ANNUAL RATE	35-64 COMBINED # OF EVENTS	CRUDE	AGE-ADJUSTED	65-94 COMBINED # OF EVENTS	CRUDE	AGE-ADJUSTED
NO	989	9	5	32	0	0	246	6	6	71	7	7
YES	25	0	0	7			4	6	7	2	7	7
UNKNOWN	3	0	0	0			0	0	0	0	0	0
TOTAL	1017	9	4	39			250	6	6	73	7	7

LOGISTIC REGRESSIONS

AGE AT EXAM		COEFFICIENT OF RISK FACTOR	SE COEFFICIENT	P-VALUE	STANDARDIZED COEFFICIENT	STD SE·
35-44		1.147	1.032	0.266	0.143	0.129
45-54		0.687	0.723	0.342	0.077	0.082
55-64		-1.259	1.006	0.211	-0.188	0.151
65-74		0.096	0.725	0.895	0.016	0.123
75-84		-	-	-	-	-
85-94						
35-64	UNIVARIATE	-0.093	0.507	0.854	-0.012	0.066
65-94	UNIVARIATE	-0.062	0.722	0.910	-0.014	0.122
35-64	BIVARIATE	-0.225	0.508	0.658	-0.030	0.067
65-94	BIVARIATE	-0.078·	0.722	0.914	-0.013	0.122
35-64	MULTIVARIATE	-0.887·	0.590	0.133	-0.116	0.077
65-94	MULTIVARIATE	-0.238	0.728	0.744	-0.040	0.124

SEX: FEMALE
EVENT: ANGINA PECTORIS
RISK FACTOR: ALBUMIN IN URINE, DEFINITE
PERSONS AT RISK: PERSONS FREE OF CORONARY HEART DISEASE

RISK FACTOR	35-44 AT EXAM PERSON EXAMS	# OF EVENTS	ANNUAL RATE	45-54 AT EXAM PERSON EXAMS	# OF EVENTS	ANNUAL RATE	55-64 AT EXAM PERSON EXAMS	# OF EVENTS	ANNUAL RATE	65-74 AT EXAM PERSON EXAMS	# OF EVENTS	ANNUAL RATE
NO	5353	4	0	9656	43	2	11263	136	6	6118	89	7
YES ·	225	1	2	278	3	5	225	4	9	120	2	8
UNKNOWN	38	0	0	25	1	20	25	0	0	10	0	0
TOTAL	5616	5	0	9959	47	2	11513	140	6	6248	91	7

RISK FACTOR	75-84 AT EXAM PERSON EXAMS	# OF EVENTS.RATE	ANNUAL	85-94 AT EXAM PERSON EXAMS	# OF EVENTS	ANNUAL RATE	35-64 COMBINED # OF EVENTS	CRUDE	AGE-ADJUSTED	65-94 COMBINED # OF EVENTS	CRUDE	AGE-ADJUSTED
NO	1904	31	8	120	3		183	3	3	121	7	7
YES ·	38	0	0	3			8	5	6	2	6	6
UNKNOWN	9	0	0	0			1	6	7	0	0	0
TOTAL	1951	31	8	123			192	4	4	123	7	7

LOGISTIC REGRESSIONS

AGE AT EXAM		COEFFICIENT OF RISK FACTOR	SE COEFFICIENT	P-VALUE	STANDARDIZED COEFFICIENT	STD SE
35-44		1.787	1.120	0.111	0.352	0.220
45-54		0.892	0.600	0.137	0.147	0.099
55-64		0.393	0.512	0.443	0.054	0.071
65-74		0.138	0.721	0.848	0.019	0.099
75-84		-	-	-	-	-
85-94						
35-64	UNIVARIATE	0.460	0.363	0.205	0.075	0.059
65-94	UNIVARIATE	-0.182	0.717	0.800	-0.025	0.099
35-64	BIVARIATE	0.673	0.365	0.065	0.109	0.059
65-94	BIVARIATE	-0.182·	0.717	0.800	-0.025	0.099
35-64	MULTIVARIATE	0.293·	0.388	0.450	0.048	0.063
65-94	MULTIVARIATE	-0.461	0.728	0.527	-0.064	0.100

RISK FACTOR	35-44 AT EXAM PERSON EXAMS	# OF EVENTS	ANNUAL RATE	45-54 AT EXAM PERSON EXAMS	# OF EVENTS	ANNUAL RATE	55-64 AT EXAM PERSON EXAMS	# OF EVENTS	ANNUAL RATE
NO	4063	16	2	6383	60	5	5967	103	9
MILD	185	4	11	650	11	8	962	21	11
DEFINITE	219	1	2	687	8	6	1119	25	11
UNKNOWN	5	0	0	4	1	125	1	0	0
TOTAL	4472	21	2	7724	80	5	8049	149	9

RISK FACTOR	75-84 AT EXAM PERSON EXAMS	# OF EVENTS	ANNUAL RATE	85-94 AT EXAM PERSON EXAMS	# OF EVENTS	ANNUAL RATE	35-64 COMBINED # OF EVENTS	CRUDE	AGE-ADJUSTED
NO	589	6	5	29			179	5	6
MILD	178	1	3	5			36	10	10
DEFINITE	250	2	4	5			34	8	7
UNKNOWN	0	0	-	0			1	50	48
TOTAL	1017	9	4	39			250	6	6

LOGISTIC REGRESSIONS

AGE AT EXAM	COEFFICIENT OF RISK FACTOR	SE COEFFICIENT	P-VALUE	STANDARDIZED COEFFICIENT
35-44	0.468	0.334	0.162	0.218
45-54	0.192	0.165	0.245	0.117
55-64	0.145	0.107	0.178	0.104
65-74	0.261	0.148	0.078	0.207
75-84	-0.165	0.419	0.693	-0.140
85-94				
35-64 UNIVARIATE	0.273	0.086	0.001	0.174
65-94 UNIVARIATE	0.185	0.138	0.178	0.150
35-64 BIVARIATE	0.162	0.087	0.064	0.103
65-94 BIVARIATE	0.194	0.138	0.159	0.157
35-64 MULTIVARIATE	0.107	0.093	0.251	0.068
65-94 MULTIVARIATE	0.160	0.145	0.269	0.129

SEX: FEMALE
EVENT: ANGINA PECTORIS
RISK FACTOR: HEART ENLARGEMENT BY X-RAY
PERSONS AT RISK: PERSONS FREE OF CORONARY HEART DISEASE

RISK FACTOR	35-44 AT EXAM PERSON EXAMS	# OF EVENTS	ANNUAL RATE	45-54 AT EXAM PERSON EXAMS	# OF EVENTS	ANNUAL RATE	55-64 AT EXAM PERSON EXAMS	# OF EVENTS	ANNUAL RATE	PE EX
NO	5164	4	0	8017	31	2	7522	67	4	4
MILD	199	0	0	943	5	3	1769	26	7	
DEFINITE	245	1	2	993	11	6	2218	47	11	
UNKNOWN	8	0	0	6	0	0	4	0	0	
TOTAL	5616	5	8	9959	47	2	11513	140	6	

RISK FACTOR	75-84 AT EXAM PERSON EXAMS	# OF EVENTS	ANNUAL RATE	85-94 AT EXAM PERSON EXAMS	# OF EVENTS	ANNUAL RATE	35-64 COMBINED # OF EVENTS	CRUDE	AGE-ADJUSTED
NO	702	7	5	32			102	2	3
MILD	373	6	8	30			51	5	4
DEFINITE	876	18	10	61			59	9	7
UNKNOWN	0	0	-	0			0	0	0
TOTAL	1951	31	8	123			192	4	4

LOGISTIC REGRESSIONS

AGE AT EXAM	COEFFICIENT OF RISK FACTOR	SE COEFFICIENT	P-VALUE	STANDARDIZED COEFFICIENT
35-44	0.766	0.590	0.194	0.338
45-54	0.515	0.176	0.003	0.328
55-64	0.442	0.095	0.001	0.352
65-74	0.414	0.118	0.001	0.364
75-84	0.356	0.217	0.100	0.319
85-94				
35-64 UNIVARIATE	0.634	0.081	0.001	0.442
65-94 UNIVARIATE	0.401	0.103	0.001	0.358
35-64 BIVARIATE	0.448	0.084	0.001	0.312
65-94 BIVARIATE	0.423	0.104	0.001	0.378
35-64 MULTIVARIATE	0.297	0.092	0.001	0.207
65-94 MULTIVARIATE	0.371	0.109	0.001	0.331

TABLE 7-16   DATE=05APR85

AVERAGE ANNUAL INCIDENCE RATE FOR EVENT PER 1000 PERSONS AT RISK AT EXAM
BY AGE AND LEVEL OF RISK FACTOR AT EXAM: FRAMINGHAM STUDY 30-YEAR FOLLOWUP

SEX: MALE
EVENT: ANGINA PECTORIS
RISK FACTOR: LEFT VENTRICULAR HYPERTROPHY BY ELECTROCARDIOGRAM
PERSONS AT RISK: PERSONS FREE OF CORONARY HEART DISEASE

RISK FACTOR	35-44 AT EXAM PERSON EXAMS	# OF EVENTS	ANNUAL RATE	45-54 AT EXAM PERSON EXAMS	# OF EVENTS	ANNUAL RATE	55-64 AT EXAM PERSON EXAMS	# OF EVENTS	ANNUAL RATE	65-74 AT EXAM PERSON EXAMS	# OF EVENTS	ANNUAL RATE
NO	4366	21	2	7497	74	5	7693	141	9	3703	60	8
BORDER	78	0	0	132	3	11	210	1	2	90	2	11
DEFINITE	28	0	0	95	3	16	146	7	24	120	0	0
UNKNOWN	0	0	-	0	0	-	0	0	-	0	0	-
TOTAL	4472	21	2	7724	80	5	8049	149	9	3913	62	8

RISK FACTOR	75-84 AT EXAM PERSON EXAMS	# OF EVENTS	ANNUAL RATE	85-94 AT EXAM PERSON EXAMS	# OF EVENTS	ANNUAL RATE	35-64 COMBINED # OF EVENTS	CRUDE	AGE-ADJUSTED	65-94 COMBINED # OF EVENTS	CRUDE	AGE-ADJUSTED
NO	949	9	5	34			236	6	6	71	8	8
BORDER	37	0	0	2			10	5	5	2	8	9
DEFINITE	31	0	0	3			10	19	16	0	0	0
UNKNOWN	0	0	-	0			0	-	-	0	-	-
TOTAL	1017	9	4	39			250	6	6	73	7	7

LOGISTIC REGRESSIONS

AGE AT EXAM		COEFFICIENT OF RISK FACTOR	SE COEFFICIENT	P-VALUE	STANDARDIZED COEFFICIENT	STD SE
35-44		-	-	-	-	-
45-54		0.636	0.266	0.017	0.162	0.068
55-64		0.319	0.210	0.130	0.098	0.065
65-74		-0.603	0.568	0.288	-0.225	0.211
75-84		-	-	-	-	-
85-94		-	-	-	-	-
35-64	UNIVARIATE	0.466	0.165	0.005	0.125	0.044
65-94	UNIVARIATE	-0.750	0.578	0.195	-0.283	0.218
35-64	BIVARIATE	0.374	0.166	0.024	0.100	0.044
65-94	BIVARIATE	-0.744	0.577	0.197	-0.281	0.218
35-64	MULTIVARIATE	0.172	0.178	0.333	0.046	0.048
65-94	MULTIVARIATE	-0.972	0.583	0.095	-0.367	0.220

SEX: FEMALE
EVENT: ANGINA PECTORIS
RISK FACTOR: LEFT VENTRICULAR HYPERTROPHY BY ELECTROCARDIOGRAM
PERSONS AT RISK: PERSONS FREE OF CORONARY HEART DISEASE

RISK FACTOR	35-44 AT EXAM PERSON EXAMS	# OF EVENTS	ANNUAL RATE	45-54 AT EXAM PERSON EXAMS	# OF EVENTS	ANNUAL RATE	55-64 AT EXAM PERSON EXAMS	# OF EVENTS	ANNUAL RATE	65-74 AT EXAM PERSON EXAMS	# OF EVENTS	ANNUAL RATE
NO	5563	5	0	9798	44	2	11137	127	6	5966	79	7
BORDER	38	0	0	107	1	5	226	5	11	134	3	11
DEFINITE	15	0	0	54	2	19	150	8	27	148	9	30
UNKNOWN	0	0	-	0	0	-	0	0	-	0	0	-
TOTAL	5616	5	0	9959	47	2	11513	140	6	6248	91	7

RISK FACTOR	75-84 AT EXAM PERSON EXAMS	# OF EVENTS	ANNUAL RATE	85-94 AT EXAM PERSON EXAMS	# OF EVENTS	ANNUAL RATE	35-64 COMBINED # OF EVENTS	CRUDE	AGE-ADJUSTED	65-94 COMBINED # OF EVENTS	CRUDE	AGE-ADJUSTED
NO	1833	29	8	111			176	3	3	107	7	7
BORDER	49	1	10	6			6	8	7	4	11	11
DEFINITE	69	1	7	6			10	23	18	10	22	24
UNKNOWN	0	0	-	0			0	-	-	0	-	-
TOTAL	1951	31	8	123			192	4	3	123	7	7

LOGISTIC REGRESSIONS

AGE AT EXAM		COEFFICIENT OF RISK FACTOR	SE COEFFICIENT	P-VALUE	STANDARDIZED COEFFICIENT	STD SE
35-44		-	-	-	-	-
45-54		1.029	0.356	0.004	0.184	0.064
55-64		0.777	0.177	0.001	0.205	0.047
65-74		0.767	0.178	0.001	0.256	0.059
75-84		0.006	0.454	0.990	0.002	0.180
85-94						
35-64	UNIVARIATE	0.973	0.157	0.001	0.207	0.033
65-94	UNIVARIATE	0.591	0.144	0.001	0.208	0.058
35-64	BIVARIATE	0.804	0.160	0.001	0.171	0.034
65-94	BIVARIATE	0.599	0.144	0.001	0.211	0.058
35-64	MULTIVARIATE	0.449	0.185	0.015	0.095	0.039
65-94	MULTIVARIATE	0.544	0.172	0.002	0.192	0.061

TABLE 7-17 · DATE=12DEC85
AVERAGE ANNUAL INCIDENCE RATE FOR EVENT PER 1000 PERSONS AT RISK AT EXAM
BY AGE AND LEVEL OF RISK FACTOR AT EXAM: FRAMINGHAM STUDY 30-YEAR FOLLOWUP

SEX: MALE
EVENT: ANGINA PECTORIS
RISK FACTOR: INTRAVENTRICULAR CONDUCTION DEFECT (WITH QRS INTERVAL MORE THAN 0.11 SECOND)
PERSONS AT RISK: PERSONS FREE OF CORONARY HEART DISEASE

RISK FACTOR	35-44 AT EXAM PERSON EXAMS	# OF EVENTS	ANNUAL RATE	45-54 AT EXAM PERSON EXAMS	# OF EVENTS	ANNUAL RATE	55-64 AT EXAM PERSON EXAMS	# OF EVENTS	ANNUAL RATE	65-74 AT EXAM PERSON EXAMS	# OF EVENTS	ANNUAL RATE
NO	4161	19	2	7213	74	5	7386	131	9	3566	48	7
INCOMPLETE	275	2	4	446	6	7	456	14	15	183	8	22
COMPLETE	36	0	0	65	0	0	207	4	10	164	6	18
UNKNOWN	0	0	-	0	0	-	0	0	-	0	0	-
TOTAL	4472	21	2	7724	80	5	8049	149	9	3913	62	8

RISK FACTOR	75-84 AT EXAM PERSON EXAMS	# OF EVENTS	ANNUAL RATE	85-94 AT EXAM PERSON EXAMS	# OF EVENTS	ANNUAL RATE	35-64 COMBINED # OF EVENTS	CRUDE	AGE-ADJUSTED	65-94 COMBINED # OF EVENTS	CRUDE	AGE-ADJUSTED
NO	918	8	4	31			224	6	6	57	6	6
INCOMPLETE	21	0	0	2			22	9	9	8	19	17
COMPLETE	78	1	6	6			4	6	4	8	16	16
UNKNOWN	0	0	-	0			0	-	-	0	-	-
TOTAL	1017	9	4	39			250	6	6	73	7	7

LOGISTIC REGRESSIONS

AGE AT EXAM	COEFFICIENT OF RISK FACTOR	SE COEFFICIENT	P-VALUE	STANDARDIZED COEFFICIENT	STD SE
35-44	0.181	0.663	0.785	0.054	0.196
45-54	0.005	0.382	0.990	0.001	0.112
55-64	0.232	0.184	0.207	0.089	0.071
65-74	0.622	0.190	0.001	0.276	0.084
75-84	0.146	0.550	0.791	0.080	0.300
85-94					
35-64 UNIVARIATE	0.243	0.163	0.135	0.081	0.054
65-94 UNIVARIATE	0.563	0.171	0.001	0.265	0.080
35-64 BIVARIATE	0.162	0.160	0.311	0.054	0.053
65-94 BIVARIATE	0.578	0.172	0.001	0.272	0.081
35-64 MULTIVARIATE	0.049	0.177	0.781	0.016	0.059
65-94 MULTIVARIATE	0.576	0.174	0.001	0.271	0.082

SEX: FEMALE
EVENT: ANGINA PECTORIS
RISK FACTOR: INTRAVENTRICULAR CONDUCTION DEFECT (WITH QRS INTERVAL MORE THAN 0.11 SECOND)
PERSONS AT RISK: PERSONS FREE OF CORONARY HEART DISEASE

RISK FACTOR	35-44 AT EXAM PERSON EXAMS	# OF EVENTS	ANNUAL RATE	45-54 AT EXAM PERSON EXAMS	# OF EVENTS	ANNUAL RATE	55-64 AT EXAM PERSON EXAMS	# OF EVENTS	ANNUAL RATE	65-74 AT EXAM PERSON EXAMS	# OF EVENTS	ANNUAL RATE
NO	5529	3	0	9771	44	2	11183	130	6	6027	87	7
INCOMPLETE	75	2	13	144	1	3	187	2	5	115	1	4
COMPLETE	12	0	0	44	2	23	143	8	28	106	3	14
UNKNOWN	0	0	-	0	0	-	0	0	-	0	0	-
TOTAL	5616	5	0	9959	47	2	11513	140	6	6248	91	7

RISK FACTOR	75-84 AT EXAM PERSON EXAMS	# OF EVENTS	ANNUAL RATE	85-94 AT EXAM PERSON EXAMS	# OF EVENTS	ANNUAL RATE	35-64 COMBINED # OF EVENTS	CRUDE	AGE-ADJUSTED	65-94 COMBINED # OF EVENTS	CRUDE	AGE-ADJUSTED
NO	1847	29	8	117			177	3	3	116	7	7
INCOMPLETE	36	1	14	1			5	6	6	3	10	14
COMPLETE	68	1	7	5			10	25	20	4	11	12
UNKNOWN	0	0	-	0			0	-	-	0	-	-
TOTAL	1951	31	8	123			192	4	4	123	7	7

LOGISTIC REGRESSIONS

AGE AT EXAM	COEFFICIENT OF RISK FACTOR	SE COEFFICIENT	P-VALUE	STANDARDIZED COEFFICIENT	STD SE
35-44	2.293	0.623	0.001	0.337	0.092
45-54	1.059	0.367	0.004	0.188	0.065
55-64	0.727	0.189	0.001	0.184	0.048
65-74	0.243	0.300	0.417	0.070	0.087
75-84	0.056	0.446	0.901	0.022	0.173
85-94					
35-64 UNIVARIATE	0.976	0.162	0.001	0.204	0.034
65-94 UNIVARIATE	0.233	0.237	0.325	0.074	0.075
35-64 BIVARIATE	0.814	0.162	0.001	0.170	0.034
65-94 BIVARIATE	0.242	0.237	0.304	0.077	0.075
35-64 MULTIVARIATE	0.711	0.178	0.001	0.148	0.037
65-94 MULTIVARIATE	0.217	0.239	0.364	0.069	0.076

TABLE 7-18 DATE=27NOV85

AVERAGE ANNUAL INCIDENCE RATE FOR EVENT PER 1000 PERSONS AT RISK AT EXAM
BY AGE AND LEVEL OF RISK FACTOR AT EXAM: FRAMINGHAM STUDY 30-YEAR FOLLOWUP

SEX: MALE
EVENT: ANGINA PECTORIS
RISK FACTOR: DEFINITE NONSPECIFIC T-WAVE OR ST-SEGMENT ABNORMALITY BY ELECTROCARDIOGRAM
PERSONS AT RISK: PERSONS FREE OF CORONARY HEART DISEASE

	35-44 AT EXAM			45-54 AT EXAM			55-64 AT EXAM			65-74 AT EXAM		
	PERSON EXAMS	# OF EVENTS	ANNUAL RATE	PERSON EXAMS	# OF EVENTS	ANNUAL RATE	PERSON EXAMS	# OF EVENTS	ANNUAL RATE	PERSON EXAMS	# OF EVENTS	ANNUAL RATE
	4362	18	2	7352	71	5	7462	130	9	3472	51	7
	110	3	14	372	9	12	587	19	16	441	11	12
N	0	0	-	0	0	-	0	0	-	0	0	-
	4472	21	2	7724	80	5	8049	149	9	3913	62	8

	75-84 AT EXAM			85-94 AT EXAM			35-64 COMBINED			63-94 COMBINED		
	PERSON EXAMS	# OF EVENTS	ANNUAL RATE	PERSON EXAMS	# OF EVENTS	ANNUAL RATE	# OF EVENTS	CRUDE	AGE-ADJUSTED	# OF EVENTS	CRUDE	AGE-ADJUSTED
	830	9	5	28	11		219	6	6	60	7	7
	187	0	0				31	14	14	13	10	11
	0	0	-	0			0	-		0	-	
	1017	9	4	39			250	6	6	73	7	7

LOGISTIC REGRESSIONS

	COEFFICIENT OF RISK FACTOR	SE COEFFICIENT	P-VALUE	STANDARDIZED COEFFICIENT	STD. SE
	1.912	0.631	0.002	0.296	0.098
	0.933	0.358	0.009	0.200	0.077
	0.635	0.249	0.011	0.165	0.065
	0.540	0.336	0.108	0.171	0.106
	-				
UNIVARIATE	0.950	0.195	0.001	0.212	0.044
UNIVARIATE	0.391	0.309	0.206	0.131	0.103
BIVARIATE	0.768	0.196	0.001	0.172	0.044
BIVARIATE	0.414	0.311	0.183	0.139	0.104
MULTIVARIATE	0.533	0.203	0.009	0.119	0.045
MULTIVARIATE	0.302	0.314	0.337	0.101	0.105

SEX: FEMALE
EVENT: ANGINA PECTORIS
RISK FACTOR: DEFINITE NONSPECIFIC T-WAVE OR ST-SEGMENT ABNORMALITY BY ELECTROCARDIOGRAM
PERSONS AT RISK: PERSONS FREE OF CORONARY HEART DISEASE

	35-44 AT EXAM			45-54 AT EXAM			55-64 AT EXAM			65-74 AT EXAM		
	PERSON EXAMS	# OF EVENTS	ANNUAL RATE	PERSON EXAMS	# OF EVENTS	ANNUAL RATE	PERSON EXAMS	# OF EVENTS	ANNUAL RATE	PERSON EXAMS	# OF EVENTS	ANNUAL RATE
	5517	5	0	9543	44	2	10647	123	6	5528	71	6
	99	0	0	416	3	4	866	17	10	720	20	14
N	0	0	-	0	0	-	0	0	-	0	0	-
	5616	5	0	9959	47	2	11513	140	6	6248	91	7

	75-84 AT EXAM			85-94 AT EXAM			35-64 COMBINED			65-94 COMBINED		
	PERSON EXAMS	# OF EVENTS	ANNUAL RATE	PERSON EXAMS	# OF EVENTS	ANNUAL RATE	# OF EVENTS	CRUDE	AGE-ADJUSTED	# OF EVENTS	CRUDE	AGE-ADJUSTED
	1632	23	7	97			172	3	3	95	7	7
	319	8	13	26			20	7	5	28	13	13
IN	0	0	-	0			0			0	-	
	1951	31	8	123			192	4	4	123	7	7

LOGISTIC REGRESSIONS

	COEFFICIENT OF RISK FACTOR	SE COEFFICIENT	P-VALUE	STANDARDIZED COEFFICIENT	STD SE
	0.450	0.599	0.452	0.090	0.120
	0.538	0.261	0.039	0.142	0.069
	0.787	0.256	0.002	0.251	0.082
	0.588	0.415	0.157	0.217	0.154
UNIVARIATE	0.780	0.238	0.001	0.172	0.052
UNIVARIATE	0.711	0.218	0.001	0.237	0.073
BIVARIATE	0.481	0.240	0.045	0.106	0.053
BIVARIATE	0.729	0.219	0.001	0.243	0.073
MULTIVARIATE	0.315	0.248	0.203	0.069	0.055
MULTIVARIATE	0.683	0.220	0.002	0.229	0.073

TABLE 8-1   DATE=13MAY85

AVERAGE ANNUAL INCIDENCE RATE FOR EVENT PER 1000 PERSONS AT RISK AT EXAM
BY AGE AND LEVEL OF RISK FACTOR AT EXAM: FRAMINGHAM STUDY 30-YEAR FOLLOWUP

SEX: MALE
EVENT: ANGINA PECTORIS UNCOMPLICATED
RISK FACTOR: SYSTOLIC BLOOD PRESSURE, FIRST EXAMINER (MM HG)
PERSONS AT RISK: PERSONS FREE OF CORONARY HEART DISEASE

35-44 AT EXAM			45-54 AT EXAM			55-64 AT EXAM			65-74 AT EXAM		
PERSON EXAMS	# OF EVENTS	ANNUAL RATE	PERSON EXAMS	# OF EVENTS	ANNUAL RATE	PERSON EXAMS	# OF EVENTS	ANNUAL RATE	PERSON EXAMS	# OF EVENTS	ANNUAL RATE
1223	1	0	1769	6	2	1440	13	5	548	5	5
2165	3	1	3417	18	3	3166	35	6	1341	12	5
841	4	2	1732	13	4	2090	30	7	1194	15	6
191	0	0	567	6	5	916	20	11	552	7	6
52	0	0	239	3	6	437	11	13	278	3	5
0	0	-	0	0	-	0	0	-	0	0	-
4472	8	1	7724	46	3	8049	109	7	3913	42	5

75-84 AT EXAM			85-94 AT EXAM			35-64 COMBINED			65-94 COMBINED		
PERSON EXAMS	# OF EVENTS	ANNUAL RATE	PERSON EXAMS	# OF EVENTS	ANNUAL RATE	# OF EVENTS	CRUDE	AGE-ADJUSTED	# OF EVENTS	CRUDE	AGE-ADJUSTED
120	0	0	7			20	2	3	5	4	4
345	3	4	15			56	3	4	15	4	4
315	4	5	11			47	5	5	18	6	6
153	1	5	4			26	8	8	8	6	5
84	0	0	2			14	10	7	3	4	4
0	0	-	0			0	-		0	-	
1017	7	3	39			163	4	4	49	5	5

LOGISTIC REGRESSIONS

	COEFFICIENT OF RISK FACTOR	SE COEFFICIENT	P-VALUE	STANDARDIZED COEFFICIENT	STD SE
	0.017	0.018	0.327	0.291	0.297
	0.017	0.006	0.008	0.330	0.125
	0.015	0.004	0.001	0.323	0.083
	0.007	0.006	0.260	0.163	0.145
	-0.002	0.017	0.895	-0.051	0.385
NIVARIATE	0.019	0.003	0.001	0.394	0.063
NIVARIATE	0.006	0.006	0.321	0.135	0.136
IVARIATE	0.015	0.003	0.001	0.305	0.065
IVARIATE	0.006	0.006	0.317	0.136	0.136
ULTIVARIATE	0.012	0.003	0.001	0.241	0.071
ULTIVARIATE	0.007	0.006	0.235	0.170	0.143

SEX: FEMALE
EVENT: ANGINA PECTORIS UNCOMPLICATED
RISK FACTOR: SYSTOLIC BLOOD PRESSURE, FIRST EXAMINER (MM HG)
PERSONS AT RISK: PERSONS FREE OF CORONARY HEART DISEASE

35-44 AT EXAM			45-54 AT EXAM			55-64 AT EXAM			65-74 AT XAM		
PERSON EXAMS	# OF EVENTS	ANNUAL RATE	PERSON EXAMS	# OF EVENTS	ANNUAL RATE	PERSON EXAMS	# OF EVENTS	ANNUAL RATE	PERSON EXAMS	# OFE EVENTS	ANNUAL RATE
2493	0	0	2714	7	1	1979	11	3	647	3	2
2244	2	1	3982	10	2	4056	29	4	1747	19	5
673	2	1	2053	10	2	3025	32	5	2013	26	6
156	0	0	768	6	4	1560	29	9	1155	9	4
50	0	0	442	7	8	893	17	10	686	15	11
0	0	-	0	0	-	0	0	-	0	0	-
5616	5	0	9959	40	2	11513	118	5	6248	72	6

75-84 AT EXAM			85-94 AT EXAM			35-64 COMBINED			65-94 COMBINED		
PERSON EXAMS	# OF EVENTS	ANNUAL RATE	PERSON EXAMS	# OF EVENTS	ANNUAL RATE	# OF EVENTS	CRUDE	AGE-ADJUSTED	# OF EVENTS	CRUDE	AGE-ADJUSTED
133	1	4	8			18	1	2	4	3	2
465	7	8	38			42	2	4	26	7	6
626	9	7	29			44	4	5	35	7	5
451	9	10	31			35	7	5	18	5	10
274	4	7	17			24	9	7	20	10	10
0	0	-	0			0	-		0	-	
1951	30	8	123			163	3	3	103	6	6

TABLE 8-2   DATE-13MAY85
.AVERAGE ANNUAL INCIDENCE RATE FOR EVENT PER 1000 PERSONS AT RISK AT EXAM
BY AGE AND LEVEL OF RISK FACTOR AT EXAM: FRAMINGHAM STUDY 30-YEAR FOLLOWUP

SEX: MALE
EVENT: ANGINA PECTORIS UNCOMPLICATED
RISK FACTOR: DIASTOLIC BLOOD PRESSURE, FIRST EXAMINER (MM HG)
PERSONS AT RISK: PERSONS FREE OF CORONARY HEART DISEASE

RISK FACTOR	35-44 AT EXAM			45-54 AT EXAM			55-64 AT EXAM			65-74 AT EXAM		
	PERSON EXAMS	# OF EVENTS	ANNUAL RATE	PERSON EXAMS	# OF EVENTS	ANNUAL RATE	PERSON EXAMS	# OF EVENTS	ANNUAL RATE	PERSON EXAMS	# OF EVENTS	ANNUAL RATE
20-74	1012	0	0	1478	8	3	1847	20	5	1225	9	4
75-84	1604	2	1	2766	12	2	2845	26	5	1323	17	6
85-94	1187	2	1	2062	11	3	1964	31	8	893	9	5
95-104	499	2	2	1008	10	5	982	22	11	343	6	9
105-160	170	2	6	410	5	6	411	10	12	129	1	4
UNKNOWN	0	0	-	0	0	-	0	0	-	0	0	-
TOTAL	4472	8	1	7724	46	3	8049	109	7	3913	42	5

RISK FACTOR	75-84 AT EXAM			85-94 AT EXAM			35-64 CUMBINED			65-94 COMBINED		
	PERSON EXAMS	# OF EVENTS	ANNUAL RATE	PERSON EXAMS	# OF EVENTS	ANNUAL RATE	# OF EVENTS	CRUDE	AGE-ADJUSTED	# OF EVENTS	CRUDE	AGE-ADJUSTED
20-74	422	3	4	22			28	3	3	12	4	4
75-84	331	3	5	12			40	3	3	20	6	6
85-94	204	0	0	5			44	4	5	10	5	4
95-104	44	0	0	0			34	7	7	6	8	7
105-160	16	0	0	0			17	9	8	1	3	3
UNKNOWN	0	0	-	0			0	-		0	-	
TOTAL	1017	7	3	39			163	4	4	49	5	5

LOGISTIC REGRESSIONS

AGE AT EXAM		COEFFICIENT OF RISK FACTOR	SE COEFFICIENT	P-VALUE	STANDARDIZED COEFFICIENT	STD SE
35-44		0.067	0.024	0.004	0.760	0.267
45-54		0.023	0.012	0.052	0.267	0.137
55-64		0.024	0.007	0.001	0.296	0.088
65-74		0.017	0.012	0.147	0.210	0.145
75-84		-0.013	0.033	0.683	-0.157	0.384
85-94						
35-64	UNIVARIATE	0.027	0.006	0.001	0.320	0.071
65-94	UNIVARIATE	0.015	0.011	0.171	0.186	0.136
35-64	BIVARIATE	0.027	0.006	0.001	0.315	0.070
65-94	BIVARIATE	0.015	0.011	0.198	0.178	0.138
35-64	MULTIVARIATE	0.022	0.006	0.001	0.264	0.075
65-94	MULTIVARIATE	0.016	0.012	0.174	0.198	0.145

SEX: FEMALE
EVENT: ANGINA PECTORIS UNCOMPLICATED
RISK FACTOR: DIASTOLIC BLOOD PRESSURE, FIRST EXAMINER (MM HG)
PERSONS AT RISK: PERSONS FREE OF CORONARY HEART DISEASE

RISK FACTOR	35-44 AT EXAM			45-54 AT EXAM			55-64 AT EXAM			65-74 AT EXAM		
	PERSON EXAMS	# OF EVENTS	ANNUAL RATE	PERSON EXAMS	# OF EVENTS	ANNUAL RATE	PERSON EXAMS	# OF EVENTS	ANNUAL RATE	PERSON EXAMS	# OF EVENTS	ANNUAL RATE
20-74	2164	1	0	2639	9	2	2969	24	4	1831	20	5
75-84	2030	1	0	3541	9	1	3853	33	4	2135	22	5
85-94	987	1	1	2361	11	2	2840	33	6	1439	20	7
95-104	324	2	3	926	6	3	1272	13	5	600	4	3
105-160	111	0	0	492	5	5	579	15	13	243	6	12
UNKNOWN	0	0	-	0	0	-	0	0	-	0	0	-
TOTAL	5616	5	0	9959	40	2	11513	118	5	6248	72	6

RISK FACTOR	75-84 AT EXAM			85-94 AT EXAM			35-64 COMBINED			65-94 COMBINED		
	PERSON EXAMS	# OF EVENTS	ANNUAL RATE	PERSON EXAMS	# OF EVENTS	ANNUAL RATE	# OF EVENTS	CRUDE	AGE-ADJUSTED	# OF EVENTS	CRUDE	AGE-ADJUSTED
20-74	724	7	5	53			34	2	2	27	5	5
75-84	596	12	10	43			43	2	2	35	6	6
85-94	443	10	11	20			45	4	3	30	8	8
95-104	146	1	3	7			21	4	4	5	3	3
105-160	42	0	0	0			20	8	7	6	11	9
UNKNOWN	0	0	-	0			0	-		0	-	
TOTAL	1951	30	8	123			163	3	3	103	6	6

LOGISTIC REGRESSIONS

AGE AT EXAM		COEFFICIENT OF RISK FACTOR	SE COEFFICIENT	P-VALUE	STANDARDIZED COEFFICIENT	STD SE
35-44		0.052	0.028	0.067	0.574	0.313
45-54		0.029	0.011	0.007	0.359	0.133
55-64		0.028	0.007	0.001	0.348	0.082
65-74		0.007	0.010	0.465	0.085	0.116
75-84		0.011	0.015	0.470	0.130	0.180
85-94						
35-64	UNIVARIATE	0.034	0.005	0.001	0.410	0.066
65-94	UNIVARIATE	0.008	0.008	0.343	0.092	0.097
35-64	BIVARIATE	0.030	0.006	0.001	0.361	0.068
65-94	BIVARIATE	0.008	0.008	0.333	0.094	0.098
35-64	MULTIVARIATE	0.021	0.006	0.001	0.257	0.074
65-94	MULTIVARIATE	0.004	0.008	0.628	0.046	0.096

TABLE 8-3A   DATE=04SEP85
AVERAGE ANNUAL INCIDENCE RATE FOR EVENT PER 1000 PERSONS AT RISK AT EXAM
BY AGE AND LEVEL OF RISK FACTOR AT EXAM: FRAMINGHAM STUDY 30-YEAR FOLLOWUP

SEX: MALE
EVENT: ANGINA PECTORIS UNCOMPLICATED
RISK FACTOR: HYPERTENSION WITH ANTIHYPERTENSIVE TREATMENT
PERSONS AT RISK: PERSONS FREE OF CORONARY HEART DISEASE

RISK FACTOR	35-44 AT EXAM			45-54 AT EXAM			55-64 AT EXAM			65-74 AT EXAM		
	PERSON EXAMS	# OF EVENTS	ANNUAL RATE	PERSON EXAMS	# OF EVENTS	ANNUAL RATE	PERSON EXAMS	# OF EVENTS	ANNUAL RATE	PERSON EXAMS	# OF EVENTS	ANNUAL RATE
NO	2651	4	1	4122	17	2	3782	34	4	1516	13	4
BORDER	1301	2	1	2205	16	4	2299	36	8	1252	14	6
DEFINITE	500	2	2	1139	10	4	1266	23	9	552	8	7
TREATED	20	0	0	258	3	6	702	16	11	593	7	6
UNKNOWN	0	0	-	0	0	-	0	0	-	0	0	-
TOTAL	4472	8	1	7724	46	3	8049	109	7	3913	42	5

RISK FACTOR	75-84 AT EXAM			85-94 AT EXAM			35-64 COMBINED			65-94 COMBINED		
	PERSON EXAMS	# OF EVENTS	ANNUAL RATE	PERSON EXAMS	# OF EVENTS	ANNUAL RATE	# OF EVENTS	CRUDE	AGE-ADJUSTED	# OF EVENTS	CRUDE	AGE-ADJUSTED
NO	373	2	3	18			55	3	3	15	4	4
BORDER	333	4	6	11			54	5	5	18	6	6
DEFINITE	123	0	0	3			35	6	6	8	6	6
TREATED	188	1	3	7			19	10	7	8	5	5
UNKNOWN	0	0	-	0			0	-		0	-	
TOTAL	1017	7	3	39			163	4	4	49	5	5

LOGISTIC REGRESSIONS

AGE AT EXAM		COEFFICIENT OF RISK FACTOR	SE COEFFICIENT	P-VALUE	STANDARDIZED COEFFICIENT	STD SE
35-44		0.417	0.458	0.362	0.290	0.318
45-54		0.417	0.180	0.020	0.320	0.138
55-64		0.398	0.115	0.001	0.324	0.094
65-74		0.213	0.189	0.259	0.174	0.155
75-84		-0.124	0.468	0.791	-0.102	0.383
85-94						
35-64	UNIVARIATE	0.495	0.095	0.001	0.385	0.074
65-94	UNIVARIATE	0.163	0.175	0.350	0.134	0.143
35-64	BIVARIATE	0.395	0.095	0.001	0.307	0.074
65-94	BIVARIATE	0.165	0.175	0.344	0.135	0.143
35-64	MULTIVARIATE	0.325	0.100	0.001	0.253	0.078
65-94	MULTIVARIATE	0.184	0.177	0.300	0.151	0.145

SEX: FEMALE
EVENT: ANGINA PECTORIS UNCOMPLICATED
RISK FACTOR: HYPERTENSION WITH ANTIHYPERTENSIVE TREATMENT
PERSONS AT RISK: PERSONS FREE OF CORONARY HEART DISEASE

RISK FACTOR	35-44 AT EXAM			45-54 AT EXAM			55-64 AT EXAM			65-74 AT EXAM		
	PERSON EXAMS	# OF EVENTS	ANNUAL RATE	PERSON EXAMS	# OF EVENTS	ANNUAL RATE	PERSON EXAMS	# OF EVENTS	ANNUAL RATE	PERSON EXAMS	# OF EVENTS	ANNUAL RATE
NO	4250	2	0	5622	11	1	4888	27	3	1767	11	3
BORDER	1001	2	1	2579	12	2	3282	41	6	1903	18	5
DEFINITE	316	1	2	1166	10	4	1835	29	8	1035	13	6
TREATED	49	0	0	592	7	6	1508	21	7	1543	30	10
UNKNOWN	0	0	-	0	0	-	0	0	-	0	0	-
TOTAL	5616	5	0	9959	40	2	11513	118	5	6248	72	6

RISK FACTOR	75-84 AT EXAM			85-94 AT EXAM			35-64 COMBINED			65-94 COMBINED		
	PERSON EXAMS	# OF EVENTS	ANNUAL RATE	PERSON EXAMS	# OF EVENTS	ANNUAL RATE	# OF EVENTS	CRUDE	AGE-ADJUSTED	# OF EVENTS	CRUDE	AGE-ADJUSTED
NO	402	6	7	31			40	1	2	17	4	4
BORDER	575	9	8	28			55	4	3	27	5	5
DEFINITE	371	9	8	21			40	6	5	19	7	6
TREATED	603	9	7	43			28	7	5	40	9	9
UNKNOWN	0	0	-	0			0	-		0	-	
TOTAL	1951	30	8	123			163	3	3	103	6	6

LOGISTIC REGRESSIONS

AGE AT EXAM		COEFFICIENT OF RISK FACTOR	SE COEFFICIENT	P-VALUE	STANDARDIZED COEFFICIENT	STD SE
35-44		0.944	0.542	0.082	0.553	0.318
45-54		0.799	0.191	0.001	0.614	0.147
55-64		0.473	0.112	0.001	0.395	0.094
65-74		0.514	0.161	0.001	0.423	0.133
75-84		0.011	0.234	0.962	0.009	0.184
85-94						
35-64	UNIVARIATE	0.742	0.094	0.001	0.589	0.075
65-94	UNIVARIATE	0.386	0.132	0.003	0.316	0.108
35-64	BIVARIATE	0.556	0.097	0.001	0.441	0.077
65-94	BIVARIATE	0.388	0.133	0.003	0.318	0.109
35-64	MULTIVARIATE	0.430	0.103	0.001	0.341	0.082
65-94	MULTIVARIATE	0.301	0.134	0.025	0.247	0.110

TABLE 5-3B   DATE=27NOV85
AVERAGE ANNUAL INCIDENCE RATE FOR EVENT PER 1000 PERSONS AT RISK AT EXAM
BY AGE AND LEVEL OF RISK FACTOR AT EXAM: FRAMINGHAM STUDY 30-YEAR FOLLOWUP

SEX: MALE
EVENT: ANGINA PECTORIS UNCOMPLICATED
RISK FACTOR: HYPERTENSION IGNORING TREATMENT
PERSONS AT RISK: PERSONS FREE OF CORONARY HEART DISEASE

RISK FACTOR	35-44 AT EXAM			45-54 AT EXAM			55-64 AT EXAM			65-74 AT EXAM		
	PERSON EXAMS	# OF EVENTS	ANNUAL RATE	PERSON EXAMS	# OF EVENTS	ANNUAL RATE	PERSON EXAMS	# OF EVENTS	ANNUAL RATE	PERSON EXAMS	# OF EVENTS	ANNUAL RATE
NO	2653	4	1	4154	17	2	3909	36	5	1631	13	4
MILD	1310	2	1	2284	18	4	2583	42	8	1499	18	6
DEFINITE	509	2	2	1286	11	4	1557	31	10	783	11	7
UNKNOWN	0	0	-	0	0	-	0	0	-	0	0	-
TOTAL	4472	8	1	7724	46	3	8049	109	7	3913	42	5

RISK FACTOR	75-84 AT EXAM			85-94 AT EXAM			35-64 COMBINED			65-94 COMBINED		
	PERSON EXAMS	# OF EVENTS	ANNUAL RATE	PERSON EXAMS	# OF EVENTS	ANNUAL RATE	# OF EVENTS	CRUDE	AGE-ADJUSTED	# OF EVENTS	CRUDE	AGE-ADJUSTED
NO	412	2	2	21			57	3	3	15	4	4
MILD	418	4	5	15			62	5	5	22	6	6
DEFINITE	187	1	3	3			44	7	6	12	6	6
UNKNOWN	0	0	-	0			0	-		0	-	
TOTAL	1017	7	3	39			163	4	4	49	3	5

LOGISTIC REGRESSIONS

AGE AT EXAM		COEFFICIENT OF RISK FACTOR	SE COEFFICIENT	P-VALUE	STANDARDIZED COEFFICIENT	STD SE
35-44		0.426	0.459	0.353	0.294	0.317
45-54		0.393	0.183	0.032	0.296	0.137
55-64		0.396	0.119	0.001	0.305	0.092
65-74		0.291	0.200	0.147	0.220	0.151
75-84		0.144	0.508	0.777	0.106	0.373
85-94						
35-64	UNIVARIATE	0.464	0.097	0.001	0.348	0.073
65-94	UNIVARIATE	0.276	0.187	0.140	0.207	0.140
35-64	BIVARIATE	0.392	0.098	0.001	0.294	0.073
65-94	BIVARIATE	0.275	0.187	0.141	0.206	0.140
35-64	MULTIVARIATE	0.310	0.104	0.003	0.233	0.078
65-94	MULTIVARIATE	0.307	0.190	0.107	0.231	0.143

SEX: FEMALE
EVENT: ANGINA PECTORIS UNCOMPLICATED
RISK FACTOR: HYPERTENSION IGNORING TREATMENT
PERSONS AT RISK: PERSONS FREE OF CORONARY HEART DISEASE

RISK FACTOR	35-44 AT EXAM			45-54 AT EXAM			55-64 AT EXAM			65-74 AT EXAM		
	PERSON EXAMS	# OF EVENTS	ANNUAL RATE	PERSON EXAMS	# OF EVENTS	ANNUAL RATE	PERSON EXAMS	# OF EVENTS	ANNUAL RATE	PERSON EXAMS	# OF EVENTS	ANNUAL RATE
NO	4272	2	0	5804	14	1	5245	30	3	2035	16	4
MILD	1016	2	1	2763	13	2	3860	48	6	2549	31	6
DEFINITE	328	1	2	1392	13	3	2408	40	8	1664	25	8
UNKNOWN	0	0	-	0	0	-	0	0	-	0	0	-
TOTAL	5616	5	0	9959	40	2	11513	118	5	6248	72	6

RISK FACTOR	75-84 AT EXAM			85-94 AT EXAM			35-64 COMBINED			65-94 COMBINED		
	PERSON EXAMS	# OF EVENTS	ANNUAL RATE	PERSON EXAMS	# OF EVENTS	ANNUAL RATE	# OF EVENTS	CRUDE	AGE-ADJUSTED	# OF EVENTS	CRUDE	AGE-ADJUSTED
NO	507	6	6	37			46	2	2	22	4	4
MILD	827	15	9	45			63	4	4	46	7	7
DEFINITE	617	9	7	41			54	7	6	35	8	8
UNKNOWN	0	0	-	0			0	-		0	-	
TOTAL	1951	30	8	123			163	3	3	103	6	6

LOGISTIC REGRESSIONS

AGE AT EXAM		COEFFICIENT OF RISK FACTOR	SE COEFFICIENT	P-VALUE	STANDARDIZED COEFFICIENT	STD SE
35-44		0.995	0.548	0.069	0.568	0.313
45-54		0.680	0.193	0.001	0.493	0.140
55-64		0.526	0.116	0.001	0.409	0.090
65-74		0.317	0.156	0.042	0.243	0.119
75-84		0.078	0.244	0.751	0.059	0.185
85-94						
35-64	UNIVARIATE	0.734	0.096	0.001	0.543	0.071
65-94	UNIVARIATE	0.272	0.130	0.037	0.208	0.100
35-64	BIVARIATE	0.568	0.099	0.001	0.420	0.073
65-94	BIVARIATE	0.272	0.131	0.038	0.208	0.100
35-64	MULTIVARIATE	0.421	0.106	0.001	0.311	0.079
65-94	MULTIVARIATE	0.181	0.133	0.174	0.139	0.102

TABLE 8-4   DATE=13MAY85

AVERAGE ANNUAL INCIDENCE RATE FOR EVENT PER 1000 PERSONS AT RISK AT EXAM
BY AGE AND LEVEL OF RISK FACTOR AT EXAM: FRAMINGHAM STUDY 30-YEAR FOLLOWUP

SEX: MALE
EVENT: ANGINA PECTORIS UNCOMPLICATED
RISK FACTOR: SERUM CHOLESTEROL
PERSONS AT RISK: PERSONS FREE OF CORONARY HEART DISEASE

35-44 AT EXAM			45-54 AT EXAM			55-64 AT EXAM			65-74 AT EXAM		
PERSON EXAMS	# OF EVENTS	ANNUAL RATE	PERSON EXAMS	# OF EVENTS	ANNUAL RATE	PERSON EXAMS	# OF EVENTS	ANNUAL RATE	PERSON EXAMS	# OF EVENTS	ANNUAL RATE
1253	1	0	1767	1	0	2255	20	4	1234	7	3
1247	1	0	2306	17	4	2405	28	6	1171	13	6
872	1	1	1745	14	4	1889	30	8	873	11	6
475	2	2	1001	3	1	918	18	10	428	5	6
291	1	2	557	8	7	454	10	11	191	6	16
334	2	3	348	3	4	128	3	12	16	0	0
4472	8	1	7724	46	3	8049	109	7	3913	42	5

75-84 AT EXAM			85-94 AT EXAM			35-64 COMBINED			65-94 COMBINED		
PERSON EXAMS	# OF EVENTS	ANNUAL RATE	PERSON EXAMS	# OF EVENTS	ANNUAL RATE	# OF EVENTS	CRUDE	AGE-ADJUSTED	# OF EVENTS	CRUDE	AGE-ADJUSTED
390	2	3	17			22	2	2	9	3	3
332	2	5	11			46	4	4	16	5	6
177	2	6	5			45	5	5	13	6	6
96	0	0	6			23	5	5	5	5	5
22	0	0	0			19	7	7	6	14	13
0	0	-	0			8	5	7	0	0	0
1017	7	3	39			163	4	4	49	5	5

LOGISTIC REGRESSIONS

	COEFFICIENT OF RISK FACTOR	SE COEFFICIENT	P-VALUE	STANDARDIZED COEFFICIENT	STD SE
	0.010	0.006	0.127	0.437	0.287
	0.008	0.003	0.003	0.341	0.115
	0.007	0.002	0.002	0.273	0.089
	0.010	0.004	0.009	0.385	0.147
	-0.002	0.010	0.872	-0.062	0.383
...RIATE	0.007	0.002	0.001	0.284	0.068
...RIATE	0.009	0.003	0.012	0.341	0.136
...RIATE	0.007	0.002	0.001	0.315	0.070
...RIATE	0.008	0.003	0.014	0.336	0.137
...VARIATE	0.007	0.002	0.001	0.288	0.071
...VARIATE	0.008	0.003	0.015	0.336	0.137

SEX: FEMALE
EVENT: ANGINA PECTORIS UNCOMPLICATED
RISK FACTOR: SERUM CHOLESTEROL
PERSONS AT RISK: PERSONS FREE OF CORONARY HEART DISEASE

35-44 AT EXAM			45-54 AT EXAM			55-64 AT EXAM			65-74 AT EXAM		
PERSON EXAMS	# OF EVENTS	ANNUAL RATE	PERSON EXAMS	# OF EVENTS	ANNUAL RATE	PERSON EXAMS	# OF EVENTS	ANNUAL RATE	PERSON EXAMS	# OF EVENTS	ANNUAL RATE
2350	0	0	2094	8	2	1625	11	3	789	9	6
1417	2	1	2558	11	2	2615	23	4	1401	11	5
806	0	0	2259	7	2	2938	25	4	1703	17	5
372	2	3	1473	6	2	2240	22	5	1287	15	6
171	1	3	1101	3	1	1866	31	8	1031	20	10
500	0	0	474	5	5	229	6	13	37	0	0
5616	5	0	9959	40	2	11513	118	5	6248	72	6

75-84 AT EXAM			85-94 AT EXAM			35-64 COMBINED			65-94 COMBINED		
PERSON EXAMS	# OF EVENTS	ANNUAL RATE	PERSON EXAMS	# OF EVENTS	ANNUAL RATE	# OF EVENTS	CRUDE	AGE-ADJUSTED	# OF EVENTS	CRUDE	AGE-ADJUSTED
361	3	4	46			19	2	2	12	5	5
486	8	8	28			36	3	3	20	5	6
542	12	11	35			32	3	3	29	6	7
350	6	9	9			30	4	4	21	6	8
205	1	2	5			35	6	4	21	8	8
7	0	0	0			11	5	7	0	0	0
1951	30	8	123			163	3	3	103	6	6

LOGISTIC REGRESSIONS

	COEFFICIENT OF RISK FACTOR	SE COEFFICIENT	P-VALUE	STANDARDIZED COEFFICIENT	STD SE
	0.023	0.005	0.001	0.921	0.196
	-0.002	0.004	0.594	-0.095	0.177
	0.006	0.002	0.001	0.294	0.086
	0.006	0.003	0.022	0.259	0.113
	-0.001	0.004	0.859	-0.033	0.185
...RIATE	0.007	0.001	0.001	0.324	0.062
...RIATE	0.004	0.002	0.079	0.169	0.096
...IATE	0.005	0.001	0.001	0.239	0.066
...IATE	0.004	0.002	0.073	0.174	0.097
...VARIATE	0.004	0.001	0.004	0.198	0.068
...VARIATE	0.004	0.002	0.064	0.182	0.098

TABLE 8-5   DATE=27NOV85
AVERAGE ANNUAL INCIDENCE RATE FOR EVENT PER 1000 PERSONS AT RISK AT EXAM
BY AGE AND LEVEL OF RISK FACTOR AT EXAM: FRAMINGHAM STUDY 30-YEAR FOLLOWUP

SEX: MALE
EVENT: ANGINA PECTORIS UNCOMPLICATED
RISK FACTOR: HEMATOCRIT
PERSONS AT RISK: PERSONS FREE OF CORONARY HEART DISEASE

RISK FACTOR	35-44 AT EXAM PERSON EXAMS	# OF EVENTS	ANNUAL RATE	45-54 AT EXAM PERSON EXAMS	# OF EVENTS	ANNUAL RATE	55-64 AT EXAM PERSON EXAMS	# OF EVENTS	ANNUAL RATE	65-74 AT EXAM PERSON EXAMS	# OF EVENTS	ANNUAL RATE
25-42	623	0	0	942	3	2	1018	14	7	450	4	4
43-44	546	0	0	967	6	3	985	16	8	493	5	5
45-46	983	0	0	1854	10	3	1953	23	6	963	12	6
47-48	775	0	0	1510	4	1	1599	21	7	861	13	8
49-70	1175	5	2	2066	19	3	2338	32	7	1127	8	4
UNKNOWN	370	3	4	385	4	5	156	3	10	19	0	0
TOTAL	4472	8	1	7724	46	3	8049	109	7	3913	42	5

RISK FACTOR	75-84 AT EXAM PERSON EXAMS	# OF EVENTS	ANNUAL RATE	85-94 AT EXAM PERSON EXAMS	# OF EVENTS	ANNUAL RATE	35-64 COMBINED # OF EVENTS	CRUDE	AGE-ADJUSTED	65-94 COMBINED # OF EVENTS	CRUDE	AGE-ADJUSTED
25-42	214	1	2	16			17	3	3	5	4	4
43-44	119	1	4	5			22	4	4	6	5	5
45-46	240	4	8	7			33	3	3	16	7	7
47-48	197	0	0	7			25	3	3	13	6	6
49-70	247	1	2	4			56	5	5	9	3	3
UNKNOWN	0	0		0			10	5	7	0	0	0
TOTAL	1017	7	3	39			163	4	4	49	5	5

LOGISTIC REGRESSIONS

AGE AT EXAM		COEFFICIENT OF RISK FACTOR	SE COEFFICIENT	P-VALUE	STANDARDIZED COEFFICIENT	STD SE
35-44		0.268	0.111	0.015	0.985	0.407
45-54		0.061	0.043	0.156	0.219	0.154
55-64		0.005	0.026	0.849	0.019	0.098
65-74		-0.009	0.042	0.829	-0.033	0.155
75-84		-0.040	0.089	0.653	-0.167	0.371
85-94						
35-64	UNIVARIATE	0.030	0.022	0.181	0.109	0.082
65-94	UNIVARIATE	-0.008	0.037	0.820	-0.033	0.143
35-64	BIVARIATE	0.027	0.022	0.213	0.101	0.081
65-94	BIVARIATE	-0.012	0.038	0.753	-0.046	0.145
35-64	MULTIVARIATE	0.011	0.022	0.613	0.042	0.082
65-94	MULTIVARIATE	-0.025	0.039	0.520	-0.096	0.150

SEX: FEMALE
EVENT: ANGINA PECTORIS UNCOMPLICATED
RISK FACTOR: HEMATOCRIT
PERSONS AT RISK: PERSONS FREE OF CORONARY HEART DISEASE

RISK FACTOR	35-44 AT EXAM PERSON EXAMS	# OF EVENTS	ANNUAL RATE	45-54 AT EXAM PERSON EXAMS	# OF EVENTS	ANNUAL RATE	55-64 AT EXAM PERSON EXAMS	# OF EVENTS	ANNUAL RATE	65-74 AT EXAM PERSON EXAMS	# OF EVENTS	ANNUAL RATE
21-39	1697	2	1	1843	5	1	1553	13	4	646	8	6
40-41	967	0	0	1741	9	3	2061	18	4	1069	9	4
42-43	1043	2	1	1955	7	2	2260	14	3	1256	14	6
44-45	856	1	1	2194	8	2	2847	34	6	1609	22	7
46-65	487	0	0	1681	6	2	2521	33	7	1626	19	6
UNKNOWN	566	0	0	545	5	5	271	6	11	42	0	0
TOTAL	5616	5	0	9959	40	2	11513	118	5	6248	72	6

RISK FACTOR	75-84 AT EXAM PERSON EXAMS	# OF EVENTS	ANNUAL RATE	85-94 AT EXAM PERSON EXAMS	# OF EVENTS	ANNUAL RATE	35-64 COMBINED # OF EVENTS	CRUDE	AGE-ADJUSTED	65-94 COMBINED # OF EVENTS	CRUDE	AGE-ADJUSTED
21-39	297	4	7	24			20	2	2	12	6	6
40-41	373	7	9	31			27	3	3	16	5	5
42-43	353	2	3	19			23	2	2	16	5	5
44-45	486	11	11	26			43	4	4	33	8	8
46-65	436	6	7	23			39	4	3	26	6	6
UNKNOWN	6	0	0	0			11	4	6	0	0	0
TOTAL	1951	30	8	123			163	3	3	103	6	6

LOGISTIC REGRESSIONS

AGE AT EXAM		COEFFICIENT OF RISK FACTOR	SE COEFFICIENT	P-VALUE	STANDARDIZED COEFFICIENT	STD SE
35-44		-0.065	0.118	0.586	-0.232	0.426
45-54		0.026	0.048	0.593	0.092	0.171
55-64		0.046	0.028	0.092	0.161	0.096
65-74		0.024	0.034	0.481	0.084	0.120
75-84		0.001	0.047	0.984	0.004	0.184
85-94						
35-64	UNIVARIATE	0.066	0.023	0.004	0.237	0.083
65-94	UNIVARIATE	0.017	0.028	0.540	0.061	0.100
35-64	BIVARIATE	0.034	0.024	0.153	0.121	0.085
65-94	BIVARIATE	0.017	0.028	0.532	0.062	0.100
35-64	MULTIVARIATE	0.024	0.024	0.303	0.088	0.085
65-94	MULTIVARIATE	0.014	0.028	0.605	0.051	0.100

TABLE 8-6   DATE=12DEC85
AVERAGE ANNUAL INCIDENCE RATE FOR EVENT PER 1000 PERSONS AT RISK AT EXAM
BY AGE AND LEVEL OF RISK FACTOR AT EXAM: FRAMINGHAM STUDY 30-YEAR FOLLOWUP

SEX: MALE
EVENT: ANGINA PECTORIS UNCOMPLICATED
RISK FACTOR: BLOOD GLUCOSE (MG/DL)
PERSONS AT RISK: PERSONS FREE OF CORONARY HEART DISEASE

35-44 AT EXAM			45-54 AT EXAM			55-64 AT EXAM			65-74 AT EXAM		
PERSON EXAMS	# OF EVENTS	ANNUAL RATE	PERSON EXAMS	# OF EVENTS	ANNUAL RATE	PERSON EXAMS	# OF EVENTS	ANNUAL RATE	PERSON EXAMS	# OF EVENTS	ANNUAL RATE
992	2	1	1490	9	3	1087	16	7	421	4	5
2501	5	1	4172	21	3	4056	55	7	1761	15	4
626	0	0	1348	11	4	1866	25	7	1046	14	7
105	0	0	305	1	2	537	6	6	335	4	6
48	0	0	193	1	3	408	7	9	334	5	7
200	1	3	216	3	7	95	0	0	16	0	0
4472	8	1	7724	46	3	8049	109	7	3913	42	5

75-84 AT EXAM			85-94 AT EXAM			35-64 COMBINED			65-94 COMBINED		
PERSON EXAMS	# OF EVENTS	ANNUAL RATE	PERSON EXAMS	# OF EVENTS	ANNUAL RATE	# OF EVENTS	CRUDE	AGE-ADJUSTED	# OF EVENTS	CRUDE	AGE-ADJUSTED
62	0	0	1	11		27	4	4	4	4	4
377	5	7	11			81	5	4	20	5	4
328	1	2	14			36	5	4	15	5	6
125	0	0	4			7	4	3	4	4	5
125	1	4	9			8	6	4	6	6	7
0	0	-	0			4	4	3	0	0	0
1017	7	3	39			163	4	4	49	5	5

LOGISTIC REGRESSIONS

	COEFFICIENT OF RISK FACTOR	SE COEFFICIENT	P-VALUE	STANDARDIZED COEFFICIENT	STD SE
	-	-		-	-
	-0.006	0.009	0.491	-0.138	0.201
	0.001	0.003	0.836	0.019	0.093
	0.001	0.004	0.892	0.020	0.149
	0.009	0.008	0.223	0.302	0.248
VARIATE	0.002	0.003	0.393	0.058	0.068
VARIATE	0.002	0.004	0.635	0.061	0.123
ARIATE	-0.001	0.003	0.692	-0.031	0.079
ARIATE	0.002	0.004	0.592	0.068	0.126
TIVARIATE	-0.001	0.003	0.640	-0.036	0.078
TIVARIATE	0.002	0.004	0.554	0.073	0.124

SEX: FEMALE
EVENT: ANGINA PECTORIS UNCOMPLICATED
RISK FACTOR: BLOOD GLUCOSE (MG/DL)
PERSONS AT RISK: PERSONS FREE OF CORONARY HEART DISEASE

35-44 AT EXAM			45-54 AT EXAM			55-64 AT EXAM			65-74 AT EXAM		
PERSON EXAMS	# OF EVENTS	ANNUAL RATE	PERSON EXAMS	# OF EVENTS	ANNUAL RATE	PERSON EXAMS	# OF EVENTS	ANNUAL RATE	PERSON EXAMS	# OF EVENTS	ANNUAL RATE
1066	2	1	1797	3	1	1369	15	5	521	7	7
3412	2	0	3744	24	2	6098	52	4	3102	33	5
672	1	1	1625	4	1	2712	24	4	1732	19	5
96	0	0	501	2	3	730	10	7	501	7	7
39	0	0	159	3	9	438	11	13	360	6	8
331	0	0	333	4	6	166	6	18	32	0	0
5616	5	0	9959	40	2	11513	118	5	6248	72	6

75-84 AT EXAM			85-94 AT EXAM			35-64 COMBINED			65-94 COMBINED		
PERSON EXAMS	# OF EVENTS	ANNUAL RATE	PERSON EXAMS	# OF EVENTS	ANNUAL RATE	# OF EVENTS	CRUDE	AGE-ADJUSTED	# OF EVENTS	CRUDE	AGE-ADJUSTED
105	2	10	3			20	2	3	9	7	7
823	8	5	44			78	3	3	41	5	5
640	11	9	51			29	3	2	31	6	6
219	4	9	13			12	5	4	4	7	7
163	5	15	12			14	11	9	11	10	10
1	0	0	0			10	6	10	0	0	0
1951	30	8	123			163	3	3	103	6	6

LOGISTIC REGRESSIONS

	COEFFICIENT OF RISK FACTOR	SE COEFFICIENT	P-VALUE	STANDARDIZED COEFFICIENT	STD SE
	-0.015	0.039	0.692	-0.234	0.590
	0.015	0.003	0.001	0.300	0.060
	0.007	0.002	0.006	0.061	0.061
	0.001	0.004	0.717	0.040	0.109
	0.005	0.004	0.127	0.182	0.119
ARIATE	0.011	0.002	0.001	0.234	0.038
ARIATE	0.004	0.003	0.184	0.104	0.079
RIATE	0.008	0.002	0.001	0.183	0.042
RIATE	0.004	0.003	0.187	0.104	0.079
IVARIATE	0.007	0.002	0.001	0.145	0.043
IVARIATE	0.003	0.003	0.341	0.077	0.081

TABLE 8-7   DATE=27NOV85
AVERAGE ANNUAL INCIDENCE RATE FOR EVENT PER 1000 PERSONS AT RISK AT EXAM
BY AGE AND LEVEL OF RISK FACTOR AT EXAM: FRAMINGHAM STUDY 30-YEAR FOLLOWUP

SEX: MALE
EVENT: ANGINA PECTORIS UNCOMPLICATED
RISK FACTOR: DIABETES MELLITUS
PERSONS AT RISK: PERSONS FREE OF CORONARY HEART DISEASE

RISK FACTOR	35-44 AT EXAM PERSON EXAMS	# OF EVENTS	ANNUAL RATE	45-54 AT EXAM PERSON EXAMS	# OF EVENTS	ANNUAL RATE	55-64 AT EXAM PERSON EXAMS	# OF EVENTS	ANNUAL RATE	65-74 AT EXAM PERSON EXAMS	# OF EVENTS	ANNUAL RATE
NO	4424	8	1	7486	43	3	7549	95	6	3504	38	5
YES	48	0	0	238	3	6	500	14	14	409	4	5
UNKNOWN	0	0	-	0	0	-	0	0	-	0	0	-
TOTAL	4472	8	1	7724	46	3	8049	109	7	3913	42	5

RISK FACTOR	75-84 AT EXAM PERSON EXAMS	# OF EVENTS	ANNUAL RATE	85-94 AT EXAM PERSON EXAMS	# OF EVENTS	ANNUAL RATE	35-64 COMBINED # OF EVENTS	CRUDE	AGE-ADJUSTED	65-94 COMBINED # OF EVENTS	CRUDE	AGE-ADJUSTED
NO	900	6	3	34		5	146	4	4	44	5	5
YES	117	1	4			5	17	11	8	5	5	5
UNKNOWN	0	0	-	0			0	-		0	-	
TOTAL	1017	7	3	39			163	4	4	49	5	5

LOGISTIC REGRESSIONS

AGE AT EXAM		COEFFICIENT OF RISK FACTOR	SE COEFFICIENT	P-VALUE	STANDARDIZED COEFFICIENT	STD SE
35-44		-	-	-	-	-
45-54		0.793	0.601	0.187	0.137	0.104
55-64		0.815	0.290	0.005	0.197	0.070
65-74		-0.104	0.528	0.843	-0.032	0.162
75-84		0.250	1.085	0.817	0.080	0.346
85-94						
35-64	UNIVARIATE	1.073	0.259	0.001	0.207	0.050
65-94	UNIVARIATE	-0.052	0.474	0.913	-0.016	0.147
35-64	BIVARIATE	0.765	0.262	0.003	0.148	0.051
65-94	BIVARIATE	-0.043	0.474	0.927	-0.013	0.147
35-64	MULTIVARIATE	0.701	0.268	0.009	0.135	0.052
65-94	MULTIVARIATE	-0.019	0.479	0.968	-0.006	0.148

SEX: FEMALE
EVENT: ANGINA PECTORIS UNCOMPLICATED
RISK FACTOR: DIABETES MELLITUS
PERSONS AT RISK: PERSONS FREE OF CORONARY HEART DISEASE

RISK FACTOR	35-44 AT EXAM PERSON EXAMS	# OF EVENTS	ANNUAL RATE	45-54 AT EXAM PERSON EXAMS	# OF EVENTS	ANNUAL RATE	55-64 AT EXAM PERSON EXAMS	# OF EVENTS	ANNUAL RATE	65-74 AT EXAM PERSON EXAMS	# OF EVENTS	ANNUAL RATE
NO	5573	5	0	9760	37	2	10985	109	5	5780	64	6
YES	43	0	0	199	3	8	528	9	9	468	8	9
UNKNOWN	0	0	-	0	0	-	0	0	-	0	0	-
TOTAL	5616	5	0	9959	40	2	11513	118	5	6248	72	6

RISK FACTOR	75-84 AT EXAM PERSON EXAMS	# OF EVENTS	ANNUAL RATE	85-94 AT EXAM PERSON EXAMS	# OF EVENTS	ANNUAL RATE	35-64 COMBINED # OF EVENTS	CRUDE	AGE-ADJUSTED	65-94 COMBINED # OF EVENTS	CRUDE	AGE-ADJUSTED
NO	1738	25	7	106		17	151	3	3	90	6	6
YES	213	5	12			17	12	8	6	13	9	9
UNKNOWN	0	0	-	0			0	-		0	-	
TOTAL	1951	30	8	123			163	3	3	103	6	6

LOGISTIC REGRESSIONS

AGE AT EXAM		COEFFICIENT OF RISK FACTOR	SE COEFFICIENT	P-VALUE	STANDARDIZED COEFFICIENT	STD SE
35-44		-	-	-	-	-
45-54		1.392	0.605	0.021	0.195	0.085
55-64		0.548	0.350	0.117	0.115	0.073
65-74		0.440	0.378	0.244	0.116	0.100
75-84		0.499	0.495	0.314	0.156	0.155
85-94						
35-64	UNIVARIATE	1.009	0.302	0.001	0.168	0.050
65-94	UNIVARIATE	0.463	0.299	0.122	0.128	0.083
35-64	BIVARIATE	0.671	0.304	0.027	0.111	0.051
65-94	BIVARIATE	0.462	0.300	0.124	0.128	0.083
35-64	MULTIVARIATE	0.472	0.311	0.128	0.079	0.052
65-94	MULTIVARIATE	0.322	0.307	0.295	0.089	0.085

TABLE 8-8   DATE=27NOV85

AVERAGE ANNUAL INCIDENCE RATE FOR EVENT PER 1000 PERSONS AT RISK AT EXAM
BY AGE AND LEVEL OF RISK FACTOR AT EXAM: FRAMINGHAM STUDY 30-YEAR FOLLOWUP

SEX: MALE
EVENT: ANGINA PECTORIS UNCOMPLICATED
RISK FACTOR: GLUCOSE IN URINE, DEFINITE OR TRACE
PERSONS AT RISK: PERSONS FREE OF CORONARY HEART DISEASE

	35-44 AT EXAM			45-54 AT EXAM			55-64 AT EXAM			65-74 AT EXAM		
OR	PERSON EXAMS	# OF EVENTS	ANNUAL RATE	PERSON EXAMS	# OF EVENTS	ANNUAL RATE	PERSON EXAMS	# OF EVENTS	ANNUAL RATE	PERSON EXAMS	# OF EVENTS	ANNUAL RATE
	4380	8	1	7535	45	3	7815	107	7	3767	40	5
	60	0	0	168	1	3	210	2	5	140	2	7
OWN	32	0	0	21	0	0	24	0	0	6	0	0
L	4472	8	1	7724	46	3	8049	109	7	3913	42	5

	75-84 AT EXAM			85-94 AT EXAM			35-64 COMBINED			65-94 COMBINED		
OR	PERSON EXAMS	# OF EVENTS	ANNUAL RATE	PERSON EXAMS	# OF EVENTS	ANNUAL RATE	# OF EVENTS	CRUDE	AGE-ADJUSTED	# OF EVENTS	CRUDE	AGE-ADJUSTED
	986	6	3	38	1		160	4	4	46	5	5
	28	1	18				3	3	3	3	9	9
OWN	3	0	0	0			0	0	0	0	0	0
L	1017	7	3	39			163	4	4	49	5	5

LOGISTIC REGRESSIONS

AT		COEFFICIENT OF RISK FACTOR	SE COEFFICIENT	P-VALUE	STANDARDIZED COEFFICIENT	STD SE
4		-0.003	1.014	0.997	-0.000	0.148
4		-0.367	0.717	0.609	-0.059	0.114
4		0.300	0.730	0.681	0.056	0.136
4		1.800	1.098	0.101	0.295	0.180
4						
4	UNIVARIATE	-0.170	0.585	0.771	-0.025	0.085
4	UNIVARIATE	0.623	0.601	0.300	0.113	0.109
4	BIVARIATE	-0.515	0.586	0.590	-0.046	0.085
4	BIVARIATE	0.615	0.601	0.306	0.112	0.109
4	MULTIVARIATE	-0.788	0.716	0.271	-0.115	0.104
4	MULTIVARIATE	0.637	0.605	0.292	0.116	0.110

SEX: FEMALE
EVENT: ANGINA PECTORIS UNCOMPLICATED
RISK FACTOR: GLUCOSE IN URINE, DEFINITE OR TRACE
PERSONS AT RISK: PERSONS FREE OF CORONARY HEART DISEASE

	35-44 AT EXAM			45-54 AT EXAM			55-64 AT EXAM			65-74 AT EXAM		
FOR	PERSON EXAMS	# OF EVENTS	ANNUAL RATE	PERSON EXAMS	# OF EVENTS	ANNUAL RATE	PERSON EXAMS	# OF EVENTS	ANNUAL RATE	PERSON EXAMS	# OF EVENTS	ANNUAL RATE
	5537	5	0	9841	38	2	11346	116	5	6131	69	6
	41	0	0	94	1	5	144	2	7	107	5	14
OWN	58	0	0	24	1	21	23	0	0	10	0	0
L	5616	5	0	9959	40	2	11513	118	5	6248	72	6

	75-84 AT EXAM			85-94 AT EXAM			35-64 COMBINED			65-94 COMBINED		
FOR	PERSON EXAMS	# OF EVENTS	ANNUAL RATE	PERSON EXAMS	# OF EVENTS	ANNUAL RATE	# OF EVENTS	CRUDE	AGE-ADJUSTED	# OF EVENTS	CRUDE	AGE-ADJUSTED
	1919	28	7	120	3		159	3	3	98	6	6
	23	2	43				3	3	3	5	19	21
OWN	9	0	0	0			1	6	8	0	0	0
L	1951	30	8	123			163	3	3	103	6	6

LOGISTIC REGRESSIONS

AT		COEFFICIENT OF RISK FACTOR	SE COEFFICIENT	P-VALUE	STANDARDIZED COEFFICIENT	ST SED
4		1.020	1.018	0.316	0.099	0.099
4		0.310	0.718	0.666	0.034	0.080
4		0.930	0.598	0.120	0.121	0.073
4		1.861	0.764	0.015	0.201	0.083
4						
4	UNIVARIATE	0.597	0.586	0.309	0.060	0.059
4	UNIVARIATE	1.169	0.467	0.012	0.147	0.059
4	BIVARIATE	0.481	0.587	0.412	0.049	0.059
4	BIVARIATE	1.170	0.467	0.012	0.147	0.059
4	MULTIVARIATE	0.194	0.596	0.745	0.020	0.060
4	MULTIVARIATE	0.942	0.480	0.050	0.118	0.060

SEX: MALE
EVENT: ANGINA PECTORIS UNCOMPLICATED
RISK FACTOR: GLUCOSE INTOLERANCE
PERSONS AT RISK: PERSONS FREE OF CORONARY HEART DISEASE

RISK FACTOR	35-44 AT EXAM			45-54 AT EXAM			55-64 AT EXAM			65-74 AT EXAM		
	PERSON EXAMS	# OF EVENTS	ANNUAL RATE	PERSON EXAMS	# OF EVENTS	ANNUAL RATE	PERSON EXAMS	# OF EVENTS	ANNUAL RATE	PERSON EXAMS	# OF EVENTS	ANNUAL RATE
NO	4335	8	1	7221	42	3	7145	91	6	3228	33	5
YES	137	0	0	503	4	4	904	18	10	685	9	7
UNKNOWN	0	0	-	0	0	-	0	0	-	0	0	-
TOTAL	4472	8	1	7724	46	3	8049	109	7	3913	42	5

RISK FACTOR	75-84 AT EXAM			85-94 AT EXAM			35-64 COMBINED			65-94 COMBINED		
	PERSON EXAMS	# OF EVENTS	ANNUAL RATE	PERSON EXAMS	# OF EVENTS	ANNUAL RATE	# OF EVENTS	CRUDE	AGE-ADJUSTED	# OF EVENTS	CRUDE	AGE-ADJUSTED
NO	802	6	4	27			141	4	4	39	5	5
YES	215	1	2	12			22	7	5	10	5	6
UNKNOWN	0	0	-	0			0	-	-	0	-	-
TOTAL	1017	7	3	39			163	4	4	49	5	5

LOGISTIC REGRESSIONS

AGE AT EXAM		COEFFICIENT OF RISK FACTOR	SE COEFFICIENT	P-VALUE	STANDARDIZED COEFFICIENT	STD SE
35-44		-	-	-	-	-
45-54		0.315	0.525	0.549	0.078	0.130
55-64		0.454	0.260	0.081	0.143	0.082
65-74		0.254	0.378	0.502	0.096	0.144
75-84		-0.478	1.083	0.659	-0.195	0.442
85-94						
35-64	UNIVARIATE	0.643	0.231	0.005	0.171	0.061
65-94	UNIVARIATE	0.133	0.356	0.709	0.051	0.138
35-64	BIVARIATE	0.382	0.233	0.101	0.101	0.062
65-94	BIVARIATE	0.149	0.357	0.677	0.058	0.138
35-64	MULTIVARIATE	0.295	0.241	0.221	0.078	0.064
65-94	MULTIVARIATE	0.187	0.360	0.602	0.073	0.139

SEX: FEMALE
EVENT: ANGINA PECTORIS UNCOMPLICATED
RISK FACTOR: GLUCOSE INTOLERANCE
PERSONS AT RISK: PERSONS FREE OF CORONARY HEART DISEASE

RISK FACTOR	35-44 AT EXAM			45-54 AT EXAM			55-64 AT EXAM			65-74 AT EXAM		
	PERSON EXAMS	# OF EVENTS	ANNUAL RATE	PERSON EXAMS	# OF EVENTS	ANNUAL RATE	PERSON EXAMS	# OF EVENTS	ANNUAL RATE	PERSON EXAMS	# OF EVENTS	ANNUAL RATE
NO	5522	5	0	9588	37	2	10503	101	5	5398	59	5
YES	94	0	0	371	3	4	1010	17	8	850	13	8
UNKNOWN	0	0	-	0	0	-	0	0	-	0	0	-
TOTAL	5616	5	0	9959	40	2	11513	118	5	6248	72	6

RISK FACTOR	75-84 AT EXAM			85-94 AT EXAM			35-64 COMBINED			65-94 COMBINED		
	PERSON EXAMS	# OF EVENTS	ANNUAL RATE	PERSON EXAMS	# OF EVENTS	ANNUAL RATE	# OF EVENTS	CRUDE	AGE-ADJUSTED	# OF EVENTS	CRUDE	AGE-ADJUSTED
NO	1581	19	6	100			143	3	3	79	6	6
YES	370	11	15	23			20	7	5	24	10	9
UNKNOWN	0	0	-	0			0	-	-	0	-	-
TOTAL	1951	30	8	123			163	3	3	103	6	6

LOGISTIC REGRESSIONS

AGE AT EXAM		COEFFICIENT OF RISK FACTOR	SE COEFFICIENT	P-VALUE	STANDARDIZED COEFFICIENT	STD SE
35-44		-	-	-	-	-
45-54		0.744	0.603	0.217	0.141	0.114
55-64		0.567	0.264	0.032	0.160	0.075
65-74		0.340	0.309	0.270	0.117	0.106
75-84		0.924	0.383	0.016	0.362	0.150
85-94						
35-64	UNIVARIATE	0.895	0.240	0.001	0.203	0.055
65-94	UNIVARIATE	0.556	0.235	0.018	0.198	0.084
35-64	BIVARIATE	0.557	0.243	0.022	0.126	0.055
65-94	BIVARIATE	0.557	0.236	0.018	0.199	0.084
35-64	MULTIVARIATE	0.489	0.246	0.047	0.111	0.056
65-94	MULTIVARIATE	0.487	0.240	0.042	0.174	0.086

TABLE 8-10   DATE=27NOV85
AVERAGE ANNUAL INCIDENCE RATE FOR EVENT PER 1000 PERSONS AT RISK AT EXAM
BY AGE AND LEVEL OF RISK FACTOR AT EXAM: FRAMINGHAM STUDY 30-YEAR FOLLOWUP

SEX: MALE
EVENT: ANGINA PECTORIS UNCOMPLICATED
RISK FACTOR: METROPOLITAN RELATIVE WEIGHT (PERCENT)
PERSONS AT RISK: PERSONS FREE OF CORONARY HEART DISEASE

RISK FACTOR	35-44 AT EXAM			45-54 AT EXAM			55-64 AT EXAM			65-74 AT EXAM		
	PERSON EXAMS	# OF EVENTS	ANNUAL RATE	PERSON EXAMS	# OF EVENTS	ANNUAL RATE	PERSON EXAMS	# OF EVENTS	ANNUAL RATE	PERSON EXAMS	# OF EVENTS	ANNUAL RATE
54-104	723	0	0	1131	2	1	1155	11	5	686	5	4
105-114	965	1	1	1540	9	3	1655	18	5	851	6	4
115-124	1122	1	0	2037	10	2	2175	28	6	1069	14	7
125-134	915	2	1	1645	13	4	1674	25	7	766	12	8
135-272	747	4	3	1370	12	4	1389	27	10	541	5	5
UNKNOWN	0	0	-	1	0	0	1	0	0	0	0	-
TOTAL	4472	8	1	7724	46	3	8049	109	7	3913	42	5

RISK FACTOR	75-84 AT EXAM			85-94 AT EXAM			35-64 COMBINED			65-94 COMBINED		
	PERSON EXAMS	# OF EVENTS	ANNUAL RATE	PERSON EXAMS	# OF EVENTS	ANNUAL RATE	# OF EVENTS	CRUDE	AGE-ADJUSTED	# OF EVENTS	CRUDE	AGE-ADJUS
54-104	224	2	4	18	.	.	13	2	2	7	4	4
105-114	252	2	4	9			28	3	3	8	4	4
115-124	296	1	2	12			39	4	4	15	5	6
125-134	155	2	6	0			40	5	5	14	8	7
135-272	90	0	0	0			43	6	6	5	4	4
UNKNOWN	0	.	-	.	0		0	0	0	0	-	
TOTAL	1017	7	3	39			163	4	4	49	5	5

LOGISTIC REGRESSIONS

AGE AT EXAM		COEFFICIENT OF RISK FACTOR	SE COEFFICIENT	P-VALUE	STANDARDIZED COEFFICIENT	STD SE
35-44		0.050	0.018	0.005	0.810	0.291
45-54		0.021	0.008	0.010	0.347	0.134
55-64		0.015	0.006	0.008	0.243	0.091
65-74		0.009	0.010	0.334	0.146	0.151
75-84		-0.012	0.026	0.646	-0.179	0.390
85-94						
35-64	UNIVARIATE	0.019	0.005	0.001	0.308	0.073
65-94	UNIVARIATE	0.008	0.009	0.384	0.122	0.140
35-64	BIVARIATE	0.019	0.005	0.001	0.310	0.073
65-94	BIVARIATE	0.007	0.009	0.419	0.114	0.141
35-64	MULTIVARIATE	0.017	0.005	0.001	0.276	0.078
65-94	MULTIVARIATE	0.005	0.009	0.618	0.073	0.147

SEX: FEMALE
EVENT: ANGINA PECTORIS UNCOMPLICATED
RISK FACTOR: METROPOLITAN RELATIVE WEIGHT (PERCENT)
PERSONS AT RISK: PERSONS FREE OF CORONARY HEART DISEASE

RISK FACTOR	35-44 AT EXAM			45-54 AT EXAM			55-64 AT EXAM			65-74 AT EXAM		
	PERSON EXAMS	# OF EVENTS	ANNUAL RATE	PERSON EXAMS	# OF EVENTS	ANNUAL RATE	PERSON EXAMS	# OF EVENTS	ANNUAL RATE	PERSON EXAMS	# OF EVENTS	ANNUAL RATE
54-104	1694	0	0	2095	10	2	2045	16	4	1216	14	6
105-114	1430	2	1	2331	3	1	2409	14	3	1145	12	5
115-124	1058	2	1	2107	10	2	2479	19	4	1395	13	5
125-134	566	0	0	1372	4	1	1743	21	6	982	15	8
135-272	860	1	1	2054	13	3	2837	48	8	1510	18	6
UNKNOWN	8	0	0	0	0	-	0	0	-	0	0	-
TOTAL	5616	5	0	9959	40	2	11513	118	5	6248	72	6

RISK FACTOR	75-84 AT EXAM			85-94 AT EXAM			35-64 COMBINED			65-94 COMBINED		
	PERSON EXAMS	# OF EVENTS	ANNUAL RATE	PERSON EXAMS	# OF EVENTS	ANNUAL RATE	# OF EVENTS	CRUDE	AGE-ADJUSTED	# OF EVENTS	CRUDE	AGE-ADJUS
54-104	517	7	7	41			26	2	3	21	6	6
105-114	375	2	3	28			19	2	3	14	5	5
115-124	399	9	11	21			31	3	3	22	6	6
125-134	266	5	9	16			25	3	3	20	8	8
135-272	394	7	9	17			62	5	5	26	7	7
UNKNOWN	0	0	-	0			0	0	0	0	-	
TOTAL	1951	30	8	123			163	3	3	103	6	6

LOGISTIC REGRESSIONS

AGE AT EXAM		COEFFICIENT OF RISK FACTOR	SE COEFFICIENT	P-VALUE	STANDARDIZED COEFFICIENT	STD SE
35-44		0.020	0.016	0.201	0.406	0.317
45-54		0.013	0.006	0.026	0.282	0.126
55-64		0.014	0.003	0.001	0.306	0.075
65-74		0.001	0.005	0.917	0.012	0.118
75-84		0.007	0.008	0.428	0.139	0.175
85-94						
35-64	UNIVARIATE	0.016	0.003	0.001	0.349	0.060
65-94	UNIVARIATE	0.003	0.004	0.547	0.058	0.096
35-64	BIVARIATE	0.014	0.003	0.001	0.301	0.063
65-94	BIVARIATE	0.003	0.004	0.535	0.060	0.097
35-64	MULTIVARIATE	0.006	0.003	0.072	0.133	0.074
65-94	MULTIVARIATE	0.000	0.005	0.936	0.008	0.099

SEX: MALE
EVENT: ANGINA PECTORIS UNCOMPLICATED
RISK FACTOR: VITAL CAPACITY - HEIGHT INDEX (ML/INCH)
PERSONS AT RISK: PERSONS FREE OF CORONARY HEART DISEASE

RISK FACTOR	35-44 AT EXAM			45-54 AT EXAM			55-64 AT EXAM			65-74 AT EXAM		
	PERSON EXAMS	# OF EVENTS	ANNUAL RATE	PERSON EXAMS	# OF EVENTS	ANNUAL RATE	PERSON EXAMS	# OF EVENTS	ANNUAL RATE	PERSON EXAMS	# OF EVENTS	ANNUAL RATE
8-44	269	0	0	1062	10	5	2045	29	7	1752	21	6
45-49	411	0	0	1125	8	4	1480	16	5	749	8	5
50-54	747	1	1	1658	10	3	1835	31	8	712	5	4
55-59	1022	2	1	1739	7	2	1416	25	9	442	6	7
60-85	2010	3	1	2135	11	3	1255	8	3	246	2	4
UNKNOWN	13	1	38	5	0	0	18	0	0	12	0	0
TOTAL	4472	8	1	7724	46	3	8049	109	7	3913	42	5

RISK FACTOR	75-84 AT EXAM			85-94 AT EXAM			35-64 COMBINED			65-94 COMBINED		
	PERSON EXAMS	# OF EVENTS	ANNUAL RATE	PERSON EXAMS	# OF EVENTS	ANNUAL RATE	# OF EVENTS	CRUDE	AGE-ADJUSTED	# OF EVENTS	CRUDE	AGE-ADJUSTED
8-44	649	4	3	24			39	6	5	25	5	5
45-49	172	2	6	6			25	4	4	10	5	5
50-54	97	0	0	7			42	5	5	5	3	7
55-59	65	1	8	2			34	4	4	6	6	7
60-85	26	0	0	0			22	2	2	2	4	3
UNKNOWN	8	0	0	0			1	14	8	0	0	0
TOTAL	1017	7	3	39			163	4	4	49	5	5

LOGISTIC REGRESSIONS

AGE AT EXAM		COEFFICIENT OF RISK FACTOR	SE COEFFICIENT	P-VALUE	STANDARDIZED COEFFICIENT	STD SE
35-44		0.001	0.043	0.991	0.004	0.379
45-54		-0.030	0.016	0.052	-0.278	0.143
55-64		-0.004	0.010	0.713	-0.035	0.096
65-74		-0.006	0.016	0.723	-0.055	0.194
75-84		0.025	0.040	0.537	0.233	0.376
85-94						
35-64	UNIVARIATE	-0.028	0.008	0.001	-0.275	0.076
65-94	UNIVARIATE	0.002	0.015	0.906	0.017	0.144
35-64	BIVARIATE	-0.009	0.008	0.265	-0.091	0.082
65-94	BIVARIATE	-0.000	0.015	0.993	-0.001	0.147
35-64	MULTIVARIATE	-0.005	0.009	0.615	-0.044	0.087
65-94	MULTIVARIATE	-0.001	0.016	0.937	-0.012	0.150

SEX: FEMALE
EVENT: ANGINA PECTORIS UNCOMPLICATED
RISK FACTOR: VITAL CAPACITY - HEIGHT INDEX (ML/INCH)
PERSONS AT RISK: PERSONS FREE OF CORONARY HEART DISEASE

RISK FACTOR	35-44 AT EXAM			45-54 AT EXAM			55-64 AT EXAM			65-74 AT EXAM		
	PERSON EXAMS	# OF EVENTS	ANNUAL RATE	PERSON EXAMS	# OF EVENTS	ANNUAL RATE	PERSON EXAMS	# OF EVENTS	ANNUAL RATE	PERSON EXAMS	# OF EVENTS	ANNUAL RATE
4-29	122	2	8	534	4	4	1461	25	9	1747	34	10
30-34	289	1	2	1217	5	2	2414	30	6	1715	15	4
35-39	769	0	0	2077	16	4	3048	32	5	1529	10	3
40-44	1427	1	0	2804	9	2	2706	22	4	854	9	5
45-92	2988	1	0	3310	6	1	1859	9	2	397	4	5
UNKNOWN	21	0	0	17	0	0	21	0	0	6	0	0
TOTAL	5616	5	0	9959	40	2	11513	118	5	6248	72	6

RISK FACTOR	75-84 AT EXAM			85-94 AT EXAM			35-64 COMBINED			65-94 COMBINED		
	PERSON EXAMS	# OF EVENTS	ANNUAL RATE	PERSON EXAMS	# OF EVENTS	ANNUAL RATE	# OF EVENTS	CRUDE	AGE-ADJUSTED	# OF EVENTS	CRUDE	AGE-ADJUSTED
4-29	930	13	7	80			31	7	7	48	9	9
30-34	569	9	8	29			36	5	4	24	5	5
35-39	314	7	11	13			48	4	4	17	5	5
40-44	111	1	5	1			32	2	2	10	5	5
45-92	25	0	0	0			16	1	1	4	5	5
UNKNOWN	2	0	0	0			0	0	0	0	0	0
TOTAL	1951	30	8	123			163	3	3	103	6	6

LOGISTIC REGRESSIONS

AGE AT EXAM		COEFFICIENT OF RISK FACTOR	SE COEFFICIENT	P-VALUE	STANDARDIZED COEFFICIENT	STD SE
35-44		-0.134	0.049	0.006	-0.980	0.359
45-54		-0.062	0.029	0.002	-0.455	0.150
55-64		-0.043	0.012	0.001	-0.326	0.089
65-74		-0.039	0.016	0.012	-0.291	0.116
75-84		0.017	0.026	0.516	0.121	0.186
85-94						
35-64	UNIVARIATE	-0.071	0.009	0.001	-0.566	0.074
65-94	UNIVARIATE	-0.028	0.013	0.030	-0.212	0.097
35-64	BIVARIATE	-0.049	0.010	0.001	-0.388	0.081
65-94	BIVARIATE	-0.030	0.013	0.026	-0.225	0.101
35-64	MULTIVARIATE	-0.038	0.011	0.001	-0.303	0.087
65-94	MULTIVARIATE	-0.024	0.014	0.080	-0.184	0.105

TABLE 8-12   DATE=27NOV85

AVERAGE ANNUAL INCIDENCE RATE FOR EVENT PER 1000 PERSONS AT RISK AT EXAM
BY AGE AND LEVEL OF RISK FACTOR AT EXAM: FRAMINGHAM STUDY 30-YEAR FOLLOWUP

SEX: MALE
EVENT: ANGINA PECTORIS UNCOMPLICATED
RISK FACTOR: HEART RATE (PER MINUTE)
PERSONS AT RISK: PERSONS FREE OF CORONARY HEART DISEASE

RISK FACTOR	35-44 AT EXAM PERSON EXAMS	# OF EVENTS	ANNUAL RATE	45-54 AT EXAM PERSON EXAMS	# OF EVENTS	ANNUAL RATE	55-64 AT EXAM PERSON EXAMS	# OF EVENTS	ANNUAL RATE	65-74 AT EXAM PERSON EXAMS	# OF EVENTS	ANNUAL RATE
30-67	1284	0	0	1949	14	4	2309	28	6	1183	14	6
68-75	1492	5	2	2490	14	3	2490	35	7	1159	10	4
76-83	722	1	1	1394	13	5	1374	21	8	624	7	6
84-91	579	2	2	1097	4	2	1064	14	7	487	7	7
92-220	395	0	0	794	1	1	812	11	7	460	4	4
UNKNOWN	0	0	-	0	0	-	0	0	-	0	0	-
TOTAL	4472	8	1	7724	46	3	8049	109	7	3913	42	5

RISK FACTOR	75-84 AT EXAM PERSON EXAMS	# OF EVENTS	ANNUAL RATE	85-94 AT EXAM PERSON EXAMS	# OF EVENTS	ANNUAL RATE	35-64 COMBINED # OF EVENTS	CRUDE	AGE-ADJUSTED	65-94 COMBINED # OF EVENTS	CRUDE	AGE-ADJUSTED
30-67	301	0	0	11			42	4	4	14	5	5
68-75	296	3	5	12			54	4	4	13	4	4
76-83	192	3	8	10			35	4	5	10	6	6
84-91	123	1	4	5			20	4	4	8	7	6
92-220	105	0	0	1			12	3	3	4	4	3
UNKNOWN	0	0	-	0			0	-	-	0	-	-
TOTAL	1017	7	5	39			163	4	4	49	5	5

LOGISTIC REGRESSIONS

AGE AT EXAM		COEFFICIENT OF RISK FACTOR	SE COEFFICIENT	P-VALUE	STANDARDIZED COEFFICIENT	STD SE
35-44		0.015	0.027	0.588	0.181	0.334
45-54		-0.016	0.012	0.209	-0.195	0.155
55-64		0.004	0.007	0.629	0.046	0.095
65-74		0.000	0.011	0.971	0.006	0.155
75-84		0.007	0.027	0.796	0.095	0.366
85-94						
35-64	UNIVARIATE	-0.001	0.006	0.879	-0.012	0.079
65-94	UNIVARIATE	0.002	0.011	0.886	0.020	0.143
35-64	BIVARIATE	-0.001	0.006	0.902	-0.010	0.078
65-94	BIVARIATE	0.002	0.011	0.884	0.021	0.143
35-64	MULTIVARIATE	-0.008	0.007	0.211	-0.105	0.084
65-94	MULTIVARIATE	-0.001	0.011	0.936	-0.012	0.147

SEX: FEMALE
EVENT: ANGINA PECTORIS UNCOMPLICATED
RISK FACTOR: HEART RATE (PER MINUTE)
PERSONS AT RISK: PERSONS FREE OF CORONARY HEART DISEASE

RISK FACTOR	35-44 AT EXAM PERSON EXAMS	# OF EVENTS	ANNUAL RATE	45-54 AT EXAM PERSON EXAMS	# OF EVENTS	ANNUAL RATE	55-64 AT EXAM PERSON EXAMS	# OF EVENTS	ANNUAL RATE	65-74 AT EXAM PERSON EXAMS	# OF EVENTS	ANNUAL RATE
30-67	1092	0	0	1897	9	2	2305	25	5	1103	11	5
68-75	1803	2	1	3244	11	2	3468	35	5	1668	25	7
76-83	1051	2	1	1911	6	2	2351	23	5	1358	16	6
84-91	939	1	1	1604	9	3	1780	17	5	1089	8	4
92-220	731	0	0	1303	5	2	1609	18	6	1030	12	6
UNKNOWN	0	0	-	0	0	-	0	0	-	0	0	-
TOTAL	5616	4	0	9959	40	2	11513	118	5	6248	72	6

RISK FACTOR	75-84 AT EXAM PERSON EXAMS	# OF EVENTS	ANNUAL RATE	85-94 AT EXAM PERSON EXAMS	# OF EVENTS	ANNUAL RATE	35-64 COMBINED # OF EVENTS	CRUDE	AGE-ADJUSTED	65-94 COMBINED # OF EVENTS	CRUDE	AGE-ADJUSTED
30-67	360	5	7	26			34	5	3	16	5	5
68-75	462	8	9	32			48	3	3	33	8	8
76-83	428	6	7	24			31	3	3	22	6	6
84-91	330	8	12	23			27	3	3	16	6	6
92-220	371	3	4	18			23	3	3	16	6	6
UNKNOWN	0	0	-	0			0	-	-	0	-	-
TOTAL	1951	30	8	123			163	3	3	103	6	6

LOGISTIC REGRESSIONS

AGE AT EXAM		COEFFICIENT OF RISK FACTOR	SE COEFFICIENT	P-VALUE	STANDARDIZED COEFFICIENT	STD SE
35-44		0.010	0.033	0.763	0.126	0.416
45-54		0.006	0.012	0.638	0.073	0.154
55-64		-0.002	0.007	0.750	-0.030	0.094
65-74		-0.006	0.009	0.524	-0.078	0.122
75-84		-0.007	0.014	0.588	-0.104	0.191
85-94						
35-64	UNIVARIATE	0.000	0.006	0.979	0.002	0.079
65-94	UNIVARIATE	-0.005	0.008	0.523	-0.065	0.101
35-64	BIVARIATE	-0.000	0.006	0.988	-0.001	0.078
65-94	BIVARIATE	-0.005	0.008	0.521	-0.065	0.101
35-64	MULTIVARIATE	-0.006	0.007	0.351	-0.078	0.083
65-94	MULTIVARIATE	-0.007	0.008	0.329	-0.100	0.102

AVERAGE ANNUAL INCIDENCE RATE FOR EVENT PER 1000 PERSONS AT RISK AT EXAM
BY AGE AND LEVEL OF RISK FACTOR AT EXAM: FRAMINGHAM STUDY 30-YEAR FOLLOWUP

SEX: MALE
EVENT: ANGINA PECTORIS UNCOMPLICATED
RISK FACTOR: CIGARETTES SMOKED (NUMBER PER DAY)
PERSONS AT RISK: PERSONS FREE OF CORONARY HEART DISEASE

RISK FACTOR	35-44 AT EXAM			45-54 AT EXAM			55-64 AT EXAM			65-74 AT EXAM		
	PERSON EXAMS	# OF EVENTS	ANNUAL RATE	PERSON EXAMS	# OF EVENTS	ANNUAL RATE	PERSON EXAMS	# OF EVENTS	ANNUAL RATE	PERSON EXAMS	# OF EVENTS	ANNUAL RATE
NONE	1383	2	1	3068	20	3	4295	60	7	2611	32	6
1-10	433	0	0	716	2	1	769	8	5	374	2	3
11-20	1485	2	1	2051	12	3	1682	21	6	526	4	4
21-40	987	2	1	1602	11	3	1119	18	8	369	4	5
41-90	158	2	6	252	1	2	162	2	6	27	0	0
UNKNOWN	26	0	0	35	0	0	22	0	0	6	0	0
TOTAL	4472	8	1	7724	46	3	8049	109	7	3913	42	5

RISK FACTOR	75-84 AT EXAM			85-94 AT EXAM			35-64 COMBINED			65-94 COMBINED		
	PERSON EXAMS	# OF EVENTS	ANNUAL RATE	PERSON EXAMS	# OF EVENTS	ANNUAL RATE	# OF EVENTS	CRUDE	AGE-ADJUSTED	# OF EVENTS	CRUDE	AGE-ADJUSTED
NONE	797	7	4	27			82	5	4	39	6	6
1-10	92	0	0	6			10	3	2	2	2	2
11-20	77	0	0	5			35	3	4	4	3	3
21-40	46	0	0	1			31	4	5	4	5	4
41-90	5	0	0	0			5	4	4	0	0	0
UNKNOWN	0	0	-	0			0	0	0	0	0	0
TOTAL	1017	7	3	39			163	4	4	49	5	5

LOGISTIC REGRESSIONS

AGE AT EXAM		COEFFICIENT OF RISK FACTOR	SE COEFFICIENT	P-VALUE	STANDARDIZED COEFFICIENT	STD SE
35-44		0.030	0.023	0.189	0.428	0.326
45-54		0.002	0.010	0.810	0.035	0.146
55-64		0.001	0.007	0.875	0.015	0.096
65-74		-0.012	0.015	0.442	-0.154	0.174
75-84		-	-	-	-	.
85-94						
35-64	UNIVARIATE	-0.004	0.006	0.496	-0.055	0.081
65-94	UNIVARIATE	-0.014	0.015	0.344	-0.158	0.167
35-64	BIVARIATE	0.004	0.005	0.478	0.056	0.079
65-94	BIVARIATE	-0.015	0.015	0.306	-0.173	0.168
35-64	MULTIVARIATE	0.006	0.006	0.284	0.086	0.081
65-94	MULTIVARIATE	-0.016	0.015	0.285	-0.179	0.167

SEX: FEMALE
EVENT: ANGINA PECTORIS UNCOMPLICATED
RISK FACTOR: CIGARETTES SMOKED (NUMBER PER DAY)
PERSONS AT RISK: PERSONS FREE OF CORONARY HEART DISEASE

RISK FACTOR	35-44 AT EXAM			45-54 AT EXAM			55-64 AT EXAM			65-74 AT EXAM		
	PERSON EXAMS	# OF EVENTS	ANNUAL RATE	PERSON EXAMS	# OF EVENTS	ANNUAL RATE	PERSON EXAMS	# OF EVENTS	ANNUAL RATE	PERSON EXAMS	# OF EVENTS	ANNUAL RATE
NONE	2582	2	0	5528	27	2	7595	91	6	4863	57	6
1-10	1314	1	0	1726	3	1	1566	11	4	626	5	4
11-20	1371	2	1	2009	6	1	1696	13	4	568	9	8
21-40	298	0	0	622	2	2	602	3	2	172	1	3
41-90	22	0	0	37	0	0	16	0	0	0	0	-
UNKNOWN	29	0	0	37	0	0	38	1	0	19	0	0
TOTAL	5616	5	0	9959	40	2	11513	118	5	6248	72	6

RISK FACTOR	75-84 AT EXAM			85-94 AT EXAM			35-64 COMBINED			65-94 COMBINED		
	PERSON EXAMS	# OF EVENTS	ANNUAL RATE	PERSON EXAMS	# OF EVENTS	ANNUAL RATE	# OF EVENTS	CRUDE	AGE-ADJUSTED	# OF EVENTS	CRUDE	AGE-ADJUSTED
NONE	1710	27	8	117			120	4	3	85	6	6
1-10	137	3	11	4			17	2	2	8	5	6
11-20	84	0	0	2			21	2	2	9	7	6
21-40	16	0	0	0			5	2	2	1	3	2
41-90	0	0	-	0			0	0	0	0	-	-
UNKNOWN	4	0	0	0			0	0	0	0	0	0
TOTAL	1951	30	8	123			163	3	3	103	6	6

LOGISTIC REGRESSIONS

AGE AT EXAM		COEFFICIENT OF RISK FACTOR	SE COEFFICIENT	P-VALUE	STANDARDIZED COEFFICIENT	STD SE
35-44		0.017	0.041	0.680	0.167	0.406
45-54		-0.017	0.018	0.340	-0.172	0.180
55-64		-0.027	0.012	0.031	-0.253	0.117
65-74		-0.001	0.016	0.946	-0.008	0.120
75-84		-0.064	0.066	0.336	-0.322	0.335
85-94						
35-64	UNIVARIATE	-0.030	0.010	0.003	-0.296	0.099
65-94	UNIVARIATE	-0.009	0.015	0.565	-0.062	0.108
35-64	BIVARIATE	-0.020	0.010	0.035	-0.203	0.096
65-94	BIVARIATE	-0.009	0.015	0.575	-0.061	0.109
35-64	MULTIVARIATE	-0.015	0.010	0.125	-0.149	0.097
65-94	MULTIVARIATE	-0.004	0.015	0.779	-0.031	0.109

TABLE 8-14   DATE=27NOV85
AVERAGE ANNUAL INCIDENCE RATE FOR EVENT PER 1000 PERSONS AT RISK AT EXAM
BY AGE AND LEVEL OF RISK FACTOR AT EXAM: FRAMINGHAM STUDY 30-YEAR FOLLOWUP

SEX: MALE
EVENT: ANGINA PECTORIS UNCOMPLICATED
RISK FACTOR: ALBUMIN IN URINE, DEFINITE
PERSONS AT RISK: PERSONS FREE OF CORONARY HEART DISEASE

RISK FACTOR	35-44 AT EXAM			45-54 AT EXAM			55-64 AT EXAM			65-74 AT EXAM		
	PERSON EXAMS	# OF EVENTS	ANNUAL RATE	PERSON EXAMS	# OF EVENTS	ANNUAL RATE	PERSON EXAMS	# OF EVENTS	ANNUAL RATE	PERSON EXAMS	# OF EVENTS	ANNUAL RATE
NO	4366	7	1	7599	45	3	7838	108	7	3792	41	5
YES	70	1	7	99	1	5	184	1	3	115	1	4
UNKNOWN	36	0	0	26	0	0	27	0	0	6	0	0
TOTAL	4472	8	1	7724	46	3	8049	109	7	3913	42	5

RISK FACTOR	75-84 AT EXAM			85-94 AT EXAM			35-64 COMBINED			65-94 COMBINED		
	PERSON EXAMS	# OF EVENTS	ANNUAL RATE	PERSON EXAMS	# OF EVENTS	ANNUAL RATE	# OF EVENTS	CRUDE	AGE-ADJUSTED	# OF EVENTS	CRUDE	AGE-ADJUSTED
NO	989	7	4	32			160	4	4	48	5	5
YES	25	0	0	7			3	4	5	1	3	3
UNKNOWN	3	0	0	0			0	0	0	0	0	0
TOTAL	1017	7	5	39			163	4	4	49	5	5

LOGISTIC REGRESSIONS

AGE AT EXAM		COEFFICIENT OF RISK FACTOR	SE COEFFICIENT	P-VALUE	STANDARDIZED COEFFICIENT	STD SE
35-44		2.200	1.076	0.041	0.274	0.134
45-54		0.533	1.016	0.596	0.061	0.114
55-64		-0.939	1.007	0.351	-0.141	0.151
65-74		-0.220	1.017	0.829	-0.037	0.172
75-84		-	-	-	-	-
85-94						
35-64	UNIVARIATE	0.051	0.585	0.931	0.007	0.077
65-94	UNIVARIATE	-0.386	1.014	0.704	-0.065	0.172
35-64	BIVARIATE	-0.112	0.586	0.848	-0.015	0.077
65-94	BIVARIATE	-0.381	1.014	0.707	-0.065	0.172
35-64	MULTIVARIATE	-0.904	0.722	0.210	-0.119	0.095
65-94	MULTIVARIATE	-0.438	1.019	0.667	-0.074	0.173

SEX: FEMALE
EVENT: ANGINA PECTORIS UNCOMPLICATED
RISK FACTOR: ALBUMIN IN URINE, DEFINITE
PERSONS AT RISK: PERSONS FREE OF CORONARY HEART DISEASE

RISK FACTOR	35-44 AT EXAM			45-54 AT EXAM			55-64 AT EXAM			65-74 AT EXAM		
	PERSON EXAMS	# OF EVENTS	ANNUAL RATE	PERSON EXAMS	# OF EVENTS	ANNUAL RATE	PERSON EXAMS	# OF EVENTS	ANNUAL RATE	PERSON EXAMS	# OF EVENTS	ANNUAL RATE
NO	5353	4	0	9656	36	2	11263	115	5	6118	70	6
YES	225	1	2	278	5	5	225	5	7	120	2	8
UNKNOWN	38	0	0	25	1	20	25	0	0	10	0	0
TOTAL	5616	5	0	9959	40	2	11513	118	5	6248	72	6

RISK FACTOR	75-84 AT EXAM			85-94 AT EXAM			35-64 COMBINED			65-94 COMBINED		
	PERSON EXAMS	# OF EVENTS	ANNUAL RATE	PERSON EXAMS	# OF EVENTS	ANNUAL RATE	# OF EVENTS	CRUDE	AGE-ADJUSTED	# OF EVENTS	CRUDE	AGE-ADJUSTED
NO	1904	30	8	120	3		155	5	5	101	6	6
YES	38	0	0	3			7	5	5	2	6	6
UNKNOWN	9	0	0	0			1	6	7	0	0	0
TOTAL	1951	30	8	123			163	5	5	103	6	6

LOGISTIC REGRESSIONS

AGE AT EXAM	COEFFICIENT OF RISK FACTOR	SE COEFFICIENT	P-VALUE	STANDARDIZED COEFFICIENT	STD SE

TABLE 8-15   DATE=27NOV85

AVERAGE ANNUAL INCIDENCE RATE FOR EVENT PER 1000 PERSONS AT RISK AT EXAM
BY AGE AND LEVEL OF RISK FACTOR AT EXAM: FRAMINGHAM STUDY 30-YEAR FOLLOWUP

SEX: MALE
EVENT: ANGINA PECTORIS UNCOMPLICATED
RISK FACTOR: HEART ENLARGEMENT BY X-RAY
PERSONS AT RISK: PERSONS FREE OF CORONARY HEART DISEASE

RISK FACTOR	35-44 AT EXAM			45-54 AT EXAM			55-64 AT EXAM			65-74 AT EXAM		
	PERSON EXAMS	# OF EVENTS	ANNUAL RATE	PERSON EXAMS	# OF EVENTS	ANNUAL RATE	PERSON EXAMS	# OF EVENTS	ANNUAL RATE	PERSON EXAMS	# OF EVENTS	ANNUAL RATE
NO	4063	7	1	6383	31	2	5967	73	6	2584	26	5
MILD	185	1	3	650	8	6	962	17	9	578	5	4
DEFINITE	219	0	0	687	6	4	1119	19	8	751	11	7
UNKNOWN	5	0	0	4	1	125	1	0	0	0	0	-
TOTAL	4472	8	1	7724	46	3	8049	109	7	3913	42	5

RISK FACTOR	75-84 AT EXAM			85-94 AT EXAM			35-64 COMBINED			65-94 COMBINED		
	PERSON EXAMS	# OF EVENTS	ANNUAL RATE	PERSON EXAMS	# OF EVENTS	ANNUAL RATE	# OF EVENTS	CRUDE	AGE-ADJUSTED	# OF EVENTS	CRUDE	AGE-ADJUSTED
NO	589	6	5	29		5	111	3	4	32	5	5
MILD	178	0	0			5	26	7	6	5	3	3
DEFINITE	250	1	2			5	25	6	5	12	6	6
UNKNOWN	0	0	-				1	50	48	0	-	
TOTAL	1017	7	5	39			163	4	4	49	5	5

LOGISTIC REGRESSIONS

AGE AT EXAM		COEFFICIENT OF RISK FACTOR	SE COEFFICIENT	P-VALUE	STANDARDIZED COEFFICIENT	STD SE
35-44		-0.070	0.800	0.930	-0.033	0.374
45-54		0.388	0.196	0.048	0.236	0.120
55-64		0.191	0.122	0.120	0.137	0.088
65-74		0.167	0.184	0.364	0.133	0.146
75-84		-0.706	0.622	0.257	-0.597	0.526
85-94						
35-64	UNIVARIATE	0.361	0.101	0.001	0.230	0.065
65-94	UNIVARIATE	0.051	0.175	0.769	0.041	0.141
35-64	BIVARIATE	0.230	0.103	0.026	0.146	0.066
65-94	BIVARIATE	0.061	0.175	0.726	0.050	0.141
35-64	MULTIVARIATE	0.182	0.110	0.098	0.116	0.070
65-94	MULTIVARIATE	0.056	0.183	0.759	0.045	0.148

SEX: FEMALE
EVENT: ANGINA PECTORIS UNCOMPLICATED
RISK FACTOR: HEART ENLARGEMENT BY X-RAY
PERSONS AT RISK: PERSONS FREE OF CORONARY HEART DISEASE

RISK FACTOR	35-44 AT EXAM			45-54 AT EXAM			55-64 AT EXAM			65-74 AT EXAM		
	PERSON EXAMS	# OF EVENTS	ANNUAL RATE	PERSON EXAMS	# OF EVENTS	ANNUAL RATE	PERSON EXAMS	# OF EVENTS	ANNUAL RATE	PERSON EXAMS	# OF EVENTS	ANNUAL RATE
NO	5164	4	0	8017	27	2	7522	58	4	3191	25	4
MILD	199	0	0	943	3	2	1769	21	6	1150	11	5
DEFINITE	245	1	2	993	10	5	2218	39	9	1907	36	9
UNKNOWN	8	0	0	6	0	0	4	0	0	0	0	-
TOTAL	5616	5	0	9959	40	2	11513	118	5	6248	72	6

RISK FACTOR	75-84 AT EXAM			85-94 AT EXAM			35-64 COMBINED			65-94 COMBINED		
	PERSON EXAMS	# OF EVENTS	ANNUAL RATE	PERSON EXAMS	# OF EVENTS	ANNUAL RATE	# OF EVENTS	CRUDE	AGE-ADJUSTED	# OF EVENTS	CRUDE	AGE-ADJUSTED
NO	702	7	5	32			89	2	2	32	4	4
MILD	373	6	8	30			24	4	3	17	5	5
DEFINITE	876	17	10	61			50	7	6	54	5	9
UNKNOWN	0	0	-	0			0	0	0	0	-	
TOTAL	1951	30	8	123			163	3	3	103	6	6

LOGISTIC REGRESSIONS

AGE AT EXAM		COEFFICIENT OF RISK FACTOR	SE COEFFICIENT	P-VALUE	STANDARDIZED COEFFICIENT	STD SE
35-44		0.766	0.590	0.194	0.338	0.261
45-54		0.515	0.191	0.007	0.329	0.122
55-64		0.418	0.104	0.001	0.333	0.083
65-74		0.453	0.133	0.001	0.398	0.117
75-84		0.325	0.218	0.137	0.291	0.195
85-94						
35-64	UNIVARIATE	0.614	0.088	0.001	0.428	0.061
65-94	UNIVARIATE	0.434	0.113	0.001	0.388	0.101
35-64	BIVARIATE	0.428	0.091	0.001	0.299	0.064
65-94	BIVARIATE	0.445	0.115	0.001	0.397	0.102
35-64	MULTIVARIATE	0.273	0.100	0.006	0.191	0.070
65-94	MULTIVARIATE	0.376	0.120	0.002	0.335	0.107

TABLE 8-16   DATE=13MAY85

AVERAGE ANNUAL INCIDENCE RATE FOR EVENT PER 1000 PERSONS AT RISK AT EXAM
BY AGE AND LEVEL OF RISK FACTOR AT EXAM: FRAMINGHAM STUDY 30-YEAR FOLLOWUP

SEX: MALE
EVENT: ANGINA PECTORIS UNCOMPLICATED
RISK FACTOR: LEFT VENTRICULAR HYPERTROPHY BY ELECTROCARDIOGRAM
PERSONS AT RISK: PERSONS FREE OF CORONARY HEART DISEASE

RISK FACTOR	35-44 AT EXAM			45-54 AT EXAM			55-64 AT EXAM			65-74 AT EXAM		
	PERSON EXAMS	# OF EVENTS	ANNUAL RATE	PERSON EXAMS	# OF EVENTS	ANNUAL RATE	PERSON EXAMS	# OF EVENTS	ANNUAL RATE	PERSON EXAMS	# OF EVENTS	ANNUAL RATE
NO	4366	8	1	7497	42	3	7693	103	7	3703	41	6
BORDER	78	0	0	132	1	4	210	1	2	90	1	6
DEFINITE	28	0	0	95	3	16	146	5	17	120	0	0
UNKNOWN	0	0	-	0	0	-	0	0	-	0	0	-
TOTAL	4472	8	1	7724	46	3	8049	109	7	3913	42	5

RISK FACTOR	75-84 AT EXAM			85-94 AT EXAM			35-64 COMBINED			65-94 COMBINED		
	PERSON EXAMS	# OF EVENTS	ANNUAL RATE	PERSON EXAMS	# OF EVENTS	ANNUAL RATE	# OF EVENTS	CRUDE	AGE- ADJUSTED	# OF EVENTS	CRUDE	AGE- ADJUSTED
NO	949	7	4	54			153	4	4	48	5	6
BORDER	37	0	0	2			2	2	2	1	4	5
DEFINITE	31	0	0	3			8	15	13	0	0	0
UNKNOWN	0	0	-	0			0	-	-	0	-	-
TOTAL	1017	7	3	39			163	4	4	49	5	5

LOGISTIC REGRESSIONS

AGE AT EXAM		COEFFICIENT OF RISK FACTOR	SE COEFFICIENT	P-VALUE	STANDARDIZED COEFFICIENT	STD SE
35-44		-	-	-	-	-
45-54		0.822	0.302	0.006	0.209	0.077
55-64		0.318	0.245	0.193	0.098	0.075
65-74		-0.799	0.812	0.325	-0.297	0.302
75-84		-	-	-	-	-
85-94						
35-64	UNIVARIATE	0.553	0.190	0.004	0.148	0.051
65-94	UNIVARIATE	-0.947	0.828	0.253	-0.357	0.312
35-64	BIVARIATE	0.443	0.191	0.020	0.119	0.051
65-94	BIVARIATE	-0.939	0.827	0.256	-0.354	0.312
35-64	MULTIVARIATE	0.256	0.205	0.213	0.068	0.055
65-94	MULTIVARIATE	-1.080	0.833	0.195	-0.407	0.314

SEX: FEMALE
EVENT: ANGINA PECTORIS UNCOMPLICATED
RISK FACTOR: LEFT VENTRICULAR HYPERTROPHY BY ELECTROCARDIOGRAM
PERSONS AT RISK: PERSONS FREE OF CORONARY HEART DISEASE

RISK FACTOR	35-44 AT EXAM			45-54 AT EXAM			55-64 AT EXAM			65-74 AT EXAM		
	PERSON EXAMS	# OF EVENTS	ANNUAL RATE	PERSON EXAMS	# OF EVENTS	ANNUAL RATE	PERSON EXAMS	# OF EVENTS	ANNUAL RATE	PERSON EXAMS	# OF EVENTS	ANNUAL RATE
NO	5563	5	0	9798	37	2	11137	107	5	5966	61	5
BORDER	38	0	0	107	1	5	226	4	9	134	2	7
DEFINITE	15	0	0	54	2	19	150	7	23	148	9	30
UNKNOWN	0	0	-	0	0	-	0	0	-	0	0	-
TOTAL	5616	5	0	9959	40	2	11513	118	5	6248	72	6

RISK FACTOR	75-84 AT EXAM			85-94 AT EXAM			35-64 COMBINED			65-94 COMBINED		
	PERSON EXAMS	# OF EVENTS	ANNUAL RATE	PERSON EXAMS	# OF EVENTS	ANNUAL RATE	# OF EVENTS	CRUDE	AGE- ADJUSTED	# OF EVENTS	CRUDE	AGE- ADJUSTED
NO	1833	28	8	111			149	3	3	90	6	6
BORDER	49	1	10	6			5	7	6	3	4	4
DEFINITE	69	1	7	6			9	21	17	10	22	24
UNKNOWN	0	0	-	0			0	-	-	0	-	-
TOTAL	1951	30	8	123			163	3	5	103	6	6

LOGISTIC REGRESSIONS

AGE AT EXAM		COEFFICIENT OF RISK FACTOR	SE COEFFICIENT	P-VALUE	STANDARDIZED COEFFICIENT	STD SE
35-44		-	-	-	-	-
45-54		1.129	0.356	0.002	0.202	0.064
55-64		0.784	0.191	0.001	0.207	0.050
65-74		0.885	0.183	0.001	0.295	0.061
75-84		0.026	0.454	0.954	0.010	0.180
85-94						
35-64	UNIVARIATE	0.996	0.167	0.001	0.212	0.035
65-94	UNIVARIATE	0.675	0.168	0.001	0.238	0.059
35-64	BIVARIATE	0.830	0.169	0.001	0.176	0.036
65-94	BIVARIATE	0.676	0.169	0.001	0.238	0.059
35-64	MULTIVARIATE	0.451	0.199	0.023	0.096	0.042
65-94	MULTIVARIATE	0.591	0.178	0.001	0.208	0.063

TABLE 8-17   DATE=12DEC85
AVERAGE ANNUAL INCIDENCE RATE FOR EVENT PER 1000 PERSONS AT RISK AT EXAM
BY AGE AND LEVEL OF RISK FACTOR AT EXAM: FRAMINGHAM STUDY 30-YEAR FOLLOWUP

SEX: MALE
EVENT: ANGINA PECTORIS UNCOMPLICATED
RISK FACTOR: INTRAVENTRICULAR CONDUCTION DEFECT (WITH QRS INTERVAL MORE THAN (
PERSONS AT RISK: PERSONS FREE OF CORONARY HEART DISEASE

RISK FACTOR	35-44 AT EXAM PERSON EXAMS	# OF EVENTS	ANNUAL RATE	45-54 AT EXAM PERSON EXAMS	# OF EVENTS	ANNUAL RATE	55-64 AT EXAM PERSON EXAMS	# OF EVENTS	ANNUAL RATE	65-74 AT EX PERSON EXAMS	# OF EVENTS
NO	4161	7	1	7213	41	3	7386	99	7	3566	30
INCOMPLETE	275	1	2	444	5	6	456	6	7	183	7
COMPLETE	36	0	0	65	0	0	207	4	10	164	5
UNKNOWN	0	0	0	0	0	-	0	0	-	0	0
TOTAL	4472	8	1	7724	46	3	8049	109	7	3913	42

RISK FACTOR	75-84 AT EXAM PERSON EXAMS	# OF EVENTS	ANNUAL RATE	85-94 AT EXAM PERSON EXAMS	# OF EVENTS	ANNUAL RATE	35-64 COMBINED # OF EVENTS	CRUDE	AGE-ADJUSTED	65-94 COM # OF EVENTS	CRU
NO	918	6	3	31			147	4	4	36	4
INCOMPLETE	21	0	0	2			12	5	5	7	17
COMPLETE	78	1	6	6			4	6	4	6	12
UNKNOWN	0	0	-	0			0	-	-	0	-
TOTAL	1017	7	3	39			163	4	4	49	5

LOGISTIC REGRESSIONS

AGE AT EXAM	COEFFICIENT OF RISK FACTOR	SE COEFFICIENT	P-VALUE	STANDARDIZED COEFFICIENT	STD SE
35-44	0.417	0.930	0.654	0.124	0.275
45-54	0.326	0.414	0.431	0.095	0.121
55-64	0.127	0.229	0.579	0.049	0.088
65-74	0.781	0.209	0.001	0.347	0.093
75-84	0.299	0.558	0.592	0.163	0.304
85-94					
35-64 UNIVARIATE	0.259	0.199	0.192	0.087	0.066
65-94 UNIVARIATE	0.675	0.194	0.001	0.317	0.091
35-64 BIVARIATE	0.159	0.195	0.415	0.053	0.065
65-94 BIVARIATE	0.694	0.196	0.001	0.327	0.092
35-64 MULTIVARIATE	0.084	0.207	0.683	0.028	0.069
65-94 MULTIVARIATE	0.710	0.199	0.001	0.334	0.093

SEX: FEMALE
EVENT: ANGINA PECTORIS UNCOMPLICATED
RISK FACTOR: INTRAVENTRICULAR CONDUCTION DEFECT (WITH QRS INTERVAL MORE THAN 0.1
PERSONS AT RISK: PERSONS FREE OF CORONARY HEART DISEASE

RISK FACTOR	35-44 AT EXAM PERSON EXAMS	# OF EVENTS	ANNUAL RATE	45-54 AT EXAM PERSON EXAMS	# OF EVENTS	ANNUAL RATE	55-64 AT EXAM PERSON EXAMS	# OF EVENTS	ANNUAL RATE	65-74 AT EXAM PERSON EXAMS	# OF EVENTS R
NO	5529	3	0	9771	37	2	11183	111	5	6027	69
INCOMPLETE	75	2	13	144	1	3	187	1	3	115	1
COMPLETE	12	0	0	44	2	23	143	6	21	106	2
UNKNOWN	0	0	-	0	0	-	0	0	-	0	0
TOTAL	5616	5	0	9959	40	2	11513	118	5	6248	72

RISK FACTOR	75-84 AT EXAM PERSON EXAMS	# OF EVENTS	ANNUAL RATE	85-94 AT EXAM PERSON EXAMS	# OF EVENTS	ANNUAL RATE	35-64 COMBINED # OF EVENTS	CRUDE	AGE-ADJUSTED	65-94 COMBI # OF EVENTS	CRUDE
NO	1847	28	4	117			151	3	3	97	6
INCOMPLETE	36	1	14	1	5		4	5	3	3	10
COMPLETE	68	1	7	5			8	20	17	3	8
UNKNOWN	0	0	-	0			0	-	-	0	-
TOTAL	1951	30	8	123			163	3	3	103	6

LOGISTIC REGRESSIONS

AGE AT EXAM	COEFFICIENT OF RISK FACTOR	SE COEFFICIENT	P-VALUE	STANDARDIZED COEFFICIENT	STD SE
35-44	2.293	0.623	0.001	0.337	0.092
45-54	1.166	0.367	0.001	0.207	0.065
55-64	0.626	0.221	0.005	0.159	0.056
65-74	0.178	0.354	0.615	0.052	0.102
75-84	0.075	0.446	0.866	0.029	0.173
85-94					
35-64 UNIVARIATE	0.935	0.180	0.001	0.195	0.038
65-94 UNIVARIATE	0.218	0.261	0.404	0.069	0.083
35-64 BIVARIATE	0.773	0.180	0.001	0.161	0.038
65-94 BIVARIATE	0.216	0.262	0.408	0.069	0.083
35-64 MULTIVARIATE	0.632	0.202	0.002	0.132	0.042
65-94 MULTIVARIATE	0.191	0.264	0.469	0.061	0.084

TABLE 8-18   DATE=27NOV85
AVERAGE ANNUAL INCIDENCE RATE FOR EVENT PER 1000 PERSONS AT RISK AT EXAM
BY AGE AND LEVEL OF RISK FACTOR AT EXAM: FRAMINGHAM STUDY 30-YEAR FOLLOWUP

SEX: MALE
EVENT: ANGINA PECTORIS UNCOMPLICATED
RISK FACTOR: DEFINITE NONSPECIFIC T-WAVE OR ST-SEGMENT ABNORMALITY BY ELECTROCARDIOGRAM
PERSONS AT RISK: PERSONS FREE OF CORONARY HEART DISEASE

35-44 AT EXAM			45-54 AT EXAM			55-64 AT EXAM			65-74 AT EXAM		
SON MS	# OF EVENTS	ANNUAL RATE	PERSON EXAMS	# OF EVENTS	ANNUAL RATE	PERSON EXAMS	# OF EVENTS	ANNUAL RATE	PERSON EXAMS	# OF EVENTS	ANNUAL RATE
362	8	1	7352	39	3	7462	94	6	3472	37	5
110	0	0	372	7	9	587	15	13	441	5	6
0	0	-	0	0	-	0	0	-	0	0	-
472	8	1	7724	46	3	8049	109	7	3913	42	5

75-84 AT EXAM			85-94 AT EXAM			35-64 COMBINED			65-94 COMBINED		
SON MS	# OF EVENTS	ANNUAL RATE	PERSON EXAMS	# OF EVENTS	ANNUAL RATE	# OF EVENTS	CRUDE	AGE-ADJUSTED	# OF EVENTS	CRUDE	AGE-ADJUSTED
830	7	4	28	0	0	141	4	4	44	5	5
187	0	0	11	0	-	22	10	9	5	4	4
0	0	-	0	0	-	0	-	-	0	-	-
017	7	3	39	0	-	163	4	4	49	5	5

LOGISTIC REGRESSIONS

	COEFFICIENT OF RISK FACTOR	SE COEFFICIENT	P-VALUE	STANDARDIZED COEFFICIENT	STD SE
	1.280	0.414	0.002	0.274	0.089
	0.721	0.281	0.010	0.187	0.073
	0.063	0.479	0.896	0.020	0.152
ATE	1.043	0.251	0.001	0.233	0.052
ATE	-0.264	0.474	0.578	-0.088	0.159
TE	0.822	0.233	0.001	0.184	0.052
TE	-0.242	0.476	0.612	-0.081	0.159
RIATE	0.585	0.243	0.016	0.131	0.054
RIATE	-0.319	0.479	0.506	-0.107	0.160

SEX: FEMALE
EVENT: ANGINA PECTORIS UNCOMPLICATED
RISK FACTOR: DEFINITE NONSPECIFIC T-WAVE OR ST-SEGMENT ABNORMALITY BY ELECTROCARDIOGRAM
PERSONS AT RISK: PERSONS FREE OF CORONARY HEART DISEASE

35-44 AT EXAM			45-54 AT EXAM			55-64 AT EXAM			65-74 AT EXAM		
SON MS	# OF EVENTS	ANNUAL RATE	PERSON EXAMS	# OF EVENTS	ANNUAL RATE	PERSON EXAMS	# OF EVENTS	ANNUAL RATE	PERSON EXAMS	# OF EVENTS	ANNUAL RATE
517	5	0	9543	38	2	10647	104	5	5528	57	5
99	0	0	416	2	2	866	14	8	720	15	10
0	0	-	0	0	-	0	0	-	0	0	-
616	5	0	9959	40	2	11513	118	5	6248	72	6

75-84 AT EXAM			85-94 AT EXAM			35-64 COMBINED			65-94 COMBINED		
SON MS	# OF EVENTS	ANNUAL RATE	PERSON EXAMS	# OF EVENTS	ANNUAL RATE	# OF EVENTS	CRUDE	AGE-ADJUSTED	# OF EVENTS	CRUDE	AGE-ADJUSTED
632	22	7	97	0	0	147	3	3	80	6	6
319	8	13	26	0	-	16	6	4	23	11	11
0	0	-	0	0	-	0	-	-	0	-	-
951	30	8	123	0	-	163	3	3	103	6	6

LOGISTIC REGRESSIONS

	COEFFICIENT OF RISK FACTOR	SE COEFFICIENT	P-VALUE	STANDARDIZED COEFFICIENT	STD SE
	0.189	0.727	0.795	0.038	0.145
	0.510	0.287	0.075	0.135	0.076
	0.714	0.293	0.015	0.228	0.094
	0.633	0.417	0.130	0.234	0.154

AVERAGE ANNUAL INCIDENCE RATE FOR EVENT PER 1000 PERSONS AT RISK AT EXAM
BY AGE AND LEVEL OF RISK FACTOR AT EXAM: FRAMINGHAM STUDY 30-YEAR FOLLOWUP

SEX: MALE
EVENT: SUDDEN DEATH FROM CORONARY HEART DISEASE
RISK FACTOR: SYSTOLIC BLOOD PRESSURE, FIRST EXAMINER (MM HG)
PERSONS AT RISK: PERSONS FREE OF CORONARY HEART DISEASE

RISK FACTOR	35-44 AT EXAM PERSON EXAMS	# OF EVENTS	ANNUAL RATE	45-54 AT EXAM PERSON EXAMS	# OF EVENTS	ANNUAL RATE	55-64 AT EXAM PERSON EXAMS	# OF EVENTS	ANNUAL RATE	65-74 AT EXAM PERSON EXAMS	# OF EVENTS	ANNUAL RATE
74-119	1223			1769	1	0	1440	1	0	548	1	1
120-139	2165			3417	10	1	3166	14	2	1341	5	2
140-159	841			1732	3	1	2090	8	2	1194	10	4
160-179	191			567	2	2	916	4	2	552	6	5
180-300	52			239	0	0	437	5	6	278	4	7
UNKNOWN	0			0	0	-	0	0	-	0	0	-
TOTAL	4472			7724	16	1	8049	32	2	3913	26	3

RISK FACTOR	75-84 AT EXAM PERSON EXAMS	# OF EVENTS	ANNUAL RATE	85-94 AT EXAM PERSON EXAMS	# OF EVENTS	ANNUAL RATE	35-64 COMBINED # OF EVENTS	CRUDE	AGE-ADJUSTED	65-94 COMBINED # OF EVENTS	CRUDE	AGE-ADJUSTED
74-119	120	2	8	7			4	0	0	3	2	2
120-139	345	2	3	15			25	1	1	7	2	2
140-159	315	2	3	11			11	1	1	12	4	4
160-179	153	2	7	4			6	2	2	8	6	6
180-300	84	3	18	2			5	3	2	7	10	9
UNKNOWN	0	0	-	0			0	-	-	0	-	-
TOTAL	1017	11	5	39			51	1	1	37	4	4

LOGISTIC REGRESSIONS

AGE AT EXAM		COEFFICIENT OF RISK FACTOR	SE COEFFICIENT	P-VALUE	STANDARDIZED COEFFICIENT	STD SE
35-44						
45-54		0.010	0.012	0.399	0.191	0.227
55-64		0.020	0.006	0.002	0.439	0.141
65-74		0.022	0.007	0.001	0.494	0.156
75-84		0.017	0.012	0.166	0.383	0.276
85-94						
35-64	UNIVARIATE	0.019	0.005	0.001	0.388	0.112
65-94	UNIVARIATE	0.020	0.006	0.001	0.467	0.135
35-64	BIVARIATE	0.015	0.006	0.009	0.300	0.115
65-94	BIVARIATE	0.021	0.006	0.001	0.479	0.138
35-64	MULTIVARIATE	0.009	0.006	0.150	0.182	0.126
65-94	MULTIVARIATE	0.016	0.006	0.013	0.360	0.146

SEX: FEMALE
EVENT: SUDDEN DEATH FROM CORONARY HEART DISEASE
RISK FACTOR: SYSTOLIC BLOOD PRESSURE, FIRST EXAMINER (MM HG)
PERSONS AT RISK: PERSONS FREE OF CORONARY HEART DISEASE

RISK FACTOR	35-44 AT EXAM PERSON EXAMS	# OF EVENTS	ANNUAL RATE	45-54 AT EXAM PERSON EXAMS	# OF EVENTS	ANNUAL RATE	55-64 AT EXAM PERSON EXAMS	# OF EVENTS	ANNUAL RATE	65-74 AT EXAM PERSON EXAMS	# OF EVENTS	ANNUAL RATE
74-119	2493			2714	1	0	1979	2	1	647	1	1
120-139	2244			3982	2	0	4056	1	0	1747	4	1
140-159	673			2053	0	0	3025	9	1	2013	3	1
160-179	156			768	0	0	1560	1	0	1155	3	1
180-300	50			442	1	1	893	0	0	686	3	2
UNKNOWN	0			0	0	-	0	0	-	0	0	-
TOTAL	5616			9959	5	0	11513	13	1	6248	14	1

RISK FACTOR	75-84 AT EXAM PERSON EXAMS	# OF EVENTS	ANNUAL RATE	85-94 AT EXAM PERSON EXAMS	# OF EVENTS	ANNUAL RATE	35-64 COMBINED # OF EVENTS	CRUDE	AGE-ADJUSTED	65-94 COMBINED # OF EVENTS	CRUDE	AGE-ADJUSTED
74-119	133	0	0	8			3	0	0	1	1	1
120-139	465	2	2	38			3	0	0	7	2	2
140-159	626	3	2	29			10	1	1	6	1	1
160-179	451	7	8	31			1	0	0	11	3	3
180-300	276	3	5	17			1	0	0	6	3	3
UNKNOWN	0	0	-	0			0	-	-	0	-	-
TOTAL	1951	15	4	123			18	0	0	31	2	2

LOGISTIC REGRESSIONS

AGE AT EXAM		COEFFICIENT OF RISK FACTOR	SE COEFFICIENT	P-VALUE	STANDARDIZED COEFFICIENT	STD SE
35-44						
45-54		0.019	0.013	0.139	0.444	0.300
55-64		0.000	0.011	0.978	0.008	0.277
65-74		0.009	0.010	0.377	0.220	0.249
75-84		0.011	0.009	0.264	0.264	0.236
85-94						
35-64	UNIVARIATE	0.013	0.008	0.118	0.301	0.193
65-94	UNIVARIATE	0.011	0.007	0.098	0.272	0.164
35-64	BIVARIATE	0.005	0.009	0.608	0.111	0.216
65-94	BIVARIATE	0.009	0.007	0.181	0.223	0.167
35-64	MULTIVARIATE	0.006	0.010	0.513	0.147	0.225
65-94	MULTIVARIATE	0.005	0.007	0.507	0.117	0.176

AVERAGE ANNUAL INCIDENCE RATE FOR EVENT PER 1000 PERSONS AT RISK AT EXAM
BY AGE AND LEVEL OF RISK FACTOR AT EXAM: FRAMINGHAM STUDY 30-YEAR FOLLOWUP

SEX: MALE
EVENT: SUDDEN DEATH FROM CORONARY HEART DISEASE
RISK FACTOR: DIASTOLIC BLOOD PRESSURE, FIRST EXAMINER (MM HG)
PERSONS AT RISK: PERSONS FREE OF CORONARY HEART DISEASE

RISK FACTOR	35-44 AT EXAM PERSON EXAMS	# OF EVENTS	ANNUAL RATE	45-54 AT EXAM PERSON EXAMS	# OF EVENTS	ANNUAL RATE	55-64 AT EXAM PERSON EXAMS	# OF EVENTS	ANNUAL RATE	65-74 AT EXAM PERSON EXAMS	# OF EVENTS	ANNUAL RATE
20-74	1012			1478	1	0	1847	8	2	1225	4	2
75-84	1604			2766	7	1	2845	7	1	1323	5	2
85-94	1187			2062	6	1	1964	8	2	893	6	3
95-104	499			1008	2	1	982	6	3	343	8	12
105-160	170			410	0	0	411	3	4	129	3	12
UNKNOWN	0			0	0	-	0	0	-	0	0	-
TOTAL	4472			7724	16	1	8049	32	2	3913	26	3

RISK FACTOR	75-84 AT EXAM PERSON EXAMS	# OF EVENTS	ANNUAL RATE	85-94 AT EXAM PERSON EXAMS	# OF EVENTS	ANNUAL RATE	35-64 COMBINED # OF EVENTS	CRUDE	AGE-ADJUSTED	65-94 COMBINED # OF EVENTS	CRUDE	AGE-ADJUSTED
20-74	422	3	4	22			9	1	1	7	2	2
75-84	331	3	5	12			17	1	1	8	2	2
85-94	204	3	7	5			14	1	1	9	4	4
95-104	44	2	23	0			8	2	2	10	13	14
105-160	16	0	0	0			5	2	1	3	10	9
UNKNOWN	0	0	-	0			0	-	-	0	-	-
TOTAL	1017	11	5	39			51	1	1	37	4	4

LOGISTIC REGRESSIONS

AGE AT EXAM		COEFFICIENT OF RISK FACTOR	SE COEFFICIENT	P-VALUE	STANDARDIZED COEFFICIENT	STD SE
35-44						
45-54		0.008	0.021	0.704	0.093	0.244
55-64		0.022	0.013	0.094	0.271	0.162
65-74		0.050	0.013	0.001	0.609	0.156
75-84		0.038	0.024	0.119	0.445	0.285
85-94						
35-64	UNIVARIATE	0.016	0.011	0.142	0.194	0.132
65-94	UNIVARIATE	0.045	0.011	0.001	0.545	0.137
35-64	BIVARIATE	0.016	0.011	0.136	0.193	0.130
65-94	BIVARIATE	0.051	0.012	0.001	0.624	0.141
35-64	MULTIVARIATE	0.008	0.011	0.499	0.089	0.132
65-94	MULTIVARIATE	0.044	0.012	0.001	0.538	0.143

SEX: FEMALE
EVENT: SUDDEN DEATH FROM CORONARY HEART DISEASE
RISK FACTOR: DIASTOLIC BLOOD PRESSURE, FIRST EXAMINER (MM HG)
PERSONS AT RISK: PERSONS FREE OF CORONARY HEART DISEASE

RISK FACTOR	35-44 AT EXAM PERSON EXAMS	# OF EVENTS	ANNUAL RATE	45-54 AT EXAM PERSON EXAMS	# OF EVENTS	ANNUAL RATE	55-64 AT EXAM PERSON EXAMS	# OF EVENTS	ANNUAL RATE	65-74 AT EXAM PERSON EXAMS	# OF EVENTS	ANNUAL RATE
20-74	2164			2639	1	0	2969	3	1	1831	4	1
75-84	2030			3541	2	0	3853	3	0	2135	6	1
85-94	987			2361	1	0	2840	5	1	1439	1	0
95-104	324			926	0	0	1272	1	0	600	2	2
105-160	111			492	1	1	579	1	1	243	1	2
UNKNOWN	0			0	0	-	0	0	-	0	0	-
TOTAL	5616			9959	5	0	11513	13	1	6248	14	1

RISK FACTOR	75-84 AT EXAM PERSON EXAMS	# OF EVENTS	ANNUAL RATE	85-94 AT EXAM PERSON EXAMS	# OF EVENTS	ANNUAL RATE	35-64 COMBINED # OF EVENTS	CRUDE	AGE-ADJUSTED	65-94 COMBINED # OF EVENTS	CRUDE	AGE ADJUSTED
20-74	724	5	3	53			4	0	0	11	2	2
75-84	596	4	3	43			5	0	0	10	2	1
85-94	443	4	5	20			6	0	0	5	1	1
95-104	146	2	7	7			1	0	0	4	3	3
105-160	42	0	0	0			2	1	1	1	2	2
UNKNOWN	0	0	-	0			0	-	-	0	-	-
TOTAL	1951	15	4	123			18	0	0	31	2	2

LOGISTIC REGRESSIONS

AGE AT EXAM		COEFFICIENT OF RISK FACTOR	SE COEFFICIENT	P-VALUE	STANDARDIZED COEFFICIENT	STD SE
35-44						
45-54		0.021	0.032	0.507	0.263	0.396
55-64		0.012	0.021	0.571	0.150	0.264
65-74		0.003	0.022	0.907	0.031	0.265
75-84		0.012	0.021	0.573	0.142	0.252
85-94						
35-64	UNIVARIATE	0.021	0.017	0.231	0.254	0.212
65-94	UNIVARIATE	-0.002	0.015	0.880	-0.027	0.181
35-64	BIVARIATE	0.015	0.018	0.398	0.185	0.219
65-94	BIVARIATE	0.004	0.015	0.798	0.046	0.181
35-64	MULTIVARIATE	0.018	0.018	0.325	0.222	0.226
65-94	MULTIVARIATE	0.003	0.015	0.862	0.031	0.181

TABLE 9-3A   DATE=05JAN86

AVERAGE ANNUAL INCIDENCE RATE FOR EVENT PER 1000 PERSONS AT RISK AT EXAM
BY AGE AND LEVEL OF RISK FACTOR AT EXAM: FRAMINGHAM STUDY 30-YEAR FOLLOWUP

SEX: MALE
EVENT: SUDDEN DEATH FROM CORONARY HEART DISEASE
RISK FACTOR: HYPERTENSION WITH ANTIHYPERTENSIVE TREATMENT
PERSONS AT RISK: PERSONS FREE OF CORONARY HEART DISEASE

RISK FACTOR	35-44 AT EXAM PERSON EXAMS	# OF EVENTS	ANNUAL RATE	45-54 AT EXAM PERSON EXAMS	# OF EVENTS	ANNUAL RATE	55-64 AT EXAM PERSON EXAMS	# OF EVENTS	ANNUAL RATE	65-74 AT EXAM PERSON EXAMS	# OF EVENTS	ANNUAL RATE
NO	2651			4122	6	1	3773	7	1	1512	1	0
MILD	1301			2205	7	2	2301	13	3	1255	9	4
DEFINITE	500			1139	3	1	1268	8	3	552	5	5
TREATED	20			258	0	0	707	4	3	594	11	9
UNKNOWN	0			0	0	-	0	0	-	0	0	-
TOTAL	4472			7724	16	1	8049	32	2	3913	26	3

RISK FACTOR	75-84 AT EXAM PERSON EXAMS	# OF EVENTS	ANNUAL RATE	85-94 AT EXAM PERSON EXAMS	# OF EVENTS	ANNUAL RATE	35-64 COMBINED # OF EVENTS	CRUDE	AGE-ADJUSTED	65-94 COMBINED # OF EVENTS	CRUDE	AGE-ADJUSTED
NO	370	3	4	18			15	1	1	4	1	1
MILD	333	2	3	11			21	2	2	11	3	3
DEFINITE	125	2	8	3			11	2	2	7	5	5
TREATED	189	4	11	7			4	2	1	15	9	9
UNKNOWN	0	0	-	0			0	-		0	-	
TOTAL	1017	11	5	39			51	1	1	37	4	4

LOGISTIC REGRESSIONS

AGE AT EXAM		COEFFICIENT OF RISK FACTOR	SE COEFFICIENT	P-VALUE	STANDARDIZED COEFFICIENT	STD SE
35-44						
45-54		0.264	0.309	0.392	0.203	0.237
55-64		0.550	0.214	0.010	0.448	0.175
65-74		1.155	0.300	0.001	0.930	0.246
75-84		0.505	0.386	0.191	0.414	0.316
85-94						
35-64	UNIVARIATE	0.500	0.169	0.003	0.389	0.132
65-94	UNIVARIATE	0.930	0.234	0.001	0.762	0.191
35-64	BIVARIATE	0.400	0.169	0.018	0.311	0.132
65-94	BIVARIATE	0.926	0.233	0.001	0.759	0.191
35-64	MULTIVARIATE	0.275	0.179	0.124	0.214	0.139
65-94	MULTIVARIATE	0.849	0.238	0.001	0.695	0.195

SEX: FEMALE
EVENT: SUDDEN DEATH FROM CORONARY HEART DISEASE
RISK FACTOR: HYPERTENSION WITH ANTIHYPERTENSIVE TREATMENT
PERSONS AT RISK: PERSONS FREE OF CORONARY HEART DISEASE

RISK FACTOR	35-44 AT EXAM PERSON EXAMS	# OF EVENTS	ANNUAL RATE	45-54 AT EXAM PERSON EXAMS	# OF EVENTS	ANNUAL RATE	55-64 AT EXAM PERSON EXAMS	# OF EVENTS	ANNUAL RATE	65-74 AT EXAM PERSON EXAMS	# OF EVENTS	ANNUAL RATE
NO	4250			5622	2	0	4863	3	0	1751	3	1
MILD	1001			2579	1	0	3290	7	1	1911	2	1
DEFINITE	316			1166	1	0	1835	2	1	1036	3	1
TREATED	49			592	1	1	1525	1	0	1550	6	2
UNKNOWN	0			0	0	-	0	0	-	0	0	-
TOTAL	5616			9959	5	0	11513	13	1	6248	14	1

RISK FACTOR	75-84 AT EXAM PERSON EXAMS	# OF EVENTS	ANNUAL RATE	85-94 AT EXAM PERSON EXAMS	# OF EVENTS	ANNUAL RATE	35-64 COMBINED # OF EVENTS	CRUDE	AGE-ADJUSTED	65-94 COMBINED # OF EVENTS	CRUDE	AGE-ADJUSTED
NO	395	1	1	31			5	0	0	5	1	1
MILD	585	3	3	28			8	1	1	5	1	1
DEFINITE	375	3	4	24			3	0	0	7	2	2
TREATED	596	8	7	40			2	0	0	14	3	3
UNKNOWN	0	0	-	0			0	-		0	-	
TOTAL	1951	15	4	123			18	0	0	31	2	2

LOGISTIC REGRESSIONS

AGE AT EXAM		COEFFICIENT OF RISK FACTOR	SE COEFFICIENT	P-VALUE	STANDARDIZED COEFFICIENT	STD SE
35-44						
45-54		0.582	0.531	0.274	0.447	0.408
55-64		0.185	0.329	0.574	0.155	0.275
65-74		0.476	0.361	0.188	0.391	0.297
75-84		0.771	0.437	0.078	0.604	0.342
85-94						
35-64	UNIVARIATE	0.495	0.277	0.074	0.393	0.220
65-94	UNIVARIATE	0.586	0.258	0.023	0.479	0.210
35-64	BIVARIATE	0.224	0.284	0.431	0.178	0.226
65-94	BIVARIATE	0.513	0.259	0.047	0.419	0.211
35-64	MULTIVARIATE	0.258	0.287	0.368	0.205	0.228
65-94	MULTIVARIATE	0.391	0.263	0.138	0.319	0.215

TABLE 9-3B DATE=05JAN86
AVERAGE ANNUAL INCIDENCE RATE FOR EVENT PER 1000 PERSONS AT RISK AT EXAM
BY AGE AND LEVEL OF RISK FACTOR AT EXAM: FRAMINGHAM STUDY 30-YEAR FOLLOWUP

SEX: MALE
EVENT: SUDDEN DEATH FROM CORONARY HEART DISEASE
RISK FACTOR: HYPERTENSION IGNORING TREATMENT
PERSONS AT RISK: PERSONS FREE OF CORONARY HEART DISEASE

RISK FACTOR	35-44 AT EXAM PERSON EXAMS	# OF EVENTS	ANNUAL RATE	45-54 AT EXAM PERSON EXAMS	# OF EVENTS	ANNUAL RATE	55-64 AT EXAM PERSON EXAMS	# OF EVENTS	ANNUAL RATE	65-74 AT EXAM PERSON EXAMS	# OF EVENTS	ANNUAL RATE
NO	2653			4154	6	1	3909	9	1	1631	3	1
MILD	1310			2284	7	2	2583	14	3	1499	14	5
DEFINITE	509			1286	3	1	1557	9	3	783	9	6
UNKNOWN	0			0	0	-	0	0	-	0	0	-
TOTAL	4472			7724	16	1	8049	32	2	3913	26	3

RISK FACTOR	75-84 AT EXAM PERSON EXAMS	# OF EVENTS	ANNUAL RATE	85-94 AT EXAM PERSON EXAMS	# OF EVENTS	ANNUAL RATE	35-64 COMBINED # OF EVENTS	CRUDE	AGE ADJUSTED	65-94 COMBINED # OF EVENTS	CRUDE	AGE ADJUSTED
NO	412	4	5	21			17	1	1	7	2	2
MILD	418	2	2	15			22	2	2	16	4	4
DEFINITE	187	5	13	3			12	2	2	14	7	7
UNKNOWN	0	0	-	0			0	-	-	0	-	-
TOTAL	1017	11	5	39			51	1	1	37	4	4

LOGISTIC REGRESSIONS

AGE AT EXAM	COEFFICIENT OF RISK FACTOR	SE COEFFICIENT	P-VALUE	STANDARDIZED COEFFICIENT	STD SE
35-44					
45-54	0.303	0.312	0.332	0.228	0.235
55-64	0.462	0.219	0.035	0.356	0.169
65-74	0.764	0.262	0.004	0.577	0.198
75-84	0.562	0.404	0.164	0.413	0.297
85-94					
35-64 UNIVARIATE	0.433	0.174	0.013	0.325	0.130
65-94 UNIVARIATE	0.704	0.219	0.001	0.529	0.164
35-64 BIVARIATE	0.361	0.174	0.038	0.271	0.131
65-94 BIVARIATE	0.719	0.219	0.001	0.540	0.165
35-64 MULTIVARIATE	0.208	0.185	0.261 *	0.156	0.139
65-94 MULTIVARIATE	0.600	0.226	0.008	0.450	0.170

SEX: FEMALE
EVENT: SUDDEN DEATH FROM CORONARY HEART DISEASE
RISK FACTOR: HYPERTENSION IGNORING TREATMENT
PERSONS AT RISK: PERSONS FREE OF CORONARY HEART DISEASE

RISK FACTOR	35-44 AT EXAM PERSON EXAMS	# OF EVENTS	ANNUAL RATE	45-54 AT EXAM PERSON EXAMS	# OF EVENTS	ANNUAL RATE	55-64 AT EXAM PERSON EXAMS	# OF EVENTS	ANNUAL RATE	65-74 AT EXAM PERSON EXAMS	# OF EVENTS	ANNUAL RATE
NO	4272			5804	2	0	5245	3	0	2035	4	1
MILD	1016			2763	2	0	3860	8	1	2549	4	1
DEFINITE	328			1392	1	0	2408	2	0	1664	6	2
UNKNOWN	0			0	0	-	0	0	-	0	0	-
TOTAL	5616			9959	5	0	11513	13	1	6248	14	1

RISK FACTOR	75-84 AT EXAM PERSON EXAMS	# OF EVENTS	ANNUAL RATE	85-94 AT EXAM PERSON EXAMS	# OF EVENTS	ANNUAL RATE	35-64 COMBINED # OF EVENTS	CRUDE	AGE ADJUSTED	65-94 COMBINED # OF EVENTS	CRUDE	AGE ADJUSTED
NO	507	2	2	37			5	0	0	7	1	1
MILD	827	6	4	45			10	1	1	10	1	1
DEFINITE	617	7	6	41			3	0	0	14	3	3
UNKNOWN	0	0	-	0			0	-	-	0	-	-
TOTAL	1951	15	4	123			18	0	0	31	2	2

LOGISTIC REGRESSIONS

AGE AT EXAM	COEFFICIENT OF RISK FACTOR	SE COEFFICIENT	P-VALUE	STANDARDIZED COEFFICIENT	STD SE
35-44					
45-54	0.412	0.558	0.460	0.299	0.405
55-64	0.271	0.346	0.434	0.210	0.269
65-74	0.345	0.352	0.327	0.265	0.270
75-84	0.511	0.365	0.161	0.387	0.276
85-94					
35-64 UNIVARIATE	0.491	0.289	0.089	0.363	0.214
65-94 UNIVARIATE	0.445	0.241	0.065	0.341	0.185
35-64 BIVARIATE	0.252	0.297	0.397	0.187	0.220
65-94 BIVARIATE	0.390	0.243	0.108	0.299	0.186
35-64 MULTIVARIATE	0.286	0.302	0.342	0.212	0.223
65-94 MULTIVARIATE	0.258	0.249	0.299	0.198	0.191

TABLE 9-4   DATE=05JAN86

AVERAGE ANNUAL INCIDENCE RATE FOR EVENT PER 1000 PERSONS AT RISK AT EXAM
BY AGE AND LEVEL OF RISK FACTOR AT EXAM: FRAMINGHAM STUDY 30-YEAR FOLLOWUP

SEX: MALE
EVENT: SUDDEN DEATH FROM CORONARY HEART DISEASE
RISK FACTOR: SERUM CHOLESTEROL
PERSONS AT RISK: PERSONS FREE OF CORONARY HEART DISEASE

RISK FACTOR	35-44 AT EXAM PERSON EXAMS	# OF EVENTS	ANNUAL RATE	45-54 AT EXAM PERSON EXAMS	# OF EVENTS	ANNUAL RATE	55-64 AT EXAM PERSON EXAMS	# OF EVENTS	ANNUAL RATE	65-74 AT EXAM PERSON EXAMS	# OF EVENTS	ANNUAL RATE
84-204	1253			1767	2	1	2255	1	0	1234	11	4
205-234	1247			2306	5	1	2405	12	2	1171	6	3
235-264	872			1745	4	1	1889	11	3	873	6	3
265-294	475			1001	1	0	918	5	3	428	2	2
295-1124	291			557	4	4	454	3	3	191	1	3
UNKNOWN	334			348	0	0	128	0	0	16	0	0
TOTAL	4472			7724	16	1	8049	32	2	3913	26	3

RISK FACTOR	75-84 AT EXAM PERSON EXAMS	# OF EVENTS	ANNUAL RATE	85-94 AT EXAM PERSON EXAMS	# OF EVENTS	ANNUAL RATE	35-64 COMBINED # OF EVENTS	CRUDE	AGE-ADJUSTED	65-94 COMBINED # OF EVENTS	CRUDE	AGE-ADJUSTED
84-204	390	4	5	17			3	0	0	15	5	5
205-234	332	4	6	11			17	1	1	10	3	3
235-264	177	1	3	5			17	2	2	7	3	3
265-294	96	2	10	6			6	1	1	4	2	4
295-1124	22	0	0	0			8	3	3	1	2	2
UNKNOWN	0	0	-	0			0	0	0	0	0	0
TOTAL	1017	11	5	39			51	1	1	37	4	4

LOGISTIC REGRESSIONS

AGE AT EXAM		COEFFICIENT OF RISK FACTOR	SE COEFFICIENT	P-VALUE	STANDARDIZED COEFFICIENT	STD SE
35-44						
45-54		0.011	0.003	0.001	0.480	0.144
55-64		0.009	0.004	0.009	0.382	0.147
65-74		-0.004	0.005	0.372	-0.181	0.202
75-84		0.002	0.008	0.788	0.081	0.299
85-94						
35-64	UNIVARIATE	0.010	0.002	0.001	0.421	0.099
65-94	UNIVARIATE	-0.003	0.004	0.478	-0.119	0.168
35-64	BIVARIATE	0.011	0.002	0.001	0.454	0.101
65-94	BIVARIATE	-0.002	0.004	0.620	-0.084	0.168
35-64	MULTIVARIATE	0.010	0.002	0.001	0.438	0.104
65-94	MULTIVARIATE	-0.002	0.004	0.618	-0.084	0.168

SEX: FEMALE
EVENT: SUDDEN DEATH FROM CORONARY HEART DISEASE
RISK FACTOR: SERUM CHOLESTEROL
PERSONS AT RISK: PERSONS FREE OF CORONARY HEART DISEASE

RISK FACTOR	35-44 AT EXAM PERSON EXAMS	# OF EVENTS	ANNUAL RATE	45-54 AT EXAM PERSON EXAMS	# OF EVENTS	ANNUAL RATE	55-64 AT EXAM PERSON EXAMS	# OF EVENTS	ANNUAL RATE	65-74 AT EXAM PERSON EXAMS	# OF EVENTS	ANNUAL RATE
84-204	2550			2994	0	0	1625	1	0	789	1	1
205-234	1417			2558	1	0	2615	2	0	1401	2	1
235-264	806			2259	2	0	2938	2	0	1703	3	1
265-294	372			1473	0	0	2240	6	1	1287	3	1
295-1124	171			1101	2	1	1866	2	1	1031	4	2
UNKNOWN	500			474	0	0	229	0	0	37	1	14
TOTAL	5616			9959	5	0	11513	13	1	6248	14	1

RISK FACTOR	75-84 AT EXAM PERSON EXAMS	# OF EVENTS	ANNUAL RATE	85-94 AT EXAM PERSON EXAMS	# OF EVENTS	ANNUAL RATE	35-64 COMBINED # OF EVENTS	CRUDE	AGE-ADJUSTED	65-94 COMBINED # OF EVENTS	CRUDE	AGE-ADJUSTED
84-204	361	1	1	46			1	0	0	4	2	1
205-234	486	5	5	28			3	0	0	7	2	2
235-264	542	6	6	35			4	0	0	9	2	2
265-294	350	3	0	9			6	1	1	6	2	2
295-1124	205	0	0	5			4	1	1	4	2	1
UNKNOWN	7	0	0	0			0	0	0	1	11	10
TOTAL	1951	15	4	123			18	0	0	31	2	2

LOGISTIC REGRESSIONS

AGE AT EXAM		COEFFICIENT OF RISK FACTOR	SE COEFFICIENT	P-VALUE	STANDARDIZED COEFFICIENT	STD SE
35-44						
45-54		0.006	0.004	0.105	0.275	0.170
55-64		0.006	0.006	0.314	0.257	0.255
65-74		0.009	0.006	0.156	0.382	0.256
75-84		-0.001	0.006	0.848	-0.050	0.261
85-94						
35-64	UNIVARIATE	0.007	0.002	0.006	0.314	0.114
65-94	UNIVARIATE	-0.001	0.004	0.887	-0.026	0.184
35-64	BIVARIATE	0.006	0.004	0.087	0.282	0.165
65-94	BIVARIATE	0.002	0.004	0.655	0.082	0.183
35-64	MULTIVARIATE	0.006	0.004	0.113	0.285	0.180
65-94	MULTIVARIATE	0.003	0.004	0.549	0.111	0.185

TABLE 9-5 DATE=05JAN86
AVERAGE ANNUAL INCIDENCE RATE FOR EVENT PER 1000 PERSONS AT RISK AT EXAM
BY AGE AND LEVEL OF RISK FACTOR AT EXAM: FRAMINGHAM STUDY 30-YEAR FOLLOWUP .

SEX: MALE
EVENT: SUDDEN DEATH FROM CORONARY HEART DISEASE
RISK FACTOR: HEMATOCRIT
PERSONS AT RISK: PERSONS FREE OF CORONARY HEART DISEASE

RISK FACTOR	35-44 AT EXAM			45-54 AT EXAM			55-64 AT EXAM			65-74 AT EXAM		
	PERSON EXAMS	# OF EVENTS	ANNUAL RATE	PERSON EXAMS	# OF EVENTS	ANNUAL RATE	PERSON EXAMS	# OF EVENTS	ANNUAL RATE	PERSON EXAMS	# OF EVENTS	ANNUAL RATE
25-42	623		:	942	3	2	1018	3	1	450	2	2
43-44	546			967	2	1	985	5	3	493	2	2
45-46	983			1854	4	1	1953	6	2	963	5	3
47-48	775			1510	2	1	1599	4	1	861	2	1
49-70	1175			2066	5	1	2338	14	3	1127	15	7
UNKNOWN	370			385	0	0	156	0	0	19	0	0
TOTAL	4472			7724	16	1	8049	32	2	3913	26	3

RISK FACTOR	75-86 AT EXAM			85-94 AT EXAM			35-64 COMBINED			65-94 COMBINED		
	PERSON EXAMS	# OF EVENTS	ANNUAL RATE	PERSON EXAMS	# OF EVENTS	ANNUAL RATE	# OF EVENTS	CRUDE	AGE-ADJUSTED	# OF EVENTS	CRUDE	AGE-ADJUSTED
25-42	214	3	7	16			7	1	1	5	4	3
43-44	119	2	8	5			7	1	1	4	3	3
45-46	240	1	2	7			10	1	1	6	2	2
47-48	197	2	5	7			6	1	1	4	2	2
49-70	247	3	6	4			21	2	2	18	7	6
UNKNOWN	0	0	-	0			0	0	0	0	0	0
TOTAL	1017	11	5	39			51	1	1	37	4	4

LOGISTIC REGRESSIONS

AGE AT EXAM		COEFFICIENT OF RISK FACTOR	SE COEFFICIENT	P-VALUE	STANDARDIZED COEFFICIENT	STD SE
35-44						
45-54		-0.031	0.070	0.658	-0.110	0.249
55-64		0.064	0.047	0.175	0.243	0.180
65-74		0.122	0.053	0.022	0.451	0.198
75-84		0.047	0.074	0.528	0.195	0.310
85-94						
35-64	UNIVARIATE	0.041	0.038	0.288	0.150	0.141
65-94	UNIVARIATE	0.093	0.044	0.034	0.357	0.169
35-64	BIVARIATE	0.038	0.038	0.312	0.142	0.140
65-94	BIVARIATE	0.105	0.044	0.016	0.401	0.167
35-64	MULTIVARIATE	0.009	0.038	0.811	0.034	0.142
65-94	MULTIVARIATE	0.100	0.044	0.022	0.384	0.168

SEX: FEMALE
EVENT: SUDDEN DEATH FROM CORONARY HEART DISEASE
RISK FACTOR: HEMATOCRIT
PERSONS AT RISK: PERSONS FREE OF CORONARY HEART DISEASE

RISK FACTOR	35-44 AT EXAM			45-54 AT EXAM			55-64 AT EXAM			65-74 AT EXAM		
	PERSON EXAMS	# OF EVENTS	ANNUAL RATE	PERSON EXAMS	# OF EVENTS	ANNUAL RATE	PERSON EXAMS	# OF EVENTS	ANNUAL RATE	PERSON EXAMS	# OF EVENTS	ANNUAL RATE
21-39	1697			1843	1	0	1553	2	1	646	1	1
40-41	947			1741	0	0	2061	1	0	1069	0	0
42-43	1043			1955	0	0	2260	1	0	1256	2	1
44-45	856			2194	2	0	2847	4	1	1609	6	2
46-65	487			1681	2	1	2521	5	1	1626	4	1
UNKNOWN	566			545	0	0	271	0	0	42	1	12
TOTAL	5616			9959	5	0	11513	13	1	6248	14	1

RISK FACTOR	75-84 AT EXAM			85-94 AT EXAM			35-64 COMBINED			65-94 COMBINED		
	PERSON EXAMS	# OF EVENTS	ANNUAL RATE	PERSON EXAMS	# OF EVENTS	ANNUAL RATE	# OF EVENTS	CRUDE	AGE-ADJUSTED	# OF EVENTS	CRUDE	AGE-ADJUSTED
21-39	297	1	2	24			3	0	0	3	2	1
40-41	373	3	4	31			1	0	0	3	1	1
42-43	353	1	1	19			1	0	0	4	1	1
44-45	486	2	2	26			6	1	1	8	2	2
46-65	436	8	9	23			7	1	1	12	3	3
UNKNOWN	6	0	0	0			0	0	0	1	10	9
TOTAL	1951	15	4	123			18	0	0	31	2	2

LOGISTIC REGRESSIONS

AGE AT EXAM		COEFFICIENT OF RISK FACTOR	SE COEFFICIENT	P-VALUE	STANDARDIZED COEFFICIENT	STD SE
35-44						
45-54		0.113	0.128	0.380	0.401	0.457
55-64		0.108	0.080	0.177	0.374	0.277
65-74		0.112	0.081	0.163	0.393	0.281
75-84		0.142	0.062	0.021	0.553	0.240
85-94						
35-64	UNIVARIATE	0.137	0.066	0.039	0.493	0.239
65-94	UNIVARIATE	0.103	0.051	0.044	0.371	0.184
35-64	BIVARIATE	0.101	0.068	0.140	0.362	0.245
65-94	BIVARIATE	0.110	0.049	0.024	0.397	0.175
35-64	MULTIVARIATE	0.071	0.070	0.307	0.257	0.251
65-94	MULTIVARIATE	0.106	0.049	0.031	0.382	0.177

TABLE 9-6   DATE=05JAN86
AVERAGE ANNUAL INCIDENCE RATE FOR EVENT PER 1000 PERSONS AT RISK AT EXAM
BY AGE AND LEVEL OF RISK FACTOR AT EXAM: FRAMINGHAM STUDY 30-YEAR FOLLOWUP

SEX: MALE
EVENT: SUDDEN DEATH FROM CORONARY HEART DISEASE
RISK FACTOR: BLOOD GLUCOSE (MG/DL)
PERSONS AT RISK: PERSONS FREE OF CORONARY HEART DISEASE

RISK FACTOR	35-44 AT EXAM PERSON EXAMS	# OF EVENTS	ANNUAL RATE	45-54 AT EXAM PERSON EXAMS	# OF EVENTS	ANNUAL RATE	55-64 AT EXAM PERSON EXAMS	# OF EVENTS	ANNUAL RATE	65-74 AT EXAM PERSON EXAMS	# OF EVENTS	ANNUAL RATE
29-69	992			1490	5	2	1087	4	2	421	2	2
70-89	2501			4172	7	1	4056	19	2	1761	10	3
90-109	626			1348	2	1	1866	6	2	1046	8	4
110-129	105			305	0	0	537	3	3	335	2	3
130-524	48			193	2	5	408	0	0	334	4	6
UNKNOWN	200			216	0	0	95	0	0	16	0	0
TOTAL	4472			7724	16	1	8049	32	2	3913	26	3

RISK FACTOR	75-84 AT EXAM PERSON EXAMS	# OF EVENTS	ANNUAL RATE	85-94 AT EXAM PERSON EXAMS	# OF EVENTS	ANNUAL RATE	35-64 COMBINED # OF EVENTS	CRUDE	AGE-ADJUSTED	65-94 COMBINED # OF EVENTS	CRUDE	AGE-ADJUSTED
29-69	62	0	0				9	1	1	2	2	2
70-89	377	6	8	11			28	1	1	16	4	4
90-109	328	4	6	14			9	1	1	12	4	4
110-129	125	1	4	4			3	2	1	3	3	3
130-524	125	0	0	9			2	2	2	4	4	5
UNKNOWN	0	0	-	0			0	0	0	0	0	0
TOTAL	1017	11	5	39			51	1	1	37	4	4

LOGISTIC REGRESSIONS

AGE AT EXAM		COEFFICIENT OF RISK FACTOR	SE COEFFICIENT	P-VALUE	STANDARDIZED COEFFICIENT	STD SE
35-44						
45-54		0.007	0.006	0.174	0.177	0.130
55-64		-0.010	0.009	0.285	-0.280	0.261
65-74		0.006	0.003	0.066	0.216	0.118
75-84		-0.012	0.014	0.388	-0.381	0.442
85-94						
35-64	UNIVARIATE	0.002	0.005	0.720	0.045	0.125
65-94	UNIVARIATE	0.004	0.004	0.225	0.147	0.122
35-64	BIVARIATE	-0.002	0.006	0.768	-0.042	0.143
65-94	BIVARIATE	0.004	0.004	0.508	0.132	0.129
35-64	MULTIVARIATE	-0.001	0.006	0.801	-0.035	0.140
65-94	MULTIVARIATE	0.003	0.004	0.434	0.103	0.131

SEX: FEMALE
EVENT: SUDDEN DEATH FROM CORONARY HEART DISEASE
RISK FACTOR: BLOOD GLUCOSE (MG/DL)
PERSONS AT RISK: PERSONS FREE OF CORONARY HEART DISEASE

RISK FACTOR	35-44 AT EXAM PERSON EXAMS	# OF EVENTS	ANNUAL RATE	45-54 AT EXAM PERSON EXAMS	# OF EVENTS	ANNUAL RATE	55-64 AT EXAM PERSON EXAMS	# OF EVENTS	ANNUAL RATE	65-74 AT EXAM PERSON EXAMS	# OF EVENTS	ANNUAL RATE
29-69	1066			1797	1	0	1369	5	2	521	1	1
70-89	3412			5744	3	0	6098	6	0	3102	3	0
90-109	672			1625	0	0	2712	2	0	1732	5	1
110-129	96			301	0	0	730	0	0	501	1	1
130-524	39			159	1	3	438	0	0	360	3	4
UNKNOWN	331			333	0	0	166	0	0	32	1	16
TOTAL	5616			9959	5	0	11513	13	1	6248	14	1

RISK FACTOR	75-84 AT EXAM PERSON EXAMS	# OF EVENTS	ANNUAL RATE	85-94 AT EXAM PERSON EXAMS	# OF EVENTS	ANNUAL RATE	35-64 COMBINED # OF EVENTS	CRUDE	AGE-ADJUSTED	65-94 COMBINED # OF EVENTS	CRUDE	AGE-ADJUSTED
29-69	105	1	5	3			6	1	1	2	2	2
70-89	823	5	3	44			9	0	0	9	1	1
90-109	640	3	2	51			2	0	0	9	2	2
110-129	219	4	9	13			0	0	0	5	3	3
130-524	163	2	6	12			1	1	1	5	5	5
UNKNOWN	1	0	0	0			0	0	0	1	15	12
TOTAL	1951	15	4	123			18	0	0	31	2	2

LOGISTIC REGRESSIONS

AGE AT EXAM		COEFFICIENT OF RISK FACTOR	SE COEFFICIENT	P-VALUE	STANDARDIZED COEFFICIENT	STD SE
35-44						
45-54		0.015	0.008	0.063	0.291	0.156
55-64		-0.048	0.022	0.030	-1.218	0.562
65-74		0.008	0.006	0.169	0.224	0.162
75-84		0.007	0.004	0.075	0.247	0.139
85-94						
35-64	UNIVARIATE	-0.003	0.012	0.812	-0.064	0.270
65-94	UNIVARIATE	0.008	0.003	0.012	0.245	0.098
35-64	BIVARIATE	-0.012	0.014	0.420	-0.252	0.313
65-94	BIVARIATE	0.007	0.003	0.046	0.205	0.103
35-64	MULTIVARIATE	-0.011	0.014	0.425	-0.246	0.308
65-94	MULTIVARIATE	0.006	0.004	0.080	0.188	0.107

TABLE 9-7   DATE=05JAN86
AVERAGE ANNUAL INCIDENCE RATE FOR EVENT PER 1000 PERSONS AT RISK AT EXAM
BY AGE AND LEVEL OF RISK FACTOR AT EXAM: FRAMINGHAM STUDY 30-YEAR FOLLOWUP

SEX: MALE
EVENT: SUDDEN DEATH FROM CORONARY HEART DISEASE
RISK FACTOR: DIABETES MELLITUS
PERSONS AT RISK: PERSONS FREE OF CORONARY HEART DISEASE

RISK FACTOR	35-44 AT EXAM			45-54 AT EXAM			55-64 AT EXAM			65-74 AT EXAM		
	PERSON EXAMS	# OF EVENTS	ANNUAL RATE	PERSON EXAMS	# OF EVENTS	ANNUAL RATE	PERSON EXAMS	# OF EVENTS	ANNUAL RATE	PERSON EXAMS	# OF EVENTS	ANNUAL RATE
NO	4424			7486	14	1	7549	32	2	3504	22	3
YES	48			238	2	4	500	0	0	409	4	5
UNKNOWN	0			0	0	-	0	0	-	0	0	-
TOTAL	4472			7724	16	1	8049	32	2	3913	26	3

RISK FACTOR	75-84 AT EXAM			85-94 AT EXAM			35-64 COMBINED			6-94 COMBINED		
	PERSON EXAMS	# OF EVENTS	ANNUAL RATE	PERSON EXAMS	# OF EVENTS	ANNUAL RATE	# OF EVENTS	CRUDE	AGE-ADJUSTED	# OF EVENTS	CRUDE	AGE-ADJUSTED
NO	900	11	6	34		5	49	1	1	33	4	4
YES	117	0	0	5			2	1	2	4	4	4
UNKNOWN	0	0	-	0			0	-		0	-	
TOTAL	1017	11	5	39			51	1	1	37	4	4

LOGISTIC REGRESSIONS

AGE AT EXAM		COEFFICIENT OF RISK FACTOR	SE COEFFICIENT	P-VALUE	STANDARDIZED COEFFICIENT	STD SE
35-44		-	-	-	-	-
45-54		1.509	0.759	0.047	0.261	0.131
55-64		-	-	-	-	-
65-74		0.447	0.546	0.413	0.137	0.167
75-84		-	-	-	-	-
85-94		-	-	-	-	-
35-64	UNIVARIATE	0.010	0.722	0.988	0.002	0.140
65-94	UNIVARIATE	0.013	0.531	0.980	0.004	0.164
35-64	BIVARIATE	-0.307	0.725	0.672	-0.059	0.140
65-94	BIVARIATE	-0.016	0.532	0.976	-0.005	0.164
35-64	MULTIVARIATE	-0.437	0.734	0.552	-0.084	0.142
65-94	MULTIVARIATE	-0.064	0.537	0.905	-0.020	0.166

SEX: FEMALE
EVENT: SUDDEN DEATH FROM CORONARY HEART DISEASE
RISK FACTOR: DIABETES MELLITUS
PERSONS AT RISK: PERSONS FREE OF CORONARY HEART DISEASE

RISK FACTOR	35-44 AT EXAM			45-54 AT EXAM			55-64 AT EXAM			65-74 AT EXAM		
	PERSON EXAMS	# OF EVENTS	ANNUAL RATE	PERSON EXAMS	# OF EVENTS	ANNUAL RATE	PERSON EXAMS	# OF EVENTS	ANNUAL RATE	PERSON EXAMS	# OF EVENTS	ANNUAL RATE
NO	5573			9760	4	0	10985	12	1	5780	12	1
YES	43			199	1	3	528	1	1	468	2	2
UNKNOWN	0			0	0	-	0	0	-	0	0	-
TOTAL	5616			9959	5	0	11513	13	1	6248	14	1

RISK FACTOR	75-84 AT EXAM			85-94 AT EXAM			35-64 COMBINED			65-94 COMBINED		
	PERSON EXAMS	# OF EVENTS	ANNUAL RATE	PERSON EXAMS	# OF EVENTS	ANNUAL RATE	# OF EVENTS	CRUDE	AGE-ADJUSTED	# OF EVENTS	CRUDE	AGE-ADJUSTED
NO	1738	11	1	106			16	0	0	25	2	2
YES	213	4	9	17			2	1	1	6	4	4
UNKNOWN	0	0	-	0			0	-		0	-	
TOTAL	1951	15	4	123			18	-	0	31	2	2

LOGISTIC REGRESSIONS

AGE AT EXAM		COEFFICIENT OF RISK FACTOR	SE COEFFICIENT	P-VALUE	STANDARDIZED COEFFICIENT	STD SE
35-44		-	-	-	-	-
45-54		2.511	1.120	0.025	0.351	0.157
55-64		0.551	1.042	0.597	0.115	0.218
65-74		0.724	0.765	0.344	0.191	0.201
75-84		1.100	0.588	0.062	0.343	0.184
85-94		-	-	-	-	-
35-64	UNIVARIATE	1.454	0.751	0.053	0.242	0.125
65-94	UNIVARIATE	0.969	0.456	0.034	0.269	0.127
35-64	BIVARIATE	1.028	0.755	0.173	0.171	0.125
65-94	BIVARIATE	0.825	0.459	0.072	0.229	0.127
35-64	MULTIVARIATE	0.928	0.765	0.225	0.154	0.127
65-94	MULTIVARIATE	0.697	0.478	0.145	0.193	0.132

TABLE 7-6   DATE=27NOV85

AVERAGE ANNUAL INCIDENCE RATE FOR EVENT PER 1000 PERSONS AT RISK AT EXAM
BY AGE AND LEVEL OF RISK FACTOR AT EXAM: FRAMINGHAM STUDY 30-YEAR FOLLOWUP

SEX: MALE
EVENT: ANGINA PECTORIS
RISK FACTOR: BLOOD GLUCOSE (MG/DL)
PERSONS AT RISK: PERSONS FREE OF CORONARY HEART DISEASE

RISK FACTOR	35-44 AT EXAM			45-54 AT EXAM			55-64 AT EXAM			65-74 AT EXAM		
	PERSON EXAMS	# OF EVENTS	ANNUAL RATE	PERSON EXAMS	# OF EVENTS	ANNUAL RATE	PERSON EXAMS	# OF EVENTS	ANNUAL RATE	PERSON EXAMS	# OF EVENTS	ANNUAL RATE
29-69	992	4	2	1490	15	5	1087	22	10	421	4	5
70-89	2501	11	2	4172	35	4	4056	78	10	1761	23	7
90-109	626	3	2	1348	19	7	1866	30	8	1046	21	10
110-129	105	1	5	305	3	5	537	8	7	535	6	9
130-524	48	0	0	193	4	10	408	8	10	334	8	12
UNKNOWN	200	2	5	216	4	9	95	5	16	16	0	0
TOTAL	4472	21	2	7724	80	5	8049	149	9	3913	62	8

RISK FACTOR	75-84 AT EXAM			85-94 AT EXAM			35-64 COMBINED			65-94 COMBINED		
	PERSON EXAMS	# OF EVENTS	ANNUAL RATE	PERSON EXAMS	# OF EVENTS	ANNUAL RATE	# OF EVENTS	CRUDE	AGE-ADJUSTED	# OF EVENTS	CRUDE	AGE-ADJUSTED
29-69	62	0	0				41	6	6	4	4	4
70-89	377	5	7	11			124	6	6	28	7	6
90-109	328	2	3	14			52	7	6	24	9	9
110-129	125	0	0	4			12	6	6	6	6	7
130-524	125	2	8	9			12	9	8	11	12	12
UNKNOWN	0	0	-	0			9	9	11	0	0	0
TOTAL	1017	9	4	39			250	6	6	73	7	7

LOGISTIC REGRESSIONS

AGE AT EXAM		COEFFICIENT OF RISK FACTOR	SE COEFFICIENT	P-VALUE	STANDARDIZED COEFFICIENT	STD SE
35-44		0.002	0.012	0.844	0.039	0.200
45-54		0.004	0.004	0.336	0.086	0.089
55-64		-0.002	0.003	0.557	-0.054	0.092
65-74		0.005	0.003	0.095	0.153	0.092
75-84		0.012	0.006	0.050	0.385	0.196
85-94						
35-64	UNIVARIATE	0.003	0.002	0.262	0.062	0.055
65-94	UNIVARIATE	0.006	0.002	0.018	0.189	0.080
35-64	BIVARIATE	-0.000	0.003	0.865	-0.011	0.062
65-94	BIVARIATE	0.006	0.002	0.016	0.192	0.079
35-64	MULTIVARIATE	-0.001	0.003	0.819	-0.014	0.061
65-94	MULTIVARIATE	0.006	0.002	0.020	0.187	0.081

SEX: FEMALE
EVENT: ANGINA PECTORIS
RISK FACTOR: BLOOD GLUCOSE (MG/DL)
PERSONS AT RISK: PERSONS FREE OF CORONARY HEART DISEASE

RISK FACTOR	35-44 AT EXAM			45-54 AT EXAM			55-64 AT EXAM			65-74 AT EXAM		
	PERSON EXAMS	# OF EVENTS	ANNUAL RATE	PERSON EXAMS	# OF EVENTS	ANNUAL RATE	PERSON EXAMS	# OF EVENTS	ANNUAL RATE	PERSON EXAMS	# OF EVENTS	ANNUAL RATE
29-69	1066	2	1	1797	4	1	1369	21	8	521	9	9
70-89	3412	2	0	5744	27	2	6098	58	5	3102	44	7
90-109	672	1	1	1625	4	1	2712	30	6	1732	24	7
110-129	96	0	0	301	2	3	730	12	8	501	8	8
130-524	39	0	0	159	4	13	438	13	15	360	8	8
UNKNOWN	331	0	0	333	6	9	166	6	18	32	0	0
TOTAL	5616	5	0	9959	47	2	11513	140	6	6248	91	7

RISK FACTOR	75-84 AT EXAM			85-94 AT EXAM			35-64 COMBINED			65-94 COMBINED		
	PERSON EXAMS	# OF EVENTS	ANNUAL RATE	PERSON EXAMS	# OF EVENTS	ANNUAL RATE	# OF EVENTS	CRUDE	AGE-ADJUSTED	# OF EVENTS	CRUDE	AGE-ADJUSTED
29-69	105	2	10	3			27	3	4	11	9	9
70-89	823	9	5	44			87	3	3	53	7	7
90-109	640	11	9	51			35	3	3	36	7	7
110-129	219	4	9	13			14	6	5	12	8	8
130-524	163	5	15	12			17	13	11	11	10	10
UNKNOWN	1	0	0	0			12	7	11	0	0	0
TOTAL	1951	31	8	123			192	4	4	123	7	7

LOGISTIC REGRESSIONS

AGE AT EXAM		COEFFICIENT OF RISK FACTOR	SE COEFFICIENT	P-VALUE	STANDARDIZED COEFFICIENT	STD SE
35-44		-0.015	0.039	0.692	-0.234	0.590
45-54		0.015	0.003	0.001	0.294	0.058
55-64		0.006	0.002	0.005	0.160	0.057
65-74		-0.001	0.004	0.825	-0.024	0.111
75-84		0.005	0.004	0.160	0.170	0.121
85-94						
35-64	UNIVARIATE	0.011	0.002	0.001	0.230	0.036
65-94	UNIVARIATE	0.002	0.003	0.520	0.052	0.081
35-64	BIVARIATE	0.008	0.002	0.001	0.176	0.039
65-94	BIVARIATE	0.002	0.003	0.473	0.058	0.081
35-64	MULTIVARIATE	0.006	0.002	0.001	0.137	0.041
65-94	MULTIVARIATE	0.001	0.003	0.651	0.037	0.082

TABLE 9-8   DATE=05JAN86
AVERAGE ANNUAL INCIDENCE RATE FOR EVENT PER 1000 PERSONS AT RISK AT EXAM
BY AGE AND LEVEL OF RISK FACTOR AT EXAM: FRAMINGHAM STUDY 30-YEAR FOLLOWUP

SEX: MALE
EVENT: SUDDEN DEATH FROM CORONARY HEART DISEASE
RISK FACTOR: GLUCOSE IN URINE, DEFINITE OR TRACE
PERSONS AT RISK: PERSONS FREE OF CORONARY HEART DISEASE

RISK FACTOR	35-44 AT EXAM			45-54 AT EXAM			55-64 AT EXAM			65-74 AT EXAM		
	PERSON EXAMS	# OF EVENTS	ANNUAL RATE	PERSON EXAMS	# OF EVENTS	ANNUAL RATE	PERSON EXAMS	# OF EVENTS	ANNUAL RATE	PERSON EXAMS	# OF EVENTS	ANNUAL RATE
NO	4380			7535	14	1	7815	32	2	3767	22	3
YES	60			168	2	6	210	0	0	140	4	14
UNKNOWN	32			21	0	0	24	0	0	6	0	0
TOTAL	4472			7724	16	1	8049	32	2	3913	26	3

RISK FACTOR	75-84 AT EXAM			85-94 AT EXAM			35-64 COMBINED			65-94 COMBINED		
	PERSON EXAMS	# OF EVENTS	ANNUAL RATE	PERSON EXAMS	# OF EVENTS	ANNUAL RATE	# OF EVENTS	CRUDE	AGE-ADJUSTED	# OF EVENTS	CRUDE	AGE-ADJUSTED
NO	986	11	6	38			49	1	1	33	3	3
YES	28	0	0	1			2	2	2	4	12	11
UNKNOWN	3	0	0	0			0	0	0	0	0	0
TOTAL	1017	11	5	39			51	1	1	37	4	4

LOGISTIC REGRESSIONS

AGE AT EXAM		COEFFICIENT OF RISK FACTOR	SE COEFFICIENT	P-VALUE	STANDARDIZED COEFFICIENT	STD SE
35-44		-	-	-	-	-
45-54		1.868	0.760	0.014	0.273	0.111
55-64		-	-	-	-	-
65-74		1.611	0.551	0.003	0.299	0.102
75-84		-	-	-	-	-
85-94		-	-	-	-	-
35-64	UNIVARIATE	0.611	0.725	0.398	0.089	0.105
65-94	UNIVARIATE	1.251	0.555	0.019	0.227	0.097
35-64	BIVARIATE	0.470	0.724	0.516	0.069	0.106
65-94	BIVARIATE	1.297	0.537	0.016	0.235	0.097
35-64	MULTIVARIATE	0.387	0.729	0.595	0.056	0.106
65-94	MULTIVARIATE	1.170	0.546	0.032	0.212	0.099

SEX: FEMALE
EVENT: SUDDEN DEATH FROM CORONARY HEART DISEASE
RISK FACTOR: GLUCOSE IN URINE, DEFINITE OR TRACE
PERSONS AT RISK: PERSONS FREE OF CORONARY HEART DISEASE

RISK FACTOR	35-44 AT EXAM			45-54 AT EXAM			55-64 AT EXAM			65-74 AT EXAM		
	PERSON EXAMS	# OF EVENTS	ANNUAL RATE	PERSON EXAMS	# OF EVENTS	ANNUAL RATE	PERSON EXAMS	# OF EVENTS	ANNUAL RATE	PERSON EXAMS	# OF EVENTS	ANNUAL RATE
NO	5537			9841	4	0	11346	13	1	6131	13	1
YES	41			94	1	5	144	0	0	107	1	5
UNKNOWN	38			24	0	0	23	0	0	10	0	0
TOTAL	5616			9959	5	0	11513	13	1	6248	14	1

RISK FACTOR	75-84 AT EXAM			85-94 AT EXAM			35-64 COMBINED			65-94 COMBINED		
	PERSON EXAMS	# OF EVENTS	ANNUAL RATE	PERSON EXAMS	# OF EVENTS	ANNUAL RATE	# OF EVENTS	CRUDE	AGE-ADJUSTED	# OF EVENTS	CRUDE	AGE-ADJUSTED
NO	1919	14	4	120			17	0	0	29	2	2
YES	23	1	22	3			1	2	2	3	3	9
UNKNOWN	9	0	0	0			0	0	0	0	0	0
TOTAL	1951	15	4	123			18	0	0	31	2	2

LOGISTIC REGRESSIONS

AGE AT EXAM	COEFFICIENT OF RISK FACTOR	SE COEFFICIENT	P-VALUE	STANDARDIZED COEFFICIENT	STD SE

AVERAGE ANNUAL INCIDENCE RATE FOR EVENT PER 1000 PERSONS AT RISK AT EXAM
BY AGE AND LEVEL OF RISK FACTOR AT EXAM: FRAMINGHAM STUDY 30-YEAR FOLLOWUP

SEX: MALE
EVENT: SUDDEN DEATH FROM CORONARY HEART DISEASE
RISK FACTOR: GLUCOSE INTOLERANCE
PERSONS AT RISK: PERSONS FREE OF CORONARY HEART DISEASE

RISK FACTOR	35-44 AT EXAM PERSON EXAMS	# OF EVENTS	ANNUAL RATE	45-54 AT EXAM PERSON EXAMS	# OF EVENTS	ANNUAL RATE	55-64 AT EXAM PERSON EXAMS	# OF EVENTS	ANNUAL RATE	65-74 AT EXAM PERSON EXAMS	# OF EVENTS	ANNUAL RATE
NO	4355			7221	14	1	7145	32	2	3228	20	3
YES	137			503	2	2	904	0	0	685	6	4
UNKNOWN	0			0	0	-	0	0	-	0	0	-
TOTAL	4472			7724	16	1	8049	32	2	3913	26	3

RISK FACTOR	75-84 AT EXAM PERSON EXAMS	# OF EVENTS	ANNUAL RATE	85-94 AT EXAM PERSON EXAMS	# OF EVENTS	ANNUAL RATE	35-64 COMBINED # OF EVENTS	CRUDE	AGE-ADJUSTED	65-94 COMBINED # OF EVENTS	CRUDE	AGE-ADJUSTED
NO	802	10	6	27			49	1	1	30	4	4
YES	215	1	2	12			2	1	1	7	4	4
UNKNOWN	0	0	-	0			0	-		0	-	
TOTAL	1017	11	5	39			51	1	1	37	4	4

LOGISTIC REGRESSIONS

AGE AT EXAM		COEFFICIENT OF RISK FACTOR	SE COEFFICIENT	P-VALUE	STANDARDIZED COEFFICIENT	STD SE
35-44						
45-54		0.720	0.757	0.342	0.178	0.187
55-64		-				
65-74		0.349	0.467	0.455	0.133	0.178
75-84		-0.994	1.052	0.345	-0.406	0.430
85-94						
35-64	UNIVARIATE	-0.706	0.722	0.328	-0.187	0.192
65-94	UNIVARIATE	0.038	0.421	0.929	0.015	0.163
35-64	BIVARIATE	-0.974	0.724	0.178	-0.259	0.192
65-94	BIVARIATE	-0.020	0.422	0.962	-0.008	0.164
35-64	MULTIVARIATE	-1.045	0.727	0.151	-0.277	0.193
65-94	MULTIVARIATE	-0.091	0.425	0.830	-0.035	0.164

SEX: FEMALE
EVENT: SUDDEN DEATH FROM CORONARY HEART DISEASE
RISK FACTOR: GLUCOSE INTOLERANCE
PERSONS AT RISK: PERSONS FREE OF CORONARY HEART DISEASE

RISK FACTOR	35-44 AT EXAM PERSON EXAMS	# OF EVENTS	ANNUAL RATE	45-54 AT EXAM PERSON EXAMS	# OF EVENTS	ANNUAL RATE	55-64 AT EXAM PERSON EXAMS	# OF EVENTS	ANNUAL RATE	65-74 AT EXAM PERSON EXAMS	# OF EVENTS	ANNUAL RATE
NO	5522			9588	4	0	10503	12	1	5398	10	1
YES	94			371	1	1	1010	1	0	850	4	2
UNKNOWN	0		-	0	0	-	0	0	-	0	0	-
TOTAL	5616			9959	5	0	11513	13	1	6248	14	1

RISK FACTOR	75-84 AT EXAM PERSON EXAMS	# OF EVENTS	ANNUAL RATE	85-94 AT EXAM PERSON EXAMS	# OF EVENTS	ANNUAL RATE	35-64 COMBINED # OF EVENTS	CRUDE	AGE-ADJUSTED	65-94 COMBINED # OF EVENTS	CRUDE	AGE-ADJUSTED
NO	1581	11	3	100			16	0	0	23	2	2
YES	370	4	3	23			2	1	1	8	3	3
UNKNOWN	0	0	-	0			0	-		0	-	
TOTAL	1951	15	4	123			18	0	0	31	2	2

LOGISTIC REGRESSIONS

AGE AT EXAM		COEFFICIENT OF RISK FACTOR	SE COEFFICIENT	P-VALUE	STANDARDIZED COEFFICIENT	STD SE
35-44						
45-54		1.868	1.119	0.095	0.354	0.212
55-64		-0.143	1.041	0.891	-0.041	0.295
65-74		0.935	0.593	0.115	0.321	0.203
75-84		0.445	0.587	0.449	0.174	0.230
85-94						
35-64	UNIVARIATE	0.776	0.750	0.301	0.176	0.170
65-94	UNIVARIATE	0.687	0.412	0.095	0.245	0.147
35-64	BIVARIATE	0.341	0.755	0.651	0.077	0.171
65-94	BIVARIATE	0.554	0.414	0.180	0.198	0.148
35-64	MULTIVARIATE	0.291	0.758	0.701	0.066	0.172
65-94	MULTIVARIATE	0.482	0.426	0.258	0.172	0.152

188

TABLE 9-10   DATE=05JAN86
AVERAGE ANNUAL INCIDENCE RATE FOR EVENT PER 1000 PERSONS AT RISK AT EXAM
BY AGE AND LEVEL OF RISK FACTOR AT EXAM: FRAMINGHAM STUDY 30-YEAR FOLLOWUP

SEX: MALE
EVENT: SUDDEN DEATH FROM CORONARY HEART DISEASE
RISK FACTOR: METROPOLITAN RELATIVE WEIGHT (PERCENT)
PERSONS AT RISK: PERSONS FREE OF CORONARY HEART DISEASE

RISK FACTOR	35-44 AT EXAM PERSON EXAMS	# OF EVENTS	ANNUAL RATE	45-54 AT EXAM PERSON EXAMS	# OF EVENTS	ANNUAL RATE	55-64 AT EXAM PERSON EXAMS	# OF EVENTS	ANNUAL RATE	65-74 AT EXAM PERSON EXAMS	# OF EVENTS	ANNUAL RATE
54-104	723			1131	2	1	1155	6	3	686	3	2
105-114	965			1540	4	1	1655	4	1	851	2	1
115-124	1122			2037	4	1	2175	6	1	1069	9	4
125-134	915			1645	2	1	1674	5	1	766	7	5
135-272	747			1370	4	1	1389	11	4	541	5	5
UNKNOWN	0			1	0	0	1	0	0	0	0	-
TOTAL	4472			7724	16	1	8049	32	2	3913	26	3

RISK FACTOR	75-84 AT EXAM PERSON EXAMS	# OF EVENTS	ANNUAL RATE	85-94 AT EXAM PERSON EXAMS	# OF EVENTS	ANNUAL RATE	35-64 COMBINED # OF EVENTS	CRUDE	AGE-ADJUSTED	65-94 COMBINED # OF EVENTS	CRUDE	AGE-ADJUSTED
54-104	224	4	9	18			9	1	1	7	4	4
105-114	252	3	6	9			8	1	1	5	2	2
115-124	296	2	3	12			11	1	1	11	4	4
125-134	155	0	0	0			8	1	1	7	5	5
135-272	90	2	11	0			15	2	2	7	6	6
UNKNOWN	0	0	-	0			0	0	0	0	-	-
TOTAL	1017	11	5	39			51	1	1	37	4	4

LOGISTIC REGRESSIONS

AGE AT EXAM		COEFFICIENT OF RISK FACTOR	SE COEFFICIENT	P-VALUE	STANDARDIZED COEFFICIENT	STD SE
35-44						
45-54		0.000	0.015	0.975	0.008	0.250
55-64		0.013	0.011	0.231	0.202	0.169
65-74		0.020	0.012	0.085	0.316	0.184
75-84		-0.032	0.021	0.133	-0.475	0.316
85-94						
35-64	UNIVARIATE	0.008	0.008	0.336	0.131	0.136
65-94	UNIVARIATE	0.006	0.010	0.544	0.098	0.162
35-64	BIVARIATE	0.008	0.008	0.342	0.130	0.137
65-94	BIVARIATE	0.009	0.011	0.375	0.145	0.163
35-64	MULTIVARIATE	0.007	0.009	0.439	0.110	0.142
65-94	MULTIVARIATE	0.005	0.011	0.644	0.077	0.166

SEX: FEMALE
EVENT: SUDDEN DEATH FROM CORONARY HEART DISEASE
RISK FACTOR: METROPOLITAN RELATIVE WEIGHT (PERCENT)
PERSONS AT RISK: PERSONS FREE OF CORONARY HEART DISEASE

RISK FACTOR	35-44 AT EXAM PERSON EXAMS	# OF EVENTS	ANNUAL RATE	45-54 AT EXAM PERSON EXAMS	# OF EVENTS	ANNUAL RATE	55-64 AT EXAM PERSON EXAMS	# OF EVENTS	ANNUAL RATE	65-74 AT EXAM PERSON EXAMS	# OF EVENTS	ANNUAL RATE
54-104	1694			2095	1	0	2045	1	0	1216	4	2
105-114	1430			2331	3	1	2409	3	1	1145	1	0
115-124	1058			2107	0	0	2479	5	1	1395	2	1
125-134	566			1372	1	0	1743	1	0	982	4	2
135-272	860			2054	3	1	2837	3	1	1510	3	1
UNKNOWN	8			0	0	-	0	0	-	0	0	-
TOTAL	5616			9959	5	0	11513	13	1	6248	14	1

RISK FACTOR	75-84 AT EXAM PERSON EXAMS	# OF EVENTS	ANNUAL RATE	85-94 AT EXAM PERSON EXAMS	# OF EVENTS	ANNUAL RATE	35-64 COMBINED # OF EVENTS	CRUDE	AGE-ADJUSTED	65-94 COMBINED # OF EVENTS	CRUDE	AGE-ADJUSTED
54-104	517	3	3	41			2	0	0	9	3	2
105-114	375	2	3	28			3	0	0	3	1	2
115-124	399	4	5	21			5	0	0	6	2	2
125-134	266	4	8	16			2	0	0	8	3	3
135-272	394	2	3	17			6	1	0	5	1	1
UNKNOWN	0	0	-	0			0	0	0	0	-	-
TOTAL	1951	15	4	123			18	0	0	31	2	2

LOGISTIC REGRESSIONS

AGE AT EXAM		COEFFICIENT OF RISK FACTOR	SE COEFFICIENT	P-VALUE	STANDARDIZED COEFFICIENT	STD SE
35-44						
45-54		0.021	0.014	0.128	0.454	0.298
55-64		-0.004	0.013	0.744	-0.096	0.293
65-74		-0.006	0.013	0.649	-0.130	0.285
75-84		0.003	0.012	0.831	0.054	0.254
85-94						
35-64	UNIVARIATE	0.008	0.010	0.397	0.176	0.208
65-94	UNIVARIATE	-0.009	0.009	0.325	-0.192	0.196
35-64	BIVARIATE	0.004	0.010	0.655	0.098	0.218
65-94	BIVARIATE	-0.005	0.009	0.567	-0.111	0.193
35-64	MULTIVARIATE	0.006	0.011	0.568	0.133	0.233
65-94	MULTIVARIATE	-0.006	0.009	0.525	-0.126	0.198

TABLE 9-11   DATE=05JAN86
AVERAGE ANNUAL INCIDENCE RATE FOR EVENT PER 1000 PERSONS AT RISK AT EXAM
BY AGE AND LEVEL OF RISK FACTOR AT EXAM: FRAMINGHAM STUDY 30-YEAR FOLLOWUP

SEX: MALE
EVENT: SUDDEN DEATH FROM CORONARY HEART DISEASE
RISK FACTOR: VITAL CAPACITY - HEIGHT INDEX (ML/INCH)
PERSONS AT RISK: PERSONS FREE OF CORONARY HEART DISEASE

RISK FACTOR	35-44 AT EXAM			45-54 AT EXAM			55-64 AT EXAM			65-74 AT EXAM		
	PERSON EXAMS	# OF EVENTS	ANNUAL RATE	PERSON EXAMS	# OF EVENTS	ANNUAL RATE	PERSON EXAMS	# OF EVENTS	ANNUAL RATE	PERSON EXAMS	# OF EVENTS	ANNUAL RATE
8-44	269			1062	1	0	2045	12	3	1752	11	3
45-49	411			1125	2	1	1480	5	2	749	7	5
50-54	747			1658	6	2	1835	9	2	712	6	4
55-59	1022			1739	2	1	1416	2	1	442	2	2
60-85	2010			2135	5	1	1255	4	2	246	0	0
UNKNOWN	13			5	0	0	18	0	0	12	0	0
TOTAL	4472			7724	16	1	8049	32	2	3913	26	3

RISK FACTOR	75-84 AT EXAM			85-94 AT EXAM			35-64 COMBINED			65-94 COMBINED		
	PERSON EXAMS	# OF EVENTS	ANNUAL RATE	PERSON EXAMS	# OF EVENTS	ANNUAL RATE	# OF EVENTS	CRUDE	AGE-ADJUSTED	# OF EVENTS	CRUDE	AGE-ADJUSTED
8-44	649	8	6	24			13	2	1	19	4	4
45-49	172	1	3	6			8	1	1	8	4	4
50-54	97	0	0	7			15	2	2	6	4	3
55-59	65	2	15	2			6	1	1	4	4	5
60-85	26	0	0	0			9	1	1	0	0	0
UNKNOWN	8	0	0	0			0	0	0	0	0	0
TOTAL	1017	11	5	39			51	1	1	37	4	4

LOGISTIC REGRESSIONS

AGE AT EXAM		COEFFICIENT OF RISK FACTOR	SE COEFFICIENT	P-VALUE	STANDARDIZED COEFFICIENT	STD SE
35-44						
45-54		0.008	0.028	0.765	0.075	0.252
55-64		-0.018	0.019	0.340	-0.165	0.173
65-74		-0.013	0.021	0.516	-0.126	0.195
75-84		-0.001	0.033	0.983	-0.007	0.303
85-94						
35-64	UNIVARIATE	-0.027	0.014	0.055	-0.260	0.135
65-94	UNIVARIATE	-0.013	0.017	0.436	-0.128	0.164
35-64	BIVARIATE	-0.008	0.015	0.574	-0.082	0.146
65-94	BIVARIATE	-0.006	0.018	0.745	-0.055	0.169
35-64	MULTIVARIATE	0.001	0.016	0.931	0.013	0.151
65-94	MULTIVARIATE	0.002	0.018	0.931	0.015	0.172

SEX: FEMALE
EVENT: SUDDEN DEATH FROM CORONARY HEART DISEASE
RISK FACTOR: VITAL CAPACITY - HEIGHT INDEX (ML/INCH)
PERSONS AT RISK: PERSONS FREE OF CORONARY HEART DISEASE

RISK FACTOR	35-44 AT EXAM			45-54 AT EXAM			55-64 AT EXAM			65-74 AT EXAM		
	PERSON EXAMS	# OF EVENTS	ANNUAL RATE	PERSON EXAMS	# OF EVENTS	ANNUAL RATE	PERSON EXAMS	# OF EVENTS	ANNUAL RATE	PERSON EXAMS	# OF EVENTS	ANNUAL RATE
4-29	122			534	0	0	1461	4	1	1747	7	2
30-34	289			1217	0	0	2418	3	1	1715	2	1
35-39	769			2077	2	0	3048	3	0	1529	2	1
40-44	1427			2804	2	0	2706	2	0	854	3	2
45-92	2988			3310	1	0	1859	1	0	397	0	0
UNKNOWN	21			17	0	0	21	0	0	6	0	0
TOTAL	5616			9959	5	0	11513	13	1	6248	14	1

RISK FACTOR	75-84 AT EXAM			85-94 AT EXAM			35-64 COMBINED			65-94 COMBINED		
	PERSON EXAMS	# OF EVENTS	ANNUAL RATE	PERSON EXAMS	# OF EVENTS	ANNUAL RATE	# OF EVENTS	CRUDE	AGE-ADJUSTED	# OF EVENTS	CRUDE	AGE-ADJUSTED
4-29	930	5	3	80			4	0	1	14	3	2
30-34	569	8	7	29	1		3	0	0	10	2	2
35-39	314	2	3	13			5	0	0	4	1	1
40-44	111	0	0	1			4	0	0	3	2	1
45-92	25	0	0	0			2	0	0	0	0	0
UNKNOWN	2	0	0	0			0	0	0	0	0	0
TOTAL	1951	15	4	123			18	0	0	31	2	2

LOGISTIC REGRESSIONS

AGE AT EXAM		COEFFICIENT OF RISK FACTOR	SE COEFFICIENT	P-VALUE	STANDARDIZED COEFFICIENT	STD SE
35-44						
45-54		-0.012	0.060	0.841	-0.089	0.444
55-64		-0.069	0.034	0.044	-0.521	0.258
65-74		-0.047	0.035	0.178	-0.349	0.259
75-84		-0.015	0.036	0.685	-0.104	0.256
85-94						
35-64	UNIVARIATE	-0.075	0.028	0.006	-0.600	0.220
65-94	UNIVARIATE	-0.052	0.023	0.023	-0.394	0.173
35-64	BIVARIATE	-0.044	0.031	0.148	-0.353	0.244
65-94	BIVARIATE	-0.032	0.025	0.193	-0.242	0.186
35-64	MULTIVARIATE	-0.043	0.032	0.171	-0.346	0.252
65-94	MULTIVARIATE	-0.033	0.026	0.195	-0.252	0.194

TABLE 9-12   DATE=05JAN86
AVERAGE ANNUAL INCIDENCE RATE FOR EVENT PER 1000 PERSONS AT RISK AT EXAM
BY AGE AND LEVEL OF RISK FACTOR AT EXAM: FRAMINGHAM STUDY 30-YEAR FOLLOWUP

SEX: MALE
EVENT: SUDDEN DEATH FROM CORONARY HEART DISEASE
RISK FACTOR: HEART RATE (PER MINUTE)
PERSONS AT RISK: PERSONS FREE OF CORONARY HEART DISEASE

RISK FACTOR	35-44 AT EXAM PERSON EXAMS	# OF EVENTS	ANNUAL RATE	45-54 AT EXAM PERSON EXAMS	# OF EVENTS	ANNUAL RATE	55-64 AT EXAM PERSON EXAMS	# OF EVENTS	ANNUAL RATE	65-74 AT EXAM PERSON EXAMS	# OF EVENTS	ANNUAL RATE
30-67	1284			1949	1	0	2309	6	1	1183	3	1
68-75	1492			2490	3	1	2490	3	2	1159	4	3
76-83	722			1394	1	1	1374	3	1	624	4	3
84-91	379			1097	3	1	1064	4	2	487	7	7
92-220	595			794	3	2	812	11	7	460	6	7
UNKNOWN	0	0	-	0	0	-	0	0	-	0	0	-
TOTAL	4472			7724	16	1	8049	32	2	3913	26	3

RISK FACTOR	75-84 AT EXAM PERSON EXAMS	# OF EVENTS	ANNUAL RATE	85-94 AT EXAM PERSON EXAMS	# OF EVENTS	ANNUAL RATE	35-64 COMBINED # OF EVENTS	CRUDE	AGE-ADJUSTED	65-94 COMBINED # OF EVENTS	CRUDE	AGE-ADJUSTED
30-67	301	2	3	11			7	1	1	5	2	3
68-75	296	3	8	12			13	1	1	9	3	4
76-83	192	3	8	10			8	1	1	7	4	4
84-91	123	2	8	5			9	2	2	9	7	7
92-220	105	1	3	1			14	3	3	7	6	6
UNKNOWN	0	0	-	0			0	-	-	0	-	-
TOTAL	1017	11	5	39			51	1	1	37	4	4

LOGISTIC REGRESSIONS

AGE AT EXAM		COEFFICIENT OF RISK FACTOR	SE COEFFICIENT	P-VALUE	STANDARDIZED COEFFICIENT	STD SE
35-44						
45-54		0.036	0.018	0.041	0.451	0.220
55-64		0.037	0.011	0.001	0.480	0.147
65-74		0.034	0.012	0.005	0.461	0.163
75-84		0.021	0.020	0.309	0.276	0.271
85-94						
35-64	UNIVARIATE	0.038	0.009	0.001	0.481	0.118
65-94	UNIVARIATE	0.030	0.010	0.003	0.409	0.140
35-64	BIVARIATE	0.037	0.009	0.001	0.470	0.116
65-94	BIVARIATE	0.030	0.010	0.003	0.410	0.140
35-64	MULTIVARIATE	0.027	0.010	0.005	0.347	0.124
65-94	MULTIVARIATE	0.029	0.011	0.005	0.396	0.142

SEX: FEMALE
EVENT: SUDDEN DEATH FROM CORONARY HEART DISEASE
RISK FACTOR: HEART RATE (PER MINUTE)
PERSONS AT RISK: PERSONS FREE OF CORONARY HEART DISEASE

RISK FACTOR	35-44 AT EXAM PERSON EXAMS	# OF EVENTS	ANNUAL RATE	45-54 AT EXAM PERSON EXAMS	# OF EVENTS	ANNUAL RATE	55-64 AT EXAM PERSON EXAMS	# OF EVENTS	ANNUAL RATE	65-74 AT EXAM PERSON EXAMS	# OF EVENTS	ANNUAL RATE
30-67	1092			1897	0	0	2505	1	0	1103	1	0
68-75	1803			3244	2	0	3468	4	1	1668	3	1
76-83	1051			1911	0	0	2551	2	1	1358	1	1
84-91	939			1604	1	0	1780	5	1	1089	1	0
92-220	731			1303	2	1	1409	1	0	1030	6	3
UNKNOWN	0	0	-	0	0	-	0	0	-	0	0	-
TOTAL	5616			9959	5	0	11513	13	1	6248	14	1

RISK FACTOR	75-84 AT EXAM PERSON EXAMS	# OF EVENTS	ANNUAL RATE	85-94 AT EXAM PERSON EXAMS	# OF EVENTS	ANNUAL RATE	35-64 COMBINED # OF EVENTS	CRUDE	AGE-ADJUSTED	65-94 COMBINED # OF EVENTS	CRUDE	AGE-ADJUSTED
30-67	360	3	4	26			1	0	0	5	2	2
68-75	462	4	4	32			6	0	0	7	2	2
76-83	428	0	0	24			2	0	0	3	1	1
84-91	330	2	3	23			6	1	1	4	1	1
92-220	371	6	8	18			3	0	0	12	4	4
UNKNOWN	0	0	-	0			0	-	-	0	-	-
TOTAL	1951	15	4	123			18	0	0	31	2	2

LOGISTIC REGRESSIONS

AGE AT EXAM		COEFFICIENT OF RISK FACTOR	SE COEFFICIENT	P-VALUE	STANDARDIZED COEFFICIENT	STD SE
35-44						
45-54		0.031	0.031	0.329	0.578	0.587
55-64		0.013	0.020	0.505	0.170	0.255
65-74		0.033	0.015	0.028	0.438	0.199
75-84		0.009	0.017	0.615	0.123	0.244
85-94						
35-64	UNIVARIATE	0.018	0.017	0.273	0.229	0.209
65-94	UNIVARIATE	0.020	0.012	0.094	0.261	0.156
35-64	BIVARIATE	0.017	0.016	0.294	0.217	0.207
65-94	BIVARIATE	0.019	0.012	0.109	0.250	0.156
35-64	MULTIVARIATE	0.014	0.017	0.422	0.175	0.218
65-94	MULTIVARIATE	0.013	0.012	0.285	0.176	0.164

TABLE 9-13   DATE=08JAN86
AVERAGE ANNUAL INCIDENCE RATE FOR EVENT PER 1000 PERSONS AT RISK AT EXAM
BY AGE AND LEVEL OF RISK FACTOR AT EXAM: FRAMINGHAM STUDY 30-YEAR FOLLOWUP

SEX: MALE
EVENT: SUDDEN DEATH FROM CORONARY HEART DISEASE
RISK FACTOR: CIGARETTES SMOKED (NUMBER PER DAY)
PERSONS AT RISK: PERSONS FREE OF CORONARY HEART DISEASE

RISK FACTOR	35-44 AT EXAM			45-54 AT EXAM			55-64 AT EXAM			65-74 AT EXAM		
	PERSON EXAMS	# OF EVENTS	ANNUAL RATE	PERSON EXAMS	# OF EVENTS	ANNUAL RATE	PERSON EXAMS	# OF EVENTS	ANNUAL RATE	PERSON EXAMS	# OF EVENTS	ANNUAL RATE
NONE	1383			3068	2	0	4295	11	1	2611	15	3
1-10	433			716	3	2	769	4	3	374	3	4
11-20	1485			2051	5	1	1682	9	3	526	2	2
21-40	987			1602	4	1	1119	6	3	369	5	7
41-90	158			252	2	4	162	2	6	27	1	19
UNKNOWN	26			35	0	0	22	0	0	6	0	0
TOTAL	4472			7724	16	1	8049	32	2	3913	26	3

RISK FACTOR	75-84 AT EXAM			85-94 AT EXAM			35-64 COMBINED			65-94 COMBINED		
	PERSON EXAMS	# OF EVENTS	ANNUAL RATE	PERSON EXAMS	# OF EVENTS	ANNUAL RATE	# OF EVENTS	CRUDE	AGE-ADJUSTED	# OF EVENTS	CRUDE	AGE-ADJUSTED
NONE	797	10	6	27			13	1	1	25	4	4
1-10	92	1	5	6			8	2	2	4	4	4
11-20	77	0	0	5			14	1	2	2	2	1
21-40	46	0	0	1			12	2	2	5	6	5
41-90	5	0	0	0			4	3	4	1	16	15
UNKNOWN	0	0	-	0			0	0	0	0	0	0
TOTAL	1017	11	5	39			51	1	1	37	4	4

LOGISTIC REGRESSIONS

AGE AT EXAM		COEFFICIENT OF RISK FACTOR	SE COEFFICIENT	P-VALUE	STANDARDIZED COEFFICIENT	STD SE
35-44						
45-54		0.027	0.014	0.060	0.409	0.217
55-64		0.021	0.011	0.049	0.294	0.149
65-74		0.018	0.014	0.203	0.212	0.166
75-84		*	*	*	*	*
85-94						
35-64	UNIVARIATE	0.019	0.009	0.031	0.271	0.126
65-94	UNIVARIATE	0.005	0.014	0.697	0.061	0.156
35-64	BIVARIATE	0.025	0.008	0.003	0.365	0.122
65-94	BIVARIATE	0.010	0.014	0.486	0.109	0.156
35-64	MULTIVARIATE	0.025	0.008	0.003	0.364	0.123
65-94	MULTIVARIATE	0.014	0.014	0.319	0.157	0.158

SEX: FEMALE
EVENT: SUDDEN DEATH FROM CORONARY HEART DISEASE
RISK FACTOR: CIGARETTES SMOKED (NUMBER PER DAY)
PERSONS AT RISK: PERSONS FREE OF CORONARY HEART DISEASE

RISK FACTOR	35-44 AT EXAM			45-54 AT EXAM			55-64 AT EXAM			65-74 AT EXAM		
	PERSON EXAMS	# OF EVENTS	ANNUAL RATE	PERSON EXAMS	# OF EVENTS	ANNUAL RATE	PERSON EXAMS	# OF EVENTS	ANNUAL RATE	PERSON EXAMS	# OF EVENTS	ANNUAL RATE
NONE	2582			5528	1	0	7595	8	1	4863	10	1
1-10	1314			1726	1	0	1566	5	1	626	1	1
11-20	1371			2009	1	0	1696	2	1	568	2	2
21-40	298			622	1	1	602	0	0	172	0	0
41-90	22			37	1	14	16	0	0	0	0	-
UNKNOWN	29			37	0	0	38	0	0	19	1	26
TOTAL	5616			9959	5	0	11513	13	1	6248	14	1

RISK FACTOR	75-84 AT EXAM			85-94 AT EXAM			35-64 COMBINED			65-94 COMBINED		
	PERSON EXAMS	# OF EVENTS	ANNUAL RATE	PERSON EXAMS	# OF EVENTS	ANNUAL RATE	# OF EVENTS	CRUDE	AGE-ADJUSTED	# OF EVENTS	CRUDE	AGE-ADJUSTED
NONE	1710	13	4	117			9	0	0	25	2	2
1-10	137	1	4	4			4	0	1	2	1	3
11-20	84	1	6	2			3	0	0	3	2	1
21-40	16	0	0	0			1	0	0	0	0	0
41-90	0	0	-	0			1	7	5	0	0	0
UNKNOWN	4	0	0	0			0	0	0	1	22	20
TOTAL	1951	15	4	123			18	0	0	31	2	-2

LOGISTIC REGRESSIONS

AGE AT EXAM		COEFFICIENT OF RISK FACTOR	SE COEFFICIENT	P-VALUE	STANDARDIZED COEFFICIENT	STD SE
35-44						
45-54		0.077	0.028	0.007	0.789	0.292
55-64		-0.009	0.032	0.777	-0.085	0.300
65-74		0.010	0.033	0.764	0.076	0.252
75-84		-0.004	0.053	0.945	-0.019	0.269
85-94						
35-64	UNIVARIATE	0.022	0.020	0.273	0.220	0.201
65-94	UNIVARIATE	-0.009	0.028	0.756	-0.062	0.200
35-64	BIVARIATE	0.031	0.019	0.114	0.306	0.193
65-94	BIVARIATE	0.006	0.028	0.825	0.044	0.199
35-64	MULTIVARIATE	0.031	0.019	0.107	0.311	0.193
65-94	MULTIVARIATE	0.011	0.028	0.698	0.077	0.199

SEX: MALE
EVENT: SUDDEN DEATH FROM CORONARY HEART DISEASE
RISK FACTOR: ALBUMIN IN URINE, DEFINITE
PERSONS AT RISK: PERSONS FREE OF CORONARY HEART DISEASE

35-44 AT EXAM RSON AMS	# OF EVENTS	ANNUAL RATE	45-54 AT EXAM PERSON EXAMS	# OF EVENTS	ANNUAL RATE	55-64 AT EXAM PERSON EXAMS	# OF EVENTS	ANNUAL RATE	65-74 AT EXAM PERSON EXAMS	# OF EVENTS	ANNUAL RATE
4366			7599	15	1	7838	31	2	3792	22	3
70			99	1	5	184	1	3	115	4	17
36			26	0	0	27	0	0	6	0	0
4472			7724	16	1	8049	32	2	3913	26	3

75-84 AT EXAM RSON AMS	# OF EVENTS	ANNUAL RATE	85-94 AT EXAM PERSON EXAMS	# OF EVENTS	ANNUAL RATE	35-64 COMBINED # OF EVENTS	CRUDE	AGE-ADJUSTED	65-94 COMBINED # OF EVENTS	CRUDE	AGE-ADJUSTED
989	11	6	32			49	1	1	33	3	3
25	0	0	7			2	3	3	4	14	14
3	0	0	0			0	0	0	0	0	0
1017	11	5	39			51	1	1	37	4	4

LOGISTIC REGRESSIONS

	COEFFICIENT OF RISK FACTOR	SE COEFFICIENT	P-VALUE	STANDARDIZED COEFFICIENT	STD SE
	1.641	1.038	0.114	0.185	0.117
	0.319	1.019	0.754	0.048	0.153
	1.821	0.552	0.001	0.308	0.093
	-	-	-	-	-
[ATE	0.832	0.723	0.250	0.109	0.095
[ATE	1.399	0.536	0.009	0.237	0.091
\TE	0.674	0.725	0.352	0.088	0.095
\TE	1.353	0.539	0.012	0.229	0.091
\RIATE	0.186	0.743	0.803	0.024	0.097
\RIATE	0.983	0.559	0.079	0.167	0.095

SEX: FEMALE
EVENT: SUDDEN DEATH FROM CORONARY HEART DISEASE
RISK FACTOR: ALBUMIN IN URINE, DEFINITE
PERSONS AT RISK: PERSONS FREE OF CORONARY HEART DISEASE

35-44 AT EXAM RSON AMS	# OF EVENTS	ANNUAL RATE	45-54 AT EXAM PERSON EXAMS	# OF EVENTS	ANNUAL RATE	55-64 AT EXAM PERSON EXAMS	# OF EVENTS	ANNUAL RATE	65-74 AT EXAM PERSON EXAMS	# OF EVENTS	ANNUAL RATE
3353			9656	5	0	11263	12	1	6118	13	1
225			278	0	0	225	1	2	120	1	4
38			25	0	0	25	0	0	10	0	0
616			9959	5	0	11513	13	1	6248	14	1

75-84 AT EXAM SON MS	# OF EVENTS	ANNUAL RATE	85-94 AT EXAM PERSON EXAMS	# OF EVENTS	ANNUAL RATE	35-64 COMBINED # OF EVENTS	CRUDE	AGE-ADJUSTED	65-94 COMBINED # OF EVENTS	CRUDE	AGE-ADJUSTED
904	15	4	120			17	0	0	30	2	2
38	0	0	3			1	1	1	1	3	3
9	0	0	0			0	0	0	0	0	0
951	15	4	123			18	0	0	31	2	2

LOGISTIC REGRESSIONS

	COEFFICIENT OF RISK FACTOR	SE COEFFICIENT	P-VALUE	STANDARDIZED COEFFICIENT	STD SE
	1.432	1.043	0.170	0.198	0.145
	1.373	1.042	0.188	0.189	0.143
	-	-	-	-	-
\TE	0.753	1.030	0.464	0.122	0.167
\TE	0.525	1.020	0.607	0.072	0.141
'E	1.017	1.032	0.324	0.165	0.167
'E	0.489	1.022	0.632	0.067	0.141
:IATE	1.054	1.057	0.319	0.171	0.171
:IATE	0.170	1.042	0.870	0.023	0.144
	-				-

SEX: MALE
EVENT: SUDDEN DEATH FROM CORONARY HEART DISEASE
RISK FACTOR: HEART ENLARGEMENT BY X-RAY
PERSONS AT RISK: PERSONS FREE OF CORONARY HEART DISEASE

RISK FACTOR	35-44 AT EXAM			45-54 AT EXAM			55-64 AT EXAM			65-74 AT EXAM		
	PERSON EXAMS	# OF EVENTS	ANNUAL RATE	PERSON EXAMS	# OF EVENTS	ANNUAL RATE	PERSON EXAMS	# OF EVENTS	ANNUAL RATE	PERSON EXAMS	# OF EVENTS	ANNUAL RATE
NO	4063			6383	12	1	5967	20	2	2584	12	2
MILD	185			650	2	2	962	3	2	578	4	3
DEFINITE	219			687	2	1	1119	9	4	751	10	7
UNKNOWN	5			4	0	0	1	0	0	0	0	-
TOTAL	4472			7724	16	1	8049	32	2	3913	26	3

RISK FACTOR	75-84 AT EXAM			85-94 AT EXAM			35-64 COMBINED			65-94 COMBINED		
	PERSON EXAMS	# OF EVENTS	ANNUAL RATE	PERSON EXAMS	# OF EVENTS	ANNUAL RATE	# OF EVENTS	CRUDE	AGE-ADJUSTED	# OF EVENTS	CRUDE	AGE-ADJUSTED
NO	589	4	3	29			35	1	1	16	2	3
MILD	178	3	8	5			5	1	1	7	5	4
DEFINITE	250	4	8	5			11	3	2	14	7	7
UNKNOWN	0	0	-	0			0	0	0	0		
TOTAL	1017	11	5	39			51	1	1	37	2	4

LOGISTIC REGRESSIONS

AGE AT EXAM		COEFFICIENT OF RISK FACTOR	SE COEFFICIENT	P-VALUE	STANDARDIZED COEFFICIENT	STD SE
35-44						
45-54		0.258	0.351	0.462	0.157	0.214
55-64		0.413	0.207	0.046	0.297	0.149
65-74		0.529	0.217	0.015	0.421	0.173
75-84		0.434	0.337	0.198	0.367	0.285
85-94						
35-64	UNIVARIATE	0.454	0.172	0.008	0.289	0.110
65-94	UNIVARIATE	0.518	0.182	0.004	0.418	0.147
35-64	BIVARIATE	0.329	0.176	0.061	0.210	0.112
65-94	BIVARIATE	0.490	0.183	0.007	0.395	0.148
35-64	MULTIVARIATE	0.267	0.188	0.157	0.170	0.120
65-94	MULTIVARIATE	0.327	0.197	0.097	0.264	0.159

SEX: FEMALE
EVENT: SUDDEN DEATH FROM CORONARY HEART DISEASE
RISK FACTOR: HEART ENLARGEMENT BY X-RAY
PERSONS AT RISK: PERSONS FREE OF CORONARY HEART DISEASE

RISK FACTOR	35-44 AT EXAM			45-54 AT EXAM			55-64 AT EXAM			65-74 AT EXAM		
	PERSON EXAMS	# OF EVENTS	ANNUAL RATE	PERSON EXAMS	# OF EVENTS	ANNUAL RATE	PERSON EXAMS	# OF EVENTS	ANNUAL RATE	PERSON EXAMS	# OF EVENTS	ANNUAL RATE
NO	5164			8017	3	0	7522	5	0	3191	4	1
MILD	199			943	1	1	1769	3	1	1150	4	2
DEFINITE	245			993	1	1	2218	5	1	1907	6	2
UNKNOWN	8			6	0	0	4	0	0	0	0	-
TOTAL	5616			9959	5	0	11513	13	1	6248	14	1

RISK FACTOR	75-84 AT EXAM			85-94 AT EXAM			35-64 COMBINED			65-94 COMBINED		
	PERSON EXAMS	# OF EVENTS	ANNUAL RATE	PERSON EXAMS	# OF EVENTS	ANNUAL RATE	# OF EVENTS	CRUDE	AGE-ADJUSTED	# OF EVENTS	CRUDE	AGE-ADJUSTED
NO	702	0	0	32			8	0	0	5	1	1
MILD	373	2	3	30			4	1	1	6	2	2
DEFINITE	876	13	7	61			6	1	1	20	4	3
UNKNOWN	0	0	-	0			0	0	0	0	-	-
TOTAL	1951	15	4	123			18	0	0	31	2	2

LOGISTIC REGRESSIONS

AGE AT EXAM		COEFFICIENT OF RISK FACTOR	SE COEFFICIENT	P-VALUE	STANDARDIZED COEFFICIENT	STD SE
35-44						
45-54		0.548	0.531	0.302	0.350	0.339
55-64		0.611	0.306	0.046	0.487	0.244
65-74		0.436	0.300	0.147	0.383	0.264
75-84		1.620	0.622	0.009	1.450	0.557
85-94						
35-64	UNIVARIATE	0.765	0.257	0.003	0.534	0.179
65-94	UNIVARIATE	0.820	0.231	0.001	0.732	0.206
35-64	BIVARIATE	0.527	0.266	0.048	0.367	0.185
65-94	BIVARIATE	0.722	0.235	0.002	0.644	0.209
35-64	MULTIVARIATE	0.612	0.276	0.027	0.427	0.193
65-94	MULTIVARIATE	0.640	0.243	0.008	0.572	0.217

TABLE 9-16   DATE=05JAN86
AVERAGE ANNUAL INCIDENCE RATE FOR EVENT PER 1000 PERSONS AT RISK AT EXAM
BY AGE AND LEVEL OF RISK FACTOR AT EXAM: FRAMINGHAM STUDY 30-YEAR FOLLOWUP

SEX: MALE
EVENT: SUDDEN DEATH FROM CORONARY HEART DISEASE
RISK FACTOR: LEFT VENTRICULAR HYPERTROPHY BY ELECTROCARDIOGRAM
PERSONS AT RISK: PERSONS FREE OF CORONARY HEART DISEASE

RISK FACTOR	35-44 AT EXAM PERSON EXAMS	# OF EVENTS	ANNUAL RATE	45-54 AT EXAM PERSON EXAMS	# OF EVENTS	ANNUAL RATE	55-64 AT EXAM PERSON EXAMS	# OF EVENTS	ANNUAL RATE	65-74 AT EXAM PERSON EXAMS	# OF EVENTS	ANNUAL RATE
NO	4366			7497	14	1	7693	28	2	3703	19	3
MILD	78			132	1	4	210	0	0	90	3	17
DEFINITE	28			95	1	5	146	4	14	120	4	17
UNKNOWN	0			0	0	-	0	0	-	0	0	-
TOTAL	4472			7724	16	1	8049	32	2	3913	26	3

RISK FACTOR	75-84 AT EXAM PERSON EXAMS	# OF EVENTS	ANNUAL RATE	85-94 AT EXAM PERSON EXAMS	# OF EVENTS	ANNUAL RATE	35-64 COMBINED # OF EVENTS	CRUDE	AGE-ADJUSTED	65-94 COMBINED # OF EVENTS	CRUDE	AGE-ADJUSTED
NO	949	10	5	34	1	-	45	1	1	29	3	3
MILD	37	0	0	2	-	-	1	1	1	3	12	13
DEFINITE	31	1	16	3	-	-	5	9	7	5	16	16
UNKNOWN	0	0	-	0	-	-	0	-	-	0	-	-
TOTAL	1017	11	5	39	-	-	51	1	1	37	4	4

LOGISTIC REGRESSIONS

AGE AT EXAM		COEFFICIENT OF RISK FACTOR	SE COEFFICIENT	P-VALUE	STANDARDIZED COEFFICIENT	STD SE
35-44						
45-54		0.946	0.459	0.039	0.240	0.117
55-64		0.906	0.289	0.002	0.279	0.089
65-74		1.027	0.254	0.001	0.382	0.095
75-84		0.413	0.577	0.474	0.159	0.223
85-94						
35-64	UNIVARIATE	0.978	0.243	0.001	0.262	0.065
65-94	UNIVARIATE	0.885	0.230	0.001	0.334	0.087
35-64	BIVARIATE	0.873	0.245	0.001	0.234	0.066
65-94	BIVARIATE	0.872	0.231	0.001	0.329	0.087
35-64	MULTIVARIATE	0.794	0.272	0.004	0.212	0.073
65-94	MULTIVARIATE	0.654	0.259	0.011	0.247	0.098

SEX: FEMALE
EVENT: SUDDEN DEATH FROM CORONARY HEART DISEASE
RISK FACTOR: LEFT VENTRICULAR HYPERTROPHY BY ELECTROCARDIOGRAM
PERSONS AT RISK: PERSONS FREE OF CORONARY HEART DISEASE

RISK FACTOR	35-44 AT EXAM PERSON EXAMS	# OF EVENTS	ANNUAL RATE	45-54 AT EXAM PERSON EXAMS	# OF EVENTS	ANNUAL RATE	55-64 AT EXAM PERSON EXAMS	# OF EVENTS	ANNUAL RATE	65-74 AT EXAM PERSON EXAMS	# OF EVENTS	ANNUAL RATE
NO	5563			9798	5	0	11137	13	1	5966	13	1
MILD	78			107	0	0	226	0	0	134	0	0
DEFINITE	15			54	0	0	150	0	0	148	1	3
UNKNOWN	0			0	0	0	0	0	-	0	0	-
TOTAL	5616			9959	5	0	11513	13	1	6248	14	1

RISK FACTOR	75-84 AT EXAM PERSON EXAMS	# OF EVENTS	ANNUAL RATE	85-94 AT EXAM PERSON EXAMS	# OF EVENTS	ANNUAL RATE	35-64 COMBINED # OF EVENTS	CRUDE	AGE-ADJUSTED	65-94 COMBINED # OF EVENTS	CRUDE	AGE-ADJUSTED
NO	1833	11	3	111			18	0	0	26	2	2
MILD	49	2	20	6			0	0	0	2	5	5
DEFINITE	69	2	14	6			0	0	0	3	7	6
UNKNOWN	0	0	-	0			0	-	-	0	-	-
TOTAL	1951	15	4	123			18	0	0	31	2	2

LOGISTIC REGRESSIONS

AGE AT EXAM		COEFFICIENT OF RISK FACTOR	SE COEFFICIENT	P-VALUE	STANDARDIZED COEFFICIENT	STD SE
35-44		-	-	-	-	-
45-54		-	-	-	-	-
55-64		-	-	-	-	-
65-74		0.449	0.555	0.418	0.150	0.185
75-84		0.911	0.342	0.008	0.361	0.136
85-94						
35-64	UNIVARIATE	-	-	-	-	-
65-94	UNIVARIATE	0.757	0.284	0.008	0.267	0.100
35-64	BIVARIATE	-	-	-	-	-
65-94	BIVARIATE	0.682	0.286	0.017	0.240	0.101
35-64	MULTIVARIATE	-	-	-	-	-
65-94	MULTIVARIATE	0.614	0.303	0.043	0.216	0.107

TABLE 9-17   DATE=05JAN86
AVERAGE ANNUAL INCIDENCE RATE FOR EVENT PER 1000 PERSONS AT RISK AT EXAM
BY AGE AND LEVEL OF RISK FACTOR AT EXAM: FRAMINGHAM STUDY 30-YEAR FOLLOWUP

SEX: MALE
EVENT: SUDDEN DEATH FROM CORONARY HEART DISEASE
RISK FACTOR: INTRAVENTRICULAR CONDUCTION DEFECT (WITH QRS INTERVAL MORE THAN 0.11 SECOND)
PERSONS AT RISK: PERSONS FREE OF CORONARY HEART DISEASE

RISK FACTOR	35-44 AT EXAM PERSON EXAMS	# OF EVENTS	ANNUAL RATE	45-54 AT EXAM PERSON EXAMS	# OF EVENTS	ANNUAL RATE	55-64 AT EXAM PERSON EXAMS	# OF EVENTS	ANNUAL RATE	65-74 AT EXAM PERSON EXAMS	# OF EVENTS	ANNUAL RATE
NO	4161			7213	15	1	7386	27	2	3566	23	3
INCOMPLETE	275			446	1	1	456	2	2	183	1	3
COMPLETE	36			65	0	0	207	3	7	164	2	6
UNKNOWN	0			0	0	-	0	0	-	0	0	-
TOTAL	4472			7724	16	1	8049	32	2	3913	26	3

RISK FACTOR	75-84 AT EXAM PERSON EXAMS	# OF EVENTS	ANNUAL RATE	85-94 AT EXAM PERSON EXAMS	# OF EVENTS	ANNUAL RATE	35-64 COMBINED # OF EVENTS	CRUDE	AGE- ADJUSTED	65-94 COMBINED # OF EVENTS	CRUDE	AGE- ADJUSTED
NO	918	11	6	31			45	1	1	34	4	4
INCOMPLETE	21	0	0	2			3	1	1	1	2	2
COMPLETE	78	0	0	6			3	5	3	2	4	5
UNKNOWN	0	0	-	0			0	-	-	0	-	-
TOTAL	1017	11	5	39			51	1	1	37	4	4

LOGISTIC REGRESSIONS

AGE AT EXAM		COEFFICIENT OF RISK FACTOR	SE COEFFICIENT	P-VALUE	STANDARDIZED COEFFICIENT	STD SE
35-44						
45-54		-0.155	0.938	0.869	-0.045	0.275
55-64		0.611	0.303	0.044	0.235	0.117
65-74		0.258	0.366	0.481	0.115	0.163
75-84		-				
85-94						
35-64	UNIVARIATE	0.541	0.294	0.066	0.180	0.098
65-94	UNIVARIATE	-0.029	0.359	0.936	-0.013	0.169
35-64	BIVARIATE	0.431	0.288	0.135	0.144	0.096
65-94	BIVARIATE	-0.074	0.355	0.835	-0.035	0.167
35-64	MULTIVARIATE	0.499	0.297	0.092	0.166	0.099
65-94	MULTIVARIATE	-0.057	0.362	0.874	-0.027	0.170

SEX: FEMALE
EVENT: SUDDEN DEATH FROM CORONARY HEART DISEASE
RISK FACTOR: INTRAVENTRICULAR CONDUCTION DEFECT (WITH QRS INTERVAL MORE THAN 0.11 SECOND)
PERSONS AT RISK: PERSONS FREE OF CORONARY HEART DISEASE

RISK FACTOR	35-44 AT EXAM PERSON EXAMS	# OF EVENTS	ANNUAL RATE	45-54 AT EXAM PERSON EXAMS	# OF EVENTS	ANNUAL RATE	55-64 AT EXAM PERSON EXAMS	# OF EVENTS	ANNUAL RATE	65-74 AT EXAM PERSON EXAMS	# OF EVENTS	ANNUAL RATE
NO	5529			9771	5	0	11183	12	1	6027	13	1
INCOMPLETE	75			144	0	0	187	0	0	115	0	0
COMPLETE	12			44	0	0	143	1	3	106	1	5
UNKNOWN	0			0	0	-	0	0	-	0	0	-
TOTAL	5616			9959	5	0	11513	13	1	6248	14	1

RISK FACTOR	75-84 AT EXAM PERSON EXAMS	# OF EVENTS	ANNUAL RATE	85-94 AT EXAM PERSON EXAMS	# OF EVENTS	ANNUAL RATE	35-64 COMBINED # OF EVENTS	CRUDE	AGE- ADJUSTED	65-94 COMBINED # OF EVENTS	CRUDE	AGE- ADJUSTED
NO	1847	14	4	117			17	0	0	29	2	2
INCOMPLETE	36	0	0	1			0	0	0	0	0	0
COMPLETE	68	1	7	5			1	3	1	2	6	5
UNKNOWN	0	0	-	0			0	-	-	0	-	-
TOTAL	1951	15	4	123			18	0	0	31	2	2

LOGISTIC REGRESSIONS

AGE AT EXAM		COEFFICIENT OF RISK FACTOR	SE COEFFICIENT	P-VALUE	STANDARDIZED COEFFICIENT	STD SE
35-44						
45-54		-				
55-64		0.822	0.556	0.140	0.208	0.141
65-74		0.621	0.556	0.264	0.179	0.161
75-84		0.246	0.546	0.653	0.095	0.212
85-94						
35-64	UNIVARIATE	0.859	0.562	0.126	0.179	0.117
65-94	UNIVARIATE	0.452	0.390	0.246	0.143	0.124
35-64	BIVARIATE	0.649	0.558	0.245	0.135	0.116
65-94	BIVARIATE	0.356	0.389	0.359	0.113	0.123
35-64	MULTIVARIATE	0.688	0.560	0.219	0.143	0.117
65-94	MULTIVARIATE	0.376	0.391	0.336	0.119	0.124

TABLE 9-18 DATE=05JAN86
AVERAGE ANNUAL INCIDENCE RATE FOR EVENT PER 1000 PERSONS AT RISK AT EXAM
BY AGE AND LEVEL OF RISK FACTOR AT EXAM: FRAMINGHAM STUDY 30-YEAR FOLLOWUP

SEX: MALE
EVENT: SUDDEN DEATH FROM CORONARY HEART DISEASE
RISK FACTOR: DEFINITE NONSPECIFIC T-WAVE OR ST-SEGMENT ABNORMALITY BY ELECTROCARDIOGRAM
PERSONS AT RISK: PERSONS FREE OF CORONARY HEART DISEASE

35-44 AT EXAM			45-54 AT EXAM			55-64 AT EXAM			65-74 AT EXAM		
PERSON EXAMS	# OF EVENTS	ANNUAL RATE	PERSON EXAMS	# OF EVENTS	ANNUAL RATE	PERSON EXAMS	# OF EVENTS	ANNUAL RATE	PERSON EXAMS	# OF EVENTS	ANNUAL RATE
4362		-	7352	14	1	7462	29	2	3472	16	2
110			372	2	3	587	3	3	441	10	11
0			0	0	-	0	0	-	0	0	-
4472			7724	16	1	8049	32	2	3913	26	3

75-84 AT EXAM			85-94 AT EXAM			35-64 COMBINED			65-94 COMBINED		
PERSON EXAMS	# OF EVENTS	ANNUAL RATE	PERSON EXAMS	# OF EVENTS	ANNUAL RATE	# OF EVENTS	CRUDE	AGE-ADJUSTED	# OF EVENTS	CRUDE	AGE-ADJUSTED
830	10	6	28			46	1	1	26	3	3
187	1	3	11			5	2	2	11	9	9
0	0	-	0			0	-		0	-	
1017	11	5	39			51	1	1	37	4	4

LOGISTIC REGRESSIONS

	COEFFICIENT OF RISK FACTOR	SE COEFFICIENT	P-VALUE	STANDARDIZED COEFFICIENT	STD SE
	1.041	0.758	0.169	0.223	0.162
	0.275	0.608	0.651	0.072	0.158
	1.612	0.406	0.001	0.510	0.129
	-0.819	1.052	0.436	-0.317	0.408
:VARIATE	0.670	0.472	0.156	0.150	0.106
:VARIATE	1.065	0.362	0.003	0.356	0.121
/ARIATE	0.448	0.474	0.345	0.100	0.106
/ARIATE	0.981	0.366	0.007	0.328	0.122
.TIVARIATE	0.083	0.484	0.863	0.019	0.108
.TIVARIATE	0.941	0.369	0.011	0.315	0.124

SEX: FEMALE
EVENT: SUDDEN DEATH FROM CORONARY HEART DISEASE
RISK FACTOR: DEFINITE NONSPECIFIC T-WAVE OR ST-SEGMENT ABNORMALITY BY ELECTROCARDIOGRAM
PERSONS AT RISK: PERSONS FREE OF CORONARY HEART DISEASE

35-44 AT EXAM			45-54 AT EXAM			55-64 AT EXAM			65-74 AT EXAM		
PERSON EXAMS	# OF EVENTS	ANNUAL RATE	PERSON EXAMS	# OF EVENTS	ANNUAL RATE	PERSON EXAMS	# OF EVENTS	ANNUAL RATE	PERSON EXAMS	# OF EVENTS	ANNUAL RATE
5517			9543	4	0	10647	10	0	5528	12	1
99			416	1	1	866	3	2	720	2	1
0			0	0	-	0	0	-	0	0	-
5616			9959	5	0	11513	13	1	6248	14	1

75-84 AT EXAM			85-94 AT EXAM			35-64 COMBINED			65-94 COMBINED		
PERSON EXAMS	# OF EVENTS	ANNUAL RATE	PERSON EXAMS	# OF EVENTS	ANNUAL RATE	# OF EVENTS	CRUDE	AGE-ADJUSTED	# OF EVENTS	CRUDE	AGE-ADJUSTED
1632	13	4	97			14	0	0	27	2	2
319	2	3	26			4	1	1	4	2	2
0	0	-	0			0	-	0	0	-	2
1951	15	4	123			18	0	0	31	2	2

LOGISTIC REGRESSIONS

	COEFFICIENT OF RISK FACTOR	SE COEFFICIENT	P-VALUE	STANDARDIZED COEFFICIENT	STD SE
	1.749	1.119	0.118	0.350	0.224
	1.308	0.659	0.047	0.345	0.174
	0.247	0.765	0.747	0.079	0.244
	-0.241	0.762	0.752	-0.089	0.282
'ARIATE	1.674	0.568	0.003	0.368	0.125
'ARIATE	0.009	0.537	0.986	0.003	0.179
RIATE	1.305	0.572	0.023	0.287	0.126
RIATE	-0.133	0.539	0.805	-0.044	0.180
IVARIATE	1.328	0.576	0.021	0.292	0.127
IVARIATE	-0.136	0.541	0.802	-0.045	0.181

TABLE 10-1   DATE=05JAN86
AVERAGE ANNUAL INCIDENCE RATE FOR EVENT PER 1000 PERSONS AT RISK AT EXAM
BY AGE AND LEVEL OF RISK FACTOR AT EXAM: FRAMINGHAM STUDY 30-YEAR FOLLOWUP

SEX: MALE
EVENT: SUDDEN DEATH FROM CORONARY HEART DISEASE
RISK FACTOR: SYSTOLIC BLOOD PRESSURE, FIRST EXAMINER (MM HG)
PERSONS AT RISK: ALL PERSONS ALIVE

RISK FACTOR	35-44 AT EXAM PERSON EXAMS	# OF EVENTS	ANNUAL RATE	45-54 AT EXAM PERSON EXAMS	# OF EVENTS	ANNUAL RATE	55-64 AT EXAM PERSON EXAMS	# OF EVENTS	ANNUAL RATE	65-74 AT EXAM PERSON EXAMS	# OF EVENTS	ANNUAL RATE
74-119	1226			1842	4	1	1603	6	2	659	2	2
120-139	2195			3542	12	2	3525	22	3	1636	11	3
140-159	858			1838	6	2	2371	16	3	1473	19	6
160-179	197			598	4	3	1062	9	4	696	9	6
180-300	53			247	0	0	507	7	7	370	8	11
UNKNOWN	0			0	0	-	0	0	-	0	0	-
TOTAL	4519			8067	26	2	9070	60	3	4834	49	5

RISK FACTOR	75-84 AT EXAM PERSON EXAMS	# OF EVENTS	ANNUAL RATE	85-94 AT EXAM PERSON EXAMS	# OF EVENTS	ANNUAL RATE	35-64 COMBINED # OF EVENTS	CRUDE	AGE-ADJUSTED	65-94 COMBINED # OF EVENTS	CRUDE	AGE-ADJUSTED
74-119	144	3	10	8			13	1	1	5	3	3
120-139	451	3	3	17			35	2	2	14	3	3
140-159	411	6	7	17			22	2	2	25	7	7
160-179	218	4	9	6			13	4	3	13	7	7
180-300	121	5	21	2			7	4	3	13	13	13
UNKNOWN	0	0	-	0			0	-	-	0	-	-
TOTAL	1345	21	8	50			90	2	2	70	6	6

LOGISTIC REGRESSIONS

AGE AT EXAM		COEFFICIENT OF RISK FACTOR	SE COEFFICIENT	P-VALUE	STANDARDIZED COEFFICIENT	STD SE
35-44						
45-54		0.009	0.009	0.341	0.171	0.180
55-64		0.013	0.005	0.014	0.278	0.114
65-74		0.017	0.005	0.001	0.387	0.122
75-84		0.016	0.008	0.055	0.374	0.195
85-94						
35-64	UNIVARIATE	0.014	0.004	0.001	0.292	0.090
65-94	UNIVARIATE	0.017	0.004	0.001	0.388	0.103
35-64	BIVARIATE	0.009	0.005	0.046	0.184	0.092
65-94	BIVARIATE	0.017	0.004	0.001	0.388	0.104
35-64	MULTIVARIATE	-0.001	0.005	0.858	-0.018	0.100
65-94	MULTIVARIATE	0.012	0.005	0.013	0.277	0.111

SEX: FEMALE
EVENT: SUDDEN DEATH FROM CORONARY HEART DISEASE
RISK FACTOR: SYSTOLIC BLOOD PRESSURE, FIRST EXAMINER (MM HG)
PERSONS AT RISK: ALL PERSONS ALIVE

RISK FACTOR	35-44 AT EXAM PERSON EXAMS	# OF EVENTS	ANNUAL RATE	45-54 AT EXAM PERSON EXAMS	# OF EVENTS	ANNUAL RATE	55-64 AT EXAM PERSON EXAMS	# OF EVENTS	ANNUAL RATE	65-74 AT EXAM PERSON EXAMS	# OF EVENTS	ANNUAL RATE
74-119	2493			2724	1	0	2039	2	0	683	2	1
120-139	2245			4027	2	0	4199	1	0	1918	6	2
140-159	674			2086	1	0	3197	9	1	2283	6	1
160-179	156			785	0	0	1682	3	1	1360	7	3
180-300	51			450	1	1	986	0	0	859	7	4
UNKNOWN	0			0	0	-	0	0	-	0	0	-
TOTAL	5619			10072	5	0	12103	15	1	7103	28	2

RISK FACTOR	75-84 AT EXAM PERSON EXAMS	# OF EVENTS	ANNUAL RATE	85-94 AT EXAM PERSON EXAMS	# OF EVENTS	ANNUAL RATE	35-64 COMBINED # OF EVENTS	CRUDE	AGE-ADJUSTED	65-94 COMBINED # OF EVENTS	CRUDE	AGE-ADJUSTED
74-119	158	1	3	10			3	0	0	3	2	2
120-139	561	4	4	51			3	0	0	11	2	2
140-159	759	5	3	40			10	1	1	11	2	2
160-179	561	9	8	36			3	1	0	17	4	4
180-300	380	4	5	26			1	0	0	11	4	4
UNKNOWN	0	0	-	0			0	-	-	0	-	-
TOTAL	2419	23	5	163			20	0	0	53	3	3

LOGISTIC REGRESSIONS

AGE AT EXAM		COEFFICIENT OF RISK FACTOR	SE COEFFICIENT	P-VALUE	STANDARDIZED COEFFICIENT	STD SE
35-44						
45-54		0.019	0.013	0.144	0.441	0.302
55-64		0.005	0.010	0.602	0.127	0.243
65-74		0.014	0.007	0.039	0.349	0.169
75-84		0.004	0.008	0.643	0.094	0.203
85-94						
35-64	UNIVARIATE	0.015	0.007	0.041	0.357	0.175
65-94	UNIVARIATE	0.010	0.005	0.042	0.258	0.127
35-64	BIVARIATE	0.007	0.008	0.400	0.167	0.198
65-94	BIVARIATE	0.009	0.005	0.094	0.215	0.128
35-64	MULTIVARIATE	0.009	0.009	0.286	0.220	0.206
65-94	MULTIVARIATE	0.005	0.005	0.385	0.116	0.134

AVERAGE ANNUAL INCIDENCE RATE FOR EVENT PER 1000 PERSONS AT RISK AT EXAM
BY AGE AND LEVEL OF RISK FACTOR AT EXAM: FRAMINGHAM STUDY 30-YEAR FOLLOWUP

SEX: MALE
EVENT: SUDDEN DEATH FROM CORONARY HEART DISEASE
RISK FACTOR: DIASTOLIC BLOOD PRESSURE, FIRST EXAMINER (MM HG)
PERSONS AT RISK: ALL PERSONS ALIVE

RISK FACTOR	35-44 AT EXAM			45-54 AT EXAM			55-64 AT EXAM			65-74 AT EXAM		
	PERSON EXAMS	# OF EVENTS	ANNUAL RATE	PERSON EXAMS	# OF EVENTS	ANNUAL RATE	PERSON EXAMS	# OF EVENTS	ANNUAL RATE	PERSON EXAMS	# OF EVENTS	ANNUAL RATE
20-74	1015			1543	5	2	2092	18	4	1524	9	3
75-84	1617			2861	7	1	3165	13	2	1643	12	4
85-94	1203			2151	8	2	2228	16	4	1072	13	6
95-104	509			1073	4	2	1114	9	4	425	12	14
105-160	175			439	2	2	471	4	4	170	3	9
UNKNOWN	0	0	-	0	0	-	0	0	-	0	0	-
TOTAL	4519			8067	26	2	9070	60	3	4834	49	5

RISK FACTOR	75-84 AT EXAM			85-94 AT EXAM			35-64 COMBINED			65-94 COMBINED		
	PERSON EXAMS	# OF EVENTS	ANNUAL RATE	PERSON EXAMS	# OF EVENTS	ANNUAL RATE	# OF EVENTS	CRUDE	AGE-ADJUSTED	# OF EVENTS	CRUDE	AGE-ADJUSTED
20-74	560	7	6	30			24	3	3	16	5	4
75-84	440	7	8	15			23	2	2	19	5	6
85-94	244	3	6	5			24	2	2	16	6	6
95-104	71	2	14	0			13	2	2	14	14	14
105-160	30	2	33	0			6	3	3	5	13	14
UNKNOWN	0	0	-	0			0	-		0	-	
TOTAL	1345	21	8	50			90	2	2	70	6	6

LOGISTIC REGRESSIONS

AGE AT EXAM		COEFFICIENT OF RISK FACTOR	SE COEFFICIENT	P-VALUE	STANDARDIZED COEFFICIENT	STD SE
35-44						
45-54		0.014	0.016	0.385	0.164	0.188
55-64		0.008	0.010	0.425	0.100	0.126
65-74		0.010	0.010	0.001	0.425	0.124
75-84		0.034	0.017	0.160	0.293	0.209
85-94		0.024				
35-64	UNIVARIATE	0.008	0.009	0.351	0.096	0.103
65-94	UNIVARIATE	0.030	0.009	0.001	0.373	0.107
35-64	BIVARIATE	0.009	0.008	0.310	0.102	0.101
65-94	BIVARIATE	0.035	0.009	0.001	0.429	0.108
35-64	MULTIVARIATE	-0.003	0.008	0.717	-0.036	0.098
65-94	MULTIVARIATE	0.028	0.009	0.002	0.346	0.110

SEX: FEMALE
EVENT: SUDDEN DEATH FROM CORONARY HEART DISEASE
RISK FACTOR: DIASTOLIC BLOOD PRESSURE, FIRST EXAMINER (MM HG)
PERSONS AT RISK: ALL PERSONS ALIVE

RISK FACTOR	35-44 AT EXAM			45-54 AT EXAM			55-64 AT EXAM			65-74 AT EXAM		
	PERSON EXAMS	# OF EVENTS	ANNUAL RATE	PERSON EXAMS	# OF EVENTS	ANNUAL RATE	PERSON EXAMS	# OF EVENTS	ANNUAL RATE	PERSON EXAMS	# OF EVENTS	ANNUAL RATE
20-74	2164			2657	1	0	3110	3	0	2069	9	2
75-84	2031			3573	2	0	4005	5	1	2416	11	2
85-94	988			2395	1	0	3000	5	1	1622	2	1
95-104	324			941	0	0	1348	1	0	685	3	2
105-160	112			506	1	0	640	1	1	311	3	5
UNKNOWN	0			0	0	-	0	0	-	0	0	-
TOTAL	5619			10072	5	0	12103	15	1	7103	28	2

RISK FACTOR	75-84 AT EXAM			85-94 AT EXAM			35-64 COMBINED			65-94 COMBINED		
	PERSON EXAMS	# OF EVENTS	ANNUAL RATE	PERSON EXAMS	# OF EVENTS	ANNUAL RATE	# OF EVENTS	CRUDE	AGE-ADJUSTED	# OF EVENTS	CRUDE	AGE-ADJUSTED
20-74	916	10	5	73			4	0	0	21	3	3
75-84	722	3	3	56			7	0	0	16	3	2
85-94	535	5	5	25			6	0	0	7	2	3
95-104	178	2	6	7			1	0	0	5	3	3
105-160	68	1	7	2			2	1	1	4	5	5
UNKNOWN	0	0	-	0			0	-		0	-	
TOTAL	2419	23	5	163			20	-	0	53	3	3

LOGISTIC REGRESSIONS

AGE AT EXAM		COEFFICIENT OF RISK FACTOR	SE COEFFICIENT	P-VALUE	STANDARDIZED COEFFICIENT	STD SE
35-44						
45-54		0.021	0.032	0.516	0.257	0.396
55-64		0.007	0.020	0.729	0.087	0.251
65-74		0.004	0.015	0.772	0.054	0.187
75-84		-0.011	0.017	0.528	-0.136	0.215
85-94						
35-64	UNIVARIATE	0.017	0.017	0.310	0.208	0.205
65-94	UNIVARIATE	-0.008	0.011	0.478	-0.100	0.141
35-64	BIVARIATE	0.011	0.017	0.533	0.131	0.211
65-94	BIVARIATE	-0.002	0.011	0.856	-0.025	0.140
35-64	MULTIVARIATE	0.014	0.018	0.433	0.170	0.217
65-94	MULTIVARIATE	-0.002	0.011	0.882	-0.021	0.138

TABLE 10-3A   DATE:05JAN86
AVERAGE ANNUAL INCIDENCE RATE FOR EVENT PER 1000 PERSONS AT RISK AT EXAM
BY AGE AND LEVEL OF RISK FACTOR AT EXAM: FRAMINGHAM STUDY 30-YEAR FOLLOWUP

SEX: MALE
EVENT: SUDDEN DEATH FROM CORONARY HEART DISEASE
RISK FACTOR: HYPERTENSION WITH ANTIHYPERTENSIVE TREATMENT
PERSONS AT RISK: ALL PERSONS ALIVE

RISK FACTOR	35-44 AT EXAM PERSON EXAMS	# OF EVENTS	ANNUAL RATE	45-54 AT EXAM PERSON EXAMS	# OF EVENTS	ANNUAL RATE	55-64 AT EXAM PERSON EXAMS	# OF EVENTS	ANNUAL RATE	65-74 AT EXAM PERSON EXAMS	# OF EVENTS	ANNUAL RATE
NO	2668			4263	9	1	4163	16	2	1803	7	2
MILD	1316			2287	9	2	2573	17	3	1520	16	5
DEFINITE	511			1216	5	2	1431	11	4	698	10	7
TREATED	24			301	3	5	903	16	9	813	16	10
UNKNOWN	0	0	-	0	0	-	0	0	-	0	0	-
TOTAL	4519			8067	26	2	9070	60	3	4834	49	5

RISK FACTOR	75-84 AT EXAM PERSON EXAMS	# OF EVENTS	ANNUAL RATE	85-94 AT EXAM PERSON EXAMS	# OF EVENTS	ANNUAL RATE	35-64 COMBINED # OF EVENTS	CRUDE	AGE-ADJUSTED	65-94 COMBINED # OF EVENTS	CRUDE	AGE-ADJUSTED
NO	465	4	4	19			27			11	2	2
MILD	424	5	6	16			27	2	2	21	5	5
DEFINITE	194	5	13	5			16	3	2	15	8	8
TREATED	262	7	13	10			20	8	10	23	11	11
UNKNOWN	0	0	-	0			0	-		0	-	
TOTAL	1345	21	8	50			90	2	2	70	6	6

LOGISTIC REGRESSIONS

AGE AT EXAM		COEFFICIENT OF RISK FACTOR	SE COEFFICIENT	P-VALUE	STANDARDIZED COEFFICIENT	STD SE
35-44						
45-54		0.463	0.237	0.051	0.358	0.184
55-64		0.556	0.157	0.001	0.457	0.129
65-74		0.697	0.190	0.001	0.576	0.157
75-84		0.609	0.290	0.036	0.504	0.240
85-94						
35-64	UNIVARIATE	0.612	0.128	0.001	0.481	0.100
65-94	UNIVARIATE	0.677	0.159	0.001	0.560	0.151
35-64	BIVARIATE	0.494	0.128	0.001	0.389	0.101
65-94	BIVARIATE	0.670	0.159	0.001	0.554	0.131
35-64	MULTIVARIATE	0.319	0.135	0.019	0.251	0.106
65-94	MULTIVARIATE	0.583	0.165	0.001	0.482	0.136

SEX: FEMALE
EVENT: SUDDEN DEATH FROM CORONARY HEART DISEASE
RISK FACTOR: HYPERTENSION WITH ANTIHYPERTENSIVE TREATMENT
PERSONS AT RISK: ALL PERSONS ALIVE

RISK FACTOR	35-44 AT EXAM PERSON EXAMS	# OF EVENTS	ANNUAL RATE	45-54 AT EXAM PERSON EXAMS	# OF EVENTS	ANNUAL RATE	55-64 AT EXAM PERSON EXAMS	# OF EVENTS	ANNUAL RATE	65-74 AT EXAM PERSON EXAMS	# OF EVENTS	ANNUAL RATE
NO	4251			5660	2	0	4992	3	0	1871	4	1
MILD	1002			2611	1	0	3449	8	1	2116	4	1
DEFINITE	316			1184	1	0	1950	2	1	1198	5	2
TREATED	50			617	1	1	1712	2	1	1918	15	4
UNKNOWN	0	0	-	0	0	-	0	0	-	0	0	-
TOTAL	5619			10072	5	0	12103	15	1	7103	28	2

RISK FACTOR	75-84 AT EXAM PERSON EXAMS	# OF EVENTS	ANNUAL RATE	85-94 AT EXAM PERSON EXAMS	# OF EVENTS	ANNUAL RATE	35-64 COMBINED # OF EVENTS	CRUDE	AGE-ADJUSTED	65-94 COMBINED # OF EVENTS	CRUDE	AGE-ADJUSTED
NO	476	3	3	40			5	0	0	8	2	2
MILD	694	4	3	37			9	1	1	8	1	1
DEFINITE	469	4	4	35			3	0	0	10	3	3
TREATED	780	12	8	51			3	1	1	27	5	5
UNKNOWN	0	0	-	0			0	-		0	-	
TOTAL	2419	23	5	163			20	0	0	53	3	3

LOGISTIC REGRESSIONS

AGE AT EXAM		COEFFICIENT OF RISK FACTOR	SE COEFFICIENT	P-VALUE	STANDARDIZED COEFFICIENT	STD SE
35-44						
45-54		0.573	0.531	0.281	0.442	0.409
55-64		0.249	0.307	0.417	0.209	0.257
65-74		0.698	0.284	0.014	0.572	0.232
75-84		0.466	0.313	0.137	0.364	0.245
85-94						
35-64	UNIVARIATE	0.539	0.263	0.041	0.431	0.210
65-94	UNIVARIATE	0.587	0.202	0.004	0.477	0.164
35-64	BIVARIATE	0.251	0.270	0.352	0.201	0.216
65-94	BIVARIATE	0.524	0.202	0.009	0.426	0.164
35-64	MULTIVARIATE	0.299	0.272	0.272	0.239	0.218
65-94	MULTIVARIATE	0.408	0.205	0.046	0.332	0.167

TABLE 10-3B   DATE=05JAN86
AVERAGE ANNUAL INCIDENCE RATE FOR EVENT PER 1000 PERSONS AT RISK AT EXAM
BY AGE AND LEVEL OF RISK FACTOR AT EXAM: FRAMINGHAM STUDY 30-YEAR FOLLOWUP

SEX: MALE
EVENT: SUDDEN DEATH FROM CORONARY HEART DISEASE
RISK FACTOR: HYPERTENSION IGNORING TREATMENT
PERSONS AT RISK: ALL PERSONS ALIVE

RISK FACTOR	35-44 AT EXAM			45-54 AT EXAM			55-64 AT EXAM			65-74 AT EXAM		
	PERSON EXAMS	# OF EVENTS	ANNUAL RATE	PERSON EXAMS	# OF EVENTS	ANNUAL RATE	PERSON EXAMS	# OF EVENTS	ANNUAL RATE	PERSON EXAMS	# OF EVENTS	ANNUAL RATE
NO	2672			4306	19	1	4347	21	2	1980	9	2
MILD	1326			2383	9	2	2922	22	4	1852	25	7
DEFINITE	521			1378	7	3	1801	17	5	1002	15	7
UNKNOWN	0			0	0	-	0	0	-	0	0	-
TOTAL	4519			8067	26	2	9070	60	3	4834	49	5

RISK FACTOR	75-84 AT EXAM			85-94 AT EXAM			35-64 COMBINED			65-94 COMBINED		
	PERSON EXAMS	# OF EVENTS	ANNUAL RATE	PERSON EXAMS	# OF EVENTS	ANNUAL RATE	# OF EVENTS	CRUDE	AGE-ADJUSTED	# OF EVENTS	CRUDE	AGE-ADJUSTED
NO	526	6	6	24			34	2	2	15	3	3
MILD	543	6	6	21			32	2	2	31	6	6
DEFINITE	276	9	16	5			24	3	3	24	9	9
UNKNOWN	0	0	-	0			0	-	-	0	-	-
TOTAL	1345	21	8	50			90	2	2	70	6	6

LOGISTIC REGRESSIONS

AGE AT EXAM		COEFFICIENT OF RISK FACTOR	SE COEFFICIENT	P-VALUE	STANDARDIZED COEFFICIENT	STD SE
35-44						
45-54		0.398	0.242	0.099	0.302	0.183
55-64		0.341	0.160	0.033	0.264	0.124
65-74		0.551	0.187	0.003	0.418	0.142
75-84		0.579	0.292	0.047	0.434	0.219
85-94						
35-64	UNIVARIATE	0.392	0.131	0.003	0.296	0.099
65-94	UNIVARIATE	0.563	0.157	0.001	0.426	0.119
35-64	BIVARIATE	0.310	0.132	0.019	0.234	0.099
65-94	BIVARIATE	0.567	0.157	0.001	0.429	0.119
35-64	MULTIVARIATE	0.089	0.140	0.525	0.067	0.106
65-94	MULTIVARIATE	0.448	0.164	0.006	0.339	0.124

SEX: FEMALE
EVENT: SUDDEN DEATH FROM CORONARY HEART DISEASE
RISK FACTOR: HYPERTENSION IGNORING TREATMENT
PERSONS AT RISK: ALL PERSONS ALIVE

RISK FACTOR	35-44 AT EXAM			45-54 AT EXAM			55-64 AT EXAM			65-74 AT EXAM		
	PERSON EXAMS	# OF EVENTS	ANNUAL RATE	PERSON EXAMS	# OF EVENTS	ANNUAL RATE	PERSON EXAMS	# OF EVENTS	ANNUAL RATE	PERSON EXAMS	# OF EVENTS	ANNUAL RATE
NO	4273			5846	2	0	5415	3	0	2205	6	1
MILD	1017			2803	2	0	4079	10	1	2907	9	2
DEFINITE	329			1423	1	0	2609	2	0	1991	13	3
UNKNOWN	0			0	0	-	0	0	-	0	0	-
TOTAL	5619			10072	5	0	12103	15	1	7103	28	2

RISK FACTOR	75-84 AT EXAM			85-94 AT EXAM			35-64 COMBINED			65-94 COMBINED		
	PERSON EXAMS	# OF EVENTS	ANNUAL RATE	PERSON EXAMS	# OF EVENTS	ANNUAL RATE	# OF EVENTS	CRUDE	AGE-ADJUSTED	# OF EVENTS	CRUDE	AGE-ADJUSTED
NO	613	4	3	52			5	0	0	11	2	2
MILD	1003	10	5	57			12	1	1	19	2	2
DEFINITE	803	9	4	54			3	0	0	23	4	4
UNKNOWN	0	0	-	0			0	-	-	0	-	-
TOTAL	2419	23	5	163			20	0	0	53	3	3

LOGISTIC REGRESSIONS

AGE AT EXAM		COEFFICIENT OF RISK FACTOR	SE COEFFICIENT	P-VALUE	STANDARDIZED COEFFICIENT	STD SE
35-44						
45-54		0.405	0.558	0.468	0.295	0.406
55-64		0.262	0.322	0.415	0.205	0.251
65-74		0.487	0.255	0.057	0.374	0.196
75-84		0.247	0.282	0.381	0.188	0.214
85-94						
35-64	UNIVARIATE	0.487	0.274	0.075	0.363	0.204
65-94	UNIVARIATE	0.397	0.185	0.031	0.305	0.142
35-64	BIVARIATE	0.232	0.282	0.411	0.173	0.210
65-94	BIVARIATE	0.350	0.185	0.059	0.269	0.142
35-64	MULTIVARIATE	0.280	0.286	0.327	0.209	0.213
65-94	MULTIVARIATE	0.240	0.189	0.206	0.184	0.145

TABLE 10-4   DATE=05JAN86
AVERAGE ANNUAL INCIDENCE RATE FOR EVENT PER 1000 PERSONS AT RISK AT EXAM
BY AGE AND LEVEL OF RISK FACTOR AT EXAM: FRAMINGHAM STUDY 30-YEAR FOLLOWUP

SEX: MALE
EVENT: SUDDEN DEATH FROM CORONARY HEART DISEASE
RISK FACTOR: SERUM CHOLESTEROL
PERSONS AT RISK: ALL PERSONS ALIVE

35-44 AT EXAM			45-54 AT EXAM			55-64 AT EXAM			65-74 AT EXAM		
PERSON EXAMS	# OF EVENTS	ANNUAL RATE	PERSON EXAMS	# OF EVENTS	ANNUAL RATE	PERSON EXAMS	# OF EVENTS	ANNUAL RATE	PERSON EXAMS	# OF EVENTS	ANNUAL RATE
1253			1807	3	1	2484	7	1	1530	17	6
1258			2390	7	1	2695	22	4	1457	11	4
883			1843	6	2	2128	16	4	1059	12	6
488			1057	5	2	1075	10	5	525	6	6
303			619	5	4	556	5	4	246	2	4
334			351	0	0	132	0	0	17	1	29
4519			8067	26	2	9070	60	3	4834	49	5

75-84 AT EXAM			85-94 AT EXAM			35-64 COMBINED			65-94 COMBINED		
PERSON EXAMS	# OF EVENTS	ANNUAL RATE	PERSON EXAMS	# OF EVENTS	ANNUAL RATE	# OF EVENTS	CRUDE	AGE-ADJUSTED	# OF EVENTS	CRUDE	AGE-ADJUSTED
537	7	7	23			10	1	1	24	6	6
422	8	9	14			29	2	2	19	5	5
232	4	9	7			24	3	3	16	6	6
130	2	8	6			15	4	4	8	6	6
24	0	0	0			12	4	4	2	4	3
0	0	-	0			0	0	0	1	29	23
1343	21	8	50			90	2	2	70	6	6

LOGISTIC REGRESSIONS

	COEFFICIENT OF RISK FACTOR	SE COEFFICIENT.	P-VALUE	STANDARDIZED COEFFICIENT	STD SE
	0.010	0.003	0.001	0.420	0.128
	0.006	0.003	0.022	0.264	0.115
	-0.001	0.004	0.850	-0.028	0.146
	-0.001	0.006	0.880	-0.033	0.221
IVARIATE	0.008	0.002	0.001	0.346	0.082
IVARIATE	-0.001	0.003	0.704	-0.046	0.122
VARIATE	0.009	0.002	0.001	0.375	0.084
VARIATE	-0.000	0.003	0.938	-0.009	0.123
LTIVARIATE	0.009	0.002	0.001	0.385	0.086
LTIVARIATE	-0.000	0.003	0.981	-0.003	0.123

SEX: FEMALE
EVENT: SUDDEN DEATH FROM CORONARY HEART DISEASE
RISK FACTOR: SERUM CHOLESTEROL
PERSONS AT RISK: ALL PERSONS ALIVE

35-44 AT EXAM			45-54 AT EXAM			55-64 AT EXAM			65-74 AT EXAM		
PERSON EXAMS	# OF EVENTS	ANNUAL RATE	PERSON EXAMS	# OF EVENTS	ANNUAL RATE	PERSON EXAMS	# OF EVENTS	ANNUAL RATE	PERSON EXAMS	# OF EVENTS	ANNUAL RATE
2350			2120	0	0	1714	2	1	897	2	1
1419			2590	1	0	2741	2	0	1570	5	2
807			2268	2	0	3052	2	0	1912	7	2
372			1490	0	0	2364	7	1	1472	6	2
171			1124	2	1	1999	2	1	1215	7	3
500			480	0	0	233	0	0	37	1	14
5619			10072	5	0	12103	15	1	7103	28	2

75-84 AT EXAM			85-94 AT EXAM			35-64 COMBINED			65-94 COMBINED		
PERSON EXAMS	# OF EVENTS	ANNUAL RATE	PERSON EXAMS	# OF EVENTS	ANNUAL RATE	# OF EVENTS	CRUDE	AGE-ADJUSTED	# OF EVENTS	CRUDE	AGE-ADJUSTED
463	3	3	60			2	0	0	7	2	2
588	8	7	39			3	0	0	13	3	3
660	6	5	42			4	0	0	13	2	2
419	5	6	12			7	1	1	11	3	3
282	1	2	10			4	1	1	8	3	3
7	0	0	0			0	0	0	1	11	10
2419	23	5	163			20	0	0	53	3	3

LOGISTIC REGRESSIONS

	COEFFICIENT OF RISK FACTOR	SE COEFFICIENT	P-VALUE	STANDARDIZED COEFFICIENT	STD SE
	0.006	0.004	0.113	0.277	0.174
	0.003	0.005	0.584	0.136	0.248
	0.005	0.004	0.221	0.222	0.181
	-0.003	0.005	0.526	-0.136	0.214
VARIATE	0.006	0.003	0.022	0.287	0.125
VARIATE	-0.001	0.003	0.756	-0.044	0.140
ARIATE	0.005	0.004	0.273	0.213	0.195
ARIATE	0.001	0.003	0.726	0.049	0.140
TIVARIATE	0.004	0.004	0.341	0.198	0.208
TIVARIATE	0.002	0.003	0.619	0.068	0.137

TABLE 10-5   DATE=05JAN86
AVERAGE ANNUAL INCIDENCE RATE FOR EVENT PER 1000 PERSONS AT RISK AT EXAM
BY AGE AND LEVEL OF RISK FACTOR AT EXAM: FRAMINGHAM STUDY 30-YEAR FOLLOWUP

SEX: MALE
EVENT: SUDDEN DEATH FROM CORONARY HEART DISEASE
RISK FACTOR: HEMATOCRIT
PERSONS AT RISK: ALL PERSONS ALIVE

RISK FACTOR	35-44 AT EXAM PERSON EXAMS	# OF EVENTS	ANNUAL RATE	45-54 AT EXAM PERSON EXAMS	# OF EVENTS	ANNUAL RATE	55-64 AT EXAM PERSON EXAMS	# OF EVENTS	ANNUAL RATE	65-74 AT EXAM PERSON EXAMS	# OF EVENTS	ANNUAL RATE
25-42	630			961	3	2	1121	6	3	558	5	4
43-44	551			996	4	2	1094	6	3	611	4	3
45-46	990			1942	8	2	2215	14	3	1194	7	3
47-48	786			1585	3	1	1825	11	3	1069	9	4
49-70	1192			2195	8	2	2656	23	4	1382	23	8
UNKNOWN	370			390	0	0	159	0	0	20	1	25
TOTAL	4519			8067	26	2	9070	60	3	4834	49	5

RISK FACTOR	75-84 AT EXAM PERSON EXAMS	# OF EVENTS	ANNUAL RATE	85-94 AT EXAM PERSON EXAMS	# OF EVENTS	ANNUAL RATE	35-64 COMBINED # OF EVENTS	CRUDE	AGE-ADJUSTED	65-94 COMBINED # OF EVENTS	CRUDE	AGE-ADJUSTED
25-42	283	4	7	19			10	2	2	9	5	5
43-44	169	5	15	8			10	2	2	9	6	6
45-46	313	1	2	7			23	2	2	8	3	3
47-48	266	6	11	9			14	2	2	15	6	6
49-70	314	5	8	7			33	3	3	28	8	8
UNKNOWN	0	0	-	0			0	0	0	1	25	19
TOTAL	1345	21	8	50			90	2	2	70	6	6

LOGISTIC REGRESSIONS

AGE AT EXAM	COEFFICIENT OF RISK FACTOR	SE COEFFICIENT	P-VALUE	STANDARDIZED COEFFICIENT	STD SE
35-44					
45-54	-9.014	0.055	0.792	-0.052	0.196
55-64	0.037	0.055	0.281	0.141	0.130
65-74	0.086	0.040	0.051	0.315	0.146
75-84	0.034	0.054	0.530	0.140	0.223
85-94					
35-64 UNIVARIATE	0.026	0.029	0.369	0.095	0.106
65-94 UNIVARIATE	0.064	0.032	0.049	0.242	0.123
35-64 BIVARIATE	0.023	0.029	0.411	0.087	0.105
65-94 BIVARIATE	0.075	0.032	0.020	0.283	0.122
35-64 MULTIVARIATE	0.004	0.029	0.890	0.015	0.106
65-94 MULTIVARIATE	0.072	0.033	0.029	0.272	0.124

SEX: FEMALE
EVENT: SUDDEN DEATH FROM CORONARY HEART DISEASE
RISK FACTOR: HEMATOCRIT
PERSONS AT RISK: ALL PERSONS ALIVE

RISK FACTOR	35-44 AT EXAM PERSON EXAMS	# OF EVENTS	ANNUAL RATE	45-54 AT EXAM PERSON EXAMS	# OF EVENTS	ANNUAL RATE	55-64 AT EXAM PERSON EXAMS	# OF EVENTS	ANNUAL RATE	65-74 AT EXAM PERSON EXAMS	# OF EVENTS	ANNUAL RATE
21-39	1698			1858	1	0	1609	2	1	751	4	3
40-41	967			1761	0	0	2144	1	0	1180	3	1
42-43	1043			1978	0	0	2371	1	1	1422	4	1
44-45	858			2222	2	0	3003	5	1	1835	9	2
46-65	487			1702	2	1	2700	6	1	1873	7	2
UNKNOWN	566			551	0	0	276	0	0	42	1	12
TOTAL	5619			10072	5	0	12103	15	1	7103	28	2

RISK FACTOR	75-84 AT EXAM PERSON EXAMS	# OF EVENTS	ANNUAL RATE	85-94 AT EXAM PERSON EXAMS	# OF EVENTS	ANNUAL RATE	35-64 COMBINED # OF EVENTS	CRUDE	AGE-ADJUSTED	65-94 COMBINED # OF EVENTS	CRUDE	AGE-ADJUSTED
21-39	375	3	4	33			4	0	0	8	3	3
40-41	449	4	4	39			1	0	0	7	2	2
42-43	437	2	2	26			1	0	0	7	2	2
44-45	607	5	4	34			7	1	1	14	3	3
46-65	545	9	8	31			8	1	1	16	3	3
UNKNOWN	6	0	0	0			0	0	0	1	10	9
TOTAL	2419	23	5	163			20	0	0	53	3	3

LOGISTIC REGRESSIONS

AGE AT EXAM	COEFFICIENT OF RISK FACTOR	SE COEFFICIENT	P-VALUE	STANDARDIZED COEFFICIENT	STD SE
35-44					
45-54	0.112	0.128	0.382	0.400	0.458
55-64	0.115	0.074	0.120	0.399	0.256
65-74	0.033	0.055	0.555	0.115	0.195
75-84	0.053	0.055	0.333	0.206	0.213
85-94					
35-64 UNIVARIATE	0.143	0.063	0.022	0.515	0.226
65-94 UNIVARIATE	0.025	0.039	0.515	0.092	0.140
35-64 BIVARIATE	0.104	0.064	0.106	0.374	0.231
65-94 BIVARIATE	0.038	0.038	0.321	0.137	0.138
35-64 MULTIVARIATE	0.078	0.066	0.237	0.281	0.238
65-94 MULTIVARIATE	0.031	0.038	0.408	0.113	0.136

TABLE 10-6   DATE=05JAN86
AVERAGE ANNUAL INCIDENCE RATE FOR EVENT PER 1000 PERSONS AT RISK AT EXAM
BY AGE AND LEVEL OF RISK FACTOR AT EXAM: FRAMINGHAM STUDY 30-YEAR FOLLOWUP

SEX: MALE
EVENT: SUDDEN DEATH FROM CORONARY HEART DISEASE
RISK FACTOR: BLOOD GLUCOSE (MG/DL)
PERSONS AT RISK: ALL PERSONS ALIVE

RISK FACTOR	35-44 AT EXAM			45-54 AT EXAM			55-64 AT EXAM			65-74 AT EXAM		
	PERSON EXAMS	# OF EVENTS	ANNUAL RATE	PERSON EXAMS	# OF EVENTS	ANNUAL RATE	PERSON EXAMS	# OF EVENTS	ANNUAL RATE	PERSON EXAMS	# OF EVENTS	ANNUAL RATE
29-69	1004			1562	6	2	1208	5	2	501	4	4
70-89	2531			4344	12	1	4526	28	3	2142	20	5
90-109	628			1402	5	2	2136	15	4	1296	13	5
110-129	107			331	1	2	608	8	7	428	5	6
130-524	49			204	2	5	491	4	4	450	6	7
UNKNOWN	200			224	0	0	101	0	0	17	1	29
TOTAL	4519			8067	26	2	9070	60	3	4834	49	5

RISK FACTOR	75-84 AT EXAM			85-94 AT EXAM			35-64 COMBINED			65-94 COMBINED		
	PERSON EXAMS	# OF EVENTS	ANNUAL RATE	PERSON EXAMS	# OF EVENTS	ANNUAL RATE	# OF EVENTS	CRUDE	AGE-ADJUSTED	# OF EVENTS	CRUDE	AGE-ADJUSTED
29-69	86	0	0	1			11	1	2	4	3	3
70-89	479	8	8	17			43	2	2	28	5	5
90-109	438	6	7	15			21	3	3	19	5	6
110-129	170	3	9	7			9	4	3	8	7	6
130-524	172	4	12	10			6	4	4	10	8	8
UNKNOWN	0	0	-	0			0	0	0	1	29	23
TOTAL	1345	21	8	50			90	2	2	70	6	6

LOGISTIC REGRESSIONS

AGE AT EXAM	COEFFICIENT OF RISK FACTOR	SE COEFFICIENT	P-VALUE	STANDARDIZED COEFFICIENT	STD SE
35-44					
45-54	0.006	0.005	0.244	0.142	0.122
55-64	0.006	0.003	0.031	0.180	0.083
65-74	0.005	0.003	0.051	0.183	0.094
75-84	0.002	0.006	0.696	0.077	0.198
85-94					
35-64 UNIVARIATE	0.008	0.002	0.001	0.193	0.058
65-94 UNIVARIATE	0.005	0.002	0.050	0.165	0.084
35-64 BIVARIATE	0.005	0.003	0.033	0.138	0.066
65-94 BIVARIATE	0.004	0.002	0.077	0.155	0.088
35-64 MULTIVARIATE	0.005	0.003	0.032	0.140	0.065
65-94 MULTIVARIATE	0.004	0.002	0.119	0.138	0.089

SEX: FEMALE
EVENT: SUDDEN DEATH FROM CORONARY HEART DISEASE
RISK FACTOR: BLOOD GLUCOSE (MG/DL)
PERSONS AT RISK: ALL PERSONS ALIVE

RISK FACTOR	35-44 AT EXAM			45-54 AT EXAM			55-64 AT EXAM			65-74 AT EXAM		
	PERSON EXAMS	# OF EVENTS	ANNUAL RATE	PERSON EXAMS	# OF EVENTS	ANNUAL RATE	PERSON EXAMS	# OF EVENTS	ANNUAL RATE	PERSON EXAMS	# OF EVENTS	ANNUAL RATE
29-69	1067			1806	1	0	1411	5	2	597	4	3
70-89	3414			5611	3	0	6371	7	1	3483	9	1
90-109	672			1646	0	0	2877	3	1	1982	7	2
110-129	96			303	0	0	777	0	0	582	2	2
130-524	39			167	1	3	496	0	0	427	5	6
UNKNOWN	331			339	0	0	171	0	0	32	1	16
TOTAL	5619			10072	5	0	12103	15	1	7103	28	2

RISK FACTOR	75-84 AT EXAM			85-94 AT EXAM			35-64 COMBINED			65-94 COMBINED		
	PERSON EXAMS	# OF EVENTS	ANNUAL RATE	PERSON EXAMS	# OF EVENTS	ANNUAL RATE	# OF EVENTS	CRUDE	AGE-ADJUSTED	# OF EVENTS	CRUDE	AGE-ADJUSTED
29-69	129	1	4	5			6	1	1	5	3	3
70-89	976	5	3	68			10	0	0	15	2	2
90-109	803	6	4	63			3	0	0	14	2	2
110-129	286	6	10	19			0	0	0	8	5	4
130-524	224	5	11	16			1	1	1	10	7	7
UNKNOWN	1	0	0	0			0	0	0	1	15	11
TOTAL	2419	23	5	163			20	0	0	53	3	3

LOGISTIC REGRESSIONS

AGE AT EXAM	COEFFICIENT OF RISK FACTOR	SE COEFFICIENT	P-VALUE	STANDARDIZED COEFFICIENT	STD SE
35-44					
45-54	0.015	0.008	0.065	0.291	0.158
55-64	-0.043	0.020	0.035	-1.162	0.552
65-74	0.007	0.004	0.074	0.206	0.115
75-84	0.009	0.003	0.002	0.318	0.103
85-94					
35-64 UNIVARIATE	-0.003	0.012	0.768	-0.078	0.266
65-94 UNIVARIATE	0.009	0.002	0.001	0.268	0.072
35-64 BIVARIATE	-0.012	0.014	0.358	-0.286	0.311
65-94 BIVARIATE	0.008	0.002	0.001	0.242	0.076
35-64 MULTIVARIATE	-0.012	0.013	0.356	-0.285	0.309
65-94 MULTIVARIATE	0.007	0.002	0.004	0.223	0.078

TABLE 10-7   DATE=05JAN86

AVERAGE ANNUAL INCIDENCE RATE FOR EVENT PER 1000 PERSONS AT RISK AT EXAM
BY AGE AND LEVEL OF RISK FACTOR AT EXAM: FRAMINGHAM STUDY 30-YEAR FOLLOWUP

SEX: MALE
EVENT: SUDDEN DEATH FROM CORONARY HEART DISEASE
RISK FACTOR: DIABETES MELLITUS
PERSONS AT RISK: ALL PERSONS ALIVE

RISK FACTOR	35-44 AT EXAM PERSON EXAMS	# OF EVENTS	ANNUAL RATE	45-54 AT EXAM PERSON EXAMS	# OF EVENTS	ANNUAL RATE	55-64 AT EXAM PERSON EXAMS	# OF EVENTS	ANNUAL RATE	65-74 AT EXAM PERSON EXAMS	# OF EVENTS	ANNUAL RATE
NO	4470			7812	24	2	8424	54	3	4269	42	5
YES	49			255	2	4	646	6	5	565	7	6
UNKNOWN	0			0	0	-	0	0	-	0	0	-
TOTAL	4519			8067	26	2	9070	60	3	4834	49	-

RISK FACTOR	75-84 AT EXAM PERSON EXAMS	# OF EVENTS	ANNUAL RATE	85-94 AT EXAM PERSON EXAMS	# OF EVENTS	ANNUAL RATE	35-64 COMBINED # OF EVENTS	CRUDE	AGE-ADJUSTED	65-94 COMBINED # OF EVENTS	CRUDE	AGE-ADJUSTED
NO	1170	19	8	44			82	2	2	61	6	6
YES	175	2	6	6			8	4	3	9	6	6
UNKNOWN	0	0	-	0			0	-		0	-	
TOTAL	1345	21	8	50			90	2	2	70	6	6

LOGISTIC REGRESSIONS

AGE AT EXAM		COEFFICIENT OF RISK FACTOR	SE COEFFICIENT	P-VALUE	STANDARDIZED COEFFICIENT	STD SE
35-44						
45-54		0.942	0.739	0.202	0.165	0.129
55-64		0.374	0.432	0.387	0.096	0.111
65-74		0.233	0.411	0.570	0.075	0.132
75-84		-0.356	0.748	0.634	-0.120	0.252
85-94						
35-64	UNIVARIATE	0.759	0.372	0.041	0.155	0.076
65-94	UNIVARIATE	0.082	0.359	0.819	0.027	0.117
35-64	BIVARIATE	0.390	0.375	0.298	0.080	0.077
65-94	BIVARIATE	0.059	0.360	0.870	0.019	0.117
35-64	MULTIVARIATE	0.131	0.588	0.735	0.027	0.079
65-94	MULTIVARIATE	-0.010	0.363	0.978	-0.003	0.118

SEX: FEMALE
EVENT: SUDDEN DEATH FROM CORONARY HEART DISEASE
RISK FACTOR: DIABETES MELLITUS
PERSONS AT RISK: ALL PERSONS ALIVE

RISK FACTOR	35-44 AT EXAM PERSON EXAMS	# OF EVENTS	ANNUAL RATE	45-54 AT EXAM PERSON EXAMS	# OF EVENTS	ANNUAL RATE	55-64 AT EXAM PERSON EXAMS	# OF EVENTS	ANNUAL RATE	65-74 AT EXAM PERSON EXAMS	# OF EVENTS	ANNUAL RATE
NO	5576			9859	4	0	11497	14	1	6499	20	2
YES	43			213	1	2	606	1	1	604	8	7
UNKNOWN	0			0	0	-	0	0	-	0	0	-
TOTAL	5619			10072	5	0	12103	15	1	7103	28	2

RISK FACTOR	75-84 AT EXAM PERSON EXAMS	# OF EVENTS	ANNUAL RATE	85-94 AT EXAM PERSON EXAMS	# OF EVENTS	ANNUAL RATE	35-64 COMBINED # OF EVENTS	CRUDE	AGE-ADJUSTED	65-94 COMBINED # OF EVENTS	CRUDE	AGE-ADJUSTED
NO	2106	15	7	139			18	0	0	37	2	2
YES	313	8	13	24			2	1	.1	16	9	8
UNKNOWN	0	0	-	0			0	-	0	0	-	
TOTAL	2419	23	5	163			20	0	0	53	3	3

LOGISTIC REGRESSIONS

AGE AT EXAM		COEFFICIENT OF RISK FACTOR	SE COEFFICIENT	P-VALUE	STANDARDIZED COEFFICIENT	STD SE
35-44						
45-54		2.453	1.120	0.029	0.353	0.161
55-64		0.304	1.036	0.769	0.066	0.226
65-74		1.470	0.421	0.001	0.410	0.117
75-84		1.296	0.442	0.003	0.435	0.148
85-94						
35-64	UNIVARIATE	1.246	0.746	0.095	0.216	0.129
65-94	UNIVARIATE	1.404	0.301	0.001	0.416	0.089
35-64	BIVARIATE	0.787	0.750	0.294	0.136	0.130
65-94	BIVARIATE	1.271	0.304	0.001	0.377	0.090
35-64	MULTIVARIATE	0.714	0.754	0.343	0.124	0.131
65-94	MULTIVARIATE	1.174	0.313	0.001	0.348	0.093

TABLE 10-8　　　　　　　　DATE=05JAN86
AVERAGE ANNUAL INCIDENCE RATE FOR EVENT PER 1000 PERSONS AT RISK AT EXAM
BY AGE AND LEVEL OF RISK FACTOR AT EXAM: FRAMINGHAM STUDY 30-YEAR FOLLOWUP

SEX: MALE
EVENT: SUDDEN DEATH FROM CORONARY HEART DISEASE
RISK FACTOR: GLUCOSE IN URINE, DEFINITE OR TRACE
PERSONS AT RISK: ALL PERSONS ALIVE

RISK FACTOR	35-44 AT EXAM PERSON EXAMS	# OF EVENTS	ANNUAL RATE	45-54 AT EXAM PERSON EXAMS	# OF EVENTS	ANNUAL RATE	55-64 AT EXAM PERSON EXAMS	# OF EVENTS	ANNUAL RATE	65-74 AT EXAM PERSON EXAMS	# OF EVENTS	ANNUAL RATE
NO	4426			7869	24	2	8786	59	3	4661	44	5
YES	60			177	2	6	260	1	2	167	5	15
UNKNOWN	33			21	0	0	24	0	0	6	0	0
TOTAL	4519			8067	26	2	9070	60	3	4834	49	5

RISK FACTOR	75-84 AT EXAM PERSON EXAMS	# OF EVENTS	ANNUAL RATE	85-94 AT EXAM PERSON EXAMS	# OF EVENTS	ANNUAL RATE	35-64 COMBINED # OF EVENTS	CRUDE	AGE-ADJUSTED	65-94 COMBINED # OF EVENTS	CRUDE	AGE-ADJUSTED
NO	1303	20	8	49			87	2	2	64	5	5
YES	39	1	13	1			3	3	3	6	14	14
UNKNOWN	3	0	0	0			0	0	0	0	0	0
TOTAL	1345	21	8	50			90	2	2	70	6	6

LOGISTIC REGRESSIONS

AGE AT EXAM		COEFFICIENT OF RISK FACTOR	SE COEFFICIENT	P-VALUE	STANDARDIZED COEFFICIENT	STD SE
35-44						
45-54		1.318	0.740	0.075	0.193	0.109
55-64		-0.560	1.010	0.579	-0.094	0.169
65-74		1.175	0.479	0.014	0.215	0.087
75-84		0.524	1.038	0.614	0.088	0.174
85-94						
35-64	UNIVARIATE	0.382	0.589	0.516	0.057	0.088
65-94	UNIVARIATE	1.021	0.433	0.018	0.183	0.078
35-64	BIVARIATE	0.207	0.590	0.726	0.031	0.089
65-94	BIVARIATE	1.052	0.434	0.015	0.189	0.078
35-64	MULTIVARIATE	-0.001	0.598	0.999	-0.000	0.090
65-94	MULTIVARIATE	0.983	0.439	0.025	0.176	0.079

SEX: FEMALE
EVENT: SUDDEN DEATH FROM CORONARY HEART DISEASE
RISK FACTOR: GLUCOSE IN URINE, DEFINITE OR TRACE
PERSONS AT RISK: ALL PERSONS ALIVE

RISK FACTOR	35-44 AT EXAM PERSON EXAMS	# OF EVENTS	ANNUAL RATE	45-54 AT EXAM PERSON EXAMS	# OF EVENTS	ANNUAL RATE	55-64 AT EXAM PERSON EXAMS	# OF EVENTS	ANNUAL RATE	65-74 AT EXAM PERSON EXAMS	# OF EVENTS	ANNUAL RATE
NO	5540			9947	4	0	11903	15	1	6957	25	2
YES	41			101	1	5	174	0	0	136	3	11
UNKNOWN	38			24	0	0	26	0	0	10	0	0
TOTAL	5619			10072	5	0	12103	15	1	7+03	28	2

RISK FACTOR	75-84 AT EXAM PERSON EXAMS	# OF EVENTS	ANNUAL RATE	85-94 AT EXAM PERSON EXAMS	# OF EVENTS	ANNUAL RATE	35-64 COMBINED # OF EVENTS	CRUDE	AGE-ADJUSTED	65-94 COMBINED # OF EVENTS	CRUDE	AGE-ADJUSTED
NO	2375	22	5	160	3		19	0	0	49	3	3
YES	35	1	14	3			1	2	2	4	11	12
UNKNOWN	9	0	0	0			0	0	0	0	0	0
TOTAL	2419	23	5	163			20	0	0	53	3	3

LOGISTIC REGRESSIONS

AGE AT EXAM		COEFFICIENT OF RISK FACTOR	SE COEFFICIENT	P-VALUE	STANDARDIZED COEFFICIENT	STD SE
35-44						
45-54		3.213	1.123	0.004	0.321	0.112
55-64						
65-74		1.833	0.617	0.003	0.251	0.085
75-84		1.146	1.037	0.269	0.137	0.124
85-94						
35-64	UNIVARIATE	1.520	1.028	0.139	0.161	0.109
65-94	UNIVARIATE	1.512	0.526	0.004	0.201	0.070
35-64	BIVARIATE	1.315	1.029	0.201	0.140	0.109
65-94	BIVARIATE	1.587	0.528	0.003	0.211	0.070
35-64	MULTIVARIATE	1.217	1.038	0.241	0.129	0.110
65-94	MULTIVARIATE	1.352	0.545	0.013	0.180	0.072

TABLE 10-9   DATE=01MAR86
AVERAGE ANNUAL INCIDENCE RATE FOR EVENT PER 1000 PERSONS AT RISK AT EXAM
BY AGE AND LEVEL OF RISK FACTOR AT EXAM: FRAMINGHAM STUDY 30-YEAR FOLLOWUP

SEX: MALE
EVENT: SUDDEN DEATH FROM CORONARY HEART DISEASE
RISK FACTOR: GLUCOSE INTOLERANCE
PERSONS AT RISK: ALL PERSONS ALIVE.

RISK FACTOR	35-44 AT EXAM			45-54 AT EXAM			55-64 AT EXAM			65-74 AT EXAM		
	PERSON EXAMS	# OF EVENTS	ANNUAL RATE	PERSON EXAMS	# OF EVENTS	ANNUAL RATE	PERSON EXAMS	# OF EVENTS	ANNUAL RATE	PERSON EXAMS	# OF EVENTS	ANNUAL RATE
NO	4381			7528	23	2	7973	50	3	3918	39	5
YES	138			539	3	3	1097	10	5	916	10	5
UNKNOWN	0			0	0	-	0	0	-	0	0	-
TOTAL	4519			8067	26	2	9070	60	3	4834	49	5

RISK FACTOR	75-84 AT EXAM			85-94 AT EXAM			35-64 COMBINED			65-94 COMBINED		
	PERSON EXAMS	# OF EVENTS	ANNUAL RATE	PERSON EXAMS	# OF EVENTS	ANNUAL RATE	# OF EVENTS	CRUDE	AGE-ADJUSTED	# OF EVENTS	CRUDE	AGE-ADJUSTED
NO	1035	16	8	36			77	2	2	55	6	6
YES	310	5	8	.14			13	4	3	15	6	6
UNKNOWN	0	0	-	0			0	-	-	0	-	-
TOTAL	1345	21	8	50			90	2	2	70	6	6

LOGISTIC REGRESSIONS

AGE AT EXAM	COEFFICIENT OF RISK FACTOR	SE COEFFICIENT	P-VALUE	STANDARDIZED COEFFICIENT	STD SE
35-44					
45-54	0.602	0.615	0.328	0.150	0.154
55-64	0.377	0.348	0.279	0.123	0.113
65-74	0.093	0.356	0.793	0.037	0.140
75-84	0.043	0.516	0.933	0.018	0.218
85-94					
35-64 UNIVARIATE	0.641	0.301	0.033	0.176	0.083
65-94 UNIVARIATE	0.094	0.293	0.749	0.038	0.117
35-64 BIVARIATE	0.343	0.303	0.259	0.094	0.083
65-94 BIVARIATE	0.048	0.294	0.869	0.019	0.117
35-64 MULTIVARIATE	0.185	0.310	0.551	0.051	0.085
65-94 MULTIVARIATE	-0.002	0.296	0.995	-0.001	0.118

SEX: FEMALE
EVENT: SUDDEN DEATH FROM CORONARY HEART DISEASE
RISK FACTOR: GLUCOSE INTOLERANCE
PERSONS AT RISK: ALL PERSONS ALIVE

RISK FACTOR	35-44 AT EXAM			45-54 AT EXAM			55-64 AT EXAM			65-74 AT EXAM		
	PERSON EXAMS	# OF EVENTS	ANNUAL RATE	PERSON EXAMS	# OF EVENTS	ANNUAL RATE	PERSON EXAMS	# OF EVENTS	ANNUAL RATE	PERSON EXAMS	# OF EVENTS	ANNUAL RATE
NO	5525			9685	4	0	10988	14	1	6066	18	1
YES	94			387	1	1	1115	1	0	1037	10	5
UNKNOWN	0			0	0	-	0	0	-	0	0	-
TOTAL	5619			10072	5	0	12103	15	1	7103	28	2

RISK FACTOR	75-84 AT EXAM			85-94 AT EXAM			35-64 COMBINED			65-94 COMBINED		
	PERSON EXAMS	# OF EVENTS	ANNUAL RATE	PERSON EXAMS	# OF EVENTS	ANNUAL RATE	# OF EVENTS	CRUDE	AGE-ADJUSTED	# OF EVENTS	CRUDE	AGE-ADJUSTED
NO	1905	14	4	131			18	0	0	34	2	2
YES	514	9	9	32			2	1	1	19	6	6
UNKNOWN	0	0	-	0			0	-	-	0	-	-
TOTAL	2419	23	5	163			20	0	0	53	3	3

LOGISTIC REGRESSIONS

AGE AT EXAM	COEFFICIENT OF RISK FACTOR	SE COEFFICIENT	P-VALUE	STANDARDIZED COEFFICIENT	STD SE
35-44					
45-54	1.836	1.119	0.101	0.353	0.215
55-64	-0.551	1.036	0.734	-0.102	0.500
65-74	1.185	0.396	0.003	0.419	0.140
75-84	0.878	0.430	0.041	0.359	0.176
85-94					
35-64 UNIVARIATE	0.602	0.746	0.420	0.140	0.174
65-94 UNIVARIATE	1.059	0.288	0.001	0.392	0.106
35-64 BIVARIATE	0.142	0.749	0.850	0.033	0.174
65-94 BIVARIATE	0.934	0.290	0.001	0.345	0.107
35-64 MULTIVARIATE	0.110	0.750	0.884	0.026	0.174
65-94 MULTIVARIATE	0.865	0.297	0.004	0.320	0.110

TABLE 10-10   DATE=05JAN86
AVERAGE ANNUAL INCIDENCE RATE FOR EVENT PER 1000 PERSONS AT RISK AT EXAM
BY AGE AND LEVEL OF RISK FACTOR AT EXAM: FRAMINGHAM STUDY 30-YEAR FOLLOWUP

SEX: MALE
EVENT: SUDDEN DEATH FROM CORONARY HEART DISEASE
RISK FACTOR: METROPOLITAN RELATIVE WEIGHT (PERCENT)
PERSONS AT RISK: ALL PERSONS ALIVE

RISK FACTOR	35-44 AT EXAM			45-54 AT EXAM			55-64 AT EXAM			65-74 AT EXAM		
	PERSON EXAMS	# OF EVENTS	ANNUAL RATE	PERSON EXAMS	# OF EVENTS	ANNUAL RATE	PERSON EXAMS	# OF EVENTS	ANNUAL RATE	PERSON EXAMS	# OF EVENTS	ANNUAL RATE
54-104	723			1149	2	1	1272	8	3	831	12	7
105-114	979			1635	5	2	1858	10	3	1052	6	3
115-124	1133			2139	9	2	2467	17	3	1322	14	5
125-134	923			1691	4	1	1851	9	2	956	9	5
135-272	761			1452	6	2	1621	16	5	673	8	6
UNKNOWN	0			1	0	0	1	0	0	0	0	-
TOTAL	4519			8067	26	2	9070	60	3	4834	49	5

RISK FACTOR	75-84 AT EXAM			85-94 AT EXAM			35-64 COMBINED			65-94 COMBINED		
	PERSON EXAMS	# OF EVENTS	ANNUAL RATE	PERSON EXAMS	# OF EVENTS	ANNUAL RATE	# OF EVENTS	CRUDE	AGE-ADJUSTED	# OF EVENTS	CRUDE	AGE-ADJUSTED
54-104	277	6	11	19			11	2	2	18	8	8
105-114	357	7	10	15			15	2	2	13	5	4
115-124	399	4	5	14			28	2	2	18	5	5
125-134	196	1	3	2			14	2	2	10	4	4
135-272	116	3	13	0			22	3	3	11	7	7
UNKNOWN	0	0	-	0			0	0	0	0	-	
TOTAL	1345	21	8	50			90	2	2	70	6	6

LOGISTIC REGRESSIONS

AGE AT EXAM		COEFFICIENT OF RISK FACTOR	SE COEFFICIENT	P-VALUE	STANDARDIZED COEFFICIENT	STD SE
35-44						
45-54		0.005	0.012	0.704	0.073	0.193
55-64		0.012	0.008	0.112	0.196	0.124
65-74		-0.002	0.009	0.802	-0.036	0.144
75-84		-0.016	0.015	0.302	-0.235	0.228
85-94						
35-64	UNIVARIATE	0.010	0.006	0.125	0.156	0.102
65-94	UNIVARIATE	-0.007	0.008	0.399	-0.103	0.122
35-64	BIVARIATE	0.010	0.006	0.132	0.154	0.102
65-94	BIVARIATE	-0.005	0.008	0.570	-0.070	0.123
35-64	MULTIVARIATE	0.008	0.007	0.205	0.136	0.107
65-94	MULTIVARIATE	-0.012	0.008	0.161	-0.179	0.127

SEX: FEMALE
EVENT: SUDDEN DEATH FROM CORONARY HEART DISEASE
RISK FACTOR: METROPOLITAN RELATIVE WEIGHT (PERCENT)
PERSONS AT RISK: ALL PERSONS ALIVE

RISK FACTOR	35-44 AT EXAM			45-54 AT EXAM			55-64 AT EXAM			65-74 AT EXAM		
	PERSON EXAMS	# OF EVENTS	ANNUAL RATE	PERSON EXAMS	# OF EVENTS	ANNUAL RATE	PERSON EXAMS	# OF EVENTS	ANNUAL RATE	PERSON EXAMS	# OF EVENTS	ANNUAL RATE
54-104	1694			2117	1	0	2140	1	0	319	9	3
105-114	1431			2564	0	0	2541	3	1	321	3	1
115-124	1059			2128	0	0	2556	5	1	536	4	1
125-134	566			1383	1	0	1831	1	0	140	6	3
135-272	861			2080	3	1	3035	5	1	787	6	2
UNKNOWN	8			0	0	-	0	0	-	0	0	-
TOTAL	5619			10072	5	0	12103	15	1	7103	28	2

RISK FACTOR	75-84 AT EXAM			85-94 AT EXAM			35-64 COMBINED			65-94 COMBINED		
	PERSON EXAMS	# OF EVENTS	ANNUAL RATE	PERSON EXAMS	# OF EVENTS	ANNUAL RATE	# OF EVENTS	CRUDE	AGE-ADJUSTED	# OF EVENTS	CRUDE	AGE-ADJUSTED
54-104	632	5	4	51			2	0	0	16	4	4
105-114	457	2	2	33			3	0	0	5	1	1
115-124	487	7	7	31			5	0	0	11	3	3
125-134	342	5	7	26			2	0	0	11	4	4
135-272	501	4	4	22			8	1	1	10	2	2
UNKNOWN	0	0	-	0			0	0	0	0	-	
TOTAL	2419	23	5	163			20	0	0	53	3	3

LOGISTIC REGRESSIONS

AGE AT EXAM		COEFFICIENT OF RISK FACTOR	SE COEFFICIENT	P-VALUE	STANDARDIZED COEFFICIENT	STD SE
35-44						
45-54		0.021	0.014	0.131	0.453	0.300
55-64		0.002	0.011	0.885	0.036	0.253
65-74		-0.015	0.010	0.121	-0.344	0.222
75-84		0.006	0.009	0.475	0.138	0.194
85-94						
35-64	UNIVARIATE	0.011	0.009	0.213	0.234	0.188
65-94	UNIVARIATE	-0.009	0.007	0.187	-0.200	0.151
35-64	BIVARIATE	0.007	0.009	0.429	0.157	0.198
65-94	BIVARIATE	-0.006	0.007	0.398	-0.126	0.149
35-64	MULTIVARIATE	0.008	0.010	0.413	0.174	0.213
65-94	MULTIVARIATE	-0.006	0.007	0.345	-0.140	0.148

SEX: MALE
EVENT: SUDDEN DEATH FROM CORONARY HEART DISEASE
RISK FACTOR: VITAL CAPACITY - HEIGHT INDEX (ML/INCH)
PERSONS AT RISK: ALL PERSONS ALIVE

RISK FACTOR	35-44 AT EXAM			45-54 AT EXAM			55-64 AT EXAM			65-74 AT EXAM		
	PERSON EXAMS	# OF EVENTS	ANNUAL RATE	PERSON EXAMS	# OF EVENTS	ANNUAL RATE	PERSON EXAMS	# OF EVENTS	ANNUAL RATE	PERSON EXAMS	# OF EVENTS	ANNUAL RATE
8-44	271			1108	1	0	2574	25	5	2179	19	4
45-49	416			1188	4	2	1692	12	4	974	15	8
50-54	791			1726	8	2	2061	13	3	857	6	4
55-59	1034			1823	7	2	1569	4	1	528	5	5
60-85	2034			2216	6	1	1356	6	2	284	4	7
UNKNOWN	13			6	0	0	18	0	0	12	0	0
TOTAL	4519			8067	26	2	9070	60	3	4834	49	5

RISK FACTOR	75-84 AT EXAM			85-94 AT EXAM			35-64 COMBINED			65-94 COMBINED		
	PERSON EXAMS	# OF EVENTS	ANNUAL RATE	PERSON EXAMS	# OF EVENTS	ANNUAL RATE	# OF EVENTS	CRUDE	AGE-ADJUSTED	# OF EVENTS	CRUDE	AGE-ADJUSTED
8-44	853	13	8	34			26	3	2	32	5	5
45-49	228	2	4	7			17	3	2	17	7	7
50-54	142	2	7	7			21	2		8	4	4
55-59	81	4	25	2			13	1	1	9	7	9
60-85	33	0	0	0			13	1	1	4	6	5
UNKNOWN	8	0	0	0			0	0	0	0	0	0
TOTAL	1345	21	8	50			90	2	2	70	6	6

LOGISTIC REGRESSIONS

AGE AT EXAM		COEFFICIENT OF RISK FACTOR	SE COEFFICIENT	P-VALUE	STANDARDIZED COEFFICIENT	STD SE
35-44						
45-54		0.006	0.022	0.773	0.057	0.198
55-64		-0.030	0.013	0.024	-0.282	0.125
65-74		0.010	0.016	0.516	0.094	0.145
75-84		0.018	0.024	0.456	0.164	0.220
85-94						
35-64	UNIVARIATE	-0.036	0.010	0.001	-0.344	0.101
65-94	UNIVARIATE	0.008	0.013	0.508	0.080	0.121
35-64	BIVARIATE	-0.015	0.011	0.182	-0.146	0.109
65-94	BIVARIATE	0.016	0.013	0.224	0.151	0.124
35-64	MULTIVARIATE	-0.006	0.012	0.629	-0.055	0.113
65-94	MULTIVARIATE	0.020	0.013	0.140	0.188	0.127

SEX: FEMALE
EVENT: SUDDEN DEATH FROM CORONARY HEART DISEASE
RISK FACTOR: VITAL CAPACITY - HEIGHT INDEX (ML/INCH)
PERSONS AT RISK: ALL PERSONS ALIVE

RISK FACTOR	35-44 AT EXAM			45-54 AT EXAM			55-64 AT EXAM			65-74 AT EXAM		
	PERSON EXAMS	# OF EVENTS	ANNUAL RATE	PERSON EXAMS	# OF EVENTS	ANNUAL RATE	PERSON EXAMS	# OF EVENTS	ANNUAL RATE	PERSON EXAMS	# OF EVENTS	ANNUAL RATE
4-29	122			546	0	0	1572	5	2	2068	14	3
30-34	291			1241	0	0	2547	3	1	1946	6	2
35-39	769			2103	2	0	3213	4	1	1728	5	1
40-44	1427			2838	2	0	2836	2	0	933	3	2
45-92	2989			3327	1	0	1914	1	0	422	0	0
UNKNOWN	21			17	0	0	21	0	0	6	0	0
TOTAL	5619			10072	5	0	12103	15	1	7103	28	2

RISK FACTOR	75-84 AT EXAM			85-94 AT EXAM			35-64 COMBINED			65-94 COMBINED		
	PERSON EXAMS	# OF EVENTS	ANNUAL RATE	PERSON EXAMS	# OF EVENTS	ANNUAL RATE	# OF EVENTS	CRUDE	AGE-ADJUSTED	# OF EVENTS	CRUDE	AGE-ADJUSTED
4-29	1212	10	4	109			5	1	1	26	3	3
30-34	686	10	7	37			3	0	0	16	3	3
35-39	370	3	4	16			6	0	0	8	3	2
40-44	123	0	0	1			4	0	0	3	1	1
45-92	26	0	0	0			2	0	0	0	0	0
UNKNOWN	2	0	0	0			0	0	0	0	0	0
TOTAL	2419	23	5	163			20	0	0	53	3	3

LOGISTIC REGRESSIONS

AGE AT EXAM		COEFFICIENT OF RISK FACTOR	SE COEFFICIENT	P-VALUE	STANDARDIZED COEFFICIENT	STD SE
35-44						
45-54		-0.011	0.060	0.851	-0.083	0.444
55-64		-0.073	0.032	0.022	-0.547	0.238
65-74		-0.064	0.025	0.009	-0.472	0.182
75-84		-0.026	0.029	0.378	-0.181	0.206
85-94						
35-64	UNIVARIATE	-0.080	0.026	0.002	-0.635	0.207
65-94	UNIVARIATE	-0.059	0.018	0.001	-0.447	0.133
35-64	BIVARIATE	-0.048	0.029	0.098	-0.381	0.230
65-94	BIVARIATE	-0.041	0.019	0.028	-0.312	0.142
35-64	MULTIVARIATE	-0.047	0.030	0.115	-0.375	0.238
65-94	MULTIVARIATE	-0.037	0.020	0.067	-0.276	0.151

AVERAGE ANNUAL INCIDENCE RATE FOR EVENT PER 1000 PERSONS AT RISK AT EXAM
BY AGE AND LEVEL OF RISK FACTOR AT EXAM: FRAMINGHAM STUDY 30-YEAR FOLLOWUP

SEX: MALE
EVENT: SUDDEN DEATH FROM CORONARY HEART DISEASE
RISK FACTOR: HEART RATE (PER MINUTE)
PERSONS AT RISK: ALL PERSONS ALIVE

RISK FACTOR	35-44 AT EXAM			45-54 AT EXAM			55-64 AT EXAM			65-74 AT EXAM		
	PERSON EXAMS	# OF EVENTS	ANNUAL RATE	PERSON EXAMS	# OF EVENTS	ANNUAL RATE	PERSON EXAMS	# OF EVENTS	ANNUAL RATE	PERSON EXAMS	# OF EVENTS	ANNUAL RATE
30-67	1295			2016	1	0	2618	13	2	1425	10	4
68-75	1506			2594	8	2	2763	12	2	1447	9	3
76-83	728			1473	6	2	1552	8	3	802	10	6
84-91	586			1145	5	2	1211	11	5	607	14	12
92-220	404			839	6	4	926	16	9	553	6	5
UNKNOWN	0			0	0	-	0	0	-	0	0	-
TOTAL	4519			8067	26	2	9070	60	3	4834	49	5

RISK FACTOR	75-84 AT EXAM			85-94 AT EXAM			35-64 COMBINED			65-94 COMBINED		
	PERSON EXAMS	# OF EVENTS	ANNUAL RATE	PERSON EXAMS	# OF EVENTS	ANNUAL RATE	# OF EVENTS	CRUDE	AGE-ADJUSTED	# OF EVENTS	CRUDE	AGE-ADJUSTED
30-67	388	4	5	17			14	1	1	14	4	4
68-75	362	5	7	14			20	1	1	14	4	4
76-83	271	5	9	13			15	2	2	15	7	7
84-91	174	5	14	5			18	3	3	19	12	12
92-220	150	2	7	1			23	5	5	8	6	6
UNKNOWN	0	0	-	0			0	-	-	0	-	-
TOTAL	1345	21	8	50			90	2	2	70	6	6

LOGISTIC REGRESSIONS

AGE AT EXAM		COEFFICIENT OF RISK FACTOR	SE COEFFICIENT	P-VALUE	STANDARDIZED COEFFICIENT	STD SE
35-44						
45-54		0.044	0.013	0.001	0.550	0.168
55-64		0.029	0.009	0.001	0.378	0.113
65-74		0.018	0.010	0.067	0.240	0.131
75-84		0.020	0.015	0.167	0.277	0.200
85-94						
35-64	UNIVARIATE	0.035	0.007	0.001	0.449	0.090
65-94	UNIVARIATE	0.019	0.008	0.020	0.254	0.109
35-64	BIVARIATE	0.034	0.007	0.001	0.438	0.089
65-94	BIVARIATE	0.019	0.008	0.022	0.252	0.110
35-64	MULTIVARIATE	0.027	0.007	0.001	0.342	0.092
65-94	MULTIVARIATE	0.018	0.008	0.033	0.238	0.111

SEX: FEMALE
EVENT: SUDDEN DEATH FROM CORONARY HEART DISEASE
RISK FACTOR: HEART RATE (PER MINUTE)
PERSONS AT RISK: ALL PERSONS ALIVE

RISK FACTOR	35-44 AT EXAM			45-54 AT EXAM			55-64 AT EXAM			65-74 AT EXAM		
	PERSON EXAMS	# OF EVENTS	ANNUAL RATE	PERSON EXAMS	# OF EVENTS	ANNUAL RATE	PERSON EXAMS	# OF EVENTS	ANNUAL RATE	PERSON EXAMS	# OF EVENTS	ANNUAL RATE
30-67	1092			1918	0	0	2450	2	0	1285	7	3
68-75	1806			3279	2	0	3641	4	1	1907	7	2
76-83	1051			1938	0	0	2442	2	0	1542	4	1
84-91	939			1620	1	0	1888	6	2	1196	1	0
92-220	731			1317	2	1	1682	1	0	1173	9	4
UNKNOWN	0			0	0	-	0	0	-	0	0	-
TOTAL	5619			10072	5	0	12103	15	1	7103	28	2

RISK FACTOR	75-84 AT EXAM			85-94 AT EXAM			35-64 COMBINED			65-94 COMBINED		
	PERSON EXAMS	# OF EVENTS	ANNUAL RATE	PERSON EXAMS	# OF EVENTS	ANNUAL RATE	# OF EVENTS	CRUDE	AGE-ADJUSTED	# OF EVENTS	CRUDE	AGE-ADJUSTED
30-67	459	5	5	36			2	0	0	13	4	4
68-75	586	5	4	41			6	0	0	12	2	2
76-83	532	0	0	30			2	0	0	4	1	1
84-91	410	5	6	29			7	1	1	7	2	2
92-220	432	8	9	27			3	0	0	17	5	5
UNKNOWN	0	0	-	0			0	-	-	0	-	-
TOTAL	2419	23	5	163			20	0	0	53	3	3

LOGISTIC REGRESSIONS

AGE AT EXAM		COEFFICIENT OF RISK FACTOR	SE COEFFICIENT	P-VALUE	STANDARDIZED COEFFICIENT	STD SE
35-44						
45-54		0.031	0.031	0.329	0.378	0.387
55-64		0.009	0.019	0.634	0.117	0.245
65-74		0.005	0.014	0.713	0.068	0.184
75-84		0.012	0.014	0.396	0.164	0.193
85-94						
35-64	UNIVARIATE	0.015	0.016	0.366	0.185	0.204
65-94	UNIVARIATE	0.007	0.010	0.452	0.099	0.132
35-64	BIVARIATE	0.014	0.016	0.386	0.175	0.201
65-94	BIVARIATE	0.007	0.010	0.492	0.090	0.131
35-64	MULTIVARIATE	0.010	0.017	0.555	0.125	0.211
65-94	MULTIVARIATE	0.000	0.010	0.964	0.006	0.135

SEX: MALE
EVENT: SUDDEN DEATH FROM CORONARY HEART DISEASE
RISK FACTOR: CIGARETTES SMOKED (NUMBER PER DAY)
PERSONS AT RISK: ALL PERSONS ALIVE

RISK FACTOR	35-44 AT EXAM			45-54 AT EXAM			55-64 AT EXAM			65-74 AT EXAM		
	PERSON EXAMS	# OF EVENTS	ANNUAL RATE	PERSON EXAMS	# OF EVENTS	ANNUAL RATE	PERSON EXAMS	# OF EVENTS	ANNUAL RATE	PERSON EXAMS	# OF EVENTS	ANNUAL RATE
NONE	1402			3220	4	1	4888	27	3	3235	30	5
1-10	437			746	5	3	856	7	4	463	6	6
11-20	1499			2136	9	2	1889	16	4	668	6	4
21-40	993			1672	6	2	1235	8	3	427	5	6
41-90	162			255	2	4	178	2	6	54	1	15
UNKNOWN	26			38	0	0	24	0	0	7	1	71
TOTAL	4519			8067	26	2	9070	60	3	4834	49	5

RISK FACTOR	75-84 AT EXAM			85-94 AT EXAM			35-64 COMBINED			65-94 COMBINED		
	PERSON EXAMS	# OF EVENTS	ANNUAL RATE	PERSON EXAMS	# OF EVENTS	ANNUAL RATE	# OF EVENTS	CRUDE	AGE-ADJUSTED	# OF EVENTS	CRUDE	AGE-ADJUSTED
NONE	1067	18	8	38			31	2	1	48	6	5
1-10	125	2	8	6			13	3	3	8	7	7
11-20	88	1	6	5			26	2	3	7	5	5
21-40	60	0	0	1			16	2	2	5	5	5
41-90	5	0	0	0			4	3	4	1	13	11
UNKNOWN	0	0	-	0			0	0	0	1	71	55
TOTAL	1345	21	8	50			90	2	2	70	6	6

LOGISTIC REGRESSIONS

AGE AT EXAM		COEFFICIENT OF RISK FACTOR	SE COEFFICIENT	P-VALUE	STANDARDIZED COEFFICIENT	STD SE
35-44						
45-54		0.019	0.012	0.115	0.280	0.178
55-64		0.007	0.009	0.414	0.099	0.122
65-74		0.005	0.012	0.700	0.053	0.139
75-84		-0.043	0.043	0.309	-0.381	0.374
85-94						
35-64	UNIVARIATE	0.006	0.007	0.418	0.083	0.102
65-94	UNIVARIATE	-0.004	0.011	0.755	-0.039	0.126
35-64	BIVARIATE	0.014	0.007	0.040	0.204	0.099
65-94	BIVARIATE	0.001	0.011	0.959	0.007	0.126
35-64	MULTIVARIATE	0.016	0.007	0.021	0.231	0.100
65-94	MULTIVARIATE	0.004	0.011	0.708	0.047	0.127

SEX: FEMALE
EVENT: SUDDEN DEATH FROM CORONARY HEART DISEASE
RISK FACTOR: CIGARETTES SMOKED (NUMBER PER DAY)
PERSONS AT RISK: ALL PERSONS ALIVE

RISK FACTOR	35-44 AT EXAM			45-54 AT EXAM			55-64 AT EXAM			65-74 AT EXAM		
	PERSON EXAMS	# OF EVENTS	ANNUAL RATE	PERSON EXAMS	# OF EVENTS	ANNUAL RATE	PERSON EXAMS	# OF EVENTS	ANNUAL RATE	PERSON EXAMS	# OF EVENTS	ANNUAL RATE
NONE	2583			5599	1	0	8010	9	1	5562	20	2
1-10	1314			1732	1	0	1646	4	1	680	2	1
11-20	1373			2037	1	0	1767	2	1	650	3	4
21-40	298			629	1	1	624	0	0	191	0	0
41-90	22			38	1	13	18	0	0	1	0	0
UNKNOWN	29			37	0	0	38	0	0	19	1	26
TOTAL	5619			10072	5	0	12103	15	1	7103	28	2

RISK FACTOR	75-84 AT EXAM			85-94 AT EXAM			35-64 COMBINED			65-94 COMBINED		
	PERSON EXAMS	# OF EVENTS	ANNUAL RATE	PERSON EXAMS	# OF EVENTS	ANNUAL RATE	# OF EVENTS	CRUDE	AGE-ADJUSTED	# OF EVENTS	CRUDE	AGE-ADJUSTED
NONE	2139	21	5	154			10	0	0	43	5	3
1-10	161	1	3	7			5	1	1	3	2	2
11-20	99	1	5	2			3	0	0	6	4	4
21-40	16	0	0	0			1	0	0	0	0	0
41-90	0	0	-	0			1	6	5	0	0	0
UNKNOWN	4	0	0	0			0	0	0	1	22	19
TOTAL	2419	23	5	163			20	0	0	53	3	3

LOGISTIC REGRESSIONS

AGE AT EXAM		COEFFICIENT OF RISK FACTOR	SE COEFFICIENT	P-VALUE	STANDARDIZED COEFFICIENT	STD SE
35-44						
45-54		0.077	0.028	0.007	0.790	0.293
55-64		-0.008	0.029	0.791	-0.073	0.277
65-74		0.018	0.021	0.391	0.138	0.161
75-84		-0.026	0.056	0.641	-0.126	0.270
85-94						
35-64	UNIVARIATE	0.019	0.020	0.320	0.193	0.194
65-94	UNIVARIATE	0.000	0.020	0.981	0.003	0.158
35-64	BIVARIATE	0.029	0.019	0.122	0.288	0.186
65-94	BIVARIATE	0.014	0.020	0.472	0.100	0.139
35-64	MULTIVARIATE	0.029	0.019	0.115	0.291	0.185
65-94	MULTIVARIATE	0.018	0.020	0.372	0.124	0.139

TABLE 10-14   DATE=05JAN86
AVERAGE ANNUAL INCIDENCE RATE FOR EVENT PER 1000 PERSONS AT RISK AT EXAM
BY AGE AND LEVEL OF RISK FACTOR AT EXAM: FRAMINGHAM STUDY 30-YEAR FOLLOWUP

SEX: MALE
EVENT: SUDDEN DEATH FROM CORONARY HEART DISEASE
RISK FACTOR: ALBUMIN IN URINE, DEFINITE
PERSONS AT RISK: ALL PERSONS ALIVE

RISK FACTOR	35-44 AT EXAM PERSON EXAMS	# OF EVENTS	ANNUAL RATE	45-54 AT EXAM PERSON EXAMS	# OF EVENTS	ANNUAL RATE	55-64 AT EXAM PERSON EXAMS	# OF EVENTS	ANNUAL RATE	65-74 AT EXAM PERSON EXAMS	# OF EVENTS	ANNUAL RATE
NO	4411			7936	25	2	8828	58	3	4671	45	5
YES	71			105	1	5	215	2	5	157	4	13
UNKNOWN	37			26	0	0	27	0	0	6	0	0
TOTAL	4519			8067	26	2	9070	60	3	4834	49	5

RISK FACTOR	75-84 AT EXAM PERSON EXAMS	# OF EVENTS	ANNUAL RATE	85-94 AT EXAM PERSON EXAMS	# OF EVENTS	ANNUAL RATE	35-64 COMBINED # OF EVENTS	CRUDE	AGE-ADJUSTED	65-94 COMBINED # OF EVENTS	CRUDE	AGE-ADJUSTED
NO	1308	21	8	43			87	2	2	66	5	5
YES	34	0	0	7			3	4	4	4	10	10
UNKNOWN	3	0	0	0			0	0	0	0	0	0
TOTAL	1345	21	8	50			90	2	2	70	6	6

LOGISTIC REGRESSIONS

AGE AT EXAM		COEFFICIENT OF RISK FACTOR	SE COEFFICIENT	P-VALUE	STANDARDIZED COEFFICIENT	STD SE
35-44						
45-54		1.113	1.025	0.277	0.126	0.116
55-64		0.351	0.723	0.628	0.053	0.110
65-74		0.989	0.528	0.061	0.175	0.094
75-84		-	-	-	-	-
85-94						
35-64	UNIVARIATE	0.628	0.589	0.287	0.084	0.079
65-94	UNIVARIATE	0.621	0.520	0.232	0.109	0.091
35-64	BIVARIATE	0.441	0.591	0.456	0.059	0.079
65-94	BIVARIATE	0.600	0.521	0.249	0.105	0.091
35-64	MULTIVARIATE	0.038	0.604	0.950	0.005	0.081
65-94	MULTIVARIATE	0.340	0.530	0.522	0.060	0.093

SEX: FEMALE
EVENT: SUDDEN DEATH FROM CORONARY HEART DISEASE
RISK FACTOR: ALBUMIN IN URINE, DEFINITE
PERSONS AT RISK: ALL PERSONS ALIVE

RISK FACTOR	35-44 AT EXAM PERSON EXAMS	# OF EVENTS	ANNUAL RATE	45-54 AT EXAM PERSON EXAMS	# OF EVENTS	ANNUAL RATE	55-64 AT EXAM PERSON EXAMS	# OF EVENTS	ANNUAL RATE	65-74 AT EXAM PERSON EXAMS	# OF EVENTS	ANNUAL RATE
NO	5356			9766	5	0	11835	14	1	6955	25	2
YES	225			281	0	0	240	1	2	138	3	11
UNKNOWN	38			25	0	0	28	0	0	10	0	0
TOTAL	5619			10072	5	0	12103	15	1	7103	28	2

RISK FACTOR	75-84 AT EXAM PERSON EXAMS	# OF EVENTS	ANNUAL RATE	85-94 AT EXAM PERSON EXAMS	# OF EVENTS	ANNUAL RATE	35-64 COMBINED # OF EVENTS	CRUDE	AGE-ADJUSTED	65-94 COMBINED # OF EVENTS	CRUDE	AGE-ADJUSTED
NO	2360	23	5	156			19	0	0	50	3	3
YES	50	0	0	7			1	1	1	3	8	8
UNKNOWN	9	0	0	0			0	0	0	0	0	0
TOTAL	2419	23	5	163			20	0	0	53	3	3

LOGISTIC REGRESSIONS

AGE AT EXAM		COEFFICIENT OF RISK FACTOR	SE COEFFICIENT	P-VALUE	STANDARDIZED COEFFICIENT	STD SE
35-44						
45-54		-	-	-	-	-
55-64		1.262	1.037	0.224	0.176	0.145
65-74		1.818	0.617	0.003	0.251	0.085
75-84		-	-	-	-	-
85-94						
35-64	UNIVARIATE	0.643	1.027	0.531	0.104	0.166
65-94	UNIVARIATE	1.080	0.599	0.071	0.152	0.084
35-64	BIVARIATE	0.901	1.028	0.381	0.146	0.166
65-94	BIVARIATE	0.986	0.602	0.102	0.139	0.085
35-64	MULTIVARIATE	0.958	1.040	0.357	0.155	0.168
65-94	MULTIVARIATE	0.598	0.617	0.333	0.084	0.087

TABLE 10-15   DATE=05JAN86

AVERAGE ANNUAL INCIDENCE RATE FOR EVENT PER 1000 PERSONS AT RISK AT EXAM
BY AGE AND LEVEL OF RISK FACTOR AT EXAM: FRAMINGHAM STUDY 30-YEAR FOLLOWUP

SEX: MALE
EVENT: SUDDEN DEATH FROM CORONARY HEART DISEASE
RISK FACTOR: HEART ENLARGEMENT BY X-RAY
PERSONS AT RISK: ALL PERSONS ALIVE

RISK FACTOR	35-44 AT EXAM PERSON EXAMS	# OF EVENTS	ANNUAL RATE	45-54 AT EXAM PERSON EXAMS	# OF EVENTS	ANNUAL RATE	55-64 AT EXAM PERSON EXAMS	# OF EVENTS	ANNUAL RATE	65-74 AT EXAM PERSON EXAMS	# OF EVENTS	ANNUAL RATE
NO	4101			6613	17	1	6543	31	2	3075	21	3
MILD	189			708	2	1	1126	6	3	703	5	4
DEFINITE	224			742	7	5	1400	23	3	1056	23	11
UNKNOWN	5			4	0	0	1	0	0	0	0	-
TOTAL	4519			8067	26	2	9070	60	3	4834	49	5

RISK FACTOR	75-84 AT EXAM PERSON EXAMS	# OF EVENTS	ANNUAL RATE	85-94 AT EXAM PERSON EXAMS	# OF EVENTS	ANNUAL RATE	35-64 COMBINED # OF EVENTS	CRUDE	AGE-ADJUSTED	65-94 COMBINED # OF EVENTS	CRUDE	AGE-ADJUSTED
NO	753	9	6	34			51	1	2	30	4	4
MILD	226	4	9	6			8	2	2	9	5	5
DEFINITE	366	8	11	10			31	7	6	31	11	11
UNKNOWN	0	0	-	0			0	0	0	0	-	
TOTAL	1345	21	8	50			90	2	2	70	6	6

LOGISTIC REGRESSIONS

AGE AT EXAM	COEFFICIENT OF RISK FACTOR	SE COEFFICIENT	P-VALUE	STANDARDIZED COEFFICIENT	STD SE
35-44					
45-54	0.625	0.231	0.007	0.386	0.143
55-64	0.617	0.143	0.001	0.460	0.106
65-74	0.590	0.157	0.001	0.487	0.130
75-84	0.308	0.243	0.205	0.267	0.210
85-94					
35-64 UNIVARIATE	0.736	0.117	0.001	0.485	0.077
65-94 UNIVARIATE	0.520	0.132	0.001	0.434	0.110
35-64 BIVARIATE	0.597	0.120	0.001	0.393	0.079
65-94 BIVARIATE	0.498	0.132	0.001	0.416	0.111
35-64 MULTIVARIATE	0.539	0.130	0.001	0.355	0.085
65-94 MULTIVARIATE	0.377	0.143	0.008	0.315	0.120

SEX: FEMALE
EVENT: SUDDEN DEATH FROM CORONARY HEART DISEASE
RISK FACTOR: HEART ENLARGEMENT BY X-RAY
PERSONS AT RISK: ALL PERSONS ALIVE

RISK FACTOR	35-44 AT EXAM PERSON EXAMS	# OF EVENTS	ANNUAL RATE	45-54 AT EXAM PERSON EXAMS	# OF EVENTS	ANNUAL RATE	55-64 AT EXAM PERSON EXAMS	# OF EVENTS	ANNUAL RATE	65-74 AT EXAM PERSON EXAMS	# OF EVENTS	ANNUAL RATE
NO	5166			8085	3	0	7836	5	0	3511	6	1
MILD	199			959	1	1	1879	3	1	1293	6	2
DEFINITE	246			1022	1	0	2384	7	1	2299	16	3
UNKNOWN	8			6	0	0	4	0	0	0	0	-
TOTAL	5619			10072	5	0	12103	15	1	7103	28	2

RISK FACTOR	75-84 AT EXAM PERSON EXAMS	# OF EVENTS	ANNUAL RATE	85-94 AT EXAM PERSON EXAMS	# OF EVENTS	ANNUAL RATE	35-64 COMBINED # OF EVENTS	CRUDE	AGE-ADJUSTED	65-94 COMBINED # OF EVENTS	CRUDE	AGE-ADJUSTED
NO	830	1	1	40			8	0	0	8	1	1
MILD	455	4	4	36			4	1	1	10	3	3
DEFINITE	1134	18	8	87			8	1	1	35	5	5
UNKNOWN	0	0	-	0			0	0	0	0	-	
TOTAL	2419	23	5	163			20	0	0	53	5	5

LOGISTIC REGRESSIONS

AGE AT EXAM	COEFFICIENT OF RISK FACTOR	SE COEFFICIENT	P-VALUE	STANDARDIZED COEFFICIENT	STD SE
35-44					
45-54	0.538	0.531	0.311	0.346	0.341
55-64	0.760	0.288	0.008	0.608	0.231
65-74	0.675	0.226	0.003	0.600	0.201
75-84	1.055	0.359	0.003	0.942	0.320
85-94					
35-64 UNIVARIATE	0.878	0.242	0.001	0.619	0.171
65-94 UNIVARIATE	0.810	0.180	0.001	0.728	0.162
35-64 BIVARIATE	0.634	0.251	0.011	0.447	0.177
65-94 BIVARIATE	0.720	0.182	0.001	0.647	0.164
35-64 MULTIVARIATE	0.706	0.260	0.007	0.498	0.184
65-94 MULTIVARIATE	0.639	0.188	0.001	0.575	0.169

TABLE 10-16   DATE=05JAN86
AVERAGE ANNUAL INCIDENCE RATE FOR EVENT PER 1000 PERSONS AT RISK AT EXAM
BY AGE AND LEVEL OF RISK FACTOR AT EXAM' FRAMINGHAM STUDY 30-YEAR FOLLOWUP

SEX: MALE
EVENT: SUDDEN DEATH FROM CORONARY HEART DISEASE
RISK FACTOR: LEFT VENTRICULAR HYPERTROPHY BY ELECTROCARDIOGRAM
PERSONS AT RISK: ALL PERSONS ALIVE

RISK FACTOR	35-44 AT EXAM			45-54 AT EXAM			55-64 AT EXAM			65-74 AT EXAM		
	PERSON EXAMS	# OF EVENTS	ANNUAL RATE	PERSON EXAMS	# OF EVENTS	ANNUAL RATE	PERSON EXAMS	# OF EVENTS	ANNUAL RATE	PERSON EXAMS	# OF EVENTS	ANNUAL RATE
NO	4408			7812	24	2	8611	45	3	4527	36	4
MILD	79			147	1	3	240	3	6	121	4	17
DEFINITE	32			108	1	5	219	12	27	186	9	24
UNKNOWN	0		'	0	0	-	0	0	-	0	0	-
TOTAL	4519			8067	26	2	9070	60	3	4834	49	5

RISK FACTOR	75-84 AT EXAM			85-94 AT EXAM			35-64 COMBINED			65-94 COMBINED		
	PERSON EXAMS	# OF EVENTS	ANNUAL RATE	PERSON EXAMS	# OF EVENTS	ANNUAL RATE	# OF EVENTS	CRUDE	AGE-ADJUSTED	# OF EVENTS	CRUDE	AGE-ADJUSTED
NO	1250	20	8	45			73	2	2	56	5	5
MILD	44	0	0	2			4	4	4	12	13
DEFINITE	51	1	10	3			13	18	13	10	21	21
UNKNOWN	0	0	-	0			0	-	-	0	-	
TOTAL	1345	21	8	50			90	2	2	70	6	6

LOGISTIC REGRESSIONS

AGE AT EXAM		COEFFICIENT OF RISK FACTOR	SE COEFFICIENT	P-VALUE	STANDARDIZED COEFFICIENT	STD SE
35-44						
45-54		0.596	0.457	0.192	0.157	0.121
55-64		1.184	0.166	0.001	0.406	0.057
65-74		0.955	0.182	0.001	0.392	0.075
75-84		-0.084	0.565	0.882	-0.035	0.235
85-94						
35-64	UNIVARIATE	1.165	0.151	0.001	0.339	0.044
65-94	UNIVARIATE	0.762	0.169	0.001	0.314	0.070
35-64	BIVARIATE	1.030	0.153	0.001	0.300	0.045
65-94	BIVARIATE	0.759	0.169	0.001	0.313	0.070
35-64	MULTIVARIATE	1.093	0.172	0.001	0.318	0.050
65-94	MULTIVARIATE	0.584	0.189	0.002	0.241	0.078

SEX: FEMALE
EVENT: SUDDEN DEATH FROM CORONARY HEART DISEASE
RISK FACTOR: LEFT VENTRICULAR HYPERTROPHY BY ELECTROCARDIOGRAM
PERSONS AT RISK: ALL PERSONS ALIVE

RISK FACTOR	35-44 AT EXAM			45-54 AT EXAM			55-64 AT EXAM			65-74 AT EXAM		
	PERSON EXAMS	# OF EVENTS	ANNUAL RATE	PERSON EXAMS	# OF EVENTS	ANNUAL RATE	PERSON EXAMS	# OF EVENTS	ANNUAL RATE	PERSON EXAMS	# OF EVENTS	ANNUAL RATE
NO	5566			9905	5	0	11479	15	1	6686	22	2
MILD	38			111	0	0	250	0	0	182	3	8
DEFINITE	15			56	0	0	174	0	0	235	3	6
UNKNOWN	0			0	0	-	0	0	-	0	0	-
TOTAL	5619			10072	5	0	12103	15	1	7103	28	2

RISK FACTOR	75-84 AT EXAM			85-94 AT EXAM			35-64 COMBINED			65-94 COMBINED		
	PERSON EXAMS	# OF EVENTS	ANNUAL RATE	PERSON EXAMS	# OF EVENTS	ANNUAL RATE	# OF EVENTS	CRUDE	AGE-ADJUSTED	# OF EVENTS	CRUDE	AGE-ADJUSTED
NO	2217	18	4	142			20	0	0	42	2	2
MILD	77	2	13	10			0	0	0	5	9	9
DEFINITE	125	3	12	11			0	0	0	6	8	8
UNKNOWN	0	0	-	0			0	-	-	0	-	
TOTAL	2419	23	5	163			20	0	0	53	3	3

LOGISTIC REGRESSIONS

AGE AT EXAM		COEFFICIENT OF RISK FACTOR	SE COEFFICIENT	P-VALUE	STANDARDIZED COEFFICIENT	STD SE
35-44		-	-	-	-	-
45-54		-	-	-	-	-
55-64		-	-	-	-	-
65-74		0.787	0.272	0.004	0.304	0.105
75-84		0.605	0.291	0.038	0.284	0.137
85-94						
35-64	UNIVARIATE	-	-	-	-	-
65-94	UNIVARIATE	0.709	0.198	0.001	0.292	0.082
35-64	BIVARIATE	-	-	-	-	-
65-94	BIVARIATE	0.632	0.200	0.002	0.261	0.082
35-64	MULTIVARIATE	-	-	-	-	-
65-94	MULTIVARIATE	0.492	0.212	0.020	0.203	0.087

TABLE 10-17   DATE=05JAN86
AVERAGE ANNUAL ·INCIDENCE RATE FOR EVENT PER 1000 PERSONS AT RISK AT EXAM
BY AGE AND LEVEL OF RISK FACTOR AT EXAM: FRAMINGHAM STUDY 30-YEAR FOLLOWUP

SEX: MALE
EVENT: SUDDEN DEATH FROM CORONARY HEART DISEASE
RISK FACTOR: INTRAVENTRICULAR CONDUCTION DEFECT (WITH QRS INTERVAL MORE THAN 0.11 SECOND)
PERSONS AT RISK: ALL PERSONS ALIVE

RISK FACTOR	35-44 AT EXAM			45-54 AT EXAM			55-64 AT EXAM			65-74 AT EXAM		
	PERSON EXAMS	# OF EVENTS	ANNUAL RATE	PERSON EXAMS	# OF EVENTS	ANNUAL RATE	PERSON EXAMS	# OF EVENTS	ANNUAL RATE	PERSON EXAMS	# OF EVENTS	ANNUAL RATE
NO	4201			7487	22	1	8247	52	3	4323	37	4
INCOMPLETE	280			491	1	1	541	5	5	250	4	8
COMPLETE	38			89	3	17	282	3	5	261	8	15
UNKNOWN	0			0	0	-	0	0	-	0	0	-
TOTAL	4519			8067	26	2	9070	60	3	4834	49	5

RISK FACTOR	75-84 AT EXAM			85-94 AT EXAM			35-64 COMBINED			65-94 COMBINED		
	PERSON EXAMS	# OF EVENTS	ANNUAL RATE	PERSON EXAMS	# OF EVENTS	ANNUAL RATE	# OF EVENTS	CRUDE	AGE-ADJUSTED	# OF EVENTS	CRUDE	AGE-ADJUSTED
NO	1180	19	8	41			78	2	2	56	5	5
INCOMPLETE	57	0	0	3			6	2	2	4	7	6
COMPLETE	128	2	8	6			6	7	9	10	13	14
UNKNOWN	0	0	-	0			0	-	-	0	-	-
TOTAL	1345	21	8	50			90	2	2	70	6	6

LOGISTIC REGRESSIONS

AGE AT EXAM	COEFFICIENT OF RISK FACTOR	SE COEFFICIENT	P-VALUE	STANDARDIZED COEFFICIENT	STD SE
35-44					
45-54	0.976	0.345	0.005	0.306	0.108
55-64	0.295	0.256	0.249	0.121	0.105
65-74	0.648	0.192	0.001	0.319	0.095
75-84	-0.082	0.389	0.833	-0.049	0.234
85-94					
35-64 UNIVARIATE	0.546	0.206	0.008	0.194	0.073
65-94 UNIVARIATE	0.455	0.170	0.008	0.236	0.089
35-64 BIVARIATE	0.415	0.203	0.041	0.148	0.072
65-94 BIVARIATE	0.416	0.171	0.015	0.216	0.089
35-64 MULTIVARIATE	0.389	0.211	0.066	0.138	0.075
65-94 MULTIVARIATE	0.431	0.173	0.013	0.224	0.090

SEX: FEMALE
EVENT: SUDDEN DEATH FROM CORONARY HEART DISEASE
RISK FACTOR: INTRAVENTRICULAR CONDUCTION DEFECT (WITH QRS INTERVAL MORE THAN 0.11 SECOND)
PERSONS AT RISK: ALL PERSONS ALIVE

RISK FACTOR	35-44 AT EXAM			45-54 AT EXAM			55-64 AT EXAM			65-74 AT EXAM		
	PERSON EXAMS	# OF EVENTS	ANNUAL RATE	PERSON EXAMS	# OF EVENTS	ANNUAL RATE	PERSON EXAMS	# OF EVENTS	ANNUAL RATE	PERSON EXAMS	# OF EVENTS	ANNUAL RATE
NO	5532			9873	5	0	11731	14	1	6794	23	2
INCOMPLETE	75			151	0	0	195	0	0	134	0	0
COMPLETE	12			48	0	0	177	1	3	175	5	14
UNKNOWN	0			0	0	-	0	0	-	0	0	-
TOTAL	5619			10072	5	0	12103	15	1	7103	28	2

RISK FACTOR	75-84 AT EXAM			85-94 AT EXAM			35-64 COMBINED			65-94 COMBINED		
	PERSON EXAMS	# OF EVENTS	ANNUAL RATE	PERSON EXAMS	# OF EVENTS	ANNUAL RATE	# OF EVENTS	CRUDE	AGE-ADJUSTED	# OF EVENTS	CRUDE	AGE-ADJUSTED
NO	2261	19	4	153			19	0	0	44	2	2
INCOMPLETE	51	0	0	3			0	0	0	0	0	0
COMPLETE	107	4	19	7			1	2	1	9	16	15
UNKNOWN	0	0	-	0			0	-	-	0	-	-
TOTAL	2419	23	5	163			20	0	0	53	3	3

LOGISTIC REGRESSIONS

AGE AT EXAM	COEFFICIENT OF RISK FACTOR	SE COEFFICIENT	P-VALUE	STANDARDIZED COEFFICIENT	STD SE
35-44					
45-54					
55-64	0.662	0.552	0.231	0.178	0.149
65-74	1.026	0.259	0.001	0.344	0.087
75-84	0.714	0.288	0.013	0.308	0.124
85-94					
35-64 UNIVARIATE	0.729	0.560	0.193	0.160	0.123
65-94 UNIVARIATE	0.898	0.192	0.001	0.327	0.070
35-64 BIVARIATE	0.501	0.555	0.366	0.110	0.122
65-94 BIVARIATE	0.823	0.193	0.001	0.300	0.070
35-64 MULTIVARIATE	0.537	0.554	0.333	0.118	0.122
65-94 MULTIVARIATE	0.794	0.196	0.001	0.289	0.071

AVERAGE ANNUAL INCIDENCE RATE FOR EVENT PER 1000 PERSONS AT RISK AT EXAM
BY AGE AND LEVEL OF RISK FACTOR AT EXAM: FRAMINGHAM STUDY 30-YEAR FOLLOWUP

SEX: MALE
EVENT: SUDDEN DEATH FROM CORONARY HEART DISEASE
RISK FACTOR: DEFINITE NONSPECIFIC T-WAVE OR ST-SEGMENT ABNORMALITY BY ELECTROCARDIOGRAM
PERSONS AT RISK: ALL PERSONS ALIVE

RISK FACTOR	35-44 AT EXAM PERSON EXAMS	# OF EVENTS	ANNUAL RATE	45-54 AT EXAM PERSON EXAMS	# OF EVENTS	ANNUAL RATE	55-64 AT EXAM PERSON EXAMS	# OF EVENTS	ANNUAL RATE	65-74 AT EXAM PERSON EXAMS	# OF EVENTS	ANNUAL RATE
NO	4406			7641	21	1	8246	46	3	4170	35	4
YES	113			426	5	6	824	14	8	664	14	11
UNKNOWN	0			0	0	-	0	0	-	0	0	-
TOTAL	4519			8067	26	2	9070	60	3	4834	49	5

RISK FACTOR	75-84 AT EXAM PERSON EXAMS	# OF EVENTS	ANNUAL RATE	85-94 AT EXAM PERSON EXAMS	# OF EVENTS	ANNUAL RATE	35-64 COMBINED # OF EVENTS	CRUDE	AGE-ADJUSTED	65-94 COMBINED # OF EVENTS	CRUDE	AGE-ADJUSTED
NO	1063	15	7	31			71	2	2	50	5	5
YES	282	6	11	19			19	7	6	20	10	10
UNKNOWN	0	0	-	0			0	-		0	-	
TOTAL	1345	21	8	50			90	2	2	70	6	6

LOGISTIC REGRESSIONS

AGE AT EXAM		COEFFICIENT OF RISK FACTOR	SE COEFFICIENT	P-VALUE	STANDARDIZED COEFFICIENT	STD SE
35-44						
45-54		1.461	0.500	0.003	0.327	0.112
55-64		1.125	0.307	0.001	0.323	0.088
65-74		0.934	0.319	0.003	0.322	0.110
75-84		0.418	0.488	0.392	0.170	0.199
85-94						
35-64	UNIVARIATE	1.393	0.260	0.001	0.338	0.063
65-94	UNIVARIATE	0.792	0.267	0.003	0.286	0.097
35-64	BIVARIATE	1.121	0.263	0.001	0.272	0.064
65-94	BIVARIATE	0.723	0.269	0.007	0.262	0.097
35-64	MULTIVARIATE	0.928	0.266	0.001	0.225	0.065
65-94	MULTIVARIATE	0.723	0.274	0.008	0.262	0.099

SEX: FEMALE
EVENT: SUDDEN DEATH FROM CORONARY HEART DISEASE
RISK FACTOR: DEFINITE NONSPECIFIC T-WAVE OR ST-SEGMENT ABNORMALITY BY ELECTROCARDIOGRAM
PERSONS AT RISK: ALL PERSONS ALIVE

RISK FACTOR	35-44 AT EXAM PERSON EXAMS	# OF EVENTS	ANNUAL RATE	45-54 AT EXAM PERSON EXAMS	# OF EVENTS	ANNUAL RATE	55-64 AT EXAM PERSON EXAMS	# OF EVENTS	ANNUAL RATE	65-74 AT EXAM PERSON EXAMS	# OF EVENTS	ANNUAL RATE
NO	5520			9642	4	0	11111	12	1	6163	18	1
YES	99			430	1	1	992	3	2	940	10	5
UNKNOWN	0			0	0	-	0	0	-	0	0	-
TOTAL	5619			10072	5	0	12103	15	1	7103	28	2

RISK FACTOR	75-84 AT EXAM PERSON EXAMS	# OF EVENTS	ANNUAL RATE	85-94 AT EXAM PERSON EXAMS	# OF EVENTS	ANNUAL RATE	35-64 COMBINED # OF EVENTS	CRUDE	AGE-ADJUSTED	65-94 COMBINED # OF EVENTS	CRUDE	AGE-ADJUSTED
NO	1945	18	5	124			16	0	0	38	2	2
YES	474	5	5	39			4	1	1	15	5	5
UNKNOWN	0	0	-	0			0	-		0	-	
TOTAL	2419	23	5	163			20	0	0	53	3	3

LOGISTIC REGRESSIONS

AGE AT EXAM		COEFFICIENT OF RISK FACTOR	SE COEFFICIENT	P-VALUE	STANDARDIZED COEFFICIENT	STD SE
35-44						
45-54		1.726	1.119	0.123	0.349	0.226
55-64		1.032	0.646	0.110	0.283	0.177
65-74		1.300	0.396	0.001	0.441	0.134
75-84		0.132	0.508	0.795	0.052	0.202
85-94						
35-64	UNIVARIATE	1.465	0.560	0.009	0.333	0.127
65-94	UNIVARIATE	0.811	0.306	0.008	0.289	0.109
35-64	BIVARIATE	1.055	0.564	0.061	0.240	0.128
65-94	BIVARIATE	0.669	0.309	0.030	0.239	0.110
35-64	MULTIVARIATE	1.082	0.569	0.057	0.246	0.129
65-94	MULTIVARIATE	0.640	0.311	0.040	0.229	0.111

TABLE 11-1   DATE=12APR85
AVERAGE ANNUAL INCIDENCE RATE FOR EVENT PER 1000 PERSONS AT RISK AT EXAM
BY AGE AND LEVEL OF RISK FACTOR AT EXAM: FRAMINGHAM STUDY 30-YEAR FOLLOWUP

SEX: MALE
EVENT: CORONARY HEART DISEASE DEATH
RISK FACTOR: SYSTOLIC BLOOD PRESSURE, FIRST EXAMINER (MM HG)
PERSONS AT RISK: PERSONS FREE OF CORONARY HEART DISEASE

RISK FACTOR	35-44 AT EXAM			45-54 AT EXAM			55-64 AT EXAM			65-74 AT EXAM		
	PERSON EXAMS	# OF EVENTS	ANNUAL RATE	PERSON EXAMS	# OF EVENTS	ANNUAL RATE	PERSON EXAMS	# OF EVENTS	ANNUAL RATE	PERSON EXAMS	# OF EVENTS	ANNUAL RATE
74-119	1223	2	1	1769	3	1	1440	4	1	548	2	2
120-139	2165	3	1	3417	12	2	3166	20	3	1341	14	5
140-159	841	0	0	1732	6	2	2090	18	4	1194	15	6
160-179	191	0	0	567	4	4	916	10	5	552	12	11
180-300	52	0	0	239	3	6	437	8	9	278	11	20
UNKNOWN	0	0	-	0	0	-	0	0	-	0	0	-
TOTAL	4472	5	1	7724	28	2	8049	60	4	3913	54	7

RISK FACTOR	75-84 AT EXAM			85-94 AT EXAM			35-64 COMBINED			65-94 COMBINED		
	PERSON EXAMS	# OF EVENTS	ANNUAL RATE	PERSON EXAMS	# OF EVENTS	ANNUAL RATE	# OF EVENTS	CRUDE	AGE-ADJUSTED	# OF EVENTS	CRUDE	AGE-ADJUSTED
74-119	120	3	13	7	.		9	1	1	5	4	4
120-139	345	5	7	15	.		35	2	2	19	6	5
140-159	315	6	10	11	.		24	3	2	21	7	7
160-179	153	5	16	4	.		14	4	4	19	13	14
180-300	84	4	24	2	.		11	8	6	15	21	21
UNKNOWN	0	0	-	0	.		0	-		0	-	
TOTAL	1017	23	11	39	.		93	2	3	79	8	8

LOGISTIC REGRESSIONS

AGE AT EXAM		COEFFICIENT OF RISK FACTOR	SE COEFFICIENT	P-VALUE	STANDARDIZED COEFFICIENT	STD SE
35-44		-0.044	0.034	0.189	-0.743	0.565
45-54		0.023	0.008	0.003	0.450	0.149
55-64		0.020	0.005	0.001	0.446	0.103
65-74		0.022	0.005	0.001	0.495	0.110
75-84		0.012	0.009	0.183	0.263	0.198
85-94						
35-64	UNIVARIATE	0.023	0.004	0.001	0.473	0.079
65-94	UNIVARIATE	0.020	0.004	0.001	0.446	0.095
35-64	BIVARIATE	0.019	0.004	0.001	0.384	0.081
65-94	BIVARIATE	0.020	0.004	0.001	0.457	0.097
35-64	MULTIVARIATE	0.016	0.004	0.001	0.332	0.089
65-94	MULTIVARIATE	0.016	0.005	0.001	0.356	0.103

SEX: FEMALE
EVENT: CORONARY HEART DISEASE DEATH
RISK FACTOR: SYSTOLIC BLOOD PRESSURE, FIRST EXAMINER (MM HG)
PERSONS AT RISK: PERSONS FREE OF CORONARY HEART DISEASE

RISK FACTOR	35-44 AT EXAM			45-54 AT EXAM			55-64 AT EXAM			65-74 AT EXAM		
	PERSON EXAMS	# OF EVENTS	ANNUAL RATE	PERSON EXAMS	# OF EVENTS	ANNUAL RATE	PERSON EXAMS	# OF EVENTS	ANNUAL RATE	PERSON EXAMS	# OF EVENTS	ANNUAL RATE
74-119	2493			2714	1	0	1979	3	1	647	1	1
120-139	2244			3982	3	0	4056	4	0	1747	10	3
140-159	673			2053	5	1	3025	12	2	2013	13	3
160-179	156			768	0	0	1560	6	2	1155	6	3
180-300	50			442	1	1	893	6	3	686	12	9
UNKNOWN	0	.		0	0	-	0	0	-	0	0	-
TOTAL	5616			9959	10	1	11513	31	1	6248	42	3

RISK FACTOR	75-84 AT EXAM			85-94 AT EXAM			35-64 COMBINED			65-94 COMBINED		
	PERSON EXAMS	# OF EVENTS	ANNUAL RATE	PERSON EXAMS	# OF EVENTS	ANNUAL RATE	# OF EVENTS	CRUDE	AGE-ADJUSTED	# OF EVENTS	CRUDE	AGE-ADJUSTED
74-119	133	0	0	8			4	0	0	1	1	1
120-139	465	4	6	38			8	0	0	18	4	4
140-159	626	4	5	29			17	1	1	18	3	3
160-179	451	11	12	31			6	1	1	18	5	5
180-300	276	9	16	17			8	3	4	21	11	11
UNKNOWN	0	0	-	0			0	-		0	-	
TOTAL	1951	30	8	123			43	1	1	76	5	4

LOGISTIC REGRESSIONS

AGE AT EXAM		COEFFICIENT OF RISK FACTOR	SE COEFFICIENT	P-VALUE	STANDARDIZED COEFFICIENT	STD SE
35-44						
45-54		0.017	0.010	0.078	0.396	0.225
55-64		0.017	0.006	0.003	0.421	0.142
65-74		0.018	0.006	0.001	0.431	0.134
75-84		0.016	0.006	0.016	0.388	0.161
85-94						
35-64	UNIVARIATE	0.022	0.004	0.001	0.516	0.104
65-94	UNIVARIATE	0.017	0.004	0.001	0.414	0.100
35-64	BIVARIATE	0.017	0.005	0.001	0.399	0.116
65-94	BIVARIATE	0.016	0.004	0.001	0.383	0.102
35-64	MULTIVARIATE	0.013	0.006	0.019	0.305	0.130
65-94	MULTIVARIATE	0.012	0.004	0.005	0.303	0.107

TABLE 11-2   DATE=12APR85

AVERAGE ANNUAL INCIDENCE RATE FOR EVENT PER 1000 PERSONS AT RISK AT EXAM
BY AGE AND LEVEL OF RISK FACTOR AT EXAM: FRAMINGHAM STUDY 30-YEAR FOLLOWUP

SEX: MALE
EVENT: CORONARY HEART DISEASE DEATH
RISK FACTOR: DIASTOLIC BLOOD PRESSURE, FIRST EXAMINER (MM HG)
PERSONS AT RISK: PERSONS FREE OF CORONARY HEART DISEASE

RISK FACTOR	35-44 AT EXAM			45-54 AT EXAM			55-64 AT EXAM			65-74 AT EXAM		
	PERSON EXAMS	# OF EVENTS	ANNUAL RATE	PERSON EXAMS	# OF EVENTS	ANNUAL RATE	PERSON EXAMS	# OF EVENTS	ANNUAL RATE	PERSON EXAMS	# OF EVENTS	ANNUAL RATE
20-74	1012	0	0	1478	3	1	1847	12	3	1225	11	4
75-84	1604	4	1	2766	9	2	2845	14	2	1323	14	5
85-94	1187	1	0	2062	7	2	1964	19	5	893	13	7
95-104	499	0	0	1008	5	2	982	10	5	343	11	16
105-160	170	0	0	410	4	5	411	5	6	129	5	19
UNKNOWN	0	0	-	0	0	-	0	0	-	0	0	-
TOTAL	4472	5	1	7724	28	2	8049	60	4	3913	54	7

RISK FACTOR	75-84 AT EXAM			85-94 AT EXAM			35-64 COMBINED			65-94 COMBINED		
	PERSON EXAMS	# OF EVENTS	ANNUAL RATE	PERSON EXAMS	# OF EVENTS	ANNUAL RATE	# OF EVENTS	CRUDE	AGE-ADJUSTED	# OF EVENTS	CRUDE	AGE-ADJUSTED
20-74	422	8	9	22			15	2	2	20	6	5
75-84	331	7	11	12			27	2	2	21	6	6
85-94	204	5	12	5			27	3	3	19	9	9
95-104	44	2	23	0			15	3	3	13	17	17
105-160	16	1	31	0			9	5	4	6	21	21
UNKNOWN	0	0	-	0			0	-	-	0	-	-
TOTAL	1017	23	11	39			93	2	3	79	8	8

LOGISTIC REGRESSIONS

AGE AT EXAM		COEFFICIENT OF RISK FACTOR	SE COEFFICIENT	P-VALUE	STANDARDIZED COEFFICIENT	STD SE
35-44		-0.021	0.042	0.615	-0.237	0.470
45-54		0.037	0.014	0.010	0.427	0.166
55-64		0.024	0.010	0.016	0.285	0.118
65-74		0.034	0.010	0.001	0.412	0.119
75-84		0.007	0.018	0.707	0.079	0.209
85-94						
35-64	UNIVARIATE	0.027	0.008	0.001	0.313	0.094
65-94	UNIVARIATE	0.024	0.009	0.004	0.296	0.104
35-64	BIVARIATE	0.026	0.008	0.001	0.308	0.092
65-94	BIVARIATE	0.030	0.009	0.001	0.369	0.106
35-64	MULTIVARIATE	0.021	0.008	0.010	0.248	0.096
65-94	MULTIVARIATE	0.024	0.009	0.006	0.295	0.107

SEX: FEMALE
EVENT: CORONARY HEART DISEASE DEATH
RISK FACTOR: DIASTOLIC BLOOD PRESSURE, FIRST EXAMINER (MM HG)
PERSONS AT RISK: PERSONS FREE OF CORONARY HEART DISEASE

RISK FACTOR	35-44 AT EXAM			45-54 AT EXAM			55-64 AT EXAM			65-74 AT EXAM		
	PERSON EXAMS	# OF EVENTS	ANNUAL RATE	PERSON EXAMS	# OF EVENTS	ANNUAL RATE	PERSON EXAMS	# OF EVENTS	ANNUAL RATE	PERSON EXAMS	# OF EVENTS	ANNUAL RATE
20-74	2164			2639	1	0	2969	7	1	1831	11	3
75-84	2030			3541	4	1	3853	4	1	2135	16	4
85-94	987			2361	3	1	2840	11	2	1439	7	2
95-104	324			926	1	1	1272	5	2	600	4	3
105-160	111			492	1	1	579	4	3	243	4	4
UNKNOWN	0			0	0	-	0	0	-	0	0	-
TOTAL	5616			9959	10	1	11515	31	1	6248	42	3

RISK FACTOR	75-84 AT EXAM			85-94 AT EXAM			35-64 COMBINED			65-94 COMBINED		
	PERSON EXAMS	# OF EVENTS	ANNUAL RATE	PERSON EXAMS	# OF EVENTS	ANNUAL RATE	# OF EVENTS	CRUDE	AGE-ADJUSTED	# OF EVENTS	CRUDE	AGE-ADJUSTED
20-74	724	11	8	53			9	1	0	25	5	5
75-84	596	8	7	43			8	0	0	16	5	3
85-94	443	7	8	20			14	1	1	14	4	3
95-104	146	3	10	7			7	1	2	7	5	5
105-160	42	1	12	0			5	2	2	5	9	9
UNKNOWN	0	0	-	0			0	-	-	0	-	-
TOTAL	1951	30	8	123			43	1	1	76	5	4

LOGISTIC REGRESSIONS

AGE AT EXAM		COEFFICIENT OF RISK FACTOR	SE COEFFICIENT	P-VALUE	STANDARDIZED COEFFICIENT	STD SE
35-44						
45-54		0.028	0.022	0.202	0.342	0.268
55-64		0.030	0.013	0.017	0.373	0.157
65-74		0.012	0.012	0.328	0.146	0.149
75-84		0.009	0.015	0.549	0.108	0.180
85-94						
35-64	UNIVARIATE	0.032	0.010	0.002	0.392	0.128
65-94	UNIVARIATE	0.004	0.009	0.642	0.053	0.114
35-64	BIVARIATE	0.028	0.011	0.012	0.336	0.134
65-94	BIVARIATE	0.008	0.009	0.366	0.103	0.114
35-64	MULTIVARIATE	0.021	0.011	0.064	0.258	0.140
65-94	MULTIVARIATE	0.006	0.009	0.537	0.070	0.114

TABLE 11-3A   DATE=30AUG85

AVERAGE ANNUAL INCIDENCE RATE FOR EVENT PER 1000 PERSONS AT RISK AT EXAM
BY AGE AND LEVEL OF RISK FACTOR AT EXAM: FRAMINGHAM STUDY 30-YEAR FOLLOWUP

SEX: MALE
EVENT: CORONARY HEART DISEASE DEATH
RISK FACTOR: HYPERTENSION WITH ANTIHYPERTENSIVE TREATMENT
PERSONS AT RISK: PERSONS FREE OF CORONARY HEART DISEASE

35-44 AT EXAM			45-54 AT EXAM			55-64 AT EXAM			65-74 AT EXAM		
PERSON EXAMS	# OF EVENTS	ANNUAL RATE	PERSON EXAMS	# OF EVENTS	ANNUAL RATE	PERSON EXAMS	# OF EVENTS	ANNUAL RATE	PERSON EXAMS	# OF EVENTS	ANNUAL RATE
2651	3	1	4122	10	1	3782	15	2	1516	8	5
1301	2	1	2205	7	2	2299	23	5	1252	13	5
500	0	0	1139	9	4	1266	14	6	552	13	12
20	0	0	258	2	4	702	8	6	593	20	17
0	0	-	0	0	-	0	0	-	0	0	-
4472	5	1	7724	28	2	8049	60	4	3913	54	7

75-84 AT EXAM			85-94 AT EXAM			35-64 COMBINED			65-94 COMBINED		
PERSON EXAMS	# OF EVENTS	ANNUAL RATE	PERSON EXAMS	# OF EVENTS	ANNUAL RATE	# OF EVENTS	CRUDE	AGE-ADJUSTED	# OF EVENTS	CRUDE	AGE-ADJUSTED
373	7	9	18			28	1	1	15	4	4
333	7	11	11			32	3	3	20	6	6
123	3	12	3			23	4	4	17	13	13
188	6	16	7			10	5	4	27	17	17
0	0	-	0			0	-	-	0	-	-
1017	23	11	39			93	2	3	79	8	8

LOGISTIC REGRESSIONS

	COEFFICIENT OF RISK FACTOR	SE COEFFICIENT	P-VALUE	STANDARDIZED COEFFICIENT	STD SE
	-0.283	0.716	0.692	-0.197	0.497
	0.596	0.229	0.009	0.458	0.176
	0.497	0.156	0.001	0.405	0.127
	0.900	0.191	0.001	0.738	0.156
	0.226	0.259	0.382	0.185	0.212
VARIATE	0.581	0.126	0.001	0.452	0.098
VARIATE	0.716	0.150	0.001	0.586	0.123
ARIATE	0.476	0.126	0.001	0.370	0.098
ARIATE	0.713	0.150	0.001	0.584	0.123
TIVARIATE	0.414	0.132	0.002	0.322	0.103
TIVARIATE	0.629	0.163	0.001	0.515	0.126

SEX: FEMALE
EVENT: CORONARY HEART DISEASE DEATH
RISK FACTOR: HYPERTENSION WITH ANTIHYPERTENSIVE TREATMENT
PERSONS AT RISK: PERSONS FREE OF CORONARY HEART DISEASE

35-44 AT EXAM			45-54 AT EXAM			55-64 AT EXAM			65-74 AT EXAM		
PERSON EXAMS	# OF EVENTS	ANNUAL RATE	PERSON EXAMS	# OF EVENTS	ANNUAL RATE	PERSON EXAMS	# OF EVENTS	ANNUAL RATE	PERSON EXAMS	# OF EVENTS	ANNUAL RATE
4250			5622	3	0	4888	6	1	1767	8	2
1001			2579	3	1	3282	13	2	1903	10	3
316			1166	1	0	1835	9	2	1035	7	3
49			592	3	3	1508	3	1	1543	17	6
0			0	0	-	0	0	-	0	0	-
5616			9959	10	1	11513	31	1	6248	42	3

75-84 AT EXAM			85-94 AT EXAM			35-64 COMBINED			65-94 COMBINED		
PERSON EXAMS	# OF EVENTS	ANNUAL RATE	PERSON EXAMS	# OF EVENTS	ANNUAL RATE	# OF EVENTS	CRUDE	AGE-ADJUSTED	# OF EVENTS	CRUDE	AGE-ADJUSTED
402	3	4	31			10	0	0	13	3	3
375	6	5	28			16	1	1	17	3	4
371	10	13	21			11	2	1	18	6	6
603	11	9	43			6	1	2	28	6	7
0	0	-	0			0	-	-	0	-	-
1951	30	8	123			43	1	1	76	5	4

LOGISTIC REGRESSIONS

	COEFFICIENT OF RISK FACTOR	SE COEFFICIENT	P-VALUE	STANDARDIZED COEFFICIENT	STD SE
	0.724	0.378	0.055	0.556	0.290
	0.467	0.218	0.032	0.390	0.182
	0.398	0.203	0.050	0.327	0.167
	0.597	0.285	0.036	0.470	0.225
RIATE	0.725	0.182	0.001	0.575	0.144
RIATE	0.433	0.156	0.005	0.354	0.127
IATE	0.495	0.187	0.008	0.393	0.149
IATE	0.380	0.156	0.015	0.311	0.128
VARIATE	0.367	0.194	0.059	0.291	0.154
VARIATE	0.270	0.159	0.090	0.221	0.130

TABLE 11-3B   DATE=21NOV85
AVERAGE ANNUAL INCIDENCE RATE FOR EVENT PER 1000 PERSONS AT RISK AT EXAM
BY AGE AND LEVEL OF RISK FACTOR AT EXAM: FRAMINGHAM STUDY 30-YEAR FOLLOWUP

SEX: MALE
EVENT: CORONARY HEART DISEASE DEATH
RISK FACTOR: HYPERTENSION IGNORING TREATMENT
PERSONS AT RISK: PERSONS FREE OF CORONARY HEART DISEASE

RISK FACTOR	35-44 AT EXAM PERSON EXAMS	# OF EVENTS	ANNUAL RATE	45-54 AT EXAM PERSON EXAMS	# OF EVENTS	ANNUAL RATE	55-64 AT EXAM PERSON EXAMS	# OF EVENTS	ANNUAL RATE	65-74 AT EXAM PERSON EXAMS	# OF EVENTS	ANNUAL RATE
NO	2653	3	1	4154	10	1	3909	17	2	1631	11	3
MILD	1310	2	1	2284	8	2	2583	26	5	1499	21	7
DEFINITE	509	0	0	1286	10	4	1557	17	5	783	22	14
UNKNOWN	0	0	-	0	0	-	0	0	-	0	0	-
TOTAL	4472	5	1	7724	28	2	8049	60	4	3913	54	7

RISK FACTOR	75-84 AT EXAM PERSON EXAMS	# OF EVENTS	ANNUAL RATE	85-94 AT EXAM PERSON EXAMS	# OF EVENTS	ANNUAL RATE	35-64 COMBINED # OF EVENTS	CRUDE	AGE-ADJUSTED	65-94 COMBINED # OF EVENTS	CRUDE	AGE-ADJUSTED
NO	412	8	10	21	0	-	30	1	1	19	5	5
MILD	418	7	8	15	0	-	36	3	3	29	8	7
DEFINITE	187	8	21	3	0	-	27	4	4	31	16	17
UNKNOWN	0	0	-	0	0	-	0	-	-	0	-	-
TOTAL	1017	23	11	39	0	-	93	2	2	79	8	8

LOGISTIC REGRESSIONS

AGE AT EXAM	COEFFICIENT OF RISK FACTOR	SE COEFFICIENT	P-VALUE	STANDARDIZED COEFFICIENT	STD SE
35-44	-0.279	0.718	0.698	-0.193	0.496
45-54	0.589	0.232	0.011	0.443	0.174
55-64	0.464	0.160	0.004	0.358	0.123
65-74	0.722	0.181	0.001	0.545	0.137
75-84	0.405	0.280	0.148	0.298	0.206
85-94					
35-64 UNIVARIATE	0.5.5	0.128	0.001	0.401	0.096
65-94 UNIVARIATE	0.635	0.149	0.001	0.484	0.112
35-64 BIVARIATE	0.4.0	0.129	0.001	0.345	0.097
65-94 BIVARIATE	0.661	0.150	0.001	0.496	0.113
35-64 MULTIVARIATE	0.380	0.135	0.005	0.285	0.101
65-94 MULTIVARIATE	0.525	0.150	0.001	0.394	0.112

SEX: FEMALE
EVENT: CORONARY HEART DISEASE DEATH
RISK FACTOR: HYPERTENSION IGNORING TREATMENT
PERSONS AT RISK: PERSONS FREE OF CORONARY HEART DISEASE

RISK FACTOR	35-44 AT EXAM PERSON EXAMS	# OF EVENTS	ANNUAL RATE	45-54 AT EXAM PERSON EXAMS	# OF EVENTS	ANNUAL RATE	55-64 AT EXAM PERSON EXAMS	# OF EVENTS	ANNUAL RATE	65-74 AT EXAM PERSON EXAMS	# OF EVENTS	ANNUAL RATE
NO	4272			5804	3	0	5245	6	1	2035	9	2
MILD	1016			2763	5	1	3860	15	2	2549	16	3
DEFINITE	328			1392	2	1	2408	10	2	1664	17	5
UNKNOWN	0			0	0	-	0	0	-	0	0	-
TOTAL	5616			9959	10	1	11513	31	1	6248	42	3

RISK FACTOR	75-84 AT EXAM PERSON EXAMS	# OF EVENTS	ANNUAL RATE	85-94 AT EXAM PERSON EXAMS	# OF EVENTS	ANNUAL RATE	35-64 COMBINED # OF EVENTS	CRUDE	AGE-ADJUSTED	65-94 COMBINED # OF EVENTS	CRUDE	AGE-ADJUSTED
NO	507	5	5	37	0	-	10	0	0	16	3	3
MILD	827	9	5	45	0	-	20	1	1	26	4	4
DEFINITE	617	16	13	41	0	-	13	2	1	34	7	7
UNKNOWN	0	0	-	0	0	-	0	-	-	0	-	-
TOTAL	1951	30	8	123	0	-	43	1	1	76	5	5

LOGISTIC REGRESSIONS

AGE AT EXAM	COEFFICIENT OF RISK FACTOR	SE COEFFICIENT	P-VALUE	STANDARDIZED COEFFICIENT	STD SE
35-44					
45-54	0.565	0.388	0.145	0.410	0.281
55-64	0.591	0.226	0.009	0.460	0.175
65-74	0.430	0.206	0.036	0.330	0.158
75-84	0.583	0.263	0.026	0.441	0.199
85-94					
35-64 UNIVARIATE	0.761	0.186	0.001	0.563	0.138
65-94 UNIVARIATE	0.468	0.155	0.003	0.359	0.119
35-64 BIVARIATE	0.560	0.192	0.003	0.414	0.142
65-94 BIVARIATE	0.429	0.156	0.006	0.329	0.120
35-64 MULTIVARIATE	0.431	0.200	0.032	0.319	0.148
65-94 MULTIVARIATE	0.325	0.160	0.042	0.249	0.122

TABLE 11-4　　　　DATE=12APR85

AVERAGE ANNUAL INCIDENCE RATE FOR EVENT PER 1000 PERSONS AT RISK AT EXAM
BY AGE AND LEVEL OF RISK FACTOR AT EXAM: FRAMINGHAM STUDY 30-YEAR FOLLOWUP

SEX: MALE
EVENT: CORONARY HEART DISEASE DEATH
RISK FACTOR: SERUM CHOLESTEROL
PERSONS AT RISK: PERSONS FREE OF CORONARY HEART DISEASE

RISK FACTOR	35-44 AT EXAM PERSON EXAMS	# OF EVENTS	ANNUAL RATE	45-54 AT EXAM PERSON EXAMS	# OF EVENTS	ANNUAL RATE	55-64 AT EXAM PERSON EXAMS	# OF EVENTS	ANNUAL RATE	65-74 AT EXAM PERSON EXAMS	# OF EVENTS	ANNUAL RATE
84-204	1253	0	0	1767	5	1	2255	7	2	1234	17	7
205-234	1247	0	0	2306	8	2	2405	21	4	1171	17	7
235-264	872	2	1	1745	6	2	1889	18	5	873	11	6
265-294	475	1	1	1001	3	1	918	7	4	428	3	4
295-1124	291	2	3	557	5	4	454	7	8	191	6	16
UNKNOWN	334	0	0	348	1	1	128	0	0	16	0	0
TOTAL	4472	5	1	7724	28	2	8049	60	4	3913	54	7

RISK FACTOR	75-84 AT EXAM PERSON EXAMS	# OF EVENTS	ANNUAL RATE	85-94 AT EXAM PERSON EXAMS	# OF EVENTS	ANNUAL RATE	35-64 COMBINED # OF EVENTS	CRUDE	AGE-ADJUSTED	65-94 COMBINED # OF EVENTS	CRUDE	AGE-ADJUSTED
84-204	390	9	12	17			12	1	1	26	8	8
205-234	332	6	9	11			29	2	2	24	8	8
235-264	177	4	11	5			26	3	3	16	8	8
265-294	96	3	16	6			11	2	2	6	6	6
295-1124	22	1	23	0			14	5	5	7	16	17
UNKNOWN	0	0	-	0			1	1	0	0	0	0
TOTAL	1017	23	11	39			93	2	3	79	8	8

LOGISTIC REGRESSIONS

AGE AT EXAM		COEFFICIENT OF RISK FACTOR	SE COEFFICIENT	P-VALUE	STANDARDIZED COEFFICIENT	STD SE
35-44		0.017	0.005	0.001	0.735	0.217
45-54		0.009	0.003	0.003	0.396	0.131
55-64		0.007	0.003	0.017	0.278	0.117
65-74		0.001	0.003	0.658	0.060	0.136
75-84		0.005	0.005	0.364	0.186	0.205
85-94						
35-64	UNIVARIATE	0.008	0.002	0.001	0.349	0.082
65-94	UNIVARIATE	0.002	0.003	0.449	0.085	0.112
35-64	BIVARIATE	0.009	0.002	0.001	0.382	0.084
65-94	BIVARIATE	0.003	0.003	0.273	0.123	0.112
35-64	MULTIVARIATE	0.008	0.002	0.001	0.347	0.086
65-94	MULTIVARIATE	0.003	0.003	0.266	0.124	0.112

SEX: FEMALE
EVENT: CORONARY HEART DISEASE DEATH
RISK FACTOR: SERUM CHOLESTEROL
PERSONS AT RISK: PERSONS FREE OF CORONARY HEART DISEASE

RISK FACTOR	35-44 AT EXAM PERSON EXAMS	# OF EVENTS	ANNUAL RATE	45-54 AT EXAM PERSON EXAMS	# OF EVENTS	ANNUAL RATE	55-64 AT EXAM PERSON EXAMS	# OF EVENTS	ANNUAL RATE	65-74 AT EXAM PERSON EXAMS	# OF EVENTS	ANNUAL RATE
84-204	2550			2094	0	0	1625	3	1	789	1	1
205-234	1417			2558	2	0	2615	4	1	1401	7	2
235-264	806			2259	2	0	2938	5	1	1703	9	3
265-294	372			1473	2	1	2240	8	2	1287	8	3
295-1124	171			1101	4	2	1866	11	3	1031	16	8
UNKNOWN	500			474	0	0	229	0	0	37	1	14
TOTAL	5616			9959	10	1	11513	31	1	6248	42	3

RISK FACTOR	75-84 AT EXAM PERSON EXAMS	# OF EVENTS	ANNUAL RATE	85-94 AT EXAM PERSON EXAMS	# OF EVENTS	ANNUAL RATE	35-64 COMBINED # OF EVENTS	CRUDE	AGE-ADJUSTED	65-94 COMBINED # OF EVENTS	CRUDE	AGE-ADJUSTED
84-204	361	4	6	46			3	0	0	8	3	3
205-234	486	9	9	28			6	0	0	16	4	4
235-264	542	10	9	35			8	1	1	19	4	5
265-294	350	5	7	9			10	1	1	14	4	7
295-1124	205	2	5	5			16	3	3	18	-	7
UNKNOWN	7	0	0	0			0	0	0		11	11
TOTAL	1951	30	8	123			43	1	1	76	5	4

LOGISTIC REGRESSIONS

AGE AT EXAM		COEFFICIENT OF RISK FACTOR	SE COEFFICIENT	P-VALUE	STANDARDIZED COEFFICIENT	STD SE
35-44						
45-54		0.007	0.002	0.007	0.304	0.113
55-64		0.012	0.003	0.001	0.536	0.144
65-74		0.012	0.003	0.001	0.540	0.140
75-84		0.000	0.004	0.981	0.004	0.184
85-94						
35-64	UNIVARIATE	0.009	0.002	0.001	0.433	0.097
65-94	UNIVARIATE	0.005	0.003	0.044	0.224	0.111
35-64	BIVARIATE	0.009	0.002	0.001	0.398	0.085
65-94	BIVARIATE	0.007	0.003	0.008	0.295	0.112
35-64	MULTIVARIATE	0.007	0.002	0.001	0.342	0.086
65-94	MULTIVARIATE	0.007	0.003	0.007	0.305	0.112

TABLE 11-5　　　　　　　　DATE=21NOV85
AVERAGE ANNUAL INCIDENCE RATE FOR EVENT PER 1000 PERSONS AT RISK AT EXAM
BY AGE AND LEVEL OF RISK FACTOR AT EXAM: FRAMINGHAM STUDY 30-YEAR FOLLOWUP

SEX: MALE
EVENT: CORONARY HEART DISEASE DEATH
RISK FACTOR: HEMATOCRIT
PERSONS AT RISK: PERSONS FREE OF CORONARY HEART DISEASE

RISK FACTOR	35-44 AT EXAM			45-54 AT EXAM			55-64 AT EXAM			65-74 AT EXAM		
	PERSON EXAMS	# OF EVENTS	ANNUAL RATE	PERSON EXAMS	# OF EVENTS	ANNUAL RATE	PERSON EXAMS	# OF EVENTS	ANNUAL RATE	PERSON EXAMS	# OF EVENTS	ANNUAL RATE
25-42	623	1	1	942	3	2	1018	8	4	450	4	4
43-44	546	0	0	967	4	2	985	5	3	493	3	3
45-46	983	0	0	1854	8	2	1953	17	4	963	11	6
47-48	775	0	0	1510	3	1	1599	6	2	861	14	8
49-70	1175	4	2	2066	9	2	2338	24	5	1127	22	10
UNKNOWN	370	0	0	385	1	1	156	0	0	19	0	0
TOTAL	4472	5	1	7724	28	2	8049	60	4	3913	54	7

RISK FACTOR	75-84 AT EXAM			85-94 AT EXAM			35-64 COMBINED			65-94 COMBINED		
	PERSON EXAMS	# OF EVENTS	ANNUAL RATE	PERSON EXAMS	# OF EVENTS	ANNUAL RATE	# OF EVENTS	CRUDE	AGE-ADJUSTED	# OF EVENTS	CRUDE	AGE-ADJUSTED
25-42	214	7	16	16			12	2	2	12	9	7
43-44	119	3	13	5			9	2	2	6	5	3
45-46	240	5	10	7			25	3	3	16	7	7
47-48	197	2	5	7			9	1	1	16	8	7
49-70	247	6	12	4			37	3	3	29	11	11
UNKNOWN	0	0	-	0			1	1	0	0	0	0
TOTAL	1017	23	11	39			93	2	2	79	8	8

LOGISTIC REGRESSIONS

AGE AT EXAM		COEFFICIENT OF RISK FACTOR	SE COEFFICIENT	P-VALUE	STANDARDIZED COEFFICIENT	STD SE
35-44		0.199	0.118	0.091	0.730	0.432
45-54		0.018	0.054	0.745	0.063	0.193
55-64		0.023	0.034	0.502	0.088	0.131
65-74		0.077	0.038	0.040	0.285	0.139
75-84		-0.015	0.050	0.772	-0.061	0.209
85-94						
35-64	UNIVARIATE	0.032	0.029	0.256	0.120	0.105
65-94	UNIVARIATE	0.038	0.030	0.206	0.146	0.115
35-64	BIVARIATE	0.030	0.028	0.286	0.111	0.104
65-94	BIVARIATE	0.051	0.030	0.086	0.197	0.114
35-64	MULTIVARIATE	0.002	0.029	0.953	0.006	0.105
65-94	MULTIVARIATE	0.040	0.030	0.186	0.154	0.117

SEX: FEMALE
EVENT: CORONARY HEART DISEASE DEATH
RISK FACTOR: HEMATOCRIT
PERSONS AT RISK: PERSONS FREE OF CORONARY HEART DISEASE

RISK FACTOR	35-44 AT EXAM			45-54 AT EXAM			55-64 AT EXAM			65-74 AT EXAM		
	PERSON EXAMS	# OF EVENTS	ANNUAL RATE	PERSON EXAMS	# OF EVENTS	ANNUAL RATE	PERSON EXAMS	# OF EVENTS	ANNUAL RATE	PERSON EXAMS	# OF EVENTS	ANNUAL RATE
21-39	1697			1843	1	0	1553	8	3	646	3	2
40-41	967			1741	1	0	2061	5	1	1049	4	2
42-43	1043			1955	0	0	2260	3	1	1256	7	3
44-45	856			2194	2	0	2847	6	1	1609	14	4
46-65	487			1681	6	2	2521	7	1	1626	13	4
UNKNOWN	566			545	0	0	271	0	0	42	1	12
TOTAL	5616			9959	10	1	11513	31	1	6248	42	3

RISK FACTOR	75-84 AT EXAM			85-94 AT EXAM			35-64 COMBINED			65-94 COMBINED		
	PERSON EXAMS	# OF EVENTS	ANNUAL RATE	PERSON EXAMS	# OF EVENTS	ANNUAL RATE	# OF EVENTS	CRUDE	AGE-ADJUSTED	# OF EVENTS	CRUDE	AGE-ADJUSTED
21-39	297	5	8	24			10	1	1	9	5	4
40-41	373	4	5	31			6	1	1	9	3	3
42-43	153	3	4	19			5	0	0	12	4	4
44-45	486	6	6	26			9	1	1	20	5	5
46-65	436	12	14	23			13	1	1	25	6	6
UNKNOWN	6	0	0	0			0	0	0	1	10	9
TOTAL	1951	30	8	123			43	1	1	76	5	5

LOGISTIC REGRESSIONS

AGE AT EXAM		COEFFICIENT OF RISK FACTOR	SE COEFFICIENT	P-VALUE	STANDARDIZED COEFFICIENT	STD SE
35-44						
45-54		0.210	0.087	0.016	0.750	0.312
55-64		-0.027	0.051	0.595	-0.095	0.178
65-74		0.083	0.046	0.068	0.291	0.159
75-84		0.054	0.048	0.255	0.212	0.186
85-94						
35-64	UNIVARIATE	0.056	0.043	0.198	0.200	0.155
65-94	UNIVARIATE	0.049	0.033	0.132	0.177	0.118
35-64	BIVARIATE	0.019	0.044	0.671	0.067	0.158
65-94	BIVARIATE	0.057	0.032	0.074	0.206	0.115
35-64	MULTIVARIATE	-0.015	0.042	0.718	-0.055	0.152
65-94	MULTIVARIATE	0.048	0.032	0.135	0.173	0.116

TABLE 11-6   DATE=12DEC85
AVERAGE ANNUAL INCIDENCE RATE FOR EVENT PER 1000 PERSONS AT RISK AT EXAM
BY AGE AND LEVEL OF RISK FACTOR AT EXAM: FRAMINGHAM STUDY 30-YEAR FOLLOWUP

SEX: MALE
EVENT: CORONARY HEART DISEASE DEATH
RISK FACTOR: BLODO GLUCOSE (MG/DL)
PERSONS AT RISK: PERSONS FREE OF CORONARY HEART DISEASE

RISK FACTOR	35-44 AT EXAM PERSON EXAMS	# OF EVENTS	ANNUAL RATE	45-54 AT EXAM PERSON EXAMS	# OF EVENTS	ANNUAL RATE	55-64 AT EXAM PERSON EXAMS	# OF EVENTS	ANNUAL RATE	65-74 AT EXAM PERSON EXAMS	# OF EVENTS	ANNUAL RATE
29-69	992	0	0	1490	5	3	1087	7	3	421	5	6
70-89	2501	4	1	4172	12	1	4056	36	4	1761	18	5
90-109	626	1	1	1348	4	1	1866	11	3	1046	15	7
110-129	105	0	0	305	0	0	537	4	4	335	7	10
130-524	48	0	0	193	3	8	408	2	2	334	9	13
UNKNOWN	200	0	0	216	1	2	95	0	0	16	0	0
TOTAL	4472	5	1	7724	28	2	8049	60	4	3913	54	7

RISK FACTOR	75-84 AT EXAM PERSON EXAMS	# OF EVENTS	ANNUAL RATE	85-94 AT EXAM PERSON EXAMS	# OF EVENTS	ANNUAL RATE	35-64 COMBINED # OF EVENTS	CRUDE	AGE-ADJUSTED	65-94 COMBINED # OF EVENTS	CRUDE	AGE-ADJUSTED
29-69	62	2	16	1			15	2	2	7	7	8
70-89	377	10	13	11			52	2	2	28	7	7
90-109	328	5	8	14			16	2	2	21	8	7
110-129	125	3	12	4			4	2	1	10	11	11
130-524	125	3	12	9			5	4	4	13	14	14
UNKNOWN	0	0	-	0			1	1	1	0	0	0
TOTAL	1017	23	11	39			93	2	2	79	8	8

LOGISTIC REGRESSIONS

AGE AT EXAM		COEFFICIENT OF RISK FACTOR	SE COEFFICIENT	P-VALUE	STANDARDIZED COEFFICIENT	STD SE
35-44		-	-	-	-	-
45-54		0.006	0.005	0.198	0.148	0.115
55-64		-0.004	0.006	0.452	-0.120	0.159
65-74		0.009	0.002	0.001	0.303	0.070
75-84		-0.002	0.007	0.816	-0.052	0.224
85-94						
35-64	UNIVARIATE	0.003	0.004	0.467	0.064	0.088
65-94	UNIVARIATE	0.008	0.002	0.001	0.260	0.066
35-64	BIVARIATE	-0.001	0.004	0.821	-0.023	0.102
65-94	BIVARIATE	0.008	0.002	0.001	0.255	0.069
35-64	MULTIVARIATE	-0.001	0.004	0.779	-0.028	0.101
65-94	MULTIVARIATE	0.007	0.002	0.001	0.236	0.071

SEX: FEMALE
EVENT: CORONARY HEART DISEASE DEATH
RISK FACTOR: BLOOD GLUCOSE (MG/DL)
PERSONS AT RISK: PERSONS FREE OF CORONARY HEART DISEASE

RISK FACTOR	35-44 AT EXAM PERSON EXAMS	# OF EVENTS	ANNUAL RATE	45-54 AT EXAM PERSON EXAMS	# OF EVENTS	ANNUAL RATE	55-64 AT EXAM PERSON EXAMS	# OF EVENTS	ANNUAL RATE	65-74 AT EXAM PERSON EXAMS	# OF EVENTS	ANNUAL RATE
29-69	1066			1797	2	1	1369	6	2	521	5	5
70-89	3412			5744	4	0	6098	16	1	3102	10	2
90-109	672			1625	0	0	2712	4	1	1732	13	4
110-129	96			301	0	0	730	1	1	501	3	3
130-524	39			159	4	13	438	4	5	360	10	14
UNKNOWN	331			333	0	0	166	0	0	32	1	16
TOTAL	5616			9959	10	1	11513	31	1	6248	42	3

RISK FACTOR	75-84 AT EXAM PERSON EXAMS	# OF EVENTS	ANNUAL RATE	85-94 AT EXAM PERSON EXAMS	# OF EVENTS	ANNUAL RATE	35-64 COMBINED # OF EVENTS	CRUDE	AGE-ADJUSTED	65-94 COMBINED # OF EVENTS	CRUDE	AGE-ADJUSTED
29-69	105	2	10	3			8	1	1	7	6	6
70-89	823	13	8	44			22	1	1	24	3	3
90-109	640	5	4	51			4	0	0	19	4	4
110-129	219	6	14	13			1	0	0	9	6	5
130-524	163	4	12	12			8	6	7	16	15	15
UNKNOWN	1	0	0	0			0	0	0	1	15	12
TOTAL	1951	30	8	123			43	1	1	76	5	5

LOGISTIC REGRESSIONS

AGE AT EXAM		COEFFICIENT OF RISK FACTOR	SE COEFFICIENT	P-VALUE	STANDARDIZED COEFFICIENT	STD SE
35-44						
45-54		0.020	0.004	0.001	0.396	0.077
55-64		0.007	0.004	0.127	0.170	0.111
65-74		0.012	0.002	0.001	0.352	0.069
75-84		0.007	0.003	0.014	0.250	0.101
85-94						
35-64	UNIVARIATE	0.013	0.003	0.001	0.275	0.058
65-94	UNIVARIATE	0.011	0.002	0.001	0.318	0.054
35-64	BIVARIATE	0.010	0.003	0.001	0.226	0.063
65-94	BIVARIATE	0.010	0.002	0.001	0.295	0.055
35-64	MULTIVARIATE	0.008	0.003	0.007	0.177	0.065
65-94	MULTIVARIATE	0.010	0.002	0.001	0.287	0.057

TABLE 11-7 DATE=12DEC85

AVERAGE ANNUAL INCIDENCE RATE FOR EVENT PER 1000 PERSONS AT RISK AT EXAM
BY AGE AND LEVEL OF RISK FACTOR AT EXAM: FRAMINGHAM STUDY 30-YEAR FOLLOWUP

SEX: MALE
EVENT: CORONARY HEART DISEASE DEATH
RISK FACTOR: DIABETES MELLITUS
PERSONS AT RISK: PERSONS FREE OF CORONARY HEART DISEASE

RISK FACTOR	35-44 AT EXAM PERSON EXAMS	# OF EVENTS	ANNUAL RATE	45-54 AT EXAM PERSON EXAMS	# OF EVENTS	ANNUAL RATE	55-64 AT EXAM PERSON EXAMS	# OF EVENTS	ANNUAL RATE	65-74 AT EXAM PERSON EXAMS	# OF EVENTS	ANNU. RATE
NO	4424	5	1	7486	26	2	7549	56	4	3504	44	6
YES	48	0	0	238	2	4	500	4	4	409	10	12
UNKNOWN	0	0	–	0	0	–	0	0	–	0	0	–
TOTAL	4472	5	1	7724	28	2	8049	60	4	3913	54	7

RISK FACTOR	75-84 AT EXAM PERSON EXAMS	# OF EVENTS	ANNUAL RATE	85-94 AT EXAM PERSON EXAMS	# OF EVENTS	ANNUAL RATE	35-64 COMBINED # OF EVENTS	CRUDE	AGE-ADJUSTED	65-94 COMBINED # OF EVENTS	CRUDE	AGE AD.
NO	900	22	12	34			87	2	2	68	8	
YES	117	1	4	5			6	4	3	11	10	
UNKNOWN	0	0	–	0			0	–	–	0		
TOTAL	1017	23	11	39			93	2	2	79	8	

LOGISTIC REGRESSIONS

AGE AT EXAM		COEFFICIENT OF RISK FACTOR	SE COEFFICIENT	P-VALUE	STANDARDIZED COEFFICIENT	STD SE
35-44		–	–	–	–	–
45-54		0.889	0.737	0.228	0.154	0.127
55-64		0.076	0.520	0.884	0.018	0.125
65-74		0.678	0.354	0.055	0.208	0.108
75-84		-1.067	1.027	0.299	-0.341	0.328
85-94						
35-64	UNIVARIATE	0.538	0.424	0.204	0.104	0.082
65-94	UNIVARIATE	0.307	0.328	0.350	0.095	0.101
35-64	BIVARIATE	0.208	0.426	0.625	0.040	0.082
65-94	BIVARIATE	0.278	0.329	0.398	0.086	0.102
35-64	MULTIVARIATE	0.104	0.433	0.809	0.020	0.084
65-94	MULTIVARIATE	0.274	0.333	0.412	0.085	0.103

SEX: FEMALE
EVENT: CORONARY HEART DISEASE DEATH
RISK FACTOR: DIABETES MELLITUS
PERSONS AT RISK: PERSONS FREE OF CORONARY HEART DISEASE

RISK FACTOR	35-44 AT EXAM PERSON EXAMS	# OF EVENTS	ANNUAL RATE	45-54 AT EXAM PERSON EXAMS	# OF EVENTS	ANNUAL RATE	55-64 AT EXAM PERSON EXAMS	# OF EVENTS	ANNUAL RATE	65-74 AT EXAM PERSON EXAMS	# OF EVENTS	ANNUAL RATE
NO	5573			9760	6	0	10985	25	1	5780	31	3
YES	43			199	4	10	528	6	6	468	11	12
UNKNOWN	0			0	0	–	0	0	–	0	0	–
TOTAL	5616			9959	10	1	11513	31	1	6248	42	3

RISK FACTOR	75-84 AT EXAM PERSON EXAMS	# OF EVENTS	ANNUAL RATE	85-94 AT EXAM PERSON EXAMS	# OF EVENTS	ANNUAL RATE	35-64 COMBINED # OF EVENTS	CRUDE	AGE-ADJUSTED	65-94 COMBINED # OF EVENTS	CRUDE	AGE-ADJ.
NO	1738	24	7	106			32	1	1	58	4	
YES	213	6	14	17			11	7	9	18	13	
UNKNOWN	0	0	–	0			0	–	–	0	–	
TOTAL	1951	30	8	123			43	1	1	76	5	

LOGISTIC REGRESSIONS

AGE AT EXAM		COEFFICIENT OF RISK FACTOR	SE COEFFICIENT	P-VALUE	STANDARDIZED COEFFICIENT	STD SE
35-44						
45-54		3.507	0.650	0.001	0.491	0.091
55-64		1.617	0.457	0.001	0.338	0.096
65-74		1.496	0.354	0.001	0.394	0.093
75-84		0.728	0.462	0.116	0.227	0.144
85-94						
35-64	UNIVARIATE	2.477	0.351	0.001	0.412	0.058
65-94	UNIVARIATE	1.239	0.273	0.001	0.344	0.076
35-64	BIVARIATE	2.108	0.356	0.001	0.350	0.059
65-94	BIVARIATE	1.142	0.275	0.001	0.316	0.076
35-64	MULTIVARIATE	1.791	0.371	0.001	0.298	0.062
65-94	MULTIVARIATE	1.021	0.285	0.001	0.283	0.079

TABLE 11-8   DATE=22NOV85
AVERAGE ANNUAL INCIDENCE RATE FOR EVENT PER 1000 PERSONS AT RISK AT EXAM
BY AGE AND LEVEL OF RISK FACTOR AT EXAM: FRAMINGHAM STUDY 30-YEAR FOLLOWUP

SEX: MALE
EVENT: CORONARY HEART DISEASE DEATH
RISK FACTOR: GLUCOSE IN URINE, DEFINITE OR TRACE
PERSONS AT RISK: PERSONS FREE OF CORONARY HEART DISEASE

RISK FACTOR	35-44 AT EXAM PERSON EXAMS	# OF EVENTS	ANNUAL RATE	45-54 AT EXAM PERSON EXAMS	# OF EVENTS	ANNUAL RATE	55-64 AT EXAM PERSON EXAMS	# OF EVENTS	ANNUAL RATE	65-74 AT EXAM PERSON EXAMS	# OF EVENTS	ANNUAL RATE
NO	4380	5	1	7535	26	2	7815	55	4	3767	44	6
YES	60	0	0	168	2	6	210	4	10	140	10	36
UNKNOWN	32	0	0	21	0	0	24	1	21	6	0	0
TOTAL	4472	5	1	7724	28	2	8049	60	4	3913	54	7

RISK FACTOR	75-84 AT EXAM PERSON EXAMS	# OF EVENTS	ANNUAL RATE	85-94 AT EXAM PERSON EXAMS	# OF EVENTS	ANNUAL RATE	35-64 COMBINED # OF EVENTS	CRUDE	AGE-ADJUSTED	65-94 COMBINED # OF EVENTS	CRUDE	AGE-ADJUSTED
NO	986	21	11	38			86	2	2	67	7	7
YES	28	2	36	1			6	7	6	12	36	35
UNKNOWN	3	0	0	0			1	6	8	0	0	0
TOTAL	1017	23	11	39			93	2	2	79	8	8

LOGISTIC REGRESSIONS

AGE AT EXAM		COEFFICIENT OF RISK FACTOR	SE COEFFICIENT	P-VALUE	STANDARDIZED COEFFICIENT	STD SE
35-44		-	-	-	-	-
45-54		1.247	0.738	0.091	0.182	0.108
55-64		1.008	0.523	0.054	0.161	0.083
65-74		1.873	0.362	0.001	0.348	0.067
75-84		1.263	0.766	0.099	0.207	0.126
85-94						
35-64	UNIVARIATE	1.155	0.425	0.007	0.168	0.062
65-94	UNIVARIATE	1.684	0.324	0.001	0.306	0.059
35-64	BIVARIATE	1.011	0.426	0.018	0.147	0.062
65-94	BIVARIATE	1.739	0.326	0.001	0.315	0.059
35-64	MULTIVARIATE	0.949	0.432	0.028	0.138	0.063
65-94	MULTIVARIATE	1.621	0.335	0.001	0.294	0.061

SEX: FEMALE
EVENT: CORONARY HEART DISEASE DEATH
RISK FACTOR: GLUCOSE IN URINE, DEFINITE OR TRACE
PERSONS AT RISK: PERSONS FREE OF CORONARY HEART DISEASE

RISK FACTOR	35-44 AT EXAM PERSON EXAMS	# OF EVENTS	ANNUAL RATE	45-54 AT EXAM PERSON EXAMS	# OF EVENTS	ANNUAL RATE	55-64 AT EXAM PERSON EXAMS	# OF EVENTS	ANNUAL RATE	65-74 AT EXAM PERSON EXAMS	# OF EVENTS	ANNUAL RATE
NO	5537			9841	9	0	11346	28	1	6131	37	3
YES	41			94	1	3	144	3	10	107	5	23
UNKNOWN	38			24	0	0	23	0	0	10	0	0
TOTAL	5616			9959	10	1	11513	31	1	6248	42	3

RISK FACTOR	75-84 AT EXAM PERSON EXAMS	# OF EVENTS	ANNUAL RATE	85-94 AT EXAM PERSON EXAMS	# OF EVENTS	ANNUAL RATE	35-64 COMBINED # OF EVENTS	CRUDE	AGE-ADJUSTED	65-94 COMBINED # OF EVENTS	CRUDE	AGE-ADJUSTED
NO	1919	28	7	120	3		38	1	1	68	4	4
YES	23	2	43	3			5	9	9	8	30	30
UNKNOWN	9	0	0	0			0	0	0	0	0	0
TOTAL	1951	30	8	123			43	1	1	76	5	5

LOGISTIC REGRESSIONS

AGE AT EXAM		COEFFICIENT OF RISK FACTOR	SE COEFFICIENT	P-VALUE	STANDARDIZED COEFFICIENT	STD SE
35-44						
45-54		2.464	1.059	0.020	0.239	0.103
55-64		2.152	0.613	0.001	0.239	0.068
65-74		2.089	0.487	0.001	0.271	0.063
75-84		1.861	0.764	0.015	0.201	0.083
85-94						
35-64	UNIVARIATE	2.551	0.480	0.001	0.258	0.048
65-94	UNIVARIATE	2.031	0.384	0.001	0.255	0.048
35-64	BIVARIATE	2.425	0.482	0.001	0.245	0.049
65-94	BIVARIATE	2.084	0.387	0.001	0.262	0.049
35-64	MULTIVARIATE	2.049	0.500	0.001	0.207	0.051
65-94	MULTIVARIATE	1.907	0.404	0.001	0.239	0.051

TABLE 11-9   DATE=01MAR86
AVERAGE ANNUAL INCIDENCE RATE FOR EVENT PER 1000 PERSONS AT RISK AT EXAM
BY AGE AND LEVEL OF RISK FACTOR AT EXAM: FRAMINGHAM STUDY 30-YEAR FOLLOWUP

SEX: MALE
EVENT: CORONARY HEART DISEASE DEATH
RISK FACTOR: GLUCOSE INTOLERANCE
PERSONS AT RISK: PERSONS FREE OF CORONARY HEART DISEASE

RISK FACTOR	35-44 AT EXAM			45-54 AT EXAM			55-64 AT EXAM			65-74 AT EXAM		
	PERSON EXAMS	# OF EVENTS	ANNUAL RATE	PERSON EXAMS	# OF EVENTS	ANNUAL RATE	PERSON EXAMS	# OF EVENTS	ANNUAL RATE	PERSON EXAMS	# OF EVENTS	ANNUAL RATE
NO	4335	5	1	7221	25	2	7145	55	4	3228	36	6
YES	137	0	0	503	3	3	904	5	3	685	18	13
UNKNOWN	0	0	-	0	0	-	0	0	-	0	0	-
TOTAL	4472	5	1	7724	28	2	8049	60	4	3913	54	7

RISK FACTOR	75-84 AT EXAM			85-94 AT EXAM			35-64 COMBINED			65-94 COMBINED		
	PERSON EXAMS	# OF EVENTS	ANNUAL RATE	PERSON EXAMS	# OF EVENTS	ANNUAL RATE	# OF EVENTS	CRUDE	AGE-ADJUSTED	# OF EVENTS	CRUDE	AGE-ADJUSTED
NO	802	19	12	27			85	2	2	56	7	7
YES	215	4	9	12			8	3	2	23	13	13
UNKNOWN	0	0	-	0			0	-	-	0	-	-
TOTAL	1017	23	11	39			93	2	2	79	8	8

LOGISTIC REGRESSIONS

AGE AT EXAM		COEFFICIENT OF RISK FACTOR	SE COEFFICIENT	P-VALUE	STANDARDIZED COEFFICIENT	STD SE
35-44		-	-		-	-
45-54		0.546	0.613	0.373	0.135	0.151
55-64		-0.333	0.468	0.478	-0.105	0.148
65-74		0.872	0.292	0.003	0.332	0.111
75-84		-0.247	0.556	0.657	-0.101	0.227
85-94						
35-64	UNIVARIATE	0.132	0.371	0.723	0.035	0.098
65-94	UNIVARIATE	0.614	0.250	0.014	0.238	0.097
35-64	BIVARIATE	-0.147	0.373	0.693	-0.059	0.099
65-94	BIVARIATE	0.559	0.251	0.026	0.216	0.097
35-64	MULTIVARIATE	-0.203	0.376	0.590	-0.054	0.100
65-94	MULTIVARIATE	0.520	0.254	0.040	0.201	0.098

SEX: FEMALE
EVENT: CORONARY HEART DISEASE DEATH
RISK FACTOR: GLUCOSE INTOLERANCE
PERSONS AT RISK: PERSONS FREE OF CORONARY HEART DISEASE

RISK FACTOR	35-44 AT EXAM			45-54 AT EXAM			55-64 AT EXAM			65-74 AT EXAM		
	PERSON EXAMS	# OF EVENTS	ANNUAL RATE	PERSON EXAMS	# OF EVENTS	ANNUAL RATE	PERSON EXAMS	# OF EVENTS	ANNUAL RATE	PERSON EXAMS	# OF EVENTS	ANNUAL RATE
NO	5522			9588	6	0	10503	24	1	5398	27	3
YES	94			371	4	5	1010	7	3	850	15	9
UNKNOWN	0		-	0	0	-	0	0	-	0	0	-
TOTAL	5616			9959	10	1	11513	31	1	6248	42	3

RISK FACTOR	75-84 AT EXAM			85-94 AT EXAM			35-64 COMBINED			65-94 COMBINED		
	PERSON EXAMS	# OF EVENTS	ANNUAL RATE	PERSON EXAMS	# OF EVENTS	ANNUAL RATE	# OF EVENTS	CRUDE	AGE-ADJUSTED	# OF EVENTS	CRUDE	AGE-ADJUSTED
NO	1581	23	7	100			31	1	1	52	4	4
YES	370	7	9	23			12	4	5	24	10	9
UNKNOWN	0	0	-	0			0	-	-	0	-	-
TOTAL	1951	30	8	123			43	1	1	76	5	5

LOGISTIC REGRESSIONS

AGE AT EXAM		COEFFICIENT OF RISK FACTOR	SE COEFFICIENT	P-VALUE	STANDARDIZED COEFFICIENT	STD SE
35-44						
45-54		2.857	0.648	0.001	0.541	0.123
55-64		1.114	0.451	0.010	0.315	0.122
65-74		1.274	0.324	0.001	0.437	0.111
75-84		0.267	0.436	0.540	0.105	0.171
85-94						
35-64	UNIVARIATE	1.912	0.341	0.001	0.434	0.077
65-94	UNIVARIATE	0.979	0.249	0.001	0.349	0.089
35-64	BIVARIATE	1.538	0.346	0.001	0.349	0.079
65-94	BIVARIATE	0.889	0.250	0.001	0.317	0.089
35-64	MULTIVARIATE	1.345	0.354	0.001	0.305	0.080
65-94	MULTIVARIATE	0.841	0.256	0.001	0.300	0.091

TABLE 11-10   DATE=21NOV85
AVERAGE ANNUAL INCIDENCE RATE FOR EVENT PER 1000 PERSONS AT RISK AT EXAM
BY AGE AND LEVEL OF RISK·FACTOR AT EXAM! FRAMINGHAM STUDY 30-YEAR FOLLOWUP

SEX! MALE
EVENT! CORONARY HEART DISEASE DEATH
RISK FACTOR! METROPOLITAN RELATIVE WEIGHT (PERCENT)
PERSONS AT RISK! PERSONS FREE OF CORONARY HEART DISEASE

RISK FACTOR	35-44 AT EXAM PERSON EXAMS	# OF EVENTS	ANNUAL RATE	45-54 AT EXAM PERSON EXAMS	# OF EVENTS	ANNUAL RATE	55-64 AT EXAM PERSON EXAMS	# OF EVENTS	ANNUAL RATE	65-74 AT EXAM PERSON EXAMS	# OF EVENTS	ANNUAL RATE
54-104	723	1	1	1131	2	1	1155	13	6	686	9	7
105-114	965	0	0	1540	7	2	1655	10	3	851	5	3
115-124	1122	1	0	2037	9	2	2175	8	2	1069	21	10
125-134	915	2	1	1645	3	1	1674	13	4	766	11	7
135-272	747	1	1	1370	7	3	1389	16	6	541	8	7
UNKNOWN	0	0	-	1	0	0	1	0	0	0	0	-
TOTAL	4472	5	1	7724	28	2	8049	60	4	3913	54	7

RISK FACTOR	75-84 AT EXAM PERSON EXAMS	# OF EVENTS	ANNUAL RATE	85-94 AT EXAM PERSON EXAMS	# OF EVENTS	ANNUAL RATE	35-64 COMBINED # OF EVENTS	CRUDE	AGE-ADJUSTED	65-94 COMBINED # OF EVENTS	CRUDE	AGE-ADJUSTED
54-104	224	5	11	18			16	3	3	14	8	7
105-114	252	6	12	9			17	2	2	12	5	5
115-124	296	5	8	12			18	2	2	27	10	10
125-134	155	3	10	0			18	2	2	14	8	8
135-272	90	4	22	0			24	3	3	12	10	10
UNKNOWN	0	0	-	0			0	0	0	0	-	
TOTAL	1017	23	11	39			93	2	2	79	8	8

LOGISTIC REGRESSIONS

AGE AT EXAM		COEFFICIENT OF RISK FACTOR	SE COEFFICIENT	P-VALUE	STANDARDIZED COEFFICIENT	STD SE
35-44		0.019	0.026	0.458	0.309	0.417
45-54		0.004	0.012	0.723	0.066	0.187
55-64		0.005	0.008	0.510	0.084	0.127
65-74		0.007	0.009	0.413	0.110	0.135
75-84		-0.001	0.014	0.937	-0.017	0.212
85-94						
35-64	UNIVARIATE	0.006	0.006	0.341	0.097	0.102
65-94	UNIVARIATE	0.004	0.007	0.622	0.055	0.112
35-64	BIVARIATE	0.006	0.006	0.349	0.096	0.102
65-94	BIVARIATE	0.007	0.007	0.362	0.103	0.113
35-64	MULTIVARIATE	0.003	0.007	0.651	0.048	0.106
65-94	MULTIVARIATE	0.002	0.007	0.819	0.027	0.116

SEX! FEMALE
EVENT! CORONARY HEART DISEASE DEATH
RISK FACTOR! METROPOLITAN RELATIVE WEIGHT (PERCENT)
PERSONS AT RISK! PERSONS FREE OF CORONARY HEART DISEASE

RISK FACTOR	35-44 AT EXAM PERSON EXAMS	# OF EVENTS	ANNUAL RATE	45-54 AT EXAM PERSON EXAMS	# OF EVENTS	ANNUAL RATE	55-64 AT EXAM PERSON EXAMS	# OF EVENTS	ANNUAL RATE	65-74 AT EXAM PERSON EXAMS	# OF EVENTS	ANNUAL RATE
54-104	1694			2095	2	0	2045	4	1	1216	9	4
105-114	1430			2351	0	0	2409	4	1	1145	8	3
115-124	1058			2107	0	0	2479	10	2	1395	4	1
125-134	566			1372	2	1	1743	4	1	982	9	5
135-272	860			2054	6	1	2837	9	2	1510	12	4
UNKNOWN	8			0	0	-	0	0	-	0	0	-
TOTAL	5616			9959	10	1	11513	31	1	6248	42	3

RISK FACTOR	75-84 AT EXAM PERSON EXAMS	# OF EVENTS	ANNUAL RATE	85-94 AT EXAM PERSON EXAMS	# OF EVENTS	ANNUAL RATE	35-64 COMBINED # OF EVENTS	CRUDE	AGE-ADJUSTED	65-94 COMBINED # OF EVENTS	CRUDE	AGE-ADJUSTED
54-104	517	7	7	41			7	1	1	19	5	5
105-114	375	3	4	28			4	0	0	11	4	4
115-124	399	8	10	21			10	1	1	13	4	4
125-134	266	6	11	16			6	1	1	15	6	6
135-272	394	6	8	17			16	1	1	18	5	5
UNKNOWN	0	0	-	0			0	0	0	0	-	
TOTAL	1951	30	8	123			43	1	1	76	5	5

LOGISTIC REGRESSIONS

AGE AT EXAM		COEFFICIENT OF RISK FACTOR	SE COEFFICIENT	P-VALUE	STANDARDIZED COEFFICIENT	STD SE
35-44						
45-54		0.027	0.009	0.002	0.576	0.185
55-64		0.004	0.008	0.617	0.085	0.170
65-74		-0.001	0.007	0.871	-0.026	0.157
75-84		0.009	0.008	0.272	0.188	0.172
85-94						
35-64	UNIVARIATE	0.014	0.006	0.012	0.303	0.121
65-94	UNIVARIATE	-0.001	0.005	0.851	-0.022	0.116
35-64	BIVARIATE	0.011	0.006	0.056	0.243	0.127
65-94	BIVARIATE	0.001	0.005	0.792	0.031	0.116
35-64	MULTIVARIATE	0.006	0.006	0.365	0.125	0.138
65-94	MULTIVARIATE	-0.002	0.005	0.775	-0.034	0.118

TABLE 11-11   DATE=21NOV85

AVERAGE ANNUAL INCIDENCE RATE FOR EVENT PER 1000 PERSONS AT RISK AT EXAM
BY AGE AND LEVEL OF RISK FACTOR AT EXAM: FRAMINGHAM STUDY 30-YEAR FOLLOWUP

SEX: MALE
EVENT: CORONARY HEART DISEASE DEATH
RISK FACTOR: VITAL CAPACITY - HEIGHT INDEX (ML/INCH)
PERSONS AT RISK: PERSONS FREE OF CORONARY HEART DISEASE

RISK FACTOR	35-44 AT EXAM PERSON EXAMS	# OF EVENTS	ANNUAL RATE	45-54 AT EXAM PERSON EXAMS	# OF EVENTS	ANNUAL RATE	55-64 AT EXAM PERSON EXAMS	# OF EVENTS	ANNUAL RATE	65-74 AT EXAM PERSON EXAMS	# OF EVENTS	ANNUAL RATE
8-44	269	0	0	1062	3	1	2045	20	5	1752	26	7
45-49	411	1	1	1125	3	1	1480	11	4	749	15	10
50-54	747	0	0	1658	9	3	1835	14	4	712	6	6
55-59	1022	2	1	1739	4	1	1416	6	2	442	2	2
60-85	2010	2	0	2135	9	2	1255	9	4	246	2	4
UNKNOWN	13	0	0	5	0	0	18	0	0	12	1	7
TOTAL	4472	5	1	7724	28	2	8049	60	4	3913	54	7

RISK FACTOR	75-84 AT EXAM PERSON EXAMS	# OF EVENTS	ANNUAL RATE	85-94 AT EXAM PERSON EXAMS	# OF EVENTS	ANNUAL RATE	35-64 COMBINED # OF EVENTS	CRUDE	AGE-ADJUSTED	65-94 COMBINED # OF EVENTS	CRUDE	AGE-ADJUSTED
8-44	649	16	12	24			23	3	2	43	9	9
45-49	172	1	3	6			15	2	2	16	9	8
50-54	97	2	10	7			23	3	3	11	7	7
55-59	65	4	31	2			12	1	1	6	8	8
60-85	26	0	0	0			20	2	2	2	4	3
UNKNOWN	8	0	0	0			0	0	0	1	25	33
TOTAL	1017	23	11	39			93	2	2	79	8	8

LOGISTIC REGRESSIONS

AGE AT EXAM		COEFFICIENT OF RISK FACTOR	SE COEFFICIENT	P-VALUE	STANDARDIZED COEFFICIENT	STD SE
35-44		-0.015	0.049	0.759	-0.134	0.439
45-54		0.015	0.021	0.476	0.137	0.192
55-64		-0.016	0.014	0.226	-0.154	0.127
65-74		-0.028	0.014	0.051	-0.264	0.135
75-84		-0.006	0.023	0.780	-0.059	0.211
85-94						
35-64	UNIVARIATE	-0.024	0.010	0.024	-0.228	0.101
65-94	UNIVARIATE	-0.026	0.012	0.025	-0.252	0.112
35-64	BIVARIATE	-0.003	0.011	0.771	-0.032	0.109
65-94	BIVARIATE	-0.019	0.012	0.123	-0.179	0.116
35-64	MULTIVARIATE	0.006	0.012	0.603	0.059	0.113
65-94	MULTIVARIATE	-0.014	0.012	0.269	-0.132	0.119

SEX: FEMALE
EVENT: CORONARY HEART DISEASE DEATH
RISK FACTOR: VITAL CAPACITY - HEIGHT INDEX (ML/INCH)
PERSONS AT RISK: PERSONS FREE OF CORONARY HEART DISEASE

RISK FACTOR	35-44 AT EXAM PERSON EXAMS	# OF EVENTS	ANNUAL RATE	45-54 AT EXAM PERSON EXAMS	# OF EVENTS	ANNUAL RATE	55-64 AT EXAM PERSON EXAMS	# OF EVENTS	ANNUAL RATE	65-74 AT EXAM PERSON EXAMS	# OF EVENTS	ANNUAL RATE
4-29	122			534	1	1	1461	12	4	1747	22	6
30-34	289			1217	0	0	2418	7	1	1715	9	3
35-39	769			2077	2	0	3048	6	1	1529	4	1
40-44	1427			2804	4	1	2786	5	1	854	5	3
45-92	2988			3310	3	0	1859	1	0	397	2	3
UNKNOWN	21			17	0	0	21	0	0	6	0	0
TOTAL	5616			9959	10	1	11513	31	1	6248	42	3

RISK FACTOR	75-84 AT EXAM PERSON EXAMS	# OF EVENTS	ANNUAL RATE	85-94 AT EXAM PERSON EXAMS	# OF EVENTS	ANNUAL RATE	35-64 COMBINED # OF EVENTS	CRUDE	AGE-ADJUSTED	65-94 COMBINED # OF EVENTS	CRUDE	AGE-ADJUSTED
4-29	930	15	8	80			13	3	2	40	7	7
30-34	569	11	10	29			7	1	1	21	5	4
35-39	314	4	6	13			9	1	1	8	2	2
40-44	111	0	0	1			9	1	1	5	3	2
45-92	25	0	0	0			5	0	0	2	3	2
UNKNOWN	2	0	0	0			0	0	0	0	0	0
TOTAL	1951	30	8	123			43	1	1	76	5	5

LOGISTIC REGRESSIONS

AGE AT EXAM		COEFFICIENT OF RISK FACTOR	SE COEFFICIENT	P-VALUE	STANDARDIZED COEFFICIENT	STD SE
35-44						
45-54		-0.006	0.043	0.887	-0.045	0.315
55-64		-0.083	0.022	0.001	-0.621	0.164
65-74		-0.051	0.020	0.012	-0.374	0.150
75-84		-0.020	0.026	0.435	-0.141	0.181
85-94						
35-64	UNIVARIATE	-0.081	0.018	0.001	-0.643	0.141
65-94	UNIVARIATE	-0.050	0.015	0.001	-0.373	0.111
35-64	BIVARIATE	-0.055	0.020	0.005	-0.435	0.156
65-94	BIVARIATE	-0.036	0.016	0.020	-0.275	0.118
35-64	MULTIVARIATE	-0.040	0.021	0.054	-0.318	0.166
65-94	MULTIVARIATE	-0.033	0.017	0.046	-0.249	0.125

TABLE 11-12   DATE=21NOV85

AVERAGE ANNUAL INCIDENCE RATE FOR EVENT PER 1000 PERSONS AT RISK AT EXAM
BY AGE AND LEVEL OF RISK FACTOR AT EXAM: FRAMINGHAM STUDY 30-YEAR FOLLOWUP

SEX: MALE
EVENT: CORONARY HEART DISEASE DEATH
RISK FACTOR: HEART RATE (PER MINUTE)
PERSONS AT RISK: PERSONS FREE OF CORONARY HEART DISEASE

RISK FACTOR	35-44 AT EXAM PERSON EXAMS	# OF EVENTS	ANNUAL RATE	45-54 AT EXAM PERSON EXAMS	# OF EVENTS	ANNUAL RATE	55-64 AT EXAM PERSON EXAMS	# OF EVENTS	ANNUAL RATE	65-74 AT EXAM PERSON EXAMS	# OF EVENTS	ANNUAL RATE
30-67	1284	0	0	1949	3	1	2309	8	2	1183	12	5
68-75	1492	0	0	2490	6	1	2490	17	3	1159	9	4
76-83	722	1	1	1394	6	2	1374	10	4	624	10	8
84-91	579	4	3	1097	6	3	1064	8	4	487	11	11
92-220	395	0	0	794	7	4	812	17	10	460	12	13
UNKNOWN	0	0	-	0	0	-	0	0	-	0	0	-
TOTAL	4472	5	1	7724	28	2	8049	60	4	3913	54	7

RISK FACTOR	75-84 AT EXAM PERSON EXAMS	# OF EVENTS	ANNUAL RATE	85-94 AT EXAM PERSON EXAMS	# OF EVENTS	ANNUAL RATE	35-64 COMBINED # OF EVENTS	CRUDE	AGE-ADJUSTED	65-94 COMBINED # OF EVENTS	CRUDE	AGE-ADJUSTED
30-67	301	7	12	11			11	1	1	19	6	6
68-75	296	7	12	12			23	2	2	16	5	5
76-83	192	4	10	10			17	2	2	14	8	
84-91	123	3	12	5			18	3	3	16	13	13
92-220	105	2	10	1			24	6	6	14	12	12
UNKNOWN	0	0	-	0			0	-	-	0	-	-
TOTAL	1017	23	11	39			93	2	2	79	8	8

LOGISTIC REGRESSIONS

AGE AT EXAM		COEFFICIENT OF RISK FACTOR	SE COEFFICIENT	P-VALUE	STANDARDIZED COEFFICIENT	STD SE
35-44		0.050	0.027	0.068	0.619	0.339
45-54		0.043	0.013	0.001	0.535	0.163
55-64		0.036	0.008	0.001	0.461	0.109
65-74		0.024	0.009	0.006	0.333	0.121
75-84		0.002	0.016	0.904	.025	0.269
85-94						
35-64	UNIVARIATE	0.039	0.007	0.001	0.495	0.087
65-94	UNIVARIATE	0.019	0.008	0.012	0.258	0.103
35-64	BIVARIATE	0.038	0.007	0.001	0.485	0.086
65-94	BIVARIATE	0.019	0.008	0.012	0.260	0.103
35-64	MULTIVARIATE	0.030	0.007	0.001	0.375	0.092
65-94	MULTIVARIATE	0.016	0.008	0.037	0.220	0.105

SEX: FEMALE
EVENT: CORONARY HEART DISEASE DEATH
RISK FACTOR: HEART RATE (PER MINUTE)
PERSONS AT RISK: PERSONS FREE OF CORONARY HEART DISEASE

RISK FACTOR	35-44 AT EXAM PERSON EXAMS	# OF EVENTS	ANNUAL RATE	45-54 AT EXAM PERSON EXAMS	# OF EVENTS	ANNUAL RATE	55-64 AT EXAM PERSON EXAMS	# OF EVENTS	ANNUAL RATE	65-74 AT EXAM PERSON EXAMS	# OF EVENTS	ANNUAL RATE
30-67	1092			1897	1	0	2305	3	1	1103	5	2
68-75	1803			3244	4	1	3468	10	1	1668	11	3
76-83	1051			1911	0	0	2351	2	0	1358	9	3
84-91	939			1604	2	1	1780	10	3	1089	7	3
92-220	731			1303	3	1	1609	6	2	1030	10	5
UNKNOWN	0			0	0	-	0	0	-	0	0	-
TOTAL	5616			9959	10	1	11513	31	1	6248	42	3

RISK FACTOR	75-84 AT EXAM PERSON EXAMS	# OF EVENTS	ANNUAL RATE	85-94 AT EXAM PERSON EXAMS	# OF EVENTS	ANNUAL RATE	35-64 COMBINED # OF EVENTS	CRUDE	AGE-ADJUSTED	65-94 COMBINED # OF EVENTS	CRUDE	AGE-ADJUSTED
30-67	360	6	8	26			4	0	0	13	4	4
68-75	462	6	6	32			14	1	1	17	4	4
76-83	428	7	8	24			3	0	0	16	4	4
84-91	330	4	6	23			13	2	2	13	5	4
92-220	371	7	9	18			9	1	1	17	6	6
UNKNOWN	0			0			0	-	-	0	-	-
TOTAL	1951	30	8	123			43	1	1	76	5	5

LOGISTIC REGRESSIONS

AGE AT EXAM		COEFFICIENT OF RISK FACTOR	SE COEFFICIENT	P-VALUE	STANDARDIZED COEFFICIENT	STD SE
35-44						
45-54		0.019	0.023	0.414	0.237	0.290
55-64		0.020	0.012	0.102	0.257	0.157
65-74		0.017	0.011	0.104	0.226	0.139
75-84		-0.003	0.013	0.833	-0.039	0.187
85-94						
35-64	UNIVARIATE	0.020	0.010	0.053	0.256	0.133
65-94	UNIVARIATE	0.007	0.008	0.376	0.098	0.111
35-64	BIVARIATE	0.019	0.010	0.062	0.246	0.132
65-94	BIVARIATE	0.007	0.008	0.417	0.089	0.110
35-64	MULTIVARIATE	0.008	0.011	0.504	0.097	0.145
65-94	MULTIVARIATE	-0.001	0.009	0.922	-0.011	0.114

TABLE 11-13   DATE=12APR85
AVERAGE ANNUAL INCIDENCE RATE FOR EVENT PER 1000 PERSONS AT RISK AT EXAM
BY AGE AND LEVEL OF RISK FACTOR AT EXAM: FRAMINGHAM STUDY 30-YEAR FOLLOWUP

SEX: MALE
EVENT: CORONARY HEART DISEASE DEATH
RISK FACTOR: CIGARETTES SMOKED (NUMBER PER DAY)
PERSONS AT RISK: PERSONS FREE OF CORONARY HEART DISEASE

RISK FACTOR	35-44 AT EXAM			45-54 AT EXAM			55-64 AT EXAM			65-74 AT EXAM		
	PERSON EXAMS	# OF EVENTS	ANNUAL RATE	PERSON EXAMS	# OF EVENTS	ANNUAL RATE	PERSON EXAMS	# OF EVENTS	ANNUAL RATE	PERSON EXAMS	# OF EVENTS	ANNUAL RATE
NONE	1383	0	0	3068	6	1	4295	22	3	2611	31	6
1-10	433	1	1	716	3	2	769	7	5	374	6	8
11-20	1485	0	0	2051	10	2	1682	18	5	526	10	10
21-40	987	3	2	1602	6	2	1119	9	4	369	6	8
41-90	158	1	3	252	2	4	162	4	12	27	1	19
UNKNOWN	26	0	0	35	1	14	22	0	0	6	0	0
TOTAL	4472	5	1	7724	28	2	8049	60	4	3913	54	7

RISK FACTOR	75-84 AT EXAM			85-94 AT EXAM			35-64 COMBINED			65-94 COMBINED		
	PERSON EXAMS	# OF EVENTS	ANNUAL RATE	PERSON EXAMS	# OF EVENTS	ANNUAL RATE	# OF EVENTS	CRUDE	AGE-ADJUSTED	# OF EVENTS	CRUDE	AGE-ADJUSTED
NONE	797	20	13	27			28	2	2	52	8	8
1-10	92	3	16	6			11	3	3	9	10	10
11-20	77	0	0	5			28	3	3	11	9	9
21-40	46	0	0	1			18	2	2	6	7	6
41-90	5	0	0	0			7	6	7	1	16	15
UNKNOWN	0	0	-	0			1	6	5	0	0	0
TOTAL	1017	23	11	39			93	2	3	79	8	8

LOGISTIC REGRESSIONS

AGE AT EXAM		COEFFICIENT OF RISK FACTOR	SE COEFFICIENT	P-VALUE	STANDARDIZED COEFFICIENT	STD SE
35-44		0.056	0.028	0.046	0.794	0.398
45-54		0.019	0.012	0.106	0.282	0.174
55-64		0.021	0.008	0.008	0.293	0.110
65-74		0.013	0.010	0.208	0.153	0.121
75-84		-0.079	0.058	0.168	-0.723	0.525
85-94						
35-64	UNIVARIATE	0.017	0.007	0.010	0.243	0.095
65-94	UNIVARIATE	0.002	0.010	0.802	0.028	0.111
35-64	BIVARIATE	0.024	0.006	0.001	0.346	0.092
65-94	BIVARIATE	0.007	0.010	0.482	0.078	0.111
35-64	MULTIVARIATE	0.025	0.006	0.001	0.357	0.092
65-94	MULTIVARIATE	0.011	0.010	0.284	0.119	0.111

SEX: FEMALE
EVENT: CORONARY HEART DISEASE DEATH
RISK FACTOR: CIGARETTES SMOKED (NUMBER PER DAY)
PERSONS AT RISK: PERSONS FREE OF CORONARY HEART DISEASE

RISK FACTOR	35-44 AT EXAM			45-54 AT EXAM			55-64 AT EXAM			65-74 AT EXAM		
	PERSON EXAMS	# OF EVENTS	ANNUAL RATE	PERSON EXAMS	# OF EVENTS	ANNUAL RATE	PERSON EXAMS	# OF EVENTS	ANNUAL RATE	PERSON EXAMS	# OF EVENTS	ANNUAL RATE
NONE	2582			5528	4	0	7595	21	1	4863	28	3
1-10	1314			1726	2	1	1566	5	2	626	5	4
11-20	1371			2009	2	0	1696	4	1	568	8	7
21-40	298			622	1	1	602	1	1	172	0	0
41-90	22			37	1	14	16	0	0	0	0	-
UNKNOWN	29			37	0	0	38	0	0	19	1	26
TOTAL	5616			9959	10	1	11513	31	1	6248	42	3

RISK FACTOR	75-84 AT EXAM			85-94 AT EXAM			35-64 COMBINED			65-94 COMBINED		
	PERSON EXAMS	# OF EVENTS	ANNUAL RATE	PERSON EXAMS	# OF EVENTS	ANNUAL RATE	# OF EVENTS	CRUDE	AGE-ADJUSTED	# OF EVENTS	CRUDE	AGE-ADJUSTED
NONE	1710	25	7	117			25	1	0	57	4	4
1-10	137	3	11	4			7	1	1	8	5	6
11-20	84	2	12	2			7	1	0	10	8	8
21-40	16	0	0	0			3	1	1	0	0	0
41-90	0	0	-	0			1	7	5	0		
UNKNOWN	4	0	0	0			0	0	0	1	22	20
TOTAL	1951	30	8	123			43	1	1	76	5	4

LOGISTIC REGRESSIONS

AGE AT EXAM		COEFFICIENT OF RISK FACTOR	SE COEFFICIENT	P-VALUE	STANDARDIZED COEFFICIENT	STD SE
35-44						
45-54		0.036	0.024	0.140	0.366	0.249
55-64		-0.010	0.021	0.634	-0.093	0.196
65-74		0.020	0.017	0.233	0.153	0.129
75-84		0.004	0.035	0.905	0.021	0.177
85-94						
35-64	UNIVARIATE	0.009	0.014	0.523	0.091	0.142
65-94	UNIVARIATE	0.008	0.015	0.618	0.054	0.107
35-64	BIVARIATE	0.018	0.014	0.199	0.176	0.138
65-94	BIVARIATE	0.018	0.015	0.251	0.124	0.108
35-64	MULTIVARIATE	0.021	0.014	0.134	0.206	0.138
65-94	MULTIVARIATE	0.025	0.015	0.095	0.180	0.108

TABLE 11-14   DATE=21NOV85

AVERAGE ANNUAL INCIDENCE RATE FOR EVENT PER 1000 PERSONS AT RISK AT EXAM
BY AGE AND LEVEL OF RISK FACTOR AT EXAM: FRAMINGHAM STUDY 30-YEAR FOLLOWUP

SEX: MALE
EVENT: CORONARY HEART DISEASE DEATH
RISK FACTOR: ALBUMIN IN URINE, DEFINITE
PERSONS AT RISK: PERSONS FREE OF CORONARY HEART DISEASE

RISK FACTOR	35-44 AT EXAM PERSON EXAMS	# OF EVENTS	ANNUAL RATE	45-54 AT EXAM PERSON EXAMS	# OF EVENTS	ANNUAL RATE	55-64 AT EXAM PERSON EXAMS	# OF EVENTS	ANNUAL RATE	65-74 AT EXAM PERSON EXAMS	# OF EVENTS	ANNUAL RATE
NO	4366	5	1	7599	26	2	7838	56	4	3792	49	6
YES	70	0	0	99	2	10	184	3	8	115	5	22
UNKNOWN	36	0	0	26	0	0	27	1	19	6	0	0
TOTAL	4472	5	1	7724	28	2	8049	60	4	3913	54	7

RISK FACTOR	75-84 AT EXAM PERSON EXAMS	# OF EVENTS	ANNUAL RATE	85-94 AT EXAM PERSON EXAMS	# OF EVENTS	ANNUAL RATE	35-64 COMBINED # OF EVENTS	CRUDE	AGE-ADJUSTED	65-94 COMBINED # OF EVENTS	CRUDE	AGE-ADJUSTED
NO	989	22	11	32		7	87	2	2	73	8	8
YES	25	1	20			7	5	7	7	6	20	21
UNKNOWN	3	0	0	0			1	6	7	0	0	0
TOTAL	1017	23	11	39			93	2	2	79	8	8

LOGISTIC REGRESSIONS

AGE AT EXAM	COEFFICIENT OF RISK FACTOR	SE COEFFICIENT	P-VALUE	STANDARDIZED COEFFICIENT	STD SE
35-44	-	-	-	-	-
45-54	1.795	0.741	0.016	0.202	0.083
55-64	0.834	0.597	0.163	0.125	0.089
65-74	1.245	0.479	0.009	0.210	0.081
75-84	0.605	1.043	0.562	0.094	0.162
85-94					
35-64 UNIVARIATE	1.181	0.463	0.011	0.155	0.061
65-94 UNIVARIATE	1.016	0.433	0.019	0.172	0.073
35-64 BIVARIATE	1.018	0.465	0.029	0.134	0.061
65-94 BIVARIATE	0.967	0.436	0.027	0.164	0.074
35-64 MULTIVARIATE	0.501	0.480	0.297	0.066	0.063
65-94 MULTIVARIATE	0.589	0.448	0.189	0.100	0.076

SEX: FEMALE
EVENT: CORONARY HEART DISEASE DEATH
RISK FACTOR: ALBUMIN IN URINE, DEFINITE
PERSONS AT RISK: PERSONS FREE OF CORONARY HEART DISEASE

RISK FACTOR	35-44 AT EXAM PERSON EXAMS	# OF EVENTS	ANNUAL RATE	45-54 AT EXAM PERSON EXAMS	# OF EVENTS	ANNUAL RATE	55-64 AT EXAM PERSON EXAMS	# OF EVENTS	ANNUAL RATE	65-74 AT EXAM PERSON EXAMS	# OF EVENTS	ANNUAL RATE
NO	5353			9656	10	1	11263	29	1	6118	40	3
YES	225			278	0	0	225	2	4	120	2	8
UNKNOWN	38			25	0	0	25	0	0	10	0	0
TOTAL	5616			9959	10	1	11513	31	1	6248	42	3

RISK FACTOR	75-84 AT EXAM PERSON EXAMS	# OF EVENTS	ANNUAL RATE	85-94 AT EXAM PERSON EXAMS	# OF EVENTS	ANNUAL RATE	35-64 COMBINED # OF EVENTS	CRUDE	AGE-ADJUSTED	65-94 COMBINED # OF EVENTS	CRUDE	AGE-ADJUSTED
NO	1904	30	8	120		3	41	1	1	74	5	5
YES	38	0	0	3			2	1	2	2	6	6
UNKNOWN	9	0	0	0			0	0	0	0	0	0
TOTAL	1951	30	8	123			43	1	1	76	5	5

LOGISTIC REGRESSIONS

AGE AT EXAM	COEFFICIENT OF RISK FACTOR	SE COEFFICIENT	P-VALUE	STANDARDIZED COEFFICIENT	STD SE
35-44	-	-	-	-	-
45-54	-	-	-	-	-
55-64	1.245	0.734	0.090	0.173	0.102
65-74	0.946	0.731	0.195	0.130	0.100
75-84	-	-	-	-	-
85-94	-	-	-	-	-
35-64 UNIVARIATE	0.567	0.725	0.434	0.092	0.117
65-94 UNIVARIATE	0.316	0.721	0.661	0.044	0.099
35-64 BIVARIATE	0.809	0.727	0.266	0.131	0.118
65-94 BIVARIATE	0.298	0.722	0.680	0.041	0.100
35-64 MULTIVARIATE	-0.167	0.872	0.848	-0.027	0.141
65-94 MULTIVARIATE	-0.133	0.737	0.857	-0.018	0.102

TABLE 11-15   DATE=21NOV85

AVERAGE ANNUAL INCIDENCE RATE FOR EVENT PER 1000 PERSONS AT RISK AT EXAM
BY AGE AND LEVEL OF RISK FACTOR AT EXAM: FRAMINGHAM STUDY 30-YEAR FOLLOWUP

SEX: MALE
EVENT: CORONARY HEART DISEASE DEATH
RISK FACTOR: HEART ENLARGEMENT BY X-RAY
PERSONS AT RISK: PERSONS FREE OF CORONARY HEART DISEASE

RISK FACTOR	35-44 AT EXAM			45-54 AT EXAM			55-64 AT EXAM			65-74 AT EXAM		
	PERSON EXAMS	# OF EVENTS	ANNUAL RATE	PERSON EXAMS	# OF EVENTS	ANNUAL RATE	PERSON EXAMS	# OF EVENTS	ANNUAL RATE	PERSON EXAMS	# OF EVENTS	ANNUAL RATE
NO	4063	5	1	6383	18	1	5967	41	3	2584	31	6
MILD	185	0	0	650	2	2	962	8	4	578	7	6
DEFINITE	219	0	0	687	8	6	1119	11	5	751	16	11
UNKNOWN	5	0	0	4	0	0	1	0	0	0	0	-
TOTAL	4472	5	1	7724	28	2	8049	60	4	3913	54	7

RISK FACTOR	75-84 AT EXAM			85-94 AT EXAM			35-64 COMBINED			65-94 COMBINED		
	PERSON EXAMS	# OF EVENTS	ANNUAL RATE	PERSON EXAMS	# OF EVENTS	ANNUAL RATE	# OF EVENTS	CRUDE	AGE-ADJUSTED	# OF EVENTS	CRUDE	AGE-ADJUSTED
NO	589	9	8	29	0	-	64	2	2	40	6	6
MILD	178	6	17	5	0	-	10	3	2	14	9	9
DEFINITE	250	8	16	5	0	-	19	5	4	25	12	12
UNKNOWN	0	0	-	0	0	-	0	0	0	0	-	-
TOTAL	1017	23	11	39	0	-	93	2	2	79	8	8

LOGISTIC REGRESSIONS

AGE AT EXAM		COEFFICIENT OF RISK FACTOR	SE COEFFICIENT	P-VALUE	STANDARDIZED COEFFICIENT	STD SE
35-44		-	-	-	-	-
45-54		0.685	0.219	0.002	0.418	0.134
55-64		0.182	0.165	0.270	0.131	0.119
65-74		0.277	0.158	0.079	0.220	0.125
75-84		0.384	0.235	0.102	0.325	0.199
85-94						
35-64	UNIVARIATE	0.435	0.129	0.001	0.277	0.082
65-94	UNIVARIATE	0.352	0.128	0.006	0.284	0.103
35-64	BIVARIATE	0.302	0.132	0.022	0.192	0.084
65-94	BIVARIATE	0.321	0.128	0.012	0.259	0.104
35-64	MULTIVARIATE	0.221	0.140	0.116	0.141	0.089
65-94	MULTIVARIATE	0.165	0.138	0.231	0.133	0.111

SEX: FEMALE
EVENT: CORONARY HEART DISEASE DEATH
RISK FACTOR: HEART ENLARGEMENT BY X-RAY
PERSONS AT RISK: PERSONS FREE OF CORONARY HEART DISEASE

RISK FACTOR	35-44 AT EXAM			45-54 AT EXAM			55-64 AT EXAM			65-74 AT EXAM		
	PERSON EXAMS	# OF EVENTS	ANNUAL RATE	PERSON EXAMS	# OF EVENTS	ANNUAL RATE	PERSON EXAMS	# OF EVENTS	ANNUAL RATE	PERSON EXAMS	# OF EVENTS	ANNUAL RATE
NO	5164			8017	5	0	7522	10	1	3191	13	2
MILD	199			943	2	1	1769	9	3	1150	8	3
DEFINITE	245			993	3	2	2218	12	3	1907	21	6
UNKNOWN	8			6	0	0	4	0	0	0	0	-
TOTAL	5616			9959	10	1	11513	31	1	6248	42	3

RISK FACTOR	75-84 AT EXAM			85-94 AT EXAM			35-64 COMBINED			65-94 COMBINED		
	PERSON EXAMS	# OF EVENTS	ANNUAL RATE	PERSON EXAMS	# OF EVENTS	ANNUAL RATE	# OF EVENTS	CRUDE	AGE-ADJUSTED	# OF EVENTS	CRUDE	AGE-ADJUSTED
NO	702	5	4	32	0	-	16	0	0	20	3	3
MILD	373	6	8	30	0	-	11	2	2	15	5	5
DEFINITE	876	19	11	61	0	-	16	2	2	41	7	7
UNKNOWN	0	0	-	0	0	-	0	0	0	0	-	-
TOTAL	1951	30	8	123	0	-	43	1	1	76	5	5

LOGISTIC REGRESSIONS

AGE AT EXAM		COEFFICIENT OF RISK FACTOR	SE COEFFICIENT	P-VALUE	STANDARDIZED COEFFICIENT	STD SE
35-44						
45-54		0.809	0.350	0.021	0.516	0.223
55-64		0.692	0.199	0.001	0.551	0.158
65-74		0.499	0.175	0.004	0.439	0.154
75-84		0.528	0.234	0.024	0.472	0.209
85-94						
35-64	UNIVARIATE	0.897	0.165	0.001	0.626	0.115
65-94	UNIVARIATE	0.516	0.134	0.001	0.461	0.119
35-64	BIVARIATE	0.694	0.172	0.001	0.484	0.120
65-94	BIVARIATE	0.444	0.136	0.001	0.397	0.121
35-64	MULTIVARIATE	0.585	0.182	0.001	0.408	0.127
65-94	MULTIVARIATE	0.363	0.142	0.011	0.324	0.127

TABLE 11-16   DATE=12APR85

AVERAGE ANNUAL INCIDENCE RATE FOR EVENT PER 1000 PERSONS AT RISK AT EXAM
BY AGE AND LEVEL OF RISK FACTOR AT EXAM: FRAMINGHAM STUDY 30-YEAR FOLLOWUP

SEX: MALE
EVENT: CORONARY HEART DISEASE DEATH
RISK FACTOR: LEFT VENTRICULAR HYPERTROPHY BY ELECTROCARDIOGRAM
PERSONS AT RISK: PERSONS FREE OF CORONARY HEART DISEASE

RISK FACTOR	35-44 AT EXAM			45-54 AT EXAM			55-64 AT EXAM			65-74 AT EXAM		
	PERSON EXAMS	# OF EVENTS	ANNUAL RATE	PERSON EXAMS	# OF EVENTS	ANNUAL RATE	PERSON EXAMS	# OF EVENTS	ANNUAL RATE	PERSON EXAMS	# OF EVENTS	ANNUAL RATE
NO	4366	5	1	7497	24	2	7693	56	4	3703	41	6
BORDER	78	0	0	132	1	4	210	0	0	90	4	22
DEFINITE	28	0	0	95	3	16	146	4	14	120	9	38
UNKNOWN	0	0	–	0	0	–	0	0	–	0	0	–
TOTAL	4472	5	1	7724	28	2	8049	60	4	3913	54	7

RISK FACTOR	75-84 AT EXAM			85-94 AT EXAM			35-64 COMBINED			65-94 COMBINED		
	PERSON EXAMS	# OF EVENTS	ANNUAL RATE	PERSON EXAMS	# OF EVENTS	ANNUAL RATE	# OF EVENTS	CRUDE	AGE-ADJUSTED	# OF EVENTS	CRUDE	AGE-ADJUSTED
NO	949	22	12	34	0	0	85	2	3	65	7	7
BORDER	37	0	0	2	0	0	1	1	2	4	16	17
DEFINITE	31	1	16	3	0	0	7	13	12	10	32	33
UNKNOWN	0	0	–	0	0	–	0	0	–	0	0	–
TOTAL	1017	23	11	39	0	0	93	2	3	79	8	8

LOGISTIC REGRESSIONS

AGE AT EXAM	COEFFICIENT OF RISK FACTOR	SE COEFFICIENT	P-VALUE	STANDARDIZED COEFFICIENT	STD SE
35-44	–	–	–	–	–
45-54	1.136	0.306	0.001	0.289	0.078
55-64	0.497	0.285	0.082	0.153	0.088
65-74	1.019	0.182	0.001	0.379	0.068
75-84	-0.076	0.579	0.896	-0.029	0.223
85-94					
35-64 UNIVARIATE	0.794	0.208	0.001	0.212	0.056
65-94 UNIVARIATE	0.800	0.170	0.001	0.302	0.064
35-64 BIVARIATE	0.681	0.209	0.001	0.182	0.056
65-94 BIVARIATE	0.788	0.171	0.001	0.297	0.064
35-64 MULTIVARIATE	0.438	0.229	0.056	0.117	0.061
65-94 MULTIVARIATE	0.571	0.188	0.002	0.215	0.071

SEX: FEMALE
EVENT: CORONARY HEART DISEASE DEATH
RISK FACTOR: LEFT VENTRICULAR HYPERTROPHY BY ELECTROCARDIOGRAM
PERSONS AT RISK: PERSONS FREE OF CORONARY HEART DISEASE

RISK FACTOR	35-44 AT EXAM			45-54 AT EXAM			55-64 AT EXAM			65-74 AT EXAM		
	PERSON EXAMS	# OF EVENTS	ANNUAL RATE	PERSON EXAMS	# OF EVENTS	ANNUAL RATE	PERSON EXAMS	# OF EVENTS	ANNUAL RATE	PERSON EXAMS	# OF EVENTS	ANNUAL RATE
NO	5563			9798	10	1	11137	29	1	5966	36	3
BORDER	38			107	0	0	226	0	0	134	2	7
DEFINITE	15			54	0	0	150	2	7	148	4	14
UNKNOWN	0	0	–	0	0	–	0	0	–	0	0	–
TOTAL	5616			9959	10	1	11513	31	1	6248	42	3

RISK FACTOR	75-84 AT EXAM			85-94 AT EXAM			35-64 COMBINED			65-94 COMBINED		
	PERSON EXAMS	# OF EVENTS	ANNUAL RATE	PERSON EXAMS	# OF EVENTS	ANNUAL RATE	# OF EVENTS	CRUDE	AGE-ADJUSTED	# OF EVENTS	CRUDE	AGE-ADJUSTED
NO	1833	26	7	111	0	0	40	1	1	66	4	4
BORDER	49	2	20	6	0	0	0	0	0	4	11	10
DEFINITE	69	2	14	6	0	0	3	7	10	6	13	14
UNKNOWN	0	0	–	0	0	–	0	0	–	0	0	–
TOTAL	1951	30	8	123	0	0	43	1	1	76	5	4

LOGISTIC REGRESSIONS

AGE AT EXAM	COEFFICIENT OF RISK FACTOR	SE COEFFICIENT	P-VALUE	STANDARDIZED COEFFICIENT	STD SE
35-44	–	–	–	–	–
45-54					
55-64	0.666	0.398	0.095	0.176	0.105
65-74	0.775	0.254	0.002	0.259	0.085
75-84	0.460	0.327	0.159	0.183	0.130
85-94					
35-64 UNIVARIATE	0.977	0.322	0.002	0.208	0.069
65-94 UNIVARIATE	0.633	0.200	0.002	0.223	0.070
35-64 BIVARIATE	0.781	0.327	0.017	0.166	0.070
65-94 BIVARIATE	0.581	0.201	0.004	0.204	0.071
35-64 MULTIVARIATE	0.401	0.363	0.269	0.085	0.077
65-94 MULTIVARIATE	0.403	0.215	0.061	0.142	0.076

TABLE 11-17   DATE=21NOV85
AVERAGE ANNUAL INCIDENCE RATE FOR EVENT PER 1000 PERSONS AT RISK AT EXAM
BY AGE AND LEVEL OF RISK FACTOR AT EXAM: FRAMINGHAM STUDY 30-YEAR FOLLOWUP

SEX: MALE
EVENT: CORONARY HEART DISEASE DEATH
RISK FACTOR: INTRAVENTRICULAR CONDUCTION DEFECT (WITH QRS INTERVAL MORE THAN 0.11 SECOND)
PERSONS AT RISK: PERSONS FREE OF CORONARY HEART DISEASE

RISK FACTOR	35-44 AT EXAM PERSON EXAMS	# OF EVENTS	ANNUAL RATE	45-54 AT EXAM PERSON EXAMS	# OF EVENTS	ANNUAL RATE	55-64 AT EXAM PERSON EXAMS	# OF EVENTS	ANNUAL RATE	65-74 AT EXAM PERSON EXAMS	# OF EVENTS	ANNUAL RATE
NO	4161	5	1	7213	25	2	7386	51	3	3566	50	7
INCOMPLETE	275	0	0	446	2	2	456	4	4	183	2	5
COMPLETE	36	0	0	65	1	8	207	5	12	164	2	6
UNKNOWN	0	0	-	0	0	-	0	0	-	0	0	-
TOTAL	4472	5	1	7724	28	2	8049	60	4	3913	54	7

RISK FACTOR	75-84 AT EXAM PERSON EXAMS	# OF EVENTS	ANNUAL RATE	85-94 AT EXAM PERSON EXAMS	# OF EVENTS	ANNUAL RATE	35-64 COMBINED # OF EVENTS	CRUDE	AGE-ADJUSTED	65-94 COMBINED # OF EVENTS	CRUDE	AGE-ADJUSTED
NO	918	20	11	31			81	2	2	71	8	8
INCOMPLETE	21	1	24	2			6	3	3	3	7	9
COMPLETE	78	2	13	4			6	10	8	5	10	8
UNKNOWN	0	0	-	0			0	-	-	0	-	-
TOTAL	1017	23	11	39			93	2	2	79	8	8

LOGISTIC REGRESSIONS

AGE AT EXAM		COEFFICIENT OF RISK FACTOR	SE COEFFICIENT	P-VALUE	STANDARDIZED COEFFICIENT	STD SE
35-44		-	-	-	-	-
45-54		0.557	0.460	0.226	0.163	0.135
55-64		0.563	0.230	0.014	0.217	0.088
65-74		-0.108	0.335	0.746	-0.048	0.149
75-84		0.135	0.350	0.700	0.073	0.191
85-94						
35-64	UNIVARIATE	0.615	0.208	0.003	0.205	0.069
65-94	UNIVARIATE	0.099	0.224	0.658	0.047	0.105
35-64	BIVARIATE	0.495	0.205	0.016	0.165	0.068
65-94	BIVARIATE	0.050	0.223	0.823	0.023	0.105
35-64	MULTIVARIATE	0.543	0.209	0.010	0.181	0.070
65-94	MULTIVARIATE	0.091	0.226	0.687	0.043	0.106

SEX: FEMALE
EVENT: CORONARY HEART DISEASE DEATH
RISK FACTOR: INTRAVENTRICULAR CONDUCTION DEFECT (WITH QRS INTERVAL MORE THAN 0.11 SECOND)
PERSONS AT RISK: PERSONS FREE OF CORONARY HEART DISEASE

RISK FACTOR	35-44 AT EXAM PERSON EXAMS	# OF EVENTS	ANNUAL RATE	45-54 AT EXAM PERSON EXAMS	# OF EVENTS	ANNUAL RATE	55-64 AT EXAM PERSON EXAMS	# OF EVENTS	ANNUAL RATE	65-74 AT EXAM PERSON EXAMS	# OF EVENTS	ANNUAL RATE
NO	5529			9771	10	1	11183	28	1	6027	40	3
INCOMPLETE	75			144	0	0	187	1	3	115	0	0
COMPLETE	12			44	0	0	143	2	7	106	2	9
UNKNOWN	0			0	0	-	0	0	-	0	0	-
TOTAL	5616			9959	10	1	11513	31	1	6248	42	3

RISK FACTOR	75-84 AT EXAM PERSON EXAMS	# OF EVENTS	ANNUAL RATE	85-94 AT EXAM PERSON EXAMS	# OF EVENTS	ANNUAL RATE	35-64 COMBINED # OF EVENTS	CRUDE	AGE-ADJUSTED	65-94 COMBINED # OF EVENTS	CRUDE	AGE-ADJUSTED
NO	1847	28	8	117			40	1	1	72	5	5
INCOMPLETE	36	0	0	1			1	1	1	0	0	0
COMPLETE	68	2	15	5			2	5	3	4	11	11
UNKNOWN	0	0	-	0			0	-	-	0	-	-
TOTAL	1951	30	8	123			43	1	1	76	5	5

LOGISTIC REGRESSIONS

AGE AT EXAM		COEFFICIENT OF RISK FACTOR	SE COEFFICIENT	P-VALUE	STANDARDIZED COEFFICIENT	STD SE
35-44		-	-	-	-	-
45-54		-	-	-	-	-
55-64		0.855	0.354	0.016	0.217	0.090
65-74		0.374	0.394	0.343	0.108	0.114
75-84		0.248	0.389	0.523	0.096	0.131
85-94						
35-64	UNIVARIATE	0.890	0.356	0.012	0.186	0.074
65-94	UNIVARIATE	0.331	0.277	0.231	0.105	0.088
35-64	BIVARIATE	0.698	0.354	0.049	0.146	0.074
65-94	BIVARIATE	0.264	0.276	0.339	0.084	0.088
35-64	MULTIVARIATE	0.707	0.358	0.048	0.147	0.075
65-94	MULTIVARIATE	0.301	0.278	0.279	0.096	0.088

TABLE 11-18 DATE=21NOV85
AVERAGE ANNUAL INCIDENCE RATE FOR EVENT PER 1000 PERSONS AT RISK AT EXAM
BY AGE AND LEVEL OF RISK FACTOR AT EXAM: FRAMINGHAM STUDY 30-YEAR FOLLOWUP

SEX: MALE
EVENT: CORONARY HEART DISEASE DEATH
RISK FACTOR: DEFINITE NONSPECIFIC T-WAVE OR ST-SEGMENT ABNORMALITY BY ELECTROCARDIOGRAM
PERSONS AT RISK: PERSONS FREE OF CORONARY HEART DISEASE

RISK FACTOR	35-44 AT EXAM PERSON EXAMS	# OF EVENTS	ANNUAL RATE	45-54 AT EXAM PERSON EXAMS	# OF EVENTS	ANNUAL RATE	55-64 AT EXAM PERSON EXAMS	# OF EVENTS	ANNUAL RATE	65-74 AT EXAM PERSON EXAMS	# OF EVENTS	ANNUAL RATE
NO	4362	5	1	7352	25	2	7462	54	4	3472	43	6
YES	110	0	0	372	3	4	587	6	5	441	11	12
UNKNOWN	0	0	-	0	0	-	0	0	-	0	0	-
TOTAL	4472	5	1	7724	28	2	8049	60	4	3913	54	7

RISK FACTOR	75-84 AT EXAM PERSON EXAMS	# OF EVENTS	ANNUAL RATE	85-94 AT EXAM PERSON EXAMS	# OF EVENTS	ANNUAL RATE	35-64 COMBINED # OF EVENTS	CRUDE	AGE-ADJUSTED	65-94 COMBINED # OF EVENTS	CRUDE	AGE-ADJUSTED
NO	830	20	12	.28			84	2	2	63	7	7
YES	187	3	3	.11			9	4	4	16	13	12
UNKNOWN	0	0	-	0			0	-	-	0	-	-
TOTAL	1017	23	11	39			93	2	2	79	8	8

LOGISTIC REGRESSIONS

AGE AT EXAM		COEFFICIENT OF RISK FACTOR	SE COEFFICIENT	P-VALUE	STANDARDIZED COEFFICIENT	STD SE
35-44		-	-	-	-	-
45-54		0.868	0.613	0.157	0.186	0.131
55-64		0.348	0.432	0.421	0.091	0.112
65-74		0.713	0.342	0.037	-0.225	0.108
75-84		-0.415	0.625	0.506	-0.161	0.242
85-94						
35-64	UNIVARIATE	0.657	0.352	0.062	0.147	0.079
65-94	UNIVARIATE	0.554	0.283	0.051	0.185	0.095
35-64	BIVARIATE	0.422	0.354	0.233	0.094	0.079
65-94	BIVARIATE	0.457	0.286	0.110	0.153	0.096
35-64	MULTIVARIATE	0.074	0.360	0.838	0.017	0.080
65-94	MULTIVARIATE	0.370	0.289	0.201	0.124	0.097

SEX: FEMALE
EVENT: CORONARY HEART DISEASE DEATH
RISK FACTOR: DEFINITE NONSPECIFIC T-WAVE OR ST-SEGMENT ABNORMALITY BY ELECTROCARDIOGRAM
PERSONS AT RISK: PERSONS FREE OF CORONARY HEART DISEASE

RISK FACTOR	35-44 AT EXAM PERSON EXAMS	# OF EVENTS	ANNUAL RATE	45-54 AT EXAM PERSON EXAMS	# OF EVENTS	ANNUAL RATE	55-64 AT EXAM PERSON EXAMS	# OF EVENTS	ANNUAL RATE	65-74 AT EXAM PERSON EXAMS	# OF EVENTS	ANNUAL RATE
NO	5517			9543	7	0	10647	23	1	5528	32	3
YES	99			416	3	4	866	8	5	720	10	7
UNKNOWN	0			0	0	-	0	0	-	0	0	-
TOTAL	5616			9959	10	1	11513	31	1	6248	42	3

RISK FACTOR	75-84 AT EXAM PERSON EXAMS	# OF EVENTS	ANNUAL RATE	85-94 AT EXAM PERSON EXAMS	# OF EVENTS	ANNUAL RATE	35-64 COMBINED # OF EVENTS	CRUDE	AGE-ADJUSTED	65-94 COMBINED # OF EVENTS	CRUDE	AGE-ADJUSTED
NO	1632	25	8	97			32	1	1	60	4	4
YES	319	5	8	26			11	4	3	16	8	7
UNKNOWN	0	0	-	0			0	-	-	0	-	-
TOTAL	1951	30	8	123			43	1	1	76	5	5

LOGISTIC REGRESSIONS

AGE AT EXAM		COEFFICIENT OF RISK FACTOR	SE COEFFICIENT	P-VALUE	STANDARDIZED COEFFICIENT	STD SE
35-44						
45-54		2.292	0.692	0.001	0.459	0.138
55-64		1.460	0.612	0.001	0.385	0.109
65-74		0.883	0.364	0.015	0.282	0.116
75-84		0.023	0.494	0.962	0.009	0.183
85-94						
35-64	UNIVARIATE	1.863	0.351	0.001	0.410	0.077
65-94	UNIVARIATE	0.604	0.283	0.033	0.202	0.095
35-64	BIVARIATE	1.533	0.355	0.001	0.337	0.078
65-94	BIVARIATE	0.507	0.285	0.075	0.169	0.095
35-64	MULTIVARIATE	1.296	0.360	0.001	0.285	0.079
65-94	MULTIVARIATE	0.415	0.287	0.148	0.139	0.096

TABLE 12-1   DATE=19APR85

AVERAGE ANNUAL INCIDENCE RATE FOR EVENT PER 1000 PERSONS AT RISK AT EXAM,
BY AGE AND LEVEL OF RISK FACTOR AT EXAM: FRAMINGHAM STUDY 30-YEAR FOLLOWUP

SEX: MALE
EVENT: CORONARY HEART DISEASE DEATH
RISK FACTOR: SYSTOLIC BLOOD PRESSURE, FIRST EXAMINER (MM HG)
PERSONS AT RISK: ALL PERSONS ALIVE

RISK FACTOR	35-44 AT EXAM PERSON EXAMS	# OF EVENTS	ANNUAL RATE	45-54 AT EXAM PERSON EXAMS	# OF EVENTS	ANNUAL RATE	55-64 AT EXAM PERSON EXAMS	# OF EVENTS	ANNUAL RATE	65-74 AT EXAM PERSON EXAMS	# OF EVENTS	ANNUAL RATE
74-119	1226	3	-	1842	9	2	1605	14	4	659	8	6
120-139	2193	3	1	3542	18	3	3525	41	6	1636	26	8
140-159	850	0	0	1838	11	3	2371	31	7	1473	35	12
160-179	197	0	0	598	8	7	1062	19	9	696	23	17
180-300	53	0	0	247	3	6	507	15	15	370	17	23
UNKNOWN	0	0	-	0	0	-	0	0	-	0	0	-
TOTAL	4519	6	1	8067	49	3	9070	120	7	4834	109	11

RISK FACTOR	75-84 AT EXAM PERSON EXAMS	# OF EVENTS	ANNUAL RATE	85-94 AT EXAM PERSON EXAMS	# OF EVENTS	ANNUAL RATE	35-64 COMBINED # OF EVENTS	CRUDE	AGE-ADJUSTED	65-94 COMBINED # OF EVENTS	CRUDE	AGE-ADJUSTED
74-119	144	9	31	8			26	3	3	17	10	11
120-139	451	15	17	17			62	3	4	41	10	10
140-159	411	20	24	17			42	4	4	55	14	14
160-179	218	9	21	6			27	7	6	34	18	19
180-300	121	11	45	2			18	11	9	28	28	28
UNKNOWN	0	0	-	0			0	-	-	0	-	-
TOTAL	1345	64	24	50			175	4	4	175	14	14

LOGISTIC REGRESSIONS

AGE AT EXAM	COEFFICIENT OF RISK FACTOR	SE COEFFICIENT	P-VALUE	STANDARDIZED COEFFICIENT	STD SE
35-44	-0.050	0.031	0.110	-0.841	0.526
45-54	0.016	0.006	0.010	0.315	0.122
55-64	0.014	0.004	0.001	0.304	0.080
65-74	0.016	0.004	0.001	0.366	0.084
75-84	0.010	0.005	0.064	0.222	0.120
85-94					
35-64 UNIVARIATE	0.018	0.003	0.001	0.361	0.062
65-94 UNIVARIATE	0.014	0.003	0.001	0.330	0.068
35-64 BIVARIATE	0.012	0.003	0.001	0.256	0.064
65-94 BIVARIATE	0.014	0.003	0.001	0.330	-0.069
35-64 MULTIVARIATE	0.005	0.003	0.142	0.103	0.070
65-94 MULTIVARIATE	0.008	0.003	0.009	0.193	0.074

SEX: FEMALE
EVENT: CORONARY HEART DISEASE DEATH
RISK FACTOR: SYSTOLIC BLOOD PRESSURE, FIRST EXAMINER (MM HG)
PERSONS AT RISK: ALL PERSONS ALIVE

RISK FACTOR	35-44 AT EXAM PERSON EXAMS	# OF EVENTS	ANNUAL RATE	45-54 AT EXAM PERSON EXAMS	# OF EVENTS	ANNUAL RATE	55-64 AT EXAM PERSON EXAMS	# OF EVENTS	ANNUAL RATE	65-74 AT EXAM PERSON EXAMS	# OF EVENTS	ANNUAL RATE
74-119	2493			2724	1	0	2039	4	1	683	2	1
120-139	2245			4027	3	0	4199	7	1	1918	14	4
140-159	674			2086	5	1	3197	14	2	2283	20	4
160-179	156			785	0	0	1682	9	3	1360	18	7
180-300	51			450	1	1	986	8	4	859	21	12
UNKNOWN	0			0	0	-	0	0	-	0	0	-
TOTAL	5619			10072	10	0	12103	42	2	7103	75	5

RISK FACTOR	75-84 AT EXAM PERSON EXAMS	# OF EVENTS	ANNUAL RATE	85-94 AT EXAM PERSON EXAMS	# OF EVENTS	ANNUAL RATE	35-64 COMBINED # OF EVENTS	CRUDE	AGE-ADJUSTED	65-94 COMBINED # OF EVENTS	CRUDE	AGE-ADJUSTED
74-119	158	2	6	10	0	0	5	0	0	4	2	2
120-139	561	12	11	51	3	29	11	1	0	35	6	6
140-159	759	14	9	40	1	13	19	2	1	35	6	5
160-179	561	18	16	36	1	14	9	2	1	37	9	5
180-300	380	16	21	26	0	0	10	3	4	37	15	14
UNKNOWN	0	0	-	0	0	-	-			0	-	-
TOTAL	2419	62	13	163	5	15	54	1	1	142	7	7

LOGISTIC REGRESSIONS

AGE AT EXAM	COEFFICIENT OF RISK FACTOR	SE COEFFICIENT	P-VALUE	STANDARDIZED COEFFICIENT	STD SE
35-44					
45-54	0.017	0.010	0.082	0.393	0.226
55-64	0.016	0.005	0.001	0.409	0.125
65-74	0.020	0.004	0.001	0.494	0.100
75-84	0.010	0.005	0.030	0.257	0.118
85-94	-0.013	0.020	0.499	-0.335	0.495
35-64 UNIVARIATE	0.022	0.004	0.001	0.517	0.093
65-94 UNIVARIATE	0.016	0.003	0.001	0.398	0.075
35-64 BIVARIATE	0.016	0.004	0.001	0.379	0.105
65-94 BIVARIATE	0.015	0.003	0.001	0.365	0.076
35-64 MULTIVARIATE	0.013	0.005	0.008	0.312	0.117
65-94 MULTIVARIATE	0.010	0.003	0.001	0.255	0.079

TABLE 12-2   DATE=19APR85

NCE RATE FOR EVENT PER 1000 PERSONS AT RISK AT EXAM
ISK FACTOR AT EXAM: FRAMINGHAM STUDY 30-YEAR FOLLOWUP

HEART DISEASE DEATH
ASTOLIC BLOOD PRESSURE, FIRST EXAMINER (MM HG)
: ALL PERSONS ALIVE

45-54 AT EXAM			55-64 AT EXAM			65-74 AT EXAM		
PERSON EXAMS	# OF EVENTS	ANNUAL RATE	PERSON EXAMS	# OF EVENTS	ANNUAL RATE	PERSON EXAMS	# OF EVENTS	ANNUAL RATE
1543	11	4	2092	30	7	1524	26	9
2861	11	2	3165	28	4	1643	28	9
2151	11	3	2228	35	8	1072	25	12
1073	9	4	1114	17	8	425	21	25
439	7	8	471	10	11	170	9	26
0	0	-	0	0	-	0	0	-
8067	49	3	9070	120	7	4834	109	11

85-94 AT EXAM			35-64 COMBINED			65-94 COMBINED		
PERSON EXAMS	# OF EVENTS	ANNUAL RATE	# OF EVENTS	CRUDE	AGE-ADJUSTED	# OF EVENTS	CRUDE	AGE-ADJUSTED
30			42	5	4	55	13	13
15			43	3	3	47	11	12
5			47	4	4	34	13	14
0			26	5	5	24	24	24
0			17	8	8	15	38	42
0			0	-		0	-	
50			175	4	4	175	14	14

LOGISTIC REGRESSIONS

SE COEFFICIENT	P-VALUE	STANDARDIZED COEFFICIENT	STD SE
0.039	0.341	-0.421	0.442
0.011	0.020	0.306	0.132
0.007	0.046	0.174	0.087
0.007	0.001	0.319	0.088
0.010	0.935	0.010	0.128
0.006	0.006	0.199	0.072
0.006	0.015	0.178	0.073
0.006	0.004	0.201	0.070
0.006	0.001	0.256	0.074
0.006	0.203	0.090	0.071
0.006	0.027	0.166	0.075

HEART DISEASE DEATH
STOLIC BLOOD PRESSURE, FIRST EXAMINER (MM HG)
ALL PERSONS ALIVE

45-54 AT EXAM			55-64 AT EXAM			65-74 AT EXAM		
PERSON EXAMS	# OF EVENTS	ANNUAL RATE	PERSON EXAMS	# OF EVENTS	ANNUAL RATE	PERSON EXAMS	# OF EVENTS	ANNUAL RATE
2657	1	0	3110	9	1	2069	20	5
3573	4	1	4005	7	1	2416	25	5
2395	3	1	3000	14	2	1622	12	4
941	1	1	1348	7	3	685	9	7
506	1	1	640	5	4	311	9	14
0	0	-	0	0	-	0	0	-
10072	10	0	12103	42	2	7103	75	5

85-94 AT EXAM			35-64 COMBINED			65-94 COMBINED		
PERSON EXAMS	# OF EVENTS	ANNUAL RATE	# OF EVENTS	CRUDE	AGE-ADJUSTED	# OF EVENTS	CRUDE	AGE-ADJUSTED
73	4	27	11	1	0	50	8	8
56	1	9	11	1	1	39	6	6
25	0	0	17	1	1	28	6	7
7	0	0	9	2	2	14	8	9
2	0	0	6	2	2	11	14	14
0	0	-				0	-	
163	5	15	54	1	1	142	7	7

LOGISTIC REGRESSIONS

SE COEFFICIENT	P-VALUE	STANDARDIZED COEFFICIENT	STD SE
0.022	0.209	0.336	0.268
0.011	0.016	0.331	0.137
0.009	0.033	0.232	0.109
0.010	0.799	-0.033	0.130
0.033	0.121	-0.661	0.427
0.009	0.001	0.371	0.116
0.007	0.596	0.044	0.084
0.010	0.011	0.306	0.120
0.007	0.206	0.105	0.083
0.010	0.058	0.240	0.127
0.007	0.323	0.081	0.082

TABLE 12-3A   DATE:30AUG85

AVERAGE ANNUAL INCIDENCE RATE FOR EVENT PER 1000 PERSONS AT RISK AT EXAM
BY AGE AND LEVEL OF RISK FACTOR AT EXAM: FRAMINGHAM STUDY 30-YEAR FOLLOWUP

SEX: MALE
EVENT: CORONARY HEART DISEASE DEATH
RISK FACTOR: HYPERTENSION WITH ANTIHYPERTENSIVE TREATMENT
PERSONS AT RISK: ALL PERSONS ALIVE

RISK FACTOR	35-44 AT EXAM			45-54 AT EXAM			55-64 AT EXAM			65-74 AT EXAM		
	PERSON EXAMS	# OF EVENTS	ANNUAL RATE	PERSON EXAMS	# OF EVENTS	ANNUAL RATE	PERSON EXAMS	# OF EVENTS	ANNUAL RATE	PERSON EXAMS	# OF EVENTS	ANNUAL RATE
NO	2668	3	1	4263	17	2	4177	37	4	1810	21	6
BORDER	1316	2	1	2287	12	3	2570	31	6	1514	28	9
DEFINITE	511	0	0	1216	13	5	1428	22	8	698	24	17
TREATED	24	1	21	301	7	12	895	30	17	812	36	22
UNKNOWN	0	0	-	0	0	-	0	0	-	0	0	-
TOTAL	4519	6	1	8067	49	3	9070	120	7	4834	109	11

RISK FACTOR	75-84 AT EXAM			85-94 AT EXAM			35-64 COMBINED			65-94 COMBINED		
	PERSON EXAMS	# OF EVENTS	ANNUAL RATE	PERSON EXAMS	# OF EVENTS	ANNUAL RATE	# OF EVENTS	CRUDE	AGE-ADJUSTED	# OF EVENTS	CRUDE	AGE-ADJUSTED
NO	470	18	19	19			57	3	3	32	8	9
BORDER	424	20	24	16			45	4	4	48	12	12
DEFINITE	190	11	29	5			35	6	5	36	20	20
TREATED	261	15	29	10			38	16	16	52	24	24
UNKNOWN	0	0	-	0			0			0	-	
TOTAL	1345	64	24	50			175	4	4	175	14	14

LOGISTIC REGRESSIONS

AGE AT EXAM		COEFFICIENT OF RISK FACTOR	SE COEFFICIENT	P-VALUE	STANDARDIZED COEFFICIENT	STD SE
35-44		0.264	0.544	0.628	0.184	0.379
45-54		0.610	0.173	0.001	0.473	0.134
55-64		0.478	0.110	0.001	0.392	0.091
65-74		0.655	0.126	0.001	0.541	0.104
75-84		0.214	0.156	0.170	0.178	0.129
85-94						
35-64	UNIVARIATE	0.604	0.092	0.001	0.475	0.072
65-94	UNIVARIATE	0.511	0.097	0.001	0.423	0.080
35-64	BIVARIATE	0.486	0.092	0.001	0.382	0.072
65-94	BIVARIATE	0.505	0.097	0.001	0.417	0.080
35-64	MULTIVARIATE	0.356	0.097	0.001	0.280	0.076
65-94	MULTIVARIATE	0.373	0.101	0.001	0.308	0.083

SEX: FEMALE
EVENT: CORONARY HEART DISEASE DEATH
RISK FACTOR: HYPERTENSION WITH ANTIHYPERTENSIVE TREATMENT
PERSONS AT RISK: ALL PERSONS ALIVE

RISK FACTOR	35-44 AT EXAM			45-54 AT EXAM			55-64 AT EXAM			65-74 AT EXAM		
	PERSON EXAMS	# OF EVENTS	ANNUAL RATE	PERSON EXAMS	# OF EVENTS	ANNUAL RATE	PERSON EXAMS	# OF EVENTS	ANNUAL RATE	PERSON EXAMS	# OF EVENTS	ANNUAL RATE
NO	4251			5660	3	1	5022	10	1	1890	10	3
BORDER	1002			2611	3	1	3437	16	2	2106	18	4
DEFINITE	316			1184	1	0	1950	11	3	1197	16	7
TREATED	50			617	3	2	1694	5	1	1910	31	8
UNKNOWN	0	0	-	0	0	-	0	0	-	0	0	-
TOTAL	5619			10072	10	0	12103	42	2	7103	75	5

RISK FACTOR	75-84 AT EXAM			85-94 AT EXAM			35-64 COMBINED			65-94 COMBINED		
	PERSON EXAMS	# OF EVENTS	ANNUAL RATE	PERSON EXAMS	# OF EVENTS	ANNUAL RATE	# OF EVENTS	CRUDE	AGE-ADJUSTED	# OF EVENTS	CRUDE	AGE-ADJUSTED
NO	484	7	7	41	2	24	14	0	0	19	4	4
BORDER	683	14	10	37	1	14	19	1	1	33	6	6
DEFINITE	465	15	16	32	1	16	13	2	2	32	9	9
TREATED	787	26	17	53	1	9	8	2	1	58	11	10
UNKNOWN	0	0	-	0	0	-	0	-		0	-	
TOTAL	2419	62	13	163	5	15	54	1	1	142	7	7

LOGISTIC REGRESSIONS

AGE AT EXAM		COEFFICIENT OF RISK FACTOR	SE COEFFICIENT	P-VALUE	STANDARDIZED COEFFICIENT	STD SE
35-44						
45-54		0.715	0.378	0.058	0.551	0.291
55-64		0.363	0.185	0.050	0.305	0.155
65-74		0.541	0.163	0.001	0.444	0.133
75-84		0.435	0.188	0.021	0.342	0.148
85-94		-0.377	0.522	0.470	-0.316	0.438
35-64	UNIVARIATE	0.653	0.161	0.001	0.522	0.129
65-94	UNIVARIATE	0.498	0.119	0.001	0.406	0.097
35-64	BIVARIATE	0.387	0.166	0.020	0.309	0.132
65-94	BIVARIATE	0.444	0.119	0.001	0.362	0.097
35-64	MULTIVARIATE	0.296	0.174	0.089	0.237	0.139
65-94	MULTIVARIATE	0.303	0.121	0.012	0.247	0.099

TABLE 12-3B   DATE=22NOV85

AVERAGE ANNUAL INCIDENCE RATE FOR EVENT PER 1000 PERSONS AT RISK AT EXAM
BY AGE AND LEVEL OF RISK FACTOR AT EXAM: FRAMINGHAM STUDY 30-YEAR FOLLOWUP

SEX: MALE
EVENT: CORONARY HEART DISEASE DEATH
RISK FACTOR: HYPERTENSION IGNORING TREATMENT
PERSONS AT RISK: ALL PERSONS ALIVE

RISK FACTOR	35-44 AT EXAM			45-54 AT EXAM			55-64 AT EXAM			65-74 AT EXAM		
	PERSON EXAMS	# OF EVENTS	ANNUAL RATE	PERSON EXAMS	# OF EVENTS	ANNUAL RATE	PERSON EXAMS	# OF EVENTS	ANNUAL RATE	PERSON EXAMS	# OF EVENTS	ANNUAL RATE
NO	2672	4	1	4506	19	2	4347	45	5	1980	25	6
MILD	1326	2	1	2383	13	3	2922	40	7	1852	45	12
DEFINITE	521	0	0	1378	17	6	1801	35	10	1002	39	19
UNKNOWN	0	0	-	0	0	-	0	0	-	0	0	-
TOTAL	4519	6	1	8067	49	3	9070	120	7	4834	109	11

RISK FACTOR	75-84 AT EXAM			85-94 AT EXAM			35-64 COMBINED			65-94 COMBINED		
	PERSON EXAMS	# OF EVENTS	ANNUAL RATE	PERSON EXAMS	# OF EVENTS	ANNUAL RATE	# OF EVENTS	CRUDE	AGE-ADJUSTED	# OF EVENTS	CRUDE	AGE-ADJUSTED
NO	526	21	20	24			68	3	3	46	9	9
MILD	543	24	22	21			55	4	4	70	14	14
DEFINITE	276	19	34	5			52	7	6	59	23	23
UNKNOWN	0	0	-	0			0	-	-	0	-	-
TOTAL	1345	64	24	50			175	4	4	175	14	14

LOGISTIC REGRESSIONS

AGE AT EXAM		COEFFICIENT OF RISK FACTOR	SE COEFFICIENT	P-VALUE	STANDARDIZED COEFFICIENT	STD SE
35-44		-0.471	0.709	0.506	-0.327	0.491
45-54		0.512	0.175	0.003	0.388	0.133
55-64		0.318	0.114	0.005	0.246	0.088
65-74		0.569	0.126	0.001	0.432	0.096
75-84		0.283	0.168	0.092	0.212	0.126
85-94						
35-64	UNIVARIATE	0.424	0.094	0.001	0.320	0.071
65-94	UNIVARIATE	0.478	0.100	0.001	0.362	0.075
35-64	BIVARIATE	0.342	0.094	0.001	0.258	0.071
65-94	BIVARIATE	0.489	0.100	0.001	0.370	0.076
35-64	MULTIVARIATE	0.177	0.100	0.078	0.133	0.076
65-94	MULTIVARIATE	0.340	0.105	0.001	0.257	0.080

SEX: FEMALE
EVENT: CORONARY HEART DISEASE DEATH
RISK FACTOR: HYPERTENSION IGNORING TREATMENT
PERSONS AT RISK: ALL PERSONS ALIVE

RISK FACTOR	35-44 AT EXAM			45-54 AT EXAM			55-64 AT EXAM			65-74 AT EXAM		
	PERSON EXAMS	# OF EVENTS	ANNUAL RATE	PERSON EXAMS	# OF EVENTS	ANNUAL RATE	PERSON EXAMS	# OF EVENTS	ANNUAL RATE	PERSON EXAMS	# OF EVENTS	ANNUAL RATE
NO	4273			5846	3	0	5415	10	1	2205	13	3
MILD	1017			2803	5	1	4079	19	2	2907	27	5
DEFINITE	329			1423	2	1	2609	13	2	1991	35	9
UNKNOWN	0			0	0	-	0	0	-	0	0	-
TOTAL	5619			10072	10	0	12103	42	2	7103	75	5

RISK FACTOR	75-84 AT EXAM			85-94 AT EXAM			35-64 COMBINED			65-94 COMBINED		
	PERSON EXAMS	# OF EVENTS	ANNUAL RATE	PERSON EXAMS	# OF EVENTS	ANNUAL RATE	# OF EVENTS	CRUDE	AGE-ADJUSTED	# OF EVENTS	CRUDE	AGE-ADJUSTED
NO	613	11	9	52	3	29	14	0	1	27	5	5
MILD	1003	22	11	57	1	9	24	2	1	50	6	6
DEFINITE	803	29	18	54	1	9	16	2	2	65	11	11
UNKNOWN	0	0	-	0	0	-	0	-	-	0	-	-
TOTAL	2419	62	13	163	5	15	54	1	1	142	7	7

LOGISTIC REGRESSIONS

AGE AT EXAM		COEFFICIENT OF RISK FACTOR	SE COEFFICIENT	P-VALUE	STANDARDIZED COEFFICIENT	STD SE
35-44						
45-54		0.558	0.388	0.150	0.406	0.282
55-64		0.477	0.193	0.013	0.373	0.150
65-74		0.571	0.159	0.001	0.438	0.122
75-84		0.389	0.177	0.028	0.296	0.135
85-94		-0.697	0.622	0.262	-0.564	0.503
35-64	UNIVARIATE	0.692	0.166	0.001	0.515	0.124
65-94	UNIVARIATE	0.477	0.115	0.001	0.366	0.088
35-64	BIVARIATE	0.460	0.171	0.007	0.343	0.127
65-94	BIVARIATE	0.437	0.115	0.001	0.336	0.088
35-64	MULTIVARIATE	0.371	0.180	0.040	0.276	0.134
65-94	MULTIVARIATE	0.313	0.118	0.008	0.240	0.091

TABLE 12-4   DATE=19APR85

AVERAGE ANNUAL INCIDENCE RATE FOR EVENT PER 1000 PERSONS AT RISK AT EXAM
BY AGE AND LEVEL OF RISK FACTOR AT EXAM: FRAMINGHAM STUDY 30-YEAR FOLLOWUP

SEX: MALE
EVENT: CORONARY HEART DISEASE DEATH
RISK FACTOR: SERUM CHOLESTEROL
PERSONS AT RISK: ALL PERSONS ALIVE

RISK FACTOR	35-44 AT EXAM PERSON EXAMS	# OF EVENTS	ANNUAL RATE	45-54 AT EXAM PERSON EXAMS	# OF EVENTS	ANNUAL RATE	55-64 AT EXAM PERSON EXAMS	# OF EVENTS	ANNUAL RATE	65-74 AT EXAM PERSON EXAMS	# OF EVENTS	ANNUAL RATE
84-204	1253	0	0	1807	10	3	2484	17	3	1530	32	10
205-234	1258	0	0	2390	12	3	2695	42	8	1457	30	10
235-264	883	2	1	1843	11	3	2128	30	7	1059	24	11
265-294	488	1	1	1057	9	4	1073	16	7	525	9	9
295-1124	303	3	5	619	6	5	556	14	13	246	13	26
UNKNOWN	334	0	0	351	1	1	132	1	4	17	1	29
TOTAL	4519	6	1	8067	49	3	9070	120	7	4834	109	11

RISK FACTOR	75-84 AT EXAM PERSON EXAMS	# OF EVENTS	ANNUAL RATE	85-94 AT EXAM PERSON EXAMS	# OF EVENTS	ANNUAL RATE	35-64 COMBINED # OF EVENTS	CRUDE	AGE- ADJUSTED	65-94 COMBINED # OF EVENTS	CRUDE	AGE- ADJUSTED
84-204	537	23	21	23			27	2	2	55	13	12
205-234	422	21	25	14			54	4	4	52	14	13
235-264	232	13	28	7			43	4	4	38	15	15
265-294	130	5	19	6			26	3	3	14	11	11
295-1124	24	2	42	0			23	8	8	15	28	29
UNKNOWN	0	0	-	0			2	1	2	1	29	23
TOTAL	1345	64	24	50			175	4	4	175	14	14

LOGISTIC REGRESSIONS

AGE AT EXAM		COEFFICIENT OF RISK FACTOR	SE COEFFICIENT	P-VALUE	STANDARDIZED COEFFICIENT	STD SE
35-44		0.018	0.004	0.001	0.801	0.192
45-54		0.007	0.003	0.014	0.284	0.116
55-64		0.007	0.002	0.001	0.274	0.082
65-74		0.003	0.002	0.158	0.134	0.095
75-84		0.002	0.003	0.497	0.086	0.127
85-94						
35-64	UNIVARIATE	0.007	0.001	0.001	0.301	0.063
65-94	UNIVARIATE	0.002	0.002	0.266	0.084	0.076
35-64	BIVARIATE	0.008	0.002	0.001	0.329	0.065
65-94	BIVARIATE	0.003	0.002	0.067	0.140	0.077
35-64	MULTIVARIATE	0.008	0.002	0.001	0.323	0.065
65-94	MULTIVARIATE	0.004	0.002	0.035	0.163	0.077

SEX: FEMALE
EVENT: CORONARY HEART DISEASE DEATH
RISK FACTOR: SERUM CHOLESTEROL
PERSONS AT RISK: ALL PERSONS ALIVE

RISK FACTOR	35-44 AT EXAM PERSON EXAMS	# OF EVENTS	ANNUAL RATE	45-54 AT EXAM PERSON EXAMS	# OF EVENTS	ANNUAL RATE	55-64 AT EXAM PERSON EXAMS	# OF EVENTS	ANNUAL RATE	65-74 AT EXAM PERSON EXAMS	# OF EVENTS	ANNUAL RATE
84-204	2350			2120	0	0	1714	4	1	897	5	3
205-234	1419			2590	2	0	2741	5	1	1570	15	5
235-264	807			2264	2	0	3032	6	1	1912	16	4
265-294	372			1490	2	1	2364	13	3	1472	14	5
295-1124	171			1124	4	2	1999	13	3	1215	24	10
UNKNOWN	500			480	0	0	233	1	2	37	1	14
TOTAL	5619			10072	10	0	12103	42	2	7103	75	5

RISK FACTOR	75-84 AT EXAM PERSON EXAMS	# OF EVENTS	ANNUAL RATE	85-94 AT EXAM PERSON EXAMS	# OF EVENTS	ANNUAL RATE	35-64 COMBINED # OF EVENTS	CRUDE	AGE- ADJUSTED	65-94 COMBINED # OF EVENTS	CRUDE	AGE- ADJUSTED
84-204	463	11	12	60	3	25	4	0	0	19	7	7
205-234	588	16	14	39	1	13	7	1	0	32	7	7
235-264	660	16	12	42	0	0	9	1	1	32	6	6
265-294	419	12	14	12	1	42	15	2	2	27	8	8
295-1124	282	7	12	10	0	0	18	3	3	31	10	10
UNKNOWN	7	0	0	0	-	-	1	0	1	1	11	10
TOTAL	2419	62	13	163	5	15	54	1	1	142	7	7

LOGISTIC REGRESSIONS

AGE AT EXAM		COEFFICIENT OF RISK FACTOR	SE COEFFICIENT	P-VALUE	STANDARDIZED COEFFICIENT	STD SE
35-44						
45-54		0.007	0.002	0.008	0.306	0.115
55-64		0.012	0.003	0.001	0.538	0.125
65-74		0.008	0.002	0.001	0.338	0.106
75-84		0.001	0.003	0.835	0.027	0.128
85-94		-0.012	0.011	0.285	-0.545	0.510
35-64	UNIVARIATE	0.010	0.002	0.001	0.458	0.095
65-94	UNIVARIATE	0.003	0.002	0.122	0.128	0.083
35-64	BIVARIATE	0.009	0.002	0.001	0.412	0.082
65-94	BIVARIATE	0.005	0.002	0.014	0.203	0.083
35-64	MULTIVARIATE	0.008	0.002	0.001	0.361	0.082
65-94	MULTIVARIATE	0.005	0.002	0.011	0.208	0.082

240

SEX: MALE
EVENT: CORONARY HEART DISEASE DEATH
RISK FACTOR: HEMATOCRIT
PERSONS AT RISK: ALL PERSONS ALIVE

RISK FACTOR	35-44 AT EXAM PERSON EXAMS	# OF EVENTS	ANNUAL RATE	45-54 AT EXAM PERSON EXAMS	# OF EVENTS	ANNUAL RATE	55-64 AT EXAM PERSON EXAMS	# OF EVENTS	ANNUAL RATE	65-74 AT EXAM PERSON EXAMS	# OF EVENTS	ANNUAL RATE
25-42	630	1	1	961	5	3	1121	14	6	558	13	12
43-44	551	0	0	996	7	4	1094	10	5	611	11	9
45-46	990	1	1	1942	14	4	2215	35	8	1194	19	8
47-48	786	0	0	1585	7	2	1825	20	5	1069	27	13
49-70	1192	4	2	2193	15	3	2656	40	8	1382	38	14
UNKNOWN	370	0	0	390	1	1	159	1	3	20	1	25
TOTAL	4519	6	1	8067	49	3	9070	120	7	4834	109	11

RISK FACTOR	75-84 AT EXAM PERSON EXAMS	# OF EVENTS	ANNUAL RATE	85-94 AT EXAM PERSON EXAMS	# OF EVENTS	ANNUAL RATE	35-64 COMBINED # OF EVENTS	CRUDE	AGE-ADJUSTED	65-94 COMBINED # OF EVENTS	CRUDE	AGE-ADJUSTED
25-42	283	16	28	19	8		20	4	4	30	17	15
43-44	169	9	27	8	7		17	3	5	20	13	13
45-46	313	14	22	7	4		50	5	5	33	11	11
47-48	266	11	21	9	7		27	3	3	38	14	14
49-70	314	14	22	7			59	5	5	53	16	16
UNKNOWN	0	0	-	0			2	1	2	1	25	19
TOTAL	1345	64	24	50			175	4	4	175	14	14

LOGISTIC REGRESSIONS

AGE AT EXAM		COEFFICIENT OF RISK FACTOR	SE COEFFICIENT	P-VALUE	STANDARDIZED COEFFICIENT	STD SE
35-44		0.148	0.111	0.181	0.544	0.407
45-54		0.005	0.040	0.906	0.017	0.145
55-64		0.015	0.025	0.533	0.058	0.093
65-74		0.032	0.027	0.238	0.116	0.098
75-84		-0.021	0.031	0.505	-0.085	0.127
85-94						
35-64	UNIVARIATE	0.019	0.021	0.368	0.069	0.077
65-94	UNIVARIATE	-0.000	0.020	0.986	-0.001	0.077
35-64	BIVARIATE	0.016	0.021	0.426	0.061	0.076
65-94	BIVARIATE	0.016	0.020	0.426	0.061	0.077
35-64	MULTIVARIATE	-0.003	0.021	0.879	-0.012	0.074
65-94	MULTIVARIATE	0.010	0.021	0.642	0.037	0.079

SEX: FEMALE
EVENT: CORONARY HEART DISEASE DEATH
RISK FACTOR: HEMATOCRIT
PERSONS AT RISK: ALL PERSONS ALIVE

RISK FACTOR	35-44 AT EXAM PERSON EXAMS	# OF EVENTS	ANNUAL RATE	45-54 AT EXAM PERSON EXAMS	# OF EVENTS	ANNUAL RATE	55-64 AT EXAM PERSON EXAMS	# OF EVENTS	ANNUAL RATE	65-74 AT EXAM PERSON EXAMS	# OF EVENTS	ANNUAL RATE
21-39	1698			1858	1	0	1609	8	2	751	9	6
40-41	967			1761	1	0	2144	6	1	1180	9	4
42-43	1043			1978	0	0	2371	5	1	1422	12	4
44-45	858			2222	2	0	3003	9	1	1835	23	6
46-65	487			1702	6	2	2700	13	2	1873	21	6
UNKNOWN	566			551	0	0	274	1	2	42	1	12
TOTAL	5619			10072	10	0	12103	42	2	7103	75	5

RISK FACTOR	75-84 AT EXAM PERSON EXAMS	# OF EVENTS	ANNUAL RATE	85-94 AT EXAM PERSON EXAMS	# OF EVENTS	ANNUAL RATE	35-64 COMBINED # OF EVENTS	CRUDE	AGE-ADJUSTED	65-94 COMBINED # OF EVENTS	CRUDE	AGE-ADJUSTED
21-39	375	11	15	33	2	30	10	1	1	22	9	9
40-41	449	10	11	39	1	13	7	1	1	20	6	6
42-43	437	8	9	26	2	38	5	0	0	22	6	6
44-45	607	13	11	34	0	0	12	1	1	36	7	7
46-65	545	20	18	31	0	0	19	2	2	41	8	9
UNKNOWN	6	0	0	0	0	-	1	0	1	1	10	9
TOTAL	2419	62	13	163	5	15	54	1	1	142	7	7

LOGISTIC REGRESSIONS

AGE AT EXAM		COEFFICIENT OF RISK FACTOR	SE COEFFICIENT	P-VALUE	STANDARDIZED COEFFICIENT	STD SE
35-44						
45-54		0.210	0.088	0.016	0.749	0.312
55-64		0.049	0.045	0.276	0.171	0.157
65-74		0.050	0.034	0.135	0.177	0.119
75-84		0.007	0.033	0.831	0.028	0.129
85-94		-0.178	0.123	0.148	-0.694	0.480
35-64	UNIVARIATE	0.105	0.039	0.007	0.379	0.140
65-94	UNIVARIATE	0.011	0.023	0.629	0.041	0.085
35-64	BIVARIATE	0.066	0.040	0.096	0.238	0.143
65-94	BIVARIATE	0.022	0.023	0.340	0.080	0.084
35-64	MULTIVARIATE	0.032	0.039	0.420	0.113	0.141
65-94	MULTIVARIATE	0.010	0.023	0.668	0.036	0.084

TABLE 12-6   DATE=12DEC85
AVERAGE ANNUAL INCIDENCE RATE FOR EVENT PER 1000 PERSONS AT RISK AT EXAM
BY AGE AND LEVEL OF RISK FACTOR AT EXAM: FRAMINGHAM STUDY 30-YEAR FOLLOWUP

SEX: MALE
EVENT: CORONARY HEART DISEASE DEATH
RISK FACTOR: BLOOD GLUCOSE (MG/DL)
PERSONS AT RISK: ALL PERSONS ALIVE

RISK FACTOR	35-44 AT EXAM			45-54 AT EXAM			55-64 AT EXAM			65-74 AT EXAM		
	PERSON EXAMS	# OF EVENTS	ANNUAL RATE	PERSON EXAMS	# OF EVENTS	ANNUAL RATE	PERSON EXAMS	# OF EVENTS	ANNUAL RATE	PERSON EXAMS	# OF EVENTS	ANNUAL RATE
29-69	1004	0	0	1562	10	3	1208	10	4	501	9	9
70-89	2531	5	1	4344	23	3	4526	58	6	2142	37	9
90-109	628	1	1	1402	8	3	2136	31	7	1296	28	11
110-129	107	0	0	331	2	3	608	11	9	428	13	15
130-524	49	0	0	204	5	12	491	10	10	450	21	23
UNKNOWN	200	0	0	224	1	2	101	0	0	17	1	29
TOTAL	4519	6	1	8067	49	3	9070	120	7	4834	109	11

RISK FACTOR	75-84 AT EXAM			85-94 AT EXAM			35-64 COMBINED			65-94 COMBINED		
	PERSON EXAMS	# OF EVENTS	ANNUAL RATE	PERSON EXAMS	# OF EVENTS	ANNUAL RATE	# OF EVENTS	CRUDE	AGE-ADJUSTED	# OF EVENTS	CRUDE	AGE-ADJUSTED
29-69	86	4	23	1			20	3	3	13	11	12
70-89	479	17	18	17			86	4	4	54	10	11
90-109	438	18	21	15			40	5	4	47	13	13
110-129	170	10	29	7			13	6	5	23	19	18
130-524	172	15	44	10			15	10	9	37	29	28
UNKNOWN	0	0	-	0			1	1	1	1	29	23
TOTAL	1345	64	24	50			175	4	4	175	14	14

LOGISTIC REGRESSIONS

AGE AT EXAM		COEFFICIENT OF RISK FACTOR	SE COEFFICIENT	P-VALUE	STANDARDIZED COEFFICIENT	STD SE
35-44		-0.019	0.033	0.559	-0.331	0.566
45-54		0.007	0.003	0.043	0.166	0.082
55-64		0.006	0.002	0.002	0.185	0.059
65-74		0.008	0.001	0.001	0.312	0.051
75-84		0.007	0.003	0.009	0.246	0.094
85-94						
35-64	UNIVARIATE	0.008	0.002	0.001	0.205	0.041
65-94	UNIVARIATE	0.008	0.001	0.001	0.297	0.044
35-64	BIVARIATE	0.006	0.002	0.001	0.151	0.045
65-94	BIVARIATE	0.008	0.001	0.001	0.294	0.045
35-64	MULTIVARIATE	0.006	0.002	0.001	0.149	0.046
65-94	MULTIVARIATE	0.008	0.001	0.001	0.280	0.046

SEX: FEMALE
EVENT: CORONARY HEART DISEASE DEATH
RISK FACTOR: BLOOD GLUCOSE (MG/DL)
PERSONS AT RISK: ALL PERSONS ALIVE

RISK FACTOR	35-44 AT EXAM			45-54 AT EXAM			55-64 AT EXAM			65-74 AT EXAM		
	PERSON EXAMS	# OF EVENTS	ANNUAL RATE	PERSON EXAMS	# OF EVENTS	ANNUAL RATE	PERSON EXAMS	# OF EVENTS	ANNUAL RATE	PERSON EXAMS	# OF EVENTS	ANNUAL RATE
29-69	1067			1806	2	1	1411	7	2	597	10	8
70-89	3414			5811	4	0	6371	18	1	3483	23	3
90-109	672			1646	0	0	2877	8	1	1982	18	5
110-129	96			303	0	0	777	2	1	582	7	6
130-524	39			167	4	12	496	6	6	427	16	19
UNKNOWN	331			339	0	0	171	1	3	32	1	16
TOTAL	5619			10072	10	0	12103	42	2	7103	75	5

RISK FACTOR	75-84 AT EXAM			85-94 AT EXAM			35-64 COMBINED			65-94 COMBINED		
	PERSON EXAMS	# OF EVENTS	ANNUAL RATE	PERSON EXAMS	# OF EVENTS	ANNUAL RATE	# OF EVENTS	CRUDE	AGE-ADJUSTED	# OF EVENTS	CRUDE	AGE-ADJUSTED
29-69	129	3	12	5	0	0	9	1	1	13	9	9
70-89	976	18	9	60	2	17	24	1	1	43	5	5
90-109	803	17	11	63	1	8	8	1	1	36	6	6
110-129	286	11	19	19	0	0	2	1	1	18	10	9
130-524	224	13	29	16	2	63	10	7	7	31	23	22
UNKNOWN	1	0	0	0	0	-	1	1	1	15	11	
TOTAL	2419	62	13	163	5	15	54	1	1	142	7	7

LOGISTIC REGRESSIONS

AGE AT EXAM		COEFFICIENT OF RISK FACTOR	SE COEFFICIENT	P-VALUE	STANDARDIZED COEFFICIENT	STD SE
35-44						
45-54		0.020	0.004	0.001	0.397	0.078
55-64		0.006	0.003	0.082	0.164	0.094
65-74		0.011	0.002	0.001	0.318	0.055
75-84		0.008	0.002	0.001	0.285	0.074
85-94		0.005	0.009	0.568	0.185	0.325
35-64	UNIVARIATE	0.011	0.002	0.001	0.253	0.053
65-94	UNIVARIATE	0.010	0.001	0.001	0.312	0.042
35-64	BIVARIATE	0.009	0.003	0.001	0.197	0.058
65-94	BIVARIATE	0.009	0.001	0.001	0.294	0.044
35-64	MULTIVARIATE	0.007	0.003	0.007	0.161	0.060
65-94	MULTIVARIATE	0.009	0.001	0.001	0.274	0.046

AVERAGE ANNUAL INCIDENCE RATE FOR EVENT PER 1000 PERSONS AT RISK AT EXAM
BY AGE AND LEVEL OF RISK FACTOR AT EXAM: FRAMINGHAM STUDY 30-YEAR FOLLOWUP

SEX: MALE
EVENT: CORONARY HEART DISEASE DEATH
RISK FACTOR: DIABETES MELLITUS
PERSONS AT RISK: ALL PERSONS ALIVE

RISK FACTOR	35-44 AT EXAM PERSON EXAMS	# OF EVENTS	ANNUAL RATE	45-54 AT EXAM PERSON EXAMS	# OF EVENTS	ANNUAL RATE	55-64 AT EXAM PERSON EXAMS	# OF EVENTS	ANNUAL RATE	65-74 AT EXAM PERSON EXAMS	# OF EVENTS	ANNUAL RATE
NO	4470	6	1	7812	44	3	8424	104	6	4269	84	10
YES	49	0	0	255	5	10	646	16	12	565	25	22
UNKNOWN	0	0	-	0	0	-	0	0	-	0	0	-
TOTAL	4519	6		8067	49	3	9070	120	7	4834	109	11

RISK FACTOR	75-84 AT EXAM PERSON EXAMS	# OF EVENTS	ANNUAL RATE	85-94 AT EXAM PERSON EXAMS	# OF EVENTS	ANNUAL RATE	35-64 COMBINED # OF EVENTS	CRUDE	AGE-ADJUSTED	65-94 COMBINED # OF EVENTS	CRUDE	AGE-ADJUSTED
NO	1170	53	23	44			154	4	4	139	13	13
YES	175	11	31	6			21	11	9	36	24	24
UNKNOWN	0	0	-	0			0	-		0	-	
TOTAL	1345	64	24	50			175	4	4	175	14	14

LOGISTIC REGRESSIONS

AGE AT EXAM		COEFFICIENT OF RISK FACTOR	SE COEFFICIENT	P-VALUE	STANDARDIZED COEFFICIENT	STD SE
35-44		-	-	-	-	-
45-54		1.262	0.476	0.008	0.221	0.083
55-64		0.709	0.272	0.009	0.182	0.070
65-74		0.836	0.232	0.001	0.269	0.075
75-84		0.346	0.342	0.311	0.116	0.115
85-94						
35-64	UNIVARIATE	1.104	0.235	0.001	0.226	0.048
65-94	UNIVARIATE	0.668	0.191	0.001	0.217	0.062
35-64	BIVARIATE	0.738	0.238	0.002	0.151	0.049
65-94	BIVARIATE	0.640	0.192	0.001	0.208	0.062
35-64	MULTIVARIATE	0.541	0.247	0.029	0.111	0.051
65-94	MULTIVARIATE	0.596	0.196	0.002	0.194	0.064

SEX: FEMALE
EVENT: CORONARY HEART DISEASE DEATH
RISK FACTOR: DIABETES MELLITUS
PERSONS AT RISK: ALL PERSONS ALIVE

RISK FACTOR	35-44 AT EXAM PERSON EXAMS	# OF EVENTS	ANNUAL RATE	45-54 AT EXAM PERSON EXAMS	# OF EVENTS	ANNUAL RATE	55-64 AT EXAM PERSON EXAMS	# OF EVENTS	ANNUAL RATE	65-74 AT EXAM PERSON EXAMS	# OF EVENTS	ANNUAL RATE
NO	5576			9859	6	0	11497	34	1	6499	49	4
YES	43			213	4	9	606	8	7	604	26	22
UNKNOWN	0			0	0	-	0	0	-	0	0	-
TOTAL	5619			10072	10	0	12103	42	2	7103	75.	5

RISK FACTOR	75-84 AT EXAM PERSON EXAMS	# OF EVENTS	ANNUAL RATE	85-94 AT EXAM PERSON EXAMS	# OF EVENTS	ANNUAL RATE	35-64 COMBINED # OF EVENTS	CRUDE	AGE-ADJUSTED	65-94 COMBINED # OF EVENTS	CRUDE	AGE-ADJUSTED
NO	2106	43	10	139	4	14	41	1	1	96	5	6
YES	313	19	30	24	1	21	13	8	9	46	24	24
UNKNOWN	0	0	-	0	0	-	0	-		0	-	
TOTAL	2419	62	13	163	5	15	54	1	1	142	7	7

LOGISTIC REGRESSIONS

AGE AT EXAM		COEFFICIENT OF RISK FACTOR	SE COEFFICIENT	P-VALUE	STANDARDIZED COEFFICIENT	STD SE
35-44						
45-54		3.448	0.649	0.001	0.496	0.093
55-64		1.506	0.395	0.001	0.329	0.086
65-74		1.779	0.246	0.001	0.496	0.069
75-84		1.132	0.282	0.001	0.380	0.095
85-94		0.383	1.141	0.737	0.136	0.405
35-64	UNIVARIATE	2.307	0.320	0.001	0.400	0.056
65-94	UNIVARIATE	1.533	0.183	0.001	0.454	0.054
35-64	BIVARIATE	1.885	0.324	0.001	0.327	0.056
65-94	BIVARIATE	1.426	0.184	0.001	0.422	0.055
35-64	MULTIVARIATE	1.632	0.336	0.001	0.283	0.058
65-94	MULTIVARIATE	1.285	0.190	0.001	0.381	0.056

TABLE 12-8   DATE=22NOV85
AVERAGE ANNUAL INCIDENCE RATE FOR EVENT PER 1000 PERSONS AT RISK AT EXAM
BY AGE AND LEVEL OF RISK FACTOR AT EXAM: FRAMINGHAM STUDY 30-YEAR FOLLOWUP

SEX: MALE
EVENT: CORONARY HEART DISEASE DEATH
RISK FACTOR: GLUCOSE IN URINE, DEFINITE OR TRACE
PERSONS AT RISK: ALL PERSONS ALIVE

RISK FACTOR	35-44 AT EXAM PERSON EXAMS	# OF EVENTS	ANNUAL RATE	45-54 AT EXAM PERSON EXAMS	# OF EVENTS	ANNUAL RATE	55-64 AT EXAM PERSON EXAMS	# OF EVENTS	ANNUAL RATE	65-74 AT EXAM PERSON EXAMS	# OF EVENTS	ANNUAL RATE
NO	4426	6	1	7869	47	3	8786	109	6	4661	96	10
YES	60	0	0	177	2	6	260	10	19	167	13	39
UNKNOWN	33	-0	0	21	0	0	24	1	21	6	0	0
TOTAL	4519	6	1	8067	49	3	9070	120	7	4834	109	11

RISK FACTOR	75-84 AT EXAM PERSON EXAMS	# OF EVENTS	ANNUAL RATE	85-94 AT EXAM PERSON EXAMS	# OF EVENTS	ANNUAL RATE	35-64 COMBINED # OF EVENTS	CRUDE	AGE-ADJUSTED	65-94 COMBINED # OF EVENTS	CRUDE	AGE-ADJUSTED
NO	1303	58	22	49			162	4	4	156	13	13
YES	39	6	77	1			12	12	10	19	46	47
UNKNOWN	3	0	0	0			1	6	9	0	0	0
TOTAL	1345	64	24	50			175	4	4	175	14	14

LOGISTIC REGRESSIONS

AGE AT EXAM	COEFFICIENT OF RISK FACTOR	SE COEFFICIENT	P-VALUE	STANDARDIZED COEFFICIENT	STD SE
35-44	-	-		-	-
45-54	0.643	0.726	0.376	0.094	0.106
55-64	1.158	0.337	0.001	0.194	0.056
65-74	1.390	0.307	0.001	0.254	0.056
75-84	1.362	0.464	0.003	0.229	0.078
85-94					
35-64 UNIVARIATE	1.162	0.303	0.001	0.174	0.045
65-94 UNIVARIATE	1.334	0.254	0.001	0.239	0.046
35-64 BIVARIATE	0.991	0.304	0.001	0.149	0.046
65-94 BIVARIATE	1.392	0.256	0.001	0.250	0.046
35-64 MULTIVARIATE	0.857	0.311	0.006	0.129	0.047
65-94 MULTIVARIATE	1.330	0.263	0.001	0.239	0.047

SEX: FEMALE
EVENT: CORONARY HEART DISEASE DEATH
RISK FACTOR: GLUCOSE IN URINE, DEFINITE OR TRACE
PERSONS AT RISK: ALL PERSONS ALIVE

RISK FACTOR	35-44 AT EXAM PERSON EXAMS	# OF EVENTS	ANNUAL RATE	45-54 AT EXAM PERSON EXAMS	# OF EVENTS	ANNUAL RATE	55-64 AT EXAM PERSON EXAMS	# OF EVENTS	ANNUAL RATE	65-74 AT EXAM PERSON EXAMS	# OF EVENTS	ANNUAL RATE
NO	5540			9947	9	0	11903	39	2	6957	66	5
YES	41			101	1	5	174	3	9	136	9	33
UNKNOWN	38			24	0	0	26	0	0	10	0	0
TOTAL	5619			10072	10	0	12103	42	2	7103	75	5

RISK FACTOR	75-84 AT EXAM PERSON EXAMS	# OF EVENTS	ANNUAL RATE	85-94 AT EXAM PERSON EXAMS	# OF EVENTS	ANNUAL RATE	35-64 COMBINED # OF EVENTS	CRUDE	AGE-ADJUSTED	65-94 COMBINED # OF EVENTS	CRUDE	AGE-ADJUSTED
NO	2375	59	12	160	4	13	49	1	1	129	7	7
YES	35	3	43	3	1	167	5	8	8	13	37	38
UNKNOWN	9	0	0	0			0	0	0	0		
TOTAL	2419	62	13	163	5	15	54	1	1	142	7	7

LOGISTIC REGRESSIONS

AGE AT EXAM	COEFFICIENT OF RISK FACTOR	SE COEFFICIENT	P-VALUE	STANDARDIZED COEFFICIENT	STD SE
35-44					
45-54	2.402	1.059	0.023	0.240	0.106
55-64	1.675	0.604	0.006	0.200	0.072
65-74	2.001	0.366	0.001	0.274	0.050
75-84	1.303	0.618	0.035	0.156	0.074
85-94	2.970	1.325	0.025	0.400	0.179
35-64 UNIVARIATE	2.194	0.473	0.001	0.233	0.050
65-94 UNIVARIATE	1.768	0.302	0.001	0.235	0.040
35-64 BIVARIATE	2.006	0.476	0.001	0.213	0.050
65-94 BIVARIATE	1.844	0.305	0.001	0.245	0.041
35-64 MULTIVARIATE	1.702	0.489	0.001	0.181	0.052
65-94 MULTIVARIATE	1.565	0.319	0.001	0.208	0.042

· 244

TABLE 12-9   DATE=01MAR86
AVERAGE ANNUAL INCIDENCE RATE FOR EVENT PER 1000 PERSONS AT RISK AT EXAM
BY AGE AND LEVEL OF RISK FACTOR AT EXAM: FRAMINGHAM STUDY 30-YEAR FOLLOWUP

SEX: MALE
EVENT: CORONARY HEART DISEASE DEATH
RISK FACTOR: GLUCOSE INTOLERANCE
PERSONS AT RISK: ALL PERSONS ALIVE

K TOR	35-44 AT EXAM PERSON EXAMS	# OF EVENTS	ANNUAL RATE	45-54 AT EXAM PERSON EXAMS	# OF EVENTS	ANNUAL RATE	55-64 AT EXAM PERSON EXAMS	# OF EVENTS	ANNUAL RATE	65-74 AT EXAM PERSON EXAMS	# OF EVENTS	ANNUAL RATE
	4381	6	1	7528	42	5	7973	98	6	3913	73	9
	138	0	0	539	7	6	1097	22	10	916	36	20
NOWN AL	0 4519	0 6	- 1	0 8067	0 49	- 5	0 9070	0 120	- 7	0 4834	0 109	- 11

K TOR	75-84 AT EXAM PERSON EXAMS	# OF EVENTS	ANNUAL RATE	85-94 AT EXAM PERSON EXAMS	# OF EVENTS	ANNUAL RATE	35-64 COMBINED # OF EVENTS	CRUDE	AGE- ADJUSTED	65-94 COMBINED # OF EVENTS	CRUDE	AGE- ADJUSTED
	1035	42	20	36			146	4	4	116	12	12
	310	22	35	14			29	8	7	59	24	23
NOWN AL	0 1345	0 64	- 24	0 50			0 175	- 4	- 4	0 175	- 14	- 14

LOGISTIC REGRESSIONS

AT	COEFFICIENT OF RISK FACTOR	SE COEFFICIENT	P-VALUE	STANDARDIZED COEFFICIENT	STD SE
44	-	-	-	-	-
54	0.852	0.411	0.038	0.213	0.103
64	0.497	0.238	0.037	0.162	0.078
74	0.768	0.207	0.001	0.301	0.081
64	0.591	0.272	0.029	0.249	0.114
94					
64 UNIVARIATE	0.809	0.205	0.001	0.222	0.056
94 UNIVARIATE	0.741	0.163	0.001	0.296	0.065
64 BIVARIATE	0.510	0.207	0.014	0.140	0.057
94 BIVARIATE	0.683	0.164	0.001	0.273	0.065
64 MULTIVARIATE	0.390	0.212	0.066	0.107	0.058
94 MULTIVARIATE	0.644	0.166	0.001	0.257	0.066

SEX: FEMALE
EVENT: CORONARY HEART DISEASE DEATH
RISK FACTOR: GLUCOSE INTOLERANCE
PERSONS AT RISK: ALL PERSONS ALIVE

OR	35-44 AT EXAM PERSON EXAMS	# OF EVENTS	ANNUAL RATE	45-54 AT EXAM PERSON EXAMS	# OF EVENTS	ANNUAL RATE	55-64 AT EXAM PERSON EXAMS	# OF EVENTS	ANNUAL RATE	65-74 AT EXAM PERSON EXAMS	# OF EVENTS	ANNUAL RATE
	5525	-	-	9685	6	0	10988	32	1	6066	44	4
	94	-	-	387	4	5	1115	10	4	1037	31	15
OWN L	0 5619	0	-	0 10072	0 10	- 0	0 12103	0 42	- 2	0 7103	0 75	- 5

OR	75-84 AT EXAM PERSON EXAMS	# OF EVENTS	ANNUAL RATE	85-94 AT EXAM PERSON EXAMS	# OF EVENTS	ANNUAL RATE	35-64 COMBINED # OF EVENTS	CRUDE	AGE- ADJUSTED	65-94 COMBINED # OF EVENTS	CRUDE	AGE- ADJUSTED
	1905	39	10	131	3	11	39	1	1	86	5	5
	514	23	22	32	2	31	15	5	5	56	18	17
OWN	0 2419	0 62	- 13	0 163	0 5	- 15	0 54	- 1	- 1	0 142	- 7	- 7

LOGISTIC REGRESSIONS

AT	COEFFICIENT OF RISK FACTOR	SE COEFFICIENT	P-VALUE	STANDARDIZED COEFFICIENT	STD SE
	2.824	0.648	0.001	0.543	0.124
	1.131	0.364	0.002	0.327	0.105
	1.439	0.237	0.001	0.508	0.084
	0.807	0.268	0.003	0.330	0.110
	1.045	0.935	0.264	0.417	0.373
UNIVARIATE	1.851	0.305	0.001	0.431	0.071
UNIVARIATE	1.229	0.174	0.001	0.455	0.064
BIVARIATE	1.431	0.309	0.001	0.333	0.072
BIVARIATE	1.128	0.175	0.001	0.417	0.065
MULTIVARIATE	1.297	0.316	0.001	0.302	0.074
MULTIVARIATE	1.032	0.180	0.001	0.381	0.066

TABLE 12-10   DATE=22NOV85

AVERAGE ANNUAL INCIDENCE RATE FOR EVENT PER 1000 PERSONS AT RISK AT EXAM
BY AGE AND LEVEL OF RISK FACTOR AT EXAM; FRAMINGHAM STUDY 30-YEAR FOLLOWUP

SEX: MALE
EVENT: CORONARY HEART DISEASE DEATH
RISK FACTOR: METROPOLITAN RELATIVE WEIGHT (PERCENT)
PERSONS AT RISK: ALL PERSONS ALIVE

RISK FACTOR	35-44 AT EXAM PERSON EXAMS	# OF EVENTS	ANNUAL RATE	45-54 AT EXAM PERSON EXAMS	# OF EVENTS	ANNUAL RATE	55-64 AT EXAM PERSON EXAMS	# OF EVENTS	ANNUAL RATE	65-74 AT EXAM PERSON EXAMS	# OF EVENTS	ANNUAL RATE
54-104	723	1	1	1149	5	2	1272	23	9	831	24	14
105-114	979	0	0	1635	11	3	1858	22	6	1052	17	8
115-124	1153	2	1	2139	17	4	2467	26	5	1322	31	12
125-134	923	2	1	1691	5	1	1851	22	6	956	22	12
135-272	761	1	1	1452	11	4	1621	27	8	673	15	11
UNKNOWN	0	0	-	1	0	0	1	0	0	0	0	-
TOTAL	4519	6	1	8067	49	3	9070	120	7	4834	109	11

RISK FACTOR	75-84 AT EXAM PERSON EXAMS	# OF EVENTS	ANNUAL RATE	85-94 AT EXAM PERSON EXAMS	# OF EVENTS	ANNUAL RATE	35-64 COMBINED # OF EVENTS	CRUDE	AGE-ADJUSTED	65-94 COMBINED # OF EVENTS	CRUDE	AGE-ADJUSTED
54-104	277	13	23	19	1	5	29	5	5	37	16	16
105-114	357	23	32	15			33	4	4	41	14	13
115-124	399	14	18	14			45	4	4	46	13	13
125-134	196	6	15	2			29	3	3	28	12	12
135-272	116	8	34	0			39	5	5	23	15	16
UNKNOWN	0	0	-	0			0	0	0	0	0	-
TOTAL	1345	64	24	50			175	4	4	175	14	14

LOGISTIC REGRESSIONS

AGE AT EXAM		COEFFICIENT OF RISK FACTOR	SE COEFFICIENT	P-VALUE	STANDARDIZED COEFFICIENT	STD SE
35-44		0.015	0.024	0.522	0.247	0.387
45-54		-0.001	0.009	0.881	-0.021	0.144
55-64		0.002	0.006	0.760	0.028	0.091
65-74		-0.003	0.006	0.611	-0.050	0.097
75-84		-0.002	0.009	0.786	-0.035	0.129
85-94						
35-64	UNIVARIATE	0.002	0.005	0.685	0.031	0.075
65-94	UNIVARIATE	-0.005	0.005	0.361	-0.071	0.077
35-64	BIVARIATE	0.002	0.005	0.718	0.027	0.076
65-94	BIVARIATE	-0.001	0.005	0.796	-0.020	0.078
35-64	MULTIVARIATE	-0.001	0.005	0.850	-0.015	0.080
65-94	MULTIVARIATE	-0.008	0.005	0.135	-0.117	0.082

SEX: FEMALE
EVENT: CORONARY HEART DISEASE DEATH
RISK FACTOR: METROPOLITAN RELATIVE WEIGHT (PERCENT)
PERSONS AT RISK: ALL PERSONS ALIVE

RISK FACTOR	35-44 AT EXAM PERSON EXAMS	# OF EVENTS	ANNUAL RATE	45-54 AT EXAM PERSON EXAMS	# OF EVENTS	ANNUAL RATE	55-64 AT EXAM PERSON EXAMS	# OF EVENTS	ANNUAL RATE	65-74 AT EXAM PERSON EXAMS	# OF EVENTS	ANNUAL RATE
54-104	1694			2117	2	0	2140	5	1	1319	18	7
105-114	1431			2364	0	0	2541	6	1	1321	11	4
115-124	1059			2128	0	0	2556	12	2	1536	11	4
125-134	566			1383	2	1	1831	6	2	1140	12	5
135-272	861			2080	6	1	3035	13	2	1787	23	6
UNKNOWN	8			0	0	-	0	0	-	0	0	-
TOTAL	5619			10072	10	0	12103	42	2	7103	75	5

RISK FACTOR	75-84 AT EXAM PERSON EXAMS	# OF EVENTS	ANNUAL RATE	85-94 AT EXAM PERSON EXAMS	# OF EVENTS	ANNUAL RATE	35-64 COMBINED # OF EVENTS	CRUDE	AGE-ADJUSTED	65-94 COMBINED # OF EVENTS	CRUDE	AGE-ADJUSTED
54-104	632	14	11	51	4	39	8	1	1	36	9	8
105-114	457	8	9	33	0	0	6	0	1	19	5	5
115-124	487	14	14	31	1	16	12	1	1	26	6	6
125-134	342	10	15	26	0	0	8	1	1	22	7	8
135-272	501	16	16	22	0	0	20	2	2	39	8	9
UNKNOWN	0	0	-	0	0	-	0	0	0	0		
TOTAL	2419	62	13	163	5	15	54	1	1	142	7	7

LOGISTIC REGRESSIONS

AGE AT EXAM		COEFFICIENT OF RISK FACTOR	SE COEFFICIENT	P-VALUE	STANDARDIZED COEFFICIENT	STD SE
35-44						
45-54		0.027	0.009	0.002	0.576	0.186
55-64		0.004	0.006	0.500	0.098	0.145
65-74		0.002	0.005	0.708	0.042	0.113
75-84		0.010	0.005	0.040	0.232	0.113
85-94		-0.064	0.032	0.042	-1.228	0.605
35-64	UNIVARIATE	0.013	0.005	0.009	0.287	0.110
65-94	UNIVARIATE	0.003	0.004	0.473	0.059	0.082
35-64	BIVARIATE	0.010	0.005	0.060	0.217	0.115
65-94	BIVARIATE	0.005	0.004	0.160	0.113	0.081
35-64	MULTIVARIATE	0.005	0.006	0.395	0.107	0.126
65-94	MULTIVARIATE	0.003	0.004	0.455	0.060	0.080

TABLE 12-11   DATE=22NOV85
AVERAGE ANNUAL INCIDENCE RATE FOR EVENT PER 1000 PERSONS AT RISK AT EXAM
BY AGE AND LEVEL OF RISK FACTOR AT EXAM: FRAMINGHAM STUDY 30-YEAR FOLLOWUP

SEX: MALE
EVENT: CORONARY HEART DISEASE DEATH
RISK FACTOR: VITAL CAPACITY - HEIGHT INDEX (ML/INCH)
PERSONS AT RISK: ALL PERSONS ALIVE

RISK FACTOR	35-44 AT EXAM			45-54 AT EXAM			55-64 AT EXAM			65-74 AT EXAM		
	PERSON EXAMS	# OF EVENTS	ANNUAL RATE	PERSON EXAMS	# OF EVENTS	ANNUAL RATE	PERSON EXAMS	# OF EVENTS	ANNUAL RATE	PERSON EXAMS	# OF EVENTS	ANNUAL RATE
8-44	271	0	0	1108	6	3	2374	49	10	2179	53	12
45-49	416	1	1	1188	6	3	1692	23	7	974	30	15
50-54	751	0	0	1726	14	4	2061	24	6	857	11	6
55-59	1034	2	1	1823	12	3	1569	12	4	528	7	7
60-85	2034	3	1	2216	11	2	1356	12	4	284	7	12
UNKNOWN	13	0	0	6	0	0	18	0	0	12	1	42
TOTAL	4519	6	1	8067	49	3	9070	120	7	4834	109	11

RISK FACTOR	75-84 AT EXAM			85-94 AT EXAM			35-64 COMBINED			65-94 COMBINED		
	PERSON EXAMS	# OF EVENTS	ANNUAL RATE	PERSON EXAMS	# OF EVENTS	ANNUAL RATE	# OF EVENTS	CRUDE	AGE-ADJUSTED	# OF EVENTS	CRUDE	AGE-ADJUSTED
8-44	853	42	25	34			55	7	5	96	16	15
45-49	228	6	13	7			30	5	4	36	15	15
50-54	142	10	35	7			38	4	4	22	11	13
55-59	81	6	37	2			26	3	3	13	11	13
60-85	33	0	0	0			26	2	3	7	11	10
UNKNOWN	8	0	0	0			0	0	0	1	25	32
TOTAL	1345	64	24	50			175	4	4	175	44	14

LOGISTIC REGRESSIONS

AGE AT EXAM		COEFFICIENT OF RISK FACTOR	SE COEFFICIENT	P-VALUE	STANDARDIZED COEFFICIENT	STD SE
35-44		-0.006	0.046	0.889	-0.057	0.405
45-54		-0.003	0.016	0.857	-0.026	0.143
55-64		-0.031	0.009	0.001	-0.291	0.088
65-74		-0.021	0.010	0.041	-0.195	0.096
75-84		-0.004	0.014	0.762	-0.039	0.128
85-94						
35-64	UNIVARIATE	-0.039	0.007	0.001	-0.376	0.072
65-94	UNIVARIATE	-0.022	0.008	0.007	-0.206	0.076
35-64	BIVARIATE	-0.019	0.008	0.022	-0.180	0.078
65-94	BIVARIATE	-0.013	0.008	0.123	-0.121	0.079
35-64	MULTIVARIATE	-0.009	0.008	0.265	-0.091	0.082
65-94	MULTIVARIATE	-0.009	0.009	0.299	-0.084	0.081

SEX: FEMALE
EVENT: CORONARY HEART DISEASE DEATH
RISK FACTOR: VITAL CAPACITY - HEIGHT INDEX (ML/INCH)
PERSONS AT RISK: ALL PERSONS ALIVE

RISK FACTOR	35-44 AT EXAM			45-54 AT EXAM			55-64 AT EXAM			65-74 AT EXAM		
	PERSON EXAMS	# OF EVENTS	ANNUAL RATE	PERSON EXAMS	# OF EVENTS	ANNUAL RATE	PERSON EXAMS	# OF EVENTS	ANNUAL RATE	PERSON EXAMS	# OF EVENTS	ANNUAL RATE
4-29	122			546	1	1	1572	17	5	2068	36	9
30-34	291			1241	0	0	2547	8	2	1946	20	5
35-39	769			2103	2	0	3213	10	2	1728	10	3
40-44	1427			2838	4	1	2836	5	1	933	7	4
45-92	2989			3327	3	0	1914	2	1	422	2	2
UNKNOWN	21			17	0	0	21	0	0	6	0	0
TOTAL	5619			10072	10	0	12103	42	2	7103	75	5

RISK FACTOR	75-84 AT EXAM			85-94 AT EXAM			35-64 COMBINED			65-94 COMBINED		
	PERSON EXAMS	# OF EVENTS	ANNUAL RATE	PERSON EXAMS	# OF EVENTS	ANNUAL RATE	# OF EVENTS	CRUDE	AGE-ADJUSTED	# OF EVENTS	CRUDE	AGE-ADJUSTED
4-29	1212	40	17	109	4	18	18	4	3	80	12	11
30-34	686	15	11	37	1	14	8	1	1	36	7	7
35-39	370	6	8	16	0	0	13	1	1	16	4	4
40-44	123	0	0	1	0	0	9	1	1	3	3	3
45-92	26	1	19	0	0	-	6	0	0	3	3	7
UNKNOWN	2	0	0	0	0	-	0	0	0	0	0	0
TOTAL	2419	62	13	163	5	15	54	1	1	142	7	7

LOGISTIC REGRESSIONS

AGE AT EXAM		COEFFICIENT OF RISK FACTOR	SE COEFFICIENT	P-VALUE	STANDARDIZED COEFFICIENT	STD SE

TABLE 12-12 DATE=22NOV85

AVERAGE ANNUAL INCIDENCE RATE FOR EVENT PER 1000 PERSONS AT RISK AT EXAM
BY AGE AND LEVEL OF RISK FACTOR AT EXAM: FRAMINGHAM STUDY 30-YEAR FOLLOWUP

SEX: MALE
EVENT: CORONARY HEART DISEASE DEATH
RISK FACTOR: HEART RATE (PER MINUTE)
PERSONS AT RISK: ALL PERSONS ALIVE

RISK FACTOR	35-44 AT EXAM			45-54 AT EXAM			55-64 AT EXAM			65-74 AT EXAM		
	PERSON EXAMS	# OF EVENTS	ANNUAL RATE	PERSON EXAMS	# OF EVENTS	ANNUAL RATE	PERSON EXAMS	# OF EVENTS	ANNUAL RATE	PERSON EXAMS	# OF EVENTS	ANNUAL RATE
30-67	1295	0	0	2016	4	1	2618	24		1425	25	9
68-75	1506	0	0	2594	12	2	2763	29		1447	23	8
76-83	728	1	1	1473	11	4	1552	19		802	21	13
84-91	586	4	3	1145	9	4	1211	20		607	23	19
92-220	404	1	1	839	13	8	926	28	8	553	17	15
UNKNOWN	0	0	-	0	0	-	0	0	-	0	0	-
TOTAL	4519	6	1	8067	49	3	9070	120	7	4834	109	11

RISK FACTOR	75-84 AT EXAM			85-94 AT EXAM			35-64 COMBINED			65-94 COMBINED		
	PERSON EXAMS	# OF EVENTS	ANNUAL RATE	PERSON EXAMS	# OF EVENTS	ANNUAL RATE	# OF EVENTS	CRUDE	AGE-ADJUSTED	# OF EVENTS	CRUDE	AGE-ADJUSTED
30-67	388	16	21	17			28	2	2	41	11	11
68-75	362	15	21	14			41	3	3	38	10	11
76-83	271	13	24	13			31	4	4	34	16	15
84-91	174	11	32	5			33	6	6	36	23	23
92-220	150	9	30	1			42	10	9	26	18	18
UNKNOWN	0	0	-	0			0	-		0	-	
TOTAL	1345	64	24	50			175	4	4	175	14	14

LOGISTIC REGRESSIONS

AGE AT EXAM		COEFFICIENT OF RISK FACTOR	SE COEFFICIENT	P-VALUE	STANDARDIZED COEFFICIENT	STD SE
35-44		0.059	0.023	0.012	0.731	0.290
45-54		0.044	0.010	0.001	0.555	0.123
55-64		0.026	0.006	0.001	0.342	0.082
65-74		0.016	0.007	0.015	0.217	0.090
75-84		0.012	0.009	0.161	0.171	0.122
85-94						
35-64	UNIVARIATE	0.03?	0.005	0.001	0.419	0.066
65-94	UNIVARIATE	0.018	0.005	0.003	0.209	0.071
35-64	BIVARIATE	0.032	0.005	0.001	0.411	0.065
65-94	BIVARIATE	0.015	0.005	0.004	0.206	0.072
35-64	MULTIVARIATE	0.024	0.005	0.001	0.309	0.068
65-94	MULTIVARIATE	0.013	0.005	0.017	0.175	0.073

SEX: FEMALE
EVENT: CORONARY HEART DISEASE DEATH
RISK FACTOR: HEART RATE (PER MINUTE)
PERSONS AT RISK: ALL PERSONS ALIVE

RISK FACTOR	35-44 AT EXAM			45-54 AT EXAM			55-64 AT EXAM			65-74 AT EXAM		
	PERSON EXAMS	# OF EVENTS	ANNUAL RATE	PERSON EXAMS	# OF EVENTS	ANNUAL RATE	PERSON EXAMS	# OF EVENTS	ANNUAL RATE	PERSON EXAMS	# OF EVENTS	ANNUAL RATE
30-67	1892			1918	1	0	2450	6	1	1285	12	5
68-75	1806			3279	4	1	3641	12	2	1907	19	5
76-83	1051			1938	0	0	2442	3	1	1542	15	5
84-91	939			1620	2	1	1888	14	4	1196	11	5
92-220	731			1317	3	1	1682	7	2	1173	18	8
UNKNOWN	0			0	0	-	0	0	-	0	0	-
TOTAL	5619			10072	10	0	12103	42	2	7103	75	5

RISK FACTOR	75-84 AT EXAM			85-94 AT EXAM			35-64 COMBINED			65-94 COMBINED		
	PERSON EXAMS	# OF EVENTS	ANNUAL RATE	PERSON EXAMS	# OF EVENTS	ANNUAL RATE	# OF EVENTS	CRUDE	AGE-ADJUSTED	# OF EVENTS	CRUDE	AGE-ADJUSTED
30-67	459	11	12	36	2	28	7	1	1	25	7	7
68-75	586	16	14	41	0	0	16	1	1	35	7	7
76-83	532	12	11	30	1	17	4	0	0	28	7	7
84-91	410	11	13	29	2	14	17	2	2	24	7	7
92-220	432	12	14	27	0	0	10	2	1	30	9	9
UNKNOWN	0	0	-	0	0	-	0	-		0	-	
TOTAL	2419	62	13	163	5	15	54	1	1	142	7	7

LOGISTIC REGRESSIONS

AGE AT EXAM		COEFFICIENT OF RISK FACTOR	SE COEFFICIENT	P-VALUE	STANDARDIZED COEFFICIENT	STD SE
35-44						
45-54		0.019	0.023	0.413	0.237	0.290
55-64		0.017	0.011	0.104	0.225	0.138
65-74		0.011	0.008	0.186	0.145	0.109
75-84		-0.002	0.009	0.857	-0.023	0.130
85-94		-0.027	0.035	0.441	-0.383	0.497
35-64	UNIVARIATE	0.018	0.010	0.055	0.232	0.121
65-94	UNIVARIATE	0.005	0.006	0.456	0.062	0.083
35-64	BIVARIATE	0.018	0.009	0.064	0.222	0.120
65-94	BIVARIATE	0.004	0.006	0.513	0.054	0.082
35-64	MULTIVARIATE	0.006	0.010	0.597	0.070	0.132
65-94	MULTIVARIATE	-0.002	0.006	0.699	-0.032	0.084

AVERAGE ANNUAL INCIDENCE RATE FOR EVENT PER 1000 PERSONS AT RISK AT EXAM
BY AGE AND LEVEL OF RISK FACTOR AT EXAM: FRAMINGHAM STUDY 30-YEAR FOLLOWUP

SEX: MALE
EVENT: CORONARY HEART DISEASE DEATH
RISK FACTOR: CIGARETTES SMOKED (NUMBER PER DAY)
PERSONS AT RISK: ALL PERSONS ALIVE

RISK FACTOR	35-44 AT EXAM			45-54 AT EXAM			55-64 AT EXAM			65-74 AT EXAM		
	PERSON EXAMS	# OF EVENTS	ANNUAL RATE	PERSON EXAMS	# OF EVENTS	ANNUAL RATE	PERSON EXAMS	# OF EVENTS	ANNUAL RATE	PERSON EXAMS	# OF EVENTS	ANNUAL RATE
NONE	1402	0	0	3220	11	2	4888	55	6	3235	72	11
1-10	437	1	1	746	6	4	856	12	7	463	10	11
11-20	1499	1	0	2136	19	4	1889	33	9	668	17	13
21-40	993	3	2	1672	10	3	1235	16	6	427	8	9
41-90	162	1	3	255	2	4	178	4	11	34	1	15
UNKNOWN	26	0	0	38	1	13	24	0	0	7	1	71
TOTAL	4519	6	1	8067	49	3	9070	120	7	4834	109	11

RISK FACTOR	75-84 AT EXAM			85-94 AT EXAM			35-64 COMBINED			65-94 COMBINED		
	PERSON EXAMS	# OF EVENTS	ANNUAL RATE	PERSON EXAMS	# OF EVENTS	ANNUAL RATE	# OF EVENTS	CRUDE	AGE-ADJUSTED	# OF EVENTS	CRUDE	AGE-ADJUSTED
NONE	1067	53	25	38			66	3	3	126	15	14
1-10	125	7	28	6			19	5	5	17	14	15
11-20	88	3	17	5			53	4	4	21	14	15
21-40	60	1	8	1			29	4	7	9	9	9
41-90	5	0	0	0			7	6	7	1	13	12
UNKNOWN	0	0	-	0			1	6	5	1	71	55
TOTAL	1345	64	24	50			175	4	4	175	14	14

LOGISTIC REGRESSIONS

AGE AT EXAM		COEFFICIENT OF RISK FACTOR	SE COEFFICIENT	P-VALUE	STANDARDIZED COEFFICIENT	STD SE
35-44		0.050	0.026	0.052	0.707	0.364
45-54		0.012	0.009	0.179	0.182	0.135
55-64		0.011	0.006	0.085	0.145	0.084
65-74		-0.001	0.009	0.869	-0.016	0.099
75-84		-0.023	0.019	0.231	-0.202	0.169
85-94						
35-64	UNIVARIATE	0.006	0.005	0.247	0.085	0.073
65-94	UNIVARIATE	-0.009	0.008	0.219	-0.104	0.085
35-64	BIVARIATE	0.014	0.005	0.003	0.209	0.072
65-94	BIVARIATE	-0.004	0.008	0.649	-0.039	0.085
35-64	MULTIVARIATE	0.016	0.005	0.001	0.239	0.072
65-94	MULTIVARIATE	0.002	0.008	0.827	0.019	0.085

SEX: FEMALE
EVENT: CORONARY HEART DISEASE DEATH
RISK FACTOR: CIGARETTES SMOKED (NUMBER PER DAY)
PERSONS AT RISK: ALL PERSONS ALIVE

RISK FACTOR	35-44 AT EXAM			45-54 AT EXAM			55-64 AT EXAM			65-74 AT EXAM		
	PERSON EXAMS	# OF EVENTS	ANNUAL RATE	PERSON EXAMS	# OF EVENTS	ANNUAL RATE	PERSON EXAMS	# OF EVENTS	ANNUAL RATE	PERSON EXAMS	# OF EVENTS	ANNUAL RATE
NONE	2583			5599	4	0	8010	27	2	5562	50	4
1-10	1514			1732	2	1	1646	7	2	680	10	7
11-20	1373			2057	2	0	1767	7	2	650	13	10
21-40	298			629	1	1	624	1	1	191	1	3
41-90	22			38	1	13	18	0	0	1	0	0
UNKNOWN	29			37	0	0	38	0	0	19	1	26
TOTAL	5619			10072	10	0	12103	42	2	7103	75	5

RISK FACTOR	75-84 AT EXAM			85-94 AT EXAM			35-64 COMBINED			65-94 COMBINED		
	PERSON EXAMS	# OF EVENTS	ANNUAL RATE	PERSON EXAMS	# OF EVENTS	ANNUAL RATE	# OF EVENTS	CRUDE	AGE-ADJUSTED	# OF EVENTS	CRUDE	AGE-ADJUSTED
NONE	2139	55	13	154	5	16	31	1	1	110	7	6
1-10	161	1	9	7	0	0	9	1	1	13	8	7
11-20	99	4	20	2	0	-	10	1	1	17	11	12
21-40	16	0	0	0	0	-	3	1	1	2	2	2
41-90	0	0	-	0	0	-	1	6	5	0	0	0
UNKNOWN	4	0	0	0	0	-	0	0	0	1	22	19
TOTAL	2419	62	13	163	5	15	54	1	1	142	7	7

LOGISTIC REGRESSIONS

AGE AT EXAM		COEFFICIENT OF RISK FACTOR	SE COEFFICIENT	P-VALUE	STANDARDIZED COEFFICIENT	STD SE
35-44						
45-54		0.036	0.024	0.142	0.366	0.249
55-64		-0.004	0.017	0.792	-0.043	0.161
65-74		0.020	0.013	0.115	0.152	0.096
75-84		0.001	0.026	0.957	0.007	0.127
85-94		-	-	-	-	-
35-64	UNIVARIATE	0.007	0.013	0.598	0.068	0.129
65-94	UNIVARIATE	0.007	0.011	0.562	0.046	0.079
35-64	BIVARIATE	0.017	0.013	0.177	0.168	0.124
65-94	BIVARIATE	0.018	0.011	0.117	0.125	0.080
35-64	MULTIVARIATE	0.018	0.013	0.160	0.179	0.127
65-94	MULTIVARIATE	0.024	0.011	0.035	0.169	0.080

TABLE 12-14   DATE=22NOV85
AVERAGE ANNUAL INCIDENCE RATE FOR EVENT PER 1000 PERSONS AT RISK AT EXAM
BY AGE AND LEVEL OF RISK FACTOR AT EXAM: FRAMINGHAM STUDY 30-YEAR FOLLOWUP

SEX: MALE
EVENT: CORONARY HEART DISEASE DEATH
RISK FACTOR: ALBUMIN IN URINE, DEFINITE
PERSONS AT RISK: ALL PERSONS ALIVE

RISK FACTOR	35-44 AT EXAM PERSON EXAMS	# OF EVENTS	ANNUAL RATE	45-54 AT EXAM PERSON EXAMS	# OF EVENTS	ANNUAL RATE	55-64 AT EXAM PERSON EXAMS	# OF EVENTS	ANNUAL RATE	65-74 AT EXAM PERSON EXAMS	# OF EVENTS	ANNUAL RATE
NO	4411	6	1	7936	47	3	8828	114	6	4671	102	11
YES	71	0	0	105	2	10	215	5	12	157	7	22
UNKNOWN	37	0	0	26	0	0	27	1	19	6	0	0
TOTAL	4519	6	1	8067	49	3	9070	120	7	4834	109	11

RISK FACTOR	75-84 AT EXAM PERSON EXAMS	# OF EVENTS	ANNUAL RATE	85-94 AT EXAM PERSON EXAMS	# OF EVENTS	ANNUAL RATE	35-64 COMBINED # OF EVENTS	CRUDE	AGE-ADJUSTED	65-94 COMBINED # OF EVENTS	CRUDE	AGE-ADJUSTED
NO	1308	62	24	43	7		167	4	4	166	14	14
YES	34	2	29		7		7	9	8	9	23	24
UNKNOWN	3	0	0	0			1	6	8	0	0	0
TOTAL	1345	64	24	50			175	4	4	175	14	14

LOGISTIC REGRESSIONS

AGE AT EXAM		COEFFICIENT OF RISK FACTOR	SE COEFFICIENT	P-VALUE	STANDARDIZED COEFFICIENT	STD SE
35-44		-	-	-	-	-
45-54		1.181	0.729	0.105	0.134	0.083
55-64		0.599	0.462	0.195	0.091	0.070
65-74		0.737	0.399	0.065	0.131	0.071
75-84		0.228	0.740	0.758	0.036	0.116
85-94						
35-64	UNIVARIATE	0.830	0.389	0.033	0.111	0.052
65-94	UNIVARIATE	0.519	0.350	0.138	0.091	0.061
35-64	BIVARIATE	0.643	0.391	0.100	0.086	0.052
65-94	BIVARIATE	0.484	0.352	0.170	0.085	0.062
35-64	MULTIVARIATE	0.051	0.432	0.906	0.007	0.058
65-94	MULTIVARIATE	0.162	0.361	0.655	0.028	0.063

SEX: FEMALE
EVENT: CORONARY HEART DISEASE DEATH
RISK FACTOR: ALBUMIN IN URINE, DEFINITE
PERSONS AT RISK: ALL PERSONS ALIVE

RISK FACTOR	35-44 AT EXAM PERSON EXAMS	# OF EVENTS	ANNUAL RATE	45-54 AT EXAM PERSON EXAMS	# OF EVENTS	ANNUAL RATE	55-64 AT EXAM PERSON EXAMS	# OF EVENTS	ANNUAL RATE	65-74 AT EXAM PERSON EXAMS	# OF EVENTS	ANNUAL RATE
NO	5356			9766	10	1	11835	40	2	6955	70	5
YES	225			281	0	0	240	2	4	138	5	18
UNKNOWN	38			25	0	0	28	0	0	10	0	0
TOTAL	5619			10072	10	0	12103	42	2	7103	75	5

RISK FACTOR	75-84 AT EXAM. PERSON EXAMS	# OF EVENTS	ANNUAL RATE	85-94 AT EXAM PERSON EXAMS	# OF EVENTS	ANNUAL RATE	35-64 COMBINED # OF EVENTS	CRUDE	AGE-ADJUSTED	65-94 COMBINED # OF EVENTS	CRUDE	AGE-ADJUSTED
NO	2360	61	13	156	4	13	52	1	1	135	7	7
YES	50	1	10	7	1	71	2	1	2	7	18	17
UNKNOWN	9	0	0	0	0	-	0	0	0	0	0	0
TOTAL	2419	62	13	163	5	15	54	1	1	142	7	7

LOGISTIC REGRESSIONS

AGE AT EXAM		COEFFICIENT OF RISK FACTOR	SE COEFFICIENT	P-VALUE	STANDARDIZED COEFFICIENT	STD SE
35-44		-	-	-	-	-
45-54		-	-	-	-	-
55-64		0.907	0.728	0.212	0.127	0.102
65-74		1.308	0.471	0.006	0.181	0.065
75-84		-0.262	1.018	0.797	-0.037	0.145
85-94		1.846	1.193	0.122	0.375	0.243
35-64	UNIVARIATE	0.530	0.722	0.647	0.053	0.117
65-94	UNIVARIATE	0.946	0.395	0.017	0.133	0.053
35-64	BIVARIATE	0.579	0.723	0.424	0.094	0.117
65-94	BIVARIATE	0.876	0.397	0.028	0.123	0.056
35-64	MULTIVARIATE	-0.352	0.860	0.682	-0.057	0.139
65-94	MULTIVARIATE	0.395	0.411	0.337	0.056	0.058

TABLE 12-15   DATE=22NOV85
AVERAGE ANNUAL INCIDENCE RATE FOR EVENT PER 1000 PERSONS AT RISK AT EXAM
BY AGE AND LEVEL OF RISK FACTOR AT EXAM: FRAMINGHAM STUDY 30-YEAR FOLLOWUP

SEX: MALE
EVENT: CORONARY HEART DISEASE DEATH
RISK FACTOR: HEART ENLARGEMENT BY X-RAY
PERSONS AT RISK: ALL PERSONS ALIVE

RISK FACTOR	35-44 AT EXAM			45-54 AT EXAM			55-64 AT EXAM			65-74 AT EXAM		
	PERSON EXAMS	# OF EVENTS	ANNUAL RATE	PERSON EXAMS	# OF EVENTS	ANNUAL RATE	PERSON EXAMS	# OF EVENTS	ANNUAL RATE	PERSON EXAMS	# OF EVENTS	ANNUAL RATE
NO	4101	5	1	6613	26	2	6543	66	5	3075	50	8
MILD	189	0	0	708	5	4	1126	15	7	703	14	10
DEFINITE	224	1	2	742	18	12	1400	39	14	1056	45	21
UNKNOWN	5	0	0	4	0	0	1	0	0	0	0	-
TOTAL	4519	6	1	8067	49	3	9070	120	7	4834	109	11

RISK FACTOR	75-84 AT EXAM			85-94 AT EXAM			35-64 COMBINED			65-94 COMBINED		
	PERSON EXAMS	# OF EVENTS	ANNUAL RATE	PERSON EXAMS	# OF EVENTS	ANNUAL RATE	# OF EVENTS	CRUDE	AGE-ADJUSTED	# OF EVENTS	CRUDE	AGE-ADJUSTED
NO	753	27	18	34			97	3	3	77	10	10
MILD	226	10	22	6			20	5	4	25	13	13
DEFINITE	366	27	37	10			58	12	11	73	25	25
UNKNOWN	0	0	-	0			0	0	0	0		-
TOTAL	1345	64	24	50			175	4	4	175	14	14

LOGISTIC REGRESSIONS

AGE AT EXAM		COEFFICIENT OF RISK FACTOR	SE COEFFICIENT	P-VALUE	STANDARDIZED COEFFICIENT	STD SE
35-44		0.562	0.584	0.335	0.264	0.274
45-54		0.912	0.157	0.001	0.564	0.097
55-64		0.508	0.103	0.001	0.378	0.077
65-74		0.492	0.106	0.001	0.406	0.088
75-84		0.380	0.142	0.007	0.329	0.123
85-94						
35-64	UNIVARIATE	0.740	0.084	0.001	0.487	0.055
65-94	UNIVARIATE	0.484	0.084	0.001	0.405	0.070
35-64	BIVARIATE	0.599	0.086	0.001	0.394	0.057
65-94	BIVARIATE	0.454	0.085	0.001	0.379	0.071
35-64	MULTIVARIATE	0.544	0.093	0.001	0.358	0.062
65-94	MULTIVARIATE	0.315	0.091	0.001	0.263	0.076

SEX: FEMALE
EVENT: CORONARY HEART DISEASE DEATH
RISK FACTOR: HEART ENLARGEMENT BY X-RAY
PERSONS AT RISK: ALL PERSONS ALIVE

RISK FACTOR	35-44 AT EXAM			45-54 AT EXAM			55-64 AT EXAM			65-74 AT EXAM		
	PERSON EXAMS	# OF EVENTS	ANNUAL RATE	PERSON EXAMS	# OF EVENTS	ANNUAL RATE	PERSON EXAMS	# OF EVENTS	ANNUAL RATE	PERSON EXAMS	# OF EVENTS	ANNUAL RATE
NO	5166			8085	5	0	7836	13	1	3511	21	3
MILD	199			959	2	1	1879	10	3	1293	10	4
DEFINITE	246			1022	3	1	2384	19	4	2299	44	10
UNKNOWN	8			6	0	0	4	0	0	0	0	-
TOTAL	5619			10072	10	0	12103	42	2	7103	75	5

RISK FACTOR	75-84 AT EXAM			85-94 AT EXAM			35-64 COMBINED			65-94 COMBINED		
	PERSON EXAMS	# OF EVENTS	ANNUAL RATE	PERSON EXAMS	# OF EVENTS	ANNUAL RATE	# OF EVENTS	CRUDE	AGE-ADJUSTED	# OF EVENTS	CRUDE	AGE-ADJUSTED
NO	830	9	5	40	3	38	19	0	0	33	4	4
MILD	455	11	12	36	1	14	12	2	2	22	6	6
DEFINITE	1134	42	19	87	1	6	23	3	3	87	12	12
UNKNOWN	0	0	-	0	0	-	0	0	0	0		-
TOTAL	2419	62	13	163	5	15	54	1	1	142	7	7

LOGISTIC REGRESSIONS

AGE AT EXAM		COEFFICIENT OF RISK FACTOR	SE COEFFICIENT	P-VALUE	STANDARDIZED COEFFICIENT	STD SE
35-44						
45-54		0.799	0.350	0.022	0.513	0.224
55-64		0.774	0.173	0.001	0.620	0.138
65-74		0.609	0.136	0.001	0.540	0.121
75-84		0.598	0.171	0.001	0.534	0.153
85-94		-0.980	0.574	0.088	-0.820	0.480
35-64	UNIVARIATE	0.967	0.148	0.001	0.681	0.104
65-94	UNIVARIATE	0.612	0.102	0.001	0.550	0.092
35-64	BIVARIATE	0.742	0.153	0.001	0.523	0.108
65-94	BIVARIATE	0.534	0.104	0.001	0.480	0.093
35-64	MULTIVARIATE	0.675	0.164	0.001	0.476	0.116
65-94	MULTIVARIATE	0.423	0.108	0.001	0.380	0.097

TABLE 12-16   DATE=12DEC85
AVERAGE ANNUAL INCIDENCE RATE FOR EVENT PER 1000 PERSONS AT RISK AT EXAM
BY AGE AND LEVEL OF RISK FACTOR AT EXAM: FRAMINGHAM STUDY 30-YEAR FOLLOWUP

SEX: MALE
EVENT: CORONARY HEART DISEASE DEATH
RISK FACTOR: LEFT VENTRICULAR HYPERTROPHY BY ELECTROCARDIOGRAM
PERSONS AT RISK: ALL PERSONS ALIVE

| | 35-44 AT EXAM | | | 45-54 AT EXAM | | | 55-64 AT EXAM | | | 65-74 AT EXAM | | |
RISK FACTOR	PERSON EXAMS	# OF EVENTS	ANNUAL RATE	PERSON EXAMS	# OF EVENTS	ANNUAL RATE	PERSON EXAMS	# OF EVENTS	ANNUAL RATE	PERSON EXAMS	# OF EVENTS	ANNUAL RATE
NO	4408	6	1	7812	39	2	8611	101	6	4527	78	9
MILD	79	0	0	147	4	14	240	4	8	121	9	37
DEFINITE	32	0	0	108	6	28	219	15	34	186	22	59
UNKNOWN	0	0	–	0	0	–	0	0	–	0	0	–
TOTAL	4519	6	1	8067	49	3	9070	120	7	4834	109	11

| | 75-84 AT EXAM | | | 85-94 AT EXAM | | | 35-64 COMBINED | | | 65-94 COMBINED | | |
RISK FACTOR	PERSON EXAMS	# OF EVENTS	ANNUAL RATE	PERSON EXAMS	# OF EVENTS	ANNUAL RATE	# OF EVENTS	CRUDE	AGE-ADJUSTED	# OF EVENTS	CRUDE	AGE-ADJUSTED
NO	1250	58	23	45			146	4	4	138	12	12
MILD	44	1	11	2			8	9	9	10	30	31
DEFINITE	51	5	49	3			21	29	25	27	56	56
UNKNOWN	0	0	–	0			0	–		0	–	
TOTAL	1345	64	24	50			175	4	4	175	14	14

LOGISTIC REGRESSIONS

AGE AT EXAM		COEFFICIENT OF RISK FACTOR	SE COEFFICIENT	P-VALUE	STANDARDIZED COEFFICIENT	STD SE
35-44		–	–	–	–	–
45-54		1.279	0.209	0.001	0.338	0.055
55-64		0.871	0.143	0.001	0.299	0.049
65-74		1.047	0.122	0.001	0.430	0.050
75-84		0.308	0.249	0.217	0.128	0.104
85-94						
35-64	UNIVARIATE	1.072	0.117	0.001	0.312	0.034
65-94	UNIVARIATE	0.836	0.108	0.001	0.345	0.044
35-64	BIVARIATE	0.935	0.119	0.001	0.273	0.035
65-94	BIVARIATE	0.838	0.109	0.001	0.346	0.045
35-64	MULTIVARIATE	0.861	0.133	0.001	0.259	0.039
65-94	MULTIVARIATE	0.790	0.120	0.001	0.289	0.049

SEX: FEMALE
EVENT: CORONARY HEART DISEASE DEATH
RISK FACTOR: LEFT VENTRICULAR HYPERTROPHY BY ELECTROCARDIOGRAM
PERSONS AT RISK: ALL PERSONS ALIVE

| | 35-44 AT EXAM | | | 45-54 AT EXAM | | | 55-64 AT EXAM | | | 65-74 AT EXAM | | |
RISK FACTOR	PERSON EXAMS	# OF EVENTS	ANNUAL RATE	PERSON EXAMS	# OF EVENTS	ANNUAL RATE	PERSON EXAMS	# OF EVENTS	ANNUAL RATE	PERSON EXAMS	# OF EVENTS	ANNUAL RATE
NO	5566			9905	10	1	11679	39	2	6686	57	9
MILD	38			111	0	0	250	1	2	182	6	16
DEFINITE	15			56	0	0	174	2	6	235	12	26
UNKNOWN	0			0	0	–	0	0	–	0	0	–
TOTAL	5619			10072	10	0	12103	42	2	7103	75	5

| | 75-84 AT EXAM | | | 85-94 AT EXAM | | | 35-64 COMBINED | | | 65-94 COMBINED | | |
RISK FACTOR	PERSON EXAMS	# OF EVENTS	ANNUAL RATE	PERSON EXAMS	# OF EVENTS	ANNUAL RATE	# OF EVENTS	CRUDE	AGE-ADJUSTED	# OF EVENTS	CRUDE	AGE-ADJUSTED
NO	2217	49	11	142	5	18	50	1	1	11	6	6
MILD	77	4	26	10	0	0	1	1	1	10	19	19
DEFINITE	125	9	36	11	0	0	3	6	9	21	28	28
UNKNOWN	0	0	–	0	0	–	0	–		0	–	
TOTAL	2419	62	13	163	5	15	54	1	1	142	7	7

LOGISTIC REGRESSIONS

AGE AT EXAM		COEFFICIENT OF RISK FACTOR	SE COEFFICIENT	P-VALUE	STANDARDIZED COEFFICIENT	STD SE
35-44		–	–	–	–	–
45-54		–	–	–	–	–
55-64		0.564	0.356	0.113	0.155	0.098
65-74		0.950	0.154	0.001	0.367	0.060
75-84		0.636	0.180	0.001	0.299	0.084
85-94		–	–	–	–	–
35-64	UNIVARIATE	0.887	0.299	0.003	0.196	0.066
65-94	UNIVARIATE	0.813	0.117	0.001	0.335	0.048
35-64	BIVARIATE	0.657	0.303	0.030	0.145	0.067
65-94	BIVARIATE	0.753	0.118	0.001	0.310	0.049
35-64	MULTIVARIATE	0.296	0.334	0.376	0.065	0.074
65-94	MULTIVARIATE	0.542	0.127	0.001	0.223	0.052

TABLE 12-17 DATE=22NOV85
AVERAGE ANNUAL INCIDENCE RATE FOR EVENT PER 1000 PERSONS AT RISK AT EXAM
BY AGE AND LEVEL OF RISK FACTOR AT EXAM: FRAMINGHAM STUDY 30-YEAR FOLLOWUP

SEX: MALE
EVENT: CORONARY HEART DISEASE DEATH
RISK FACTOR: INTRAVENTRICULAR CONDUCTION DEFECT (WITH QRS INTERVAL MORE THAN 0.11 SECOND)
PERSONS AT RISK: ALL PERSONS ALIVE

RISK FACTOR	35-44 AT EXAM			45-54 AT EXAM			55-64 AT EXAM			65-74 AT EXAM		
	PERSON EXAMS	# OF EVENTS	ANNUAL RATE	PERSON EXAMS	# OF EVENTS	ANNUAL RATE	PERSON EXAMS	# OF EVENTS	ANNUAL RATE	PERSON EXAMS	# OF EVENTS	ANNUAL RATE
NO	4201	6	1	7487	40	3	8247	100	6	4323	91	11
INCOMPLETE	280	0	0	491	3	3	541	10	9	250	5	10
COMPLETE	38	0	0	89	6	34	282	10	18	261	13	25
UNKNOWN	0	0	-	0	0	-	0	0	-	0	0	-
TOTAL	4519	6	1	8067	49	3	9070	120	7	4834	109	11

RISK FACTOR	75-84 AT EXAM			85-94 AT EXAM			35-64 COMBINED			65-94 COMBINED		
	PERSON EXAMS	# OF EVENTS	ANNUAL RATE	PERSON EXAMS	# OF EVENTS	ANNUAL RATE	# OF EVENTS	CRUDE	AGE-ADJUSTED	# OF EVENTS	CRUDE	AGE-ADJUSTED
NO	1180	48	20	41			146	4	4	140	13	13
INCOMPLETE	37	4	54	3			13	5	5	9	16	19
COMPLETE	128	12	47	6			16	20	20	26	33	30
UNKNOWN	0	0	-	0			0	-	-	0	-	-
TOTAL	1345	64	24	50			175	4	4	175	14	14

LOGISTIC REGRESSIONS

AGE AT EXAM		COEFFICIENT OF RISK FACTOR	SE COEFFICIENT	P-VALUE	STANDARDIZED COEFFICIENT	STD SE
35-44		-	-	-	-	-
45-54		1.092	0.239	0.001	0.342	0.075
55-64		0.526	0.157	0.001	0.216	0.065
65-74		0.397	0.151	0.009	0.195	0.074
75-84		0.473	0.163	0.004	0.284	0.098
85-94						
35-64	UNIVARIATE	0.745	0.132	0.001	0.265	0.047
65-94	UNIVARIATE	0.478	0.109	0.001	0.249	0.057
35-64	BIVARIATE	0.608	0.131	0.001	0.216	0.046
65-94	BIVARIATE	0.422	0.109	0.001	0.219	0.057
35-64	MULTIVARIATE	0.595	0.136	0.001	0.212	0.048
65-94	MULTIVARIATE	0.440	0.111	0.001	0.229	0.058

SEX: FEMALE
EVENT: CORONARY HEART DISEASE DEATH
RISK FACTOR: INTRAVENTRICULAR CONDUCTION DEFECT (WITH QRS INTERVAL MORE THAN 0.11 SECOND)
PERSONS AT RISK: ALL PERSONS ALIVE

RISK FACTOR	35-44 AT EXAM			45-54 AT EXAM			55-64 AT EXAM			65-74 AT EXAM		
	PERSON EXAMS	# OF EVENTS	ANNUAL RATE	PERSON EXAMS	# OF EVENTS	ANNUAL RATE	PERSON EXAMS	# OF EVENTS	ANNUAL RATE	PERSON EXAMS	# OF EVENTS	ANNUAL RATE
NO	5532			9873	10	1	11731	38	2	6794	66	5
INCOMPLETE	75			151	0	0	195	1	3	134	0	0
COMPLETE	12			48	0	0	177	3	8	175	9	26
UNKNOWN	0			0	0	-	0	0	-	0	0	-
TOTAL	5619			10072	10	0	12103	42	2	7103	75	5

RISK FACTOR	75-84 AT EXAM			85-94 AT EXAM			35-64 COMBINED			65-94 COMBINED		
	PERSON EXAMS	# OF EVENTS	ANNUAL RATE	PERSON EXAMS	# OF EVENTS	ANNUAL RATE	# OF EVENTS	CRUDE	AGE-ADJUSTED	# OF EVENTS	CRUDE	AGE-ADJUSTED
NO	2261	53	12	153	5	16	50	1	1	124	7	7
INCOMPLETE	51	0	0	3	0	0	1	1	1	0	0	0
COMPLETE	107	9	42	7	0	0	3	6	4	18	31	29
UNKNOWN	0	0	-	0	0	-	0	-	-	0	-	-
TOTAL	2419	62	13	163	5	15	54	1	1	142	7	7

LOGISTIC REGRESSIONS

AGE AT EXAM		COEFFICIENT OF RISK FACTOR	SE COEFFICIENT	P-VALUE	STANDARDIZED COEFFICIENT	STD SE
35-44		-	-	-	-	-
45-54		-	-	-	-	-
55-64		0.802	0.298	0.007	0.216	0.080
65-74		0.779	0.190	0.001	0.261	0.064
75-84		0.613	0.193	0.002	0.265	0.083
85-94		-	-	-	-	-
35-64	UNIVARIATE	0.894	0.300	0.003	0.196	0.066
65-94	UNIVARIATE	0.722	0.135	0.001	0.263	0.049
35-64	BIVARIATE	0.673	0.298	0.024	0.148	0.066
65-94	BIVARIATE	0.659	0.156	0.001	0.240	0.049
35-64	MULTIVARIATE	0.676	0.302	0.025	0.148	0.066
65-94	MULTIVARIATE	0.636	0.139	0.001	0.232	0.051

· 253

TABLE 12-18   DATE=22NOV85

AVERAGE ANNUAL INCIDENCE RATE FOR EVENT PER 1000 PERSONS AT RISK AT EXAM
BY AGE AND LEVEL OF RISK FACTOR AT EXAM: FRAMINGHAM STUDY 30-YEAR FOLLOWUP

SEX: MALE
EVENT: CORONARY HEART DISEASE DEATH
RISK FACTOR: DEFINITE NONSPECIFIC T-WAVE OR ST-SEGMENT ABNORMALITY BY ELECTROCARDIOGRAM
PERSONS AT RISK: ALL PERSONS ALIVE

RISK FACTOR	35-44 AT EXAM PERSON EXAMS	# OF EVENTS	ANNUAL RATE	45-54 AT EXAM PERSON EXAMS	# OF EVENTS	ANNUAL RATE	55-64 AT EXAM PERSON EXAMS	# OF EVENTS	ANNUAL RATE	65-74 AT EXAM PERSON EXAMS	# OFT EVEN S	ANNUAL RATE
NO	4406	6	1	7641	41	3	8246	91	6	4170	82	10
YES	113	0	0	426	8	9	824	29	18	664	27	20
UNKNOWN	0	0	-	0	0	-	0	0	-	0	0	-
TOTAL	4519	6	1	8067	49	3	9070	120	7	4834	109	11

RISK FACTOR	75-84 AT EXAM PERSON EXAMS	# OF EVENTS	ANNUAL RATE	85-94 AT EXAM PERSON EXAMS	# OF EVENTS	ANNUAL RATE	35-64 COMBINED # OF EVENTS	CRUDE	AGE-ADJUSTED	65-94 COMBINED # OF EVENTS	CRUDE	AGE-ADJUSTED
NO	1063	44	21	31			138	3	3	126	12	12
YES	282	20	35	19			37	14	11	49	25	24
UNKNOWN	0	0	-	0			0			0		
TOTAL	1345	64	24	50			175	4	4	175	14	14

LOGISTIC REGRESSIONS

AGE AT EXAM		COEFFICIENT OF RISK FACTOR	SE COEFFICIENT	P-VALUE	STANDARDIZED COEFFICIENT	STD SE
35-44		-	-	-	-	-
45-54		1.266	0.390	0.001	0.283	0.087
55-64		1.184	0.216	0.001	0.340	0.062
65-74		0.748	0.226	0.001	0.258	0.078
75-84		0.570	0.278	0.041	0.232	0.113
85-94						
35-64	UNIVARIATE	1.405	0.187	0.001	0.341	0.045
65-94	UNIVARIATE	0.780	0.172	0.001	0.282	0.062
35-64	BIVARIATE	1.151	0.190	0.001	0.275	0.046
65-94	BIVARIATE	0.677	0.174	0.001	0.245	0.063
35-64	MULTIVARIATE	0.932	0.193	0.001	0.226	0.047
65-94	MULTIVARIATE	0.667	0.178	0.001	0.241	0.064

SEX: FEMALE
EVENT: CORONARY HEART DISEASE DEATH
RISK FACTOR: DEFINITE NONSPECIFIC T-WAVE OR ST-SEGMENT ABNORMALITY BY ELECTROCARDIOGRAM
PERSONS AT RISK: ALL PERSONS ALIVE

RISK FACTOR	35-44 AT EXAM PERSON EXAMS	# OF EVENTS	ANNUAL RATE	45-54 AT EXAM PERSON EXAMS	# OF EVENTS	ANNUAL RATE	55-64 AT EXAM PERSON EXAMS	# OF EVENTS	ANNUAL RATE	65-74 AT EXAM PERSON EXAMS	# OF EVENTS	ANNUAL RATE
NO	5520			9642	7	0	11111	33	1	6163	53	4
YES	99			430	3	3	992	9	5	940	22	12
UNKNOWN	0			0	0	-	0	0	-	0	0	-
TOTAL	5619			10072	10	0	12103	42	2	7103	75	5

RISK FACTOR	75-84 AT EXAM PERSON EXAMS	# OF EVENTS	ANNUAL RATE	85-94 AT EXAM PERSON EXAMS	# OF EVENTS	ANNUAL RATE	35-64 COMBINED # OF EVENTS	CRUDE	AGE-ADJUSTED	65-94 COMBINED # OF EVENTS	CRUDE	AGE-ADJUSTED
NO	1945	42	11	124	3	12	43	1	1	98	6	6
YES	474	20	21	39	2	26	12	4	3	44	15	14
UNKNOWN	0	0	-	0	0	-	0			0		
TOTAL	2419	62	13	163	5	15	54	1	1	142	7	7

LOGISTIC REGRESSIONS

AGE AT EXAM		COEFFICIENT OF RISK FACTOR	SE COEFFICIENT	P-VALUE	STANDARDIZED COEFFICIENT	STD SE
35-44						
45-54		2.269	0.692	0.001	0.459	0.140
55-64		1.123	0.378	0.003	0.308	0.104
65-74		1.016	0.256	0.001	0.344	0.087
75-84		0.691	0.277	0.012	0.274	0.110
85-94		0.779	0.932	0.403	0.334	0.399
35-64	UNIVARIATE	1.603	0.328	0.001	0.365	0.075
65-94	UNIVARIATE	0.952	0.184	0.001	0.340	0.066
35-64	BIVARIATE	1.214	0.332	0.001	0.276	0.075
65-94	BIVARIATE	0.837	0.185	0.001	0.299	0.066
35-64	MULTIVARIATE	0.996	0.338	0.003	0.226	0.077
65-94	MULTIVARIATE	0.771	0.188	0.001	0.275	0.067

TABLE 13-1   DATE=02NOV85

AVERAGE ANNUAL INCIDENCE RATE FOR EVENT PER 1000 PERSONS AT RISK AT EXAM
BY AGE AND LEVEL OF RISK FACTOR AT EXAM: FRAMINGHAM STUDY 30-YEAR FOLLOWUP

SEX: MALE
EVENT: STROKE AND TIA ONLY
RISK FACTOR: SYSTOLIC BLOOD PRESSURE, FIRST EXAMINER (MM HG)
PERSONS AT RISK: PERSONS FREE OF STROKE AND TIA ONLY

RISK FACTOR	35-44 AT EXAM PERSON EXAMS	# OF EVENTS	ANNUAL RATE	45-54 AT EXAM PERSON EXAMS	# OF EVENTS	ANNUAL RATE	55-64 AT EXAM PERSON EXAMS	# OF EVENTS	ANNUAL RATE	65-74 AT EXAM PERSON EXAMS	# OF EVENTS	ANNUAL RATE
74-119	1225			1837	4	1	1589	4	1	649	7	5
120-139	2191			3551	9	1	3482	16	2	1582	15	5
140-159	850			1830	9	2	2324	19	4	1394	23	8
160-179	197			589	7	6	1029	15	7	659	17	13
180-300	53			245	3	6	490	13	13	342	15	22
UNKNOWN	0			0	0	-	0	0	-	0	0	-
TOTAL	4516			8032	32	2	8914	67	4	4626	77	8

RISK FACTOR	75-84 AT EXAM PERSON EXAMS	# OF EVENTS	ANNUAL RATE	85-94 AT EXAM PERSON EXAMS	# OF EVENTS	ANNUAL RATE	35-64 COMBINED # OF EVENTS	CRUDE	AGE-ADJUSTED	65-94 COMBINED # OF EVENTS	CRUDE	AGE-ADJUSTED
74-119	136	3	11	7			9	1	1	12	8	8
120-139	420	14	17	13			25	1	1	29	7	9
140-159	379	8	11	15			28	3	3	31	9	9
160-179	193	16	41	5			22	6	5	33	19	19
180-300	106	9	42	2			16	10	8	24	27	26
UNKNOWN	0	0	-	0			0	-		0	-	
TOTAL	1234	50	20	42			100	2	2	129	11	11

LOGISTIC REGRESSIONS

AGE AT EXAM		COEFFICIENT OF RISK FACTOR	SE COEFFICIENT	P-VALUE	STANDARDIZED COEFFICIENT	STD SE
35-44						
45-54		0.027	0.007	0.001	0.537	0.134
55-64		0.028	0.004	0.001	0.625	0.091
65-74		0.019	0.004	0.001	0.431	0.096
75-84		0.018	0.005	0.001	0.426	0.127
85-94						
35-64	UNIVARIATE	0.031	0.003	0.001	0.639	0.070
65-94	UNIVARIATE	0.018	0.003	0.001	0.415	0.076
35-64	BIVARIATE	0.027	0.004	0.001	0.543	0.072
65-94	BIVARIATE	0.018	0.003	0.001	0.418	0.077
35-64	MULTIVARIATE	0.024	0.004	0.001	0.490	0.079
65-94	MULTIVARIATE	0.015	0.004	0.001	0.356	0.082

SEX: FEMALE
EVENT: STROKE AND TIA ONLY
RISK FACTOR: SYSTOLIC BLOOD PRESSURE, FIRST EXAMINER (MM HG)
PERSONS AT RISK: PERSONS FREE OF STROKE AND TIA ONLY

RISK FACTOR	35-44 AT EXAM PERSON EXAMS	# OF EVENTS	ANNUAL RATE	45-54 AT EXAM PERSON EXAMS	# OF EVENTS	ANNUAL RATE	55-64 AT EXAM PERSON EXAMS	# OF EVENTS	ANNUAL RATE	65-74 AT EXAM PERSON EXAMS	# OF EVENTS	ANNUAL RATE
74-119	2488			2717	4	1	2011	2	0	675	3	2
120-139	2245			4015	6	1	4168	13	2	1876	15	4
140-159	673			2070	9	2	3155	18	3	2212	25	6
160-179	156			783	2	1	1654	9	3	1302	24	9
180-300	51			448	3	3	962	15	8	806	33	20
UNKNOWN	0			0	0	-	0	0	-	0	0	-
TOTAL	5613			10033	24	1	11950	57	2	6871	100	7

RISK FACTOR	75-84 AT EXAM PERSON EXAMS	# OF EVENTS	ANNUAL RATE	85-94 AT EXAM PERSON EXAMS	# OF EVENTS	ANNUAL RATE	35-64 COMBINED # OF EVENTS	CRUDE	AGE-ADJUSTED	65-94 COMBINED # OF EVENTS	CRUDE	AGE-ADJUSTED
74-119	145	3	10	10	1	50	9	1	1	7	4	5
120-139	527	13	12	47	1	11	20	1	-1	29	6	7
140-159	704	16	11	34	1	15	27	2	2	42	7	7
160-179	512	13	13	32	2	31	11	2	2	39	11	10
180-300	339	12	18	25	2	40	18	6	5	47	20	20
UNKNOWN	0	0	-	0		-	0	-		0	-	
TOTAL	2227	57	13	148	7	24	85	2	2	164	9	9

LOGISTIC REGRESSIONS

AGE AT EXAM		COEFFICIENT OF RISK FACTOR	SE COEFFICIENT	P-VALUE	STANDARDIZED COEFFICIENT	STD SE
35-44						
45-54		0.019	0.006	0.003	0.425	0.142
55-64		0.025	0.004	0.001	0.621	0.094
65-74		0.024	0.003	0.001	0.582	0.085
75-84		0.007	0.005	0.144	0.185	0.126
85-94		0.014	0.015	0.352	0.346	0.371
35-64	UNIVARIATE	0.025	0.003	0.001	0.594	0.071
65-94	UNIVARIATE	0.018	0.003	0.001	0.458	0.069
35-64	BIVARIATE	0.022	0.003	0.001	0.525	0.076
65-94	BIVARIATE	0.017	0.003	0.001	0.433	0.069
35-64	MULTIVARIATE	0.017	0.004	0.001	0.400	0.086
65-94	MULTIVARIATE	0.015	0.003	0.001	0.373	0.072

TABLE 13-2   DATE=02NOV85

AVERAGE ANNUAL INCIDENCE RATE FOR EVENT PER 1000 PERSONS AT RISK AT EXAM.
BY AGE AND LEVEL OF RISK FACTOR AT EXAM, FRAMINGHAM STUDY 30-YEAR FOLLOWUP

SEX: MALE
EVENT: STROKE AND TIA ONLY
RISK FACTOR: DIASTOLIC BLOOD PRESSURE, FIRST EXAMINER (MM HG)
PERSONS AT RISK: PERSONS FREE OF STROKE AND TIA ONLY

RISK FACTOR	35-44 AT EXAM			45-54 AT EXAM			55-64 AT EXAM			65-74 AT EXAM		
	PERSON EXAMS	# OF EVENTS	ANNUAL RATE	PERSON EXAMS	# OF EVENTS	ANNUAL RATE	PERSON EXAMS	# OF EVENTS	ANNUAL RATE	PERSON EXAMS	# OF EVENTS	ANNUAL RATE
20-74	1015			1532	4	1	2056	10	2	1469	23	10
75-84	1614			2857	6	1	3127	10	2	1575	18	6
85-94	1203			2142	13	3	2192	15	3	1030	14	7
95-104	509			1065	6	3	1081	17	8	399	9	11
105-160	175			436	3	3	458	15	16	153	8	26
UNKNOWN	0			0	0	-	0	0	-	0	0	-
TOTAL	4516			8032	32	2	8914	67	4	4626	77	8

RISK FACTOR	75-84 AT EXAM			85-94 AT EXAM			35-64 COMBINED			65-94 COMBINED		
	PERSON EXAMS	# OF EVENTS	ANNUAL RATE	PERSON EXAMS	# OF EVENTS	ANNUAL RATE	# OF EVENTS	CRUDE	AGE-ADJUSTED	# OF EVENTS	CRUDE	AGE-ADJUSTED
20-74	513	20	19	22			15	2	2	49	12	12
75-84	403	14	17	15			16	2	1	33	8	8
85-94	224	7	16	5			28	3	3	21	8	9
95-104	66	8	61	0			23	4	4	17	18	22
105-160	28	1	18	0			18	8	8	9	25	24
UNKNOWN	0	0	-	0			0	-	-	0	-	-
TOTAL	1234	50	20	42			100	2	2	129	11	11

LOGISTIC REGRESSIONS

AGE AT EXAM		COEFFICIENT OF RISK FACTOR	SE COEFFICIENT	P-VALUE	STANDARDIZED COEFFICIENT	STD SE
35-44						
45-54		0.032	0.013	0.016	0.380	0.158
55-64		0.055	0.008	0.001	0.668	0.098
65-74		0.011	0.009	0.214	0.137	0.110
75-84		0.013	0.011	0.254	0.161	0.141
85-94						
35-64	UNIVARIATE	0.049	0.007	0.001	0.585	0.082
65-94	UNIVARIATE	0.007	0.007	0.345	0.082	0.087
35-64	BIVARIATE	0.048	0.007	0.001	0.572	0.081
65-94	BIVARIATE	0.013	0.007	0.060	0.164	0.088
35-64	MULTIVARIATE	0.044	0.007	0.001	0.524	0.085
65-94	MULTIVARIATE	0.010	0.007	0.136	0.129	0.087

SEX: FEMALE
EVENT: STROKE AND TIA ONLY
RISK FACTOR: DIASTOLIC BLOOD PRESSURE, FIRST EXAMINER (MM HG)
PERSONS AT RISK: PERSONS FREE OF STROKE AND TIA ONLY

RISK FACTOR	35-44 AT EXAM			45-54 AT EXAM			55-64 AT EXAM			65-74 AT EXAM		
	PERSON EXAMS	# OF EVENTS	ANNUAL RATE	PERSON EXAMS	# OF EVENTS	ANNUAL RATE	PERSON EXAMS	# OF EVENTS	ANNUAL RATE	PERSON EXAMS	# OF EVENTS	ANNUAL RATE
20-74	2161			2650	3	1	3078	9	1	2026	20	5
75-84	2028			3558	6	1	3964	12	2	2362	23	5
85-94	988			2387	7	1	2965	12	2	1559	29	9
95-104	324			933	5	3	1326	10	4	642	10	8
105-160	112			505	3	3	617	14	11	282	18	32
UNKNOWN	0			0	0	-	0	0	-	0	0	-
TOTAL	5613			10033	24	1	11950	57	2	6871	100	7

RISK FACTOR	75-84 AT EXAM			85-94 AT EXAM			35-64 COMBINED			65-94 COMBINED		
	PERSON EXAMS	# OF EVENTS	ANNUAL RATE	PERSON EXAMS	# OF EVENTS	ANNUAL RATE	# OF EVENTS	CRUDE	AGE-ADJUSTED	# OF EVENTS	CRUDE	AGE-ADJUSTED
20-74	869	20	12	66	4	30	13	1	1	44	7	7
75-84	681	22	16	52	2	19	20	1	1	47	8	8
85-94	466	12	13	21	1	24	20	2	2	42	10	10
95-104	155	2	6	7	0	0	15	3	3	12	7	7
105-160	56	1	9	2	0	0	17	7	6	19	28	26
UNKNOWN	0	0	-	0	0	-	0	-	-	0	-	-
TOTAL	2227	57	13	148	7	24	85	2	2	164	9	9

LOGISTIC REGRESSIONS

AGE AT EXAM		COEFFICIENT OF RISK FACTOR	SE COEFFICIENT	P-VALUE	STANDARDIZED COEFFICIENT	STD SE
35-44						
45-54		0.035	0.013	0.008	0.433	0.163
55-64		0.046	0.008	0.001	0.581	0.106
65-74		0.039	0.007	0.001	0.483	0.088
75-84		-0.007	0.011	0.549	-0.082	0.156
85-94		-0.019	0.029	0.508	-0.253	0.182
35-64	UNIVARIATE	0.044	0.007	0.001	0.547	0.084
65-94	UNIVARIATE	0.020	0.006	0.001	0.245	0.074
35-64	BIVARIATE	0.041	0.007	0.001	0.511	0.087
65-94	BIVARIATE	0.024	0.006	0.001	0.294	0.074
35-64	MULTIVARIATE	0.033	0.007	0.001	0.404	0.091
65-94	MULTIVARIATE	0.023	0.006	0.001	0.289	0.074

SEX: MALE
EVENT: STROKE AND TIA ONLY
RISK FACTOR: HYPERTENSION WITH ANTIHYPERTENSIVE TREATMENT
PERSONS AT RISK: PERSONS FREE OF STROKE AND TIA ONLY

RISK FACTOR	35-44 AT EXAM PERSON EXAMS	# OF EVENTS	ANNUAL RATE	45-54 AT EXAM PERSON EXAMS	# OF EVENTS	ANNUAL RATE	55-64 AT EXAM PERSON EXAMS	# OF EVENTS	ANNUAL RATE	65-74 AT EXAM PERSON EXAMS	# OF EVENTS	ANNUAL RATE
NO	2665			4248	8	1	4128	8	1	1764	15	4
MILD	1316			2280	10	2	2534	17	3	1474	22	7
DEFINITE	511			1212	9	4	1402	26	9	675	14	10
TREATED	24			292	5	9	850	16	9	715	26	18
UNKNOWN	0			0	0	-	0	0	-	0	0	-
TOTAL	4516			8032	32	2	8914	67	4	4626	77	8

RISK FACTOR	75-84 AT EXAM PERSON EXAMS	# OF EVENTS	ANNUAL RATE	85-94 AT EXAM PERSON EXAMS	# OF EVENTS	ANNUAL RATE	35-64 COMBINED # OF EVENTS	CRUDE	AGE-ADJUSTED	65-94 COMBINED # OF EVENTS	CRUDE	AGE-ADJUSTED
NO	440	12	14	15			17	1	1	29	7	7
MILD	395	9	11	14			27	2	2	31	8	8
DEFINITE	182	14	38	4			35	6	5	28	16	16
TREATED	217	15	35	9			21	9	7	41	22	22
UNKNOWN	0	0	-	0			0	-	-	0	-	-
TOTAL	1234	50	20	42			100	2	2	129	11	11

LOGISTIC REGRESSIONS

AGE AT EXAM		COEFFICIENT OF RISK FACTOR	SE COEFFICIENT	P-VALUE	STANDARDIZED COEFFICIENT	STD SE
35-44						
45-54		0.799	0.218	0.001	0.618	0.169
55-64		1.114	0.173	0.001	0.913	0.141
65-74		0.629	0.148	0.001	0.516	0.122
75-84		0.597	0.188	0.001	0.492	0.155
85-94						
35-64	UNIVARIATE	0.834	0.094	0.001	0.751	0.085
65-94	UNIVARIATE	0.441	0.078	0.001	0.473	0.084
35-64	BIVARIATE	0.686	0.093	0.001	0.617	0.084
65-94	BIVARIATE	0.432	0.078	0.001	0.464	0.084
35-64	MULTIVARIATE	0.648	0.098	0.001	0.583	0.088
65-94	MULTIVARIATE	0.391	0.081	0.001	0.419	0.087

SEX: FEMALE
EVENT: STROKE AND TIA ONLY
RISK FACTOR: HYPERTENSION WITH ANTIHYPERTENSIVE TREATMENT
PERSONS AT RISK: PERSONS FREE OF STROKE AND TIA ONLY

RISK FACTOR	35-44 AT EXAM PERSON EXAMS	# OF EVENTS	ANNUAL RATE	45-54 AT EXAM PERSON EXAMS	# OF EVENTS	ANNUAL RATE	55-64 AT EXAM PERSON EXAMS	# OF EVENTS	ANNUAL RATE	65-74 AT EXAM PERSON EXAMS	# OF EVENTS	ANNUAL RATE
NO	4247			5644	8	1	4955	9	1	1847	9	2
MILD	1001			2597	7	1	3409	13	2	2075	20	5
DEFINITE	316			1178	6	3	1924	17	4	1167	25	11
TREATED	49			614	3	2	1662	18	5	1782	46	13
UNKNOWN	0			0	0	-	0	0	-	0	0	-
TOTAL	5613			10033	24	1	11950	57	2	6871	100	7

RISK FACTOR	75-84 AT EXAM PERSON EXAMS	# OF EVENTS	ANNUAL RATE	85-94 AT EXAM PERSON EXAMS	# OF EVENTS	ANNUAL RATE	35-64 COMBINED # OF EVENTS	CRUDE	AGE-ADJUSTED	65-94 COMBINED # OF EVENTS	CRUDE	AGE-ADJUSTED
NO	450	7	8	38	1	13	21	1	1	17	4	4
MILD	654	14	11	32	1	16	20	1	1	35	6	6
DEFINITE	431	5	6	31	2	32	23	3	3	32	10	10
TREATED	692	31	22	47	3	32	21	5	3	80	16	15
UNKNOWN	0	0	-	0	0	-	0	-	-	0	-	-
TOTAL	2227	57	13	148	7	24	85	2	2	164	9	9

LOGISTIC REGRESSIONS

AGE AT EXAM		COEFFICIENT OF RISK FACTOR	SE COEFFICIENT	P-VALUE	STANDARDIZED COEFFICIENT	STD SE
35-44						
45-54		0.634	0.243	0.009	0.489	0.187
55-64		0.864	0.179	0.001	0.724	0.150
65-74		0.847	0.159	0.001	0.694	0.130
75-84		0.381	0.192	0.046	0.299	0.150
85-94		0.515	0.545	0.345	0.436	0.461
35-64	UNIVARIATE	0.643	0.094	0.001	0.623	0.091
65-94	UNIVARIATE	0.487	0.074	0.001	0.555	0.085
35-64	BIVARIATE	0.522	0.099	0.001	0.506	0.095
65-94	BIVARIATE	0.461	0.075	0.001	0.526	0.085
35-64	MULTIVARIATE	0.434	0.104	0.001	0.420	0.101
65-94	MULTIVARIATE	0.409	0.076	0.001	0.466	0.087

TABLE 13-3B   DATE=02NOV85

AVERAGE ANNUAL INCIDENCE RATE FOR EVENT PER 1000 PERSONS AT RISK AT EXAM
BY AGE AND LEVEL OF RISK FACTOR AT EXAM: FRAMINGHAM STUDY 30-YEAR FOLLOWUP

SEX: MALE
EVENT: STROKE AND TIA ONLY
RISK FACTOR: HYPERTENSION IGNORING TREATMENT
PERSONS AT RISK: PERSONS FREE OF STROKE AND TIA ONLY

RISK FACTOR	35-44 AT EXAM PERSON EXAMS	# OF EVENTS	ANNUAL RATE	45-54 AT EXAM PERSON EXAMS	# OF EVENTS	ANNUAL RATE	55-64 AT EXAM PERSON EXAMS	# OF EVENTS	ANNUAL RATE	65-74 AT EXAM PERSON EXAMS	# OF EVENTS	ANNUAL RATE
NO	2669			4291	8	1	4304	13	2	1928	19	5
MILD	1326			2375	12	3	2863	21	4	1756	28	8
DEFINITE	521			1366	12	4	1747	33	9	942	30	16
UNKNOWN	0			0	0	-	0	0	-	0	0	-
TOTAL	4516			8032	32	2	8914	67	4	4626	77	8

RISK FACTOR	75-84 AT EXAM PERSON EXAMS	# OF EVENTS	ANNUAL RATE	85-94 AT EXAM PERSON EXAMS	# OF EVENTS	ANNUAL RATE	35-64 COMBINED # OF EVENTS	CRUDE	AGE-ADJUSTED	65-94 COMBINED # OF EVENTS	CRUDE	AGE-ADJUSTED
NO	490	15	15	20			22	1	1	36	7	7
MILD	503	14	14	18			33	3	2	42	9	9
DEFINITE	241	21	44	4			45	6	6	51	21	22
UNKNOWN	0	0	-	0			0	-	-	0	0	-
TOTAL	1234	50	20	42			100	2	2	129	11	11

LOGISTIC REGRESSIONS

AGE AT EXAM		COEFFICIENT OF RISK FACTOR	SE COEFFICIENT	P-VALUE	STANDARDIZED COEFFICIENT	STD SE
35-44						
45-54		0.763	0.218	0.001	0.577	0.165
55-64		0.930	0.160	0.001	0.718	0.123
65-74		0.607	0.150	0.001	0.460	0.114
75-84		0.590	0.193	0.002	0.439	0.143
85-94						
35-64	UNIVARIATE	0.926	0.127	0.001	0.698	0.096
65-94	UNIVARIATE	0.569	0.116	0.001	0.429	0.088
35-64	BIVARIATE	0.844	0.128	0.001	0.636	0.096
65-94	BIVARIATE	0.585	0.117	0.001	0.441	0.088
35-64	MULTIVARIATE	0.761	0.133	0.001	0.573	0.100
65-94	MULTIVARIATE	0.517	0.121	0.001	0.390	0.091

SEX: FEMALE
EVENT: STROKE AND TIA ONLY
RISK FACTOR: HYPERTENSION IGNORING TREATMENT
PERSONS AT RISK: PERSONS FREE OF STROKE AND TIA ONLY

RISK FACTOR	35-44 AT EXAM PERSON EXAMS	# OF EVENTS	ANNUAL RATE	45-54 AT EXAM PERSON EXAMS	# OF EVENTS	ANNUAL RATE	55-64 AT EXAM PERSON EXAMS	# OF EVENTS	ANNUAL RATE	65-74 AT EXAM PERSON EXAMS	# OF EVENTS	ANNUAL RATE
NO	4268			5829	8	1	7360	14	1	2168	13	3
MILD	1016			2787	9	2	6931	16	2	2816	32	6
DEFINITE	329			1417	7	2	2315	27	5	1887	55	15
UNKNOWN	0			0	0	-	0	0	-	0	0	-
TOTAL	5613			10033	24	1	11950	57	2	6871	100	7

RISK FACTOR	75-84 AT EXAM PERSON EXAMS	# OF EVENTS	ANNUAL RATE	85-94 AT EXAM PERSON EXAMS	# OF EVENTS	ANNUAL RATE	35-64 COMBINED # OF EVENTS	CRUDE	AGE-ADJUSTED	65-94 COMBINED # OF EVENTS	CRUDE	AGE-ADJUSTED
NO	575	13	11	48	2	21	26	1	1	28	5	5
MILD	932	22	12	51	1	10	25	2	1	55	7	7
DEFINITE	720	22	15	49	4	41	34	4	3	81	15	15
UNKNOWN	0	0	-	0	0	-	0	-	-	0	-	-
TOTAL	2227	57	13	148	7	24	85	2	2	164	9	9

LOGISTIC REGRESSIONS

AGE AT EXAM		COEFFICIENT OF RISK FACTOR	SE COEFFICIENT	P-VALUE	STANDARDIZED COEFFICIENT	STD SE
35-44						
45-54		0.647	0.249	0.009	0.471	0.181
55-64		0.733	0.170	0.001	0.571	0.133
65-74		0.847	0.146	0.001	0.650	0.112
75-84		0.167	0.179	0.351	0.127	0.136
85-94		0.462	0.502	0.358	0.375	0.408
35-64	UNIVARIATE	0.782	0.133	0.001	0.581	0.099
65-94	UNIVARIATE	0.607	0.109	0.001	0.466	0.084
35-64	BIVARIATE	0.641	0.137	0.001	0.477	0.102
65-94	BIVARIATE	0.578	0.110	0.001	0.443	0.084
35-64	MULTIVARIATE	0.516	0.146	0.001	0.383	0.108
65-94	MULTIVARIATE	0.518	0.112	0.001	0.397	0.086

TABLE 13-4   DATE=02NOV85

**AVERAGE ANNUAL INCIDENCE RATE FOR EVENT PER 1000 PERSONS AT RISK AT EXAM
BY AGE AND LEVEL OF RISK FACTOR AT EXAM: FRAMINGHAM STUDY 30-YEAR FOLLOWUP**

SEX: MALE
EVENT: STROKE AND TIA ONLY
RISK FACTOR: SERUM CHOLESTEROL
PERSONS AT RISK: PERSONS FREE OF STROKE AND TIA ONLY

RISK FACTOR	35-44 AT EXAM			45-54 AT EXAM			55-64 AT EXAM			65-74 AT EXAM		
	PERSON EXAMS	# OF EVENTS	ANNUAL RATE	PERSON EXAMS	# OF EVENTS	ANNUAL RATE	PERSON EXAMS	# OF EVENTS	ANNUAL RATE	PERSON EXAMS	# OF EVENTS	ANNUAL RATE
84-204	1253			1796	6	2	2431	19	4	1471	29	10
205-234	1257			2382	9	2	2649	21	4	1389	20	7
235-264	882			1835	5	1	2099	10	2	1007	14	7
265-294	487			1050	9	4	1059	15	7	505	6	6
295-1124	303			618	3	2	544	1	1	237	8	17
UNKNOWN	334			351	0	0	132	1	4	17	0	0
TOTAL	4516			8032	32	2	8914	67	4	4626	77	8

RISK FACTOR	75-84 AT EXAM			85-94 AT EXAM			35-64 COMBINED			65-94 COMBINED		
	PERSON EXAMS	# OF EVENTS	ANNUAL RATE	PERSON EXAMS	# OF EVENTS	ANNUAL RATE	# OF EVENTS	CRUDE	AGE-ADJUSTED	# OF EVENTS	CRUDE	AGE-ADJUSTED
84-204	480	27	28	18			25	2	2	56	14	14
205-234	389	16	21	11			31	2	1	38	11	11
235-264	221	5	11	7			15	2	2	19	8	8
265-294	121	2	8	6			24	5	5	8	6	6
295-1124	23	0	0	0			4	1	1	8	15	13
UNKNOWN	0	0	~	0			1	1	2	0	0	0
TOTAL	1234	50	20	42			100	2	2	129	11	11

LOGISTIC REGRESSIONS

AGE AT EXAM		COEFFICIENT OF RISK FACTOR	SE COEFFICIENT	P-VALUE	STANDARDIZED COEFFICIENT	STD SE
35-44						
45-54		0.006	0.003	0.075	0.259	0.145
55-64		-0.003	0.003	0.351	-0.119	0.127
65-74		-0.001	0.003	0.776	-0.033	0.116
75-84		-0.012	0.004	0.002	-0.461	0.152
85-94						
35-64	UNIVARIATE	0.000	0.002	0.836	0.021	0.100
65-94	UNIVARIATE	-0.006	0.002	0.014	-0.226	0.092
35-64	BIVARIATE	0.001	0.002	0.684	0.041	0.102
65-94	BIVARIATE	-0.004	0.002	0.057	-0.177	0.093
35-64	MULTIVARIATE	-0.000	0.002	0.959	-0.005	0.103
65-94	MULTIVARIATE	-0.004	0.002	0.071	-0.168	0.093

SEX: FEMALE
EVENT: STROKE AND TIA ONLY
RISK FACTOR: SERUM CHOLESTEROL
PERSONS AT RISK: PERSONS FREE OF STROKE AND TIA ONLY

RISK FACTOR	35-44 AT EXAM			45-54 AT EXAM			55-64 AT EXAM			65-74 AT EXAM		
	PERSON EXAMS	# OF EVENTS	ANNUAL RATE	PERSON EXAMS	# OF EVENTS	ANNUAL RATE	PERSON EXAMS	# OF EVENTS	ANNUAL RATE	PERSON EXAMS	# OF EVENTS	ANNUAL RATE
84-204	2349			2113	4	1	1681	11	3	869	25	14
205-234	1415			2578	8	2	2704	11	2	1510	17	6
235-264	806			2262	2	0	3017	16	3	1867	24	6
265-294	372			1481	3	1	2339	8	2	1417	15	5
295-1124	171			1119	6	3	1976	11	3	1171	18	8
UNKNOWN	500			480	1	1	233	0	0	37	1	14
TOTAL	5613			10033	24	1	11950	57	2	6871	100	7

RISK FACTOR	75-84 AT EXAM			85-94 AT EXAM			35-64 COMBINED			65-94 COMBINED		
	PERSON EXAMS	# OF EVENTS	ANNUAL RATE	PERSON EXAMS	# OF EVENTS	ANNUAL RATE	# OF EVENTS	CRUDE	AGE-ADJUSTED	# OF EVENTS	CRUDE	AGE-ADJUSTED
84-204	407	15	18	53	3	28	17	1	2	43	16	16
205-234	537	16	15	35	1	14	20	1	2	34	8	8
235-264	619	14	11	41	2	24	19	2	1	40	8	7
265-294	382	8	10	9	1	56	11	1	1	24	7	7
295-1124	275	4	7	10	0	0	17	3	2	22	8	7
UNKNOWN	7	0	0	0	0	~	1	0	0	7	11	10
TOTAL	2227	57	13	144	7	24	85	2	2	164	9	9

LOGISTIC REGRESSIONS

AGE AT EXAM		COEFFICIENT OF RISK FACTOR	SE COEFFICIENT	P-VALUE	STANDARDIZED COEFFICIENT	STD SE
35-44						
45-54		0.003	0.004	0.446	0.130	0.171
55-64		-0.001	0.003	0.856	-0.024	0.134
65-74		-0.005	0.002	0.031	-0.227	0.105
75-84		-0.007	0.003	0.024	-0.318	0.141
85-94		-0.005	0.009	0.557	-0.240	0.408
35-64	UNIVARIATE	0.003	0.002	0.134	0.145	0.097
65-94	UNIVARIATE	-0.007	0.002	0.001	-0.300	0.083
35-64	BIVARIATE	0.000	0.002	0.930	0.010	0.112
65-94	BIVARIATE	-0.006	0.002	0.003	-0.250	0.083
35-64	MULTIVARIATE	-0.001	0.002	0.700	-0.044	0.114
65-94	MULTIVARIATE	-0.006	0.002	0.003	-0.251	0.083

TABLE 13-5　　　　　　　DATE=02NOV85
AVERAGE ANNUAL INCIDENCE RATE FOR EVENT PER 1000 PERSONS AT RISK AT EXAM
BY AGE AND LEVEL OF RISK FACTOR AT EXAM: FRAMINGHAM STUDY 30-YEAR FOLLOWUP

SEX: MALE
EVENT: STROKE AND TIA ONLY
RISK FACTOR: HEMATOCRIT
PERSONS AT RISK: PERSONS FREE OF STROKE AND TIA ONLY

RISK FACTOR	35-44 AT EXAM			45-54 AT EXAM			55-64 AT EXAM			65-74 AT EXAM		
	PERSON EXAMS	# OF EVENTS	ANNUAL RATE	PERSON EXAMS	# OF EVENTS	ANNUAL RATE	PERSON EXAMS	# OF EVENTS	ANNUAL RATE	PERSON EXAMS	# OF EVENTS	ANNUAL RATE
25-42	630			961	3	2	1117	3	1	541	6	6
43-44	551			993	1	1	1082	4	2	582	9	8
45-46	990			1935	10	3	2192	12	3	1151	16	7
47-48	785			1573	5	2	1783	18	5	1033	20	10
49-70	1190			2180	13	3	2581	29	6	1299	26	10
UNKNOWN	370			390	0	0	159	1	3	20	0	0
TOTAL	4516			8032	32	2	8914	67	4	4626	77	8

RISK FACTOR	75-84 AT EXAM			85-94 AT EXAM			35-64 COMBINED			65-94 COMBINED		
	PERSON EXAMS	# OF EVENTS	ANNUAL RATE	PERSON EXAMS	# OF EVENTS	ANNUAL RATE	# OF EVENTS	CRUDE	AGE-ADJUSTED	# OF EVENTS	CRUDE	AGE-ADJUSTED
25-42	258	8	16	14			6	1	1	15	9	8
43-44	150	9	30	7			5	1	1	18	12	12
45-46	292	13	22	6			23	2	2	29	10	10
47-48	248	7	14	8			23	3	3	27	10	11
49-70	286	13	23	7			42	4	3	40	13	13
UNKNOWN	0	0	-	0			1	1	1	0	0	0
TOTAL	1234	50	20	42			100	2	2	129	11	11

LOGISTIC REGRESSIONS

AGE AT EXAM		COEFFICIENT OF RISK FACTOR	SE COEFFICIENT	P-VALUE	STANDARDIZED COEFFICIENT	STD SE
35-44						
45-54		0.070	0.049	0.153	0.252	0.176
55-64		0.125	0.032	0.001	0.472	0.121
65-74		0.036	0.032	0.263	0.130	0.116
75-84		0.001	0.036	0.978	0.004	0.145
85-94						
35-64	UNIVARIATE	0.110	0.027	0.001	0.404	0.100
65-94	UNIVARIATE	0.008	0.024	0.736	0.030	0.089
35-64	BIVARIATE	0.107	0.027	0.001	0.392	0.099
65-94	BIVARIATE	0.025	0.024	0.296	0.093	0.089
35-64	MULTIVARIATE	0.089	0.027	0.001	0.327	0.100
65-94	MULTIVARIATE	0.022	0.024	0.366	0.082	0.090

SEX: FEMALE
EVENT: STROKE AND TIA ONLY
RISK FACTOR: HEMATOCRIT
PERSONS AT RISK: PERSONS FREE OF STROKE AND TIA ONLY

RISK FACTOR	35-44 AT EXAM			45-54 AT EXAM			55-64 AT EXAM			65-74 AT EXAM		
	PERSON EXAMS	# OF EVENTS	ANNUAL RATE	PERSON EXAMS	# OF EVENTS	ANNUAL RATE	PERSON EXAMS	# OF EVENTS	ANNUAL RATE	PERSON EXAMS	# OF EVENTS	ANNUAL RATE
21-39	1697			1853	4	1	1585	9	3	719	14	10
40-41	966			1760	1	0	2127	10	2	1132	20	9
42-43	1041			1972	4	1	2353	10	2	1388	11	4
44-45	857			2206	5	1	2969	10	2	1782	25	7
46-65	486			1694	7	2	2646	17	3	1810	29	8
UNKNOWN	566			548	3	3	270	1	2	40	1	13
TOTAL	5613			10033	24	1	11950	57	2	6871	100	7

RISK FACTOR	75-84 AT EXAM			85-94 AT EXAM			35-64 COMBINED			65-94 COMBINED		
	PERSON EXAMS	# OF EVENTS	ANNUAL RATE	PERSON EXAMS	# OF EVENTS	ANNUAL RATE	# OF EVENTS	CRUDE	AGE-ADJUSTED	# OF EVENTS	CRUDE	AGE-ADJUSTED
21-39	324	11	17	28	1	18	14	1	2	26	12	12
40-41	427	8	9	36	3	42	11	1	1	31	10	9
42-43	418	10	12	26	2	38	15	1	1	23	6	6
44-45	567	12	11	29	1	17	16	1	1	38	8	8
46-65	485	16	16	29	0	0	25	3	3	45	10	10
UNKNOWN	6	0	0	0	0	-	4	1	2	1	11	9
TOTAL	2227	57	13	148	7	24	85	2	2	164	9	9

LOGISTIC REGRESSIONS

AGE AT EXAM		COEFFICIENT OF RISK FACTOR	SE COEFFICIENT	P-VALUE	STANDARDIZED COEFFICIENT	STD SE
35-44						
45-54		0.116	0.063	0.063	0.415	0.223
55-64		0.036	0.039	0.358	0.124	0.135
65-74		-0.017	0.029	0.543	-0.061	0.100
75-84		-0.012	0.035	0.739	-0.044	0.133
85-94		-0.063	0.101	0.534	-0.245	0.394
35-64	UNIVARIATE	0.085	0.032	0.007	0.306	0.113
65-94	UNIVARIATE	-0.025	0.022	0.254	-0.089	0.078
35-64	BIVARIATE	0.057	0.032	0.075	0.206	0.116
65-94	BIVARIATE	-0.016	0.022	0.458	-0.058	0.078
35-64	MULTIVARIATE	0.039	0.032	0.215	0.141	0.114
65-94	MULTIVARIATE	-0.022	0.021	0.294	-0.078	0.075

TABLE 13-7   DATE=02NOV85
AVERAGE ANNUAL INCIDENCE RATE FOR EVENT PER 1000 PERSONS AT RISK AT EXAM
BY AGE AND LEVEL OF RISK FACTOR AT EXAM: FRAMINGHAM STUDY 30-YEAR FOLLOWUP

SEX: MALE
EVENT: STROKE AND TIA ONLY
RISK FACTOR: DIABETES MELLITUS
PERSONS AT RISK: PERSONS FREE OF STROKE AND TIA ONLY

RISK FACTOR	35-44 AT EXAM PERSON EXAMS	# OF EVENTS	ANNUAL RATE	45-54 AT EXAM PERSON EXAMS	# OF EVENTS	ANNUAL RATE	55-64 AT EXAM PERSON EXAMS	# OF EVENTS	ANNUAL RATE	65-74 AT EXAM PERSON EXAMS	# OF EVENTS	ANNUAL RATE
NO	4467			7780	28	2	8292	57	3	4101	63	8
YES	49			252	4	8	622	10	8	525	14	13
UNKNOWN	0			0	0	-	0	0	-	0	0	-
TOTAL	4516			8032	32	2	8914	67	4	4626	77	8

RISK FACTOR	75-84 AT EXAM PERSON EXAMS	# OF EVENTS	ANNUAL RATE	85-94 AT EXAM PERSON EXAMS	# OF EVENTS	ANNUAL RATE	35-64 COMBINED # OF EVENTS	CRUDE	AGE-ADJUSTED	65-94 COMBINED # OF EVENTS	CRUDE	AGE-ADJUSTED
NO	1076	40	19	36			86	2	2	104	10	10
YES	158	10	32	6			14	8	6	25	18	18
UNKNOWN	0	0	-	0			0	-	-	0	-	-
TOTAL	1234	50	20	42			100	2	2	129	11	11

LOGISTIC REGRESSIONS

AGE AT EXAM		COEFFICIENT OF RISK FACTOR	SE COEFFICIENT	P-VALUE	STANDARDIZED COEFFICIENT	STD SE
35-44						
45-54		1.496	0.538	0.005	0.261	0.094
55-64		0.859	0.345	0.013	0.219	0.088
65-74		0.563	0.299	0.060	0.179	0.095
75-84		0.560	0.364	0.125	0.187	0.122
85-94						
35-64	UNIVARIATE	1.298	0.290	0.001	0.263	0.059
65-94	UNIVARIATE	0.615	0.227	0.007	0.197	0.073
35-64	BIVARIATE	0.905	0.294	0.002	0.184	0.060
65-94	BIVARIATE	0.578	0.227	0.011	0.186	0.073
35-64	MULTIVARIATE	0.625	0.304	0.040	0.127	0.062
65-94	MULTIVARIATE	0.518	0.231	0.025	0.166	0.074

SEX: FEMALE
EVENT: STROKE AND TIA ONLY
RISK FACTOR: DIABETES MELLITUS
PERSONS AT RISK: PERSONS FREE OF STROKE AND TIA ONLY

RISK FACTOR	35-44 AT EXAM PERSON EXAMS	# OF EVENTS	ANNUAL RATE	45-54 AT EXAM PERSON EXAMS	# OF EVENTS	ANNUAL RATE	55-64 AT EXAM PERSON EXAMS	# OF EVENTS	ANNUAL RATE	65-74 AT EXAM PERSON EXAMS	# OF EVENTS	ANNUAL RATE
NO	5570			9820	24	1	11350	48	2	6289	40	6
YES	43			213	0	0	600	9	8	582	20	17
UNKNOWN	0			0	0	-	0	0	-	0	0	-
TOTAL	5613			10033	24	1	11950	57	2	6871	100	7

RISK FACTOR	75-84 AT EXAM PERSON EXAMS	# OF EVENTS	ANNUAL RATE	85-94 AT EXAM PERSON EXAMS	# OF EVENTS	ANNUAL RATE	35-64 COMBINED # OF EVENTS	CRUDE	AGE-ADJUSTED	65-94 COMBINED # OF EVENTS	CRUDE	AGE-ADJUSTED
NO	1942	43	11	127	5	20	76	1	1	128	8	8
YES	285	14	25	21	2	48	9	5	3	36	20	19
UNKNOWN	0	0	-	0	0	-	0	-	-	0	-	-
TOTAL	2227	57	13	148	7	24	85	2	2	164	9	9

LOGISTIC REGRESSIONS

AGE AT EXAM		COEFFICIENT OF RISK FACTOR	SE COEFFICIENT	P-VALUE	STANDARDIZED COEFFICIENT	STD SE
35-44		-	-	-	-	-
45-54		-	-	-	-	-
55-64		1.277	0.366	0.001	0.279	0.080
65-74		1.016	0.254	0.001	0.283	0.071
75-84		0.825	0.314	0.009	0.276	0.105
85-94		0.943	0.872	0.280	0.330	0.305
35-64	UNIVARIATE	1.316	0.354	0.001	0.228	0.061
65-94	UNIVARIATE	0.999	0.192	0.001	0.295	0.057
35-64	BIVARIATE	1.010	0.358	0.005	0.175	0.062
65-94	BIVARIATE	0.912	0.193	0.001	0.269	0.057
35-64	MULTIVARIATE	0.636	0.369	0.085	0.110	0.064
65-94	MULTIVARIATE	0.625	0.202	0.002	0.184	0.060

TABLE 13-6   DATE=02NOV85
AVERAGE ANNUAL INCIDENCE RATE FOR EVENT PER 1000 PERSONS AT RISK AT EXAM
BY AGE AND LEVEL OF RISK FACTOR AT EXAM: FRAMINGHAM STUDY 30-YEAR FOLLOWUP

SEX: MALE
EVENT: STROKE AND TIA ONLY
RISK FACTOR: BLOOD GLUCOSE (MG/DL)
PERSONS AT RISK: PERSONS FREE OF STROKE AND TIA ONLY

RISK FACTOR	35-44 AT EXAM			45-54 AT EXAM			55-64 AT EXAM			65-74 AT EXAM		
	PERSON EXAMS	# OF EVENTS	ANNUAL RATE	PERSON EXAMS	# OF EVENTS	ANNUAL RATE	PERSON EXAMS	# OF EVENTS	ANNUAL RATE	PERSON EXAMS	# OF EVENTS	ANNUAL RATE
29-69	1004			1554	3	1	1179	8	5	484	7	7
70-89	2528			4326	18	2	4458	36	4	2050	32	8
90-109	628			1397	5	2	2105	16	4	1242	23	9
110-129	107			329	4	6	599	2	2	412	6	7
130-524	49			202	2	5	472	5	5	421	9	11
KNKNOWN	200			224	0	0	101	0	0	17	0	0
TOTAL	4516			8032	32	2	8914	67	4	4626	77	8

RISK FACTOR	75-84 AT EXAM			85-94 AT EXAM			35-64 COMBINED			65-94 COMBINED		
	PERSON EXAMS	# OF EVENTS	ANNUAL RATE	PERSON EXAMS	# OF EVENTS	ANNUAL RATE	# OF EVENTS	CRUDE	AGE-ADJUSTED	# OF EVENTS	CRUDE	AGE-ADJUSTED
29-69	82	3	18	1			11	1	2	10	9	9
70-89	449	19	21	16			54	2	2	51	10	11
90-109	394	15	19	10			22	3	3	38	12	11
110-129	151	8	26	5			6	3	3	14	12	11
130-524	158	5	16	10			7	5	4	16	14	12
UNKNOWN	0	0	-	0			0	0	0	0	0	0
TOTAL	1234	50	20	42			100	2	2	129	11	11

LOGISTIC REGRESSIONS

AGE AT EXAM		COEFFICIENT OF RISK FACTOR	SE COEFFICIENT	P-VALUE	STANDARDIZED COEFFICIENT	STD SE
35-44						
45-54		0.009	0.004	0.017	0.203	0.085
55-64		-0.000	0.004	0.941	-0.009	0.125
65-74		0.005	0.002	0.017	0.183	0.076
75-84		-0.003	0.005	0.551	-0.097	0.163
85-94						
35-64	UNIVARIATE	0.006	0.003	0.026	0.145	0.065
65-94	UNIVARIATE	0.004	0.002	0.030	0.145	0.067
35-64	BIVARIATE	0.003	0.003	0.358	0.069	0.075
65-94	BIVARIATE	0.004	0.002	0.080	0.126	0.072
35-64	MULTIVARIATE	0.002	0.003	0.625	0.039	0.080
65-94	MULTIVARIATE	0.003	0.002	0.130	0.112	0.074

SEX: FEMALE
EVENT: STROKE AND TIA ONLY
RISK FACTOR: BLOOD GLUCOSE (MG/DL)
PERSONS AT RISK: PERSONS FREE OF STROKE AND TIA ONLY

RISK FACTOR	35-44 AT EXAM			45-54 AT EXAM			55-64 AT EXAM			65-74 AT EXAM		
	PERSON EXAMS	# OF EVENTS	ANNUAL RATE	PERSON EXAMS	# OF EVENTS	ANNUAL RATE	PERSON EXAMS	# OF EVENTS	ANNUAL RATE	PERSON EXAMS	# OF EVENTS	ANNUAL RATE
29-69	1067			1803	3	1	1403	7	2	571	11	10
70-89	3408			5786	12	1	6298	24	2	3369	45	7
90-109	672			1640	4	1	2833	13	2	1933	17	4
110-129	96			302	2	3	762	7	5	563	9	8
130-524	39			166	1	3	489	5	5	405	17	21
UNKNOWN	331			336	2	3	165	1	3	30	1	17
TOTAL	5613			10033	24	1	11950	57	2	6871	100	7

RISK FACTOR	75-84 AT EXAM			85-94 AT EXAM			35-64 COMBINED			65-94 COMBINED		
	PERSON EXAMS	# OF EVENTS	ANNUAL RATE	PERSON EXAMS	# OF EVENTS	ANNUAL RATE	# OF EVENTS	CRUDE	AGE-ADJUSTED	# OF EVENTS	CRUDE	AGE-ADJUSTED
29-69	120	5	21	4	0	0	10	1	1	16	12	12
70-89	891	19	11	55	2	18	39	1	1	66	8	8
90-109	744	19	13	58	2	17	18	2	2	38	7	7
110-129	263	5	10	17	1	29	9	4	3	15	9	9
130-524	208	9	22	14	2	71	6	4	3	28	22	22
UNKNOWN	1	0	0	0	0	-	3	2	2	1	16	12
TOTAL	2227	57	13	148	7	24	85	2	2	164	9	9

LOGISTIC REGRESSIONS

AGE AT EXAM		COEFFICIENT OF RISK FACTOR	SE COEFFICIENT	P-VALUE	STANDARDIZED COEFFICIENT	STD SE
35-44						
45-54		0.008	0.007	0.226	0.158	0.130
55-64		0.008	0.002	0.001	0.225	0.065
65-74		0.006	0.002	0.004	0.187	0.065
75-84		0.004	0.003	0.223	0.127	0.104
85-94		0.009	0.009	0.337	0.257	0.268
35-64	UNIVARIATE	0.010	0.002	0.001	0.224	0.048
65-94	UNIVARIATE	0.006	0.002	0.001	0.183	0.052
35-64	BIVARIATE	0.008	0.002	0.001	0.182	0.052
65-94	BIVARIATE	0.005	0.002	0.003	0.161	0.055
35-64	MULTIVARIATE	0.006	0.002	0.017	0.130	0.054
65-94	MULTIVARIATE	0.003	0.002	0.063	0.106	0.057

TABLE 13-8   DATE=02NOV85

AVERAGE ANNUAL INCIDENCE RATE FOR EVENT PER 1000 PERSONS AT RISK AT EXAM
BY AGE AND LEVEL OF RISK FACTOR AT EXAM: FRAMINGHAM STUDY 30-YEAR FOLLOWUP

SEX: MALE
EVENT: STROKE AND TIA ONLY
RISK FACTOR: GLUCOSE IN URINE, DEFINITE OR TRACE
PERSONS AT RISK: PERSONS FREE OF STROKE AND TIA ONLY

RISK FACTOR	35-44 AT EXAM PERSON EXAMS	# OF EVENTS	ANNUAL RATE	45-54 AT EXAM PERSON EXAMS	# OF EVENTS	ANNUAL RATE	55-64 AT EXAM PERSON EXAMS	# OF EVENTS	ANNUAL RATE	65-74 AT EXAM PERSON EXAMS	# OF EVENTS	ANNUAL RATE
NO	4423			7834	50	2	8639	65	4	4460	73	8
YES	60			177	2	6	251	2	4	160	4	13
UNKNOWN	33			21	0	0	24	0	0	6	0	0
TOTAL	4516			8032	32	2	8914	67	4	4626	77	8

RISK FACTOR	75-84 AT EXAM PERSON EXAMS	# OF EVENTS	ANNUAL RATE	85-94 AT EXAM PERSON EXAMS	# OF EVENTS	ANNUAL RATE	35-64 COMBINED # OF EVENTS	CRUDE	AGE-ADJUSTED	65-94 COMBINED # OF EVENTS	CRUDE	AGE-ADJUSTED
NO	1199	50	21	41			96	2	2	125	11	11
YES	32	0	0	1			4	4	4	4	10	10
UNKNOWN	3	0	0	0			0	0	0	0	0	0
TOTAL	1234	50	20	42			100	2	2	129	11	11

LOGISTIC REGRESSIONS.

AGE AT EXAM		COEFFICIENT OF RISK FACTOR	SE COEFFICIENT	P-VALUE	STANDARDIZED COEFFICIENT	STD SE
35-44						
45-54		1.325	0.660	0.045	0.195	0.097
55-64		0.058	0.721	0.936	0.010	0.119
65-74		0.432	0.520	0.406	0.079	0.095
75-84		-	-	-	-	-
85-94						
35-64	UNIVARIATE	0.583	0.512	0.256	0.087	0.077
65-94	UNIVARIATE	-0.058	0.513	0.910	-0.010	0.091
35-64	BIVARIATE	0.597	0.514	0.440	0.059	0.077
65-94	BIVARIATE	0.010	0.515	0.984	0.002	0.092
35-64	MULTIVARIATE	-0.089	0.595	0.881	-0.013	0.089
65-94	MULTIVARIATE	-0.051	0.517	0.921	-0.009	0.092

SEX: FEMALE
EVENT: STROKE AND TIA ONLY
RISK FACTOR: GLUCOSE IN URINE, DEFINITE OR TRACE
PERSONS AT RISK: PERSONS FREE OF STROKE AND TIA ONLY

RISK FACTOR	35-44 AT EXAM PERSON EXAMS	# OF EVENTS	ANNUAL RATE	45-54 AT EXAM PERSON EXAMS	# OF EVENTS	ANNUAL RATE	55-64 AT EXAM PERSON EXAMS	# OF EVENTS	ANNUAL RATE	65-74 AT EXAM PERSON EXAMS	# OF EVENTS	ANNUAL RATE
NO	5534			9908	23	1	11753	53	2	6733	94	7
YES	41			101	1	5	171	4	12	128	6	23
UNKNOWN	38			24	0	0	26	0	0	10	0	0
TOTAL	5613			10033	24	1	11950	57	2	6871	100	7

RISK FACTOR	75-84 AT EXAM PERSON EXAMS	# OF EVENTS	ANNUAL RATE	85-94 AT EXAM PERSON EXAMS	# OF EVENTS	ANNUAL RATE	35-64 COMBINED # OF EVENTS	CRUDE	AGE-ADJUSTED	65-94 COMBINED # OF EVENTS	CRUDE	AGE-ADJUSTED
NO	2183	54	12	145	7	24	80	1	1	155	9	9
YES	35	3	43	3	0	0	5	8	7	9	27	28
UNKNOWN	9	0	0	0			0	0	0	0	0	0
TOTAL	2227	57	13	148	7	24	85	2	2	164	9	9

LOGISTIC REGRESSIONS

AGE AT EXAM		COEFFICIENT OF RISK FACTOR	SE COEFFICIENT	P-VALUE	STANDARDIZED COEFFICIENT	STD SE
35-44						
45-54		1.458	1.026	0.155	0.146	0.103
55-64		1.665	0.524	0.001	0.198	0.062
65-74		1.245	0.431	0.004	0.168	0.058
75-84		1.307	0.619	0.035	0.163	0.077
85-94		-				
35-64	UNIVARIATE	1.705	0.465	0.001	0.181	0.049
65-94	UNIVARIATE	1.192	0.352	0.001	0.158	0.047
35-64	BIVARIATE	1.575	0.466	0.001	0.167	0.049
65-94	BIVARIATE	1.227	0.354	0.001	0.163	0.047
35-64	MULTIVARIATE	1.097	0.485	0.024	0.116	0.051
65-94	MULTIVARIATE	0.829	0.367	0.024	0.110	0.049

TABLE 13-9   DATE=05MAR86
AVERAGE ANNUAL INCIDENCE RATE FOR EVENT PER 1000 PERSONS AT RISK AT EXAM
BY AGE AND LEVEL OF RISK FACTOR AT EXAM: FRAMINGHAM STUDY 30-YEAR FOLLOWUP

SEX: MALE
EVENT: STROKE AND TIA ONLY
RISK FACTOR: GLUCOSE INTOLERANCE
PERSONS AT RISK: PERSONS FREE OF STROKE AND TIA ONLY

RISK FACTOR	35-44 AT EXAM			45-54 AT EXAM			55-64 AT EXAM			65-74 AT EXAM		
	PERSON EXAMS	# OF EVENTS	ANNUAL RATE	PERSON EXAMS	# OF EVENTS	ANNUAL RATE	PERSON EXAMS	# OF EVENTS	ANNUAL RATE	PERSON EXAMS	# OF EVENTS	ANNUAL RATE
NO	4378			7497	27	2	7847	54	3	3766	58	8
YES	138			535	5	.5	1067	13	6	860	19	11
UNKNOWN	0			0	0		0	0		0	0	
TOTAL	4516			8032	32	2	8914	67	4	4626	77	8

RISK FACTOR	75-84 AT EXAM			85-94 AT EXAM			35-64 COMBINED			65-94 COMBINED		
	PERSON EXAMS	# OF EVENTS	ANNUAL RATE	PERSON EXAMS	# OF EVENTS	ANNUAL RATE	# OF EVENTS	CRUDE	AGE-ADJUSTED	# OF EVENTS	CRUDE	AGE-ADJUSTED
NO	952	36	19	29			82	2	2	94	10	10
YES	282	14	25	13			18	5	4	35	15	14
UNKNOWN	0	0	-	0			0	-	-	0	-	-
TOTAL	1234	50	20	42			100	2	2	129	11	11

LOGISTIC REGRESSIONS

AGE AT EXAM		COEFFICIENT OF RISK FACTOR	SE COEFFICIENT	P-VALUE	STANDARDIZED COEFFICIENT	STD SE
35-44						
45-54		0.959	0.489	0.050	0.239	0.122
55-64		0.577	0.311	0.063	0.187	0.101
65-74		0.368	0.267	0.169	0.143	0.104
75-84		0.285	0.323	0.378	0.120	0.135
85-94						
35-64	UNIVARIATE	0.918	0.262	0.001	0.250	0.071
65-94	UNIVARIATE	0.436	0.201	0.030	0.173	0.080
35-64	BIVARIATE	0.596	0.264	0.024	0.163	0.072
65-94	BIVARIATE	0.365	0.202	0.070	0.145	0.080
35-64	MULTIVARIATE	0.346	0.275	0.209	0.094	0.075
65-94	MULTIVARIATE	0.313	0.204	0.124	0.124	0.081

SEX: FEMALE
EVENT: STROKE AND TIA ONLY
RISK FACTOR: GLUCOSE INTOLERANCE
PERSONS AT RISK: PERSONS FREE OF STROKE AND TIA ONLY

RISK FACTOR	35-44 AT EXAM			45-54 AT EXAM			55-64 AT EXAM			65-74 AT EXAM		
	PERSON EXAMS	# OF EVENTS	ANNUAL RATE	PERSON EXAMS	# OF EVENTS	ANNUAL RATE	PERSON EXAMS	# OF EVENTS	ANNUAL RATE	PERSON EXAMS	# OF EVENTS	ANNUAL RATE
NO	5519			9648	22	1	10848	47	2	5877	68	6
YES	94			385	2	3	1102	10	5	994	32	16
UNKNOWN	0			0	0	-	0	0	-	0	0	-
TOTAL	5613			10033	24	1	11950	57	2	6871	100	7

RISK FACTOR	75-84 AT EXAM			85-94 AT EXAM			35-64 COMBINED			65-94 COMBINED		
	PERSON EXAMS	# OF EVENTS	ANNUAL RATE	PERSON EXAMS	# OF EVENTS	ANNUAL RATE	# OF EVENTS	CRUDE	AGE-ADJUSTED	# OF EVENTS	CRUDE	AGE-ADJUSTED
NO	1752	41	12	120	4	17	73	1	1	113	7	7
YES	475	16	17	28	3	54	12	4	3	51	17	17
UNKNOWN	0	0	-	0	0	-	0	-	-	0	-	-
TOTAL	2227	57	13	148	7	24	85	2	2	164	9	9

LOGISTIC REGRESSIONS

AGE AT EXAM		COEFFICIENT OF RISK FACTOR	SE COEFFICIENT	P-VALUE	STANDARDIZED COEFFICIENT	STD SE
35-44						
45-54		0.826	0.740	0.264	0.159	0.142
55-64		0.744	0.350	0.033	0.215	0.101
65-74		1.044	0.217	0.001	0.367	0.076
75-84		0.375	0.299	0.211	0.154	0.125
85-94		1.247	0.795	0.117	0.490	0.312
35-64	UNIVARIATE	1.000	0.313	0.001	0.232	0.073
65-94	UNIVARIATE	0.869	0.171	0.001	0.320	0.063
35-64	BIVARIATE	0.698	0.316	0.027	0.162	0.074
65-94	BIVARIATE	0.785	0.172	0.001	0.289	0.064
35-64	MULTIVARIATE	0.467	0.324	0.149	0.199	0.075
65-94	MULTIVARIATE	0.596	0.178	0.001	0.220	0.066

TABLE 13-10 DATE=02NOV85
AVERAGE ANNUAL INCIDENCE RATE FOR EVENT PER 1000 PERSONS AT RISK AT EXAM
BY AGE AND LEVEL OF RISK FACTOR AT EXAM: FRAMINGHAM STUDY 30-YEAR FOLLOWUP

SEX: MALE
EVENT: STROKE AND TIA ONLY
RISK FACTOR: METROPOLITAN RELATIVE WEIGHT (PERCENT)
PERSONS AT RISK: PERSONS FREE OF STROKE AND TIA ONLY

RISK FACTOR	35-44 AT EXAM PERSON EXAMS	# OF EVENTS	ANNUAL RATE	45-54 AT EXAM PERSON EXAMS	# OF EVENTS	ANNUAL RATE	55-64 AT EXAM PERSON EXAMS	# OF EVENTS	ANNUAL RATE	65-74 AT EXAM PERSON EXAMS	# OF EVENTS	ANNUAL RATE
54-104	720			1145	3	1	1251	8	3	795	18	11
105-114	979			1631	4	1	1830	13	4	998	11	6
115-124	1133			2131	9	2	2428	17	4	1272	25	10
125-134	923			1681	8	2	1819	14	4	918	13	7
135-272	761			1443	8	3	1585	15	5	643	10	8
UNKNOWN	0			1	0	0	1	0	0	0	0	
TOTAL	4516			8032	32	2	8914	67	4	4626	77	8

RISK FACTOR	75-84 AT EXAM PERSON EXAMS	# OF EVENTS	ANNUAL RATE	85-94 AT EXAM PERSON EXAMS	# OF EVENTS	ANNUAL RATE	35-64 COMBINED # OF EVENTS	CRUDE	AGE-ADJUSTED	65-94 COMBINED # OF EVENTS	CRUDE	AGE-ADJUSTED
54-104	247	8	16	15			12	2	2	28	13	13
105-114	323	16	25	12			17	2	2	27	10	9
115-124	377	15	20	13			26	2	2	40	12	12
125-134	184	5	14	2			22	2	2	18	8	8
135-272	103	6	29	0			23	3	3	16	11	12
UNKNOWN	0	0		0			0	0	0	0		
TOTAL	1234	50	20	42			100	2	2	129	11	11

LOGISTIC REGRESSIONS

AGE AT EXAM		COEFFICIENT OF RISK FACTOR	SE COEFFICIENT	P-VALUE	STANDARDIZED COEFFICIENT	STD SE
35-44						
45-54		0.011	0.010	0.291	0.179	0.169
55-64		0.010	0.007	0.158	0.166	0.118
65-74		-0.007	0.008	0.341	-0.111	0.116
75-84		0.000	0.010	0.991	0.002	0.144
85-94						
35-64	UNIVARIATE	0.010	0.006	0.101	0.158	0.096
65-94	UNIVARIATE	-0.008	0.006	0.182	-0.121	0.090
35-64	BIVARIATE	0.010	0.006	0.103	0.158	0.097
65-94	BIVARIATE	-0.005	0.006	0.442	-0.071	0.092
35-64	MULTIVARIATE	0.004	0.006	0.572	0.057	0.100
65-94	MULTIVARIATE	-0.010	0.006	0.098	-0.155	0.094

SEX: FEMALE
EVENT: STROKE AND TIA ONLY
RISK FACTOR: METROPOLITAN RELATIVE WEIGHT (PERCENT)
PERSONS AT RISK: PERSONS FREE OF STROKE AND TIA ONLY

RISK FACTOR	35-44 AT EXAM PERSON EXAMS	# OF EVENTS	ANNUAL RATE	45-54 AT EXAM PERSON EXAMS	# OF EVENTS	ANNUAL RATE	55-64 AT EXAM PERSON EXAMS	# OF EVENTS	ANNUAL RATE	65-74 AT EXAM PERSON EXAMS	# OF EVENTS	ANNUAL RATE
54-104	1689			2108	5	1	2108	5	1	1275	21	8
105-114	1431			2351	4	1	2509	15	3	1289	18	7
115-124	1059			2123	5	1	2535	5	1	1485	14	5
125-134	565			1578	0	0	1813	13	4	1119	18	8
135-272	861			2073	10	2	2985	19	3	1703	29	9
UNKNOWN	8			0	0		0	0		0	0	
TOTAL	5613			10033	24	1	11950	57	2	6871	100	7

RISK FACTOR	75-84 AT EXAM PERSON EXAMS	# OF EVENTS	ANNUAL RATE	85-94 AT EXAM PERSON EXAMS	# OF EVENTS	ANNUAL RATE	35-64 COMBINED # OF EVENTS	CRUDE	AGE-ADJUSTED	65-94 COMBINED # OF EVENTS	CRUDE	AGE-ADJUSTED
54-104	592	13	11	47	3	32	11	1	1	37	10	9
105-114	405	15	19	31	1	16	19	2	1	34	10	10
115-124	448	8	9	28	2	36	12	1	1	24	6	6
125-134	325	8	12	24	0	0	14	2	2	26	9	9
135-272	457	13	14	18	1	28	29	2	2	43	10	10
UNKNOWN	0	0		0	0		0	0	0	0		
TOTAL	2227	57	13	148	7	24	85	2	2	164	9	9

LOGISTIC REGRESSIONS

AGE AT EXAM		COEFFICIENT OF RISK FACTOR	SE COEFFICIENT	P-VALUE	STANDARDIZED COEFFICIENT	STD SE
35-44						
45-54		0.017	0.007	0.018	0.359	0.151
55-64		0.009	0.005	0.069	0.210	0.116
65-74		0.001	0.004	0.750	0.032	0.099
75-84		0.001	0.006	0.917	0.014	0.134
85-94		-0.011	0.021	0.590	-0.215	0.399
35-64	UNIVARIATE	0.013	0.004	0.001	0.294	0.087
65-94	UNIVARIATE	-0.001	0.004	0.816	-0.019	0.079
35-64	BIVARIATE	0.011	0.004	0.007	0.246	0.091
65-94	BIVARIATE	0.001	0.004	0.744	0.026	0.079
35-64	MULTIVARIATE	0.003	0.004	0.437	0.075	0.097
65-94	MULTIVARIATE	-0.001	0.003	0.809	-0.018	0.076

TABLE 13-11　　　　　　DATE=02NOV85
AVERAGE ANNUAL INCIDENCE RATE FOR EVENT PER 1000 PERSONS AT RISK AT EXAM
BY AGE AND LEVEL OF RISK FACTOR AT EXAM: FRAMINGHAM STUDY 30-YEAR FOLLOWUP

SEX: MALE
EVENT: STROKE AND TIA ONLY
RISK FACTOR: VITAL CAPACITY - HEIGHT INDEX (ML/INCH)
PERSONS AT RISK: PERSONS FREE OF STROKE AND TIA ONLY

RISK FACTOR	35-44 AT EXAM			45-54 AT EXAM			55-64 AT EXAM			65-74 AT EXAM		
	PERSON EXAMS	# OF EVENTS	ANNUAL RATE	PERSON EXAMS	# OF EVENTS	ANNUAL RATE	PERSON EXAMS	# OF EVENTS	ANNUAL RATE	PERSON EXAMS	# OF EVENTS	ANNUAL RATE
8-44	268			1100	7	3	2319	26	6	2063	44	11
45-49	416			1182	6	3	1667	12	4	950	13	7
50-54	751			1721	5	1	2024	12	3	824	10	6
55-59	1034			1809	13	4	1540	9	3	509	8	8
60-85	2034			2214	1	0	1346	8	3	268	2	4
UNKNOWN	13			6	0	0	18	0	0	12	0	0
TOTAL	4516			8032	32	2	8914	67	4	4626	77	8

RISK FACTOR	75-84 AT EXAM			85-94 AT EXAM			35-64 COMBINED			65-94 COMBINED		
	PERSON EXAMS	# OF EVENTS	ANNUAL RATE	PERSON EXAMS	# OF EVENTS	ANNUAL RATE	# OF EVENTS	CRUDE	AGE-ADJUSTED	# OF EVENTS	CRUDE	AGE-ADJUSTED
8-44	778	35	22	28			33	4	4	81	14	13
45-49	207	7	17	5			19	3	3	20	9	9
50-54	130	7	27	7			17	2	2	17	9	10
55-59	79	0	0	2			22	3	3	8	7	6
60-85	32	1	16	0			9	1	1	3	5	6
UNKNOWN	8	0	0	0			0	0	0	0	0	0
TOTAL	1234	50	20	42			100	2	2	129	11	11

LOGISTIC REGRESSIONS

AGE AT EXAM		COEFFICIENT OF RISK FACTOR	SE COEFFICIENT	P-VALUE	STANDARDIZED COEFFICIENT	STD SE
35-44						
45-54		-0.033	0.019	0.080	-0.299	0.171
55-64		-0.025	0.013	0.050	-0.233	0.119
65-74		-0.023	0.012	0.053	-0.218	0.113
75-84		-0.025	0.016	0.114	-0.230	0.145
85-94						
35-64	UNIVARIATE	-0.044	0.010	0.001	-0.428	0.095
65-94	UNIVARIATE	-0.033	0.009	0.001	-0.309	0.087
35-64	BIVARIATE	-0.023	0.011	0.033	-0.218	0.102
65-94	BIVARIATE	-0.024	0.010	0.013	-0.226	0.090
35-64	MULTIVARIATE	-0.007	0.011	0.547	-0.065	0.108
65-94	MULTIVARIATE	-0.018	0.010	0.063	-0.171	0.092

SEX: FEMALE
EVENT: STROKE AND TIA ONLY
RISK FACTOR: VITAL CAPACITY - HEIGHT INDEX (ML/INCH)
PERSONS AT RISK: PERSONS FREE OF STROKE AND TIA ONLY

RISK FACTOR	35-44 AT EXAM			45-54 AT EXAM			55-64 AT EXAM			65-74 AT EXAM		
	PERSON EXAMS	# OF EVENTS	ANNUAL RATE	PERSON EXAMS	# OF EVENTS	ANNUAL RATE	PERSON EXAMS	# OF EVENTS	ANNUAL RATE	PERSON EXAMS	# OF EVENTS	ANNUAL RATE
4-29	121			539	3	3	1530	14	5	1948	32	8
30-34	291			1240	4	2	2520	18	4	1888	36	10
35-39	769			2096	7	2	3181	10	2	1694	24	7
40-44	1427			2825	4	1	2809	8	1	919	6	3
45-92	2984			3316	6	1	1889	7	2	416	2	2
UNKNOWN	21			17	0	0	21	0	0	6	0	0
TOTAL	5613			10033	24	1	11950	57	2	6871	100	7

RISK FACTOR	75-84 AT EXAM			85-94 AT EXAM			35-64 COMBINED			65-94 COMBINED		
	PERSON EXAMS	# OF EVENTS	ANNUAL RATE	PERSON EXAMS	# OF EVENTS	ANNUAL RATE	# OF EVENTS	CRUDE	AGE-ADJUSTED	# OF EVENTS	CRUDE	AGE-ADJUSTED
4-29	1090	31	14	100	4	20	18	4	4	67	11	10
30-34	645	21	16	33	3	45	22	3	2	60	12	12
35-39	351	3	4	14	0	0	17	1	1	27	7	6
40-44	113	1	4	1	0	0	14	1	1	7	3	3
45-92	26	1	19	0	0	-	14	1	1	3	3	6
UNKNOWN	2	0	0	0	0	-	0	0	0	0	0	0
TOTAL	2227	57	13	148	7	24	85	2	2	164	9	9

LOGISTIC REGRESSIONS

AGE AT EXAM		COEFFICIENT OF RISK FACTOR	SE COEFFICIENT	P-VALUE	STANDARDIZED COEFFICIENT	STD SE
35-44						
45-54		-0.044	0.027	0.097	-0.327	0.197
55-64		-0.059	0.017	0.001	-0.440	0.125
65-74		-0.039	0.013	0.004	-0.282	0.099
75-84		-0.035	0.019	0.058	-0.248	0.131
85-94		0.006	0.059	0.918	0.040	0.392
35-64	UNIVARIATE	-0.071	0.013	0.001	-0.562	0.102
65-94	UNIVARIATE	-0.044	0.010	0.001	-0.331	0.077
35-64	BIVARIATE	-0.052	0.014	0.001	-0.417	0.112
65-94	BIVARIATE	-0.032	0.011	0.003	-0.243	0.082
35-64	MULTIVARIATE	-0.031	0.015	0.037	-0.250	0.120
65-94	MULTIVARIATE	-0.016	0.012	0.156	-0.122	0.086

TABLE 13-12   DATE=02NOV85
AVERAGE ANNUAL INCIDENCE RATE FOR EVENT PER 1000 PERSONS AT RISK AT EXAM
BY AGE AND LEVEL OF RISK FACTOR AT EXAM: FRAMINGHAM STUDY 30-YEAR FOLLOWUP

SEX: MALE
EVENT: STROKE AND TIA ONLY
RISK FACTOR: HEART RATE (PER MINUTE)
PERSONS AT RISK: PERSONS FREE OF STROKE AND TIA ONLY

RISK FACTOR	35-44 AT EXAM			45-54 AT EXAM			55-64 AT EXAM			65-74 AT EXAM		
	PERSON EXAMS	# OF EVENTS	ANNUAL RATE	PERSON EXAMS	# OF EVENTS	ANNUAL RATE	PERSON EXAMS	# OF EVENTS	ANNUAL RATE	PERSON EXAMS	# OF EVENTS	ANNUAL RATE
30-67	1295			2013	5	1	2590	16	3	1365	23	8
68-75	1503			2579	9	2	2709	12	2	1403	17	6
76-83	728			1465	7	2	1520	16	5	767	14	9
84-91	586			1139	7	3	1188	12	5	570	13	11
92-220	404			836	4	2	907	11	6	521	10	10
UNKNOWN	0			0	0	-	0	0	-	0	0	-
TOTAL	4516			8032	32	2	8914	67	4	4626	77	8

RISK FACTOR	75-84 AT EXAM			85-94 AT EXAM			55-64 COMBINED			65-94 COMBINED		
	PERSON EXAMS	# OF EVENTS	ANNUAL RATE	PERSON EXAMS	# OF EVENTS	ANNUAL RATE	# OF EVENTS	CRUDE	AGE-ADJUSTED	# OF EVENTS	CRUDE	AGE-ADJUSTED
30-67	353	13	18	13			21	2	2	36	10	10
68-75	339	11	16	13			21	2	3	29	8	8
76-83	246	9	18	11			23	3	3	23	11	11
84-91	163	6	18	4			19	3	3	19	13	13
92-220	133	11	41	1			16	4	4	22	17	20
UNKNOWN	0	0	-	0			0			0	-	
TOTAL	1234	50	20	42			100	2	2	129	11	11

LOGISTIC REGRESSIONS

AGE AT EXAM		COEFFICIENT OF RISK FACTOR	SE COEFFICIENT	P-VALUE	STANDARDIZED COEFFICIENT	STD SE
35-44						
45-54		0.019	0.013	0.152	0.238	0.166
55-64		0.021	0.009	0.013	0.275	0.111
65-74		0.010	0.008	0.244	0.128	0.110
75-84		0.016	0.010	0.121	0.211	0.136
85-94						
35-64	UNIVARIATE	0.022	0.007	0.002	0.280	0.092
65-94	UNIVARIATE	0.013	0.006	0.038	0.174	0.084
35-64	BIVARIATE	0.022	0.007	0.002	0.277	0.090
65-94	BIVARIATE	0.013	0.006	0.044	0.170	0.084
35-64	MULTIVARIATE	0.010	0.007	0.189	0.123	0.094
65-94	MULTIVARIATE	0.009	0.006	0.145	0.125	0.086

SEX: FEMALE
EVENT: STROKE AND TIA ONLY
RISK FACTOR: HEART RATE (PER MINUTE)
PERSONS AT RISK: PERSONS FREE OF STROKE AND TIA ONLY

RISK FACTOR	35-44 AT EXAM			45-54 AT EXAM			55-64 AT EXAM			65-74 AT EXAM		
	PERSON EXAMS	# OF EVENTS	ANNUAL RATE	PERSON EXAMS	# OF EVENTS	ANNUAL RATE	PERSON EXAMS	# OF EVENTS	ANNUAL RATE	PERSON EXAMS	# OF EVENTS	ANNUAL RATE
30-67	1090			1908	7	2	2402	7	1	1242	13	5
68-75	1804			3269	4	1	3602	11	2	1853	26	7
76-83	1050			1930	4	1	2412	16	3	1498	22	7
84-91	939			1612	5	2	1867	12	3	1154	18	8
92-220	730			1314	4	2	1667	11	3	1124	14	4
UNKNOWN	0			0	0	-	0	0	-	0	0	-
TOTAL	5613			10033	24	1	11950	57	2	6871	100	7

RISK FACTOR	75-84 AT EXAM			85-94 AT EXAM			55-64 COMBINED			65-94 COMBINED		
	PERSON EXAMS	# OF EVENTS	ANNUAL RATE	PERSON EXAMS	# OF EVENTS	ANNUAL RATE	# OF EVENTS	CRUDE	AGE-ADJUSTED	# OF EVENTS	CRUDE	AGE-ADJUSTED
30-67	429	14	16	30	3	50	14	1	1	30	9	9
68-75	535	11	10	38	2	26	17	1	1	59	8	8
76-83	484	15	15	27	1	19	20	2	2	38	9	9
84-91	375	11	15	29	1	17	19	2	2	33	11	11
92-220	404	6	7	24	0	0	15	2	2	24	8	8
UNKNOWN	0	0	-	0			0			0	-	
TOTAL	2227	57	13	148	7	24	85	2	2	164	9	9

LOGISTIC REGRESSIONS

AGE AT EXAM		COEFFICIENT OF RISK FACTOR	SE COEFFICIENT	P-VALUE	STANDARDIZED COEFFICIENT	STD SE
35-44						
45-54		0.007	0.016	0.684	0.081	0.199
55-64		0.020	0.009	0.024	0.261	0.116
65-74		0.015	0.007	0.034	0.196	0.093
75-84		-0.013	0.010	0.188	-0.188	0.143
85-94		-0.057	0.034	0.099	-0.786	0.476
35-64	UNIVARIATE	0.016	0.008	0.036	0.205	0.098
65-94	UNIVARIATE	0.003	0.006	0.592	0.042	0.078
35-64	BIVARIATE	0.016	0.008	0.041	0.200	0.098
65-94	BIVARIATE	0.002	0.006	0.673	0.033	0.077
35-64	MULTIVARIATE	0.008	0.008	0.324	0.102	0.104
65-94	MULTIVARIATE	-0.000	0.006	0.953	-0.005	0.077

TABLE 13-13   DATE=02NOV85
AVERAGE ANNUAL INCIDENCE RATE FOR EVENT PER 1000 PERSONS AT RISK AT EXAM
BY AGE AND LEVEL OF RISK FACTOR AT EXAM: FRAMINGHAM STUDY 30-YEAR FOLLOWUP

SEX: MALE
EVENT: STROKE AND TIA ONLY
RISK FACTOR: CIGARETTES SMOKED (NUMBER PER DAY)
PERSONS AT RISK: PERSONS FREE OF STROKE AND TIA ONLY

RISK FACTOR	35-44 AT EXAM PERSON EXAMS	# OF EVENTS	ANNUAL RATE	45-54 AT EXAM PERSON EXAMS	# OF EVENTS	ANNUAL RATE	55-64 AT EXAM PERSON EXAMS	# OF EVENTS	ANNUAL RATE	65-74 AT EXAM PERSON EXAMS	# OF EVENTS	ANNUAL RATE
NONE	1402			3211	6	1	4815	28	3	3084	47	8
1-10	437			742	5	3	842	4	2	449	6	7
11-20	1497			2127	11	3	1848	28	8	633	17	13
21-40	993			1664	6	2	1217	7	3	421	7	8
41-90	161			250	4	8	168	0	0	32	0	0
UNKNOWN	26			38	0	0	24	0	0	7	0	0
TOTAL	4516			8032	32	2	8914	67	4	4626	77	8

RISK FACTOR	75-84 AT EXAM PERSON EXAMS	# OF EVENTS	ANNUAL RATE	85-94 AT EXAM PERSON EXAMS	# OF EVENTS	ANNUAL RATE	35-64 COMBINED # OF EVENTS	CRUDE	AGE-ADJUSTED	65-94 COMBINED # OF EVENTS	CRUDE	AGE-ADJUSTED
NONE	963	40	21	32			34	2	2	88	11	10
1-10	122	4	16	4			9	2	2	10	9	9
11-20	86	3	17	5			39	2	2	20	14	14
21-40	58	3	26	1			14	2	2	11	11	15
41-90	5	0	0	0			4	3	3	0	0	0
UNKNOWN	0	0	-	0			0	0	0	0	0	0
TOTAL	1234	50	20	42			100	2	2	129	11	11

LOGISTIC REGRESSIONS

AGE AT EXAM		COEFFICIENT OF RISK FACTOR	SE COEFFICIENT	P-VALUE	STANDARDIZED COEFFICIENT	STD SE
35-44						
45-54		0.028	0.010	0.006	0.419	0.154
55-64		0.008	0.008	0.345	0.108	0.115
65-74		0.006	0.009	0.551	0.065	0.109
75-84		-0.002	0.017	0.887	-0.021	0.149
85-94						
35-64	UNIVARIATE	0.010	0.007	0.126	0.144	0.094
65-94	UNIVARIATE	0.001	0.008	0.914	0.010	0.088
35-64	BIVARIATE	0.019	0.006	0.002	0.280	0.092
65-94	BIVARIATE	0.007	0.008	0.383	0.077	0.088
35-64	MULTIVARIATE	0.022	0.006	0.001	0.323	0.093
65-94	MULTIVARIATE	0.010	0.008	0.200	0.114	0.089

SEX: FEMALE
EVENT: STROKE AND TIA ONLY
RISK FACTOR: CIGARETTES SMOKED (NUMBER PER DAY)
PERSONS AT RISK: PERSONS FREE OF STROKE AND TIA ONLY

RISK FACTOR	35-44 AT EXAM PERSON EXAMS	# OF EVENTS	ANNUAL RATE	45-54 AT EXAM PERSON EXAMS	# OF EVENTS	ANNUAL RATE	55-64 AT EXAM PERSON EXAMS	# OF EVENTS	ANNUAL RATE	65-74 AT EXAM PERSON EXAMS	# OF EVENTS	ANNUAL RATE
NONE	2580			5582	12	1	7896	41	3	5364	71	7
1-10	1312			1723	4	1	1634	5	2	674	12	9
11-20	1372			2027	7	2	1743	9	3	627	14	11
21-40	298			626	1	1	621	2	2	186	3	8
41-90	22			38	0	0	18	0	0	1	0	0
UNKNOWN	29			57	0	0	38	0	0	19	0	0
TOTAL	5613			10033	24	1	11950	57	2	6871	100	7

RISK FACTOR	75-84 AT EXAM PERSON EXAMS	# OF EVENTS	ANNUAL RATE	85-94 AT EXAM PERSON EXAMS	# OF EVENTS	ANNUAL RATE	35-64 COMBINED # OF EVENTS	CRUDE	AGE-ADJUSTED	65-94 COMBINED # OF EVENTS	CRUDE	AGE-ADJUSTED
NONE	1959	49	13	139	7	25	54	2	2	127	9	8
1-10	158	4	13	7	0	0	10	1	1	16	-0	10
11-20	93	4	22	2	0	0	17	1	1	18	12	13
21-40	13	0	0	0	0	-	4	1	1	3	8	6
41-90	0	0	-	0	0	-	0	0	0	0	0	0
UNKNOWN	4	0	0	0	0	-	0	0	0	0	0	0
TOTAL	2227	57	13	148	7	24	85	2	2	164	9	9

LOGISTIC REGRESSIONS

AGE AT EXAM		COEFFICIENT OF RISK FACTOR	SE COEFFICIENT	P-VALUE	STANDARDIZED COEFFICIENT	STD SE
35-44						
45-54		0.009	0.019	0.614	0.096	0.191
55-64		-0.003	0.014	0.825	-0.030	0.137
65-74		0.018	0.011	0.102	0.138	0.084
75-84		0.018	0.023	0.452	0.085	0.113
85-94						
35-64	UNIVARIATE	-0.001	0.011	0.904	-0.013	0.110
65-94	UNIVARIATE	0.012	0.010	0.242	0.082	0.070
35-64	BIVARIATE	0.006	0.011	0.609	0.055	0.107
65-94	BIVARIATE	0.020	0.010	0.044	0.143	0.071
35-64	MULTIVARIATE	0.010	0.011	0.375	0.094	0.106
65-94	MULTIVARIATE	0.024	0.010	0.017	0.172	0.072

TABLE 13-14   DATE=02NOV85
AVERAGE ANNUAL INCIDENCE RATE FOR EVENT PER 1000 PERSONS AT RISK AT EXAM
BY AGE AND LEVEL OF RISK FACTOR AT EXAM: FRAMINGHAM STUDY 30-YEAR FOLLOWUP

SEX: MALE
EVENT: STROKE AND TIA ONLY
RISK FACTOR: ALBUMIN IN URINE, DEFINITE
PERSONS AT RISK: PERSONS FREE OF STROKE AND TIA ONLY

RISK FACTOR	35-44 AT EXAM PERSON EXAMS	# OF EVENTS	ANNUAL RATE	45-54 AT EXAM PERSON EXAMS	# OF EVENTS	ANNUAL RATE	55-64 AT EXAM PERSON EXAMS	# OF EVENTS	ANNUAL RATE	65-74 AT EXAM PERSON EXAMS	# OF EVENTS	ANNUAL RATE
NO	4408			7903	30	2	8682	62	4	4476	75	8
YES	71			103	2	10	205	5	12	144	2	7
UNKNOWN	37			26	0	0	27	0	0	6	0	0
TOTAL	4516			8032	32	2	8914	67	4	4626	77	8

RISK FACTOR	75-84 AT EXAM PERSON EXAMS	# OF EVENTS	ANNUAL RATE	85-94 AT EXAM PERSON EXAMS	# OF EVENTS	ANNUAL RATE	35-64 COMBINED # OF EVENTS	CRUDE	AGE-ADJUSTED	65-94 COMBINED # OF EVENTS	CRUDE	AGE-ADJUSTED
NO	1199	47	20	37			93	2	2	123	11	11
YES	32	3	47	5			7	9	9	6	17	16
UNKNOWN	3	0	0	0			0	0	0	0	0	0
TOTAL	1234	50	20	42			100	2	2	129	11	11

LOGISTIC REGRESSIONS

AGE AT EXAM		COEFFICIENT OF RISK FACTOR	SE COEFFICIENT	P-VALUE	STANDARDIZED COEFFICIENT	STD SE
35-44						
45-54		1.648	0.737	0.025	0.186	0.083
55-64		1.246	0.470	0.008	0.187	0.071
65-74		-0.191	0.722	0.792	-0.033	0.125
75-84		0.930	0.624	0.136	0.148	0.099
85-94						
35-64	UNIVARIATE	1.442	0.395	0.001	0.190	0.052
65-94	UNIVARIATE	0.443	0.425	0.297	0.077	0.073
35-64	BIVARIATE	1.246	0.398	0.002	0.164	0.052
65-94	BIVARIATE	0.393	0.428	0.358	0.068	0.074
35-64	MULTIVARIATE	0.732	0.413	0.077	0.097	0.055
65-94	MULTIVARIATE	0.133	0.434	0.759	0.023	0.075

SEX: FEMALE
EVENT: STROKE AND TIA ONLY
RISK FACTOR: ALBUMIN IN URINE, DEFINITE
PERSONS AT RISK: PERSONS FREE OF STROKE AND TIA ONLY

RISK FACTOR	35-44 AT EXAM PERSON EXAMS	# OF EVENTS	ANNUAL RATE	45-54 AT EXAM PERSON EXAMS	# OF EVENTS	ANNUAL RATE	55-64 AT EXAM PERSON EXAMS	# OF EVENTS	ANNUAL RATE	65-74 AT EXAM PERSON EXAMS	# OF EVENTS	ANNUAL RATE
NO	5350			9728	23	1	11684	55	2	6729	96	7
YES	225			280	1	2	238	2	4	132	4	15
UNKNOWN	38			25	0	0	28	0	0	10	0	0
TOTAL	5613			10033	24	1	11950	57	2	6871	100	7

RISK FACTOR	75-84 AT EXAM PERSON EXAMS	# OF EVENTS	ANNUAL RATE	85-94 AT EXAM PERSON EXAMS	# OF EVENTS	ANNUAL RATE	35-64 COMBINED # OF EVENTS	CRUDE	AGE-ADJUSTED	65-94 COMBINED # OF EVENTS	CRUDE	AGE-ADJUSTED
NO	2173	54	12	141	6	21	82	2	2	156	9	9
YES	45	3	33	7	1	71	3	2	2	8	22	20
UNKNOWN	9	0	0	0	0	-	0	0	0	0	0	0
TOTAL	2227	57	13	148	7	24	85	2	2	164	9	9

LOGISTIC REGRESSIONS

AGE AT EXAM		COEFFICIENT OF RISK FACTOR	SE COEFFICIENT	P-VALUE	STANDARDIZED COEFFICIENT	STD SE
35-44						
45-54		0.414	1.023	0.686	0.068	0.169
55-64		0.585	0.723	0.420	0.082	0.101
65-74		0.770	0.518	0.137	0.106	0.071
75-84		1.031	0.613	0.093	0.145	0.086
85-94		1.322	1.158	0.254	0.282	0.247
35-64	UNIVARIATE	0.277	0.589	0.638	0.045	0.095
65-94	UNIVARIATE	0.951	0.370	0.010	0.133	0.052
35-64	BIVARIATE	0.452	0.590	0.444	0.073	0.096
65-94	BIVARIATE	0.906	0.372	0.015	0.127	0.052
35-64	MULTIVARIATE	-0.177	0.613	0.773	-0.029	0.099
65-94	MULTIVARIATE	0.476	0.386	0.218	0.067	0.054

TABLE 13-15   DATE=02NOV85
AVERAGE ANNUAL INCIDENCE RATE FOR EVENT PER 1000 PERSONS AT RISK AT EXAM
BY AGE AND LEVEL OF RISK FACTOR AT EXAM: FRAMINGHAM STUDY 30-YEAR FOLLOWUP

SEX: MALE
EVENT: STROKE AND TIA ONLY
RISK FACTOR: HEART ENLARGEMENT BY X-RAY
PERSONS AT RISK: PERSONS FREE OF STROKE AND TIA ONLY

RISK FACTOR	35-44 AT EXAM			45-54 AT EXAM			55-64 AT EXAM			65-74 AT EXAM		
	PERSON EXAMS	# OF EVENTS	ANNUAL RATE	PERSON EXAMS	# OF EVENTS	ANNUAL RATE	PERSON EXAMS	# OF EVENTS	ANNUAL RATE	PERSON EXAMS	# OF EVENTS	ANNUAL RATE
NO	4098			6590	22	2	6455	42	3	2971	39	7
MILD	189			702	3	2	1102	10	5	662	11	8
DEFINITE	224			736	7	5	1356	15	6	993	27	14
UNKNOWN	5			4	0	0	1	0	0	0	0	-
TOTAL	4516	.		8032	32	2	8914	67	4	4626	77	8

RISK FACTOR	75-84 AT EXAM			85-94 AT EXAM			35-64 COMBINED			65-94 COMBINED		
	PERSON EXAMS	# OF EVENTS	ANNUAL RATE	PERSON EXAMS	# OF EVENTS	ANNUAL RATE	# OF EVENTS	CRUDE	AGE-ADJUSTED	# OF EVENTS	CRUDE	AGE-ADJUSTED
NO	693	24	17	28			65	2	2	64	9	9
MILD	205	11	27	6			13	3	3	22	13	12
DEFINITE	336	15	22	8			22	5	4	43	16	16
UNKNOWN	0	0	-	0			0	0	0	0	-	-
TOTAL	1234	50	20	42			100	2	2	129	11	11

LOGISTIC REGRESSIONS

AGE AT EXAM		COEFFICIENT OF RISK FACTOR	SE COEFFICIENT	P-VALUE	STANDARDIZED COEFFICIENT	STD SE
35-44						
45-54		0.506	0.218	0.020	0.312	0.135
55-64		0.273	0.147	0.063	0.203	0.109
65-74		0.367	0.127	0.004	0.302	0.105
75-84		0.147	0.162	0.363	0.128	0.140
85-94						
35-64	UNIVARIATE	0.469	0.121	0.001	0.307	0.079
65-94	UNIVARIATE	0.319	0.099	0.001	0.265	0.082
35-64	BIVARIATE	0.303	0.123	0.014	0.198	0.080
65-94	BIVARIATE	0.281	0.100	0.005	0.234	0.083
35-64	MULTIVARIATE	0.082	0.133	0.541	0.053	0.087
65-94	MULTIVARIATE	0.119	0.108	0.269	0.099	0.090

SEX: FEMALE
EVENT: STROKE AND TIA ONLY
RISK FACTOR: HEART ENLARGEMENT BY X-RAY
PERSONS AT RISK: PERSONS FREE OF STROKE AND TIA ONLY

RISK FACTOR	35-44 AT EXAM			45-54 AT EXAM			55-64 AT EXAM			65-74 AT EXAM		
	PERSON EXAMS	# OF EVENTS	ANNUAL RATE	PERSON EXAMS	# OF EVENTS	ANNUAL RATE	PERSON EXAMS	# OF EVENTS	ANNUAL RATE	PERSON EXAMS	# OF EVENTS	ANNUAL RATE
NO	5161			8054	19	1	7762	21	1	3444	35	5
MILD	198			954	2	1	1850	8	2	1254	13	5
DEFINITE	246			1019	3	1	2334	28	6	2173	52	12
UNKNOWN	8			6	0	0	4	0	0	0	0	-
TOTAL	5613			10033	24	1	11950	57	2	6871	100	7

RISK FACTOR	75-84 AT EXAM			85-94 AT EXAM			35-64 COMBINED			65-94 COMBINED		
	PERSON EXAMS	# OF EVENTS	ANNUAL RATE	PERSON EXAMS	# OF EVENTS	ANNUAL RATE	# OF EVENTS	CRUDE	AGE-ADJUSTED	# OF EVENTS	CRUDE	AGE-ADJUSTED
NO	778	14	9	38	2	26	44	1	1	51	6	6
MILD	435	9	10	30	1	17	10	2	3	23	7	7
DEFINITE	1014	34	17	80	4	25	31	4	3	90	14	13
UNKNOWN	0	0	-	0	0	-	0	0	0	0	-	-
TOTAL	2227	57	13	148	7	24	85	2	2	164	9	9

LOGISTIC REGRESSIONS

AGE AT EXAM		COEFFICIENT OF RISK FACTOR	SE COEFFICIENT	P-VALUE	STANDARDIZED COEFFICIENT	STD SE
35-44						
45-54		0.081	0.303	0.790	0.052	0.195
55-64		0.756	0.148	0.001	0.604	0.118
65-74		0.448	0.114	0.001	0.397	0.100
75-84		0.334	0.161	0.038	0.297	0.143
85-94		0.003	0.458	0.995	0.002	0.389
35-64	UNIVARIATE	0.704	0.119	0.001	0.495	0.084
65-94	UNIVARIATE	0.439	0.090	0.001	0.394	0.081
35-64	BIVARIATE	0.553	0.124	0.001	0.388	0.087
65-94	BIVARIATE	0.378	0.092	0.001	0.339	0.082
35-64	MULTIVARIATE	0.295	0.137	0.032	0.207	0.097
65-94	MULTIVARIATE	0.206	0.097	0.033	0.185	0.087

TABLE 13-16   DATE=05NOV85

AVERAGE ANNUAL INCIDENCE RATE FOR EVENT PER 1000 PERSONS AT RISK AT EXAM
BY AGE AND LEVEL OF RISK FACTOR AT EXAM: FRAMINGHAM STUDY 30-YEAR FOLLOWUP

SEX: MALE
EVENT: STROKE AND TIA ONLY
RISK FACTOR: LEFT VENTRICULAR HYPERTROPHY BY ELECTROCARDIOGRAM
PERSONS AT RISK: PERSONS FREE OF STROKE AND TIA ONLY

RISK FACTOR	35-44 AT EXAM PERSON EXAMS	# OF EVENTS	ANNUAL RATE	45-54 AT EXAM PERSON EXAMS	# OF EVENTS	ANNUAL RATE	55-64 AT EXAM PERSON EXAMS	# OF EVENTS	ANNUAL RATE	65-74 AT EXAM PERSON EXAMS	# OF EVENTS	ANNUAL RATE
NO	4405			7780	30	2	8471	54	3	4347	64	7
MILD	79			145			235	4	9	114	4	18
DEFINITE	32			107	2	9	208	9	22	165	9	27
UNKNOWN	0			0	0	-	0	0	-	0	0	-
TOTAL	4516			8032	32	2	8914	67	4	4626	77	8

RISK FACTOR	75-84 AT EXAM PERSON EXAMS	# OF EVENTS	ANNUAL RATE	85-94 AT EXAM PERSON EXAMS	# OF EVENTS	ANNUAL RATE	35-64 COMBINED # OF EVENTS	CRUDE	AGE-ADJUSTED	65-94 COMBINED # OF EVENTS	CRUDE	AGE-ADJUSTED
NO	1148	43	19	38	2		85	2	2	109	10	10
MILD	39	2	26	2			4	4	4	6	19	19
DEFINITE	47	5	53	2			11	16	12	14	33	32
UNKNOWN	0	0	-	0			0	-		0		
TOTAL	1234	50	20	42			100	2	2	129	11	11

LOGISTIC REGRESSIONS

AGE AT EXAM	COEFFICIENT OF RISK FACTOR	SE COEFFICIENT	P-VALUE	STANDARDIZED COEFFICIENT	STD SE
35-44					
45-54	0.651	0.396	0.100	0.171	0.104
55-64	0.978	0.178	0.001	0.331	0.060
65-74	0.692	0.175	0.001	0.275	0.070
75-84	0.537	0.243	0.027	0.223	0.101
85-94					
35-64 UNIVARIATE	1.011	0.160	0.001	0.292	0.046
65-94 UNIVARIATE	0.630	0.141	0.001	0.253	0.057
35-64 BIVARIATE	0.865	0.161	0.001	0.249	0.047
65-94 BIVARIATE	0.622	0.142	0.001	0.250	0.057
35-64 MULTIVARIATE	0.470	0.181	0.009	0.136	0.052
65-94 MULTIVARIATE	0.359	0.157	0.022	0.144	0.063

SEX: FEMALE
EVENT: STROKE AND TIA ONLY
RISK FACTOR: LEFT VENTRICULAR HYPERTROPHY BY ELECTROCARDIOGRAM
PERSONS AT RISK: PERSONS FREE OF STROKE AND TIA ONLY

RISK FACTOR	35-44 AT EXAM PERSON EXAMS	# OF EVENTS	ANNUAL RATE	45-54 AT EXAM PERSON EXAMS	# OF EVENTS	ANNUAL RATE	55-64 AT EXAM PERSON EXAMS	# OF EVENTS	ANNUAL RATE	65-74 AT EXAM PERSON EXAMS	# OF EVENTS	ANNUAL RATE
NO	5560			9866	24	1	11542	41	2	6504	76	6
MILD	38			111	0	0	242	8	17	174	7	20
DEFINITE	15			56	0	0	166	8	24	193	17	44
UNKNOWN	0			0	0	-	0	0	-	0	0	-
TOTAL	5613			10033	24	1	11950	57	2	6871	100	7

RISK FACTOR	75-84 AT EXAM PERSON EXAMS	# OF EVENTS	ANNUAL RATE	85-94 AT EXAM PERSON EXAMS	# OF EVENTS	ANNUAL RATE	35-64 COMBINED # OF EVENTS	CRUDE	AGE-ADJUSTED	65-94 COMBINED # OF EVENTS	CRUDE	AGE-ADJUSTED
NO	2059	48	12	129	6	25	69	1	1	130	7	7
MILD	65	2	15	8	0	0	8	10	7	9	18	19
DEFINITE	103	7	34	11	1	45	8	17	10	25	41	42
UNKNOWN	0	0	-	0	0	-	0	-		0	-	
TOTAL	2227	57	13	148	7	24	85	2	2	164	9	9

LOGISTIC REGRESSIONS

AGE AT EXAM	COEFFICIENT OF RISK FACTOR	SE COEFFICIENT	P-VALUE	STANDARDIZED COEFFICIENT	STD SE
35-44	-			-	
45-54	-			-	
55-64	1.426	0.175	0.001	0.387	0.048
65-74	1.065	0.135	0.001	0.386	0.049
75-84	0.540	0.286	0.009	0.241	0.092
85-94	0.240	0.599	0.688	0.134	0.335
35-64 UNIVARIATE	1.404	0.167	0.001	0.306	0.036
65-94 UNIVARIATE	0.884	0.110	0.001	0.343	0.043
35-64 BIVARIATE	1.265	0.170	0.001	0.276	0.037
65-94 BIVARIATE	0.834	0.111	0.001	0.324	0.043
35-64 MULTIVARIATE	0.837	0.203	0.001	0.182	0.044
65-94 MULTIVARIATE	0.583	0.121	0.001	0.226	0.047

TABLE 13-17   DATE=05NOV85
AVERAGE ANNUAL INCIDENCE RATE FOR EVENT PER 1000 PERSONS AT RISK AT EXAM
BY AGE AND LEVEL OF RISK FACTOR AT EXAM: FRAMINGHAM STUDY 30-YEAR FOLLOWUP

SEX: MALE
EVENT: STROKE AND TIA ONLY
RISK FACTOR: INTRAVENTRICULAR CONDUCTION DEFECT (WITH QRS INTERVAL MORE THAN 0.11 SECOND)
PERSONS AT RISK: PERSONS FREE OF STROKE AND TIA ONLY

RISK FACTOR	35-44 AT EXAM PERSON EXAMS	# OF EVENTS	ANNUAL RATE	45-54 AT EXAM PERSON EXAMS	# OF EVENTS	ANNUAL RATE	55-64 AT EXAM PERSON EXAMS	# OF EVENTS	ANNUAL RATE	65-74 AT EXAM PERSON EXAMS	# OF EVENTS	ANNUAL RATE
NO	4198			7452	30	2	8107	58	4	4150	62	7
INCOMPLETE	280			491	1	1	534	8	7	231	6	13
COMPLETE	38			89	1	6	273	1	2	245	9	18
UNKNOWN	0			0			0			0		
TOTAL	4516			8032	32	2	8914	67	4	4626	77	8

RISK FACTOR	75-84 AT EXAM PERSON EXAMS	# OF EVENTS	ANNUAL RATE	85-94 AT EXAM PERSON EXAMS	# OF EVENTS	ANNUAL RATE	35-64 COMBINED # OF EVENTS	CRUDE	AGE-ADJUSTED	65-94 COMBINED # OF EVENTS	CRUDE	AGE-ADJUSTED
NO	1083	44	20	35			89	2	2	107	10	10
INCOMPLETE	35	2	29	3			9	3	3	8	15	16
COMPLETE	116	4	17	4			2	3	3	14	19	19
UNKNOWN	0	0	-	0			0	-	-	0	-	-
TOTAL	1234	50	20	42			100	2	2	129	11	11

LOGISTIC REGRESSIONS

AGE AT EXAM	COEFFICIENT OF RISK FACTOR	SE COEFFICIENT	P-VALUE	STANDARDIZED COEFFICIENT	STD SE
35-44					
45-54	0.100	0.530	0.850	0.031	0.166
55-64	0.151	0.269	0.574	0.062	0.110
65-74	0.473	0.172	0.006	0.231	0.084
75-84	-0.049	0.250	0.843	-0.030	0.150
85-94					
35-64 UNIVARIATE	0.219	0.244	0.369	0.078	0.086
65-94 UNIVARIATE	0.334	0.138	0.016	0.172	0.071
35-64 BIVARIATE	0.086	0.239	0.720	0.030	0.085
65-94 BIVARIATE	0.273	0.138	0.048	0.140	0.071
35-64 MULTIVARIATE	-0.005	0.246	0.983	-0.002	0.087
65-94 MULTIVARIATE	0.264	0.139	0.058	0.136	0.072

SEX: FEMALE
EVENT: STROKE AND TIA ONLY
RISK FACTOR: INTRAVENTRICULAR CONDUCTION DEFECT (WITH QRS INTERVAL MORE THAN 0.11 SECOND)
PERSONS AT RISK: PERSONS FREE OF STROKE AND TIA ONLY

RISK FACTOR	35-44 AT EXAM PERSON EXAMS	# OF EVENTS	ANNUAL RATE	45-54 AT EXAM PERSON EXAMS	# OF EVENTS	ANNUAL RATE	55-64 AT EXAM PERSON EXAMS	# OF EVENTS	ANNUAL RATE	65-74 AT EXAM PERSON EXAMS	# OF EVENTS	ANNUAL RATE
NO	5526			9834	24	1	11582	56	2	6577	92	7
INCOMPLETE	75			151	0	0	195	0	0	130	2	8
COMPLETE	12			48	0	0	173	1	3	164	6	18
UNKNOWN	0			0		-	0		-	0		-
TOTAL	5613			10033	24	1	11950	57	2	6871	100	7

RISK FACTOR	75-84 AT EXAM PERSON EXAMS	# OF EVENTS	ANNUAL RATE	85-94 AT EXAM PERSON EXAMS	# OF EVENTS	ANNUAL RATE	35-64 COMBINED # OF EVENTS	CRUDE	AGE-ADJUSTED	65-94 COMBINED # OF EVENTS	CRUDE	AGE-ADJUSTED
NO	2088	52	12	138	6	22	84	2	2	150	9	9
INCOMPLETE	49	1	10	3	0	0	0	0	-	3	8	8
COMPLETE	90	4	22	7	1	71	1	2	-1	11	21	20
UNKNOWN	0	0	-	0	0	-	0	-	-	0	-	-
TOTAL	2227	57	13	148	7	24	85	2	2	164	9	9

LOGISTIC REGRESSIONS

AGE AT EXAM	COEFFICIENT OF RISK FACTOR	SE COEFFICIENT	P-VALUE	STANDARDIZED COEFFICIENT	STD SE
35-44					
45-54					
55-64	-0.162	0.566	0.775	-0.043	0.152
65-74	0.457	0.211	0.030	0.152	0.070
75-84	0.264	0.263	0.316	0.110	0.110
85-94	0.591	0.595	0.319	0.263	0.264
35-64 UNIVARIATE	-0.213	0.588	0.718	-0.047	0.129
65-94 UNIVARIATE	0.427	0.158	0.007	0.152	0.056
35-64 BIVARIATE	-0.346	0.581	0.551	-0.076	0.127
65-94 BIVARIATE	0.381	0.159	0.016	0.136	0.056
35-64 MULTIVARIATE	-0.334	0.588	0.571	-0.073	0.129
65-94 MULTIVARIATE	0.351	0.162	0.030	0.125	0.058

TABLE 13-18   DATE=02NOV85

AVERAGE ANNUAL INCIDENCE RATE FOR EVENT PER 1000 PERSONS AT RISK AT EXAM
BY AGE AND LEVEL OF RISK FACTOR AT EXAM: FRAMINGHAM STUDY 30-YEAR FOLLOWUP

SEX: MALE
EVENT: STROKE AND TIA ONLY
RISK FACTOR: DEFINITE NONSPECIFIC T-WAVE OR ST-SEGMENT ABNORMALITY BY ELECTROCARDIOGRAM
PERSONS AT RISK: PERSONS FREE OF STROKE AND TIA ONLY

RISK FACTOR	35-44 AT EXAM PERSON EXAMS	# OF EVENTS	ANNUAL RATE	45-54 AT EXAM PERSON EXAMS	# OF EVENTS	ANNUAL RATE	55-64 AT EXAM PERSON EXAMS	# OF EVENTS	ANNUAL RATE	65-74 AT EXAM PERSON EXAMS	# OF EVENTS	ANNUAL RATE
NO	4403			7609	26	2	8128	54	3	4007	59	7
YES	113			423	6	7	786	13	8	619	18	15
UNKNOWN	0			0			0			0		
TOTAL	4516			8032	32	2	8914	67	4	4626	77	8

RISK FACTOR	75-84 AT EXAM PERSON EXAMS	# OF EVENTS	ANNUAL RATE	85-94 AT EXAM PERSON EXAMS	# OF EVENTS	ANNUAL RATE	35-64 COMBINED # OF EVENTS	CRUDE	AGE-ADJUSTED	65-94 COMBINED # OF EVENTS	CRUDE	AGE-ADJUSTED
NO	988	36	18	24			81	2	2	96	10	10
YES	246	14	28	18			19	7	6	33	19	18
UNKNOWN	0	0	–	0			0	–		0	–	
TOTAL	1234	50	20	42			100	2	2	129	11	11

LOGISTIC REGRESSIONS

AGE AT EXAM	COEFFICIENT OF RISK FACTOR	SE COEFFICIENT	P-VALUE	STANDARDIZED COEFFICIENT	STD SE
35-44					
45-54	1.434	0.456	0.002	0.320	0.102
55-64	0.922	0.311	0.003	0.261	0.088
65-74	0.695	0.273	0.011	0.237	0.093
75-84	0.467	0.323	0.148	0.187	0.129
85-94					
35-64 UNIVARIATE	1.284	0.257	0.001	0.309	0.062
65-94 UNIVARIATE	0.689	0.205	0.001	0.246	0.073
35-64 BIVARIATE	0.984	0.259	0.001	0.237	0.062
65-94 BIVARIATE	0.571	0.208	0.006	0.204	0.074
35-64 MULTIVARIATE	0.752	0.262	0.004	0.181	0.063
65-94 MULTIVARIATE	0.500	0.210	0.017	0.178	0.075

SEX: FEMALE
EVENT: STROKE AND TIA ONLY
RISK FACTOR: DEFINITE NONSPECIFIC T-WAVE OR ST-SEGMENT ABNORMALITY BY ELECTROCARDIOGRAM
PERSONS AT RISK: PERSONS FREE OF STROKE AND TIA ONLY

RISK FACTOR	35-44 AT EXAM PERSON EXAMS	# OF EVENTS	ANNUAL RATE	45-54 AT EXAM PERSON EXAMS	# OF EVENTS	ANNUAL RATE	55-64 AT EXAM PERSON EXAMS	# OF EVENTS	ANNUAL RATE	65-74 AT EXAM PERSON EXAMS	# OF EVENTS	ANNUAL RATE
NO	5514			9603	20	1	10984	48	2	5994	78	7
YES	99			430	4	5	966	9	5	877	22	13
UNKNOWN	0			0			0			0		
TOTAL	5613			10033	24	1	11950	57	2	6871	100	7

RISK FACTOR	75-84 AT EXAM PERSON EXAMS	# OF EVENTS	ANNUAL RATE	85-94 AT EXAM PERSON EXAMS	# OF EVENTS	ANNUAL RATE	35-64 COMBINED # OF EVENTS	CRUDE	AGE-ADJUSTED	65-94 COMBINED # OF EVENTS	CRUDE	AGE-ADJUSTED
NO	1811	37	10	114	6	26	72	1	1	121	8	8
YES	416	20	24	34	1	15	13	4	4	43	16	15
UNKNOWN	0	0	–	0			0	–		0	–	
TOTAL	2227	57	13	148	7	24	85	2	2	164	9	9

LOGISTIC REGRESSIONS

AGE AT EXAM	COEFFICIENT OF RISK FACTOR	SE COEFFICIENT	P-VALUE	STANDARDIZED COEFFICIENT	STD SE
35-44					
45-54	1.504	0.550	0.006	0.305	0.111
55-64	0.762	0.365	0.037	0.208	0.099
65-74	0.669	0.244	0.006	0.223	0.081
75-84	0.884	0.283	0.002	0.345	0.110
85-94	-0.606	1.098	0.581	-0.256	0.464
35-64 UNIVARIATE	1.154	0.303	0.001	0.261	0.068
65-94 UNIVARIATE	0.769	0.180	0.001	0.270	0.063
35-64 BIVARIATE	0.886	0.306	0.004	0.201	0.069
65-94 BIVARIATE	0.681	0.182	0.001	0.239	0.064
35-64 MULTIVARIATE	0.717	0.309	0.020	0.162	0.070
65-94 MULTIVARIATE	0.627	0.184	0.001	0.220	0.064

TABLE 14-1   DATE=06NOV85

AVERAGE ANNUAL INCIDENCE RATE FOR EVENT PER 1000 PERSONS AT RISK AT EXAM
BY AGE AND LEVEL OF RISK FACTOR AT EXAM: FRAMINGHAM STUDY 30-YEAR FOLLOWUP

SEX: MALE
EVENT: ATHEROTHROMBOTIC BRAIN INFARCTION
RISK FACTOR: SYSTOLIC BLOOD PRESSURE, FIRST EXAMINER (MM HG)
PERSONS AT RISK: PERSONS FREE OF STROKE AND TIA ONLY

RISK FACTOR	35-44 AT EXAM			45-54 AT EXAM			55-64 AT EXAM			65-74 AT EXAM		
	PERSON EXAMS	# OF EVENTS	ANNUAL RATE	PERSON EXAMS	# OF EVENTS	ANNUAL RATE	PERSON EXAMS	# OF EVENTS	ANNUAL RATE	PERSON EXAMS	# OF EVENTS	ANNUAL RATE
74-119	1225			1837	3	1	1589	1	0	649	2	2
120-139	2191			3531	4	1	3482	5	1	1582	8	3
140-159	850			1830	3	1	2324	10	2	1394	13	5
160-179	197			589	4	3	1029	9	4	659	10	8
180-300	53			245	2	4	490	10	10	342	8	12
UNKNOWN	0			0	0	-	0	0	-	0	0	-
TOTAL	4516			8032	16	1	8914	35	2	4626	41	4

RISK FACTOR	75-84 AT EXAM			85-94 AT EXAM			35-64 COMBINED			65-94 COMBINED		
	PERSON EXAMS	# OF EVENTS	ANNUAL RATE	PERSON EXAMS	# OF EVENTS	ANNUAL RATE	# OF EVENTS	CRUDE	AGE-ADJUSTED	# OF EVENTS	CRUDE	AGE-ADJUSTED
74-119	136	1	4	7			4	0	0	3	2	2
120-139	420	8	10	13			9	1	1	16	4	4
140-159	379	4	5	15			13	1	1	17	5	5
160-179	193	8	21	5			13	4	3	18	11	10
180-300	106	5	24	2			12	8	6	13	14	14
UNKNOWN	0	0	-	0			0	-	-	0	-	-
TOTAL	1234	26	11	42			51	1	1	67	6	6

LOGISTIC REGRESSIONS

AGE AT EXAM	COEFFICIENT OF RISK FACTOR	SE COEFFICIENT	P-VALUE	STANDARDIZED COEFFICIENT	STD SE
35-44					
45-54	0.029	0.009	0.002	0.574	0.184
55-64	0.057	0.005	0.001	0.824	0.114
65-74	0.023	0.005	0.001	0.524	0.124
75-84	0.021	0.007	0.004	0.485	0.168
85-94					
35-64 UNIVARIATE	0.039	0.004	0.001	0.789	0.089
65-94 UNIVARIATE	0.022	0.004	0.001	0.516	0.099
35-64 BIVARIATE	0.034	0.005	0.001	0.695	0.092
65-94 BIVARIATE	0.023	0.004	0.001	0.520	0.100
35-64 MULTIVARIATE	0.032	0.005	0.001	0.646	0.102
65-94 MULTIVARIATE	0.020	0.005	0.001	0.470	0.109

SEX: FEMALE
EVENT: ATHEROTHROMBOTIC BRAIN INFARCTION
RISK FACTOR: SYSTOLIC BLOOD PRESSURE, FIRST EXAMINER (MM HG)
PERSONS AT RISK: PERSONS FREE OF STROKE AND TIA ONLY

RISK FACTOR	35-44 AT EXAM			45-54 AT EXAM			55-64 AT EXAM			65-74 AT EXAM		
	PERSON EXAMS	# OF EVENTS	ANNUAL RATE	PERSON EXAMS	# OF EVENTS	ANNUAL RATE	PERSON EXAMS	# OF EVENTS	ANNUAL RATE	PERSON EXAMS	# OF EVENTS	ANNUAL RATE
74-119	2488			2717	1	0	2011	0	0	675	0	0
120-139	2245			4015	3	0	4168	3	0	1876	7	2
140-159	673			2070	6	1	3155	11	2	2212	11	3
160-179	156			783	1	1	1654	7	2	1302	11	4
180-300	51			448	3	3	962	10	5	806	24	15
UNKNOWN	0			0	0	-	0	0	-	0	0	-
TOTAL	5613			10033	14	1	11950	31	1	6871	53	4

RISK FACTOR	75-84 AT EXAM			85-94 AT EXAM			35-64 COMBINED			65-94 COMBINED		
	PERSON EXAMS	# OF EVENTS	ANNUAL RATE	PERSON EXAMS	# OF EVENTS	ANNUAL RATE	# OF EVENTS	CRUDE	AGE-ADJUSTED	# OF EVENTS	CRUDE	AGE-ADJUSTED
74-119	145	1	3	10			1	0	0	1	1	1
120-139	527	4	4	47			6	0	0	11	2	2
140-159	704	8	6	34			17	1	1	20	3	3
160-179	512	6	6	32			8	2	1	17	5	5
180-300	339	8	12	25			13	4	3	32	14	14
UNKNOWN	0	0	-	0			0	-	-	0	-	-
TOTAL	2227	27	6	148			45	1	1	81	4	4

LOGISTIC REGRESSIONS

AGE AT EXAM	COEFFICIENT OF RISK FACTOR	SE COEFFICIENT	P-VALUE	STANDARDIZED COEFFICIENT	STD SE
35-44					
45-54	0.026	0.007	0.001	0.589	0.153
55-64	0.032	0.005	0.001	0.796	0.113
65-74	0.031	0.004	0.001	0.752	0.110
75-84	0.015	0.007	0.021	0.392	0.170
85-94					
35-64 UNIVARIATE	0.032	0.004	0.001	0.766	0.083
65-94 UNIVARIATE	0.026	0.004	0.001	0.636	0.091
35-64 BIVARIATE	0.030	0.004	0.001	0.717	0.089
65-94 BIVARIATE	0.025	0.004	0.001	0.623	0.092
35-64 MULTIVARIATE	0.026	0.004	0.001	0.624	0.104
65-94 MULTIVARIATE	0.022	0.004	0.001	0.550	0.097

TABLE 14-2   DATE=06NOV85
AVERAGE ANNUAL INCIDENCE RATE FOR EVENT PER 1000 PERSONS AT RISK AT EXAM
BY AGE AND LEVEL OF RISK FACTOR AT EXAM: FRAMINGHAM STUDY 30-YEAR FOLLOWUP

SEX: MALE
EVENT: ATHEROTHROMBOTIC BRAIN INFARCTION
RISK FACTOR: DIASTOLIC BLOOD PRESSURE, FIRST EXAMINER (MM HG)
PERSONS AT RISK: PERSONS FREE OF STROKE AND TIA ONLY

RISK FACTOR	35-44 AT EXAM			45-54 AT EXAM			55-64 AT EXAM			65-74 AT EXAM		
	PERSON EXAMS	# OF EVENTS	ANNUAL RATE	PERSON EXAMS	# OF EVENTS	ANNUAL RATE	PERSON EXAMS	# OF EVENTS	ANNUAL RATE	PERSON EXAMS	# OF EVENTS	ANNUAL RATE
20-74	1015			1532	2	1	2056	4	1	1469	13	4
75-84	1614			2857	4	1	3127	3	0	1575	9	3
85-94	1203			2142	4	1	2192	8	2	1030	12	6
95-104	509			1065	5	2	1081	11	5	399	4	5
105-160	175			436	1	1	458	9	10	153	3	10
UNKNOWN	0	-	-	0	0	-	0	0	-	0	0	-
TOTAL	4516			8032	16	1	8914	35	2	4626	41	4

RISK FACTOR	75-84 AT EXAM			85-94 AT EXAM			35-64 COMBINED			65-94 COMBINED		
	PERSON EXAMS	# OF EVENTS	ANNUAL RATE	PERSON EXAMS	# OF EVENTS	ANNUAL RATE	# OF EVENTS	CRUDE	AGE- ADJUSTED	# OF EVENTS	CRUDE	AGE- ADJUSTED
20-74	513	8	8	22			6	1	1	21	5	5
75-84	403	8	10	15			7	0	0	17	4	4
85-94	224	4	9	5			12	1	1	16	6	6
95-104	66	5	38	0			16	3	3	9	10	12
105-160	28	1	18	0			10	5	5	4	11	11
UNKNOWN	0	0	-	0			0	-	-	0	-	-
TOTAL	1234	26	11	42			51	1	1	67	6	6

LOGISTIC REGRESSIONS

AGE AT EXAM		COEFFICIENT OF RISK FACTOR	SE COEFFICIENT	P-VALUE	STANDARDIZED COEFFICIENT	STD SE
35-44						
45-54		0.037	0.019	0.049	0.431	0.219
55-64		0.066	0.010	0.001	0.806	0.127
65-74		0.014	0.012	0.264	0.166	0.149
75-84		0.029	0.015	0.058	0.354	0.187
85-94						
35-64	UNIVARIATE	0.061	0.009	0.001	0.717	0.108
65-94	UNIVARIATE	0.016	0.009	0.093	0.195	0.116
35-64	BIVARIATE	0.059	0.009	0.001	0.697	0.106
65-94	BIVARIATE	0.022	0.009	0.022	0.267	0.117
35-64	MULTIVARIATE	0.054	0.010	0.001	0.634	0.114
63-94	MULTIVARIATE	0.018	0.009	0.052	0.226	0.116

SEX: FEMALE
EVENT: ATHEROTHROMBOTIC BRAIN INFARCTION
RISK FACTOR: DIASTOLIC BLOOD PRESSURE, FIRST EXAMINER (MM HG)
PERSONS AT RISK: PERSONS FREE OF STROKE AND TIA ONLY

RISK FACTOR	35-44 AT EXAM			45-54 AT EXAM			55-64 AT EXAM			65-74 AT EXAM		
	PERSON EXAMS	# OF EVENTS	ANNUAL RATE	PERSON EXAMS	# OF EVENTS	ANNUAL RATE	PERSON EXAMS	# OF EVENTS	ANNUAL RATE	PERSON EXAMS	# OF EVENTS	ANNUAL RATE
20-74	2161			2650	1	0	3078	2	0	2026	6	1
75-84	2028			3558	2	0	3964	4	1	2362	11	2
85-94	988			2587	4	1	2965	7	1	1559	19	6
95-104	324			933	4	2	1326	8	3	642	5	4
105-160	112			505	3	3	617	10	8	282	12	21
UNKNOWN	0			0	0	-	0	0	-	0	0	-
TOTAL	5613			10053	14	1	11950	31	1	6871	53	4

RISK FACTOR	75-84 AT EXAM			85-94 AT EXAM			35-64 COMBINED			65-94 COMBINED		
	PERSON EXAMS	# OF EVENTS	ANNUAL RATE	PERSON EXAMS	# OF EVENTS	ANNUAL RATE	# OF EVENTS	CRUDE	AGE- ADJUSTED	# OF EVENTS	CRUDE	AGE- ADJUSTED
20-74	869	8	5	66			3	0	0	14	2	2
75-84	681	11	8	52			6	0	0	23	4	4
85-94	466	5	5	21			11	1	1	24	4	4
95-104	155	2	6	7			12	2	2	7	4	4
105-160	56	1	9	2			13	5	5	13	19	18
UNKNOWN	0	0	-	0			0	-	-	0	-	-
TOTAL	2227	27	6	148			45	1	1	81	4	4

LOGISTIC REGRESSIONS

AGE AT EXAM		COEFFICIENT OF RISK FACTOR	SE COEFFICIENT	P-VALUE	STANDARDIZED COEFFICIENT	STD SE
35-44						
45-54		0.054	0.015	0.001	0.672	0.182
55-64		0.064	0.010	0.001	0.805	0.130
65-74		0.055	0.009	0.001	0.674	0.112
75-84		0.007	0.016	0.670	0.081	0.191
85-94						
35-64	UNIVARIATE	0.064	0.008	0.001	0.784	0.100
65-94	UNIVARIATE	0.040	0.008	0.001	0.488	0.098
35-64	BIVARIATE	0.062	0.009	0.001	0.768	0.105
65-94	BIVARIATE	0.042	0.008	0.001	0.522	0.098
35-64	MULTIVARIATE	0.054	0.009	0.001	0.665	0.114
65-94	MULTIVARIATE	0.040	0.008	0.001	0.493	0.098

TABLE 14-3A   DATE=06NOV85
AVERAGE ANNUAL INCIDENCE RATE FOR EVENT PER 1000 PERSONS AT RISK AT EXAM
BY AGE AND LEVEL OF RISK FACTOR AT EXAM: FRAMINGHAM STUDY 30-YEAR FOLLOWUP

SEX: MALE
EVENT: ATHEROTHROMBOTIC BRAIN INFARCTION
RISK FACTOR: HYPERTENSION WITH ANTIHYPERTENSIVE TREATMENT
PERSONS AT RISK: PERSONS FREE OF STROKE AND TIA ONLY

RISK FACTOR	35-44 AT EXAM PERSON EXAMS	# OF EVENTS	ANNUAL RATE	45-54 AT EXAM PERSON EXAMS	# OF EVENTS	ANNUAL RATE	55-64 AT EXAM PERSON EXAMS	# OF EVENTS	ANNUAL RATE	65-74 AT EXAM PERSON EXAMS	# OF EVENTS	ANNUAL RATE
NO	2665			4248	4	0	4128	2	0	1764	8	2
MILD	1316			2280	4	1	2534	7	1	1474	13	4
DEFINITE	511			1212	6	2	1402	18	6	675	6	4
TREATED	24			292	2	3	850	8	5	713	14	10
UNKNOWN	0			0	0	-	0	0	-	0	0	-
TOTAL	4516			8032	16	1	8914	35	2	4626	41	4

RISK FACTOR	75-84 AT EXAM PERSON EXAMS	# OF EVENTS	ANNUAL RATE	85-94 AT EXAM PERSON EXAMS	# OF EVENTS	ANNUAL RATE	35-64 COMBINED # OF EVENTS	CRUDE	AGE-ADJUSTED	65-94 COMBINED # OF EVENTS	CRUDE	AGE-ADJUSTED
NO	440	6	7	15			6	0	0	14	3	3
MILD	395	5	6	14			11	1	1	18	5	5
DEFINITE	182	9	25	4			24	4	1	15	9	9
TREATED	217	6	14	9			10	4	3	20	11	11
UNKNOWN	0	0	-	0			0	-	-	0	-	-
TOTAL	1234	26	11	42			51	1	1	67	6	6

LOGISTIC REGRESSIONS

AGE AT EXAM	COEFFICIENT OF RISK FACTOR	SE COEFFICIENT	P-VALUE	STANDARDIZED COEFFICIENT	STD SE
35-44					
45-54	0.893	0.313	0.004	0.691	0.242
55-64	1.532	0.288	0.001	1.255	0.236
65-74	0.570	0.200	0.004	0.468	0.164
75-84	0.596	0.258	0.021	0.491	0.213
85-94					
35-64 UNIVARIATE	0.960	0.133	0.001	0.864	0.120
65-94 UNIVARIATE	0.423	0.108	0.001	0.453	0.116
35-64 BIVARIATE	0.792	0.132	0.001	0.712	0.119
65-94 BIVARIATE	0.414	0.108	0.001	0.444	0.116
35-64 MULTIVARIATE	0.753	0.139	0.001	0.678	0.125
65-94 MULTIVARIATE	0.370	0.111	0.001	0.397	0.119

SEX: FEMALE
EVENT: ATHEROTHROMBOTIC BRAIN INFARCTION
RISK FACTOR: HYPERTENSION WITH ANTIHYPERTENSIVE TREATMENT
PERSONS AT RISK: PERSONS FREE OF STROKE AND TIA ONLY

RISK FACTOR	35-44 AT EXAM PERSON EXAMS	# OF EVENTS	ANNUAL RATE	45-54 AT EXAM PERSON EXAMS	# OF EVENTS	ANNUAL RATE	55-64 AT EXAM PERSON EXAMS	# OF EVENTS	ANNUAL RATE	65-74 AT EXAM PERSON EXAMS	# OF EVENTS	ANNUAL RATE
NO	4247			5644	3	0	4955	2	0	1847	2	1
MILD	1001			2597	5	1	3409	9	1	2075	9	2
DEFINITE	316			1178	5	2	1924	12	3	1167	16	7
TREATED	49			614	1	1	1662	8	2	1782	26	7
UNKNOWN	0			0	0	-	0	0	-	0	0	-
TOTAL	5613			10033	14	1	11950	31	1	6871	53	4

RISK FACTOR	75-84 AT EXAM PERSON EXAMS	# OF EVENTS	ANNUAL RATE	85-94 AT EXAM PERSON EXAMS	# OF EVENTS	ANNUAL RATE	35-64 COMBINED # OF EVENTS	CRUDE	AGE-ADJUSTED	65-94 COMBINED # OF EVENTS	CRUDE	AGE-ADJUSTED
NO	450	5	3	38			5	0	0	5	1	1
MILD	654	8	6	32			14	1	1	18	3	3
DEFINITE	431	1	1	31			17	2	2	17	5	5
TREATED	692	15	11	47			9	2	1	41	8	8
UNKNOWN	0	0	-	0			0	-	-	0	-	-
TOTAL	2227	27	6	148			45	1	1	81	4	4

LOGISTIC REGRESSIONS

AGE AT EXAM	COEFFICIENT OF RISK FACTOR	SE COEFFICIENT	P-VALUE	STANDARDIZED COEFFICIENT	STD SE
35-44					
45-54	0.882	0.326	0.007	0.680	0.251
55-64	1.113	0.268	0.001	0.932	0.225
65-74	1.250	0.266	0.001	1.025	0.218
75-84	0.321	0.271	0.236	0.252	0.213
85-94					
35-64 UNIVARIATE	0.753	0.130	0.001	0.730	0.126
65-94 UNIVARIATE	0.569	0.109	0.001	0.649	0.124
35-64 BIVARIATE	0.624	0.136	0.001	0.604	0.132
65-94 BIVARIATE	0.553	0.109	0.001	0.631	0.125
35-64 MULTIVARIATE	0.518	0.142	0.001	0.502	0.138
65-94 MULTIVARIATE	0.488	0.112	0.001	0.556	0.127

TABLE 14-3B   DATE=06NOV85
AVERAGE ANNUAL INCIDENCE RATE FOR EVENT PER 1000 PERSONS AT RISK AT EXAM
BY AGE AND LEVEL OF RISK FACTOR AT EXAM: FRAMINGHAM STUDY 30-YEAR FOLLOWUP

SEX: MALE
EVENT: ATHEROTHROMBOTIC BRAIN INFARCTION
RISK FACTOR: HYPERTENSION IGNORING TREATMENT
PERSONS AT RISK: PERSONS FREE OF STROKE AND TIA ONLY

RISK FACTOR	35-44 AT EXAM			45-54 AT EXAM			55-64 AT EXAM			65-74 AT EXAM		
	PERSON EXAMS	# OF EVENTS	ANNUAL RATE	PERSON EXAMS	# OF EVENTS	ANNUAL RATE	PERSON EXAMS	# OF EVENTS	ANNUAL RATE	PERSON EXAMS	# OF EVENTS	ANNUAL RATE
NO	2669			4291	4	0	4304	3	0	1928		2
MILD	1326			2375	5	1	2863	10	2	1756	1	5
DEFINITE	521			1366	7	3	1747	22	6	942	15	8
UNKNOWN	0			0	0	-	0	0	-	0	0	-
TOTAL	4516			8032	16	1	8914	35	2	4626	41	4

RISK FACTOR	75-84 AT EXAM			85-94 AT EXAM			35-64 COMBINED			65-94 COMBINED		
	PERSON EXAMS	# OF EVENTS	ANNUAL RATE	PERSON EXAMS	# OF EVENTS	ANNUAL RATE	# OF EVENTS	CRUDE	AGE-ADJUSTED	# OF EVENTS	CRUDE	AGE-ADJUSTED
NO	490	7	7	20			7	0	0	15	3	3
MILD	503	5	5	18			15	1	1	23	5	5
DEFINITE	241	14	29	4			29	4	4	29	12	12
UNKNOWN	0	0	-	0			0	-	-	0	-	-
TOTAL	1234	26	11	42			51	1	1	67	6	6

LOGISTIC REGRESSIONS

AGE AT EXAM		COEFFICIENT OF RISK FACTOR	SE COEFFICIENT	P-VALUE	STANDARDIZED COEFFICIENT	STD SE
35-44						
45-54		0.858	0.312	0.006	0.649	0.236
55-64		1.398	0.254	0.001	1.080	0.197
65-74		0.650	0.205	0.002	0.493	0.156
75-84		0.855	0.273	0.002	0.635	0.203
85-94						
35-64	UNIVARIATE	1.274	0.194	0.001	0.960	0.146
65-94	UNIVARIATE	0.724	0.163	0.001	0.546	0.123
35-64	BIVARIATE	1.187	0.195	0.001	0.895	0.147
65-94	BIVARIATE	0.737	0.164	0.001	0.556	0.124
35-64	MULTIVARIATE	1.092	0.201	0.001	0.823	0.151
65-94	MULTIVARIATE	0.675	0.168	0.001	0.509	0.127

SEX: FEMALE
EVENT: ATHEROTHROMBOTIC BRAIN INFARCTION
RISK FACTOR: HYPERTENSION IGNORING TREATMENT
PERSONS AT RISK: PERSONS FREE OF STROKE AND TIA ONLY

RISK FACTOR	35-44 AT EXAM			45-54 AT EXAM			55-64 AT EXAM			65-74 AT EXAM		
	PERSON EXAMS	# OF EVENTS	ANNUAL RATE	PERSON EXAMS	# OF EVENTS	ANNUAL RATE	PERSON EXAMS	# OF EVENTS	ANNUAL RATE	PERSON EXAMS	# OF EVENTS	ANNUAL RATE
NO	4268			5829	3	0	5360	2	0	2168	3	1
MILD	1016			2787	5	1	4031	10	1	2816	17	3
DEFINITE	329			1417	6	2	2559	19	4	1887	33	9
UNKNOWN	0			0	0	-	0	0	-	0	0	-
TOTAL	5613			10033	14	1	11950	31	1	6871	53	4

RISK FACTOR	75-84 AT EXAM			85-94 AT EXAM			35-64 COMBINED			65-94 COMBINED		
	PERSON EXAMS	# OF EVENTS	ANNUAL RATE	PERSON EXAMS	# OF EVENTS	ANNUAL RATE	# OF EVENTS	CRUDE	AGE-ADJUSTED	# OF EVENTS	CRUDE	AGE-ADJUSTED
NO	575	4	3	48			5	0	0	7	1	1
MILD	932	10	5	51			15	1	1	28	4	4
DEFINITE	720	13	9	49			25	3	2	46	9	9
UNKNOWN	0	0	-	0			0	-	-	0	-	-
TOTAL	2227	27	6	148			45	1	1	81	4	4

LOGISTIC REGRESSIONS

AGE AT EXAM		COEFFICIENT OF RISK FACTOR	SE COEFFICIENT	P-VALUE	STANDARDIZED COEFFICIENT	STD SE
35-44						
45-54		1.033	0.335	0.002	0.751	0.244
55-64		1.340	0.273	0.001	1.045	0.213
65-74		1.183	0.222	0.001	0.907	0.170
75-84		0.493	0.271	0.069	0.375	0.206
85-94						
35-64	UNIVARIATE	1.356	0.206	0.001	1.008	0.153
65-94	UNIVARIATE	0.933	0.168	0.001	0.716	0.129
35-64	BIVARIATE	1.231	0.211	0.001	0.915	0.157
65-94	BIVARIATE	0.915	0.169	0.001	0.702	0.129
35-64	MULTIVARIATE	1.099	0.217	0.001	0.817	0.162
65-94	MULTIVARIATE	0.843	0.172	0.001	0.647	0.132

TABLE 14-4   DATE=06NOV85
AVERAGE ANNUAL INCIDENCE RATE FOR EVENT PER 1000 PERSONS AT RISK AT EXAM
BY AGE AND LEVEL OF RISK FACTOR AT EXAM: FRAMINGHAM STUDY 30-YEAR FOLLOWUP

SEX: MALE
EVENT: ATHEROTHROMBOTIC BRAIN INFARCTION
RISK FACTOR: SERUM CHOLESTEROL
PERSONS AT RISK: PERSONS FREE OF STROKE AND TIA ONLY

RISK FACTOR	35-44 AT EXAM PERSON EXAMS	# OF EVENTS	ANNUAL RATE	45-54 AT EXAM PERSON EXAMS	# OF EVENTS	ANNUAL RATE	55-64 AT EXAM PERSON EXAMS	# OF EVENTS	ANNUAL RATE	65-74 AT EXAM PERSON EXAMS	# OF EVENTS	ANNUAL RATE
84-204	1253			1796			2431	9	2	1471	13	4
205-234	1257			2382	3	1	2649	10	2	1389	16	6
235-264	882			1835	3	1	2099	6	1	1007	5	2
265-294	487			1050	6	3	1059	10	5	505	2	2
295-1124	303			618	1	1	544	0	0	237	5	11
UNKNOWN	334			351	0	0	132	0	0	17	0	0
TOTAL	4516			8032	16	1	8914	35	2	4626	41	4

RISK FACTOR	75-84 AT EXAM PERSON EXAMS	# OF EVENTS	ANNUAL RATE	85-94 AT EXAM PERSON EXAMS	# OF EVENTS	ANNUAL RATE	35-64 COMBINED # OF EVENTS	CRUDE	AGE-ADJUSTED	65-94 COMBINED # OF EVENTS	CRUDE	AGE-ADJUSTED
84-204	480	15	16	18			12	1	1	28	7	7
205-234	389	6	8	11			13	1	1	22	6	6
235-264	221	5	7	7			9	1	1	8	3	3
265-294	121	2	8	6			16	3	3	4	3	3
295-1124	23	0	0	0			1	0	0	5	10	8
UNKNOWN	0	0	-	0			0	0	0	0	0	0
TOTAL	1234	26	11	42			51	1	1	67	6	6

LOGISTIC REGRESSIONS

AGE AT EXAM		COEFFICIENT OF RISK FACTOR	SE COEFFICIENT	P-VALUE	STANDARDIZED COEFFICIENT	STD SE
35-44						
45-54		0.008	0.004	0.039	0.362	0.175
55-64		-0.000	0.004	0.973	-0.006	0.170
65-74		0.001	0.004	0.844	0.031	0.156
75-84		-0.012	0.005	0.023	-0.473	0.208
85-94						
35-64	UNIVARIATE	0.003	0.003	0.291	0.138	0.131
65-94	UNIVARIATE	-0.005	0.003	0.137	-0.188	0.127
35-64	BIVARIATE	0.004	0.003	0.212	0.167	0.134
65-94	BIVARIATE	-0.004	0.003	0.254	-0.145	0.127
35-64	MULTIVARIATE	0.003	0.003	0.423	0.111	0.138
65-94	MULTIVARIATE	-0.003	0.003	0.272	-0.140	0.128

SEX: FEMALE
EVENT: ATHEROTHROMBOTIC BRAIN INFARCTION
RISK FACTOR: SERUM CHOLESTEROL
PERSONS AT RISK: PERSONS FREE OF STROKE AND TIA ONLY

RISK FACTOR	35-44 AT EXAM PERSON EXAMS	# OF EVENTS	ANNUAL RATE	45-54 AT EXAM PERSON EXAMS	# OF EVENTS	ANNUAL RATE	55-64 AT EXAM PERSON EXAMS	# OF EVENTS	ANNUAL RATE	65-74 AT EXAM PERSON EXAMS	# OF EVENTS	ANNUAL RATE
84-204	2349			2113	2	1	1681	5	1	869	14	3
205-234	1415			2578	5	1	2704	6	1	1510	9	3
235-264	806			2262	1	0	3017	7	1	1867	13	3
265-294	372			1481	2	1	2339	5	1	1417	5	2
295-1124	171			1119	4	2	1976	8	2	1171	11	3
UNKNOWN	500			480	0	0	233	0	0	37	1	14
TOTAL	5613			10033	14	1	11950	31	1	6871	53	4

RISK FACTOR	75-84 AT EXAM PERSON EXAMS	# OF EVENTS	ANNUAL RATE	85-94 AT EXAM PERSON EXAMS	# OF EVENTS	ANNUAL RATE	35-64 COMBINED # OF EVENTS	CRUDE	AGE-ADJUSTED	65-94 COMBINED # OF EVENTS	CRUDE	AGE-ADJUSTED
84-204	407	9	11	53			7	1	1	23	9	9
205-234	537	6	6	35			11	1	1	15	4	4
235-264	619	4	5	41			8	1	1	20	4	3
265-294	382	4	5	9			7	1	1	9	2	3
295-1124	275	2	4	10			12	1	1	13	4	4
UNKNOWN	7	0	0	0			0	0	0	1	11	10
TOTAL	2227	27	6	148			45	1	1	81	4	4

LOGISTIC REGRESSIONS

AGE AT EXAM		COEFFICIENT OF RISK FACTOR	SE COEFFICIENT	P-VALUE	STANDARDIZED COEFFICIENT	STD SE
35-44						
45-54		0.004	0.004	0.265	0.195	0.173
55-64		0.004	0.004	0.316	0.171	0.171
65-74		-0.005	0.003	0.099	-0.239	0.145
75-84		-0.009	0.005	0.068	-0.373	0.204
85-94						
35-64	UNIVARIATE	0.006	0.002	0.005	0.268	0.096
65-94	UNIVARIATE	-0.006	0.003	0.015	-0.285	0.118
35-64	BIVARIATE	0.004	0.003	0.125	0.197	0.128
65-94	BIVARIATE	-0.006	0.003	0.053	-0.252	0.118
35-64	MULTIVARIATE	0.003	0.003	0.374	0.124	0.139
65-94	MULTIVARIATE	-0.006	0.003	0.025	-0.262	0.117

TABLE 14-5
DATE=06NOV85
AVERAGE ANNUAL INCIDENCE RATE FOR EVENT PER 1000 PERSONS AT RISK AT EXAM
BY AGE AND LEVEL OF RISK FACTOR AT EXAM: FRAMINGHAM STUDY 30-YEAR FOLLOWUP

SEX: MALE
EVENT: ATHEROTHROMBOTIC BRAIN INFARCTION
RISK FACTOR: HEMATOCRIT
PERSONS AT RISK: PERSONS FREE OF STROKE AND TIA ONLY

RISK FACTOR	35-44 AT EXAM PERSON EXAMS	# OF EVENTS	ANNUAL RATE	45-54 AT EXAM PERSON EXAMS	# OF EVENTS	ANNUAL RATE	55-64 AT EXAM PERSON EXAMS	# OF EVENTS	ANNUAL RATE	65-74 AT EXAM PERSON EXAMS	# OF EVENTS	ANNUAL RATE
25-42	630			961	2	1	1117	1	0	541	3	3
43-44	551			993	0	0	1082	3	1	582	6	5
45-46	990			1935	5	1	2192	7	2	1151	8	3
47-48	785			1573	2	1	1783	7	2	1033	11	5
49-70	1190			2180	7	2	2581	17	3	1299	13	5
UNKNOWN	370			390	0	0	159	0	0	20	0	0
TOTAL	4516			8032	16	1	8914	35	2	4626	41	4

RISK FACTOR	75-84 AT EXAM PERSON EXAMS	# OF EVENTS	ANNUAL RATE	85-94 AT EXAM PERSON EXAMS	# OF EVENTS	ANNUAL RATE	35-64 COMBINED # OF EVENTS	CRUDE	AGE-ADJUSTED	65-94 COMBINED # OF EVENTS	CRUDE	AGE-ADJUSTED
25-42	258	4	8	14			3	1	1	7	4	4
43-44	150	5	17	7			3	1	1	11	7	8
45-46	292	5	9	6			12	1	1	13	4	5
47-48	248	5	10	8			9	1	1	16	6	6
49-70	286	7	12	7			24	2	2	20	6	6
UNKNOWN	0	0		0			0	0	0	0	0	0
TOTAL	1234	26	11	42			51	1	1	67	6	6

LOGISTIC REGRESSIONS

AGE AT EXAM		COEFFICIENT OF RISK FACTOR	SE COEFFICIENT	P-VALUE	STANDARDIZED COEFFICIENT	STD SE
35-44						
45-54		0.092	0.069	0.180	0.331	0.247
55-64		0.124	0.044	0.005	0.468	0.166
65-74		0.031	0.044	0.480	0.112	0.158
75-84		0.003	0.049	0.945	0.014	0.199
85-94						
35-64	UNIVARIATE	0.118	0.038	0.002	0.434	0.138
65-94	UNIVARIATE	0.008	0.033	0.807	0.030	0.123
35-64	BIVARIATE	0.114	0.037	0.002	0.420	0.137
65-94	BIVARIATE	0.023	0.033	0.490	0.085	0.122
35-64	MULTIVARIATE	0.087	0.038	0.021	0.320	0.139
65-94	MULTIVARIATE	0.018	0.033	0.585	0.068	0.125

SEX: FEMALE
EVENT: ATHEROTHROMBOTIC BRAIN INFARCTION
RISK FACTOR: HEMATOCRIT
PERSONS AT RISK: PERSONS FREE OF STROKE AND TIA ONLY

RISK FACTOR	35-44 AT EXAM PERSON EXAMS	# OF EVENTS	ANNUAL RATE	45-54 AT EXAM PERSON EXAMS	# OF EVENTS	ANNUAL RATE	55-64 AT EXAM PERSON EXAMS	# OF EVENTS	ANNUAL RATE	65-74 AT EXAM PERSON EXAMS	# OF EVENTS	ANNUAL RATE
21-39	1697			1853	2	1	1585	5	2	719	6	4
40-41	966			1760	0	0	2127	4	1	1132	10	4
42-43	1041			1972	2	1	2333	6	1	1388	6	2
44-45	857			2206	5	1	2969	7	1	1782	15	4
46-65	486			1694	3	1	2646	8	2	1810	15	4
UNKNOWN	566			548	2	2	270	31	1	40	1	13
TOTAL	5613			10033	14	1	11950	31	1	6871	53	4

RISK FACTOR	75-84 AT EXAM PERSON EXAMS	# OF EVENTS	ANNUAL RATE	85-94 AT EXAM PERSON EXAMS	# OF EVENTS	ANNUAL RATE	35-64 COMBINED # OF EVENTS	CRUDE	AGE-ADJUSTED	65-94 COMBINED # OF EVENTS	CRUDE	AGE-ADJUSTED
21-39	324	5	8	28			7	1	1	11	5	5
40-41	427	5	6	36			4	0	0	16	5	5
42-43	418	4	4	26			8	1	1	10	3	3
44-45	567	4	4	29			12	1	1	19	4	4
46-65	485	9	9	29			11	1	1	24	5	5
UNKNOWN	6	0	0	0			3	1	1	1	11	9
TOTAL	2227	27	6	148			45	1	1	81	4	4

LOGISTIC REGRESSIONS

AGE AT EXAM		COEFFICIENT OF RISK FACTOR	SE COEFFICIENT	P-VALUE	STANDARDIZED COEFFICIENT	STD SE
35-44						
45-54		0.119	0.083	0.149	0.426	0.295
55-64		0.037	0.053	0.490	0.127	0.184
65-74		-0.018	0.039	0.652	-0.062	0.138
75-84		0.036	0.052	0.493	0.134	0.196
85-94						
35-64	UNIVARIATE	0.090	0.044	0.039	0.324	0.157
65-94	UNIVARIATE	-0.003	0.031	0.918	-0.012	0.112
35-64	BIVARIATE	0.059	0.045	0.184	0.214	0.161
65-94	BIVARIATE	0.003	0.031	0.932	0.010	0.112
35-64	MULTIVARIATE	0.033	0.044	0.456	0.117	0.157
65-94	MULTIVARIATE	-0.011	0.029	0.699	-0.041	0.106

TABLE 14-6   DATE=06NOV85
AVERAGE ANNUAL INCIDENCE RATE FOR EVENT PER 1000 PERSONS AT RISK AT EXAM
BY AGE AND LEVEL OF RISK FACTOR AT EXAM: FRAMINGHAM STUDY 30-YEAR FOLLOWUP

SEX: MALE
EVENT: ATHEROTHROMBOTIC BRAIN INFARCTION
RISK FACTOR: BLOOD GLUCOSE (MG/DL)
PERSONS AT RISK: PERSONS FREE OF STROKE AND TIA ONLY

RISK FACTOR	35-44 AT EXAM PERSON EXAMS	# OF EVENTS	ANNUAL RATE	45-54 AT EXAM PERSON EXAMS	# OF EVENTS	ANNUAL RATE	55-64 AT EXAM PERSON EXAMS	# OF EVENTS	ANNUAL RATE	65-74 AT EXAM PERSON EXAMS	# OF EVENTS	ANNUAL RATE
29-69	1004			1554	1	0	1179	4	2	484	2	2
70-89	2528			4326	10	1	4458	20	2	2050	18	4
90-109	628			1397	2	1	2105	7	2	1242	11	4
110-129	107			329	2	3	599	1	1	412	4	5
130-524	49			202	1	2	472	3	3	421	6	7
UNKNOWN	200			224	0	0	101	0	0	17	0	0
TOTAL	4516			8032	16	1	8914	35	2	4626	41	4

RISK FACTOR	75-84 AT EXAM PERSON EXAMS	# OF EVENTS	ANNUAL RATE	85-94 AT EXAM PERSON EXAMS	# OF EVENTS	ANNUAL RATE	35-64 COMBINED # OF EVENTS	CRUDE	AGE-ADJUSTED	65-94 COMBINED # OF EVENTS	CRUDE	AGE-ADJUSTED
29-69	82	1	6	1			5	1	1	3	3	3
70-89	449	11	12	16			30	1	1	29	6	6
90-109	394	9	11	10			9	1	1	20	6	6
110-129	151	3	10	5			3	1	1	7	6	6
130-524	158	2	6	10			4	3	2	8	7	7
UNKNOWN	0	0	-	0			0	0	0	0	0	0
TOTAL	1234	26	11	42			51	1	1	67	6	6

LOGISTIC REGRESSIONS

AGE AT EXAM		COEFFICIENT OF RISK FACTOR	SE COEFFICIENT	P-VALUE	STANDARDIZED COEFFICIENT	STD SE
35-44						
45-54		0.010	0.004	0.029	0.227	0.104
55-64		0.001	0.005	0.790	0.041	0.155
65-74		0.007	0.002	0.001	0.270	0.083
75-84		-0.002	0.006	0.732	-0.075	0.218
85-94						
35-64	UNIVARIATE	0.007	0.003	0.032	0.173	0.081
65-94	UNIVARIATE	0.006	0.002	0.007	0.209	0.078
35-64	BIVARIATE	0.004	0.004	0.283	0.101	0.094
65-94	BIVARIATE	0.006	0.002	0.015	0.202	0.083
35-64	MULTIVARIATE	0.003	0.004	0.487	0.070	0.100
65-94	MULTIVARIATE	0.005	0.002	0.026	0.189	0.085

SEX: FEMALE
EVENT: ATHEROTHROMBOTIC BRAIN INFARCTION
RISK FACTOR: BLOOD GLUCOSE (MG/DL)
PERSONS AT RISK: PERSONS FREE OF STROKE AND TIA ONLY

RISK FACTOR	35-44 AT EXAM PERSON EXAMS	# OF EVENTS	ANNUAL RATE	45-54 AT EXAM PERSON EXAMS	# OF EVENTS	ANNUAL RATE	55-64 AT EXAM PERSON EXAMS	# OF EVENTS	ANNUAL RATE	65-74 AT EXAM PERSON EXAMS	# OF EVENTS	ANNUAL RATE
29-69	1067			1803	2	1	1403	2	1	571	6	5
70-89	3408			5786	6	1	6298	15	1	3369	24	4
90-109	672			1640	3	1	2833	4	1	1933	8	2
110-129	96			302	1	2	762	5	3	563	4	4
130-524	39			166	0	0	489	4	4	405	10	12
UNKNOWN	531			336	2	3	165	1	3	30	1	17
TOTAL	5613			10033	14	1	11950	31	1	6871	53	4

RISK FACTOR	75-84 AT EXAM PERSON EXAMS	# OF EVENTS	ANNUAL RATE	85-94 AT EXAM PERSON EXAMS	# OF EVENTS	ANNUAL RATE	35-64 COMBINED # OF EVENTS	CRUDE	AGE-ADJUSTED	65-94 COMBINED # OF EVENTS	CRUDE	AGE-ADJUSTED
29-69	120	0	0	4			4	0	-1	6	4	4
70-89	891	11	6	55			21	1	1	35	4	4
90-109	744	7	5	58			7	1	1	15	3	3
110-129	263	5	10	17			6	3	2	9	5	5
130-524	208	4	10	14			4	3	2	15	---	12
UNKNOWN	1	0	0	0			3	2	2	16	12	12
TOTAL	2227	27	6	148			43	1	1	81	4	4

LOGISTIC REGRESSIONS

AGE AT EXAM		COEFFICIENT OF RISK FACTOR	SE COEFFICIENT	P-VALUE	STANDARDIZED COEFFICIENT	STD SE
35-44						
45-54		0.005	0.011	0.609	0.108	0.211
55-64		0.009	0.003	0.002	0.248	0.079
65-74		0.008	0.003	0.001	0.245	0.076
75-84		0.003	0.004	0.501	0.105	0.156
85-94						
35-64	UNIVARIATE	0.011	0.003	0.001	0.242	0.061
65-94	UNIVARIATE	0.007	0.002	0.001	0.218	0.067
35-64	BIVARIATE	0.009	0.003	0.003	0.198	0.066
65-94	BIVARIATE	0.007	0.002	0.003	0.205	0.069
35-64	MULTIVARIATE	0.006	0.003	0.063	0.128	0.069
65-94	MULTIVARIATE	0.005	0.002	0.047	0.143	0.072

TABLE 14-7   DATE=06NOV85

AVERAGE ANNUAL INCIDENCE RATE FOR EVENT PER 1000 PERSONS AT RISK AT EXAM
BY AGE AND LEVEL OF RISK FACTOR AT EXAM: FRAMINGHAM STUDY 30-YEAR FOLLOWUP

SEX: MALE
EVENT: ATHEROTHROMBOTIC BRAIN INFARCTION
RISK FACTOR: DIABETES MELLITUS
PERSONS AT RISK: PERSONS FREE OF STROKE AND TIA ONLY

RISK FACTOR	35-44 AT EXAM			45-54 AT EXAM			55-64 AT EXAM			65-74 AT EXAM		
	PERSON EXAMS	# OF EVENTS	ANNUAL RATE	PERSON EXAMS	# OF EVENTS	ANNUAL RATE	PERSON EXAMS	# OF EVENTS	ANNUAL RATE	PERSON EXAMS	# OF EVENTS	ANNUAL RATE
NO	4467			7780	14	1	8292	31	2	4101	32	4
YES	49			252	2	4	622	4	3	525	9	9
UNKNOWN	0			0	0	-	0	0	-	0	0	-
TOTAL	4516			8032	16	1	8914	35	2	4626	41	4

RISK FACTOR	75-84 AT EXAM			85-94 AT EXAM			35-64 COMBINED			65-94 COMBINED		
	PERSON EXAMS	# OF EVENTS	ANNUAL RATE	PERSON EXAMS	# OF EVENTS	ANNUAL RATE	# OF EVENTS	CRUDE	AGE-ADJUSTED	# OF EVENTS	CRUDE	AGE-ADJUSTED
NO	1076	22	10	36		-	45	1	1	54	5	5
YES	158	4	13	6			6	3	3	13	9	9
UNKNOWN	0	0	-	0			0	-		0	-	
TOTAL	1234	26	11	42			51	1	1	67	6	6

LOGISTIC REGRESSIONS

AGE AT EXAM		COEFFICIENT OF RISK FACTOR	SE COEFFICIENT	P-VALUE	STANDARDIZED COEFFICIENT	STD SE
35-44						
45-54		1.490	0.759	0.050	0.260	0.132
55-64		0.545	0.533	0.306	0.139	0.136
65-74		0.797	0.580	0.036	0.253	0.121
75-84		0.219	0.550	0.691	0.073	0.184
85-94						
35-64	UNIVARIATE	1.0 2	0.436	0.012	0.222	0.088
65-94	UNIVARIATE	0.6 8	0.312	0.051	0.195	0.100
35-64	BIVARIATE	0. 6	0.440	0.133	0.134	0.089
65-94	BIVARIATE	0.574	0.312	0.066	0.184	0.100
35-64	MULTIVARIATE	0.30	0.453	0.499	0.062	0.092
65-94	MULTIVARIATE	0.504	0.317	0.111	0.162	0.102

SEX: FEMALE
EVENT: ATHEROTHROMBOTIC BRAIN INFARCTION
RISK FACTOR: DIABETES MELLITUS
PERSONS AT RISK: PERSONS FREE OF STROKE AND TIA ONLY

RISK FACTOR	35-44 AT EXAM			45-54 AT EXAM			55-64 AT EXAM			65-74 AT EXAM		
	PERSON EXAMS	# OF EVENTS	ANNUAL RATE	PERSON EXAMS	# OF EVENTS	ANNUAL RATE	PERSON EXAMS	# OF EVENTS	ANNUAL RATE	PERSON EXAMS	# OF EVENTS	ANNUAL RATE
NO	5570			9820	14	1	11350	26	1	6289	38	3
YES	43			213	0	0	600	5	4	582	15	13
UNKNOWN	0			0	0	-	0	0	-	0	0	-
TOTAL	5613			10033	14	1	11950	31	1	6871	53	4

RISK FACTOR	75-84 AT EXAM			85-94 AT EXAM			35-64 COMBINED			65-94 COMBINED		
	PERSON EXAMS	# OF EVENTS	ANNUAL RATE	PERSON EXAMS	# OF EVENTS	ANNUAL RATE	# OF EVENTS	CRUDE	AGE-ADJUSTED	# OF EVENTS	CRUDE	AGE-ADJUSTED
NO	1942	20	5	127			40	1	1	58	3	3
YES	285	7	12	21			5	3	2	23	13	13
UNKNOWN	0	0	-	0			0	-	1	0	-	
TOTAL	2227	27	6	148			45	1	1	81	4	4

LOGISTIC REGRESSIONS

AGE AT EXAM		COEFFICIENT OF RISK FACTOR	SE COEFFICIENT	P-VALUE	STANDARDIZED COEFFICIENT	STD SE
35-44		-			-	
45-54		-			-	
55-64		1.297	0.490	0.008	0.283	0.107
65-74		1.471	0.308	0.001	0.410	0.086
75-84		0.884	0.444	0.046	0.295	0.148
85-94						
35-64	UNIVARIATE	1.367	0.476	0.004	0.237	0.082
65-94	UNIVARIATE	1.336	0.249	0.001	0.394	0.073
35-64	BIVARIATE	1.031	0.480	0.032	0.179	0.083
65-94	BIVARIATE	1.284	0.251	0.001	0.378	0.074
35-64	MULTIVARIATE	0.565	0.490	0.249	0.098	0.085
65-94	MULTIVARIATE	0.939	0.265	0.001	0.277	0.078

TABLE 14-8   DATE=06NOV85
AVERAGE ANNUAL INCIDENCE RATE FOR EVENT PER 1000 PERSONS AT RISK AT EXAM
BY AGE AND LEVEL OF RISK FACTOR AT EXAM: FRAMINGHAM STUDY 30-YEAR FOLLOWUP

SEX: MALE
EVENT: ATHEROTHROMBOTIC BRAIN INFARCTION
RISK FACTOR: GLUCOSE IN URINE, DEFINITE OR TRACE
PERSONS AT RISK: PERSONS FREE OF STROKE AND TIA ONLY

RISK FACTOR	35-44 AT EXAM PERSON EXAMS	# OF EVENTS	ANNUAL RATE	45-54 AT EXAM PERSON EXAMS	# OF EVENTS	ANNUAL RATE	55-64 AT EXAM PERSON EXAMS	# OF EVENTS	ANNUAL RATE	65-74 AT EXAM PERSON EXAMS	# OF EVENTS	ANNUAL RATE
NO	4423			7834	15	1	8639	34	2	4460	39	4
YES	60			177	1	3	251	1	2	160	2	6
UNKNOWN	33			21	0	0	24	0	0	6	0	0
TOTAL	4516			8032	16	1	8914	35	2	4626	41	4

RISK FACTOR	75-84 AT EXAM PERSON EXAMS	# OF EVENTS	ANNUAL RATE	85-94 AT EXAM PERSON EXAMS	# OF EVENTS	ANNUAL RATE	35-64 COMBINED # OF EVENTS	CRUDE	AGE-ADJUSTED	65-94 COMBINED # OF EVENTS	CRUDE	AGE-ADJUSTED
NO	1199	26	11	41	1		49	1	1	65	6	6
YES	32	0	0	1			2	2	2	2	5	5
UNKNOWN	3	0	0	0			0	0	0	0	0	0
TOTAL	1234	26	11	42			51	1	1	67	6	6

LOGISTIC REGRESSIONS

AGE AT EXAM	COEFFICIENT OF RISK FACTOR	SE COEFFICIENT	P-VALUE	STANDARDIZED COEFFICIENT	STD SE
35-44					
45-54	1.325	0.928	0.153	0.195	0.136
55-64	0.012	1.017	0.990	0.002	0.168
65-74	0.361	0.730	0.621	0.066	0.133
75-84	-				
85-94	-				
35-64 UNIVARIATE	0.560	0.723	0.438	0.084	0.108
65-94 UNIVARIATE	-0.097	0.722	0.893	-0.017	0.128
35-64 BIVARIATE	0.359	0.724	0.620	0.054	0.108
65-94 BIVARIATE	-0.038	0.723	0.958	-0.007	0.129
35-64 MULTIVARIATE	0.110	0.733	0.881	0.016	0.109
65-94 MULTIVARIATE	-0.128	0.725	0.859	-0.023	0.129

SEX: FEMALE
EVENT: ATHEROTHROMBOTIC BRAIN INFARCTION
RISK FACTOR: GLUCOSE IN URINE, DEFINITE OR TRACE
PERSONS AT RISK: PERSONS FREE OF STROKE AND TIA ONLY

RISK FACTOR	35-44 AT EXAM PERSON EXAMS	# OF EVENTS	ANNUAL RATE	45-54 AT EXAM PERSON EXAMS	# OF EVENTS	ANNUAL RATE	55-64 AT EXAM PERSON EXAMS	# OF EVENTS	ANNUAL RATE	65-74 AT EXAM PERSON EXAMS	# OF EVENTS	ANNUAL RATE
NO	5534			9908	13	1	11753	29	1	6733	49	4
YES	41			101	1	5	171	2	6	128	4	16
UNKNOWN	38			24	0	0	26	0	0	10	0	0
TOTAL	5613			10033	14	1	11950	31	1	6871	53	4

RISK FACTOR	75-84 AT EXAM PERSON EXAMS	# OF EVENTS	ANNUAL RATE	85-94 AT EXAM PERSON EXAMS	# OF EVENTS	ANNUAL RATE	35-64 COMBINED # OF EVENTS	CRUDE	AGE-ADJUSTED	65-94 COMBINED # OF EVENTS	CRUDE	AGE-ADJUSTED
NO	2183	26	6	145	3		42	1	1	76	4	4
YES	35	1	14	3			3	5	4	5	15	15
UNKNOWN	9	0	0	0			0	0	0	0	0	0
TOTAL	2227	27	6	148			45	1	1	81	4	4

LOGISTIC REGRESSIONS

AGE AT EXAM	COEFFICIENT OF RISK FACTOR	SE COEFFICIENT	P-VALUE	STANDARDIZED COEFFICIENT	STD SE
35-44					
45-54	2.030	1.043	0.052	0.203	0.104
55-64	1.565	0.735	0.033	0.186	0.087
65-74	1.482	0.528	0.005	0.200	0.071
75-84	0.892	1.034	0.388	0.111	0.129
85-94					
35-64 UNIVARIATE	1.834	0.600	0.002	0.194	0.064
65-94 UNIVARIATE	1.301	0.468	0.005	0.173	0.062
35-64 BIVARIATE	1.689	0.602	0.005	0.179	0.064
65-94 BIVARIATE	1.321	0.469	0.005	0.176	0.062
35-64 MULTIVARIATE	1.104	0.621	0.075	0.117	0.066
65-94 MULTIVARIATE	0.858	0.488	0.079	0.114	0.065

TABLE 14-9   DATE=04MAR86
AVERAGE ANNUAL INCIDENCE RATE FOR EVENT PER 1000 PERSONS AT RISK AT EXAM
BY AGE AND LEVEL OF RISK FACTOR AT EXAM: FRAMINGHAM STUDY 30-YEAR FOLLOWUP

SEX: MALE
EVENT: ATHEROTHROMBOTIC BRAIN INFARCTION
RISK FACTOR: GLUCOSE INTOLERANCE
PERSONS AT RISK: PERSONS FREE OF STROKE AND TIA ONLY

RISK FACTOR	35-44 AT EXAM			45-54 AT EXAM			55-64 AT EXAM			65-74 AT EXAM		
	PERSON EXAMS	# OF EVENTS	ANNUAL RATE	PERSON EXAMS	# OF EVENTS	ANNUAL RATE	PERSON EXAMS	# OF EVENTS	ANNUAL RATE	PERSON EXAMS	# 05 EVENTS	ANNUAL RATE
NO	4378			7497	13	1	7847	30	2	3766	29	4
YES	138			535	3	3	1067	5	2	860	12	7
UNKNOWN	0	0	-	0	0	-	0	0	-	0	0	-
TOTAL	4516			8032	16	1	8914	35	2	4626	41	4

RISK FACTOR	75-84 AT EXAM			85-94 AT EXAM			35-64 COMBINED			65-94 COMBINED		
	PERSON EXAMS	# OF EVENTS	ANNUAL RATE	PERSON EXAMS	# OF EVENTS	ANNUAL RATE	# OF EVENTS	CRUDE	AGE-ADJUSTED	# OF EVENTS	CRUDE	AGE-ADJUSTED
NO	952	20	11	29			43	1	1	49	5	5
YES	282	6	11	13			8	2	2	18	8	8
UNKNOWN	0	0	-	0			0	-	-	0	-	-
TOTAL	1234	26	11	42			51	1	1	67	6	6

LOGISTIC REGRESSIONS

AGE AT EXAM		COEFFICIENT OF RISK FACTOR	SE COEFFICIENT	P-VALUE	STANDARDIZED COEFFICIENT	STD SE
35-44						
45-54		1.178	0.642	0.067	0.294	0.160
55-64		0.204	0.484	0.673	0.066	0.157
65-74		0.601	0.345	0.082	0.234	0.134
75-84		0.013	0.470	0.978	0.005	0.198
85-94						
35-64	UNIVARIATE	0.749	0.386	0.052	0.204	0.105
65-94	UNIVARIATE	0.417	0.278	0.133	0.166	0.110
35-64	BIVARIATE	0.398	0.388	0.306	0.109	0.106
65-94	BIVARIATE	0.354	0.279	0.204	0.140	0.111
35-64	MULTIVARIATE	0.180	0.395	0.649	0.049	0.108
65-94	MULTIVARIATE	0.301	0.280	0.282	0.120	0.111

SEX: FEMALE
EVENT: ATHEROTHROMBOTIC BRAIN INFARCTION
RISK FACTOR: GLUCOSE INTOLERANCE
PERSONS AT RISK: PERSONS FREE OF STROKE AND TIA ONLY

RISK FACTOR	35-44 AT EXAM			45-54 AT EXAM			55-64 AT EXAM			65-74 AT EXAM		
	PERSON EXAMS	# OF EVENTS	ANNUAL RATE	PERSON EXAMS	# OF EVENTS	ANNUAL RATE	PERSON EXAMS	# OF EVENTS	ANNUAL RATE	PERSON EXAMS	# OF EVENTS	ANNUAL RATE
NO	5519			9648	13	1	10848	26	1	5877	34	5
YES	94			385	1	1	1102	5	2	994	19	10
UNKNOWN	0			0	0	-	0	0	-	0	0	-
TOTAL	5613			10033	14	1	11950	31	1	6871	53	4

RISK FACTOR	75-84 AT EXAM			85-94 AT EXAM			35-64 COMBINED			65-94 COMBINED		
	PERSON EXAMS	# OF EVENTS	ANNUAL RATE	PERSON EXAMS	# OF EVENTS	ANNUAL RATE	# OF EVENTS	CRUDE	AGE-ADJUSTED	# OF EVENTS	CRUDE	AGE-ADJUSTED
NO	1752	18	5	120			39	1	1	52	3	3
YES	475	9	9	28			6	2	1	29	10	10
UNKNOWN	0	0	-	0			0	-	-	0	-	-
TOTAL	2227	27	6	148			45	1	1	81	4	4

LOGISTIC REGRESSIONS

AGE AT EXAM		COEFFICIENT OF RISK FACTOR	SE COEFFICIENT	P-VALUE	STANDARDIZED COEFFICIENT	STD SE
35-44						
45-54		0.658	1.039	0.527	0.126	0.200
55-64		0.640	0.489	0.191	0.185	0.142
65-74		1.209	0.289	0.001	0.425	0.101
75-84		0.621	0.412	0.131	0.254	0.169
85-94						
35-64	UNIVARIATE	0.931	0.439	0.034	0.216	0.102
65-94	UNIVARIATE	1.073	0.234	0.001	0.395	0.086
35-64	BIVARIATE	0.597	0.444	0.179	0.139	0.103
65-94	BIVARIATE	1.022	0.235	0.001	0.377	0.087
35-64	MULTIVARIATE	0.322	0.451	0.475	0.075	0.105
65-94	MULTIVARIATE	0.802	0.244	0.001	0.295	0.090

TABLE 14-10   DATE=06NOV85

AVERAGE ANNUAL INCIDENCE RATE FOR EVENT PER 1000 PERSONS AT RISK AT EXAM
BY AGE AND LEVEL OF RISK FACTOR AT EXAM: FRAMINGHAM STUDY 30-YEAR FOLLOWUP

SEX: MALE
EVENT: ATHEROTHROMBOTIC BRAIN INFARCTION
RISK FACTOR: METROPOLITAN RELATIVE WEIGHT (PERCENT)
PERSONS AT RISK: PERSONS FREE OF STROKE AND TIA ONLY

RISK FACTOR	35-44 AT EXAM PERSON EXAMS	# OF EVENTS	ANNUAL RATE	45-54 AT EXAM PERSON EXAMS	# OF EVENTS	ANNUAL RATE	55-64 AT EXAM PERSON EXAMS	# OF EVENTS	ANNUAL RATE	65-74 AT EXAM PERSON EXAMS	# OF EVENTS	ANNUAL RATE
54-104	720			1145	2	1	1251	3	1	795	10	6
105-114	979			1631	2	1	1830	6	2	998	7	4
115-124	1133			2131	4	1	2428	8	2	1272	8	3
125-134	923			1681	4	1	1819	10	3	918	10	5
135-272	761			1443	4	1	1585	8	3	643	6	5
UNKNOWN	0			1	0	0	1	0	0	0	0	-
TOTAL	4516			8032	16	1	8914	35	2	4626	41	4

RISK FACTOR	75-84 AT EXAM PERSON EXAMS	# OF EVENTS	ANNUAL RATE	85-94 AT EXAM PERSON EXAMS	# OF EVENTS	ANNUAL RATE	35-64 COMBINED # OF EVENTS	CRUDE	AGE-ADJUSTED	65-94 COMBINED # OF EVENTS	CRUDE	AGE-ADJUSTED
54-104	247	5	10	15			5	1	1	15	7	7
105-114	323	6	9	12			8	1	1	13	5	5
115-124	377	9	12	13			12	1	1	17	5	5
125-134	184	2	5	2			14	2	2	12	5	5
135-272	103	4	19	0			12	2	2	10	7	8
UNKNOWN	0	0	-	0			0	0	0	0	0	-
TOTAL	1234	26	11	42			51	1	1	67	6	6

LOGISTIC REGRESSIONS

AGE AT EXAM		COEFFICIENT OF RISK FACTOR	SE COEFFICIENT	P-VALUE	STANDARDIZED COEFFICIENT	STD SE
35-44						
45-54		0.007	0.015	0.640	0.114	0.243
55-64		0.020	0.010	0.045	0.312	0.156
65-74		-0.007	0.010	0.511	-0.104	0.159
75-84		0.005	0.013	0.700	0.075	0.194
85-94						
35-64	UNIVARIATE	0.016	0.008	0.048	0.259	0.131
65-94	UNIVARIATE	-0.004	0.008	0.605	-0.064	0.124
35-64	BIVARIATE	0.016	0.008	0.047	0.261	0.132
65-94	BIVARIATE	-0.001	0.008	0.877	-0.019	0.126
35-64	MULTIVARIATE	0.010	0.008	0.247	0.158	0.136
65-94	MULTIVARIATE	-0.008	0.008	0.314	-0.128	0.128

SEX: FEMALE
EVENT: ATHEROTHROMBOTIC BRAIN INFARCTION
RISK FACTOR: METROPOLITAN RELATIVE WEIGHT (PERCENT)
PERSONS AT RISK: PERSONS FREE OF STROKE AND TIA ONLY

RISK FACTOR	35-44 AT EXAM PERSON EXAMS	# OF EVENTS	ANNUAL RATE	45-54 AT EXAM PERSON EXAMS	# OF EVENTS	ANNUAL RATE	55-64 AT EXAM PERSON EXAMS	# OF EVENTS	ANNUAL RATE	65-74 AT EXAM PERSON EXAMS	# OF EVENTS	ANNUAL RATE
54-104	1689			2108	2	0	2108	2	0	1275	4	2
105-114	1431			2351	1	0	2509	8	2	1289	7	3
115-124	1059			2123	4	1	2535	2	0	1485	12	4
125-134	565			1378	0	0	1813	5	1	1119	11	5
135-272	861			2073	7	2	2985	14	2	1703	19	6
UNKNOWN	8			0	0	-	0	0	-	0	0	-
TOTAL	5613			10033	14	1	11950	31	1	6871	53	4

RISK FACTOR	75-84 AT EXAM PERSON EXAMS	# OF EVENTS	ANNUAL RATE	85-94 AT EXAM PERSON EXAMS	# OF EVENTS	ANNUAL RATE	35-64 COMBINED # OF EVENTS	CRUDE	AGE-ADJUSTED	65-94 COMBINED # OF EVENTS	CRUDE	AGE-ADJUSTED
54-104	592	6	5	47			4	0	0	10	3	2
105-114	405	7	9	31			9	1	1	14	4	4
115-124	448	5	6	28			6	1	1	18	5	5
125-134	325	3	5	24			5	1	1	14	5	5
135-272	457	6	7	18			21	2	2	25	6	6
UNKNOWN	0	0	-	0			0	0	0	0	0	-
TOTAL	2227	27	6	148			45	1	1	81	4	4

LOGISTIC REGRESSIONS

AGE AT EXAM		COEFFICIENT OF RISK FACTOR	SE COEFFICIENT	P-VALUE	STANDARDIZED COEFFICIENT	STD SE
35-44						
45-54		0.024	0.008	0.002	0.527	0.167
55-64		0.016	0.006	0.009	0.363	0.139
65-74		0.011	0.005	0.034	0.250	0.118
75-84		0.006	0.009	0.476	0.133	0.186
85-94						
35-64	UNIVARIATE	0.021	0.005	0.001	0.461	0.103
65-94	UNIVARIATE	0.009	0.005	0.044	0.201	0.100
35-64	BIVARIATE	0.019	0.005	0.001	0.422	0.107
65-94	BIVARIATE	0.010	0.005	0.022	0.231	0.100
35-64	MULTIVARIATE	0.008	0.006	0.176	0.166	0.122
65-94	MULTIVARIATE	0.005	0.004	0.219	0.117	0.095

TABLE 14-11   DATE=06NOV85

AVERAGE ANNUAL INCIDENCE RATE FOR EVENT PER 1000 PERSONS AT RISK AT EXAM
BY AGE AND LEVEL OF RISK FACTOR AT EXAM: FRAMINGHAM STUDY 30-YEAR FOLLOWUP

SEX: MALE
EVENT: ATHEROTHROMBOTIC BRAIN INFARCTION
RISK FACTOR: VITAL CAPACITY - HEIGHT INDEX (ML/INCH)
PERSONS AT RISK: PERSONS FREE OF STROKE AND TIA ONLY

35-44 AT EXAM			45-54 AT EXAM			55-64 AT EXAM			65-74 AT EXAM		
PERSON EXAMS	# OF EVENTS	ANNUAL RATE	PERSON EXAMS	# OF EVENTS	ANNUAL RATE	PERSON EXAMS	# OF EVENTS	ANNUAL RATE	PERSON EXAMS	# OF EVENTS	ANNUAL RATE
268			1100	3	1	2319	13	3	2063	26	6
416			1182	5	2	1667	7	2	950	4	2
751			1721	5	1	2024	9	2	824	4	2
1034			1809	5	1	1540	2	1	509	5	5
2034			2214	0	0	1346	4	1	268	2	4
13			6	0	0	18	0	0	12	0	0
4516			8032	16	1	8914	35	2	4626	41	4

75-84 AT EXAM			85-94 AT EXAM			55-64 COMBINED			65-94 COMBINED		
PERSON EXAMS	# OF EVENTS	ANNUAL RATE	PERSON EXAMS	# OF EVENTS	ANNUAL RATE	# OF EVENTS	CRUDE	AGE-ADJUSTED	# OF EVENTS	CRUDE	AGE-ADJUSTED
778	17	11	28			16	2	2	43	7	7
207	4	10	5			12	2	2	8	3	4
130	4	15	7			12	1	1	8	4	5
79	0	0	2			7	1	1	5	4	5
32	1	16	0			4	0	1	3	5	6
8	0	0	0			0	0	0	0	0	0
1234	26	11	42			51	1	1	67	6	6

LOGISTIC REGRESSIONS

	COEFFICIENT OF RISK FACTOR	SE COEFFICIENT	P-VALUE	STANDARDIZED COEFFICIENT	STD SE
	-0.043	0.026	0.101	-0.390	0.237
	-0.030	0.017	0.083	-0.283	0.163
	-0.024	0.017	0.142	-0.225	0.154
	-0.012	0.022	0.586	-0.108	0.199
NIVARIATE	-0.050	0.014	0.001	-0.484	0.131
NIVARIATE	-0.027	0.013	0.039	-0.249	0.121
IVARIATE	-0.027	0.015	0.067	-0.261	0.142
IVARIATE	-0.018	0.013	0.163	-0.174	0.125
ULTIVARIATE	-0.008	0.016	0.601	-0.079	0.152
ULTIVARIATE	-0.012	0.014	0.356	-0.117	0.127

SEX: FEMALE
EVENT: ATHEROTHROMBOTIC BRAIN INFARCTION
RISK FACTOR: VITAL CAPACITY - HEIGHT INDEX (ML/INCH)
PERSONS AT RISK: PERSONS FREE OF STROKE AND TIA ONLY

35-44 AT EXAM			45-54 AT EXAM			55-64 AT EXAM			65-74 AT EXAM		
PERSON EXAMS	# OF EVENTS	ANNUAL RATE	PERSON EXAMS	# OF EVENTS	ANNUAL RATE	PERSON EXAMS	# OF EVENTS	ANNUAL RATE	PERSON EXAMS	# OF EVENTS	ANNUAL RATE
121			539	3	3	1530	9	3	1948	18	5
291			1240	0	0	2520	9	2	1888	16	4
769			2096	5	1	3181	4	1	1694	14	4
1427			2825	3	1	2809	4	1	919	3	2
2984			3316	3	0	1889	5	1	416	2	2
21			17	0	0	21	0	0	6	0	0
5613			10033	14	1	11950	31	1	6871	53	4

75-84 AT EXAM			85-94 AT EXAM			55-64 COMBINED			65-94 COMBINED		
PERSON EXAMS	# OF EVENTS	ANNUAL RATE	PERSON EXAMS	# OF EVENTS	ANNUAL RATE	# OF EVENTS	CRUDE	AGE-ADJUSTED	# OF EVENTS	CRUDE	AGE-ADJUSTED
1090	11	5	100			12	3	2	30	5	5
645	14	11	33			9	1	1	30	6	6
351	2	3	14			9	1	1	16	4	4
113	0	0	1			7	0	1	3	1	1
26	0	0	0			8	0	0	2	2	2
2	0	0	0			0	0	0	0	0	0
2227	27	6	148			45	1	1	81	4	4

LOGISTIC REGRESSIONS

	COEFFICIENT OF RISK FACTOR	SE COEFFICIENT	P-VALUE	STANDARDIZED COEFFICIENT	STD SE
	-0.053	0.035	0.124	-0.393	0.256
	-0.066	0.022	0.003	-0.492	0.168
	-0.031	0.018	0.099	-0.223	0.135
	-0.036	0.027	0.178	-0.253	0.188
NIVARIATE	-0.079	0.017	0.001	-0.631	0.138
NIVARIATE	-0.036	0.015	0.013	-0.271	0.109
IVARIATE	-0.060	0.019	0.002	-0.476	0.152
IVARIATE	-0.029	0.015	0.059	-0.217	0.115
ULTIVARIATE	-0.031	0.021	0.132	-0.247	0.164
ULTIVARIATE	-0.006	0.017	0.717	-0.045	0.124

TABLE 14-12   DATE=06NOV85
AVERAGE ANNUAL INCIDENCE RATE FOR EVENT PER 1000 PERSONS AT RISK AT EXAM
BY AGE AND LEVEL OF RISK FACTOR AT EXAM: FRAMINGHAM STUDY 30-YEAR FOLLOWUP

SEX: MALE
EVENT: ATHEROTHROMBOTIC BRAIN INFARCTION
RISK FACTOR: HEART RATE (PER MINUTE)
PERSONS AT RISK: PERSONS FREE OF STROKE AND TIA ONLY

RISK FACTOR	35-44 AT EXAM PERSON EXAMS	# OF EVENTS	ANNUAL RATE	45-54 AT EXAM PERSON EXAMS	# OF EVENTS	ANNUAL RATE	55-64 AT EXAM PERSON EXAMS	# OF EVENTS	ANNUAL RATE	65-74 AT EXAM PERSON EXAMS	# OF EVENTS	ANNUAL RATE
30-67	1295			2013	2	0	2590	5	1	1365	14	5
68-75	1503			2579	4	1	2709	7	1	1403	8	3
76-83	728			1465	4	1	1520	10	3	767	7	5
84-91	586			1139	3	1	1188	6	3	570	7	6
92-220	404			836	3	2	907	7	4	521	5	5
UNKNOWN	0			0	0	-	0	0	-	0	0	-
TOTAL	4516			8032	16	1	8914	35	2	4626	41	4

RISK FACTOR	75-84 AT EXAM PERSON EXAMS	# OF EVENTS	ANNUAL RATE	85-94 AT EXAM PERSON EXAMS	# OF EVENTS	ANNUAL RATE	35-64 COMBINED # OF EVENTS	CRUDE	AGE-ADJUSTED	65-94 COMBINED # OF EVENTS	CRUDE	AGE-ADJUSTED
30-67	353	6	8	13			7	1	1	20	6	6
68-75	339	4	6	13			11	1	1	12	3	3
76-83	246	6	12	11			14	2	2	13	5	5
84-91	163	3	9	4			9	2	2	10	7	7
92-220	133	7	26	1			10	2	2	12	9	9
UNKNOWN	0	0	-	0			0	-	-	0	-	-
TOTAL	1234	26	11	42			51	1	1	67	6	6

LOGISTIC REGRESSIONS

AGE AT EXAM		COEFFICIENT OF RISK FACTOR	SE COEFFICIENT	P-VALUE	STANDARDIZED COEFFICIENT	STD SE
35-44						
45-54		0.032	0.018	0.069	0.407	0.223
55-64		0.033	0.011	0.002	0.434	0.143
65-74		0.007	0.011	0.532	0.095	0.152
75-84		0.022	0.014	0.109	0.290	0.181
85-94						
35-64	UNIVARIATE	0.034	0.009	0.001	0.432	0.120
65-94	UNIVARIATE	0.013	0.009	0.126	0.177	0.115
35-64	BIVARIATE	0.033	0.009	0.001	0.421	0.117
65-94	BIVARIATE	0.013	0.009	0.136	0.173	0.116
35-64	MULTIVARIATE	0.019	0.010	0.057	0.237	0.124
65-94	MULTIVARIATE	0.010	0.009	0.272	0.129	0.118

SEX: FEMALE
EVENT: ATHEROTHROMBOTIC BRAIN INFARCTION
RISK FACTOR: HEART RATE (PER MINUTE)
PERSONS AT RISK: PERSONS FREE OF STROKE AND TIA ONLY

RISK FACTOR	35-44 AT EXAM PERSON EXAMS	# OF EVENTS	ANNUAL RATE	45-54 AT EXAM PERSON EXAMS	# OF EVENTS	ANNUAL RATE	55-64 AT EXAM PERSON EXAMS	# OF EVENTS	ANNUAL RATE	65-74 AT EXAM PERSON EXAMS	# OF EVENTS	ANNUAL RATE
30-67	1090			1908	5	1	2402	4	1	1242	8	3
68-75	1804			3269	3	0	3602	6	1	1853	16	4
76-83	1050			1930	3	1	2412	8	2	1498	14	5
84-91	939			1612	1	0	1867	6	2	1154	8	3
92-220	730			1314	2	1	1667	7	2	1124	7	3
UNKNOWN	0			0	0	-	0	0	-	0	0	-
TOTAL	5613			10033	14	1	11950	31	1	6871	53	4

RISK FACTOR	75-84 AT EXAM PERSON EXAMS	# OF EVENTS	ANNUAL RATE	85-94 AT EXAM PERSON EXAMS	# OF EVENTS	ANNUAL RATE	35-64 COMBINED # OF EVENTS	CRUDE	AGE-ADJUSTED	65-94 COMBINED # OF EVENTS	CRUDE	AGE-ADJUSTED
30-67	429	7	8	50			9	1	1	15	4	4
68-75	535	5	5	38			9	1	1	22	5	5
76-83	484	6	6	27			11	1	1	20	5	5
84-91	375	6	8	29			7	1	1	14	4	4
92-220	404	3	4	24			9	1	1	10	3	3
UNKNOWN	0	0	-	0			0	-	-	0	-	-
TOTAL	2227	27	6	148			45	1	1	81	4	4

LOGISTIC REGRESSIONS

AGE AT EXAM		COEFFICIENT OF RISK FACTOR	SE COEFFICIENT	P-VALUE	STANDARDIZED COEFFICIENT	STD SE
35-44						
45-54		-0.016	0.023	0.485	-0.199	0.285
55-64		0.022	0.012	0.073	0.278	0.155
65-74		0.001	0.010	0.902	0.017	0.137
75-84		-0.018	0.015	0.231	-0.251	0.210
85-94						
35-64	UNIVARIATE	0.012	0.011	0.259	0.157	0.139
65-94	UNIVARIATE	-0.006	0.009	0.507	-0.076	0.115
35-64	BIVARIATE	0.012	0.011	0.274	0.150	0.137
65-94	BIVARIATE	-0.006	0.009	0.475	-0.082	0.114
35-64	MULTIVARIATE	-0.001	0.012	0.948	-0.010	0.148
65-94	MULTIVARIATE	-0.011	0.008	0.209	-0.143	0.113

TABLE 14-13   DATE=06NOV85

AVERAGE ANNUAL INCIDENCE RATE FOR EVENT PER 1000 PERSONS AT RISK AT EXAM
BY AGE AND LEVEL OF RISK FACTOR AT EXAM: FRAMINGHAM STUDY 30-YEAR FOLLOWUP

SEX: MALE
EVENT: ATHEROTHROMBOTIC BRAIN INFARCTION
RISK FACTOR: CIGARETTES SMOKED (NUMBER PER DAY)
PERSONS AT RISK: PERSONS FREE OF STROKE AND TIA ONLY

RISK FACTOR	35-44 AT EXAM			45-54 AT EXAM			55-64 AT EXAM			65-74 AT EXAM		
	PERSON EXAMS	# OF EVENTS	ANNUAL RATE	PERSON EXAMS	# OF EVENTS	ANNUAL RATE	PERSON EXAMS	# OF EVENTS	ANNUAL RATE	PERSON EXAMS	# OF EVENTS	ANNUAL RATE
NONE	1402			3211	2	0	4815	14	1	3084	28	5
1-10	437			742	3	2	842	2	1	449	3	3
11-20	1497			2127	5	1	1848	16	4	633	7	6
21-40	993			1664	3	1	1217	3	1	421	3	4
41-90	161			250	3	6	168	0	0	32	0	0
UNKNOWN	26			38	0	0	24	0	0	7	0	0
TOTAL	4516			8032	16	1	8914	35	2	4626	41	4

RISK FACTOR	75-84 AT EXAM			85-94 AT EXAM			35-64 COMBINED			65-94 COMBINED		
	PERSON EXAMS	# OF EVENTS	ANNUAL RATE	PERSON EXAMS	# OF EVENTS	ANNUAL RATE	# OF EVENTS	CRUDE	AGE-ADJUSTED	# OF EVENTS	CRUDE	AGE-ADJUSTED
NONE	963	19	10	32			16	1	1	47	6	6
1-10	122	2	8	4			5	1	1	5	4	4
11-20	86	2	12	5			21	2	2	9	4	7
21-40	58	3	26	1			6	1	1	6	6	8
41-90	5	0	0	0			3	3	2	0	0	0
UNKNOWN	0	0	-	0			0	0	0	0	0	0
TOTAL	1234	26	11	42			51	1	1	67	6	6

LOGISTIC REGRESSIONS

AGE AT EXAM		COEFFICIENT OF RISK FACTOR	SE COEFFICIENT	P-VALUE	STANDARDIZED COEFFICIENT	STD SE
35-44						
45-54		0.040	0.014	0.004	0.595	0.205
55-64		0.010	0.011	0.400	0.132	0.156
65-74		-0.007	0.015	0.612	-0.086	0.169
75-84		0.019	0.018	0.295	0.166	0.159
85-94						
35-64	UNIVARIATE	0.014	0.009	0.105	0.208	0.129
65-94	UNIVARIATE	-0.003	0.011	0.798	-0.032	0.127
35-64	BIVARIATE	0.024	0.009	0.005	0.352	0.124
65-94	BIVARIATE	0.002	0.011	0.836	0.026	0.127
35-64	MULTIVARIATE	0.029	0.009	0.001	0.422	0.126
65-94	MULTIVARIATE	0.005	0.011	0.652	0.058	0.128

SEX: FEMALE
EVENT: ATHEROTHROMBOTIC BRAIN INFARCTION
RISK FACTOR: CIGARETTES SMOKED (NUMBER PER DAY)
PERSONS AT RISK: PERSONS FREE OF STROKE AND TIA ONLY

RISK FACTOR	35-44 AT EXAM			45-54 AT EXAM			55-64 AT EXAM			65-74 AT EXAM		
	PERSON EXAMS	# OF EVENTS	ANNUAL RATE	PERSON EXAMS	# OF EVENTS	ANNUAL RATE	PERSON EXAMS	# OF EVENTS	ANNUAL RATE	PERSON EXAMS	# OF EVENTS	ANNUAL RATE
NONE	2580			5582	9	1	7896	24	2	5364	40	4
1-10	1312			1723	2	1	1634	0	0	674	6	4
11-20	1372			2027	3	1	1743	5	1	627	6	3
21-40	298			626	0	0	621	2	2	186	3	8
41-90	22			38	0	0	18	0	0	1	0	0
UNKNOWN	29			37	0	0	38	0	0	19	0	0
TOTAL	5613			10033	14	1	11950	31	1	6871	53	4

RISK FACTOR	75-84 AT EXAM			85-94 AT EXAM			35-64 COMBINED			65-94 COMBINED		
	PERSON EXAMS	# OF EVENTS	ANNUAL RATE	PERSON EXAMS	# OF EVENTS	ANNUAL RATE	# OF EVENTS	CRUDE	AGE-ADJUSTED	# OF EVENTS	CRUDE	AGE-ADJUSTED
NONE	1959	20	5	139			33	1	1	61	4	4
1-10	158	3	9	7			2	0	0	9	5	6
11-20	93	4	22	2			8	1	1	6	6	6
21-40	13	0	0	0			2	1	1	3	6	6
41-90	0	0	-	0			0	0	0	0	0	0
UNKNOWN	4	0	0	0			0	0	0	0	0	0
TOTAL	2227	27	6	148			45	1	1	81	4	4

LOGISTIC REGRESSIONS

AGE AT EXAM		COEFFICIENT OF RISK FACTOR	SE COEFFICIENT	P-VALUE	STANDARDIZED COEFFICIENT	STD SE
35-44						
45-54		-0.019	0.030	0.528	-0.195	0.309
55-64		0.001	0.019	0.974	0.006	0.179
65-74		0.015	0.016	0.334	0.115	0.119
75-84		0.054	0.024	0.022	0.261	0.114
85-94						
35-64	UNIVARIATE	-0.012	0.017	0.461	-0.121	0.164
65-94	UNIVARIATE	0.020	0.013	0.114	0.144	0.091
35-64	BIVARIATE	-0.004	0.016	0.805	-0.039	0.159
65-94	BIVARIATE	0.026	0.013	0.044	0.185	0.092
35-64	MULTIVARIATE	0.002	0.016	0.904	0.019	0.155
65-94	MULTIVARIATE	0.033	0.013	0.012	0.235	0.094

TABLE 14-14   DATE=06NOV85

AVERAGE ANNUAL INCIDENCE RATE FOR EVENT PER 1000 PERSONS AT RISK AT EXAM
BY AGE AND LEVEL OF RISK FACTOR AT EXAM: FRAMINGHAM STUDY 30-YEAR FOLLOWUP

SEX: MALE
EVENT: ATHEROTHROMBOTIC BRAIN INFARCTION
RISK FACTOR: ALBUMIN IN URINE, DEFINITE
PERSONS AT RISK: PERSONS FREE OF STROKE AND TIA ONLY

RISK FACTOR	35-44 AT EXAM PERSON EXAMS	# OF EVENTS	ANNUAL RATE	45-54 AT EXAM PERSON EXAMS	# OF EVENTS	ANNUAL RATE	55-64 AT EXAM PERSON EXAMS	# OF EVENTS	ANNUAL RATE	65-74 AT EXAM PERSON EXAMS	# OF EVENTS	ANNUAL RATE
NO	4408			7903	16	1	8682	31	2	4476	39	4
YES	71			103	0	0	205	4	10	144	2	7
UNKNOWN	37			26	0	0	27	0	0	6	0	0
TOTAL	4516			8032	16	1	8914	35	2	4626	41	4

RISK FACTOR	75-84 AT EXAM PERSON EXAMS	# OF EVENTS	ANNUAL RATE	85-94 AT EXAM PERSON EXAMS	# OF EVENTS	ANNUAL RATE	35-64 COMBINED # OF EVENTS	CRUDE	AGE-ADJUSTED	65-94 COMBINED # OF EVENTS	CRUDE	AGE-ADJUSTED
NO	1199	26	11	37			47	1	1	65	6	6
YES	32	0	0	5			4	5	4	2	6	5
UNKNOWN	3	0	0	0			0	0	0	0	0	0
TOTAL	1234	26	11	42			51	1	1	67	6	6

LOGISTIC REGRESSIONS

AGE AT EXAM		COEFFICIENT OF RISK FACTOR	SE COEFFICIENT	P-VALUE	STANDARDIZED COEFFICIENT	STD SE
35-44		-	-		-	-
45-54		-	-		-	-
55-64		1.714	0.536	0.001	0.257	0.080
65-74		0.471	0.730	0.518	0.082	0.127
75-84		-	-		-	-
85-94		-	-		-	-
35-64	UNIVARIATE	1.559	0.523	0.003	0.206	0.069
65-94	UNIVARIATE	-0.030	0.722	0.967	-0.005	0.125
35-64	BIVARIATE	1.340	0.526	0.011	0.177	0.069
65-94	BIVARIATE	-0.075	0.723	0.917	-0.013	0.125
35-64	MULTIVARIATE	0.615	0.548	0.261	0.081	0.072
65-94	MULTIVARIATE	-0.370	0.730	0.612	-0.064	0.126

SEX: FEMALE
EVENT: ATHEROTHROMBOTIC BRAIN INFARCTION
RISK FACTOR: ALBUMIN IN URINE, DEFINITE
PERSONS AT RISK: PERSONS FREE OF STROKE AND TIA ONLY

RISK FACTOR	35-44 AT EXAM PERSON EXAMS	# OF EVENTS	ANNUAL RATE	45-54 AT EXAM PERSON EXAMS	# OF EVENTS	ANNUAL RATE	55-64 AT EXAM PERSON EXAMS	# OF EVENTS	ANNUAL RATE	65-74 AT EXAM PERSON EXAMS	# OF EVENTS	ANNUAL RATE
NO	5350			9728	13	1	11684	31	1	6729	50	4
YES	225			280	1	2	238	0	0	132	3	11
UNKNOWN	38			25	0	0	28	0	0	10	0	0
TOTAL	5613			10033	14	1	11950	31	1	6871	53	4

RISK FACTOR	75-84 AT EXAM PERSON EXAMS	# OF EVENTS	ANNUAL RATE	85-94 AT EXAM PERSON EXAMS	# OF EVENTS	ANNUAL RATE	35-64 COMBINED # OF EVENTS	CRUDE	AGE-ADJUSTED	65-94 COMBINED # OF EVENTS	CRUDE	AGE-ADJUSTED
NO	2173	26	6	141			44	1	1	77	4	4
YES	45	1	11	7			1	1	1	4	11	11
UNKNOWN	9	0	0	0			0	0	0	0	0	0
TOTAL	2227	27	6	148			45	1	1	81	4	4

LOGISTIC REGRESSIONS

AGE AT EXAM		COEFFICIENT OF RISK FACTOR	SE COEFFICIENT	P-VALUE	STANDARDIZED COEFFICIENT	STD SE
35-44		-	-		-	-
45-54		0.985	1.039	0.343	0.162	0.171
55-64		-	-		-	-
65-74		1.134	0.601	0.059	0.156	0.083
75-84		0.630	1.030	0.541	0.089	0.145
85-94		-	-		-	-
35-64	UNIVARIATE	-0.200	1.012	0.843	-0.032	0.164
65-94	UNIVARIATE	0.951	0.518	0.067	0.133	0.072
35-64	BIVARIATE	-0.010	1.013	0.992	-0.002	0.164
65-94	BIVARIATE	0.923	0.519	0.076	0.129	0.073
35-64	MULTIVARIATE	-1.045	1.045	0.318	-0.169	0.169
65-94	MULTIVARIATE	0.388	0.538	0.470	0.054	0.075

TABLE 14-15   DATE=06NOV85
AVERAGE ANNUAL INCIDENCE RATE FOR EVENT PER 1000 PERSONS AT RISK AT EXAM
BY AGE AND LEVEL OF RISK FACTOR AT EXAM: FRAMINGHAM STUDY 30-YEAR FOLLOWUP

SEX: MALE
EVENT: ATHEROTHROMBOTIC BRAIN INFARCTION
RISK FACTOR: HEART ENLARGEMENT BY X-RAY
PERSONS AT RISK: PERSONS FREE OF STROKE AND TIA ONLY

RISK FACTOR	35-44 AT EXAM PERSON EXAMS	# OF EVENTS	ANNUAL RATE	45-54 AT EXAM PERSON EXAMS	# OF EVENTS	ANNUAL RATE	55-64 AT EXAM PERSON EXAMS	# OF EVENTS	ANNUAL RATE	65-74 AT EXAM PERSON EXAMS	# OF EVENTS	ANNUAL RATE
NO	4098			6590	11	1	6455	18	1	2971	24	4
MILD	189			702	1	1	1102	8	4	662	5	4
DEFINITE	224			736	4	3	1356	9	3	993	12	6
UNKNOWN	5			4	0	0	1	0	0	0	0	-
TOTAL	4516			8032	16	1	8914	35	2	4626	41	4

RISK FACTOR	75-84 AT EXAM PERSON EXAMS	# OF EVENTS	ANNUAL RATE	85-94 AT EXAM PERSON EXAMS	# OF EVENTS	ANNUAL RATE	35-64 COMBINED # OF EVENTS	CRUDE	AGE-ADJUSTED	65-94 COMBINED # OF EVENTS	CRUDE	AGE-ADJUSTED
NO	693	14	10	28			29	1	1	38	5	5
MILD	205	5	12	6			9	2	2	10	6	6
DEFINITE	336	7	10	8			13	3	2	19	7	7
UNKNOWN	0	0	-	0			0	0	0	0	-	-
TOTAL	1234	26	11	42			51	1	1	67	6	6

LOGISTIC REGRESSIONS

AGE AT EXAM	COEFFICIENT OF RISK FACTOR	SE COEFFICIENT	P-VALUE	STANDARDIZED COEFFICIENT	STD SE
35-44					
45-54	0.551	0.302	0.068	0.340	0.186
55-64	0.470	0.192	0.014	0.348	0.142
65-74	0.189	0.180	0.293	0.155	0.148
75-84	0.027	0.227	0.905	0.024	0.197
85-94					
35-64 UNIVARIATE	0.62	0.160	0.001	0.410	0.105
65-94 UNIVARIATE	0.166	0.141	0.255	0.133	0.117
35-64 BIVARIATE	0.452	0.163	0.005	0.296	0.107
65-94 BIVARIATE	0.124	0.141	0.380	0.103	0.118
35-64 MULTIVARIATE	0.179	0.177	0.312	0.117	0.116
65-94 MULTIVARIATE	-0.103	0.154	0.506	-0.085	0.128

SEX: FEMALE
EVENT: ATHEROTHROMBOTIC BRAIN INFARCTION
RISK FACTOR: HEART ENLARGEMENT BY X-RAY
PERSONS AT RISK: PERSONS FREE OF STROKE AND TIA ONLY

RISK FACTOR	35-44 AT EXAM PERSON EXAMS	# OF EVENTS	ANNUAL RATE	45-54 AT EXAM PERSON EXAMS	# OF EVENTS	ANNUAL RATE	55-64 AT EXAM PERSON EXAMS	# OF EVENTS	ANNUAL RATE	65-74 AT EXAM PERSON EXAMS	# OF EVENTS	ANNUAL RATE
NO	5161			8054	12	1	7762	13	1	3444	17	2
MILD	198			954	1	1	1850	1	1	1254	6	2
DEFINITE	246			1019	1	0	2334	13	3	2173	30	7
UNKNOWN	8			6	0	0	4	0	0	0	0	-
TOTAL	5613			10033	14	1	11950	31	1	6871	53	4

RISK FACTOR	75-84 AT EXAM PERSON EXAMS	# OF EVENTS	ANNUAL RATE	85-94 AT EXAM PERSON EXAMS	# OF EVENTS	ANNUAL RATE	35-64 COMBINED # OF EVENTS	CRUDE	AGE-ADJUSTED	65-94 COMBINED # OF EVENTS	CRUDE	AGE-ADJUSTED
NO	778	10	6	38			25	1	1	27	3	3
MILD	435	2	2	30			6	1	1	8	2	2
DEFINITE	1014	15	7	80			14	2	1	46	7	7
UNKNOWN	0	0	-	0			0	0	0	0	-	-
TOTAL	2227	27	6	148			45	1	1	81	4	4

LOGISTIC REGRESSIONS

AGE AT EXAM	COEFFICIENT OF RISK FACTOR	SE COEFFICIENT	P-VALUE	STANDARDIZED COEFFICIENT	STD SE
35-44					
45-54	-0.237	0.486	0.626	-0.152	0.312
55-64	0.603	0.199	0.002	0.482	0.159
65-74	0.543	0.158	0.001	0.481	0.140
75-84	0.102	0.220	0.642	0.091	0.196
85-94					
35-64 UNIVARIATE	0.590	0.167	0.001	0.414	0.117
65-94 UNIVARIATE	0.431	0.128	0.001	0.386	0.114
35-64 BIVARIATE	0.409	0.173	0.018	0.288	0.122
65-94 BIVARIATE	0.394	0.130	0.002	0.353	0.116
35-64 MULTIVARIATE	0.057	0.191	0.767	0.040	0.134
65-94 MULTIVARIATE	0.157	0.138	0.255	0.140	0.123

TABLE 14-16   DATE=06NOV85
AVERAGE ANNUAL INCIDENCE RATE FOR EVENT PER 1000 PERSONS AT RISK AT EXAM
BY AGE AND LEVEL OF RISK FACTOR AT EXAM, FRAMINGHAM STUDY 30-YEAR FOLLOWUP

SEX: MALE
EVENT: ATHEROTHROMBOTIC BRAIN INFARCTION
RISK FACTOR: LEFT VENTRICULAR HYPERTROPHY BY ELECTROCARDIOGRAM
PERSONS AT RISK: PERSONS FREE OF STROKE AND TIA ONLY

RISK FACTOR	35-44 AT EXAM			45-54 AT EXAM			55-64 AT EXAM			65-74 AT EXAM		
	PERSON EXAMS	# OF EVENTS	ANNUAL RATE	PERSON EXAMS	# OF EVENTS	ANNUAL RATE	PERSON EXAMS	# OF EVENTS	ANNUAL RATE	PERSON EXAMS	# OF EVENTS	ANNUAL RATE
NO	4405			7780	16	1	8471	25	1	4347	35	4
MILD	79			145	0	0	235	4	9	114	2	9
DEFINITE	32			107	0	0	206	6	14	165	4	12
UNKNOWN	0			0			0			0		
TOTAL	4516			8032	16	1	8914	35	2	4626	41	4

RISK FACTOR	75-84 AT EXAM			85-94 AT EXAM			35-64 COMBINED			65-94 COMBINED		
	PERSON EXAMS	# OF EVENTS	ANNUAL RATE	PERSON EXAMS	# OF EVENTS	ANNUAL RATE	# OF EVENTS	CRUDE	AGE-ADJUSTED	# OF EVENTS	CRUDE	AGE-ADJUSTED
NO	1148	21	9	38			41	1	1	56	5	5
MILD	39	2	26	2			4	4	4	4	13	12
DEFINITE	47	3	32	2			6	9	6	7	16	16
UNKNOWN	0	0	-	0			0	-		0		
TOTAL	1234	26	11	42			51	1	1	67	6	6

LOGISTIC REGRESSIONS

AGE AT EXAM		COEFFICIENT OF RISK FACTOR	SE COEFFICIENT	P-VALUE	STANDARDIZED COEFFICIENT	STD SE
35-44		-	-	-	-	-
45-54						
55-64		1.203	0.214	0.001	0.407	0.073
65-74		0.580	0.253	0.022	0.230	0.101
75-84		0.691	0.296	0.020	0.287	0.123
85-94						
35-64	UNIVARIATE	1.130	0.205	0.001	0.326	0.059
65-94	UNIVARIATE	0.631	0.191	0.001	0.254	0.077
35-64	BIVARIATE	0.970	0.207	0.001	0.280	0.060
65-94	BIVARIATE	0.622	0.192	0.001	0.250	0.077
35-64	MULTIVARIATE	0.460	0.234	0.050	0.133	0.068
65-94	MULTIVARIATE	0.259	0.215	0.228	0.104	0.086

SEX: FEMALE
EVENT: ATHEROTHROMBOTIC BRAIN INFARCTION
RISK FACTOR: LEFT VENTRICULAR HYPERTROPHY BY ELECTROCARDIOGRAM
PERSONS AT RISK: PERSONS FREE OF STROKE AND TIA ONLY

RISK FACTOR	35-44 AT EXAM			45-54 AT EXAM			55-64 AT EXAM			65-74 AT EXAM		
	PERSON EXAMS	# OF EVENTS	ANNUAL RATE	PERSON EXAMS	# OF EVENTS	ANNUAL RATE	PERSON EXAMS	# OF EVENTS	ANNUAL RATE	PERSON EXAMS	# OF EVENTS	ANNUAL RATE
NO	5560			9866	14	1	11542	23	1	6504	39	3
MILD	38			111	0	0	242	3	6	174	3	9
DEFINITE	15			56	0	0	166	5	15	193	11	28
UNKNOWN	0			0			0			0		
TOTAL	5613			10033	14	1	11950	31	1	6871	53	4

RISK FACTOR	75-84 AT EXAM			85-94 AT EXAM			35-64 COMBINED			65-94 COMBINED		
	PERSON EXAMS	# OF EVENTS	ANNUAL RATE	PERSON EXAMS	# OF EVENTS	ANNUAL RATE	# OF EVENTS	CRUDE	AGE-ADJUSTED	# OF EVENTS	CRUDE	AGE-ADJUSTED
NO	2059	23	6	129			37	1	1	63	4	4
MILD	65	0	0	8			3	4	3	3	6	6
DEFINITE	103	4	19	11			5	11	7	15	24	26
UNKNOWN	0	0	-	0			0	-		0		
TOTAL	2227	27	6	148			45	1	1	81	4	4

LOGISTIC REGRESSIONS

AGE AT EXAM		COEFFICIENT OF RISK FACTOR	SE COEFFICIENT	P-VALUE	STANDARDIZED COEFFICIENT	STD SE
35-44		-	-	-	-	-
45-54						
55-64		1.410	0.235	0.001	0.382	0.064
65-74		1.148	0.173	0.001	0.416	0.063
75-84		0.564	0.288	0.050	0.252	0.129
85-94						
35-64	UNIVARIATE	1.410	0.226	0.001	0.307	0.049
65-94	UNIVARIATE	0.954	0.147	0.001	0.371	0.057
35-64	BIVARIATE	1.256	0.229	0.001	0.274	0.050
65-94	BIVARIATE	0.923	0.148	0.001	0.358	0.058
35-64	MULTIVARIATE	0.509	0.277	0.066	0.111	0.060
65-94	MULTIVARIATE	0.534	0.165	0.001	0.208	0.064

TABLE 14-17   DATE=06NOV85

AVERAGE ANNUAL INCIDENCE RATE FOR EVENT PER 1000 PERSONS AT RISK AT EXAM
BY AGE AND LEVEL OF RISK FACTOR AT EXAM: FRAMINGHAM STUDY 30-YEAR FOLLOWUP

SEX: MALE
EVENT: ATHEROTHROMBOTIC BRAIN INFARCTION
RISK FACTOR: INTRAVENTRICULAR CONDUCTION DEFECT (WITH QRS INTERVAL MORE THAN 0.11 SECOND)
PERSONS AT RISK: PERSONS FREE OF STROKE AND TIA ONLY

RISK FACTOR	35-44 AT EXAM PERSON EXAMS	# OF EVENTS	ANNUAL RATE	45-54 AT EXAM PERSON EXAMS	# OF EVENTS	ANNUAL RATE	55-64 AT EXAM PERSON EXAMS	# OF EVENTS	ANNUAL RATE	65-74 AT EXAM PERSON EXAMS	# OF EVENTS	ANNUAL RATE
NO	4198			7452	14	1	8107	33	2	4150	32	4
INCOMPLETE	280			491	1	1	534	2	2	231	3	6
COMPLETE	38			89	1	6	273	0	0	245	6	12
UNKNOWN	0			0	0	-	0	0	~	0	0	-
TOTAL	4516			8032	16	1	8914	35	2	4626	41	4

RISK FACTOR	75-84 AT EXAM PERSON EXAMS	# OF EVENTS	ANNUAL RATE	85-94 AT EXAM PERSON EXAMS	# OF EVENTS	ANNUAL RATE	35-64 COMBINED # OF EVENTS	CRUDE	AGE-ADJUSTED	65-94 COMBINED # OF EVENTS	CRUDE	AGE-ADJUSTED
NO	1083	25	12	35			47	1	1	57	5	5
INCOMPLETE	35	1	14	3			3	1	1	4	7	8
COMPLETE	116	0	0	4			1	1	2	6	8	10
UNKNOWN	0	0	-	0			0	-	-	0	-	-
TOTAL	1234	26	11	42			51	1	1	67	6	6

LOGISTIC REGRESSIONS

AGE AT EXAM		COEFFICIENT OF RISK FACTOR	SE COEFFICIENT	P-VALUE	STANDARDIZED COEFFICIENT	STD SE
35-44						
45-54		0.673	0.523	0.198	0.211	0.164
55-64		-0.557	0.618	0.367	-0.228	0.253
65-74		0.581	0.219	0.008	0.283	0.107
75-84		-1.032	0.777	0.184	-0.617	0.465
85-94						
35-64	UNIVARIATE	-0.000	0.395	0.999	-0.000	0.140
65-94	UNIVARIATE	0.225	0.204	0.269	0.116	0.105
35-64	BIVARIATE	-0.136	0.385	0.724	-0.048	0.137
65-94	BIVARIATE	0.171	0.203	0.398	0.088	0.104
35-64	MULTIVARIATE	-0.235	0.400	0.557	-0.083	0.142
65-94	MULTIVARIATE	0.170	0.204	0.405	0.088	0.105

SEX: FEMALE
EVENT: ATHEROTHROMBOTIC BRAIN INFARCTION
RISK FACTOR: INTRAVENTRICULAR CONDUCTION DEFECT (WITH QRS INTERVAL MORE THAN 0.11 SECOND)
PERSONS AT RISK: PERSONS FREE OF STROKE AND TIA ONLY

RISK FACTOR	35-44 AT EXAM PERSON EXAMS	# OF EVENTS	ANNUAL RATE	45-54 AT EXAM PERSON EXAMS	# OF EVENTS	ANNUAL RATE	55-64 AT EXAM PERSON EXAMS	# OF EVENTS	ANNUAL RATE	65-74 AT EXAM PERSON EXAMS	# OF EVENTS	ANNUAL RATE
NO	5526			9834	14	1	11582	30	1	6577	49	4
INCOMPLETE	75			151	0	0	195	0	0	130	2	8
COMPLETE	12			48	0	0	173	1	3	164	2	6
UNKNOWN	0			0	0	-	0	0	~	0	0	-
TOTAL	5613			10033	14	1	11950	31	1	6871	53	4

RISK FACTOR	75-84 AT EXAM PERSON EXAMS	# OF EVENTS	ANNUAL RATE	85-94 AT EXAM PERSON EXAMS	# OF EVENTS	ANNUAL RATE	35-64 COMBINED # OF EVENTS	CRUDE	AGE-ADJUSTED	65-94 COMBINED # OF EVENTS	CRUDE	AGE-ADJUSTED
NO	2088	25	6	138			44	1	1	75	4	4
INCOMPLETE	49	0	0	3			0	0	0	2	5	6
COMPLETE	90	2	11	7			1	2	1	4	8	7
UNKNOWN	0	0	-	0			0	-	-	0	-	-
TOTAL	2227	27	6	148			45	1	1	81	4	4

LOGISTIC REGRESSIONS

AGE AT EXAM		COEFFICIENT OF RISK FACTOR	SE COEFFICIENT	P-VALUE	STANDARDIZED COEFFICIENT	STD SE
35-44		-				-
45-54		-				
55-64		0.221	0.556	0.691	0.059	0.149
65-74		0.323	0.319	0.312	0.107	0.106
75-84		0.221	0.390	0.572	0.092	0.162
85-94						
35-64	UNIVARIATE	0.214	0.572	0.708	0.047	0.125
65-94	UNIVARIATE	0.294	0.247	0.235	0.104	0.088
35-64	BIVARIATE	0.057	0.566	0.919	0.013	0.124
65-94	BIVARIATE	0.261	0.247	0.289	0.093	0.088
35-64	MULTIVARIATE	0.084	0.569	0.885	0.018	0.125
65-94	MULTIVARIATE	0.240	0.251	0.338	0.086	0.089

SEX: MALE
EVENT: ATHEROTHROMBOTIC BRAIN INFARCTION
RISK FACTOR: DEFINITE NONSPECIFIC T-WAVE OR ST-SEGMENT ABNORMALITY BY ELECTROCARDIOGRAM
PERSONS AT RISK: PERSONS FREE OF STROKE AND TIA ONLY

RISK FACTOR	35-44 AT EXAM			45-54 AT EXAM			55-64 AT EXAM			65-74 AT EXAM		
	PERSON EXAMS	# OF EVENTS	ANNUAL RATE	PERSON EXAMS	# OF EVENTS	ANNUAL RATE	PERSON EXAMS	# OF EVENTS	ANNUAL RATE	PERSON EXAMS	# OF EVENTS	ANNUAL RATE
NO	4403			7609	13	1	8128	27	2	4007	31	4
YES	113			423	3	4	786	8	5	619	10	8
UNKNOWN	0		-	0	0	-	0	0	-	0	0	-
TOTAL	4516			8032	16	1	8914	35	2	4626	41	4

RISK FACTOR	75-84 AT EXAM			85-94 AT EXAM			35-64 COMBINED			65-94 COMBINED		
	PERSON EXAMS	# OF EVENTS	ANNUAL RATE	PERSON EXAMS	# OF EVENTS	ANNUAL RATE	# OF EVENTS	CRUDE	AGE-ADJUSTED	# OF EVENTS	CRUDE	AGE-ADJUSTED
NO	988	21	11	24			40	1	1	52	5	5
YES	246	5	10	18			11	4	3	15	8	8
UNKNOWN	0	0	-	0			0	-		0	-	
TOTAL	1234	26	11	42			51	1	1	67	6	6

LOGISTIC REGRESSIONS

AGE AT EXAM		COEFFICIENT OF RISK FACTOR	SE COEFFICIENT	P-VALUE	STANDARDIZED COEFFICIENT	STD SE
35-44						
45-54		1.429	0.642	0.026	0.319	0.144
55-64		1.127	0.404	0.005	0.314	0.115
65-74		0.745	0.366	0.042	0.254	0.125
75-84		-0.046	0.503	0.928	-0.018	0.201
85-94						
35-64	UNIVARIATE	1.439	0.342	0.001	0.546	0.082
65-94	UNIVARIATE	0.501	0.295	0.090	0.179	0.105
35-64	BIVARIATE	1.113	0.345	0.001	0.268	0.083
65-94	BIVARIATE	0.395	0.298	0.185	0.141	0.106
35-64	MULTIVARIATE	0.826	0.348	0.018	0.199	0.084
65-94	MULTIVARIATE	0.303	0.301	0.314	0.108	0.107

SEX: FEMALE
EVENT: ATHEROTHROMBOTIC BRAIN INFARCTION
RISK FACTOR: DEFINITE NONSPECIFIC T-WAVE OR ST-SEGMENT ABNORMALITY BY ELECTROCARDIOGRAM
PERSONS AT RISK: PERSONS FREE OF STROKE AND TIA ONLY

RISK FACTOR	35-44 AT EXAM			45-54 AT EXAM			55-64 AT EXAM			65-74 AT EXAM		
	PERSON EXAMS	# OF EVENTS	ANNUAL RATE	PERSON EXAMS	# OF EVENTS	ANNUAL RATE	PERSON EXAMS	# OF EVENTS	ANNUAL RATE	PERSON EXAMS	# OF EVENTS	ANNUAL RATE
NO	5514			9603	13	1	10984	27	1	5994	43	4
YES	99			430	1	1	966	4	2	877	10	6
UNKNOWN	0		-	0	0	-	0	0	-	0	0	-
TOTAL	5613			10033	14	1	11950	31	1	6871	53	4

RISK FACTOR	75-84 AT EXAM			85-94 AT EXAM			35-64 COMBINED			65-94 COMBINED		
	PERSON EXAMS	# OF EVENTS	ANNUAL RATE	PERSON EXAMS	# OF EVENTS	ANNUAL RATE	# OF EVENTS	CRUDE	AGE-ADJUSTED	# OF EVENTS	CRUDE	AGE-ADJUSTED
NO	1811	21	6	114			40	1	1	65	4	4
YES	416	6	7	34			5	2	1	16	6	6
UNKNOWN	0	0	-	0			0	-		0	-	
TOTAL	2227	27	6	148			45	1	1	81	4	4

LOGISTIC REGRESSIONS

AGE AT EXAM		COEFFICIENT OF RISK FACTOR	SE COEFFICIENT	P-VALUE	STANDARDIZED COEFFICIENT	STD SE
35-44						
45-54		0.542	1.039	0.602	0.110	0.210
55-64		0.523	0.537	0.330	0.143	0.146
65-74		0.468	0.353	0.185	0.156	0.118
75-84		0.221	0.466	0.635	0.086	0.182
85-94						
35-64	UNIVARIATE	0.782	0.475	0.100	0.177	0.108
65-94	UNIVARIATE	0.388	0.281	0.166	0.136	0.098
35-64	BIVARIATE	0.480	0.479	0.316	0.109	0.108
65-94	BIVARIATE	0.327	0.282	0.247	0.115	0.099
35-64	MULTIVARIATE	0.252	0.484	0.603	0.057	0.109
65-94	MULTIVARIATE	0.208	0.286	0.467	0.073	0.100

TABLE 15-1   DATE=23AUG85

AVERAGE ANNUAL INCIDENCE RATE FOR EVENT PER 1000 PERSONS AT RISK AT EXAM
BY AGE AND LEVEL OF RISK FACTOR AT EXAM: FRAMINGHAM STUDY 30-YEAR FOLLOWUP

SEX: MALE
EVENT: TRANSIENT ISCHEMIC ATTACK
RISK FACTOR: SYSTOLIC BLOOD PRESSURE, FIRST EXAMINER (MM HG)
PERSONS AT RISK: PERSONS FREE OF STROKE AND TIA ONLY

RISK FACTOR	35-44 AT EXAM			45-54 AT EXAM			55-64 AT EXAM			65-74 AT EXAM		
	PERSON EXAMS	# OF EVENTS	ANNUAL RATE	PERSON EXAMS	# OF EVENTS	ANNUAL RATE	PERSON EXAMS	# OF EVENTS	ANNUAL RATE	PERSON EXAMS	# OF EVENTS	ANNUAL RATE
74-119	1225			1837	1	0	1589	1	0	649	1	1
120-139	2191			3531	2	0	3482	8	1	1582	7	2
140-159	850			1830	1	0	2324	5	1	1394	4	1
160-179	197			589	4	3	1029	2	1	659	2	2
180-300	53			245	0	0	490	2	2	342	3	4
UNKNOWN	0			0	0	-	0	0	-	0	0	-
TOTAL	4516			8032	8	0	8914	18	1	4626	17	2

RISK FACTOR	75-84 AT EXAM			85-94 AT EXAM			35-64 COMBINED			65-94 COMBINED		
	PERSON EXAMS	# OF EVENTS	ANNUAL RATE	PERSON EXAMS	# OF EVENTS	ANNUAL RATE	# OF EVENTS	CRUDE	AGE-ADJUSTED	# OF EVENTS	CRUDE	AGE-ADJUSTED
74-119	136	1	4	7			2	0	0	2	1	2
120-139	420	2	2	13			10	1	0	9	2	2
140-159	379	2	3	15			6	1	2	6	2	2
160-179	193	1	3	5			6	2	2	3	2	2
180-300	106	3	14	2			2	1	1	6	7	6
UNKNOWN	0	0	-	0			0			0	-	
TOTAL	1234	9	4	42			26	1	0	26	2	2

LOGISTIC REGRESSIONS

AGE AT EXAM		COEFFICIENT OF RISK FACTOR	SE COEFFICIENT	P-VALUE	STANDARDIZED COEFFICIENT	STD SE
35-44						
45-54		0.024	0.014	0.083	0.475	0.274
55-64		0.011	0.009	0.240	0.246	0.210
65-74		0.007	0.010	0.507	0.151	0.228
75-84		0.023	0.012	0.056	0.527	0.276
85-94						
35-64	UNIVARIATE	0.019	0.008	0.012	0.395	0.157
65-94	UNIVARIATE	0.013	0.007	0.084	0.298	0.173
35-64	BIVARIATE	0.013	0.008	0.106	0.260	0.161
65-94	BIVARIATE	0.013	0.008	0.087	0.297	0.174
35-64	MULTIVARIATE	0.013	0.008	0.117	0.269	0.172
65-94	MULTIVARIATE	0.013	0.008	0.096	0.307	0.184

SEX: FEMALE
EVENT: TRANSIENT ISCHEMIC ATTACK
RISK FACTOR: SYSTOLIC BLOOD PRESSURE, FIRST EXAMINER (MM HG)
PERSONS AT RISK: PERSONS FREE OF STROKE AND TIA ONLY

RISK FACTOR	35-44 AT EXAM			45-54 AT EXAM			55-64 AT EXAM			65-74 AT EXAM		
	PERSON EXAMS	# OF EVENTS	ANNUAL RATE	PERSON EXAMS	# OF EVENTS	ANNUAL RATE	PERSON EXAMS	# OF EVENTS	ANNUAL RATE	PERSON EXAMS	# OF EVENTS	ANNUAL RATE
74-119	2488			2717			2011	0	0	675	0	0
120-139	2245			4015			4168	3	0	1876	5	1
140-159	673			2070			3155	5	1	2212	1	0
160-179	156			783			1654	2	1	1302	6	2
180-300	51			448			962	2	1	806	3	2
UNKNOWN	0			0			0	0	-	0	0	-
TOTAL	5613			10033			11950	12	1	6871	15	1

RISK FACTOR	75-84 AT EXAM			85-94 AT EXAM			35-64 COMBINED			65-94 COMBINED		
	PERSON EXAMS	# OF EVENTS	ANNUAL RATE	PERSON EXAMS	# OF EVENTS	ANNUAL RATE	# OF EVENTS	CRUDE	AGE-ADJUSTED	# OF EVENTS	CRUDE	AGE-ADJUSTED
74-119	145	1	3	10			0	0	0	1	1	1
120-139	527	4	4	47			4	0	0	9	2	2
140-159	704	4	3	34			6	1	0	5	1	1
160-179	512	3	4	32			3	1	1	10	3	2
180-300	339	3	4	25			2	1	0	6	3	2
UNKNOWN	0	0	-	0			0	-		0	-	
TOTAL	2227	15	3	148			15	0	0	31	2	2

LOGISTIC REGRESSIONS

AGE AT EXAM		COEFFICIENT OF RISK FACTOR	SE COEFFICIENT	P-VALUE	STANDARDIZED COEFFICIENT	STD SE
35-44						
45-54						
55-64		0.020	0.009	0.022	0.497	0.217
65-74		0.023	0.009	0.007	0.573	0.213
75-84		0.003	0.010	0.739	0.084	0.252
85-94						
35-64	UNIVARIATE	0.024	0.007	0.001	0.571	0.167
65-94	UNIVARIATE	0.015	0.006	0.015	0.383	0.158
35-64	BIVARIATE	0.019	0.008	0.017	0.452	0.189
65-94	BIVARIATE	0.014	0.006	0.030	0.348	0.160
35-64	MULTIVARIATE	0.018	0.009	0.044	0.419	0.209
65-94	MULTIVARIATE	0.010	0.007	0.128	0.253	0.167

TABLE 15-2   DATE=22AUG85
AVERAGE ANNUAL INCIDENCE RATE FOR EVENT PER 1000 PERSONS AT RISK AT EXAM
BY AGE AND LEVEL OF RISK FACTOR AT EXAM: FRAMINGHAM STUDY 30-YEAR FOLLOWUP

SEX: MALE
EVENT: TRANSIENT ISCHEMIC ATTACK
RISK FACTOR: DIASTOLIC BLOOD PRESSURE, FIRST EXAMINER (MM HG)
PERSONS AT RISK: PERSONS FREE OF STROKE AND TIA ONLY

RISK FACTOR	35-44 AT EXAM			45-54 AT EXAM			55-64 AT EXAM			65-74 AT EXAM		
	PERSON EXAMS	# OF EVENTS	ANNUAL RATE	PERSON EXAMS	# OF EVENTS	ANNUAL RATE	PERSON EXAMS	# OF EVENTS	ANNUAL RATE	PERSON EXAMS	# OF EVENTS	ANNUAL RATE
20-74	1015			1532	2	1	2056	3	1	1469	8	3
75-84	1614			2857	0	0	3127	5	1	1575	4	1
85-94	1203			2142	5	1	2192	5	1	1030	2	1
95-104	509			1065	1	0	1081	5	2	399	2	3
105-160	175			436	0	0	458	0	0	153	1	3
UNKNOWN	0			0	0	-	0	0	-	0	0	-
TOTAL	4516			8032	8	0	8914	18	1	4626	17	2

RISK FACTOR	75-84 AT EXAM			85-94 AT EXAM			35-64 COMBINED			65-94 COMBINED		
	PERSON EXAMS	# OF EVENTS	ANNUAL RATE	PERSON EXAMS	# OF EVENTS	ANNUAL RATE	# OF EVENTS	CRUDE	AGE-ADJUSTED	# OF EVENTS	CRUDE	AGE-ADJUSTED
20-74	513	3	3	22			5	1	1	11	3	3
75-84	403	3	4	15			5	0	0	7	2	2
85-94	224	1	2	5			10	1	1	3	1	1
95-104	66	2	15	0			6	1	1	4	4	5
105-160	28	0	0	0			0	0	0	1	3	2
UNKNOWN	0	0	-	0			0	-	-	0	-	-
TOTAL	1234	9	4	42			26	1	0	26	2	2

LOGISTIC REGRESSIONS

AGE AT EXAM		COEFFICIENT OF RISK FACTOR	SE COEFFICIENT	P-VALUE	STANDARDIZED COEFFICIENT	STD SE
35-44						
45-54		0.011	0.029	0.700	0.132	0.342
55-64		0.016	0.018	0.387	0.192	0.222
65-74		0.000	0.020	0.989	0.003	0.243
75-84		0.028	0.026	0.264	0.350	0.314
85-94						
35-64	UNIVARIATE	0.015	0.016	0.329	0.181	0.186
65-94	UNIVARIATE	0.008	0.015	0.618	0.096	0.191
35-64	BIVARIATE	0.016	0.015	0.283	0.193	0.180
65-94	BIVARIATE	0.011	0.016	0.484	0.135	0.193
35-64	MULTIVARIATE	0.017	0.016	0.294	0.197	0.188
65-94	MULTIVARIATE	0.009	0.016	0.561	0.114	0.197

SEX: FEMALE
EVENT: TRANSIENT ISCHEMIC ATTACK
RISK FACTOR: DIASTOLIC BLOOD PRESSURE, FIRST EXAMINER (MM HG)
PERSONS AT RISK: PERSONS FREE OF STROKE AND TIA ONLY

RISK FACTOR	35-44 AT EXAM			45-54 AT EXAM			55-64 AT EXAM			65-74 AT EXAM		
	PERSON EXAMS	# OF EVENTS	ANNUAL RATE	PERSON EXAMS	# OF EVENTS	ANNUAL RATE	PERSON EXAMS	# OF EVENTS	ANNUAL RATE	PERSON EXAMS	# OF EVENTS	ANNUAL RATE
20-74	2161			2650			3078	1	0	2026	4	1
75-84	2028			3558			3964	2	0	2362	3	1
85-94	988			2387			2965	4	1	1559	3	1
95-104	324			933			1326	3	1	642	2	2
105-160	112			505			617	2	2	282	3	5
UNKNOWN	0			0			0	0	-	0	0	-
TOTAL	5613			10033			11950	12	1	6871	15	1

RISK FACTOR	75-84 AT EXAM			85-94 AT EXAM			35-64 COMBINED			65-94 COMBINED		
	PERSON EXAMS	# OF EVENTS	ANNUAL RATE	PERSON EXAMS	# OF EVENTS	ANNUAL RATE	# OF EVENTS	CRUDE	AGE-ADJUSTED	# OF EVENTS	CRUDE	AGE-ADJUSTED
20-74	869	6	3	66			1	0	0	11	2	2
75-84	681	4	3	52			2	0	0	7	1	1
85-94	466	4	4	21			7	1	1	7	2	2
95-104	155	1	3	7			3	1	0	3	2	2
105-160	56	0	0	2			2	1	1	3	4	4
UNKNOWN	0	0	-	0			0	-	-	0	-	-
TOTAL	2227	15	3	148			15	0	0	31	2	2

LOGISTIC REGRESSIONS

AGE AT EXAM		COEFFICIENT OF RISK FACTOR	SE COEFFICIENT	P-VALUE	STANDARDIZED COEFFICIENT	STD SE
35-44						
45-54						
55-64		0.046	0.018	0.011	0.580	0.228
65-74		0.042	0.018	0.020	0.509	0.219
75-84		-0.014	0.022	0.515	-0.175	0.268
85-94						
35-64	UNIVARIATE	0.049	0.016	0.002	0.598	0.191
65-94	UNIVARIATE	0.009	0.014	0.542	0.107	0.175
35-64	BIVARIATE	0.046	0.016	0.006	0.561	0.203
65-94	BIVARIATE	0.014	0.014	0.335	0.168	0.175
35-64	MULTIVARIATE	0.044	0.017	0.012	0.540	0.214
65-94	MULTIVARIATE	0.011	0.014	0.422	0.138	0.172

TABLE 15-3A   DATE=26AUG85

AVERAGE ANNUAL INCIDENCE RATE FOR EVENT PER 1000 PERSONS AT RISK AT EXAM
BY AGE AND LEVEL OF RISK FACTOR AT EXAM: FRAMINGHAM STUDY 30-YEAR FOLLOWUP

SEX: MALE
EVENT: TRANSIENT ISCHEMIC ATTACK
RISK FACTOR: HYPERTENSION WITH ANTIHYPERTENSIVE TREATMENT
PERSONS AT RISK: PERSONS FREE OF STROKE AND TIA ONLY

RISK FACTOR	35-44 AT EXAM			45-54 AT EXAM			55-64 AT EXAM			65-74 AT EXAM		
	PERSON EXAMS	# OF EVENTS	ANNUAL RATE	PERSON EXAMS	# OF EVENTS	ANNUAL RATE	PERSON EXAMS	# OF EVENTS	ANNUAL RATE	PERSON EXAMS	# OF EVENTS	ANNUAL RATE
NO	2665			4248	2	0	4142	4	0	1771	3	1
BORDER	1316			2280	3	1	2531	7	1	1469	4	1
DEFINITE	511			1212	2	1	1399	4	1	675	3	2
TREATED	24			292	1	2	842	3	2	711	7	5
UNKNOWN	0			0	0	-	0	0	-	0	0	-
TOTAL	4516			8032	8	0	8914	18	1	4626	17	2

RISK FACTOR	75-84 AT EXAM			85-94 AT EXAM			55-64 COMBINED			65-94 COMBINED		
	PERSON EXAMS	# OF EVENTS	ANNUAL RATE	PERSON EXAMS	# OF EVENTS	ANNUAL RATE	# OF EVENTS	CRUDE	AGE-ADJUSTED	# OF EVENTS	CRUDE	AGE-ADJUSTED
NO	445	1	1	15			6	0	0	4	1	1
BORDER	395	1	1	14			10	1	1	5	1	1
DEFINITE	178	2	6	4			6	1	1	5	3	3
TREATED	216	5	12	9			4	2	2	12	6	6
UNKNOWN	0	0	-	0			0	-		0	-	
TOTAL	1234	9	4	42			26	1	0	26	2	2

LOGISTIC REGRESSIONS

AGE AT EXAM		COEFFICIENT OF RISK FACTOR	SE COEFFICIENT	P-VALUE	STANDARDIZED COEFFICIENT	STD SE
35-44						
45-54		0.702	0.430	0.102	0.544	0.333
55-64		0.547	0.285	0.055	0.448	0.234
65-74		0.774	0.327	0.018	0.636	0.269
75-84		1.285	0.565	0.023	1.059	0.465
85-94						
35-64	UNIVARIATE	0.700	0.239	0.003	0.549	0.188
65-94	UNIVARIATE	0.931	0.280	0.001	0.765	0.230
35-64	BIVARIATE	0.546	0.238	0.022	0.428	0.187
65-94	BIVARIATE	0.925	0.280	0.001	0.761	0.230
35-64	MULTIVARIATE	0.550	0.244	0.024	0.431	0.192
65-94	MULTIVARIATE	0.949	0.284	0.001	0.781	0.233

SEX: FEMALE
EVENT: TRANSIENT ISCHEMIC ATTACK
RISK FACTOR: HYPERTENSION WITH ANTIHYPERTENSIVE TREATMENT
PERSONS AT RISK: PERSONS FREE OF STROKE AND TIA ONLY

RISK FACTOR	35-44 AT EXAM			45-54 AT EXAM			55-64 AT EXAM			65-74 AT EXAM		
	PERSON EXAMS	# OF EVENTS	ANNUAL RATE	PERSON EXAMS	# OF EVENTS	ANNUAL RATE	PERSON EXAMS	# OF EVENTS	ANNUAL RATE	PERSON EXAMS	# OF EVENTS	ANNUAL RATE
NO	4247			5644			4983	2	0	1866	2	1
BORDER	1001			2597			3397	2	0	2066	3	1
DEFINITE	316			1178			1924	3	1	1166	1	0
TREATED	49			614			1646	5	2	1773	9	3
UNKNOWN	0			0			0	0	-	0	0	-
TOTAL	5613			10033			11950	12	1	6871	15	1

RISK FACTOR	75-84 AT EXAM			85-94 AT EXAM			55-64 COMBINED			65-94 COMBINED		
	PERSON EXAMS	# OF EVENTS	ANNUAL RATE	PERSON EXAMS	# OF EVENTS	ANNUAL RATE	# OF EVENTS	CRUDE	AGE-ADJUSTED	# OF EVENTS	CRUDE	AGE-ADJUSTED
NO	458	1	1	39			2	0	0	3	1	1
BORDER	643	3	2	32			4	0	0	4	1	1
DEFINITE	429	1	1	28			4	1	0	2	1	0
TREATED	697	10	7	49			5	1	1	20	4	4
UNKNOWN	0	0	-	0			0	-		0	-	
TOTAL	2227	15	3	148			15	0	0	31	2	2

LOGISTIC REGRESSIONS

AGE AT EXAM		COEFFICIENT OF RISK FACTOR	SE COEFFICIENT	P-VALUE	STANDARDIZED COEFFICIENT	STD SE
35-44						
45-54						
55-64		0.947	0.402	0.018	0.794	0.337
65-74		0.644	0.374	0.086	0.529	0.308
75-84		0.750	0.433	0.084	0.591	0.342
85-94						
35-64	UNIVARIATE	1.182	0.348	0.001	0.944	0.278
65-94	UNIVARIATE	0.769	0.281	0.006	0.628	0.229
35-64	BIVARIATE	0.947	0.357	0.008	0.755	0.284
65-94	BIVARIATE	0.717	0.282	0.011	0.585	0.230
35-64	MULTIVARIATE	0.903	0.362	0.013	0.721	0.289
65-94	MULTIVARIATE	0.605	0.284	0.033	0.494	0.232

TABLE 15-3B   DATE=07NOV85
AVERAGE ANNUAL INCIDENCE RATE FOR EVENT PER 1000 PERSONS AT RISK AT EXAM
BY AGE AND LEVEL OF RISK FACTOR AT EXAM: FRAMINGHAM STUDY 30-YEAR FOLLOWUP

SEX: MALE
EVENT: TRANSIENT ISCHEMIC ATTACK
RISK FACTOR: HYPERTENSION IGNORING TREATMENT
PERSONS AT RISK: PERSONS FREE OF STROKE AND TIA ONLY

RISK FACTOR	35-44 AT EXAM			45-54 AT EXAM			55-64 AT EXAM			65-74 AT EXAM		
	PERSON EXAMS	# OF EVENTS	ANNUAL RATE	PERSON EXAMS	# OF EVENTS	ANNUAL RATE	PERSON EXAMS	# OF EVENTS	ANNUAL RATE	PERSON EXAMS	# OF EVENTS	ANNUAL RATE
NO	2669			4291	2	0	4304	5	1	1928	7	2
MILD	1326			2375	3	1	2863	8	1	1756	5	1
DEFINITE	521			1366	3	1	1747	5	1	942	5	3
UNKNOWN	0			0	0	-	0	0	-	0	0	-
TOTAL	4516			8032	8	0	8914	18	1	4626	17	2

RISK FACTOR	75-84 AT EXAM			85-94 AT EXAM			35-64 COMBINED			65-94 COMBINED		
	PERSON EXAMS	# OF EVENTS	ANNUAL RATE	PERSON EXAMS	# OF EVENTS	ANNUAL RATE	# OF EVENTS	CRUDE	AGE-ADJUSTED	# OF EVENTS	CRUDE	AGE-ADJUSTED
NO	490	3	3	20			7	0	0	10	2	2
MILD	503	3	3	18			11	1	1	8	2	2
DEFINITE	241	3	6	4			8	1	1	8	3	3
UNKNOWN	0	0	-	0			0	-	-	0	-	-
TOTAL	1234	9	4	42			26	1	1	26	2	2

LOGISTIC REGRESSIONS

AGE AT EXAM		COEFFICIENT OF RISK FACTOR	SE COEFFICIENT	P-VALUE	STANDARDIZED COEFFICIENT	STD SE
35-44						
45-54		0.761	0.436	0.081	0.575	0.330
55-64		0.452	0.291	0.121	0.349	0.225
65-74		0.163	0.315	0.604	0.124	0.239
75-84		0.358	0.440	0.417	0.266	0.327
85-94						
35-64	UNIVARIATE	0.625	0.242	0.010	0.471	0.182
65-94	UNIVARIATE	0.232	0.255	0.362	0.175	0.192
35-64	BIVARIATE	0.524	0.243	0.031	0.395	0.183
65-94	BIVARIATE	0.235	0.255	0.357	0.177	0.193
35-64	MULTIVARIATE	0.516	0.249	0.039	0.389	0.188
65-94	MULTIVARIATE	0.223	0.261	0.394	0.168	0.197

SEX: FEMALE
EVENT: TRANSIENT ISCHEMIC ATTACK
RISK FACTOR: HYPERTENSION IGNORING TREATMENT
PERSONS AT RISK: PERSONS FREE OF STROKE AND TIA ONLY

RISK FACTOR	35-44 AT EXAM			45-54 AT EXAM			55-64 AT EXAM			65-74 AT EXAM		
	PERSON EXAMS	# OF EVENTS	ANNUAL RATE	PERSON EXAMS	# OF EVENTS	ANNUAL RATE	PERSON EXAMS	# OF EVENTS	ANNUAL RATE	PERSON EXAMS	# OF EVENTS	ANNUAL RATE
NO	4268			5829			5360	3	0	2168	5	1
MILD	1016			2787			4031	4	0	2816	3	1
DEFINITE	329			1417			2559	5	1	1887	7	2
UNKNOWN	0			0			0	0	-	0	0	-
TOTAL	5613			10033			11950	12	1	6871	15	1

RISK FACTOR	75-84 AT EXAM			85-94 AT EXAM			35-64 COMBINED			65-94 COMBINED		
	PERSON EXAMS	# OF EVENTS	ANNUAL RATE	PERSON EXAMS	# OF EVENTS	ANNUAL RATE	# OF EVENTS	CRUDE	AGE-ADJUSTED	# OF EVENTS	CRUDE	AGE-ADJUSTED
NO	575	3	3	48			3	0	0	8	1	1
MILD	932	7	4	51			6	0	0	10	1	1
DEFINITE	720	5	3	49			6	1	1	13	2	2
UNKNOWN	0	0	-	0			0	-	-	0	-	-
TOTAL	2227	15	3	148			15	0	0	31	2	2

LOGISTIC REGRESSIONS

AGE AT EXAM		COEFFICIENT OF RISK FACTOR	SE COEFFICIENT	P-VALUE	STANDARDIZED COEFFICIENT	STD SE
35-44						
45-54						
55-64		0.630	0.365	0.084	0.492	0.284
65-74		0.298	0.340	0.384	0.228	0.261
75-84		0.120	0.344	0.727	0.091	0.261
85-94						
35-64	UNIVARIATE	0.940	0.322	0.003	0.699	0.239
65-94	UNIVARIATE	0.302	0.238	0.204	0.232	0.183
35-64	BIVARIATE	0.727	0.331	0.028	0.540	0.246
65-94	BIVARIATE	0.262	0.239	0.273	0.201	0.183
35-64	MULTIVARIATE	0.682	0.341	0.045	0.507	0.254
65-94	MULTIVARIATE	0.149	0.243	0.540	0.114	0.186

TABLE 15-4   DATE=23AUG85
AVERAGE ANNUAL INCIDENCE RATE FOR EVENT PER 1000 PERSONS AT RISK AT EXAM
BY AGE AND LEVEL OF RISK FACTOR AT EXAM: FRAMINGHAM STUDY 30-YEAR FOLLOWUP

SEX: MALE
EVENT: TRANSIENT ISCHEMIC ATTACK
RISK FACTOR: SERUM CHOLESTEROL
PERSONS AT RISK: PERSONS FREE OF STROKE AND TIA ONLY

RISK FACTOR	35-44 AT EXAM PERSON EXAMS	# OF EVENTS	ANNUAL RATE	45-54 AT EXAM PERSON EXAMS	# OF EVENTS	ANNUAL RATE	55-64 AT EXAM PERSON EXAMS	# OF EVENTS	ANNUAL RATE	65-74 AT EXAM PERSON EXAMS	# OF EVENTS	ANNUAL RATE
84-204	1253			1796	0	0	2431	7	1	1471	5	2
205-234	1257			2382	2	0	2649	5	1	1389	5	2
235-264	882			1835	1	0	2099	1	0	1007	4	2
265-294	487			1050	3	1	1059	5	2	505	2	2
295-1124	303			618	2	2	544	0	0	237	1	2
UNKNOWN	334			351	0	0	132	0	0	17	0	0
TOTAL	4516			8032	8	0	8914	18	1	4626	17	2

RISK FACTOR	75-84 AT EXAM PERSON EXAMS	# OF EVENTS	ANNUAL RATE	85-94 AT EXAM PERSON EXAMS	# OF EVENTS	ANNUAL RATE	35-64 COMBINED # OF EVENTS	CRUDE	AGE-ADJUSTED	65-94 COMBINED # OF EVENTS	CRUDE	AGE-ADJUSTED
84-204	480	4	4	18			7	1	0	9	2	2
205-234	389	2	3	11			7	1	0	7	2	2
235-264	221	2	5	7			2	0	0	6	2	3
265-294	121	1	4	6			8	2	1	3	2	2
295-1124	23	0	0	0			2	1	1	1	2	2
UNKNOWN	0	0	-	0			0	0	0	0	0	0
TOTAL	1234	9	4	42			26	1	0	26	2	2

LOGISTIC REGRESSIONS

AGE AT EXAM		COEFFICIENT OF RISK FACTOR	SE COEFFICIENT	P-VALUE	STANDARDIZED COEFFICIENT	STD SE
35-44						
45-54		0.014	0.004	0.001	0.600	0.155
55-64		-0.005	0.006	0.386	-0.215	0.248
65-74		0.004	0.006	0.482	0.165	0.235
75-84		0.001	0.009	0.889	0.046	0.333
85-94						
35-64	UNIVARIATE	0.006	0.004	0.160	0.239	0.170
65-94	UNIVARIATE	0.002	0.005	0.611	0.098	0.193
35-64	BIVARIATE	0.006	0.004	0.112	0.278	0.175
65-94	BIVARIATE	0.003	0.005	0.522	0.124	0.194
35-64	MULTIVARIATE	0.006	0.004	0.149	0.253	0.176
65-94	MULTIVARIATE	0.003	0.005	0.576	0.108	0.194

SEX: FEMALE
EVENT: TRANSIENT ISCHEMIC ATTACK
RISK FACTOR: SERUM CHOLESTEROL
PERSONS AT RISK: PERSONS FREE OF STROKE AND TIA ONLY

RISK FACTOR	35-44 AT EXAM PERSON EXAMS	# OF EVENTS	ANNUAL RATE	45-54 AT EXAM PERSON EXAMS	# OF EVENTS	ANNUAL RATE	55-64 AT EXAM PERSON EXAMS	# OF EVENTS	ANNUAL RATE	65-74 AT EXAM PERSON EXAMS	# OF EVENTS	ANNUAL RATE
84-204	2349			2113			1681	0	0	869	3	2
205-234	1415			2578			2704	3	1	1510	1	0
235-264	806			2262			3017	5	1	1867	2	1
265-294	372			1481			2339	1	0	1417	3	1
295-1124	171			1119			1976	3	1	1171	6	3
UNKNOWN	500			480			233	0	0	37	0	0
TOTAL	5613			10033			11950	12	1	6871	15	1

RISK FACTOR	75-84 AT EXAM PERSON EXAMS	# OF EVENTS	ANNUAL RATE	85-94 AT EXAM PERSON EXAMS	# OF EVENTS	ANNUAL RATE	35-64 COMBINED # OF EVENTS	CRUDE	AGE-ADJUSTED	65-94 COMBINED # OF EVENTS	CRUDE	AGE-ADJUSTED
84-204	407	2	2	53			1	0	0	5	2	2
205-234	537	6	6	35			4	0	0	8	2	2
235-264	619	3	2	41			5	0	0	5	1	1
265-294	382	3	4	9			2	0	0	6	2	2
295-1124	275	1	2	10			3	0	0	7	2	3
UNKNOWN	7	0	0	0			0	0	0	0	0	0
TOTAL	2227	15	3	148			15	0	0	31	2	2

LOGISTIC REGRESSIONS

AGE AT EXAM		COEFFICIENT OF RISK FACTOR	SE COEFFICIENT	P-VALUE	STANDARDIZED COEFFICIENT	STD SE
35-44						
45-54						
55-64		0.005	0.006	0.672	0.118	0.279
65-74		0.009	0.005	0.087	0.383	0.223
75-84		-0.003	0.006	0.668	-0.113	0.264
85-94						
35-64	UNIVARIATE	0.004	0.004	0.353	0.191	0.206
65-94	UNIVARIATE	0.002	0.004	0.637	0.083	0.177
35-64	BIVARIATE	0.000	0.006	0.937	0.021	0.263
65-94	BIVARIATE	0.003	0.004	0.398	0.149	0.176
35-64	MULTIVARIATE	-0.001	0.006	0.881	-0.040	0.266
65-94	MULTIVARIATE	0.003	0.004	0.431	0.136	0.173

AVERAGE ANNUAL INCIDENCE RATE FOR EVENT PER 1000 PERSONS AT RISK AT EXAM
BY AGE AND LEVEL OF RISK FACTOR AT EXAM: FRAMINGHAM STUDY 30-YEAR FOLLOWUP

SEX: MALE
EVENT: TRANSIENT ISCHEMIC ATTACK
RISK FACTOR: HEMATOCRIT
PERSONS AT RISK: PERSONS FREE OF STROKE AND TIA ONLY

RISK FACTOR	35-44 AT EXAM PERSON EXAMS	# OF EVENTS	ANNUAL RATE	45-54 AT EXAM PERSON EXAMS	# OF EVENTS	ANNUAL RATE	55-64 AT EXAM PERSON EXAMS	# OF EVENTS	ANNUAL RATE	65-74 AT EXAM PERSON EXAMS	# OF EVENTS	ANNUAL RATE
25-42	630			961	1	1	1117	1	0	541	1	1
43-44	551			993	0	0	1082	0	0	582	1	1
45-46	990			1935	4	1	2192	3	1	1151	5	2
47-48	785			1573	0	0	1783	6	2	1033	3	1
49-70	1190			2180	3	1	2581	8	2	1299	7	3
UNKNOWN	370			390	0	0	159	0	0	20	0	0
TOTAL	4516			8032	8	0	8914	18	1	4626	17	2

RISK FACTOR	75-84 AT EXAM PERSON EXAMS	# OF EVENTS	ANNUAL RATE	85-94 AT EXAM PERSON EXAMS	# OF EVENTS	ANNUAL RATE	35-64 COMBINED # OF EVENTS	CRUDE	AGE-ADJUSTED	65-94 COMBINED # OF EVENTS	CRUDE	AGE-ADJUSTED
25-42	258	2	4	14			2	0	0	3	2	2
43-44	150	2	7	7			0	0	0	3	2	2
45-46	292	3	5	6			7	1	1	4	2	3
47-48	248	1	2	8			6	1	1	4	2	2
49-70	286	1	2	7			11	1	1	8	3	2
UNKNOWN	0	0	-	0			0	0	0	0	0	0
TOTAL	1234	9	4	42			26	1	1	26	2	2

LOGISTIC REGRESSIONS

AGE AT EXAM		COEFFICIENT OF RISK FACTOR	SE COEFFICIENT	P-VALUE	STANDARDIZED COEFFICIENT	STD SE
35-44						
45-54		0.012	0.099	0.901	0.044	0.354
55-64		0.097	0.062	0.119	0.365	0.234
65-74		0.078	0.068	0.249	0.282	0.245
75-84		-0.076	0.078	0.332	-0.307	0.317
85-94						
35-64	UNIVARIATE	0.075	0.053	0.160	0.276	0.196
65-94	UNIVARIATE	0.007	0.053	0.901	0.024	0.197
35-64	BIVARIATE	0.072	0.053	0.174	0.264	0.194
65-94	BIVARIATE	0.015	0.053	0.781	0.055	0.197
35-64	MULTIVARIATE	0.047	0.053	0.381	0.172	0.196
65-94	MULTIVARIATE	0.001	0.053	0.980	0.005	0.201

SEX: FEMALE
EVENT: TRANSIENT ISCHEMIC ATTACK
RISK FACTOR: HEMATOCRIT
PERSONS AT RISK: PERSONS FREE OF STROKE AND TIA ONLY

RISK FACTOR	35-44 AT EXAM PERSON EXAMS	# OF EVENTS	ANNUAL RATE	45-54 AT EXAM PERSON EXAMS	# OF EVENTS	ANNUAL RATE	55-64 AT EXAM PERSON EXAMS	# OF EVENTS	ANNUAL RATE	65-74 AT EXAM PERSON EXAMS	# OF EVENTS	ANNUAL RATE
21-39	1697			1853			1585	2	1	719	3	2
40-41	966			1760			2127	2	0	1132	0	0
42-43	1041			1972			2353	2	0	1388	1	0
44-45	857			2206			2969	4	1	1782	3	1
46-65	486			1694			2646	2	0	1810	8	2
UNKNOWN	566			548			270	0	0	40	0	0
TOTAL	5613			10033			11950	12	1	6871	15	1

RISK FACTOR	75-84 AT EXAM PERSON EXAMS	# OF EVENTS	ANNUAL RATE	85-94 AT EXAM PERSON EXAMS	# OF EVENTS	ANNUAL RATE	35-64 COMBINED # OF EVENTS	CRUDE	AGE-ADJUSTED	65-94 COMBINED # OF EVENTS	CRUDE	AGE-ADJUSTED
21-39	324	2	3	28			3	0	0	6	3	3
40-41	427	0	0	36			3	0	0	0	0	0
42-43	418	5	6	26			2	0	0	6	2	2
44-45	567	3	3	29			4	0	0	6	1	1
46-65	485	5	5	29			3	0	0	13	3	3
UNKNOWN	6	0	0	0			0	0	0	0	0	0
TOTAL	2227	15	3	148			15	0	0	31	2	2

LOGISTIC REGRESSIONS

AGE AT EXAM		COEFFICIENT OF RISK FACTOR	SE COEFFICIENT	P-VALUE	STANDARDIZED COEFFICIENT	STD SE
35-44						
45-54						
55-64		-0.037	0.082	0.658	-0.127	0.286
65-74		0.077	0.075	0.303	0.270	0.262
75-84		0.068	0.069	0.323	0.257	0.260
85-94						
35-64	UNIVARIATE	0.002	0.072	0.973	0.009	0.259
65-94	UNIVARIATE	0.051	0.051	0.314	0.184	0.182
35-64	BIVARIATE	-0.041	0.073	0.576	-0.147	0.262
65-94	BIVARIATE	0.061	0.050	0.224	0.218	0.179
35-64	MULTIVARIATE	-0.046	0.072	0.521	-0.167	0.260
65-94	MULTIVARIATE	0.055	0.050	0.271	0.197	0.179

TABLE 15-6   DATE=07NOV85
AVERAGE ANNUAL INCIDENCE RATE FOR EVENT PER 1000 PERSONS AT RISK AT EXAM
BY AGE AND LEVEL OF RISK FACTOR AT EXAM: FRAMINGHAM STUDY 30-YEAR FOLLOWUP

SEX: MALE
EVENT: TRANSIENT ISCHEMIC ATTACK
RISK FACTOR: BLOOD GLUCOSE (MG/DL)
PERSONS AT RISK: PERSONS FREE OF STROKE AND TIA ONLY

RISK FACTOR	35-44 AT EXAM			45-54 AT EXAM			55-64 AT EXAM			65-74 AT EXAM		
	PERSON EXAMS	# OF EVENTS	ANNUAL RATE	PERSON EXAMS	# OF EVENTS	ANNUAL RATE	PERSON EXAMS	# OF EVENTS	ANNUAL RATE	PERSON EXAMS	# OF EVENTS	ANNUAL RATE
29-69	1004			1554	1	0	1179	1	0	484	3	3
70-89	2528			4326	4	0	4458	8	1	2050	6	1
90-109	628			1397	2	1	2105	6	1	1242	5	2
110-129	107			329	1	2	599	2	2	412	1	1
130-524	49			202	0	0	472	1	1	421	2	2
UNKNOWN	200			224	0	0	101	0	0	17	0	0
TOTAL	4516			8032	8	0	8914	18	1	4626	17	2

RISK FACTOR	75-84 AT EXAM			85-94 AT EXAM			35-64 COMBINED			65-94 COMBINED		
	PERSON EXAMS	# OF EVENTS	ANNUAL RATE	PERSON EXAMS	# OF EVENTS	ANNUAL RATE	# OF EVENTS	CRUDE	AGE-ADJUSTED	# OF EVENTS	CRUDE	AGE-ADJUSTED
29-69	82	0	0	1			2	0	0	3	3	2
70-89	449	5	6	16			12	1	1	11	2	2
90-109	394	2	3	10			8	1	1	7	2	2
110-129	151	2	7	5			3	1	1	3	2	2
130-524	158	0	0	10			1	1	0	2	2	2
UNKNOWN	0	0	-	0			0	0	0	0	0	0
TOTAL	1234	9	4	42			26	1	1	26	2	2

LOGISTIC REGRESSIONS

AGE AT EXAM		COEFFICIENT OF RISK FACTOR	SE COEFFICIENT	P-VALUE	STANDARDIZED COEFFICIENT	STD SE
35-44						
45-54		0.000	0.015	0.976	0.010	0.344
55-64		0.002	0.007	0.788	0.056	0.208
65-74		-0.001	0.007	0.882	-0.039	0.262
75-84		-0.011	0.015	0.451	-0.375	0.498
85-94						
35-64	UNIVARIATE	0.004	0.006	0.435	0.111	0.142
65-94	UNIVARIATE	-0.003	0.007	0.659	-0.103	0.234
35-64	BIVARIATE	0.000	0.007	0.974	0.006	0.171
65-94	BIVARIATE	-0.004	0.007	0.585	-0.134	0.245
35-64	MULTIVARIATE	0.001	0.007	0.902	0.021	0.167
65-94	MULTIVARIATE	-0.004	0.007	0.591	-0.132	0.246

SEX: FEMALE
EVENT: TRANSIENT ISCHEMIC ATTACK
RISK FACTOR: BLOOD GLUCOSE (MG/DL)
PERSONS AT RISK: PERSONS FREE OF STROKE AND TIA ONLY

RISK FACTOR	35-44 AT EXAM			45-54 AT EXAM			55-64 AT EXAM			65-74 AT EXAM		
	PERSON EXAMS	# OF EVENTS	ANNUAL RATE	PERSON EXAMS	# OF EVENTS	ANNUAL RATE	PERSON EXAMS	# OF EVENTS	ANNUAL RATE	PERSON EXAMS	# OF EVENTS	ANNUAL RATE
29-69	1067			1803			1403	1	0	571	3	3
70-89	3408			5786			6298	6	0	3369	3	1
90-109	672			1640			2833	3	1	1933	3	1
110-129	96			302			762	1	1	563	3	3
130-524	39			166			489	1	1	405	3	4
UNKNOWN	331			336			165	0	0	30	0	0
TOTAL	5613			10033			11950	12	1	6871	15	1

RISK FACTOR	75-84 AT EXAM			85-94 AT EXAM			35-64 COMBINED			65-94 COMBINED		
	PERSON EXAMS	# OF EVENTS	ANNUAL RATE	PERSON EXAMS	# OF EVENTS	ANNUAL RATE	# OF EVENTS	CRUDE	AGE-ADJUSTED	# OF EVENTS	CRUDE	AGE-ADJUSTED
29-69	120	3	13	4			1	0	0	6	4	5
70-89	891	5	3	55			7	0	0	8	1	1
90-109	744	2	1	58			5	0	0	6	1	1
110-129	263	2	4	17			1	0	0	5	3	3
130-524	208	3	7	14			1	1	0	6	5	4
UNKNOWN	1	0	0	0			0	0	0	0	0	0
TOTAL	2227	15	5	148			15	0	0	31	2	2

LOGISTIC REGRESSIONS

AGE AT EXAM		COEFFICIENT OF RISK FACTOR	SE COEFFICIENT	P-VALUE	STANDARDIZED COEFFICIENT	STD SE
35-44						
45-54						
55-64		0.002	0.009	0.812	0.059	0.246
65-74		0.007	0.005	0.123	0.224	0.145
75-84		0.007	0.004	0.100	0.244	0.149
85-94						
35-64	UNIVARIATE	0.007	0.007	0.307	0.154	0.151
65-94	UNIVARIATE	0.008	0.003	0.009	0.251	0.096
35-64	BIVARIATE	0.003	0.008	0.704	0.068	0.180
65-94	BIVARIATE	0.007	0.003	0.021	0.231	0.100
35-64	MULTIVARIATE	0.001	0.008	0.881	0.027	0.179
65-94	MULTIVARIATE	0.006	0.003	0.060	0.196	0.105

TABLE 15-7   DATE=07NOV85
AVERAGE ANNUAL INCIDENCE RATE FOR EVENT PER 1000 PERSONS AT RISK AT EXAM
BY AGE AND LEVEL OF RISK FACTOR AT EXAM: FRAMINGHAM STUDY 30-YEAR FOLLOWUP

SEX: MALE
EVENT: TRANSIENT ISCHEMIC ATTACK
RISK FACTOR: DIABETES MELLITUS
PERSONS AT RISK: PERSONS FREE OF STROKE AND TIA ONLY

RISK FACTOR	35-44 AT EXAM			45-54 AT EXAM			55-64 AT EXAM			65-74 AT EXAM		
	PERSON EXAMS	# OF EVENTS	ANNUAL RATE	PERSON EXAMS	# OF EVENTS	ANNUAL RATE	PERSON EXAMS	# OF EVENTS	ANNUAL RATE	PERSON EXAMS	# OF EVENTS	ANNUAL RATE
NO	4467			7780	8	1	8292	16	1	4101	16	2
YES	49			252	0	0	622	2	2	525	1	1
UNKNOWN	0			0	0	-	0	0	-	0	0	-
TOTAL	4516			8032	8	0	8914	18	1	4626	17	2

RISK FACTOR	75-84 AT EXAM			85-94 AT EXAM			35-64 COMBINED			65-94 COMBINED		
	PERSON EXAMS	# OF EVENTS	ANNUAL RATE	PERSON EXAMS	# OF EVENTS	ANNUAL RATE	# OF EVENTS	CRUDE	AGE-ADJUSTED	# OF EVENTS	CRUDE	AGE-ADJUSTED
NO	1076	7	3	36			24	1	1	23	2	2
YES	158	2	6	6			2	1	1	3	2	2
UNKNOWN	0	0	-	0			0	-		0	-	
TOTAL	1234	9	4	42			26	1	1	26	2	2

LOGISTIC REGRESSIONS

AGE AT EXAM		COEFFICIENT OF RISK FACTOR	SE COEFFICIENT	P-VALUE	STANDARDIZED COEFFICIENT	STD SE
35-44		-	-	-	-	-
45-54						
55-64		0.512	0.751	0.496	0.130	0.191
65-74		-0.719	1.032	0.486	-0.228	0.327
75-84		0.672	0.806	0.405	0.225	0.270
85-94						
35-64	UNIVARIATE	0.619	0.737	0.401	0.125	0.149
65-94	UNIVARIATE	-0.013	0.615	0.983	-0.004	0.198
35-64	BIVARIATE	0.141	0.740	0.849	0.029	0.150
65-94	BIVARIATE	-0.034	0.616	0.956	-0.011	0.198
35-64	MULTIVARIATE	0.161	0.747	0.829	0.033	0.152
65-94	MULTIVARIATE	-0.007	0.621	0.991	-0.002	0.200

SEX: FEMALE
EVENT: TRANSIENT ISCHEMIC ATTACK
RISK FACTOR: DIABETES MELLITUS
PERSONS AT RISK: PERSONS FREE OF STROKE AND TIA ONLY

RISK FACTOR	35-44 AT EXAM			45-54 AT EXAM			55-64 AT EXAM			65-74 AT EXAM		
	PERSON EXAMS	# OF EVENTS	ANNUAL RATE	PERSON EXAMS	# OF EVENTS	ANNUAL RATE	PERSON EXAMS	# OF EVENTS	ANNUAL RATE	PERSON EXAMS	# OF EVENTS	ANNUAL RATE
NO	5570			9820			11350	10	0	6289	12	1
YES	43			213			600	2	2	582	3	3
UNKNOWN	0			0			0	0	-	0	0	-
TOTAL	5613			10033			11950	12	1	6871	15	1

RISK FACTOR	75-84 AT EXAM			85-94 AT EXAM			35-64 COMBINED			65-94 COMBINED		
	PERSON EXAMS	# OF EVENTS	ANNUAL RATE	PERSON EXAMS	# OF EVENTS	ANNUAL RATE	# OF EVENTS	CRUDE	AGE-ADJUSTED	# OF EVENTS	CRUDE	AGE-ADJUSTED
NO	1942	11	3	127			13	0	0	24	1	1
YES	285	4	7	21			2	1	1	7	4	4
UNKNOWN	0	0	-	0			0	-	0	0	-	
TOTAL	2227	15	3	148			15	0	0	31	2	2

LOGISTIC REGRESSIONS

AGE AT EXAM		COEFFICIENT OF RISK FACTOR	SE COEFFICIENT	P-VALUE	STANDARDIZED COEFFICIENT	STD SE
35-44						
45-54						
55-64		1.333	0.776	0.086	0.291	0.169
65-74		0.997	0.647	0.123	0.278	0.180
75-84		0.916	0.587	0.119	0.306	0.196
85-94						
35-64	UNIVARIATE	1.572	0.760	0.039	0.272	0.132
65-94	UNIVARIATE	1.015	0.431	0.019	0.299	0.127
35-64	BIVARIATE	1.133	0.766	0.139	0.196	0.133
65-94	BIVARIATE	0.900	0.434	0.038	0.265	0.128
35-64	MULTIVARIATE	0.882	0.779	0.258	0.153	0.135
65-94	MULTIVARIATE	0.699	0.447	0.118	0.206	0.132

TABLE 15-8   DATE=07NOV85
AVERAGE ANNUAL INCIDENCE RATE FOR EVENT PER 1000 PERSONS AT RISK AT EXAM
BY AGE AND LEVEL OF RISK FACTOR AT EXAM: FRAMINGHAM STUDY 30-YEAR FOLLOWUP

SEX: MALE
EVENT: TRANSIENT ISCHEMIC ATTACK
RISK FACTOR: GLUCOSE IN URINE, DEFINITE OR TRACE
PERSONS AT RISK: PERSONS FREE OF STROKE AND TIA ONLY

RISK FACTOR	35-44 AT EXAM			45-54 AT EXAM			55-64 AT EXAM			65-74 AT EXAM		
	PERSON EXAMS	# OF EVENTS	ANNUAL RATE	PERSON EXAMS	# OF EVENTS	ANNUAL RATE	PERSON EXAMS	# OF EVENTS	ANNUAL RATE	PERSON EXAMS	# OF EVENTS	ANNUAL RATE
NO	4423			7834	8	1	8639	18	1	4460	16	2
YES	60			177	0	0	251	0	0	160	1	3
UNKNOWN	33			21	0	0	24	0	0	6	0	0
TOTAL	4516			8032	8	0	8914	18	1	4626	17	2

RISK FACTOR	75-84 AT EXAM			85-94 AT EXAM			35-64 COMBINED			65-94 COMBINED		
	PERSON EXAMS	# OF EVENTS	ANNUAL RATE	PERSON EXAMS	# OF EVENTS	ANNUAL RATE	# OF EVENTS	CRUDE	AGE-ADJUSTED	# OF EVENTS	CRUDE	AGE-ADJUSTED
NO	1199	9	4	41			26	1	1	25	2	2
YES	32	0	0	1			0	0	0	1	3	2
UNKNOWN	3	0	0	0			0	0	0	0	0	0
TOTAL	1234	9	4	42			26	1	1	26	2	2

LOGISTIC REGRESSIONS

AGE AT EXAM		COEFFICIENT OF RISK FACTOR	SE COEFFICIENT	P-VALUE	STANDARDIZED COEFFICIENT	STD SE
35-44		-	-	-		-
45-54		-	-	-	-	-
55-64		-	-	-	-	-
65-74		0.558	1.034	0.590	0.102	0.189
75-84		-	-	-	-	-
85-94		-	-	-	-	-
35-64	UNIVARIATE	-	-	-	-	-
65-94	UNIVARIATE	0.167	1.022	0.870	0.030	0.182
35-64	BIVARIATE	-	-	-	-	-
65-94	BIVARIATE	0.198	1.023	0.847	0.035	0.182
35-64	MULTIVARIATE	-	-	-	-	-
65-94	MULTIVARIATE	0.160	1.026	0.876	0.029	0.183

SEX: FEMALE
EVENT: TRANSIENT ISCHEMIC ATTACK
RISK FACTOR: GLUCOSE IN URINE, DEFINITE OR TRACE
PERSONS AT RISK: PERSONS FREE OF STROKE AND TIA ONLY

RISK FACTOR	35-44 AT EXAM			45-54 AT EXAM			55-64 AT EXAM			65-74 AT EXAM		
	PERSON EXAMS	# OF EVENTS	ANNUAL RATE	PERSON EXAMS	# OF EVENTS	ANNUAL RATE	PERSON EXAMS	# OF EVENTS	ANNUAL RATE	PERSON EXAMS	# OF EVENTS	ANNUAL RATE
NO	5534			9908			11753	12	1	6733	13	1
YES	41			101			171	0	0	128	2	8
UNKNOWN	38			24			26	0	0	10	0	0
TOTAL	5613			10033			11950	12	1	6871	15	1

RISK FACTOR	75-84 AT EXAM			85-94 AT EXAM			35-64 COMBINED			65-94 COMBINED		
	PERSON EXAMS	# OF EVENTS	ANNUAL RATE	PERSON EXAMS	# OF EVENTS	ANNUAL RATE	# OF EVENTS	CRUDE	AGE-ADJUSTED	# OF EVENTS	CRUDE	AGE-ADJUSTED
NO	2183	14	5	145	3		15	0	0	28	2	2
YES	35	1	14	3			0	0	0	3	9	9
UNKNOWN	9	0	0	0			0	0	0	0	0	0
TOTAL	2227	15	3	148			15	0	0	31	2	2

LOGISTIC REGRESSIONS

AGE AT EXAM		COEFFICIENT OF RISK FACTOR	SE COEFFICIENT	P-VALUE	STANDARDIZED COEFFICIENT	STD SE
35-44						
45-54						
55-64		-	-	-		
65-74		2.105	0.765	0.006	0.285	0.103
75-84		1.517	1.049	0.148	0.189	0.131
85-94						
35-64	UNIVARIATE	-	-	-		-
65-94	UNIVARIATE	1.781	0.613	0.004	0.237	0.081
35-64	BIVARIATE	-	-	-	-	-
65-94	BIVARIATE	1.822	0.614	0.003	0.242	0.082
35-64	MULTIVARIATE	-	-	-	-	-
65-94	MULTIVARIATE	1.511	0.643	0.019	0.201	0.086

SEX: MALE
EVENT: TRANSIENT ISCHEMIC ATTACK
RISK FACTOR: GLUCOSE INTOLERANCE
PERSONS AT RISK: PERSONS FREE OF STROKE AND TIA ONLY

RISK FACTOR	35-44 AT EXAM PERSON EXAMS	# OF EVENTS	ANNUAL RATE	45-54 AT EXAM PERSON EXAMS	# OF EVENTS	ANNUAL RATE	55-64 AT EXAM PERSON EXAMS	# OF EVENTS	ANNUAL RATE
NO	4378			7497	8	1	7847	15	1
YES	138			535	0	0	1067	3	1
UNKNOWN	0			0	0	-	0	0	-
TOTAL	4516			8032	8	0	8914	18	1

RISK FACTOR	75-84 AT EXAM PERSON EXAMS	# OF EVENTS	ANNUAL RATE	85-94 AT EXAM PERSON EXAMS	# OF EVENTS	ANNUAL RATE	35-64 COMBINED # OF EVENTS	CRUDE	AGE-ADJUSTED
NO	952	7	4	29			23	1	1
YES	282	2	4	13			3	1	1
UNKNOWN	0	0	-	0			0	-	
TOTAL	1234	9	4	42			26	1	1

LOGISTIC REGRESSIONS

AGE AT EXAM		COEFFICIENT OF RISK FACTOR	SE COEFFICIENT	P-VALUE	STANDARDIZE COEFFICIENT
35-44		-	-	-	-
45-54					
55-64		0.387	0.633	0.541	0.126
65-74		-0.540	0.754	0.474	-0.210
75-84		-0.036	0.805	0.964	-0.015
85-94					
35-64	UNIVARIATE	0.392	0.614	0.524	0.107
65-94	UNIVARIATE	-0.293	0.545	0.591	-0.116
35-64	BIVARIATE	0.008	0.617	0.990	0.002
65-94	BIVARIATE	-0.332	0.546	0.543	-0.132
35-64	MULTIVARIATE	0.021	0.620	0.973	0.006
65-94	MULTIVARIATE	-0.325	0.547	0.553	-0.129

SEX: FEMALE
EVENT: TRANSIENT ISCHEMIC ATTACK
RISK FACTOR: GLUCOSE INTOLERANCE
PERSONS AT RISK: PERSONS FREE OF STROKE AND TIA ONLY

RISK FACTOR	35-44 AT EXAM PERSON EXAMS	# OF EVENTS	ANNUAL RATE	45-54 AT EXAM PERSON EXAMS	# OF EVENTS	ANNUAL RATE	55-64 AT EXAM PERSON EXAMS	# OF EVENTS	ANNUAL RATE	P E
NO	5519			9648			10848	10	0	
YES	94			385			1102	2	1	
UNKNOWN	0			0			0	0	-	
TOTAL	5613			10033			11950	12	1	

RISK FACTOR	75-84 AT EXAM PERSON EXAMS	# OF EVENTS	ANNUAL RATE	85-94 AT EXAM PERSON EXAMS	# OF EVENTS	ANNUAL RATE	35-64 COMBINED # OF EVENTS	CRUDE	AGE-ADJUSTED
NO	1752	11	3	120			13	0	0
YES	475	4	4	28			2	1	0
UNKNOWN	0	0	-	0			0	-	
TOTAL	2227	15	3	148			15	0	0

LOGISTIC REGRESSIONS

AGE AT EXAM		COEFFICIENT OF RISK FACTOR	SE COEFFICIENT	P-VALUE	STANDARDIZED COEFFICIENT
35-44					
45-54				-	
55-64		0.678	0.775	0.382	0.196
65-74		1.087	0.549	0.048	0.382
75-84		0.296	0.586	0.614	0.121
85-94					
35-64	UNIVARIATE	0.930	0.760	0.221	0.216
65-94	UNIVARIATE	0.753	0.397	0.058	0.278
35-64	BIVARIATE	0.490	0.765	0.522	0.114
65-94	BIVARIATE	0.643	0.399	0.108	0.237
55-64	MULTIVARIATE	0.338	0.772	0.661	0.079
65-94	MULTIVARIATE	0.494	0.410	0.228	0.182

TABLE 15-10   DATE=07NOV85
AVERAGE ANNUAL INCIDENCE RATE FOR EVENT PER 1000 PERSONS AT RISK AT EXAM
BY AGE AND LEVEL OF RISK FACTOR AT EXAM: FRAMINGHAM STUDY 30-YEAR FOLLOWUP

SEX: MALE
EVENT: TRANSIENT ISCHEMIC ATTACK
RISK FACTOR: METROPOLITAN RELATIVE WEIGHT (PERCENT)
PERSONS AT RISK: PERSONS FREE OF STROKE AND TIA ONLY

RISK FACTOR	35-44 AT EXAM			45-54 AT EXAM			55-64 AT EXAM			65-74 AT EXAM		
	PERSON EXAMS	# OF EVENTS	ANNUAL RATE	PERSON EXAMS	# OF EVENTS	ANNUAL RATE	PERSON EXAMS	# OF EVENTS	ANNUAL RATE	PERSON EXAMS	# OF EVENTS	ANNUAL RATE
54-104	720			1145	0	0	1251	5	2	795	2	1
105-114	979			1631	2	1	1830	5	1	998	2	1
115-124	1133			2131	0	0	2428	3	1	1272	10	4
125-134	923			1681	5	1	1819	1	0	918	3	2
135-272	761			1443	1	0	1585	4	1	643	0	0
UNKNOWN	0			1	0	0	1	0	0	0	0	-
TOTAL	4516			8032	8	0	8914	18	1	4626	17	2

RISK FACTOR	75-84 AT EXAM			85-94 AT EXAM			35-64 COMBINED			65-94 COMBINED		
	PERSON EXAMS	# OF EVENTS	ANNUAL RATE	PERSON EXAMS	# OF EVENTS	ANNUAL RATE	# OF EVENTS	CRUDE	AGE-ADJUSTED	# OF EVENTS	CRUDE	AGE-ADJUSTED
54-104	247	2	4	15			5	1	1	4	2	2
105-114	323	5	8	12			7	1	1	7	3	2
115-124	377	2	3	13			3	0	0	12	4	4
125-134	184	0	0	2			6	1	1	3	1	1
135-272	103	0	0	0			5	1	1	0	0	0
UNKNOWN	0	0	-	0			0	0	0	0	-	
TOTAL	1234	9	4	42			26	1	1	26	2	2

LOGISTIC REGRESSIONS

AGE AT EXAM		COEFFICIENT OF RISK FACTOR	SE COEFFICIENT	P-VALUE	STANDARDIZED COEFFICIENT	STD SE
35-44						
45-54		0.017	0.020	0.409	0.279	0.327
55-64		-0.021	0.016	0.186	-0.329	0.249
65-74		-0.003	0.016	0.847	-0.047	0.244
75-84		-0.041	0.023	0.075	-0.609	0.343
85-94						
35-64	UNIVARIATE	-0.007	0.012	0.563	-0.116	0.201
65-94	UNIVARIATE	-0.016	0.013	0.217	-0.249	0.202
35-64	BIVARIATE	-0.007	0.013	0.563	-0.117	0.202
65-94	BIVARIATE	-0.015	0.013	0.261	-0.229	0.204
35-64	MULTIVARIATE	-0.009	0.013	0.498	-0.141	0.207
65-94	MULTIVARIATE	-0.019	0.014	0.168	-0.288	0.209

SEX: FEMALE
EVENT: TRANSIENT ISCHEMIC ATTACK
RISK FACTOR: METROPOLITAN RELATIVE WEIGHT (PERCENT)
PERSONS AT RISK: PERSONS FREE OF STROKE AND TIA ONLY

RISK FACTOR	35-44 AT EXAM			45-54 AT EXAM			55-64 AT EXAM			65-74 AT EXAM		
	PERSON EXAMS	# OF EVENTS	ANNUAL RATE	PERSON EXAMS	# OF EVENTS	ANNUAL RATE	PERSON EXAMS	# OF EVENTS	ANNUAL RATE	PERSON EXAMS	# OF EVENTS	ANNUAL RATE
54-104	1689			2108			2108	1	0	1275	6	2
105-114	1431			2351			2509	1	0	1289	4	2
115-124	1059			2123			2535	1	0	1485	1	0
125-134	565			1378			1813	7	2	1119	2	1
135-272	861			2073			2985	2	0	1703	2	1
UNKNOWN	8			0			0	0	-	0	0	-
TOTAL	5613			10053			11950	12	1	6871	15	1

RISK FACTOR	75-84 AT EXAM			85-94 AT EXAM			35-64 COMBINED			65-94 COMBINED		
	PERSON EXAMS	# OF EVENTS	ANNUAL RATE	PERSON EXAMS	# OF EVENTS	ANNUAL RATE	# OF EVENTS	CRUDE	AGE-ADJUSTED	# OF EVENTS	CRUDE	AGE-ADJUSTED
54-104	592	1	1	47			1	0	0	8	2	2
105-114	405	7	9	51			1	0	0	11	3	3
115-124	448	1	1	28			2	0	0	2	1	1
125-134	325	3	5	24			7	1	1	5	2	2
135-272	457	3	3	18			4	0	0	5	1	1
UNKNOWN	0	0	-	0			0	0	0	0	-	
TOTAL	2227	15	3	148			15	0	0	31	2	2

LOGISTIC REGRESSIONS

AGE AT EXAM		COEFFICIENT OF RISK FACTOR	SE COEFFICIENT	P-VALUE	STANDARDIZED COEFFICIENT	STD SE
35-44						
45-54						
55-64		0.012	0.011	0.280	0.261	0.241
65-74		-0.025	0.014	0.082	-0.558	0.321
75-84		0.002	0.012	0.863	0.044	0.256
85-94						
35-64	UNIVARIATE	0.018	0.009	0.037	0.395	0.189
65-94	UNIVARIATE	-0.016	0.009	0.088	-0.351	0.206
35-64	BIVARIATE	0.015	0.009	0.089	0.338	0.199
65-94	BIVARIATE	-0.013	0.009	0.158	-0.287	0.203
35-64	MULTIVARIATE	0.008	0.010	0.410	0.180	0.219
65-94	MULTIVARIATE	-0.016	0.009	0.088	-0.351	0.206

TABLE 15-11   DATE=07NOV85
AVERAOE ANNUAL INCIDENCE RATE FOR EVENT PER 1000 PERSONS AT RISK AT EXAM
BY AGE AND LEVEL OF RISK FACTOR AT EXAM: FRAMINGHAM STUDY 30-YEAR FOLLOWUP

SEX: MALE
EVENT: TRANSIENT ISCHEMIC ATTACK
RISK FACTOR: VITAL CAPACITY - HEIGHT INDEX (ML/INCH)
PERSONS AT RISK: PERSONS FREE OF STROKE AND TIA ONLY

RISK FACTOR	35-44 AT EXAM PERSON EXAMS	# OF EVENTS	ANNUAL RATE	45-54 AT EXAM PERSON EXAMS	# OF EVENTS	ANNUAL RATE	55-64 AT EXAM PERSON EXAMS	# OF EVENTS	ANNUAL RATE	65-74 AT EXAM PERSON EXAMS	# OF EVENTS	ANNUAL RATE
8-44	268			1100	2	1	2319	9	2	2063	7	2
45-49	416			1182	3	1	1667	2	1	950	4	2
50-54	751			1721	0	0	2024	1	0	824	3	2
55-59	1034			1809	3	1	1540	2	1	509	3	3
60-85	2034			2214	0	0	1346	4	1	268	0	0
UNKNOWN	13			6	0	0	18	0	0	12	0	0
TOTAL	4516			8032	8	0	8914	18	1	4626	17	2

RISK FACTOR	75-84 AT EXAM PERSON EXAMS	# OF EVENTS	ANNUAL RATE	85-94 AT EXAM PERSON EXAMS	# OF EVENTS	ANNUAL RATE	35-64 COMBINED # OF EVENTS	CRUDE	AGE-ADJUSTED	65-94 COMBINED # OF EVENTS	CRUDE	AGE-ADJUSTED
8-44	778	7	4	28			11	1	1	14	2	2
45-49	207	0	0	5			5	1	1	4	2	2
50-54	130	2	8	7			1	0	0	3	3	3
55-59	79	0	0	2			5	1	1	3	3	2
60-85	32	0	0	0			4	0	1	0	0	0
UNKNOWN	8	0	0	0			0	0	0	0	0	0
TOTAL	1234	9	4	42			26	1	1	26	2	2

LOGISTIC REGRESSIONS

AGE AT EXAM		COEFFICIENT OF RISK FACTOR	SE COEFFICIENT	P-VALUE	STANDARDIZED COEFFICIENT	STD SE
35-44						
45-54		-0.054	0.036	0.138	-0.489	0.329
55-64		-0.036	0.024	0.133	-0.338	0.225
65-74		-0.000	0.026	0.999	-0.000	0.243
75-84		-0.026	0.036	0.477	-0.238	0.336
85-94						
35-64	UNIVARIATE	-0.057	0.019	0.002	-0.543	0.181
65-94	UNIVARIATE	-0.014	0.021	0.489	-0.135	0.195
35-64	BIVARIATE	-0.032	0.020	0.116	-0.309	0.197
65-94	BIVARIATE	-0.010	0.021	0.648	-0.091	0.200
35-64	MULTIVARIATE	-0.024	0.021	0.261	-0.228	0.203
65-94	MULTIVARIATE	-0.007	0.022	0.744	-0.066	0.203

SEX: FEMALE
EVENT: TRANSIENT ISCHEMIC ATTACK
RISK FACTOR: VITAL CAPACITY - HEIGHT INDEX (ML/INCH)
PERSONS AT RISK: PERSONS FREE OF STROKE AND TIA ONLY

RISK FACTOR	35-44 AT EXAM PERSON EXAMS	# OF EVENTS	ANNUAL RATE	45-54 AT EXAM PERSON EXAMS	# OF EVENTS	ANNUAL RATE	55-64 AT EXAM PERSON EXAMS	# OF EVENTS	ANNUAL RATE	65-74 AT EXAM PERSON EXAMS	# OF EVENTS	ANNUAL RATE
4-29	121			539			1530	2	1	1948	3	1
30-34	291			1240			2520	3	1	1888	8	2
35-39	769			2096			3181	3	1	1694	1	0
40-44	1427			2825			2809	4	1	919	3	2
45-92	2984			3316			1889	0	0	416	0	0
UNKNOWN	21			17			21	0	0	6	0	0
TOTAL	5613			10053			11950	12	1	6871	15	1

RISK FACTOR	75-84 AT EXAM PERSON EXAMS	# OF EVENTS	ANNUAL RATE	85-94 AT EXAM PERSON EXAMS	# OF EVENTS	ANNUAL RATE	35-64 COMBINED # OF EVENTS	CRUDE	AGE-ADJUSTED	65-94 COMBINED # OF EVENTS	CRUDE	AGE-ADJUSTED
4-29	1090	9	4	100			3	1	1	12	2	2
30-34	645	4	3	33			3	0	0	13	3	3
35-39	351	1	1	14			3	0	0	2	0	1
40-44	113	0	0	1			5	0	0	3	1	1
45-92	26	1	19	0			1	0	0	1	1	5
UNKNOWN	2	0	0	0			0	0	0	0	0	0
TOTAL	2227	15	3	148			15	0	0	31	2	2

LOGISTIC REGRESSIONS

AGE AT EXAM		COEFFICIENT OF RISK FACTOR	SE COEFFICIENT	P-VALUE	STANDARDIZED COEFFICIENT	STD SE
35-44						
45-54						
55-64		-0.055	0.036	0.133	-0.410	0.273
65-74		-0.016	0.035	0.654	-0.115	0.256
75-84		-0.039	0.036	0.272	-0.275	0.251
85-94						
35-64	UNIVARIATE	-0.077	0.030	0.011	-0.611	0.240
65-94	UNIVARIATE	-0.041	0.023	0.081	-0.305	0.175
35-64	BIVARIATE	-0.046	0.034	0.166	-0.369	0.267
65-94	BIVARIATE	-0.024	0.025	0.336	-0.180	0.188
35-64	MULTIVARIATE	-0.033	0.035	0.341	-0.265	0.279
65-94	MULTIVARIATE	-0.015	0.026	0.571	-0.112	0.198

TABLE 15-12   DATE=07NOV85
AVERAGE ANNUAL INCIDENCE RATE FOR EVENT PER 1000 PERSONS AT RISK AT EXAM
BY AGE AND LEVEL OF RISK FACTOR AT EXAM: FRAMINGHAM STUDY 30-YEAR FOLLOWUP

SEX: MALE
EVENT: TRANSIENT ISCHEMIC ATTACK
RISK FACTOR: HEART RATE (PER MINUTE)
PERSONS AT RISK: PERSONS FREE OF STROKE AND TIA ONLY

RISK FACTOR	35-44 AT EXAM			45-54 AT EXAM			55-64 AT EXAM			65-74 AT EXAM		
	PERSON EXAMS	# OF EVENTS	ANNUAL RATE	PERSON EXAMS	# OF EVENTS	ANNUAL RATE	PERSON EXAMS	# OF EVENTS	ANNUAL RATE	PERSON EXAMS	# OF EVENTS	ANNUAL RATE
30-67	1295			2013	0	0	2590	6	1	365	4	1
68-75	1503			2579	2	0	2709	3	1	403	3	1
76-83	728			1465	3	1	1520	3	1	767	4	3
84-91	586			1139	2	1	1188	2	1	570	5	4
92-220	404			836	1	1	907	4	2	521	1	1
UNKNOWN	0			0	0	-	0	0	-	0	0	-
TOTAL	4516			8032	8	0	8914	18	1	4626	17	2

RISK FACTOR	75-84 AT EXAM			85-94 AT EXAM			35-64 COMBINED			65-94 COMBINED		
	PERSON EXAMS	# OF EVENTS	ANNUAL RATE	PERSON EXAMS	# OF EVENTS	ANNUAL RATE	# OF EVENTS	CRUDE	AGE-ADJUSTED	# OF EVENTS	CRUDE	AGE-ADJUSTED
30-67	353	3	4	13			6	1	0	7	2	2
68-75	339	2	3	13			5	0	0	5	1	1
76-83	246	2	4	11			6	1	1	6	3	3
84-91	163	0	0	4			4	1	1	5	3	3
92-220	133	2	8	1			5	1	1	3	2	2
UNKNOWN	0	0	-	0			0	-	-	0	-	
TOTAL	1234	9	4	42			26	1	1	26	2	2

LOGISTIC REGRESSIONS

AGE AT EXAM		COEFFICIENT OF RISK FACTOR	SE COEFFICIENT	P-VALUE	STANDARDIZED COEFFICIENT	STD SE
35-44						
45-54		0.041	0.024	0.090	0.517	0.305
55-64		0.022	0.016	0.173	0.289	0.212
65-74		0.016	0.017	0.333	0.216	0.223
75-84		-0.000	0.025	0.999	-0.000	0.335
85-94						
35-64	UNIVARIATE	0.029	0.014	0.036	0.363	0.173
65-94	UNIVARIATE	0.011	0.014	0.422	0.149	0.186
35-64	BIVARIATE	0.028	0.013	0.034	0.355	0.168
65-94	BIVARIATE	0.011	0.014	0.432	0.147	0.186
35-64	MULTIVARIATE	0.019	0.014	0.178	0.240	0.178
65-94	MULTIVARIATE	0.008	0.014	0.567	0.110	0.192

SEX: FEMALE
EVENT: TRANSIENT ISCHEMIC ATTACK
RISK FACTOR: HEART RATE (PER MINUTE)
PERSONS AT RISK: PERSONS FREE OF STROKE AND TIA ONLY

RISK FACTOR	35-44 AT EXAM			45-54 AT EXAM			55-64 AT EXAM			65-74 AT EXAM		
	PERSON EXAMS	# OF EVENTS	ANNUAL RATE	PERSON EXAMS	# OF EVENTS	ANNUAL RATE	PERSON EXAMS	# OF EVENTS	ANNUAL RATE	PERSON EXAMS	# OF EVENTS	ANNUAL RATE
30-67	1090			1908			2402	2	0	1242	0	0
68-75	1804			3269			3602	3	0	1853	4	1
76-83	1050			1930			2412	4	1	1498	2	1
84-91	939			1612			1867	2	1	1154	4	2
92-220	730			1314			1667	1	0	1124	5	2
UNKNOWN	0			0			0	0	-	0	0	-
TOTAL	5613			10033			11950	12	1	6871	15	1

RISK FACTOR	75-84 AT EXAM			85-94 AT EXAM			35-64 COMBINED			65-94 COMBINED		
	PERSON EXAMS	# OF EVENTS	ANNUAL RATE	PERSON EXAMS	# OF EVENTS	ANNUAL RATE	# OF EVENTS	CRUDE	AGE-ADJUSTED	# OF EVENTS	CRUDE	AGE-ADJUSTED
30-67	429	6	7	30			3	0	0	7	2	2
68-75	535	3	3	38			4	0	0	5	1	1
76-83	484	3	3	27			2	0	0	7	2	2
84-91	375	3	4	29			1	0	0	5	2	2
92-220	404	0	0	24								
UNKNOWN	0	0	-	0			0	-	-	0	-	
TOTAL	2227	15	3	148			15	0	0	31	2	2

LOGISTIC REGRESSIONS

AGE AT EXAM		COEFFICIENT OF RISK FACTOR	SE COEFFICIENT	P-VALUE	STANDARDIZED COEFFICIENT	STD SE
35-44						
45-54						
55-64		-0.010	0.023	0.684	-0.123	0.303
65-74		0.036	0.014	0.012	0.472	0.188
75-84		-0.056	0.022	0.012	-0.781	0.310
85-94						
35-64	UNIVARIATE	-0.016	0.022	0.459	-0.205	0.277
65-94	UNIVARIATE	-0.004	0.014	0.781	-0.051	0.183
35-64	BIVARIATE	-0.016	0.022	0.455	-0.204	0.273
65-94	BIVARIATE	-0.005	0.014	0.738	-0.061	0.182
35-64	MULTIVARIATE	-0.024	0.022	0.274	-0.306	0.279
65-94	MULTIVARIATE	-0.007	0.013	0.583	-0.099	0.181

TABLE 15-13   DATE=23AUG85

AVERAGE ANNUAL INCIDENCE RATE FOR EVENT PER 1000 PERSONS AT RISK AT EXAM
BY AGE AND LEVEL OF RISK FACTOR AT EXAM: FRAMINGHAM STUDY 30-YEAR FOLLOWUP

SEX: MALE
EVENT: TRANSIENT ISCHEMIC ATTACK
RISK FACTOR: CIGARETTES SMOKED (NUMBER PER DAY)
PERSONS AT RISK: PERSONS FREE OF STROKE AND TIA ONLY

RISK FACTOR	35-44 AT EXAM PERSON EXAMS	# OF EVENTS	ANNUAL RATE	45-54 AT EXAM PERSON EXAMS	# OF EVENTS	ANNUAL RATE	55-64 AT EXAM PERSON EXAMS	# OF EVENTS	ANNUAL RATE	65-74 AT EXAM PERSON EXAMS	# OF EVENTS	ANNUAL RATE
NONE	1402			3211	0	0	4815	7	1	3084	8	1
1-10	437			742	1	1	842	2	1	449	2	2
11-20	1497			2127	2	0	1848	8	2	633	6	5
21-40	993			1664	5	2	1217	1	0	421	1	1
41-90	161			250	0	0	168	0	0	32	0	0
UNKNOWN	26			38	0	0	24	0	0	7	0	0
TOTAL	4516			8032	8	0	8914	18	1	4626	17	2

RISK FACTOR	75-84 AT EXAM PERSON EXAMS	# OF EVENTS	ANNUAL RATE	85-94 AT EXAM PERSON EXAMS	# OF EVENTS	ANNUAL RATE	35-64 COMBINED # OF EVENTS	CRUDE	AGE-ADJUSTED	65-94 COMBINED # OF EVENTS	CRUDE	AGE-ADJUSTED
NONE	963	7	4	32			7	0	0	15	2	2
1-10	122	1	4	4			3	1	1	3	3	2
11-20	86	1	6	5			10	1	1	7	5	5
21-40	58	0	0	1			6	1	1	1	1	1
41-90	5	0	0	0			0	0	0	0	0	0
UNKNOWN	0	0	-	0			0	0	0	0	0	0
TOTAL	1234	9	4	42			26	1	0	26	2	2

LOGISTIC REGRESSIONS

AGE AT EXAM		COEFFICIENT OF RISK FACTOR	SE COEFFICIENT	P-VALUE	STANDARDIZED COEFFICIENT	STD SE
35-44	—					
45-54		0.050	0.018	0.007	0.745	0.275
55-64		0.008	0.016	0.641	0.103	0.221
65-74		0.016	0.018	0.366	0.189	-0.209
75-84		-0.020	0.048	0.677	-0.179	0.429
85-94						
35-64	UNIVARIATE	0.017	0.012	0.161	0.248	0.177
65-94	UNIVARIATE	0.007	0.017	0.694	0.072	0.184
35-64	BIVARIATE	0.028	0.012	0.018	0.404	0.170
65-94	BIVARIATE	0.009	0.017	0.569	0.105	0.185
35-64	MULTIVARIATE	0.029	0.012	0.014	0.420	0.171
65-94	MULTIVARIATE	0.009	0.017	0.585	0.101	0.184

SEX: FEMALE
EVENT: TRANSIENT ISCHEMIC ATTACK
RISK FACTOR: CIGARETTES SMOKED (NUMBER PER DAY)
PERSONS AT RISK: PERSONS FREE OF STROKE AND TIA ONLY

RISK FACTOR	35-44 AT EXAM PERSON EXAMS	# OF EVENTS	ANNUAL RATE	45-54 AT EXAM PERSON EXAMS	# OF EVENTS	ANNUAL RATE	55-64 AT EXAM PERSON EXAMS	# OF EVENTS	ANNUAL RATE	65-74 AT EXAM PERSON EXAMS	# OF EVENTS	ANNUAL RATE
NONE	2580			5582			7896	9	1	5364	11	1
1-10	1312			1723			1634	1	0	674	2	1
11-20	1372			2027			1743	2	1	627	2	2
21-40	298			626			621	0	0	186	0	0
41-90	22			38			18	0	0	1	0	0
UNKNOWN	29			37			38	0	0	19	0	0
TOTAL	5613			10033			11950	12	1	6871	15	1

RISK FACTOR	75-84 AT EXAM PERSON EXAMS	# OF EVENTS	ANNUAL RATE	85-94 AT EXAM PERSON EXAMS	# OF EVENTS	ANNUAL RATE	35-64 COMBINED # OF EVENTS	CRUDE	AGE-ADJUSTED	65-94 COMBINED # OF EVENTS	CRUDE	AGE-ADJUSTED
NONE	1959	15	4	139			11	0	0	27	2	2
1-10	158	0	0	7			2	0	0	2	1	1
11-20	93	0	0	2			2	0	0	2	1	1
21-40	13	0	0	0			0	0	0	0	0	0
41-90	0	0	-	0			0	0	0	0	0	0
UNKNOWN	4	0	0	0			0	0	0	0	0	0
TOTAL	2227	15	3	148			15	0	0	31	2	2

LOGISTIC REGRESSIONS

AGE AT EXAM		COEFFICIENT OF RISK FACTOR	SE COEFFICIENT	P-VALUE	STANDARDIZED COEFFICIENT	STD SE
35-44						
45-54		*				
55-64		-0.015	0.035	0.672	-0.139	0.329
65-74		0.008	0.031	0.811	0.057	0.240
75-84		-			-	
85-94						
35-64	UNIVARIATE	-0.032	0.034	0.345	-0.314	0.332
65-94	UNIVARIATE	-0.024	0.033	0.458	-0.172	0.231
35-64	BIVARIATE	-0.019	0.032	0.560	-0.186	0.319
65-94	BIVARIATE	-0.013	0.033	0.660	-0.095	0.231
35-64	MULTIVARIATE	-0.016	0.032	0.621	-0.156	0.315
65-94	MULTIVARIATE	-0.008	0.033	0.797	-0.059	0.230

TABLE 15-14   DATE=08NOV85
AVERAGE ANNUAL INCIDENCE RATE FOR EVENT PER 1000 PERSONS AT RISK AT EXAM
BY AGE AND LEVEL OF RISK FACTOR AT EXAM: FRAMINGHAM STUDY 30-YEAR FOLLOWUP

SEX: MALE
EVENT: TRANSIENT ISCHEMIC ATTACK
RISK FACTOR: ALBUMIN IN URINE, DEFINITE
PERSONS AT RISK: PERSONS FREE OF STROKE AND TIA ONLY

RISK FACTOR	35-44 AT EXAM			45-54 AT EXAM			55-64 AT EXAM			65-74 AT EXAM		
	PERSON EXAMS	# OF EVENTS	ANNUAL RATE	PERSON EXAMS	# OF EVENTS	ANNUAL RATE	PERSON EXAMS	# OF EVENTS	ANNUAL RATE	PERSON EXAMS	# OF EVENTS	ANNUAL RATE
NO	4408			7903	8	1	8682	17	1	4476	17	2
YES	71			103	0	0	205	1	2	144	0	0
UNKNOWN	37		.	26	0	0	27	0	0	6	0	0
TOTAL	4516			8032	8	0	8914	18	1	4626	17	2

RISK FACTOR	75-84 AT EXAM			85-94 AT EXAM			35-64 COMBINED			65-94 COMBINED		
	PERSON EXAMS	# OF EVENTS	ANNUAL RATE	PERSON EXAMS	# OF EVENTS	ANNUAL RATE	# OF EVENTS	CRUDE	AGE-ADJUSTED	# OF EVENTS	CRUDE	AGE-ADJUSTED
NO	1199	7	3	37			25	1	1	24	2	2
YES	32	2	31	5			1	1	1	2	6	7
UNKNOWN	3	0	0	0			0	0	0	0	0	0
TOTAL	1234	9	4	42			26	1	1	26	2	2

LOGISTIC REGRESSIONS

AGE AT EXAM		COEFFICIENT OF RISK FACTOR	SE COEFFICIENT	P-VALUE	STANDARDIZED COEFFICIENT	STD SE
35-44		-	-		-	-
45-54		-	-		-	-
55-64		0.916	1.031	0.375	0.137	0.155
65-74		-	-		-	-
75-84		2.429	0.823	0.003	0.387	0.131
85-94						
35-64	UNIVARIATE	0.797	1.021	0.435	0.105	0.135
65-94	UNIVARIATE	0.974	0.740	0.188	0.168	0.128
35-64	BIVARIATE	0.547	1.023	0.593	0.072	0.135
65-94	BIVARIATE	0.957	0.741	0.196	0.165	0.128
35-64	MULTIVARIATE	0.251	1.037	0.809	0.033	0.137
65-94	MULTIVARIATE	0.858	0.750	0.253	0.148	0.129

SEX: FEMALE
EVENT: TRANSIENT ISCHEMIC ATTACK
RISK FACTOR: ALBUMIN IN URINE, DEFINITE
PERSONS AT RISK: PERSONS FREE OF STROKE AND TIA ONLY

RISK FACTOR	35-44 AT EXAM			45-54 AT EXAM			55-64 AT EXAM			65-74 AT EXAM		
	PERSON EXAMS	# OF EVENTS	ANNUAL RATE	PERSON EXAMS	# OF EVENTS	ANNUAL RATE	PERSON EXAMS	# OF EVENTS	ANNUAL RATE	PERSON EXAMS	# OF EVENTS	ANNUAL RATE
NO	5350			9728			11684	12	1	6729	15	1
YES	225			280			238	0	0	132	0	0
UNKNOWN	38			25			28	0	0	10	0	0
TOTAL	5613			10033			11950	12	1	6871	15	1

RISK FACTOR	75-84 AT EXAM			85-94 AT EXAM			35-64 COMBINED			65-94 COMBINED		
	PERSON EXAMS	# OF EVENTS	ANNUAL RATE	PERSON EXAMS	# OF EVENTS	ANNUAL RATE	# OF EVENTS	CRUDE	AGE-ADJUSTED	# OF EVENTS	CRUDE	AGE-ADJUSTED
NO	2173	15	3	141	7		15	0	0	31	2	2
YES	45	0	0	7			0	0	0	0	0	0
UNKNOWN	9	0	0	0			0	0	0	0	0	0
TOTAL	2227	15	3	148			15	0	0	31	2	2

LOGISTIC REGRESSIONS

AGE AT EXAM		COEFFICIENT OF RISK FACTOR	SE COEFFICIENT	P-VALUE	STANDARDIZED COEFFICIENT	STD SE
35-44						
45-54						
55-64						
65-74						
75-84						
85-94						
35-64	UNIVARIATE	-				
65-94	UNIVARIATE	-	-		-	-
35-64	BIVARIATE	-				
65-94	BIVARIATE	-	-			
35-64	MULTIVARIATE	-	.			
65-94	MULTIVARIATE	-				

TABLE 15-15   DATE=07NOV85
AVERAGE ANNUAL INCIDENCE RATE FOR EVENT PER 1000 PERSONS AT RISK AT EXAM
BY AGE AND LEVEL OF RISK FACTOR AT EXAM: FRAMINGHAM STUDY 30-YEAR FOLLOWUP

SEX: MALE
EVENT: TRANSIENT ISCHEMIC ATTACK
RISK FACTOR: HEART ENLARGEMENT BY X-RAY
PERSONS AT RISK: PERSONS FREE OF STROKE AND TIA ONLY

RISK FACTOR	35-44 AT EXAM PERSON EXAMS	# OF EVENTS	ANNUAL RATE	45-54 AT EXAM PERSON EXAMS	# OF EVENTS	ANNUAL RATE	55-64 AT EXAM PERSON EXAMS	# OF EVENTS	ANNUAL RATE	65-74 AT EXAM PERSON EXAMS	# OF EVENTS	ANNUAL RATE
NO	4098			6590	6	0	6455	13	1	2971	10	2
MILD	189			702	0	0	1102	2	1	662	3	2
DEFINITE	224			736	2	1	1356	3	1	993	4	2
UNKNOWN	5			4	0	0	1	0	0	0	0	-
TOTAL	4516			8032	8	0	8914	18	1	4626	17	2

RISK FACTOR	75-84 AT EXAM PERSON EXAMS	# OF EVENTS	ANNUAL RATE	85-94 AT EXAM PERSON EXAMS	# OF EVENTS	ANNUAL RATE	35-64 COMBINED # OF EVENTS	CRUDE	AGE-ADJUSTED	65-94 COMBINED # OF EVENTS	CRUDE	AGE-ADJUSTED
NO	693	5	4	28			19	1	1	15	2	2
MILD	205	2	5	6			2	1	0	3	2	3
DEFINITE	336	2	3	8			5	1	1	6	2	2
UNKNOWN	0	0	-	0			0	0	0	0	0	-
TOTAL	1234	9	4	42			26	1	1	26	2	2

LOGISTIC REGRESSIONS

AGE AT EXAM		COEFFICIENT OF RISK FACTOR	SE COEFFICIENT	P-VALUE	STANDARDIZED COEFFICIENT	STD SE
35-44						
45-54		0.455	0.444	0.305	0.281	0.274
55-64		0.030	0.314	0.925	0.022	0.232
65-74		0.107	0.285	0.708	0.088	0.234
75-84		-0.060	0.393	0.878	-0.052	0.340
85-94						
35-64	UNIVARIATE	0.298	0.254	0.240	0.196	0.166
65-94	UNIVARIATE	0.075	0.230	0.746	0.062	0.192
35-64	BIVARIATE	0.096	0.258	0.709	0.063	0.169
65-94	BIVARIATE	0.054	0.231	-0.816	0.045	0.193
35-64	MULTIVARIATE	0.085	0.270	0.754	0.055	0.177
65-94	MULTIVARIATE	-0.004	0.245	0.986	-0.004	0.204

SEX: FEMALE
EVENT: TRANSIENT ISCHEMIC ATTACK
RISK FACTOR: HEART ENLARGEMENT BY X-RAY
PERSONS AT RISK: PERSONS FREE OF STROKE AND TIA ONLY

RISK FACTOR	35-44 AT EXAM PERSON EXAMS	# OF EVENTS	ANNUAL RATE	45-54 AT EXAM PERSON EXAMS	# OF EVENTS	ANNUAL RATE	55-64 AT EXAM PERSON EXAMS	# OF EVENTS	ANNUAL RATE	65-74 AT EXAM PERSON EXAMS	# OF EVENTS	ANNUAL RATE
NO	5161			8054			7762	5	0	3444	6	1
MILD	198			954			1850	3	1	1254	4	2
DEFINITE	246			1019			2334	4	1	2173	5	1
UNKNOWN	8			6			4	0	0	0	0	-
TOTAL	5613			10033			11950	12	1	6871	15	1

RISK FACTOR	75-84 AT EXAM PERSON EXAMS	# OF EVENTS	ANNUAL RATE	85-94 AT EXAM PERSON EXAMS	# OF EVENTS	ANNUAL RATE	35-64 COMBINED # OF EVENTS	CRUDE	AGE-ADJUSTED	65-94 COMBINED # OF EVENTS	CRUDE	AGE-ADJUSTED
NO	778	1	1	38			7	0	0	8	1	1
MILD	435	5	6	30			3	0	0	9	3	3
DEFINITE	1014	9	4	80			5	1	1	14	2	2
UNKNOWN	0	0	-	0			0	0	0	0	0	-
TOTAL	2227	15	3	148			15	0	0	31	2	2

LOGISTIC REGRESSIONS

AGE AT EXAM		COEFFICIENT OF RISK FACTOR	SE COEFFICIENT	P-VALUE	STANDARDIZED COEFFICIENT	STD SE
35-44						
45-54						
55-64		0.500	0.320	0.118	0.399	0.256
65-74		0.149	0.287	0.605	0.131	0.254
75-84		0.620	0.347	0.074	0.552	0.309
85-94						
35-64	UNIVARIATE	0.725	0.282	0.010	0.509	0.199
65-94	UNIVARIATE	0.375	0.204	0.065	0.336	0.183
35-64	BIVARIATE	0.485	0.292	0.097	0.341	0.205
65-94	BIVARIATE	0.290	0.208	0.162	0.260	0.186
35-64	MULTIVARIATE	0.173	0.310	0.294	0.229	0.218
65-94	MULTIVARIATE	0.173	0.214	0.420	0.155	0.192

TABLE 15-16   DATE=23AUG85
AVERAGE ANNUAL INCIDENCE RATE FOR EVENT PER 1000 PERSONS AT RISK AT EXAM
BY AGE AND LEVEL OF RISK FACTOR AT EXAM: FRAMINGHAM STUDY 30-YEAR FOLLOWUP

SEX: MALE
EVENT: TRANSIENT ISCHEMIC ATTACK
RISK FACTOR: LEFT VENTRICULAR HYPERTROPHY BY ELECTROCARDIOGRAM
PERSONS AT RISK: PERSONS FREE OF STROKE AND TIA ONLY

RISK FACTOR	35-44 AT EXAM			45-54 AT EXAM			55-64 AT EXAM			65-74 AT EXAM		
	PERSON EXAMS	# OF EVENTS	ANNUAL RATE	PERSON EXAMS	# OF EVENTS	ANNUAL RATE	PERSON EXAMS	# OF EVENTS	ANNUAL RATE	PERSON EXAMS	# OF EVENTS	ANNUAL RATE
NO	4405			7780	8	1	8471	17	1	4347	16	2
BORDER	79			145	0	0	235	0	0	114	1	4
DEFINITE	32			107	0	0	208	1	2	165	0	0
UNKNOWN	0			0	0	-	0	0	-	0	0	-
TOTAL	4516			8032	8	0	8914	18	1	4626	17	2

RISK FACTOR	75-84 AT EXAM			85-94 AT EXAM			35-64 COMBINED			65-94 COMBINED		
	PERSON EXAMS	# OF EVENTS	ANNUAL RATE	PERSON EXAMS	# OF EVENTS	ANNUAL RATE	# OF EVENTS	CRUDE	AGE-ADJUSTED	# OF EVENTS	CRUDE	AGE-ADJUSTED
NO	1148	8	3	38			25	1	1	24	2	2
BORDER	39	0	0	2			0	0	0	1	3	3
DEFINITE	47	1	11	2			1	1	1	1	2	2
UNKNOWN	0			0			0			0	-	
TOTAL	1234	9	4	42			26	1	0	26	2	2

LOGISTIC REGRESSIONS

AGE AT EXAM		COEFFICIENT OF RISK FACTOR	SE COEFFICIENT	P-VALUE	STANDARDIZED COEFFICIENT	STD SE
35-44		-	-	-		
45-54						
55-64		0.267	0.562	0.635	0.090	0.190
65-74		-0.300	0.786	0.703	-0.119	0.312
75-84		0.459	0.569	0.419	0.191	0.236
85-94						
35-64	UNIVARIATE	0.230	0.563	0.683	0.066	0.162
65-94	UNIVARIATE	0.095	0.452	0.833	0.038	0.182
35-64	BIVARIATE	0.044	0.565	0.938	0.013	0.163
65-94	BIVARIATE	0.087	0.453	0.847	0.035	0.182
35-64	MULTIVARIATE	-0.139	0.594	0.815	-0.040	0.171
65-94	MULTIVARIATE	-0.105	0.479	0.826	-0.042	0.193

SEX: FEMALE
EVENT: TRANSIENT ISCHEMIC ATTACK
RISK FACTOR: LEFT VENTRICULAR HYPERTROPHY BY ELECTROCARDIOGRAM
PERSONS AT RISK: PERSONS FREE OF STROKE AND TIA ONLY

RISK FACTOR	35-44 AT EXAM			45-54 AT EXAM			55-64 AT EXAM			65-74 AT EXAM		
	PERSON EXAMS	# OF EVENTS	ANNUAL RATE	PERSON EXAMS	# OF EVENTS	ANNUAL RATE	PERSON EXAMS	# OFE EVENTS	ANNUAL RATE	PERSON EXAMS	# OF EVENTS	ANNUAL RATE
NO	5560			9866			11542	11	0	6504	12	1
BORDER	38			111			242	0	0	174	1	3
DEFINITE	15			56			166	1	3	193	2	5
UNKNOWN	0			0			0	0	-	0	0	-
TOTAL	5613			10053			11950	12	1	6871	15	1

RISK FACTOR	75-84 AT EXAM			85-94 AT EXAM			35-64 COMBINED			65-94 COMBINED		
	PERSON EXAMS	# OF EVENTS	ANNUAL RATE	PERSON EXAMS	# OF EVENTS	ANNUAL RATE	# OF EVENTS	CRUDE	AGE-ADJUSTED	# OF EVENTS	CRUDE	AGE-ADJUSTED
NO	2059	12	5	129			14	0	0	25	1	2
BORDER	65	1	8	8			0	0	0	2	4	4
DEFINITE	103	2	10	11			1	2	1	4	7	6
UNKNOWN	0	0	-	0			0	-		0	-	
TOTAL	2227	15	3	148			15	0	0	31	2	2

LOGISTIC REGRESSIONS

AGE AT EXAM		COEFFICIENT OF RISK FACTOR	SE COEFFICIENT	P-VALUE	STANDARDIZED COEFFICIENT	STD SE
35-44						
45-54						
55-64		0.792	0.563	0.159	0.215	0.153
65-74		0.889	0.366	0.015	0.322	0.132
75-84		0.637	0.366	0.081	0.285	0.163
85-94						
35-64	UNIVARIATE	0.915	0.557	0.100	0.199	0.121
65-94	UNIVARIATE	0.783	0.258	0.002	0.304	0.100
35-64	BIVARIATE	0.696	0.565	0.218	0.152	0.123
65-94	BIVARIATE	0.712	0.260	0.006	0.277	0.101
35-64	MULTIVARIATE	0.212	0.627	0.735	0.046	0.137
65-94	MULTIVARIATE	0.531	0.278	0.056	0.206	0.108

TABLE 15-17   DATE=07NOV85
AVERAGE ANNUAL INCIDENCE RATE FOR EVENT PER 1000 PERSONS AT RISK AT EXAM
BY AGE AND LEVEL OF RISK FACTOR AT EXAM: FRAMINGHAM STUDY 30-YEAR FOLLOWUP

SEX: MALE
EVENT: TRANSIENT ISCHEMIC ATTACK
RISK FACTOR: INTRAVENTRICULAR CONDUCTION DEFECT (WITH QRS INTERVAL MORE THAN 0.11 SECOND)
PERSONS AT RISK: PERSONS FREE OF STROKE AND TIA ONLY

RISK FACTOR	35-44 AT EXAM PERSON EXAMS	# OF EVENTS	ANNUAL RATE	45-54 AT EXAM PERSON EXAMS	# OF EVENTS	ANNUAL RATE	55-64 AT EXAM PERSON EXAMS	# OF EVENTS	ANNUAL RATE	65-74 AT EXAM PERSON EXAMS	# OF EVENTS	ANNUAL RATE
NO	4198			7452	7	0	8107	15	1	4150	14	2
INCOMPLETE	280			491	0	0	534	3	3	231	1	2
COMPLETE	38			69	1	6	273	0	0	245	2	4
UNKNOWN	0			0	0	-	0	0	-	0	0	-
TOTAL	4516			8032	8	0	8914	18	1	4626	17	2

RISK FACTOR	75-84 AT EXAM PERSON EXAMS	# OF EVENTS	ANNUAL RATE	85-94 AT EXAM PERSON EXAMS	# OF EVENTS	ANNUAL RATE	35-64 COMBINED # OF EVENTS	CRUDE	AGE-ADJUSTED	65-94 COMBINED # OF EVENTS	CRUDE	AGE-ADJUSTED
NO	1083	8	4	35			22	1	1	22	2	2
INCOMPLETE	35	0	0	3			3	1	1	1	2	2
COMPLETE	116	1	4	4			1	1	2	3	4	4
UNKNOWN	0	0	-	0			0	-	-	0	-	
TOTAL	1234	9	4	42			26	1	1	26	2	2

LOGISTIC REGRESSIONS

AGE AT EXAM		COEFFICIENT OF RISK FACTOR	SE COEFFICIENT	P-VALUE	STANDARDIZED COEFFICIENT	STD SE
35-44						
45-54		0.907	0.641	0.157	0.285	0.201
55-64		0.230	0.489	0.639	0.094	0.200
65-74		0.424	0.370	0.252	0.207	0.180
75-84		0.016	0.553	0.976	0.010	0.331
85-94						
35-64	UNIVARIATE	0.518	0.391	0.185	0.184	0.138
65-94	UNIVARIATE	0.302	0.308	0.326	0.155	0.158
35-64	BIVARIATE	0.341	0.381	0.371	0.121	0.135
65-94	BIVARIATE	0.272	0.308	0.376	0.140	0.158
35-64	MULTIVARIATE	0.368	0.386	0.340	0.131	0.137
65-94	MULTIVARIATE	0.287	0.309	0.352	0.148	0.159

SEX: FEMALE
EVENT: TRANSIENT ISCHEMIC ATTACK
RISK FACTOR: INTRAVENTRICULAR CONDUCTION DEFECT (WITH QRS INTERVAL MORE THAN 0.11 SECOND)
PERSONS AT RISK: PERSONS FREE OF STROKE AND TIA ONLY

RISK FACTOR	35-44 AT EXAM PERSON EXAMS	# OF EVENTS	ANNUAL RATE	45-54 AT EXAM PERSON EXAMS	# OF EVENTS	ANNUAL RATE	55-64 AT EXAM PERSON EXAMS	# OF EVENTS	ANNUAL RATE	65-74 AT EXAM PERSON EXAMS	# OF EVENTS	ANNUAL RATE
NO	5526			9834			11582	11	0	6577	14	1
INCOMPLETE	75			151			195	0	0	130	0	0
COMPLETE	12			48			173	1	3	164	1	3
UNKNOWN	0			0			0	0	-	0	0	-
TOTAL	5613			10033			11950	12	1	6871	15	1

RISK FACTOR	75-84 AT EXAM PERSON EXAMS	# OF EVENTS	ANNUAL RATE	85-94 AT EXAM PERSON EXAMS	# OF EVENTS	ANNUAL RATE	35-64 COMBINED # OF EVENTS	CRUDE	AGE-ADJUSTED	65-94 COMBINED # OF EVENTS	CRUDE	AGE-ADJUSTED
NO	2088	15	4	138			14	0	0	30	2	2
INCOMPLETE	49	0	0	3			0	0	0	0	0	0
COMPLETE	90	0	0	7			1	2	1	2	2	2
UNKNOWN	0	0	-	0			0	-	-	0	-	
TOTAL	2227	15	3	148			15	0	0	31	2	2

LOGISTIC REGRESSIONS

AGE AT EXAM		COEFFICIENT OF RISK FACTOR	SE COEFFICIENT	P-VALUE	STANDARDIZED COEFFICIENT	STD SE
35-44						
45-54						
55-64		0.801	0.555	0.149	0.215	0.149
65-74		0.419	0.550	0.446	0.139	0.182
75-84		-	-	-	-	-
85-94						
35-64	UNIVARIATE	0.912	0.560	0.103	0.200	0.123
65-94	UNIVARIATE	-0.100	0.550	0.856	-0.036	0.196
35-64	BIVARIATE	0.691	0.556	0.214	0.151	0.122
65-94	BIVARIATE	-0.160	0.549	0.771	-0.057	0.195
35-64	MULTIVARIATE	0.673	0.556	0.227	0.147	0.122
65-94	MULTIVARIATE	-0.223	0.557	0.689	-0.079	0.198

TABLE 15-18   DATE=07NOV85
AVERAGE ANNUAL INCIDENCE RATE FOR EVENT PER 1000 PERSONS AT RISK AT EXAM
BY AGE AND LEVEL OF RISK FACTOR AT EXAM: FRAMINGHAM STUDY 30-YEAR FOLLOWUP

SEX: MALE
EVENT: TRANSIENT ISCHEMIC ATTACK
RISK FACTOR: DEFINITE NONSPECIFIC T-WAVE OR ST-SEGMENT ABNORMALITY BY ELECTROCARDIOGRAM
PERSONS AT RISK: PERSONS FREE OF STROKE AND TIA ONLY

RISK FACTOR	35-44 AT EXAM PERSON EXAMS	# OF EVENTS	ANNUAL RATE	45-54 AT EXAM PERSON EXAMS	# OF EVENTS	ANNUAL RATE	55-64 AT EXAM PERSON EXAMS	# OF EVENTS	ANNUAL RATE	65-74 AT EXAM PERSON EXAMS	# OF EVENTS	ANNUAL RATE
NO	4403			7609	5	0	8128	17	1	4007	13	2
YES	113			423	3	4	786	1	1	619	4	3
UNKNOWN	0			0	0	-	0	0	-	0	0	-
TOTAL	4516			8032	8	0	8914	18	1	4626	17	2

RISK FACTOR	75-84 AT EXAM PERSON EXAMS	# OF EVENTS	ANNUAL RATE	85-94 AT EXAM PERSON EXAMS	# OF EVENTS	ANNUAL RATE	35-64 COMBINED- # OF EVENTS	CRUDE	AGE-ADJUSTED	65-94 COMBINED- # OF EVENTS	CRUDE	AGE-ADJUSTED
NO	988	8	4	24			22	1	1	21	2	2
YES	246	1	2	18			4	2	2	5	3	3
UNKNOWN	0	0	-	0			0	-	-	0	-	-
TOTAL	1234	9	4	42			26	1	1	26	2	2

LOGISTIC REGRESSIONS

AGE AT EXAM		COEFFICIENT OF RISK FACTOR	SE COEFFICIENT	P-VALUE	STANDARDIZED COEFFICIENT	STD SE
35-44						
45-54		2.385	0.732	0.001	0.533	0.164
55-64		-0.498	1.030	0.629	-0.141	0.292
65-74		0.692	0.573	0.227	0.236	0.195
75-84		-0.693	1.063	0.514	-0.277	0.425
85-94						
35-64	UNIVARIATE	1.021	0.544	0.061	0.245	0.131
65-94	UNIVARIATE	0.304	0.499	0.542	0.108	0.178
35-64	BIVARIATE	0.658	0.547	0.229	0.158	0.132
65-94	BIVARIATE	0.246	0.502	0.625	0.088	0.179
35-64	MULTIVARIATE	0.460	0.554	0.407	0.111	0.133
65-94	MULTIVARIATE	0.173	0.505	0.732	0.062	0.180

SEX: FEMALE
EVENT: TRANSIENT ISCHEMIC ATTACK
RISK FACTOR: DEFINITE NONSPECIFIC T-WAVE OR ST-SEGMENT ABNORMALITY BY ELECTROCARDIOGRAM
PERSONS AT RISK: PERSONS FREE OF STROKE AND TIA ONLY

RISK FACTOR	35-44 AT EXAM PERSON EXAMS	# OF EVENTS	ANNUAL RATE	45-54 AT EXAM PERSON EXAMS	# OF EVENTS	ANNUAL RATE	55-64 AT EXAM PERSON EXAMS	# OF EVENTS	ANNUAL RATE	65-74 AT EXAM PERSON EXAMS	# OF EVENTS	ANNUAL RATE
NO	5514			9603			10984	11	1	5994	11	1
YES	99			430			966	1	1	877	4	2
UNKNOWN	0			0			0	0	-	0	0	-
TOTAL	5613			10033			11950	12	1	6871	15	1

RISK FACTOR	75-84 AT EXAM PERSON EXAMS	# OF EVENTS	ANNUAL RATE	85-94 AT EXAM PERSON EXAMS	# OF EVENTS	ANNUAL RATE	35-64 COMBINED # OF EVENTS	CRUDE	AGE-ADJUSTED	65-94 COMBINED # OF EVENTS	CRUDE	AGE-ADJUSTED
NO	1811	9	2	114			13	0	0	21	1	1
YES	416	6	7	34			2	1	1	10	4	3
UNKNOWN	0	0	-	0			0	-	-	0	-	-
TOTAL	2227	15	3	148			15	0	0	31	2	2

LOGISTIC REGRESSIONS

AGE AT EXAM		COEFFICIENT OF RISK FACTOR	SE COEFFICIENT	P-VALUE	STANDARDIZED COEFFICIENT	STD SE
35-44						
45-54						
55-64		0.033	1.045	0.975	0.009	0.285
65-74		0.913	0.585	0.119	0.305	0.195
75-84		1.075	0.530	0.042	0.419	0.207
85-94						
35-64	UNIVARIATE	0.989	0.760	0.193	0.224	0.172
65-94	UNIVARIATE	1.049	0.385	0.006	0.368	0.135
35-64	BIVARIATE	0.593	0.765	0.438	0.134	0.173
65-94	BIVARIATE	0.938	0.388	0.016	0.329	0.136
35-64	MULTIVARIATE	0.414	0.771	0.592	0.094	0.175
65-94	MULTIVARIATE	0.924	0.391	0.018	0.324	0.137

TABLE 16-1   DATE=14DEC85

AVERAGE ANNUAL INCIDENCE RATE FOR EVENT PER 1000 PERSONS AT RISK AT EXAM
BY AGE AND LEVEL OF RISK FACTOR AT EXAM: FRAMINGHAM STUDY 30-YEAR FOLLOWUP

SEX: MALE
EVENT: DEATH FROM STROKE
RISK FACTOR: SYSTOLIC BLOOD PRESSURE, FIRST EXAMINER (MM HG)
PERSONS AT RISK: PERSONS FREE OF STROKE AND TIA ONLY

RISK FACTOR	35-44 AT EXAM PERSON EXAMS	# OF EVENTS	ANNUAL RATE	45-54 AT EXAM PERSON EXAMS	# OF EVENTS	ANNUAL RATE	55-64 AT EXAM PERSON EXAMS	# OF EVENTS	ANNUAL RATE	65-74 AT EXAM PERSON EXAMS	# OF EVENTS	ANNUAL RATE
74-119	1225			1837			1589	0	0	649	2	2
120-139	2191			3531			3482	2	0	1582	1	0
140-159	850			1830			2524	2	0	1394	6	2
160-179	197			589			1029	4	2	659	3	2
180-300	53			245			490	3	3	342	2	3
UNKNOWN	0			0			0	0	-	0	0	-
TOTAL	4516			8032			8914	11	1	4626	14	2

RISK FACTOR	75-84 AT EXAM PERSON EXAMS	# OF EVENTS	ANNUAL RATE	85-94 AT EXAM PERSON EXAMS	# OF EVENTS	ANNUAL RATE	35-64 COMBINED # OF EVENTS	CRUDE	AGE-ADJUSTED	65-94 COMBINED # OF EVENTS	CRUDE	AGE-ADJUSTED
74-119	136	0	0	7			0	0	0	4	3	2
120-139	420	6	7	13			2	0	0	7	2	2
140-159	379	3	4	15			5	0	0	9	3	3
160-179	193	6	16	5			4	1	1	9	5	5
180-300	106	5	24	2			3	2	1	7	8	7
UNKNOWN	0	0	-	0			0	-	-	0	-	-
TOTAL	1234	20	8	42			14	0	0	36	3	3

LOGISTIC REGRESSIONS

AGE AT EXAM		COEFFICIENT OF RISK FACTOR	SE COEFFICIENT	P-VALUE	STANDARDIZED COEFFICIENT	STD SE
35-44						
45-54						
55-64		0.040	0.009	0.001	0.877	0.189
65-74		0.011	0.010	0.266	0.264	0.238
75-84		0.026	0.008	0.001	0.591	0.185
85-94						
35-64	UNIVARIATE	0.043	0.008	0.001	0.870	0.156
65-94	UNIVARIATE	0.017	0.006	0.005	0.398	0.140
35-64	BIVARIATE	0.037	0.008	0.001	0.764	0.163
65-94	BIVARIATE	0.018	0.006	0.004	0.411	0.144
35-64	MULTIVARIATE	0.039	0.009	0.001	0.794	0.187
65-94	MULTIVARIATE	0.017	0.007	0.011	0.386	0.152

SEX: FEMALE
EVENT: DEATH FROM STROKE
RISK FACTOR: SYSTOLIC BLOOD PRESSURE, FIRST EXAMINER (MM HG)
PERSONS AT RISK: PERSONS FREE OF STROKE AND TIA ONLY

RISK FACTOR	35-44 AT EXAM PERSON EXAMS	# OF EVENTS	ANNUAL RATE	45-54 AT EXAM PERSON EXAMS	# OF EVENTS	ANNUAL RATE	55-64 AT EXAM PERSON EXAMS	# OF EVENTS	ANNUAL RATE	65-74 AT EXAM PERSON EXAMS	# OF EVENTS	ANNUAL RATE
74-119	2488			2717	0	0	2011	0	0	675	1	1
120-139	2245			4015	2	0	4168	4	0	1876	1	0
140-159	673			2070	2	0	3155	3	0	2212	8	2
160-179	156			783	0	0	1654	0	0	1302	4	2
180-300	51			448	1	1	962	3	2	806	5	3
UNKNOWN	0			0	0	-	0	0	-	0	0	-
TOTAL	5613			10033	5	0	11950	10	0	6871	21	2

RISK FACTOR	75-84 AT EXAM PERSON EXAMS	# OF EVENTS	ANNUAL RATE	85-94 AT EXAM PERSON EXAMS	# OF EVENTS	ANNUAL RATE	35-64 COMBINED # OF EVENTS	CRUDE	AGE-ADJUSTED	65-94 COMBINED # OF EVENTS	CRUDE	AGE-ADJUSTED
74-119	145	0	0	10			0	0	0	2	1	1
120-139	527	5	5	47			6	0	0	8	2	2
140-159	704	6	4	34			5	0	0	15	3	3
160-179	512	5	5	32			0	0	0	10	3	3
180-300	339	2	3	25			4	1	1	8	3	3
UNKNOWN	0	0	-	0			0	-	-	0	-	-
TOTAL	2227	18	4	148			15	0	0	43	2	2

LOGISTIC REGRESSIONS

AGE AT EXAM		COEFFICIENT OF RISK FACTOR	SE COEFFICIENT	P-VALUE	STANDARDIZED COEFFICIENT	STD SE
35-44						
45-54						
55-64		0.020	0.013	0.128	0.453	0.298
65-74		0.024	0.009	0.007	0.596	0.222
75-84		0.021	0.008	0.006	0.509	0.184
85-94		0.010	0.009	0.250	0.250	0.217
35-64	UNIVARIATE	0.026	0.007	0.001	0.610	0.161
65-94	UNIVARIATE	0.016	0.005	0.003	0.402	0.134
35-64	BIVARIATE	0.023	0.007	0.002	0.540	0.176
65-94	BIVARIATE	0.014	0.005	0.010	0.352	0.156
35-64	MULTIVARIATE	0.015	0.008	0.070	0.364	0.201
65-94	MULTIVARIATE	0.011	0.006	0.052	0.279	0.144

TABLE 16-2   DATE=14DEC85

AVERAGE ANNUAL INCIDENCE RATE FOR EVENT PER 1000 PERSONS AT RISK AT EXAM
BY AGE AND LEVEL OF RISK FACTOR AT EXAM: FRAMINGHAM STUDY 30-YEAR FOLLOWUP

SEX: MALE
EVENT: DEATH FROM STROKE
RISK FACTOR: DIASTOLIC BLOOD PRESSURE, FIRST EXAMINER (MM HG)
PERSONS AT RISK: PERSONS FREE OF STROKE AND TIA ONLY

RISK FACTOR	35-44 AT EXAM			45-54 AT EXAM			55-64 AT EXAM			65-74 AT EXAM		
	PERSON EXAMS	# OF EVENTS	ANNUAL RATE	PERSON EXAMS	# OF EVENTS	ANNUAL RATE	PERSON EXAMS	# OF EVENTS	ANNUAL RATE	PERSON EXAMS	# OF EVENTS	ANNUAL RATE
20-74	1015			1532			2056	0	0	1469	7	2
75-84	1614			2857			3127	0	0	1575	2	1
85-94	1203			2142			2192	4	1	1030	1	0
95-104	509			1065			1081	1	0	399	1	1
105-160	175			436			458	6	7	153	3	10
UNKNOWN	0			0				0	-		0	-
TOTAL	4516			8032			8914	11	1	4626	14	2

RISK FACTOR	75-84 AT EXAM			85-94 AT EXAM			35-64 COMBINED			65-94 COMBINED		
	PERSON EXAMS	# OF EVENTS	ANNUAL RATE	PERSON EXAMS	# OF EVENTS	ANNUAL RATE	# OF EVENTS	CRUDE	AGE-ADJUSTED	# OF EVENTS	CRUDE	AGE-ADJUSTED
20-74	513	6	6	22			0	0	0	14	3	3
75-84	403	6	7	15			0	0	0	9	2	2
85-94	224	4	9	5			6	1	1	5	2	2
95-104	66	4	30	0			2	0	0	5	5	7
105-160	28	0	0	0			6	3	3	3	8	8
UNKNOWN	0	0	-	0			0	-		0	-	
TOTAL	1234	20	8	42			14	0	0	36	3	3

LOGISTIC REGRESSIONS

AGE AT EXAM		COEFFICIENT OF RISK FACTOR	SE COEFFICIENT	P-VALUE	STANDARDIZED COEFFICIENT	STD SE
35-44						
45-54						
55-64		0.097	0.017	0.001	1.173	0.203
65-74		0.005	0.021	0.798	0.067	0.262
75-84		0.025	0.017	0.145	0.312	0.214
85-94						
35-64	UNIVARIATE	0.092	0.015	0.001	1.086	0.174
65-94	UNIVARIATE	0.004	0.013	0.768	0.049	0.165
35-64	BIVARIATE	0.090	0.015	0.001	1.060	0.175
65-94	BIVARIATE	0.016	0.013	0.222	0.201	0.165
35-64	MULTIVARIATE	0.091	0.018	0.001	1.071	0.208
65-94	MULTIVARIATE	0.018	0.013	0.154	0.228	0.160

SEX: FEMALE
EVENT: DEATH FROM STROKE
RISK FACTOR: DIASTOLIC BLOOD PRESSURE, FIRST EXAMINER (MM HG)
PERSONS AT RISK: PERSONS FREE OF STROKE AND TIA ONLY

RISK FACTOR	35-44 AT EXAM			45-54 AT EXAM			55-64 AT EXAM			65-74 AT EXAM		
	PERSON EXAMS	# OF EVENTS	ANNUAL RATE	PERSON EXAMS	# OF EVENTS	ANNUAL RATE	PERSON EXAMS	# OF EVENTS	ANNUAL RATE	PERSON EXAMS	# OF EVENTS	ANNUAL RATE
20-74	2161			2650	0	0	3078	2	0	2026	6	1
75-84	2028			3558	2	0	3964	3	0	2362	5	1
85-94	988			2387	0	0	2965	0	0	1559	7	2
95-104	324			933	2	1	1326	2	1	642	1	1
105-160	112			505	1	1	617	3	2	282	2	4
UNKNOWN	0			0			0	0	-	0	0	-
TOTAL	5613			10033	5	0	11950	10	0	6871	21	2

RISK FACTOR	75-84 AT EXAM			85-94 AT EXAM			35-64 COMBINED			65-94 COMBINED		
	PERSON EXAMS	# OF EVENTS	ANNUAL RATE	PERSON EXAMS	# OF EVENTS	ANNUAL RATE	# OF EVENTS	CRUDE	AGE-ADJUSTED	# OF EVENTS	CRUDE	AGE-ADJUSTED
20-74	869	5	3	66			2	0	0	12	2	2
75-84	681	7	5	52			5	0	0	14	2	2
85-94	466	5	5	21			0	0	0	13	3	3
95-104	155	1	3	7			4	1	1	2	1	1
105-160	56	0	0	2			4	2	1	2	3	3
UNKNOWN	0	0	-	0			0	-		0	-	
TOTAL	2227	18	4	148			15	0	0	43	2	2

LOGISTIC REGRESSIONS

AGE AT EXAM		COEFFICIENT OF RISK FACTOR	SE COEFFICIENT	P-VALUE	STANDARDIZED COEFFICIENT	STD SE
35-44						
45-54		0.044	0.027	0.100	0.544	0.331
55-64		0.037	0.021	0.084	0.457	0.265
65-74		0.012	0.017	0.470	0.152	0.210
75-84		0.012	0.019	0.511	0.151	0.230
85-94						
35-64	UNIVARIATE	0.044	0.016	0.007	0.537	0.198
65-94	UNIVARIATE	0.006	0.012	0.606	0.077	0.150
35-64	BIVARIATE	0.041	0.017	0.016	0.499	0.206
65-94	BIVARIATE	0.014	0.012	0.243	0.174	0.149
35-64	MULTIVARIATE	0.029	0.018	0.101	0.355	0.216
65-94	MULTIVARIATE	0.016	0.012	0.189	0.194	0.148

SEX: MALE
EVENT: DEATH FROM STROKE
RISK FACTOR: HYPERTENSION WITH ANTIHYPERTENSIVE TREATMENT
PERSONS AT RISK: PERSONS FREE OF STROKE AND TIA ONLY

RISK FACTOR	35-44 AT EXAM			45-54 AT EXAM			55-64 AT EXAM			65-74 AT EXAM		
	PERSON EXAMS	# OF EVENTS	ANNUAL RATE	PERSON EXAMS	# OF EVENTS	ANNUAL RATE	PERSON EXAMS	# OF EVENTS	ANNUAL RATE	PERSON EXAMS	# OF EVENTS	ANNUAL RATE
NO	2665			4248			4128	0	0	1764	3	1
MILD	1316			2280			2534	2	0	1474	4	1
DEFINITE	511			1212			1402	5	2	675	4	5
TREATED	24			292			850	4	2	713	3	2
UNKNOWN	0			0			0	0	-	0	0	-
TOTAL	4516			8032			8914	11	1	4626	14	2

RISK FACTOR	75-84 AT EXAM			85-94 AT EXAM			35-64 COMBINED			65-94 COMBINED		
	PERSON EXAMS	# OF EVENTS	ANNUAL RATE	PERSON EXAMS	# OF EVENTS	ANNUAL RATE	# OF EVENTS	CRUDE	AGE-ADJUSTED	# OF EVENTS	CRUDE	AGE-ADJUSTED
NO	440	4	5	15			0	0	0	9	2	2
MILD	395	5	6	14			4	0	0	9	2	2
DEFINITE	182	6	16	4			6	1	1	10	6	6
TREATED	217	5	12	9			4	2	1	8	4	4
UNKNOWN	0	0	-	0			0	-		0	-	
TOTAL	1234	20	8	42			14	0	0	36	3	3

LOGISTIC REGRESSIONS

AGE AT EXAM		COEFFICIENT OF RISK FACTOR	SE COEFFICIENT	P-VALUE	STANDARDIZED COEFFICIENT	STD SE
35-44						
45-54						
55-64		2.018	0.668	0.003	1.654	0.548
65-74		0.555	0.341	0.103	0.456	0.280
75-84		0.599	0.294	0.042	0.494	0.243
85-94						
35-64	UNIVARIATE	1.810	0.480	0.001	1.420	0.377
65-94	UNIVARIATE	0.484	0.210	0.021	0.398	0.173
35-64	BIVARIATE	1.629	0.477	0.001	1.278	0.374
65-94	BIVARIATE	0.477	0.211	0.023	0.393	0.173
35-64	MULTIVARIATE	1.746	0.553	0.002	1.370	0.434
65-94	MULTIVARIATE	0.444	0.216	0.039	0.365	0.177

SEX: FEMALE
EVENT: DEATH FROM STROKE
RISK FACTOR: HYPERTENSION WITH ANTIHYPERTENSIVE TREATMENT
PERSONS AT RISK: PERSONS FREE OF STROKE AND TIA ONLY

RISK FACTOR	35-44 AT EXAM			45-54 AT EXAM			55-64 AT EXAM			65-74 AT EXAM		
	PERSON EXAMS	# OF EVENTS	ANNUAL RATE	PERSON EXAMS	# OF EVENTS	ANNUAL RATE	PERSON EXAMS	# OF EVENTS	ANNUAL RATE	PERSON EXAMS	# OF EVENTS	ANNUAL RATE
NO	4247			5644	2	0	4955	2	0	1847	2	1
MILD	1001			2597	1	0	3409	0	0	2075	8	2
DEFINITE	316			1178	1	0	1924	2	1	1167	4	2
TREATED	49			614	1	1	1662	6	2	1782	7	2
UNKNOWN	0			0	0	-	0	0	-	0	0	-
TOTAL	5613			10033	5	0	11950	10	0	6871	21	2

RISK FACTOR	75-84 AT EXAM			85-94 AT EXAM			35-64 COMBINED			65-94 COMBINED		
	PERSON EXAMS	# OF EVENTS	ANNUAL RATE	PERSON EXAMS	# OF EVENTS	ANNUAL RATE	# OF EVENTS	CRUDE	AGE-ADJUSTED	# OF EVENTS	CRUDE	AGE-ADJUSTED
NO	450	3	5	38			4	0	0	5	1	1
MILD	654	4	3	32			1	0	0	13	2	2
DEFINITE	431	1	1	31			0	0	0	6	2	2
TREATED	692	10	7	47			7	2	1	19	4	4
UNKNOWN	0	0	-	0			0	-		0	-	
TOTAL	2227	18	4	148			15	0	0	43	2	2

LOGISTIC REGRESSIONS

AGE AT EXAM		COEFFICIENT OF RISK FACTOR	SE COEFFICIENT	P-VALUE	STANDARDIZED COEFFICIENT	STD SE
35-44						
45-54		0.574	0.531	0.283	0.442	0.409
55-64		1.155	0.482	0.017	0.967	0.403
65-74		0.435	0.294	0.140	0.357	0.241
75-84		0.249	0.323	0.442	0.195	0.254
85-94						
35-64	UNIVARIATE	1.064	0.334	0.001	0.849	0.267
65-94	UNIVARIATE	0.449	0.211	0.033	0.366	0.171
35-64	BIVARIATE	0.916	0.346	0.008	0.732	0.276
65-94	BIVARIATE	0.370	0.211	0.080	0.301	0.172
35-64	MULTIVARIATE	0.742	0.357	0.038	0.592	0.285
65-94	MULTIVARIATE	0.296	0.217	0.172	0.241	0.176

TABLE 16-3B　　　　　　　　DATE=08NOV85
AVERAGE ANNUAL INCIDENCE RATE FOR EVENT PER 1000 PERSONS AT RISK AT EXAM
BY AGE AND LEVEL OF RISK FACTOR AT EXAM: FRAMINGHAM STUDY 30-YEAR FOLLOWUP

SEX: MALE
EVENT: DEATH FROM STROKE
RISK FACTOR: HYPERTENSION IGNORING TREATMENT
PERSONS AT RISK: PERSONS FREE OF STROKE AND TIA ONLY

35-44 AT EXAM			45-54 AT EXAM			55-64 AT EXAM			65-74 AT EXAM		
PERSON EXAMS	# OF EVENTS	ANNUAL RATE	PERSON EXAMS	# OF EVENTS	ANNUAL RATE	PERSON EXAMS	# OF EVENTS	ANNUAL RATE	PERSON EXAMS	# OF EVENTS	ANNUAL RATE
2669			4291			4304	0	0	1928	3	1
1326			2375			2863	3	1	1756	6	2
521			1366			1747	8	2	942	5	3
0			0			0	0	-	0	0	-
4516			8032			8914	11	1	4626	14	2

75-84 AT EXAM			85-94 AT EXAM			35-64 COMBINED			65-94 COMBINED		
PERSON EXAMS	# OF EVENTS	ANNUAL RATE	PERSON EXAMS	# OF EVENTS	ANNUAL RATE	# OF EVENTS	CRUDE	AGE-ADJUSTED	# OF EVENTS	CRUDE	AGE-ADJUSTED
490	4	4	20			0	0	0	9	2	2
503	7	7	18			5	0	0	13	3	3
241	9	19	4			9	1	1	14	6	6
0	0	-	0			0	-	-	0	-	-
1234	20	8	42			14	0	0	36	3	3

LOGISTIC REGRESSIONS

	COEFFICIENT OF RISK FACTOR	SE COEFFICIENT	P-VALUE	STANDARDIZED COEFFICIENT	STD SE
	1.905	0.567	0.001	1.472	0.438
	0.598	0.348	0.086	0.453	0.264
	0.813	0.308	0.008	0.604	0.229
IVARIATE	1.749	0.444	0.001	1.318	0.334
IVARIATE	0.597	0.219	0.006	0.451	0.165
VARIATE	1.651	0.445	0.001	1.244	0.336
VARIATE	0.641	0.221	0.004	0.484	0.167
LTIVARIATE	1.706	0.501	0.001	1.285	0.577
LTIVARIATE	0.615	0.226	0.007	0.464	0.171

SEX: FEMALE
EVENT: DEATH FROM STROKE AND TIA
RISK FACTOR: HYPERTENSION ... TREATMENT
PERSONS AT RISK: PERSONS ... OF STROKE AND TIA ONLY

35-44 AT EXAM			45-54 AT EXAM			55-64 AT EXAM			65-74 AT EXAM		
PERSON EXAMS	# OF EVENTS	ANNUAL RATE	PERSON EXAMS	# OF EVENTS	ANNUAL RATE	PERSON EXAMS	# OF EVENTS	ANNUAL RATE	PERSON EXAMS	# OF EVENTS	ANNUAL RATE
4268			5829	2	0	5360	4	0	2168	2	0
1016			2787	1	0	4031	1	0	2816	10	2
329			1417	2	1	2559	5	1	1887	9	2
0			0	0	-	0	0	-	0	0	-
5613			10033	5	0	11950	10	0	6871	21	2

75-84 AT EXAM			85-94 AT EXAM			35-64 COMBINED			65-94 COMBINED		
PERSON EXAMS	# OF EVENTS	ANNUAL RATE	PERSON EXAMS	# OF EVENTS	ANNUAL RATE	# OF EVENTS	CRUDE	AGE-ADJUSTED	# OF EVENTS	CRUDE	AGE-ADJUSTED
575	4	3	48			6	0	0	7	1	1
932	7	4	51			2	0	0	19	3	3
720	7	5	49			7	1	1	17	3	3
0	0	-	0			0	-	-	0	-	-
2227	18	4	148			15	0	0	43	2	2

LOGISTIC REGRESSIONS

	COEFFICIENT OF RISK FACTOR	SE COEFFICIENT	P-VALUE	STANDARDIZED COEFFICIENT	STD SE
	0.708	0.545	0.194	0.515	0.396
	0.525	0.395	0.184	0.409	0.308
	0.662	0.304	0.029	0.508	0.233
	0.179	0.316	0.570	0.136	0.240
VARIATE	0.738	0.315	0.019	0.549	0.234
VARIATE	0.430	0.205	0.036	0.330	0.158
ARIATE	0.585	0.325	0.072	0.435	0.242
ARIATE	0.372	0.206	0.071	0.285	0.158
TIVARIATE	0.367	0.346	0.289	0.273	0.257
TIVARIATE	0.294	0.212	0.165	0.226	0.163

TABLE 16-4   DATE=14DEC85

AVERAGE ANNUAL INCIDENCE RATE FOR EVENT PER 1000 PERSONS AT RISK AT EXAM
BY AGE AND LEVEL OF RISK FACTOR AT EXAM: FRAMINGHAM STUDY 30-YEAR FOLLOWUP

SEX: MALE
EVENT: DEATH FROM STROKE
RISK FACTOR: SERUM CHOLESTEROL
PERSONS AT RISK: PERSONS FREE OF STROKE AND TIA ONLY

RISK FACTOR	35-44 AT EXAM			45-54 AT EXAM			55-64 AT EXAM			65-74 AT EXAM		
	PERSON EXAMS	# OF EVENTS	ANNUAL RATE	PERSON EXAMS	# OF EVENTS	ANNUAL RATE	PERSON EXAMS	# OF EVENTS	ANNUAL RATE	PERSON EXAMS	# OF EVENTS	ANNUAL RATE
84-204	1253			1796			2431	5	1	1471	11	4
205-234	1257			2382			2649	2	0	1389	1	0
235-264	882			1835			2099	2	0	1007	1	0
265-294	487			1050			1059	1	0	505	1	1
295-1124	303			618			544	0	0	237	0	0
UNKNOWN	334			351			132	1	4	17	0	0
TOTAL	4516			8032			8914	11	1	4626	14	2

RISK FACTOR	75-84 AT EXAM			85-94 AT EXAM			35-64 COMBINED			65-94 COMBINED		
	PERSON EXAMS	# OF EVENTS	ANNUAL RATE	PERSON EXAMS	# OF EVENTS	ANNUAL RATE	# OF EVENTS	CRUDE	AGE-ADJUSTED	# OF EVENTS	CRUDE	AGE-ADJUSTED
84-204	480	12	13	18			5	0	0	23	6	6
205-234	389	6	8	11			3	0	0	9	3	3
235-264	221	2	5	7			3	0	0	3	1	1
265-294	121	0	0	6			2	0	0	1	1	1
295-1124	23	0	0	0			0	0	0	0	0	0
UNKNOWN	0	0	-	0			1	1	2	0	0	0
TOTAL	1234	20	8	42			14	0	0	36	3	3

LOGISTIC REGRESSIONS

AGE AT EXAM		COEFFICIENT OF RISK FACTOR	SE COEFFICIENT	P-VALUE	STANDARDIZED COEFFICIENT	STD SE
35-44						
45-54						
55-64		-0.014	0.008	0.096	-0.585	0.352
65-74		-0.023	0.008	0.003	-0.932	0.309
75-84		-0.017	0.006	0.007	-0.646	0.239
85-94						
35-64	UNIVARIATE	-0.008	0.007	0.243	-0.356	0.305
65-94	UNIVARIATE	-0.020	0.005	0.001	-0.804	0.187
35-64	BIVARIATE	-0.008	0.007	0.282	-0.330	0.307
65-94	BIVARIATE	-0.018	0.005	0.001	-0.711	0.186
35-64	MULTIVARIATE	-0.010	0.007	0.167	-0.440	0.318
65-94	MULTIVARIATE	-0.017	0.005	0.001	-0.699	0.189

SEX: FEMALE
EVENT: DEATH FROM STROKE
RISK FACTOR: SERUM CHOLESTEROL
PERSONS AT RISK: PERSONS FREE OF STROKE AND TIA ONLY

RISK FACTOR	35-44 AT EXAM			45-54 AT EXAM			55-64 AT EXAM			65-74 AT EXAM		
	PERSON EXAMS	# OF EVENTS	ANNUAL RATE	PERSON EXAMS	# OF EVENTS	ANNUAL RATE	PERSON EXAMS	# OF EVENTS	ANNUAL RATE	PERSON EXAMS	# OF EVENTS	ANNUAL RATE
84-204	2349			2113	0	0	1681	2	1	869	8	5
205-234	1415			2578	0	0	2704	1	0	1510	3	1
235-264	806			2262	0	0	3017	4	1	1867	4	1
265-294	372			1481	1	0	2339	1	0	1417	2	1
295-1124	171			1119	3	1	1976	2	1	1171	3	1
UNKNOWN	500			480	0	0	233	0	0	37	1	14
TOTAL	5613			10053	5	0	11950	10	0	6871	21	2

RISK FACTOR	75-84 AT EXAM			85-94 AT EXAM			35-64 COMBINED			65-94 COMBINED		
	PERSON EXAMS	# OF EVENTS	ANNUAL RATE	PERSON EXAMS	# OF EVENTS	ANNUAL RATE	# OF EVENTS	CRUDE	AGE-ADJUSTED	# OF EVENTS	CRUDE	AGE-ADJUSTED
84-204	407	4	5	53			2	0	0	14	5	5
205-234	537	5	5	35			1	0	0	8	2	2
235-264	619	2	2	41			5	0	0	8	2	2
265-294	382	5	7	9			2	0	0	7	2	2
295-1124	275	2	4	10			5	1	1	5	2	2
UNKNOWN	7	0	0	0			0	0	0	1	11	10
TOTAL	2227	18	4	148			15	0	0	43	2	2

LOGISTIC REGRESSIONS

AGE AT EXAM		COEFFICIENT OF RISK FACTOR	SE COEFFICIENT	P-VALUE	STANDARDIZED COEFFICIENT	STD SE
35-44						
45-54		0.007	0.003	0.014	0.326	0.132
55-64		-0.002	0.007	0.827	-0.070	0.322
65-74		-0.013	0.005	0.020	-0.568	0.243
75-84		0.000	0.005	0.976	0.007	0.236
85-94						
35-64	UNIVARIATE	0.006	0.003	0.031	0.294	0.136
65-94	UNIVARIATE	-0.010	0.004	0.009	-0.428	0.165
35-64	BIVARIATE	0.005	0.004	0.155	0.258	0.181
65-94	BIVARIATE	-0.007	0.004	0.060	-0.309	0.164
35-64	MULTIVARIATE	0.004	0.004	0.253	0.211	0.184
65-94	MULTIVARIATE	-0.007	0.004	0.077	-0.291	0.165

TABLE 16-5   DATE=08NOV85
AVERAGE ANNUAL INCIDENCE RATE FOR EVENT PER 1000 PERSONS AT RISK AT EXAM
BY AGE AND LEVEL OF RISK FACTOR AT EXAM: FRAMINGHAM STUDY 30-YEAR FOLLOWUP

SEX: MALE
EVENT: DEATH FROM STROKE
RISK FACTOR: HEMATOCRIT
PERSONS AT RISK: PERSONS FREE OF STROKE AND TIA ONLY

RISK FACTOR	35-44 AT EXAM			45-54 AT EXAM			55-64 AT EXAM			65-74 AT EXAM		
	PERSON EXAMS	# OF EVENTS	ANNUAL RATE	PERSON EXAMS	# OF EVENTS	ANNUAL RATE	PERSON EXAMS	# OF EVENTS	ANNUAL RATE	PERSON EXAMS	# OF EVENTS	ANNUAL RATE
25-42	630			961			1117	1	0	541	2	2
43-44	551			993			1082	0	0	582	0	0
45-46	990			1935			2192	3	1	1151	3	1
47-48	785			1573			1783	1	0	1033	5	2
49-70	1190			2180			2581	5	1	1299	4	2
UNKNOWN	370			390			159	1	3	20	0	0
TOTAL	4516			8032			8914	11	1	4626	14	2

RISK FACTOR	75-84 AT EXAM			85-94 AT EXAM			35-64 COMBINED			65-94 COMBINED		
	PERSON EXAMS	# OF EVENTS	ANNUAL RATE	PERSON EXAMS	# OF EVENTS	ANNUAL RATE	# OF EVENTS	CRUDE	AGE-ADJUSTED	# OF EVENTS	CRUDE	AGE-ADJUSTED
25-42	258	3	6	14			2	0	0	6	4	3
43-44	150	5	17	7			1	0	0	5	3	3
45-46	292	6	10	6			2	0	0	9	3	3
47-48	248	3	6	8			2	0	0	8	3	3
49-70	286	3	5	7			5	0	0	8	3	3
UNKNOWN	0	0	-	0			1	1	1	0	0	0
TOTAL	1234	20	8	42			14	0	0	36	3	3

LOGISTIC REGRESSIONS

AGE AT EXAM		COEFFICIENT OF RISK FACTOR	SE COEFFICIENT	P-VALUE	STANDARDIZED COEFFICIENT	STD SE
35-44						
45-54						
55-64		0.122	0.082	0.137	0.459	0.309
65-74		0.043	0.075	0.563	0.156	0.270
75-84		-0.039	0.054	0.468	-0.159	0.220
85-94						
35-64	UNIVARIATE	0.059	0.076	0.438	0.216	0.278
65-94	UNIVARIATE	-0.036	0.044	0.406	-0.136	0.164
35-64	BIVARIATE	0.055	0.075	0.460	0.203	0.275
65-94	BIVARIATE	-0.005	0.043	0.900	-0.020	0.161
35-64	MULTIVARIATE	0.036	0.075	0.631	0.133	0.277
65-94	MULTIVARIATE	0.002	0.043	0.955	0.009	0.162

SEX: FEMALE
EVENT: DEATH FROM STROKE
RISK FACTOR: HEMATOCRIT
PERSONS AT RISK: PERSONS FREE OF STROKE AND TIA ONLY

RISK FACTOR	35-44 AT EXAM			45-54 AT EXAM			55-64 AT EXAM			65-74 AT EXAM		
	PERSON EXAMS	# OF EVENTS	ANNUAL RATE	PERSON EXAMS	# OF EVENTS	ANNUAL RATE	PERSON EXAMS	# OF EVENTS	ANNUAL RATE	PERSON EXAMS	# OF EVENTS	ANNUAL RATE
21-39	1697			1853	1	0	1585	1	0	719	5	3
40-41	966			1760	0	0	2127	3	1	1132	4	2
42-43	1041			1972	1	0	2353	2	0	1388	2	1
44-45	857			2206	0	0	2969	1	0	1782	4	1
46-65	486			1694	3	1	2646	3	1	1810	5	1
UNKNOWN	566			548	0	0	270	0	0	40	1	13
TOTAL	5613			10033	5	0	11950	10	0	6871	21	2

RISK FACTOR	75-84 AT EXAM			85-94 AT EXAM			35-64 COMBINED			65-94 COMBINED		
	PERSON EXAMS	# OF EVENTS	ANNUAL RATE	PERSON EXAMS	# OF EVENTS	ANNUAL RATE	# OF EVENTS	CRUDE	AGE-ADJUSTED	# OF EVENTS	CRUDE	AGE-ADJUSTED
21-39	324	5	8	28			2	0	0	10	5	4
40-41	427	5	6	36			3	0	0	11	3	3
42-43	418	0	0	26			3	0	0	3	1	1
44-45	567	3	3	29			1	0	0	7	1	1
46-65	485	5	5	29			6	1	1	11	2	3
UNKNOWN	6	0	0	0			0	0	0	1	11	9
TOTAL	2227	18	4	148			15	0	0	43	2	2

LOGISTIC REGRESSIONS

AGE AT EXAM		COEFFICIENT OF RISK FACTOR	SE COEFFICIENT	P-VALUE	STANDARDIZED COEFFICIENT	STD SE
35-44						
45-54		0.211	0.124	0.088	0.751	0.441
55-64		0.024	0.092	0.792	0.084	0.318
65-74		-0.061	0.061	0.323	-0.213	0.216
75-84		-0.079	0.057	0.163	-0.299	0.214
85-94						
35-64	UNIVARIATE	0.114	0.073	0.117	0.411	0.262
65-94	UNIVARIATE	-0.074	0.040	0.068	-0.265	0.145
35-64	BIVARIATE	0.087	0.075	0.241	0.314	0.268
65-94	BIVARIATE	-0.056	0.041	0.172	-0.199	0.146
35-64	MULTIVARIATE	0.027	0.070	0.703	0.096	0.252
65-94	MULTIVARIATE	-0.059	0.038	0.125	-0.210	0.137

TABLE 16-6   DATE=08NOV85
AVERAGE ANNUAL INCIDENCE RATE FOR EVENT PER 1000 PERSONS AT RISK AT EXAM
BY AGE AND LEVEL OF RISK FACTOR AT EXAM: FRAMINGHAM STUDY 30-YEAR FOLLOWUP

SEX: MALE
EVENT: DEATH FROM STROKE
RISK FACTOR: BLOOD GLUCOSE (MG/DL)
PERSONS AT RISK: PERSONS FREE OF STROKE AND TIA ONLY

RISK FACTOR	35-44 AT EXAM			45-54 AT EXAM			55-64 AT EXAM			65-74 AT EXAM		
	PERSON EXAMS	# OF EVENTS	ANNUAL RATE	PERSON EXAMS	# OF EVENTS	ANNUAL RATE	PERSON EXAMS	# OF EVENTS	ANNUAL RATE	PERSON EXAMS	# OF EVENTS	ANNUAL RATE
29-69	1004			1554			1179	2	1	484	0	0
70-89	2528			4326			4458	5	1	2050	3	1
90-109	628			1397			2105	4	1	1242	8	3
110-129	107			329			599	0	0	412	1	1
130-524	49			202			472	0	0	421	2	2
UNKNOWN	200			224			101	0	0	17	0	0
TOTAL	4516			8032			8914	11	1	4626	14	2

RISK FACTOR	75-84 AT EXAM			85-94 AT EXAM			35-64 COMBINED			65-94 COMBINED		
	PERSON EXAMS	# OF EVENTS	ANNUAL RATE	PERSON EXAMS	# OF EVENTS	ANNUAL RATE	# OF EVENTS	CRUDE	AGE-ADJUSTED	# OF EVENTS	CRUDE	AGE-ADJUSTED
29-69	82	2	12	1			2	0	0	2	2	3
70-89	449	9	10	16			8	0	0	12	2	3
90-109	394	5	6	10			4	0	0	13	2	4
110-129	151	1	3	5			0	0	0	2	2	2
130-524	158	3	9	10			0	0	0	7	6	5
UNKNOWN	0	0	-	0			0	0	0	0	0	0
TOTAL	1234	20	8	42			14	0	0	36	3	3

LOGISTIC REGRESSIONS

AGE AT EXAM		COEFFICIENT OF RISK FACTOR	SE COEFFICIENT	P-VALUE	STANDARDIZED COEFFICIENT	STD SE
35-44						
45-54						
55-64		-0.019	0.019	0.304	-0.558	0.543
65-74		0.006	0.004	0.131	0.230	0.152
75-84		-0.002	0.007	0.745	-0.082	0.250
85-94						
35-64	UNIVARIATE	-0.015	0.017	0.389	-0.373	0.433
65-94	UNIVARIATE	0.005	0.003	0.094	0.186	0.111
35-64	BIVARIATE	-0.025	0.018	0.167	-0.636	0.461
65-94	BIVARIATE	0.005	0.004	0.194	0.165	0.127
35-64	MULTIVARIATE	-0.029	0.019	0.124	-0.727	0.473
65-94	MULTIVARIATE	0.004	0.004	0.267	0.151	0.136

SEX: FEMALE
EVENT: DEATH FROM STROKE
RISK FACTOR: BLOOD GLUCOSE (MG/DL)
PERSONS AT RISK: PERSONS FREE OF STROKE AND TIA ONLY

RISK FACTOR	35-44 AT EXAM			45-54 AT EXAM			55-64 AT EXAM			65-74 AT EXAM		
	PERSON EXAMS	# OF EVENTS	ANNUAL RATE	PERSON EXAMS	# OF EVENTS	ANNUAL RATE	PERSON EXAMS	# OF EVENTS	ANNUAL RATE	PERSON EXAMS	# OF EVENTS	ANNUAL RATE
29-69	1067			1803	0	0	1403	2	1	571	0	0
70-89	3408			5786	5	0	6298	3	0	3369	11	2
90-109	672			1640	0	0	2833	1	0	1933	3	1
110-129	96			302	0	0	762	3	2	563	2	2
130-524	39			166	0	0	489	1	1	405	4	5
UNKNOWN	331			336	0	0	165	0	0	30	1	17
TOTAL	5613			10033	5	0	11950	10	0	6871	21	2

RISK FACTOR	75-84 AT EXAM			85-94 AT EXAM			35-64 COMBINED			65-94 COMBINED		
	PERSON EXAMS	# OF EVENTS	ANNUAL RATE	PERSON EXAMS	# OF EVENTS	ANNUAL RATE	# OF EVENTS	CRUDE	AGE-ADJUSTED	# OF EVENTS	CRUDE	AGE-ADJUSTED
29-69	120	1	4	4			2	0	0	1	1	1
70-89	891	6	3	55			8	0	0	19	2	2
90-109	744	6	4	58			1	0	0	9	2	3
110-129	263	3	6	17			3	1	1	6	4	5
130-524	208	2	5	14			1	1	0	7	6	5
UNKNOWN	1	0	0	0			0	0	0	1	16	12
TOTAL	2227	18	4	148			15	0	0	43	2	2

LOGISTIC REGRESSIONS

AGE AT EXAM		COEFFICIENT OF RISK FACTOR	SE COEFFICIENT	P-VALUE	STANDARDIZED COEFFICIENT	STD SE
35-44						
45-54		0.003	0.019	0.857	0.067	0.372
55-64		0.012	0.004	0.001	0.325	0.099
65-74		0.009	0.004	0.015	0.269	0.111
75-84		-0.000	0.007	0.989	-0.003	0.238
85-94						
35-64	UNIVARIATE	0.013	0.003	0.001	0.289	0.080
65-94	UNIVARIATE	0.006	0.003	0.031	0.203	0.094
35-64	BIVARIATE	0.011	0.004	0.002	0.257	0.085
65-94	BIVARIATE	0.005	0.003	0.116	0.162	0.103
35-64	MULTIVARIATE	0.008	0.004	0.037	0.190	0.091
65-94	MULTIVARIATE	0.004	0.003	0.232	0.128	0.107

TABLE 16-7   DATE=08NOV85
AVERAGE ANNUAL INCIDENCE RATE FOR EVENT PER 1000 PERSONS AT RISK AT EXAM
BY AGE AND LEVEL OF RISK FACTOR AT EXAM: FRAMINGHAM STUDY 30-YEAR FOLLOWUP

SEX: MALE
EVENT: DEATH FROM STROKE
RISK FACTOR: DIABETES MELLITUS
PERSONS AT RISK: PERSONS FREE OF STROKE AND TIA ONLY

RISK FACTOR	35-44 AT EXAM			45-54 AT EXAM			55-64 AT EXAM			65-74 AT EXAM		
	PERSON EXAMS	# OF EVENTS	ANNUAL RATE	PERSON EXAMS	# OF EVENTS	ANNUAL RATE	PERSON EXAMS	# OF EVENTS	ANNUAL RATE	PERSON EXAMS	# OF EVENTS	ANNUAL RATE
NO	4467			7780			8292	10	1	4101	10	1
YES	49			252			622	1	1	525	4	4
UNKNOWN	0			0			0	0	-	0	0	-
TOTAL	4516			8032			8914	11	1	4626	14	2

RISK FACTOR	75-84 AT EXAM			85-94 AT EXAM			35-64 COMBINED			65-94 COMBINED		
	PERSON EXAMS	# OF EVENTS	ANNUAL RATE	PERSON EXAMS	# OF EVENTS	ANNUAL RATE	# OF EVENTS	CRUDE	AGE-ADJUSTED	# OF EVENTS	CRUDE	AGE-ADJUSTED
NO	1076	16	7	36			13	0	0	27	5	3
YES	158	4	13	6			1	1	0	9	7	6
UNKNOWN	0	0	-	0			0	-	-	0	-	-
TOTAL	1234	20	8	42			14	0	0	36	3	3

LOGISTIC REGRESSIONS

AGE AT EXAM		COEFFICIENT OF RISK FACTOR	SE COEFFICIENT	P-VALUE	STANDARDIZED COEFFICIENT	STD SE
35-44						
45-54						
55-64		0.288	1.050	0.784	0.073	0.267
65-74		1.145	0.593	0.054	0.363	0.188
75-84		0.343	0.566	0.337	0.181	0.189
85-94						
35-64	UNIVARIATE	0.538	1.038	0.604	0.109	0.211
65-94	UNIVARIATE	0.933	0.387	0.016	0.300	0.124
35-64	BIVARIATE	0.027	1.042	0.980	0.005	0.211
65-94	BIVARIATE	0.875	0.389	0.024	0.281	0.125
35-64	MULTIVARIATE	-0.497	1.067	0.641	-0.101	0.216
65-94	MULTIVARIATE	0.853	0.399	0.032	0.274	0.128

SEX: FEMALE
EVENT: DEATH FROM STROKE
RISK FACTOR: DIABETES MELLITUS
PERSONS AT RISK: PERSONS FREE OF STROKE AND TIA ONLY

RISK FACTOR	35-44 AT EXAM			45-54 AT EXAM			55-64 AT EXAM			65-74 AT EXAM		
	PERSON EXAMS	# OF EVENTS	ANNUAL RATE	PERSON EXAMS	# OF EVENTS	ANNUAL RATE	PERSON EXAMS	# OF EVENTS	ANNUAL RATE	PERSON EXAMS	# OF EVENTS	ANNUAL RATE
NO	5570			9820	5	0	11350	8	0	6289	17	1
YES	43			213	0	0	600	2	2	582	4	3
UNKNOWN	0			0	0	-	0	0	-	0	0	-
TOTAL	5613			10033	5	0	11950	10	0	6871	21	2

RISK FACTOR	75-84 AT EXAM			85-94 AT EXAM			35-64 COMBINED			65-94 COMBINED		
	PERSON EXAMS	# OF EVENTS	ANNUAL RATE	PERSON EXAMS	# OF EVENTS	ANNUAL RATE	# OF EVENTS	CRUDE	AGE-ADJUSTED	# OF EVENTS	CRUDE	AGE-ADJUSTED
NO	1942	15	4	127			13	0	0	35	2	2
YES	285	3	5	21			2	1	1	8	5	4
UNKNOWN	0	0	-	0			0	-	-	0	-	-
TOTAL	2227	18	4	148			15	0	0	43	2	2

LOGISTIC REGRESSIONS

AGE AT EXAM		COEFFICIENT OF RISK FACTOR	SE COEFFICIENT	P-VALUE	STANDARDIZED COEFFICIENT	STD SE
35-44		-	-	-	-	-
45-54		-	-	-	-	-
55-64		1.556	0.792	0.049	0.340	0.173
65-74		0.937	0.557	0.093	0.261	0.155
75-84		0.312	0.636	0.623	0.104	0.212
85-94						
35-64	UNIVARIATE	1.572	0.760	0.039	0.272	0.132
65-94	UNIVARIATE	0.771	0.393	0.050	0.227	0.116
35-64	BIVARIATE	1.255	0.769	0.102	0.218	0.133
65-94	BIVARIATE	0.595	0.396	0.133	0.175	0.117
35-64	MULTIVARIATE	0.806	0.788	0.306	0.140	0.137
65-94	MULTIVARIATE	0.197	0.431	0.648	0.058	0.127

TABLE 16-8　　　　　DATE=14DEC85

AVERAGE ANNUAL INCIDENCE RATE FOR EVENT PER 1000 PERSONS AT RISK AT EXAM
BY AGE AND LEVEL OF RISK FACTOR AT EXAM: FRAMINGHAM STUDY 30-YEAR FOLLOWUP

SEX: MALE
EVENT: DEATH FROM STROKE
RISK FACTOR: GLUCOSE IN URINE, DEFINITE OR TRACE
PERSONS AT RISK: PERSONS FREE OF STROKE AND TIA ONLY

RISK FACTOR	35-44 AT EXAM PERSON EXAMS	# OF EVENTS	ANNUAL RATE	45-54 AT EXAM PERSON EXAMS	# OF EVENTS	ANNUAL RATE	55-64 AT EXAM PERSON EXAMS	# OF EVENTS	ANNUAL RATE	65-74 AT EXAM PERSON EXAMS	# OF EVENTS	ANNUAL RATE
NO	4423			7834			8639	10	1	4460	14	2
YES	60			177			251	1	2	160	0	0
UNKNOWN	53			21			24	0	0	6	0	0
TOTAL	4516			8032			8914	11	1	4626	14	2

RISK FACTOR	75-84 AT EXAM PERSON EXAMS	# OF EVENTS	ANNUAL RATE	85-94 AT EXAM PERSON EXAMS	# OF EVENTS	ANNUAL RATE	35-64 COMBINED # OF EVENTS	CRUDE	AGE-ADJUSTED	65-94 COMBINED # OF EVENTS	CRUDE	AGE-ADJUSTED
NO	1199	20	8	41		1	13	0	0	36	3	3
YES	32	0	0	1			1	1	1	0	0	0
UNKNOWN	3	0	0	0			0	0	0	0	0	0
TOTAL	1234	20	8	42			14	0	0	36	3	3

LOGISTIC REGRESSIONS

AGE AT EXAM		COEFFICIENT OF RISK FACTOR	SE COEFFICIENT	P-VALUE	STANDARDIZED COEFFICIENT	STD SE
35-44						
45-54						
55-64		1.239	1.051	0.238	0.205	0.174
65-74		-	-	-	-	-
75-84		-	-	-	-	-
85-94		-	-		-	-
35-64	UNIVARIATE	1.193	1.039	0.251	0.178	0.155
65-94	UNIVARIATE	-	-	-	-	-
35-64	BIVARIATE	0.962	1.040	0.355	0.144	0.155
65-94	BIVARIATE	-	-	-	-	-
35-64	MULTIVARIATE	-	-	-		
65-94	MULTIVARIATE	-	-	-		

SEX: FEMALE
EVENT: DEATH FROM STROKE
RISK FACTOR: GLUCOSE IN URINE, DEFINITE OR TRACE
PERSONS AT RISK: PERSONS FREE OF STROKE AND TIA ONLY

RISK FACTOR	35-44 AT EXAM PERSON EXAMS	# OF EVENTS	ANNUAL RATE	45-54 AT EXAM PERSON EXAMS	# OF EVENTS	ANNUAL RATE	55-64 AT EXAM PERSON EXAMS	# OF EVENTS	ANNUAL RATE	65-74 AT EXAM PERSON EXAMS	# OF EVENTS	ANNUAL RATE
NO	5534			9908	4	0	11753	8	0	6733	20	1
YES	41			101	1	5	171	2	6	128	1	4
UNKNOWN	38			24	0	0	26	0	0	10	0	0
TOTAL	5613			10033	5	0	11950	10	0	6871	21	2

RISK FACTOR	75-84 AT EXAM PERSON EXAMS	# OF EVENTS	ANNUAL RATE	85-94 AT EXAM PERSON EXAMS	# OF EVENTS	ANNUAL RATE	35-64 COMBINED # OF EVENTS	CRUDE	AGE-ADJUSTED	65-94 COMBINED # OF EVENTS	CRUDE	AGE-ADJUSTED
NO	2183	17	4	145		3	12	0	0	41	2	2
YES	35	1	14	3			3	5	4	2	6	6
UNKNOWN	9	0	0	0			0	0	0	0	0	0
TOTAL	2227	18	4	148			15	0	0	43	2	2

LOGISTIC REGRESSIONS

AGE AT EXAM		COEFFICIENT OF RISK FACTOR	SE COEFFICIENT	P-VALUE	STANDARDIZED COEFFICIENT	STD SE
35-44						
45-54		3.209	1.123	0.004	0.321	0.112
55-64		2.855	0.794	0.001	0.339	0.094
65-74		0.972	1.029	0.345	0.132	0.139
75-84		1.321	1.043	0.205	0.165	0.130
85-94						
35-64	UNIVARIATE	.087	0.648	0.001	0.327	0.069
65-94	UNIVARIATE	0.987	0.728	0.175	0.131	0.097
35-64	BIVARIATE	2.955	0.651	0.001	0.313	0.069
65-94	BIVARIATE	1.046	0.732	0.153	0.139	0.097
35-64	MULTIVARIATE	2.424	0.687	0.001	0.257	0.073
65-94	MULTIVARIATE	0.665	0.745	0.372	0.088	0.099

TABLE 16-9   DATE=04MAR86
AVERAGE ANNUAL INCIDENCE RATE FOR EVENT PER 1000 PERSONS AT RISK AT EXAM
BY AGE AND LEVEL OF RISK FACTOR AT EXAM: FRAMINGHAM STUDY 30-YEAR FOLLOWUP

SEX: MALE
EVENT: DEATH FROM STROKE
RISK FACTOR: GLUCOSE INTOLERANCE
PERSONS AT RISK: PERSONS FREE OF STROKE AND TIA ONLY

RISK FACTOR	35-44 AT EXAM PERSON EXAMS	# OF EVENTS	ANNUAL RATE	45-54 AT EXAM PERSON EXAMS	# OF EVENTS	ANNUAL RATE	55-64 AT EXAM PERSON EXAMS	# OF EVENTS	ANNUAL RATE	65-74 AT EXAM PERSON EXAMS	# OF EVENTS	ANNUAL RATE
NO	4378			7497			7847	9	1	3766	9	1
YES	138			535			1067	2	1	860	5	3
UNKNOWN	0			0			0	0	-	0	0	-
TOTAL	4516			8032			8914	11	1	4626	14	2

RISK FACTOR	75-84 AT EXAM PERSON EXAMS	# OF EVENTS	ANNUAL RATE	85-94 AT EXAM PERSON EXAMS	# OF EVENTS	ANNUAL RATE	35-64 COMBINED # OF EVENTS	CRUDE	AGE-ADJUSTED	65-94 COMBINED # OF EVENTS	CRUDE	AGE-ADJUSTED
NO	952	15	8	29			12	0	0	24	3	3
YES	282	5	9	13			2	1	0	12	5	5
UNKNOWN	0	0	-	0			0	-		0	-	
TOTAL	1234	20	8	42			14	0	0	36	3	3

LOGISTIC REGRESSIONS

AGE AT EXAM		COEFFICIENT OF RISK FACTOR	SE COEFFICIENT	P-VALUE	STANDARDIZED COEFFICIENT	STD SE
35-44						
45-54						
55-64		0.492	0.782	0.530	0.160	0.254
65-74		0.892	0.559	0.110	0.347	0.218
75-84		0.120	0.521	0.818	0.050	0.219
85-94						
35-64	UNIVARIATE	0.637	0.764	0.405	0.174	0.209
65-94	UNIVARIATE	0.726	0.355	0.041	0.288	0.141
35-64	BIVARIATE	0.230	0.767	0.764	0.063	0.209
65-94	BIVARIATE	0.608	0.357	0.088	0.241	0.142
35-64	MULTIVARIATE	-0.869	1.055	0.410	-0.237	0.288
65-94	MULTIVARIATE	0.564	0.362	0.120	0.224	0.144

SEX: FEMALE
EVENT: DEATH FROM STROKE
RISK FACTOR: GLUCOSE INTOLERANCE
PERSONS AT RISK: PERSONS FREE OF STROKE AND TIA ONLY

RISK FACTOR	35-44 AT EXAM PERSON EXAMS	# OF EVENTS	ANNUAL RATE	45-54 AT EXAM PERSON EXAMS	# OF EVENTS	ANNUAL RATE	55-64 AT EXAM PERSON EXAMS	# OF EVENTS	ANNUAL RATE	65-74 AT EXAM PERSON EXAMS	# OF EVENTS	ANNUAL RATE
NO	5519			9648	4	0	10848	7	0	5877	14	1
YES	94			385	1	1	1102	3	1	994	7	4
UNKNOWN	0			0	0	-	0	0	-	0	0	-
TOTAL	5613			10033	5	0	11950	10	0	6871	21	2

RISK FACTOR	75-84 AT EXAM PERSON EXAMS	# OF EVENTS	ANNUAL RATE	85-94 AT EXAM PERSON EXAMS	# OF EVENTS	ANNUAL RATE	35-64 COMBINED # OF EVENTS	CRUDE	AGE-ADJUSTED	65-94 COMBINED # OF EVENTS	CRUDE	AGE-ADJUSTED
NO	1752	14	4	120			11	0	0	30	2	2
YES	475	4	4	28			4	1	1	13	4	4
UNKNOWN	0	0	-	0			0	-		0	-	
TOTAL	2227	18	4	148			15	0	0	43	2	2

LOGISTIC REGRESSIONS

AGE AT EXAM		COEFFICIENT OF RISK FACTOR	SE COEFFICIENT	P-VALUE	STANDARDIZED COEFFICIENT	STD SE
35-44						
45-54		1.837	1.119	0.101	0.353	0.215
55-64		1.442	0.691	0.037	0.417	0.200
65-74		1.089	0.464	0.019	0.383	0.163
75-84		0.053	0.569	0.926	0.022	0.233
85-94						
35-64	UNIVARIATE	1.791	0.584	0.002	0.416	0.136
65-94	UNIVARIATE	0.813	0.333	0.015	0.299	0.123
35-64	BIVARIATE	1.498	0.596	0.012	0.348	0.139
65-94	BIVARIATE	0.648	0.336	0.053	0.239	0.124
35-64	MULTIVARIATE	1.260	0.607	0.038	0.293	0.141
65-94	MULTIVARIATE	0.395	0.354	0.265	0.145	0.130

TABLE 16-10   DATE=06NOV85

AVERAGE ANNUAL INCIDENCE RATE FOR EVENT PER 1000 PERSONS AT RISK AT EXAM
BY AGE AND LEVEL OF RISK FACTOR AT EXAM, FRAMINGHAM STUDY 30-YEAR FOLLOWUP

SEX: MALE
EVENT: DEATH FROM STROKE
RISK FACTOR: METROPOLITAN RELATIVE WEIGHT (PERCENT)
PERSONS AT RISK: PERSONS FREE OF STROKE AND TIA ONLY

RISK FACTOR	35-44 AT EXAM PERSON EXAMS	# OF EVENTS	ANNUAL RATE	45-54 AT EXAM PERSON EXAMS	# OF EVENTS	ANNUAL RATE	55-64 AT EXAM PERSON EXAMS	# OF EVENTS	ANNUAL RATE	65-74 AT EXAM PERSON EXAMS	# OF EVENTS	ANNUAL RATE
54-104	720			1145			1251	1	0	795	4	3
105-114	979			1631			1830	3	1	998	2	1
115-124	1133			2131			2428	4	1	1272	5	2
125-134	923			1681			1819	1	0	918	1	1
135-272	761			1443			1585	2	1	643	2	2
UNKNOWN	0			1			1	0	0	0	0	-
TOTAL	4516			8032			8914	11	1	4626	14	2

RISK FACTOR	75-84 AT EXAM PERSON EXAMS	# OF EVENTS	ANNUAL RATE	85-94 AT EXAM PERSON EXAMS	# OF EVENTS	ANNUAL RATE	35-64 COMBINED # OF EVENTS	CRUDE	AGE-ADJUSTED	65-94 COMBINED # OF EVENTS	CRUDE	AGE-ADJUSTED
54-104	247	2	4	15			3	0	0	8	4	3
105-114	323	5	7	12			3	0	0	7	3	2
115-124	377	7	9	13			5	0	0	12	4	3
125-134	184	4	11	2			1	0	0	5	2	3
135-272	103	2	10	0			2	0	0	4	3	3
UNKNOWN	0	0	-	0			0	0	0	0	-	
TOTAL	1234	20	8	42			14	0	0	36	3	3

LOGISTIC REGRESSIONS

AGE AT EXAM		COEFFICIENT OF RISK FACTOR	SE COEFFICIENT	P-VALUE	STANDARDIZED COEFFICIENT	STD SE
35-44						
45-54						
55-64		-0.007	0.019	0.712	-0.114	0.308
65-74		-0.019	0.018	0.295	-0.287	0.274
75-84		0.006	0.015	0.677	0.092	0.219
85-94						
35-64	UNIVARIATE	-0.022	0.018	0.217	-0.350	0.284
65-94	UNIVARIATE	-0.014	0.011	0.214	-0.212	0.171
35-64	BIVARIATE	-0.022	0.018	0.217	-0.350	0.284
65-94	BIVARIATE	-0.008	0.012	0.508	-0.117	0.177
35-64	MULTIVARIATE	-0.037	0.019	0.059	-0.591	0.313
65-94	MULTIVARIATE	-0.011	0.012	0.329	-0.173	0.177

SEX: FEMALE
EVENT: DEATH FROM STROKE
RISK FACTOR: METROPOLITAN RELATIVE WEIGHT (PERCENT)
PERSONS AT RISK: PERSONS FREE OF STROKE AND TIA ONLY

RISK FACTOR	35-44 AT EXAM PERSON EXAMS	# OF EVENTS	ANNUAL RATE	45-54 AT EXAM PERSON EXAMS	# OF EVENTS	ANNUAL RATE	55-64 AT EXAM PERSON EXAMS	# OF EVENTS	ANNUAL RATE	65-74 AT EXAM PERSON EXAMS	# OF EVENTS	ANNUAL RATE
54-104	1689			2108	2	0	2108	0	0	1275	4	2
105-114	1431			2351	2	0	2509	5	1	1289	1	1
115-124	1059			2123	0	0	2535	1	0	1485	5	2
125-134	565			1378	0	0	1813	1	0	1119	3	1
135-272	861			2073	2	0	2985	3	1	1703	6	2
UNKNOWN	8			0	0	-	0	0	-	0	0	-
TOTAL	5613			10033	5	0	11950	10	0	6871	21	2

RISK FACTOR	75-84 AT EXAM PERSON EXAMS	# OF EVENTS	ANNUAL RATE	85-94 AT EXAM PERSON EXAMS	# OF EVENTS	ANNUAL RATE	35-64 COMBINED # OF EVENTS	CRUDE	AGE-ADJUSTED	65-94 COMBINED # OF EVENTS	CRUDE	AGE-ADJUSTED
54-104	592	7	6	47			2	0	0	11	3	3
105-114	405	4	5	31			6	0	1	9	3	3
115-124	448	1	1	28			1	0	0	8	2	2
125-134	325	0	0	24			1	0	0	5	1	1
135-272	457	6	7	18			5	0	0	12	3	3
UNKNOWN	0	0	-	0			0	0	0	0	-	
TOTAL	2227	18	4	148			15	0	0	43	2	2

LOGISTIC REGRESSIONS

AGE AT EXAM		COEFFICIENT OF RISK FACTOR	SE COEFFICIENT	P-VALUE	STANDARDIZED COEFFICIENT	STD SE
35-44						
45-54		0.008	0.018	0.639	0.183	0.389
55-64		0.005	0.013	0.716	0.108	0.296
65-74		-0.001	0.010	0.897	-0.029	0.222
75-84		-0.012	0.012	0.306	-0.259	0.253
85-94						
35-64	UNIVARIATE	0.009	0.010	0.391	0.193	0.225
65-94	UNIVARIATE	-0.009	0.008	0.225	-0.202	0.166
35-64	BIVARIATE	0.006	0.011	0.566	0.134	0.234
65-94	BIVARIATE	-0.005	0.007	0.526	-0.104	0.164
35-64	MULTIVARIATE	-0.001	0.011	0.962	-0.011	0.235
65-94	MULTIVARIATE	-0.004	0.007	0.617	-0.080	0.160

TABLE 16-11   DATE=08NOV85
AVERAGE ANNUAL INCIDENCE RATE FOR EVENT PER 1000 PERSONS AT RISK AT EXAM
BY AGE AND LEVEL OF RISK FACTOR AT EXAM, FRAMINGHAM STUDY 30-YEAR FOLLOWUP

SEX: MALE
EVENT: DEATH FROM STROKE
RISK FACTOR: VITAL CAPACITY - HEIGHT INDEX (ML/INCH)
PERSONS AT RISK: PERSONS FREE OF STROKE AND TIA ONLY

RISK FACTOR	35-44 AT EXAM PERSON EXAMS	# OF EVENTS	ANNUAL RATE	45-54 AT EXAM PERSON EXAMS	# OF EVENTS	ANNUAL RATE	55-64 AT EXAM PERSON EXAMS	# OF EVENTS	ANNUAL RATE	65-74 AT EXAM PERSON EXAMS	# OF EVENTS	ANNUAL RATE
8-44	268			1100			2319	4	1	2063	8	2
45-49	416			1182			1667	2	1	950	3	3
50-54	751			1721			2024	1	0	824	3	2
55-59	1034			1809			1540	3	1	509	0	0
60-85	2034			2214			1346	1	0	268	0	0
UNKNOWN	13			6			18	0	0	12	0	0
TOTAL	4516			8032			8914	11	1	4626	14	2

RISK FACTOR	75-84 AT EXAM PERSON EXAMS	# OF EVENTS	ANNUAL RATE	85-94 AT EXAM PERSON EXAMS	# OF EVENTS	ANNUAL RATE	35-64 COMBINED # OF EVENTS	CRUDE	AGE-ADJUSTED	65-94 COMBINED # OF EVENTS	CRUDE	AGE-ADJUSTED
8-44	778	14	9	28			4	1	0	24	4	4
45-49	207	2	5	5			2	0	0	5	2	2
50-54	130	3	12	7			3	0	0	6	3	4
55-59	79	0	0	2			3	0	0	0	0	0
60-85	32	1	16	0			2	0	0	1	2	3
UNKNOWN	8	0	0	0			0	0	0	0	0	0
TOTAL	1234	20	8	42			14	0	0	36	3	3

LOGISTIC REGRESSIONS

AGE AT EXAM		COEFFICIENT OF RISK FACTOR	SE COEFFICIENT	P-VALUE	STANDARDIZED COEFFICIENT	STD SE
35-44						
45-54						
55-64		-0.017	0.032	0.590	-0.159	0.296
65-74		-0.030	0.028	0.287	-0.277	0.260
75-84		-0.015	0.024	0.547	-0.136	0.226
85-94						
35-64	UNIVARIATE	-0.032	0.027	0.227	-0.310	0.257
65-94	UNIVARIATE	-0.043	0.017	0.013	-0.402	0.162
35-64	BIVARIATE	-0.000	0.029	0.990	-0.003	0.280
65-94	BIVARIATE	-0.027	0.018	0.136	-0.255	0.171
35-64	MULTIVARIATE	0.032	0.032	0.321	0.506	0.308
65-94	MULTIVARIATE	-0.015	0.018	0.409	-0.142	0.172

SEX: FEMALE
EVENT: DEATH FROM STROKE
RISK FACTOR: VITAL CAPACITY - HEIGHT INDEX (ML/INCH)
PERSONS AT RISK: PERSONS FREE OF STROKE AND TIA ONLY

RISK FACTOR	35-44 AT EXAM PERSON EXAMS	# OF EVENTS	ANNUAL RATE	45-54 AT EXAM PERSON EXAMS	# OF EVENTS	ANNUAL RATE	55-64 AT EXAM PERSON EXAMS	# OF EVENTS	ANNUAL RATE	65-74 AT EXAM PERSON EXAMS	# OF EVENTS	ANNUAL RATE
4-29	121			539	0	0	1530	2	1	1948	5	1
30-34	291			1240	2	1	2520	4	1	1888	9	2
35-39	769			2096	1	0	3181	1	0	1694	6	2
40-44	1427			2825	1	0	2809	2	0	919	1	1
45-92	2984			3316	1	0	1889	1	0	416	0	0
UNKNOWN	21			17	0	0	21	0	0	6	0	0
TOTAL	5613			10033	5	0	11950	10	0	6871	21	2

RISK FACTOR	75-84 AT EXAM PERSON EXAMS	# OF EVENTS	ANNUAL RATE	85-94 AT EXAM PERSON EXAMS	# OF EVENTS	ANNUAL RATE	35-64 COMBINED # OF EVENTS	CRUDE	AGE-ADJUSTED	65-94 COMBINED # OF EVENTS	CRUDE	AGE-ADJUSTED
4-29	1090	10	5	100			2	0	0	17	3	2
30-34	645	6	5	33			6	1	1	17	3	2
35-39	351	2	3	14			2	0	0	8	2	2
40-44	113	0	0	1			3	0	0	1	0	0
45-92	26	0	0	0			2	0	0	0	0	0
UNKNOWN	2	0	0	0			0	0	0	0	0	0
TOTAL	2227	18	4	148			15	0	0	43	2	2

LOGISTIC REGRESSIONS

AGE AT EXAM		COEFFICIENT OF RISK FACTOR	SE COEFFICIENT	P-VALUE	STANDARDIZED COEFFICIENT	STD SE
35-44						
45-54		-0.057	0.058	0.325	-0.419	0.426
55-64		-0.062	0.040	0.120	-0.461	0.296
65-74		-0.022	0.029	0.459	-0.160	0.216
75-84		-0.029	0.033	0.378	-0.204	0.231
85-94						
35-64	UNIVARIATE	-0.076	0.030	0.011	-0.608	0.240
65-94	UNIVARIATE	-0.039	0.020	0.051	-0.291	0.149
35-64	BIVARIATE	-0.058	0.033	0.080	-0.460	0.263
65-94	BIVARIATE	-0.010	0.022	0.638	-0.076	0.162
35-64	MULTIVARIATE	-0.032	0.036	0.377	-0.251	0.285
65-94	MULTIVARIATE	0.007	0.023	0.747	0.056	0.173

- 323 -

TABLE 16-12   DATE=08NOV85

AVERAGE ANNUAL INCIDENCE RATE FOR EVENT PER 1000 PERSONS AT RISK AT EXAM
BY AGE AND LEVEL OF RISK FACTOR AT EXAM: FRAMINGHAM STUDY 30-YEAR FOLLOWUP

SEX: MALE
EVENT: DEATH FROM STROKE
RISK FACTOR: HEART RATE (PER MINUTE)
PERSONS AT RISK: PERSONS FREE OF STROKE AND TIA ONLY

RISK FACTOR	35-44 AT EXAM PERSON EXAMS	# OF EVENTS	ANNUAL RATE	45-54 AT EXAM PERSON EXAMS	# OF EVENTS	ANNUAL RATE	55-64 AT EXAM PERSON EXAMS	# OF EVENTS	ANNUAL RATE	65-74 AT EXAM PERSON EXAMS	# OF EVENTS	ANNUAL RATE
30-67	1295			2013			2590	3	1	1365	3	1
68-75	1503			2579			2709	1	0	1403	7	2
76-83	728			1465			1520	3	1	767	2	1
84-91	586			1139			1188	2	1	570	1	1
92-220	404			836			907	2	1	521	1	1
UNKNOWN	0			0			0			0		
TOTAL	4516			8032			8914	11	1	4626	14	2

RISK FACTOR	75-84 AT EXAM PERSON EXAMS	# OF EVENTS	ANNUAL RATE	85-94 AT EXAM PERSON EXAMS	# OF EVENTS	ANNUAL RATE	35-64 COMBINED # OF EVENTS	CRUDE	AGE-ADJUSTED	65-94 COMBINED # OF EVENTS	CRUDE	AGE-ADJUSTED
30-67	335	6	8	13			5	0	0	9	3	3
68-75	339	7	10	13			1	0	0	15	4	4
76-83	246	3	6	11			3	0	0	5	2	2
84-91	163	1	3	4			2	0	1	2	1	1
92-220	133	3	11	1			3	1	1	5	4	7
UNKNOWN	0	0	-	0			0	-	-	0	-	-
TOTAL	1234	20	8	42			14	0	0	36	3	3

LOGISTIC REGRESSIONS

AGE AT EXAM		COEFFICIENT OF RISK FACTOR	SE COEFFICIENT	P-VALUE	STANDARDIZED COEFFICIENT	STD SE
35-44						
45-54						
55-64		0.023	0.021	0.262	0.302	0.270
65-74		-0.012	0.021	0.566	-0.162	0.283
75-84		-0.010	0.018	0.560	-0.137	0.235
85-94						
35-64	UNIVARIATE	0.016	0.020	0.423	0.201	0.251
65-94	UNIVARIATE	-0.006	0.013	0.642	-0.080	0.172
35-64	BIVARIATE	0.016	0.019	0.406	0.202	0.243
65-94	BIVARIATE	-0.007	0.013	0.603	-0.090	0.174
35-64	MULTIVARIATE	0.005	0.020	0.809	0.062	0.256
65-94	MULTIVARIATE	-0.013	0.013	0.320	-0.171	0.172

SEX: FEMALE
EVENT: DEATH FROM STROKE
RISK FACTOR: HEART RATE (PER MINUTE)
PERSONS AT RISK: PERSONS FREE OF STROKE AND TIA ONLY

RISK FACTOR	35-44 AT EXAM PERSON EXAMS	# OF EVENTS	ANNUAL RATE	45-54 AT EXAM PERSON EXAMS	# OF EVENTS	ANNUAL RATE	55-64 AT EXAM PERSON EXAMS	# OF EVENTS	ANNUAL RATE	65-74 AT EXAM PERSON EXAMS	# OF EVENTS	ANNUAL RATE
30-67	1090			1908	1	0	2402	0	0	1242	3	1
68-75	1804			3269	0	0	3602	1	0	1853	8	2
76-83	1050			1930	1	0	2412	1	1	1498	2	1
84-91	939			1612	2	1	1867	2	1	1154	6	3
92-220	730			1314	1	0	1667	2	1	1124	2	1
UNKNOWN	0			0	0	-	0	0	-	0	0	-
TOTAL	5613			10033	5	0	11950	10	0	6871	21	2

RISK FACTOR	75-84 AT EXAM PERSON EXAMS	# OF EVENTS	ANNUAL RATE	85-94 AT EXAM PERSON EXAMS	# OF EVENTS	ANNUAL RATE	35-64 COMBINED # OF EVENTS	CRUDE	AGE-ADJUSTED	65-94 COMBINED # OF EVENTS	CRUDE	AGE-ADJUSTED
30-67	429	2	2	30			1	0	0	7	2	2
68-75	535	5	5	38			1	0	0	14	3	3
76-83	484	3	3	27			6	1	1	5	1	1
84-91	375	4	5	29			3	0	0	10	3	3
92-220	404	4	5	24			3	0	0	7	2	2
UNKNOWN	0	0	-	0			0	-	-	0	-	-
TOTAL	2227	18	4	148			15	0	0	43	2	2

LOGISTIC REGRESSIONS

AGE AT EXAM		COEFFICIENT OF RISK FACTOR	SE COEFFICIENT	P-VALUE	STANDARDIZED COEFFICIENT	STD SE
35-44						
45-54		0.047	0.029	0.097	0.588	0.354
55-64		0.038	0.017	0.026	0.493	0.222
65-74		-0.000	0.017	0.978	-0.006	0.219
75-84		0.020	0.014	0.167	0.276	0.199
85-94						
35-64	UNIVARIATE	0.039	0.013	0.003	0.490	0.166
65-94	UNIVARIATE	0.008	0.011	0.440	0.112	0.146
35-64	BIVARIATE	0.039	0.014	0.005	0.493	0.174
65-94	BIVARIATE	0.007	0.011	0.507	0.096	0.145
35-64	MULTIVARIATE	0.028	0.017	0.095	0.350	0.210
65-94	MULTIVARIATE	0.006	0.011	0.575	0.081	0.145

TABLE 16-13   DATE=14DEC85
AVERAGE ANNUAL INCIDENCE RATE FOR EVENT PER 1000 PERSONS AT RISK AT EXAM
BY AGE AND LEVEL OF RISK FACTOR AT EXAM: FRAMINGHAM STUDY 30-YEAR FOLLOWUP

SEX: MALE
EVENT: DEATH FROM STROKE
RISK FACTOR: CIGARETTES SMOKED (NUMBER PER DAY)
PERSONS AT RISK: PERSONS FREE OF STROKE AND TIA ONLY

RISK FACTOR	35-44 AT EXAM PERSON EXAMS	# OF EVENTS	ANNUAL RATE	45-54 AT EXAM PERSON EXAMS	# OF EVENTS	ANNUAL RATE	55-64 AT EXAM PERSON EXAMS	# OF EVENTS	ANNUAL RATE	65-74 AT EXAM PERSON EXAMS	# OF EVENTS	ANNUAL RATE
NONE	1402			3211			4815	2	0	3084	7	1
1-10	437			742			842	1	1	449	1	1
11-20	1497			2127			1848	5	1	633	4	3
21-40	993			1664			1217	3	1	421	2	2
41-90	161			250			168	0	0	32	0	0
UNKNOWN	26			38			24	0	0	7	0	0
TOTAL	4516			8032			8914	11	1	4626	14	2

RISK FACTOR	75-84 AT EXAM PERSON EXAMS	# OF EVENTS	ANNUAL RATE	85-94 AT EXAM PERSON EXAMS	# OF EVENTS	ANNUAL RATE	35-64 COMBINED # OF EVENTS	CRUDE	AGE-ADJUSTED	65-94 COMBINED # OF EVENTS	CRUDE	AGE-ADJUSTED
NONE	963	15	8	32			3	0	0	23	3	3
1-10	122	1	4	4			1	0	0	2	2	2
11-20	86	2	12	5			7	1	1	6	4	5
21-40	58	2	17	1			3	0	1	5	5	9
41-90	5	0	0	0			0	0	0	0	0	0
UNKNOWN	0	0	-	0			0	0	0	0	0	0
TOTAL	1234	20	8	42			14	0	0	36	3	3

LOGISTIC REGRESSIONS

AGE AT EXAM		COEFFICIENT OF RISK FACTOR	SE COEFFICIENT	P-VALUE	STANDARDIZED COEFFICIENT	STD SE
35-44						
45-54						
55-64		0.027	0.018	0.137	0.363	0.244
65-74		0.020	0.019	0.300	0.231	0.223
75-84		0.016	0.021	0.438	0.144	0.186
85-94						
35-64	UNIVARIATE	0.012	0.017	0.471	0.179	0.248
65-94	UNIVARIATE	0.013	0.013	0.313	0.148	0.146
35-64	BIVARIATE	0.024	0.016	0.138	0.353	0.238
65-94	BIVARIATE	0.024	0.013	0.067	0.266	0.145
35-64	MULTIVARIATE	0.027	0.018	0.139	0.388	0.262
65-94	MULTIVARIATE	0.030	0.014	0.026	0.337	0.152

SEX: FEMALE
EVENT: DEATH FROM STROKE
RISK FACTOR: CIGARETTES SMOKED (NUMBER PER DAY)
PERSONS AT RISK: PERSONS FREE OF STROKE AND TIA ONLY

RISK FACTOR	35-44 AT EXAM PERSON EXAMS	# OF EVENTS	ANNUAL RATE	45-54 AT EXAM PERSON EXAMS	# OF EVENTS	ANNUAL RATE	55-64 AT EXAM PERSON EXAMS	# OF EVENTS	ANNUAL RATE	65-74 AT EXAM PERSON EXAMS	# OF EVENTS	ANNUAL RATE
NONE	2580			5582	0	0	7896	5	0	5364	11	1
1-10	1312			1723	1	0	1634	2	1	674	6	4
11-20	1372			2027	3	1	1743	3	1	627	4	3
21-40	298			626	1	1	621	0	0	186	0	0
41-90	22			38	0	0	18	0	0	1	0	0
UNKNOWN	29			37	0	0	38	0	0	19	0	0
TOTAL	5613			10033	5	0	11950	10	0	6871	21	2

RISK FACTOR	75-84 AT EXAM PERSON EXAMS	# OF EVENTS	ANNUAL RATE	85-94 AT EXAM PERSON EXAMS	# OF EVENTS	ANNUAL RATE	35-64 COMBINED # OF EVENTS	CRUDE	AGE-ADJUSTED	65-94 COMBINED # OF EVENTS	CRUDE	AGE-ADJUSTED
NONE	1959	13	3	139			5	0	0	28	2	2
1-10	158	1	3	7			3	0	0	7	4	4
11-20	93	4	22	2			1	0	1	8	6	8
21-40	13	0	0	0			1	0	0	0	0	0
41-90	0	0	-	0			0	0	0	0	0	0
UNKNOWN	4	0	0	0			0	0	0	0	0	0
TOTAL	2227	18	4	148			15	0	0	43	2	2

LOGISTIC REGRESSIONS

AGE AT EXAM		COEFFICIENT OF RISK FACTOR	SE COEFFICIENT	P-VALUE	STANDARDIZED COEFFICIENT	STD SE
35-44						
45-54		0.072	0.029	0.013	0.739	0.297
55-64		0.020	0.028	0.474	0.191	0.267
65-74		0.031	0.021	0.137	0.239	0.160
75-84		0.070	0.025	0.004	0.338	0.119
85-94						
35-64	UNIVARIATE	0.038	0.020	0.055	0.380	0.198
65-94	UNIVARIATE	0.031	0.016	0.050	0.213	0.111
35-64	BIVARIATE	0.043	0.019	0.026	0.450	0.193
65-94	BIVARIATE	0.049	0.016	0.002	0.347	0.114
35-64	MULTIVARIATE	0.047	0.020	0.016	0.469	0.194
65-94	MULTIVARIATE	0.044	0.017	0.001	0.387	0.117

TABLE 16-14   DATE=08NOV85

AVERAGE ANNUAL INCIDENCE RATE FOR EVENT PER 1000 PERSONS AT RISK AT EXAM
BY AGE AND LEVEL OF RISK FACTOR AT EXAM: FRAMINGHAM STUDY 30-YEAR FOLLOWUP

SEX: MALE
EVENT: DEATH FROM STROKE
RISK FACTOR: ALBUMIN IN URINE, DEFINITE
PERSONS AT RISK: PERSONS FREE OF STROKE AND TIA ONLY

RISK FACTOR	35-44 AT EXAM PERSON EXAMS	# OF EVENTS	ANNUAL RATE	45-54 AT EXAM PERSON EXAMS	# OF EVENTS	ANNUAL RATE	55-64 AT EXAM PERSON EXAMS	# OF EVENTS	ANNUAL RATE	65-74 AT EXAM PERSON EXAMS	# OF EVENTS	ANNUAL RATE
NO	4408			7903			8682	11	1	4476	14	2
YES	71			103			205	0	0	144	0	0
UNKNOWN	37			26			27	0	0	6	0	0
TOTAL	4516			8032			8914	11	1	4626	14	2

RISK FACTOR	75-84 AT EXAM PERSON EXAMS	# OF EVENTS	ANNUAL RATE	85-94 AT EXAM PERSON EXAMS	# OF EVENTS	ANNUAL RATE	35-64 COMBINED # OF EVENTS	CRUDE	AGE-ADJUSTED	65-94 COMBINED # OF EVENTS	CRUDE	AGE-ADJUSTED
NO	1199	20	8	37			13	0	0	35	3	3
YES	32	0	0	5			1	1	2	1	3	1
UNKNOWN	3	0	0	0			0	0	0	0	0	0
TOTAL	1234	20	8	42			14	0	0	36	3	3

LOGISTIC REGRESSIONS

AGE AT EXAM	COEFFICIENT OF RISK FACTOR	SE COEFFICIENT	P-VALUE	STANDARDIZED COEFFICIENT	STD SE
35-44					
45-54					
55-64					
65-74					
75-84	-	-	-	-	-
85-94					
35-64 UNIVARIATE	1.451	1.039	0.162	0.192	0.137
65-94 UNIVARIATE	-0.104	1.017	0.918	-0.018	0.175
35-64 BIVARIATE	1.181	1.042	0.257	0.156	0.138
65-94 BIVARIATE	-0.265	1.023	0.796	-0.046	0.177
35-64 MULTIVARIATE	0.614	1.078	0.569	0.081	0.142
65-94 MULTIVARIATE	-0.496	1.030	0.630	-0.086	0.178

SEX: FEMALE
EVENT: DEATH FROM STROKE
RISK FACTOR: ALBUMIN IN URINE, DEFINITE
PERSONS AT RISK: PERSONS FREE OF STROKE AND TIA ONLY

RISK FACTOR	35-44 AT EXAM PERSON EXAMS	# OF EVENTS	ANNUAL RATE	45-54 AT EXAM PERSON EXAMS	# OF EVENTS	ANNUAL RATE	55-64 AT EXAM PERSON EXAMS	# OF EVENTS	ANNUAL RATE	65-74 AT EXAM PERSON EXAMS	# OF EVENTS	ANNUAL RATE
NO	5350			9728	5	0	11684	10	0	6729	21	2
YES	225			280	0	0	238	0	0	132	0	0
UNKNOWN	38			25	0	0	28	0	0	10	0	0
TOTAL	5613			10033	5	0	11950	10	0	6871	21	2

RISK FACTOR	75-84 AT EXAM PERSON EXAMS	# OF EVENTS	ANNUAL RATE	85-94 AT EXAM PERSON EXAMS	# OF EVENTS	ANNUAL RATE	35-64 COMBINED # OF EVENTS	CRUDE	AGE-ADJUSTED	65-94 COMBINED # OF EVENTS	CRUDE	AGE-ADJUSTED
NO	2173	16	4	141			15	0	0	41	2	2
YES	45	2	22	7			0	0	0	2	5	5
UNKNOWN	9	0	0	0			0	0	0	0	0	0
TOTAL	2227	18	4	148			15	0	0	43	2	2

LOGISTIC REGRESSIONS

AGE AT EXAM	COEFFICIENT OF RISK FACTOR	SE COEFFICIENT	P-VALUE	STANDARDIZED COEFFICIENT	STD SE
35-44					
45-54					
55-64	-	-	-	-	-
65-74					
75-84	1.856	0.766	0.016	0.259	0.108
85-94					
35-64 UNIVARIATE	-	-	-	-	-
65-94 UNIVARIATE	0.881	0.728	0.226	0.123	0.102
35-64 BIVARIATE	-	-	-	-	-
65-94 BIVARIATE	0.743	0.733	0.311	0.104	0.103
35-64 MULTIVARIATE	-	-	-	-	-
65-94 MULTIVARIATE	0.317	0.751	0.673	0.044	0.105

TABLE 16-15   DATE=08NOV85
AVERAGE ANNUAL INCIDENCE RATE FOR EVENT PER 1000 PERSONS AT RISK AT EXAM
BY AGE AND LEVEL OF RISK FACTOR AT EXAM: FRAMINGHAM STUDY 30-YEAR FOLLOWUP

SEX: MALE
EVENT: DEATH FROM STROKE
RISK FACTOR: HEART ENLARGEMENT BY X-RAY
PERSONS AT RISK: PERSONS FREE OF STROKE AND TIA ONLY

RISK FACTOR	35-44 AT EXAM			45-54 AT EXAM			55-64 AT EXAM			65-74 AT EXAM		
	PERSON EXAMS	# OF EVENTS	ANNUAL RATE	PERSON EXAMS	# OF EVENTS	ANNUAL RATE	PERSON EXAMS	# OF EVENTS	ANNUAL RATE	PERSON EXAMS	# OF EVENTS	ANNUAL RATE
NO	4098			6590			6455	8	1	2971	4	1
MILD	189			702			1102	0	0	662	4	3
DEFINITE	224			736			1356	3	1	993	6	3
UNKNOWN	5			4			1	0	0	0	0	-
TOTAL	4516			8032			8914	11	1	4626	14	2

RISK FACTOR	75-84 AT EXAM			85-94 AT EXAM			35-64 COMBINED			65-94 COMBINED		
	PERSON EXAMS	# OF EVENTS	ANNUAL RATE	PERSON EXAMS	# OF EVENTS	ANNUAL RATE	# OF EVENTS	CRUDE	AGE-ADJUSTED	# OF EVENTS	CRUDE	AGE-ADJUSTED
NO	693	9	6	28			11	0	0	14	2	2
MILD	205	5	12	6			0	0	0	9	5	5
DEFINITE	336	6	9	8			3	1	0	13	5	5
UNKNOWN	0	0	-	0			0	0	0	0	-	
TOTAL	1234	20	8	42			14	0	0	36	3	3

LOGISTIC REGRESSIONS

AGE AT EXAM		COEFFICIENT OF RISK FACTOR	SE COEFFICIENT	P-VALUE	STANDARDIZED COEFFICIENT	STD SE
35-44						
45-54						
55-64		0.195	0.372	0.600	0.145	0.276
65-74		0.722	0.296	0.015	0.593	0.243
75-84		0.182	0.251	0.470	0.157	0.218
85-94						
35-64	UNIVARIATE	0.242	0.356	0.497	0.158	0.233
65-94	UNIVARIATE	0.478	0.183	0.009	0.398	0.152
35-64	BIVARIATE	0.024	0.361	0.947	0.016	0.237
65-94	BIVARIATE	0.416	0.184	0.024	0.347	0.153
35-64	MULTIVARIATE	-0.442	0.401	0.270	-0.290	0.263
65-94	MULTIVARIATE	0.262	0.199	0.189	0.218	0.166

SEX: FEMALE
EVENT: DEATH FROM STROKE
RISK FACTOR: HEART ENLARGEMENT BY X-RAY
PERSONS AT RISK: PERSONS FREE OF STROKE AND TIA ONLY

RISK FACTOR	35-44 AT EXAM			45-54 AT EXAM			55-64 AT EXAM			65-74 AT EXAM		
	PERSON EXAMS	# OF EVENTS	ANNUAL RATE	PERSON EXAMS	# OF EVENTS	ANNUAL RATE	PERSON EXAMS	# OF EVENTS	ANNUAL RATE	PERSON EXAMS	# OF EVENTS	ANNUAL RATE
NO	5161			8054	4	0	7762	3	0	3444	13	2
MILD	198			954	1	1	1850	3	1	1254	1	0
DEFINITE	246			1019	0	0	2334	4	1	2173	7	2
UNKNOWN	8			6	0	0	4	0	0	0	0	-
TOTAL	5613			10033	5	0	11950	10	0	6871	21	2

RISK FACTOR	75-84 AT EXAM			85-94 AT EXAM			35-64 COMBINED			65-94 COMBINED		
	PERSON EXAMS	# OF EVENTS	ANNUAL RATE	PERSON EXAMS	# OF EVENTS	ANNUAL RATE	# OF EVENTS	CRUDE	AGE-ADJUSTED	# OF EVENTS	CRUDE	AGE-ADJUSTED
NO	778	5	5	38			7	0	0	18	2	2
MILD	435	3	3	30			4	1	1	5	1	1
DEFINITE	1014	10	5	80			4	1	0	20	3	3
UNKNOWN	0	0	-	0			0	0	0	0	-	
TOTAL	2227	18	4	148			15	0	0	43	2	2

LOGISTIC REGRESSIONS

AGE AT EXAM		COEFFICIENT OF RISK FACTOR	SE COEFFICIENT	P-VALUE	STANDARDIZED COEFFICIENT	STD SE
35-44						
45-54		-0.285	0.840	0.734	-0.183	0.540
55-64		0.724	0.351	0.039	0.578	0.281
65-74		-0.132	0.254	0.602	-0.117	0.225
75-84		0.225	0.276	0.414	0.201	0.246
85-94						
35-64	UNIVARIATE	0.644	0.286	0.024	0.453	0.201
65-94	UNIVARIATE	0.191	0.170	0.261	0.171	0.152
35-64	BIVARIATE	0.478	0.297	0.108	0.336	0.209
65-94	BIVARIATE	0.046	0.174	0.790	0.042	0.156
35-64	MULTIVARIATE	0.239	0.326	0.464	0.168	0.229
65-94	MULTIVARIATE	-0.154	0.187	0.411	-0.138	0.168

TABLE 16-16   DATE=14DEC85
AVERAGE ANNUAL INCIDENCE RATE FOR EVENT PER 1000 PERSONS AT RISK AT EXAM
BY AGE AND LEVEL OF RISK FACTOR AT EXAM: FRAMINGHAM STUDY 30-YEAR FOLLOWUP

SEX: MALE
EVENT: DEATH FROM STROKE
RISK FACTOR: LEFT VENTRICULAR HYPERTROPHY BY ELECTROCARDIOGRAM
PERSONS AT RISK: PERSONS FREE OF STROKE AND TIA ONLY

RISK FACTOR	35-44 AT EXAM			45-54 AT EXAM			55-64 AT EXAM			65-74 AT EXAM		
	PERSON EXAMS	# OF EVENTS	ANNUAL RATE	PERSON EXAMS	# OF EVENTS	ANNUAL RATE	PERSON EXAMS	# OF EVENTS	ANNUAL RATE	PERSON EXAMS	# OF EVENTS	ANNUAL RATE
NO	4405			7780			8471	8	0	4347	12	1
MILD	79			145			235	0	0	114	0	0
DEFINITE	32			107			208	3	7	165	2	6
UNKNOWN	0			0			0	0	-	0	0	-
TOTAL	4516			8032			8914	11	1	4626	14	2

RISK FACTOR	75-84 AT EXAM			85-94 AT EXAM			35-64 COMBINED			65-94 COMBINED		
	PERSON EXAMS	# OF EVENTS	ANNUAL RATE	PERSON EXAMS	# OF EVENTS	ANNUAL RATE	# OF EVENTS	CRUDE	AGE-ADJUSTED	# OF EVENTS	CRUDE	AGE-ADJUSTED
NO	1148	17	7	38	2		11	0	0	31	3	3
MILD	39	1	13	2			0	0	0	1	3	3
DEFINITE	47	2	21	2			3	4	3	4	9	9
UNKNOWN	0	0	-	0			0	-		0	-	
TOTAL	1234	20	8	42			14	0	0	36	3	3

LOGISTIC REGRESSIONS

AGE AT EXAM		COEFFICIENT OF RISK FACTOR	SE COEFFICIENT	P-VALUE	STANDARDIZED COEFFICIENT	STD SE
35-44						
45-54						
55-64		1.324	0.357	0.001	0.448	0.121
65-74		0.675	0.402	0.093	0.268	0.160
75-84		0.544	0.366	0.137	0.226	0.152
85-94						
35-64	UNIVARIATE	1.337	0.343	0.001	0.386	0.099
65-94	UNIVARIATE	0.579	0.267	0.030	0.233	0.107
35-64	BIVARIATE	1.148	0.346	0.001	0.331	0.100
65-94	BIVARIATE	0.561	0.270	0.038	0.226	0.109
35-64	MULTIVARIATE	0.562	0.402	0.162	0.162	0.116
65-94	MULTIVARIATE	0.299	0.298	0.316	0.120	0.120

SEX: FEMALE
EVENT: DEATH FROM STROKE
RISK FACTOR: LEFT VENTRICULAR HYPERTROPHY BY ELECTROCARDIOGRAM
PERSONS AT RISK: PERSONS FREE OF STROKE AND TIA ONLY

RISK FACTOR	35-44 AT EXAM			45-54 AT EXAM			55-64 AT EXAM			65-74 AT EXAM		
	PERSON EXAMS	# OF EVENTS	ANNUAL RATE	PERSON EXAMS	# OF EVENTS	ANNUAL RATE	PERSON EXAMS	# OF EVENTS	ANNUAL RATE	PERSON EXAMS	# OF EVENTS	ANNUAL RATE
NO	5560			9866	5	0	11542	7	0	6504	16	1
MILD	38			111	0	0	242	1	2	174	1	3
DEFINITE	15			56	0	0	166	2	6	193	4	10
UNKNOWN	0			0	0	-	0	0	-	0	0	-
TOTAL	5613			10033	5	0	11950	10	0	6871	21	2

RISK FACTOR	75-84 AT EXAM			85-94 AT EXAM			35-64 COMBINED			65-94 COMBINED		
	PERSON EXAMS	# OF EVENTS	ANNUAL RATE	PERSON EXAMS	# OF EVENTS	ANNUAL RATE	# OF EVENTS	CRUDE	AGE-ADJUSTED	# OF EVENTS	CRUDE	AGE-ADJUSTED
NO	2059	14	3	129	8		12	0	0	33	2	2
MILD	65	1	8	8			1	1	1	2	4	4
DEFINITE	103	3	15	11			2	4	3	8	13	12
UNKNOWN	0	0	-	0			0	-		0	-	
TOTAL	2227	18	4	148			15	0	0	43	2	2

LOGISTIC REGRESSIONS

AGE AT EXAM		COEFFICIENT OF RISK FACTOR	SE COEFFICIENT	P-VALUE	STANDARDIZED COEFFICIENT	STD SE
35-44		-	-	-	-	-
45-54		-	-	-	-	-
55-64		1.527	0.385	0.001	0.414	0.104
65-74		1.064	0.280	0.001	0.385	0.102
75-84		0.744	0.315	0.018	0.332	0.141
85-94						
35-64	UNIVARIATE	1.497	0.365	0.001	0.326	0.080
65-94	UNIVARIATE	0.964	0.198	0.001	0.374	0.077
35-64	BIVARIATE	1.353	0.371	0.001	0.295	0.081
65-94	BIVARIATE	0.859	0.201	0.001	0.334	0.078
35-64	MULTIVARIATE	0.930	0.441	0.035	0.203	0.096
65-94	MULTIVARIATE	0.602	0.224	0.007	0.234	0.087

TABLE 16-17   DATE=08NOV85
AVERAGE ANNUAL INCIDENCE RATE FOR EVENT PER 1000 PERSONS AT RISK AT EXAM
BY AGE AND LEVEL OF RISK FACTOR AT EXAM: FRAMINGHAM STUDY 30-YEAR FOLLOWUP

SEX: MALE
EVENT: DEATH FROM STROKE
RISK FACTOR: INTRAVENTRICULAR CONDUCTION DEFECT (WITH QRS INTERVAL MORE THAN 0.11 SECOND)
PERSONS AT RISK: PERSONS FREE OF STROKE AND TIA ONLY

RISK FACTOR	35-44 AT EXAM PERSON EXAMS	# OF EVENTS	ANNUAL RATE	45-54 AT EXAM PERSON EXAMS	# OF EVENTS	ANNUAL RATE	55-64 AT EXAM PERSON EXAMS	# OF EVENTS	ANNUAL RATE	65-74 AT EXAM PERSON EXAMS	# OF EVENTS	ANNUAL RATE
NO	4198			7452			8107	9	1	4150	11	1
INCOMPLETE	280			491			534	2	2	231	3	6
COMPLETE	38			89			273	0	0	245	0	0
UNKNOWN	0			0			0	0	-	0	0	-
TOTAL	4516			8032			8914	11	1	4626	14	2

RISK FACTOR	75-84 AT EXAM PERSON EXAMS	# OF EVENTS	ANNUAL RATE	85-94 AT EXAM PERSON EXAMS	# OF EVENTS	ANNUAL RATE	35-64 COMBINED # OF EVENTS	CRUDE	AGE-ADJUSTED	65-94 COMBINED # OF EVENTS	CRUDE	AGE-ADJUSTED
NO	1083	20	9	35	3		12	0	0	32	3	3
INCOMPLETE	35	0	0	3	0		2	1	1	3	6	5
COMPLETE	116	0	0	4			0	0	0	1	1	1
UNKNOWN	0	0	-	0			0	-		0	-	
TOTAL	1234	20	8	42			14	0	0	36	3	3

LOGISTIC REGRESSIONS

AGE AT EXAM		COEFFICIENT OF RISK FACTOR	SE COEFFICIENT	P-VALUE	STANDARDIZED COEFFICIENT	STD SE
35-44						
45-54						
55-64		0.292	0.599	0.626	0.120	0.245
65-74		0.210	0.471	0.655	0.103	0.230
75-84		-	-	-	-	-
85-94						
35-64	UNIVARIATE	0.290	0.619	0.639	0.103	0.219
65-94	UNIVARIATE	-0.127	0.358	0.723	-0.065	0.184
35-64	BIVARIATE	0.111	0.601	0.853	0.039	0.213
65-94	BIVARIATE	-0.223	0.352	0.527	-0.115	0.181
35-64	MULTIVARIATE	0.077	0.632	0.903	0.027	0.224
65-94	MULTIVARIATE	-0.303	0.358	0.398	-0.156	0.184

SEX: FEMALE
EVENT: DEATH FROM STROKE
RISK FACTOR: INTRAVENTRICULAR CONDUCTION DEFECT (WITH QRS INTERVAL MORE THAN 0.11 SECOND)
PERSONS AT RISK: PERSONS FREE OF STROKE AND TIA ONLY

RISK FACTOR	35-44 AT EXAM PERSON EXAMS	# OF EVENTS	ANNUAL RATE	45-54 AT EXAM PERSON EXAMS	# OF EVENTS	ANNUAL RATE	55-64 AT EXAM PERSON EXAMS	# OF EVENTS	ANNUAL RATE	65-74 AT EXAM PERSON EXAMS	# OF EVENTS	ANNUAL RATE
NO	5526			9834	5	0	11582	10	0	6577	20	2
INCOMPLETE	75			151	0	0	195	0	0	130	0	0
COMPLETE	12			48	0	0	173	0	0	164	1	3
UNKNOWN	0			0			0	0	-	0	0	-
TOTAL	5613			10033	5	0	11950	10	0	6871	21	2

RISK FACTOR	75-84 AT EXAM PERSON EXAMS	# OF EVENTS	ANNUAL RATE	85-94 AT EXAM PERSON EXAMS	# OF EVENTS	ANNUAL RATE	35-64 COMBINED # OF EVENTS	CRUDE	AGE-ADJUSTED	65-94 COMBINED # OF EVENTS	CRUDE	AGE-ADJUSTED
NO	2088	16	4	138	3		15	0	0	40	2	2
INCOMPLETE	49	1	10	3	0		0	0	0	1	3	2
COMPLETE	90	1	6	7			0	0	0	2	4	4
UNKNOWN	0	0	-	0			0	-		0	-	
TOTAL	2227	18	4	148			15	0	0	43	2	2

LOGISTIC REGRESSIONS

AGE AT EXAM		COEFFICIENT OF RISK FACTOR	SE COEFFICIENT	P-VALUE	STANDARDIZED COEFFICIENT	STD SE
35-44						
45-54		-	-	-	-	-
55-64		-	-	-	-	-
65-74		0.216	0.550	0.694	0.072	0.182
75-84		0.291	0.451	0.518	0.121	0.187
85-94						
35-64	UNIVARIATE	-	-	-	-	-
65-94	UNIVARIATE	0.255	0.347	0.463	0.091	0.124
35-64	BIVARIATE	-	-	-	-	-
65-94	BIVARIATE	0.164	0.347	0.636	0.059	0.124
35-64	MULTIVARIATE	-	-	-	-	-
65-94	MULTIVARIATE	0.144	0.352	0.681	0.051	0.125

TABLE 16-18   DATE=08NOV85
AVERAGE ANNUAL INCIDENCE RATE FOR EVENT PER 1000 PERSONS AT RISK AT EXAM
BY AGE AND LEVEL OF RISK FACTOR AT EXAM: FRAMINGHAM STUDY 30-YEAR FOLLOWUP

SEX: MALE
EVENT: DEATH FROM STROKE
RISK FACTOR: DEFINITE NONSPECIFIC T-WAVE OR ST-SEGMENT ABNORMALITY BY ELECTROCARDIOGRAM
PERSONS AT RISK: PERSONS FREE OF STROKE AND TIA ONLY

RISK FACTOR	35-44 AT EXAM PERSON EXAMS	# OF EVENTS	ANNUAL RATE	45-54 AT EXAM PERSON EXAMS	# OF EVENTS	ANNUAL RATE	55-64 AT EXAM PERSON EXAMS	# OF EVENTS	ANNUAL RATE	65-74 AT EXAM PERSON EXAMS	# OF EVENTS	ANNUAL RATE
NO	4403			7609			8128	9	1	4007	12	1
YES	113			423			786	2	1	619	2	2
UNKNOWN	0			0			0	0	-	0	0	-
TOTAL	4516			8032			8914	11	1	4626	14	2

RISK FACTOR	75-84 AT EXAM PERSON EXAMS	# OF EVENTS	ANNUAL RATE	85-94 AT EXAM PERSON EXAMS	# OF EVENTS	ANNUAL RATE	35-64 COMBINED # OF EVENTS	CRUDE	AGE-ADJUSTED	65-94 COMBINED # OF EVENTS	CRUDE	AGE-ADJUSTED
NO	988	15	8	24			12	0	0	28	3	3
YES	246	5	10	18			2	1	1	8	5	4
UNKNOWN	0	0	-	0			0	-		0	-	
TOTAL	1234	20	8	42			14	0	0	36	3	3

LOGISTIC REGRESSIONS

AGE AT EXAM		COEFFICIENT OF RISK FACTOR	SE COEFFICIENT	P-VALUE	STANDARDIZED COEFFICIENT	STD SE
35-44						
45-54						
55-64		0.833	0.783	0.287	0.236	0.222
65-74		0.076	0.765	0.921	0.026	0.260
75-84		0.297	0.521	0.569	0.119	0.208
85-94						
35-64	UNIVARIATE	0.933	0.764	0.222	0.224	0.184
65-94	UNIVARIATE	0.488	0.403	0.225	0.174	0.144
35-64	BIVARIATE	0.542	0.768	0.480	0.130	0.185
65-94	BIVARIATE	0.255	0.408	0.532	0.091	0.146
35-64	MULTIVARIATE	0.474	0.776	0.542	0.114	0.187
65-94	MULTIVARIATE	0.195	0.410	0.634	0.070	0.146

SEX: FEMALE
EVENT: DEATH FROM STROKE
RISK FACTOR: DEFINITE NONSPECIFIC T-WAVE OR ST-SEGMENT ABNORMALITY BY ELECTROCARDIOGRAM
PERSONS AT RISK: PERSONS FREE OF STROKE AND TIA ONLY

RISK FACTOR	35-44 AT EXAM PERSON EXAMS	# OF EVENTS	ANNUAL RATE	45-54 AT EXAM PERSON EXAMS	# OF EVENTS	ANNUAL RATE	55-64 AT EXAM PERSON EXAMS	# OF EVENTS	ANNUAL RATE	65-74 AT EXAM PERSON EXAMS	# OF EVENTS	ANNUAL RATE
NO	5514			9603	4	0	10984	9	0	5994	16	1
YES	99			430	1	1	966	1	1	877	5	3
UNKNOWN	0			0			0			0		
TOTAL	5613			10033	5	0	11950	10	0	6871	21	2

RISK FACTOR	75-84 AT EXAM PERSON EXAMS	# OF EVENTS	ANNUAL RATE	85-94 AT EXAM PERSON EXAMS	# OF EVENTS	ANNUAL RATE	35-64 COMBINED # OF EVENTS	CRUDE	AGE-ADJUSTED	65-94 COMBINED # OF EVENTS	CRUDE	AGE-ADJUSTED
NO	1811	13	4	114			13	0	0	33	2	2
YES	416	5	6	34			2	1	1	10	4	4
UNKNOWN	0	0	-	0			0	-		0	-	
TOTAL	2227	18	4	148			15	0	0	43	2	2

LOGISTIC REGRESSIONS

AGE AT EXAM		COEFFICIENT OF RISK FACTOR	SE COEFFICIENT	P-VALUE	STANDARDIZED COEFFICIENT	STD SE
35-44						
45-54		1.722	1.119	0.124	0.349	0.227
55-64		0.234	1.055	0.824	0.064	0.287
65-74		0.762	0.514	0.138	0.254	0.171
75-84		0.520	0.529	0.325	0.203	0.206
85-94						
35-64	UNIVARIATE	0.989	0.760	0.193	0.224	0.172
65-94	UNIVARIATE	0.596	0.362	0.100	0.209	0.127
35-64	BIVARIATE	0.704	0.767	0.359	0.159	0.174
65-94	BIVARIATE	0.417	0.365	0.254	0.146	0.128
35-64	MULTIVARIATE	0.413	0.776	0.594	0.094	0.176
65-94	MULTIVARIATE	0.397	0.369	0.283	0.139	0.130

TABLE 17-1   DATE=14DEC85

AVERAGE ANNUAL INCIDENCE RATE FOR EVENT PER 1000 PERSONS AT RISK AT EXAM
BY AGE AND LEVEL OF RISK FACTOR AT EXAM: FRAMINGHAM STUDY 30-YEAR FOLLOWUP

SEX: MALE
EVENT: DEATH FROM STROKE
RISK FACTOR: SYSTOLIC BLOOD PRESSURE, FIRST EXAMINER (MM HG)
PERSONS AT RISK: ALL PERSONS ALIVE

RISK FACTOR	35-44 AT EXAM PERSON EXAMS	# OF EVENTS	ANNUAL RATE	45-54 AT EXAM PERSON EXAMS	# OF EVENTS	ANNUAL RATE	55-64 AT EXAM PERSON EXAMS	# OF EVENTS	ANNUAL RATE	65-74 AT EXAM PERSON EXAMS	# OF EVENTS	ANNUAL RATE
74-119	1226			1842			1605	0	0	659	3	2
120-139	2193			3542			3525	4	1	1636	2	1
140-159	850			1838			2371	2	0	1473	9	3
160-179	197			598			1062	7	3	696	4	3
180-300	53			247			507	4	4	370	6	8
UNKNOWN	0			0			0	0	-	0	0	-
TOTAL	4519			8067			9070	17	1	4834	24	2

RISK FACTOR	75-84 AT EXAM PERSON EXAMS	# OF EVENTS	ANNUAL RATE	85-94 AT EXAM PERSON EXAMS	# OF EVENTS	ANNUAL RATE	35-64 COMBINED # OF EVENTS	CRUDE	AGE-ADJUSTED	65-94 COMBINED # OF EVENTS	CRUDE	AGE-ADJUSTED
74-119	144	0	0	8			0	0	0	6	4	3
120-139	451	7	8	17			5	0	0	9	2	2
140-159	411	4	5	17			5	0	0	13	3	3
160-179	218	6	14	6			7	2	1	10	5	5
180-300	121	6	25	2			4	2	2	12	12	12
UNKNOWN	0	0	-	0			0	-	-	0	-	-
TOTAL	1345	23	9	50			21	0	0	50	4	4

LOGISTIC REGRESSIONS

AGE AT EXAM		COEFFICIENT OF RISK FACTOR	SE COEFFICIENT	P-VALUE	STANDARDIZED COEFFICIENT	STD SE
35-44						
45-54						
55-64		0.035	0.007	0.001	0.771	0.163
65-74		0.018	0.007	0.012	0.427	0.169
75-84		0.024	0.008	0.002	0.553	0.178
85-94						
35-64	UNIVARIATE	0.038	0.007	0.001	0.771	0.136
65-94	UNIVARIATE	0.018	0.005	0.001	0.414	0.120
35-64	BIVARIATE	0.033	0.007	0.001	0.674	0.142
65-94	BIVARIATE	0.018	0.005	0.001	0.423	0.123
35-64	MULTIVARIATE	0.035	0.008	0.001	0.710	0.159
65-94	MULTIVARIATE	0.016	0.006	0.004	0.376	0.129

SEX: FEMALE
EVENT: DEATH FROM STROKE
RISK FACTOR: SYSTOLIC BLOOD PRESSURE, FIRST EXAMINER (MM HG)
PERSONS AT RISK: ALL PERSONS ALIVE

RISK FACTOR	35-44 AT EXAM PERSON EXAMS	# OF EVENTS	ANNUAL RATE	45-54 AT EXAM PERSON EXAMS	# OF EVENTS	ANNUAL RATE	55-64 AT EXAM PERSON EXAMS	# OF EVENTS	ANNUAL RATE	65-74 AT EXAM PERSON EXAMS	# OF EVENTS	ANNUAL RATE
74-119	2493			2724	1	0	2039	0	0	683	1	1
120-139	2245			4027	2	0	4199	5	1	1918	5	1
140-159	674			2086	2	0	3197	3	0	2283	9	2
160-179	156			785	0	0	1682	0	0	1360	6	2
180-300	51			450	2	2	986	3	2	859	7	4
UNKNOWN	0			0	0	-	0	0	-	0	0	-
TOTAL	5619			10072	7	0	12103	11	0	7103	28	2

RISK FACTOR	75-84 AT EXAM PERSON EXAMS	# OF EVENTS	ANNUAL RATE	85-94 AT EXAM PERSON EXAMS	# OF EVENTS	ANNUAL RATE	35-64 COMBINED # OF EVENTS	CRUDE	AGE-ADJUSTED	65-94 COMBINED # OF EVENTS	CRUDE	AGE-ADJUSTED
74-119	158	2	6	10	1	50	1	0	0	4	2	3
120-139	561	7	6	51	0	0	7	0	0	12	2	3
140-159	759	8	5	40	3	38	5	0	0	20	3	3
160-179	561	9	8	36	1	14	0	0	0	16	4	4
180-300	380	6	8	26	1	19	5	2	1	14	6	5
UNKNOWN	0	0	-	0	0	-	0	-	-	0	-	-
TOTAL	2419	32	7	163	6	18	18	0	0	66	3	3

LOGISTIC REGRESSIONS

AGE AT EXAM		COEFFICIENT OF RISK FACTOR	SE COEFFICIENT	P-VALUE	STANDARDIZED COEFFICIENT	STD SE
35-44						
45-54		0.021	0.011	0.043	0.489	0.242
55-64		0.021	0.009	0.020	0.517	0.222
65-74		0.021	0.006	0.001	0.517	0.160
75-84		0.008	0.006	0.206	0.209	0.166
85-94		-0.001	0.017	0.936	-0.034	0.420
35-64	UNIVARIATE	0.024	0.006	0.001	0.569	0.152
65-94	UNIVARIATE	0.015	0.004	0.001	0.367	0.109
35-64	BIVARIATE	0.021	0.007	0.003	0.500	0.166
65-94	BIVARIATE	0.013	0.004	0.005	0.315	0.112
35-64	MULTIVARIATE	0.013	0.008	0.108	0.306	0.190
65-94	MULTIVARIATE	0.011	0.005	0.019	0.273	0.117

TABLE 17-2   DATE=14DEC85
AVERAGE ANNUAL INCIDENCE RATE FOR EVENT PER 1000 PERSONS AT RISK AT EXAM
BY AGE AND LEVEL OF RISK FACTOR AT EXAM: FRAMINGHAM STUDY 30-YEAR FOLLOWUP

SEX: MALE
EVENT: DEATH FROM STROKE
RISK FACTOR: DIASTOLIC BLOOD PRESSURE, FIRST EXAMINER (MM HG)
PERSONS AT RISK: ALL PERSONS ALIVE

RISK FACTOR	35-44 AT EXAM PERSON EXAMS	# OF EVENTS	ANNUAL RATE	45-54 AT EXAM PERSON EXAMS	# OF EVENTS	ANNUAL RATE	55-64 AT EXAM PERSON EXAMS	# OF EVENTS	ANNUAL RATE	65-74 AT EXAM PERSON EXAMS	# OF EVENTS	ANNUAL RATE
20-74	1015			1543			2092	0	0	1524	10	3
75-84	1617			2861			3165	1	0	1643	4	1
85-94	1203			2151			2228	5	1	1072	3	1
95-104	509			1073			1114	3	1	425	1	1
105-160	175			439			471	8	8	170	6	18
UNKNOWN	0			0			0	0	-	0	0	-
TOTAL	4519			8067			9070	17	1	4834	24	2

RISK FACTOR	75-84 AT EXAM PERSON EXAMS	# OF EVENTS	ANNUAL RATE	85-94 AT EXAM PERSON EXAMS	# OF EVENTS	ANNUAL RATE	35-64 COMBINED # OF EVENTS	CRUDE	AGE-ADJUSTED	65-94 COMBINED # OF EVENTS	CRUDE	AGE-ADJUSTED
20-74	560	7	6	30			1	0	0	19	4	4
75-84	440	4	8	13			1	0	0	12	3	3
85-94	244	5	10	0			7	1	1	5	3	3
95-104	71	4	28	0			4	1	1	5	5	7
105-160	30	0	0	0			8	4	4	6	15	14
UNKNOWN	0	0	-	0			0	-		0	-	
TOTAL	1345	23	9	50			21	0	0	50	4	4

LOGISTIC REGRESSIONS

AGE AT EXAM		COEFFICIENT OF RISK FACTOR	SE COEFFICIENT	P-VALUE	STANDARDIZED COEFFICIENT	STD SE
35-44						
45-54						
55-64		1.089(	0.014	0.001	1.083	0.169
65-74		1.019(	0.015	0.225	0.230	0.189
75-84		1.024(	0.016	0.150	0.288	0.200
85-94						
35-64	UNIVARIATE	0.082	0.013	0.001	0.978	0.150
65-94	UNIVARIATE	0.010	0.011	0.374	0.122	0.137
35-64	BIVARIATE	0.080	0.013	0.001	0.949	0.149
65-94	BIVARIATE	0.020	0.011	0.069	0.249	0.137
35-64	MULTIVARIATE	0.085	0.014	0.001	1.008	0.171
65-94	MULTIVARIATE	0.019	0.011	0.073	0.239	0.133

SEX: FEMALE
EVENT: DEATH FROM STROKE
RISK FACTOR: DIASTOLIC BLOOD PRESSURE, FIRST EXAMINER (MM HG)
PERSONS AT RISK: ALL PERSONS ALIVE

RISK FACTOR	35-44 AT EXAM PERSON EXAMS	# OF EVENTS	ANNUAL RATE	45-54 AT EXAM PERSON EXAMS	# OF EVENTS	ANNUAL RATE	55-64 AT EXAM PERSON EXAMS	# OF EVENTS	ANNUAL RATE	65-74 AT EXAM PERSON EXAMS	# OF EVENTS	ANNUAL RATE
20-74	2164			2657	1	0	3110	2	0	2069	8	2
75-84	2031			3573	2	0	4005	4	0	2416	6	1
85-94	988			2395	0	0	3000	0	0	1622	9	3
95-104	324			941	3	2	1348	2	1	685	3	2
105-160	112			506	1	1	640	3	2	311	2	3
UNKNOWN	0			0	0	-	0	0	-	0	0	-
TOTAL	5619			10072	7	0	12103	11	0	7103	28	2

RISK FACTOR	75-84 AT EXAM PERSON EXAMS	# OF EVENTS	ANNUAL RATE	85-94 AT EXAM PERSON EXAMS	# OF EVENTS	ANNUAL RATE	35-64 COMBINED # OF EVENTS	CRUDE	AGE-ADJUSTED	65-94 COMBINED # OF EVENTS	CRUDE	AGE-ADJUSTED
20-74	916	11	6	73	2	14	3	0	0	21	3	3
75-84	722	11	8	56	3	27	6	0	0	20	3	3
85-94	535	6	6	25	1	20	0	0	0	16	4	4
95-104	178	4	11	7	0	0	5	1	1	7	4	4
105-160	68	0	0	2	0	0	4	2	1	2	4	2
UNKNOWN	0	0	-	0	0	-	0	-		0	-	
TOTAL	2419	32	7	163	6	18	18	0	0	66	3	3

LOGISTIC REGRESSIONS

AGE AT EXAM		COEFFICIENT OF RISK FACTOR	SE COEFFICIENT	P-VALUE	STANDARDIZED COEFFICIENT	STD SE
35-44						
45-54		0.036	0.024	0.141	0.441	0.300
55-64		0.032	0.021	0.122	0.400	0.259
65-74		0.013	0.015	0.376	0.160	0.181
75-84		0.003	0.014	0.856	0.032	0.177
85-94		0.002	0.032	0.954	0.024	0.418
35-64	UNIVARIATE	0.038	0.015	0.013	0.470	0.189
65-94	UNIVARIATE	0.000	0.010	0.973	0.004	0.123
35-64	BIVARIATE	0.035	0.016	0.029	0.427	0.195
65-94	BIVARIATE	0.009	0.010	0.365	0.111	0.122
35-64	MULTIVARIATE	0.022	0.016	0.182	0.271	0.203
65-94	MULTIVARIATE	0.011	0.010	0.248	0.140	0.121

TABLE 17-3A   DATE=16JAN86
AVERAGE ANNUAL INCIDENCE RATE FOR EVENT PER 1000 PERSONS AT RISK AT EXAM
BY AGE AND LEVEL OF RISK FACTOR AT EXAM: FRAMINGHAM STUDY 30-YEAR FOLLOWUP

SEX: MALE
EVENT: DEATH FROM STROKE
RISK FACTOR: HYPERTENSION WITH ANTIHYPERTENSIVE TREATMENT
PERSONS AT RISK: ALL PERSONS ALIVE

RISK FACTOR	35-44 AT EXAM			45-54 AT EXAM			55-64 AT EXAM			65-74 AT EXAM		
	PERSON EXAMS	# OF EVENTS	ANNUAL RATE	PERSON EXAMS	# OF EVENTS	ANNUAL RATE	PERSON EXAMS	# OF EVENTS	ANNUAL RATE	PERSON EXAMS	# OF EVENTS	ANNUAL RATE
NO	2668			4263			4163	1	0	1803	4	1
MILD	1316			2287			2573	3	1	1520	4	1
DEFINITE	511			1216			1431	9	3	698	7	5
TREATED	24			301			903	4	2	813	9	6
UNKNOWN	0			0			0	0	-	0	0	-
TOTAL	4519			8067			9070	17	1	4834	24	2

RISK FACTOR	75-84 AT EXAM			85-94 AT EXAM			35-64 COMBINED			65-94 COMBINED		
	PERSON EXAMS	# OF EVENTS	ANNUAL RATE	PERSON EXAMS	# OF EVENTS	ANNUAL RATE	# OF EVENTS	CRUDE	AGE-ADJUSTED	# OF EVENTS	CRUDE	AGE-ADJUSTED
NO	465	5	5	19			2	0	0	12	3	3
MILD	424	6	7	16			5	0	0	10	3	3
DEFINITE	194	6	15	5			10	2	1	15	7	7
TREATED	262	6	11	10			4	2	1	15	7	7
UNKNOWN	0	0	-	0			0	-	-	0	-	-
TOTAL	1345	23	9	50			21	0	0	50	4	4

LOGISTIC REGRESSIONS

AGE AT EXAM		COEFFICIENT OF RISK FACTOR	SE COEFFICIENT	P-VALUE	STANDARDIZED COEFFICIENT	STD SE
35-44						
45-54						
55-64		1.572	0.425	0.001	1.292	0.350
65-74		0.905	0.291	0.002	0.747	0.240
75-84		0.479	0.268	0.074	0.396	0.222
85-94						
35-64	UNIVARIATE	1.419	0.328	0.001	1.116	0.258
65-94	UNIVARIATE	0.563	0.182	0.002	0.465	0.151
35-64	BIVARIATE	1.269	0.327	0.001	0.998	0.257
65-94	BIVARIATE	0.554	0.182	0.002	0.458	0.151
35-64	MULTIVARIATE	1.304	0.351	0.001	1.026	0.276
65-94	MULTIVARIATE	0.509	0.187	0.006	0.421	0.154

SEX: FEMALE
EVENT: DEATH FROM STROKE
RISK FACTOR: HYPERTENSION WITH ANTIHYPERTENSIVE TREATMENT
PERSONS AT RISK: ALL PERSONS ALIVE

RISK FACTOR	35-44 AT EXAM			45-54 AT EXAM			55-64 AT EXAM			65-74 AT EXAM		
	PERSON EXAMS	# OF EVENTS	ANNUAL RATE	PERSON EXAMS	# OF EVENTS	ANNUAL RATE	PERSON EXAMS	# OF EVENTS	ANNUAL RATE	PERSON EXAMS	# OF EVENTS	ANNUAL RATE
NO	4251			5660	3	0	4992	2	0	1871	3	1
MILD	1002			2611	1	0	3449	1	0	2116	9	2
DEFINITE	316			1184	2	1	1950	2	1	1198	5	2
TREATED	50			617	1	1	1712	6	2	1918	11	3
UNKNOWN	0			0	0	-	0	0	-	0	0	-
TOTAL	5619			10072	7	0	12103	11	0	7103	28	2

RISK FACTOR	75-84 AT EXAM			85-94 AT EXAM			35-64 COMBINED			65-94 COMBINED		
	PERSON EXAMS	# OF EVENTS	ANNUAL RATE	PERSON EXAMS	# OF EVENTS	ANNUAL RATE	# OF EVENTS	CRUDE	AGE-ADJUSTED	# OF EVENTS	CRUDE	AGE-ADJUSTED
NO	476	6	4	40	0	0	5	0	0	9	2	2
MILD	694	6	4	37	3	41	2	0	0	18	3	3
DEFINITE	469	3	3	35	1	14	4	1	1	9	3	3
TREATED	780	17	11	51	2	20	7	1	1	30	5	5
UNKNOWN	0	0	-	0	0	-	0	-	-	0	-	-
TOTAL	2419	32	7	163	6	18	18	0	0	66	3	3

LOGISTIC REGRESSIONS

AGE AT EXAM		COEFFICIENT OF RISK FACTOR	SE COEFFICIENT	P-VALUE	STANDARDIZED COEFFICIENT	STD SE
35-44						
45-54		0.573	0.449	0.202	0.442	0.346
55-64		1.027	0.436	0.018	0.861	0.365
65-74		0.477	0.260	0.067	0.391	0.213
75-84		0.207	0.241	0.392	0.162	0.189
85-94		0.360	0.558	0.519	0.301	0.466
35-64	UNIVARIATE	0.950	0.296	0.001	0.760	0.236
65-94	UNIVARIATE	0.402	0.169	0.017	0.327	0.137
35-64	BIVARIATE	0.814	0.307	0.008	0.651	0.246
65-94	BIVARIATE	0.316	0.169	0.062	0.257	0.137
35-64	MULTIVARIATE	0.622	0.319	0.051	0.497	0.255
65-94	MULTIVARIATE	0.277	0.173	0.110	0.225	0.141

TABLE 17-5B   DATE=09NOV85
AVERAGE ANNUAL INCIDENCE RATE FOR EVENT PER 1000 PERSONS AT RISK AT EXAM
BY AGE AND LEVEL OF RISK FACTOR AT EXAM: FRAMINGHAM STUDY 30-YEAR FOLLOWUP

SEX: MALE
EVENT: DEATH FROM STROKE
RISK FACTOR: HYPERTENSION IGNORING TREATMENT
PERSONS AT RISK: ALL PERSONS ALIVE

RISK FACTOR	35-44 AT EXAM PERSON EXAMS	# OF EVENTS	ANNUAL RATE	45-54 AT EXAM PERSON EXAMS	# OF EVENTS	ANNUAL RATE	55-64 AT EXAM PERSON EXAMS	# OF EVENTS	ANNUAL RATE	65-74 AT EXAM PERSON EXAMS	# OF EVENTS	ANNUAL RATE
NO	2672			4306			4347	1	0	1980	4	1
MILD	1326			2383			2922	4	1	1852	10	3
DEFINITE	521			1378			1801	12	3	1002	10	5
UNKNOWN	0			0			0	0	-	0	0	-
TOTAL	4519			8067			9070	17	1	4834	24	2

RISK FACTOR	75-84 AT EXAM PERSON EXAMS	# OF EVENTS	ANNUAL RATE	85-94 AT EXAM PERSON EXAMS	# OF EVENTS	ANNUAL RATE	35-64 COMBINED # OF EVENTS	CRUDE	AGE-ADJUSTED	65-94 COMBINED # OF EVENTS	CRUDE	AGE-ADJUSTED
NO	526	5	5	24			2	0	0	12	2	2
MILD	543	8	7	21			6	0	0	18	4	4
DEFINITE	276	10	18	5			13	2	2	20	8	8
UNKNOWN	0	0	-	0			0	-		0	-	
TOTAL	1345	23	9	50			21	0	0	50	4	4

LOGISTIC REGRESSIONS

AGE AT EXAM		COEFFICIENT OF RISK FACTOR	SE COEFFICIENT	P-VALUE	STANDARDIZED COEFFICIENT	STD SE
35-44						
45-54						
55-64		1.645	0.406	0.001	1.273	0.314
65-74		0.769	0.273	0.005	0.584	0.207
75-84		0.715	0.283	0.012	0.536	0.212
85-94						
35-64	UNIVARIATE	1.453	0.322	0.001	1.097	0.243
65-94	UNIVARIATE	0.615	0.186	0.001	0.465	0.141
35-64	BIVARIATE	1.363	0.323	0.001	1.029	0.244
65-94	BIVARIATE	0.640	0.188	0.001	0.484	0.142
35-64	MULTIVARIATE	1.392	0.345	0.001	1.051	0.260
65-94	MULTIVARIATE	0.594	0.192	0.002	0.449	0.145

SEX: FEMALE
EVENT: DEATH FROM STROKE
RISK FACTOR: HYPERTENSION IGNORING TREATMENT
PERSONS AT RISK: ALL PERSONS ALIVE

RISK FACTOR	35-44 AT EXAM PERSON EXAMS	# OF EVENTS	ANNUAL RATE	45-54 AT EXAM PERSON EXAMS	# OF EVENTS	ANNUAL RATE	55-64 AT EXAM PERSON EXAMS	# OF EVENTS	ANNUAL RATE	65-74 AT EXAM PERSON EXAMS	# OF EVENTS	ANNUAL RATE
NO	4273			5846	3	0	5415	4	0	2205	4	1
MILD	1017			2803	1	0	4079	2	0	2907	11	2
DEFINITE	329			1423	3	1	2609	5	1	1991	13	3
UNKNOWN	0			0		-	0	0	-	0	0	-
TOTAL	.5619			10072	7	0	12103	11	0	7103	28	2

RISK FACTOR	75-84 AT EXAM PERSON EXAMS	# OF EVENTS	ANNUAL RATE	85-94 AT EXAM PERSON EXAMS	# OF EVENTS	ANNUAL RATE	35-64 COMBINED # OF EVENTS	CRUDE	AGE-ADJUSTED	65-94 COMBINED # OF EVENTS	CRUDE	AGE-ADJUSTED
NO	613	8	7	52	1	10	7	0	0	13	2	2
MILD	1003	11	5	57	4	35	3	0	0	26	3	3
DEFINITE	803	13	8	54	1	9	8	1	1	27	5	5
UNKNOWN	0	0	-	0	0	-	0	-		0	-	
TOTAL	2419	32	7	163	6	18	18	0	0	66	3	3

LOGISTIC REGRESSIONS

AGE AT EXAM		COEFFICIENT OF RISK FACTOR	SE COEFFICIENT	P-VALUE	STANDARDIZED COEFFICIENT	STD SE
35-44						
45-54		0.707	0.461	0.125	0.514	0.335
55-64		0.506	0.376	0.179	0.395	0.294
65-74		0.620	0.262	0.018	0.476	0.201
75-84		0.137	0.236	0.561	0.104	0.180
85-94		-0.020	0.516	0.970	-0.016	0.417
35-64	UNIVARIATE	0.718	0.287	0.012	0.535	0.214
65-94	UNIVARIATE	0.372	0.165	0.024	0.286	0.127
35-64	BIVARIATE	0.578	0.297	0.051	0.431	0.221
65-94	BIVARIATE	0.309	0.166	0.062	0.237	0.127
35-64	MULTIVARIATE	0.345	0.317	0.277	0.257	0.236
65-94	MULTIVARIATE	0.279	0.170	0.101	0.214	0.131

TABLE 17-4   DATE=14DEC85
AVERAGE ANNUAL INCIDENCE RATE FOR EVENT PER 1000 PERSONS AT RISK AT EXAM
BY AGE AND LEVEL OF RISK FACTOR AT EXAM: FRAMINGHAM STUDY 30-YEAR FOLLOWUP

SEX: MALE
EVENT: DEATH FROM STROKE
RISK FACTOR: SERUM CHOLESTEROL
PERSONS AT RISK: ALL PERSONS ALIVE

RISK FACTOR	35-44 AT EXAM			45-54 AT EXAM			55-64 AT EXAM			65-74 AT EXAM		
	PERSON EXAMS	# OF EVENTS	ANNUAL RATE	PERSON EXAMS	# OF EVENTS	ANNUAL RATE	PERSON EXAMS	# OF EVENTS	ANNUAL RATE	PERSON EXAMS	# OF EVENTS	ANNUAL RATE
84-204	1253			1807			2484	7	1	1530	14	5
205-234	1258			2390			2695	4	1	1457	3	2
235-264	883			1843			2128	4	1	1059	2	1
265-294	488			1057			1075	1	0	525	2	2
295-1124	303			619			556	0	0	246	0	0
UNKNOWN	334			351			132	1	4	17	0	0
TOTAL	4519			8067			9070	17	1	4834	24	2

RISK FACTOR	75-84 AT EXAM			85-94 AT EXAM			35-64 COMBINED			65-94 COMBINED		
	PERSON EXAMS	# OF EVENTS	ANNUAL RATE	PERSON EXAMS	# OF EVENTS	ANNUAL RATE	# OF EVENTS	CRUDE	AGE-ADJUSTED	# OF EVENTS	CRUDE	AGE-ADJUSTED
84-204	537	13	12	23			7	1	1	28	7	6
205-234	422	8	9	14			5	0	0	15	4	4
235-264	232	2	4	7			6	1	1	5	2	2
265-294	130	0	0	6			2	0	0	2	2	1
295-1124	24	0	0	0			0	0	0	0	0	0
UNKNOWN	0	0	-	0			1	1	2	0	0	0
TOTAL	1345	23	9	50			21	0	0	50	4	4

LOGISTIC REGRESSIONS

AGE AT EXAM		COEFFICIENT OF RISK FACTOR	SE COEFFICIENT	P-VALUE	STANDARDIZED COEFFICIENT	STD SE
35-44						
45-54						
55-64		-0.013	0.007	0.054	-0.534	0.277
65-74		-0.014	0.006	0.011	-0.573	0.225
75-84		-0.015	0.006	0.011	-0.570	0.223
85-94						
35-64	UNIVARIATE	-0.009	0.006	0.132	-0.371	0.247
65-94	UNIVARIATE	-0.016	0.004	0.001	-0.655	0.156
35-64	BIVARIATE	-0.008	0.006	0.155	-0.353	0.248
65-94	BIVARIATE	-0.014	0.004	0.001	-0.567	0.155
35-64	MULTIVARIATE	-0.010	0.006	0.086	-0.433	0.253
65-94	MULTIVARIATE	-0.014	0.004	0.001	-0.556	0.156

SEX: FEMALE
EVENT: DEATH FROM STROKE AND TIA ONLY
RISK FACTOR: SERUM CHOLESTEROL
PERSONS AT RISK: ALL PERSONS ALIVE

RISK FACTOR	35-44 AT EXAM			45-54 AT EXAM			55-64 AT EXAM			65-74 AT EXAM		
	PERSON EXAMS	# OF EVENTS	ANNUAL RATE	PERSON EXAMS	# OF EVENTS	ANNUAL RATE	PERSON EXAMS	# OF EVENTS	ANNUAL RATE	PERSON EXAMS	# OF EVENTS	ANNUAL RATE
84-204	2350			2120	0	0	1714	2	1	897	9	5
205-234	1419			2590	1	0	2741	1	0	1570	4	1
235-264	807			2268	1	0	3052	5	1	1912	6	2
265-294	372			1490	1	0	2364	1	0	1472	4	1
295-1124	171			1124	4	2	1999	2	1	1215	4	2
UNKNOWN	500			480	0	0	233	0	0	37	1	14
TOTAL	5619			10072	7	0	12103	11	0	7103	28	2

RISK FACTOR	75-84 AT EXAM			85-94 AT EXAM			35-64 COMBINED			65-94 COMBINED		
	PERSON EXAMS	# OF EVENTS	ANNUAL RATE	PERSON EXAMS	# OF EVENTS	ANNUAL RATE	# OF EVENTS	CRUDE	AGE-ADJUSTED	# OF EVENTS	CRUDE	AGE-ADJUSTED
84-204	463	9	10	60	4	33	2	0	0	22	8	7
205-234	588	8	7	39	0	0	2	0	0	12	3	3
235-264	660	7	5	42	2	24	6	0	0	15	3	2
265-294	419	5	6	12	0	0	2	0	0	9	2	1
295-1124	282	3	5	10	0	0	6	1	1	7	2	3
UNKNOWN	7	0	0	0		-	0	0	0	1	11	10
TOTAL	2419	32	7	163	6	18	18	0	0	66	3	3

LOGISTIC REGRESSIONS

AGE AT EXAM		COEFFICIENT OF RISK FACTOR	SE COEFFICIENT	P-VALUE	STANDARDIZED COEFFICIENT	STD SE
35-44						
45-54		0.007	0.003	0.007	0.322	0.120
55-64		-0.001	0.007	0.881	-0.046	0.305
65-74		-0.010	0.005	0.024	-0.467	0.207
75-84		-0.005	0.004	0.261	-0.206	0.184
85-94		-0.022	0.011	0.050	-1.007	0.513
35-64	UNIVARIATE	0.006	0.003	0.013	0.303	0.122
65-94	UNIVARIATE	-0.011	0.003	0.001	-0.495	0.134
35-64	BIVARIATE	0.006	0.003	0.073	0.275	0.153
65-94	BIVARIATE	-0.008	0.003	0.008	-0.352	0.133
35-64	MULTIVARIATE	0.005	0.003	0.140	0.224	0.152
65-94	MULTIVARIATE	-0.008	0.003	0.009	-0.351	0.135

TABLE 17-5   DATE=09NOV85
AVERAGE ANNUAL INCIDENCE RATE FOR EVENT PER 1000 PERSONS AT RISK AT EXAM,
BY AGE AND LEVEL OF RISK FACTOR AT EXAM, FRAMINGHAM STUDY 30-YEAR FOLLOWUP

SEX: MALE
EVENT: DEATH FROM STROKE
RISK FACTOR: HEMATOCRIT
PERSONS AT RISK: ALL PERSONS ALIVE

RISK FACTOR	35-44 AT EXAM PERSON EXAMS	# OF EVENTS	ANNUAL RATE	45-54 AT EXAM PERSON EXAMS	# OF EVENTS	ANNUAL RATE	55-64 AT EXAM PERSON EXAMS	# OF EVENTS	ANNUAL RATE	65-74 AT EXAM PERSON EXAMS	# OF EVENTS	ANNUAL RATE
25-42	630			961			1121	2	1	558	3	3
43-44	551			996			1094	0	0	611	2	2
45-46	990			1942			2215	4	1	1194	5	2
47-48	786			1585			1625	2	1	1069	6	3
49-70	1192			2193			2656	8	2	1382	8	3
UNKNOWN	370			390			159	1	3	20	0	0
TOTAL	4519			8067			9070	17	1	4834	24	2

RISK FACTOR	75-84 AT EXAM PERSON EXAMS	# OF EVENTS	ANNUAL RATE	85-94 AT EXAM PERSON EXAMS	# OF EVENTS	ANNUAL RATE	35-64 COMBINED # OF EVENTS	CRUDE	AGE-ADJUSTED	65-94 COMBINED # OF EVENTS	CRUDE	AGE-ADJUSTED
25-42	283	3	5	19			3	1	1	7	4	4
43-44	169	5	15	8			1	0	0	7	4	4
45-46	313	6	10	7			5	0	0	12	4	4
47-48	266	3	6	9			3	0	0	9	3	3
49-70	314	6	10	7			8	1	1	15	4	5
UNKNOWN	0	0	-	0			1	1	1	0	0	0
TOTAL	1345	23	9	50			21	0	0	50	4	4

LOGISTIC REGRESSIONS

AGE AT EXAM	COEFFICIENT OF RISK FACTOR	SE COEFFICIENT	P-VALUE	STANDARDIZED COEFFICIENT	STD SE
35-44					
45-54					
55-64	0.126	0.064	0.050	0.476	0.243
65-74	0.039	0.057	0.487	0.143	0.206
75-84	-0.005	0.051	0.915	-0.023	0.210
85-94					
35-64 UNIVARIATE	0.079	0.061	0.191	0.292	0.223
65-94 UNIVARIATE	-0.006	0.037	0.875	-0.022	0.142
35-64 BIVARIATE	0.075	0.060	0.211	0.276	0.221
65-94 BIVARIATE	0.021	0.037	0.376	0.078	0.139
35-64 MULTIVARIATE	0.060	0.060	0.311	0.223	0.220
65-94 MULTIVARIATE	0.024	0.037	0.518	0.091	0.140

SEX: FEMALE
EVENT: DEATH FROM STROKE
RISK FACTOR: HEMATOCRIT
PERSONS AT RISK: ALL PERSONS ALIVE

RISK FACTOR	35-44 AT EXAM PERSON EXAMS	# OF EVENTS	ANNUAL RATE	45-54 AT EXAM PERSON EXAMS	# OF EVENTS	ANNUAL RATE	55-64 AT EXAM PERSON EXAMS	# OF EVENTS	ANNUAL RATE	65-74 AT EXAM PERSON EXAMS	# OF EVENTS	ANNUAL RATE
21-39	1698			1858	1	0	1609	1	0	751	6	4
40-41	967			1761	0	0	2144	1	1	1180	4	2
42-43	1043			1978	1	0	2371	3	1	1422	3	1
44-45	858			2222	2	0	3003	1	0	1835	3	2
46-65	487			1702	3	1	2700	3	1	1873	6	2
UNKNOWN	566			551	0	0	276	0	0	42	1	12
TOTAL	5619			10072	7	0	12103	11	0	7103	28	2

RISK FACTOR	75-84 AT EXAM PERSON EXAMS	# OF EVENTS	ANNUAL RATE	85-94 AT EXAM PERSON EXAMS	# OF EVENTS	ANNUAL RATE	35-64 COMBINED # OF EVENTS	CRUDE	AGE-ADJUSTED	65-94 COMBINED # OF EVENTS	CRUDE	AGE-ADJUSTED
21-39	375	10	13	33	2	30	2	0	0	18	8	7
40-41	449	7	8	39	2	26	3	0	0	13	4	4
42-43	437	2	2	26	1	19	4	0	0	6	2	2
44-45	607	6	5	34	0	0	3	0	0	14	3	3
46-65	545	7	6	31	1	16	6	1	1	14	3	3
UNKNOWN	6	0	0	0	0	-	0	0	0	1	10	9
TOTAL	2419	32	7	163	6	18	18	0	0	66	3	3

LOGISTIC REGRESSIONS

AGE AT EXAM	COEFFICIENT OF RISK FACTOR	SE COEFFICIENT	P-VALUE	STANDARDIZED COEFFICIENT	STD SE
35-44					
45-54	0.196	0.106	0.064	0.697	0.376
55-64	0.019	0.087	0.826	0.067	0.303
65-74	-0.041	0.053	0.446	-0.144	0.189
75-84	-0.105	0.040	0.009	-0.407	0.156
85-94	-0.079	0.109	0.467	-0.309	0.424
35-64 UNIVARIATE	0.112	0.066	0.091	0.404	0.239
65-94 UNIVARIATE	-0.099	0.031	0.001	-0.360	0.113
35-64 BIVARIATE	0.088	0.068	0.196	0.316	0.244
65-94 BIVARIATE	-0.079	0.031	0.012	-0.285	0.114
35-64 MULTIVARIATE	0.037	0.064	0.567	0.133	0.232
65-94 MULTIVARIATE	-0.072	0.030	0.016	-0.261	0.108

TABLE 17-6   DATE=09NOV85
AVERAGE ANNUAL INCIDENCE RATE FOR EVENT PER 1000 PERSONS AT RISK AT EXAM
BY AGE AND LEVEL OF RISK FACTOR AT EXAM: FRAMINGHAM STUDY 30-YEAR FOLLOWUP

SEX: MALE
EVENT: DEATH FROM STROKE
RISK FACTOR: BLOOD GLUCOSE (MG/DL)
PERSONS AT RISK: ALL PERSONS ALIVE

RISK FACTOR	35-44 AT EXAM PERSON EXAMS	# OF EVENTS	ANNUAL RATE	45-54 AT EXAM PERSON EXAMS	# OF EVENTS	ANNUAL RATE	55-64 AT EXAM PERSON EXAMS	# OF EVENTS	ANNUAL RATE	65-74 AT EXAM PERSON EXAMS	# OF EVENTS	ANNUAL RATE
29-69	1004			1562			1208	5	2	501	2	2
70-89	2531			4344			4526	6	1	2142	7	2
90-109	628			1402			2136	4	1	1296	10	4
110-129	107			331			608	0	0	428	1	1
130-524	49			204			491	2	2	450	4	4
UNKNOWN	200			224			101	0	0	17	0	0
TOTAL	4519			8067			9070	17	1	4834	24	2

RISK FACTOR	75-84 AT EXAM PERSON EXAMS	# OF EVENTS	ANNUAL RATE	85-94 AT EXAM PERSON EXAMS	# OF EVENTS	ANNUAL RATE	35-64 COMBINED # OF EVENTS	CRUDE	AGE-ADJUSTED	65-94 COMBINED # OF EVENTS	CRUDE	AGE-ADJUSTED
29-69	86	2	12	1			5	1	1	4	3	4
70-89	479	11	11	17			10	0	0	18	3	4
90-109	438	6	7	15			4	0	0	17	5	5
110-129	170	1	7	7			0	0	0	2	2	2
130-524	172	3	9	10			2	1	1	9	7	6
UNKNOWN	0	0	-	0			0	0	0	0	0	0
TOTAL	1345	23	9	50			21	0	0	50	4	4

LOGISTIC REGRESSIONS

AGE AT EXAM		COEFFICIENT OF RISK FACTOR	SE COEFFICIENT	P-VALUE	STANDARDIZED COEFFICIENT	STD SE
35-44						
45-54						
55-64		0.003	0.007	0.687	0.081	0.201
65-74		0.004	0.004	0.352	0.137	0.148
75-84		-0.004	0.007	0.552	-0.150	0.253
85-94						
35-64	UNIVARIATE	0.004	0.006	0.469	0.114	0.157
65-94	UNIVARIATE	0.003	0.003	0.278	0.119	0.110
35-64	BIVARIATE	0.001	0.007	0.926	0.017	0.186
65-94	BIVARIATE	0.002	0.003	0.488	0.086	0.124
35-64	MULTIVARIATE	-0.002	0.008	0.837	-0.041	0.200
65-94	MULTIVARIATE	0.002	0.004	0.566	0.075	0.130

SEX: FEMALE
EVENT: DEATH FROM STROKE
RISK FACTOR: BLOOD GLUCOSE (MG/DL)
PERSONS AT RISK: ALL PERSONS ALIVE

RISK FACTOR	35-44 AT EXAM PERSON EXAMS	# OF EVENTS	ANNUAL RATE	45-54 AT EXAM PERSON EXAMS	# OF EVENTS	ANNUAL RATE	55-64 AT EXAM PERSON EXAMS	# OF EVENTS	ANNUAL RATE	65-74 AT EXAM PERSON EXAMS	# OF EVENTS	ANNUAL RATE
29-69	1067			1806	1	0	1411	2	1	597	1	1
70-89	3414			5811	6	1	6371	3	0	3483	13	2
90-109	672			1646	0	0	2877	1	0	1982	4	1
110-129	96			303	0	0	777	4	3	582	3	3
130-524	39			167	0	0	496	1	1	427	6	7
UNKNOWN	331			339	0	0	171	0	0	32	1	16
TOTAL	5619			10072	7	0	12103	11	0	7103	28	2

RISK FACTOR	75-84 AT EXAM PERSON EXAMS	# OF EVENTS	ANNUAL RATE	85-94 AT EXAM PERSON EXAMS	# OF EVENTS	ANNUAL RATE	35-64 COMBINED # OF EVENTS	CRUDE	AGE-ADJUSTED	65-94 COMBINED # OF EVENTS	CRUDE	AGE-ADJUSTED
29-69	129	2	8	5	1	100	3	0	0	4	3	3
70-89	976	13	7	60	2	17	9	0	0	28	3	2
90-109	803	9	6	63	1	8	1	0	0	14	2	2
110-129	286	4	7	19	1	26	4	2	0	8	5	4
130-524	224	4	9	16	1	31	1	1	0	11	8	8
UNKNOWN	1	0	0	0	0	-	0	0	0	1	15	11
TOTAL	2419	32	7	163	6	18	18	0	0	66	3	3

LOGISTIC REGRESSIONS

AGE AT EXAM		COEFFICIENT OF RISK FACTOR	SE COEFFICIENT	P-VALUE	STANDARDIZED COEFFICIENT	STD SE
35-44						
45-54		-0.003	0.022	0.877	-0.067	0.435
55-64		0.012	0.003	0.001	0.327	0.094
65-74		0.009	0.003	0.008	0.262	0.098
75-84		-0.002	0.006	0.747	-0.065	0.200
85-94		-0.002	0.013	0.900	-0.058	0.460
35-64	UNIVARIATE	0.012	0.003	0.001	0.277	0.078
65-94	UNIVARIATE	0.005	0.003	0.048	0.167	0.084
35-64	BIVARIATE	0.011	0.004	0.003	0.246	0.082
65-94	BIVARIATE	0.004	0.003	0.236	0.111	0.094
35-64	MULTIVARIATE	0.008	0.004	0.038	0.184	0.089
65-94	MULTIVARIATE	0.002	0.003	0.451	0.073	0.097

TABLE 17-7   DATE=09NOV85
AVERAGE ANNUAL INCIDENCE RATE FOR EVENT PER 1000 PERSONS AT RISK AT EXAM
BY AGE AND LEVEL OF RISK FACTOR AT EXAM: FRAMINGHAM STUDY 30-YEAR FOLLOWUP

SEX: MALE
EVENT: DEATH FROM STROKE
RISK FACTOR: DIABETES MELLITUS
PERSONS AT RISK: ALL PERSONS ALIVE

RISK FACTOR	35-44 AT EXAM			45-54 AT EXAM			55-64 AT EXAM			65-74 AT EXAM		
	PERSON EXAMS	# OF EVENTS	ANNUAL RATE	PERSON EXAMS	# OF EVENTS	ANNUAL RATE	PERSON EXAMS	# OF EVENTS	ANNUAL RATE	PERSON EXAMS	# OF EVENTS	ANNUAL RATE
NO	4470			7812			8424	14	1	4269	16	2
YES	49			255			646	3	2	565	8	7
UNKNOWN	0			0			0	0	-	0	0	-
TOTAL	4519			8067			9070	17	1	4834	24	2

RISK FACTOR	75-84 AT EXAM			85-94 AT EXAM			35-64 COMBINED			65-94 COMBINED		
	PERSON EXAMS	# OF EVENTS	ANNUAL RATE	PERSON EXAMS	# OF EVENTS	ANNUAL RATE	# OF EVENTS	CRUDE	AGE-ADJUSTED	# OF EVENTS	CRUDE	AGE-ADJUSTED
NO	1170	19	8	44			18	0	0	37	3	3
YES	175	4	11	6			3	2	1	13	9	9
UNKNOWN	0	0	-	0			0	-	-	0	-	-
TOTAL	1345	23	9	50			21	0	0	50	4	4

LOGISTIC REGRESSIONS

AGE AT EXAM		COEFFICIENT OF RISK FACTOR	SE COEFFICIENT	P-VALUE	STANDARDIZED COEFFICIENT	STD SE
35-44						
45-54						
55-64		1.031	0.638	0.106	0.265	0.164
65-74		1.340	0.435	0.002	0.430	0.140
75-84		0.349	0.556	0.531	0.117	0.187
85-94						
35-64	UNIVARIATE	1.292	0.624	0.039	0.265	0.128
65-94	UNIVARIATE	0.960	0.325	0.003	0.312	0.105
35-64	BIVARIATE	0.855	0.630	0.175	0.175	0.129
65-94	BIVARIATE	0.920	0.326	0.005	0.299	0.106
35-64	MULTIVARIATE	0.478	0.651	0.463	0.098	0.133
65-94	MULTIVARIATE	0.911	0.333	0.006	0.296	0.108

SEX: FEMALE
EVENT: DEATH FROM STROKE
RISK FACTOR: DIABETES MELLITUS
PERSONS AT RISK: ALL PERSONS ALIVE

RISK FACTOR	35-44 AT EXAM			45-54 AT EXAM			55-64 AT EXAM			65-74 AT EXAM		
	PERSON EXAMS	# OF EVENTS	ANNUAL RATE	PERSON EXAMS	# OF EVENTS	ANNUAL RATE	PERSON EXAMS	# OF EVENTS	ANNUAL RATE	PERSON EXAMS	# OF EVENTS	ANNUAL RATE
NO	5576			9859	7	0	11497	8	0	6499	23	2
YES	43			213	0	0	606	3	2	604	5	4
UNKNOWN	0			0	0	-	0	0	-	0	0	-
TOTAL	5619			10072	7	0	12103	11	0	7103	28	2

RISK FACTOR	75-84 AT EXAM			85-94 AT EXAM			35-64 COMBINED			65-94 COMBINED		
	PERSON EXAMS	# OF EVENTS	ANNUAL RATE	PERSON EXAMS	# OF EVENTS	ANNUAL RATE	# OF EVENTS	CRUDE	AGE-ADJUSTED	# OF EVENTS	CRUDE	AGE-ADJUSTED
NO	2106	24	6	139	4	14	15	0	0	51	3	3
YES	313	8	13	24	2	42	3	2	1	15	8	7
UNKNOWN	0	0	-	0	0	-	0	-	-	0	-	-
TOTAL	2419	32	7	163	6	18	18	0	0	66	3	3

LOGISTIC REGRESSIONS

AGE AT EXAM		COEFFICIENT OF RISK FACTOR	SE COEFFICIENT	P-VALUE	STANDARDIZED COEFFICIENT	STD SE
35-44		-	-	-	-	-
45-54		-	-	-	-	-
55-64		1.966	0.678	0.004	0.429	0.148
65-74		0.855	0.495	0.084	0.238	0.138
75-84		0.822	0.413	0.046	0.276	0.139
85-94		1.121	0.896	0.211	0.398	0.318
35-64	UNIVARIATE	1.835	0.633	0.004	0.318	0.110
65-94	UNIVARIATE	1.016	0.296	0.001	0.301	0.088
35-64	BIVARIATE	1.553	0.643	0.016	0.269	0.111
65-94	BIVARIATE	0.827	0.298	0.006	0.245	0.088
35-64	MULTIVARIATE	1.134	0.663	0.087	0.197	0.115
65-94	MULTIVARIATE	0.554	0.313	0.078	0.164	0.093

TABLE 17-8   DATE=14DEC85
AVERAGE ANNUAL INCIDENCE RATE FOR EVENT PER 1000 PERSONS AT RISK AT EXAM
BY AGE AND LEVEL OF RISK FACTOR AT EXAM: FRAMINGHAM STUDY 30-YEAR FOLLOWUP

SEX: MALE
EVENT: DEATH FROM STROKE
RISK FACTOR: GLUCOSE IN URINE, DEFINITE OR TRACE
PERSONS AT RISK: ALL PERSONS ALIVE

RISK FACTOR	35-44 AT EXAM			45-54 AT EXAM			55-64 AT EXAM			65-74 AT EXAM		
	PERSON EXAMS	# OF EVENTS	ANNUAL RATE	PERSON EXAMS	# OF EVENTS	ANNUAL RATE	PERSON EXAMS	# OF EVENTS	ANNUAL RATE	PERSON EXAMS	# OF EVENTS	ANNUAL RATE
NO	4426			7869			8786	15	1	4661	24	3
YES	60			177			260	2	4	167	0	0
UNKNOWN	33			21			24	0	0	6	0	0
TOTAL	4519			8067			9070	17	1	4834	24	2

RISK FACTOR	75-84 AT EXAM			85-94 AT EXAM			35-64 COMBINED			65-94 COMBINED		
	PERSON EXAMS	# OF EVENTS	ANNUAL RATE	PERSON EXAMS	# OF EVENTS	ANNUAL RATE	# OF EVENTS	CRUDE	AGE-ADJUSTED	# OF EVENTS	CRUDE	AGE-ADJUSTED
NO	1303	23	9	49			19	0	0	50	4	4
YES	39	0	0	1			2	2	2	0	0	0
UNKNOWN	3	0	0	0			0	0	0	0	0	0
TOTAL	1345	23	9	50			21	0	0	50	4	4

LOGISTIC REGRESSIONS

AGE AT EXAM		COEFFICIENT OF RISK FACTOR	SE COEFFICIENT	P-VALUE	STANDARDIZED COEFFICIENT	STD SE
35-44						
45-54						
55-64		1.511	0.755	0.045	0.253	0.126
65-74		-	-	-	-	-
75-84		-	-	-	-	-
85-94						
35-64	UNIVARIATE	1.499	0.745	0.044	0.225	0.112
65-94	UNIVARIATE	-	-	-	-	-
35-64	BIVARIATE	1.292	0.746	0.083	0.194	0.112
65-94	BIVARIATE	-	-	-	-	-
35-64	MULTIVARIATE	0.277	1.049	0.792	0.042	0.157
65-94	MULTIVARIATE	-	-	-	-	-

SEX: FEMALE
EVENT: DEATH FROM STROKE
RISK FACTOR: GLUCOSE IN URINE, DEFINITE OR TRACE
PERSONS AT RISK: ALL PERSONS ALIVE

RISK FACTOR	35-44 AT EXAM			45-54 AT EXAM			55-64 AT EXAM			65-74 AT EXAM		
	PERSON EXAMS	# OF EVENTS	ANNUAL RATE	PERSON EXAMS	# OF EVENTS	ANNUAL RATE	PERSON EXAMS	# OF EVENTS	ANNUAL RATE	PERSON EXAMS	# OF EVENTS	ANNUAL RATE
NO	5540			9947	6	0	11903	9	0	6957	26	2
YES	41			101	1	5	174	2	6	136	2	7
UNKNOWN	38			24	0	0	26	0	0	10	0	0
TOTAL	5619			10072	7	0	12103	11	0	7103	28	2

RISK FACTOR	75-84 AT EXAM			85-94 AT EXAM			35-64 COMBINED			65-94 COMBINED		
	PERSON EXAMS	# OF EVENTS	ANNUAL RATE	PERSON EXAMS	# OF EVENTS	ANNUAL RATE	# OF EVENTS	CRUDE	AGE-ADJUSTED	# OF EVENTS	CRUDE	AGE-ADJUSTED
NO	2375	31	7	160	6	19	15	0	0	63	3	3
YES	35	1	14	3	0	0	3	5	4	3	9	9
UNKNOWN	9	0	0	0	0	-	0	0	0	0	0	0
TOTAL	2419	32	7	163	6	18	18	0	0	66	3	3

LOGISTIC REGRESSIONS

AGE AT EXAM		COEFFICIENT OF RISK FACTOR	SE COEFFICIENT	P-VALUE	STANDARDIZED COEFFICIENT	STD SE
35-44						
45-54		2.807	1.085	0.010	0.280	0.108
55-64		2.732	0.785	0.001	0.326	0.094
65-74		1.381	0.739	0.062	0.189	0.101
75-84		0.799	1.031	0.438	0.096	0.123
85-94		-	-	-	-	-
35-64	UNIVARIATE	2.862	0.635	0.001	0.304	0.067
65-94	UNIVARIATE	0.965	0.596	0.105	0.128	0.079
35-64	BIVARIATE	2.738	0.637	0.001	0.291	0.068
65-94	BIVARIATE	1.069	0.601	0.075	0.142	0.080
35-64	MULTIVARIATE	2.234	0.671	0.001	0.237	0.071
65-94	MULTIVARIATE	0.759	0.611	0.214	0.101	0.081

TABLE 17-9   DATE=04MAR86
AVERAGE ANNUAL INCIDENCE RATE FOR EVENT PER 1000 PERSONS AT RISK AT EXAM
BY AGE AND LEVEL OF RISK FACTOR AT EXAM: FRAMINGHAM STUDY 30-YEAR FOLLOWUP

SEX: MALE
EVENT: DEATH FROM STROKE
RISK FACTOR: GLUCOSE INTOLERANCE
PERSONS AT RISK: ALL PERSONS ALIVE

RISK FACTOR	35-44 AT EXAM			45-54 AT EXAM			55-64 AT EXAM			65-74 AT EXAM		
	PERSON EXAMS	# OF EVENTS	ANNUAL RATE	PERSON EXAMS	# OF EVENTS	ANNUAL RATE	PERSON EXAMS	# OF EVENTS	ANNUAL RATE	PERSON EXAMS	# OF EVENTS	ANNUAL RATE
NO	4381			7528			7973	13	1	3918	15	2
YES	138			539			1097	4	2	916	9	5
UNKNOWN	0			0			0	0	-	0	0	-
TOTAL	4519			8067			9070	17	1	4834	24	2

RISK FACTOR	75-84 AT EXAM			85-94 AT EXAM			35-64 COMBINED			65-94 COMBINED		
	PERSON EXAMS	# OF EVENTS	ANNUAL RATE	PERSON EXAMS	# OF EVENTS	ANNUAL RATE	# OF EVENTS	CRUDE	AGE-ADJUSTED	# OF EVENTS	CRUDE	AGE-ADJUSTED
NO	1035	18	9	36			17	0	0	34	3	3
YES	310	5	8	14			4	1	1	16	6	6
UNKNOWN	0	0	-	0			0	-		0		
TOTAL	1345	23	9	50			21	0	0	50	4	4

LOGISTIC REGRESSIONS

AGE AT EXAM		COEFFICIENT OF RISK FACTOR	SE COEFFICIENT	P-VALUE	STANDARDIZED COEFFICIENT	STD SE
35-64						
45-54						
55-64		0.807	0.573	0.159	0.263	0.187
65-74		0.949	0.423	0.025	0.372	0.166
75-84		-0.077	0.510	0.880	-0.032	0.215
85-94						
35-64	UNIVARIATE	0.971	0.556	0.081	0.266	0.153
65-94	UNIVARIATE	0.645	0.305	0.034	0.257	0.122
35-64	BIVARIATE	0.617	0.560	0.271	0.169	0.154
65-94	BIVARIATE	0.552	0.306	0.072	0.220	0.122
35-64	MULTIVARIATE	0.028	0.640	0.965	0.008	0.176
65-94	MULTIVARIATE	0.510	0.310	0.100	0.204	0.124

SEX: FEMALE
EVENT: DEATH FROM STROKE
RISK FACTOR: GLUCOSE INTOLERANCE
PERSONS AT RISK: ALL PERSONS ALIVE

RISK FACTOR	35-44 AT EXAM			45-54 AT EXAM			55-64 AT EXAM			65-74 AT EXAM		
	PERSON EXAMS	# OF EVENTS	ANNUAL RATE	PERSON EXAMS	# OF EVENTS	ANNUAL RATE	PERSON EXAMS	# OF EVENTS	ANNUAL RATE	PERSON EXAMS	# OF EVENTS	ANNUAL RATE
NO	5525			9685	6	0	10988	7	0	6066	19	2
YES	94			387	1	3	1115	4	2	1037	9	4
UNKNOWN	0			0	0	-	0	0	-	0	0	-
TOTAL	5619			10072	7	0	12103	11	0	7103	28	2

RISK FACTOR	75-84 AT EXAM			85-94 AT EXAM			35-64 COMBINED			65-94 COMBINED		
	PERSON EXAMS	# OF EVENTS	ANNUAL RATE	PERSON EXAMS	# OF EVENTS	ANNUAL RATE	# OF EVENTS	CRUDE	AGE-ADJUSTED	# OF EVENTS	CRUDE	AGE-ADJUSTED
NO	1905	22	6	131	3	11	13	0	0	44	3	3
YES	514	10	10	32	3	47	5	2	1	22	7	6
UNKNOWN	0	0	-	0	0	-	0	-	0	0	-	
TOTAL	2419	32	7	163	6	18	18	0	0	66	3	3

LOGISTIC REGRESSIONS

AGE AT EXAM		COEFFICIENT OF RISK FACTOR	SE COEFFICIENT	P-VALUE	STANDARDIZED COEFFICIENT	STD SE
35-44						
45-54		1.430	1.081	0.186	0.275	0.208
55-64		1.731	0.628	0.006	0.501	0.182
65-74		1.025	0.406	0.012	0.362	0.143
75-84		0.530	0.385	0.169	0.217	0.157
85-94		1.485	0.842	0.078	0.592	0.335
35-64	UNIVARIATE	1.845	0.527	0.001	0.429	0.123
65-94	UNIVARIATE	0.948	0.263	0.001	0.351	0.097
35-64	BIVARIATE	1.584	0.539	0.003	0.368	0.125
65-94	BIVARIATE	0.774	0.265	0.003	0.287	0.098
35-64	MULTIVARIATE	1.355	0.549	0.014	0.315	0.128
65-94	MULTIVARIATE	0.591	0.274	0.031	0.218	0.101

TABLE 17-10   DATE=09NOV85
AVERAGE ANNUAL INCIDENCE RATE FOR EVENT PER 1000 PERSONS AT RISK AT EXAM
BY AGE AND LEVEL OF RISK FACTOR AT EXAM: FRAMINGHAM STUDY 30-YEAR FOLLOWUP

SEX: MALE
EVENT: DEATH FROM STROKE
RISK FACTOR: METROPOLITAN RELATIVE WEIGHT (PERCENT)
PERSONS AT RISK: ALL PERSONS ALIVE

RISK FACTOR	35-44 AT EXAM			45-54 AT EXAM			55-64 AT EXAM			65-74 AT EXAM		
	PERSON EXAMS	# OF EVENTS	ANNUAL RATE	PERSON EXAMS	# OF EVENTS	ANNUAL RATE	PERSON EXAMS	# OF EVENTS	ANNUAL RATE	PERSON EXAMS	# OF EVENTS	ANNUAL RATE
54-104	723			1149			1272	1	0	831	6	4
105-114	979			1635			1858	4	1	1052	4	2
115-124	1133			2139			2467	5	1	1322	6	2
125-134	923			1691			1851	3	1	956	2	1
135-272	761			1452			1621	4	1	673	6	4
UNKNOWN	0			1			1	0	0	0	0	-
TOTAL	4519			8067			9070	17	1	4834	24	2

RISK FACTOR	75-84 AT EXAM			85-94 AT EXAM			35-64 COMBINED			65-94 COMBINED		
	PERSON EXAMS	# OF EVENTS	ANNUAL RATE	PERSON EXAMS	# OF EVENTS	ANNUAL RATE	# OF EVENTS	CRUDE	AGE-ADJUSTED	# OF EVENTS	CRUDE	AGE-ADJUSTED
54-104	277	3	5	19			4	1	1	12	5	5
105-114	357	5	7	15			4	0	0	9	3	3
115-124	399	8	10	14			6	1	1	14	4	4
125-134	196	5	13	2			3	0	0	7	3	4
135-272	116	2	9	0			4	1	1	8	5	5
UNKNOWN	0	0	-	0			0	0	0	0	-	
TOTAL	1345	23	9	50			21	0	0	50	4	4

LOGISTIC REGRESSIONS

AGE AT EXAM	COEFFICIENT OF RISK FACTOR	SE COEFFICIENT	P-VALUE	STANDARDIZED COEFFICIENT	STD SE
35-44					
45-54					
55-64	0.013	0.014	0.386	0.200	0.231
65-74	0.005	0.013	0.677	0.084	0.202
75-84	0.007	0.014	0.637	0.097	0.205
85-94					
35-64 UNIVARIATE	-0.007	0.014	0.593	-0.119	0.224
65-94 UNIVARIATE	-0.005	0.009	0.581	-0.079	0.144
35-64 BIVARIATE	-0.008	0.014	0.586	-0.122	0.225
65-94 BIVARIATE	0.001	0.009	0.951	0.009	0.146
35-64 MULTIVARIATE	-0.018	0.015	0.222	-0.288	0.236
65-94 MULTIVARIATE	-0.004	0.010	0.661	-0.064	0.147

SEX: FEMALE
EVENT: DEATH FROM STROKE
RISK FACTOR: METROPOLITAN RELATIVE WEIGHT (PERCENT)
PERSONS AT RISK: ALL PERSONS ALIVE

RISK FACTOR	35-44 AT EXAM			45-54 AT EXAM			55-64 AT EXAM			65-74 AT EXAM		
	PERSON EXAMS	# OF EVENTS	ANNUAL RATE	PERSON EXAMS	# OF EVENTS	ANNUAL RATE	PERSON EXAMS	# OF EVENTS	ANNUAL RATE	PERSON EXAMS	# OF EVENTS	ANNUAL RATE
54-104	1694			2117	2	0	2140	0	0	1319	5	2
105-114	1431			2364	3	1	2541	6	1	1321	4	2
115-124	1059			2128	0	0	2556	1	0	1536	8	3
125-134	566			1383	0	0	1831	1	0	1140	3	1
135-272	861			2080	2	0	3035	3	0	1787	8	2
UNKNOWN	8			0	0	-	0	0	-	0	0	-
TOTAL	5619			10072	7	0	12103	11	0	7103	28	2

RISK FACTOR	75-84 AT EXAM			85-94 AT EXAM			35-64 COMBINED			65-94 COMBINED		
	PERSON EXAMS	# OF EVENTS	ANNUAL RATE	PERSON EXAMS	# OF EVENTS	ANNUAL RATE	# OF EVENTS	CRUDE	AGE-ADJUSTED	# OF EVENTS	CRUDE	AGE-ADJUSTED
54-104	632	11	9	51	0	0	2	0	0	16	4	4
105-114	457	9	10	33	2	30	9	1	1	15	4	4
115-124	487	3	3	31	3	48	1	0	0	14	3	3
125-134	342	1	1	26	1	19	1	0	0	5	2	2
135-272	501	8	8	22	0	0	5	0	0	16	3	4
UNKNOWN	0	0	-	0	0	-	0	0	0	0	-	
TOTAL	2419	32	7	163	6	18	18	0	0	66	3	3

LOGISTIC REGRESSIONS

AGE AT EXAM	COEFFICIENT OF RISK FACTOR	SE COEFFICIENT	P-VALUE	STANDARDIZED COEFFICIENT	STD SE
35-44					
45-54	-0.000	0.018	0.986	-0.007	0.380
55-64	0.002	0.013	0.874	0.046	0.293
65-74	0.000	0.008	0.959	0.010	0.188
75-84	-0.014	0.009	0.111	-0.319	0.200
85-94	0.005	0.022	0.807	0.101	0.414
35-64 UNIVARIATE	0.004	0.010	0.690	0.088	0.222
65-94 UNIVARIATE	-0.010	0.006	0.101	-0.224	0.137
35-64 BIVARIATE	0.001	0.010	0.909	0.026	0.230
65-94 BIVARIATE	-0.005	0.006	0.390	-0.115	0.134
35-64 MULTIVARIATE	-0.005	0.010	0.601	-0.119	0.228
65-94 MULTIVARIATE	-0.004	0.006	0.437	-0.101	0.129

TABLE 17-11   DATE=09NOV85

AVERAGE ANNUAL INCIDENCE RATE FOR EVENT PER 1000 PERSONS AT RISK AT EXAM
BY AGE AND LEVEL OF RISK FACTOR AT EXAM: FRAMINGHAM STUDY 30-YEAR FOLLOWUP

SEX: MALE
EVENT: DEATH FROM STROKE
RISK FACTOR: VITAL CAPACITY - HEIGHT INDEX (ML/INCH)
PERSONS AT RISK: ALL PERSONS ALIVE

RISK FACTOR	35-44 AT EXAM			45-54 AT EXAM			55-64 AT EXAM			65-74 AT EXAM		
	PERSON EXAMS	# OF EVENTS	ANNUAL RATE	PERSON EXAMS	# OF EVENTS	ANNUAL RATE	PERSON EXAMS	# OF EVENTS	ANNUAL RATE	PERSON EXAMS	# OF EVENTS	ANNUAL RATE
8-44	271			1108			2374	6	1	2179	14	3
45-49	416			1188			1692	3	1	974	4	2
50-54	751			1726			2061	3	1	857	4	2
55-59	1034			1823			1569	4	1	528	2	2
60-85	2034			2216			1356	1	0	284	0	0
UNKNOWN	13			6			18	0	0	12	0	0
TOTAL	4519			8067			9070	17	1	4834	24	2

RISK FACTOR	75-84 AT EXAM			85-94 AT EXAM			35-64 COMBINED			65-94 COMBINED		
	PERSON EXAMS	# OF EVENTS	ANNUAL RATE	PERSON EXAMS	# OF EVENTS	ANNUAL RATE	# OF EVENTS	CRUDE	AGE-ADJUSTED	# OF EVENTS	CRUDE	AGE-ADJUSTED
8-44	853	15	9	34			7	1	1	31	5	5
45-49	228	4	9	7			3	0	0	9	4	4
50-54	142	3	11	7			5	1	1	7	3	4
55-59	81	0	0	2			4	0	1	2	2	1
60-85	33	1	15	0			2	0	0	1	2	3
UNKNOWN	8	0	0	0			0	0	0	0	0	0
TOTAL	1345	23	9	50			21	0	0	50	4	4

LOGISTIC REGRESSIONS

AGE AT EXAM		COEFFICIENT OF RISK FACTOR	SE COEFFICIENT	P-VALUE	STANDARDIZED COEFFICIENT	STD SE
35-44						
45-54						
55-64		-0.019	0.025	0.461	-0.175	0.237
65-74		-0.037	0.021	0.085	-0.341	0.198
75-84		-0.016	0.023	0.484	-0.147	0.210
85-94						
35-64	UNIVARIATE	-0.044	0.021	0.039	-0.423	0.206
65-94	UNIVARIATE	-0.041	0.015	0.005	-0.388	0.138
35-64	BIVARIATE	-0.019	0.023	0.404	-0.186	0.223
65-94	BIVARIATE	-0.028	0.015	0.066	-0.266	0.145
35-64	MULTIVARIATE	0.005	0.025	0.850	0.045	0.240
65-94	MULTIVARIATE	-0.017	0.015	0.261	-0.164	0.146

SEX: FEMALE
EVENT: DEATH FROM STROKE
RISK FACTOR: VITAL CAPACITY - HEIGHT INDEX (ML/INCH)
PERSONS AT RISK: ALL PERSONS ALIVE

RISK FACTOR	35-44 AT EXAM			45-54 AT EXAM			55-64 AT EXAM			65-74 AT EXAM		
	PERSON EXAMS	# OF EVENTS	ANNUAL RATE	PERSON EXAMS	# OF EVENTS	ANNUAL RATE	PERSON EXAMS	# OF EVENTS	ANNUAL RATE	PERSON EXAMS	# OF EVENTS	ANNUAL RATE
4-29	122			546	1	1	1572	3	1	2068	8	2
30-34	291			1241	2	1	2547	4	1	1946	13	2
35-39	769			2103	2	0	3213	1	0	1728	6	2
40-44	1427			2838	1	0	2836	2	0	933	1	1
45-92	2989			3327	1	0	1914	1	0	422	0	0
UNKNOWN	21			17	0	0	21	0	0	6	0	0
TOTAL	5619			10072	7	0	12103	11	0	7103	28	2

RISK FACTOR	75-84 AT EXAM			85-94 AT EXAM			35-64 COMBINED			65-94 COMBINED		
	PERSON EXAMS	# OF EVENTS	ANNUAL RATE	PERSON EXAMS	# OF EVENTS	ANNUAL RATE	# OF EVENTS	CRUDE	AGE-ADJUSTED	# OF EVENTS	CRUDE	AGE-ADJUSTED
4-29	1212	19	8	109	2	9	4	1	1	29	4	4
30-34	686	8	6	37	4	54	6	1	1	25	5	5
35-39	370	4	5	16	0	0	3	0	0	10	2	3
40-44	123	1	4	1	0	0	3	0	0	2	1	1
45-92	26	0	0	0	0	-	2	0	0	0	0	0
UNKNOWN	2	0	0	0	0	0	0	0	0	0	0	0
TOTAL	2419	32	7	163	6	18	18	0	0	66	3	3

LOGISTIC REGRESSIONS

AGE AT EXAM		COEFFICIENT OF RISK FACTOR	SE COEFFICIENT	P-VALUE	STANDARDIZED COEFFICIENT	STD SE
35-44						
45-54		-0.082	0.047	0.083	-0.604	0.349
55-64		-0.087	0.036	0.017	-0.651	0.273
65-74		-0.035	0.025	0.168	-0.256	0.185
75-84		-0.016	0.025	0.531	-0.110	0.176
85-94		0.105	0.073	0.149	0.691	0.479
35-64	UNIVARIATE	-0.095	0.027	0.001	-0.755	0.214
65-94	UNIVARIATE	-0.040	0.016	0.012	-0.304	0.121
35-64	BIVARIATE	-0.082	0.029	0.005	-0.653	0.232
65-94	BIVARIATE	-0.009	0.018	0.620	-0.066	0.132
35-64	MULTIVARIATE	-0.061	0.032	0.052	-0.487	0.251
65-94	MULTIVARIATE	0.008	0.018	0.665	0.060	0.139

TABLE 17-12   DATE=09NOV85
AVERAGE ANNUAL INCIDENCE RATE FOR EVENT PER 1000 PERSONS AT RISK AT EXAM
BY AGE AND LEVEL OF RISK FACTOR AT EXAM: FRAMINGHAM STUDY 30-YEAR FOLLOWUP

SEX: MALE
EVENT: DEATH FROM STROKE
RISK FACTOR: HEART RATE (PER MINUTE)
PERSONS AT RISK: ALL PERSONS ALIVE

RISK FACTOR	35-44 AT EXAM			45-54 AT EXAM			55-64 AT EXAM			65-74 AT EXAM		
	PERSON EXAMS	# OF EVENTS	ANNUAL RATE	PERSON EXAMS	# OF EVENTS	ANNUAL RATE	PERSON EXAMS	# OF EVENTS	ANNUAL RATE	PERSON EXAMS	# OF EVENTS	ANNUAL RATE
30-67	1295			2016			2618	4	1	1425	5	2
68-75	1506			2594			2763	1	0	1447	11	4
76-83	728			1473			1552	4	1	802	5	3
84-91	586			1145			1211	5	2	607	2	2
92-220	404			839			926	3	2	553	1	1
UNKNOWN	0			0			0	0	-	0	0	-
TOTAL	4519			8067			9070	17	1	4834	24	2

RISK FACTOR	75-84 AT EXAM			85-94 AT EXAM			35-64 COMBINED			65-94 COMBINED		
	PERSON EXAMS	# OF EVENTS	ANNUAL RATE	PERSON EXAMS	# OF EVENTS	ANNUAL RATE	# OF EVENTS	CRUDE	AGE-ADJUSTED	# OF EVENTS	CRUDE	AGE-ADJUSTED
30-67	388	6	8	17			6	1	1	11	3	3
68-75	362	8	11	14			1	0	0	21	6	4
76-83	271	3	6	13			1	1	1	8	4	4
84-91	174	2	6	5			5	1	1	4	3	3
92-220	150	4	13	1			4	1	1	6	4	8
UNKNOWN	0	0	-	0			0	-	-	0	-	-
TOTAL	1345	23	9	50			21	0	0	50	4	4

LDOISTIC REGRESSIONS

AGE AT EXAM		COEFFICIENT OF RISK FACTOR	SE COEFFICIENT	P-VALUE	STANDARDIZED COEFFICIENT	STD SE
35-44						
45-54						
55-64		0.033	0.016	0.037	0.428	0.205
65-74		-0.009	0.016	0.567	-0.122	0.213
75-84		0.003	0.015	0.823	0.046	0.207
85-94						
35-64	UNIVARIATE	0.027	0.015	0.071	0.349	0.194
65-94	UNIVARIATE	0.000	0.011	0.979	0.004	0.142
35-64	BIVARIATE	0.027	0.015	0.071	0.340	0.188
65-94	BIVARIATE	-0.000	0.011	0.997	-0.000	0.142
35-64	MULTIVARIATE	0.017	0.016	0.290	0.210	0.199
65-94	MULTIVARIATE	-0.005	0.011	0.614	-0.072	0.143

SEX: FEMALE
EVENT: DEATH FROM STROKE
RISK FACTOR: HEART RATE (PER MINUTE)
PERSONS AT RISK: ALL PERSONS ALIVE

RISK FACTOR	35-44 AT EXAM			45-54 AT EXAM			55-64 AT EXAM			65-74 AT EXAM		
	PERSON EXAMS	# OF EVENTS	ANNUAL RATE	PERSON EXAMS	# OF EVENTS	ANNUAL RATE	PERSON EXAMS	# OF EVENTS	ANNUAL RATE	PERSON EXAMS	# OF EVENTS	ANNUAL RATE
30-67	1092			1918	1	0	2450	0	0	1285	5	2
68-75	1806			3279	1	0	3641	1	0	1907	10	3
76-83	1051			1938	2	1	2642	5	1	1542	2	1
84-91	939			1620	2	1	1888	2	1	1196	6	3
92-220	731			1317	1	0	1682	3	1	1173	5	2
UNKNOWN	0			0	0	-	0	0	-	0	0	-
TOTAL	5619			10072	7	0	12103	11	0	7103	28	2

RISK FACTOR	75-84 AT EXAM			85-94 AT EXAM			35-64 COMBINED			65-94 COMBINED		
	PERSON EXAMS	# OF EVENTS	ANNUAL RATE	PERSON EXAMS	# OF EVENTS	ANNUAL RATE	# OF EVENTS	CRUDE	AGE-ADJUSTED	# OF EVENTS	CRUDE	AGE-ADJUSTED
30-67	459	3	3	36	2	28	1	0	0	10	3	3
68-75	586	9	8	41	2	24	2	0	0	21	4	2
76-83	532	7	7	30	0	0	7	1	1	9	2	2
84-91	410	6	7	29	0	0	4	1	0	12	4	4
92-220	432	7	8	27	2	37	4	1	1	14	4	4
UNKNOWN	0	0	-	0	0	-	0	-	-	0	-	-
TOTAL	2419	32	7	163	6	18	18	0	0	66	3	3

LOGISTIC REGRESSIONS

AGE AT EXAM		COEFFICIENT OF RISK FACTOR	SE COEFFICIENT	P-VALUE	STANDARDIZED COEFFICIENT	STD SE
35-44						
45-54		0.032	0.026	0.217	0.401	0.324
55-64		0.041	0.016	0.011	0.524	0.205
65-74		0.003	0.014	0.856	0.034	0.187
75-84		0.014	0.012	0.225	0.195	0.161
85-94		-0.001	0.030	0.961	-0.021	0.419
35-64	UNIVARIATE	0.037	0.013	0.004	0.463	0.160
65-94	UNIVARIATE	0.009	0.009	0.309	0.119	0.117
35-64	BIVARIATE	0.037	0.013	0.005	0.464	0.166
65-94	BIVARIATE	0.008	0.009	0.351	0.109	0.117
35-64	MULTIVARIATE	0.025	0.015	0.101	0.322	0.196
65-94	MULTIVARIATE	0.007	0.009	0.400	0.099	0.117

TABLE 17-13   DATE=14DEC85
AVERAGE ANNUAL INCIDENCE RATE FOR EVENT PER 1000 PERSONS AT RISK AT EXAM
BY AGE AND LEVEL OF RISK FACTOR AT EXAM: FRAMINGHAM STUDY 30-YEAR FOLLOWUP

SEX: MALE
EVENT: DEATH FROM STROKE
RISK FACTOR: CIGARETTES SMOKED (NUMBER PER DAY)
PERSONS AT RISK: ALL PERSONS ALIVE

RISK FACTOR	35-44 AT EXAM PERSON EXAMS	# OF EVENTS	ANNUAL RATE	45-54 AT EXAM PERSON EXAMS	# OF EVENTS	ANNUAL RATE	55-64 AT EXAM PERSON EXAMS	# OF EVENTS	ANNUAL RATE	65-74 AT EXAM PERSON EXAMS	# OF EVENTS	ANNUAL RATE
NONE	1402			3220			4888	3	0	3235	11	2
1-10	437			746			856	2	1	463	3	3
11-20	1499			2136			1889	9	2	668	7	5
21-40	993			1672			1235	3	1	427	3	4
41-90	162			255			178	0	0	34	0	0
UNKNOWN	26			38			24	0	0	7	0	0
TOTAL	4519			8067			9070	17	1	4834	24	2

RISK FACTOR	75-84 AT EXAM PERSON EXAMS	# OF EVENTS	ANNUAL RATE	85-94 AT EXAM PERSON EXAMS	# OF EVENTS	ANNUAL RATE	35-64 COMBINED # OF EVENTS	CRUDE	AGE-ADJUSTED	65-94 COMBINED # OF EVENTS	CRUDE	AGE-ADJUSTED
NONE	1067	17	8	38			4	0	0	30	3	3
1-10	125	2	8	6			2	0	0	5	4	4
11-20	88	2	11	5			11	1	1	9	6	7
21-40	60	2	17	1			4	1	1	6	6	10
41-90	5	0	0	0			0	0	0	0	0	0
UNKNOWN	0	0	-	0			0	0	0	0	0	0
TOTAL	1345	23	9	50			21	0	0	50	4	4

LOGISTIC REGRESSIONS

AGE AT EXAM		COEFFICIENT OF RISK FACTOR	SE COEFFICIENT	P-VALUE	STANDARDIZED COEFFICIENT	STD SE
35-44						
45-54						
55-64		0.024	0.014	0.095	0.333	0.199
65-74		0.019	0.015	0.213	0.214	0.172
75-84		0.015	0.020	0.450	0.132	0.175
85-94						
35-64	UNIVARIATE	0.016	0.014	0.245	0.231	0.198
65-94	UNIVARIATE	0.013	0.011	0.265	0.139	0.125
35-64	BIVARIATE	0.025	0.013	0.053	0.370	0.191
65-94	BIVARIATE	0.022	0.011	0.050	0.244	0.124
35-64	MULTIVARIATE	0.028	0.014	0.061	0.412	0.201
65-94	MULTIVARIATE	0.029	0.012	0.014	0.316	0.129

SEX: FEMALE
EVENT: DEATH FROM STROKE
RISK FACTOR: CIGARETTES SMOKED (NUMBER PER DAY)
PERSONS AT RISK: ALL PERSONS ALIVE

RISK FACTOR	35-44 AT EXAM PERSON EXAMS	# OF EVENTS	ANNUAL RATE	45-54 AT EXAM PERSON EXAMS	# OF EVENTS	ANNUAL RATE	55-64 AT EXAM PERSON EXAMS	# OF EVENTS	ANNUAL RATE	65-74 AT EXAM PERSON EXAMS	# OF EVENTS	ANNUAL RATE
NONE	2583			5599	1	0	8010	6	0	5562	18	2
1-10	1314			1732	2	1	1646	2	1	680	6	4
11-20	1373			2037	3	1	1767	3	1	650	4	3
21-40	298			629	1	1	624	0	0	191	0	0
41-90	22			38	0	0	18	0	0	1	0	0
UNKNOWN	29			37	0	0	38	0	0	19	0	0
TOTAL	5619			10072	7	0	12103	11	0	7103	28	2

RISK FACTOR	75-84 AT EXAM PERSON EXAMS	# OF EVENTS	ANNUAL RATE	85-94 AT EXAM PERSON EXAMS	# OF EVENTS	ANNUAL RATE	35-64 COMBINED # OF EVENTS	CRUDE	AGE-ADJUSTED	65-94 COMBINED # OF EVENTS	CRUDE	AGE-ADJUSTED
NONE	2139	26	6	154	6	19	7	0	0	50	3	3
1-10	161	1	3	7	0	0	4	0	0	7	4	4
11-20	99	5	25	2	0	0	6	1	1	9	6	9
21-40	16	0	0	0	0	-	1	0	0	0	0	0
41-90	0	0	-	0	0	-	0	0	0	0	0	0
UNKNOWN	4	0	9	0	0	-	0	0	0	0	0	0
TOTAL	2419	32	7	163	6	18	18	0	0	66	3	3

LOGISTIC REGRESSIONS

AGE AT EXAM		COEFFICIENT OF RISK FACTOR	SE COEFFICIENT	P-VALUE	STANDARDIZED COEFFICIENT	STD SE
35-44						
45-54		0.054	0.026	0.042	0.550	0.270
55-64		0.015	0.028	0.601	0.159	0.266
65-74		0.016	0.021	0.444	0.123	0.161
75-84		0.060	0.020	0.003	0.289	0.099
85-94		-		-	-	-
35-64	UNIVARIATE	0.030	0.019	0.120	0.296	0.190
65-94	UNIVARIATE	0.013	0.015	0.384	0.094	0.108
35-64	BIVARIATE	0.035	0.019	0.062	0.547	0.186
65-94	BIVARIATE	0.033	0.015	0.033	0.231	0.109
35-64	MULTIVARIATE	0.039	0.019	0.041	0.382	0.188
65-94	MULTIVARIATE	0.037	0.016	0.018	0.262	0.110

TABLE 17-14   DATE=13NOV85
AVERAGE ANNUAL INCIDENCE RATE FOR EVENT PER 1000 PERSONS AT RISK AT EXAM
BY AGE AND LEVEL OF RISK FACTOR AT EXAM: FRAMINGHAM STUDY 30-YEAR FOLLOWUP

SEX: MALE
EVENT: DEATH FROM STROKE
RISK FACTOR: ALBUMIN IN URINE, DEFINITE
PERSONS AT RISK: ALL PERSONS ALIVE

RISK FACTOR	35-44 AT EXAM PERSON EXAMS	# OF EVENTS	ANNUAL RATE	45-54 AT EXAM PERSON EXAMS	# OF EVENTS	ANNUAL RATE	55-64 AT EXAM PERSON EXAMS	# OF EVENTS	ANNUAL RATE	65-74 AT EXAM PERSON EXAMS	# OF EVENTS	ANNUAL RATE
NO	4411			7936			8828	15	1	4671	24	3
YES	71			105			215	2	5	157	0	0
UNKNOWN	37			26			27	0	0	6	0	0
TOTAL	4519			8067			9070	17	1	4834	24	2

RISK FACTOR	75-84 AT EXAM PERSON EXAMS	# OF EVENTS	ANNUAL RATE	85-94 AT EXAM PERSON EXAMS	# OF EVENTS	ANNUAL RATE	35-64 COMBINED # OF EVENTS	CRUDE	AGE-ADJUSTED	65-94 COMBINED # OF EVENTS	CRUDE	AGE-ADJUSTED
NO	1308	23	9	43			18	0	0	49	4	4
YES	34	0	0	7			3	4	4	1	3	1
UNKNOWN	3	0	0	0			0	0	0	0	0	0
TOTAL	1345	23	9	50			21	0	0	50	4	4

LOGISTIC REGRESSIONS

AGE AT EXAM		COEFFICIENT OF RISK FACTOR	SE COEFFICIENT	P-VALUE	STANDARDIZED COEFFICIENT	STD SE
35-44						
45-54						
55-64		1.708	0.756	0.024	0.260	0.115
65-74		-	-	-	-	-
75-84		-	-	-	-	-
85-94						
35-64	UNIVARIATE	2.207	0.626	0.001	0.294	0.083
65-94	UNIVARIATE	-0.480	1.013	0.636	-0.084	0.178
35-64	BIVARIATE	1.981	0.629	0.002	0.264	0.084
65-94	BIVARIATE	-0.586	1.016	0.564	-0.103	0.178
35-64	MULTIVARIATE	1.547	0.659	0.019	0.206	0.088
65-94	MULTIVARIATE	-0.949	1.025	0.355	-0.167	0.180

SEX: FEMALE
EVENT: DEATH FROM STROKE
RISK FACTOR: ALBUMIN IN URINE, DEFINITE
PERSONS AT RISK: ALL PERSONS ALIVE

RISK FACTOR	35-44 AT EXAM PERSON EXAMS	# OF EVENTS	ANNUAL RATE	45-54 AT EXAM PERSON EXAMS	# OF EVENTS	ANNUAL RATE	55-64 AT EXAM PERSON EXAMS	# OF EVENTS	ANNUAL RATE	65-74 AT EXAM PERSON EXAMS	# OF EVENTS	ANNUAL RATE
NO	5356			9766	7	0	11835	11	0	6955	28	2
YES	225			281	0	0	240	0	0	138	0	0
UNKNOWN	38			25	0	0	28	0	0	10	0	0
TOTAL	5619			10072	7	0	12103	11	0	7103	28	2

RISK FACTOR	75-84 AT EXAM PERSON EXAMS	# OF EVENTS	ANNUAL RATE	85-94 AT EXAM PERSON EXAMS	# OF EVENTS	ANNUAL RATE	35-64 COMBINED # OF EVENTS	CRUDE	AGE-ADJUSTED	65-94 COMBINED # OF EVENTS	CRUDE	AGE-ADJUSTED
NO	2360	29	6	156	6	19	18	0	0	63	3	3
YES	50	3	30	7	0	0	0	0	0	3	8	7
UNKNOWN	9	0	0	0	0	-	0	0	0	0	0	0
TOTAL	2419	32	7	163	6	18	18	0	0	66	3	3

LOGISTIC REGRESSIONS

AGE AT EXAM		COEFFICIENT OF RISK FACTOR	SE COEFFICIENT	P-VALUE	STANDARDIZED COEFFICIENT	STD SE
35-44						
45-54		-	-	-	-	-
55-64		-	-	-	-	-
65-74		-	-	-	-	-
75-84		1.635	0.624	0.009	0.233	0.089
85-94		-	-	-	-	-
35-64	UNIVARIATE	-	-	-	-	-
65-94	UNIVARIATE	0.847	0.595	0.155	0.119	0.084
35-64	BIVARIATE	-	-	-	-	-
65-94	BIVARIATE	0.691	0.602	0.251	0.097	0.085
35-64	MULTIVARIATE	-	-	-	-	-
65-94	MULTIVARIATE	0.295	0.617	0.632	0.042	0.087

AVERAGE ANNUAL INCIDENCE RATE FOR EVENT PER 1000 PERSONS AT RISK AT EXAM
BY AGE AND LEVEL OF RISK FACTOR AT EXAM: FRAMINGHAM STUDY 30-YEAR FOLLOWUP

SEX: MALE
EVENT: DEATH FROM STROKE
RISK FACTOR: HEART ENLARGEMENT BY X-RAY
PERSONS AT RISK: ALL PERSONS ALIVE

RISK FACTOR	35-44 AT EXAM			45-54 AT EXAM			55-64 AT EXAM			65-74 AT EXAM		
	PERSON EXAMS	# OF EVENTS	ANNUAL RATE	PERSON EXAMS	# OF EVENTS	ANNUAL RATE	PERSON EXAMS	# OF EVENTS	ANNUAL RATE	PERSON EXAMS	# OF EVENTS	ANNUAL RATE
NO	4101			6613			6543	12	1.	3075	8	1
MILD	189			708			1126	1	0	703	4	3
DEFINITE	224			742			1400	4	1	1056	12	6
UNKNOWN	5			4			1	0	0	0	0	-
TOTAL	4519			8067			9070	17	1	4834	24	2

RISK FACTOR	75-84 AT EXAM			85-94 AT EXAM			35-64 COMBINED			65-94 COMBINED		
	PERSON EXAMS	# OF EVENTS	ANNUAL RATE	PERSON EXAMS	# OF EVENTS	ANNUAL RATE	# OF EVENTS	CRUDE	AGE-ADJUSTED	# OF EVENTS	CRUDE	AGE-ADJUSTED
NO	753	11	7	34			16	0	0	21	3	3
MILD	226	6	13	6			1	0	0	10	5	5
DEFINITE	366	6	8	10			4	1	1	19	7	7
UNKNOWN	0	0	-	0			0	0	0	0	-	-
TOTAL	1345	23	9	50			21	0	0	50	4	4

LOGISTIC REGRESSIONS

AGE AT EXAM		COEFFICIENT OF RISK FACTOR	SE COEFFICIENT	P-VALUE	STANDARDIZED COEFFICIENT	STD SE
35-44						
45-54						
55-64		0.161	0.303	0.594	0.120	0.225
65-74		0.740	0.227	0.001	0.610	0.187
75-84		0.094	0.238	0.695	0.081	0.206
85-94						
35-64	UNIVARIATE	0.234	0.290	0.420	0.154	0.191
65-94	UNIVARIATE	0.452	0.156	0.004	0.378	0.130
35-64	BIVARIATE	0.042	0.295	0.886	0.028	0.194
65-94	BIVARIATE	0.404	0.157	0.010	0.337	0.131
35-64	MULTIVARIATE	-0.278	0.320	.385	-0.183	0.210
65-94	MULTIVARIATE	0.243	0.169	0.152	0.203	0.141

SEX: FEMALE
EVENT: DEATH FROM STROKE
RISK FACTOR: HEART ENLARGEMENT BY X-RAY
PERSONS AT RISK: ALL PERSONS ALIVE

RISK FACTOR	35-44 AT EXAM			45-54 AT EXAM			55-64 AT EXAM			65-74 AT EXAM		
	PERSON EXAMS	# OF EVENTS	ANNUAL RATE	PERSON EXAMS	# OF EVENTS	ANNUAL RATE	PERSON EXAMS	# OF EVENTS	ANNUAL RATE	PERSON EXAMS	# OF EVENTS	ANNUAL RATE
NO	5166			8085	5	0	7836	3	0	3511	14	2
MILD	199			959	1	1	1879	3	1	1293	3	1
DEFINITE	246			1022	1	0	2384	5	1	2299	11	2
UNKNOWN	8			6	0	0	4	0	0	0	0	-
TOTAL	5619			10072	7	0	12103	11	0	7103	28	2

RISK FACTOR	75-84 AT EXAM			85-94 AT EXAM			35-64 COMBINED			65-94 COMBINED		
	PERSON EXAMS	# OF EVENTS	ANNUAL RATE	PERSON EXAMS	# OF EVENTS	ANNUAL RATE	# OF EVENTS	CRUDE	AGE-ADJUSTED	# OF EVENTS	CRUDE	AGE-ADJUSTED
NO	830	10	6	40	1	13	8	0	0	25	3	3
MILD	455	4	4	36	2	28	4	1	1	9	3	2
DEFINITE	1134	18	8	87	3	17	6	1	1	32	5	4
UNKNOWN	0	0	-	0	0	-	0	0	0	0	-	-
TOTAL	2419	32	7	163	6	18	18	0	0	66	3	3

LOGISTIC REGRESSIONS

AGE AT EXAM		COEFFICIENT OF RISK FACTOR	SE COEFFICIENT	P-VALUE	STANDARDIZED COEFFICIENT	STD SE
35-44						
45-54		0.269	0.505	0.595	0.173	0.324
55-64		0.820	0.339	0.016	0.657	0.272
65-74		0.080	0.211	0.705	0.071	0.188
75-84		0.162	0.205	0.429	0.145	0.183
85-94		0.068	0.507	0.493	0.057	0.424
35-64	UNIVARIATE	0.746	0.257	0.004	0.526	0.181
65-94	UNIVARIATE	0.242	0.138	0.079	0.217	0.124
35-64	BIVARIATE	0.605	0.268	0.024	0.427	0.189
65-94	BIVARIATE	0.083	0.141	0.558	0.074	0.127
35-64	MULTIVARIATE	0.392	0.294	0.182	0.276	0.207
65-94	MULTIVARIATE	-0.084	0.150	0.576	-0.075	0.135

TABLE 17-16 · DATE=14DEC85
AVERAGE ANNUAL INCIDENCE RATE FOR EVENT PER 1000 PERSONS AT RISK AT EXAM
BY AGE AND LEVEL OF RISK FACTOR AT EXAM: FRAMINGHAM STUDY 30-YEAR FOLLOWUP

SEX: MALE
EVENT: DEATH FROM STROKE
RISK FACTOR: LEFT VENTRICULAR HYPERTROPHY BY ELECTROCARDIOGRAM
PERSONS AT RISK: ALL PERSONS ALIVE

RISK FACTOR	35-44 AT EXAM PERSON EXAMS	# OF EVENTS	ANNUAL RATE	45-54 AT EXAM PERSON EXAMS	# OF EVENTS	ANNUAL RATE	55-64 AT EXAM PERSON EXAMS	# OF EVENTS	ANNUAL RATE	65-74 AT EXAM PERSON EXAMS	# OF EVENTS	ANNUAL RATE
NO	4408			7812			8611	14	1	4527	20	2
MILD	79			147			240	0	0	121	1	4
DEFINITE	32			108			219	3	7	186	3	8
UNKNOWN	0			0			0	0	-	0	0	-
TOTAL	4519			8067			9070	17	1	4834	24	2

RISK FACTOR	75-84 AT EXAM PERSON EXAMS	# OF EVENTS	ANNUAL RATE	85-94 AT EXAM PERSON EXAMS	# OF EVENTS	ANNUAL RATE	35-64 COMBINED # OF EVENTS	CRUDE	AGE-ADJUSTED	65-94 COMBINED # OF EVENTS	CRUDE	AGE-ADJUSTED
NO	1250	19	8	45			18	0	0	42	4	4
MILD	44	1	11	2			0	0	0	2	6	6
DEFINITE	51	3	29	3			3	4	3	6	13	13
UNKNOWN	0	0	-	0			0	-		0	-	
TOTAL	1345	23	9	50			21	0	0	50	4	4

LOGISTIC REGRESSIONS

AGE AT EXAM		COEFFICIENT OF RISK FACTOR	SE COEFFICIENT	P-VALUE	STANDARDIZED COEFFICIENT	STD SE
35-44						
45-54						
55-64		0.996	0.337	0.003	0.341	0.116
65-74		0.652	0.305	0.032	0.268	0.125
75-84		0.677	0.315	0.032	0.281	0.131
85-94						
35-64	UNIVARIATE	1.051	0.331	0.002	0.306	0.096
65-94	UNIVARIATE	0.622	0.216	0.004	0.257	0.089
35-64	BIVARIATE	0.879	0.333	0.008	0.256	0.097
65-94	BIVARIATE	0.616	0.218	0.005	0.254	0.090
35-64	MULTIVARIATE	0.294	0.377	0.436	0.086	0.110
65-94	MULTIVARIATE	0.383	0.241	0.112	0.158	0.099

SEX: FEMALE
EVENT: DEATH FROM STROKE
RISK FACTOR: LEFT VENTRICULAR HYPERTROPHY BY ELECTROCARDIOGRAM
PERSONS AT RISK: ALL PERSONS ALIVE

RISK FACTOR	35-44 AT EXAM PERSON EXAMS	# OF EVENTS	ANNUAL RATE	45-54 AT EXAM PERSON EXAMS	# OF EVENTS	ANNUAL RATE	55-64 AT EXAM PERSON EXAMS	# OF EVENTS	ANNUAL RATE	65-74 AT EXAM PERSON EXAMS	# OF EVENTS	ANNUAL RATE
NO	5566			9905	7	0	11679	7	0	6686	22	2
MILD	38			111	0	0	250	2	4	182	1	3
DEFINITE	15			56	0	0	174	2	6	235	5	11
UNKNOWN	0		-	0	0	-	0	0	-	0	0	-
TOTAL	5619			10072	7	0	12103	11	0	7103	28	2

RISK FACTOR	75-84 AT EXAM PERSON EXAMS	# OF EVENTS	ANNUAL RATE	85-94 AT EXAM PERSON EXAMS	# OF EVENTS	ANNUAL RATE	35-64 COMBINED # OF EVENTS	CRUDE	AGE-ADJUSTED	65-94 COMBINED # OF EVENTS	CRUDE	AGE-ADJUSTED
NO	2217	26	6	142	5	18	14	0	0	53	3	3
MILD	77	2	13	10	0	0	2	3	2	3	6	5
DEFINITE	125	4	16	11	1	45	2	4	3	10	13	13
UNKNOWN	0	0	-	0	0	-	0	-		0	-	
TOTAL	2419	32	7	163	6	18	18	0	0	66	3	3

LOGISTIC REGRESSIONS

AGE AT EXAM		COEFFICIENT OF RISK FACTOR	SE COEFFICIENT	P-VALUE	STANDARDIZED COEFFICIENT	STD SE
35-44		-	-	-	-	-
45-54						
55-64		1.575	0.357	0.001	0.433	0.098
65-74		0.923	0.251	0.001	0.357	0.097
75-84		0.536	0.259	0.039	0.252	0.122
85-94		0.380	0.611	0.534	0.206	0.332
35-64	UNIVARIATE	1.485	0.334	0.001	0.327	0.074
65-94	UNIVARIATE	0.766	0.172	0.001	0.316	0.071
35-64	BIVARIATE	1.347	0.340	0.001	0.297	0.075
65-94	BIVARIATE	0.662	0.174	0.001	0.273	0.072
35-64	MULTIVARIATE	0.982	0.405	0.015	0.217	0.089
65-94	MULTIVARIATE	0.422	0.192	0.028	0.174	0.079

TABLE 17-17   DATE=13NOV85
AVERAGE ANNUAL INCIDENCE RATE FOR EVENT PER 1000 PERSONS AT RISK AT EXAM
BY AGE AND LEVEL OF RISK FACTOR AT EXAM: FRAMINGHAM STUDY 30-YEAR FOLLOWUP

SEX: MALE
EVENT: DEATH FROM STROKE
RISK FACTOR: INTRAVENTRICULAR CONDUCTION DEFECT (WITH QRS INTERVAL MORE THAN 0.11 SECOND)
PERSONS AT RISK: ALL PERSONS ALIVE

RISK FACTOR	35-44 AT EXAM PERSON EXAMS	# OF EVENTS	ANNUAL RATE	45-54 AT EXAM PERSON EXAMS	# OF EVENTS	ANNUAL RATE	55-64 AT EXAM PERSON EXAMS	# OF EVENTS	ANNUAL RATE	65-74 AT EXAM PERSON EXAMS	# OF EVENTS	ANNUAL RATE
NO	4201			7487			8247	15	1	4323	19	2
INCOMPLETE	280			491			541	2	2	250	4	8
COMPLETE	38			89			282	0	0	261	1	2
UNKNOWN	0			0			0	0	-	0	0	-
TOTAL	4519			8067			9070	17	1	4834	24	2

RISK FACTOR	75-84 AT EXAM PERSON EXAMS	# OF EVENTS	ANNUAL RATE	85-94 AT EXAM PERSON EXAMS	# OF EVENTS	ANNUAL RATE	35-64 COMBINED # OF EVENTS	CRUDE	AGE-ADJUSTED	65-94 COMBINED # OF EVENTS	CRUDE	AGE-ADJUSTED
NO	1180	23	10	41			19	0	0	44	4	4
INCOMPLETE	37	0	0	3			2	1	1	4	7	6
COMPLETE	128	0	0	6			0	0	0	2	3	2
UNKNOWN	0	0	-	0			0	-	-	0	-	-
TOTAL	1345	23	9	50			21	0	0	50	4	4

LOGISTIC REGRESSIONS

AGE AT EXAM		COEFFICIENT OF RISK FACTOR	SE COEFFICIENT	P-VALUE	STANDARDIZED COEFFICIENT	STD SE
35-44		-				
45-54		-				
55-64		-0.025	0.601	0.967	-0.010	0.247
65-74		0.299	0.336	0.373	0.147	0.165
75-84		-				
85-94		-				
35-64	UNIVARIATE	-0.025	0.625	0.968	-0.009	0.222
65-94	UNIVARIATE	-0.052	0.284	0.855	-0.027	0.148
35-64	BIVARIATE	-0.161	0.608	0.791	-0.057	0.216
65-94	BIVARIATE	-0.137	0.280	0.624	-0.071	0.146
35-64	MULTIVARIATE	-0.212	0.623	0.734	-0.075	0.221
65-94	MULTIVARIATE	-0.219	0.285	0.443	-0.114	0.148

SEX: FEMALE
EVENT: DEATH FROM STROKE
RISK FACTOR: INTRAVENTRICULAR CONDUCTION DEFECT (WITH QRS INTERVAL MORE THAN 0.11 SECOND)
PERSONS AT RISK: ALL PERSONS ALIVE

RISK FACTOR	35-44 AT EXAM PERSON EXAMS	# OF EVENTS	ANNUAL RATE	45-54 AT EXAM PERSON EXAMS	# OF EVENTS	ANNUAL RATE	55-64 AT EXAM PERSON EXAMS	# OF EVENTS	ANNUAL RATE	65-74 AT EXAM PERSON EXAMS	# OF EVENTS	ANNUAL RATE
NO	5532			9873	7	0	11731	10	0	6794	27	2
INCOMPLETE	75			151	0	0	195	0	0	134	0	0
COMPLETE	12			48	0	0	177	1	5	175	1	3
UNKNOWN	0			0	0	-	0	0	-	0	0	-
TOTAL	5619			10072	7	0	12103	11	0	7103	28	2

RISK FACTOR	75-84 AT EXAM PERSON EXAMS	# OF EVENTS	ANNUAL RATE	85-94 AT EXAM PERSON EXAMS	# OF EVENTS	ANNUAL RATE	35-64 COMBINED # OF EVENTS	CRUDE	AGE-ADJUSTED	65-94 COMBINED # OF EVENTS	CRUDE	AGE-ADJUSTED
NO	2261	29	6	153	6	20	17	0	0	62	3	3
INCOMPLETE	51	1	10	3	0	0	0	0	0	1	3	2
COMPLETE	107	2	9	7	0	0	1	2	1	3	5	4
UNKNOWN	0	0	-	0	0	-	0	-	-	0	-	-
TOTAL	2419	32	7	163	6	18	18	0	0	66	3	3

LOGISTIC REGRESSIONS

AGE AT EXAM		COEFFICIENT OF RISK FACTOR	SE COEFFICIENT	P-VALUE	STANDARDIZED COEFFICIENT	STD SE
35-44		-				
45-54		-				
55-64		0.851	0.555	0.125	0.229	0.150
65-74		0.029	0.550	0.959	0.010	0.165
75-84		0.213	0.348	0.540	0.092	0.150
85-94		-				
35-64	UNIVARIATE	0.794	0.559	0.156	0.174	0.123
65-94	UNIVARIATE	0.176	0.293	0.548	0.064	0.107
35-64	BIVARIATE	0.648	0.557	0.245	0.142	0.122
65-94	BIVARIATE	0.068	0.293	0.816	0.025	0.107
35-64	MULTIVARIATE	0.716	0.570	0.209	0.157	0.125
65-94	MULTIVARIATE	0.041	0.296	0.890	0.015	0.108

TABLE 17-18   DATE=09NOV85

AVERAGE ANNUAL INCIDENCE RATE FOR EVENT PER 1000 PERSONS AT RISK AT EXAM
BY AGE AND LEVEL OF RISK FACTOR AT EXAM: FRAMINGHAM STUDY 30-YEAR FOLLOWUP

SEX: MALE
EVENT: DEATH FROM STROKE
RISK FACTOR: DEFINITE NONSPECIFIC T-WAVE OR ST-SEGMENT ABNORMALITY BY ELECTROCARDIOGRAM
PERSONS AT RISK: ALL PERSONS ALIVE

RISK FACTOR	35-44 AT EXAM			45-54 AT EXAM			55-64 AT EXAM			65-74 AT EXAM		
	PERSON EXAMS	# OF EVENTS	ANNUAL RATE	PERSON EXAMS	# OF EVENTS	ANNUAL RATE	PERSON EXAMS	# OF EVENTS	ANNUAL RATE	PERSON EXAMS	# OF EVENTS	ANNUAL RATE
NO	4406			7641			8246	13	1	4170	20	2
YES	113			426			824	4	2	664	4	1
UNKNOWN	0			0			0	0	-	0	0	-
TOTAL	4519			8067			9070	17	1	4834	24	2

RISK FACTOR	75-84 AT EXAM			85-94 AT EXAM			35-64 COMBINED			65-94 COMBINED		
	PERSON EXAMS	# OF EVENTS	ANNUAL RATE	PERSON EXAMS	# OF EVENTS	ANNUAL RATE	# OF EVENTS	CRUDE	AGE-ADJUSTED	# OF EVENTS	CRUDE	AGE-ADJUSTED
NO	1065	18	8	31			17	0	0	40	4	4
YES	282	5	9	19			4	1	1	10	5	4
UNKNOWN	0	0	-	0			0	-		0	-	
TOTAL	1345	23	9	50			21	0	0	50	4	4

LOGISTIC REGRESSIONS

AGE AT EXAM		COEFFICIENT OF RISK FACTOR	SE COEFFICIENT	P-VALUE	STANDARDIZED COEFFICIENT	STD SE
35-44						
45-54						
55-64		1.128	0.573	0.049	0.324	0.165
65-74		0.229	0.549	0.677	0.079	0.189
75-84		0.047	0.510	0.927	0.019	0.208
85-94						
35-64	UNIVARIATE	1.256	0.556	0.024	0.305	0.135
65-94	UNIVARIATE	0.313	0.355	0.378	0.113	0.129
35-64	BIVARIATE	0.919	0.560	0.101	0.223	0.136
65-94	BIVARIATE	0.123	0.359	0.731	0.045	0.130
35-64	MULTIVARIATE	0.773	0.566	0.172	0.188	0.137
65-94	MULTIVARIATE	0.065	0.361	0.857	0.024	0.131

SEX: FEMALE
EVENT: DEATH FROM STROKE
RISK FACTOR: DEFINITE NONSPECIFIC T-WAVE OR ST-SEGMENT ABNORMALITY BY ELECTROCARDIOGRAM
PERSONS AT RISK: ALL PERSONS ALIVE

RISK FACTOR	35-44 AT EXAM			45-54 AT EXAM			55-64 AT EXAM			65-74 AT EXAM		
	PERSON EXAMS	# OF EVENTS	ANNUAL RATE	PERSON EXAMS	# OF EVENTS	ANNUAL RATE	PERSON EXAMS	# OF EVENTS	ANNUAL RATE	PERSON EXAMS	# OF EVENTS	ANNUAL RATE
NO	5520			9642	6	0	11111	9	0	6163	21	2
YES	99			430	1	1	992	2	1	940	7	4
UNKNOWN	0			0	0	-	0	0	-	0	0	-
TOTAL	5619			10072	7	0	12103	11	0	7103	28	2

RISK FACTOR	75-84 AT EXAM			85-94 AT EXAM			35-64 COMBINED			65-94 COMBINED		
	PERSON EXAMS	# OF EVENTS	ANNUAL RATE	PERSON EXAMS	# OF EVENTS	ANNUAL RATE	# OF EVENTS	CRUDE	AGE-ADJUSTED	# OF EVENTS	CRUDE	AGE-ADJUSTED
NO	1945	23	6	124	5	20	15	0	0	49	3	3
YES	474	9	9	39	1	13	3	1	1	17	6	5
UNKNOWN	0	0	-	0	0	-	0	-		0	-	
TOTAL	2419	32	7	163	6	18	18	0	0	66	3	3

LOGISTIC REGRESSIONS

AGE AT EXAM		COEFFICIENT OF RISK FACTOR	SE COEFFICIENT	P-VALUE	STANDARDIZED COEFFICIENT	STD SE
35-44						
45-54		1.320	1.081	0.222	0.267	0.219
55-64		0.913	0.782	0.243	0.250	0.215
65-74		0.786	0.438	0.073	0.266	0.148
75-84		0.481	0.397	0.225	0.191	0.157
85-94		-0.468	1.111	0.674	-0.200	0.476
35-64	UNIVARIATE	1.241	0.633	0.050	0.282	0.144
65-94	UNIVARIATE	0.682	0.283	0.016	0.243	0.101
35-64	BIVARIATE	0.983	0.640	0.125	0.223	0.146
65-94	BIVARIATE	0.484	0.285	0.090	0.173	0.102
35-64	MULTIVARIATE	0.707	0.648	0.275	0.161	0.147
65-94	MULTIVARIATE	0.467	0.288	0.105	0.167	0.103

- 349 -

TABLE 18-1   DATE=15MAY85

AVERAGE ANNUAL INCIDENCE RATE FOR EVENT PER 1000 PERSONS AT RISK AT EXAM
BY AGE AND LEVEL OF RISK FACTOR AT EXAM: FRAMINGHAM STUDY 30-YEAR FOLLOWUP

SEX: MALE
EVENT: INTERMITTENT CLAUDICATION
RISK FACTOR: SYSTOLIC BLOOD PRESSURE, FIRST EXAMINER (MM HG)
PERSONS AT RISK: PERSONS FREE OF INTERMITTENT CLAUDICATION

RISK FACTOR	35-44 AT EXAM			45-54 AT EXAM			55-64 AT EXAM			65-74 AT EXAM		
	PERSON EXAMS	# OF EVENTS	ANNUAL RATE	PERSON EXAMS	# OF EVENTS	ANNUAL RATE	PERSON EXAMS	# OF EVENTS	ANNUAL RATE	PERSON EXAMS	# OF EVENTS	ANNUAL RATE
74-119	1226			1839	5	1	1570	12	4	634	6	5
120-139	2188			3519	9	1	3453	17	2	1559	18	6
140-159	850			1816	10	3	2276	26	8	1268	19	7
160-179	197			581	3	3	1015	16	8	630	13	10
180-300	53			243	2	4	486	10	10	325	9	14
UNKNOWN	0			0	0	-	0	0	-	0	0	-
TOTAL	4514			7998	29	2	8800	91	5	4516	65	7

RISK FACTOR	75-84 AT EXAM			85-94 AT EXAM			35-64 COMBINED			65-94 COMBINED		
	PERSON EXAMS	# OF EVENTS	ANNUAL RATE	PERSON EXAMS	# OF EVENTS	ANNUAL RATE	# OF EVENTS	CRUDE	AGE-ADJUSTED	# OF EVENTS	CRUDE	AGE-ADJUSTED
74-119	136	2	7	7			17	2	2	8	5	5
120-139	413	3	6	16			28	2	1	23	6	6
140-159	380	3	6	17			47	5	3	22	6	6
160-179	193	4	10	4			19	5	4	17	10	10
180-300	101	0	0	2			12	8	6	9	11	11
UNKNOWN	0	0	-	0			0	-	-	0	-	-
TOTAL	1223	14	6	46			123	3	3	79	7	7

LOGISTIC REGRESSIONS

AGE AT EXAM		COEFFICIENT OF RISK FACTOR	SE COEFFICIENT	P-VALUE	STANDARDIZED COEFFICIENT	STD SE
35-44						
45-54		0.016	0.008	0.050	0.311	0.158
55-64		0.016	0.004	0.001	0.351	0.089
65-74		0.015	0.005	0.001	0.350	0.108
75-84		-0.005	0.012	0.709	-0.107	0.279
85-94						
35-64	UNIVARIATE	0.021	0.003	0.001	0.425	0.071
65-94	UNIVARIATE	0.012	0.004	0.006	0.281	0.101
35-64	BIVARIATE	0.016	0.004	0.001	0.321	0.073
65-94	BIVARIATE	0.012	0.004	0.005	0.282	0.101
35-64	MULTIVARIATE	0.014	0.004	0.001	0.286	0.080
65-94	MULTIVARIATE	0.012	0.005	0.013	0.265	0.107

SEX: FEMALE
EVENT: INTERMITTENT CLAUDICATION
RISK FACTOR: SYSTOLIC BLOOD PRESSURE, FIRST EXAMINER (MM HG)
PERSONS AT RISK: PERSONS FREE OF INTERMITTENT CLAUDICATION

RISK FACTOR	35-44 AT EXAM			45-54 AT EXAM			55-64 AT EXAM			65-74 AT EXAM		
	PERSON EXAMS	# OF EVENTS	ANNUAL RATE	PERSON EXAMS	# OF EVENTS	ANNUAL RATE	PERSON EXAMS	# OF EVENTS	ANNUAL RATE	PERSON EXAMS	# OF EVENTS	ANNUAL RATE
74-119	2491			2722	2	0	2024	7	2	669	4	3
120-139	2240			4018	5	1	4156	14	2	1880	8	2
140-159	672			2076	2	0	3163	18	3	2225	13	3
160-179	156			777	3	2	1649	13	4	1316	11	4
180-300	51			449	3	3	954	11	6	811	11	7
UNKNOWN	0			0	0	-	0	0	-	0	0	-
TOTAL	5610			10042	15	1	11946	63	3	6901	47	3

RISK FACTOR	75-84 AT EXAM			85-94 AT EXAM			35-64 COMBINED			65-94 COMBINED		
	PERSON EXAMS	# OF EVENTS	ANNUAL RATE	PERSON EXAMS	# OF EVENTS	ANNUAL RATE	# OF EVENTS	CRUDE	AGE-ADJUSTED	# OF EVENTS	CRUDE	AGE-ADJUSTED
74-119	150	0	0	9			10	1	1	4	2	2
120-139	530	4	4	46			19	1	1	12	2	2
140-159	736	6	4	38			21	2	2	19	3	3
160-179	531	6	4	34			16	2	2	15	4	4
180-300	349	3	4	26			14	3	4	14	6	6
UNKNOWN	0	0	-	0			0	-	-	0	-	-
TOTAL	2296	17	4	153			80	1	2	64	3	3

LOGISTIC REGRESSIONS

AGE AT EXAM		COEFFICIENT OF RISK FACTOR	SE COEFFICIENT	P-VALUE	STANDARDIZED COEFFICIENT	STD SE
35-44						
45-54		0.025	0.007	0.001	0.563	0.152
55-64		0.016	0.004	0.001	0.390	0.102
65-74		0.013	0.005	0.020	0.309	0.133
75-84		0.011	0.009	0.226	0.271	0.224
85-94						
35-64	UNIVARIATE	0.022	0.003	0.001	0.519	0.077
65-94	UNIVARIATE	0.012	0.005	0.009	0.300	0.114
35-64	BIVARIATE	0.017	0.004	0.001	0.395	0.086
65-94	BIVARIATE	0.012	0.005	0.008	0.304	0.115
35-64	MULTIVARIATE	0.014	0.004	0.001	0.339	0.095
65-94	MULTIVARIATE	0.012	0.005	0.016	0.288	0.119

TABLE 18-2   DATE=15MAY85

AVERAGE ANNUAL INCIDENCE RATE FOR EVENT PER 1000 PERSONS AT RISK AT EXAM
BY AGE AND LEVEL OF RISK FACTOR AT EXAM: FRAMINGHAM STUDY 30-YEAR FOLLOWUP

SEX: MALE
EVENT: INTERMITTENT CLAUDICATION
RISK FACTOR: DIASTOLIC BLOOD PRESSURE, FIRST EXAMINER (MM HG)
PERSONS AT RISK: PERSONS FREE OF INTERMITTENT CLAUDICATION

RISK FACTOR	35-44 AT EXAM PERSON EXAMS	# OF EVENTS	ANNUAL RATE	45-54 AT EXAM PERSON EXAMS	# OF EVENTS	ANNUAL RATE	55-64 AT EXAM PERSON EXAMS	# OF EVENTS	ANNUAL RATE	65-74 AT EXAM PERSON EXAMS	# OF EVENTS	ANNUAL RATE
20-74	1014			1535	4	1	2009	18	4	1392	19	7
75-84	1615			2845	9	2	5088	26	4	1548	23	7
85-94	1202			2127	5	1	2152	24	6	1019	11	5
95-104	508			1055	7	3	1084	17	8	400	8	10
105-160	175			436	4	5	667	6	6	157	4	13
UNKNOWN	0			0	0	-	0	0	-	0		
TOTAL	4514			7998	29	-	8800	91	5	4516	65	7

RISK FACTOR	75-84 AT EXAM PERSON EXAMS	# OF EVENTS	ANNUAL RATE	85-94 AT EXAM PERSON EXAMS	# OF EVENTS	ANNUAL RATE	35-64 COMBINED # OF EVENTS	CRUDE	AGE-ADJUSTED	65-94 COMBINED # OF EVENTS	CRUDE	AGE-ADJUSTED
20-74	496	7	7	28			22	2	2	26	7	7
75-84	408	6	7	13			33	3	3	29	7	7
85-94	228	1	2	5			30	3	3	12	5	4
95-104	61	0	0	0			26	5	5	8	9	8
105-160	30	0	0	0			10	5	4	4	11	10
UNKNOWN	0	0	-	0			0	-	-	0	-	-
TOTAL	1223	14	6	46			123	3	3	79	7	7

LOGISTIC REGRESSIONS

AGE AT EXAM		COEFFICIENT OF RISK FACTOR	SE COEFFICIENT	P-VALUE	STANDARDIZED COEFFICIENT	STD SE
35-44						
45-54		0.021	0.015	0.164	0.242	0.174
55-64		0.011	0.008	0.206	0.128	0.101
65-74		0.005	0.010	0.624	0.060	0.123
75-84		-0.023	0.023	-0.305	-0.286	0.278
85-94						
35-64	UNIVARIATE	0.015	0.007	0.036	0.180	0.086
65-94	UNIVARIATE	0.001	0.009	0.874	0.018	0.113
35-64	BIVARIATE	0.015	0.007	0.031	0.180	0.084
65-94	BIVARIATE	0.000	0.009	0.959	0.006	0.113
35-64	MULTIVARIATE	0.013	0.007	0.076	0.156	0.088
65-94	MULTIVARIATE	-0.002	0.010	0.829	-0.025	0.118

SEX: FEMALE
EVENT: INTERMITTENT CLAUDICATION
RISK FACTOR: DIASTOLIC BLOOD PRESSURE, FIRST EXAMINER (MM HG)
PERSONS AT RISK: PERSONS FREE OF INTERMITTENT CLAUDICATION

RISK FACTOR	35-44 AT EXAM PERSON EXAMS	# OF EVENTS	ANNUAL RATE	45-54 AT EXAM PERSON EXAMS	# OF EVENTS	ANNUAL RATE	55-64 AT EXAM PERSON EXAMS	# OF EVENTS	ANNUAL RATE	65-74 AT EXAM PERSON EXAMS	# OF EVENTS	ANNUAL RATE
20-74	2161			2654	4	1	3066	12	4	2004	15	4
75-84	2027			3559	7	1	3971	18	2	2348	9	2
85-94	988			2387	1	0	2959	15	3	1580	17	5
95-104	322			937	1	1	1125	12	5	669	4	3
105-160	112			505	2	2	625	0	-	300	2	3
UNKNOWN	0			0	0	-	0	0	-	0	0	-
TOTAL	5610			10042	15	1	11946	63	3	6901	47	3

RISK FACTOR	75-84 AT EXAM PERSON EXAMS	# OF EVENTS	ANNUAL RATE	85-94 AT EXAM PERSON EXAMS	# OF EVENTS	ANNUAL RATE	35-64 COMBINED # OF EVENTS	CRUDE	AGE-ADJUSTED	65-94 COMBINED # OF EVENTS	CRUDE	AGE-ADJUSTED
20-74	868	8	5	71			17	1	1	23	4	4
75-84	688	9	7	48			25	1	1	18	3	3
85-94	511	0	0	25			16	1	1	17	4	4
95-104	162	0	0	7			14	3	3	4	2	2
105-160	67	0	0	2			8	3	3	2	3	2
UNKNOWN	0	0	-	0			0	-	-	0	-	-
TOTAL	2296	17	4	153			80	1	2	64	3	3

LOGISTIC REGRESSIONS

AGE AT EXAM		COEFFICIENT OF RISK FACTOR	SE COEFFICIENT	P-VALUE	STANDARDIZED COEFFICIENT	STD SE
35-44						
45-54		0.020	0.019	0.278	0.250	0.230
55-64		0.022	0.009	0.014	0.270	0.115
65-74		0.004	0.012	0.700	0.056	0.144
75-84		-0.037	0.021	0.087	-0.454	0.265
85-94						
35-64	UNIVARIATE	0.027	0.008	0.001	0.332	0.097
65-94	UNIVARIATE	-0.005	0.010	0.612	-0.065	0.127
35-64	BIVARIATE	0.022	0.008	0.008	0.268	0.100
65-94	BIVARIATE	-0.005	0.010	0.594	-0.368	0.128
35-64	MULTIVARIATE	0.018	0.009	0.037	0.219	0.105
65-94	MULTIVARIATE	-0.007	0.010	0.506	-0.083	0.127

TABLE 18-3A   DATE=04SEP83

AVERAGE ANNUAL INCIDENCE RATE FOR EVENT PER 1000 PERSONS AT RISK AT EXAM
BY AGE AND LEVEL OF RISK FACTOR AT EXAM: FRAMINGHAM STUDY 30-YEAR FOLLOWUP

SEX: MALE
EVENT: INTERMITTENT CLAUDICATION
RISK FACTOR: HYPERTENSION WITH ANTIHYPERTENSIVE TREATMENT
PERSONS AT RISK: PERSONS FREE OF INTERMITTENT CLAUDICATION

RISK FACTOR	35-44 AT EXAM			45-54 AT EXAM			55-64 AT EXAM			65-74 AT EXAM		
	PERSON EXAMS	# OF EVENTS	ANNUAL RATE	PERSON EXAMS	# OF EVENTS	ANNUAL RATE	PERSON EXAMS	# OF EVENTS	ANNUAL RATE	PERSON EXAMS	# OF EVENTS	ANNUAL RATE
NO	2665			4254	7	1	4106	24	3	1751	19	5
BORDER	1315			2262	10	2	2477	34	7	1411	17	6
DEFINITE	510			1193	7	3	1380	22	8	633	14	11
TREATED	24			289	5	9	837	11	7	721	15	10
UNKNOWN	0			0	0	-	0	0	-	0	0	-
TOTAL	4514			7998	29	2	8800	91	5	4516	65	7

RISK FACTOR	75-84 AT EXAM			85-94 AT EXAM			35-64 COMBINED			65-94 COMBINED		
	PERSON EXAMS	# OF EVENTS	ANNUAL RATE	PERSON EXAMS	# OF EVENTS	ANNUAL RATE	# OF EVENTS	CRUDE	AGE-ADJUSTED	# OF EVENTS	CRUDE	AGE-ADJUSTED
NO	438	5	6	17			31	1	2	24	5	5
BORDER	398	4	5	15			46	4	4	21	6	6
DEFINITE	164	0	0	4			30	5	5	14	9	9
TREATED	223	5	11	10			16	7	6	20	10	10
UNKNOWN	0	0	-	0			0	-		0	-	
TOTAL	1223	14	6	46			123	3	3	79	7	7

LOGISTIC REGRESSIONS

AGE AT EXAM		COEFFICIENT OF RISK FACTOR	SE COEFFICIENT	P-VALUE	STANDARDIZED COEFFICIENT	STD SE
35-44						
45-54		0.784	0.228	0.001	0.606	0.176
55-64		0.453	0.126	0.001	0.372	0.103
65-74		0.360	0.153	0.019	0.297	0.126
75-84		0.063	0.327	0.848	0.051	0.269
85-94						
35-64	UNIVARIATE	0.660	0.110	0.001	0.517	0.086
65-94	UNIVARIATE	0.305	0.138	0.027	0.251	0.114
35-64	BIVARIATE	0.541	0.110	0.001	0.424	0.086
65-94	BIVARIATE	0.308	0.138	0.026	0.254	0.114
35-64	MULTIVARIATE	0.499	0.114	0.001	0.391	0.089
65-94	MULTIVARIATE	0.271	0.142	0.056	0.223	0.117

SEX: FEMALE
EVENT: INTERMITTENT CLAUDICATION
RISK FACTOR: HYPERTENSION WITH ANTIHYPERTENSIVE TREATMENT
PERSONS AT RISK: PERSONS FREE OF INTERMITTENT CLAUDICATION

RISK FACTOR	35-44 AT EXAM			45-54 AT EXAM			55-64 AT EXAM			65-74 AT EXAM		
	PERSON EXAMS	# OF EVENTS	ANNUAL RATE	PERSON EXAMS	# OF EVENTS	ANNUAL RATE	PERSON EXAMS	# OF EVENTS	ANNUAL RATE	PERSON EXAMS	# OF EVENTS	ANNUAL RATE
NO	4244			5655	6	1	4980	15	2	1856	8	2
BORDER	1002			2601	1	0	3416	10	1	2075	10	3
DEFINITE	314			1180	3	1	1913	16	4	1176	6	3
TREATED	50			606	5	4	1637	22	7	1794	23	6
UNKNOWN	0			0	0	-	0	0	-	0	0	-
TOTAL	5610			10042	15	1	11946	63	3	6901	47	3

RISK FACTOR	75-84 AT EXAM			85-94 AT EXAM			35-64 COMBINED			65-94 COMBINED		
	PERSON EXAMS	# OF EVENTS	ANNUAL RATE	PERSON EXAMS	# OF EVENTS	ANNUAL RATE	# OF EVENTS	CRUDE	AGE-ADJUSTED	# OF EVENTS	CRUDE	AGE-ADJUSTED
NO	460	2	2	35			22	1	1	10	2	2
BORDER	662	5	4	35			11	1	0	15	3	2
DEFINITE	453	2	2	32			20	3	3	8	2	3
TREATED	721	8	6	51			27	6	4	31	6	6
UNKNOWN	0	0	-	0			0	-		0	-	
TOTAL	2296	17	4	153			80	1	2	64	3	3

LOGISTIC REGRESSIONS

AGE AT EXAM		COEFFICIENT OF RISK FACTOR	SE COEFFICIENT	P-VALUE	STANDARDIZED COEFFICIENT	STD SE
35-44						
45-54		0.767	0.310	0.013	0.590	0.239
55-64		0.715	0.162	0.001	0.598	0.136
65-74		0.468	0.199	0.019	0.384	0.163
75-84		0.282	0.337	0.403	0.221	0.265
85-94						
35-64	UNIVARIATE	0.931	0.139	0.001	0.742	0.111
65-94	UNIVARIATE	0.420	0.171	0.014	0.342	0.139
35-64	BIVARIATE	0.705	0.143	0.001	0.562	0.114
65-94	BIVARIATE	0.427	0.172	0.013	0.348	0.140
35-64	MULTIVARIATE	0.650	0.147	0.001	0.518	0.117
65-94	MULTIVARIATE	0.405	0.175	0.020	0.330	0.142

TABLE 18-3B   DATE=13NOV85
ANNUAL INCIDENCE RATE FOR EVENT PER 1000 PERSONS AT RISK AT EXAM
AND LEVEL OF RISK FACTOR AT EXAM: FRAMINGHAM STUDY 30-YEAR FOLLOWUP

SEX: MALE
EVENT: INTERMITTENT CLAUDICATION
RISK FACTOR: HYPERTENSION IGNORING TREATMENT
PERSONS AT RISK: PERSONS FREE OF INTERMITTENT CLAUDICATION

AT EXAM # OF EVENTS	ANNUAL RATE	45-54 AT EXAM PERSON EXAMS	# OF EVENTS	ANNUAL RATE	55-64 AT EXAM PERSON EXAMS	# OF EVENTS	ANNUAL RATE	65-74 AT EXAM PERSON EXAMS	# OF EVENTS	ANNUAL RATE
		4294	9	1	4267	24	3	1896	22	6
		2351	10	2	2800	19	7	1709	23	7
		1353	10	4	1733	28	8	911	20	11
		0	0	-	0	0	-	0	0	-
		7998	29	2	8800	91	5	4516	65	7

AT EXAM # OF EVENTS	ANNUAL RATE	85-94 AT EXAM PERSON EXAMS	# OF EVENTS	ANNUAL RATE	35-64 COMBINED # OF EVENTS	CRUDE	AGE-ADJUSTED	65-94 COMBINED # OF EVENTS	CRUDE	AGE-ADJUSTED
6	6	22			33	1	2	28	6	6
6	6	20			51	4	4	29	7	7
2	4	4			39	5	5	22	10	9
0	-	0			0	-	-	0	-	-
14	6	46			123	3	3	79	7	7

LOGISTIC REGRESSIONS

COEFFICIENT RISK FACTOR	SE COEFFICIENT	P-VALUE	STANDARDIZED COEFFICIENT	STD SE
0.632	0.227	0.005	0.477	0.172
0.523	0.130	0.001	0.405	0.101
0.321	0.161	0.046	0.243	0.122
-0.154	0.371	0.678	-0.114	0.275
0.646	0.111	0.001	0.487	0.084
0.244	0.147	0.097	0.184	0.111
0.565	0.112	0.001	0.426	0.084
0.244	0.147	0.096	0.184	0.111
0.517	0.116	0.001	0.390	0.088
0.206	0.152	0.174	0.155	0.114

SEX: FEMALE
EVENT: INTERMITTENT CLAUDICATION
RISK FACTOR: HYPERTENSION IGNORING TREATMENT
PERSONS AT RISK: PERSONS FREE OF INTERMITTENT CLAUDICATION

AT EXAM # OF EVENTS	ANNUAL RATE	45-54 AT EXAM PERSON EXAMS	# OF EVENTS	ANNUAL RATE	55-64 AT EXAM PERSON EXAMS	# OF EVENTS	ANNUAL RATE	65-74 AT EXAM PERSON EXAMS	# OF EVENTS	ANNUAL RATE
		5836	6	1	5369	17	2	2164	11	5
		2792	3	1	4034	18	2	2829	16	5
		1414	6	2	2543	28	6	1908	20	5
		0	0	-	0	0	-	0	0	-
		10042	15	1	11946	63	3	6901	47	5

AT EXAM # OF EVENTS	ANNUAL RATE	85-94 AT EXAM PERSON EXAMS	# OF EVENTS	ANNUAL RATE	35-64 COMBINED # OF EVENTS	CRUDE	AGE-ADJUSTED	65-94 COMBINED # OF EVENTS	CRUDE	AGE-ADJUSTED
3	5	46			24	1	1	14	3	3
9	5	54			21	1	1	25	3	3
5	3	53			35	4	3	25	5	5
0	-	0			0	-	-	0	-	-
17	4	153			80	1	1	64	3	3

LOGISTIC REGRESSIONS

COEFFICIENT RISK FACTOR	SE COEFFICIENT	P-VALUE	STANDARDIZED COEFFICIENT	STD SE
0.710	0.315	0.024	0.516	0.229
0.651	0.160	0.001	0.507	0.125
0.395	0.195	0.042	0.303	0.149
0.084	0.323	0.794	0.064	0.245
0.850	0.138	0.001	0.632	0.102
0.314	0.166	0.059	0.241	0.128
0.649	0.142	0.001	0.482	0.106
0.318	0.167	0.056	0.244	0.128
0.603	0.147	0.001	0.448	0.109
0.293	0.170	0.084	0.225	0.130

TABLE 18-4   DATE=15MAY85

AVERAGE ANNUAL INCIDENCE RATE FOR EVENT PER 1000 PERSONS AT RISK AT EXAM
BY AGE AND LEVEL OF RISK FACTOR AT EXAM, FRAMINGHAM STUDY 30-YEAR FOLLOWUP

SEX: MALE
EVENT: INTERMITTENT CLAUDICATION
RISK FACTOR: SERUM CHOLESTEROL
PERSONS AT RISK: PERSONS FREE OF INTERMITTENT CLAUDICATION

35-44 AT EXAM			45-54 AT EXAM			55-64 AT EXAM			65-74 AT EXAM		
PERSON EXAMS	# OF EVENTS	ANNUAL RATE	PERSON EXAMS	# OF EVENTS	ANNUAL RATE	PERSON EXAMS	# OF EVENTS	ANNUAL RATE	PERSON EXAMS	# OF EVENTS	ANNUAL RATE
1253			1793	6	2	2410	19	4	1423	17	6
1255			2375	8	2	2618	24	5	1370	17	6
882			1830	6	2	2075	22	5	1001	15	7
487			1047	4	2	1039	15	7	485	12	12
303			602	5	4	529	10	9	220	4	9
334			351	0	0	129	1	4	17	0	0
4514			7998	29	2	8800	91	5	4516	65	7

75-84 AT EXAM			85-94 AT EXAM			35-64 COMBINED			65-94 COMBINED		
PERSON EXAMS	# OF EVENTS	ANNUAL RATE	PERSON EXAMS	# OF EVENTS	ANNUAL RATE	# OF EVENTS	CRUDE	AGE-ADJUSTED	# OF EVENTS	CRUDE	AGE-ADJUSTED
477	5	5	21	0	0	25	2	2	22	6	6
401	2	2	13	0	0	32	3	3	19	5	5
205	3	7	6			28	3	3	18	7	7
116	4	17	6			20	4	4	16	13	13
24	0	0	0			16	6	6	4	8	7
0	0	-	0			2	1	2	0	0	0
1223	14	6	46			123	3	3	79	7	7

LOGISTIC REGRESSIONS

	COEFFICIENT OF RISK FACTOR	SE COEFFICIENT	P-VALUE	STANDARDIZED COEFFICIENT	STD SE
	0.006	0.004	0.072	0.273	0.152
	0.006	0.002	0.015	0.235	0.096
	0.005	0.003	0.091	0.203	0.120
	0.012	0.007	0.071	0.456	0.252
:VARIATE	0.006	0.002	0.001	0.254	0.079
:VARIATE	0.006	0.003	0.018	0.256	0.108
/ARIATE	0.007	0.002	0.001	0.283	0.080
/ARIATE	0.006	0.003	0.021	0.251	0.109
.TIVARIATE	0.006	0.002	0.002	0.252	0.080
.TIVARIATE	0.006	0.003	0.016	0.259	0.108

SEX: FEMALE
EVENT: INTERMITTENT CLAUDICATION
RISK FACTOR: SERUM CHOLESTEROL
PERSONS AT RISK: PERSONS FREE OF INTERMITTENT CLAUDICATION

35-44 AT EXAM			45-54 AT EXAM			55-64 AT EXAM			65-74 AT EXAM		
PERSON EXAMS	# OF EVENTS	ANNUAL RATE	PERSON EXAMS	# OF EVENTS	ANNUAL RATE	PERSON EXAMS	# OF EVENTS	ANNUAL RATE	PERSON EXAMS	# OF EVENTS	ANNUAL RATE
2350			2118	2	0	1701	5	1	868	10	6
1414			2579	4	1	2705	13	2	1535	5	2
807			2262	4	1	3015	11	2	1869	9	2
369			1484	3	1	2325	16	3	1426	12	4
170			1119	2	1	1967	17	4	1166	11	5
500			480	0	0	233	1	2	37	0	0
5610			10042	15	1	11944	63	3	6901	47	3

75-84 AT EXAM			85-94 AT EXAM			35-64 COMBINED			65-94 COMBINED		
PERSON EXAMS	# OF EVENTS	ANNUAL RATE	PERSON EXAMS	# OF EVENTS	ANNUAL RATE	# OF EVENTS	CRUDE	AGE-ADJUSTED	# OF EVENTS	CRUDE	AGE-ADJUSTED
436	1	1	57			8	1	1	11	4	3
555	5	5	37			18	1	1	10	2	3
623	4	3	38			15	1	1	13	3	2
402	4	5	11			19	2	2	16	4	4
273	3	5	10			19	3	2	14	5	5
7	0	0	0			1	0	1	0	0	0
2294	17	4	153			80	1	2	64	3	3

LOGISTIC REGRESSIONS

	COEFFICIENT OF RISK FACTOR	SE COEFFICIENT	P-VALUE	STANDARDIZED COEFFICIENT	STD SE
	0.004	0.003	0.196	0.204	0.158
	0.007	0.002	0.006	0.314	0.115
	0.003	0.003	0.284	0.151	0.141
	0.004	0.005	0.497	0.161	0.237
VARIATE	0.007	0.002	0.001	0.326	0.074
VARIATE	0.004	0.003	0.189	0.159	0.121
ARIATE	0.006	0.002	0.003	0.262	0.087
ARIATE	0.004	0.003	0.195	0.158	0.122
TIVARIATE	0.005	0.002	0.010	0.220	0.086
TIVARIATE	0.003	0.003	0.220	0.150	0.122

TABLE 18-5   DATE=13NOV85
AVERAGE ANNUAL INCIDENCE RATE FOR EVENT PER 1000 PERSONS AT RISK AT EXAM
BY AGE AND LEVEL OF RISK FACTOR AT EXAM: FRAMINGHAM STUDY 30-YEAR FOLLOWUP

SEX: MALE
EVENT: INTERMITTENT CLAUDICATION
RISK FACTOR: HEMATOCRIT
PERSONS AT RISK: PERSONS FREE OF INTERMITTENT CLAUDICATION

RISK FACTOR	35-44 AT EXAM PERSON EXAMS	# OF EVENTS	ANNUAL RATE	45-54 AT EXAM PERSON EXAMS	# OF EVENTS	ANNUAL RATE	55-64 AT EXAM PERSON EXAMS	# OF EVENTS	ANNUAL RATE	65-74 AT EXAM PERSON EXAMS	# OF EVENTS	ANNUAL RATE
25-42	630			951	2	1	1093	8	4	516	5	5
43-44	551			990	2	1	1068	8	4	584	6	5
45-46	990			1928	8	2	2145	16	4	1117	14	6
47-48	783			1574	5	2	1774	22	6	998	17	9
49-70	1190			2165	12	3	2564	36	7	1281	23	9
UNKNOWN	370			390	0	0	156	1	3	20	0	0
TOTAL	4514			7998	29	2	8800	91	5	4516	65	7

RISK FACTOR	75-84 AT EXAM PERSON EXAMS	# OF EVENTS	ANNUAL RATE	85-94 AT EXAM PERSON EXAMS	# OF EVENTS	ANNUAL RATE	35-64 COMBINED # OF EVENTS	CRUDE	AGE-ADJUSTED	65-94 COMBINED # OF EVENTS	CRUDE	AGE-ADJUSTED
25-42	262	2	4	17			10	2	2	7	4	5
43-44	154	0	0	7			10	2	2	6	4	4
45-46	286	3	5	7			24	2	2	17	6	6
47-48	236	7	15	9			28	3	3	24	10	10
49-70	285	2	4	6			49	4	4	25	8	8
UNKNOWN	0	0	-	0			2	1	2	0	0	0
TOTAL	1223	14	6	46			123	3	3	79	7	7

LOGISTIC REGRESSIONS

AGE AT EXAM		COEFFICIENT OF RISK FACTOR	SE COEFFICIENT	P-VALUE	STANDARDIZED COEFFICIENT	STD SE
35-44						
45-54		0.110	0.051	0.031	0.395	0.183
55-64		0.082	0.028	0.003	0.310	0.106
65-74		0.077	0.035	0.027	0.279	0.126
75-84		0.054	0.066	0.415	0.223	0.274
85-94						
35-64	UNIVARIATE	0.094	0.025	0.001	0.346	0.091
65-94	UNIVARIATE	0.074	0.030	0.015	0.281	0.115
35-64	BIVARIATE	0.091	0.024	0.001	0.334	0.090
65-94	BIVARIATE	0.073	0.031	0.017	0.276	0.116
35-64	MULTIVARIATE	0.064	0.025	0.010	0.234	0.091
65-94	MULTIVARIATE	0.062	0.031	0.049	0.234	0.119

SEX: FEMALE
EVENT: INTERMITTENT CLAUDICATION
RISK FACTOR: HEMATOCRIT
PERSONS AT RISK: PERSONS FREE OF INTERMITTENT CLAUDICATION

RISK FACTOR	35-44 AT EXAM PERSON EXAMS	# OF EVENTS	ANNUAL RATE	45-54 AT EXAM PERSON EXAMS	# OF EVENTS	ANNUAL RATE	55-64 AT EXAM PERSON EXAMS	# OF EVENTS	ANNUAL RATE	65-74 AT EXAM PERSON EXAMS	# OF EVENTS	ANNUAL RATE
21-39	1697			1854	6	2	1591	8	3	729	7	5
40-41	967			1758	2	1	2124	10	2	1159	5	2
42-43	1041			1973	1	0	2352	8	2	1388	8	3
44-45	856			2213	3	1	2960	15	3	1778	13	4
46-65	483			1693	3	1	2643	21	4	1805	14	4
UNKNOWN	566			551	0	0	276	1	2	42	0	0
TOTAL	5610			10042	15	1	11946	63	3	6901	47	3

RISK FACTOR	75-84 AT EXAM PERSON EXAMS	# OF EVENTS	ANNUAL RATE	85-94 AT EXAM PERSON EXAMS	# OF EVENTS	ANNUAL RATE	35-64 COMBINED # OF EVENTS	CRUDE	AGE-ADJUSTED	65-94 COMBINED # OF EVENTS	CRUDE	AGE-ADJUSTED
21-39	551	3	4	30			14	1	2	10	5	5
40-41	428	4	5	38			12	1	1	9	3	3
42-43	422	3	4	23			9	1	1	11	3	3
44-45	577	5	4	53			18	1	1	18	4	4
46-65	512	2	2	29			26	3	2	16	3	3
UNKNOWN	6	0	0	0			1	0	1	0	0	0
TOTAL	2296	17	4	153			80	1	1	64	3	3

LOGISTIC REGRESSIONS

AGE AT EXAM		COEFFICIENT OF RISK FACTOR	SE COEFFICIENT	P-VALUE	STANDARDIZED COEFFICIENT	STD SE
35-44						
45-54		-0.047	0.070	0.500	-0.168	0.249
55-64		0.067	0.037	0.068	0.233	0.128
65-74		-0.001	0.042	0.981	-0.003	0.146
75-84		0.012	0.064	0.855	0.045	0.246
85-94						
35-64	UNIVARIATE	0.086	0.032	0.007	0.311	0.115
65-94	UNIVARIATE	0.003	0.035	0.922	0.012	0.126
35-64	BIVARIATE	0.049	0.033	0.131	0.177	0.117
65-94	BIVARIATE	0.003	0.035	0.935	0.010	0.126
35-64	MULTIVARIATE	-0.002	0.031	0.956	-0.006	0.112
65-94	MULTIVARIATE	-0.016	0.055	0.639	-0.059	0.125

SEX: MALE
EVENT: INTERMITTENT CLAUDICATION
RISK FACTOR: BLOOD GLUCOSE (MG/DL)
PERSONS AT RISK: PERSONS FREE OF INTERMITTENT CLAUDICATION

RISK FACTOR	35-44 AT EXAM PERSON EXAMS	# OF EVENTS	ANNUAL RATE	45-54 AT EXAM PERSON EXAMS	# OF EVENTS	ANNUAL RATE	55-64 AT EXAM PERSON EXAMS	# OF EVENTS	ANNUAL RATE	65-74 AT EXAM PERSON EXAMS	# OF EVENTS	ANNUAL RATE
29-69	1003			1554	4	1	1175	7	3	476	7	7
70-89	2527			4298	15	2	4435	35	4	2021	22	5
90-109	628			1397	6	2	2049	27	7	1215	17	7
110-129	107			327	2	3	584	10	9	395	7	9
130-524	49			198	2	5	459	11	12	392	12	15
UNKNOWN	200			224	0	0	98	1	5	17	0	0
TOTAL	4514			7998	29	2	8800	91	5	4516	65	7

RISK FACTOR	75-84 AT EXAM PERSON EXAMS	# OF EVENTS	ANNUAL RATE	85-94 AT EXAM PERSON EXAMS	# OF EVENTS	ANNUAL RATE	35-64 COMBINED # OF EVENTS	CRUDE	AGE-ADJUSTED	65-94 COMBINED # OF EVENTS	CRUDE	AGE-ADJUSTED
29-69	81	0	0	1			11	1	2	7	6	6
70-89	440	3	3	16			52	2	2	25	5	5
90-109	402	9	11	15			33	4	4	26	8	8
110-129	150	0	0	5			12	6	5	7	6	7
130-524	150	2	7	9			13	9	7	14	13	13
UNKNOWN	0	0	-	0			2	2	3	0	0	0
TOTAL	1223	14	6	46			123	3	3	79	7	7

LOGISTIC REGRESSIONS

AGE AT EXAM		COEFFICIENT OF RISK FACTOR	SE COEFFICIENT	P-VALUE	STANDARDIZED COEFFICIENT	STD SE
35-44						
45-54		0.008	0.004	0.040	0.191	0.093
55-64		0.008	0.002	0.001	0.227	0.061
65-74		0.005	0.002	0.036	0.172	0.082
75-84		0.007	0.006	0.235	0.225	0.189
85-94						
35-64	UNIVARIATE	0.010	0.002	0.001	0.239	0.043
65-94	UNIVARIATE	0.005	0.002	0.020	0.176	0.076
35-64	BIVARIATE	0.008	0.002	0.001	0.193	0.047
65-94	BIVARIATE	0.005	0.002	0.017	0.179	0.075
35-64	MULTIVARIATE	0.008	0.002	0.001	0.198	0.047
65-94	MULTIVARIATE	0.005	0.002	0.018	0.177	0.075

SEX: FEMALE
EVENT: INTERMITTENT CLAUDICATION
RISK FACTOR: BLOOD GLUCOSE (MG/DL)
PERSONS AT RISK: PERSONS FREE OF INTERMITTENT CLAUDICATION

RISK FACTOR	35-44 AT EXAM PERSON EXAMS	# OF EVENTS	ANNUAL RATE	45-54 AT EXAM PERSON EXAMS	# OF EVENTS	ANNUAL RATE	55-64 AT EXAM PERSON EXAMS	# OF EVENTS	ANNUAL RATE	65-74 AT EXAM PERSON EXAMS	# OF EVENTS	ANNUAL RATE
29-69	1064			1803	2	1	1398	4	1	584	4	3
70-89	3469			5796	7	1	6312	27	2	3403	22	3
90-109	671			1643	1	0	2837	17	3	1920	10	3
110-129	96			302	1	2	764	4	3	563	5	4
130-524	39			159	4	13	464	11	12	399	6	8
UNKNOWN	331			339	0	0	171	0	0	32	0	0
TOTAL	5610			10042	15	1	11946	63	3	6901	47	3

RISK FACTOR	75-84 AT EXAM PERSON EXAMS	# OF EVENTS	ANNUAL RATE	85-94 AT EXAM PERSON EXAMS	# OF EVENTS	ANNUAL RATE	35-64 COMBINED # OF EVENTS	CRUDE	AGE-ADJUSTED	65-94 COMBINED # OF EVENTS	CRUDE	AGE-ADJUSTED
29-69	123	1	4	4			7	1	1	5	4	4
70-89	930	5	3	58			35	1	1	27	3	3
90-109	753	7	5	57			18	2	1	17	3	3
110-129	281	3	5	18			5	2	2	8	5	5
130-524	208	1	2	16			15	11	10	7	6	6
UNKNOWN	1	0	0	0			0	0	0	0	0	0
TOTAL	2296	17	4	153			80	1	1	64	3	3

LOGISTIC REGRESSIONS

AGE AT EXAM		COEFFICIENT OF RISK FACTOR	SE COEFFICIENT	P-VALUE	STANDARDIZED COEFFICIENT	STD SE
35-44						
45-54		0.020	0.004	0.001	0.383	0.067
55-64		0.012	0.002	0.001	0.313	0.048
65-74		0.006	0.003	0.082	0.171	0.098
75-84		0.002	0.006	0.799	0.055	0.217
85-94						
35-64	UNIVARIATE	0.015	0.002	0.001	0.329	0.035
65-94	UNIVARIATE	0.005	0.003	0.121	0.141	0.091
35-64	BIVARIATE	0.013	0.002	0.001	0.284	0.037
65-94	BIVARIATE	0.005	0.003	0.112	0.145	0.091
35-64	MULTIVARIATE	0.012	0.002	0.001	0.254	0.039
65-94	MULTIVARIATE	0.004	0.003	0.142	0.134	0.091

TABLE 18-7   DATE=13DEC85
AVERAGE ANNUAL INCIDENCE RATE FOR EVENT PER 1000 PERSONS AT RISK AT EXAM
BY AGE AND LEVEL OF RISK FACTOR AT EXAM: FRAMINGHAM STUDY 30-YEAR FOLLOWUP

SEX: MALE
EVENT: INTERMITTENT CLAUDICATION
RISK FACTOR: DIABETES MELLITUS
PERSONS AT RISK: PERSONS FREE OF INTERMITTENT CLAUDICATION

RISK FACTOR	35-44 AT EXAM			45-54 AT EXAM			55-64 AT EXAM			65-74 AT EXAM		
	PERSON EXAMS	# OF EVENTS	ANNUAL RATE	PERSON EXAMS	# OF EVENTS	ANNUAL RATE	PERSON EXAMS	# OF EVENTS	ANNUAL RATE	PERSON EXAMS	# OF EVENTS	ANNUAL RATE
NO	4465			7753	25	2	8219	73	4	4030	50	6
YES	49			245	4	8	581	18	15	486	15	15
UNKNOWN	0			0	0	-	0	0	-	0	0	-
TOTAL	4514			7998	29	2	8800	91	5	4516	65	7

RISK FACTOR	75-84 AT EXAM			85-94 AT EXAM			35-64 COMBINED			65-94 COMBINED		
	PERSON EXAMS	# OF EVENTS	ANNUAL RATE	PERSON EXAMS	# OF EVENTS	ANNUAL RATE	# OF EVENTS	CRUDE	AGE-ADJUSTED	# OF EVENTS	CRUDE	AGE-ADJUSTED
NO	1074	11	5	40			101	2	3	61	6	6
YES	149	3	10	6			22	13	9	18	14	14
UNKNOWN	0	0	-	0			0	-		0	-	
TOTAL	1223	14	6	46			123	3	3	79	7	7

LOGISTIC REGRESSIONS

AGE AT EXAM		COEFFICIENT OF RISK FACTOR	SE COEFFICIENT	P-VALUE	STANDARDIZED COEFFICIENT	STD SE
35-44						
45-54		1.635	0.542	0.003	0.282	0.093
55-64		1.272	0.267	0.001	0.316	0.066
65-74		0.930	0.298	0.002	0.288	0.092
75-84		0.686	0.657	0.297	0.224	0.215
85-94						
35-64	UNIVARIATE	1.647	0.238	0.001	0.327	0.047
65-94	UNIVARIATE	0.879	0.272	0.001	0.276	0.085
35-64	BIVARIATE	1.295	0.241	0.001	0.257	0.048
65-94	BIVARIATE	0.888	0.272	0.001	0.279	0.085
35-64	MULTIVARIATE	1.334	0.250	0.001	0.265	0.050
65-94	MULTIVARIATE	0.936	0.277	0.001	0.294	0.087

SEX: FEMALE
EVENT: INTERMITTENT CLAUDICATION
RISK FACTOR: DIABETES MELLITUS
PERSONS AT RISK: PERSONS FREE OF INTERMITTENT CLAUDICATION

RISK FACTOR	35-44 AT EXAM			45-54 AT EXAM			55-64 AT EXAM			65-74 AT EXAM		
	PERSON EXAMS	# OF EVENTS	ANNUAL RATE	PERSON EXAMS	# OF EVENTS	ANNUAL RATE	PERSON EXAMS	# OF EVENTS	ANNUAL RATE	PERSON EXAMS	# OF EVENTS	ANNUAL RATE
NO	5567			9841	11	1	11388	48	2	6348	36	3
YES	43			201	4	10	558	15	13	553	11	10
UNKNOWN	0			0	0	-	0	0	-	0	0	-
TOTAL	5610			10042	15	1	11946	63	3	6901	47	3

RISK FACTOR	75-84 AT EXAM			85-94 AT EXAM			35-64 COMBINED			65-94 COMBINED		
	PERSON EXAMS	# OF EVENTS	ANNUAL RATE	PERSON EXAMS	# OF EVENTS	ANNUAL RATE	# OF EVENTS	CRUDE	AGE-ADJUSTED	# OF EVENTS	CRUDE	AGE-ADJUSTED
NO	2014	15	4	291			61	1	1	51	3	3
YES	282	2	4	24			19	12	9	13	8	8
UNKNOWN	0	0	-	0			0	-		0	-	
TOTAL	2296	17	4	531			80	1	1	64	3	3

LOGISTIC REGRESSIONS

AGE AT EXAM		COEFFICIENT OF RISK FACTOR	SE COEFFICIENT	P-VALUE	STANDARDIZED COEFFICIENT	STD SE
35-44						
45-54		2.898	0.588	0.001	0.406	0.082
55-64		1.876	0.299	0.001	0.396	0.063
65-74		1.269	0.347	0.001	0.345	0.094
75-84		-0.049	0.755	0.948	-0.016	0.248
85-94						
35-64	UNIVARIATE	2.364	0.265	0.001	0.397	0.045
65-94	UNIVARIATE	0.933	0.313	0.003	0.270	0.090
35-64	BIVARIATE	1.985	0.269	0.001	0.333	0.045
65-94	BIVARIATE	0.950	0.315	0.003	0.274	0.091
35-64	MULTIVARIATE	1.796	0.275	0.001	0.302	0.046
65-94	MULTIVARIATE	0.878	0.320	0.006	0.254	0.092

TABLE 18-8   DATE=14NOV85
AVERAGE ANNUAL INCIDENCE RATE FOR EVENT PER 1000 PERSONS AT RISK AT EXAM
BY AGE AND LEVEL OF RISK FACTOR AT EXAM, FRAMINGHAM STUDY 30-YEAR FOLLOWUP

SEX: MALE
EVENT: INTERMITTENT CLAUDICATION
RISK FACTOR: GLUCOSE IN URINE, DEFINITE OR TRACE
PERSONS AT RISK: PERSONS FREE OF INTERMITTENT CLAUDICATION

RISK FACTOR	35-44 AT EXAM			45-54 AT EXAM			55-64 AT EXAM			65-74 AT EXAM		
	PERSON EXAMS	# OF EVENTS	ANNUAL RATE	PERSON EXAMS	# OF EVENTS	ANNUAL RATE	PERSON EXAMS	# OF EVENTS	ANNUAL RATE	PERSON EXAMS	# OF EVENTS	ANNUAL RATE
NO	4421			7806	26	2	8545	82	5	4366	61	7
YES	60			171	3	9.	231	9	19.	144	4	14
UNKNOWN	33			21	0	0	24	0	0	6	0	0
TOTAL	4514			7998	29	2	8800	91	5	4516	65	7

RISK FACTOR	75-84 AT EXAM			85-94 AT EXAM			35-64 COMBINED			65-94 COMBINED		
	PERSON EXAMS	# OF EVENTS	ANNUAL RATE	PERSON EXAMS	# OF EVENTS	ANNUAL RATE	# OF EVENTS	CRUDE	AGE-ADJUSTED	# OF EVENTS	CRUDE	AGE-ADJUSTED
NO	1183	14	6	45			111	3	3	75	7	7
YES	37	0	0	1			12	13	11	4	11	11
UNKNOWN	3	0	0	0			0	0	0	0	0	0
TOTAL	1223	14	6	46			123	3	3	79	7	7

LOGISTIC REGRESSIONS

AGE AT EXAM		COEFFICIENT OF RISK FACTOR	SE COEFFICIENT	P-VALUE	STANDARDIZED COEFFICIENT	STD SE
35-44						
45-54		1.676	0.615	0.006	0.243	0.089
55-64		1.431	0.358	0.001	0.229	0.057
65-74		0.701	0.523	0.180	0.123	0.092
75-84		-	-	-	-	-
85-94						
35-64	UNIVARIATE	1.602	0.308	0.001	0.234	0.045
65-94	UNIVARIATE	0.503	0.519	0.332	0.088	0.091
35-64	BIVARIATE	1.450	0.309	0.001	0.212	0.045
65-94	BIVARIATE	1.502	0.519	0.334	0.088	0.091
35-64	MULTIVARIATE	1.415	0.314	0.001	0.206	0.046
65-94	MULTIVARIATE	0.478	0.521	0.360	0.083	0.091

SEX: FEMALE
EVENT: INTERMITTENT CLAUDICATION
RISK FACTOR: GLUCOSE IN URINE, DEFINITE OR TRACE
PERSONS AT RISK: PERSONS FREE OF INTERMITTENT CLAUDICATION

RISK FACTOR	35-44 AT EXAM			45-54 AT EXAM			55-64 AT EXAM			65-74 AT EXAM		
	PERSON EXAMS	# OF EVENTS	ANNUAL RATE	PERSON EXAMS	# OF EVENTS	ANNUAL RATE	PERSON EXAMS	# OF EVENTS	ANNUAL RATE	PERSON EXAMS	# OF EVENTS	ANNUAL RATE
NO	5531			9925	13	1	11764	55	2	6772	45	3
YES	41			93	2	11	156	8	26	119	2	8
UNKNOWN	38			24	0	0	26	0	0	10	0	0
TOTAL	5610			10042	15	1	11946	63	3	6901	47	3

RISK FACTOR	75-84 AT EXAM			85-94 AT EXAM			35-64 COMBINED			65-94 COMBINED		
	PERSON EXAMS	# OF EVENTS	ANNUAL RATE	PERSON EXAMS	# OF EVENTS	ANNUAL RATE	# OF EVENTS	CRUDE	AGE-ADJUSTED	# OF EVENTS	CRUDE	AGE-ADJUSTED
NO	2255	17	4	150			70	1	1	62	3	3
YES	32	0	0	3			10	17	15	2	6	6
UNKNOWN	9	0	0	0			0	0	0	0	0	0
TOTAL	2296	17	4	153			80	1	1	64	3	3

LOGISTIC REGRESSIONS

AGE AT EXAM		COEFFICIENT OF RISK FACTOR	SE COEFFICIENT	P-VALUE	STANDARDIZED COEFFICIENT	STD SE
35-44						
45-54		2.819	0.767	0.001	0.270	0.074
55-64		2.443	0.387	0.001	0.278	0.044
65-74		0.938	0.729	0.198	0.122	0.095
75-84		-	-	-	-	-
85-94						
35-64	UNIVARIATE	0.503	0.519	0.332	0.051	0.053
65-94	UNIVARIATE	0.660	0.723	0.361	0.084	0.092
35-64	BIVARIATE	2.475	0.347	0.001	0.253	0.035
65-94	BIVARIATE	0.658	0.723	0.363	0.084	0.092
35-64	MULTIVARIATE	2.198	0.360	0.001	0.224	0.037
65-94	MULTIVARIATE	0.455	0.730	0.534	0.058	0.093

TABLE 18-9   DATE=04MAR86
AVERAGE ANNUAL INCIDENCE RATE FOR EVENT PER 1000 PERSONS AT RISK AT EXAM
BY AGE AND LEVEL OF RISK FACTOR AT EXAM: FRAMINGHAM STUDY 30-YEAR FOLLOWUP

SEX: MALE
EVENT: INTERMITTENT CLAUDICATION
RISK FACTOR: GLUCOSE INTOLERANCE
PERSONS AT RISK: PERSONS FREE OF INTERMITTENT CLAUDICATION

RISK FACTOR	35-44 AT EXAM			45-54 AT EXAM			55-64 AT EXAM			65-74 AT EXAM		
	PERSON EXAMS	# OF EVENTS	ANNUAL RATE	PERSON EXAMS	# OF EVENTS	ANNUAL RATE	PERSON EXAMS	# OF EVENTS	ANNUAL RATE	PERSON EXAMS	# OF EVENTS	ANNUAL RATE
NO	4376			7473	25	2	7787	69	4	3700	45	6
YES	138			525	4	4	1013	22	11	816	20	12
UNKNOWN	0			0	0	-	0	0	-	0	0	-
TOTAL	4514			7998	29	2	8800	91	5	4516	65	7

RISK FACTOR	75-84 AT EXAM			85-94 AT EXAM			35-64 COMBINED			65-94 COMBINED		
	PERSON EXAMS	# OF EVENTS	ANNUAL RATE	PERSON EXAMS	# OF EVENTS	ANNUAL RATE	# OF EVENTS	CRUDE	AGE-ADJUSTED	# OF EVENTS	CRUDE	AGE-ADJUST
NO	949	10	5	35			97	2	3	55	6	6
YES	274	4	7	11			26	8	6	24	11	11
UNKNOWN	0	0	-	0			0	-	-	0	-	-
TOTAL	1223	14	6	46			123	3	3	79	7	7

LOGISTIC REGRESSIONS

AGE AT EXAM		COEFFICIENT OF RISK FACTOR	SE COEFFICIENT	P-VALUE	STANDARDIZED COEFFICIENT	STD SE
35-44						
45-54		0.827	0.540	0.126	0.205	0.134
55-64		0.910	0.247	0.001	0.290	0.079
65-74		0.713	0.272	0.009	0.274	0.105
75-84		0.330	0.596	0.579	0.138	0.248
85-94						
35-64	UNIVARIATE	1.155	0.222	0.001	0.311	0.060
65-94	UNIVARIATE	0.629	0.247	0.011	0.247	0.097
35-64	BIVARIATE	0.862	0.225	0.001	0.232	0.060
65-94	BIVARIATE	0.646	0.248	0.009	0.254	0.097
35-64	MULTIVARIATE	0.880	0.229	0.001	0.237	0.062
65-94	MULTIVARIATE	0.664	0.249	0.008	0.261	0.098

SEX: FEMALE
EVENT: INTERMITTENT CLAUDICATION
RISK FACTOR: GLUCOSE INTOLERANCE
PERSONS AT RISK: PERSONS FREE OF INTERMITTENT CLAUDICATION

RISK FACTOR	35-44 AT EXAM			45-54 AT EXAM			55-64 AT EXAM			65-74 AT EXAM		
	PERSON EXAMS	# OF EVENTS	ANNUAL RATE	PERSON EXAMS	# OF EVENTS	ANNUAL RATE	PERSON EXAMS	# OF EVENTS	ANNUAL RATE	PERSON EXAMS	# OF EVENTS	ANNUAL RATE
NO	5516			9669	10	1	10885	45	2	5923	34	3
YES	94			373	5	7	1061	18	8	978	13	7
UNKNOWN	0			0	0	-	0	0	-	0	0	-
TOTAL	5610			10042	15	1	11946	63	3	6901	47	3

RISK FACTOR	75-84 AT EXAM			85-94 AT EXAM			35-64 COMBINED			65-94 COMBINED		
	PERSON EXAMS	# OF EVENTS	ANNUAL RATE	PERSON EXAMS	# OF EVENTS	ANNUAL RATE	# OF EVENTS	CRUDE	AGE-ADJUSTED	# OF EVENTS	CRUDE	AGE-ADJUST
NO	1820	15	4	121			57	1	1	49	3	3
YES	476	2	2	32			23	8	6	15	5	5
UNKNOWN	0	0	-	0			0	-	-	0	-	-
TOTAL	2296	17	4	153			80	1	1	64	3	3

LOGISTIC REGRESSIONS

AGE AT EXAM		COEFFICIENT OF RISK FACTOR	SE COEFFICIENT	P-VALUE	STANDARDIZED COEFFICIENT	STD SE
35-44						
45-54		2.574	0.550	0.001	0.487	0.104
55-64		1.425	0.281	0.001	0.405	0.080
65-74		0.847	0.328	0.010	0.296	0.114
75-84		-0.678	0.755	0.369	-0.275	0.306
85-94						
35-64	UNIVARIATE	1.942	0.248	0.001	0.444	0.057
65-94	UNIVARIATE	0.486	0.296	0.101	0.178	0.108
35-64	BIVARIATE	1.562	0.252	0.001	0.357	0.058
65-94	BIVARIATE	0.499	0.298	0.095	0.182	0.109
35-64	MULTIVARIATE	1.475	0.255	0.001	0.337	0.058
65-94	MULTIVARIATE	0.448	0.301	0.136	0.164	0.110

TABLE 18-10   DATE=13NOV85
AVERAGE ANNUAL INCIDENCE RATE FOR EVENT PER 1000 PERSONS AT RISK AT EXAM
BY AGE AND LEVEL OF RISK FACTOR AT EXAM: FRAMINGHAM STUDY 30-YEAR FOLLOWUP

SEX: MALE
EVENT: INTERMITTENT CLAUDICATION
RISK FACTOR: METROPOLITAN RELATIVE WEIGHT (PERCENT)
PERSONS AT RISK: PERSONS FREE OF INTERMITTENT CLAUDICATION

RISK FACTOR	35-44 AT EXAM			45-54 AT EXAM			55-64 AT EXAM			65-74 AT EXAM		
	PERSON EXAMS	# OF EVENTS	ANNUAL RATE	PERSON EXAMS	# OF EVENTS	ANNUAL RATE	PERSON EXAMS	# OF EVENTS	ANNUAL RATE	PERSON EXAMS	# OF EVENTS	ANNUAL RATE
54-104	723			1144	5	2	1204	18	7	741	11	7
105-114	979			1609	9	3	1805	18	5	985	14	7
115-124	1133			2124	7	2	2398	26	5	1251	16	6
125-134	923			1677	5	1	1812	16	4	886	15	8
135-272	756			1443	3	1	1580	13	4	653	9	7
UNKNOWN	0			1	0	2	1	0	0	0	0	-
TOTAL	4514			7998	29	2	8800	91	5	4516	65	7

RISK FACTOR	75-84 AT EXAM			85-94 AT EXAM			35-64 COMBINED			65-94 COMBINED		
	PERSON EXAMS	# OF EVENTS	ANNUAL RATE	PERSON EXAMS	# OF EVENTS	ANNUAL RATE	# OF EVENTS	CRUDE	AGE-ADJUSTED	# OF EVENTS	CRUDE	AGE-ADJUSTED
54-104	247	1	2	18			23	4	4	12	6	6
105-114	327	7	11	14			27	3	3	21	8	8
115-124	360	3	4	13			33	3	3	19	6	6
125-134	178	2	6	1			22	2	2	17	8	8
135-272	111	1	5	0			18	2	2	10	7	6
UNKNOWN	0	0	-	0			0	0	0	0	0	-
TOTAL	1223	14	6	46			123	3	3	79	7	7

LOGISTIC REGRESSIONS

AGE AT EXAM		COEFFICIENT OF RISK FACTOR	SE COEFFICIENT	P-VALUE	STANDARDIZED COEFFICIENT	STD SE
35-44						
45-54		-0.013	0.012	0.287	-0.206	0.193
55-64		-0.015	0.007	0.032	-0.236	0.110
65-74		-0.003	0.008	0.755	-0.039	0.126
75-84		0.002	0.018	0.927	0.025	0.267
85-94						
35-64	UNIVARIATE	-0.010	0.006	0.077	-0.165	0.093
65-94	UNIVARIATE	-0.001	0.007	0.895	-0.015	0.114
35-64	BIVARIATE	-0.011	0.006	0.062	-0.175	0.094
65-94	BIVARIATE	-0.002	0.007	0.836	-0.024	0.114
35-64	MULTIVARIATE	-0.016	0.006	0.007	-0.265	0.099
65-94	MULTIVARIATE	-0.006	0.008	0.401	-0.099	0.118

SEX: FEMALE
EVENT: INTERMITTENT CLAUDICATION
RISK FACTOR: METROPOLITAN RELATIVE WEIGHT (PERCENT)
PERSONS AT RISK: PERSONS FREE OF INTERMITTENT CLAUDICATION

RISK FACTOR	35-44 AT EXAM			45-54 AT EXAM			55-64 AT EXAM			65-74 AT EXAM		
	PERSON EXAMS	# OF EVENTS	ANNUAL RATE	PERSON EXAMS	# OF EVENTS	ANNUAL RATE	PERSON EXAMS	# OF EVENTS	ANNUAL RATE	PERSON EXAMS	# OF EVENTS	ANNUAL RATE
54-104	1694			2116	2	0	2108	21	5	1271	10	4
105-114	1430			2362	0	0	2516	12	2	1284	8	3
115-124	1057			2121	3	1	2518	8	2	1496	10	3
125-134	565			1378	4	1	1812	6	2	1113	5	2
135-272	856			2065	6	1	2992	16	3	1737	14	4
UNKNOWN	8			0	0	-	0	0	-	0	0	-
TOTAL	5610			10042	15	1	11946	63	3	6901	47	3

RISK FACTOR	75-84 AT EXAM			85-94 AT EXAM			35-64 COMBINED			65-94 COMBINED		
	PERSON EXAMS	# OF EVENTS	ANNUAL RATE	PERSON EXAMS	# OF EVENTS	ANNUAL RATE	# OF EVENTS	CRUDE	AGE-ADJUSTED	# OF EVENTS	CRUDE	AGE-ADJUSTED
54-104	589	3	3	49			23	2	2	13	3	4
105-114	436	4	5	32			13	1	1	12	3	3
115-124	464	6	6	28			11	1	1	16	4	3
125-134	323	1	2	23			10	1	1	6	2	2
135-272	484	3	3	21			23	2	2	17	4	4
UNKNOWN	0	0	-	0			0	0	0	0	0	-
TOTAL	2296	17	4	153			80	1	1	64	3	3

LOGISTIC REGRESSIONS

AGE AT EXAM		COEFFICIENT OF RISK FACTOR	SE COEFFICIENT	P-VALUE	STANDARDIZED COEFFICIENT	STD SE
35-44						
45-54		0.023	0.008	0.003	0.494	0.166
55-64		-0.002	0.006	0.765	-0.039	0.129
65-74		0.001	0.006	0.857	0.026	0.144
75-84		-0.003	0.011	0.780	-0.070	0.252
85-94						
35-64	UNIVARIATE	0.009	0.004	0.050	0.191	0.098
65-94	UNIVARIATE	0.000	0.006	0.970	0.005	0.125
35-64	BIVARIATE	0.005	0.005	0.257	0.116	0.102
65-94	BIVARIATE	0.000	0.006	0.986	0.002	0.126
35-64	MULTIVARIATE	0.001	0.005	0.893	0.014	0.107
65-94	MULTIVARIATE	-0.002	0.006	0.777	-0.036	0.128

AVERAGE ANNUAL INCIDENCE RATE FOR EVENT PER 1000 PERSONS AT RISK AT EXAM
BY AGE AND LEVEL OF RISK FACTOR AT EXAM: FRAMINGHAM STUDY 30-YEAR FOLLOWUP

SEX: MALE
EVENT: INTERMITTENT CLAUDICATION
RISK FACTOR: VITAL CAPACITY - HEIGHT INDEX (ML/INCH)
PERSONS AT RISK: PERSONS FREE OF INTERMITTENT CLAUDICATION

35-44 AT EXAM			45-54 AT EXAM			55-64 AT EXAM			65-74 AT EXAM		
PERSON EXAMS	# OF EVENTS	ANNUAL RATE	PERSON EXAMS	# OF EVENTS	ANNUAL RATE	PERSON EXAMS	# OF EVENTS	ANNUAL RATE	PERSON EXAMS	# OF EVENTS	ANNUAL RATE
270			1093	8	4	2262	37	8	2003	37	9
413			1172	4	2	1642	12	4	909	11	6
750			1713	4	1	2012	16	4	820	9	5
1034			1813	8	2	1529	21	7	501	5	5
2034			2201	5	1	1337	5	2	271	3	6
13			6	0	0	18	0	0	12	0	0
4514			7998	29	2	8800	91	5	4516	65	7

75-84 AT EXAM			85-94 AT EXAM			35-64 COMBINED			65-94 COMBINED		
PERSON EXAMS	# OF EVENTS	ANNUAL RATE	PERSON EXAMS	# OF EVENTS	ANNUAL RATE	# OF EVENTS	CRUDE	AGE-ADJUSTED	# OF EVENTS	CRUDE	AGE-ADJUSTED
757	10	7	30			45	6	5	47	8	9
214	2	5	7			16	2	2	13	6	6
136	0	0	7			21	3	4	9	5	4
77	1	6	2			30	3	4	6	5	5
31	1	16	0			11	1	1	4	7	8
8	0	0	0			0	0	0	0	0	0
1223	14	6	46			123	3	3	79	7	7

LOGISTIC REGRESSIONS

	COEFFICIENT OF RISK FACTOR	SE COEFFICIENT	P-VALUE	STANDARDIZED COEFFICIENT	STD SE
	-0.040	0.020	0.039	-0.365	0.177
	-0.031	0.011	0.004	-0.291	0.101
	-0.019	0.013	0.141	-0.180	0.123
	0.021	0.029	0.461	0.198	0.269
VARIATE	-0.050	0.009	0.001	-0.481	0.085
VARIATE	-0.010	0.012	0.398	-0.095	0.112
ARIATE	-0.031	0.009	0.001	-0.299	0.091
ARIATE	-0.012	0.012	0.313	-0.116	0.115
TIVARIATE	-0.018	0.010	0.071	-0.172	0.095
TIVARIATE	-0.009	0.013	0.472	-0.085	0.118

SEX: FEMALE
EVENT: INTERMITTENT CLAUDICATION
RISK FACTOR: VITAL CAPACITY - HEIGHT INDEX (ML/INCH)
PERSONS AT RISK: PERSONS FREE OF INTERMITTENT CLAUDICATION

35-44 AT EXAM			45-54 AT EXAM			55-64 AT EXAM			65-74 AT EXAM		
PERSON EXAMS	# OF EVENTS	ANNUAL RATE	PERSON EXAMS	# OF EVENTS	ANNUAL RATE	PERSON EXAMS	# OF EVENTS	ANNUAL RATE	PERSON EXAMS	# OF EVENTS	ANNUAL RATE
122			543	2	2	1549	13	4	1995	17	4
291			1237	1	0	2510	17	3	1888	12	3
766			2094	4	1	3173	14	2	1680	13	4
1423			2831	4	1	2803	9	2	915	4	2
2987			3320	4	1	1890	10	3	417	1	1
21			17	0	0	21	0	0	6	0	0
5610			10042	15	1	11946	63	3	6901	47	3

75-84 AT EXAM			85-94 AT EXAM			35-64 COMBINED			65-94 COMBINED		
PERSON EXAMS	# OF EVENTS	ANNUAL RATE	PERSON EXAMS	# OF EVENTS	ANNUAL RATE	# OF EVENTS	CRUDE	AGE-ADJUSTED	# OF EVENTS	CRUDE	AGE-ADJUSTED
1134	8	4	102			15	2	2	25	4	4
652	7	5	34			18	2	2	19	4	4
364	1	1	16			18	1	1	14	3	3
119	1	4	1			14	1	1	5	2	1
25	0	0	0			15	1	1	1	1	1
2	0	0	0			0	0	0	0	0	0
2296	17	4	153			80	1	1	64	3	3

LOGISTIC REGRESSIONS

	COEFFICIENT OF RISK FACTOR	SE COEFFICIENT	P-VALUE	STANDARDIZED COEFFICIENT	STD SE
	-0.039	0.034	0.249	-0.289	0.251
	-0.027	0.016	0.099	-0.204	0.124
	-0.025	0.020	0.212	-0.180	0.144
	0.014	0.035	0.695	0.096	0.245
ARIATE	-0.055	0.013	0.001	-0.435	0.107
ARIATE	-0.014	0.017	0.408	-0.103	0.125
RIATE	-0.022	0.015	0.142	-0.174	0.119
RIATE	-0.016	0.017	0.351	-0.121	0.130
IVARIATE	-0.004	0.016	0.794	-0.033	0.126
IVARIATE	-0.009	0.018	0.611	-0.068	0.134

AVERAGE ANNUAL INCIDENCE RATE FOR EVENT PER 1000 PERSONS AT RISK AT EXAM
BY AGE AND LEVEL OF RISK FACTOR AT EXAM: FRAMINGHAM STUDY 30-YEAR FOLLOWUP

SEX: MALE
EVENT: INTERMITTENT CLAUDICATION
RISK FACTOR: HEART RATE (PER MINUTE)
PERSONS AT RISK: PERSONS FREE OF INTERMITTENT CLAUDICATION

RISK FACTOR	35-44 AT EXAM			45-54 AT EXAM			55-64 AT EXAM			65-74 AT EXAM		
	PERSON EXAMS	# OF EVENTS	ANNUAL RATE	PERSON EXAMS	# OF EVENTS	ANNUAL RATE	PERSON EXAMS	# OF EVENTS	ANNUAL RATE	PERSON EXAMS	# OF EVENTS	ANNUAL RATE
30-67	1295			2002	4	1	2562	17	3	1338	25	9
68-75	1502			2581	7	1	2688	25	5	1358	12	4
76-83	728			1452	8	3	1497	22	7	744	12	8
84-91	586			1136	5	2	1164	10	4	562	7	6
92-220	403			827	5	3	889	17	10	514	9	9
UNKNOWN	0			0	0	-	0	0	-	0	0	-
TOTAL	4514			7998	29	2	8800	91	5	4516	65	7

RISK FACTOR	75-84 AT EXAM			85-94 AT EXAM			35-64 COMBINED			65-94 COMBINED		
	PERSON EXAMS	# OF EVENTS	ANNUAL RATE	PERSON EXAMS	# OF EVENTS	ANNUAL RATE	# OF EVENTS	CRUDE	AGE-ADJUSTED	# OF EVENTS	CRUDE	AGE-ADJUSTED
30-67	351	5	7	17			21	2	2	30	9	9
68-75	334	1	1	13			33	2	2	13	4	4
76-83	240	3	6	12			30	4	4	15	8	8
84-91	161	5	16	4			15	3	3	12	8	8
92-220	137	0	0	0			24	6	6	9	7	7
UNKNOWN	0	0	-	0			0	-		0	-	
TOTAL	1223	14	6	46			123	3	3	79	7	7

LOGISTIC REGRESSIONS

AGE AT EXAM		COEFFICIENT OF RISK FACTOR	SE COEFFICIENT	P-VALUE	STANDARDIZED COEFFICIENT	STD SE
35-44						
45-54		0.026	0.014	0.052	0.330	0.170
55-64		0.019	0.007	0.009	0.251	0.097
65-74		-0.004	0.010	0.696	-0.050	0.127
75-84		-0.007	0.020	0.739	-0.092	0.277
85-94						
35-64	UNIVARIATE	0.023	0.006	0.001	0.288	0.082
65-94	UNIVARIATE	-0.004	0.009	0.630	-0.056	0.115
35-64	BIVARIATE	0.022	0.006	0.001	0.285	0.081
65-94	BIVARIATE	-0.004	0.009	0.638	-0.054	0.116
35-64	MULTIVARIATE	0.009	0.007	0.188	0.115	0.087
65-94	MULTIVARIATE	-0.010	0.009	0.243	-0.139	0.119

SEX: FEMALE
EVENT: INTERMITTENT CLAUDICATION
RISK FACTOR: HEART RATE (PER MINUTE)
PERSONS AT RISK: PERSONS FREE OF INTERMITTENT CLAUDICATION

RISK FACTOR	35-44 AT EXAM			45-54 AT EXAM			55-64 AT EXAM			65-74 AT EXAM		
	PERSON EXAMS	# OF EVENTS	ANNUAL RATE	PERSON EXAMS	# OF EVENTS	ANNUAL RATE	PERSON EXAMS	# OF EVENTS	ANNUAL RATE	PERSON EXAMS	# OF EVENTS	ANNUAL RATE
30-67	1092			1916	1	0	2426	12	2	1247	9	4
68-75	1803			3275	3	0	3598	17	2	1845	13	4
76-83	1047			1929	3	1	2412	11	2	1507	11	4
84-91	939			1613	5	2	1857	14	4	1167	8	3
92-220	729			1309	3	1	1653	9	3	1135	6	3
UNKNOWN	0			0	0	-	0	0	-	0	0	-
TOTAL	5610			10042	15	1	11946	63	3	6901	47	3

RISK FACTOR	75-84 AT EXAM			85-94 AT EXAM			35-64 COMBINED			65-94 COMBINED		
	PERSON EXAMS	# OF EVENTS	ANNUAL RATE	PERSON EXAMS	# OF EVENTS	ANNUAL RATE	# OF EVENTS	CRUDE	AGE-ADJUSTED	# OF EVENTS	CRUDE	AGE-ADJUSTED
30-67	430	3	3	35			13	1	1	12	4	4
68-75	554	5	5	36			20	1	1	18	4	4
76-83	507	3	3	29			14	1	1	14	3	3
84-91	390	4	5	28			20	2	2	12	4	4
92-220	415	2	2	25			13	2	2	8	3	3
UNKNOWN	0	0	-	0			0	-		0	-	
TOTAL	2296	17	4	153			80	1	1	64	3	3

LOGISTIC REGRESSIONS

AGE AT EXAM		COEFFICIENT OF RISK FACTOR	SE COEFFICIENT	P-VALUE	STANDARDIZED COEFFICIENT	STD SE
35-44						
45-54		0.028	0.018	0.120	0.351	0.226
55-64		0.005	0.010	0.603	0.064	0.123
65-74		-0.005	0.011	0.662	-0.065	0.150
75-84		-0.009	0.018	0.636	-0.121	0.255
85-94						
35-64	UNIVARIATE	0.011	0.008	0.169	0.144	0.105
65-94	UNIVARIATE	-0.006	0.010	0.541	-0.079	0.129
35-64	BIVARIATE	0.011	0.008	0.182	0.138	0.103
65-94	BIVARIATE	-0.006	0.010	0.544	-0.078	0.129
35-64	MULTIVARIATE	-0.004	0.009	0.673	-0.047	0.112
65-94	MULTIVARIATE	-0.009	0.010	0.341	-0.122	0.128

TABLE 18-13   DATE=15MAY85

AVERAGE ANNUAL INCIDENCE RATE FOR EVENT PER 1000 PERSONS AT RISK AT EXAM
BY AGE AND LEVEL OF RISK FACTOR AT EXAM: FRAMINGHAM STUDY 30-YEAR FOLLOWUP

SEX: MALE
EVENT: INTERMITTENT CLAUDICATION
RISK FACTOR: CIGARETTES SMOKED (NUMBER PER DAY)
PERSONS AT RISK: PERSONS FREE OF INTERMITTENT CLAUDICATION

RISK FACTOR	35-44 AT EXAM			45-54 AT EXAM			55-64 AT EXAM			65-74 AT EXAM		
	PERSON EXAMS	# OF EVENTS	ANNUAL RATE	PERSON EXAMS	# OF EVENTS	ANNUAL RATE	PERSON EXAMS	# OF EVENTS	ANNUAL RATE	PERSON EXAMS	# OF EVENTS	ANNUAL RATE
NONE	1398			3194	7	1	4777	29	3	3049	41	7
1-10	437			739	1	1	828	7	4	429	8	9
11-20	1498			2115	12	3	1809	26	7	608	13	11
21-40	993			1658	9	3	1198	20	8	391	3	4
41-90	162			254	0	0	167	8	24	32	0	0
UNKNOWN	26			38	0	0	21	1	24	7	0	0
TOTAL	4514			7998	29	2	8800	91	5	4516	65	7

RISK FACTOR	75-84 AT EXAM			85-94 AT EXAM			35-64 COMBINED			65-94 COMBINED		
	PERSON EXAMS	# OF EVENTS	ANNUAL RATE	PERSON EXAMS	# OF EVENTS	ANNUAL RATE	# OF EVENTS	CRUDE	AGE-ADJUSTED	# OF EVENTS	CRUDE	AGE-ADJUSTED
NONE	972	9	5	36			38	2	2	50	6	7
1-10	112	3	13	6			9	2	2	11	10	10
11-20	80	1	6	4			38	4	4	14	10	10
21-40	57	0	0	0			29	4	4	3	3	3
41-90	2	1	250	0			8	7	10	1	15	53
UNKNOWN	0	0	-	0			1	6	10	0	0	0
TOTAL	1223	14	6	46			123	3	3	79	7	7

LOGISTIC REGRESSIONS

AGE AT EXAM		COEFFICIENT OF RISK FACTOR	SE COEFFICIENT	P-VALUE	STANDARDIZED COEFFICIENT	STD SE
35-44						
45-54		0.016	0.011	0.152	0.244	0.170
55-64		0.034	0.006	0.001	0.463	0.083
65-74		-0.003	0.011	0.792	-0.034	0.129
75-84		0.025	0.023	0.286	0.215	0.202
85-94						
35-64	UNIVARIATE	0.021	0.006	0.001	0.305	0.080
65-94	UNIVARIATE	0.002	0.010	0.824	0.025	0.111
35-64	BIVARIATE	0.029	0.005	0.001	0.424	0.078
65-94	BIVARIATE	0.001	0.010	0.902	0.014	0.112
35-64	MULTIVARIATE	0.033	0.005	0.001	0.476	0.078
65-94	MULTIVARIATE	0.003	0.010	0.784	0.051	0.112

SEX: FEMALE
EVENT: INTERMITTENT CLAUDICATION
RISK FACTOR: CIGARETTES SMOKED (NUMBER PER DAY)
PERSONS AT RISK: PERSONS FREE OF INTERMITTENT CLAUDICATION

RISK FACTOR	35-44 AT EXAM			45-54 AT EXAM			55-64 AT EXAM			65-74 AT EXAM		
	PERSON EXAMS	# OF EVENTS	ANNUAL RATE	PERSON EXAMS	# OF EVENTS	ANNUAL RATE	PERSON EXAMS	# OF EVENTS	ANNUAL RATE	PERSON EXAMS	# OF EVENTS	ANNUAL RATE
NONE	2581			5582	9	1	7918	30	2	5437	29	3
1-10	1314			1732	1	0	1627	9	3	649	6	5
11-20	1368			2032	2	0	1742	16	5	616	10	8
21-40	298			621	3	2	605	8	7	182	2	5
41-90	20			38	0	0	16	0	0	0	0	-
UNKNOWN	29			37	0	0	38	0	0	19	0	0
TOTAL	5610			10042	15	1	11946	63	3	6901	47	3

RISK FACTOR	75-84 AT EXAM			85-94 AT EXAM			35-64 COMBINED			65-94 COMBINED		
	PERSON EXAMS	# OF EVENTS	ANNUAL RATE	PERSON EXAMS	# OF EVENTS	ANNUAL RATE	# OF EVENTS	CRUDE	AGE-ADJUSTED	# OF EVENTS	CRUDE	AGE-ADJUSTED
NONE	2034	15	4	144			39	1	1	44	3	3
1-10	152	1	3	7			10	1	1	7	4	4
11-20	91	1	5	2			19	2	2	11	8	7
21-40	15	0	0	0			11	4	4	2	5	4
41-90	0	0	-	0			1	7	5	0	-	-
UNKNOWN	4	0	0	0			0	0	0	0	0	0
TOTAL	2296	17	4	153			80	1	2	64	3	3

LOGISTIC REGRESSIONS

AGE AT EXAM		COEFFICIENT OF RISK FACTOR	SE COEFFICIENT	P-VALUE	STANDARDIZED COEFFICIENT	STD SE
35-44						
45-54		0.021	0.022	0.332	0.216	0.223
55-64		0.039	0.010	0.001	0.365	0.093
65-74		0.036	0.014	0.010	0.268	0.104
75-84		-0.002	0.052	0.965	-0.011	0.249
85-94						
35-4	UNIVARIATE	0.033	0.009	0.001	0.324	0.089
65-64	UNIVARIATE	0.032	0.013	0.016	0.220	0.091
3-64	BIVARIATE	0.041	0.009	0.001	0.402	0.086
65-94	BIVARIATE	0.032	0.013	0.017	0.222	0.092
5-4	MULTIVARIATE	0.043	0.009	0.001	0.428	0.087
65-64	MULTIVARIATE	0.036	0.013	0.007	0.248	0.092

AVERAGE ANNUAL INCIDENCE RATE FOR EVENT PER 1000 PERSONS AT RISK AT EXAM
BY AGE AND LEVEL OF RISK FACTOR AT EXAM: FRAMINGHAM STUDY 30-YEAR FOLLOWUP

SEX: MALE
EVENT: INTERMITTENT CLAUDICATION
RISK FACTOR: ALBUMIN IN URINE, DEFINITE
PERSONS AT RISK: PERSONS FREE OF INTERMITTENT CLAUDICATION

RISK FACTOR	35-44 AT EXAM			45-54 AT EXAM			55-64 AT EXAM			65-74 AT EXAM		
	PERSON EXAMS	# OF EVENTS	ANNUAL RATE	PERSON EXAMS	# OF EVENTS	ANNUAL RATE	PERSON EXAMS	# OF EVENTS	ANNUAL RATE	PERSON EXAMS	# OF EVENTS	ANNUAL RATE
NO	4407			7870	28	2	8572	91	5	4367	62	7
YES	70			102	1	5	201	0	0	143	3	10
UNKNOWN	37			26	0	0	27	0	0	6	0	0
TOTAL	4514			7998	29	2	8800	91	5	4516	65	7

RISK FACTOR	75-84 AT EXAM			85-94 AT EXAM			35-64 COMBINED			65-94 COMBINED		
	PERSON EXAMS	# OF EVENTS	ANNUAL RATE	PERSON EXAMS	# OF EVENTS	ANNUAL RATE	# OF EVENTS	CRUDE	AGE-ADJUSTED	# OF EVENTS	CRUDE	AGE-ADJUSTED
NO	1189	13	5	41			121	3	3	75	7	7
YES	31	1	16	5			2	3	3	4	11	12
UNKNOWN	3	0	0	0			0	0	0	0	0	0
TOTAL	1223	14	6	46			123	3	3	79	7	7

LOGISTIC REGRESSIONS

AGE AT EXAM		COEFFICIENT OF RISK FACTOR	SE COEFFICIENT	P-VALUE	STANDARDIZED COEFFICIENT	STD SE
35-44						
45-54		1.020	1.023	0.319	0.115	0.115
55-64		-				
65-74		0.397	0.597	0.506	0.070	0.105
75-84		1.104	1.054	0.295	0.174	0.166
85-94						
35-64	UNIVARIATE	-0.080	0.715	0.911	-0.010	0.094
65-94	UNIVARIATE	0.521	0.519	0.316	0.090	0.090
35-64	BIVARIATE	-0.268	0.716	0.709	-0.035	0.094
65-94	BIVARIATE	0.522	0.519	0.314	0.090	0.090
35-64	MULTIVARIATE	-1.467	1.011	0.147	-0.193	0.133
65-94	MULTIVARIATE	0.250	0.528	0.636	0.043	0.091

SEX: FEMALE
EVENT: INTERMITTENT CLAUDICATION
RISK FACTOR: ALBUMIN IN URINE, DEFINITE
PERSONS AT RISK: PERSONS FREE OF INTERMITTENT CLAUDICATION

RISK FACTOR	35-44 AT EXAM			45-54 AT EXAM			55-64 AT EXAM			65-74 AT EXAM		
	PERSON EXAMS	# OF EVENTS	ANNUAL RATE	PERSON EXAMS	# OF EVENTS	ANNUAL RATE	PERSON EXAMS	# OF EVENTS	ANNUAL RATE	PERSON EXAMS	# OF EVENTS	ANNUAL RATE
NO	5347			9737	14	1	11686	62	5	6759	47	3
YES	225			280	1	2	232	1	2	132	0	0
UNKNOWN	38			25	0	0	28	0	0	10	0	0
TOTAL	5610			10042	15	1	11946	63	3	6901	47	3

RISK FACTOR	75-84 AT EXAM			85-94 AT EXAM			35-64 COMBINED			65-94 COMBINED		
	PERSON EXAMS	# OF EVENTS	ANNUAL RATE	PERSON EXAMS	# OF EVENTS	ANNUAL RATE	# OF EVENTS	CRUDE	AGE-ADJUSTED	# OF EVENTS	CRUDE	AGE-ADJUSTED
NO	2241	16	4	461			78	1	1	63	3	3
YES	46	1	11	7			2	1	2	1	3	3
UNKNOWN	9	0	0	0			0	0	0	0	0	0
TOTAL	2296	17	4	531			80	1	1	64	3	3

LOGISTIC REGRESSIONS

AGE AT EXAM		COEFFICIENT OF RISK FACTOR	SE COEFFICIENT	P-VALUE	STANDARDIZED COEFFICIENT	STD SE
35-44						
45-54		0.912	1.037	0.379	0.150	0.171
55-64		-0.209	1.010	0.836	-0.029	0.140
65-74		-				
75-84		1.128	1.042	0.279	0.158	0.146
85-94						
35-64	UNIVARIATE	-0.071	0.717	0.921	-0.012	0.116
65-94	UNIVARIATE	-0.244	1.011	0.809	-0.034	0.141
35-64	BIVARIATE	0.170	0.718	0.813	0.028	0.116
65-94	BIVARIATE	-0.242	1.011	0.811	-0.034	0.141
35-64	MULTIVARIATE	-0.274	0.748	0.714	-0.044	0.121
65-94	MULTIVARIATE	-0.474	1.018	0.641	-0.066	0.142

TABLE 18-15   DATE=13NOV85

AVERAGE ANNUAL INCIDENCE RATE FOR EVENT PER 1000 PERSONS AT RISK AT EXAM
BY AGE AND LEVEL OF RISK FACTOR AT EXAM: FRAMINGHAM STUDY 30-YEAR FOLLOWUP

SEX: MALE
EVENT: INTERMITTENT CLAUDICATION
RISK FACTOR: HEART ENLARGEMENT BY X-RAY
PERSONS AT RISK: PERSONS FREE OF INTERMITTENT CLAUDICATION

RISK FACTOR	35-44 AT EXAM			45-54 AT EXAM			55-64 AT EXAM			65-74 AT EXAM		
	PERSON EXAMS	# OF EVENTS	ANNUAL RATE	PERSON EXAMS	# OF EVENTS	ANNUAL RATE	PERSON EXAMS	# OF EVENTS	ANNUAL RATE	PERSON EXAMS	# OF EVENTS	ANNUAL RATE
NO	4096			6565	22	2	6375	67	3	2898	44	8
MILD	189			694	0	0	1081	13	6	661	4	3
DEFINITE	224			735	7	5	1343	11	4	957	17	9
UNKNOWN	5			4	0	0	1	0	0	0	0	-
TOTAL	4514			7998	29	2	8800	91	5	4516	65	7

RISK FACTOR	75-84 AT EXAM			85-94 AT EXAM			35-64 COMBINED			65-94 COMBINED		
	PERSON EXAMS	# OF EVENTS	ANNUAL RATE	PERSON EXAMS	# OF EVENTS	ANNUAL RATE	# OF EVENTS	CRUDE	AGE-ADJUSTED	# OF EVENTS	CRUDE	AGE-ADJUSTED
NO	686	8	6	32			92	3	3	52	7	7
MILD	204	2	5	6			13	3	2	6	3	3
DEFINITE	333	4	6	8			18	4	3	21	8	8
UNKNOWN	0	0	-	0			0	0	0	0	-	-
TOTAL	1223	14	6	46			123	3	3	79	7	7

LOGISTIC REGRESSIONS

AGE AT EXAM		COEFFICIENT OF RISK FACTOR	SE COEFFICIENT	P-VALUE	STANDARDIZED COEFFICIENT	STD SE
35-44						
45-54		0.429	0.237	0.070	0.265	0.146
55-64		-0.083	0.148	0.574	-0.062	0.110
65-74		0.022	0.152	0.886	0.018	0.124
75-84		0.004	0.310	0.990	0.003	0.269
85-94						
35-64	UNIVARIATE	0.188	0.124	0.128	0.124	0.081
65-94	UNIVARIATE	0.012	0.136	0.929	0.010	0.113
35-64	BIVARIATE	0.021	0.126	0.867	0.014	0.083
65-94	BIVARIATE	0.020	0.136	0.883	0.017	0.113
35-64	MULTIVARIATE	-0.037	0.153	0.784	-0.024	0.087
65-94	MULTIVARIATE	-0.079	0.145	0.585	-0.066	0.120

SEX: FEMALE
EVENT: INTERMITTENT CLAUDICATION
RISK FACTOR: HEART ENLARGEMENT BY X-RAY
PERSONS AT RISK: PERSONS FREE OF INTERMITTENT CLAUDICATION

RISK FACTOR	35-44 AT EXAM			45-54 AT EXAM			55-64 AT EXAM			65-74 AT EXAM		
	PERSON EXAMS	# OF EVENTS	ANNUAL RATE	PERSON EXAMS	# OF EVENTS	ANNUAL RATE	PERSON EXAMS	# OF EVENTS	ANNUAL RATE	PERSON EXAMS	# OF EVENTS	ANNUAL RATE
NO	5157			8066	9	1	7738	40	3	3419	17	2
MILD	199			951	3	2	1854	11	3	1260	7	3
DEFINITE	246			1019	3	1	2350	12	3	2222	23	5
UNKNOWN	8			6	0	0	4	0	0	0	0	-
TOTAL	5610			10042	15	1	11946	63	3	6901	47	3

RISK FACTOR	75-84 AT EXAM			85-94 AT EXAM			35-64 COMBINED			65-94 COMBINED		
	PERSON EXAMS	# OF EVENTS	ANNUAL RATE	PERSON EXAMS	# OF EVENTS	ANNUAL RATE	# OF EVENTS	CRUDE	AGE-ADJUSTED	# OF EVENTS	CRUDE	AGE-ADJUSTED
NO	794	1	1	55			51	1	1	18	2	2
MILD	439	5	6	36			14	2	2	12	3	3
DEFINITE	1063	11	5	82			15	2	2	34	5	5
UNKNOWN	0	0	-	0			0	0	0	0	-	-
TOTAL	2296	17	4	153			80	1	1	64	3	3

LOGISTIC REGRESSIONS

AGE AT EXAM		COEFFICIENT OF RISK FACTOR	SE COEFFICIENT	P-VALUE	STANDARDIZED COEFFICIENT	STD SE
35-44						
45-54		0.540	0.307	0.078	0.346	0.197
55-64		0.011	0.157	0.947	0.008	0.126
65-74		0.376	0.164	0.022	0.334	0.145
75-84		0.707	0.342	0.039	0.630	0.305
85-94						
35-64	UNIVARIATE	0.308	0.137	0.024	0.217	0.096
65-94	UNIVARIATE	0.434	0.144	0.003	0.389	0.129
35-64	BIVARIATE	0.057	0.142	0.689	0.040	0.100
65-94	BIVARIATE	0.453	0.146	0.002	0.406	0.131
35-64	MULTIVARIATE	-0.045	0.150	0.762	-0.032	0.106
65-94	MULTIVARIATE	0.420	0.152	0.006	0.377	0.137

AVERAGE ANNUAL INCIDENCE RATE FOR EVENT PER 1000 PERSONS AT RISK AT EXAM
BY AGE AND LEVEL OF RISK FACTOR AT EXAM: FRAMINGHAM STUDY 30-YEAR FOLLOWUP

SEX: MALE
EVENT: INTERMITTENT CLAUDICATION
RISK FACTOR: LEFT VENTRICULAR HYPERTROPHY BY ELECTROCARDIOGRAM
PERSONS AT RISK: PERSONS FREE OF INTERMITTENT CLAUDICATION

35-44 AT EXAM			45-54 AT EXAM			55-64 AT EXAM			65-74 AT EXAM		
PERSON EXAMS	# OF EVENTS	ANNUAL RATE	PERSON EXAMS	# OF EVENTS	ANNUAL RATE	PERSON EXAMS	# OF EVENTS	ANNUAL RATE	PERSON EXAMS	# OF EVENTS	ANNUAL RATE
4404			7749	26	2	8366	86	5	4255	60	7
79			141	3	11	236	1	2	115	1	4
31			108	0	0	198	4	10	146	4	14
0			0	0	-	0	0	-	0	0	-
4514			7998	29	2	8800	91	5	4516	65	7

75-84 AT EXAM			85-94 AT EXAM			35-64 COMBINED			65-94 COMBINED		
PERSON EXAMS	# OF EVENTS	ANNUAL RATE	PERSON EXAMS	# OF EVENTS	ANNUAL RATE	# OF EVENTS	CRUDE	AGE-ADJUSTED	# OF EVENTS	CRUDE	AGE-ADJUSTED
1138	13	6	42			113	3	3	73	7	7
40	0	0	2			5	6	6	1	3	3
45	1	11	2			5	7	8	5	13	13
0	0	-	0			0	-	-	0	-	-
1223	14	6	46			123	3	3	79	7	7

LOGISTIC REGRESSIONS

	COEFFICIENT OF RISK FACTOR	SE COEFFICIENT	P-VALUE	STANDARDIZED COEFFICIENT	STD SE
	0.530	0.456	0.245	0.140	0.120
	0.207	0.268	0.439	0.069	0.089
	0.268	0.264	0.510	0.103	0.101
	0.189	0.566	0.739	0.078	0.232
ARIATE	0.534	0.206	0.010	0.153	0.059
ARIATE	0.248	0.239	0.301	0.096	0.093
RIATE	0.391	0.207	0.059	0.112	0.059
RIATE	0.250	0.239	0.296	0.097	0.093
IVARIATE	0.036	0.239	0.881	0.010	0.068
IVARIATE	0.056	0.253	0.824	0.022	0.098

SEX: FEMALE
EVENT: INTERMITTENT CLAUDICATION
RISK FACTOR: LEFT VENTRICULAR HYPERTROPHY BY ELECTROCARDIOGRAM
PERSONS AT RISK: PERSONS FREE OF INTERMITTENT CLAUDICATION

35-44 AT EXAM			45-54 AT EXAM			55-64 AT EXAM			65-74 AT EXAM		
PERSON EXAMS	# OF EVENTS	ANNUAL RATE	PERSON EXAMS	# OF EVENTS	ANNUAL RATE	PERSON EXAMS	# OF EVENTS	ANNUAL RATE	PERSON EXAMS	# OF EVENTS	ANNUAL RATE
5557			9877	14	1	11539	59	3	6512	41	3
38			110	0	0	248	1	2	177	2	6
15			55	1	9	159	3	9	212	4	9
0			0	0	-	0	0	-	0	0	-
5610			10042	15	1	11946	63	3	6901	47	3

75-84 AT EXAM			85-94 AT EXAM			35-64 COMBINED			65-94 COMBINED		
PERSON EXAMS	# OF EVENTS	ANNUAL RATE	PERSON EXAMS	# OF EVENTS	ANNUAL RATE	# OF EVENTS	CRUDE	AGE-ADJUSTED	# OF EVENTS	CRUDE	AGE-ADJUSTED
2110	16	4	133			75	1	2	57	3	3
71	0	0	9			1	1	1	2	4	4
115	1	4	11			4	9	7	5	7	8
0	0	-	0			0	-	-	0	-	-
2296	17	4	153			80	1	2	64	3	3

LOGISTIC REGRESSIONS

	COEFFICIENT OF RISK FACTOR	SE COEFFICIENT	P-VALUE	STANDARDIZED COEFFICIENT	STD SE
	1.152	0.558	0.039	0.207	0.100
	0.554	0.303	0.068	0.148	0.081
	0.559	0.253	0.027	0.210	0.095
	-0.067	0.555	0.904	-0.031	0.257
ARIATE	0.827	0.264	0.002	0.178	0.057
ARIATE	0.393	0.229	0.087	0.158	0.092
RIATE	0.617	0.269	0.022	0.133	0.058
RIATE	0.399	0.230	0.083	0.161	0.093
IVARIATE	0.068	0.328	0.836	0.015	0.071
IVARIATE	0.268	0.240	0.265	0.108	0.097

TABLE 18-17   DATE=13NOV85
AVERAGE ANNUAL INCIDENCE RATE FOR EVENT PER 1000 PERSONS AT RISK AT EXAM
BY AGE AND LEVEL OF RISK FACTOR AT EXAM: FRAMINGHAM STUDY 30-YEAR FOLLOWUP

SEX: MALE
EVENT: INTERMITTENT CLAUDICATION
RISK FACTOR: INTRAVENTRICULAR CONDUCTION DEFECT (WITH QRS INTERVAL MORE THAN 0.11 SECOND)
PERSONS AT RISK: PERSONS FREE OF INTERMITTENT CLAUDICATION

RISK FACTOR	35-44 AT EXAM			45-54 AT EXAM			55-64 AT EXAM			65-74 AT EXAM		
	PERSON EXAMS	# OF EVENTS	ANNUAL RATE	PERSON EXAMS	# OF EVENTS	ANNUAL RATE	PERSON EXAMS	# OF EVENTS	ANNUAL RATE	PERSON EXAMS	# OF EVENTS	ANNUAL RATE
COMPLETE	4196			7427	27	2	8016	81	5	4043	58	7
MPLETE	280			484	1	1	530	4	4	238	4	8
	38			85	1	6	254	6	12	235	3	6
KNOWN	0			0	0	-	0	0	-	0	0	-
TAL	4514			7998	29	2	8800	91	5	4516	65	7

RISK FACTOR	75-84 AT EXAM			85-94 AT EXAM			35-64 COMBINED			65-94 COMBINED		
	PERSON EXAMS	# OF EVENTS	ANNUAL RATE	PERSON EXAMS	# OF EVENTS	ANNUAL RATE	# OF EVENTS	CRUDE	AGE-ADJUSTED	# OF EVENTS	CRUDE	AGE-ADJUSTED
COMPLETE	1069	14	7	38			110	3	3	72	7	7
MPLETE	36	0	0	5			6	2	2	4	7	7
	118	0	0	5			7	9	7	3	4	5
KNOWN	0	0	-	0			0	-		0	-	
TAL	1223	14	6	46			123	3	3	79	7	7

LOGISTIC REGRESSIONS

E AT AM		COEFFICIENT OF RISK FACTOR	SE COEFFICIENT	P-VALUE	STANDARDIZED COEFFICIENT	STD SE
-44						
-54		0.196	0.529	0.711	0.061	0.165
-64		0.293	0.214	0.171	0.118	0.086
-74		-0.013	0.259	0.961	-0.006	0.126
-84		-.				
-94						
-64	UNIVARIATE	0.410	0.197	0.037	0.143	0.069
-94	UNIVARIATE	-0.198	0.255	0.436	-0.102	0.132
-64	BIVARIATE	0.283	0.193	0.144	0.099	0.068
-94	BIVARIATE	-0.189	0.256	0.460	-0.098	0.132
-64	MULTIVARIATE	0.264	0.197	0.180	0.092	0.069
-94	MULTIVARIATE	-0.206	0.257	0.424	-0.106	0.133

SEX: FEMALE
EVENT: INTERMITTENT CLAUDICATION
RISK FACTOR: INTRAVENTRICULAR CONDUCTION DEFECT (WITH QRS INTERVAL MORE THAN 0.11 SECOND)
PERSONS AT RISK: PERSONS FREE OF INTERMITTENT CLAUDICATION

RISK FACTOR	35-44 AT EXAM			45-54 AT EXAM			55-64 AT EXAM			65-74 AT EXAM		
	PERSON EXAMS	# OF EVENTS	ANNUAL RATE	PERSON EXAMS	# OF EVENTS	ANNUAL RATE	PERSON EXAMS	# OF EVENTS	ANNUAL RATE	PERSON EXAMS	# OF EVENTS	ANNUAL RATE
COMPLETE	5523			9843	15	1	11577	61	3	6600	44	3
MPLETE	75			151	0	0	193	1	3	127	0	0
	12			48	0	0	176	1	3	174	3	9
KNOWN	0			0	0	-	0	0	-	0	0	-
TAL	5610			10042	15	1	11946	63	3	6901	47	3

RISK FACTOR	75-84 AT EXAM			85-94 AT EXAM			35-64 COMBINED			65-94 COMBINED		
	PERSON EXAMS	# OF EVENTS	ANNUAL RATE	PERSON EXAMS	# OF EVENTS	ANNUAL RATE	# OF EVENTS	CRUDE	AGE-ADJUSTED	# OF EVENTS	CRUDE	AGE-ADJUSTED
COMPLETE	2150	14	3	143			78	1	1	58	3	3
MPLETE	45	0	0	3			1	1	1	0	0	0
	101	3	15	7			1	2	1	6	11	10
KNOWN	0	0	-	0			0	-		0	-	
TAL	2296	17	4	153			80	1	1	44	3	3

LOGISTIC REGRESSIONS

AT M		COEFFICIENT OF RISK FACTOR	SE COEFFICIENT	P-VALUE	STANDARDIZED COEFFICIENT	STD SE
44		-			-	-
54		-			-	-
64		0.027	0.457	0.953	0.007	0.124
74		0.373	0.318	0.240	0.126	0.108
84		0.728	0.332	0.028	0.312	0.142
94						
64	UNIVARIATE	0.100	0.470	0.832	0.022	0.103
94	UNIVARIATE	0.520	0.226	0.022	0.190	0.083
64	BIVARIATE	-0.090	0.463	0.846	-0.020	0.102
94	BIVARIATE	0.526	0.227	0.021	0.192	0.083
64	MULTIVARIATE	-0.034	0.462	0.942	-0.007	0.102
94	MULTIVARIATE	0.529	0.228	0.020	0.193	0.083

TABLE 18-18   DATE=15NOV85
AVERAGE ANNUAL INCIDENCE RATE FOR EVENT PER 1000 PERSONS AT RISK AT EXAM
BY AGE AND LEVEL OF RISK FACTOR AT EXAM: FRAMINGHAM STUDY 30-YEAR FOLLOWUP

SEX: MALE
EVENT: INTERMITTENT CLAUDICATION
RISK FACTOR: DEFINITE NONSPECIFIC T-WAVE OR ST-SEGMENT ABNORMALITY BY ELECTROCARDIOGRAM
PERSONS AT RISK: PERSONS FREE OF INTERMITTENT CLAUDICATION

RISK FACTOR	35-44 AT EXAM PERSON EXAMS	# OF EVENTS	ANNUAL RATE	45-54 AT EXAM PERSON EXAMS	# OF EVENTS	ANNUAL RATE	55-64 AT EXAM PERSON EXAMS	# OF EVENTS	ANNUAL RATE	65-74 AT EXAM PERSON EXAMS	# OF EVENTS	ANNUAL RATE
NO	4401			7584	23	2	8033	78	5	3929	51	6
YES	113			414	6	7	767	13	8	587	14	12
UNKNOWN	0			0	0		0	0	-	0	0	-
TOTAL	4514			7998	29	2	8800	91	5	4516	65	7

RISK FACTOR	75-84 AT EXAM PERSON EXAMS	# OF EVENTS	ANNUAL RATE	85-94 AT EXAM PERSON EXAMS	# OF EVENTS	ANNUAL RATE	35-64 COMBINED # OF EVENTS	CRUDE	AGE-ADJUSTED	65-94 COMBINED # OF EVENTS	CRUDE	AGE-ADJUSTED
NO	988	12	6	30			104	3	3	63	6	6
YES	235	2	4	16			19	7	6	16	10	10
UNKNOWN	0	0	-	0			0	-		0	-	
TOTAL	1223	14	6	46			123	5	3	79	7	7

LOGISTIC REGRESSIONS

AGE AT EXAM	COEFFICIENT OF RISK FACTOR	SE COEFFICIENT	P-VALUE	STANDARDIZED COEFFICIENT	STD SE
35-44					
45-54	1.576	0.461	0.001	0.349	0.102
55-64	0.564	0.302	0.062	0.159	0.085
65-74	0.619	0.305	0.042	0.208	0.103
75-84	-0.359	0.767	0.640	-0.142	0.302
85-94					
35-64 UNIVARIATE	1.049	0.251	0.001	0.250	0.060
65-94 UNIVARIATE	0.411	0.282	0.145	0.145	0.099
35-64 BIVARIATE	0.765	0.253	0.003	0.183	0.060
65-94 BIVARIATE	0.434	0.284	0.126	0.153	0.100
35-64 MULTIVARIATE	0.516	0.257	0.044	0.125	0.061
65-94 MULTIVARIATE	0.338	0.286	0.237	0.119	0.101

SEX: FEMALE
EVENT: INTERMITTENT CLAUDICATION
RISK FACTOR: DEFINITE NONSPECIFIC T-WAVE OR ST-SEGMENT ABNORMALITY BY ELECTROCARDIOGRAM
PERSONS AT RISK: PERSONS FREE OF INTERMITTENT CLAUDICATION

RISK FACTOR	35-44 AT EXAM PERSON EXAMS	# OF EVENTS	ANNUAL RATE	45-54 AT EXAM PERSON EXAMS	# OF EVENTS	ANNUAL RATE	55-64 AT EXAM PERSON EXAMS	# OF EVENTS	ANNUAL RATE	65-74 AT EXAM PERSON EXAMS	# OF EVENTS	ANNUAL RATE
NO	5512			9616	14	1	10987	49	2	6004	40	3
YES	98			426	1	1	959	14	7	897	7	4
UNKNOWN	0			0	0	-	0	0	-	0	0	-
TOTAL	5610			10042	15	1	11946	63	3	6901	47	3

RISK FACTOR	75-84 AT EXAM PERSON EXAMS	# OF EVENTS	ANNUAL RATE	85-94 AT EXAM PERSON EXAMS	# OF EVENTS	ANNUAL RATE	35-64 COMBINED # OF EVENTS	CRUDE	AGE-ADJUSTED	65-94 COMBINED # OF EVENTS	CRUDE	AGE-ADJUSTED
NO	1854	15	4	120			65	1	1	55	3	3
YES	442	2	2	33			15	5	4	9	5	3
UNKNOWN	0	0	-	0			0	-		0	-	
TOTAL	2296	17	4	153			80	1	1	64	3	3

LOGISTIC REGRESSIONS

AGE AT EXAM	COEFFICIENT OF RISK FACTOR	SE COEFFICIENT	P-VALUE	STANDARDIZED COEFFICIENT	STD SE
35-44					
45-54	0.479	1.036	0.644	0.096	0.209
55-64	1.196	0.305	0.001	0.325	0.083
65-74	0.159	0.411	0.699	0.054	0.138
75-84	-0.585	0.755	0.438	-0.231	0.298
85-94					
35-64 UNIVARIATE	1.410	0.288	0.001	0.318	0.065
65-94 UNIVARIATE	-0.050	0.361	0.890	-0.018	0.128
35-64 BIVARIATE	1.049	0.291	0.001	0.237	0.066
65-94 BIVARIATE	-0.043	0.362	0.905	-0.015	0.128
35-64 MULTIVARIATE	0.779	0.297	0.009	0.176	0.067
65-94 MULTIVARIATE	-0.141	0.364	0.698	-0.050	0.129

TABLE 19-1   DATE=16NOV85

AVERAGE ANNUAL INCIDENCE RATE FOR EVENT PER 1000 PERSONS AT RISK AT EXAM
BY AGE AND LEVEL OF RISK FACTOR AT EXAM: FRAMINGHAM STUDY 30-YEAR FOLLOWUP

SEX: MALE
EVENT: CONGESTIVE HEART FAILURE
RISK FACTOR: SYSTOLIC BLOOD PRESSURE, FIRST EXAMINER (MM HG)
PERSONS AT RISK: PERSONS FREE OF CONGESTIVE HEART FAILURE

RISK FACTOR	35-44 AT EXAM PERSON EXAMS	45-54 AT EXAM PERSON EXAMS	# OF EVENTS	ANNUAL RATE	55-64 AT EXAM PERSON EXAMS	# OF EVENTS	ANNUAL RATE	65-74 AT EX: PERSON EXAMS	# OF EVENTS
74-119	1226	1842	5	1	1591	11	3	640	9
120-139	2188	3533	4	1	3490	22	3	1592	13
140-159	850	1830	10	3	2336	15	3	1436	25
160-179	197	594	4	3	1055	13	6	664	18
180-300	51	240	7	15	492	13	13	346	12
UNKNOWN	0	0	0	-	0	0	-	0	0
TOTAL	4512	8039	30	2	8964	74	4	4678	77

RISK FACTOR	75-84 AT EXAM PERSON EXAMS	# OF EVENTS	ANNUAL RATE	85-94 AT EXAM PERSON EXAMS	# OF EVENTS	ANNUAL RATE	35-64 COMBINED # OF EVENTS	CRUDE	AGE-ADJUSTED	65-94 COMB: # OF EVENTS	CRUDE
74-119	135	5	19	8	0	-	16	2	2	14	9
120-139	434	4	5	15	1	33	27	1	2	18	4
140-159	398	12	15	15	2	67	26	3	2	39	11
160-179	204	9	22	6	1	83	18	5	4	28	16
180-300	110	5	23	2	1	250	21	13	13	18	20
UNKNOWN	0	0	-	0	0	-	0	-	-	0	-
TOTAL	1281	35	14	46	5	54	108	3	3	117	10

LOGISTIC REGRESSIONS

AGE AT EXAM		COEFFICIENT OF RISK FACTOR	SE COEFFICIENT	P-VALUE	STANDARDIZED COEFFICIENT	STD SE
35-44						
45-54		0.037	0.007	0.001	0.721	0.128
55-64		0.017	0.004	0.001	0.370	0.098
65-74		0.018	0.004	0.001	0.423	0.096
75-84		0.013	0.007	0.051	0.303	0.155
85-94		0.061	0.029	0.037	1.134	0.544
35-64	UNIVARIATE	0.028	0.003	0.001	0.563	0.071
65-94	UNIVARIATE	0.017	0.003	0.001	0.399	0.080
35-64	BIVARIATE	0.023	0.004	0.001	0.474	0.073
65-94	BIVARIATE	0.017	0.004	0.001	0.402	0.081
35-64	MULTIVARIATE	0.012	0.004	0.003	0.250	0.084
65-94	MULTIVARIATE	0.011	0.004	0.004	0.249	0.087

SEX: FEMALE
EVENT: CONGESTIVE HEART FAILURE
RISK FACTOR: SYSTOLIC BLOOD PRESSURE, FIRST EXAMINER (MM HG)
PERSONS AT RISK: PERSONS FREE OF CONGESTIVE HEART FAILURE

RISK FACTOR	35-44 AT EXAM PERSON EXAMS	45-54 AT EXAM PERSON EXAMS	# OF EVENTS	ANNUAL RATE	55-64 AT EXAM PERSON EXAMS	# OF EVENTS	ANNUAL RATE	65-74 AT EXAM PERSON EXAMS	# OF EVENTS
74-119	2493	2713	2	0	2019	4	1	670	3
120-139	2245	4017	7	1	4166	14	2	1896	13
140-159	674	2085	5	1	3168	13	2	2228	28
160-179	156	780	2	1	1657	12	4	1321	15
180-300	51	448	3	3	960	19	10	805	17
UNKNOWN	0	0	0	-	0	0	-	0	0
TOTAL	5619	10043	19	1	11970	62	3	6930	76

RISK FACTOR	75-84 AT EXAM PERSON EXAMS	# OF EVENTS	ANNUAL RATE	85-94 AT EXAM PERSON EXAMS	# OF EVENTS	ANNUAL RATE	35-64 COMBINED # OF EVENTS	CRUDE	AGE-ADJUSTED	65-94 COMB: # OF EVENTS	CRUDE
74-119	148	0	0	10	1	50	8	1	1	4	2
120-139	525	15	14	45	1	11	22	1	1	29	6
140-159	728	13	9	37	4	54	19	2	1	45	7
160-179	536	12	11	34	4	59	14	3	2	31	8
180-300	360	18	25	25	0	0	22	8	6	35	15
UNKNOWN	0	0	-	0	0	-	0	-	-	0	-
TOTAL	2297	58	13	151	10	33	85	2	2	144	8

LOGISTIC REGRESSIONS

AGE AT EXAM		COEFFICIENT OF RISK FACTOR	SE COEFFICIENT	P-VALUE	STANDARDIZED COEFFICIENT	STD SE
35-44						
45-54		0.018	0.007	0.012	0.407	0.162
55-64		0.025	0.004	0.001	0.610	0.090
65-74		0.015	0.004	0.001	0.372	0.103
75-84		0.013	0.005	0.007	0.327	0.121
85-94		0.001	0.013	0.962	0.016	0.327
35-64	UNIVARIATE	0.025	0.003	0.001	0.603	0.070
65-94	UNIVARIATE	0.015	0.003	0.001	0.562	0.075
35-64	BIVARIATE	0.022	0.003	0.001	0.512	0.077
65-94	BIVARIATE	0.013	0.003	0.001	0.326	0.076
35-64	MULTIVARIATE	0.013	0.004	0.001	0.317	0.087
65-94	MULTIVARIATE	0.007	0.003	0.036	0.170	0.081

TABLE 19-2   DATE=15NOV85
AVERAGE ANNUAL INCIDENCE RATE FOR EVENT PER 1000 PERSONS AT RISK AT EXAM
BY AGE AND LEVEL OF RISK FACTOR AT EXAM: FRAMINGHAM STUDY 30-YEAR FOLLOWUP

SEX: MALE
EVENT: CONGESTIVE HEART FAILURE
RISK FACTOR: DIASTOLIC BLOOD PRESSURE, FIRST EXAMINER (MM HG)
PERSONS AT RISK: PERSONS FREE OF CONGESTIVE HEART FAILURE

RISK FACTOR	35-44 AT EXAM PERSON EXAMS	# OF EVENTS	ANNUAL RATE	45-54 AT EXAM PERSON EXAMS	# OF EVENTS	ANNUAL RATE	55-64 AT EXAM PERSON EXAMS	# OF EVENTS	ANNUAL RATE	65-74 AT EXAM PERSON EXAMS	# OF EVENTS	ANNUAL RATE
20-74	1014			1338	6	2	2066	17	4	1484	20	7
75-84	1615			2855	4	1	3128	21	3	1583	22	7
85-94	1202			2143	5	1	2205	16	4	1043	18	9
95-104	508			1069	8	4	1101	12	5	408	12	15
105-160	173			434	7	8	464	8	9	160	5	16
UNKNOWN	0	-	-	0	0	-	0	0	-	0	0	-
TOTAL	4512			8039	30	2	8964	74	4	4678	77	8

RISK FACTOR	75-84 AT EXAM PERSON EXAMS	# OF EVENTS	ANNUAL RATE	85-94 AT EXAM PERSON EXAMS	# OF EVENTS	ANNUAL RATE	35-64 COMBINED # OF EVENTS	CRUDE	AGE-ADJUSTED	65-94 COMBINED # OF EVENTS	CRUDE	AGE-ADJUSTED
20-74	534	16	15	26	4	77	23	2	2	40	10	9
75-84	424	11	13	15	0	0	25	2	2	33	8	8
85-94	238	3	6	5	1	100	21	2	2	22	9	9
95-104	58	3	26	0	0	-	22	4	4	15	16	17
105-160	27	2	37	0	0	-	17	8	8	7	19	20
UNKNOWN	0	0	-	0	0	-	0	-		0	-	
TOTAL	1281	35	14	46	5	54	108	3	3	117	10	10

LOGISTIC REGRESSIONS

AGE AT EXAM		COEFFICIENT OF RISK FACTOR	SE COEFFICIENT	P-VALUE	STANDARDIZED COEFFICIENT	STD SE
35-44						
45-54		0.040	0.014	0.003	0.469	0.158
55-64		0.013	0.009	0.165	0.154	0.111
65-74		0.021	0.009	0.017	0.253	0.106
75-84		0.002	0.014	0.901	0.021	0.171
85-94		-0.019	0.044	0.672	-0.207	0.489
35-64	UNIVARIATE	0.026	0.007	0.001	0.308	0.088
65-94	UNIVARIATE	0.010	0.007	0.181	0.121	0.090
35-64	BIVARIATE	0.026	0.007	0.001	0.303	0.086
65-94	BIVARIATE	0.016	0.007	0.035	0.192	0.091
35-64	MULTIVARIATE	0.010	0.007	0.195	0.115	0.088
65-94	MULTIVARIATE	0.007	0.007	0.335	0.088	0.091

SEX: FEMALE
EVENT: CONGESTIVE HEART FAILURE
RISK FACTOR: DIASTOLIC BLOOD PRESSURE, FIRST EXAMINER (MM HG)
PERSONS AT RISK: PERSONS FREE OF CONGESTIVE HEART FAILURE

RISK FACTOR	35-44 AT EXAM PERSON EXAMS	# OF EVENTS	ANNUAL RATE	45-54 AT EXAM PERSON EXAMS	# OF EVENTS	ANNUAL RATE	55-64 AT EXAM PERSON EXAMS	# OF EVENTS	ANNUAL RATE	65-74 AT EXAM PERSON EXAMS	# OF EVENTS	ANNUAL RATE
20-74	2164			2638	4	1	3066	11	2	2011	30	7
75-84	2031			3570	5	1	3966	12	2	2374	14	3
85-94	988			2392	6	1	2981	15	3	1593	20	6
95-104	324			940	1	1	1336	11	4	663	6	5
105-160	112			503	3	3	621	13	10	289	6	10
UNKNOWN	0	-	-	0	0	-	0	0	-	0	0	-
TOTAL	5619			10043	19	1	11970	62	3	6930	76	5

RISK FACTOR	75-84 AT EXAM PERSON EXAMS	# OF EVENTS	ANNUAL RATE	85-94 AT EXAM PERSON EXAMS	# OF EVENTS	ANNUAL RATE	35-64 COMBINED # OF EVENTS	CRUDE	AGE-ADJUSTED	65-94 COMBINED # OF EVENTS	CRUDE	AGE-ADJUSTED
20-74	854	24	14	67	4	30	18	1	1	58	10	9
75-84	694	10	7	51	5	49	17	1	1	29	5	5
85-94	518	16	15	24	1	21	21	2	2	37	9	9
95-104	168	6	18	7	0	0	13	3	3	12	7	8
105-160	63	2	16	2	0	0	16	6	6	8	11	12
UNKNOWN	0	0	-	0	0	-	0	-		0	-	
TOTAL	2297	58	13	151	10	33	85	2	2	144	8	8

LOGISTIC REGRESSIONS

AGE AT EXAM		COEFFICIENT OF RISK FACTOR	SE COEFFICIENT	P-VALUE	STANDARDIZED COEFFICIENT	STD SE
35-44						
45-54		0.033	0.015	0.032	0.402	0.187
55-64		0.040	0.008	0.001	0.503	0.105
65-74		-0.000	0.009	0.972	-0.004	0.115
75-84		0.001	0.011	0.893	0.018	0.133
85-94		-0.002	0.025	0.934	-0.027	0.328
35-64	UNIVARIATE	0.040	0.007	0.001	0.493	0.086
65-94	UNIVARIATE	-0.004	0.007	0.524	-0.054	0.085
35-64	BIVARIATE	0.036	0.007	0.001	0.446	0.090
65-94	BIVARIATE	0.000	0.007	1.000	0.000	0.085
35-64	MULTIVARIATE	0.024	0.008	0.002	0.295	0.094
65-94	MULTIVARIATE	-0.006	0.007	0.406	-0.069	0.083

TABLE 19-3A   DATE=16NOV85
AVERAGE ANNUAL INCIDENCE RATE FOR EVENT PER 1000 PERSONS AT RISK AT EXAM
BY AGE AND LEVEL OF RISK FACTOR AT EXAM: FRAMINGHAM STUDY 30-YEAR FOLLOWUP

SEX: MALE
EVENT: CONGESTIVE HEART FAILURE
RISK FACTOR: HYPERTENSION WITH ANTIHYPERTENSIVE TREATMENT
PERSONS AT RISK: PERSONS FREE OF CONGESTIVE HEART FAILURE

35-44 AT EXAM			45-54 AT EXAM			55-64 AT EXAM			65-74 AT EXAM		
PERSON EXAMS	# OF EVENTS	ANNUAL RATE	PERSON EXAMS	# OF EVENTS	ANNUAL RATE	PERSON EXAMS	# OF EVENTS	ANNUAL RATE	PERSON EXAMS	# OF EVENTS	ANNUAL RATE
2665			4261	8	1	4137	20	2	1772	14	4
1315			2279	6	1	2544	14	3	1496	22	7
508			1213	11	5	1416	23	8	670	18	13
24		.	286	5	9	867	17	10	740	23	16
0			0	0	-	0	0	-	0	0	-
4512			8039	30	2	8964	74	4	4678	77	8

75-84 AT EXAM			85-94 AT EXAM			35-64 COMBINED			65-94 COMBINED		
PERSON EXAMS	# OF EVENTS	ANNUAL RATE	PERSON EXAMS	# OF EVENTS	ANNUAL RATE	# OF EVENTS	CRUDE	AGE-ADJUSTED	# OF EVENTS	CRUDE	AGE-ADJUSTED
450	5	6	17	0	0	28	1	1	19	4	4
419	12	14	14	2	71	21	2	2	36	9	9
176	7	20	5	1	100	37	6	6	26	15	15
236	11	23	10	2	100	22	9	7	36	18	18
0	0	-	0	0	-	0	-	-	0	-	-
1281	35	14	46	5	54	108	3	3	117	10	10

LOGISTIC REGRESSIONS

	COEFFICIENT OF RISK FACTOR	SE COEFFICIENT	P-VALUE	STANDARDIZED COEFFICIENT	STD SE
	0.924	0.230	0.001	0.715	0.178
	0.697	0.145	0.001	0.572	0.119
	0.668	0.150	0.001	0.549	0.123
	0.645	0.227	0.005	0.529	0.186
	1.215	0.732	0.097	1.023	0.617
UNIVARIATE	0.904	0.122	0.001	0.710	0.096
UNIVARIATE	0.685	0.123	0.001	0.563	0.101
BIVARIATE	0.796	0.122	0.001	0.625	0.096
BIVARIATE	0.680	0.123	0.001	0.559	0.101
MULTIVARIATE	0.584	0.129	0.001	0.458	0.101
MULTIVARIATE	0.545	0.127	0.001	0.448	0.104

SEX: FEMALE
EVENT: CONGESTIVE HEART FAILURE
RISK FACTOR: HYPERTENSION WITH ANTIHYPERTENSIVE TREATMENT
PERSONS AT RISK: PERSONS FREE OF CONGESTIVE HEART FAILURE

35-44 AT EXAM			45-54 AT EXAM			55-64 AT EXAM			65-74 AT EXAM		
PERSON EXAMS	# OF EVENTS	ANNUAL RATE	PERSON EXAMS	# OF EVENTS	ANNUAL RATE	PERSON EXAMS	# OF EVENTS	ANNUAL RATE	PERSON EXAMS	# OF EVENTS	ANNUAL RATE
4251			5650	6	1	4962	12	1	1853	9	2
1002			2609	5	1	3423	11	2	2090	17	4
316			1180	2	1	1925	19	5	1169	13	6
50			604	6	5	1660	20	6	1818	37	10
0			0	0	-	0	0	-	0	0	-
5619			10043	19	1	11970	62	3	6930	76	5

75-84 AT EXAM			85-94 AT EXAM			35-64 COMBINED			65-94 COMBINED		
PERSON EXAMS	# OF EVENTS	ANNUAL RATE	PERSON EXAMS	# OF EVENTS	ANNUAL RATE	# OF EVENTS	CRUDE	AGE-ADJUSTED	# OF EVENTS	CRUDE	AGE-ADJUSTED
447	9	10	36	1	14	21	1	1	19	4	4
663	12	9	35	4	57	17	1	1	33	6	6
450	15	17	34	1	15	21	3	2	29	9	8
737	22	15	46	4	43	26	6	4	63	12	12
0	0	-	0	0	-	0	-	-	0	-	-
2297	58	13	151	10	33	85	2	2	144	8	8

LOGISTIC REGRESSIONS

	COEFFICIENT OF RISK FACTOR	SE COEFFICIENT	P-VALUE	STANDARDIZED COEFFICIENT	STD SE
	0.727	0.275	0.008	0.560	0.211
	0.829	0.170	0.001	0.694	0.142
	0.652	0.168	0.001	0.534	0.138
	0.293	0.186	0.114	0.229	0.145
	0.178	0.414	0.668	0.147	0.343
UNIVARIATE	0.918	0.135	0.001	0.732	0.108
UNIVARIATE	0.524	0.119	0.001	0.426	0.097
BIVARIATE	0.719	0.139	0.001	0.573	0.111
BIVARIATE	0.469	0.120	0.001	0.381	0.097
MULTIVARIATE	0.555	0.147	0.001	0.443	0.117
MULTIVARIATE	0.318	0.122	0.009	0.259	0.099

TABLE 19-3B   DATE=15NOV85
AVERAGE ANNUAL INCIDENCE RATE FOR EVENT PER 1000 PERSONS AT RISK AT EXAM
BY AGE AND LEVEL OF RISK FACTOR AT EXAM: FRAMINGHAM STUDY 30-YEAR FOLLOWUP

SEX: MALE
EVENT: CONGESTIVE HEART FAILURE
RISK FACTOR: HYPERTENSION IGNORING TREATMENT
PERSONS AT RISK: PERSONS FREE OF CONGESTIVE HEART FAILURE

RISK FACTOR	35-44 AT EXAM PERSON EXAMS	# OF EVENTS	ANNUAL RATE	45-54 AT EXAM PERSON EXAMS	# OF EVENTS	ANNUAL RATE	55-64 AT EXAM PERSON EXAMS	# OF EVENTS	ANNUAL RATE	65-74 AT EXAM PERSON EXAMS	# OF EVENTS	ANNUAL RATE
NO	2669			4300	8	1	4308	27	3	1928	16	4
MILD	1325			2374	7	1	2882	18	3	1803	29	8
DEFINITE	518			1365	15	5	1774	29	8	947	32	17
UNKNOWN	0			0	0	-	0	0	-	0	0	-
TOTAL	4512			8039	30	2	8964	74	4	4678	77	8

RISK FACTOR	75-84 AT EXAM PERSON EXAMS	# OF EVENTS	ANNUAL RATE	85-94 AT EXAM PERSON EXAMS	# OF EVENTS	ANNUAL RATE	35-64 COMBINED # OF EVENTS	CRUDE	AGE-ADJUSTED	65-94 COMBINED # OF EVENTS	CRUDE	AGE-ADJUSTED
NO	503	7	7	22	1	23	35	2	2	24	5	5
MILD	531	16	15	19	3	79	26	2	2	48	10	10
DEFINITE	247	12	24	5	1	100	47	6	6	45	19	19
UNKNOWN	0	0	-	0	0	-	0	-	-	0	-	-
TOTAL	1281	35	14	46	5	54	108	3	3	117	10	10

LOGISTIC REGRESSIONS

AGE AT EXAM		COEFFICIENT OF RISK FACTOR	SE COEFFICIENT	P-VALUE	STANDARDIZED COEFFICIENT	STD SE
35-44						
45-54		0.933	0.231	0.001	0.705	0.174
55-64		0.489	0.144	0.001	0.379	0.112
65-74		0.721	0.152	0.001	0.545	0.115
75-84		0.627	0.230	0.006	0.463	0.170
85-94		0.859	0.682	0.208	0.583	0.463
35-64	UNIVARIATE	0.736	0.120	0.001	0.555	0.090
65-94	UNIVARIATE	0.679	0.124	0.001	0.510	0.093
35-64	BIVARIATE	0.661	0.120	0.001	0.499	0.091
65-94	BIVARIATE	0.692	0.125	0.001	0.520	0.094
35-64	MULTIVARIATE	0.412	0.129	0.001	0.310	0.097
65-94	MULTIVARIATE	0.541	0.130	0.001	0.406	0.097

SEX: FEMALE
EVENT: CONGESTIVE HEART FAILURE
RISK FACTOR: HYPERTENSION IGNORING TREATMENT
PERSONS AT RISK: PERSONS FREE OF CONGESTIVE HEART FAILURE

RISK FACTOR	35-44 AT EXAM PERSON EXAMS	# OF EVENTS	ANNUAL RATE	45-54 AT EXAM PERSON EXAMS	# OF EVENTS	ANNUAL RATE	55-64 AT EXAM PERSON EXAMS	# OF EVENTS	ANNUAL RATE	65-74 AT EXAM PERSON EXAMS	# OF EVENTS	ANNUAL RATE
NO	4273			5827	8	1	5369	12	1	2172	14	3
MILD	1017			2800	6	1	4038	19	2	2856	33	6
DEFINITE	329			1416	5	2	2563	31	6	1902	29	8
UNKNOWN	0			0	0	-	0	0	-	0	0	-
TOTAL	5619			10043	19	1	11970	62	3	6930	76	5

RISK FACTOR	75-84 AT EXAM PERSON EXAMS	# OF EVENTS	ANNUAL RATE	85-94 AT EXAM PERSON EXAMS	# OF EVENTS	ANNUAL RATE	35-64 COMBINED # OF EVENTS	CRUDE	AGE-ADJUSTED	65-94 COMBINED # OF EVENTS	CRUDE	AGE-ADJUSTED
NO	574	11	10	47	2	21	23	1	1	27	5	5
MILD	960	20	10	53	5	47	26	2	2	58	7	8
DEFINITE	763	27	18	51	3	29	36	4	3	59	11	10
UNKNOWN	0	0	-	0	0	-	0	-	-	0	-	-
TOTAL	2297	58	13	151	10	33	85	2	2	144	8	8

LOGISTIC REGRESSIONS

AGE AT EXAM		COEFFICIENT OF RISK FACTOR	SE COEFFICIENT	P-VALUE	STANDARDIZED COEFFICIENT	STD SE
35-44						
45-54		0.471	0.284	0.097	0.343	0.207
55-64		0.866	0.167	0.001	0.675	0.130
65-74		0.412	0.154	0.007	0.315	0.118
75-84		0.357	0.182	0.051	0.270	0.138
85-94		0.122	0.409	0.765	0.099	0.330
35-64	UNIVARIATE	0.871	0.134	0.001	0.648	0.100
65-94	UNIVARIATE	0.406	0.113	0.001	0.311	0.086
35-64	BIVARIATE	0.694	0.138	0.001	0.516	0.103
65-94	BIVARIATE	0.362	0.113	0.001	0.278	0.087
35-64	MULTIVARIATE	0.511	0.148	0.001	0.380	0.110
65-94	MULTIVARIATE	0.211	0.117	0.072	0.162	0.090

TABLE 19-4　　　　DATE=15NOV85
AVERAGE ANNUAL INCIDENCE RATE FOR EVENT PER 1000 PERSONS AT RISK AT EXAM
BY AGE AND LEVEL OF RISK FACTOR AT EXAM: FRAMINGHAM STUDY 30-YEAR FOLLOWUP

SEX: MALE
EVENT: CONGESTIVE HEART FAILURE
RISK FACTOR: SERUM CHOLESTEROL
PERSONS AT RISK: PERSONS FREE OF CONGESTIVE HEART FAILURE

	35-44 AT EXAM			45-54 AT EXAM			55-64 AT EXAM			65-74 AT EXAM		
RISK FACTOR	PERSON EXAMS	# OF EVENTS	ANNUAL RATE	PERSON EXAMS	# OF EVENTS	ANNUAL RATE	PERSON EXAMS	# OF EVENTS	ANNUAL RATE	PERSON EXAMS	# OF EVENTS	ANNUAL RATE
84-204	1252			1794	4	1	2450	19	4	1459	21	7
205-234	1255			2385	8	2	2670	17	3	1424	25	9
235-264	881			1838	8	2	2108	17	4	1042	11	5
265-294	487			1054	5	2	1058	14	7	506	13	13
295-1124	303			617	3	2	546	7	6	230	7	15
UNKNOWN	334			351	2	3	132	0	0	17	0	0
TOTAL	4512			8039	30	2	8964	74	4	4678	77	8

	75-84 AT EXAM			85-94 AT EXAM			35-64 COMBINED			65-94 COMBINED		
RISK FACTOR	PERSON EXAMS	# OF EVENTS	ANNUAL RATE	PERSON EXAMS	# OF EVENTS	ANNUAL RATE	# OF EVENTS	CRUDE	AGE-ADJUSTED	# OF EVENTS	CRUDE	AGE-ADJUSTED
84-204	498	17	17	20	3	75	24	2	2	41	10	10
205-234	407	10	12	13	2	77	26	3	2	37	10	10
235-264	228	6	13	7	0	0	25	3	2	17	7	7
265-294	125	1	4	6	0	0	19	4	4	14	11	11
295-1124	23	1	22	0	0	-	11	4	4	8	16	16
UNKNOWN	0	0	-	0	0	-	3	2	1	0	0	0
TOTAL	1281	35	14	46	5	54	108	3	3	117	10	10

LOGISTIC REGRESSIONS

AGE AT EXAM		COEFFICIENT OF RISK FACTOR	SE COEFFICIENT	P-VALUE	STANDARDIZED COEFFICIENT	STD SE
35-44						
45-54		0.004	0.004	0.331	0.166	0.171
55-64		0.005	0.003	0.074	0.193	0.108
65-74		0.004	0.003	0.124	0.172	0.112
75-84		-0.006	0.005	0.212	-0.219	0.176
85-94		-0.009	0.012	0.433	-0.389	0.496
35-64	UNIVARIATE	0.004	0.002	0.033	0.189	0.088
65-94	UNIVARIATE	0.000	0.002	0.936	0.007	0.093
35-64	BIVARIATE	0.005	0.002	0.018	0.212 *	0.090
65-94	BIVARIATE	0.001	0.002	0.580	0.052	0.094
35-64	MULTIVARIATE	0.004	0.002	0.035	0.192	0.091
65-94	MULTIVARIATE	0.002	0.002	0.425	0.076	0.095

SEX: FEMALE
EVENT: CONGESTIVE HEART FAILURE
RISK FACTOR: SERUM CHOLESTEROL
PERSONS AT RISK: PERSONS FREE OF CONGESTIVE HEART FAILURE

	35-44 AT EXAM			45-54 AT EXAM			55-64 AT EXAM			65-74 AT EXAM		
RISK FACTOR	PERSON EXAMS	# OF EVENTS	ANNUAL RATE	PERSON EXAMS	# OF EVENTS	ANNUAL RATE	PERSON EXAMS	# OF EVENTS	ANNUAL RATE	PERSON EXAMS	# OF EVENTS	ANNUAL RATE
84-204	2350			2113	5	1	1681	14	4	859	12	7
205-234	1419			2582	1	0	2695	13	2	1532	15	5
235-264	807			2263	5	1	3035	12	2	1874	15	4
265-294	372			1486	1	1	2343	10	2	1436	14	5
295-1124	171			1119	5	2	1983	12	3	1192	20	8
UNKNOWN	500			480	0	0	233	1	2	37	0	0
TOTAL	5619			10043	19	1	11970	62	3	6930	76	5

	75-84 AT EXAM			85-94 AT EXAM			35-64 COMBINED			65-94 COMBINED		
RISK FACTOR	PERSON EXAMS	# OF EVENTS	ANNUAL RATE	PERSON EXAMS	# OF EVENTS	ANNUAL RATE	# OF EVENTS	CRUDE	AGE-ADJUSTED	# OF EVENTS	CRUDE	AGE-ADJUSTED
84-204	439	16	18	54	4	37	21	2	2	32	12	10
205-234	552	8	7	37	2	27	14	1	1	25	6	6
235-264	629	19	15	40	1	13	18	1	1	35	7	7
265-294	402	7	9	12	3	125	14	2	2	24	6	8
295-1124	268	7	13	8	0	0	17	3	2	27	9	9
UNKNOWN	7	1	71	0	0	-	0	0	0	1	11	17
TOTAL	2297	58	13	151	10	33	85	2	2	144	8	8

LOGISTIC REGRESSIONS

AGE AT EXAM		COEFFICIENT OF RISK FACTOR	SE COEFFICIENT	P-VALUE	STANDARDIZED COEFFICIENT	STD SE
35-44						
45-54		0.005	1.0026	0.028	0.253	0.115
55-64		-0.003	1.0030	0.373	-0.118	0.132
65-74		0.003	1.0030	0.290	0.119	0.112
75-84		-0.002	1.0030	0.530	-0.085	0.136
85-94		-0.001	1.0060	0.845	-0.065	0.333
35-64	UNIVARIATE	0.004	1.0020	0.076	0.166	0.094
65-94	UNIVARIATE	-0.001	1.0020	0.577	-0.045	0.085
35-64	BIVARIATE	0.000	1.0020	0.946	0.008	0.112
65-94	BIVARIATE	0.001	1.0020	0.776	0.024	0.085
35-64	MULTIVARIATE	-0.001	1.0020	0.738	-0.038	0.113
65-94	MULTIVARIATE	0.001	1.0020	0.559	0.049	0.084

SEX: MALE
EVENT: CONGESTIVE HEART FAILURE
RISK FACTOR: HEMATOCRIT
PERSONS AT RISK: PERSONS FREE OF CONGESTIVE HEART FAILURE

RISK FACTOR	35-44 AT EXAM			45-54 AT EXAM			55-64 AT EXAM			65-74 AT EXAM		
	PERSON EXAMS	# OF EVENTS	ANNUAL RATE	PERSON EXAMS	# OF EVENTS	ANNUAL RATE	PERSON EXAMS	# OF EVENTS	ANNUAL RATE	PERSON EXAMS	# OF EVENTS	ANNUAL RATE
25-42	628			959	3	2	1115	13	6	538	13	12
43-44	551			991	3	2	1083	5	2	594	8	7
45-46	990			1938	6	2	2189	17	4	1167	17	7
47-48	783			1578	8	3	1798	16	4	1041	15	7
49-70	1190			2183	7	2	2620	23	4	1318	24	9
UNKNOWN	370			390	3	4	159	0	0	20	0	0
TOTAL	4512			8039	30	2	8964	74	4	4678	77	8

RISK FACTOR	75-84 AT EXAM			85-94 AT EXAM			35-64 COMBINED			65-94 COMBINED		
	PERSON EXAMS	# OF EVENTS	ANNUAL RATE	PERSON EXAMS	# OF EVENTS	ANNUAL RATE	# OF EVENTS	CRUDE	AGE-ADJUSTED	# OF EVENTS	CRUDE	AGE-ADJUSTED
25-42	273	10	18	17	2	59	16	3	3	25	15	14
43-44	157	2	6	8	1	63	9	2	3	11	7	7
45-46	298	10	17	7	2	143	23	2	3	29	10	10
47-48	250	7	14	7	0	0	26	3	3	22	8	9
49-70	303	6	10	7	0	0	30	3	2	30	9	9
UNKNOWN	0	0	-	0	0	-	4	2	2	0	0	
TOTAL	1281	35	14	46	5	54	108	3	3	117	10	10

LOGISTIC REGRESSIONS

AGE AT EXAM		COEFFICIENT OF RISK FACTOR	SE COEFFICIENT	P-VALUE	STANDARDIZED COEFFICIENT	STD SE
35-44						
45-54		-0.003	0.054	0.950	-0.012	0.193
55-64		-0.006	0.031	0.840	-0.024	0.117
65-74		-0.030	0.032	0.350	-0.106	0.114
75-84		-0.049	0.041	0.227	-0.202	0.168
85-94		-0.025	0.097	0.801	-0.120	0.474
35-64	UNIVARIATE	-0.004	0.027	0.875	-0.016	0.098
65-94	UNIVARIATE	-0.049	0.024	0.043	-0.185	0.091
35-64	BIVARIATE	-0.006	0.026	0.821	-0.022	0.097
65-94	BIVARIATE	-0.034	0.024	0.158	-0.129	0.091
35-64	MULTIVARIATE	-0.023	0.026	0.375	-0.086	0.096
65-94	MULTIVARIATE	-0.047	0.025	0.064	-0.176	0.095

SEX: FEMALE
EVENT: CONGESTIVE HEART FAILURE
RISK FACTOR: HEMATOCRIT
PERSONS AT RISK: PERSONS FREE OF CONGESTIVE HEART FAILURE

RISK FACTOR	35-44 AT EXAM			45-54 AT EXAM			55-64 AT EXAM			65-74 AT EXAM		
	PERSON EXAMS	# OF EVENTS	ANNUAL RATE	PERSON EXAMS	# OF EVENTS	ANNUAL RATE	PERSON EXAMS	# OF EVENTS	ANNUAL RATE	PERSON EXAMS	# OF EVENTS	ANNUAL RATE
21-39	1698			1855	3	1	1597	8	3	736	15	10
40-41	967			1757	3	1	2125	12	3	1156	10	4
42-43	1043			1975	1	0	2356	6	1	1388	10	4
44-45	858			2215	0	0	2980	14	2	1791	20	6
46-65	487			1693	11	3	2637	21	4	1817	21	6
UNKNOWN	566			548	1	1	275	1	2	42	0	0
TOTAL	5619			10043	19	1	11970	62	3	6930	76	5

RISK FACTOR	75-84 AT EXAM			85-94 AT EXAM			35-64 COMBINED			65-94 COMBINED		
	PERSON EXAMS	# OF EVENTS	ANNUAL RATE	PERSON EXAMS	# OF EVENTS	ANNUAL RATE	# OF EVENTS	CRUDE	AGE-ADJUSTED	# OF EVENTS	CRUDE	AGE-ADJUSTED
21-39	356	9	13	29	1	17	12	1	1	25	11	11
40-41	433	16	18	33	3	43	15	2	2	29	9	8
42-43	415	12	14	23	2	40	8	1	1	24	7	7
44-45	576	7	6	33	3	45	15	1	1	30	6	6
46-65	511	13	13	29	1	17	32	3	3	35	7	8
UNKNOWN	6	1	83	0	0	-	3	1	1	1	10	20
TOTAL	2297	58	13	151	10	33	85	2	2	144	8	8

LOGISTIC REGRESSIONS

AGE AT EXAM		COEFFICIENT OF RISK FACTOR	SE COEFFICIENT	P-VALUE	STANDARDIZED COEFFICIENT	STD SE
35-44						
45-54		0.188	0.066	0.005	0.669	0.236
55-64		0.089	0.037	0.017	0.307	0.129
65-74		-0.032	0.032	0.320	-0.113	0.113
75-84		-0.062	0.034	0.066	-0.236	0.129
85-94		0.006	0.083	0.938	0.026	0.328
35-64	UNIVARIATE	0.136	0.051	0.001	0.489	0.112
65-94	UNIVARIATE	-0.054	0.023	0.017	-0.195	0.082
35-64	BIVARIATE	0.104	0.032	0.001	0.373	0.115
65-94	BIVARIATE	-0.043	0.023	0.054	-0.157	0.081
35-64	MULTIVARIATE	0.085	0.032	0.007	0.307	0.113
65-94	MULTIVARIATE	-0.050	0.022	0.025	-0.179	0.080

TABLE 19-6   DATE=15NOV85

AVERAGE ANNUAL INCIDENCE RATE FOR EVENT PER 1000 PERSONS AT RISK AT EXAM
BY AGE AND LEVEL OF RISK FACTOR AT EXAM: FRAMINGHAM STUDY 30-YEAR FOLLOWUP

SEX: MALE
EVENT: CONGESTIVE HEART FAILURE
RISK FACTOR: BLOOD GLUCOSE (MG/DL)
PERSONS AT RISK: PERSONS FREE OF CONGESTIVE HEART FAILURE

RISK FACTOR	35-44 AT EXAM			45-54 AT EXAM			55-64 AT EXAM			65-74 AT EXAM		
	PERSON EXAMS	# OF EVENTS	ANNUAL RATE	PERSON EXAMS	# OF EVENTS	ANNUAL RATE	PERSON EXAMS	# OF EVENTS	ANNUAL RATE	PERSON EXAMS	# OF EVENTS	ANNUAL RATE
29-69	1003			1554	5	2	1204	6	2	488	7	7
70-89	2527			4335	13	1	4477	33	4	2068	31	7
90-109	628			1396	8	3	2103	19	5	1267	15	6
110-129	106			330	2	3	602	4	5	412	8	10
130-524	48			200	2	5	477	12	13	426	16	19
UNKNOWN	200			224	0	0	101	0	0	17	0	0
TOTAL	4512			8039	30	2	8964	74	4	4678	77	8

RISK FACTOR	75-84 AT EXAM			85-94 AT EXAM			35-64 COMBINED			65-94 COMBINED		
	PERSON EXAMS	# OF EVENTS	ANNUAL RATE	PERSON EXAMS	# OF EVENTS	ANNUAL RATE	# OF EVENTS	CRUDE	AGE-ADJUSTED	# OF EVENTS	CRUDE	AGE-ADJUSTED
29-69	84	2	12	1	0	0	11	1	2	9	8	8
70-89	461	11	12	16	0	0	47	2	2	42	8	8
90-109	416	13	16	15	1	33	27	3	3	29	9	8
110-129	162	3	9	5	2	200	6	3	3	13	11	11
130-524	158	6	19	9	2	111	16	11	11	24	20	20
UNKNOWN	0	0	-	0	0	-	1	1	1	0	0	0
TOTAL	1281	35	14	46	5	54	108	3	3	117	10	10

LOGISTIC REGRESSIONS

AGE AT EXAM		COEFFICIENT OF RISK FACTOR	SE COEFFICIENT	P-VALUE	STANDARDIZED COEFFICIENT	STD SE
35-44						
45-54		0.009	0.004	0.011	0.213	0.084
55-64		0.009	0.002	0.001	0.273	0.058
65-74		0.007	0.002	0.001	0.258	0.066
75-84		0.009	0.003	0.012	0.288	0.114
85-94		0.024	0.013	0.067	0.770	0.421
35-64	UNIVARIATE	0.012	0.002	0.001	0.294	0.039
65-94	UNIVARIATE	0.008	0.002	0.001	0.275	0.054
35-64	BIVARIATE	0.010	0.002	0.001	0.253	0.044
65-94	BIVARIATE	0.008	0.002	0.001	0.274	0.056
35-64	MULTIVARIATE	0.010	0.002	0.001	0.257	0.043
65-94	MULTIVARIATE	0.007	0.002	0.001	0.256	0.057

SEX: FEMALE
EVENT: CONGESTIVE HEART FAILURE
RISK FACTOR: BLOOD GLUCOSE (MG/DL)
PERSONS AT RISK: PERSONS FREE OF CONGESTIVE HEART FAILURE

RISK FACTOR	35-44 AT EXAM			45-54 AT EXAM			55-64 AT EXAM			65-74 AT EXAM		
	PERSON EXAMS	# OF EVENTS	ANNUAL RATE	PERSON EXAMS	# OF EVENTS	ANNUAL RATE	PERSON EXAMS	# OF EVENTS	ANNUAL RATE	PERSON EXAMS	# OF EVENTS	ANNUAL RATE
29-69	1067			1802	3	1	1403	7	2	575	5	4
70-89	3414			5802	7	1	6308	31	2	3419	30	4
90-109	672			1642	2	1	2842	13	2	1931	21	5
110-129	96			302	1	2	771	5	3	567	6	5
130-524	39			159	6	19	476	5	5	406	14	17
UNKNOWN	531			336	0	0	170	1	3	32	0	0
TOTAL	5619			10043	19	1	11970	62	3	6930	76	5

RISK FACTOR	75-84 AT EXAM			85-94 AT EXAM			35-64 COMBINED			65-94 COMBINED		
	PERSON EXAMS	# OF EVENTS	ANNUAL RATE	PERSON EXAMS	# OF EVENTS	ANNUAL RATE	# OF EVENTS	CRUDE	AGE-ADJUSTED	# OF EVENTS	CRUDE	AGE-ADJUSTED
29-69	124	5	20	5	0	0	10	1	1	10	7	8
70-89	954	17	9	56	3	27	40	1	1	50	6	6
90-109	745	18	12	60	2	17	15	1	1	41	7	7
110-129	269	7	13	18	4	111	6	3	2	17	10	9
130-524	204	10	25	12	1	42	12	9	12	25	20	19
UNKNOWN	1	1	500	0	0	-	2	1	2	1	15	122
TOTAL	2297	58	13	151	10	33	85	2	2	144	8	8

LOGISTIC REGRESSIONS

AGE AT EXAM		COEFFICIENT OF RISK FACTOR	SE COEFFICIENT	P-VALUE	STANDARDIZED COEFFICIENT	STD SE
35-44						
45-54		0.020	0.003	0.001	0.373	0.065
55-64		0.007	0.003	0.006	0.196	0.072
65-74		0.011	0.002	0.001	0.326	0.053
75-84		0.005	0.003	0.054	0.178	0.093
85-94		0.004	0.009	0.614	0.130	0.259
35-64	UNIVARIATE	0.011	0.002	0.001	0.286	0.040
65-94	UNIVARIATE	0.009	0.001	0.001	0.281	0.044
35-64	BIVARIATE	0.011	0.002	0.001	0.237	0.042
65-94	BIVARIATE	0.008	0.001	0.001	0.263	0.046
35-64	MULTIVARIATE	0.008	0.002	0.001	0.188	0.046
65-94	MULTIVARIATE	0.008	0.002	0.001	0.236	0.048

TABLE 19-7   DATE=15NOV85
AVERAGE ANNUAL INCIDENCE RATE FOR EVENT PER 1000 PERSONS AT RISK AT EXAM
BY AGE AND LEVEL OF RISK FACTOR AT EXAM: FRAMINGHAM STUDY 30-YEAR FOLLOWUP

SEX: MALE
EVENT: CONGESTIVE HEART FAILURE
RISK FACTOR: DIABETES MELLITUS
PERSONS AT RISK: PERSONS FREE OF CONGESTIVE HEART FAILURE

RISK FACTOR	35-44 AT EXAM PERSON EXAMS	# OF EVENTS	ANNUAL RATE	45-54 AT EXAM PERSON EXAMS	# OF EVENTS	ANNUAL RATE	55-64 AT EXAM PERSON EXAMS	# OF EVENTS	ANNUAL RATE	65-74 AT EXAM PERSON EXAMS	# OF EVENTS	ANNUAL RATE
NO	4465			7794	28	2	8344	60	4	4145	59	7
YES	47			245	2	4	620	14	11	533	18	17
UNKNOWN	0						0	0	-			
TOTAL	4512			8039	30	2	8964	74	4	4678	77	8

RISK FACTOR	75-84 AT EXAM PERSON EXAMS	# OF EVENTS	ANNUAL RATE	85-94 AT EXAM PERSON EXAMS	# OF EVENTS	ANNUAL RATE	35-64 COMBINED # EVENTS	CRUDE	AGE-ADJUSTED	65-94 COMBINED # EVENTS	CRUDE	AGE-ADJUSTED
NO	1128	29	13	41	4	49	90	2	2	92	9	9
YES	133	6	20	5	1	100	18	10	11	25	18	18
UNKNOWN	0	0	-	0	0	-	0	-		0	-	
TOTAL	1281	35	14	46	5	54	108	3	3	117	10	10

LOGISTIC REGRESSIONS

AGE AT EXAM		COEFFICIENT OF RISK FACTOR	SE COEFFICIENT	P-VALUE	STANDARDIZED COEFFICIENT	STD SE
35-44						
45-54		0.825	0.735	0.261	0.142	0.126
55-64		1.160	0.300	0.001	0.294	0.076
65-74		0.884	0.273	0.001	0.281	0.087
75-84		0.436	0.457	0.340	0.142	0.148
85-94		0.838	1.236	0.498	0.264	0.389
35-64	UNIVARIATE	1.524	0.260	0.001	0.307	0.052
65-94	UNIVARIATE	0.756	0.229	0.001	0.241	0.073
35-64	BIVARIATE	1.189	0.265	0.001	0.240	0.053
65-94	BIVARIATE	0.745	0.230	0.001	0.238	0.073
35-64	MULTIVARIATE	0.867	0.278	0.002	0.175	0.056
65-94	MULTIVARIATE	0.693	0.234	0.003	0.221	0.075

SEX: FEMALE
EVENT: CONGESTIVE HEART FAILURE
RISK FACTOR: DIABETES MELLITUS
PERSONS AT RISK: PERSONS FREE OF CONGESTIVE HEART FAILURE

RISK FACTOR	35-44 AT EXAM PERSON EXAMS	# OF EVENTS	ANNUAL RATE	45-54 AT EXAM PERSON EXAMS	# OF EVENTS	ANNUAL RATE	55-64 AT EXAM PERSON EXAMS	# OF EVENTS	ANNUAL RATE	65-74 AT EXAM PERSON EXAMS	# OF EVENTS	ANNUAL RATE
NO	5576			9837	15	1	11390	51	2	6383	53	4
YES	43			206	4	10	580	11	9	547	23	21
UNKNOWN	0						0	0	-			
TOTAL	5619			10043	19	1	11970	62	3	6930	76	5

RISK FACTOR	75-84 AT EXAM PERSON EXAMS	# OF EVENTS	ANNUAL RATE	85-94 AT EXAM PERSON EXAMS	# OF EVENTS	ANNUAL RATE	35-64 COMBINED # EVENTS	CRUDE	AGE-ADJUSTED	65-94 COMBINED # EVENTS	CRUDE	AGE-ADJUSTED
NO	2016	43	11	130	7	27	69	1	1	103	4	4
YES	281	15	27	21	3	71	16	10	10	41	24	23
UNKNOWN	0	0	-	0	0	-	0	-		0	-	
TOTAL	2297	58	13	151	10	33	85	2	2	144	8	8

LOGISTIC REGRESSIONS

AGE AT EXAM		COEFFICIENT OF RISK FACTOR	SE COEFFICIENT	P-VALUE	STANDARDIZED COEFFICIENT	STD SE
35-44						
45-54		2.562	0.567	0.001	0.363	0.080
55-64		1.458	0.335	0.001	0.313	0.072
65-74		1.657	0.254	0.001	0.447	0.068
75-84		0.951	0.307	0.002	0.312	0.101
85-94		1.075	0.735	0.144	0.373	0.255
35-64	UNIVARIATE	2.031	0.280	0.001	0.347	0.048
65-94	UNIVARIATE	1.423	0.188	0.001	0.408	0.054
35-64	BIVARIATE	1.679	0.283	0.001	0.286	0.048
65-94	BIVARIATE	1.316	0.190	0.001	0.378	0.055
35-64	MULTIVARIATE	1.374	0.296	0.001	0.234	0.050
65-94	MULTIVARIATE	1.166	0.196	0.001	0.335	0.056

TABLE 19-8   DATE=15NOV85

AVERAGE ANNUAL INCIDENCE RATE FOR EVENT PER 1000 PERSONS AT RISK AT EXAM
BY AGE AND LEVEL OF RISK FACTOR AT EXAM: FRAMINGHAM STUDY 30-YEAR FOLLOWUP

SEX: MALE
EVENT: CONGESTIVE HEART FAILURE
RISK FACTOR: GLUCOSE IN URINE, DEFINITE OR TRACE
PERSONS AT RISK: PERSONS FREE OF CONGESTIVE HEART FAILURE

RISK FACTOR	35-44 AT EXAM			45-54 AT EXAM			55-64 AT EXAM			65-74 AT EXAM		
	PERSON EXAMS	# OF EVENTS	ANNUAL RATE	PERSON EXAMS	# OF EVENTS	ANNUAL RATE	PERSON -EXAMS	# OF EVENTS	ANNUAL RATE	PERSON EXAMS	# OF EVENTS	ANNUAL RATE
NO	4421			7843	29	2	8688	66	4	4516	70	8
YES	58			175	1	3	252	8	16	157	6	19
UNKNOWN	33			21	0	0	24	0	0	5	1	100
TOTAL	4512			8039	30	2	8964	74	4	4678	77	8

RISK FACTOR	75-84 AT EXAM			85-94 AT EXAM			35-64 COMBINED			65-94 COMBINED		
	PERSON EXAMS	# OF EVENTS	ANNUAL RATE	PERSON EXAMS	# OF EVENTS	ANNUAL RATE	# OF EVENTS	CRUDE	AGE-ADJUSTED	# OF EVENTS	CRUDE	AGE-ADJUSTED
NO	1248	33	13	45	4	44	97	2	2	107	9	9
YES	33	2	30	1	1	500	10	10	9	9	24	25
UNKNOWN	0	0	-	0	0	-	1	6	3	1	100	78
TOTAL	1281	35	14	46	5	54	108	3	3	117	10	10

LOGISTIC REGRESSIONS

AGE AT EXAM		COEFFICIENT OF RISK FACTOR	SE COEFFICIENT	P-VALUE	STANDARDIZED COEFFICIENT	STD SE
35-44		-	-		-	-
45-54		0.437	1.020	0.668	0.064	0.149
55-64		1.455	0.380	0.001	0.241	0.063
65-74		0.926	0.433	0.033	0.167	0.078
75-84		0.865	0.751	0.249	0.137	0.119
85-94		-	-		-	-
35-64	UNIVARIATE	1.510	0.335	0.001	0.225	0.050
65-94	UNIVARIATE	0.969	0.355	0.006	0.170	0.062
35-64	BIVARIATE	1.345	0.337	0.001	0.200	0.050
65-94	BIVARIATE	1.029	0.357	0.004	0.181	0.063
35-64	MULTIVARIATE	1.214	0.346	0.001	0.181	0.051
65-94	MULTIVARIATE	0.949	0.363	0.009	0.167	0.064

SEX: FEMALE
EVENT: CONGESTIVE HEART FAILURE
RISK FACTOR: GLUCOSE IN URINE, DEFINITE OR TRACE
PERSONS AT RISK: PERSONS FREE OF CONGESTIVE HEART FAILURE

RISK FACTOR	35-44 AT EXAM			45-54 AT EXAM			55-64 AT EXAM			65-74 AT EXAM		
	PERSON EXAMS	# OF EVENTS	ANNUAL RATE	PERSON EXAMS	# OF EVENTS	ANNUAL RATE	PERSON EXAMS	# OF EVENTS	ANNUAL RATE	PERSON EXAMS	# OF EVENTS	ANNUAL RATE
NO	5540			9922	16	1	11777	62	3	6794	68	5
YES	41			97	3	15	167	0	0	126	8	32
UNKNOWN	38			24	0	0	26	0	0	10	0	0
TOTAL	5619			10043	19	1	11970	62	3	6930	76	5

RISK FACTOR	75-84 AT EXAM			85-94 AT EXAM			35-64 COMBINED			65-94 COMBINED		
	PERSON EXAMS	# OF EVENTS	ANNUAL RATE	PERSON EXAMS	# OF EVENTS	ANNUAL RATE	# OF EVENTS	CRUDE	AGE-ADJUSTED	# OF EVENTS	CRUDE	AGE-ADJUSTED
NO	2261	55	12	150	9	30	81	1	1	132	7	7
YES	27	3	56	1	1	500	4	7	8	12	39	45
UNKNOWN	9	0	0	0	0	-	0	0	0	0	0	0
TOTAL	2297	58	13	151	10	33	85	2	2	144	8	8

LOGISTIC REGRESSIONS

AGE AT EXAM		COEFFICIENT OF RISK FACTOR	SE COEFFICIENT	P-VALUE	STANDARDIZED COEFFICIENT	STD SE
35-44		-	-		-	-
45-54		2.984	0.638	0.001	0.292	0.062
55-64		-	-		-	-
65-74		1.903	0.385	0.001	0.254	0.051
75-84		1.612	0.627	0.010	0.174	0.068
85-94		-	-		-	-
35-64	UNIVARIATE	1.494	0.515	0.004	0.156	0.054
65-94	UNIVARIATE	1.759	0.313	0.001	0.224	0.040
35-64	BIVARIATE	1.342	0.517	0.009	0.140	0.054
65-94	BIVARIATE	1.873	0.316	0.001	0.238	0.040
35-64	MULTIVARIATE	0.818	0.542	0.131	0.086	0.057
65-94	MULTIVARIATE	1.554	0.332	0.001	0.198	0.042

TABLE 19-9   DATE=08MAR86
AVERAGE ANNUAL INCIDENCE RATE FOR EVENT PER 1000 PERSONS AT RISK AT EXAM
BY AGE AND LEVEL OF RISK FACTOR AT EXAM: FRAMINGHAM STUDY 30-YEAR FOLLOWUP

SEX: MALE
EVENT: CONGESTIVE HEART FAILURE
RISK FACTOR: GLUCOSE INTOLERANCE
PERSONS AT RISK: PERSONS FREE OF CONGESTIVE HEART FAILURE

RISK FACTOR	35-44 AT EXAM			45-54 AT EXAM			55-64 AT EXAM			65-74 AT EXAM		
	PERSON EXAMS	# OF EVENTS	ANNUAL RATE	PERSON EXAMS	# OF EVENTS	ANNUAL RATE	PERSON EXAMS	# OF EVENTS	ANNUAL RATE	PERSON EXAMS	# OF EVENTS	ANNUAL RATE
NO	4376			7510	26	2	7895	57	4	3802	53	7
YES	136			529	4	4	1069	17	8	876	24	14
UNKNOWN	0			0	0	-	0	0	-	0	0	-
TOTAL	4512			8039	30	2	8964	74	4	4678	77	8

RISK FACTOR	75-84 AT EXAM			85-94 AT EXAM			35-64 COMBINED			65-94 COMBINED		
	PERSON EXAMS	# OF EVENTS	ANNUAL RATE	PERSON EXAMS	# OF EVENTS	ANNUAL RATE	# OF EVENTS	CRUDE	AGE-ADJUSTED	# OF EVENTS	CRUDE	AGE-ADJUSTED
NO	996	27	14	35	2	29	85	2	2	82	8	9
YES	285	8	14	11	3	136.	23	7	6	35	15	15
UNKNOWN	0	0	-	0	0	-	0	-	-	0	-	-
TOTAL	1281	35	14	46	5	54	108	3	3	117	10	10

LOGISTIC REGRESSIONS

AGE AT EXAM		COEFFICIENT OF RISK FACTOR	SE COEFFICIENT	P-VALUE	STANDARDIZED COEFFICIENT	STD SE
35-44						
45-54		0.785	0.539	0.145	0.195	0.134
55-64		0.798	0.278	0.004	0.259	0.090
65-74		0.689	0.249	0.006	0.269	0.097
75-84		0.036	0.408	0.930	0.015	0.170
85-94		1.823	0.994	0.067	0.786	0.429
35-64	UNIVARIATE	1.136	0.236	0.001	0.309	0.064
65-94	UNIVARIATE	0.579	0.205	0.005	0.229	0.081
35-64	BIVARIATE	0.861	0.239	0.001	0.234	0.065
65-94	BIVARIATE	0.534	0.205	0.009	0.211	0.081
35-64	MULTIVARIATE	0.673	0.247	0.006	0.183	0.067
65-94	MULTIVARIATE	0.476	0.207	0.022	0.189	0.082

SEX: FEMALE
EVENT: CONGESTIVE HEART FAILURE
RISK FACTOR: GLUCOSE INTOLERANCE
PERSONS AT RISK: PERSONS FREE OF CONGESTIVE HEART FAILURE

RISK FACTOR	35-44 AT EXAM			45-54 AT EXAM			55-64 AT EXAM			65-74 AT EXAM		
	PERSON EXAMS	# OF EVENTS	ANNUAL RATE	PERSON EXAMS	# OF EVENTS	ANNUAL RATE	PERSON EXAMS	# OF EVENTS	ANNUAL RATE	PERSON EXAMS	# OF EVENTS	ANNUAL RATE
NO	5525			9665	13	1	10885	49	2	5958	51	4
YES	94			378	6	8	1085	13	6	972	25	13
UNKNOWN	0			0	0	-	0	0	-	0	0	-
TOTAL	5619			10043	19	1	11970	62	3	6930	76	5

RISK FACTOR	75-84 AT EXAM			85-94 AT EXAM			35-64 COMBINED			65-94 COMBINED		
	PERSON EXAMS	# OF EVENTS	ANNUAL RATE	PERSON EXAMS	# OF EVENTS	ANNUAL RATE	# OF EVENTS	CRUDE	AGE-ADJUSTED	# OF EVENTS	CRUDE	AGE-ADJUSTED
NO	1829	39	11	123	6	24	65	1	1	96	6	6
YES	468	19	20	28	4	71	20	6	7	48	16	16
UNKNOWN	0	0	-	0	0	-	0	-	-	0	-	-
TOTAL	2297	58	13	151	10	33	85	2	2	144	8	8

LOGISTIC REGRESSIONS

AGE AT EXAM		COEFFICIENT OF RISK FACTOR	SE COEFFICIENT	P-VALUE	STANDARDIZED COEFFICIENT	STD SE
35-44						
45-54		2.483	0.496	0.001	0.473	0.094
55-64		0.986	0.314	0.002	0.283	0.090
65-74		1.118	0.247	0.001	0.388	0.086
75-84		0.664	0.285	0.020	0.267	0.115
85-94		1.179	0.683	0.085	0.460	0.266
35-64	UNIVARIATE	1.650	0.257	0.001	0.380	0.059
65-94	UNIVARIATE	1.012	0.179	0.001	0.368	0.065
35-64	BIVARIATE	1.300	0.261	0.001	0.300	0.060
65-94	BIVARIATE	0.913	0.181	0.001	0.332	0.066
35-64	MULTIVARIATE	1.105	0.269	0.001	0.255	0.062
65-94	MULTIVARIATE	0.783	0.185	0.001	0.285	0.067

TABLE 19-10   DATE=15NOV85
AVERAGE ANNUAL INCIDENCE RATE FOR EVENT PER 1000 PERSONS AT RISK AT EXAM
BY AGE AND LEVEL OF RISK FACTOR AT EXAM: FRAMINGHAM STUDY 30-YEAR FOLLOWUP

SEX: MALE
EVENT: CONGESTIVE HEART FAILURE
RISK FACTOR: METROPOLITAN RELATIVE WEIGHT (PERCENT)
PERSONS AT RISK: PERSONS FREE OF CONGESTIVE HEART FAILURE

RISK FACTOR	35-44 AT EXAM			45-54 AT EXAM			55-64 AT EXAM			65-74 AT EXAM		
	PERSON EXAMS	# OF EVENTS	ANNUAL RATE	PERSON EXAMS	# OF EVENTS	ANNUAL RATE	PERSON -EXAMS	# OF EVENTS	ANNUAL RATE	PERSON EXAMS	# OF EVENTS	ANNUAL RATE
54-104	722			1146	1	0	1258	13	5	781	17	11
105-114	979			1630	9	3	1841	15	4	1037	16	8
115-124	1133			2132	9	2	2440	18	4	1290	16	6
125-134	922			1690	3	1	1837	11	3	930	14	8
135-272	756			1440	6	2	1587	17	5	640	14	11
UNKNOWN	0			1	0	0	0	0	0	0	0	-
TOTAL	4512			8039	30	2	-8964	74	4	4678	77	8

RISK FACTOR	75-84 AT EXAM			85-94 AT EXAM			35-64 COMBINED			65-94 COMBINED		
	PERSON EXAMS	# OF EVENTS	ANNUAL RATE	PERSON EXAMS	# OF EVENTS	ANNUAL RATE	# OF EVENTS	CRUDE	AGE-ADJUSTED	# OF EVENTS	CRUDE	AGE-ADJUSTED
54-104	270	5	9	19	1	26	14	3	3	23	11	11
105-114	348	12	18	13	3	115	25	3	2	31	11	11
115-124	385	7	9	12	1	42	27	2	2	24	7	7
125-134	187	7	19	2	0	0	16	2	2	21	9	10
135-272	99	4	20	0	0	-	26	3	3	18	12	13
UNKNOWN	0	0	-	0	0	-	0	0	0	0	0	-
TOTAL	1281	35	14	46	5	54	108	3	3	117	10	10

LOGISTIC REGRESSIONS

AGE AT EXAM		COEFFICIENT OF RISK FACTOR	SE COEFFICIENT	P-VALUE	STANDARDIZED COEFFICIENT	STD SE
35-44						
45-54		0.012	0.011	0.251	0.199	0.174
55-64		0.000	0.007	0.997	0.000	0.117
65-74		0.004	0.008	0.590	0.062	0.114
75-84		0.020	0.011	0.063	0.291	0.156
85-94		0.008	0.041	0.854	0.089	0.484
35-64	UNIVARIATE	0.007	0.006	0.258	0.111	0.094
65-94	UNIVARIATE	0.006	0.006	0.353	0.086	0.092
35-64	BIVARIATE	0.007	0.006	0.252	0.108	0.095
65-94	BIVARIATE	0.009	0.006	0.137	0.139	0.095
35-64	MULTIVARIATE	-0.000	0.006	0.968	-0.004	0.100
65-94	MULTIVARIATE	0.003	0.006	0.591	0.052	0.097

SEX: FEMALE
EVENT: CONGESTIVE HEART FAILURE
RISK FACTOR: METROPOLITAN RELATIVE WEIGHT (PERCENT)
PERSONS AT RISK: PERSONS FREE OF CONGESTIVE HEART FAILURE

RISK FACTOR	35-44 AT EXAM			45-54 AT EXAM			55-64 AT EXAM			65-74 AT EXAM		
	PERSON EXAMS	# OF EVENTS	ANNUAL RATE	PERSON EXAMS	# OF EVENTS	ANNUAL RATE	PERSON EXAMS	# OF EVENTS	ANNUAL RATE	PERSON EXAMS	# OF EVENTS	ANNUAL RATE
54-104	1694			2106	5	1	2117	9	2	1291	16	6
105-114	1431			2355	2	0	2508	7	1	1300	9	3
115-124	1059			2125	4	1	2534	7	1	1498	11	4
125-134	566			1382	0	0	1825	11	3	1115	11	5
135-272	861			2075	8	2	2986	28	5	1726	29	8
UNKNOWN	8			0	0	-	0	0	-	0	0	-
TOTAL	5619			10043	19	1	11970	62	3	6930	76	5

RISK FACTOR	75-84 AT EXAM			85-94 AT EXAM			35-64 COMBINED			65-94 COMBINED		
	PERSON EXAMS	# OF EVENTS	ANNUAL RATE	PERSON EXAMS	# OF EVENTS	ANNUAL RATE	# OF EVENTS	CRUDE	AGE-ADJUSTED	# OF EVENTS	CRUDE	AGE-ADJUSTED
54-104	606	13	11	48	3	31	15	1	1	32	8	8
105-114	436	4	5	30	2	33	10	1	1	15	4	4
115-124	436	16	18	26	3	58	12	1	1	30	4	7
125-134	330	9	14	25	1	20	11	1	1	21	7	7
135-272	469	16	17	22	1	23	37	3	3	46	10	11
UNKNOWN	0	0	-	0	0	-	0	0	0	0	0	-
TOTAL	2297	58	13	151	10	33	85	2	2	144	8	8

LOGISTIC REGRESSIONS

AGE AT EXAM		COEFFICIENT OF RISK FACTOR	SE COEFFICIENT	P-VALUE	STANDARDIZED COEFFICIENT	STD SE
35-44						
45-54		0.015	0.008	0.067	0.322	0.176
55-64		0.023	0.004	0.001	0.521	0.087
65-74		0.008	0.005	0.105	0.172	0.106
75-84		0.010	0.006	0.069	0.225	0.124
85-94		-0.001	0.017	0.962	-0.016	0.329
35-64	UNIVARIATE	0.023	0.003	0.001	0.495	0.073
65-94	UNIVARIATE	0.006	0.004	0.105	0.129	0.079
35-64	BIVARIATE	0.021	0.004	0.001	0.459	0.077
65-94	BIVARIATE	0.008	0.004	0.021	0.183	0.079
35-64	MULTIVARIATE	0.015	0.004	0.001	0.323	0.083
65-94	MULTIVARIATE	0.007	0.004	0.037	0.162	0.077

TABLE 19-11   DATE=15NOV85

AVERAGE ANNUAL INCIDENCE RATE FOR EVENT PER 1000 PERSONS AT RISK AT EXAM
BY AGE AND LEVEL OF RISK FACTOR AT EXAM: FRAMINGHAM STUDY 30-YEAR FOLLOWUP

SEX: MALE
EVENT: CONGESTIVE HEART FAILURE
RISK FACTOR: VITAL CAPACITY - HEIGHT INDEX (ML/INCH)
PERSONS AT RISK: PERSONS FREE OF CONGESTIVE HEART FAILURE

RISK FACTOR	35-44 AT EXAM			45-54 AT EXAM			55-64 AT EXAM			65-74 AT EXAM		
	PERSON EXAMS	# OF EVENTS	ANNUAL RATE	PERSON EXAMS	# OF EVENTS	ANNUAL RATE	PERSON EXAMS	# OF EVENTS	ANNUAL RATE	PERSON EXAMS	# OF EVENTS	ANNUAL RATE
8-44	268			1095	7	3	2315	35	8	2077	51	12
45-49	413			1181	7	3	1676	14	4	940	16	9
50-54	750			1721	5	1	2050	12	3	846	5	3
55-59	1034			1821	8	2	1556	10	3	525	3	3
60-85	2034			2215	3	1	1349	3	1	278	2	4
UNKNOWN	13			6	0	0	18	0	0	12	0	0
TOTAL	4512			8039	30	2	8964	74	4	4678	77	8

RISK FACTOR	75-84 AT EXAM			85-94 AT EXAM			35-64 COMBINED			65-94 COMBINED		
	PERSON EXAMS	# OF EVENTS	ANNUAL RATE	PERSON EXAMS	# OF EVENTS	ANNUAL RATE	# OF EVENTS	CRUDE	AGE-ADJUSTED	# OF EVENTS	CRUDE	AGE-ADJUSTED
8-44	797	28	18	30	5	83	43	6	5	84	14	14
45-49	226	3	7	7	0	0	22	3	3	19	8	8
50-54	137	3	11	7	0	0	18	2	2	8	4	5
55-59	80	0	0	2	0	0	19	1	1	3	2	2
60-85	33	1	15	0	0	-	6	1	1	3	5	6
UNKNOWN	8	0	0	0	0	-	0	0	0	0	0	0
TOTAL	1281	35	14	46	5	54	108	3	3	117	10	10

LOGISTIC REGRESSIONS

AGE AT EXAM		COEFFICIENT OF RISK FACTOR	SE COEFFICIENT	P-VALUE	STANDARDIZED COEFFICIENT	STD SE
35-44						
45-54		-0.054	0.019	0.004	-0.489	0.171
55-64		-0.051	0.012	0.001	-0.477	0.109
65-74		-0.054	0.012	0.001	-0.496	0.110
75-84		-0.063	0.019	0.001	-0.571	0.173
85-94		-0.063	0.064	0.324	-0.490	0.497
35-64	UNIVARIATE	-0.066	0.009	0.001	-0.637	0.088
65-94	UNIVARIATE	-0.059	0.010	0.001	-0.553	0.091
35-64	BIVARIATE	-0.052	0.010	0.001	-0.499	0.095
65-94	BIVARIATE	-0.054	0.010	0.001	-0.505	0.093
35-64	MULTIVARIATE	-0.041	0.011	0.001	-0.395	0.101
65-94	MULTIVARIATE	-0.051	0.010	0.001	-0.482	0.097

SEX: FEMALE
EVENT: CONGESTIVE HEART FAILURE
RISK FACTOR: VITAL CAPACITY - HEIGHT INDEX (ML/INCH)
PERSONS AT RISK: PERSONS FREE OF CONGESTIVE HEART FAILURE

RISK FACTOR	35-44 AT EXAM			45-54 AT EXAM			55-64 AT EXAM			65-74 AT EXAM		
	PERSON EXAMS	# OF EVENTS	ANNUAL RATE	PERSON EXAMS	# OF EVENTS	ANNUAL RATE	PERSON EXAMS	# OF EVENTS	ANNUAL RATE	PERSON EXAMS	# OF EVENTS	ANNUAL RATE
4-29	122			539	5	5	1512	24	8	1970	35	9
30-34	291			1238	3	1	2520	18	4	1899	27	7
35-39	769			2093	7	2	3184	12	2	1705	6	2
40-44	1427			2831	1	0	2827	6	1	929	8	4
45-92	2989			3325	3	0	1906	2	1	421	0	0
UNKNOWN	21			17	0	0	21	0	0	6	0	0
TOTAL	5619			10043	19	1	11970	62	3	6930	76	5

RISK FACTOR	75-84 AT EXAM			85-94 AT EXAM			35-64 COMBINED			65-94 COMBINED		
	PERSON EXAMS	# OF EVENTS	ANNUAL RATE	PERSON EXAMS	# OF EVENTS	ANNUAL RATE	# OF EVENTS	CRUDE	AGE-ADJUSTED	# OF EVENTS	CRUDE	AGE-ADJUSTED
4-29	1126	39	17	97	7	36	30	7	6	81	13	11
30-34	660	12	9	37	2	27	22	3	2	41	8	8
35-39	361	4	6	16	1	31	19	2	1	11	3	3
40-44	123	1	4	1	0	0	8	1	1	9	4	4
45-92	26	1	19	0	0	-	6	0	0	1	1	5
UNKNOWN	1	1	500	0	0	-	0	0	0	1	71	122
TOTAL	2297	58	13	151	10	33	85	2	2	144	8	8

LOGISTIC REGRESSIONS

AGE AT EXAM		COEFFICIENT OF RISK FACTOR	SE COEFFICIENT	P-VALUE	STANDARDIZED COEFFICIENT	STD SE
35-44						
45-54		-0.099	0.028	0.001	-0.729	0.208
55-64		-0.108	0.015	0.001	-0.807	0.113
65-74		-0.066	0.015	0.001	-0.487	0.111
75-84		-0.062	0.018	0.001	-0.433	0.128
85-94		-0.026	0.048	0.584	-0.173	0.315
35-64	UNIVARIATE	-0.119	0.012	0.001	-0.941	0.097
65-94	UNIVARIATE	-0.074	0.011	0.001	-0.552	0.081
35-64	BIVARIATE	-0.103	0.013	0.001	-0.817	0.105
65-94	BIVARIATE	-0.062	0.011	0.001	-0.468	0.085
35-64	MULTIVARIATE	-0.087	0.014	0.001	-0.691	0.113
65-94	MULTIVARIATE	-0.053	0.012	0.001	-0.395	0.091

TABLE 19-12   DATE=15NOV85

AVERAGE ANNUAL INCIDENCE RATE FOR EVENT PER 1000 PERSONS AT RISK AT EXAM
BY AGE AND LEVEL OF RISK FACTOR AT EXAM: FRAMINGHAM STUDY 30-YEAR FOLLOWUP

SEX: MALE
EVENT: CONGESTIVE HEART FAILURE
RISK FACTOR: HEART RATE (PER MINUTE)
PERSONS AT RISK: PERSONS FREE OF CONGESTIVE HEART FAILURE

RISK FACTOR	35-44 AT EXAM PERSON EXAMS	# OF EVENTS	ANNUAL RATE	45-54 AT EXAM PERSON EXAMS	# OF EVENTS	ANNUAL RATE	55-64 AT EXAM PERSON EXAMS	# OF EVENTS	ANNUAL RATE	65-74 AT EXAM PERSON EXAMS	# OF EVENTS	ANNUAL RATE
30-67	1295			2012	5	1	2596	10	2	1402	14	5
68-75	1502			2587	8	2	2744	18	3	1406	24	9
76-83	727			1467	4	1	1536	8	3	763	15	10
84-91	586			1141	6	3	1191	17	7	580	10	9
92-220	402			832	7	4	897	21	12	527	14	13
UNKNOWN	0			0	0	-	0	0	-	0	0	-
TOTAL	4512			8039	30	2	8964	74	4	4678	77	8

RISK FACTOR	75-84 AT EXAM PERSON EXAMS	# OF EVENTS	ANNUAL RATE	85-94 AT EXAM PERSON EXAMS	# OF EVENTS	ANNUAL RATE	35-64 COMBINED # OF EVENTS	CRUDE	AGE-ADJUSTED	65-94 COMBINED # OF EVENTS	CRUDE	AGE-ADJUSTED
30-67	380	5	7	16	2	63	15	1	1	21	6	6
68-75	350	8	11	13	1	38	28	2	2	33	9	9
76-83	253	11	22	11	1	45	13	2	2	27	13	13
84-91	161	4	12	5	1	100	23	4	4	15	10	10
92-220	137	7	26	1	0	0	29	7	7	21	16	16
UNKNOWN	0	0	-	0	0	-	0	-	-	0	-	-
TOTAL	1281	35	14	46	5	54	108	3	3	117	10	10

LOGISTIC REGRESSIONS

AGE AT EXAM		COEFFICIENT OF RISK FACTOR	SE COEFFICIENT	P-VALUE	STANDARDIZED COEFFICIENT	STD SE
35-44		--				
45-54		0.026	0.013	0.052	0.326	0.167
55-64		0.041	0.007	0.001	0.531	0.095
65-74		0.020	0.008	0.009	0.271	0.104
75-84		0.024	0.011	0.037	0.321	0.154
85-94		-0.019	0.044	0.675	-0.204	0.488
35-64	UNIVARIATE	0.038	0.006	0.001	0.480	0.082
65-94	UNIVARIATE	0.020	0.006	0.002	0.264	0.085
35-64	BIVARIATE	0.037	0.006	0.001	0.470	0.080
65-94	BIVARIATE	0.020	0.006	0.002	0.263	0.085
35-64	MULTIVARIATE	0.028	0.007	0.001	0.553	0.085
65-94	MULTIVARIATE	0.018	0.006	0.005	0.241	0.087

SEX: FEMALE
EVENT: CONGESTIVE HEART FAILURE
RISK FACTOR: HEART RATE (PER MINUTE)
PERSONS AT RISK: PERSONS FREE OF CONGESTIVE HEART FAILURE

RISK FACTOR	35-44 AT EXAM PERSON EXAMS	# OF EVENTS	ANNUAL RATE	45-54 AT EXAM PERSON EXAMS	# OF EVENTS	ANNUAL RATE	55-64 AT EXAM PERSON EXAMS	# OF EVENTS	ANNUAL RATE	65-74 AT EXAM PERSON EXAMS	# OF EVENTS	ANNUAL RATE
30-67	1092			1914	2	1	2434	9	2	1256	6	2
68-75	1806			3274	4	1	3615	9	1	1863	19	5
76-83	1051			1928	5	1	2409	19	4	1506	20	7
84-91	939			1612	3	2	1855	14	4	1161	16	7
92-220	731			1315	5	2	1657	11	3	1144	15	7
UNKNOWN	0			0	0	-	0	0	-	0	0	-
TOTAL	5619			10043	19	1	11970	62	5	6930	76	5

RISK FACTOR	75-84 AT EXAM PERSON EXAMS	# OF EVENTS	ANNUAL RATE	85-94 AT EXAM PERSON EXAMS	# OF EVENTS	ANNUAL RATE	35-64 COMBINED # OF EVENTS	CRUDE	AGE-ADJUSTED	65-94 COMBINED # OF EVENTS	CRUDE	AGE-ADJUSTED
30-67	446	12	13	34	4	59	11	1	1	22	6	6
68-75	557	10	9	36	0	0	14	1	1	29	6	6
76-83	501	17	17	30	1	17	26	2	2	38	9	9
84-91	383	10	13	26	0	0	20	2	2	26	8	8
92-220	410	9	11	25	5	100	14	2	2	29	9	9
UNKNOWN	0	0	-	0	0	-	0	-	-	0	-	-
TOTAL	2297	58	13	151	10	33	85	2	2	144	8	8

LOGISTIC REGRESSIONS

AGE AT EXAM		COEFFICIENT OF RISK FACTOR	SE COEFFICIENT	P-VALUE	STANDARDIZED COEFFICIENT	STD SE
35-44						
45-54		0.022	0.017	0.182	0.277	0.207
55-64		0.019	0.009	0.033	0.243	0.114
65-74		0.014	0.008	0.089	0.183	0.107
75-84		-0.002	0.010	0.816	-0.031	0.135
85-94		0.014	0.022	0.508	0.207	0.314
35-64	UNIVARIATE	0.019	0.008	0.012	0.241	0.096
65-94	UNIVARIATE	0.008	0.006	0.201	0.103	0.081
35-64	BIVARIATE	0.019	0.008	0.014	0.237	0.096
65-94	BIVARIATE	0.007	0.006	0.233	0.095	0.080
35-64	MULTIVARIATE	0.009	0.008	0.253	0.118	0.103
65-94	MULTIVARIATE	0.003	0.006	0.669	0.035	0.081

TABLE 19-13   DATE=15NOV85
AVERAGE ANNUAL INCIDENCE RATE FOR EVENT PER 1000 PERSONS AT RISK AT EXAM
BY AGE AND LEVEL OF RISK FACTOR AT EXAM: FRAMINGHAM STUDY 30-YEAR FOLLOWUP

SEX: MALE
EVENT: CONGESTIVE HEART FAILURE
RISK FACTOR: CIGARETTES SMOKED (NUMBER PER DAY)
PERSONS AT RISK: PERSONS FREE OF CONGESTIVE HEART FAILURE

RISK FACTOR	35-44 AT EXAM			45-54 AT EXAM			55-64 AT EXAM			65-74 AT EXAM		
	PERSON EXAMS	# OF EVENTS	ANNUAL RATE	PERSON EXAMS	# OF EVENTS	ANNUAL RATE	PERSON EXAMS	# OF EVENTS	ANNUAL RATE	PERSON EXAMS	# OF EVENTS	ANNUAL RATE
NONE	1397			3207	13	2	4816	28	3	3132	52	8
1-10	437			742	1	1	849	5	3	454	7	8
11-20	1498			2128	10	2	1880	28	7	641	12	9
21-40	993			1670	5	1	1218	12	5	411	6	7
41-90	161			254	1	2	177	1	3	33	0	0
UNKNOWN	26			38	0	0	24	0	0	7	0	0
TOTAL	4512			8039	30	2	8964	74	4	4678	77	8

RISK FACTOR	75-84 AT EXAM			85-94 AT EXAM			35-64 COMBINED			65-94 COMBINED		
	PERSON EXAMS	# OF EVENTS	ANNUAL RATE	PERSON EXAMS	# OF EVENTS	ANNUAL RATE	# OF EVENTS	CRUDE	AGE-ADJUSTED	# OF EVENTS	CRUDE	AGE-ADJUSTED
NONE	1013	28	14	34	4	59	43	2	2	84	10	10
1-10	117	3	13	6	0	0	6	1	1	10	9	9
11-20	87	3	17	5	1	100	39	4	4	16	11	12
21-40	59	1	8	1	0	0	17	2	3	7	7	7
41-90	5	0	0	0	0	-	3	3	3	0	0	0
UNKNOWN	0	0	-	0	0	-	0	0	0	0	0	0
TOTAL	1281	35	14	46	5	54	108	3	3	117	10	10

LOGISTIC REGRESSIONS

AGE AT EXAM		COEFFICIENT OF RISK FACTOR	SE COEFFICIENT	P-VALUE	STANDARDIZED COEFFICIENT	STD SE
35-44						
45-54		-0.003	0.012	0.781	-0.052	0.187
55-64		0.016	0.007	0.029	0.224	0.102
65-74		-0.004	0.010	0.720	-0.045	0.119
75-84		-0.005	0.021	0.805	-0.045	0.182
85-94		-0.008	0.064	0.902	-0.064	0.515
35-64	UNIVARIATE	0.004	0.006	0.542	0.057	0.094
65-94	UNIVARIATE	-0.007	0.009	0.431	-0.079	0.101
35-64	BIVARIATE	0.012	0.006	0.060	0.173	0.092
65-94	BIVARIATE	-0.002	0.009	0.817	-0.023	0.101
35-64	MULTIVARIATE	0.018	0.006	0.005	0.260	0.092
65-94	MULTIVARIATE	0.003	0.009	0.715	0.037	0.101

SEX: FEMALE
EVENT: CONGESTIVE HEART FAILURE
RISK FACTOR: CIGARETTES SMOKED (NUMBER PER DAY)
PERSONS AT RISK: PERSONS FREE OF CONGESTIVE HEART FAILURE

RISK FACTOR	35-44 AT EXAM			45-54 AT EXAM			55-64 AT EXAM			65-74 AT EXAM		
	PERSON EXAMS	# OF EVENTS	ANNUAL RATE	PERSON EXAMS	# OF EVENTS	ANNUAL RATE	PERSON EXAMS	# OF EVENTS	ANNUAL RATE	PERSON EXAMS	# OF EVENTS	ANNUAL RATE
NONE	2583			5575	12	1	7921	39	2	5424	54	5
1-10	1314			1731	2	1	1624	6	2	666	8	6
11-20	1373			2034	4	1	1753	10	3	629	11	9
21-40	298			628	1	1	616	7	6	191	3	8
41-90	22			38	0	0	18	0	0	1	0	0
UNKNOWN	29			37	0	0	38	0	0	19	0	0
TOTAL	5619			10043	19	1	11970	62	3	6930	76	5

RISK FACTOR	75-84 AT EXAM			85-94 AT EXAM			35-64 COMBINED			65-94 COMBINED		
	PERSON EXAMS	# OF EVENTS	ANNUAL RATE	PERSON EXAMS	# OF EVENTS	ANNUAL RATE	# OF EVENTS	CRUDE	AGE-ADJUSTED	# OF EVENTS	CRUDE	AGE-ADJUSTED
NONE	2037	50	12	143	9	31	52	2	2	113	7	7
1-10	151	4	13	6	1	83	10	2	1	13	8	9
11-20	89	4	22	2	0	0	15	1	2	15	10	12
21-40	16	0	0	0	0	-	8	3	3	3	7	6
41-90	0	0	-	0	0	-	0	0	0	0	0	0
UNKNOWN	4	0	0	0	0	-	0	0	0	0	0	0
TOTAL	2297	58	13	151	10	33	85	2	2	144	8	8

LOGISTIC REGRESSIONS

AGE AT EXAM		COEFFICIENT OF RISK FACTOR	SE COEFFICIENT	P-VALUE	STANDARDIZED COEFFICIENT	STD SE
35-44						
45-54		-0.017	0.025	0.506	-0.174	0.261
55-64		0.017	0.012	0.150	0.159	0.111
65-74		0.022	0.012	0.067	0.170	0.093
75-84		0.010	0.025	0.689	0.048	0.120
85-94		0.054	0.089	0.543	0.141	0.232
35-64	UNIVARIATE	0.003	0.011	0.803	0.027	0.106
65-94	UNIVARIATE	0.010	0.011	0.345	0.071	0.076
35-64	BIVARIATE	0.011	0.010	0.285	0.110	0.103
65-94	BIVARIATE	0.021	0.011	0.054	0.147	0.076
35-64	MULTIVARIATE	0.014	0.011	0.173	0.142	0.104
65-94	MULTIVARIATE	0.023	0.011	0.038	0.162	0.078

TABLE 19-14   DATE=15NOV85
AVERAGE ANNUAL INCIDENCE RATE FOR EVENT PER 1000 PERSONS AT RISK AT EXAM
BY AGE AND LEVEL OF RISK FACTOR AT EXAM: FRAMINGHAM STUDY 30-YEAR FOLLOWUP

SEX: MALE
EVENT: CONGESTIVE HEART FAILURE
RISK FACTOR: ALBUMIN IN URINE, DEFINITE
PERSONS AT RISK: PERSONS FREE OF CONGESTIVE HEART FAILURE

RISK FACTOR	35-44 AT EXAM			45-54 AT EXAM			55-64 AT EXAM			65-74 AT EXAM		
	PERSON EXAMS	# OF EVENTS	ANNUAL RATE	PERSON EXAMS	# OF EVENTS	ANNUAL RATE	PERSON EXAMS	# OF EVENTS	ANNUAL RATE	PERSON EXAMS	# OF EVENTS	ANNUAL RATE
NO	4405	-	-	7911	30	2	8728	65	4	4528	69	8
YES	70			102	0	0	209	9	22	145	7	24
UNKNOWN	37			26	0	0	27	0	0	5	1	100
TOTAL	4512			8039	30	2	8964	74	4	4678	77	8

RISK FACTOR	75-84 AT EXAM			85-94 AT EXAM			35-64 COMBINED			65-94 COMBINED		
	PERSON EXAMS	# OF EVENTS	ANNUAL RATE	PERSON EXAMS	# OF EVENTS	ANNUAL RATE	# OF EVENTS	CRUDE	AGE-ADJUSTED	# OF EVENTS	CRUDE	AGE-ADJUSTED
NO	1248	33	13	40	5	63	97	2	2	107	9	9
YES	33	2	30	6	0	0	10	13	10	9	24	25
UNKNOWN	0	0	-	0	0	-	1	6	3	1	100	78
TOTAL	1281	35	14	46	5	54	108	3	3	117	10	10

LOGISTIC REGRESSIONS

AGE AT EXAM		COEFFICIENT OF RISK FACTOR	SE COEFFICIENT	P-VALUE	STANDARDIZED COEFFICIENT	STD SE
35-44		-	-		-	-
45-54		-	-		-	-
55-64		1.791	0.563	0.001	0.271	0.055
65-74		1.187	0.406	0.003	0.206	0.070
75-84		0.865	0.751	0.249	0.137	0.119
85-94		-	-		-	-
35-64	UNIVARIATE	1.761	0.536	0.001	0.235	0.044
65-94	UNIVARIATE	1.009	0.355	0.005	0.174	0.061
35-64	BIVARIATE	1.590	0.338	0.001	0.210	0.045
65-94	BIVARIATE	0.977	0.357	0.006	0.169	0.062
35-64	MULTIVARIATE	0.951	0.375	0.011	0.126	0.050
65-94	MULTIVARIATE	0.666	0.367	0.069	0.115	0.063

SEX: FEMALE
EVENT: CONGESTIVE HEART FAILURE
RISK FACTOR: ALBUMIN IN URINE, DEFINITE
PERSONS AT RISK: PERSONS FREE OF CONGESTIVE HEART FAILURE

RISK FACTOR	35-44 AT EXAM			45-54 AT EXAM			55-64 AT EXAM			65-74 AT EXAM		
	PERSON EXAMS	# OF EVENTS	ANNUAL RATE	PERSON EXAMS	# OF EVENTS	ANNUAL RATE	PERSON EXAMS	# OF EVENTS	ANNUAL RATE	PERSON EXAMS	# OF EVENTS	ANNUAL RATE
NO	5356			9739	19	1	11707	60	5	6795	73	5
YES	225			279	0	0	235	2	4	125	3	12
UNKNOWN	38			25	0	0	28	0	0	10	0	0
TOTAL	5619			10043	19	1	11970	62	5	6930	76	5

RISK FACTOR	75-84 AT EXAM			85-94 AT EXAM			35-64 COMBINED			65-94 COMBINED		
	PERSON EXAMS	# OF EVENTS	ANNUAL RATE	PERSON EXAMS	# OF EVENTS	ANNUAL RATE	# OF EVENTS	CRUDE	AGE-ADJUSTED	# OF EVENTS	CRUDE	AGE-ADJUSTED
NO	2243	55	12	144	8	28	83	2	2	136	7	7
YES	45	3	33	7	2	143	2	1	2	8	23	19
UNKNOWN	9	0	0	0	0	-	0	0	0	0	0	0
TOTAL	2297	58	13	151	10	33	85	2	2	144	8	8

LOGISTIC REGRESSIONS

AGE AT EXAM		COEFFICIENT OF RISK FACTOR	SE COEFFICIENT	P-VALUE	STANDARDIZED COEFFICIENT	STD SE
35-44		-	-		-	-
45-54		-	-		-	-
55-64		0.511	0.722	0.479	0.071	0.100
65-74		0.817	0.596	0.170	0.109	0.079
75-84		1.044	0.613	0.088	0.145	0.085
85-94		1.917	0.912	0.036	0.404	0.192
35-64	UNIVARIATE	-0.135	0.717	0.850	-0.022	0.114
65-94	UNIVARIATE	1.147	0.372	0.002	0.156	0.051
35-64	BIVARIATE	0.083	0.718	0.908	0.013	0.116
65-94	BIVARIATE	1.073	0.375	0.004	0.146	0.051
35-64	MULTIVARIATE	-0.742	0.746	0.320	-0.120	0.121
65-94	MULTIVARIATE	0.607	0.391	0.120	0.083	0.053

TABLE 19-15   DATE=15NOV85
AVERAGE ANNUAL INCIDENCE RATE FOR EVENT PER 1000 PERSONS AT RISK AT EXAM
BY AGE AND LEVEL OF RISK FACTOR AT EXAM: FRAMINGHAM STUDY 30-YEAR FOLLOWUP

SEX: MALE
EVENT: CONGESTIVE HEART FAILURE
RISK FACTOR: HEART ENLARGEMENT BY X-RAY
PERSONS AT RISK: PERSONS FREE OF CONGESTIVE HEART FAILURE

RISK FACTOR	35-44 AT EXAM PERSON EXAMS	# OF EVENTS	ANNUAL RATE	45-54 AT EXAM PERSON EXAMS	# OF EVENTS	ANNUAL RATE	55-64 AT EXAM PERSON EXAMS	# OF EVENTS	ANNUAL RATE	65-74 AT EXAM PERSON EXAMS	# OF EVENTS	ANNUAL RATE
NO	4096			6604	14	1	6510	29	2	3022	27	4
MILD	188			704	5	4	1102	14	6	679	10	7
DEFINITE	223			727	11	8	1351	31	11	977	40	20
UNKNOWN	5			4	0	0	1	0	0	0	0	-
TOTAL	4512			8039	30	2	8964	74	4	4678	77	8

RISK FACTOR	75-84 AT EXAM PERSON EXAMS	# OF EVENTS	ANNUAL RATE	85-94 AT EXAM PERSON EXAMS	# OF EVENTS	ANNUAL RATE	35-64 COMBINED # OF EVENTS	CRUDE	AGE-ADJUSTED	65-94 COMBINED # OF EVENTS	CRUDE	AGE-ADJUSTED
NO	728	9	6	32	2	31	46	1	1	38	5	5
MILD	223	8	18	8	2	167	19	5	8	20	11	11
DEFINITE	330	18	27	6	1	63	43	9	8	59	22	22
UNKNOWN	0	0	-	0	0	-	0	0	0	0	0	-
TOTAL	1281	35	14	46	5	54	108	3	3	117	10	10

LOGISTIC REGRESSIONS

AGE AT EXAM		COEFFICIENT OF RISK FACTOR	SE COEFFICIENT	P-VALUE	STANDARDIZED COEFFICIENT	STD SE
35-44						
45-54		0.994	0.199	0.001	0.610	0.122
55-64		0.828	0.128	0.001	0.612	0.094
65-74		0.785	0.128	0.001	0.640	0.104
75-84		0.742	0.198	0.001	0.634	0.169
85-94		0.521	0.542	0.336	0.407	0.424
35-64	UNIVARIATE	0.982	0.104	0.001	0.642	0.068
65-94	UNIVARIATE	0.765	0.105	0.001	0.631	0.086
35-64	BIVARIATE	0.865	0.107	0.001	0.565	0.070
65-94	BIVARIATE	0.741	0.105	0.001	0.611	0.087
35-64	MULTIVARIATE	0.725	0.117	0.001	0.474	0.076
65-94	MULTIVARIATE	0.604	0.112	0.001	0.498	0.092

SEX: FEMALE
EVENT: CONGESTIVE HEART FAILURE
RISK FACTOR: HEART ENLARGEMENT BY X-RAY
PERSONS AT RISK: PERSONS FREE OF CONGESTIVE HEART FAILURE

RISK FACTOR	35-44 AT EXAM PERSON EXAMS	# OF EVENTS	ANNUAL RATE	45-54 AT EXAM PERSON EXAMS	# OF EVENTS	ANNUAL RATE	55-64 AT EXAM PERSON EXAMS	# OF EVENTS	ANNUAL RATE	65-74 AT EXAM PERSON EXAMS	# OF EVENTS	ANNUAL RATE
NO	5166			8073	8	0	7795	18	1	3474	19	3
MILD	199			957	4	2	1855	12	3	1264	13	5
DEFINITE	246			1007	7	3	2316	32	7	2192	44	10
UNKNOWN	8			6	0	0	4	0	0	0	0	-
TOTAL	5619			10043	19	1	11970	62	3	6930	76	5

RISK FACTOR	75-84 AT EXAM PERSON EXAMS	# OF EVENTS	ANNUAL RATE	85-94 AT EXAM PERSON EXAMS	# OF EVENTS	ANNUAL RATE	35-64 COMBINED # OF EVENTS	CRUDE	AGE-ADJUSTED	65-94 COMBINED # OF EVENTS	CRUDE	AGE-ADJUSTED
NO	820	8	5	39	1	13	27	1	1	28	3	3
MILD	438	9	10	36	3	42	17	3	3	25	7	7
DEFINITE	1039	41	20	76	6	39	41	6	5	91	14	13
UNKNOWN	0	0	-	0	0	-	0	0	0	0	0	-
TOTAL	2297	58	13	151	10	33	85	2	2	144	8	8

LOGISTIC REGRESSIONS

AGE AT EXAM		COEFFICIENT OF RISK FACTOR	SE COEFFICIENT	P-VALUE	STANDARDIZED COEFFICIENT	STD SE
35-44						
45-54		0.983	0.249	0.001	0.628	0.159
55-64		0.893	0.145	0.001	0.712	0.115
65-74		0.659	0.136	0.001	0.583	0.120
75-84		0.707	0.183	0.001	0.633	0.164
85-94		0.432	0.441	0.327	0.363	0.370
35-64	UNIVARIATE	1.088	0.120	0.001	0.762	0.084
65-94	UNIVARIATE	0.727	0.105	0.001	0.651	0.094
35-64	BIVARIATE	0.922	0.124	0.001	0.646	0.087
65-94	BIVARIATE	0.660	0.106	0.001	0.591	0.095
35-64	MULTIVARIATE	0.721	0.135	0.001	0.505	0.095
65-94	MULTIVARIATE	0.544	0.110	0.001	0.488	0.099

TABLE 19-16   DATE=16NOV85

L INCIDENCE RATE FOR EVENT PER 1000 PERSONS AT RISK AT EXAM
VEL OF RISK FACTOR AT EXAM: FRAMINGHAM STUDY 30-YEAR FOLLOWUP

LE
CONGESTIVE HEART FAILURE
CTOR: LEFT VENTRICULAR HYPERTROPHY BY ELECTROCARDIOGRAM
AT RISK: PERSONS FREE OF CONGESTIVE HEART FAILURE

M ANNUAL RATE	45-54 AT EXAM			55-64 AT EXAM			65-74 AT EXAM		
	PERSON EXAMS	# OF EVENTS	ANNUAL RATE	PERSON EXAMS	# OF EVENTS	ANNUAL RATE	PERSON EXAMS	# OF EVENTS	ANNUAL RATE
	7798	19	1	8531	62	4	4401	61	7
	143	2	7	233	2	4	114	2	9
	98	9	46	200	10	25	163	14	43
	0	0	-	0	0	-	0	0	-
	8039	30	2	8964	74	4	4678	77	8

M ANNUAL RATE	85-94 AT EXAM			35-64 COMBINED			65-94 COMBINED		
	PERSON EXAMS	# OF EVENTS	ANNUAL RATE	# OF EVENTS	CRUDE	AGE-ADJUSTED	# OF EVENTS	CRUDE	AGE-ADJUSTED
12	42	4	48	83	2	2	93	8	8
0	2	1	250	4	4	4	3	10	9
83	2	0	0	21	32	35	21	51	51
-	0	0	-	0	-	-	0	-	-
14	46	5	54	108	3	3	117	10	10

LOGISTIC REGRESSIONS

ICIENT FACTOR	SE COEFFICIENT	P-VALUE	STANDARDIZED COEFFICIENT	STD SE
8	0.209	0.001	0.473	1.0530
4	0.177	0.001	0.310	1.0590
1	0.155	0.001	0.362	1.0610
0	0.237	0.001	0.383	1.0920
3	0.888	0.716	0.146	1.4020
8	0.127	0.001	0.391	1.0360
9	0.128	0.001	0.362	1.0500
4	0.128	0.001	0.356	1.0360
5	0.129	0.001	0.364	1.0510
2	0.154	0.001	0.282	1.0430
4	0.144	0.001	0.293	1.0570

MALE
CONGESTIVE HEART FAILURE
CTOR: LEFT VENTRICULAR HYPERTROPHY BY ELECTROCARDIOGRAM
AT RISK: PERSONS FREE OF CONGESTIVE HEART FAILURE

1 ANNUAL RATE	45-54 AT EXAM			55-64 AT EXAM			65-74 AT EXAM		
	PERSON EXAMS	# OF EVENTS	ANNUAL RATE	PERSON EXAMS	# OF EVENTS	ANNUAL RATE	PERSON EXAMS	# OF EVENTS	ANNUAL RATE
	9879	17	1	11580	45	2	6568	61	5
	108	1	5	236	7	15	167	2	6
	56	1	9	154	10	32	195	13	33
	0	0	-	0	0	-	0	0	-
	10043	19	1	11970	62	3	6930	76	5

1 ANNUAL RATE	85-94 AT EXAM			35-64 COMBINED			65-94 COMBINED		
	PERSON EXAMS	# OF EVENTS	ANNUAL RATE	# OF EVENTS	CRUDE	AGE-ADJUSTED	# OF EVENTS	CRUDE	AGE-ADJUSTED
10	132	6	23	66	1	1	109	6	6
38	9	2	111	8	10	8	9	19	15
55	10	2	100	11	24	17	26	43	40
-	0	0	-	0	-	-	0	-	-
13	151	10	33	85	2	2	144	8	8

LOGISTIC REGRESSIONS

CIENT FACTOR	SE COEFFICIENT	P-VALUE	STANDARDIZED COEFFICIENT	STD SE
	0.457	0.006	0.226	0.083
	0.167	0.001	0.393	0.044
	0.159	0.001	0.354	0.057
	0.170	0.001	0.409	0.074
	0.409	0.024	0.499	0.220
	0.153	0.001	0.338	0.033
	0.110	0.001	0.389	0.042
	0.156	0.001	0.305	0.033
	0.112	0.001	0.367	0.043
	0.189	0.001	0.229	0.040
	0.121	0.001	0.313	0.047

TABLE 19-17   DATE=15NOV85
AVERAGE ANNUAL INCIDENCE RATE FOR EVENT PER 1000 PERSONS AT RISK AT EXAM
BY AGE AND LEVEL OF RISK FACTOR AT EXAM: FRAMINGHAM STUDY 30-YEAR FOLLOWUP

SEX: MALE
EVENT: CONGESTIVE HEART FAILURE
RISK FACTOR: INTRAVENTRICULAR CONDUCTION DEFECT (WITH QRS INTERVAL MORE THAN 0.11 SECOND)
PERSONS AT RISK: PERSONS FREE OF CONGESTIVE HEART FAILURE

RISK FACTOR	35-44 AT EXAM PERSON EXAMS	# OF EVENTS	ANNUAL RATE	45-54 AT EXAM PERSON EXAMS	# OF EVENTS	ANNUAL RATE	55-64 AT EXAM PERSON EXAMS	# OF EVENTS	ANNUAL RATE	65-74 AT EXAM PERSON EXAMS	# OF EVENTS	ANNUAL RATE
NO	4194			7464	25	2	8159	62	4	4212	59	7
INCOMPLETE	280			489	1	1	532	6	6	234	7	15
COMPLETE	38			86	4	23	273	6	11	232	11	24
UNKNOWN	0			0	0	-	0	0	-	0	0	-
TOTAL	4512			8039	30	2	8964	74	4	4678	77	8

RISK FACTOR	75-84 AT EXAM PERSON EXAMS	# OF EVENTS	ANNUAL RATE	85-94 AT EXAM PERSON EXAMS	# OF EVENTS	ANNUAL RATE	35-64 COMBINED # OF EVENTS	CRUDE	AGE-ADJUSTED	65-94 COMBINED # OF EVENTS	CRUDE	AGE-ADJUSTED
NO	1134	26	11	38	4	53	90	2	2	89	8	8
INCOMPLETE	32	2	31	0	0	0	8	3	3	9	17	18
COMPLETE	115	7	30	5	1	100	10	13	13	19	27	26
UNKNOWN	0	0	-	0	0	-	0	-		0	-	
TOTAL	1281	35	14	46	5	54	108	3	3	117	10	10

LOGISTIC REGRESSIONS

AGE AT EXAM		COEFFICIENT OF RISK FACTOR	SE COEFFICIENT	P-VALUE	STANDARDIZED COEFFICIENT	STD SE
35-44						
45-54		1.080	0.307	0.001	0.336	0.095
55-64		0.512	0.202	0.011	0.209	0.082
65-74		0.642	0.160	0.001	0.305	0.076
75-84		0.529	0.212	0.013	0.309	0.124
85-94		0.274	0.645	0.671	0.180	0.423
35-64	UNIVARIATE	0.757	0.167	0.001	0.267	0.059
65-94	UNIVARIATE	0.620	0.125	0.001	0.312	0.063
35-64	BIVARIATE	0.629	0.165	0.001	0.222	0.058
65-94	BIVARIATE	0.568	0.126	0.001	0.286	0.063
35-64	MULTIVARIATE	0.505	0.180	0.005	0.178	0.064
65-94	MULTIVARIATE	0.585	0.128	0.001	0.294	0.064

SEX: FEMALE
EVENT: CONGESTIVE HEART FAILURE
RISK FACTOR: INTRAVENTRICULAR CONDUCTION DEFECT (WITH QRS INTERVAL MORE THAN 0.11 SECOND)
PERSONS AT RISK: PERSONS FREE OF CONGESTIVE HEART FAILURE

RISK FACTOR	35-44 AT EXAM PERSON EXAMS	# OF EVENTS	ANNUAL RATE	45-54 AT EXAM PERSON EXAMS	# OF EVENTS	ANNUAL RATE	55-64 AT EXAM PERSON EXAMS	# OF EVENTS	ANNUAL RATE	65-74 AT EXAM PERSON EXAMS	# OF EVENTS	ANNUAL RATE
NO	5532			9845	18	1	11615	58	2	6651	65	5
INCOMPLETE	75			150	1	3	189	1	3	127	1	4
COMPLETE	12			48	0	0	166	3	9	152	10	33
UNKNOWN	0			0	0	-	0	0	-	0	0	-
TOTAL	5619			10043	19	1	11970	62	3	6930	76	5

RISK FACTOR	75-84 AT EXAM PERSON EXAMS	# OF EVENTS	ANNUAL RATE	85-94 AT EXAM PERSON EXAMS	# OF EVENTS	ANNUAL RATE	35-64 COMBINED # OF EVENTS	CRUDE	AGE-ADJUSTED	65-94 COMBINED # OF EVENTS	CRUDE	AGE-ADJUSTED
NO	2167	51	12	144	8	28	80	1	1	124	7	7
INCOMPLETE	51	2	20	3	1	167	2	2	2	4	11	10
COMPLETE	79	5	32	4	1	125	3	7	4	16	34	34
UNKNOWN	0	0	-	0	0	-	0	-		0	-	
TOTAL	2297	58	13	151	10	33	85	2	2	144	8	8

LOGISTIC REGRESSIONS

AGE AT EXAM		COEFFICIENT OF RISK FACTOR	SE COEFFICIENT	P-VALUE	STANDARDIZED COEFFICIENT	STD SE
35-44						
45-54		0.540	0.827	0.513	0.099	0.151
55-64		0.591	0.297	0.046	0.156	0.078
65-74		0.933	0.178	0.001	0.298	0.057
75-84		0.516	0.233	0.027	0.201	0.091
85-94		1.023	0.545	0.061	0.356	0.190
35-64	UNIVARIATE	0.713	0.281	0.011	0.154	0.061
65-94	UNIVARIATE	0.803	0.136	0.001	0.272	0.046
35-64	BIVARIATE	0.536	0.280	0.055	0.116	0.060
65-94	BIVARIATE	0.762	0.137	0.001	0.258	0.046
35-64	MULTIVARIATE	0.417	0.328	0.204	0.090	0.071
65-94	MULTIVARIATE	0.735	0.140	0.001	0.249	0.047

AVERAGE ANNUAL INCIDENCE RATE FOR EVENT PER 1000 PERSONS AT RISK AT EXAM BY AGE AND LEVEL OF RISK FACTOR AT EXAM: FRAMINGHAM STUDY 30-YEAR FOLLOWUP

SEX: MALE
EVENT: CONGESTIVE HEART FAILURE
RISK FACTOR: DEFINITE NONSPECIFIC T-WAVE OR ST-SEGMENT ABNORMALITY BY ELECTROCARDIOGRAM
PERSONS AT RISK: PERSONS FREE OF CONGESTIVE HEART FAILURE

RISK FACTOR	35-44 AT EXAM PERSON EXAMS	# OF EVENTS	ANNUAL RATE	45-54 AT EXAM PERSON EXAMS	# OF EVENTS	ANNUAL RATE	55-64 AT EXAM PERSON EXAMS	# OF EVENTS	ANNUAL RATE	65-74 AT EXAM PERSON EXAMS	# OF EVENTS	ANNUAL RATE
NO	4399			7618	25	2	8175	50	3	4067	61	7
YES	113			421	5	6	789	24	15	611	16	13
UNKNOWN	0			0			0			0		
TOTAL	4512			8039	30	2	8964	74	4	4678	77	8

RISK FACTOR	75-84 AT EXAM PERSON EXAMS	# OF EVENTS	ANNUAL RATE	85-94 AT EXAM PERSON EXAMS	# OF EVENTS	ANNUAL RATE	35-64 COMBINED # OF EVENTS	CRUDE	AGE-ADJUSTED	65-94 COMBINED # OF EVENTS	CRUDE	AGE-ADJUSTED
NO	1021	21	10	30	1	17	79	2	2	83	8	8
YES	260	14	27	16	4	125	29	11	9	34	19	17
UNKNOWN	0	0	-	0	0	-	0			0		
TOTAL	1281	35	14	46	5	94	108	3	3	117	10	10

LOGISTIC REGRESSIONS

AGE AT EXAM		COEFFICIENT OF RISK FACTOR	SE COEFFICIENT	P-VALUE	STANDARDIZED COEFFICIENT	STD SE
35-44						
45-54		1.295	0.492	0.009	0.288	0.110
55-64		1.629	0.251	0.001	0.462	0.071
65-74		0.569	0.284	0.045	0.192	0.096
75-84		0.997	0.352	0.005	0.401	0.142
85-94		2.269	1.170	0.052	1.092	0.563
35-64	UNIVARIATE	1.741	0.219	0.001	0.418	0.053
65-94	UNIVARIATE	0.883	0.207	0.001	0.313	0.073
35-64	BIVARIATE	1.498	0.222	0.001	0.360	0.053
65-94	BIVARIATE	0.793	0.209	0.001	0.282	0.074
35-64	MULTIVARIATE	1.266	0.226	0.001	0.304	0.054
65-94	MULTIVARIATE	0.774	0.213	0.001	0.275	0.076

SEX: FEMALE
EVENT: CONGESTIVE HEART FAILURE
RISK FACTOR: DEFINITE NONSPECIFIC T-WAVE OR ST-SEGMENT ABNORMALITY BY ELECTROCARDIOGRAM
PERSONS AT RISK: PERSONS FREE OF CONGESTIVE HEART FAILURE

RISK FACTOR	35-44 AT EXAM PERSON EXAMS	# OF EVENTS	ANNUAL RATE	45-54 AT EXAM PERSON EXAMS	# OF EVENTS	ANNUAL RATE	55-64 AT EXAM PERSON EXAMS	# OF EVENTS	ANNUAL RATE	65-74 AT EXAM PERSON EXAMS	# OF EVENTS	ANNUAL RATE
NO	5520			9622	15	1	11020	51	2	6033	58	5
YES	99			421	4	5	950	11	6	897	18	10
UNKNOWN	0			0			0			0		
TOTAL	5619			10043	19	1	11970	62	3	6930	76	5

RISK FACTOR	75-84 AT EXAM PERSON EXAMS	# OF EVENTS	ANNUAL RATE	85-94 AT EXAM PERSON EXAMS	# OF EVENTS	ANNUAL RATE	35-64 COMBINED # OF EVENTS	CRUDE	AGE-ADJUSTED	65-94 COMBINED # OF EVENTS	CRUDE	AGE-ADJUSTED
NO	1864	41	11	117	9	38	70	1	1	108	7	7
YES	433	17	20	34	1	15	15	5	4	36	13	12
UNKNOWN	0	0	-	0	0	-	0			0		
TOTAL	2297	58	13	151	10	33	85	2	2	144	8	8

LOGISTIC REGRESSIONS

AGE AT EXAM		COEFFICIENT OF RISK FACTOR	SE COEFFICIENT	P-VALUE	STANDARDIZED COEFFICIENT	STD SE
35-44						
45-54		1.815	0.565	0.001	0.364	0.113
55-64		0.924	0.334	0.006	0.250	0.090
65-74		0.746	0.272	0.006	0.251	0.091
75-84		0.597	0.294	0.042	0.234	0.115
85-94		-1.012	1.073	0.346	-0.424	0.450
35-64	UNIVARIATE	1.346	0.286	0.001	0.302	0.064
65-94	UNIVARIATE	0.685	0.195	0.001	0.242	0.069
35-64	BIVARIATE	1.023	0.289	0.001	0.230	0.065
65-94	BIVARIATE	0.575	0.196	0.003	0.203	0.069
35-64	MULTIVARIATE	0.837	0.295	0.004	0.188	0.066
65-94	MULTIVARIATE	0.530	0.199	0.008	0.187	0.070

TABLE 20-1   DATE=22MAY85

AVERAGE ANNUAL INCIDENCE RATE FOR EVENT PER 1000 PERSONS AT RISK AT EXAM
BY AGE AND LEVEL OF RISK FACTOR AT EXAM: FRAMINGHAM STUDY 30-YEAR FOLLOWUP

SEX: MALE
EVENT: CARDIOVASCULAR DISEASE
RISK FACTOR: SYSTOLIC BLOOD PRESSURE, FIRST EXAMINER (MM HG)
PERSONS AT RISK: PERSONS FREE OF CARDIOVASCULAR DISEASE

RISK FACTOR	35-44 AT EXAM			45-54 AT EXAM			55-64 AT EXAM			65-74 AT EXAM		
	PERSON EXAMS	# OF EVENTS	ANNUAL RATE	PERSON EXAMS	# OF EVENTS	ANNUAL RATE	PERSON EXAMS	# OF EVENTS	ANNUAL RATE	PERSON EXAMS	# OF EVENTS	ANNUAL RATE
74-119	1222	8	3	1761	27	8	1408	44	16	525	17	16
120-139	2163	21	5	3394	77	11	3089	113	18	1264	59	23
140-159	841	11	7	1708	66	19	1989	122	31	1085	81	37
160-179	191	5	13	544	32	29	863	74	43	494	51	52
180-300	50	1	10	228	16	35	410	51	62	238	37	78
UNKNOWN	0	0	-	0	0	-	0	0	-	0	0	-
TOTAL	4467	46	5	7635	218	14	7759	404	26	3606	245	34

RISK FACTOR	75-84 AT EXAM			85-94 AT EXAM			35-64 COMBINED			65-94 COMBINED		
	PERSON EXAMS	# OF EVENTS	ANNUAL RATE	PERSON EXAMS	# OF EVENTS	ANNUAL RATE	# OF EVENTS	CRUDE	AGE-ADJUSTED	# OF EVENTS	CRUDE	AGE-ADJUSTED
74-119	107	7	33	5	1	100	79	9	10	25	20	20
120-139	301	23	38	10	0	0	211	12	12	82	26	26
140-159	266	26	49	10	1	50	199	22	21	108	40	39
160-179	127	23	91	3	2	333	111	35	31	76	61	61
180-300	58	13	112	2	1	250	68	49	40	51	86	86
UNKNOWN	0	0	-	0	0	-	0	-		0	-	
TOTAL	859	92	54	30	5	83	668	17	17	342	38	38

LOGISTIC REGRESSIONS

AGE AT EXAM		COEFFICIENT OF RISK FACTOR	SE COEFFICIENT	P-VALUE	STANDARDIZED COEFFICIENT	STD SE
35-44		0.023	0.007	0.002	0.378	0.123
45-54		0.022	0.003	0.001	0.427	0.057
55-64		0.020	0.002	0.001	0.438	0.044
65-74		0.020	0.003	0.001	0.459	0.058
75-84		0.021	0.005	0.001	0.444	0.106
85-94		0.051	0.028	0.069	0.970	0.534
35-64	UNIVARIATE	0.024	0.002	0.001	0.484	0.032
65-94	UNIVARIATE	0.020	0.002	0.001	0.454	0.050
35-64	BIVARIATE	0.020	0.002	0.001	0.406	0.032
65-94	BIVARIATE	0.021	0.002	0.001	0.463	0.051
35-64	MULTIVARIATE	0.017	0.002	0.001	0.341	0.035
65-94	MULTIVARIATE	0.018	0.002	0.001	0.410	0.053

SEX: FEMALE
EVENT: CARDIOVASCULAR DISEASE
RISK FACTOR: SYSTOLIC BLOOD PRESSURE, FIRST EXAMINER (MM HG)
PERSONS AT RISK: PERSONS FREE OF CARDIOVASCULAR DISEASE

RISK FACTOR	35-44 AT EXAM			45-54 AT EXAM			55-64 AT EXAM			65-74 AT EXAM		
	PERSON EXAMS	# OF EVENTS	ANNUAL RATE	PERSON EXAMS	# OF EVENTS	ANNUAL RATE	PERSON EXAMS	# OF EVENTS	ANNUAL RATE	PERSON EXAMS	# OF EVENTS	ANNUAL RATE
74-119	2486	6	1	2694	18	3	1937	25	9	625	15	12
120-139	2239	6	1	3951	36	5	3979	73	9	1680	56	17
140-159	670	4	3	2028	36	9	2944	93	16	1896	85	22
160-179	156	0	0	758	14	9	1501	71	24	1070	42	20
180-300	50	1	10	437	14	16	842	61	36	615	55	45
UNKNOWN	0	0	-	0	0	-	0	0	-	0	0	-
TOTAL	5601	17	2	9868	118	6	11203	323	14	5886	253	21

RISK FACTOR	75-84 AT EXAM			85-94 AT EXAM			35-64 COMBINED			65-94 COMBINED		
	PERSON EXAMS	# OF EVENTS	ANNUAL RATE	PERSON EXAMS	# OF EVENTS	ANNUAL RATE	# OF EVENTS	CRUDE	AGE-ADJUSTED	# OF EVENTS	CRUDE	AGE-ADJUSTED
74-119	118	4	17	8	1	63	49	3	4	20	13	14
120-139	406	34	42	31	4	65	115	6	6	94	22	23
140-159	569	37	33	21	4	95	133	12	11	126	25	25
160-179	394	37	47	24	3	63	85	18	13	82	28	27
180-300	252	26	52	17	4	118	76	29	23	85	48	48
UNKNOWN	0	0	-	0	0	-	0	-		0	-	
TOTAL	1739	138	40	101	16	79	458	9	9	407	26	26

LOGISTIC REGRESSIONS

AGE AT EXAM		COEFFICIENT OF RISK FACTOR	SE COEFFICIENT	P-VALUE	STANDARDIZED COEFFICIENT	STD SE
35-44		0.011	0.012	0.363	0.194	0.215
45-54		0.017	0.003	0.001	0.397	0.068
55-64		0.020	0.002	0.001	0.500	0.046
65-74		0.014	0.002	0.001	0.340	0.059
75-84		0.008	0.003	0.025	0.193	0.085
85-94		0.006	0.010	0.589	0.145	0.269
35-64	UNIVARIATE	0.023	0.001	0.001	0.544	0.034
65-94	UNIVARIATE	0.013	0.002	0.001	0.310	0.047
35-64	BIVARIATE	0.019	0.002	0.001	0.442	0.037
65-94	BIVARIATE	0.012	0.002	0.001	0.282	0.048
35-64	MULTIVARIATE	0.015	0.002	0.001	0.361	0.041
65-94	MULTIVARIATE	0.008	0.002	0.001	0.207	0.050

TABLE 20-2   DATE=21MAY85

AVERAGE ANNUAL INCIDENCE RATE FOR EVENT PER 1000 PERSONS AT RISK AT EXAM
BY AGE AND LEVEL OF RISK FACTOR AT EXAM: FRAMINGHAM STUDY 30-YEAR FOLLOWUP

SEX: MALE
EVENT: CARDIOVASCULAR DISEASE
RISK FACTOR: DIASTOLIC BLOOD PRESSURE, FIRST EXAMINER (MM HG)
PERSONS AT RISK: PERSONS FREE OF CARDIOVASCULAR DISEASE

RISK FACTOR	35-44 AT EXAM			45-54 AT EXAM			55-64 AT EXAM			65-74 AT EXAM		
	PERSON EXAMS	# OF EVENTS	ANNUAL RATE	PERSON EXAMS	# OF EVENTS	ANNUAL RATE	PERSON EXAMS	# OF EVENTS	ANNUAL RATE	PERSON EXAMS	# OF EVENTS	ANNUAL RATE
20-74	1012	3	1	1457	31	11	1769	70	20	1121	63	28
75-84	1601	14	4	2747	56	10	2766	104	19	1225	67	27
85-94	1187	12	5	2039	61	15	1885	115	31	833	61	37
95-104	499	12	12	990	44	22	940	68	36	311	36	58
105-160	168	5	15	402	26	32	399	47	59	116	18	78
UNKNOWN	0	0	-	0	0	-	0	0	-	0	0	-
TOTAL	4467	46	5	7635	218	14	7759	404	26	3606	245	34

RISK FACTOR	75-84 AT EXAM			85-94 AT EXAM			35-64 COMBINED			65-94 COMBINED		
	PERSON EXAMS	# OF EVENTS	ANNUAL RATE	PERSON EXAMS	# OF EVENTS	ANNUAL RATE	# OF EVENTS	CRUDE	AGE-ADJUSTED	# OF EVENTS	CRUDE	AGE-ADJUSTED
20-74	351	31	44	14	3	107	104	12	12	97	33	32
75-84	289	31	54	11	0	0	174	12	12	98	32	32
85-94	175	18	51	5	2	200	188	18	19	81	40	41
95-104	29	7	121	0	0	-	124	26	25	43	63	70
105-160	15	5	167	0	0	-	78	40	39	23	88	94
UNKNOWN	0	0	-	0	0	-	0	-	-	0	-	-
TOTAL	859	92	54	30	5	83	668	17	17	342	38	38

LOGISTIC REGRESSIONS

AGE AT EXAM		COEFFICIENT OF RISK FACTOR	SE COEFFICIENT	P-VALUE	STANDARDIZED COEFFICIENT	STD SE
35-44		0.051	0.011	0.001	0.571	0.127
45-54		0.027	0.005	0.001	0.315	0.063
55-64		0.028	0.004	0.001	0.342	0.047
65-74		0.023	0.005	0.001	0.284	0.062
75-84		0.023	0.009	0.014	0.269	0.110
85-94		0.005	0.041	0.907	0.058	0.499
35-64	UNIVARIATE	0.030	0.003	0.001	0.355	0.036
65-94	UNIVARIATE	0.020	0.004	0.001	0.244	0.053
35-64	BIVARIATE	0.030	0.003	0.001	0.351	0.036
65-94	BIVARIATE	0.025	0.004	0.001	0.300	0.054
35-64	MULTIVARIATE	0.026	0.003	0.001	*.302	0.038
65-94	MULTIVARIATE	0.021	0.005	0.001	0.259	0.055

SEX: FEMALE
EVENT: CARDIOVASCULAR DISEASE
RISK FACTOR: DIASTOLIC BLOOD PRESSURE, FIRST EXAMINER (MM HG)
PERSONS AT RISK: PERSONS FREE OF CARDIOVASCULAR DISEASE

RISK FACTOR	35-44 AT EXAM			45-54 AT EXAM			55-64 AT EXAM			65-74 AT EXAM		
	PERSON EXAMS	# OF EVENTS	ANNUAL RATE	PERSON EXAMS	# OF EVENTS	ANNUAL RATE	PERSON EXAMS	# OF EVENTS	ANNUAL RATE	PERSON EXAMS	# OF EVENTS	ANNUAL RATE
20-74	2158	7	2	2610	25	5	2888	57	10	1732	69	20
75-84	2023	3	1	3511	31	4	3772	82	11	2035	64	16
85-94	987	2	1	2344	35	7	2770	91	16	1348	80	30
95-104	322	5	8	914	13	7	1228	51	21	550	21	19
105-160	111	0	0	489	14	14	545	42	39	221	19	19
UNKNOWN	0	0	-	0	0	-	0	0	-	0	0	-
TOTAL	5601	17	2	9868	118	6	11203	323	14	5886	253	21

RISK FACTOR	75-84 AT EXAM			85-94 AT EXAM			35-64 COMBINED			65-94 COMBINED		
	PERSON EXAMS	# OF EVENTS	ANNUAL RATE	PERSON EXAMS	# OF EVENTS	ANNUAL RATE	# OF EVENTS	CRUDE	AGE-ADJUSTED	# OF EVENTS	CRUDE	AGE-ADJUSTED
20-74	650	49	38	44	6	68	89	6	6	124	26	25
75-84	546	46	42	34	7	103	116	6	6	117	22	23
85-94	379	30	40	16	3	94	128	10	10	113	32	33
95-104	125	9	36	7	0	0	69	14	13	30	22	23
105-160	39	4	51	0	0	-	56	24	22	23	44	44
UNKNOWN	0	0	-	0	0	-	0	-	-	0	-	-
TOTAL	1739	138	40	101	16	79	458	9	9	407	26	26

LOGISTIC REGRESSIONS

AGE AT EXAM		COEFFICIENT OF RISK FACTOR	SE COEFFICIENT	P-VALUE	STANDARDIZED COEFFICIENT	STD SE
35-44		0.015	0.020	0.459	0.168	0.227
45-54		0.025	0.007	0.001	0.305	0.081
55-64		0.033	0.004	0.001	0.401	0.050
65-74		0.016	0.005	0.002	0.190	0.062
75-84		-0.001	0.008	0.938	-0.007	0.089
85-94		0.005	0.022	0.815	0.065	0.277
35-64	UNIVARIATE	0.034	0.003	0.001	0.415	0.040
65-94	UNIVARIATE	0.007	0.004	0.079	0.088	0.050
35-64	BIVARIATE	0.030	0.003	0.001	0.368	0.041
65-94	BIVARIATE	0.010	0.004	0.015	0.122	0.050
35-64	MULTIVARIATE	0.024	0.004	0.001	0.288	0.044
65-94	MULTIVARIATE	0.007	0.004	0.080	0.089	0.051

TABLE 20-3A   DATE=29AUG85

AVERAGE ANNUAL INCIDENCE RATE FOR EVENT PER 1000 PERSONS AT RISK AT EXAM
BY AGE AND LEVEL OF RISK FACTOR AT EXAM: FRAMINGHAM STUDY 30-YEAR FOLLOWUP

SEX: MALE
EVENT: CARDIOVASCULAR DISEASE
RISK FACTOR: HYPERTENSION WITH ANTIHYPERTENSIVE TREATMENT
PERSONS AT RISK: PERSONS FREE OF CARDIOVASCULAR DISEASE

RISK FACTOR	35-44 AT EXAM			45-54 AT EXAM			55-64 AT EXAM			65-74 AT EXAM		
	PERSON EXAMS	# OF EVENTS	ANNUAL RATE	PERSON EXAMS	# OF EVENTS	ANNUAL RATE	PERSON EXAMS	# OF EVENTS	ANNUAL RATE	PERSON EXAMS	# OF EVENTS	ANNUAL RATE
NO	2648	20	4	4101	71	9	3715	116	16	1457	49	17
BORDER	1501	12	5	2176	71	16	2203	135	31	1163	34	36
DEFINITE	498	13	13	1119	58	26	1212	105	43	511	59	58
TREATED	20	1	25	239	18	38	629	48	38	475	53	56
UNKNOWN	0	0	-	0	0	-	0	0	-	0	0	-
TOTAL	4467	46	5	7635	218	14	7759	404	26	3606	245	34

RISK FACTOR	75-84 AT EXAM			85-94 AT EXAM			35-64 COMBINED			65-94 COMBINED		
	PERSON EXAMS	# OF EVENTS	ANNUAL RATE	PERSON EXAMS	# OF EVENTS	ANNUAL RATE	# OF EVENTS	CRUDE	AGE-ADJUSTED	# OF EVENTS	CRUDE	AGE-ADJUSTED
NO	334	29	30	12	1	42	207	10	11	70	19	20
BORDER	296	27	46	9	1	56	218	19	19	112	38	38
DEFINITE	95	16	84	3	2	333	176	31	30	77	63	65
TREATED	134	29	108	6	1	83	67	38	35	83	67	66
UNKNOWN	0	0	-	0	0	-	0	-	-	0	-	-
TOTAL	859	92	54	30	5	83	668	17	17	342	38	38

LOGISTIC REGRESSIONS

AGE AT EXAM		COEFFICIENT OF RISK FACTOR	SE COEFFICIENT	P-VALUE	STANDARDIZED COEFFICIENT	STD SE
35-44		0.620	0.187	0.001	0.430	0.130
45-54		0.607	0.084	0.001	0.464	0.064
55-64		0.511	0.062	0.001	0.414	0.050
65-74		0.633	0.084	0.001	0.514	0.068
75-84		0.698	0.144	0.001	0.559	0.115
85-94		0.918	0.655	0.161	0.776	0.554
35-64	UNIVARIATE	0.620	0.048	0.001	0.480	0.037
65-94	UNIVARIATE	0.650	0.072	0.001	0.527	0.058
35-64	BIVARIATE	0.536	0.048	0.001	0.414	0.037
65-94	BIVARIATE	0.654	0.072	0.001	0.530	0.059
35-64	MULTIVARIATE	0.471	0.051	0.001	0.364	0.039
65-94	MULTIVARIATE	0.600	0.074	0.001	0.486	0.060

SEX: FEMALE
EVENT: CARDIOVASCULAR DISEASE
RISK FACTOR: HYPERTENSION WITH ANTIHYPERTENSIVE TREATMENT
PERSONS AT RISK: PERSONS FREE OF CARDIOVASCULAR DISEASE

RISK FACTOR	35-44 AT EXAM			45-54 AT EXAM			55-64 AT EXAM			65-74 AT EXAM		
	PERSON EXAMS	# OF EVENTS	ANNUAL RATE	PERSON EXAMS	# OF EVENTS	ANNUAL RATE	PERSON EXAMS	# OF EVENTS	ANNUAL RATE	PERSON EXAMS	# OF EVENTS	ANNUAL RATE
NO	4239	11	1	5591	41	4	4806	66	7	1714	45	13
BORDER	1000	3	2	2554	37	7	3221	91	14	1840	65	18
DEFINITE	314	3	5	1154	20	9	1770	87	25	988	52	26
TREATED	48	0	0	569	20	18	1406	79	28	1344	91	34
UNKNOWN	0	0	-	0	0	-	0	0	-	0	0	-
TOTAL	5601	17	2	9868	118	6	11203	323	14	5886	253	26

RISK FACTOR	75-84 AT EXAM			85-94 AT EXAM			35-64 COMBINED			65-94 COMBINED		
	PERSON EXAMS	# OF EVENTS	ANNUAL RATE	PERSON EXAMS	# OF EVENTS	ANNUAL RATE	# OF EVENTS	CRUDE	AGE-ADJUSTED	# OF EVENTS	CRUDE	AGE-ADJUSTED
NO	357	22	31	26	3	58	118	4	5	70	11	14
BORDER	528	35	33	22	3	68	131	10	9	103	22	18
DEFINITE	341	32	47	18	3	83	110	17	15	87	32	31
TREATED	513	49	48	35	7	100	99	24	18	147	39	38
UNKNOWN	0	0	-	0	0	-	0	-	-	0	-	-
TOTAL	1739	138	40	101	16	79	458	9	9	407	26	26

LOGISTIC REGRESSIONS

AGE AT EXAM		COEFFICIENT OF RISK FACTOR	SE COEFFICIENT	P-VALUE	STANDARDIZED COEFFICIENT	STD SE
35-44		0.508	0.332	0.126	0.297	0.194
45-54		0.585	0.110	0.001	0.448	0.085
55-64		0.682	0.071	0.001	0.567	0.059
65-74		0.466	0.085	0.001	0.383	0.070
75-84		0.265	0.120	0.028	0.208	0.094
85-94		0.302	0.345	0.382	0.256	0.293
35-64	UNIVARIATE	0.808	0.057	0.001	0.638	0.045
65-94	UNIVARIATE	0.432	0.068	0.001	0.353	0.055
35-64	BIVARIATE	0.622	0.059	0.001	0.491	0.046
65-94	BIVARIATE	0.394	0.068	0.001	0.322	0.056
35-64	MULTIVARIATE	0.540	0.061	0.001	0.426	0.048
65-94	MULTIVARIATE	0.318	0.069	0.001	0.260	0.057

TABLE 29-3B   DATE=14NOV85

AVERAGE ANNUAL INCIDENCE RATE FOR EVENT PER 1000 PERSONS AT RISK AT EXAM
BY AGE AND LEVEL OF RISK FACTOR AT EXAM: FRAMINGHAM STUDY 30-YEAR FOLLOWUP

SEX: MALE
EVENT: CARDIOVASCULAR DISEASE
RISK FACTOR: HYPERTENSION IGNORING TREATMENT
PERSONS AT RISK: PERSONS FREE OF CARDIOVASCULAR DISEASE

RISK FACTOR	35-44 AT EXAM PERSON EXAMS	# OF EVENTS	ANNUAL RATE	45-54 AT EXAM PERSON EXAMS	# OF EVENTS	ANNUAL RATE	55-64 AT EXAM PERSON EXAMS	# OF EVENTS	ANNUAL RATE	65-74 AT EXAM PERSON EXAMS	# OF EVENTS	ANNUAL RATE
NO	2650	20	4	4132	73	9	3831	124	16	1552	59	19
MILD	1310	12	5	2252	77	17	2453	150	31	1358	101	37
DEFINITE	507	14	14	1251	68	27	1475	130	44	696	85	61
UNKNOWN	0	0	-	0	0	-	0	0	-	0	0	-
TOTAL	4467	46	5	7635	218	14	7759	404	26	3606	245	34

RISK FACTOR	75-84 AT EXAM PERSON EXAMS	# OF EVENTS	ANNUAL RATE	85-94 AT EXAM PERSON EXAMS	# OF EVENTS	ANNUAL RATE	35-64 COMBINED # OF EVENTS	CRUDE	AGE-ADJUSTED	65-94 COMBINED # OF EVENTS	CRUDE	AGE-ADJUSTED
NO	360	25	35	15	1	33	217	10	11	85	22	22
MILD	363	36	59	12	2	83	239	20	20	139	40	40
DEFINITE	136	31	114	3	2	333	212	33	31	118	71	73
UNKNOWN	0	0	-	0	0	-	0	-	-	0	-	-
TOTAL	859	92	54	30	5	83	668	17	17	342	38	38

LOGISTIC REGRESSIONS

AGE AT EXAM	COEFFICIENT OF RISK FACTOR	SE COEFFICIENT	P-VALUE	STANDARDIZED COEFFICIENT	STD SE
35-44	0.630	0.188	0.001	0.435	0.130
45-54	0.583	0.085	0.001	0.437	0.063
55-64	0.531	0.063	0.001	0.408	0.049
65-74	0.625	0.087	0.001	0.471	0.065
75-84	0.691	0.153	0.001	0.493	0.109
85-94	1.641	0.836	0.050	1.107	0.564
35-64 UNIVARIATE	0.607	0.049	0.001	0.454	0.036
65-94 UNIVARIATE	0.636	0.074	0.001	0.474	0.055
35-64 BIVARIATE	0.548	0.049	0.001	0.409	0.037
65-94 BIVARIATE	0.653	0.075	0.001	0.487	0.056
35-64 MULTIVARIATE	0.470	0.052	0.001	0.352	0.039
65-94 MULTIVARIATE	0.590	0.077	0.001	0.440	0.057

SEX: FEMALE
EVENT: CARDIOVASCULAR DISEASE
RISK FACTOR: HYPERTENSION IGNORING TREATMENT
PERSONS AT RISK: PERSONS FREE OF CARDIOVASCULAR DISEASE

RISK FACTOR	35-44 AT EXAM PERSON EXAMS	# OF EVENTS	ANNUAL RATE	45-54 AT EXAM PERSON EXAMS	# OF EVENTS	ANNUAL RATE	55-64 AT EXAM PERSON EXAMS	# OF EVENTS	ANNUAL RATE	65-74 AT EXAM PERSON EXAMS	# OF EVENTS	ANNUAL RATE
NO	4260	11	1	5758	46	4	5137	77	7	1964	57	15
MILD	1015	3	3	2734	42	8	3766	120	16	2405	103	21
DEFINITE	326	3	5	1376	30	11	2300	126	27	1517	93	31
UNKNOWN	0	0	-	0	0	-	0	0	-	0	0	-
TOTAL	5601	17	2	9868	118	6	11203	323	14	5886	253	21

RISK FACTOR	75-84 AT EXAM PERSON EXAMS	# OF EVENTS	ANNUAL RATE	85-94 AT EXAM PERSON EXAMS	# OF EVENTS	ANNUAL RATE	35-64 COMBINED # OF EVENTS	CRUDE	AGE-ADJUSTED	65-94 COMBINED # OF EVENTS	CRUDE	AGE-ADJUSTED
NO	448	28	31	31	4	65	134	4	5	89	18	19
MILD	747	58	39	35	6	86	165	11	10	167	26	26
DEFINITE	544	52	48	35	6	86	159	20	17	151	36	35
UNKNOWN	0	0	-	0	0	-	0	-	-	0	-	-
TOTAL	1739	138	40	101	16	79	458	9	9	407	26	26

LOGISTIC REGRESSIONS

AGE AT EXAM	COEFFICIENT OF RISK FACTOR	SE COEFFICIENT	P-VALUE	STANDARDIZED COEFFICIENT	STD SE
35-44	0.549	0.335	0.101	0.315	0.191
45-54	0.520	0.114	0.001	0.377	0.083
55-64	0.660	0.071	0.001	0.511	0.055
65-74	0.390	0.085	0.001	0.298	0.065
75-84	0.230	0.120	0.054	0.173	0.090
85-94	0.157	0.340	0.645	0.127	0.276
35-64 UNIVARIATE	0.764	0.058	0.001	0.563	0.043
65-94 UNIVARIATE	0.358	0.067	0.001	0.274	0.052
35-64 BIVARIATE	0.594	0.060	0.001	0.440	0.044
65-94 BIVARIATE	0.327	0.068	0.001	0.250	0.052
35-64 MULTIVARIATE	0.500	0.063	0.001	0.369	0.046
65-94 MULTIVARIATE	0.251	0.069	0.001	0.192	0.053

TABLE 20-4   DATE=22MAY85

AVERAGE ANNUAL INCIDENCE RATE FOR EVENT PER 1000 PERSONS AT RISK AT EXAM
BY AGE AND LEVEL OF RISK FACTOR AT EXAM: FRAMINGHAM STUDY 30-YEAR FOLLOWUP

SEX: MALE
EVENT: CARDIOVASCULAR DISEASE
RISK FACTOR: SERUM CHOLESTEROL
PERSONS AT RISK: PERSONS FREE OF CARDIOVASCULAR DISEASE

RISK FACTOR	35-44 AT EXAM PERSON EXAMS	# OF EVENTS	ANNUAL RATE	45-54 AT EXAM PERSON EXAMS	# OF EVENTS	ANNUAL RATE	55-64 AT EXAM PERSON EXAMS	# OF EVENTS	ANNUAL RATE	65-74 AT EXAM PERSON EXAMS	# OF EVENTS	ANNUAL RATE
84-204	1252	5	2	1737	31	9	2169	79	18	1145	73	32
205-234	1246	9	4	2280	61	13	2319	121	26	1089	70	32
235-264	870	11	6	1733	50	14	1832	106	29	797	51	32
265-294	474	10	11	985	36	18	879	60	34	390	34	44
295-1124	291	7	12	552	32	29	435	29	33	169	16	47
UNKNOWN	334	4	6	348	8	11	125	9	36	16	1	31
TOTAL	4467	46	5	7635	218	14	7759	404	26	3606	245	34

RISK FACTOR	75-84 AT EXAM PERSON EXAMS	# OF EVENTS	ANNUAL RATE	85-94 AT EXAM PERSON EXAMS	# OF EVENTS	ANNUAL RATE	35-64 COMBINED # OF EVENTS	CRUDE	AGE-ADJUSTED	65-94 COMBINED # OF EVENTS	CRUDE	AGE-ADJUSTED
84-204	314	35	56	12	2	83	115	11	11	110	37	37
205-234	295	30	51	8	2	125	191	16	16	102	37	36
235-264	153	17	56	4	1	125	167	19	18	69	36	37
265-294	76	8	53	0	0	0	106	23	23	42	44	45
295-1124	21	2	48	0	0	-	68	27	27	18	47	47
UNKNOWN	0	0	-	0	0	-	21	13	20	1	31	25
TOTAL	859	92	54	30	5	83	668	17	17	342	38	38

LOGISTIC REGRESSIONS

AGE AT EXAM		COEFFICIENT OF RISK FACTOR	SE COEFFICIENT	P-VALUE	STANDARDIZED COEFFICIENT	STD SE
35-44		0.010	0.003	0.001	0.465	0.112
45-54		0.007	0.001	0.001	0.312	0.059
55-64		0.005	0.001	0.001	0.184	0.049
65-74		0.002	0.002	0.195	0.083	0.066
75-84		0.000	0.003	0.916	0.012	0.110
85-94		-0.001	0.012	0.944	-0.035	0.498
35-64	UNIVARIATE	0.006	0.001	0.001	0.243	0.036
65-94	UNIVARIATE	0.001	0.001	0.403	0.047	0.056
35-64	BIVARIATE	0.006	0.001	0.001	0.269	0.057
65-94	BIVARIATE	0.002	0.001	0.200	0.072	0.056
35-64	MULTIVARIATE	0.005	0.001	0.001	0.230	0.038
65-94	MULTIVARIATE	0.002	0.001	0.112	0.091	0.057

SEX: FEMALE
EVENT: CARDIOVASCULAR DISEASE
RISK FACTOR: SERUM CHOLESTEROL
PERSONS AT RISK: PERSONS FREE OF CARDIOVASCULAR DISEASE

RISK FACTOR	35-44 AT EXAM PERSON EXAMS	# OF EVENTS	ANNUAL RATE	45-54 AT EXAM PERSON EXAMS	# OF EVENTS	ANNUAL RATE	55-64 AT EXAM PERSON EXAMS	# OF EVENTS	ANNUAL RATE	65-74 AT EXAM PERSON EXAMS	# OF EVENTS	ANNUAL RATE
84-204	2349	5	1	2079	18	4	1579	34	11	741	36	24
205-234	1408	4	1	2532	30	6	2511	63	12	1320	43	16
235-264	805	3	2	2242	18	4	2866	65	11	1620	57	18
265-294	369	3	4	1454	19	7	2173	77	18	1184	53	22
295-1124	170	2	6	1087	24	11	1823	77	21	984	62	32
UNKNOWN	500	0	0	474	9	9	229	7	15	37	2	27
TOTAL	5601	17	2	9868	118	6	11203	323	14	5886	253	21

RISK FACTOR	75-84 AT EXAM PERSON EXAMS	# OF EVENTS	ANNUAL RATE	85-94 AT EXAM PERSON EXAMS	# OF EVENTS	ANNUAL RATE	35-64 COMBINED # OF EVENTS	CRUDE	AGE-ADJUSTED	65-94 COMBINED # OF EVENTS	CRUDE	AGE-ADJUSTED
84-204	302	27	45	38	6	79	57	5	6	69	32	29
205-234	419	38	45	23	3	65	97	7	7	84	24	23
235-264	495	42	42	29	4	69	86	7	7	103	24	24
265-294	318	22	35	6	3	250	99	12	11	78	26	28
295-1124	198	8	20	5	0	0	103	17	14	70	29	29
UNKNOWN	7	1	71	0	0	-	16	7	10	3	34	37
TOTAL	1739	138	40	101	16	79	458	9	9	407	26	26

LOGISTIC REGRESSIONS

AGE AT EXAM		COEFFICIENT OF RISK FACTOR	SE COEFFICIENT	P-VALUE	STANDARDIZED COEFFICIENT	STD SE
35-44		0.015	0.004	0.001	0.589	0.152
45-54		0.005	0.002	0.002	0.213	0.069
55-64		0.006	0.001	0.001	0.252	0.053
65-74		0.005	0.001	0.017	0.151	0.063
75-84		-0.005	0.002	0.023	-0.209	0.092
85-94		-0.003	0.006	0.625	-0.137	0.281
35-64	UNIVARIATE	0.008	0.001	0.001	0.370	0.042
65-94	UNIVARIATE	-0.000	0.001	0.771	-0.015	0.051
35-64	BIVARIATE	0.005	0.001	0.001	0.249	0.042
65-94	BIVARIATE	0.001	0.001	0.641	0.024	0.052
35-64	MULTIVARIATE	0.004	0.001	0.001	0.202	0.043
65-94	MULTIVARIATE	0.001	0.001	0.446	0.040	0.052

TABLE 20-5   DATE=14NOV85
AVERAGE ANNUAL INCIDENCE RATE FOR EVENT PER 1000 PERSONS AT RISK AT EXAM
BY AGE AND LEVEL OF RISK FACTOR AT EXAM: FRAMINGHAM STUDY 30-YEAR FOLLOWUP

SEX: MALE
EVENT: CARDIOVASCULAR DISEASE
RISK FACTOR: HEMATOCRIT
PERSONS AT RISK: PERSONS FREE OF CARDIOVASCULAR DISEASE

RISK FACTOR	35-44 AT EXAM PERSON EXAMS	# OF EVENTS	ANNUAL RATE	45-54 AT EXAM PERSON EXAMS	# OF EVENTS	ANNUAL RATE	55-64 AT EXAM PERSON EXAMS	# OF EVENTS	ANNUAL RATE	65-74 AT EXAM PERSON EXAMS	# OF EVENTS	ANNUAL RATE
25-42	621	7	6	933	18	10	990	52	26	416	24	29
43-44	546	1	1	956	25	13	960	43	22	461	29	31
45-46	983	4	2	1836	49	13	1883	86	23	905	54	30
47-48	774	5	3	1469	42	14	1540	77	25	800	52	33
49-70	1173	24	10	2036	73	18	2233	137	31	1005	85	42
UNKNOWN	370	5	7	385	11	.14	153	9	29	19	1	26
TOTAL	4467	46	5	7635	218	14	7759	404	26	3606	245	34

RISK FACTOR	75-84 AT EXAM PERSON EXAMS	# OF EVENTS	ANNUAL RATE	85-94 AT EXAM PERSON EXAMS	# OF EVENTS	ANNUAL RATE	35-64 COMBINED # OF EVENTS	CRUDE	AGE-ADJUSTED	65-94 COMBINED # OF EVENTS	CRUDE	AGE-ADJUSTED
25-42	185	19	51	11	1	45	77	15	15	44	36	33
43-44	100	14	70	4	0	0	69	14	14	43	38	39
45-46	205	23	56	6	2	167	139	15	15	79	35	36
47-48	165	14	42	6	0	0	124	16	16	66	34	34
49-70	204	22	54	3	2	333	234	21	21	109	45	46
UNKNOWN	0	0	-	0	0	-	25	14	19	1	26	21
TOTAL	859	92	54	30	5	83	668	17	17	342	38	38

LOGISTIC REGRESSIONS

AGE AT EXAM		COEFFICIENT OF RISK FACTOR	SE COEFFICIENT	P-VALUE	STANDARDIZED COEFFICIENT	STD SE
35-44		0.130	0.043	0.003	0.476	0.158
45-54		0.047	0.020	0.018	0.167	0.071
55-64		0.032	0.014	0.020	0.122	0.052
65-74		0.038	0.019	0.042	0.136	0.067
75-84		-0.005	0.027	0.856	-0.020	0.110
85-94		0.208	0.154	0.175	0.904	0.666
35-64	UNIVARIATE	0.044	0.011	0.001	0.161	0.040
65-94	UNIVARIATE	0.019	0.015	0.211	0.071	0.057
35-64	BIVARIATE	0.042	0.011	0.001	0.156	0.040
65-94	BIVARIATE	0.029	0.015	0.053	0.110	0.057
35-64	MULTIVARIATE	0.019	0.011	0.082	0.072	0.041
65-94	MULTIVARIATE	0.018	0.016	0.248	0.068	0.059

SEX: FEMALE
EVENT: CARDIOVASCULAR DISEASE
RISK FACTOR: HEMATOCRIT
PERSONS AT RISK: PERSONS FREE OF CARDIOVASCULAR DISEASE

RISK FACTOR	35-44 AT EXAM PERSON EXAMS	# OF EVENTS	ANNUAL RATE	45-54 AT EXAM PERSON EXAMS	# OF EVENTS	ANNUAL RATE	55-64 AT EXAM PERSON EXAMS	# OF EVENTS	ANNUAL RATE	65-74 AT EXAM PERSON EXAMS	# OF EVENTS	ANNUAL RATE
21-39	1695	5	1	1832	19	5	1512	43	14	604	31	26
40-41	966	0	0	1735	16	5	2022	54	13	998	43	22
42-43	1039	4	2	1941	19	5	2223	46	10	1190	35	15
44-45	853	4	2	2164	20	5	2779	79	14	1525	70	23
46-65	482	3	3	1657	32	10	2402	92	19	1531	72	24
UNKNOWN	566	1	1	539	12	11	265	9	17	40	2	25
TOTAL	5601	17	2	9868	118	6	11203	323	14	5886	253	21

RISK FACTOR	75-84 AT EXAM PERSON EXAMS	# OF EVENTS	ANNUAL RATE	85-94 AT EXAM PERSON EXAMS	# OF EVENTS	ANNUAL RATE	35-64 COMBINED # OF EVENTS	CRUDE	AGE-ADJUSTED	65-94 COMBINED # OF EVENTS	CRUDE	AGE-ADJUSTED
21-39	255	24	47	17	4	118	67	7	8	59	34	32
40-41	332	30	45	25	4	80	70	7	7	77	28	28
42-43	327	19	29	18	4	111	69	7	7	58	19	19
44-45	439	27	31	21	2	48	103	9	8	99	25	25
46-65	380	37	49	20	2	50	127	14	12	111	29	30
UNKNOWN	6	1	83	0	0	-	22	8	11	3	33	38
TOTAL	1739	138	40	101	16	79	458	9	9	407	26	26

LOGISTIC REGRESSIONS

AGE AT EXAM		COEFFICIENT OF RISK FACTOR	SE COEFFICIENT	P-VALUE	STANDARDIZED COEFFICIENT	STD SE
35-44		0.085	0.072	0.236	0.308	0.260
45-54		0.074	0.028	0.008	0.265	0.100
55-64		0.047	0.017	0.005	0.162	0.058
65-74		0.007	0.019	0.716	0.024	0.065
75-84		-0.013	0.024	0.585	-0.048	0.088
85-94		-0.070	0.069	0.311	-0.282	0.278
35-64	UNIVARIATE	0.082	0.014	0.001	0.295	0.049
65-94	UNIVARIATE	-0.011	0.014	0.430	-0.040	0.051
35-64	BIVARIATE	0.050	0.014	0.001	0.180	0.051
65-94	BIVARIATE	-0.005	0.014	0.745	-0.017	0.051
35-64	MULTIVARIATE	0.020	0.014	0.147	0.073	0.050
65-94	MULTIVARIATE	-0.011	0.014	0.439	-0.039	0.051

TABLE 20-6   DATE=14NOV85

AVERAGE ANNUAL INCIDENCE RATE FOR EVENT PER 1000 PERSONS AT RISK AT EXAM
BY AGE AND LEVEL OF RISK FACTOR AT EXAM: FRAMINGHAM STUDY 30-YEAR FOLLOWUP

SEX: MALE
EVENT: CARDIOVASCULAR DISEASE
RISK FACTOR: BLOOD GLUCOSE (MG/DL)
PERSONS AT RISK: PERSONS FREE OF CARDIOVASCULAR DISEASE

RISK FACTOR	35-44 AT EXAM			45-54 AT EXAM			55-64 AT EXAM			65-74 AT EXAM		
	PERSON EXAMS	# OF EVENTS	ANNUAL RATE	PERSON EXAMS	# OF EVENTS	ANNUAL RATE	PERSON EXAMS	# OF EVENTS	ANNUAL RATE	PERSON EXAMS	# OF EVENTS	ANNUAL RATE
29-69	992	9	5	1475	43	15	1043	51	24	399	21	26
70-89	2498	25	5	4127	98	12	3928	205	26	1635	101	31
90-109	626	6	5	1334	42	16	1795	90	25	966	70	36
110-129	104	1	5	298	16	27	515	27	26	308	25	41
130-524	47	2	21	185	13	35	386	27	35	282	27	48
UNKNOWN	200	3	8	216	6	14	92	4	22	16	1	31
TOTAL	4467	46	5	7635	218	14	7759	404	26	3606	245	34

RISK FACTOR	75-84 AT EXAM			85-94 AT EXAM			35-64 COMBINED			65-94 COMBINED		
	PERSON EXAMS	# OF EVENTS	ANNUAL RATE	PERSON EXAMS	# OF EVENTS	ANNUAL RATE	# OF EVENTS	CRUDE	AGE-ADJUSTED	# OF EVENTS	CRUDE	AGE-ADJUSTED
29-69	59	5	42	10	0	0	103	15	16	26	28	29
70-89	330	31	47	10	1	50	328	16	16	132	33	34
90-109	273	35	64	10	1	0	138	18	17	106	42	42
110-129	97	8	41	1	0	0	138	24	22	33	41	40
130-524	100	13	65	8	4	250	42	34	32	44	56	52
UNKNOWN	0	0	-	0	0	-	13	13	16	1	31	25
TOTAL	859	92	54	30	5	83	668	17	17	342	38	38

LOGISTIC REGRESSIONS

AGE AT EXAM		COEFFICIENT OF RISK FACTOR	SE COEFFICIENT	P-VALUE	STANDARDIZED COEFFICIENT	STD SE
35-44		0.015	0.005	0.002	0.241	0.079
45-54		0.007	0.002	0.001	0.163	0.044
55-64		0.002	0.002	0.189	0.060	0.046
65-74		0.005	0.002	0.003	0.153	0.051
75-84		0.006	0.003	0.035	0.195	0.092
85-94		0.038	0.016	0.015	1.444	0.594
35-64	UNIVARIATE	0.007	0.001	0.001	0.157	0.027
65-94	UNIVARIATE	0.006	0.001	0.001	0.182	0.043
35-64	BIVARIATE	0.004	0.001	0.001	0.097	0.030
65-94	BIVARIATE	0.005	0.001	0.001	0.171	0.044
35-64	MULTIVARIATE	0.004	0.001	0.005	0.087	0.031
65-94	MULTIVARIATE	0.004	0.001	0.001	0.146	0.045

SEX: FEMALE
EVENT: CARDIOVASCULAR DISEASE
RISK FACTOR: BLOOD GLUCOSE (MG/DL)
PERSONS AT RISK: PERSONS FREE OF CARDIOVASCULAR DISEASE

RISK FACTOR	35-44 AT EXAM			45-54 AT EXAM			55-64 AT EXAM			65-74 AT EXAM		
	PERSON EXAMS	# OF EVENTS	ANNUAL RATE	PERSON EXAMS	# OF EVENTS	ANNUAL RATE	PERSON EXAMS	# OF EVENTS	ANNUAL RATE	PERSON EXAMS	# OF EVENTS	ANNUAL RATE
29-69	1063	3	1	1787	16	4	1346	45	17	487	18	18
70-89	3401	10	1	5699	60	5	5963	140	12	2951	118	20
90-109	671	2	1	1612	12	4	2616	74	14	1624	64	20
110-129	96	0	0	298	6	10	705	30	21	473	21	22
130-524	39	1	13	145	15	52	413	26	31	321	30	47
UNKNOWN	331	1	2	327	9	14	160	8	25	30	2	33
TOTAL	5601	17	2	9868	118	6	11203	323	14	5886	253	21

RISK FACTOR	75-84 AT EXAM			85-94 AT EXAM			35-64 COMBINED			65-94 COMBINED		
	PERSON EXAMS	# OF EVENTS	ANNUAL RATE	PERSON EXAMS	# OF EVENTS	ANNUAL RATE	# OF EVENTS	CRUDE	AGE-ADJUSTED	# OF EVENTS	CRUDE	AGE-ADJUSTED
29-69	96	12	63	3	0	0	64	8	9	30	26	28
70-89	748	45	30	40	5	63	210	7	7	168	22	23
90-109	559	40	36	40	6	75	88	9	9	110	25	24
110-129	193	20	52	9	2	111	36	16	13	43	32	30
130-524	142	20	70	9	3	167	42	35	35	53	56	54
UNKNOWN	1	1	500	0	0	-	18	11	16	3	48	138
TOTAL	1739	138	40	101	16	79	458	9	9	407	26	26

LOGISTIC REGRESSIONS

AGE AT EXAM		COEFFICIENT OF RISK FACTOR	SE COEFFICIENT	P-VALUE	STANDARDIZED COEFFICIENT	STD SE
35-44		0.017	0.005	0.001	0.264	0.082
45-54		0.017	0.002	0.001	0.310	0.041
55-64		0.008	0.001	0.001	0.208	0.035
65-74		0.007	0.002	0.001	0.192	0.045
75-84		0.007	0.002	0.001	0.228	0.062
85-94		0.012	0.010	0.209	0.295	0.235
35-64	UNIVARIATE	0.013	0.001	0.001	0.266	0.023
65-94	UNIVARIATE	0.008	0.001	0.001	0.219	0.034
35-64	BIVARIATE	0.010	0.001	0.001	0.213	0.025
65-94	BIVARIATE	0.007	0.001	0.001	0.200	0.035
35-64	MULTIVARIATE	0.008	0.001	0.001	0.176	0.026
65-94	MULTIVARIATE	0.006	0.001	0.001	0.173	0.036

TABLE 20-7   DATE=14NOV85
AVERAGE ANNUAL INCIDENCE RATE FOR EVENT PER 1000 PERSONS AT RISK AT EXAM
BY AGE AND LEVEL OF RISK FACTOR AT EXAM: FRAMINGHAM STUDY 30-YEAR FOLLOWUP

SEX: MALE
EVENT: CARDIOVASCULAR DISEASE
RISK FACTOR: DIABETES MELLITUS
PERSONS AT RISK: PERSONS FREE OF CARDIOVASCULAR DISEASE

RISK FACTOR	35-44 AT EXAM			45-54 AT EXAM			55-64 AT EXAM			65-74 AT EXAM		
	PERSON EXAMS	# OF EVENTS	ANNUAL RATE	PERSON EXAMS	# OF EVENTS	ANNUAL RATE	PERSON EXAMS	# OF EVENTS	ANNUAL RATE	PERSON EXAMS	# OF EVENTS	ANNUAL RATE
NO	4421	44	5	7411	202	14	7310	360	25	3267	204	31
YES	46	2	22	224	16	36	449	44	49	339	41	60
UNKNOWN	0	0	-	0	0	-	0	0	-	0	0	-
TOTAL	4467	46	5	7635	218	14	7759	404	26	3606	245	34

RISK FACTOR	75-84 AT EXAM			85-94 AT EXAM			35-64 COMBINED			65-94 COMBINED		
	PERSON EXAMS	# OF EVENTS	ANNUAL RATE	PERSON EXAMS	# OF EVENTS	ANNUAL RATE	# OF EVENTS	CRUDE	AGE-ADJUSTED	# OF EVENTS	CRUDE	AGE-ADJUSTED
NO	769	78	51	25	3	64	606	16	16	285	35	35
YES	90	14	78	5	2	200	62	43	38	57	66	65
UNKNOWN	0	0	-	0	0	-	0	-		0	-	
TOTAL	859	92	54	30	5	83	668	17	17	342	38	38

LOGISTIC REGRESSIONS

AGE AT EXAM		COEFFICIENT OF RISK FACTOR	SE COEFFICIENT	P-VALUE	STANDARDIZED COEFFICIENT	STD SE
35-44		1.509	0.739	0.041	0.152	0.075
45-54		1.010	0.269	0.001	0.170	0.045
55-64		0.741	0.168	0.001	0.173	0.039
65-74		0.726	0.182	0.001	0.212	0.053
75-84		0.490	0.314	0.119	0.150	0.096
85-94		1.587	1.101	0.149	0.602	0.417
35-64	UNIVARIATE	1.060	0.139	0.001	0.198	0.026
65-94	UNIVARIATE	0.695	0.155	0.001	0.205	0.046
35-64	BIVARIATE	0.779	0.141	0.001	0.145	0.026
65-94	BIVARIATE	0.672	0.155	0.001	0.198	0.046
35-64	MULTIVARIATE	0.694	0.147	0.001	0.130	0.027
65-94	MULTIVARIATE	0.637	0.160	0.001	0.188	0.047

SEX: FEMALE
EVENT: CARDIOVASCULAR DISEASE
RISK FACTOR: DIABETES MELLITUS
PERSONS AT RISK: PERSONS FREE OF CARDIOVASCULAR DISEASE

RISK FACTOR	35-44 AT EXAM			45-54 AT EXAM			55-64 AT EXAM			65-74 AT EXAM		
	PERSON EXAMS	# OF EVENTS	ANNUAL RATE	PERSON EXAMS	# OF EVENTS	ANNUAL RATE	PERSON EXAMS	# OF EVENTS	ANNUAL RATE	PERSON EXAMS	# OF EVENTS	ANNUAL RATE
NO	5558	15	1	9683	107	6	10713	291	14	5469	220	20
YES	43	2	23	185	11	30	490	32	33	417	33	40
UNKNOWN	0	0	-	0	0	-	0	0	-	0	0	-
TOTAL	5601	17	2	9868	118	6	11203	323	14	5886	253	21

RISK FACTOR	75-84 AT EXAM			85-94 AT EXAM			35-64 COMBINED			65-94 COMBINED		
	PERSON EXAMS	# OF EVENTS	ANNUAL RATE	PERSON EXAMS	# OF EVENTS	ANNUAL RATE	# OF EVENTS	CRUDE	AGE-ADJUSTED	# OF EVENTS	CRUDE	AGE-ADJUSTED
NO	1544	112	36	89	12	67	413	8	8	344	24	24
YES	191	26	68	12	4	167	45	31	30	63	51	48
UNKNOWN	0	0	-	0	0	-	0	-		0	-	
TOTAL	1739	138	40	101	16	79	458	9	9	407	26	26

LOGISTIC REGRESSIONS

AGE AT EXAM		COEFFICIENT OF RISK FACTOR	SE COEFFICIENT	P-VALUE	STANDARDIZED COEFFICIENT	STD SE
35-44		2.892	0.769	0.001	0.252	0.067
45-54		1.733	0.326	0.001	0.235	0.044
55-64		0.917	0.192	0.001	0.188	0.039
65-74		0.718	0.194	0.001	0.184	0.050
75-84		0.703	0.233	0.003	0.220	0.073
85-94		1.166	0.687	0.090	0.379	0.223
35-64	UNIVARIATE	1.420	0.162	0.001	0.230	0.026
65-94	UNIVARIATE	0.799	0.144	0.001	0.217	0.039
35-64	BIVARIATE	1.084	0.164	0.001	0.176	0.027
65-94	BIVARIATE	0.725	0.145	0.001	0.197	0.039
35-64	MULTIVARIATE	0.826	0.169	0.001	0.154	0.027
65-94	MULTIVARIATE	0.533	0.152	0.001	0.145	0.041

TABLE 20-8   DATE=14NOV85
AVERAGE ANNUAL INCIDENCE RATE FOR EVENT PER 1000 PERSONS AT RISK AT EXAM
BY AGE AND LEVEL OF RISK FACTOR AT EXAM: FRAMINGHAM STUDY 30-YEAR FOLLOWUP

SEX: MALE
EVENT: CARDIOVASCULAR DISEASE
RISK FACTOR: GLUCOSE IN URINE, DEFINITE OR TRACE
PERSONS AT RISK: PERSONS FREE OF CARDIOVASCULAR DISEASE

RISK FACTOR	35-44 AT EXAM			45-54 AT EXAM			55-64 AT EXAM			65-74 AT EXAM		
	PERSON EXAMS	# OF EVENTS	ANNUAL RATE	PERSON EXAMS	# OF EVENTS	ANNUAL RATE	PERSON EXAMS	# OF EVENTS	ANNUAL RATE	PERSON EXAMS	# OF EVENTS	ANNUAL RATE
NO	4377	44	5	7453	207	14	7546	381	25	3487	229	33
YES	58	1	9	161	11	34	189	22	58	114	15	66
UNKNOWN	32	1	16	21	0	0	24	1	21	5	1	100
TOTAL	4467	46	5	7635	218	14	7759	404	26	3606	245	34

RISK FACTOR	75-84 AT EXAM			85-94 AT EXAM			35-64 COMBINED			65-94 COMBINED		
	PERSON EXAMS	# OF EVENTS	ANNUAL RATE	PERSON EXAMS	# OF EVENTS	ANNUAL RATE	# OF EVENTS	CRUDE	AGE-ADJUSTED	# OF EVENTS	CRUDE	AGE-ADJUSTED
NO	839	88	52	29	4	69	632	16	16	321	37	37
YES	20	4	100	1	1	500	34	42	38	20	74	75
UNKNOWN	0	0	-	0	0	-	2	13	12	1	100	80
TOTAL	859	92	54	30	5	83	668	17	17	342	38	38

LOGISTIC REGRESSIONS

AGE AT EXAM		COEFFICIENT OF RISK FACTOR	SE COEFFICIENT	P-VALUE	STANDARDIZED COEFFICIENT	STD SE
35-44		0.547	1.020	0.592	0.062	0.116
45-54		0.943	0.320	0.003	0.136	0.046
55-64		0.907	0.233	0.001	0.140	0.036
65-74		0.768	0.285	0.007	0.134	0.050
75-84		0.758	0.570	0.184	0.114	0.086
85-94		-	-	-	-	-
35-64	UNIVARIATE	0.992	0.184	0.001	0.141	0.026
65-94	UNIVARIATE	0.782	0.249	0.002	0.134	0.043
35-64	BIVARIATE	0.880	0.186	0.001	0.125	0.026
65-94	BIVARIATE	0.810	0.250	0.001	0.138	0.043
35-64	MULTIVARIATE	0.748	0.199	0.001	0.106	0.028
65-94	MULTIVARIATE	0.793	0.256	0.002	0.135	0.044

SEX: FEMALE
EVENT: CARDIOVASCULAR DISEASE
RISK FACTOR: GLUCOSE IN URINE, DEFINITE OR TRACE
PERSONS AT RISK: PERSONS FREE OF CARDIOVASCULAR DISEASE

RISK FACTOR	35-44 AT EXAM			45-54 AT EXAM			55-64 AT EXAM			65-74 AT EXAM		
	PERSON EXAMS	# OF EVENTS	ANNUAL RATE	PERSON EXAMS	# OF EVENTS	ANNUAL RATE	PERSON EXAMS	# OF EVENTS	ANNUAL RATE	PERSON EXAMS	# OF EVENTS	ANNUAL RATE
NO	5522	15	1	9758	111	6	11045	309	14	5785	242	21
YES	41	2	24	86	6	35	135	14	52	91	11	60
UNKNOWN	38	0	0	24	1	21	23	0	0	10	0	0
TOTAL	5601	17	2	9868	118	6	11203	323	14	5886	253	21

RISK FACTOR	75-84 AT EXAM			85-94 AT EXAM			35-64 COMBINED			65-94 COMBINED		
	PERSON EXAMS	# OF EVENTS	ANNUAL RATE	PERSON EXAMS	# OF EVENTS	ANNUAL RATE	# OF EVENTS	CRUDE	AGE-ADJUSTED	# OF EVENTS	CRUDE	AGE-ADJUSTED
NO	1710	131	38	100	15	75	435	8	8	388	26	26
YES	20	7	175	1	1	500	22	42	40	19	85	92
UNKNOWN	9	0	0	0	0	-	1	6	8	0	0	0
TOTAL	1739	138	40	101	16	79	458	9	9	407	26	26

LOGISTIC REGRESSIONS

AGE AT EXAM		COEFFICIENT OF RISK FACTOR	SE COEFFICIENT	P-VALUE	STANDARDIZED COEFFICIENT	STD SE
35-44		2.935	0.770	0.001	0.251	0.066
45-54		1.875	0.434	0.001	0.174	0.040
55-64		1.391	0.288	0.001	0.152	0.031
65-74		1.147	0.328	0.001	0.142	0.041
75-84		1.870	0.478	0.001	0.200	0.051
85-94		-	-	-	-	-
35-64	UNIVARIATE	1.697	0.228	0.001	0.168	0.023
65-94	UNIVARIATE	1.334	0.257	0.001	0.160	0.031
35-64	BIVARIATE	1.608	0.232	0.001	0.159	0.023
65-94	BIVARIATE	1.384	0.259	0.001	0.166	0.031
35-64	MULTIVARIATE	1.258	0.244	0.001	0.124	0.024
65-94	MULTIVARIATE	1.139	0.271	0.001	0.136	0.032

TABLE 20-9   DATE=10MAR86
AVERAGE ANNUAL INCIDENCE RATE FOR EVENT PER 1000 PERSONS AT RISK AT EXAM
BY AGE AND LEVEL OF RISK FACTOR AT EXAM: FRAMINGHAM STUDY 30-YEAR FOLLOWUP

SEX: MALE
EVENT: CARDIOVASCULAR DISEASE
RISK FACTOR: GLUCOSE INTOLERANCE
PERSONS AT RISK: PERSONS FREE OF CARDIOVASCULAR DISEASE

35-44 AT EXAM			45-54 AT EXAM			55-64 AT EXAM			65-74 AT EXAM		
RSON AMS	# OF EVENTS	ANNUAL RATE	PERSON EXAMS	# OF EVENTS	ANNUAL RATE	PERSON EXAMS	# OF EVENTS	ANNUAL RATE	PERSON EXAMS	# OF EVENTS	ANNUAL RATE
4332	43	5	7150	190	13	6919	341	25	3019	187	31
135	3	11	485	28	29	840	63	38	587	58	49
0	0	-	0	0	-	0	0	-	0	0	-
4467	46	5	7635	218	14	7759	404	26	3606	245	34

75-84 AT EXAM			85-94 AT EXAM			35-64 COMBINED			65-94 COMBINED		
RSON AMS	# OF EVENTS	ANNUAL RATE	PERSON EXAMS	# OF EVENTS	ANNUAL RATE	# OF EVENTS	CRUDE	AGE-ADJUSTED	# OF EVENTS	CRUDE	AGE-ADJUSTED
686	70	51	22	1	23	574	16	16	258	35	35
173	22	64	8	4	250	94	32	28	84	55	53
0	0	-	0	0	-	0	-	-	0	-	-
859	92	54	30	5	83	668	17	17	342	38	38

LOGISTIC REGRESSIONS

	COEFFICIENT OF RISK FACTOR	SE COEFFICIENT	P-VALUE	STANDARDIZED COEFFICIENT	STD SE
	0.818	0.604	0.175	0.140	0.103
	0.808	0.208	0.001	0.197	0.051
	0.447	0.142	0.002	0.139	0.044
	0.507	0.158	0.001	0.187	0.058
	0.249	0.261	0.341	0.100	0.105
	3.045	1.244	0.014	1.369	0.560
IATE IATE	0.759	0.115	0.001	0.198	0.030
	0.502	0.132	0.001	0.189	0.050
ITE ITE	0.522	0.116	0.001	0.136	0.030
	0.461	0.133	0.001	0.174	0.050
IRIATE IRIATE	0.439	0.121	0.001	0.114	0.032
	0.424	0.136	0.002	0.160	0.051 *

SEX: FEMALE
EVENT: CARDIOVASCULAR DISEASE
RISK FACTOR: GLUCOSE INTOLERANCE
PERSONS AT RISK: PERSONS FREE OF CARDIOVASCULAR DISEASE

35-44 AT EXAM			45-54 AT EXAM			55-64 AT EXAM			65-74 AT EXAM		
SON MS	# OF EVENTS	ANNUAL RATE	PERSON EXAMS	# OF EVENTS	ANNUAL RATE	PERSON EXAMS	# OF EVENTS	ANNUAL RATE	PERSON EXAMS	# OF EVENTS	ANNUAL RATE
507	15	1	9517	100	5	10242	277	14	5112	200	20
94	2	11	351	18	26	961	46	24	774	53	34
0	0	-	0	0	-	0	0	-	0	0	-
601	17	2	9868	118	6	11203	323	14	5886	253	21

75-84 AT EXAM			85-94 AT EXAM			35-64 COMBINED			65-94 COMBINED		
SON MS	# OF EVENTS	ANNUAL RATE	PERSON EXAMS	# OF EVENTS	ANNUAL RATE	# OF EVENTS	CRUDE	AGE-ADJUSTED	# OF EVENTS	CRUDE	AGE-ADJUSTED
412	101	36	83	9	54	392	8	8	310	23	24
327	37	57	18	7	194	66	23	22	97	43	41
0	0	-	0	0	-	0	-	-	0	-	-
739	138	40	101	16	79	458	9	9	407	26	26

LOGISTIC REGRESSIONS

	COEFFICIENT OF RISK FACTOR	SE COEFFICIENT	P-VALUE	STANDARDIZED COEFFICIENT	STD SE
	2.074	0.760	0.006	0.266	0.098
	1.627	0.262	0.001	0.301	0.049
	0.593	0.163	0.001	0.166	0.046
	0.591	0.160	0.001	0.200	0.054
	0.504	0.203	0.013	0.197	0.079
	1.655	0.599	0.006	0.636	0.230
ATE ATE	1.140	0.136	0.001	0.255	0.030
	0.656	0.121	0.001	0.231	0.043
TE TE	0.796	0.138	0.001	0.178	0.031
	0.591	0.122	0.001	0.208	0.043
RIATE RIATE	0.678	0.141	0.001	0.152	0.032
	0.474	0.126	0.001	0.167	0.044

TABLE 20-10 : DATE=14NOV85
AVERAGE ANNUAL INCIDENCE RATE FOR EVENT PER 1000 PERSONS AT RISK AT EXAM
BY AGE AND LEVEL OF RISK FACTOR AT EXAM: FRAMINGHAM STUDY 30-YEAR FOLLOWUP

SEX: MALE
EVENT: CARDIOVASCULAR DISEASE
RISK FACTOR: METROPOLITAN RELATIVE WEIGHT (PERCENT)
PERSONS AT RISK: PERSONS FREE OF CARDIOVASCULAR -DISEASE

RISK FACTOR	35-44 AT EXAM PERSON EXAMS	# OF EVENTS	ANNUAL RATE	45-54 AT EXAM PERSON EXAMS	# OF EVENTS	ANNUAL RATE	55-64 AT EXAM PERSON EXAMS	# OF EVENTS	ANNUAL RATE	65-74 AT EXAM PERSON EXAMS	# OF EVENTS	ANNUAL RATE
54-104	719	3	2	1119	19	8	1096	59	27	610	37	30
105-114	965	9	5	1519	42	14	1594	71	22	780	40	26
115-124	1122	8	4	2017	58	14	2105	102	24	1008	80	40
125-134	914	13	7	1621	51	16	1618	83	26	703	48	34
135-272	747	13	9	1358	48	18	1345	89	33	505	40	40
UNKNOWN	0	0	-	1	0	0	1	0	0	0	0	-
TOTAL	4467	46	5	7635	218	14	7759	404	26	3606	245	34

RISK FACTOR	75-84 AT EXAM PERSON EXAMS	# OF EVENTS	ANNUAL RATE	85-94 AT EXAM PERSON EXAMS	# OF EVENTS	ANNUAL RATE	35-64 COMBINED # OF EVENTS	CRUDE	AGE-ADJUSTED	65-94 COMBINED # OF EVENTS	CRUDE	AGE-ADJUSTED
54-104	188	17	45	14	2	71	81	14	14	56	34	33
105-114	208	22	53	6	2	167	122	15	15	64	32	32
115-124	256	23	45	10	1	50	168	16	16	104	41	41
125-134	135	18	67	0	0	-	147	18	18	66	39	40
135-272	72	12	83	0	0	-	150	22	22	52	45	48
UNKNOWN	0	0	-	0	0	-	0	0	0	0	-	
TOTAL	859	92	54	30	5	83	668	17	17	342	38	38

LOGISTIC REGRESSIONS

AGE AT EXAM		COEFFICIENT OF RISK FACTOR	SE COEFFICIENT	P-VALUE	STANDARDIZED COEFFICIENT	STD SE
35-44		0.020	0.009	0.017	0.330	0.138
45-54		0.011	0.004	0.007	0.179	0.066
55-64		0.006	0.003	0.058	0.096	0.050
65-74		0.007	0.004	0.084	0.113	0.066
75-84		0.013	0.007	0.072	0.192	0.107
85-94		-0.019	0.046	0.688	-0.200	0.496
35-64	UNIVARIATE	0.009	0.002	0.001	0.146	0.038
65-94	UNIVARIATE	0.007	0.004	0.058	0.106	0.056
35-64	BIVARIATE	0.009	0.002	0.001	0.145	0.039
65-94	BIVARIATE	0.009	0.004	0.013	0.141	0.056
35-64	MULTIVARIATE	0.005	0.003	0.049	0.080	0.041
65-94	MULTIVARIATE	0.003	0.004	0.447	0.044	0.058

SEX: FEMALE
EVENT: CARDIOVASCULAR DISEASE
RISK FACTOR: METROPOLITAN RELATIVE WEIGHT (PERCENT)
PERSONS AT RISK: PERSONS FREE OF CARDIOVASCULAR DISEASE

RISK FACTOR	35-44 AT EXAM PERSON EXAMS	# OF EVENTS	ANNUAL RATE	45-54 AT EXAM PERSON EXAMS	# OF EVENTS	ANNUAL RATE	55-64 AT EXAM PERSON EXAMS	# OF EVENTS	ANNUAL RATE	65-74 AT EXAM PERSON EXAMS	# OF EVENTS	ANNUAL RATE
54-104	1689	3	1	2075	24	6	1981	52	13	1138	47	21
105-114	1429	4	1	2308	15	3	2351	51	11	1101	49	22
115-124	1056	5	2	2092	26	6	2415	54	11	1313	46	18
125-134	564	1	1	1362	9	3	1711	50	15	938	43	23
135-272	855	4	2	2031	44	11	2745	116	21	1396	68	24
UNKNOWN	8	0	-	0	0	-	0	0	-	0	0	-
TOTAL	5601	17	2	9868	118	6	11203	323	14	5886	253	21

RISK FACTOR	75-84 AT EXAM PERSON EXAMS	# OF EVENTS	ANNUAL RATE	85-94 AT EXAM PERSON EXAMS	# OF EVENTS	ANNUAL RATE	35-64 COMBINED # OF EVENTS	CRUDE	AGE-ADJUSTED	65-94 COMBINED # OF EVENTS	CRUDE	AGE-ADJUSTED
54-104	458	33	36	34	7	103	79	7	8	87	27	25
105-114	326	17	26	27	2	37	70	6	6	68	23	23
115-124	351	36	51	14	5	179	85	8	7	87	26	27
125-134	242	20	41	13	0	0	60	8	8	63	26	27
135-272	362	32	44	13	2	77	164	15	13	102	29	30
UNKNOWN	0	0	-	0	0	-	0	0	0	0	-	
TOTAL	1739	138	40	101	16	79	458	9	9	407	26	26

LOGISTIC REGRESSIONS

AGE AT EXAM		COEFFICIENT OF RISK FACTOR	SE COEFFICIENT	P-VALUE	STANDARDIZED COEFFICIENT	STD SE
35-44		0.015	0.009	0.112	0.303	0.190
45-54		0.016	0.003	0.001	0.337	0.071
55-64		0.010	0.002	0.001	0.230	0.049
65-74		0.004	0.003	0.159	0.087	0.062
75-84		0.005	0.004	0.232	0.103	0.086
85-94		-0.013	0.015	0.599	-0.240	0.234
35-64	UNIVARIATE	0.014	0.002	0.001	0.318	0.038
65-94	UNIVARIATE	0.002	0.002	0.337	0.048	0.050
35-64	BIVARIATE	0.012	0.002	0.001	0.260	0.040
65-94	BIVARIATE	0.004	0.002	0.104	0.081	0.050
35-64	MULTIVARIATE	0.006	0.002	0.002	0.134	0.044
65-94	MULTIVARIATE	0.002	0.002	0.305	0.052	0.051

SEX: MALE
EVENT: CARDIOVASCULAR DISEASE
RISK FACTOR: VITAL CAPACITY - HEIGHT INDEX (ML/INCH)
PERSONS AT RISK: PERSONS FREE OF CARDIOVASCULAR DISEASE

RISK FACTOR	35-44 AT EXAM PERSON EXAMS	# OF EVENTS	ANNUAL RATE	45-54 AT EXAM PERSON EXAMS	# OF EVENTS	ANNUAL RATE	55-64 AT EXAM PERSON EXAMS	# OF EVENTS	ANNUAL RATE	65-74 PERSON EXAMS
8-44	264	3	6	1035	46	22	1920	135	35	1578
45-49	411	5	6	1108	38	17	1435	61	21	702
50-54	747	1	.1	1647	48	15	1780	88	25	674
55-59	1022	11	5	1716	47	14	1367	78	29	415
60-85	2010	25	6	2124	39	9	1239	42	17	225
UNKNOWN	13	1	.38	5	0	0	18	0	0	12
TOTAL	4467	46	5	7635	218	14	7759	404	26	3606

RISK FACTOR	75-84 AT EXAM PERSON EXAMS	# OF EVENTS	ANNUAL RATE	85-94 AT EXAM PERSON EXAMS	# OF EVENTS	ANNUAL RATE	35-64 COMBINED # OF EVENTS	CRUDE	AGE- ADJUSTED	65- # OF EVENTS
8-44	530	65	61	17	4	118	184	29	24	190
45-49	143	9	31	4	0	0	104	18	16	67
50-54	91	11	60	7	1	71	137	16	15	47
55-59	63	6	48	2	0	0	156	17	18	25
60-85	24	1	21	0	0	-	106	10	12	12
UNKNOWN	8	0	0	0	0	-	1	14	9	1
TOTAL	859	92	54	30	5	83	668	17	17	342

LOGISTIC REGRESSIONS

AGE AT EXAM		COEFFICIENT OF RISK FACTOR	SE COEFFICIENT	P-VALUE	STANDARDIZED COEFFICIENT	S S
35-44		0.000	0.017	0.998	0.000	0
45-54		-0.032	0.007	0.001	-0.292	0
55-64		-0.019	0.005	0.001	-0.181	0
65-74		-0.018	0.007	0.010	-0.169	0
75-84		-0.012	0.012	0.326	-0.109	0
85-94		-0.089	0.062	0.151	-0.797	0
35-64	UNIVARIATE	-0.036	0.004	0.001	-0.342	0
65-94	UNIVARIATE	-0.021	0.006	0.001	-0.197	0
35-64	BIVARIATE	-0.020	0.004	0.001	-0.188	0
65-94	BIVARIATE	-0.015	0.006	0.011	-0.145	0
35-64	MULTIVARIATE	-0.009	0.005	0.039	-0.089	0
65-94	MULTIVARIATE	-0.012	0.006	0.066	-0.109	0

SEX: FEMALE
EVENT: CARDIOVASCULAR DISEASE
RISK FACTOR: VITAL CAPACITY - HEIGHT INDEX (ML/INCH)
PERSONS AT RISK: PERSONS FREE OF CARDIOVASCULAR DISEASE

RISK FACTOR	35-44 AT EXAM PERSON EXAMS	# OF EVENTS	ANNUAL RATE	45-54 AT EXAM PERSON EXAMS	# OF EVENTS	ANNUAL RATE	55-64 AT EXAM PERSON EXAMS	# OF EVENTS	ANNUAL RATE	65-74 PERSON EXAMS
4-29	121	4	17	518	15	14	1373	67	24	1573
30-34	289	2	3	1211	17	7	2359	87	18	1619
35-39	766	1	1	2054	36	9	2991	74	12	1475
40-44	1423	5	2	2778	26	5	2647	68	13	827
45-92	2981	5	1	3290	24	4	1812	27	7	386
UNKNOWN	21	0	0	17	0	0	21	0	0	6
TOTAL	5601	17	2	9868	118	6	11203	323	14	5886

RISK FACTOR	75-84 AT EXAM PERSON EXAMS	# OF EVENTS	ANNUAL RATE	85-94 AT EXAM PERSON EXAMS	# OF EVENTS	ANNUAL RATE	35-64 COMBINED # OF EVENTS	CRUDE	AGE- ADJUSTED	65- # OF EVENTS
4-29	803	68	42	63	11	87	86	21	19	178
30-34	512	50	49	25	5	100	106	14	11	122
35-39	296	16	27	12	0	0	111	10	9	63
40-44	103	2	10	1	0	0	99	7	7	29
45-92	24	1	21	0	0	-	56	3	5	14
UNKNOWN	1	1	500	0	0	-	0	0	0	1
TOTAL	1739	138	40	101	16	79	458	9	9	407

LOGISTIC REGRESSIONS

AGE AT EXAM		COEFFICIENT OF RISK FACTOR	SE COEFFICIENT	P-VALUE	STANDARDIZED COEFFICIENT	S S
35-44		-0.091	0.029	0.002	-0.667	0
45-54		-0.055	0.012	0.001	-0.407	0
55-64		-0.045	0.007	0.001	-0.339	0
65-74		-0.037	0.009	0.001	-0.273	0
75-84		-0.026	0.012	0.034	-0.186	0
85-94		-0.042	0.039	0.281	-0.280	0
35-64	UNIVARIATE	-0.070	0.006	0.001	-0.550	0
65-94	UNIVARIATE	-0.043	0.007	0.001	-0.319	0
35-64	BIVARIATE	-0.045	0.006	0.001	-0.360	0
65-94	BIVARIATE	-0.034	0.007	0.001	-0.256	0
35-64	MULTIVARIATE	-0.032	0.007	0.001	-0.252	0
65-94	MULTIVARIATE	-0.029	0.007	0.001	-0.216	0

SEX: MALE
EVENT: CARDIOVASCULAR DISEASE
RISK FACTOR: HEART RATE (PER MINUTE)
PERSONS AT RISK: PERSONS FREE OF CARDIOVASCULAR DISEASE

RISK FACTOR	35-44 AT EXAM			45-54 AT EXAM			55-64 AT EXAM			65-74 AT EXAM		
	PERSON EXAMS	# OF EVENTS	ANNUAL RATE	PERSON EXAMS	# OF EVENTS	ANNUAL RATE	PERSON EXAMS	# OF EVENTS	ANNUAL RATE	PERSON EXAMS	# OF EVENTS	ANNUAL RATE
30-67	1284	6	2	1934	46	12	2256	88	20	1099	72	33
68-75	1489	12	4	2465	58	12	2405	123	26	1094	56	26
76-83	721	9	6	1374	43	16	1315	75	29	574	41	36
84-91	579	14	12	1084	39	18	1019	53	26	432	38	44
92-220	394	5	6	778	32	21	764	65	43	407	38	47
UNKNOWN	0	0	-	0	0	-	0	0	-	0	0	-
TOTAL	4467	46	5	7635	218	14	7759	404	26	3606	245	34

RISK FACTOR	75-84 AT EXAM			85-94 AT EXAM			35-64 COMBINED			65-94 COMBINED		
	PERSON EXAMS	# OF EVENTS	ANNUAL RATE	PERSON EXAMS	# OF EVENTS	ANNUAL RATE	# OF EVENTS	CRUDE	AGE-ADJUSTED	# OF EVENTS	CRUDE	AGE-ADJUSTED
30-67	257	24	47	9	0	0	140	13	13	96	35	35
68-75	255	26	51	10	2	100	193	15	15	84	31	31
76-83	155	13	42	8	1	63	127	19	19	57	37	37
84-91	111	17	77	3	2	333	106	20	20	57	52	52
92-220	81	12	74	0	0	-	102	26	26	50	51	52
UNKNOWN	0	0	-	0	0	-	0	-		0	-	
TOTAL	859	92	54	30	5	83	668	17	17	342	38	38

LOGISTIC REGRESSIONS

AGE AT EXAM		COEFFICIENT OF RISK FACTOR	SE COEFFICIENT	P-VALUE	STANDARDIZED COEFFICIENT	STD SE
35-44		0.032	0.010	0.002	0.395	0.129
45-54		0.016	0.005	0.002	0.203	0.065
55-64		0.017	0.004	0.001	0.222	0.048
65-74		0.012	0.005	0.010	0.162	0.063
75-84		0.016	0.008	0.059	0.200	0.106
85-94		0.107	0.069	0.124	0.988	0.642
35-64	UNIVARIATE	0.018	0.003	0.001	0.228	0.037
65-94	UNIVARIATE	0.013	0.004	0.001	0.171	0.053
35-64	BIVARIATE	0.018	0.003	0.001	0.230	0.037
65-94	BIVARIATE	0.013	0.004	0.001	0.174	0.054
35-64	MULTIVARIATE	0.006	0.003	0.043	0.081	0.040
65-94	MULTIVARIATE	0.010	0.004	0.024	0.126	0.056

SEX: FEMALE
EVENT: CARDIOVASCULAR DISEASE
RISK FACTOR: HEART RATE (PER MINUTE)
PERSONS AT RISK: PERSONS FREE OF CARDIOVASCULAR DISEASE

RISK FACTOR	35-44 AT EXAM			45-54 AT EXAM			55-64 AT EXAM			65-74 AT EXAM		
	PERSON EXAMS	# OF EVENTS	ANNUAL RATE	PERSON EXAMS	# OF EVENTS	ANNUAL RATE	PERSON EXAMS	# OF EVENTS	ANNUAL RATE	PERSON EXAMS	# OF EVENTS	ANNUAL RATE
30-67	1090	0	0	1881	25	7	2247	50	11	1038	38	18
68-75	1798	5	1	3227	33	5	3394	84	12	1572	73	23
76-83	1046	5	2	1886	18	5	2278	73	16	1297	60	23
84-91	939	6	3	1582	25	8	1723	59	17	1021	42	21
92-220	728	1	1	1292	17	7	1561	57	18	958	40	21
UNKNOWN	0	0	-	0	0	-	0	0	-	0	0	-
TOTAL	5601	17	2	9868	118	6	11203	323	14	5886	253	21

RISK FACTOR	75-84 AT EXAM			85-94 AT EXAM			35-64 COMBINED			65-94 COMBINED		
	PERSON EXAMS	# OF EVENTS	ANNUAL RATE	PERSON EXAMS	# OF EVENTS	ANNUAL RATE	# OF EVENTS	CRUDE	AGE-ADJUSTED	# OF EVENTS	CRUDE	AGE-ADJUSTED
30-67	321	27	42	22	7	159	75	7	7	72	26	25
68-75	419	29	35	25	2	40	122	7	7	104	26	26
76-83	372	32	43	21	0	0	96	9	9	92	27	27
84-91	300	26	43	20	4	100	90	11	11	72	27	27
92-220	327	24	37	13	3	115	75	10	10	67	26	26
UNKNOWN	0	0	-	0	0	-	-	-		0	-	
TOTAL	1739	138	40	101	16	79	458	9	9	407	26	26

LOGISTIC REGRESSIONS

AGE AT EXAM		COEFFICIENT OF RISK FACTOR	SE COEFFICIENT	P-VALUE	STANDARDIZED COEFFICIENT	STD SE
35-44		0.016	0.017	0.352	0.197	0.212
45-54		0.006	0.007	0.420	0.073	0.090
55-64		0.012	0.004	0.005	0.150	0.053
65-74		0.001	0.005	0.780	0.018	0.064
75-84		-0.000	0.006	0.938	-0.007	0.089
85-94		-0.011	0.021	0.609	-0.144	0.282
35-64	UNIVARIATE	0.011	0.004	0.003	0.133	0.045
65-94	UNIVARIATE	0.001	0.004	0.869	0.008	0.051
35-64	BIVARIATE	0.010	0.004	0.003	0.130	0.044
65-94	BIVARIATE	0.000	0.004	0.972	0.002	0.051
35-64	MULTIVARIATE	0.002	0.004	0.669	0.020	0.048
65-94	MULTIVARIATE	-0.003	0.004	0.417	-0.042	0.052

TABLE 20-13   DATE=22MAY85

AVERAGE ANNUAL INCIDENCE RATE FOR EVENT PER 1000 PERSONS AT RISK AT EXAM
BY AGE AND LEVEL OF RISK FACTOR AT EXAM: FRAMINGHAM STUDY 30-YEAR FOLLOWUP

SEX: MALE
EVENT: CARDIOVASCULAR DISEASE
RISK FACTOR: CIGARETTES SMOKED (NUMBER PER DAY)
PERSONS AT RISK: PERSONS FREE OF CARDIOVASCULAR DISEASE

RISK FACTOR	35-44 AT EXAM			45-54 AT EXAM			55-64 AT EXAM			65-74 AT EXAM		
	PERSON EXAMS	# OF EVENTS	ANNUAL RATE	PERSON EXAMS	# OF EVENTS	ANNUAL RATE	PERSON EXAMS	# OF EVENTS	ANNUAL RATE	PERSON EXAMS	# OF EVENTS	ANNUAL RATE
NONE	1382	6	2	3046	57	9	4185	173	21	2406	159	33
1-10	433	5	6	704	16	11	736	26	18	352	24	34
11-20	1483	16	5	2021	76	19	1597	120	38	473	36	38
21-40	987	13	7	1584	58	18	1076	72	33	345	23	33
41-90	156	6	19	245	9	18	146	12	41	24	2	42
UNKNOWN	26	0		35	2	29	19	1	26	6	1	83
TOTAL	4467	46	5	7635	218	14	7759	404	26	3606	245	34

RISK FACTOR	75-84 AT EXAM			85-94 AT EXAM			35-64 COMBINED			65-94 COMBINED		
	PERSON EXAMS	# OF EVENTS	ANNUAL RATE	PERSON EXAMS	# OF EVENTS	ANNUAL RATE	# OF EVENTS	CRUDE	AGE-ADJUSTED	# OF EVENTS	CRUDE	AGE-ADJUSTED
NONE	659	74	56	22	4	91	236	14	12	237	38	38
1-10	83	8	48	4	0	0	47	13	13	32	36	36
11-20	71	6	42	4	1	125	212	21	23	43	39	39
21-40	44	3	34	0	0	-	163	20	21	26	33	33
41-90	2	1	250	0	0	-	27	25	27	3	58	81
UNKNOWN	0	0	-	0	0	-	3	19	21	1	83	67
TOTAL	859	92	54	30	5	83	668	17	17	342	38	38

LOGISTIC REGRESSIONS

AGE AT EXAM	COEFFICIENT OF RISK FACTOR	SE COEFFICIENT	P-VALUE	STANDARDIZED COEFFICIENT	STD SE
35-44	0.032	0.010	0.001	0.446	0.137
45-54	0.018	0.004	0.001	0.275	0.063
55-64	0.017	0.003	0.001	0.232	0.046
65-74	0.002	0.006	0.738	0.022	0.065
75-84	-0.011	0.013	0.434	-0.095	0.122
85-94	-0.003	0.082	0.968	-0.020	0.505
35-64 UNIVARIATE	0.013	0.003	0.001	0.186	0.037
65-94 UNIVARIATE	-0.002	0.005	0.657	-0.026	0.057
35-64 BIVARIATE	0.020	0.003	0.001	0.289	0.037
65-94 BIVARIATE	0.001	0.005	0.854	0.011	0.058
35-64 MULTIVARIATE	0.023	0.003	0.001	0.333	0.037
65-94 MULTIVARIATE	0.004	0.005	0.437	0.045	0.058

SEX: FEMALE
EVENT: CARDIOVASCULAR DISEASE
RISK FACTOR: CIGARETTES SMOKED (NUMBER PER DAY)
PERSONS AT RISK: PERSONS FREE OF CARDIOVASCULAR DISEASE

RISK FACTOR	35-44 AT EXAM			45-54 AT EXAM			55-64 AT EXAM			65-74 AT EXAM		
	PERSON EXAMS	# OF EVENTS	ANNUAL RATE	PERSON EXAMS	# OF EVENTS	ANNUAL RATE	PERSON EXAMS	# OF EVENTS	ANNUAL RATE	PERSON EXAMS	# OF EVENTS	ANNUAL RATE
NONE	2577	4	1	5476	68	6	7386	204	14	4586	191	21
1-10	1312	4	2	1716	17	5	1535	39	13	593	23	19
11-20	1365	6	2	1992	23	6	1652	57	17	527	33	31
21-40	298	2	3	610	9	7	578	22	19	161	5	16
41-90	20	1	25	37	1	14	14	1	36	0	0	-
UNKNOWN	29	0	0	37	0	0	38	0	0	19	1	26
TOTAL	5601	17	2	9868	118	6	11203	323	14	5886	253	21

RISK FACTOR	74-84 AT EXAM			65-94 AT EXAM			35-64 COMBINED			65-94 COMBINED		
	PERSON EXAMS	# OF EVENTS	ANNUAL RATE	PERSON EXAMS	# OF EVENTS	ANNUAL RATE	# OF EVENTS	CRUDE	AGE-ADJUSTED	# OF EVENTS	CRUDE	AGE-ADJUSTED
NONE	1523	119	39	95	16	84	276	9	8	326	26	26
1-10	127	12	47	4	0	0	60	7	8	35	24	25
11-20	73	7	48	2	0	0	86	9	10	40	33	34
21-40	12	0	0	0	0	-	33	11	11	5	14	12
41-90	0	0	-	0	0	-	3	21	26	0	-	
UNKNOWN	4	0	0	0	0	-	0	0	0	1	22	20
TOTAL	1739	138	40	101	16	79	458	9	9	407	26	26

LOGISTIC REGRESSIONS

AGE AT EXAM	COEFFICIENT OF RISK FACTOR	SE COEFFICIENT	P-VALUE	STANDARDIZED COEFFICIENT	STD SE
35-44	0.055	0.019	0.003	0.543	0.184
45-54	0.005	0.009	0.602	0.047	0.090
55-64	0.012	0.005	0.025	0.115	0.051
65-74	0.009	0.008	0.257	0.067	0.059
75-84	-0.000	0.018	0.990	-0.001	0.089
85-94	-	-	-	-	-
35-64 UNIVARIATE	0.006	0.005	0.182	0.060	0.045
65-94 UNIVARIATE	0.001	0.007	0.917	0.005	0.051
35-64 BIVARIATE	0.014	0.004	0.002	0.138	0.044
65-94 BIVARIATE	0.007	0.007	0.305	0.052	0.051
35-64 MULTIVARIATE	0.019	0.005	0.001	0.183	0.045
65-94 MULTIVARIATE	0.012	0.007	0.107	0.083	0.052

TABLE 20-14   DATE=14NOV85
AVERAGE ANNUAL INCIDENCE RATE FOR EVENT PER 1000 PERSONS AT RISK AT EXAM
BY AGE AND LEVEL OF RISK FACTOR AT EXAM: FRAMINGHAM STUDY 30-YEAR FOLLOWUP

SEX: MALE
EVENT: CARDIOVASCULAR DISEASE
RISK FACTOR: ALBUMIN IN URINE, DEFINITE
PERSONS AT RISK: PERSONS FREE OF CARDIOVASCULAR DISEASE

RISK FACTOR	35-44 AT EXAM PERSON EXAMS	# OF EVENTS	ANNUAL RATE	45-54 AT EXAM PERSON EXAMS	# OF EVENTS	ANNUAL RATE	55-64 AT EXAM PERSON EXAMS	# OF EVENTS	ANNUAL RATE	65-74 AT EXAM PERSON EXAMS	# OF EVENTS	ANNUAL RATE
NO	4361	44	5	7515	212	14	7565	390	26	3508	232	33
YES	70	1	7	94	6	32	167	13	39	93	12	65
UNKNOWN	36	1	14	26	0	0	27	1	19	5	1	100
TOTAL	4467	46	5	7635	218	14	7759	404	26	3606	245	34

RISK FACTOR	75-84 AT EXAM PERSON EXAMS	# OF EVENTS	ANNUAL RATE	85-94 AT EXAM PERSON EXAMS	# OF EVENTS	ANNUAL RATE	35-64 COMBINED # OF EVENTS	CRUDE	AGE-ADJUSTED	65-94 COMBINED # OF EVENTS	CRUDE	AGE-ADJUSTED
NO	835	89	53	27	5	93	646	17	17	326	37	37
YES	24	3	63	3	0	0	20	30	29	15	63	64
UNKNOWN	0	0	-	0	0	-	2	11	10	1	100	80
TOTAL	859	92	54	30	5	83	668	17	17	342	38	38

LOGISTIC REGRESSIONS

AGE AT EXAM		COEFFICIENT OF RISK FACTOR	SE COEFFICIENT	P-VALUE	STANDARDIZED COEFFICIENT	STD SE
35-44		0.352	1.019	0.730	0.044	0.127
45-54		0.854	0.428	0.046	0.094	0.047
55-64		0.440	0.293	0.134	0.064	0.043
65-74		0.738	0.317	0.020	0.117	0.050
75-84		0.180	0.627	0.774	0.030	0.103
85-94		-	-	-	-	-
35-64	UNIVARIATE	0.626	0.234	0.007	0.080	0.030
65-94	UNIVARIATE	0.572	0.282	0.042	0.092	0.045
35-64	BIVARIATE	0.499	0.236	0.035	0.064	0.030
65-94	BIVARIATE	0.534	0.284	0.060	0.086	0.046
35-64	MULTIVARIATE	-0.104	0.258	0.689	-0.013	0.033
65-94	MULTIVARIATE	0.237	0.292	0.419	0.038	0.047

SEX: FEMALE
EVENT: CARDIOVASCULAR DISEASE
RISK FACTOR: ALBUMIN IN URINE, DEFINITE
PERSONS AT RISK: PERSONS FREE OF CARDIOVASCULAR DISEASE

RISK FACTOR	35-44 AT EXAM PERSON EXAMS	# OF EVENTS	ANNUAL RATE	45-54 AT EXAM PERSON EXAMS	# OF EVENTS	ANNUAL RATE	55-64 AT EXAM PERSON EXAMS	# OF EVENTS	ANNUAL RATE	65-74 AT EXAM PERSON EXAMS	# OF EVENTS	ANNUAL RATE
NO	5338	16	1	9569	113	6	10960	314	14	5767	246	21
YES	225	1	2	274	4	7	218	9	21	109	7	32
UNKNOWN	38	0	0	25	1	20	25	0	0	10	0	0
TOTAL	5601	17	2	9868	118	6	11203	323	14	5886	253	21

RISK FACTOR	75-84 AT EXAM PERSON EXAMS	# OF EVENTS	ANNUAL RATE	85-94 AT EXAM PERSON EXAMS	# OF EVENTS	ANNUAL RATE	35-64 COMBINED # OF EVENTS	CRUDE	AGE-ADJUSTED	65-94 COMBINED # OF EVENTS	CRUDE	AGE-ADJUSTED
NO	1697	135	40	98	15	77	443	9	9	396	26	26
YES	33	3	45	3	1	167	14	10	12	11	38	37
UNKNOWN	9	0	0	0	0	-	1	6	7	0	0	0
TOTAL	1739	138	10	101	16	79	458	9	9	407	26	26

LOGISTIC REGRESSIONS

AGE AT EXAM		COEFFICIENT OF RISK FACTOR	SE COEFFICIENT	P-VALUE	STANDARDIZED COEFFICIENT	STD SE
35-44		0.395	1.033	0.702	0.078	0.204
45-54		0.215	0.515	0.675	0.035	0.084
55-64		0.378	0.345	0.273	0.052	0.048
65-74		0.432	0.396	0.276	0.058	0.053
75-84		0.146	0.612	0.812	0.020	0.084
85-94		1.018	1.256	0.418	0.174	0.214
35-64	UNIVARIATE	0.134	0.274	0.626	0.022	0.044
65-94	UNIVARIATE	0.396	0.318	0.213	0.054	0.043
35-64	BIVARIATE	0.356	0.276	0.198	0.058	0.045
65-94	BIVARIATE	0.380	0.319	0.234	0.052	0.043
35-64	MULTIVARIATE	-0.088	0.294	0.764	-0.014	0.048
65-94	MULTIVARIATE	0.014	0.334	0.966	0.002	0.045

TABLE 20-15   DATE=14NOV85

AVERAGE ANNUAL INCIDENCE RATE FOR EVENT PER 1000 PERSONS AT RISK AT EXAM
BY AGE AND LEVEL OF RISK FACTOR AT EXAM: FRAMINGHAM STUDY 30-YEAR FOLLOWUP

SEX: MALE
EVENT: CARDIOVASCULAR DISEASE
RISK FACTOR: HEART ENLARGEMENT BY X-RAY
PERSONS AT RISK: PERSONS FREE OF CARDIOVASCULAR DISEASE

RISK FACTOR	35-44 AT EXAM			45-54 AT EXAM			55-64 AT EXAM			65-74 AT EXAM		
	PERSON EXAMS	# OF EVENTS	ANNUAL RATE	PERSON EXAMS	# OF EVENTS	ANNUAL RATE	PERSON EXAMS	# OF EVENTS	ANNUAL RATE	PERSON EXAMS	# OF EVENTS	ANNUAL RATE
NO	4060	40	5	6325	160	13	5776	276	24	2412	144	30
MILD	184	4	11	637	22	17	922	51	28	529	27	26
DEFINITE	218	2	5	669	35	26	1060	77	36	665	74	56
UNKNOWN	5	0	0	4	1	125	1	0	0	0	0	-
TOTAL	4467	46	5	7635	218	14	7759	404	26	3606	245	34

RISK FACTOR	75-84 AT EXAM			85-94 AT EXAM			35-64 COMBINED			65-94 COMBINED		
	PERSON EXAMS	# OF EVENTS	ANNUAL RATE	PERSON EXAMS	# OF EVENTS	ANNUAL RATE	# OF EVENTS	CRUDE	AGE-ADJUSTED	# OF EVENTS	CRUDE	AGE-ADJUSTED
NO	503	41	41	23	2	43	476	15	15	187	32	32
MILD	153	20	65	5	2	200	77	22	20	49	36	34
DEFINITE	203	31	76	2	1	250	114	29	25	106	61	61
UNKNOWN	0	0	-	0	0	-	1	50	48	0	-	-
TOTAL	859	92	54	30	5	83	668	17	17	342	38	38

LOGISTIC REGRESSIONS

AGE AT EXAM	COEFFICIENT OF RISK FACTOR	SE COEFFICIENT	P-VALUE	STANDARDIZED COEFFICIENT	STD SE
35-44	0.145	0.285	0.610	0.068	0.133
45-54	0.371	0.093	0.001	0.225	0.056
55-64	0.216	0.065	0.001	0.154	0.047
65-74	0.312	0.077	0.001	0.245	0.060
75-84	0.361	0.125	0.004	0.302	0.105
85-94	1.368	0.743	0.066	0.815	0.443
35-64 UNIVARIATE	0.365	0.052	0.001	0.231	0.033
65-94 UNIVARIATE	0.344	0.065	0.001	0.274	0.052
35-64 BIVARIATE	0.245	0.053	0.001	0.155	0.033
65-94 BIVARIATE	0.322	0.065	0.001	0.257	0.052
35-64 MULTIVARIATE	0.139	0.057	0.015	0.088	0.036
65-94 MULTIVARIATE	0.175	0.070	0.012	0.140	0.056

SEX: FEMALE
EVENT: CARDIOVASCULAR DISEASE
RISK FACTOR: HEART ENLARGEMENT BY X-RAY
PERSONS AT RISK: PERSONS FREE OF CARDIOVASCULAR DISEASE

RISK FACTOR	35-44 AT EXAM			45-54 AT EXAM			55-64 AT EXAM			65-74 AT EXAM		
	PERSON EXAMS	# OF EVENTS	ANNUAL RATE	PERSON EXAMS	# OF EVENTS	ANNUAL RATE	PERSON EXAMS	# OF EVENTS	ANNUAL RATE	PERSON EXAMS	# OF EVENTS	ANNUAL RATE
NO	5150	12	1	7958	78	5	7362	159	11	3054	104	17
MILD	198	1	3	929	13	7	1723	58	17	1088	39	18
DEFINITE	245	4	8	975	27	14	2114	106	25	1744	110	32
UNKNOWN	8	0	0	6	0	0	0	0	-	0	0	-
TOTAL	5601	17	2	9868	118	6	11203	323	14	5886	253	21

RISK FACTOR	75-84 AT EXAM			85-94 AT EXAM			35-64 COMBINED			65-94 COMBINED		
	PERSON EXAMS	# OF EVENTS	ANNUAL RATE	PERSON EXAMS	# OF EVENTS	ANNUAL RATE	# OF EVENTS	CRUDE	AGE-ADJUSTED	# OF EVENTS	CRUDE	AGE-ADJUSTED
NO	643	24	19	27	4	74	249	6	7	132	18	18
MILD	339	30	44	24	4	83	72	13	10	73	25	25
DEFINITE	757	84	55	50	8	80	137	21	17	202	40	38
UNKNOWN	0	0	-	0	0	-	0	0	0	0	-	-
TOTAL	1739	138	40	101	16	79	458	9	9	407	26	26

LOGISTIC REGRESSIONS

AGE AT EXAM	COEFFICIENT OF RISK FACTOR	SE COEFFICIENT	P-VALUE	STANDARDIZED COEFFICIENT	STD SE
35-44	0.972	0.290	0.001	0.430	0.128
45-54	0.516	0.112	0.001	0.328	0.071
55-64	0.437	0.064	0.001	0.346	0.050
65-74	0.324	0.072	0.001	0.284	0.063
75-84	0.542	0.110	0.001	0.485	0.099
85-94	0.037	0.325	0.908	0.032	0.275
35-64 UNIVARIATE	0.629	0.053	0.001	0.436	0.037
65-94 UNIVARIATE	0.427	0.058	0.001	0.380	0.051
35-64 BIVARIATE	0.439	0.055	0.001	0.304	0.038
65-94 BIVARIATE	0.380	0.059	0.001	0.338	0.052
35-64 MULTIVARIATE	0.308	0.060	0.001	0.214	0.042
65-94 MULTIVARIATE	0.290	0.061	0.001	0.257	0.054

TABLE 20-16   DATE=22MAY85

AVERAGE ANNUAL INCIDENCE RATE FOR EVENT PER 1000 PERSONS AT RISK AT EXAM
BY AGE AND LEVEL OF RISK FACTOR AT EXAM: FRAMINGHAM STUDY 30-YEAR FOLLOWUP

SEX: MALE
EVENT: CARDIOVASCULAR DISEASE
RISK FACTOR: LEFT VENTRICULAR HYPERTROPHY BY ELECTROCARDIOGRAM
PERSONS AT RISK: PERSONS FREE OF CARDIOVASCULAR DISEASE

RISK FACTOR	35-44 AT EXAM PERSON EXAMS	# OF EVENTS	ANNUAL RATE	45-54 AT EXAM PERSON EXAMS	# OF EVENTS	ANNUAL RATE	55-64 AT EXAM PERSON EXAMS	# OF EVENTS	ANNUAL RATE	65-74 AT EXAM PERSON EXAMS	# OF EVENTS	ANNUAL RATE
NO	4363	44	5	7423	197	13	7439	366	25	3433	215	31
BORDER	78	1	9	127	9	35	200	12	30	85	12	71
DEFINITE	26	1	19	85	12	71	120	26	108	88	18	102
UNKNOWN	0	0	-	0	0	-	0	0	-	0	0	-
TOTAL	4467	46	5	7635	218	14	7759	404	26	3606	245	34

RISK FACTOR	75-84 AT EXAM PERSON EXAMS	# OF EVENTS	ANNUAL RATE	85-94 AT EXAM PERSON EXAMS	# OF EVENTS	ANNUAL RATE	35-64 COMBINED # OF EVENTS	CRUDE	AGE-ADJUSTED	65-94 COMBINED # OF EVENTS	CRUDE	AGE-ADJUSTED
NO	809	83	51	27	4	74	607	16	16	302	35	35
BORDER	29	2	34	2	1	250	22	27	27	15	65	65
DEFINITE	21	7	167	1	0	0	39	84	74	25	114	114
UNKNOWN	0	0	-	0	0	-	0	-	-	0	-	-
TOTAL	859	92	54	30	5	83	668	17	17	342	38	38

LOGISTIC REGRESSIONS

AGE AT EXAM	COEFFICIENT OF RISK FACTOR	SE COEFFICIENT	P-VALUE	STANDARDIZED COEFFICIENT	STD SE
35-44	0.576	0.482	0.232	0.115	0.096
45-54	0.918	0.147	0.001	0.224	0.036
55-64	0.746	0.110	0.001	0.217	0.032
65-74	0.703	0.127	0.001	0.240	0.043
75-84	0.585	0.232	0.012	0.207	0.082
85-94	0.374	0.991	0.706	0.162	0.430
35-64 UNIVARIATE	0.863	0.086	0.001	0.220	0.022
65-94 UNIVARIATE	0.675	0.111	0.001	0.232	0.038
35-64 BIVARIATE	0.779	0.087	0.001	0.199	0.022
65-94 BIVARIATE	0.672	0.111	0.001	0.231	0.038
35-64 MULTIVARIATE	0.476	0.099	0.001	0.121	0.025
65-94 MULTIVARIATE	0.413	0.121	0.001	0.142	0.042

SEX: FEMALE
EVENT: CARDIOVASCULAR DISEASE
RISK FACTOR: LEFT VENTRICULAR HYPERTROPHY BY ELECTROCARDIOGRAM
PERSONS AT RISK: PERSONS FREE OF CARDIOVASCULAR DISEASE

RISK FACTOR	35-44 AT EXAM PERSON EXAMS	# OF EVENTS	ANNUAL RATE	45-54 AT EXAM PERSON EXAMS	# OF EVENTS	ANNUAL RATE	55-64 AT EXAM PERSON EXAMS	# OF EVENTS	ANNUAL RATE	65-74 AT EXAM PERSON EXAMS	# OF EVENTS	ANNUAL RATE
NO	5548	16	1	9711	113	6	10867	287	13	5661	218	19
BORDER	38	0	0	104	2	10	211	17	40	125	12	48
DEFINITE	15	1	33	53	3	28	125	19	76	100	23	115
UNKNOWN	0	0	-	0	0	-	0	0	-	0	0	-
TOTAL	5601	17	2	9868	118	6	11203	323	14	5886	253	21

RISK FACTOR	75-84 AT EXAM PERSON EXAMS	# OF EVENTS	ANNUAL RATE	85-94 AT EXAM PERSON EXAMS	# OF EVENTS	ANNUAL RATE	35-64 COMBINED # OF EVENTS	CRUDE	AGE-ADJUSTED	65-94 COMBINED # OF EVENTS	CRUDE	AGE-ADJUSTED
NO	1651	122	37	92	13	71	416	8	8	353	24	24
BORDER	40	6	75	3	0	0	19	27	21	18	54	53
DEFINITE	48	10	104	6	3	250	23	60	49	36	117	114
UNKNOWN	0	0	-	0	0	-	0	-	-	0	-	-
TOTAL	1739	138	40	101	16	79	458	9	9	407	26	26

LOGISTIC REGRESSIONS

AGE AT EXAM	COEFFICIENT OF RISK FACTOR	SE COEFFICIENT	P-VALUE	STANDARDIZED COEFFICIENT	STD SE
35-44	1.470	0.563	0.009	0.194	0.074
45-54	0.764	0.285	0.007	0.136	0.051
55-64	0.982	0.116	0.001	0.244	0.029
65-74	1.001	0.117	0.001	0.294	0.034
75-84	0.620	0.172	0.001	0.221	0.061
85-94	0.790	0.426	0.064	0.393	0.213
35-64 UNIVARIATE	1.095	0.104	0.001	0.222	0.021
65-94 UNIVARIATE	0.900	0.094	0.001	0.281	0.029
35-64 BIVARIATE	0.950	0.106	0.001	0.193	0.022
65-94 BIVARIATE	0.865	0.095	0.001	0.270	0.030
35-64 MULTIVARIATE	0.550	0.122	0.001	0.112	0.025
65-94 MULTIVARIATE	0.734	0.100	0.001	0.229	0.031

TABLE 20-17   DATE=14NOV85

AVERAGE ANNUAL INCIDENCE RATE FOR EVENT PER 1000 PERSONS AT RISK AT EXAM
BY AGE AND LEVEL OF RISK FACTOR AT EXAM: FRAMINGHAM STUDY 30-YEAR FOLLOWUP

SEX: MALE
EVENT: CARDIOVASCULAR DISEASE
RISK FACTOR: INTRAVENTRICULAR CONDUCTION DEFECT (WITH QRS INTERVAL MORE THAN 0.11 SECOND)
PERSONS AT RISK: PERSONS FREE OF CARDIOVASCULAR DISEASE

RISK FACTOR	35-44 AT EXAM PERSON EXAMS	# OF EVENTS	ANNUAL RATE	45-54 AT EXAM PERSON EXAMS	# OF EVENTS	ANNUAL RATE	55-64 AT EXAM PERSON EXAMS	# OF EVENTS	ANNUAL RATE	65-74 AT EXAM PERSON EXAMS	# OF EVENTS	ANNUAL RATE
NO	4156	43	5	7129	200	14	7124	361	25	3299	213	32
INCOMPLETE	275	3	5	443	13	15	446	28	31	161	17	53
COMPLETE	36	0	0	63	5	40	189	15	40	146	15	51
UNKNOWN	0	0	-	0	0	-	0	0	-	0	0	-
TOTAL	4467	46	5	7635	218	14	7759	404	26	3606	245	34

RISK FACTOR	75-84 AT EXAM PERSON EXAMS	# OF EVENTS	ANNUAL RATE	85-94 AT EXAM PERSON EXAMS	# OF EVENTS	ANNUAL RATE	35-64 COMBINED # OF EVENTS	CRUDE	AGE-ADJUSTED	65-94 COMBINED # OF EVENTS	CRUDE	AGE-ADJUSTED
NO	779	84	54	25	4	80	604	16	16	301	37	37
INCOMPLETE	18	1	28	2	0	0	44	19	19	18	50	48
COMPLETE	62	7	56	3	1	167	20	35	31	23	55	53
UNKNOWN	0	0	-	0	0	-	0	-	-	0	-	-
TOTAL	859	92	54	30	5	83	668	17	17	342	38	38

LOGISTIC REGRESSIONS

AGE AT EXAM	COEFFICIENT OF RISK FACTOR	SE COEFFICIENT	P-VALUE	STANDARDIZED COEFFICIENT	STD SE
35-44	-0.157	0.548	0.775	-0.047	0.162
45-54	0.311	0.197	0.115	0.091	0.058
55-64	0.236	0.116	0.043	0.089	0.044
65-74	0.305	0.125	0.015	0.133	0.055
75-84	-0.009	0.209	0.965	-0.005	0.111
85-94	0.352	0.686	0.608	0.225	0.439
35-64 UNIVARIATE	0.300	0.099	0.003	0.099	0.033
65-94 UNIVARIATE	0.234	0.106	0.027	0.107	0.049
35-64 BIVARIATE	0.213	0.098	0.030	0.070	0.032
65-94 BIVARIATE	0.205	0.106	0.053	0.094	0.049
35-64 MULTIVARIATE	0.148	0.105	0.160	0.049	0.035
65-94 MULTIVARIATE	0.210	0.108	0.052	0.096	0.050

SEX: FEMALE
EVENT: CARDIOVASCULAR DISEASE
RISK FACTOR: INTRAVENTRICULAR CONDUCTION DEFECT (WITH QRS INTERVAL MORE THAN 0.11 SECOND)
PERSONS AT RISK: PERSONS FREE OF CARDIOVASCULAR DISEASE.

RISK FACTOR	35-44 AT EXAM PERSON EXAMS	# OF EVENTS	ANNUAL RATE	45-54 AT EXAM PERSON EXAMS	# OF EVENTS	ANNUAL RATE	55-64 AT EXAM PERSON EXAMS	# OF EVENTS	ANNUAL RATE	65-74 AT EXAM PERSON EXAMS	# OF EVENTS	ANNUAL RATE
NO	5514	15	1	9680	115	6	10882	308	14	5687	242	21
INCOMPLETE	75	2	13	144	1	3	183	5	14	108	2	9
COMPLETE	12	0	0	44	2	23	138	10	36	91	9	49
UNKNOWN	0	0	-	0	0	-	0	0	-	0	0	-
TOTAL	5601	17	2	9868	118	6	11203	323	14	5886	253	21

RISK FACTOR	75-84 AT EXAM PERSON EXAMS	# OF EVENTS	ANNUAL RATE	85-94 AT EXAM PERSON EXAMS	# OF EVENTS	ANNUAL RATE	35-64 COMBINED # OF EVENTS	CRUDE	AGE-ADJUSTED	65-94 COMBINED # OF EVENTS	CRUDE	AGE-ADJUSTED
NO	1652	127	38	98	15	77	438	8	8	384	26	26
INCOMPLETE	35	3	43	1	1	500	8	10	10	6	21	23
COMPLETE	52	8	77	2	0	0	12	31	24	17	59	55
UNKNOWN	0	0	-	0	0	-	0	-	-	0	-	-
TOTAL	1739	138	40	101	16	79	458	9	9	407	26	26

LOGISTIC REGRESSIONS

AGE AT EXAM	COEFFICIENT OF RISK FACTOR	SE COEFFICIENT	P-VALUE	STANDARDIZED COEFFICIENT	STD SE
35-44	1.428	0.599	0.017	0.210	0.088
45-54	0.429	0.377	0.256	0.077	0.067
55-64	0.417	0.162	0.010	0.106	0.041
65-74	0.313	0.183	0.086	0.087	0.051
75-84	0.364	0.191	0.057	0.133	0.070
85-94	0.162	0.844	0.848	0.048	0.250
35-64 UNIVARIATE	0.584	0.145	0.001	0.121	0.030
65-94 UNIVARIATE	0.370	0.130	0.004	0.111	0.039
35-64 BIVARIATE	0.422	0.145	0.004	0.088	0.030
65-94 BIVARIATE	0.336	0.130	0.010	0.101	0.039
35-64 MULTIVARIATE	0.362	0.157	0.021	0.075	0.033
65-94 MULTIVARIATE	0.320	0.132	0.016	0.096	0.040

TABLE 20-18   DATE=14NOV85
AVERAGE ANNUAL INCIDENCE RATE FOR EVENT PER 1000 PERSONS AT RISK AT EXAM
BY AGE AND LEVEL OF RISK FACTOR AT EXAM: FRAMINGHAM STUDY 30-YEAR FOLLOWUP

SEX: MALE
EVENT: CARDIOVASCULAR DISEASE
RISK FACTOR: DEFINITE NONSPECIFIC T-WAVE OR ST-SEGMENT ABNORMALITY BY ELECTROCARDIOGRAM
PERSONS AT RISK: PERSONS FREE OF CARDIOVASCULAR DISEASE

RISK FACTOR	35-44 AT EXAM			45-54 AT EXAM			55-64 AT EXAM			65-74 AT EXAM		
	PERSON EXAMS	# OF EVENTS	ANNUAL RATE	PERSON EXAMS	# OF EVENTS	ANNUAL RATE	PERSON EXAMS	# OF EVENTS	ANNUAL RATE	PERSON EXAMS	# OF EVENTS	ANNUAL RATE
NO	4357	41	5	7273	196	13	7231	363	25	3226	200	31
YES	110	5	23	362	22	30	528	41	39	380	45	59
UNKNOWN	0	0	-	0	0	-	0	0	-	0	0	-
TOTAL	4467	46	5	7635	218	14	7759	404	26	3606	245	34

RISK FACTOR	75-84 AT EXAM			85-94 AT EXAM			35-64 COMBINED			65-94 COMBINED		
	PERSON EXAMS	# OF EVENTS	ANNUAL RATE	PERSON EXAMS	# OF EVENTS	ANNUAL RATE	# OF EVENTS	CRUDE	AGE-ADJUSTED	# OF EVENTS	CRUDE	AGE-ADJUSTED
NO	727	71	49	21	1	24	600	16	16	272	34	34
YES	132	21	80	9	4	222	68	34	32	70	67	64
UNKNOWN	0	0	-	0	0	-				0	-	
TOTAL	859	92	54	30	5	83	668	17	17	342	38	38

LOGISTIC REGRESSIONS

AGE AT EXAM		COEFFICIENT OF RISK FACTOR	SE COEFFICIENT	P-VALUE	STANDARDIZED COEFFICIENT	STD SE
35-44		1.612	0.484	0.001	0.250	0.075
45-54		0.849	0.232	0.001	0.180	0.049
55-64		0.466	0.171	0.007	0.117	0.043
65-74		0.709	0.175	0.001	0.218	0.054
75-84		0.558	0.269	0.038	0.202	0.097
85-94		2.773	1.225	0.024	1.292	0.571
35-64	UNIVARIATE	0.79	0.132	0.001	0.174	0.029
65-94	UNIVARIATE	0.748	0.143	0.001	0.239	0.046
35-64	BIVARIATE	0.651	0.134	0.001	0.131	0.029
65-94	BIVARIATE	0.695	0.144	0.001	0.223	0.046
35-64	MULTIVARIATE	0.2	0.140	0.088	0.052	0.031
65-94	MULTIVARIATE	0.586	0.147	0.001	0.187	0.047

SEX: FEMALE
EVENT: CARDIOVASCULAR DISEASE
RISK FACTOR: DEFINITE NONSPECIFIC T-WAVE OR ST-SEGMENT ABNORMALITY BY ELECTROCARDIOGRAM
PERSONS AT RISK: PERSONS FREE OF CARDIOVASCULAR DISEASE

RISK FACTOR	35-44 AT EXAM			45-54 AT EXAM			55-64 AT EXAM			65-74 AT EXAM		
	PERSON EXAMS	# OF EVENTS	ANNUAL RATE	PERSON EXAMS	# OF EVENTS	ANNUAL RATE	PERSON EXAMS	# OF EVENTS	ANNUAL RATE	PERSON EXAMS	# OF EVENTS	ANNUAL RATE
NO	5503	17	2	9464	106	6	10401	276	13	5239	213	20
YES	98	0	0	404	12	15	802	47	29	647	40	31
UNKNOWN	0	0	-	0	0	-	0	0	-	0	0	-
TOTAL	5601	17	2	9868	118	6	11203	323	14	5886	253	21

RISK FACTOR	75-84 AT EXAM			85-94 AT EXAM			35-64 COMBINED			65-94 COMBINED		
	PERSON EXAMS	# OF EVENTS	ANNUAL RATE	PERSON EXAMS	# OF EVENTS	ANNUAL RATE	# OF EVENTS	CRUDE	AGE-ADJUSTED	# OF EVENTS	CRUDE	AGE-ADJUSTED
NO	1462	101	35	82	15	91	399	8	8	529	24	24
YES	277	37	67	19	1	26	59	23	18	78	41	39
UNKNOWN	0	0	-	0	0	-				0	-	
TOTAL	1739	138	40	101	16	79	458	9	9	407	26	26

LOGISTIC REGRESSIONS

AGE AT EXAM		COEFFICIENT OF RISK FACTOR	SE COEFFICIENT	P-VALUE	STANDARDIZED COEFFICIENT	STD SE
35-44		-	-	-	-	-
45-54		0.994	0.309	0.001	0.197	0.061
55-64		0.826	0.162	0.001	0.213	0.042
65-74		0.441	0.178	0.013	0.138	0.056
75-84		0.731	0.205	0.001	0.268	0.075
85-94		-1.394	1.066	0.191	-0.547	0.419
35-64	UNIVARIATE	1.087	0.142	0.001	0.234	0.031
65-94	UNIVARIATE	0.570	0.131	0.001	0.187	0.043
35-64	BIVARIATE	0.797	0.144	0.001	0.172	0.031
65-94	BIVARIATE	0.505	0.132	0.001	0.165	0.043
35-64	MULTIVARIATE	0.6	0.148	0.001	0.130	0.032
65-94	MULTIVARIATE	0.429	0.133	0.001	0.147	0.044

AVERAGE ANNUAL INCIDENCE RATE FOR EVENT PER 1000 PERSONS AT RISK AT EXAM
BY AGE AND LEVEL OF RISK FACTOR AT EXAM: FRAMINGHAM STUDY 30-YEAR FOLLOWUP

SEX: MALE
EVENT: DEATH FROM CARDIOVASCULAR DISEASE
RISK FACTOR: SYSTOLIC BLOOD PRESSURE, FIRST EXAMINER (MM HG)
PERSONS AT RISK: PERSONS FREE OF CARDIOVASCULAR DISEASE

RISK FACTOR	35-44 AT EXAM PERSON EXAMS	# OF EVENTS	ANNUAL RATE	45-54 AT EXAM PERSON EXAMS	# OF EVENTS	ANNUAL RATE	55-64 AT EXAM PERSON EXAMS	# OF EVENTS	ANNUAL RATE	65-74 AT EXAM PERSON EXAMS	# OF EVENTS	ANNUAL RATE
74-119	1222	2	1	1761	4	1	1408	6	2	525	7	7
120-139	2163	4	1	3394	14	2	3089	22	4	1264	15	6
140-159	841	0	0	1708	9	3	1989	19	5	1085	18	8
160-179	191	1	3	544	4	4	863	13	8	494	18	18
180-300	50	0	0	228	3	7	410	14	17	238	13	27
UNKNOWN	0	0	-	0	0	-	0	0	-	0	0	-
TOTAL	4467	7	1	7635	34	2	7759	74	5	3606	71	10

RISK FACTOR	75-84 AT EXAM PERSON EXAMS	# OF EVENTS	ANNUAL RATE	85-94 AT EXAM PERSON EXAMS	# OF EVENTS	ANNUAL RATE	35-64 COMBINED # OF EVENTS	CRUDE	AGE-ADJUSTED	65-94 COMBINED # OF EVENTS	CRUDE	AGE-ADJUSTED
74-119	107	4	19	5	1	12	1	1	12	9	10
120-139	301	7	12	10			40	2	2	22	7	7
140-159	266	11	21	10			28	3	3	29	11	11
160-179	127	7	28	3			18	6	5	27	22	22
180-300	58	6	52	2			17	12	9	20	34	33
UNKNOWN	0	0	-	0			0	-		0	-	
TOTAL	859	35	20	30			115	3	3	110	12	12

LOGISTIC REGRESSIONS

AGE AT EXAM	COEFFICIENT OF RISK FACTOR	SE COEFFICIENT	P-VALUE	STANDARDIZED COEFFICIENT	STD SE
35-44	-0.003	0.023	0.883	-0.057	0.387
45-54	0.021	0.007	0.004	0.403	0.139
55-64	0.025	0.004	0.001	0.549	0.089
65-74	0.020	0.004	0.001	0.455	0.098
75-84	0.019	0.008	0.013	0.400	0.161
85-94					
35-64 UNIVARIATE	0.027	0.003	0.001	0.540	0.069
65-94 UNIVARIATE	0.019	0.004	0.001	0.435	0.082
35-64 BIVARIATE	0.022	0.003	0.001	0.449	0.070
65-94 BIVARIATE	0.020	0.004	0.001	0.454	0.084
35-64 MULTIVARIATE	0.018	0.004	0.001	0.364	0.078
65-94 MULTIVARIATE	0.017	0.004	0.001	0.380	0.088

SEX: FEMALE
EVENT: DEATH FROM CARDIOVASCULAR DISEASE
RISK FACTOR: SYSTOLIC BLOOD PRESSURE, FIRST EXAMINER (MM HG)
PERSONS AT RISK: PERSONS FREE OF CARDIOVASCULAR DISEASE

RISK FACTOR	35-44 AT EXAM PERSON EXAMS	# OF EVENTS	ANNUAL RATE	45-54 AT EXAM PERSON EXAMS	# OF EVENTS	ANNUAL RATE	55-64 AT EXAM PERSON EXAMS	# OF EVENTS	ANNUAL RATE	65-74 AT EXAM PERSON EXAMS	# OF EVENTS	ANNUAL RATE
74-119	2486	2	0	2694	4	1	1937	6	2	625	2	2
120-139	2239	2	0	3951	5	1	3979	9	1	1680	16	5
140-159	670	0	0	2028	7	2	2944	12	2	1896	23	6
160-179	156	0	0	758	1	1	1501	6	2	1070	9	4
180-300	50	1	10	437	4	5	842	10	6	615	19	15
UNKNOWN	0	0	-	0	0	-	0	0	-	0	0	-
TOTAL	5601	5	0	9868	21	1	11203	43	2	5886	69	6

RISK FACTOR	75-84 AT EXAM PERSON EXAMS	# OF EVENTS	ANNUAL RATE	85-94 AT EXAM PERSON EXAMS	# OF EVENTS	ANNUAL RATE	35-64 COMBINED # OF EVENTS	CRUDE	AGE-ADJUSTED	65-94 COMBINED # OF EVENTS	CRUDE	AGE-ADJUSTED
74-119	118	1	4	8	1	63	12	1	1	4	3	3
120-139	406	10	12	31	2	32	16	1	1	28	7	7
140-159	569	10	9	21	2	48	19	2	1	35	7	7
160-179	394	13	16	24	2	42	7	1	1	24	8	7
180-300	252	11	22	17	1	29	15	6	6	31	18	17
UNKNOWN	0	0	-	0	0	-	0	-		0	-	
TOTAL	1739	45	13	101	8	40	69	1	1	122	8	8

LOGISTIC REGRESSIONS

AGE AT EXAM	COEFFICIENT OF RISK FACTOR	SE COEFFICIENT	P-VALUE	STANDARDIZED COEFFICIENT	STD SE
35-44	0.020	0.020	0.305	0.345	0.337
45-54	0.020	0.006	0.002	0.450	0.147
55-64	0.014	0.005	0.006	0.351	0.127
65-74	0.019	0.004	0.001	0.448	0.105
75-84	0.011	0.006	0.043	0.284	0.140
85-94	-0.002	0.014	0.905	-0.045	0.374
35-64 UNIVARIATE	0.019	0.004	0.001	0.451	0.087
65-94 UNIVARIATE	0.016	0.003	0.001	0.393	0.081
35-64 BIVARIATE	0.016	0.004	0.001	0.370	0.095
65-94 BIVARIATE	0.014	0.003	0.001	0.347	0.082
35-64 MULTIVARIATE	0.007	0.005	0.150	0.153	0.106
65-94 MULTIVARIATE	0.012	0.004	0.001	0.288	0.087

TABLE 21-2   DATE=17NOV85
AVERAGE ANNUAL INCIDENCE RATE FOR EVENT PER 1000 PERSONS AT RISK AT EXAM
BY AGE AND LEVEL OF RISK FACTOR AT EXAM: FRAMINGHAM STUDY 30-YEAR FOLLOWUP

SEX: MALE
EVENT: DEATH FROM CARDIOVASCULAR DISEASE
RISK FACTOR: DIASTOLIC BLOOD PRESSURE, FIRST EXAMINER (MM HG)
PERSONS AT RISK: PERSONS FREE OF CARDIOVASCULAR DISEASE

RISK FACTOR	35-44 AT EXAM			45-54 AT EXAM			55-64 AT EXAM			65-74 AT EXAM		
	PERSON EXAMS	# OF EVENTS	ANNUAL RATE	PERSON EXAMS	# OF EVENTS	ANNUAL RATE	PERSON EXAMS	# OF EVENTS	ANNUAL RATE	PERSON EXAMS	# OF EVENTS	ANNUAL RATE
20-74	1012	0	0	1457	4	1	1769	12	3	1121	17	8
75-84	1601	4	1	2747	10	2	2766	18	3	1225	15	6
85-94	1187	1	0	2039	9	2	1885	21	6	833	19	11
95-104	499	2	2	990	7	4	940	9	5	311	9	14
105-160	168	0	0	402	4	5	399	14	18	116	11	47
UNKNOWN	0	0	-	0	0	-	0	0	-	0	0	-
TOTAL	4467	7	1	7635	34	2	7759	74	5	3606	71	10

RISK FACTOR	75-84 AT EXAM			85-94 AT EXAM			35-64 COMBINED			65-94 COMBINED		
	PERSON EXAMS	# OF EVENTS	ANNUAL RATE	PERSON EXAMS	# OF EVENTS	ANNUAL RATE	# OF EVENTS	CRUDE	AGE-ADJUSTED	# OF EVENTS	CRUDE	AGE-ADJUSTED
20-74	351	10	14	14			16	2	2	29	10	9
75-84	289	13	22	11			32	2	2	28	9	9
85-94	175	7	20	5			31	3	3	28	14	14
95-104	29	3	52	0			18	4	4	12	18	21
105-160	15	2	67	0			18	9	9	13	50	51
UNKNOWN	0	0	-	0			0	-		0	-	
TOTAL	859	35	20	30			115	3	3	110	12	12

LOGISTIC REGRESSIONS

AGE AT EXAM		COEFFICIENT OF RISK FACTOR	SE COEFFICIENT	P-VALUE	STANDARDIZED COEFFICIENT	STD SE
35-44		0.021	0.032	0.518	0.231	0.357
45-54		0.034	0.013	0.009	0.396	0.152
55-64		0.037	0.008	0.001	0.447	0.100
65-74		0.036	0.009	0.001	0.434	0.103
75-84		0.020	0.014	0.166	0.235	0.170
85-94						
35-64	UNIVARIATE	0.037	0.007	0.001	0.430	0.081
65-94	UNIVARIATE	0.026	0.007	0.001	0.317	0.088
35-64	BIVARIATE	0.036	0.007	0.001	0.418	0.080
65-94	BIVARIATE	0.034	0.007	0.001	0.410	0.090
35-64	MULTIVARIATE	0.027	0.007	0.001	0.315	0.083
65-94	MULTIVARIATE	0.030	0.007	0.001	0.359	0.090

SEX: FEMALE
EVENT: DEATH FROM CARDIOVASCULAR DISEASE
RISK FACTOR: DIASTOLIC BLOOD PRESSURE, FIRST EXAMINER (MM HG)
PERSONS AT RISK: PERSONS FREE OF CARDIOVASCULAR DISEASE

RISK FACTOR	35-44 AT EXAM			45-54 AT EXAM			55-64 AT EXAM			65-74 AT EXAM		
	PERSON EXAMS	# OF EVENTS	ANNUAL RATE	PERSON EXAMS	# OF EVENTS	ANNUAL RATE	PERSON EXAMS	# OF EVENTS	ANNUAL RATE	PERSON EXAMS	# OF EVENTS	ANNUAL RATE
20-74	2158	2	0	2610	4	1	2888	10	2	1732	17	5
75-84	2023	0	0	3511	7	1	3772	9	1	2035	20	5
85-94	987	1	1	2344	3	1	2770	10	2	1348	19	7
95-104	122	2	3	914	3	2	1228	8	3	550	7	6
105-160	111	0	0	489	4	4	545	6	6	221	6	14
UNKNOWN	0	0	-	0	0	-	0	0	-	0	0	-
TOTAL	5401	5	0	9868	21	1	11203	43	2	5886	69	6

RISK FACTOR	75-84 AT EXAM			85-94 AT EXAM			35-64 COMBINED			65-94 COMBINED		
	PERSON EXAMS	# OF EVENTS	ANNUAL RATE	PERSON EXAMS	# OF EVENTS	ANNUAL RATE	# OF EVENTS	CRUDE	AGE-ADJUSTED	# OF EVENTS	CRUDE	AGE-ADJUSTED
20-74	650	15	12	44	4	45	16	1	1	36	7	7
75-84	546	12	11	34	3	44	16	1	1	35	7	7
85-94	379	12	16	11	1	31	14	1	1	32	9	9
95-104	125	5	20	7	0	0	13	3	3	12	9	9
105-160	39	7	13	0	0	-	10	4	4	7	13	13
UNKNOWN	0	0	-	0	0	-	0	-		0	-	
TOTAL	1739	45	13	101	8	40	69	1	1	122	8	8

LOGISTIC REGRESSIONS

AGE AT EXAM		COEFFICIENT OF RISK FACTOR	SE COEFFICIENT	P-VALUE	STANDARDIZED COEFFICIENT	STD SE
35-44		0.014	0.038	0.720	0.151	0.421
45-54		0.036	0.014	0.012	0.439	0.174
55-64		0.027	0.011	0.014	0.333	0.135
65-74		0.015	0.010	0.128	0.176	0.116
75-84		0.011	0.012	0.361	0.135	0.148
85-94		-0.013	0.028	0.644	-0.167	0.361
35-64	UNIVARIATE	0.032	0.008	0.001	0.395	0.101
65-94	UNIVARIATE	0.007	0.007	0.323	0.088	0.089
35-64	BIVARIATE	0.029	0.009	0.001	0.355	0.104
65-94	BIVARIATE	0.012	0.007	0.099	0.148	0.090
35-64	MULTIVARIATE	0.013	0.009	0.139	0.164	0.111
65-94	MULTIVARIATE	0.012	0.007	0.108	0.145	0.090

TABLE 21-3A DATE=17NOV85

AVERAGE ANNUAL INCIDENCE RATE FOR EVENT PER 1000 PERSONS AT RISK AT EXAM
BY AGE AND LEVEL OF RISK FACTOR AT EXAM: FRAMINGHAM STUDY 30-YEAR FOLLOWUP

SEX: MALE
EVENT: DEATH FROM CARDIOVASCULAR DISEASE
RISK FACTOR: HYPERTENSION WITH ANTIHYPERTENSIVE TREATMENT
PERSONS AT RISK: PERSONS FREE OF CARDIOVASCULAR DISEASE

RISK FACTOR	35-44 AT EXAM PERSON EXAMS	# OF EVENTS	ANNUAL RATE	45-54 AT EXAM PERSON EXAMS	# OF EVENTS	ANNUAL RATE	55-64 AT EXAM PERSON EXAMS	# OF EVENTS	ANNUAL RATE	65-74 AT EXAM PERSON EXAMS	# OF EVENTS	ANNUAL RATE
NO	2648	3	1	4101	12	1	3707	18	2	1454	14	5
MILD	1301	3	1	2176	9	2	2204	22	5	1163	21	9
DEFINITE	498	1	1	1119	11	5	1213	24	10	511	22	22
TREATED	20	0	0	239	2	4	635	10	8	478	14	15
UNKNOWN	0	0	-	0	0	-	0	0	-	0	0	-
TOTAL	4467	7	1	7635	34	2	7759	74	5	3606	71	10

RISK FACTOR	75-84 AT EXAM PERSON EXAMS	# OF EVENTS	ANNUAL RATE	85-94 AT EXAM PERSON EXAMS	# OF EVENTS	ANNUAL RATE	35-64 COMBINED # OF EVENTS	CRUDE	AGE-ADJUSTED	65-94 COMBINED # OF EVENTS	CRUDE	AGE-ADJUSTED
NO	331	9	14	12			33	2	2	24	7	7
MILD	296	13	22	9			34	3	3	34	12	11
DEFINITE	97	4	21	3			36	6	6	28	23	23
TREATED	135	9	33	6			12	7	5	24	19	19
UNKNOWN	0	0	-	0			0	-		0	-	
TOTAL	859	35	20	30			115	3	3	110	12	12

LOGISTIC REGRESSIONS

AGE AT EXAM		COEFFICIENT OF RISK FACTOR	SE COEFFICIENT	P-VALUE	STANDARDIZED COEFFICIENT	STD SE
35-44		0.358	0.495	0.470	0.248	0.343
45-54		0.599	0.208	0.004	0.458	0.159
55-64		0.668	0.143	0.001	0.542	0.116
65-74		0.685	0.155	0.001	0.557	0.126
75-84		0.369	0.216	0.087	0.296	0.173
85-94						
35-64	UNIVARIATE	0.712	0.114	0.001	0.551	0.088
65-94	UNIVARIATE	0.595	0.123	0.001	0.482	0.099
35-64	BIVARIATE	0.607	0.114	0.001	0.470	0.088
65-94	BIVARIATE	0.598	0.123	0.001	0.485	0.100
35-64	MULTIVARIATE	0.516	0.120	0.001	0.400	0.093
65-94	MULTIVARIATE	0.535	0.126	0.001	0.434	0.102

SEX: FEMALE
EVENT: DEATH FROM CARDIOVASCULAR DISEASE
RISK FACTOR: HYPERTENSION WITH ANTIHYPERTENSIVE TREATMENT
PERSONS AT RISK: PERSONS FREE OF CARDIOVASCULAR DISEASE

RISK FACTOR	35-44 AT EXAM PERSON EXAMS	# OF EVENTS	ANNUAL RATE	45-54 AT EXAM PERSON EXAMS	# OF EVENTS	ANNUAL RATE	55-64 AT EXAM PERSON EXAMS	# OF EVENTS	ANNUAL RATE	65-74 AT EXAM PERSON EXAMS	# OF EVENTS	ANNUAL RATE
NO	4239	3	0	5591	9	1	4782	11	1	1698	10	3
MILD	1000	0	0	2554	4	1	3229	12	2	1846	21	6
DEFINITE	314	2	3	1154	4	2	1770	11	3	989	14	7
TREATED	48	0	0	569	4	4	1422	9	3	1353	24	9
UNKNOWN	0	0	-	0	0	-	0	0	-	0	0	-
TOTAL	5601	5	0	9868	21	1	11203	43	2	5886	69	6

RISK FACTOR	75-84 AT EXAM PERSON EXAMS	# OF EVENTS	ANNUAL RATE	85-94 AT EXAM PERSON EXAMS	# OF EVENTS	ANNUAL RATE	35-64 COMBINED # OF EVENTS	CRUDE	AGE-ADJUSTED	65-94 COMBINED # OF EVENTS	CRUDE	AGE-ADJUSTED
NO	353	7	10	26	2	38	23	1	1	19	5	5
MILD	537	12	11	22	2	45	16	1	1	35	7	7
DEFINITE	343	13	19	21	2	48	17	3	3	29	11	10
TREATED	506	13	13	32	2	31	13	3	3	39	10	10
UNKNOWN	0	0	-	0	-		0	-		0	-	
TOTAL	1739	45	13	101	8	40	69	1	1	122	8	8

LOGISTIC REGRESSIONS

AGE AT EXAM		COEFFICIENT OF RISK FACTOR	SE COEFFICIENT	P-VALUE	STANDARDIZED COEFFICIENT	STD SE
35-44		0.947	0.543	0.081	0.554	0.318
45-54		0.523	0.260	0.044	0.401	0.499
55-64		0.505	0.186	0.007	0.420	0.155
65-74		0.479	0.162	0.003	0.394	0.133
75-84		0.244	0.206	0.235	0.191	0.161
85-94		-0.026	0.435	0.952	-0.022	0.368
35-64	UNIVARIATE	0.652	0.143	0.001	0.515	0.113
65-94	UNIVARIATE	0.411	0.122	0.001	0.335	0.099
35-64	BIVARIATE	0.510	0.148	0.001	0.403	0.117
65-94	BIVARIATE	0.344	0.123	0.005	0.281	0.100
35-64	MULTIVARIATE	0.330	0.157	0.035	0.261	0.124
65-94	MULTIVARIATE	0.269	0.125	0.032	0.220	0.102

TABLE 21-3B   DATE=17NOV85
AVERAGE ANNUAL INCIDENCE RATE FOR EVENT PER 1000 PERSONS AT RISK AT EXAM
BY AGE AND LEVEL OF RISK FACTOR AT EXAM: FRAMINGHAM STUDY 30-YEAR FOLLOWUP

SEX: MALE
EVENT: DEATH FROM CARDIOVASCULAR DISEASE
RISK FACTOR: HYPERTENSION IGNORING TREATMENT
PERSONS AT RISK: PERSONS FREE OF CARDIOVASCULAR DISEASE

RISK FACTOR	35-44 AT EXAM PERSON EXAMS	# OF EVENTS	ANNUAL RATE	45-54 AT EXAM PERSON EXAMS	# OF EVENTS	ANNUAL RATE	55-64 AT EXAM PERSON EXAMS	# OF EVENTS	ANNUAL RATE	65-74 AT EXAM PERSON EXAMS	# OF EVENTS	ANNUAL RATE
NO	2650	3	1	4132	12	1	3831	20	3	1552	17	5
MILD	1310	3	1	2252	10	2	2453	26	5	1358	25	9
DEFINITE	507	1	1	1251	12	5	1475	28	9	696	29	21
UNKNOWN	0	0	-	0	0	-	0	0	-	0	0	-
TOTAL	4467	7	1	7635	34	2	7759	74	5	3606	71	10

RISK FACTOR	75-84 AT EXAM PERSON EXAMS	# OF EVENTS	ANNUAL RATE	85-94 AT EXAM PERSON EXAMS	# OF EVENTS	ANNUAL RATE	35-64 COMBINED # OF EVENTS	CRUDE	AGE-ADJUSTED	65-94 COMBINED # OF EVENTS	CRUDE	AGE-ADJUSTED
NO	360	9	13	15			35	2	2	27	7	7
MILD	363	15	21	12			39	3	3	41	12	12
DEFINITE	136	11	40	3			41	6	6	42	25	27
UNKNOWN	0	0	-	0			0	-	-	0	-	-
TOTAL	859	35	20	30			115	3	3	110	12	12

LOGISTIC REGRESSIONS

AGE AT EXAM		COEFFICIENT OF RISK FACTOR	SE COEFFICIENT	P-VALUE	STANDARDIZED COEFFICIENT	STD SE
35-44		0.366	0.496	0.461	0.253	0.343
45-54		0.601	0.210	0.004	0.451	0.158
55-64		0.649	0.145	0.001	0.499	0.112
65-74		0.699	0.157	0.001	0.527	0.118
75-84		0.620	0.235	0.008	0.443	0.168
85-94						
35-64	UNIVARIATE	0.678	0.115	0.001	0.507	0.086
65-94	UNIVARIATE	0.668	0.127	0.001	0.498	0.095
35-64	BIVARIATE	0.605	0.116	0.001	0.452	0.087
65-94	BIVARIATE	0.699	0.129	0.001	0.521	0.096
35-64	MULTIVARIATE	0.492	0.123	0.001	0.368	0.092
65-94	MULTIVARIATE	0.623	0.132	0.001	0.464	0.098

SEX: FEMALE
EVENT: DEATH FROM CARDIOVASCULAR DISEASE
RISK FACTOR: HYPERTENSION IGNORING TREATMENT
PERSONS AT RISK: PERSONS FREE OF CARDIOVASCULAR DISEASE

RISK FACTOR	35-44 AT EXAM PERSON EXAMS	# OF EVENTS	ANNUAL RATE	45-54 AT EXAM PERSON EXAMS	# OF EVENTS	ANNUAL RATE	55-64 AT EXAM PERSON EXAMS	# OF EVENTS	ANNUAL RATE	65-74 AT EXAM PERSON EXAMS	# OF EVENTS	ANNUAL RATE
NO	4260	3	0	5758	9	1	5137	13	1	1964	13	3
MILD	1015	0	0	2734	5	1	3766	15	2	2405	28	6
DEFINITE	326	2	3	1376	7	3	2300	15	3	1517	28	9
UNKNOWN	0	0	-	0	0	-	0	0	-	0	0	-
TOTAL	5601	5	0	9868	21	1	11203	43	2	5886	69	6

RISK FACTOR	75-84 AT EXAM PERSON EXAMS	# OF EVENTS	ANNUAL RATE	85-94 AT EXAM PERSON EXAMS	# OF EVENTS	ANNUAL RATE	35-64 COMBINED # OF EVENTS	CRUDE	AGE-ADJUSTED	65-94 COMBINED # OF EVENTS	CRUDE	AGE-ADJUSTED
NO	448	8	9	31	3	48	25	1	1	24	5	5
MILD	747	17	11	35	3	43	20	1	1	48	8	8
DEFINITE	544	20	18	35	2	29	24	3	3	50	12	12
UNKNOWN	0	0	-	0	0	-	0	-	-	0	-	-
TOTAL	1739	45	13	101	8	40	69	1	1	122	8	8

LOGISTIC REGRESSIONS

AGE AT EXAM		COEFFICIENT OF RISK FACTOR	SE COEFFICIENT	P-VALUE	STANDARDIZED COEFFICIENT	STD SE
35-44		0.996	0.548	0.069	0.568	0.313
45-54		0.575	0.268	0.032	0.417	0.194
55-64		0.476	0.190	0.012	0.369	0.147
65-74		0.511	0.162	0.002	0.391	0.124
75-84		0.394	0.208	0.059	0.297	0.157
85-94		-0.275	0.460	0.550	-0.223	0.374
35-64	UNIVARIATE	0.646	0.147	0.001	0.476	0.108
65-94	UNIVARIATE	0.453	0.122	0.001	0.347	0.094
35-64	BIVARIATE	0.517	0.152	0.001	0.381	0.112
65-94	BIVARIATE	0.401	0.123	0.001	0.307	0.094
35-64	MULTIVARIATE	0.311	0.163	0.056	0.229	0.120
65-94	MULTIVARIATE	0.331	0.126	0.009	0.253	0.096

TABLE 21-4   DATE=17NOV85

AVERAGE ANNUAL INCIDENCE RATE FOR EVENT PER 1000 PERSONS AT RISK AT EXAM
BY AGE AND LEVEL OF RISK FACTOR AT EXAM: FRAMINGHAM STUDY 30-YEAR FOLLOWUP

SEX: MALE
EVENT: DEATH FROM CARDIOVASCULAR DISEASE
RISK FACTOR: SERUM CHOLESTEROL
PERSONS AT RISK: PERSONS FREE OF CARDIOVASCULAR DISEASE

RISK FACTOR	35-44 AT EXAM PERSON EXAMS	# OF EVENTS	ANNUAL RATE	45-54 AT EXAM PERSON EXAMS	# OF EVENTS	ANNUAL RATE	55-64 AT EXAM PERSON EXAMS	# OF EVENTS	ANNUAL RATE	65-74 AT EXAM PERSON EXAMS	# OF EVENTS	ANNUAL RATE
84-204	1252	0	0	1737	7	2	2169	12	3	1145	29	13
205-234	1246	1	0	2260	9	2	2319	26	6	1089	19	9
235-264	870	3	2	1733	7	2	1832	22	6	797	11	7
265-294	474	1	1	985	5	3	879	8	5	390	7	9
295-1124	291	2	3	552	5	5	435	5	6	169	5	15
UNKNOWN	334	0	0	348	1	1	125	1	4	16	0	0
TOTAL	4467	7	1	7635	34	2	7759	74	5	3606	71	10

RISK FACTOR	75-84 AT EXAM PERSON EXAMS	# OF EVENTS	ANNUAL RATE	85-94 AT EXAM PERSON EXAMS	# OF EVENTS	ANNUAL RATE	35-64 COMBINED # OF EVENTS	CRUDE	AGE-ADJUSTED	65-94 COMBINED # OF EVENTS	CRUDE	AGE-ADJUSTED
84-204	314	13	21	12			19	2	2	43	15	14
205-234	295	11	19	8			36	3	3	32	11	11
235-264	153	7	23	6			32	4	4	19	10	11
265-294	76	3	20	6	0		14	3	3	10	11	11
295-1124	21	1	24	0			12	5	5	6	16	16
UNKNOWN	0	0	-	0			2	1	2	0	0	0
TOTAL	859	35	20	30			115	3	3	110	12	12

LOGISTIC REGRESSIONS

AGE AT EXAM		COEFFICIENT OF RISK FACTOR	SE COEFFICIENT	P-VALUE	STANDARDIZED COEFFICIENT	STD SE
35-44		0.014	0.005	0.004	0.620	0.212
45-54		0.008	0.003	0.007	0.347	0.128
55-64		0.003	0.003	0.227	0.137	0.113
65-74		-0.003	0.003	0.581	-0.107	0.122
75-84		0.000	0.005	0.939	0.013	0.172
85-94						
35-64	UNIVARIATE	0.006	0.002	0.003	0.246	0.082
65-94	UNIVARIATE	-0.002	0.002	0.337	-0.094	0.098
35-64	BIVARIATE	0.007	0.002	0.001	0.276	0.084
65-94	BIVARIATE	-0.001	0.002	0.582	-0.054	0.098
35-64	MULTIVARIATE	0.005	0.002	0.007	0.232	0.086
65-94	MULTIVARIATE	-0.001	0.002	0.725	-0.035	0.099

SEX: FEMALE
EVENT: DEATH FROM CARDIOVASCULAR DISEASE
RISK FACTOR: SERUM CHOLESTEROL
PERSONS AT RISK: PERSONS FREE OF CARDIOVASCULAR DISEASE

RISK FACTOR	35-44 AT EXAM PERSON EXAMS	# OF EVENTS	ANNUAL RATE	45-54 AT EXAM PERSON EXAMS	# OF EVENTS	ANNUAL RATE	55-64 AT EXAM PERSON EXAMS	# OF EVENTS	ANNUAL RATE	65-74 AT EXAM PERSON EXAMS	# OF EVENTS	ANNUAL RATE
84-204	2349	0	0	2079	1	0	1579	4	1	741	11	7
205-234	1408	1	0	2532	3	1	2533	4	1	1320	13	5
235-264	805	1	1	2242	4	1	2866	9	2	1620	14	4
265-294	369	0	0	1454	3	1	2173	12	3	1184	13	5
295-1124	170	2	6	1087	9	4	1823	14	4	984	16	8
UNKNOWN	500	1	1	474	1	1	229	0	0	37	2	27
TOTAL	5601	5	0	9868	21	1	11203	43	2	5886	69	6

RISK FACTOR	75-84 AT EXAM PERSON EXAMS	# OF EVENTS	ANNUAL RATE	85-94 AT EXAM PERSON EXAMS	# OF EVENTS	ANNUAL RATE	35-64 COMBINED # OF EVENTS	CRUDE	AGE-ADJUSTED	65-94 COMBINED # OF EVENTS	CRUDE	AGE-ADJUSTED
84-204	302	7	12	38	5	66	5	0	1	23	11	9
205-234	419	14	17	23	0	0	8	1	1	27	8	8
235-264	495	12	12	29	2	34	14	1	1	28	7	6
265-294	318	7	11	6	1	83	15	2	2	21	7	8
295-1124	198	5	13	5	0	0	25	4	4	21	9	9
UNKNOWN	7	0	0	0		-	2	1	1	2	23	21
TOTAL	1739	45	13	101	8	40	69	1	1	122	8	8

LOGISTIC REGRESSIONS

AGE AT EXAM		COEFFICIENT OF RISK FACTOR	SE COEFFICIENT	P-VALUE	STANDARDIZED COEFFICIENT	STD SE
35-44		0.020	0.006	0.001	0.778	0.225
45-54		0.007	0.002	0.001	0.312	0.094
55-64		0.009	0.003	0.001	0.430	0.131
65-74		0.003	0.003	0.320	0.120	0.120
75-84		-0.002	0.004	0.582	-0.084	0.153
85-94		-0.018	0.010	0.071	-0.781	0.432
35-64	UNIVARIATE	0.009	0.002	0.001	0.433	0.089
65-94	UNIVARIATE	-0.002	0.002	0.446	-0.071	0.093
35-64	BIVARIATE	0.008	0.002	0.001	0.385	0.080
65-94	BIVARIATE	0.000	0.002	0.984	0.002	0.093
35-64	MULTIVARIATE	0.007	0.002	0.001	0.349	0.079
65-94	MULTIVARIATE	0.000	0.002	0.842	0.019	0.094

TABLE 21-5   DATE=17NOV85
AVERAGE ANNUAL INCIDENCE RATE FOR EVENT PER 1000 PERSONS AT RISK AT EXAM
BY AGE AND LEVEL OF RISK FACTOR AT EXAM: FRAMINGHAM STUDY 30-YEAR FOLLOWUP

SEX: MALE
EVENT: DEATH FROM CARDIOVASCULAR DISEASE
RISK FACTOR: HEMATOCRIT
PERSONS AT RISK: PERSONS FREE OF CARDIOVASCULAR DISEASE

RISK FACTOR	35-44 AT EXAM PERSON EXAMS	# OF EVENTS	ANNUAL RATE	45-54 AT EXAM PERSON EXAMS	# OF EVENTS	ANNUAL RATE	55-64 AT EXAM PERSON EXAMS	# OF EVENTS	ANNUAL RATE	65-74 AT EXAM PERSON EXAMS	# OF EVENTS	ANNUAL RATE
25-42	621	1	1	933	4	2	990	11	6	416	6	7
43-44	546	0	0	956	5	3	960	5	3	461	3	3
45-46	983	2	1	1836	8	2	1883	17	5	905	16	9
47-48	774	0	0	1489	5	2	1540	8	3	800	18	11
49-70	1173	4	2	2036	10	2	2233	32	7	1005	28	14
UNKNOWN	370	0	0	385	2	3	153	1	3	19	0	0
TOTAL	4467	7	1	7635	34	2	7759	74	5	3606	71	10

RISK FACTOR	75-84 AT EXAM PERSON EXAMS	# OF EVENTS	ANNUAL RATE	85-94 AT EXAM PERSON EXAMS	# OF EVENTS	ANNUAL RATE	35-64 COMBINED # OF EVENTS	CRUDE	AGE-ADJUSTED	65-94 COMBINED # OF EVENTS	CRUDE	AGE-ADJUSTED
25-42	185	7	19	11			16	3	3	14	11	10
43-44	100	6	30	4			10	2	2	9	8	8
45-46	205	10	24	6			27	3	3	27	12	12
47-48	165	3	9	6			13	2	2	21	11	11
49-70	204	9	22	3			46	4	4	39	16	18
UNKNOWN	0	0	-	0			3	2	2	0	0	0
TOTAL	859	35	20	30			115	3	3	110	12	12

LOGISTIC REGRESSIONS

AGE AT EXAM		COEFFICIENT OF RISK FACTOR	SE COEFFICIENT	P-VALUE	STANDARDIZED COEFFICIENT	STD SE
35-44		0.111	0.104	0.283	0.408	0.380
45-54		0.007	0.050	0.888	0.025	0.177
55-64		0.044	0.032	0.162	0.166	0.119
65-74		0.089	0.033	0.008	0.322	0.121
75-84		0.009	0.042	0.823	0.039	0.174
85-94						
35-64	UNIVARIATE	0.039	0.026	0.132	0.144	0.095
65-94	UNIVARIATE	0.050	0.026	0.055	0.188	0.098
35-64	BIVARIATE	0.037	0.026	0.148	0.137	0.095
65-94	BIVARIATE	0.066	0.026	0.011	0.249	0.097
35-64	MULTIVARIATE	0.014	0.026	0.596	0.051	0.096
65-94	MULTIVARIATE	0.059	0.027	0.026	0.222	0.100

SEX: FEMALE
EVENT: DEATH FROM CARDIOVASCULAR DISEASE
RISK FACTOR: HEMATOCRIT
PERSONS AT RISK: PERSONS FREE OF CARDIOVASCULAR DISEASE

RISK FACTOR	35-44 AT EXAM PERSON EXAMS	# OF EVENTS	ANNUAL RATE	45-54 AT EXAM PERSON EXAMS	# OF EVENTS	ANNUAL RATE	55-64 AT EXAM PERSON EXAMS	# OF EVENTS	ANNUAL RATE	65-74 AT EXAM PERSON EXAMS	# OF EVENTS	ANNUAL RATE
21-39	1695	1	0	1832	3	1	1512	6	2	604	4	7
40-41	966	0	0	1735	2	1	2022	7	2	998	12	6
42-43	1039	1	0	1941	1	0	2223	6	1	1190	5	2
44-45	853	1	1	2164	3	1	2779	9	2	1523	17	6
46-65	482	1	1	1657	11	3	2402	15	3	1531	25	8
UNKNOWN	566	1	1	539	1	1	263	0	0	40	2	25
TOTAL	5601	5	0	9868	21	1	11203	43	2	5886	69	6

RISK FACTOR	75-84 AT EXAM PERSON EXAMS	# OF EVENTS	ANNUAL RATE	85-94 AT EXAM PERSON EXAMS	# OF EVENTS	ANNUAL RATE	35-64 COMBINED # OF EVENTS	CRUDE	AGE-ADJUSTED	65-94 COMBINED # OF EVENTS	CRUDE	AGE-ADJUSTED
21-39	255	9	18	17	1	29	10	1	1	18	10	9
40-41	332	7	11	25	3	60	9	1	1	22	8	8
42-43	327	6	9	18	3	83	8	1	1	14	5	6
44-45	439	8	9	21	0	0	13	1	1	25	6	6
46-65	380	15	20	20	1	25	27	3	3	41	11	11
UNKNOWN	6	0	0	0	0	-	2	1	1	2	22	19
TOTAL	1739	45	13	101	8	40	69	1	1	122	8	8

LOGISTIC REGRESSIONS

AGE AT EXAM		COEFFICIENT OF RISK FACTOR	SE COEFFICIENT	P-VALUE	STANDARDIZED COEFFICIENT	STD SE
35-44		0.135	0.146	0.355	0.484	0.524
45-54		0.189	0.063	0.003	0.674	0.223
55-64		0.107	0.044	0.016	0.369	0.153
65-74		0.042	0.036	0.239	0.146	0.124
75-84		0.007	0.041	0.858	0.027	0.152
85-94		-0.038	0.092	0.679	-0.154	0.371
35-64	UNIVARIATE	0.152	0.035	0.001	0.545	0.124
65-94	UNIVARIATE	0.011	0.026	0.670	0.039	0.092
35-64	BIVARIATE	0.130	0.035	0.001	0.465	0.127
65-94	BIVARIATE	0.023	0.026	0.377	0.081	0.092
35-64	MULTIVARIATE	0.079	0.035	0.025	0.283	0.127
65-94	MULTIVARIATE	0.012	0.025	0.643	0.042	0.090

AVERAGE ANNUAL INCIDENCE RATE FOR EVENT PER 1000 PERSONS AT RISK AT EXAM
BY AGE AND LEVEL OF RISK FACTOR AT EXAM: FRAMINGHAM STUDY 30-YEAR FOLLOWUP

SEX: MALE
EVENT: DEATH FROM CARDIOVASCULAR DISEASE
RISK FACTOR: BLOOD GLUCOSE (MG/DL)
PERSONS AT RISK: PERSONS FREE OF CARDIOVASCULAR DISEASE

RISK FACTOR	35-44 AT EXAM PERSON EXAMS	# OF EVENTS	ANNUAL RATE	45-54 AT EXAM PERSON EXAMS	# OF EVENTS	ANNUAL RATE	55-64 AT EXAM PERSON EXAMS	# OF EVENTS	ANNUAL RATE	65-74 AT EXAM PERSON EXAMS	# OF EVENTS	ANNUAL RATE
29-69	992	0	0	1475	9	3	1043	9	4	399	5	6
70-89	2498	6	1	4127	16	2	3928	45	6	1635	27	8
90-109	626	1	1	1334	5	2	1795	13	4	966	22	11
110-129	104	0	0	298	0	0	515	5	5	308	8	13
130-524	47	0	0	185	3	8	386	2	3	282	9	16
UNKNOWN	200	0	0	216	1	2	92	0	0	16	0	0
TOTAL	4467	7	1	7635	34	2	7759	74	5	3606	71	10

RISK FACTOR	75-84 AT EXAM PERSON EXAMS	# OF EVENTS	ANNUAL RATE	85-94 AT EXAM PERSON EXAMS	# OF EVENTS	ANNUAL RATE	35-64 COMBINED # OF EVENTS	CRUDE	AGE-ADJUSTED	65-94 COMBINED # OF EVENTS	CRUDE	AGE-ADJUSTED
29-69	59	3	25	1			18	3	3	8	9	10
70-89	330	12	18	10			67	3	3	39	10	10
90-109	273	10	18	10			19	3	2	33	13	13
110-129	97	3	15	1			5	3	2	11	14	13
130-524	100	7	35	8			5	4	4	19	24	21
UNKNOWN	0	0	-	0			1	1	1	0	0	0
TOTAL	859	35	20	30			115	3	3	110	12	12

LOGISTIC REGRESSIONS

AGE AT EXAM		COEFFICIENT OF RISK FACTOR	SE COEFFICIENT	P-VALUE	STANDARDIZED COEFFICIENT	STD SE
35-44		-0.019	0.031	0.535	-0.303	0.489
45-54		0.005	0.005	0.366	0.110	0.121
55-64		-0.008	0.006	0.182	-0.212	0.158
65-74		0.008	0.002	0.001	0.249	0.068
75-84		0.007	0.004	0.124	0.205	0.134
85-94						
35-64	UNIVARIATE	0.000	0.004	0.926	0.009	0.092
65-94	UNIVARIATE	0.008	0.002	0.001	0.259	0.057
35-64	BIVARIATE	-0.004	0.004	0.329	-0.102	0.105
65-94	BIVARIATE	0.008	0.002	0.001	0.254	0.060
35-64	MULTIVARIATE	-0.006	0.004	0.212	-0.131	0.105
65-94	MULTIVARIATE	0.007	0.002	0.001	0.232	0.061

SEX: FEMALE
EVENT: DEATH FROM CARDIOVASCULAR DISEASE
RISK FACTOR: BLOOD GLUCOSE (MG/DL)
PERSONS AT RISK: PERSONS FREE OF CARDIOVASCULAR DISEASE

RISK FACTOR	35-44 AT EXAM PERSON EXAMS	# OF EVENTS	ANNUAL RATE	45-54 AT EXAM PERSON EXAMS	# OF EVENTS	ANNUAL RATE	55-64 AT EXAM PERSON EXAMS	# OF EVENTS	ANNUAL RATE	65-74 AT EXAM PERSON EXAMS	# OF EVENTS	ANNUAL RATE
29-69	1063	1	0	1787	3	1	1346	8	3	487	6	6
70-89	3401	3	0	5699	11	1	5963	22	2	2951	21	4
90-109	671	0	0	1612	1	0	2616	6	1	1624	18	6
110-129	96	0	0	298	0	0	705	4	3	473	9	10
130-524	39	0	0	145	5	17	413	3	4	321	13	20
UNKNOWN	331	1	2	327	1	2	160	0	0	30	2	33
TOTAL	5601	5	0	9868	21	1	11203	43	2	5886	69	6

RISK FACTOR	75-84 AT EXAM PERSON EXAMS	# OF EVENTS	ANNUAL RATE	85-94 AT EXAM PERSON EXAMS	# OF EVENTS	ANNUAL RATE	35-64 COMBINED # OF EVENTS	CRUDE	AGE-ADJUSTED	65-94 COMBINED # OF EVENTS	CRUDE	AGE-ADJUSTED
29-69	96	3	16	5	0	0	12	1	2	9	8	8
70-89	748	18	12	40	3	38	36	1	1	42	6	6
90-109	559	12	11	40	1	13	7	1	1	31	7	7
110-129	193	8	21	9	1	56	4	2	1	18	13	13
130-524	142	4	14	9	3	167	8	7	8	20	21	21
UNKNOWN	1	0	0	0	0	-	2	1	1	2	32	25
TOTAL	1739	45	13	101	8	40	69	1	1	122	8	8

LOGISTIC REGRESSIONS

AGE AT EXAM		COEFFICIENT OF RISK FACTOR	SE COEFFICIENT	P-VALUE	STANDARDIZED COEFFICIENT	STD SE
35-44		-	-	-	-	-
45-54		0.019	0.004	0.001	0.356	0.066
55-64		0.007	0.004	0.062	0.174	0.093
65-74		0.013	0.002	0.001	0.345	0.056
75-84		0.005	0.003	0.088	0.173	0.102
85-94		0.018	0.011	0.118	0.425	0.272
35-64	UNIVARIATE	0.012	0.002	0.001	0.256	0.050
65-94	UNIVARIATE	0.010	0.002	0.001	0.294	0.047
35-64	BIVARIATE	0.010	0.003	0.001	0.217	0.054
65-94	BIVARIATE	0.009	0.002	0.001	0.270	0.043
35-64	MULTIVARIATE	0.007	0.003	0.007	0.156	0.058
65-94	MULTIVARIATE	0.009	0.002	0.001	0.249	0.049

TABLE 21-7   DATE=17NOV85

AVERAGE ANNUAL INCIDENCE RATE FOR EVENT PER 1000 PERSONS AT RISK AT EXAM
BY AGE AND LEVEL OF RISK FACTOR AT EXAM: FRAMINGHAM STUDY 30-YEAR FOLLOWUP

SEX: MALE
EVENT: DEATH FROM CARDIOVASCULAR DISEASE
RISK FACTOR: DIABETES MELLITUS
PERSONS AT RISK: PERSONS FREE OF CARDIOVASCULAR DISEASE

RISK FACTOR	35-44 AT EXAM			45-54 AT EXAM			55-64 AT EXAM			65-74 AT EXAM		
	PERSON EXAMS	# OF EVENTS	ANNUAL RATE	PERSON EXAMS	# OF EVENTS	ANNUAL RATE	PERSON EXAMS	# OF EVENTS	ANNUAL RATE	PERSON EXAMS	# OF EVENTS	ANNUAL RATE
NO	4421	7	1	7411	32	2	7310	72	5	3267	61	9
YES	46	0	0	224	2	4	449	2	2	339	10	15
UNKNOWN	0	0	-	0	0	-	0	0	-	0	0	-
TOTAL	4467	7	1	7635	34	2	7759	74	5	3606	71	10

RISK FACTOR	75-84 AT EXAM			85-94 AT EXAM			35-64 COMBINED			65-94 COMBINED		
	PERSON EXAMS	# OF EVENTS	ANNUAL RATE	PERSON EXAMS	# OF EVENTS	ANNUAL RATE	# OF EVENTS	CRUDE	AGE-ADJUSTED	# OF EVENTS	CRUDE	AGE-ADJUSTED
NO	769	30	20	25	5		111	3	3	94	12	12
YES	90	5	28	5			4	3	3	16	18	18
UNKNOWN	0	0	-	0			0	-		0	-	
TOTAL	859	35	20	30			115	3	3	110	12	12

LOGISTIC REGRESSIONS

AGE AT EXAM		COEFFICIENT OF RISK FACTOR	SE COEFFICIENT	P-VALUE	STANDARDIZED COEFFICIENT	STD SE
35-44		-	-	-	-	-
45-54		0.731	0.732	0.318	0.123	0.124
55-64		-0.799	0.719	0.266	-0.187	0.168
65-74		0.468	0.346	0.176	0.137	0.101
75-84		0.371	0.496	0.455	0.114	0.152
85-94						
35-64	UNIVARIATE	-0.042	0.510	0.935	-0.008	0.095
65-94	UNIVARIATE	0.480	0.275	0.081	0.142	0.081
35-64	BIVARIATE	-0.386	0.512	0.451	-0.072	0.096
65-94	BIVARIATE	0.439	0.276	0.112	0.130	0.082
35-64	MULTIVARIATE	-0.692	0.522	0.185	-0.129	0.098
65-94	MULTIVARIATE	0.377	0.281	0.180	0.111	0.083

SEX: FEMALE
EVENT: DEATH FROM CARDIOVASCULAR DISEASE
RISK FACTOR: DIABETES MELLITUS
PERSONS AT RISK: PERSONS FREE OF CARDIOVASCULAR DISEASE

RISK FACTOR	35-44 AT EXAM			45-54 AT EXAM			55-64 AT EXAM			65-74 AT EXAM		
	PERSON EXAMS	# OF EVENTS	ANNUAL RATE	PERSON EXAMS	# OF EVENTS	ANNUAL RATE	PERSON EXAMS	# OF EVENTS	ANNUAL RATE	PERSON EXAMS	# OF EVENTS	ANNUAL RATE
NO	5558	4	0	9683	16	1	10713	39	2	5469	55	5
YES	43	1	12	185	5	14	490	4	4	417	14	17
UNKNOWN	0	0	-	0	0	-	0	0	-	0	0	-
TOTAL	5601	5	0	9868	21	1	11203	43	2	5886	69	6

RISK FACTOR	75-84 AT EXAM			85-94 AT EXAM			35-64 COMBINED			65-94 COMBINED		
	PERSON EXAMS	# OF EVENTS	ANNUAL RATE	PERSON EXAMS	# OF EVENTS	ANNUAL RATE	# OF EVENTS	CRUDE	AGE-ADJUSTED	# OF EVENTS	CRUDE	AGE-ADJUSTED
NO	1548	38	12	89	6	34	59	1	1	99	7	7
YES	191	7	18	12	2	83	10	7	9	23	19	18
UNKNOWN	0	0	-	0	0	-	0	-		0	-	
TOTAL	1739	45	13	101	8	40	69	1	1	122	8	8

LOGISTIC REGRESSIONS

AGE AT EXAM		COEFFICIENT OF RISK FACTOR	SE COEFFICIENT	P-VALUE	STANDARDIZED COEFFICIENT	STD SE
35-44		3.498	1.129	0.002	0.305	0.099
45-54		2.820	0.518	0.001	0.383	0.070
55-64		0.812	0.527	0.123	0.166	0.108
65-74		1.230	0.304	0.001	0.315	0.078
75-84		0.415	0.419	0.324	0.129	0.131
85-94		1.018	0.882	0.249	0.331	0.287
35-64	UNIVARIATE	1.824	0.344	0.001	0.295	0.056
65-94	UNIVARIATE	1.003	0.235	0.001	0.273	0.064
35-64	BIVARIATE	1.569	0.348	0.001	0.254	0.056
65-94	BIVARIATE	0.877	0.237	0.001	0.238	0.065
35-64	MULTIVARIATE	1.149	0.374	0.002	0.186	0.060
65-94	MULTIVARIATE	0.636	0.252	0.012	0.173	0.068

SEX: MALE
EVENT: DEATH FROM CARDIOVASCULAR DISEASE
RISK FACTOR: GLUCOSE IN URINE, DEFINITE OR TRACE
PERSONS AT RISK: PERSONS FREE OF CARDIOVASCULAR DISEASE

RISK FACTOR	35-44 AT EXAM PERSON EXAMS	# OF EVENTS	ANNUAL RATE	45-54 AT EXAM PERSON EXAMS	# OF EVENTS	ANNUAL RATE	55-64 AT EXAM PERSON EXAMS	# OF EVENTS	ANNUAL RATE	65-74 AT EXAM PERSON EXAMS	# OF EVENTS	ANNUAL RATE
NO	4377	7	1	7453	32	2	7546	69	5	3487	64	9
YES	58	0	0	161	2	6	189	4	11	114	7	31
UNKNOWN	32	0	0	21	0	0	24	1	21	5	0	0
TOTAL	4467	7	1	7635	34	2	7759	74	5	3606	71	10

RISK FACTOR	75-84 AT EXAM PERSON EXAMS	# OF EVENTS	ANNUAL RATE	85-94 AT EXAM PERSON EXAMS	# OF EVENTS	ANNUAL RATE	35-64 COMBINED # OF EVENTS	CRUDE	AGE-ADJUSTED	65-94 COMBINED # OF EVENTS	CRUDE	AGE-ADJUSTED
NO	839	32	19	29			108	3	3	100	11	11
YES	20	3	75	1			6	7	7	10	37	39
UNKNOWN	0	0	-	0			1	6	8	0	0	0
TOTAL	859	35	20	30			115	3	3	110	12	12

LOGISTIC REGRESSIONS

AGE AT EXAM		COEFFICIENT OF RISK FACTOR	SE COEFFICIENT	P-VALUE	STANDARDIZED COEFFICIENT	STD SE
35-44		-	-	-	-	-
45-54		1.071	0.733	0.144	0.154	0.106
55-64		0.851	0.520	0.101	0.131	0.080
65-74		1.252	0.410	0.002	0.219	0.072
75-84		1.493	0.652	0.022	0.225	0.098
85-94						
35-64	UNIVARIATE	0.979	0.422	0.020	0.139	0.060
65-94	UNIVARIATE	1.225	0.344	0.001	0.209	0.059
35-64	BIVARIATE	0.841	0.424	0.047	0.120	0.060
65-94	BIVARIATE	1.281-	0.346	0.001	0.219	0.059
35-64	MULTIVARIATE	0.515	0.470	0.273	0.073	0.067
65-94	MULTIVARIATE	1.278	0.332	0.001	0.218	0.060

SEX: FEMALE
EVENT: DEATH FROM CARDIOVASCULAR DISEASE
RISK FACTOR: GLUCOSE IN URINE, DEFINITE OR TRACE
PERSONS AT RISK: PERSONS FREE OF CARDIOVASCULAR DISEASE

RISK FACTOR	35-44 AT EXAM PERSON EXAMS	# OF EVENTS	ANNUAL RATE	45-54 AT EXAM PERSON EXAMS	# OF EVENTS	ANNUAL RATE	55-64 AT EXAM PERSON EXAMS	# OF EVENTS	ANNUAL RATE	65-74 AT EXAM PERSON EXAMS	# OF EVENTS	ANNUAL RATE
NO	5522	4	0	9758	18	1	11045	38	2	5785	65	6
YES	41	1	12	86	3	17	135	5	19	91	4	22
UNKNOWN	38	0	0	24	0	0	23	0	0	10	0	0
TOTAL	5601	5	0	9868	21	1	11203	43	2	5886	69	6

RISK FACTOR	75-84 AT EXAM PERSON EXAMS	# OF EVENTS	ANNUAL RATE	85-94 AT EXAM PERSON EXAMS	# OF EVENTS	ANNUAL RATE	35-64 COMBINED # OF EVENTS	CRUDE	AGE-ADJUSTED	65-94 COMBINED # OF EVENTS	CRUDE	AGE-ADJUSTED
NO	1710	43	13	100	7	35	60	1	1	115	8	8
YES	20	2	50	1	1	500	9	17	17	7	31	35
UNKNOWN	9	0	0	0	0	-	0	0	0	0	0	0
TOTAL	1739	45	13	101	8	40	69	1	1	122	8	8

LOGISTIC REGRESSIONS

AGE AT EXAM		COEFFICIENT OF RISK FACTOR	SE COEFFICIENT	P-VALUE	STANDARDIZED COEFFICIENT	STD SE
35-44		3.541	1.129	0.002	0.303	0.097
45-54		2.973	0.633	0.001	0.277	0.059
55-64		2.411	0.484	0.001	0.263	0.053
65-74		1.398	0.526	0.008	0.173	0.065
75-84		1.460	0.761	0.055	0.156	0.081
85-94		-				
35-64	UNIVARIATE	2.745	0.363	0.001	0.271	0.036
65-94	UNIVARIATE	1.467	0.402	0.001	0.176	0.048
35-64	BIVARIATE	2.665	0.365	0.001	0.263	0.036
65-94	BIVARIATE	1.556-	0.405	0.001	0.186	0.048
35-64	MULTIVARIATE	2.228	0.394	0.001	0.220	0.039
65-94	MULTIVARIATE	1.256	0.420	0.003	0.150	0.050

TABLE 21-9   DATE=07MAR86
AVERAGE ANNUAL INCIDENCE RATE FOR EVENT PER 1000 PERSONS AT RISK AT EXAM
BY AGE AND LEVEL OF RISK FACTOR AT EXAM: FRAMINGHAM STUDY 30-YEAR FOLLOWUP

SEX: MALE
EVENT: DEATH FROM CARDIOVASCULAR DISEASE
RISK FACTOR: GLUCOSE INTOLERANCE
PERSONS AT RISK: PERSONS FREE OF CARDIOVASCULAR DISEASE

RISK FACTOR	35-44 AT EXAM PERSON EXAMS	# OF EVENTS	ANNUAL RATE	45-54 AT EXAM PERSON EXAMS	# OF EVENTS	ANNUAL RATE	55-64 AT EXAM PERSON EXAMS	# OF EVENTS	ANNUAL RATE	65-74 AT EXAM PERSON EXAMS	# OF EVENTS	ANNUAL RATE
NO	4332	7	1	7150	31	2	6919	69	5	3019	54	9
YES	135	0	0	485	3	3	840	5	3	587	17	14
UNKNOWN	0	0	-	0	0	-	0	0	-	0	0	-
TOTAL	4467	7	1	7635	34	2	7759	74	5	3606	71	10

RISK FACTOR	75-84 AT EXAM PERSON EXAMS	# OF EVENTS	ANNUAL RATE	85-94 AT EXAM PERSON EXAMS	# OF EVENTS	ANNUAL RATE	35-64 COMBINED # OF EVENTS	CRUDE	AGE-ADJUSTED	65-94 COMBINED # OF EVENTS	CRUDE	AGE-ADJUSTED
NO	686	26	19	22			107	3	3	81	11	11
YES	173	9	26	8			8	3	2	29	19	18
UNKNOWN	0	0	-	0			0	-		0	-	
TOTAL	859	35	20	30			115	3	3	110	12	12

LOGISTIC REGRESSIONS

AGE AT EXAM		COEFFICIENT OF RISK FACTOR	SE COEFFICIENT	P-VALUE	STANDARDIZED COEFFICIENT	STD SE
35-44		-	-	-	-	-
45-54		0.357	0.606	0.556	0.087	0.148
55-64		-0.520	0.465	0.263	-0.162	0.144
65-74		0.493	0.282	0.080	0.182	0.104
75-84		0.332	0.396	0.403	0.133	0.159
85-94		-	-	-	-	-
35-64	UNIVARIATE	-0.060	0.368	0.871	-0.016	0.096
65-94	UNIVARIATE	0.569	0.220	0.010	0.214	0.083
35-64	BIVARIATE	-0.347	0.369	0.347	-0.091	0.096
65-94	BIVARIATE	0.504	0.221	0.023	0.190	0.083
35-64	MULTIVARIATE	-0.658	0.398	0.098	-0.172	0.104
65-94	MULTIVARIATE	0.445	0.224	0.047	0.168	0.084

SEX: FEMALE
EVENT: DEATH FROM CARDIOVASCULAR DISEASE
RISK FACTOR: GLUCOSE INTOLERANCE
PERSONS AT RISK: PERSONS FREE OF CARDIOVASCULAR DISEASE

RISK FACTOR	35-44 AT EXAM PERSON EXAMS	# OF EVENTS	ANNUAL RATE	45-54 AT EXAM PERSON EXAMS	# OF EVENTS	ANNUAL RATE	55-64 AT EXAM PERSON EXAMS	# OF EVENTS	ANNUAL RATE	65-74 AT EXAM PERSON EXAMS	# OF EVENTS	ANNUAL RATE
NO	5507	4	0	9517	15	1	10242	37	2	5112	48	5
YES	94	1	5	351	6	9	961	6	3	774	21	14
UNKNOWN	0	0	-	0	0	-	0	0	-	0	0	-
TOTAL	5601	5	0	9868	21	1	11203	43	2	5886	69	6

RISK FACTOR	75-84 AT EXAM PERSON EXAMS	# OF EVENTS	ANNUAL RATE	85-94 AT EXAM PERSON EXAMS	# OF EVENTS	ANNUAL RATE	35-64 COMBINED # OF EVENTS	CRUDE	AGE-ADJUSTED	65-94 COMBINED # OF EVENTS	CRUDE	AGE-ADJUSTED
NO	1412	37	13	83	4	24	56	1	1	89	7	7
YES	327	8	12	18	4	111	13	5	6	33	15	15
UNKNOWN	0	0	-	0	0	-	0	-		0	-	
TOTAL	1739	45	13	101	8	40	69	1	1	122	8	8

LOGISTIC REGRESSIONS

AGE AT EXAM		COEFFICIENT OF RISK FACTOR	SE COEFFICIENT	P-VALUE	STANDARDIZED COEFFICIENT	STD SE
35-44		2.694	1.123	0.016	0.346	0.144
45-54		2.399	0.486	0.001	0.444	0.090
55-64		0.550	0.441	0.213	0.154	0.124
65-74		1.079	0.265	0.001	0.365	0.089
75-84		-0.070	0.395	0.858	-0.028	0.154
85-94		1.730	0.764	0.024	0.666	0.294
35-64	UNIVARIATE	1.435	0.309	0.001	0.321	0.069
65-94	UNIVARIATE	0.800	0.206	0.001	0.282	0.073
35-64	BIVARIATE	1.179	0.314	0.001	0.263	0.070
65-94	BIVARIATE	0.688	0.208	0.001	0.242	0.073
35-64	MULTIVARIATE	0.927	0.329	0.005	0.207	0.073
65-94	MULTIVARIATE	0.538	0.216	0.013	0.189	0.076

SEX: MALE
EVENT: DEATH FROM CARDIOVASCULAR DISEASE
RISK FACTOR: METROPOLITAN RELATIVE WEIGHT   (PERCENT)
PERSONS AT RISK: PERSONS FREE OF CARDIOVASCULAR DISEASE

RISK FACTOR	35-44 AT EXAM			45-54 AT EXAM			55-64 AT EXAM			65-74 AT EXAM		
	PERSON EXAMS	# OF EVENTS	ANNUAL RATE	PERSON EXAMS	# OF EVENTS	ANNUAL RATE	PERSON EXAMS	# OF EVENTS	ANNUAL RATE	PERSON EXAMS	# OF EVENTS	ANNUAL RATE
54-104	719	1	1	1119	4	2	1096	16	7	610	17	14
105-114	965	0	0	1519	8	3	1594	15	5	780	8	5
115-124	1122	2	1	2017	11	3	2105	12	3	1008	25	12
125-134	914	3	2	1621	4	1	1618	13	4	703	10	7
135-272	747	1	2	1358	7	3	1345	18	7	505	11	11
UNKNOWN	0	0	-	1	0	0	1	0	0	0	0	-
TOTAL	4467	7	1	7635	34	2	7759	74	5	3606	71	10

RISK FACTOR	75-84 AT EXAM			85-94 AT EXAM			35-64 COMBINED			65-94 COMBINED		
	PERSON EXAMS	# OF EVENTS	ANNUAL RATE	PERSON EXAMS	# OF EVENTS	ANNUAL RATE	# OF EVENTS	CRUDE	AGE-ADJUSTED	# OF EVENTS	CRUDE	AGE-ADJUSTED
54-104	188	7	19	14			21	4	4	26	16	15
105-114	208	8	19	6			23	3	3	17	9	8
115-124	256	10	20	10			25	2	2	36	14	14
125-134	135	5	19	0			20	2	2	15	9	9
135-272	72	5	35	0			26	4	4	16	14	15
UNKNOWN	0	0	-	0			0	0	0	0	0	-
TOTAL	859	35	20	30			115	3	3	110	12	12

LOGISTIC REGRESSIONS

AGE AT EXAM	COEFFICIENT OF RISK FACTOR	SE COEFFICIENT	P-VALUE	STANDARDIZED COEFFICIENT	STD SE
35-44	0.022	0.022	0.308	0.355	0.348
45-54	-0.003	0.011	0.785	-0.047	0.174
55-64	-0.002	0.007	0.797	-0.030	0.117
65-74	-0.006	0.008	0.425	-0.096	0.121
75-84	0.001	0.012	0.901	0.021	0.172
85-94					
35-64 UNIVARIATE	-0.000	0.006	0.977	-0.003	0.094
65-94 UNIVARIATE	-0.007	0.006	0.243	-0.114	0.098
35-64 BIVARIATE	-0.000	0.006	0.942	-0.007	0.094
65-94 BIVARIATE	-0.004	0.006	0.545	-0.060	0.099
35-64 MULTIVARIATE	-0.006	0.006	0.331	-0.097	0.099
65-94 MULTIVARIATE	-0.010	0.007	0.148	-0.147	0.102

SEX: FEMALE
EVENT: DEATH FROM CARDIOVASCULAR DISEASE
RISK FACTOR: METROPOLITAN RELATIVE WEIGHT   (PERCENT)
PERSONS AT RISK: PERSONS FREE OF CARDIOVASCULAR DISEASE

RISK FACTOR	35-44 AT EXAM			45-54 AT EXAM			55-64 AT EXAM			65-74 AT EXAM		
	PERSON EXAMS	# OF EVENTS	ANNUAL RATE	PERSON EXAMS	# OF EVENTS	ANNUAL RATE	PERSON EXAMS	# OF EVENTS	ANNUAL RATE	PERSON EXAMS	# OF EVENTS	ANNUAL RATE
54-104	1689	2	1	2075	6	1	1981	6	2	1138	11	5
105-114	1429	0	0	2308	1	0	2351	10	2	1101	14	6
115-124	1056	1	0	2092	0	0	2415	9	2	1313	9	3
125-134	564	1	1	1362	4	1	1711	2	1	938	10	5
135-272	855	1	1	2031	10	2	2745	16	3	1396	25	9
UNKNOWN	8	0	0	0	0	-	0	0	-	0	0	-
TOTAL	5601	5	0	9868	21	1	11203	43	2	5886	69	6

RISK FACTOR	75-84 AT EXAM			85-94 AT EXAM			35-64 COMBINED			65-94 COMBINED		
	PERSON EXAMS	# OF EVENTS	ANNUAL RATE	PERSON EXAMS	# OF EVENTS	ANNUAL RATE	# OF EVENTS	CRUDE	AGE-ADJUSTED	# OF EVENTS	CRUDE	AGE-ADJUSTED
54-104	458	18	20	34	3	44	14	1	1	32	10	9
105-114	326	3	5	27	2	37	11	1	1	19	7	6
115-124	351	6	9	14	3	107	10	1	1	18	5	6
125-134	242	5	10	13	0	0	7	1	1	15	6	6
135-272	362	13	18	13	0	0	27	2	2	38	11	11
UNKNOWN	0	0	-	0	0	-	0	0	0	0	0	-
TOTAL	1739	45	13	101	8	40	69	1	1	122	8	8

LOGISTIC REGRESSIONS

AGE AT EXAM	COEFFICIENT OF RISK FACTOR	SE COEFFICIENT	P-VALUE	STANDARDIZED COEFFICIENT	STD SE
35-44	0.003	0.021	0.890	0.059	0.429
45-54	0.019	0.007	0.008	0.406	0.154
55-64	0.004	0.007	0.539	0.089	0.145
65-74	0.010	0.005	0.055	0.210	0.109
75-84	0.004	0.007	0.525	0.093	0.147
85-94	-0.021	0.021	0.323	-0.395	0.400
35-64 UNIVARIATE	0.011	0.005	0.016	0.244	0.101
65-94 UNIVARIATE	0.004	0.004	0.300	0.091	0.088
35-64 BIVARIATE	0.009	0.005	0.057	0.199	0.105
65-94 BIVARIATE	0.007	0.004	0.084	0.151	0.088
35-64 MULTIVARIATE	0.008	0.005	0.116	0.172	0.109
65-94 MULTIVARIATE	0.006	0.004	0.148	0.127	0.088

TABLE 21-11　　　　DATE=17NOV85
AVERAGE ANNUAL INCIDENCE RATE FOR EVENT PER 1000 PERSONS AT RISK AT EXAM
BY AGE AND LEVEL OF RISK FACTOR AT EXAM: FRAMINGHAM STUDY 30-YEAR FOLLOWUP

SEX: MALE
EVENT: DEATH FROM CARDIOVASCULAR DISEASE
RISK FACTOR: VITAL CAPACITY - HEIGHT INDEX (ML/INCH)
PERSONS AT RISK: PERSONS FREE OF CARDIOVASCULAR DISEASE

RISK FACTOR	35-44 AT EXAM			45-54 AT EXAM			55-64 AT EXAM			65-74 AT EXAM		
	PERSON EXAMS	# OF EVENTS	ANNUAL RATE	PERSON EXAMS	# OF EVENTS	ANNUAL RATE	PERSON EXAMS	# OF EVENTS	ANNUAL RATE	PERSON EXAMS	# OF EVENTS	ANNUAL RATE
8-44	264	0	0	1035	6	3	1920	24	6	1578	34	11
45-49	411	1	1	1108	4	2	1435	13	5	702	19	14
50-54	747	0	0	1647	11	3	1780	15	4	674	13	10
55-59	1022	4	2	1716	4	1	1367	11	4	415	2	2
60-85	2010	2	0	2124	9	2	1239	11	4	225	2	4
UNKNOWN	13	0	0	5	0	0	18	0	0	12	1	42
TOTAL	4467	7	1	7635	34	2	7759	74	5	3606	71	10

RISK FACTOR	75-84 AT EXAM			85-94 AT EXAM			35-64 COMBINED			65-94 COMBINED		
	PERSON EXAMS	# OF EVENTS	ANNUAL RATE	PERSON EXAMS	# OF EVENTS	ANNUAL RATE	# OF EVENTS	CRUDE	AGE-ADJUSTED	# OF EVENTS	CRUDE	AGE-ADJUSTED
8-44	530	23	22	17			30	5	4	60	14	13
45-49	143	2	7	4			18	3	3	21	12	12
50-54	91	6	33	7			26	3	3	20	13	15
55-59	63	4	32	2			19	2	2	6	8	8
60-85	24	0	0	0			22	2	3	2	4	4
UNKNOWN	8	0	0	0			0	0	0	1	25	33
TOTAL	859	35	20	30			115	3	3	110	12	12

LOGISTIC REGRESSIONS

AGE AT EXAM		COEFFICIENT OF RISK FACTOR	SE COEFFICIENT	P-VALUE	STANDARDIZED COEFFICIENT	STD SE
35-44		-0.018	0.042	0.665	-0.160	0.370
45-54		-0.010	0.019	0.591	-0.092	0.170
55-64		-0.020	0.012	0.109	-0.183	0.114
65-74		-0.039	0.013	0.002	-0.360	0.117
75-84		0.005	0.018	0.785	0.047	0.173
85-94						
35-64	UNIVARIATE	-0.032	0.009	0.001	-0.307	0.090
65-94	UNIVARIATE	-0.033	0.010	0.001	-0.314	0.095
35-64	BIVARIATE	-0.012	0.010	0.226	-0.117	0.097
65-94	BIVARIATE	-0.024	0.010	0.019	-0.230	0.098
35-64	MULTIVARIATE	-0.001	0.011	0.948	-0.007	0.102
65-94	MULTIVARIATE	-0.019	0.011	0.068	-0.183	0.100

SEX: FEMALE
EVENT: DEATH FROM CARDIOVASCULAR DISEASE
RISK FACTOR: VITAL CAPACITY - HEIGHT INDEX (ML/INCH)
PERSONS AT RISK: PERSONS FREE OF CARDIOVASCULAR DISEASE

RISK FACTOR	35-44 AT EXAM			45-54 AT EXAM			55-64 AT EXAM			65-74 AT EXAM		
	PERSON EXAMS	# OF EVENTS	ANNUAL RATE	PERSON EXAMS	# OF EVENTS	ANNUAL RATE	PERSON EXAMS	# OF EVENTS	ANNUAL RATE	PERSON EXAMS	# OF EVENTS	ANNUAL RATE
4-29	121	0	0	518	1	1	1373	10	4	1573	31	10
30-34	289	0	0	1211	5	2	2359	12	3	1619	16	5
35-39	766	2	1	2054	5	1	2991	8	1	1475	12	4
40-44	1423	0	0	2778	5	1	2647	11	2	827	8	5
45-92	2981	3	1	3290	5	1	1812	2	1	386	2	3
UNKNOWN	21	0	0	17	0	0	21	0	0	6	0	0
TOTAL	5601	5	0	9868	21	1	11203	43	2	5886	69	6

RISK FACTOR	75-84 AT EXAM			85-94 AT EXAM			35-64 COMBINED			65-94 COMBINED		
	PERSON EXAMS	# OF EVENTS	ANNUAL RATE	PERSON EXAMS	# OF EVENTS	ANNUAL RATE	# OF EVENTS	CRUDE	AGE-ADJUSTED	# OF EVENTS	CRUDE	AGE-ADJUSTED
4-29	803	23	14	63	5	40	11	3	2	59	12	11
30-34	512	16	16	25	3	60	17	2	2	35	8	8
35-39	296	5	8	12	0	0	15	1	1	17	5	5
40-44	103	1	5	1	0	0	16	1	1	9	5	5
45-92	24	0	0	0	0	-	10	1	1	2	2	2
UNKNOWN	1	0	0	0	0	-	0	0	0	4	0	0
TOTAL	1739	45	13	101	8	40	69	1	1	122	8	8

LOGISTIC REGRESSIONS

AGE AT EXAM		COEFFICIENT OF RISK FACTOR	SE COEFFICIENT	P-VALUE	STANDARDIZED COEFFICIENT	STD SE
35-44		-0.031	0.060	0.606	-0.225	0.436
45-54		-0.039	0.029	0.170	-0.291	0.212
55-64		-0.050	0.020	0.020	-0.370	0.146
65-74		-0.057	0.016	0.001	-0.420	0.116
75-84		-0.034	0.021	0.103	-0.239	0.146
85-94		0.009	0.056	0.868	0.063	0.378
35-64	UNIVARIATE	-0.059	0.015	0.001	-0.466	0.115
65-94	UNIVARIATE	-0.058	0.012	0.001	-0.430	0.087
35-64	BIVARIATE	-0.041	0.016	0.010	-0.324	0.126
65-94	BIVARIATE	-0.042	0.012	0.001	-0.315	0.093
35-64	MULTIVARIATE	-0.020	0.017	0.236	-0.161	0.136
65-94	MULTIVARIATE	-0.038	0.013	0.004	-0.282	0.098

AVERAGE ANNUAL INCIDENCE RATE FOR EVENT PER 1000 PERSONS AT RISK AT EXAM
BY AGE AND LEVEL OF RISK FACTOR AT EXAM: FRAMINGHAM STUDY 30-YEAR FOLLOWUP

SEX: MALE
EVENT: DEATH FROM CARDIOVASCULAR DISEASE
RISK FACTOR: HEART RATE (PER MINUTE)
PERSONS AT RISK: PERSONS FREE OF CARDIOVASCULAR DISEASE

RISK FACTOR	35-44 AT EXAM			45-54 AT EXAM			55-64 AT EXAM			65-74 AT EXAM		
	PERSON EXAMS	# OF EVENTS	ANNUAL RATE	PERSON EXAMS	# OF EVENTS	ANNUAL RATE	PERSON EXAMS	# OF EVENTS	ANNUAL RATE	PERSON EXAMS	# OF EVENTS	ANNUAL RATE
30-67	1284	1	0	1934	6	2	2256	11	2	1099	14	6
68-75	1489	0	0	2465	6	1	2405	20	4	1094	18	8
76-83	721	1	1	1374	7	3	1315	15	6	574	13	11
84-91	579	5	4	1084	7	3	1019	9	4	432	11	13
92-220	394	0	0	778	8	5	764	19	12	407	15	18
UNKNOWN	0	0	-	0	0	-	0	0	-	0	0	-
TOTAL	4467	7	1	7635	34	2	7759	74	5	3606	71	10

RISK FACTOR	75-84 AT EXAM			85-94 AT EXAM			35-64 COMBINED			65-94 COMBINED		
	PERSON EXAMS	# OF EVENTS	ANNUAL RATE	PERSON EXAMS	# OF EVENTS	ANNUAL RATE	# OF EVENTS	CRUDE	AGE-ADJUSTED	# OF EVENTS	CRUDE	AGE-ADJUSTED
30-67	257	12	23	9			18	2	2	26	10	10
68-75	255	10	20	10			26	2	3	29	11	11
76-83	155	2	6	8			23	3	3	16	11	11
84-91	111	5	23	3			21	4	4	18	16	17
92-220	81	6	37	0			27	7	7	21	22	22
UNKNOWN	0	0	-	0			0	-		0	-	
TOTAL	859	35	20	30			115	3	3	110	12	12

LOGISTIC REGRESSIONS

AGE AT EXAM		COEFFICIENT OF RISK FACTOR	SE COEFFICIENT	P-VALUE	STANDARDIZED COEFFICIENT	STD SE
35-44		0.037	0.025	0.143	0.462	0.315
45-54		0.037	0.012	0.002	0.460	0.152
55-64		0.034	0.008	0.001	0.432	0.100
65-74		0.024	0.008	0.002	0.326	0.106
75-84		0.006	0.013	0.630	0.081	0.168
85-94						
35-64	UNIVARIATE	0.035	0.006	0.001	0.443	0.081
65-94	UNIVARIATE	0.019	0.007	0.004	0.254	0.088
35-64	BIVARIATE	0.035	0.006	0.001	0.437	0.079
65-94	BIVARIATE	0.020	0.007	0.003	0.261	0.089
35-64	MULTIVARIATE	0.026	0.007	0.001	0.323	0.084
65-94	MULTIVARIATE	0.017	0.007	0.011	0.232	0.092

SEX: FEMALE
EVENT: DEATH FROM CARDIOVASCULAR DISEASE
RISK FACTOR: HEART RATE (PER MINUTE)
PERSONS AT RISK: PERSONS FREE OF CARDIOVASCULAR DISEASE

RISK FACTOR	35-44 AT EXAM			45-54 AT EXAM			55-64 AT EXAM			65-74 AT EXAM		
	PERSON EXAMS	# OF EVENTS	ANNUAL RATE	PERSON EXAMS	# OF EVENTS	ANNUAL RATE	PERSON EXAMS	# OF EVENTS	ANNUAL RATE	PERSON EXAMS	# OF EVENTS	ANNUAL RATE
30-67	1090	0	0	1881	3	1	2247	4	1	1038	5	2
68-75	1798	1	0	3227	5	1	3394	13	2	1572	24	8
76-83	1046	1	0	1886	2	1	2278	9	2	1297	15	6
84-91	939	1	1	1582	8	3	1723	10	3	1021	13	6
92-220	728	2	1	1292	3	1	1561	7	2	958	12	6
UNKNOWN	0	0	-	0	0	-	0	0	-	0	0	-
TOTAL	5601	5	0	9868	21	1	11203	43	2	5886	69	6

RISK FACTOR	75-84 AT EXAM			85-94 AT EXAM			35-64 COMBINED			65-94 COMBINED		
	PERSON EXAMS	# OF EVENTS	ANNUAL RATE	PERSON EXAMS	# OF EVENTS	ANNUAL RATE	# OF EVENTS	CRUDE	AGE-ADJUSTED	# OF EVENTS	CRUDE	AGE-ADJUSTED
30-67	321	4	6	22	4	91	7	1	1	13	5	4
68-75	419	14	17	25	1	20	19	1	1	39	10	10
76-83	372	10	13	21	0	0	12	1	1	25	7	7
84-91	500	7	12	20	2	50	19	2	2	22	8	8
92-220	327	10	15	13	1	38	12	2	2	23	9	9
UNKNOWN	0	0	-	0	0	-	0	-		0	-	
TOTAL	1739	45	13	101	8	40	69	1	1	122	8	8

LOGISTIC REGRESSIONS

AGE AT EXAM		COEFFICIENT OF RISK FACTOR	SE COEFFICIENT	P-VALUE	STANDARDIZED COEFFICIENT	STD SE
35-44		0.033	0.021	0.110	0.414	0.259
45-54		0.022	0.016	0.172	0.270	0.198
55-64		0.021	0.010	0.042	0.271	0.133
65-74		0.011	0.009	0.210	0.143	0.114
75-84		0.014	0.010	0.162	0.191	0.136
85-94		-0.031	0.030	0.305	-0.418	0.408
35-64	UNIVARIATE	0.023	0.008	0.005	0.287	0.103
65-94	UNIVARIATE	0.010	0.006	0.108	0.138	0.086
35-64	BIVARIATE	0.022	0.008	0.006	0.283	0.103
65-94	BIVARIATE	0.010	0.006	0.136	0.128	0.086
35-64	MULTIVARIATE	0.011	0.009	0.238	0.139	0.117
65-94	MULTIVARIATE	0.005	0.007	0.406	0.073	0.087

AVERAGE ANNUAL INCIDENCE RATE FOR EVENT PER 1000 PERSONS AT RISK AT EXAM
BY AGE AND LEVEL OF RISK FACTOR AT EXAM: FRAMINGHAM STUDY 30-YEAR FOLLOWUP

SEX: MALE
EVENT: DEATH FROM CARDIOVASCULAR DISEASE
RISK FACTOR: CIGARETTES SMOKED (NUMBER PER DAY)
PERSONS AT RISK: PERSONS FREE OF CARDIOVASCULAR DISEASE

RISK FACTOR	35-44 AT EXAM PERSON EXAMS	# OF EVENTS	ANNUAL RATE	45-54 AT EXAM PERSON EXAMS	# OF EVENTS	ANNUAL RATE	55-64 AT EXAM PERSON EXAMS	# OF EVENTS	ANNUAL RATE	65-74 AT EXAM PERSON EXAMS	# OF EVENTS	ANNUAL RATE
NONE	1382	0	0	3046	7	1	4185	29	3	2406	41	9
1-10	433	1	1	704	3	2	736	8	5	352	9	13
11-20	1483	1	0	2021	15	4	1597	23	7	473	10	11
21-40	987	4	2	1584	6	2	1076	10	5	345	9	13
41-90	156	1	3	245	2	4	146	4	14	24	2	42
UNKNOWN	26	0	0	35	1	14	19	0	0	6	0	0
TOTAL	4467	7	1	7635	34	2	7759	74	5	3606	71	10

RISK FACTOR	75-84 AT EXAM PERSON EXAMS	# OF EVENTS	ANNUAL RATE	85-94 AT EXAM PERSON EXAMS	# OF EVENTS	ANNUAL RATE	35-64 COMBINED # OF EVENTS	CRUDE	AGE-ADJUSTED	65-94 COMBINED # OF EVENTS	CRUDE	AGE-ADJUSTED
NONE	659	30	23	22			36	2	2	74	12	12
1-10	83	3	18	4			12	3	4	12	14	14
11-20	71	2	14	4			39	4	3	13	12	- 12
21-40	44	0	0	0			20	3	3	9	12	10
41-90	2	0	0	0			7	6	8	2	38	33
UNKNOWN	0	0	-	0			1	6	5	0	0	0
TOTAL	859	35	20	30			115	3	3	110	12	12

LOGISTIC REGRESSIONS

AGE AT EXAM		COEFFICIENT OF RISK FACTOR	SE COEFFICIENT	P-VALUE	STANDARDIZED COEFFICIENT	STD SE
35-44		0.052	— 0.024	-0.030	0.732	0.338
45-54		0.018	0.011	0.099	0.262	0.159
55-64		0.016	0.008	0.039	0.213	0.103
65-74		0.014	0.009	0.156-	0.158	0.106
75-84		-0.036	0.028	0.200	-0.330	0.257
85-94						
35-64	UNIVARIATE	0.012	0.006	0.043	0.177	0.087
65-94	UNIVARIATE	0.002	0.008	0.799	0.024	0.095
35-64	BIVARIATE	0.020	0.006	0.001	0.294	0.085
65-94	BIVARIATE	0.008	0.008	0.370	0.086	0.095
35-64	MULTIVARIATE	0.021	0.006	0.001	0.309	0.086
65-94	MULTIVARIATE	0.011	0.009	0.179	0.129	0.096

SEX: FEMALE
EVENT: DEATH FROM CARDIOVASCULAR DISEASE
RISK FACTOR: CIGARETTES SMOKED (NUMBER PER DAY)
PERSONS AT RISK: PERSONS FREE OF CARDIOVASCULAR DISEASE

RISK FACTOR	35-44 AT EXAM PERSON EXAMS	# OF EVENTS	ANNUAL RATE	45-54 AT EXAM PERSON EXAMS	# OF EVENTS	ANNUAL RATE	55-64 AT EXAM PERSON EXAMS	# OF EVENTS	ANNUAL RATE	65-74 AT EXAM PERSON EXAMS	# OF EVENTS	ANNUAL RATE
NONE	2577	0	0	5476	7	1	7386	23	2	4586	47	5
1-10	1312	1	0	1716	4	1	1535	6	2	593	8	7
11-20	1365	1	0	1992	6	2	1652	12	4	527	11	10
21-40	298	1	2	610	2	2	578	2	2	161	2	6
41-90	20	1	25	37	1	14	14	0	0	0	0	-
UNKNOWN	29	1	17	37	2	14	38	0	0	19	1	26
TOTAL	5601	5	0	9868	21	1	11203	43	2	5886	69	6

RISK FACTOR	75-84 AT EXAM PERSON EXAMS	# OF EVENTS	ANNUAL RATE	85-94 AT EXAM PERSON EXAMS	# OF EVENTS	ANNUAL RATE	35-64 COMBINED # OF EVENTS	CRUDE	AGE-ADJUSTED	65-94 COMBINED # OF EVENTS	CRUDE	AGE-ADJUSTED
NONE	1523	36	12	95	8	42	30	1	1	91	7	7
1-10	127	6	24	4	0	0	11	1	1	14	10	10
11-20	73	3	21	2	0	0	19	2	2	14	12	13
21-40	12	0	0	0	0	-	5	2	2	2	6	5
41-90	0	0	-	0	0	-	2	14	10	0	-	-
UNKNOWN	4	0	0	0	0	-	2	10	9	1	22	20
TOTAL	1739	45	13	101	8	40	69	1	1	122	8	8

LOGISTIC REGRESSIONS

AGE AT EXAM		COEFFICIENT OF RISK FACTOR	SE COEFFICIENT	P-VALUE	STANDARDIZED COEFFICIENT	STD SE
35-44		0.127	0.033	0.001	1.243	0.326
45-54		0.037	0.017	0.031	0.377	0.175
55-64		0.022	0.013	0.102	0.208	0.128
65-74		0.025	0.013	0.044	0.192	0.096
75-84		0.022	0.025	0.380	0.107	0.122
85-94		-	-	-	-	-
35-64	UNIVARIATE	0.031	0.010	0.002	0.311	0.098
65-94	UNIVARIATE	0.015	0.011	0.182	0.105	0.079
35-64	BIVARIATE	0.037	0.010	0.001	0.362	0.096
65-94	BIVARIATE	0.027	0.011	0.017	0.191	0.080
35-64	MULTIVARIATE	0.039	0.010	0.001	0.391	0.098
65-94	MULTIVARIATE	0.034	0.011	0.003	0.241	0.080

TABLE 21-14   DATE=17NOV85
AVERAGE ANNUAL INCIDENCE RATE FOR EVENT PER 1000 PERSONS AT RISK AT EXAM
BY AGE AND LEVEL OF RISK FACTOR AT EXAM: FRAMINGHAM STUDY 30-YEAR FOLLOWUP

SEX: MALE
EVENT: DEATH FROM CARDIOVASCULAR DISEASE
RISK FACTOR: ALBUMIN IN URINE, DEFINITE
PERSONS AT RISK: PERSONS FREE OF CARDIOVASCULAR DISEASE

RISK FACTOR	35-44 AT EXAM PERSON EXAMS	# OF EVENTS	ANNUAL RATE	45-54 AT EXAM PERSON EXAMS	# OF EVENTS	ANNUAL RATE	55-64 AT EXAM PERSON EXAMS	# OF EVENTS	ANNUAL RATE	65-74 AT EXAM PERSON EXAMS	# OF EVENTS	ANNUAL RATE
NO	4361	7	1	7515	32	2	7565	69	5	3508	66	9
YES	70	0	0	94	2	11	167	4	12	93	5	27
UNKNOWN	36	0	0	26	0	0	27	1	19	5	0	0
TOTAL	4467	7	1	7635	34	2	7759	74	5	3606	71	10

RISK FACTOR	75-84 AT EXAM PERSON EXAMS	# OF EVENTS	ANNUAL RATE	85-94 AT EXAM PERSON EXAMS	# OF EVENTS	ANNUAL RATE	35-64 COMBINED # OF EVENTS	CRUDE	AGE-ADJUSTED	65-94 COMBINED # OF EVENTS	CRUDE	AGE-ADJUSTED
NO	835	34	20	27	3		108	3	3	104	12	12
YES	24	1	21	3			6	9	9	6	25	26
UNKNOWN	0	0	-	0			1	6	7	0	0	0
TOTAL	859	35	20	30			115	3	3	110	12	12

LOGISTIC REGRESSIONS

AGE AT EXAM	COEFFICIENT OF RISK FACTOR	SE COEFFICIENT	P-VALUE	STANDARDIZED COEFFICIENT	STD SE
35-44	-	-	-	-	-
45-54	1.626	0.736	0.027	0.180	0.081
55-64	0.981	0.520	0.060	0.143	0.076
65-74	1.086	0.476	0.023	0.172	0.076
75-84	0.024	1.036	0.982	0.004	0.171
85-94					
35-64 UNIVARIATE	1.195	0.423	0.005	0.153	0.054
65-94 UNIVARIATE	0.770	0.430	0.074	0.124	0.069
35-64 BIVARIATE	1.042	0.425	0.014	0.134	0.055
65-94 BIVARIATE	0.690	0.435	0.113	0.111	0.070
35-64 MULTIVARIATE	0.478	0.446	0.283	0.061	0.057
65-94 MULTIVARIATE	0.413	0.444	0.353	0.067	0.072

SEX: FEMALE
EVENT: DEATH FROM CARDIOVASCULAR DISEASE
RISK FACTOR: ALBUMIN IN URINE, DEFINITE
PERSONS AT RISK: PERSONS FREE OF CARDIOVASCULAR DISEASE

RISK FACTOR	35-44 AT EXAM PERSON EXAMS	# OF EVENTS	ANNUAL RATE	45-54 AT EXAM PERSON EXAMS	# OF EVENTS	ANNUAL RATE	55-64 AT EXAM PERSON EXAMS	# OF EVENTS	ANNUAL RATE	65-74 AT EXAM PERSON EXAMS	# OF EVENTS	ANNUAL RATE
NO	5338	5	0	9569	21	1	10960	40	2	5767	68	6
YES	225	0	0	274	0	0	218	1	7	109	1	5
UNKNOWN	38	0	0	25	0	0	25	0	0	10	0	0
TOTAL	5601	5	0	9868	21	1	11203	43	2	5886	69	6

RISK FACTOR	75-84 AT EXAM PERSON EXAMS	# OF EVENTS	ANNUAL RATE	85-94 AT EXAM PERSON EXAMS	# OF EVENTS	ANNUAL RATE	35-64 COMBINED # OF EVENTS	CRUDE	AGE-ADJUSTED	65-94 COMBINED # OF EVENTS	CRUDE	AGE-ADJUSTED
NO	1697	45	13	98	8	41	66	1	1	121	8	8
YES	33	0	0	3	0	0	3	2	3	1	3	3
UNKNOWN	9	0	0	0	8	-	0	0	0	0	0	0
TOTAL	1739	45	13	101	8	40	69	1	1	122	8	8

LOGISTIC REGRESSIONS

AGE AT EXAM	COEFFICIENT OF RISK FACTOR	SE COEFFICIENT	P-VALUE	STANDARDIZED COEFFICIENT	STD SE
35-44	-	-	-	-	-
45-54	-	-	-	-	-
55-64	1.337	0.603	0.026	0.185	0.083
65-74	-0.254	1.012	0.802	-0.034	0.137
75-84	-	-	-	-	-
85-94	-	-	-	-	-
35-64 UNIVARIATE	0.496	0.592	0.402	0.080	0.096
65-94 UNIVARIATE	-0.851	1.008	0.398	-0.116	0.137
35-64 BIVARIATE	0.665	0.593	0.262	0.108	0.096
65-94 BIVARIATE	-0.903	1.009	0.371	-0.123	0.137
35-64 MULTIVARIATE	-0.127	0.701	0.856	-0.021	0.113
65-94 MULTIVARIATE	-1.330	1.020	0.192	-0.181	0.139

TABLE 21-15   DATE=17NOV85
AVERAGE ANNUAL INCIDENCE RATE FOR EVENT PER 1000 PERSONS AT RISK AT EXAM
BY AGE AND LEVEL OF RISK FACTOR AT EXAM: FRAMINGHAM STUDY 30-YEAR FOLLOWUP

SEX: MALE
EVENT: DEATH FROM CARDIOVASCULAR DISEASE
RISK FACTOR: HEART ENLARGEMENT BY X-RAY
PERSONS AT RISK: PERSONS FREE OF CARDIOVASCULAR DISEASE

RISK FACTOR	35-44 AT EXAM			45-54 AT EXAM			55-64 AT EXAM			65-74 AT EXAM		
	PERSON EXAMS	# OF EVENTS	ANNUAL RATE	PERSON EXAMS	# OF EVENTS	ANNUAL RATE	PERSON EXAMS	# OF EVENTS	ANNUAL RATE	PERSON EXAMS	# OF EVENTS	ANNUAL RATE
NO	4060	6	1	6325	22	2	5776	51	4	2412	41	8
MILD	184	0	0	637	2	2	922	9	5	529	12	11
DEFINITE	218	1	2	669	10	7	1060	14	7	665	18	14
UNKNOWN	5	0	0	4	0	0	1	0	0	0	0	-
TOTAL	4467	7	1	7635	34	2	7759	74	5	3606	71	10

RISK FACTOR	75-84 AT EXAM			85-94 AT EXAM			35-64 COMBINED			65-94 COMBINED		
	PERSON EXAMS	# OF EVENTS	ANNUAL RATE	PERSON EXAMS	# OF EVENTS	ANNUAL RATE	# OF EVENTS	CRUDE	AGE-ADJUSTED	# OF EVENTS	CRUDE	AGE-ADJUSTED
NO	503	14	14	23			79	2	3	56	10	10
MILD	153	10	33	5			11	3	3	24	17	17
DEFINITE	203	11	27	2			25	6	6	30	17	18
UNKNOWN	0	0	-	0			0	0	0	0		-
TOTAL	859	35	20	30			115	3	3	110	12	12

LOGISTIC REGRESSIONS

AGE AT EXAM		COEFFICIENT OF RISK FACTOR	SE COEFFICIENT	P-VALUE	STANDARDIZED COEFFICIENT	STD SE
35-44		0.468	0.577	0.418	0.218	0.269
45-54		0.701	0.199	0.001	0.425	0.121
55-64		0.193	0.149	0.194	0.138	0.106
65-74		0.242	0.140	0.084	0.190	0.110
75-84		0.363	0.194	0.061	0.304	0.162
85-94						
35-64	UNIVARIATE	0.470	0.115	0.001	0.297	0.073
65-94	UNIVARIATE	0.324	0.110	0.003	0.259	0.088
35-64	BIVARIATE	0.334	0.117	0.004	0.211	0.074
65-94	BIVARIATE	0.289-	0.111	0.009	0.230	0.089
35-64	MULTIVARIATE	0.178	0.127	0.161	0.112	0.080
65-94	MULTIVARIATE	0.123	0.120	0.302	0.098	0.095

SEX: FEMALE
EVENT: DEATH FROM CARDIOVASCULAR DISEASE
RISK FACTOR: HEART ENLARGEMENT BY X-RAY
PERSONS AT RISK: PERSONS FREE OF CARDIOVASCULAR DISEASE

RISK FACTOR	35-44 AT EXAM			45-54 AT EXAM			55-64 AT EXAM			65-74 AT EXAM		
	PERSON EXAMS	# OF EVENTS	ANNUAL RATE	PERSON EXAMS	# OF EVENTS	ANNUAL RATE	PERSON EXAMS	# OF EVENTS	ANNUAL RATE	PERSON EXAMS	# OF EVENTS	ANNUAL RATE
NO	5150	2	0	7958	9	1	7362	16	1	3054	24	4
MILD	198	1	3	929	4	2	1723	10	3	1088	10	5
DEFINITE	245	2	4	975	8	4	2114	17	4	1744	35	10
UNKNOWN	8	0	0	6	0	0	4	0	0	0	0	-
TOTAL	5601	5	0	9868	21	1	11203	43	2	5886	69	6

RISK FACTOR	75-84 AT EXAM			85-94 AT EXAM			35-64 COMBINED			65-94 COMBINED		
	PERSON EXAMS	# OF EVENTS	ANNUAL RATE	PERSON EXAMS	# OF EVENTS	ANNUAL RATE	# OF EVENTS	CRUDE	AGE-ADJUSTED	# OF EVENTS	CRUDE	AGE-ADJUSTED
NO	643	7	5	27	2	37	27	1	1	33	4	5
MILD	339	10	15	24	2	42	15	3	3	22	8	7
DEFINITE	757	28	18	50	4	40	27	4	4	67	13	12
UNKNOWN	0	0	-	0	0	-	0	0	0	0	0	-
TOTAL	1739	45	13	101	8	40	69	1	1	122	8	8

LOGISTIC REGRESSIONS

AGE AT EXAM		COEFFICIENT OF RISK FACTOR	SE COEFFICIENT	P-VALUE	STANDARDIZED COEFFICIENT	STD SE
35-44		1.527	0.468	0.001	0.675	0.207
45-54		0.998	0.237	0.001	0.635	0.150
55-64		0.655	0.169	0.001	0.519	0.134
65-74		0.485	0.137	0.001	0.424	0.120
75-84		0.566	0.192	0.003	0.506	0.172
85-94		0.034	0.440	0.938	0.029	0.372
35-64	UNIVARIATE	0.911	0.131	0.001	0.631	0.091
65-94	UNIVARIATE	0.553	0.106	0.001	0.491	0.094
35-64	BIVARIATE	0.798	0.136	0.001	0.553	0.095
65-94	BIVARIATE	0.468-	0.108	0.001	0.416	0.096
35-64	MULTIVARIATE	0.699	0.149	0.001	0.484	0.103
65-94	MULTIVARIATE	0.377	0.112	0.001	0.335	0.100

TABLE 21-16   DATE=12DEC85

AVERAGE ANNUAL INCIDENCE RATE FOR EVENT PER 1000 PERSONS AT RISK AT EXAM
BY AGE AND LEVEL OF RISK FACTOR AT EXAM: FRAMINGHAM STUDY 30-YEAR FOLLOWUP

SEX: MALE
EVENT: DEATH FROM CARDIOVASCULAR DISEASE
RISK FACTOR: LEFT VENTRICULAR HYPERTROPHY BY ELECTROCARDIOGRAM
PERSONS AT RISK: PERSONS FREE OF CARDIOVASCULAR DISEASE

RISK FACTOR	35-44 AT EXAM PERSON EXAMS	# OF EVENTS	ANNUAL RATE	45-54 AT EXAM PERSON EXAMS	# OF EVENTS	ANNUAL RATE	55-64 AT EXAM PERSON EXAMS	# OF EVENTS	ANNUAL RATE	65-74 AT EXAM PERSON EXAMS	# OF EVENTS	ANNUAL RATE
NO	4363	6	1	7423	28	2	7439	65	4	3433	58	8
MILD	78	0	0	127	1	4	200	2	5	85	4	24
DEFINITE	26	1	19	85	5	29	120	7	29	88	9	51
UNKNOWN	0	0	-	0	0	-	0	0	-	0	0	-
TOTAL	4467	7	1	7635	34	2	7759	74	5	3606	71	10

RISK FACTOR	75-84 AT EXAM PERSON EXAMS	# OF EVENTS	ANNUAL RATE	85-94 AT EXAM PERSON EXAMS	# OF EVENTS	ANNUAL RATE	35-64 COMBINED # OF EVENTS	CRUDE	AGE-ADJUSTED	65-94 COMBINED # OF EVENTS	CRUDE	AGE-ADJUSTED
NO	809	33	20	27			99	3	3	94	11	11
MILD	29	1	17	2			3	4	3	6	26	24
DEFINITE	21	1	24	1			13	28	27	10	45	46
UNKNOWN	0	0	-	0			0	-		0	-	
TOTAL	859	35	20	30			115	3	3	110	12	12

LOGISTIC REGRESSIONS

AGE AT EXAM		COEFFICIENT OF RISK FACTOR	SE COEFFICIENT	P-VALUE	STANDARDIZED COEFFICIENT	STD SE
35-44		1.561	0.590	0.008	0.312	0.118
45-54		1.364	0.251	0.001	0.333	0.061
55-64		0.895	0.207	0.001	0.260	0.060
65-74		0.955	0.181	0.001	0.325	0.062
75-84		0.025	0.480	0.958	0.009	0.170
85-94						
35-64	UNIVARIATE	1.156	0.153	0.001	0.295	0.039
65-94	UNIVARIATE	0.761	0.164	0.001	0.262	0.056
35-64	BIVARIATE	1.053	0.155	0.001	0.269	0.040
65-94	BIVARIATE	0.758	0.166	0.001	0.260	0.057
35-64	MULTIVARIATE	0.820	0.175	0.001	0.209	0.045
65-94	MULTIVARIATE	0.513	0.181	0.005	0.176	0.062

SEX: FEMALE
EVENT: DEATH FROM CARDIOVASCULAR DISEASE
RISK FACTOR: LEFT VENTRICULAR HYPERTROPHY BY ELECTROCARDIOGRAM
PERSONS AT RISK: PERSONS FREE OF CARDIOVASCULAR DISEASE

RISK FACTOR	35-44 AT EXAM PERSON EXAMS	# OF EVENTS	ANNUAL RATE	45-54 AT EXAM PERSON EXAMS	# OF EVENTS	ANNUAL RATE	55-64 AT EXAM PERSON EXAMS	# OF EVENTS	ANNUAL RATE	65-74 AT EXAM PERSON EXAMS	# OF EVENTS	ANNUAL RATE
NO	5548	3	0	9711	18	1	10867	38	2	5661	59	5
MILD	38	0	0	104	0	0	211	1	2	125	5	20
DEFINITE	15	2	67	53	3	28	125	4	16	100	5	25
UNKNOWN	0	0	-	0	0	-	0	0	-	0	0	-
TOTAL	5601	5	0	9868	21	1	11203	43	2	5886	69	6

RISK FACTOR	75-84 AT EXAM PERSON EXAMS	# OF EVENTS	ANNUAL RATE	85-94 AT EXAM PERSON EXAMS	# OF EVENTS	ANNUAL RATE	35-64 COMBINED # OF EVENTS	CRUDE	AGE-ADJUSTED	65-94 COMBINED # OF EVENTS	CRUDE	AGE-ADJUSTED
NO	1651	38	12	92	7	38	59	1	1	104	7	7
MILD	40	3	38	3	0	0	1	1	1	8	24	24
DEFINITE	48	4	42	6	1	83	9	23	31	10	32	30
UNKNOWN	0	0	-	0	0	-	0	-		0	-	
TOTAL	1739	45	13	101	8	40	69	1	1	122	8	8

LOGISTIC REGRESSIONS

AGE AT EXAM		COEFFICIENT OF RISK FACTOR	SE COEFFICIENT	P-VALUE	STANDARDIZED COEFFICIENT	STD SE
35-44		2.805	0.492	0.001	0.369	0.065
45-54		1.658	0.334	0.001	0.295	0.059
55-64		1.054	0.270	0.001	0.262	0.067
65-74		0.886	0.211	0.001	0.260	0.062
75-84		0.738	0.250	0.003	0.263	0.089
85-94		0.357	0.600	0.553	0.177	0.299
35-64	UNIVARIATE	1.473	0.188	0.001	0.300	0.038
65-94	UNIVARIATE	0.850	0.156	0.001	0.266	0.049
35-64	BIVARIATE	1.365	0.190	0.001	0.278	0.039
65-94	BIVARIATE	0.780	0.157	0.001	0.244	0.049
35-64	MULTIVARIATE	1.127	0.230	0.001	0.229	0.047
65-94	MULTIVARIATE	0.562	0.174	0.001	0.176	0.054

TABLE 21-17   DATE=12DEC85
AVERAGE ANNUAL INCIDENCE RATE FOR EVENT PER 1000 PERSONS AT RISK AT EXAM
BY AGE AND LEVEL OF RISK FACTOR AT EXAM: FRAMINGHAM STUDY 30-YEAR FOLLOWUP

SEX: MALE
EVENT: DEATH FROM CARDIOVASCULAR DISEASE
RISK FACTOR: INTRAVENTRICULAR CONDUCTION DEFECT (WITH QRS INTERVAL MORE THAN 0.11. SECOND)
PERSONS AT RISK: PERSONS FREE OF CARDIOVASCULAR DISEASE

RISK FACTOR	35-44 AT EXAM			45-54 AT EXAM			55-64 AT EXAM			65-74 AT EXAM		
	PERSON EXAMS	# OF EVENTS	ANNUAL RATE	PERSON EXAMS	# OF EVENTS	ANNUAL RATE	PERSON EXAMS	# OF EVENTS	ANNUAL RATE	PERSON EXAMS	# OF EVENTS	ANNUAL RATE
NO	4156	7	1	7129	31	2	7124	64	4	3299	66	10
INCOMPLETE	275	0	0	443	2	2	446	4	4	161	2	6
COMPLETE	36	0	0	63	1	8	189	6	16	146	3	10
UNKNOWN	0	0	-	0	0	-	0	0	-	0	0	-
TOTAL	4467	7	1	7635	34	2	7759	74	5	3606	71	10

RISK FACTOR	75-84 AT EXAM			85-94 AT EXAM			35-64 COMBINED			65-94 COMBINED		
	PERSON EXAMS	# OF EVENTS	ANNUAL RATE	PERSON EXAMS	# OF EVENTS	ANNUAL RATE	# OF EVENTS	CRUDE	AGE-ADJUSTED	# OF EVENTS	CRUDE	AGE-ADJUSTED
NO	779	32	21	25			102	3	3	101	12	12
INCOMPLETE	18	1	28	2			6	3	3	3	8	10
COMPLETE	62	2	16	3			7	12	9	6	14	12
UNKNOWN	0	0	-	0			0	-		0	-	
TOTAL	859	35	20	30			115	3	3	110	12	12

LOGISTIC REGRESSIONS

AGE AT EXAM		COEFFICIENT OF RISK FACTOR	SE COEFFICIENT	P-VALUE	STANDARDIZED COEFFICIENT	STD SE
35-44		-	-	-	-	-
45-54		0.395	0.462	0.393	0.115	0.135
55-64		0.527	0.216	0.015	0.200	0.082
65-74		-0.073	0.290	0.801	-0.032	0.127
75-84		-0.089	0.347	0.799	-0.047	0.185
85-94						
35-64	UNIVARIATE	0.545	0.199	0.006	0.180	0.066
65-94	UNIVARIATE	0.011	0.209	0.959	0.005	0.096
35-64	BIVARIATE	0.430	0.195	0.027	0.142	0.064
65-94	BIVARIATE	-0.037	0.207	0.858	-0.017	0.095
35-64	MULTIVARIATE	0.454	0.201	0.024	0.150	0.066
65-94	MULTIVARIATE	-0.040	0.210	0.848	-0.018	0.096

SEX: FEMALE
EVENT: DEATH FROM CARDIOVASCULAR DISEASE
RISK FACTOR: INTRAVENTRICULAR CONDUCTION DEFECT (WITH QRS INTERVAL MORE THAN 0.11 SECOND)
PERSONS AT RISK: PERSONS FREE OF CARDIOVASCULAR DISEASE

RISK FACTOR	35-44 AT EXAM			45-54 AT EXAM			55-64 AT EXAM			65-74 AT EXAM		
	PERSON EXAMS	# OF EVENTS	ANNUAL RATE	PERSON EXAMS	# OF EVENTS	ANNUAL RATE	PERSON EXAMS	# OF EVENTS	ANNUAL RATE	PERSON EXAMS	# OF EVENTS	ANNUAL RATE
NO	5514	5	0	9680	21	1	10882	40	2	5687	65	6
INCOMPLETE	75	0	0	144	0	0	183	2	5	108	0	0
COMPLETE	12	0	0	44	0	0	138	1	4	91	4	22
UNKNOWN	0	0	-	0	0	-	0	0	-	0	0	-
TOTAL	5601	5	0	9868	21	1	11203	43	2	5886	69	6

RISK FACTOR	75-84 AT EXAM			85-94 AT EXAM			35-64 COMBINED			65-94 COMBINED		
	PERSON EXAMS	# OF EVENTS	ANNUAL RATE	PERSON EXAMS	# OF EVENTS	ANNUAL RATE	# OF EVENTS	CRUDE	AGE-ADJUSTED	# OF EVENTS	CRUDE	AGE-ADJUSTED
NO	1652	42	13	98	8	41	66	1	1	115	8	8
INCOMPLETE	35	1	14	1	0	0	2	2	2	1	3	3
COMPLETE	52	2	19	2	0	0	1	3	2	6	21	21
UNKNOWN	0	0	-	0	0	-	0	-		0	-	
TOTAL	1739	45	13	101	8	40	69	1	1	122	8	8

LOGISTIC REGRESSIONS

AGE AT EXAM		COEFFICIENT OF RISK FACTOR	SE COEFFICIENT	P-VALUE	STANDARDIZED COEFFICIENT	STD SE
35-44		-	-	-	-	-
45-54		-	-	-	-	-
55-64		0.517	0.394	0.190	0.131	0.100
65-74		0.542	0.282	0.054	0.151	0.079
75-84		0.203	0.352	0.564	0.074	0.129
85-94		-	-	-	-	-
35-64	UNIVARIATE	0.448	0.404	0.267	0.093	0.084
65-94	UNIVARIATE	0.402	0.220	0.068	0.121	0.066
35-64	BIVARIATE	0.322	0.401	0.422	0.067	0.083
65-94	BIVARIATE	0.341	0.220	0.121	0.103	0.066
35-64	MULTIVARIATE	0.342	0.421	0.416	0.071	0.088
65-94	MULTIVARIATE	0.355	0.222	0.109	0.107	0.067

TABLE 21-18   DATE=17NOV85
AVERAGE ANNUAL INCIDENCE RATE FOR EVENT PER 1000 PERSONS AT RISK AT EXAM
BY AGE AND LEVEL OF RISK FACTOR AT EXAM: FRAMINGHAM STUDY 30-YEAR FOLLOWUP

SEX: MALE
EVENT: DEATH FROM CARDIOVASCULAR DISEASE
RISK FACTOR: DEFINITE NONSPECIFIC T-WAVE OR ST-SEGMENT ABNORMALITY BY ELECTROCARDIOGRAM
PERSONS AT RISK: PERSONS FREE OF CARDIOVASCULAR DISEASE

35-44 AT EXAM			45-54 AT EXAM			55-64 AT EXAM			65-74 AT EXAM		
PERSON EXAMS	# OF EVENTS	ANNUAL RATE	PERSON EXAMS	# OF EVENTS	ANNUAL RATE	PERSON EXAMS	# OF EVENTS	ANNUAL RATE	PERSON EXAMS	# OF EVENTS	ANNUAL RATE
4357	7	1	7273	31	2	7231	65	4	3226	62	10
110	0	0	362	3	4	528	9	9	380	9	12
0	0	-	0	0	-	0	0	-	0	0	-
4467	7	1	7635	34	2	7759	74	5	3606	71	10

75-84 AT EXAM			85-94 AT EXAM			35-64 COMBINED			65-94 COMBINED		
PERSON EXAMS	# OF EVENTS	ANNUAL RATE	PERSON EXAMS	# OF EVENTS	ANNUAL RATE	# OF EVENTS	CRUDE	AGE- ADJUSTED	# OF EVENTS	CRUDE	AGE- ADJUSTED
727	27	19	21			103	3	3	90	11	11
132	8	30	9			12	6	5	20	19	16
0	0	-	0			0	-		0	-	
859	35	20	30			115	3	3	110	12	12

LOGISTIC REGRESSIONS

	COEFFICIENT OF RISK FACTOR	SE COEFFICIENT	P-VALUE	STANDARDIZED COEFFICIENT	STD SE
	0.669	0.607	0.270	0.142	0.129
	0.648	0.359	0.071	0.163	0.090
	0.213	0.361	0.554	0.066	0.111
	0.514	0.414	0.214	0.186	0.149
NIVARIATE	0.794	0.307	0.010	0.174	0.067
NIVARIATE	0.544	0.252	0.031	0.174	0.081
IVARIATE	0.561	0.309	0.069	0.123	0.067
IVARIATE	0.448	0.254	0.078	0.143	0.081
ULTIVARIATE	0.220	0.313	0.482	0.048	0.069
ULTIVARIATE	0.358	0.256	0.162	0.114	0.082

SEX: FEMALE
EVENT: DEATH FROM CARDIOVASCULAR DISEASE
RISK FACTOR: DEFINITE NONSPECIFIC T-WAVE OR ST-SEGMENT ABNORMALITY BY ELECTROCARDIOGRAM
PERSONS AT RISK: PERSONS FREE OF CARDIOVASCULAR DISEASE

35-44 AT EXAM			45-54 AT EXAM			55-64 AT EXAM			65-74 AT EXAM		
PERSON EXAMS	# OF EVENTS	ANNUAL RATE	PERSON EXAMS	# OF EVENTS	ANNUAL RATE	PERSON EXAMS	# OF EVENTS	ANNUAL RATE	PERSON EXAMS	# OF EVENTS	ANNUAL RATE
5503	5	0	9464	17	1	10401	36	2	5239	58	6
98	0	0	404	4	5	802	7	4	647	11	9
0	0	-	0	0	-	0	0	-	0	0	-
5601	5	0	9868	21	1	11203	43	2	5886	69	6

75-84 AT EXAM			85-94 AT EXAM			35-64 COMBINED			65-94 COMBINED		
PERSON EXAMS	# OF EVENTS	ANNUAL RATE	PERSON EXAMS	# OF EVENTS	ANNUAL RATE	# OF EVENTS	CRUDE	AGE- ADJUSTED	# OF EVENTS	CRUDE	AGE- ADJUSTED
1462	38	13	82	7	43	58	1	1	103	8	8
277	7	13	19	1	26	11	4	4	19	10	10
0	0	-	0	0	-	0	-		0	-	
1739	45	13	101	8	40	69	1	1	122	8	8

LOGISTIC REGRESSIONS

	COEFFICIENT OF RISK FACTOR	SE COEFFICIENT	P-VALUE	STANDARDIZED COEFFICIENT	STD SE
	1.715	0.558	0.002	0.340	0.111
	0.930	0.415	0.025	0.240	0.107
	0.435	0.332	0.189	0.136	0.104
	-0.029	0.417	0.945	-0.011	0.153
	-0.519	1.101	0.637	-0.204	0.432
IVARIATE	1.312	0.330	0.001	0.283	0.071
IVARIATE	0.288	0.252	0.254	0.094	0.083
VARIATE	1.089	0.333	0.001	0.235	0.072
VARIATE	0.169	0.254	0.505	0.055	0.083
LTIVARIATE	0.870	0.337	0.010	0.188	0.073
LTIVARIATE	0.101	0.255	0.694	0.033	0.084

SEX: MALE
EVENT: DEATH FROM CARDIOVASCULAR DISEASE
RISK FACTOR: SYSTOLIC BLOOD PRESSURE, FIRST EXAMINER (MM HG)
PERSONS AT RISK: ALL PERSONS ALIVE

RISK FACTOR	35-44 AT EXAM			45-54 AT EXAM			55-64 AT EXAM			65-74 AT EXAM		
	PERSON EXAMS	# OF EVENTS	ANNUAL RATE	PERSON EXAMS	# OF EVENTS	ANNUAL RATE	PERSON EXAMS	# OF EVENTS	ANNUAL RATE	PERSON EXAMS	# OF EVENTS	ANNUAL RATE
74-119	1226	3	1	1842	10	3	1605	17	5	659	18	14
120-139	2193	4	1	3542	21	3	3525	54	8	1636	33	10
140-159	850	0	0	1838	19	5	2371	42	9	1473	52	18
160-179	197	1	3	598	8	7	1062	28	13	696	33	24
180-300	53	0	0	247	5	10	507	28	28	370	29	39
UNKNOWN	0	0	-	0	0	-	0	0	-	0	0	-
TOTAL	4519	8	1	8067	63	4	9070	169	9	4834	165	17

RISK FACTOR	75-84 AT EXAM			85-94 AT EXAM			35-64 COMBINED			65-94 COMBINED		
	PERSON EXAMS	# OF EVENTS	ANNUAL RATE	PERSON EXAMS	# OF EVENTS	ANNUAL RATE	# OF EVENTS	CRUDE	AGE-ADJUSTED	# OF EVENTS	CRUDE	AGE-ADJUSTED
74-119	144	12	42	8	3	188	30	3	3	33	20	21
120-139	451	24	27	17	0	0	79	4	5	57	14	14
140-159	411	32	39	17	0	0	61	6	6	84	22	22
160-179	218	17	39	6	2	167	37	10	9	52	28	28
180-300	121	19	79	2	1	250	33	20	15	49	50	49
UNKNOWN	0	0	-	0	0	-	0	-	-	0	-	-
TOTAL	1345	104	39	50	6	60	240	6	6	275	22	22

LOGISTIC REGRESSIONS

AGE AT EXAM		COEFFICIENT OF RISK FACTOR	SE COEFFICIENT	P-VALUE	STANDARDIZED COEFFICIENT	STD SE
35-44		-0.011	0.023	0.632	-0.183	0.382
45-54		0.019	0.005	0.001	0.380	0.104
55-64		0.019	0.003	0.001	0.426	0.064
65-74		0.015	0.003	0.001	0.352	0.069
75-84		0.014	0.004	0.001	0.321	0.093
85-94		0.002	0.024	0.919	0.045	0.437
35-64	UNIVARIATE	0.023	0.002	0.001	0.471	0.051
65-94	UNIVARIATE	0.015	0.002	0.001	0.340	0.055
35-64	BIVARIATE	0.018	0.003	0.001	0.367	0.052
65-94	BIVARIATE	0.015	0.002	0.001	0.342	0.056
35-64	MULTIVARIATE	0.010	0.003	0.001	0.197	0.058
65-94	MULTIVARIATE	0.009	0.003	0.001	0.218	0.060

SEX: FEMALE
EVENT: DEATH FROM CARDIOVASCULAR DISEASE
RISK FACTOR: SYSTOLIC BLOOD PRESSURE, FIRST EXAMINER (MM HG)
PERSONS AT RISK: ALL PERSONS ALIVE

RISK FACTOR	35-44 AT EXAM			45-54 AT EXAM			55-64 AT EXAM			65-74 AT EXAM		
	PERSON EXAMS	# OF EVENTS	ANNUAL RATE	PERSON EXAMS	# OF EVENTS	ANNUAL RATE	PERSON EXAMS	# OF EVENTS	ANNUAL RATE	PERSON EXAMS	# OF EVENTS	ANNUAL RATE
74-119	2493	2	0	2724	5	1	2039	9	2	683	3	2
120-139	2245	2	0	4027	7	1	4199	15	2	1918	27	7
140-159	674	0	0	2086	7	2	3197	20	4	2283	39	9
160-179	156	0	0	785	2	1	1682	14	4	1360	31	11
180-300	51	1	10	450	5	6	986	17	9	859	42	24
UNKNOWN	0	0	-	0	0	-	0	0	-	0	0	-
TOTAL	5619	5	0	10072	26	1	12103	75	3	7103	142	10

RISK FACTOR	75-84 AT EXAM			85-94 AT EXAM			35-64 COMBINED			65-94 COMBINED		
	PERSON EXAMS	# OF EVENTS	ANNUAL RATE	PERSON EXAMS	# OF EVENTS	ANNUAL RATE	# OF EVENTS	CRUDE	AGE-ADJUSTED	# OF EVENTS	CRUDE	AGE-ADJUSTED
74-119	158	6	19	10	1	50	16	1	1	10	6	7
120-139	561	26	23	51	6	59	24	1	1	59	12	12
140-159	759	26	17	40	4	50	27	2	2	69	11	11
160-179	561	29	26	36	4	56	16	3	2	64	16	16
180-300	380	28	37	26	1	19	23	8	8	71	28	27
UNKNOWN	0	0	-	0	0	-	0	-	-	0	-	-
TOTAL	2419	115	24	163	16	49	106	2	2	273	14	14

LOGISTIC REGRESSIONS

AGE AT EXAM		COEFFICIENT OF RISK FACTOR	SE COEFFICIENT	P-VALUE	STANDARDIZED COEFFICIENT	STD SE
35-44		0.020	0.020	0.306	0.345	0.337
45-54		0.019	0.006	0.001	0.443	0.134
55-64		0.016	0.004	0.001	0.409	0.093
65-74		0.021	0.003	0.001	0.529	0.073
75-84		0.008	0.004	0.016	0.216	0.090
85-94		-0.008	0.011	0.464	-0.202	0.276
35-64	UNIVARIATE	0.021	0.003	0.001	0.496	0.069
65-94	UNIVARIATE	0.016	0.002	0.001	0.394	0.055
35-64	BIVARIATE	0.016	0.003	0.001	0.388	0.076
65-94	BIVARIATE	0.014	0.002	0.001	0.359	0.056
35-64	MULTIVARIATE	0.005	0.004	0.138	0.127	0.086
65-94	MULTIVARIATE	0.011	0.002	0.001	0.263	0.058

TABLE 22-2   DATE=17NOV85
AVERAGE ANNUAL INCIDENCE RATE FOR EVENT PER 1000 PERSONS AT RISK AT EXAM
BY AGE AND LEVEL OF RISK FACTOR AT EXAM: FRAMINGHAM STUDY 30-YEAR FOLLOWUP

SEX: MALE
EVENT: DEATH FROM CARDIOVASCULAR DISEASE
RISK FACTOR: DIASTOLIC BLOOD PRESSURE, FIRST EXAMINER (MM HG)
PERSONS AT RISK: ALL PERSONS ALIVE

RISK FACTOR	35-44 AT EXAM			45-54 AT EXAM			55-64 AT EXAM			65-74 AT EXAM		
	PERSON EXAMS	# OF EVENTS	ANNUAL RATE	PERSON EXAMS	# OF EVENTS	ANNUAL RATE	PERSON EXAMS	# OF EVENTS	ANNUAL RATE	PERSON EXAMS	# OF EVENTS	ANNUAL RATE
20-74	1015	1	0	1543	16	5	2092	34	8	1524	47	15
75-84	1617	4	1	2861	12	2	3165	42	7	1643	37	11
85-94	1203	1	0	2151	14	3	2228	47	11	1072	38	18
95-104	509	2	2	1073	13	6	1114	24	11	425	24	28
105-160	175	0	0	439	8	9	471	22	23	170	19	56
UNKNOWN	0	0	-	0	0	-	0	0	-	0	0	-
TOTAL	4519	8	1	8067	63	4	9070	169	9	4834	165	17

RISK FACTOR	75-84 AT EXAM			85-94 AT EXAM			35-64 COMBINED			65-94 COMBINED		
	PERSON EXAMS	# OF EVENTS	ANNUAL RATE	PERSON EXAMS	# OF EVENTS	ANNUAL RATE	# OF EVENTS	CRUDE	AGE-ADJUSTED	# OF EVENTS	CRUDE	AGE-ADJUSTED
20-74	560	39	35	30	3	50	51	5	5	89	21	20
75-84	440	33	38	15	1	33	58	4	4	71	17	17
85-94	244	17	35	5	2	200	62	6	6	57	22	23
95-104	71	8	56	0	0	-	39	7	7	32	32	34
105-160	30	7	117	0	0	-	30	14	13	26	65	69
UNKNOWN	0	0	-	0	0	-	0	-	-	0	-	-
TOTAL	1345	104	39	50	6	60	240	6	6	275	22	22

LOGISTIC REGRESSIONS

AGE AT EXAM		COEFFICIENT OF RISK FACTOR	SE COEFFICIENT	P-VALUE	STANDARDIZED COEFFICIENT	STD SE
35-44		0.005	0.031	0.865	0.059	0.349
45-54		0.014	0.010	0.190	0.159	0.121
55-64		0.026	0.006	0.001	0.317	0.070
65-74		0.022	0.006	0.001	0.274	0.073
75-84		0.010	0.008	0.236	0.119	0.101
85-94		0.053	0.042	0.213	0.562	0.451
35-64	UNIVARIATE	0.023	0.005	0.001	0.275	0.060
65-94	UNIVARIATE	0.013	0.005	0.005	0.166	0.059
35-64	BIVARIATE	0.023	0.005	0.001	0.275	0.059
65-94	BIVARIATE	0.020	0.005	0.001	0.255	0.060
35-64	MULTIVARIATE	0.012	0.005	0.018	0.140	0.059
65-94	MULTIVARIATE	0.014	0.005	0.004	0.176	0.061

SEX: FEMALE
EVENT: DEATH FROM CARDIOVASCULAR DISEASE
RISK FACTOR: DIASTOLIC BLOOD PRESSURE, FIRST EXAMINER (MM HG)
PERSONS AT RISK: ALL PERSONS ALIVE

RISK FACTOR	35-44 AT EXAM			45-54 AT EXAM			55-64 AT EXAM			65-74 AT EXAM		
	PERSON EXAMS	# OF EVENTS	ANNUAL RATE	PERSON EXAMS	# OF EVENTS	ANNUAL RATE	PERSON EXAMS	# OF EVENTS	ANNUAL RATE	PERSON EXAMS	# OF EVENTS	ANNUAL RATE
20-74	2164	2	0	2657	5	1	3110	18	3	2069	38	9
75-84	2031	0	0	3573	8	1	4005	18	2	2416	38	8
85-94	988	0	0	2395	4	1	3000	17	3	1622	35	11
95-104	324	2	3	941	4	2	1348	11	4	685	15	11
105-160	112	0	0	506	5	5	640	11	9	311	16	26
UNKNOWN	0	0	-	0	0	-	0	0	-	0	0	-
TOTAL	5619	5	0	10072	26	1	12103	75	3	7103	142	10

RISK FACTOR	75-84 AT EXAM			85-94 AT EXAM			35-64 COMBINED			65-94 COMBINED		
	PERSON EXAMS	# OF EVENTS	ANNUAL RATE	PERSON EXAMS	# OF EVENTS	ANNUAL RATE	# OF EVENTS	CRUDE	AGE-ADJUSTED	# OF EVENTS	CRUDE	AGE-ADJUSTED
20-74	916	44	24	73	9	62	25	2	2	91	15	14
75-84	722	29	20	56	6	54	26	1	1	73	11	12
85-94	535	28	26	25	1	20	22	2	2	64	15	15
95-104	178	12	34	7	0	0	17	3	3	27	16	16
105-160	68	2	15	2	0	0	16	6	6	18	24	23
UNKNOWN	0	0	-	0	0	-	0	-	-	0	-	-
TOTAL	2419	115	24	163	16	49	106	2	2	273	14	14

LOGISTIC REGRESSIONS

AGE AT EXAM		COEFFICIENT OF RISK FACTOR	SE COEFFICIENT	P-VALUE	STANDARDIZED COEFFICIENT	STD SE
35-44		0.014	0.038	0.721	0.150	0.421
45-54		0.035	0.013	0.005	0.436	0.157
55-64		0.020	0.008	0.020	0.247	0.106
65-74		0.017	0.006	0.007	0.217	0.080
75-84		0.000	0.008	0.977	0.003	0.096
85-94		-0.028	0.020	0.167	-0.362	0.262
35-64	UNIVARIATE	0.028	0.007	0.001	0.346	0.084
65-94	UNIVARIATE	0.004	0.005	0.472	0.044	0.061
35-64	BIVARIATE	0.023	0.007	0.001	0.289	0.086
65-94	BIVARIATE	0.009	0.005	0.059	0.115	0.061
35-64	MULTIVARIATE	0.005	0.007	0.466	0.066	0.090
65-94	MULTIVARIATE	0.007	0.005	0.118	0.094	0.060

TABLE 22-3A   DATE=17NOV85
AVERAGE ANNUAL INCIDENCE RATE FOR EVENT PER 1000 PERSONS AT RISK AT EXAM
BY AGE AND LEVEL OF RISK FACTOR AT EXAM: FRAMINGHAM STUDY 30-YEAR FOLLOWUP

SEX: MALE
EVENT: DEATH FROM CARDIOVASCULAR DISEASE
RISK FACTOR: HYPERTENSION WITH ANTIHYPERTENSIVE TREATMENT
PERSONS AT RISK: ALL PERSONS ALIVE

RISK FACTOR	35-44 AT EXAM			45-54 AT EXAM			55-64 AT EXAM			65-74 AT EXAM		
	PERSON EXAMS	# OF EVENTS	ANNUAL RATE	PERSON EXAMS	# OF EVENTS	ANNUAL RATE	PERSON EXAMS	# OF EVENTS	ANNUAL RATE	PERSON EXAMS	# OF EVENTS	ANNUAL RATE
NO	2668	3	1	4263	20	2	4163	44	5	1303	31	9
MILD	1316	3	1	2287	17	4	2573	40	8	1520	40	13
DEFINITE	511	1	1	1216	16	7	1431	42	15	698	40	29
TREATED	24	1	21	301	10	17	903	43	24	813	54	33
UNKNOWN	0	0	-	0	0	-	0	0	-	0	0	-
TOTAL	4519	8	1	8067	63	4	9070	169	9	4834	165	17

RISK FACTOR	75-84 AT EXAM			85-94 AT EXAM			35-64 COMBINED			65-94 COMBINED		
	PERSON EXAMS	# OF EVENTS	ANNUAL RATE	PERSON EXAMS	# OF EVENTS	ANNUAL RATE	# OF EVENTS	CRUDE	AGE-ADJUSTED	# OF EVENTS	CRUDE	AGE-ADJUSTED
NO	465	24	26	19	3	79	67	3	3	58	13	13
MILD	424	31	37	16	0	0	60	5	5	71	18	18
DEFINITE	194	18	46	5	2	200	59	9	9	60	33	34
TREATED	262	31	59	10	1	50	54	22	21	86	40	39
UNKNOWN	0	0	-	0	0	-	0	-		0	-	
TOTAL	1345	104	39	50	6	60	240	6	6	275	22	22

LOGISTIC REGRESSIONS

AGE AT EXAM		COEFFICIENT OF RISK FACTOR	SE COEFFICIENT	P-VALUE	STANDARDIZED COEFFICIENT	STD SE
35-44		0.610	0.445	0.170	0.425	0.310
45-54		0.661	0.153	0.001	0.513	0.119
55-64		0.653	0.096	0.001	0.537	0.079
65-74		0.704	0.105	0.001	0.582	0.086
75-84		0.400	0.128	0.002	0.331	0.106
85-94		0.134	0.529	0.799	0.111	0.438
35-64	UNIVARIATE	0.754	0.080	0.001	0.593	0.063
65-94	UNIVARIATE	0.584	0.079	0.001	0.483	0.066
35-64	BIVARIATE	0.631	0.080	0.001	0.497	0.063
65-94	BIVARIATE	0.579	0.080	0.001	0.479	0.066
35-64	MULTIVARIATE	0.460	0.085	0.001	0.378	0.067
65-94	MULTIVARIATE	0.467	0.082	0.001	0.386	0.068

SEX: FEMALE
EVENT: DEATH FROM CARDIOVASCULAR DISEASE
RISK FACTOR: HYPERTENSION WITH ANTIHYPERTENSIVE TREATMENT
PERSONS AT RISK: ALL PERSONS ALIVE

RISK FACTOR	35-44 AT EXAM			45-54 AT EXAM			55-64 AT EXAM			65-74 AT EXAM		
	PERSON EXAMS	# OF EVENTS	ANNUAL RATE	PERSON EXAMS	# OF EVENTS	ANNUAL RATE	PERSON EXAMS	# OF EVENTS	ANNUAL RATE	PERSON EXAMS	# OF EVENTS	ANNUAL RATE
NO	4251	3	0	5660	10	1	4992	19	2	1871	15	4
MILD	1002	0	0	2611	1	1	3449	20	3	2116	35	8
DEFINITE	316	2	3	1184	4	3	1959	21	5	1198	30	13
TREATED	50	0	0	617	4	5	1712	15	4	1918	62	16
UNKNOWN	0	0	-	0	0	-	0	0	-	0	0	-
TOTAL	5619	5	0	10072	20	1	12103	75	3	7103	142	10

RISK FACTOR	75-84 AT EXAM			85-94 AT EXAM			35-64 COMBINED			65-94 COMBINED		
	PERSON EXAMS	# OF EVENTS	ANNUAL RATE	PERSON EXAMS	# OF EVENTS	ANNUAL RATE	# OF EVENTS	CRUDE	AGE-ADJUSTED	# OF EVENTS	CRUDE	AGE-ADJUSTED
NO	476	19	20	40	3	38	32	1	1	37	8	9
MILD	694	25	18	37	4	54	24	2	2	64	11	11
DEFINITE	469	22	23	35	2	29	29	4	4	54	16	16
TREATED	780	49	31	51	7	69	21	4	4	118	21	21
UNKNOWN	0	0	-	0	0	-	0	-		0	-	
TOTAL	2419	115	24	163	16	49	106	2	2	273	14	14

LOGISTIC REGRESSIONS

AGE AT EXAM		COEFFICIENT OF RISK FACTOR	SE COEFFICIENT	P-VALUE	STANDARDIZED COEFFICIENT	STD SE
35-44		0.943	0.542	0.082	0.553	0.318
45-54		0.683	0.234	0.003	0.527	0.180
55-64		0.482	0.141	0.001	0.404	0.119
65-74		0.643	0.123	0.001	0.527	0.101
75-84		0.242	0.131	0.063	0.189	0.102
85-94		0.153	0.327	0.639	0.128	0.273
35-64	UNIVARIATE	0.709	0.116	0.001	0.567	0.093
65-94	UNIVARIATE	0.489	0.086	0.001	0.398	0.070
35-64	BIVARIATE	0.512	0.120	0.001	0.409	0.096
65-94	BIVARIATE	0.430	0.086	0.001	0.349	0.070
35-64	MULTIVARIATE	0.288	0.129	0.026	0.230	0.104
65-94	MULTIVARIATE	0.313	0.088	0.001	0.255	0.072

SEX: MALE
EVENT: DEATH FROM CARDIOVASCULAR DISEASE
RISK FACTOR: HYPERTENSION IGNORING TREATMENT
PERSONS AT RISK: ALL PERSONS ALIVE

RISK FACTOR	35-44 AT EXAM PERSON EXAMS	# OF EVENTS	ANNUAL RATE	45-54 AT EXAM PERSON EXAMS	# OF EVENTS	ANNUAL RATE	55-64 AT EXAM PERSON EXAMS	# OF EVENTS	ANNUAL RATE	65-74 AT EXAM PERSON EXAMS	# OF EVENTS	ANNUAL RATE
NO	2672	4	1	4306	22	5	4347	55	6	1980	40	10
MILD	1326	3	1	2383	18	4	2922	53	9	1852	64	17
DEFINITE	521	1	1	1378	23	8	1801	61	17	1002	61	30
UNKNOWN	0	0	-	0	0	-	0	0	-	0	0	-
TOTAL	4519	8	1	8067	63	4	9070	169	9	4834	165	17

RISK FACTOR	75-84 AT EXAM PERSON EXAMS	# OF EVENTS	ANNUAL RATE	85-94 AT EXAM PERSON EXAMS	# OF EVENTS	ANNUAL RATE	35-64 COMBINED # OF EVENTS	CRUDE	AGE-ADJUSTED	65-94 COMBINED # OF EVENTS	CRUDE	AGE-ADJUSTED
NO	526	10	29	24	3	63	81	4	4	73	14	14
MILD	543	41	38	21	1	24	74	6	5	106	22	22
DEFINITE	276	33	60	5	2	200	85	11	10	96	37	38
UNKNOWN	0	0	-	0	0	-	0	-	-	0	-	-
TOTAL	1345	104	39	50	6	60	240	6	6	275	22	22

LOGISTIC REGRESSIONS

AGE AT EXAM		COEFFICIENT OF RISK FACTOR	SE COEFFICIENT	P-VALUE	STANDARDIZED COEFFICIENT	STD SE
35-44		0.198	0.482	0.681	0.137	0.334
45-54		0.599	0.155	0.001	0.453	0.117
55-64		0.505	0.096	0.001	0.391	0.074
65-74		0.574	0.103	0.001	0.436	0.078
75-84		0.405	0.134	0.003	0.304	0.101
85-94		0.526	0.632	0.405	0.351	0.421
35-64	UNIVARIATE	0.592	0.080	0.001	0.447	0.060
65-94	UNIVARIATE	0.504	0.080	0.001	0.381	0.061
35-64	BIVARIATE	0.509	0.081	0.001	0.384	0.061
65-94	BIVARIATE	0.518	0.081	0.001	0.392	0.061
35-64	MULTIVARIATE	0.320	0.086	0.001	0.242	0.065
65-94	MULTIVARIATE	0.388	0.085	0.001	0.293	0.064

SEX: FEMALE
EVENT: DEATH FROM CARDIOVASCULAR DISEASE
RISK FACTOR: HYPERTENSION IGNORING TREATMENT
PERSONS AT RISK: ALL PERSONS ALIVE

RISK FACTOR	35-44 AT EXAM PERSON EXAMS	# OF EVENTS	ANNUAL RATE	45-54 AT EXAM PERSON EXAMS	# OF EVENTS	ANNUAL RATE	55-64 AT EXAM PERSON EXAMS	# OF EVENTS	ANNUAL RATE	65-74 AT EXAM PERSON EXAMS	# OF EVENTS	ANNUAL RATE
NO	4273	3	0	5846	11	1	5415	21	2	2205	23	5
MILD	1017	0	0	2803	6	1	4079	27	3	2907	50	9
DEFINITE	329	2	3	1423	9	3	2609	27	5	1991	69	17
UNKNOWN	0	0	-	0	0	-	0	0	-	0	0	-
TOTAL	5619	5	0	10072	26	1	12103	75	3	7103	142	10

RISK FACTOR	75-84 AT EXAM PERSON EXAMS	# OF EVENTS	ANNUAL RATE	85-94 AT EXAM PERSON EXAMS	# OF EVENTS	ANNUAL RATE	35-64 COMBINED # OF EVENTS	CRUDE	AGE-ADJUSTED	65-94 COMBINED # OF EVENTS	CRUDE	AGE-ADJUSTED
NO	613	26	21	52	7	67	35	1	1	56	10	10
MILD	1003	39	19	57	5	44	33	2	2	94	12	12
DEFINITE	803	50	31	54	4	37	38	4	4	123	22	21
UNKNOWN	0	0	-	0	0	-	0	-	-	0	-	-
TOTAL	2419	115	24	163	16	49	106	2	2	273	14	14

LOGISTIC REGRESSIONS

AGE AT EXAM		COEFFICIENT OF RISK FACTOR	SE COEFFICIENT	P-VALUE	STANDARDIZED COEFFICIENT	STD SE
35-44		0.993	0.548	0.070	0.568	0.313
45-54		0.594	0.240	0.013	0.432	0.175
55-64		0.492	0.144	0.001	0.384	0.113
65-74		0.636	0.117	0.001	0.488	0.090
75-84		0.240	0.128	0.061	0.183	0.098
85-94		-0.345	0.333	0.300	-0.279	0.270
35-64	UNIVARIATE	0.680	0.119	0.001	0.506	0.088
65-94	UNIVARIATE	0.442	0.083	0.001	0.339	0.064
35-64	BIVARIATE	0.504	0.122	0.001	0.375	0.091
65-94	BIVARIATE	0.399	0.083	0.001	0.306	0.064
35-64	MULTIVARIATE	0.248	0.133	0.063	0.184	0.099
65-94	MULTIVARIATE	0.297	0.085	0.001	0.223	0.066

TABLE 22-4   DATE=17NOV85

AVERAGE ANNUAL INCIDENCE RATE FOR EVENT PER 1000 PERSONS AT RISK AT EXAM
BY AGE AND LEVEL OF RISK FACTOR AT EXAM: FRAMINGHAM STUDY 30-YEAR FOLLOWUP

SEX: MALE
EVENT: DEATH FROM CARDIOVASCULAR DISEASE
RISK FACTOR: SERUM CHOLESTEROL
PERSONS AT RISK: ALL PERSONS ALIVE

RISK FACTOR	35-44 AT EXAM			45-54 AT EXAM			55-64 AT EXAM			65-74 AT EXAM		
	PERSON EXAMS	# OF EVENTS	ANNUAL RATE	PERSON EXAMS	# OF EVENTS	ANNUAL RATE	PERSON EXAMS	# OF EVENTS	ANNUAL RATE	PERSON EXAMS	# OF EVENTS	ANNUAL RATE
84-204	1253	0	0	1807	15	4	2484	32	6	1530	58	19
205-234	1238	1	0	2390	13	3	2695	35	10	1457	43	15
235-264	883	3	2	1843	15	4	2128	43	10	1059	29	14
265-294	488	1	1	1057	12	6	1075	22	10	525	18	17
295-1124	303	3	5	619	7	6	556	15	13	244	16	33
UNKNOWN	334	0	0	351	1	1	132	2	8	17	1	29
TOTAL	4519	8	1	8067	63	4	9070	169	8	4834	165	17

RISK FACTOR	75-84 AT EXAM			85-94 AT EXAM			35-64 COMBINED			65-94 COMBINED		
	PERSON EXAMS	# OF EVENTS	ANNUAL RATE	PERSON EXAMS	# OF EVENTS	ANNUAL RATE	# OF EVENTS	CRUDE	AGE-ADJUSTED	# OF EVENTS	CRUDE	AGE-ADJUSTED
84-204	537	42	39	23	2	43	47	4	4	102	24	24
205-234	422	34	48	14	3	107	69	5	5	80	21	21
235-264	232	20	43	7	1	71	61	6	6	50	19	21
265-294	130	6	23	6	0	0	33	7	7	24	18	18
295-1124	24	2	42	0	0	-	25	8	9	18	33	34
UNKNOWN	0	0	-	0	0	-	3	2	4	1	29	23
TOTAL	1345	104	39	50	6	60	240	6	6	275	22	22

LOGISTIC REGRESSIONS

AGE AT EXAM		COEFFICIENT OF RISK FACTOR	SE COEFFICIENT	P-VALUE	STANDARDIZED COEFFICIENT	STD SE
35-44		0.016	0.004	0.001	0.696	0.187
45-54		0.005	0.003	0.058	0.211	0.111
55-64		0.004	0.002	0.025	0.165	0.074
65-74		0.001	0.002	0.492	-0.054	0.079
75-84		-0.001	0.003	0.601	-0.054	0.103
85-94		-0.002	0.010	0.868	-0.073	0.438
35-64	UNIVARIATE	0.005	0.001	0.001	0.199	0.059
65-94	UNIVARIATE	-0.001	0.002	0.570	-0.035	0.062
35-64	BIVARIATE	0.005	0.001	0.001	0.225	0.060
65-94	BIVARIATE	0.001	0.002	0.693	0.025	0.063
35-64	MULTIVARIATE	0.005	0.001	0.001	0.214	0.061
65-94	MULTIVARIATE	0.001	0.002	0.487	0.044	0.063

SEX: FEMALE
EVENT: DEATH FROM CARDIOVASCULAR DISEASE
RISK FACTOR: SERUM CHOLESTEROL
PERSONS AT RISK: ALL PERSONS ALIVE

RISK FACTOR	35-44 AT EXAM			45-54 AT EXAM			55-64 AT EXAM			65-74 AT EXAM		
	PERSON EXAMS	# OF EVENTS	ANNUAL RATE	PERSON EXAMS	# OF EVENTS	ANNUAL RATE	PERSON EXAMS	# OF EVENTS	ANNUAL RATE	PERSON EXAMS	# OF EVENTS	ANNUAL RATE
84-204	2350	0	0	2120	1	0	1714	9	3	897	23	13
205-234	1419	1	0	2590	5	1	2741	11	2	1570	28	9
235-264	807	1	1	2268	6	1	3052	16	3	1912	26	7
265-294	372	0	0	1490	3	1	2364	18	4	1472	30	10
295-1124	171	2	4	1124	10	4	1999	20	5	1215	33	14
UNKNOWN	500	1	1	480	1	1	233	1	2	37	2	27
TOTAL	5619	5	0	10072	26	1	12103	75	3	7103	142	10

RISK FACTOR	75-84 AT EXAM			85-94 AT EXAM			35-64 COMBINED			65-94 COMBINED		
	PERSON EXAMS	# OF EVENTS	ANNUAL RATE	PERSON EXAMS	# OF EVENTS	ANNUAL RATE	# OF EVENTS	CRUDE	AGE-ADJUSTED	# OF EVENTS	CRUDE	AGE-ADJUSTED
84-204	463	27	29	69	10	83	10	1	1	60	21	18
205-234	588	30	26	39	1	13	17	1	1	59	13	13
235-264	660	28	21	42	3	16	23	2	2	57	11	11
265-294	419	18	21	12	2	83	21	2	2	50	13	14
295-1124	282	12	21	10	0	0	32	5	5	45	15	15
UNKNOWN	7	0	-	0	0	-	3	1	2	2	23	20
TOTAL	2419	115	24	163	16	49	106	2	2	273	14	14

LOGISTIC REGRESSIONS

AGE AT EXAM		COEFFICIENT OF RISK FACTOR	SE COEFFICIENT	P-VALUE	STANDARDIZED COEFFICIENT	STD SE
35-44		0.020	0.006	0.001	0.778	0.226
45-54		0.006	0.002	0.001	0.298	0.093
55-64		0.007	0.002	0.002	0.322	0.105
65-74		0.002	0.002	0.334	0.081	0.084
75-84		-0.003	0.002	0.163	-0.136	0.098
85-94		-0.015	0.007	0.024	-0.700	0.310
35-64	UNIVARIATE	0.008	0.002	0.001	0.395	0.073
65-94	UNIVARIATE	-0.003	0.001	0.060	-0.118	0.063
35-64	BIVARIATE	0.007	0.001	0.001	0.330	0.069
65-94	BIVARIATE	-0.001	0.001	0.621	-0.031	0.063
35-64	MULTIVARIATE	0.006	0.001	0.001	0.301	0.069
65-94	MULTIVARIATE	-0.000	0.001	0.768	-0.019	0.063

TABLE 22-6   DATE=12DEC85
AVERAGE ANNUAL INCIDENCE RATE FOR EVENT PER 1000 PERSONS AT RISK AT EXAM
BY AGE AND LEVEL OF RISK FACTOR AT EXAM: FRAMINGHAM STUDY 30-YEAR FOLLOWUP

SEX: MALE
EVENT: DEATH FROM CARDIOVASCULAR DISEASE
RISK FACTOR: BLOOD GLUCOSE (MG/DL)
PERSONS AT RISK: ALL PERSONS ALIVE

RISK FACTOR	35-44 AT EXAM			45-54 AT EXAM			55-64 AT EXAM			65-74 AT EXAM		
	PERSON EXAMS	# OF EVENTS	ANNUAL RATE	PERSON EXAMS	# OF EVENTS	ANNUAL RATE	PERSON EXAMS	# OF EVENTS	ANNUAL RATE	PERSON EXAMS	# OF EVENTS	ANNUAL RATE
29-69	1004	0	0	1562	12	4	1208	15	6	501	13	13
70-89	2531	7	1	4344	31	4	4526	83	9	2142	61	14
90-109	628	1	1	1402	11	4	2136	43	10	1296	42	16
110-129	107	0	0	331	3	5	608	15	12	428	18	21
130-524	49	0	0	204	5	12	491	13	13	450	30	33
UNKNOWN	200	0	0	224	1	2	101	0	0	17	1	29
TOTAL	4519	8	1	8067	63	4	9070	169	9	4834	165	17

RISK FACTOR	75-84 AT EXAM			85-94 AT EXAM			35-64 COMBINED			65-94 COMBINED		
	PERSON EXAMS	# OF EVENTS	ANNUAL RATE	PERSON EXAMS	# OF EVENTS	ANNUAL RATE	# OF EVENTS	CRUDE	AGE-ADJUSTED	# OF EVENTS	CRUDE	AGE-ADJUSTED
29-69	86	8	47	1	0	0	27	4	4	21	18	20
70-89	479	34	33	17	0	0	121	5	5	95	18	19
90-109	438	29	33	15	2	67	55	7	6	73	21	20
110-129	170	12	35	7	0	0	18	9	7	30	25	24
130-524	172	21	61	10	4	200	18	12	10	55	44	41
UNKNOWN	0	0	-	0	0	-	1	1	1	1	29	23
TOTAL	1345	104	39	50	6	60	240	6	6	275	22	22

LOGISTIC REGRESSIONS

AGE AT EXAM		COEFFICIENT OF RISK FACTOR	SE COEFFICIENT	P-VALUE	STANDARDIZED COEFFICIENT	STD SE
35-44		-0.021	0.029	0.464	-0.364	0.497
45-54		0.006	0.003	0.078	0.142	0.080
55-64		0.005	0.002	0.006	0.151	0.055
65-74		0.008	0.001	0.001	0.280	0.046
75-84		0.006	0.002	0.019	0.193	0.082
85-94		0.023	0.011	0.042	0.792	0.389
35-64	UNIVARIATE	0.007	0.001	0.001	0.184	0.038
65-94	UNIVARIATE	0.007	0.001	0.001	0.268	0.039
35-64	BIVARIATE	0.005	0.002	0.005	0.121	0.043
65-94	BIVARIATE	0.007	0.001	0.001	0.261	0.040
35-64	MULTIVARIATE	0.004	0.002	0.012	0.112	0.045
65-94	MULTIVARIATE	0.007	0.001	0.001	0.252	0.041

SEX: FEMALE
EVENT: DEATH FROM CARDIOVASCULAR DISEASE
RISK FACTOR: BLOOD GLUCOSE (MG/DL)
PERSONS AT RISK: ALL PERSONS ALIVE

RISK FACTOR	35-44 AT EXAM			45-54 AT EXAM			55-64 AT EXAM			65-74 AT EXAM		
	PERSON EXAMS	# OF EVENTS	ANNUAL RATE	PERSON EXAMS	# OF EVENTS	ANNUAL RATE	PERSON EXAMS	# OF EVENTS	ANNUAL RATE	PERSON EXAMS	# OF EVENTS	ANNUAL RATE
29-69	1067	1	0	1806	4	1	1411	11	4	597	17	14
70-89	3414	3	0	5811	14	1	6371	33	3	3483	52	7
90-109	672	0	0	1646	1	0	2877	15	3	1982	30	8
110-129	96	0	0	303	0	0	777	7	5	582	15	13
130-524	39	0	0	167	5	15	496	8	8	427	26	30
UNKNOWN	331	1	2	339	2	3	171	1	3	32	2	31
TOTAL	5619	5	0	10072	26	1	12103	75	3	7103	142	10

RISK FACTOR	75-84 AT EXAM			85-94 AT EXAM			35-64 COMBINED			65-94 COMBINED		
	PERSON EXAMS	# OF EVENTS	ANNUAL RATE	PERSON EXAMS	# OF EVENTS	ANNUAL RATE	# OF EVENTS	CRUDE	AGE-ADJUSTED	# OF EVENTS	CRUDE	AGE-ADJUSTED
29-69	129	5	19	5	1	100	16	2	2	23	16	17
70-89	976	35	18	60	4	33	50	2	2	91	10	11
90-109	803	40	25	63	3	24	16	2	1	73	13	12
110-129	286	17	30	19	3	79	7	3	2	35	20	18
130-524	224	18	40	16	5	156	13	9	9	49	37	35
UNKNOWN	1	0	0	0	0	-	4	2	3	2	30	23
TOTAL	2419	115	24	163	14	49	106	2	2	273	14	14

LOGISTIC REGRESSIONS

AGE AT EXAM		COEFFICIENT OF RISK FACTOR	SE COEFFICIENT	P-VALUE	STANDARDIZED COEFFICIENT	STD SE
35-44		-	-	-	-	-
45-54		0.016	0.004	0.001	0.312	0.070
55-64		0.007	0.002	0.002	0.197	0.063
65-74		0.009	0.002	0.001	0.279	0.046
75-84		0.006	0.002	0.001	0.212	0.066
85-94		0.012	0.005	0.027	0.413	0.186
35-64	UNIVARIATE	0.011	0.002	0.001	0.241	0.041
65-94	UNIVARIATE	0.009	0.001	0.001	0.275	0.035
35-64	BIVARIATE	0.008	0.002	0.001	0.193	0.045
65-94	BIVARIATE	0.008	0.001	0.001	0.250	0.037
35-64	MULTIVARIATE	0.007	0.002	0.002	0.149	0.049
65-94	MULTIVARIATE	0.007	0.001	0.001	0.218	0.039

TABLE 22-5   DATE=17NOV85
AVERAGE ANNUAL INCIDENCE RATE FOR EVENT PER 1000 PERSONS AT RISK AT EXAM
BY AGE AND LEVEL OF RISK FACTOR AT EXAM: FRAMINGHAM STUDY 30-YEAR FOLLOWUP

SEX: MALE
EVENT: DEATH FROM CARDIOVASCULAR DISEASE
RISK FACTOR: HEMATOCRIT
PERSONS AT RISK: ALL PERSONS ALIVE

RISK FACTOR	35-44 AT EXAM PERSON EXAMS	# OF EVENTS	ANNUAL RATE	45-54 AT EXAM PERSON EXAMS	# OF EVENTS	ANNUAL RATE	55-64 AT EXAM PERSON EXAMS	# OF EVENTS	ANNUAL RATE	65-74 AT EXAM PERSON EXAMS	# OF EVENTS	ANNUAL RATE
25-42	630	1	1	961	6	3	1121	23	10	558	21	19
43-44	551	0	0	996	11	6	1094	11	5	611	14	11
45-46	990	3	2	1942	16	4	2215	45	10	1194	31	13
47-48	786	0	0	1585	11	3	1825	29	8	1069	39	18
49-70	1192	4	2	2193	17	4	2656	59	11	1382	59	21
UNKNOWN	370	0	0	390	2	3	159	2	6	20	1	25
TOTAL	4519	8	1	8067	63	4	9070	169	9	4834	165	17

RISK FACTOR	75-84 AT EXAM PERSON EXAMS	# OF EVENTS	ANNUAL RATE	85-94 AT EXAM PERSON EXAMS	# OF EVENTS	ANNUAL RATE	35-64 COMBINED # OF EVENTS	CRUDE	AGE-ADJUSTED	65-94 COMBINED # OF EVENTS	CRUDE	AGE-ADJUSTED
25-42	283	21	37	19	2	53	30	6	6	44	26	23
43-44	169	16	47	8	0	0	22	4	4	30	19	19
45-46	313	25	40	8	2	143	64	6	6	58	19	20
47-48	266	17	32	9	0	0	40	5	5	56	21	21
49-70	314	25	40	7	2	143	80	7	6	86	25	26
UNKNOWN	0	0	-	0	0	-	4	2	4	1	25	19
TOTAL	1345	104	39	50	6	60	240	6	6	275	22	22

LOGISTIC REGRESSIONS

AGE AT EXAM		COEFFICIENT OF RISK FACTOR	SE COEFFICIENT	P-VALUE	STANDARDIZED COEFFICIENT	STD SE
35-44		0.082	0.098	0.399	0.302	0.358
45-54		-0.000	0.036	0.997	-0.000	0.129
55-64		0.021	0.021	0.304	0.081	0.079
65-74		0.034	0.022	0.118	0.125	0.080
75-84		-0.007	0.025	0.784	-0.028	0.102
85-94		0.089	0.096	0.355	0.437	0.472
35-64	UNIVARIATE	0.020	0.018	0.266	0.073	0.066
65-94	UNIVARIATE	0.005	0.016	0.782	0.017	0.062
35-64	BIVARIATE	0.017	0.018	0.324	0.064	0.065
65-94	BIVARIATE	0.023	0.016	0.160	0.087	0.062
35-64	MULTIVARIATE	0.000	0.018	0.999	0.000	0.066
65-94	MULTIVARIATE	0.017	0.017	0.320	0.063	0.064

SEX: FEMALE
EVENT: DEATH FROM CARDIOVASCULAR DISEASE
RISK FACTOR: HEMATOCRIT
PERSONS AT RISK: ALL PERSONS ALIVE

RISK FACTOR	35-44 AT EXAM PERSON EXAMS	# OF EVENTS	ANNUAL RATE	45-54 AT EXAM PERSON EXAMS	# OF EVENTS	ANNUAL RATE	55-64 AT EXAM PERSON EXAMS	# OF EVENTS	ANNUAL RATE	65-74 AT EXAM PERSON EXAMS	# OF EVENTS	ANNUAL RATE
21-39	1698	1	0	1858	3	1	1609	11	3	751	20	13
40-41	967	0	0	1761	2	1	2144	11	3	1180	20	8
42-43	1043	1	0	1978	1	1	2371	12	3	1422	20	7
44-45	858	1	1	2222	6	1	3003	15	2	1835	37	10
46-65	487	1	1	1702	12	4	2700	25	5	1873	43	11
UNKNOWN	566	1	1	551	2	2	276	1	2	42	2	24
TOTAL	5619	5	0	10072	26	1	12103	75	3	7103	142	10

RISK FACTOR	75-84 AT EXAM PERSON EXAMS	# OF EVENTS	ANNUAL RATE	85-94 AT EXAM PERSON EXAMS	# OF EVENTS	ANNUAL RATE	35-64 COMBINED # OF EVENTS	CRUDE	AGE-ADJUSTED	65-94 COMBINED # OF EVENTS	CRUDE	AGE-ADJUSTED
21-39	375	24	32	33	4	61	15	1	2	48	21	19
40-41	449	19	21	39	5	64	13	1	1	44	13	13
42-43	437	16	18	26	3	58	14	1	1	39	10	11
44-45	607	26	21	34	1	15	22	2	2	64	13	13
46-65	545	30	28	31	3	48	38	4	4	76	16	16
UNKNOWN	6	0	0	0	0	-	4	1	2	2	21	17
TOTAL	2419	115	24	163	16	49	106	2	2	273	14	14

LOGISTIC REGRESSIONS

AGE AT EXAM		COEFFICIENT OF RISK FACTOR	SE COEFFICIENT	P-VALUE	STANDARDIZED COEFFICIENT	STD SE
35-44		0.134	0.146	0.358	0.482	0.525
45-54		0.187	0.057	0.001	0.666	0.205
55-64		0.083	0.034	0.013	0.290	0.117
65-74		0.028	0.024	0.257	0.098	0.086
75-84		-0.032	0.024	0.183	-0.124	0.093
85-94		-0.055	0.069	0.424	-0.214	0.268
35-64	UNIVARIATE	0.135	0.028	0.001	0.484	0.100
65-94	UNIVARIATE	-0.017	0.017	0.314	-0.062	0.061
35-64	BIVARIATE	0.103	0.029	0.001	0.372	0.103
65-94	BIVARIATE	-0.004	0.017	0.813	-0.014	0.061
35-64	MULTIVARIATE	0.067	0.028	0.017	0.241	0.101
65-94	MULTIVARIATE	-0.013	0.016	0.436	-0.046	0.060

AVERAGE ANNUAL INCIDENCE RATE FOR EVENT PER 1000 PERSONS AT RISK AT EXAM
BY AGE AND LEVEL OF RISK FACTOR AT EXAM: FRAMINGHAM STUDY 30-YEAR FOLLOWUP

SEX: MALE
EVENT: DEATH FROM CARDIOVASCULAR DISEASE
RISK FACTOR: DIABETES MELLITUS
PERSONS AT RISK: ALL PERSONS ALIVE

RISK FACTOR	35-44 AT EXAM PERSON EXAMS	# OF EVENTS	ANNUAL RATE	45-54 AT EXAM PERSON EXAMS	# OF EVENTS	ANNUAL RATE	55-64 AT EXAM PERSON EXAMS	# OF EVENTS	ANNUAL RATE	65-74 AT EXAM PERSON EXAMS	# OF EVENTS	ANNUAL RATE
NO	4470	8	1	7812	57	4	8424	148	9	4269	128	15
YES	49	0	0	255	6	12	646	21	16	565	37	33
UNKNOWN	0	0	-	0	0	-	0	0	-	0	0	-
TOTAL	4519	8	1	8067	63	4	9070	169	9	4834	165	17

RISK FACTOR	75-84 AT EXAM PERSON EXAMS	# OF EVENTS	ANNUAL RATE	85-94 AT EXAM PERSON EXAMS	# OF EVENTS	ANNUAL RATE	35-64 COMBINED # OF EVENTS	CRUDE	AGE-ADJUSTED	65-94 COMBINED # OF EVENTS	CRUDE	AGE-ADJUSTED
NO	1170	85	36	44	5	57	213	5	5	218	20	20
YES	175	19	54	6	1	83	27	14	11	57	38	38
UNKNOWN	0	0	-	0	0	-	0	-	-	0	-	-
TOTAL	1345	104	39	50	6	60	240	6	6	275	22	22

LOGISTIC REGRESSIONS

AGE AT EXAM		COEFFICIENT OF RISK FACTOR	SE COEFFICIENT	P-VALUE	STANDARDIZED COEFFICIENT	STD SE
35-44		-	-	-	-	-
45-54		1.187	0.434	0.006	0.208	0.076
55-64		0.631	0.237	0.008	0.162	0.061
65-74		0.818	0.192	0.001	0.263	0.062
75-84		0.441	0.268	0.099	0.149	0.090
85-94		0.445	1.194	0.710	0.146	0.392
35-64	UNIVARIATE	1.035	0.207	0.001	0.212	0.042
65-94	UNIVARIATE	0.692	0.154	0.001	0.225	0.050
35-64	BIVARIATE	0.652	0.209	0.002	0.133	0.043
65-94	BIVARIATE	0.665	0.155	0.001	0.216	0.050
35-64	MULTIVARIATE	0.347	0.220	0.115	0.071	0.045
65-94	MULTIVARIATE	0.631	0.159	0.001	0.205	0.052

SEX: FEMALE
EVENT: DEATH FROM CARDIOVASCULAR DISEASE
RISK FACTOR: DIABETES MELLITUS
PERSONS AT RISK: ALL PERSONS ALIVE

RISK FACTOR	35-44 AT EXAM PERSON EXAMS	# OF EVENTS	ANNUAL RATE	45-54 AT EXAM PERSON EXAMS	# OF EVENTS	ANNUAL RATE	55-64 AT EXAM PERSON EXAMS	# OF EVENTS	ANNUAL RATE	65-74 AT EXAM PERSON EXAMS	# OF EVENTS	ANNUAL RATE
NO	5576	4	0	9859	21	1	11497	61	3	6499	105	8
YES	43	1	12	213	5	12	606	14	12	604	37	31
UNKNOWN	0	0	-	0	0	-	0	0	-	0	0	-
TOTAL	5619	5	0	10072	26	1	12103	75	3	7103	142	10

RISK FACTOR	75-84 AT EXAM PERSON EXAMS	# OF EVENTS	ANNUAL RATE	85-94 AT EXAM PERSON EXAMS	# OF EVENTS	ANNUAL RATE	35-64 COMBINED # OF EVENTS	CRUDE	AGE-ADJUSTED	65-94 COMBINED # OF EVENTS	CRUDE	AGE-ADJUSTED
NO	2106	83	20	139	11	40	86	2	2	199	11	12
YES	313	32	51	24	5	104	20	12	12	74	39	37
UNKNOWN	0	0	-	0	0	-	0	-	-	0	-	-
TOTAL	2419	115	24	163	16	49	106	2	2	273	14	14

LOGISTIC REGRESSIONS

AGE AT EXAM		COEFFICIENT OF RISK FACTOR	SE COEFFICIENT	P-VALUE	STANDARDIZED COEFFICIENT	STD SE
35-44		3.502	1.129	0.002	0.305	0.098
45-54		2.421	0.503	0.001	0.348	0.072
55-64		1.489	0.299	0.001	0.325	0.065
65-74		1.380	0.196	0.001	0.385	0.055
75-84		1.021	0.218	0.001	0.343	0.073
85-94		1.119	0.593	0.059	0.398	0.211
35-64	UNIVARIATE	2.003	0.251	0.001	0.347	0.043
65-94	UNIVARIATE	1.299	0.141	0.001	0.385	0.042
35-64	BIVARIATE	1.670	0.254	0.001	0.289	0.044
65-94	BIVARIATE	1.179	0.142	0.001	0.349	0.042
35-64	MULTIVARIATE	1.341	0.269	0.001	0.233	0.047
65-94	MULTIVARIATE	0.978	0.148	0.001	0.290	0.044

AVERAGE ANNUAL INCIDENCE RATE FOR EVENT PER 1000 PERSONS AT RISK AT EXAM
BY AGE AND LEVEL OF RISK FACTOR AT EXAM: FRAMINGHAM STUDY 30-YEAR FOLLOWUP

SEX: MALE
EVENT: DEATH FROM CARDIOVASCULAR DISEASE
RISK FACTOR: GLUCOSE IN URINE, DEFINITE OR TRACE
PERSONS AT RISK: ALL PERSONS ALIVE

RISK FACTOR	35-44 AT EXAM PERSON EXAMS	# OF EVENTS	ANNUAL RATE	45-54 AT EXAM PERSON EXAMS	# OF EVENTS	ANNUAL RATE	55-64 AT EXAM PERSON EXAMS	# OF EVENTS	ANNUAL RATE	65-74 AT EXAM PERSON EXAMS	# OF EVENTS	ANNUAL RATE
NO	4426	8	1	7869	61	4	8786	155	9	4661	150	16
YES	60	0	0	177	2	6	260	13	25	167	15	45
UNKNOWN	33	0	0	21	0	0	24	1	21	6	0	0
TOTAL	4519	8	1	8067	63	4	9070	169	9	4834	165	17

RISK FACTOR	75-84 AT EXAM PERSON EXAMS	# OF EVENTS	ANNUAL RATE	85-94 AT EXAM PERSON EXAMS	# OF EVENTS	ANNUAL RATE	35-64 COMBINED # OF EVENTS	CRUDE	AGE-ADJUSTED	65-94 COMBINED # OF EVENTS	CRUDE	AGE-ADJUSTED
NO	1303	96	37	49	6	61	224	5	5	252	21	21
YES	39	8	103	1	0	0	15	15	13	23	56	57
UNKNOWN	3	0	0	0	0	-	1	6	9	0	0	0
TOTAL	1345	104	39	50	6	60	240	6	6	275	22	22

LOGISTIC REGRESSIONS

AGE AT EXAM		COEFFICIENT OF RISK FACTOR	SE COEFFICIENT	P-VALUE	STANDARDIZED COEFFICIENT	STD SE
35-44		-	-	-	-	-
45-54		0.380	0.723	0.599	0.056	0.106
55-64		1.075	0.296	0.001	0.180	0.049
65-74		1.088	0.283	0.001	0.199	0.052
75-84		1.177	0.410	0.004	0.198	0.069
85-94		-	-	-	-	-
35-64	UNIVARIATE	1.065	0.271	0.001	0.160	0.041
65-94	UNIVARIATE	1.053	0.230	0.001	0.188	0.041
35-64	BIVARIATE	0.887	0.272	0.001	0.133	0.041
65-94	BIVARIATE	1.117	0.233	0.001	0.200	0.042
35-64	MULTIVARIATE	0.679	0.289	0.019	0.102	0.043
65-94	MULTIVARIATE	1.074	0.238	0.001	0.193	0.043

SEX: FEMALE
EVENT: DEATH FROM CARDIOVASCULAR DISEASE
RISK FACTOR: GLUCOSE IN URINE, DEFINITE OR TRACE
PERSONS AT RISK: ALL PERSONS ALIVE

RISK FACTOR	35-44 AT EXAM PERSON EXAMS	# OF EVENTS	ANNUAL RATE	45-54 AT EXAM PERSON EXAMS	# OF EVENTS	ANNUAL RATE	55-64 AT EXAM PERSON EXAMS	# OF EVENTS	ANNUAL RATE	65-74 AT EXAM PERSON EXAMS	# OF EVENTS	ANNUAL RATE
NO	5540	4	0	9947	23	1	11903	68	3	6957	130	9
YES	41	1	12	101	3	15	174	7	20	136	12	44
UNKNOWN	38	0	0	24	0	0	26	0	0	10	0	0
TOTAL	5619	5	0	10072	26	1	12103	75	3	7103	142	10

RISK FACTOR	75-84 AT EXAM PERSON EXAMS	# OF EVENTS	ANNUAL RATE	85-94 AT EXAM PERSON EXAMS	# OF EVENTS	ANNUAL RATE	35-64 COMBINED # OF EVENTS	CRUDE	AGE-ADJUSTED	65-94 COMBINED # OF EVENTS	CRUDE	AGE-ADJUSTED
NO	2375	111	23	160	15	47	95	2	2	256	13	13
YES	35	4	57	3	1	167	11	17	17	17	49	49
UNKNOWN	9	0	0	0	0	-	0	0	0	0	0	0
TOTAL	2419	115	24	163	16	49	106	2	2	273	14	14

LOGISTIC REGRESSIONS

AGE AT EXAM		COEFFICIENT OF RISK FACTOR	SE COEFFICIENT	P-VALUE	STANDARDIZED COEFFICIENT	STD SE
35-44		3.544	1.129	0.002	0.303	0.096
45-54		2.581	0.622	0.001	0.257	0.062
55-64		1.987	0.405	0.001	0.237	0.048
65-74		1.626	0.315	0.001	0.223	0.043
75-84		0.968	0.540	0.073	0.116	0.065
85-94		1.576	1.254	0.209	0.212	0.169
35-64	UNIVARIATE	2.338	0.324	0.001	0.248	0.034
65-94	UNIVARIATE	1.363	0.263	0.001	0.181	0.035
35-64	BIVARIATE	2.195	0.326	0.001	0.233	0.035
65-94	BIVARIATE	1.455	0.267	0.001	0.193	0.036
35-64	MULTIVARIATE	1.791	0.349	0.001	0.190	0.037
65-94	MULTIVARIATE	1.106	0.280	0.001	0.147	0.037

- 434 -

TABLE 22-9   DATE=07MAR86
AVERAGE ANNUAL INCIDENCE RATE FOR EVENT PER 1000 PERSONS AT RISK AT EXAM
BY AGE AND LEVEL OF RISK FACTOR AT EXAM: FRAMINGHAM STUDY 30-YEAR FOLLOWUP

SEX: MALE
EVENT: DEATH FROM CARDIOVASCULAR DISEASE
RISK FACTOR: GLUCOSE INTOLERANCE
PERSONS AT RISK: ALL PERSONS ALIVE

RISK FACTOR	35-44 AT EXAM			45-54 AT EXAM			55-64 AT EXAM			65-74 AT EXAM		
	PERSON EXAMS	# OF EVENTS	ANNUAL RATE	PERSON EXAMS	# OF EVENTS	ANNUAL RATE	PERSON EXAMS	# OF EVENTS	ANNUAL RATE	PERSON EXAMS	# OF EVENTS	ANNUAL RATE
NO	4381	8	1	7528	55	4	7973	139	9	3918	112	14
YES	138	0	0	539	8	7	1097	30	14	916	53	29
UNKNOWN	0	0	-	0	0	-	0	0	-	0	0	-
TOTAL	4519	8	1	8067	63	4	9070	169	9	4834	165	17

RISK FACTOR	75-84 AT EXAM			85-94 AT EXAM			35-64 COMBINED			65-94 COMBINED		
	PERSON EXAMS	# OF EVENTS	ANNUAL RATE	PERSON EXAMS	# OF EVENTS	ANNUAL RATE	# OF EVENTS	CRUDE	AGE-ADJUSTED	# OF EVENTS	CRUDE	AGE-ADJUSTED
NO	1035	72	35	36	2	28	202	5	5	186	19	19
YES	310	32	52	14	4	143	38	11	8	89	36	35
UNKNOWN	0	0	-	0	0	-	0	-	-	0	-	-
TOTAL	1345	104	39	50	6	60	240	6	6	275	22	22

LOGISTIC REGRESSIONS

AGE AT EXAM		COEFFICIENT OF RISK FACTOR	SE COEFFICIENT	P-VALUE	STANDARDIZED COEFFICIENT	STD SE
35-44		-	-	-	-	-
45-54		0.716	0.381	0.060	0.179	0.095
55-64		0.460	0.204	0.024	0.150	0.067
65-74		0.736	0.171	0.001	0.288	0.067
75-84		0.432	0.223	0.053	0.182	0.094
85-94		1.917	0.938	0.041	0.869	0.425
35-64	UNIVARIATE	0.757	0.179	0.001	0.208	0.049
65-94	UNIVARIATE	0.691	0.133	0.001	0.276	0.053
35-64	BIVARIATE	0.445	0.180	0.014	0.122	0.049
65-94	BIVARIATE	0.629	0.134	0.001	0.251	0.054
35-64	MULTIVARIATE	0.234	0.188	0.213	0.064	0.052
65-94	MULTIVARIATE	0.594	0.136	0.001	0.237	0.054

SEX: FEMALE
EVENT: DEATH FROM CARDIOVASCULAR DISEASE
RISK FACTOR: GLUCOSE INTOLERANCE
PERSONS AT RISK: ALL PERSONS ALIVE

RISK FACTOR	35-44 AT EXAM			45-54 AT EXAM			55-64 AT EXAM			65-74 AT EXAM		
	PERSON EXAMS	# OF EVENTS	ANNUAL RATE	PERSON EXAMS	# OF EVENTS	ANNUAL RATE	PERSON EXAMS	# OF EVENTS	ANNUAL RATE	PERSON EXAMS	# OF EVENTS	ANNUAL RATE
NO	5525	4	0	9685	20	1	10988	57	3	6066	94	8
YES	94	1	5	387	6	8	1115	18	8	1037	48	23
UNKNOWN	0	0	-	0	0	-	0	0	-	0	0	-
TOTAL	5619	5	0	10072	26	1	12103	75	3	7103	142	10

RISK FACTOR	75-84 AT EXAM			85-94 AT EXAM			35-64 COMBINED			65-94 COMBINED		
	PERSON EXAMS	# OF EVENTS	ANNUAL RATE	PERSON EXAMS	# OF EVENTS	ANNUAL RATE	# OF EVENTS	CRUDE	AGE-ADJUSTED	# OF EVENTS	CRUDE	AGE-ADJUSTED
NO	1905	77	20	131	9	34	81	2	2	180	11	11
YES	514	38	37	32	7	109	25	8	7	93	29	28
UNKNOWN	0	0	-	0	0	-	0	-	-	0	-	-
TOTAL	2419	115	24	163	16	49	106	2	2	273	14	14

LOGISTIC REGRESSIONS

AGE AT EXAM		COEFFICIENT OF RISK FACTOR	SE COEFFICIENT	P-VALUE	STANDARDIZED COEFFICIENT	STD SE
35-44		2.697	1.123	0.016	0.346	0.144
45-54		2.029	0.468	0.001	0.390	0.090
55-64		1.146	0.272	0.001	0.332	0.079
65-74		1.126	0.181	0.001	0.398	0.064
75-84		0.639	0.205	0.002	0.262	0.084
85-94		1.334	0.550	0.015	0.531	0.219
35-64	UNIVARIATE	1.635	0.250	0.001	0.380	0.054
65-94	UNIVARIATE	1.011	0.131	0.001	0.374	0.048
35-64	BIVARIATE	1.306	0.234	0.001	0.304	0.054
65-94	BIVARIATE	0.895	0.132	0.001	0.331	0.049
35-64	MULTIVARIATE	1.119	0.243	0.001	0.260	0.056
65-94	MULTIVARIATE	0.753	0.136	0.001	0.278	0.050

TABLE 22-10   DATE=17NOV85
AVERAGE ANNUAL INCIDENCE RATE FOR EVENT PER 1000 PERSONS AT RISK AT EXAM
BY AGE AND LEVEL OF RISK FACTOR AT EXAM: FRAMINGHAM STUDY 30-YEAR FOLLOWUP

SEX: MALE
EVENT: DEATH FROM CARDIOVASCULAR DISEASE
RISK FACTOR: METROPOLITAN RELATIVE WEIGHT (PERCENT)
PERSONS AT RISK: ALL PERSONS ALIVE

RISK FACTOR	35-44 AT EXAM PERSON EXAMS	# OF EVENTS	ANNUAL RATE	45-54 AT EXAM PERSON EXAMS	# OF EVENTS	ANNUAL RATE	55-64 AT EXAM PERSON EXAMS	# OF EVENTS	ANNUAL RATE	65-74 AT EXAM PERSON EXAMS	# OF EVENTS	ANNUAL RATE
54-104	723	1	1	1149	11	5	1272	29	11	831	45	27
105-114	979	0	0	1635	14	4	1858	36	10	1052	27	13
115-124	1133	3	1	2139	20	5	2467	40	8	1322	42	16
125-134	923	3	2	1691	6	2	1851	30	8	956	28	15
135-272	761	1	1	1452	12	4	1621	34	10	673	23	17
UNKNOWN	0	0	-	1	0	0	1	0	0	b	0	-
TOTAL	4519	8	1	8067	63	4	9070	169	9	4834	165	17

RISK FACTOR	75-84 AT EXAM PERSON EXAMS	# OF EVENTS	ANNUAL RATE	85-94 AT EXAM PERSON EXAMS	# OF EVENTS	ANNUAL RATE	35-64 COMBINED # OF EVENTS	CRUDE	AGE-ADJUSTED	65-94 COMBINED # OF EVENTS	CRUDE	AGE-ADJUSTED
54-104	277	18	32	19	4	105	41	7	7	67	30	29
105-114	357	32	45	15	1	33	50	6	6	60	21	20
115-124	399	30	38	14	1	36	63	5	5	73	21	21
125-134	196	12	31	2	0	0	39	4	4	40	17	18
135-272	116	12	52	0	0	-	47	6	6	35	22	24
UNKNOWN	0	0	-	0	0	-	0	0	0	0	0	-
TOTAL	1345	104	39	50	6	60	240	6	6	275	22	22

LOGISTIC REGRESSIONS

AGE AT EXAM		COEFFICIENT OF RISK FACTOR	SE COEFFICIENT	P-VALUE	STANDARDIZED COEFFICIENT	STD SE
35-44		0.019	0.020	0.357	0.304	0.330
45-54		-0.011	0.008	0.163	-0.183	0.131
55-64		-0.001	0.005	0.866	-0.013	0.078
65-74		-0.010	0.005	0.046	-0.161	0.081
75-84		0.000	0.007	0.997	0.000	0.102
85-94		-0.091	0.043	0.032	-1.058	0.494
35-64	UNIVARIATE	-0.002	0.004	0.579	-0.036	0.065
65-94	UNIVARIATE	-0.010	0.004	0.012	-0.159	0.063
35-64	BIVARIATE	-0.002	0.004	0.542	-0.040	0.066
65-94	BIVARIATE	-0.007	0.004	0.102	-0.105	0.064
35-64	MULTIVARIATE	-0.007	0.004	0.087	-0.119	0.069
65-94	MULTIVARIATE	-0.013	0.004	0.003	-0.198	0.067

SEX: FEMALE
EVENT: DEATH FROM CARDIOVASCULAR DISEASE
RISK FACTOR: METROPOLITAN RELATIVE WEIGHT (PERCENT)
PERSONS AT RISK: ALL PERSONS ALIVE

RISK FACTOR	35-44 AT EXAM PERSON EXAMS	# OF EVENTS	ANNUAL RATE	45-54 AT EXAM PERSON EXAMS	# OF EVENTS	ANNUAL RATE	55-64 AT EXAM PERSON EXAMS	# OF EVENTS	ANNUAL RATE	65-74 AT EXAM PERSON EXAMS	# OF EVENTS	ANNUAL RATE
54-104	1694	2	1	2117	7	2	2140	15	4	1319	31	12
105-114	1431	0	0	2364	3	1	2541	16	3	1321	23	9
115-124	1059	1	0	2128	0	0	2556	13	3	1536	24	8
125-134	566	1	1	1383	4	1	1831	8	2	1140	18	8
135-272	861	1	1	2080	12	3	3035	23	4	1787	46	13
UNKNOWN	8	0	0	0	0	-	0	0	-	0	0	-
TOTAL	5619	5	0	10072	26	1	12103	75	3	7103	142	10

RISK FACTOR	75-84 AT EXAM PERSON EXAMS	# OF EVENTS	ANNUAL RATE	85-94 AT EXAM PERSON EXAMS	# OF EVENTS	ANNUAL RATE	35-64 COMBINED # OF EVENTS	CRUDE	AGE-ADJUSTED	65-94 COMBINED # OF EVENTS	CRUDE	AGE-ADJUSTED
54-104	632	35	28	51	7	69	24	2	2	73	18	17
105-114	457	19	21	33	3	45	19	1	2	45	12	12
115-124	487	18	18	31	4	65	14	1	1	46	11	11
125-134	342	13	19	26	1	19	13	2	2	32	11	11
135-272	501	30	30	22	1	23	36	3	3	77	17	17
UNKNOWN	0	0	-	0	0	-	0	0	0	0	0	-
TOTAL	2419	115	24	163	16	49	106	2	2	273	14	14

LOGISTIC REGRESSIONS

AGE AT EXAM		COEFFICIENT OF RISK FACTOR	SE COEFFICIENT	P-VALUE	STANDARDIZED COEFFICIENT	STD SE
35-44		0.003	0.021	0.893	0.058	0.429
45-54		0.016	0.007	0.019	0.347	0.147
55-64		-0.002	0.005	0.742	-0.039	0.119
65-74		0.003	0.004	0.461	0.060	0.082
75-84		0.004	0.004	0.394	0.078	0.092
85-94		-0.023	0.015	0.122	-0.437	0.282
35-64	UNIVARIATE	0.006	0.004	0.113	0.140	0.088
65-94	UNIVARIATE	-0.000	0.003	0.912	-0.007	0.062
35-64	BIVARIATE	0.003	0.004	0.439	0.071	0.092
65-94	BIVARIATE	0.003	0.003	0.337	0.059	0.061
35-64	MULTIVARIATE	0.000	0.004	0.909	0.011	0.095
65-94	MULTIVARIATE	0.001	0.003	0.637	0.028	0.060

SEX: MALE
EVENT: DEATH FROM CARDIOVASCULAR DISEASE
RISK FACTOR: VITAL CAPACITY - HEIGHT INDEX (ML/INCH)
PERSONS AT RISK: ALL PERSONS ALIVE

35-44 AT EXAM			45-54 AT EXAM			55-64 AT EXAM			65-74 AT EXAM		
PERSON EXAMS	# OF EVENTS	ANNUAL RATE	PERSON EXAMS	# OF EVENTS	ANNUAL RATE	PERSON EXAMS	# OF EVENTS	ANNUAL RATE	PERSON EXAMS	# OF EVENTS	ANNUAL RATE
271	0	0	1108	15	7	2374	72	15	2179	92	21
416	1	1	1188	8	3	1692	29	9	974	38	20
751	0	0	1726	16	5	2061	31	8	857	17	10
1034	4	2	1823	12	3	1569	21	7	523	10	9
2034	3	1	2216	12	3	1356	16	6	284	7	12
13	0	0	6	0	0	18	0	0	12	1	42
4519	8	1	8067	63	4	9070	169	9	4834	165	17

75-84 AT EXAM			85-94 AT EXAM			35-64 COMBINED			65-94 COMBINED		
PERSON EXAMS	# OF EVENTS	ANNUAL RATE	PERSON EXAMS	# OF EVENTS	ANNUAL RATE	# OF EVENTS	CRUDE	AGE-ADJUSTED	# OF EVENTS	CRUDE	AGE-ADJUSTED
853	68	40	34	4	59	87	12	9	164	27	25
228	12	26	7	1	71	38	6	5	51	21	21
142	16	56	7	1	71	47	5	5	34	17	20
81	7	43	2	0	0	37	4	4	17	14	17
33	1	15	0	0	-	31	3	4	8	13	13
8	0	0	0	0	-	0	0	0	1	25	32
1345	104	39	50	6	60	240	6	6	275	22	22

LOGISTIC REGRESSIONS

	COEFFICIENT OF RISK FACTOR	SE COEFFICIENT	P-VALUE	STANDARDIZED COEFFICIENT	STD SE
	-0.011	0.039	0.783	-0.096	0.349
	-0.032	0.015	0.018	-0.290	0.122
	-0.035	0.008	0.001	-0.330	0.074
	-0.041	0.008	0.001	-0.382	0.077
	-0.010	0.011	0.344	-0.097	0.102
	-0.079	0.062	0.206	-0.604	0.478
NIVARIATE	-0.049	0.006	0.001	-0.475	0.061
NIVARIATE	-0.037	0.006	0.001	-0.354	0.061
IVARIATE	-0.029	0.007	0.001	-0.285	0.066
IVARIATE	-0.029	0.007	0.001	-0.270	0.063
ULTIVARIATE	-0.019	0.007	0.010	-0.180	0.070
ULTIVARIATE	-0.025	0.007	0.001	-0.232	0.065

SEX: FEMALE
EVENT: DEATH FROM CARDIOVASCULAR DISEASE
RISK FACTOR: VITAL CAPACITY - HEIGHT INDEX (ML/INCH)
PERSONS AT RISK: ALL PERSONS ALIVE

35-44 AT EXAM			45-54 AT EXAM			55-64 AT EXAM			65-74 AT EXAM		
PERSON EXAMS	# OF EVENTS	ANNUAL RATE	PERSON EXAMS	# OF EVENTS	ANNUAL RATE	PERSON EXAMS	# OF EVENTS	ANNUAL RATE	PERSON EXAMS	# OF EVENTS	ANNUAL RATE
122	0	0	546	2	2	1572	33	10	2068	66	16
291	0	0	1241	6	2	2547	15	3	1946	38	10
769	2	1	2103	6	1	3213	13	2	1728	26	8
1427	0	0	2838	7	1	2836	11	2	933	10	5
2989	3	1	3327	5	1	1914	3	1	422	2	2
21	0	0	17	0	0	21	0	0	6	0	0
5619	5	0	10072	26	1	12103	75	3	7103	142	10

75-84 AT EXAM			85-94 AT EXAM			35-64 COMBINED			65-94 COMBINED		
PERSON EXAMS	# OF EVENTS	ANNUAL RATE	PERSON EXAMS	# OF EVENTS	ANNUAL RATE	# OF EVENTS	CRUDE	AGE-ADJUSTED	# OF EVENTS	CRUDE	AGE-ADJUSTED
1212	72	30	109	10	46	35	8	5	148	22	20
686	29	21	37	6	81	21	3	2	73	14	14
370	11	15	16	0	0	21	2	2	37	9	9
123	2	8	1	0	0	18	1	1	12	6	6
26	1	19	0	0	-	11	1	1	3	3	7
2	0	0	0	0	-	0	0	0	0	0	0
2419	115	24	163	16	49	106	2	2	273	14	14

LOGISTIC REGRESSIONS

	COEFFICIENT OF RISK FACTOR	SE COEFFICIENT	P-VALUE	STANDARDIZED COEFFICIENT	STD SE
	-0.031	0.060	0.606	-0.225	0.436
	-0.051	0.025	0.044	-0.379	0.188
	-0.092	0.014	0.001	-0.692	0.105
	-0.062	0.011	0.001	-0.454	0.082
	-0.042	0.013	0.002	-0.294	0.093
	-0.013	0.040	0.742	-0.086	0.260
IVARIATE	-0.094	0.011	0.001	-0.746	0.089
IVARIATE	-0.064	0.008	0.001	-0.484	0.060
VARIATE	-0.075	0.012	0.001	-0.600	0.097
VARIATE	-0.048	0.008	0.001	-0.359	0.064
LTIVARIATE	-0.061	0.013	0.001	-0.483	0.107
LTIVARIATE	-0.035	0.009	0.001	-0.265	0.068

TABLE 22-12   DATE=17NOV85
AVERAGE ANNUAL INCIDENCE RATE FOR EVENT PER 1000 PERSONS AT RISK AT EXAM
BY AGE AND LEVEL OF RISK FACTOR AT EXAM: FRAMINGHAM STUDY 30-YEAR FOLLOWUP

SEX: MALE
EVENT: DEATH FROM CARDIOVASCULAR DISEASE
RISK FACTOR: HEART RATE (PER MINUTE)
PERSONS AT RISK: ALL PERSONS ALIVE

RISK FACTOR	35-44 AT EXAM PERSON EXAMS	# OF EVENTS	ANNUAL RATE	45-54 AT EXAM PERSON EXAMS	# OF EVENTS	ANNUAL RATE	55-64 AT EXAM PERSON EXAMS	# OF EVENTS	ANNUAL RATE	65-74 AT EXAM PERSON EXAMS	# OF EVENTS	ANNUAL RATE
30-67	1295	1	0	2016	9	2	2618	32	6	1425	35	12
68-75	1506	0	0	2594	13	3	2763	41	7	1447	43	15
76-83	728	1	1	1473	13	4	1552	28	9	802	33	21
84-91	586	5	4	1145	12	5	1211	26	11	607	30	25
92-220	404	1	1	839	16	10	926	42	23	553	24	22
UNKNOWN	0	0	-	0	0	-	0	0	-	0	0	-
TOTAL	4519	8	1	8067	63	4	9070	169	9	4834	165	17

RISK FACTOR	75-84 AT EXAM PERSON EXAMS	# OF EVENTS	ANNUAL RATE	85-94 AT EXAM PERSON EXAMS	# OF EVENTS	ANNUAL RATE	35-64 COMBINED # OF EVENTS	CRUDE	AGE-ADJUSTED	65-94 COMBINED # OF EVENTS	CRUDE	AGE-ADJUSTED
30-67	388	26	34	17	0	0	42	4	3	61	17	17
68-75	362	28	39	14	2	71	54	4	4	73	20	20
76-83	271	18	33	13	1	38	42	6	6	52	24	23
84-91	174	16	46	5	2	200	43	7	7	48	31	31
92-220	150	16	53	1	1	500	59	14	13	41	29	32
UNKNOWN	0	0	-	0	0	-	0	6	-	0	-	
TOTAL	1345	104	39	50	6	60	240	6	6	275	22	22

LOGISTIC REGRESSIONS

AGE AT EXAM		COEFFICIENT OF RISK FACTOR	SE COEFFICIENT	P-VALUE	STANDARDIZED COEFFICIENT	STD SE
35-44		0.047	0.022	0.035	-0.582	0.276
45-54		0.039	0.009	0.001	0.483	0.111
55-64		0.028	0.003	0.001	0.369	0.069
65-74		0.015	0.006	0.005	0.206	0.074
75-84		0.010	0.007	0.184	0.131	0.099
85-94		0.114	0.052	0.027	1.264	0.572
35-64	UNIVARIATE	0.032	0.004	0.001	0.407	0.057
65-94	UNIVARIATE	0.014	0.004	0.001	0.189	0.058
35-64	BIVARIATE	0.031	0.004	0.001	0.401	0.056
65-94	BIVARIATE	0.014	0.004	0.001	0.186	0.058
35-64	MULTIVARIATE	0.023	0.005	0.001	0.288	0.059
65-94	MULTIVARIATE	0.011	0.004	0.014	0.147	0.060

SEX: FEMALE
EVENT: DEATH FROM CARDIOVASCULAR DISEASE
RISK FACTOR: HEART RATE (PER MINUTE)
PERSONS AT RISK: ALL PERSONS ALIVE

RISK FACTOR	35-44 AT EXAM PERSON EXAMS	# OF EVENTS	ANNUAL RATE	45-54 AT EXAM PERSON EXAMS	# OF EVENTS	ANNUAL RATE	55-64 AT EXAM PERSON EXAMS	# OF EVENTS	ANNUAL RATE	65-74 AT EXAM PERSON EXAMS	# OF EVENTS	ANNUAL RATE
30-67	1092	0	0	1918	3	1	2450	8	2	1285	18	7
68-75	1806	1	0	3279	6	1	3641	18	2	1907	42	11
76-83	1051	1	0	1938	6	1	2442	17	3	1542	28	9
84-91	939	1	1	1620	8	2	1888	21	6	1196	23	10
92-220	731	2	1	1317	5	2	1682	11	3	1173	31	13
UNKNOWN	0	0	-	0	0	-	0	0	-	0	0	-
TOTAL	5619	5	0	10072	26	1	12103	75	3	7103	142	10

RISK FACTOR	75-84 AT EXAM PERSON EXAMS	# OF EVENTS	ANNUAL RATE	85-94 AT EXAM PERSON EXAMS	# OF EVENTS	ANNUAL RATE	35-64 COMBINED # OF EVENTS	CRUDE	AGE-ADJUSTED	65-94 COMBINED # OF EVENTS	CRUDE	AGE-ADJUSTED
30-67	459	14	15	36	4	56	11	1	1	36	10	10
68-75	586	32	27	41	3	37	25	1	1	77	15	16
76-83	532	24	23	30	2	33	22	2	2	54	13	13
84-91	410	20	24	29	3	52	30	2	2	46	14	14
92-220	432	25	29	27	4	74	18	2	2	60	18	18
UNKNOWN	0	0	-	0	0	-	0			0	-	
TOTAL	2419	115	24	163	16	49	106	2	2	273	14	14

LOGISTIC REGRESSIONS

AGE AT EXAM		COEFFICIENT OF RISK FACTOR	SE COEFFICIENT	P-VALUE	STANDARDIZED COEFFICIENT	STD SE
35-44		0.033	0.021	0.110	0.414	0.259
45-54		0.025	0.014	0.080	0.307	0.175
55-64		0.021	0.008	0.007	0.273	0.101
63-74		0.013	0.006	0.027	0.174	0.079
75-84		0.007	0.007	0.259	0.103	0.092
85-94		0.006	0.018	0.744	0.085	0.260
35-64	UNIVARIATE	0.023	0.007	0.001	0.292	0.083
65-94	UNIVARIATE	0.011	0.004	0.013	0.144	0.058
35-64	BIVARIATE	0.023	0.007	0.001	0.285	0.083
65-94	BIVARIATE	0.010	0.004	0.019	0.136	0.058
35-64	MULTIVARIATE	0.011	0.007	0.122	0.145	0.094
65-94	MULTIVARIATE	0.006	0.004	0.159	0.083	0.059

TABLE 22-13 DATE=17NOV85
AVERAGE ANNUAL INCIDENCE RATE FOR EVENT PER 1000 PERSONS AT RISK AT EXAM
BY AGE AND LEVEL OF RISK FACTOR AT EXAM: FRAMINGHAM STUDY 30-YEAR FOLLOWUP

SEX: MALE
EVENT: DEATH FROM CARDIOVASCULAR DISEASE
RISK FACTOR: CIGARETTES SMOKED (NUMBER PER DAY)
PERSONS AT RISK: ALL PERSONS ALIVE

RISK FACTOR	35-44 AT EXAM			45-54 AT EXAM			55-64 AT EXAM			65-74 AT EXAM		
	PERSON EXAMS	# OF EVENTS	ANNUAL RATE	PERSON EXAMS	# OF EVENTS	ANNUAL RATE	PERSON EXAMS	# OF EVENTS	ANNUAL RATE	PERSON EXAMS	# OF EVENTS	ANNUAL RATE
NONE	1402	0	0	3220	14	2	4888	75	8	3235	101	16
1-10	437	1	1	746	6	4	856	16	9	463	17	18
11-20	1499	2	1	2136	27	6	1889	50	13	668	31	23
21-40	993	4	2	1672	12	4	1235	23	9	427	13	15
41-90	162	1	3	255	3	6	178	5	14	34	2	29
UNKNOWN	26	0	0	38	1	13	24	0	0	7	1	71
TOTAL	4519	8	1	8067	63	4	9070	169	9	4834	165	17

RISK FACTOR	75-84 AT EXAM			85-94 AT EXAM			35-64 COMBINED			65-94 COMBINED		
	PERSON EXAMS	# OF EVENTS	ANNUAL RATE	PERSON EXAMS	# OF EVENTS	ANNUAL RATE	# OF EVENTS	CRUDE	AGE-ADJUSTED	# OF EVENTS	CRUDE	AGE-ADJUSTED
NONE	1067	85	40	38	4	53	89	5	4	190	22	21
1-10	125	10	40	6	0	0	23	6	6	27	23	23
11-20	88	5	28	5	1	100	79	7	8	37	24	25
21-40	60	4	33	1	1	500	39	5	6	18	18	23
41-90	5	0	0	0	0	-	9	8	9	2	26	23
UNKNOWN	0	0	-	0	0	-	1	6	5	1	71	55
TOTAL	1345	104	39	50	6	60	240	6	6	275	22	22

LOGISTIC REGRESSIONS

AGE AT EXAM	COEFFICIENT OF RISK FACTOR	SE COEFFICIENT	P-VALUE	STANDARDIZED COEFFICIENT	STD SE
35-44	0.047	0.022	0.033	0.672	0.316
45-54	0.015	0.008	0.058	0.223	0.118
55-64	0.010	0.005	0.049	0.141	0.071
65-74	0.005	0.007	0.454	0.057	0.076
75-84	-0.009	0.013	0.481	-0.080	0.113
85-94	0.076	0.044	0.082	0.388	0.338
35-64 UNIVARIATE	0.006	0.004	0.173	0.085	0.063
65-94 UNIVARIATE	-0.002	0.006	0.711	-0.023	0.063
35-64 BIVARIATE	0.015	0.004	0.001	0.216	0.061
65-94 BIVARIATE	0.005	0.006	0.425	0.051	0.063
35-64 MULTIVARIATE	0.017	0.004	0.001	0.253	0.062
65-94 MULTIVARIATE	0.010	0.006	0.087	0.110	0.064

SEX: FEMALE
EVENT: DEATH FROM CARDIOVASCULAR DISEASE
RISK FACTOR: CIGARETTES SMOKED (NUMBER PER DAY)
PERSONS AT RISK: ALL PERSONS ALIVE

RISK FACTOR	35-44 AT EXAM			45-54 AT EXAM			55-64 AT EXAM			65-74 AT EXAM		
	PERSON EXAMS	# OF EVENTS	ANNUAL RATE	PERSON EXAMS	# OF EVENTS	ANNUAL RATE	PERSON EXAMS	# OF EVENTS	ANNUAL RATE	PERSON EXAMS	# OF EVENTS	ANNUAL RATE
NONE	2583	0	0	5599	10	1	8010	45	3	5562	95	9
1-10	1314	1	0	1732	5	1	1646	9	3	680	20	15
11-20	1373	1	0	2037	6	1	1767	19	5	650	22	17
21-40	298	1	2	629	3	2	624	2	2	191	4	10
41-90	22	1	23	38	1	13	18	0	0	1	0	0
UNKNOWN	29	1	17	37	1	14	38	0	0	19	1	26
TOTAL	5619	5	0	10072	26	1	12103	75	3	7103	142	10

RISK FACTOR	75-84 AT EXAM			85-94 AT EXAM			35-64 COMBINED			65-94 COMBINED		
	PERSON EXAMS	# OF EVENTS	ANNUAL RATE	PERSON EXAMS	# OF EVENTS	ANNUAL RATE	# OF EVENTS	CRUDE	AGE-ADJUSTED	# OF EVENTS	CRUDE	AGE-ADJUSTED
NONE	2139	100	23	154	16	52	55	2	2	211	13	13
1-10	161	6	19	7	0	0	15	2	2	26	15	15
11-20	99	9	45	2	0	0	26	3	3	31	21	24
21-40	16	0	0	0	0	-	6	2	2	4	10	8
41-90	0	0	-	0	0	-	2	13	9	0	0	0
UNKNOWN	4	0	0	0	0	-	2	10	8	1	22	19
TOTAL	2419	115	24	163	16	49	106	2	2	273	14	14

LOGISTIC REGRESSIONS

AGE AT EXAM	COEFFICIENT OF RISK FACTOR	SE COEFFICIENT	P-VALUE	STANDARDIZED COEFFICIENT	STD SE
35-44	0.126	0.033	0.001	1.241	0.327
45-54	0.032	0.016	0.044	0.326	0.161
55-64	0.010	0.011	0.359	0.097	0.106
65-74	0.023	0.009	0.009	0.178	0.068
75-84	0.017	0.017	0.313	0.083	0.082
85-94	-	-	-	-	-
35-64 UNIVARIATE	0.018	0.009	0.042	0.175	0.086
65-94 UNIVARIATE	0.011	0.008	0.149	0.079	0.055
35-64 BIVARIATE	0.025	0.008	0.003	0.249	0.084
65-94 BIVARIATE	0.024	0.008	0.002	0.171	0.056
35-64 MULTIVARIATE	0.027	0.009	0.002	0.268	0.086
65-94 MULTIVARIATE	0.029	0.008	0.001	0.204	0.056

TABLE 22-14   DATE=17NOV85

AVERAGE ANNUAL INCIDENCE RATE FOR EVENT PER 1000 PERSONS AT RISK AT EXAM
BY AGE AND LEVEL OF RISK FACTOR AT EXAM: FRAMINGHAM STUDY 30-YEAR FOLLOWUP

SEX: MALE
EVENT: DEATH FROM CARDIOVASCULAR DISEASE
RISK FACTOR: ALBUMIN IN URINE, DEFINITE
PERSONS AT RISK: ALL PERSONS ALIVE

RISK FACTOR	35-44 AT EXAM			45-54 AT EXAM			55-64 AT EXAM			65-74 AT EXAM		
	PERSON EXAMS	# OF EVENTS	ANNUAL RATE	PERSON EXAMS	# OF EVENTS	ANNUAL RATE	PERSON EXAMS	# OF EVENTS	ANNUAL RATE	PERSON EXAMS	# OF EVENTS	ANNUAL RATE
NO	4411	8	1	7936	58	4	8828	158	9	4671	153	16
YES	71	0	0	105	5	24	215	10	23	157	12	38
UNKNOWN	37	0	0	26	0	0	27	1	19	6	0	0
TOTAL	4519	8	1	8067	63	4	9070	169	9	4834	165	17

RISK FACTOR	75-84 AT EXAM			85-94 AT EXAM			35-64 COMBINED			65-94 COMBINED		
	PERSON EXAMS	# OF EVENTS	ANNUAL RATE	PERSON EXAMS	# OF EVENTS	ANNUAL RATE	# OF EVENTS	CRUDE	AGE-ADJUSTED	# OF EVENTS	CRUDE	AGE-ADJUSTED
NO	1308	101	39	43	5	58	224	5	5	259	22	22
YES	34	3	44	7	1	71	15	19	19	16	40	40
UNKNOWN	3	0	0	0	0	-	1	6	8	0	0	0
TOTAL	1345	104	39	50	6	60	240	6	6	275	22	22

LOGISTIC REGRESSIONS

AGE AT EXAM		COEFFICIENT OF RISK FACTOR	SE COEFFICIENT	P-VALUE	STANDARDIZED COEFFICIENT	STD SE
35-44		-	-	-	-	-
45-54		1.916	0.477	0.001	0.217	0.054
55-64		0.985	0.334	0.003	0.150	0.051
65-74		0.894	0.311	0.004	0.159	0.055
75-84		0.145	0.613	0.813	0.023	0.096
85-94		0.236	1.180	0.841	0.083	0.414
35-64	UNIVARIATE	1.317	0.272	0.001	0.176	0.036
65-94	UNIVARIATE	0.671	0.268	0.012	0.118	0.047
35-64	BIVARIATE	1.130	0.274	0.001	0.151	0.037
65-94	BIVARIATE	0.638	0.272	0.019	0.112	0.048
35-64	MULTIVARIATE	0.591	0.299	0.048	0.079	0.040
65-94	MULTIVARIATE	0.323	0.280	0.249	0.057	0.049

SEX: FEMALE
EVENT: DEATH FROM CARDIOVASCULAR DISEASE
RISK FACTOR: ALBUMIN IN URINE, DEFINITE
PERSONS AT RISK: ALL PERSONS ALIVE

RISK FACTOR	35-44 AT EXAM			45-54 AT EXAM			55-64 AT EXAM			65-74 AT EXAM		
	PERSON EXAMS	# OF EVENTS	ANNUAL RATE	PERSON EXAMS	# OF EVENTS	ANNUAL RATE	PERSON EXAMS	# OF EVENTS	ANNUAL RATE	PERSON EXAMS	# OF EVENTS	ANNUAL RATE
NO	5356	5	0	9766	26	1	11835	69	3	6955	136	10
YES	225	0	0	281	0	0	240	6	13	138	6	22
UNKNOWN	38	0	0	25	0	0	28	0	0	10	0	0
TOTAL	5619	5	0	10072	26	1	12103	75	3	7103	142	10

RISK FACTOR	75-84 AT EXAM			85-94 AT EXAM			35-64 COMBINED			65-94 COMBINED		
	PERSON EXAMS	# OF EVENTS	ANNUAL RATE	PERSON EXAMS	# OF EVENTS	ANNUAL RATE	# OF EVENTS	CRUDE	AGE-ADJUSTED	# OF EVENTS	CRUDE	AGE-ADJUSTED
NO	2360	111	24	156	14	45	100	2	2	261	14	14
YES	50	4	40	7	2	143	6	4	5	12	31	28
UNKNOWN	9	0	0	0	0	-	0	0	0	0	0	0
TOTAL	2419	115	24	163	16	49	106	2	2	273	14	14

LOGISTIC REGRESSIONS

AGE AT EXAM		COEFFICIENT OF RISK FACTOR	SE COEFFICIENT	P-VALUE	STANDARDIZED COEFFICIENT	STD SE
35-44		-	-	-	-	-
45-54		-	-	-	-	-
55-64		1.475	0.431	0.001	0.206	0.060
65-74		0.824	0.426	0.053	0.114	0.059
75-84		0.566	0.530	0.285	0.081	0.076
85-94		1.400	0.882	0.112	0.285	0.179
35-64	UNIVARIATE	0.778	0.422	0.065	0.126	0.068
65-94	UNIVARIATE	0.839	0.305	0.006	0.118	0.043
35-64	BIVARIATE	0.982	0.424	0.020	0.159	0.069
65-94	BIVARIATE	0.762	0.309	0.014	0.107	0.043
35-64	MULTIVARIATE	0.222	0.478	0.642	0.036	0.077
65-94	MULTIVARIATE	0.300	0.322	0.351	0.042	0.045

TABLE 22-15   DATE=17NOV85
AVERAGE ANNUAL INCIDENCE RATE FOR EVENT PER 1000 PERSONS AT RISK AT EXAM
BY AGE AND LEVEL OF RISK FACTOR AT EXAM: FRAMINGHAM STUDY 30-YEAR FOLLOWUP

SEX: MALE
EVENT: DEATH FROM CARDIOVASCULAR DISEASE
RISK FACTOR: HEART ENLARGEMENT BY X-RAY
PERSONS AT RISK: ALL PERSONS ALIVE

RISK FACTOR	35-44 AT EXAM PERSON EXAMS	# OF EVENTS	ANNUAL RATE	45-54 AT EXAM PERSON EXAMS	# OF EVENTS	ANNUAL RATE	55-64 AT EXAM PERSON EXAMS	# OF EVENTS	ANNUAL RATE	65-74 AT EXAM PERSON EXAMS	# OF EVENTS	ANNUAL RATE
NO	4101	6	1	6613	33	2	6543	91	7	3075	71	12
MILD	189	0	0	708	5	4	1126	21	9	703	22	16
DEFINITE	224	2	4	742	25	17	1400	57	20	1056	72	34
UNKNOWN	5	0	0	4	0	0	1	0	0	0	0	-
TOTAL	4519	8	1	8067	63	4	9070	169	9	4834	165	17

RISK FACTOR	75-84 AT EXAM PERSON EXAMS	# OF EVENTS	ANNUAL RATE	85-94 AT EXAM PERSON EXAMS	# OF EVENTS	ANNUAL RATE	35-64 COMBINED # OF EVENTS	CRUDE	AGE-ADJUSTED	65-94 COMBINED # OF EVENTS	CRUDE	AGE-ADJUSTED
NO	753	47	31	34	2	29	130	4	4	120	16	16
MILD	226	19	42	6	2	167	26	6	5	43	23	23
DEFINITE	366	38	52	10	2	100	84	18	16	112	39	35
UNKNOWN	0	0	-	0	0	-	0	0	0	0	0	-
TOTAL	1345	104	39	50	6	60	240	6	6	275	22	22

LOGISTIC REGRESSIONS

AGE AT EXAM	COEFFICIENT OF RISK FACTOR	SE COEFFICIENT	P-VALUE	STANDARDIZED COEFFICIENT	STD SE
35-44	0.850	0.432	0.049	0.399	0.203
45-54	0.959	0.138	0.001	0.593	0.085
55-64	0.542	0.087	0.001	0.403	-0.065
65-74	0.564	0.087	0.001	0.466	0.072
75-84	0.278	0.114	0.014	0.241	0.098
85-94	0.721	0.486	0.138	0.587	0.395
35-64 UNIVARIATE	0.782	0.072	0.001	0.515	0.047
65-94 UNIVARIATE	0.486	0.068	0.001	0.406	0.057
35-64 BIVARIATE	0.637	0.073	0.001	0.419	0.048
65-94 BIVARIATE	0.453	0.069	0.001	0.379	0.057
35-64 MULTIVARIATE	0.511	0.080	0.001	0.337	0.053
65-94 MULTIVARIATE	0.316	0.074	0.001	0.264	0.062

SEX: FEMALE
EVENT: DEATH FROM CARDIOVASCULAR DISEASE
RISK FACTOR: HEART ENLARGEMENT BY X-RAY
PERSONS AT RISK: ALL PERSONS ALIVE

RISK FACTOR	35-44 AT EXAM PERSON EXAMS	# OF EVENTS	ANNUAL RATE	45-54 AT EXAM PERSON EXAMS	# OF EVENTS	ANNUAL RATE	55-64 AT EXAM PERSON EXAMS	# OF EVENTS	ANNUAL RATE	65-74 AT EXAM PERSON EXAMS	# OF EVENTS	ANNUAL RATE
NO	5166	2	0	8085	11	1	7836	22	1	3511	45	6
MILD	199	1	3	959	4	2	1879	14	4	1295	19	7
DEFINITE	246	2	4	1022	11	5	2384	39	8	2299	78	17
UNKNOWN	8	0	0	6	0	0	4	0	0	0	0	-
TOTAL	5619	5	0	10072	26	1	12103	75	3	7103	142	10

RISK FACTOR	75-84 AT EXAM PERSON EXAMS	# OF EVENTS	ANNUAL RATE	85-94 AT EXAM PERSON EXAMS	# OF EVENTS	ANNUAL RATE	35-64 COMBINED # OF EVENTS	CRUDE	AGE-ADJUSTED	65-94 COMBINED # OF EVENTS	CRUDE	AGE-ADJUSTED
NO	830	22	13	40	4	50	35	1	1	71	8	9
MILD	455	17	19	36	4	56	19	3	3	40	11	11
DEFINITE	1134	76	34	87	8	46	52	7	6	162	23	22
UNKNOWN	0	0	-	0	0	-	0	0	0	0	0	-
TOTAL	2419	115	24	163	16	49	106	2	2	273	14	14

LOGISTIC REGRESSIONS

AGE AT EXAM	COEFFICIENT OF RISK FACTOR	SE COEFFICIENT	P-VALUE	STANDARDIZED COEFFICIENT	STD SE
35-44	1.527	0.468	0.001	0.675	0.207
45-54	1.039	0.213	0.001	0.667	0.136
55-64	0.883	0.132	0.001	0.708	0.106
65-74	0.513	0.097	0.001	0.456	0.086
75-84	0.500	0.121	0.001	0.446	0.108
85-94	-0.060	0.312	0.847	-0.051	0.261
35-64 UNIVARIATE	1.073	0.107	0.001	0.756	0.075
65-94 UNIVARIATE	0.552	0.072	0.001	0.496	0.065
35-64 BIVARIATE	0.921	0.112	0.001	0.649	0.079
65-94 BIVARIATE	0.459	0.074	0.001	0.412	0.066
35-64 MULTIVARIATE	0.813	0.122	0.001	0.573	0.086
65-94 MULTIVARIATE	0.315	0.078	0.001	0.283	0.070

TABLE 21-17   DATE=12DEC85
AVERAGE ANNUAL INCIDENCE RATE FOR EVENT PER 1000 PERSONS AT RISK AT EXAM
BY AGE AND LEVEL OF RISK FACTOR AT EXAM: FRAMINGHAM STUDY 30-YEAR FOLLOWUP

SEX: MALE
EVENT: DEATH FROM CARDIOVASCULAR DISEASE
RISK FACTOR: INTRAVENTRICULAR CONDUCTION DEFECT (WITH QRS INTERVAL MORE THAN 0.11 SECOND)
PERSONS AT RISK: ALL PERSONS ALIVE

RISK FACTOR	35-44 AT EXAM PERSON EXAMS	# OF EVENTS	ANNUAL RATE	45-54 AT EXAM PERSON EXAMS	# OF EVENTS	ANNUAL RATE	55-64 AT EXAM PERSON EXAMS	# OF EVENTS	ANNUAL RATE	65-74 AT EXAM PERSON EXAMS	# OF EVENTS	ANNUAL RATE
NO	4201	8	1	7487	53	4	8247	141	9	4323	137	16
INCOMPLETE	280	0	0	491	4	4	541	12	11	250	9	18
COMPLETE	38	0	0	89	6	34	282	16	28	261	19	36
UNKNOWN	0	0	-	0	0	-	0	0	-	0	0	-
TOTAL	4519	8	1	8067	63	4	9070	169	9	4834	165	17

RISK FACTOR	75-84 AT EXAM PERSON EXAMS	# OF EVENTS	ANNUAL RATE	85-94 AT EXAM PERSON EXAMS	# OF EVENTS	ANNUAL RATE	35-64 COMBINED # OF EVENTS	CRUDE	AGE-ADJUSTED	65-94 COMBINED # OF EVENTS	CRUDE	AGE-ADJUSTED
NO	1180	86	36	41	4	49	202	5	5	227	20	21
INCOMPLETE	37	4	54	3	0	0	16	6	6	13	22	26
COMPLETE	128	14	55	6	2	167	22	27	24	35	44	41
UNKNOWN	0	0	-	0	0	-	0	-	-	0	-	-
TOTAL	1345	104	39	50	6	60	240	6	6	275	22	22

LOGISTIC REGRESSIONS

AGE AT EXAM		COEFFICIENT OF RISK FACTOR	SE COEFFICIENT	P-VALUE	STANDARDIZED COEFFICIENT	STD SE
35-44		-	-	-	-	-
45-54		0.937	0.230	0.001	0.294	0.072
55-64		0.561	0.131	0.001	0.231	0.054
65-74		0.405	0.124	0.001	0.199	0.061
75-84		0.234	0.148	0.113	0.140	0.089
85-94		0.703	0.519	0.175	0.477	0.352
35-64	UNIVARIATE	0.727	0.115	0.001	0.259	0.041
65-94	UNIVARIATE	0.384	0.093	0.001	0.200	0.049
35-64	BIVARIATE	0.586	0.114	0.001	0.208	0.040
65-94	BIVARIATE	0.323	0.094	0.001	0.168	0.049
35-64	MULTIVARIATE	0.558	0.120	0.001	0.199	0.043
65-94	MULTIVARIATE	0.324	0.096	0.001	0.169	0.050

SEX: FEMALE
EVENT: DEATH FROM CARDIOVASCULAR DISEASE
RISK FACTOR: INTRAVENTRICULAR CONDUCTION DEFECT (WITH QRS INTERVAL MORE THAN 0.11 SECOND)
PERSONS AT RISK: ALL PERSONS ALIVE

RISK FACTOR	35-44 AT EXAM PERSON EXAMS	# OF EVENTS	ANNUAL RATE	45-54 AT EXAM PERSON EXAMS	# OF EVENTS	ANNUAL RATE	55-64 AT EXAM PERSON EXAMS	# OF EVENTS	ANNUAL RATE	65-74 AT EXAM PERSON EXAMS	# OF EVENTS	ANNUAL RATE
NO	5532	5	0	9873	26	1	11731	68	3	6794	130	10
INCOMPLETE	75	0	0	151	0	0	195	2	5	134	0	0
COMPLETE	12	0	0	48	0	0	177	5	14	175	12	34
UNKNOWN	0	0	-	0	0	-	0	0	-	0	0	-
TOTAL	5619	5	0	10072	26	1	12103	75	3	7103	142	10

RISK FACTOR	75-84 AT EXAM PERSON EXAMS	# OF EVENTS	ANNUAL RATE	85-94 AT EXAM PERSON EXAMS	# OF EVENTS	ANNUAL RATE	35-64 COMBINED # OF EVENTS	CRUDE	AGE-ADJUSTED	65-94 COMBINED # OF EVENTS	CRUDE	AGE-ADJUSTED
NO	2261	98	22	153	14	46	99	2	2	242	13	13
INCOMPLETE	51	3	29	3	0	0	5	2	2	3	8	7
COMPLETE	107	14	65	7	2	143	5	11	6	28	48	44
UNKNOWN	0	0	-	0	0	-	0	-	-	0	-	-
TOTAL	2419	115	24	163	16	49	106	2	2	273	14	14

LOGISTIC REGRESSIONS

AGE AT EXAM		COEFFICIENT OF RISK FACTOR	SE COEFFICIENT	P-VALUE	STANDARDIZED COEFFICIENT	STD SE
35-44		-	-	-	-	-
45-54		-	-	-	-	-
55-64		0.782	0.228	0.001	0.211	0.062
65-74		0.565	0.163	0.001	0.190	0.055
75-84		0.585	0.150	0.001	0.253	0.065
85-94		0.614	0.446	0.168	0.261	0.189
35-64	UNIVARIATE	0.812	0.230	0.001	0.178	0.050
65-94	UNIVARIATE	0.634	0.106	0.001	0.231	0.039
35-64	BIVARIATE	0.639	0.229	0.005	0.140	0.050
65-94	BIVARIATE	0.566	0.107	0.001	0.206	0.039
35-64	MULTIVARIATE	0.658	0.239	0.006	0.144	0.053
65-94	MULTIVARIATE	0.553	0.110	0.001	0.201	0.040

TABLE 22-16   DATE=17NOV85

AVERAGE ANNUAL INCIDENCE RATE FOR EVENT PER 1000 PERSONS AT RISK AT EXAM
BY AGE AND LEVEL OF RISK FACTOR AT EXAM: FRAMINGHAM STUDY 30-YEAR FOLLOWUP

SEX: MALE
EVENT: DEATH FROM CARDIOVASCULAR DISEASE
RISK FACTOR: LEFT VENTRICULAR HYPERTROPHY BY ELECTROCARDIOGRAM
PERSONS AT RISK: ALL PERSONS ALIVE

RISK FACTOR	35-44 AT EXAM PERSON EXAMS	# OF EVENTS	ANNUAL RATE	45-54 AT EXAM PERSON EXAMS	# OF EVENTS	ANNUAL RATE	55-64 AT EXAM PERSON EXAMS	# OF EVENTS	ANNUAL RATE	65-74 AT EXAM PERSON EXAMS	# OF EVENTS	ANNUAL RATE
NO	4408	7	1	7812	45	3	8611	139	8	4527	122	13
MILD	79	0	0	147	4	14	240	6	13	121	10	41
DEFINITE	32	1	16	108	14	65	219	24	55	186	33	89
UNKNOWN	0	0	-	0	0	-	0	0	-	0	0	-
TOTAL	4519	8	1	8067	63	4	9070	169	9	4834	165	17

RISK FACTOR	75-84 AT EXAM PERSON EXAMS	# OF EVENTS	ANNUAL RATE	85-94 AT EXAM PERSON EXAMS	# OF EVENTS	ANNUAL RATE	35-64 COMBINED # OF EVENTS	CRUDE	AGE-ADJUSTED	65-94 COMBINED # OF EVENTS	CRUDE	AGE-ADJUSTED
NO	1250	94	38	45	5	56	191	5	5	221	19	19
MILD	44	2	23	2	1	250	10	11	10	13	39	39
DEFINITE	51	8	78	3	0	0	39	54	50	41	85	86
UNKNOWN	0	0	-	0	0	-	0	-		0	-	
TOTAL	1345	104	39	50	6	60	240	6	6	275	22	22

LOGISTIC REGRESSIONS

AGE AT EXAM	COEFFICIENT OF RISK FACTOR	SE COEFFICIENT	P-VALUE	STANDARDIZED COEFFICIENT	STD SE
35-44	1.381	0.581	0.017	0.292	0.123
45-54	1.621	0.159	0.001	0.428	0.042
55-64	0.968	0.116	0.001	0.332	0.040
65-74	1.035	0.104	0.001	0.425	0.043
75-84	0.326	0.200	0.104	0.135	0.083
85-94	0.029	0.848	0.972	0.015	0.432
35-64 UNIVARIATE	1.261	0.092	0.001	0.567	0.027
65-94 UNIVARIATE	0.822	0.090	0.001	0.339	0.037
35-64 BIVARIATE	1.128	0.093	0.001	0.329	0.027
65-94 BIVARIATE	0.829	0.092	0.001	0.342	0.038
35-64 MULTIVARIATE	1.018	0.106	0.001	0.297	0.031
65-94 MULTIVARIATE	0.684	0.100	0.001	0.282	0.041

SEX: FEMALE
EVENT: DEATH FROM CARDIOVASCULAR DISEASE
RISK FACTOR: LEFT VENTRICULAR HYPERTROPHY BY ELECTROCARDIOGRAM
PERSONS AT RISK: ALL PERSONS ALIVE

RISK FACTOR	35-44 AT EXAM PERSON EXAMS	# OF EVENTS	ANNUAL RATE	45-54 AT EXAM PERSON EXAMS	# OF EVENTS	ANNUAL RATE	55-64 AT EXAM PERSON EXAMS	# OF EVENTS	ANNUAL RATE	65-74 AT EXAM PERSON EXAMS	# OF EVENTS	ANNUAL RATE
NO	5566	3	0	9905	23	1	11679	58	2	6686	106	8
MILD	38	0	0	111	0	0	250	5	10	182	11	30
DEFINITE	15	2	67	56	3	27	174	12	34	235	25	53
UNKNOWN	0	0	-	0	0	-	0	0	-	0	0	-
TOTAL	5619	5	0	10072	26	1	12103	75	3	7103	142	10

RISK FACTOR	75-84 AT EXAM PERSON EXAMS	# OF EVENTS	ANNUAL RATE	85-94 AT EXAM PERSON EXAMS	# OF EVENTS	ANNUAL RATE	35-64 COMBINED # OF EVENTS	CRUDE	AGE-ADJUSTED	65-94 COMBINED # OF EVENTS	CRUDE	AGE-ADJUSTED
NO	2217	89	20	142	13	46	84	2	2	208	11	12
MILD	77	8	52	10	1	50	5	6	4	20	37	36
DEFINITE	125	18	72	11	2	91	17	35	38	45	61	59
UNKNOWN	0	0	-	0	0	-	0	-		0	-	
TOTAL	2419	115	24	163	16	49	106	2	2	273	14	14

LOGISTIC REGRESSIONS

AGE AT EXAM	COEFFICIENT OF RISK FACTOR	SE COEFFICIENT	P-VALUE	STANDARDIZED COEFFICIENT	STD SE
35-44	2.806	0.491	0.001	0.369	0.065
45-54	1.500	0.331	0.001	0.271	0.060
55-64	1.353	0.159	0.001	0.372	0.044
65-74	1.026	0.112	0.001	0.397	0.043
75-84	0.720	0.133	0.001	0.338	0.062
85-94	0.361	0.405	0.373	0.196	0.220
35-64 UNIVARIATE	1.578	0.135	0.001	0.348	0.030
65-94 UNIVARIATE	0.909	0.083	0.001	0.375	0.034
35-64 BIVARIATE	1.418	0.138	0.001	0.313	0.030
65-94 BIVARIATE	0.849	0.085	0.001	0.350	0.035
35-64 MULTIVARIATE	1.258	0.166	0.001	0.277	0.037
65-94 MULTIVARIATE	0.661	0.091	0.001	0.272	0.038

TABLE 23-1   DATE=12JUN85
AVERAGE ANNUAL INCIDENCE RATE FOR EVENT PER 1000 PERSONS AT RISK AT EXAM
BY AGE AND LEVEL OF RISK FACTOR AT EXAM: FRAMINGHAM STUDY 30-YEAR FOLLOWUP

SEX: MALE
EVENT: DEATH
RISK FACTOR: SYSTOLIC BLOOD PRESSURE, FIRST EXAMINER (MM HG)
PERSONS AT RISK: ALL PERSONS ALIVE

RISK FACTOR	35-44 AT EXAM			45-54 AT EXAM			55-64 AT EXAM			65-74 AT EXAM		
	PERSON EXAMS	# OF EVENTS	ANNUAL RATE	PERSON EXAMS	# OF EVENTS	ANNUAL RATE	PERSON EXAMS	# OF EVENTS	ANNUAL RATE	PERSON EXAMS	# OF EVENTS	ANNUAL RATE
74-119	1226	5	2	1842	25	7	1605	39	12	659	37	28
120-139	2193	7	2	3542	50	7	3525	105	15	1636	81	25
140-159	850	3	2	1838	33	9	2371	76	16	1473	109	37
160-179	197	4	10	598	18	15	1062	40	19	696	63	45
180-300	53	1	9	247	10	20	507	36	36	370	44	59
UNKNOWN	0	0	-	0	0	-	0	0	-	0	0	-
TOTAL	4519	20	2	8067	136	8	9070	296	16	4834	334	35

RISK FACTOR	75-84 AT EXAM			85-94 AT EXAM			35-64 COMBINED			65-94 COMBINED		
	PERSON EXAMS	# OF EVENTS	ANNUAL RATE	PERSON EXAMS	# OF EVENTS	ANNUAL RATE	# OF EVENTS	CRUDE	AGE-ADJUSTED	# OF EVENTS	CRUDE	AGE-ADJUSTED
74-119	144	29	101	8	3	188	69	7	8	69	43	45
120-139	451	58	64	17	3	68	162	9	9	142	34	34
140-159	411	68	83	17	5	147	112	11	10	182	48	48
160-179	218	28	64	6	1	417	62	17	16	96	52	52
180-300	121	33	136	2	1	250	47	29	24	78	79	77
UNKNOWN	0	0	-	0	0	-	0	-	-	0	-	-
TOTAL	1345	216	80	50	17	170	452	10	10	567	46	46

LOGISTIC REGRESSIONS

AGE AT EXAM		COEFFICIENT OF RISK FACTOR	SE COEFFICIENT	P-VALUE	STANDARDIZED COEFFICIENT	STD SE
35-44		0.021	0.011	0.048	0.356	0.180
45-54		0.016	0.004	0.001	0.316	0.074
55-64		0.012	0.002	0.001	0.257	0.053
65-74		0.012	0.002	0.001	0.267	0.052
75-84		0.007	0.003	0.027	0.160	0.072
85-94		0.023	0.017	0.175	0.420	0.310
35-64	UNIVARIATE	0.017	0.002	0.001	0.347	0.040
65-94	UNIVARIATE	0.010	0.002	0.001	0.234	0.041
35-64	BIVARIATE	0.012	0.002	0.001	0.251	0.041
65-94	BIVARIATE	0.010	0.002	0.001	0.236	0.042
35-64	MULTIVARIATE	0.007	0.002	0.002	0.138	0.045
65-94	MULTIVARIATE	0.007	0.002	0.001	0.173	0.044

SEX: FEMALE
EVENT: DEATH
RISK FACTOR: SYSTOLIC BLOOD PRESSURE, FIRST EXAMINER (MM HG)
PERSONS AT RISK: ALL PERSONS ALIVE

RISK FACTOR	35-44 AT EXAM			45-54 AT EXAM			55-64 AT EXAM			65-74 AT EXAM		
	PERSON EXAMS	# OF EVENTS	ANNUAL RATE	PERSON EXAMS	# OF EVENTS	ANNUAL RATE	PERSON EXAMS	# OF EVENTS	ANNUAL RATE	PERSON EXAMS	# OF EVENTS	ANNUAL RATE
74-119	2493	8	2	2724	29	5	2039	27	7	683	19	14
120-139	2245	5	1	4027	25	3	4199	53	6	1918	59	15
140-159	674	4	3	2086	27	6	3197	43	7	2283	86	19
160-179	156	1	3	785	9	6	1682	40	12	1360	50	18
180-300	51	2	20	450	9	10	936	32	16	859	66	38
UNKNOWN	0	0	-	0	0	-	0	0	-	0	0	-
TOTAL	5619	20	2	10072	99	5	12103	195	8	7103	280	20

RISK FACTOR	75-84 AT EXAM			85-94 AT EXAM			35-64 COMBINED			65-94 COMBINED		
	PERSON EXAMS	# OF EVENTS	ANNUAL RATE	PERSON EXAMS	# OF EVENTS	ANNUAL RATE	# OF EVENTS	CRUDE	AGE-ADJUSTED	# OF EVENTS	CRUDE	AGE-ADJUSTED
74-119	158	16	51	10	3	150	64	4	5	38	22	26
120-139	561	54	48	51	11	108	83	4	4	124	25	25
140-159	759	63	42	40	10	125	74	6	6	159	26	27
160-179	561	55	49	36	4	56	50	10	8	109	28	26
180-300	380	42	55	26	5	96	43	14	15	113	45	43
UNKNOWN	0	0	-	0	0	-	0	-	-	0	-	-
TOTAL	2419	230	48	163	33	101	314	6	6	543	28	28

LOGISTIC REGRESSIONS

AGE AT EXAM		COEFFICIENT OF RISK FACTOR	SE COEFFICIENT	P-VALUE	STANDARDIZED COEFFICIENT	STD SE
35-44		0.020	0.010	0.044	0.344	0.170
45-54		0.008	0.004	0.037	0.186	0.089
55-64		0.011	0.003	0.001	0.270	0.063
65-74		0.013	0.002	0.001	0.323	0.056
75-84		0.003	0.003	0.263	0.076	0.068
85-94		-0.005	0.008	0.530	-0.125	0.199
35-64	UNIVARIATE	0.014	0.002	0.001	0.331	0.046
65-94	UNIVARIATE	0.009	0.002	0.001	0.231	0.042
35-64	BIVARIATE	0.009	0.002	0.001	0.224	0.050
65-94	BIVARIATE	0.008	0.002	0.001	0.191	0.042
35-64	MULTIVARIATE	0.006	0.002	0.011	0.136	0.054
65-94	MULTIVARIATE	0.005	0.002	0.003	0.129	0.044

TABLE 23-2   DATE=11JUN95

AVERAGE ANNUAL INCIDENCE RATE FOR EVENT PER 1000 PERSONS AT RISK AT EXAM
BY AGE AND LEVEL OF RISK FACTOR AT EXAM: FRAMINGHAM STUDY 30-YEAR FOLLOWUP

SEX: MALE
EVENT: DEATH
RISK FACTOR: DIASTOLIC BLOOD PRESSURE, FIRST EXAMINER (MM HG)
PERSONS AT RISK: ALL PERSONS ALIVE

RISK FACTOR	35-44 AT EXAM PERSON EXAMS	# OF EVENTS	ANNUAL RATE	45-54 AT EXAM PERSON EXAMS	# OF EVENTS	ANNUAL RATE	55-64 AT EXAM PERSON EXAMS	# OF EVENTS	ANNUAL RATE	65-74 AT EXAM PERSON EXAMS	# OF EVENTS	ANNUAL RATE
20-74	1015	2	1	1543	27	9	2092	70	17	1524	105	34
75-84	1617	7	2	2861	33	6	3165	82	15	1643	98	30
85-94	1203	5	2	2151	39	9	2228	80	18	1072	71	33
95-104	509	4	4	1073	21	10	1114	56	16	625	36	42
105-160	175	2	6	439	16	18	471	28	30	170	24	71
UNKNOWN	0	0	-	0	0	-	0	0	-	0	0	-
TOTAL	4519	20	2	8067	136	8	9070	296	16	4834	334	35

RISK FACTOR	75-84 AT EXAM PERSON EXAMS	# OF EVENTS	ANNUAL RATE	85-94 AT EXAM PERSON EXAMS	# OF EVENTS	ANNUAL RATE	35-64 COMBINED # OF EVENTS	CRUDE	AGE-ADJUSTED	65-94 COMBINED # OF EVENTS	CRUDE	AGE-ADJUSTED
20-74	560	91	81	30	12	200	99	11	11	208	49	45
75-84	440	65	74	15	3	100	122	8	8	166	40	40
85-94	244	34	70	5	2	200	124	11	11	107	40	42
95-104	71	17	120	0	0	-	61	11	11	53	53	59
105-160	30	9	150	0	0	-	46	21	21	33	83	87
UNKNOWN	0	0	-	0	0	-	0	-		0	-	
TOTAL	1345	216	80	50	17	170	452	10	10	567	46	46

LOGISTIC REGRESSIONS

AGE AT EXAM		COEFFICIENT OF RISK FACTOR	SE COEFFICIENT	P-VALUE	STANDARDIZED COEFFICIENT	STD SE
35-44		0.041	0.017	0.017	0.463	0.195
45-54		0.018	0.007	0.011	0.208	0.082
55-64		0.012	0.005	0.009	0.148	0.057
65-74		0.009	0.004	0.051	0.107	0.055
75-84		-0.000	0.006	0.975	-0.002	0.074
85-94		0.009	0.028	0.746	0.098	0.301
35-64	UNIVARIATE	0.016	0.004	0.001	0.185	0.045
65-94	UNIVARIATE	-0.000	0.004	0.943	-0.003	0.044
35-64	BIVARIATE	0.016	0.004	0.001	0.187	0.044
65-94	BIVARIATE	0.008	0.004	0.026	0.100	0.045
35-64	MULTIVARIATE	0.010	0.004	0.010	0.116	0.045
65-94	MULTIVARIATE	0.005	0.004	0.135	0.068	0.045

SEX: FEMALE
EVENT: DEATH
RISK FACTOR: DIASTOLIC BLOOD PRESSURE, FIRST EXAMINER (MM HG)
PERSONS AT RISK: ALL PERSONS ALIVE

RISK FACTOR	35-44 AT EXAM PERSON EXAMS	# OF EVENTS	ANNUAL RATE	45-54 AT EXAM PERSON EXAMS	# OF EVENTS	ANNUAL RATE	55-64 AT EXAM PERSON EXAMS	# OF EVENTS	ANNUAL RATE	65-74 AT EXAM PERSON EXAMS	# OF EVENTS	ANNUAL RATE
20-74	2164	7	2	2657	26	5	3110	46	7	2069	78	19
75-84	2031	2	0	3573	28	4	4005	49	6	2416	76	16
85-94	988	8	4	2395	21	4	3000	51	9	1622	72	22
95-104	324	2	3	961	16	9	1348	30	11	685	28	20
105-160	112	1	4	506	8	8	640	19	15	311	26	42
UNKNOWN	0	0	-	0	0	-	0	0	-	0	0	-
TOTAL	5619	20	2	10072	99	5	12103	195	8	7103	280	20

RISK FACTOR	75-84 AT EXAM PERSON EXAMS	# OF EVENTS	ANNUAL RATE	85-94 AT EXAM PERSON EXAMS	# OF EVENTS	ANNUAL RATE	35-64 COMBINED # OF EVENTS	CRUDE	AGE-ADJUSTED	65-94 COMBINED # OF EVENTS	CRUDE	AGE-ADJUSTED
20-74	916	87	47	73	17	116	79	5	5	182	30	28
75-84	722	59	41	56	12	107	79	4	4	147	23	24
85-94	535	60	56	25	3	60	80	4	4	135	31	31
95-104	178	21	59	7	1	71	48	9	9	50	29	31
105-160	68	3	22	2	0	0	28	11	10	29	38	36
UNKNOWN	0	0	-	0	0	-	0	-		0	-	
TOTAL	2419	230	48	163	33	101	314	6	6	543	28	28

LOGISTIC REGRESSIONS

AGE AT EXAM		COEFFICIENT OF RISK FACTOR	SE COEFFICIENT	P-VALUE	STANDARDIZED COEFFICIENT	STD SE
35-44		0.031	0.017	0.074	0.339	0.190
45-54		0.015	0.008	0.044	0.188	0.093
55-64		0.015	0.005	0.006	0.186	0.068
65-74		0.013	0.005	0.006	0.160	0.059
75-84		-0.000	0.006	0.972	-0.002	0.069
85-94		-0.024	0.015	0.114	-0.313	0.199
35-64	UNIVARIATE	0.020	0.004	0.001	0.245	0.051
65-94	UNIVARIATE	0.001	0.004	0.828	0.010	0.044
35-64	BIVARIATE	0.016	0.004	0.001	0.193	0.053
65-94	BIVARIATE	0.007	0.004	0.064	0.082	0.044
35-64	MULTIVARIATE	0.011	0.004	0.016	0.131	0.054
65-94	MULTIVARIATE	0.006	0.004	0.081	0.077	0.044

TABLE 23-3A   DATE=30AUG85

AVERAGE ANNUAL INCIDENCE RATE FOR EVENT PER 1000 PERSONS AT RISK AT EXAM
BY AGE AND LEVEL OF RISK FACTOR AT EXAM: FRAMINGHAM STUDY 30-YEAR FOLLOWUP

SEX: MALE
EVENT: DEATH
RISK FACTOR: HYPERTENSION WITH ANTIHYPERTENSIVE TREATMENT
PERSONS AT RISK: ALL PERSONS ALIVE

RISK FACTOR	35-44 AT EXAM			45-54 AT EXAM			55-64 AT EXAM			65-74 AT EXAM		
	PERSON EXAMS	# OF EVENTS	ANNUAL RATE	PERSON EXAMS	# OF EVENTS	ANNUAL RATE	PERSON EXAMS	# OF EVENTS	ANNUAL RATE	PERSON EXAMS	# OF EVENTS	ANNUAL RATE
NO	2668	7	1	4263	57	7	4177	99	12	1810	85	23
BORDER	1316	5	2	2287	36	8	2570	76	15	1514	94	31
DEFINITE	511	6	6	1216	30	12	1428	59	21	698	68	49
TREATED	24	2	42	301	13	22	895	62	35	812	87	54
UNKNOWN	0	0	-	0	0	-	0	0	-	0	0	-
TOTAL	4519	20	2	8067	136	8	9070	296	16	4834	334	35

RISK FACTOR	75-84 AT EXAM			85-94 AT EXAM			-64 COMBINED			65-94 COMBINED		
	PERSON EXAMS	# OF EVENTS	ANNUAL RATE	PERSON EXAMS	# OF EVENTS	ANNUAL RATE	# OFSS EVENTS	CRUDE	AGE- ADJUSTED	# OF EVENTS	CRUDE	AGE- ADJUSTED
NO	470	67	71	19	5	132	163	7	8	157	34	34
BORDER	424	66	78	16	6	188	117	9	10	166	42	42
DEFINITE	190	35	92	5	4	400	95	15	15	107	60	61
TREATED	261	48	92	10	2	100	77	32	32	137	63	63
UNKNOWN	0	0	-	0	0	-	0	-	-	0	-	-
TOTAL	1345	216	80	50	17	170	452	10	10	567	46	46

LOGISTIC REGRESSIONS

AGE AT EXAM		COEFFICIENT OF RISK FACTOR	SE COEFFICIENT	P-VALUE	STANDARDIZED COEFFICIENT	STD SE
35-44		0.887	0.280	0.002	0.618	0.195
45-54		0.374	0.105	0.001	0.290	0.081
55-64		0.413	0.071	0.001	0.339	0.058
65-74		0.432	0.070	0.001	0.357	0.058
75-84		0.154	0.090	0.088	0.127	0.075
85-94		0.313	0.366	0.393	0.260	0.304
35-64	UNIVARIATE	0.506	0.057	0.001	0.398	0.045
65-94	UNIVARIATE	0.332	0.054	0.001	0.275	0.045
35-64	BIVARIATE	0.400	0.057	0.001	0.315	0.045
65-94	BIVARIATE	0.330	0.055	0.001	0.273	0.045
35-64	MULTIVARIATE	0.299	0.061	0.001	0.235	0.048
65-94	MULTIVARIATE	0.275	0.056	0.001	0.227	0.047

SEX: FEMALE
EVENT: DEATH
RISK FACTOR: HYPERTENSION WITH ANTIHYPERTENSIVE TREATMENT
PERSONS AT RISK: ALL PERSONS ALIVE

RISK FACTOR	35-44 AT EXAM			45-54 AT EXAM			55-64 AT EXAM			65-74 AT EXAM		
	PERSON EXAMS	# OF EVENTS	ANNUAL RATE	PERSON EXAMS	# OF EVENTS	ANNUAL RATE	PERSON EXAMS	# OF EVENTS	ANNUAL RATE	PERSON EXAMS	# OF EVENTS	ANNUAL RATE
NO	4251	11	1	5640	45	4	5022	65	6	1890	49	13
BORDER	1002	6	3	2611	24	5	3437	45	7	2106	75	18
DEFINITE	316	3	5	1184	18	8	1950	47	12	1197	50	21
TREATED	50	0	0	617	12	10	1694	40	12	1910	106	26
UNKNOWN	0	0	-	0	0	-	0	0	-	0	0	-
TOTAL	5619	20	2	10072	99	5	12103	195	8	7103	280	20

RISK FACTOR	75-84 AT EXAM			85-94 AT EXAM			35-64 COMBINED			65-94 COMBINED		
	PERSON EXAMS	# OF EVENTS	ANNUAL RATE	PERSON EXAMS	# OF EVENTS	ANNUAL RATE	# OF EVENTS	CRUDE	AGE- ADJUSTED	# OF EVENTS	CRUDE	AGE- ADJUSTED
NO	484	45	46	41	9	110	119	4	4	103	21	23
BORDER	683	55	40	37	10	155	75	5	5	140	25	25
DEFINITE	465	45	48	32	5	78	68	10	9	100	30	29
TREATED	787	85	54	53	9	85	52	11	9	200	36	35
UNKNOWN	0	0	-	0	0	-	-	-	-	0	-	-
TOTAL	2419	230	48	163	33	101	314	6	6	543	28	28

LOGISTIC REGRESSIONS

AGE AT EXAM		COEFFICIENT OF RISK FACTOR	SE COEFFICIENT	P-VALUE	STANDARDIZED COEFFICIENT	STD SE
35-44		0.631	0.293	0.031	0.370	0.172
45-54		0.361	0.122	0.003	0.278	0.094
55-64		0.339	0.086	0.001	0.285	0.072
65-74		0.346	0.080	0.001	0.284	0.065
75-84		0.095	0.090	0.291	0.075	0.071
85-94		-0.207	0.228	0.364	-0.174	0.192
35-64	UNIVARIATE	0.482	0.067	0.001	0.386	0.053
65-94	UNIVARIATE	0.256	0.057	0.001	0.209	0.047
35-64	BIVARIATE	0.327	0.069	0.001	0.261	0.055
65-94	BIVARIATE	0.195	0.058	0.001	0.159	0.047
35-64	MULTIVARIATE	0.261	0.072	0.001	0.208	0.058
65-94	MULTIVARIATE	0.126	0.059	0.033	0.103	0.048

TABLE 23-3B   DATE=20NOV85
AVERAGE ANNUAL INCIDENCE RATE FOR EVENT PER 1000 PERSONS AT RISK AT EXAM
BY AGE AND LEVEL OF RISK FACTOR AT EXAM: FRAMINGHAM STUDY 30-YEAR FOLLOWUP

SEX: MALE
EVENT: DEATH
RISK FACTOR: HYPERTENSION IGNORING TREATMENT
PERSONS AT RISK: ALL PERSONS ALIVE

RISK FACTOR	35-44 AT EXAM PERSON EXAMS	# OF EVENTS	ANNUAL RATE	45-54 AT EXAM PERSON EXAMS	# OF EVENTS	ANNUAL RATE	55-64 AT EXAM PERSON EXAMS	# OF EVENTS	ANNUAL RATE	65-74 AT EXAM PERSON EXAMS	# OF EVENTS	ANNUAL RATE
NO	2672	8	1	4306	60	7	4347	113	13	1980	101	26
MILD	1326	5	2	2383	37	8	2922	98	17	1852	133	36
DEFINITE	521	7	7	1378	39	14	1801	85	24	1002	100	50
UNKNOWN	0	0	-	0	0	-	0	0	-	0	0	-
TOTAL	4519	20	.2	8067	136	8	9070	296	16	4834	334	35

RISK FACTOR	75-84 AT EXAM PERSON EXAMS	# OF EVENTS	ANNUAL RATE	85-94 AT EXAM PERSON EXAMS	# OF EVENTS	ANNUAL RATE	35-64 COMBINED # OF EVENTS	CRUDE	AGE-ADJUSTED	65-94 COMBINED # OF EVENTS	CRUDE	AGE-ADJUSTED
NO	526	76	72	24	6	125	181	8	8	183	36	36
MILD	543	85	78	21	7	167	140	11	10	225	47	46
DEFINITE	276	55	100	5	4	400	131	18	17	159	62	63
UNKNOWN	0	0	-	0	0	-	0	-	-	0	-	-
TOTAL	1345	216	80	50	17	170	452	10	10	567	46	46

LOGISTIC REGRESSIONS

AGE AT EXAM		COEFFICIENT OF RISK FACTOR	SE COEFFICIENT	P-VALUE	STANDARDIZED COEFFICIENT	STD SE
35-44		0.744	0.281	0.008	0.515	0.195
45-54		0.344	0.107	0.001	0.260	0.081
55-64		0.307	0.073	0.001	0.237	0.057
65-74		0.362	0.073	0.001	0.275	0.056
75-84		0.186	0.098	0.058	0.139	0.074
85-94		0.927	0.477	0.052	0.618	0.318
35-64	UNIVARIATE	0.401	0.059	0.001	0.302	0.045
65-94	UNIVARIATE	0.297	0.057	0.001	0.225	0.043
35-64	BIVARIATE	0.325	0.059	0.001	0.246	0.045
65-94	BIVARIATE	0.315	0.058	0.001	0.238	0.044
35-64	MULTIVARIATE	0.201	0.063	0.002	0.152	0.048
65-94	MULTIVARIATE	0.250	0.060	0.001	0.189	0.046

SEX: FEMALE
EVENT: DEATH
RISK FACTOR: HYPERTENSION IGNORING TREATMENT
PERSONS AT RISK: ALL PERSONS ALIVE

RISK FACTOR	35-44 AT EXAM PERSON EXAMS	# OF EVENTS	ANNUAL RATE	45-54 AT EXAM PERSON EXAMS	# OF EVENTS	ANNUAL RATE	55-64 AT EXAM PERSON EXAMS	# OF EVENTS	ANNUAL RATE	65-74 AT EXAM PERSON EXAMS	# OF EVENTS	ANNUAL RATE
NO	4273	11	1	5846	48	4	5415	73	7	2205	62	14
MILD	1017	6	3	2803	28	5	4079	57	7	2907	110	19
DEFINITE	329	3	5	1423	23	8	2609	65	12	1991	108	27
UNKNOWN	0	0	-	0	0	-	0	0	-	0	0	-
TOTAL	5619	20	2	10072	99	5	12103	195	8	7103	280	20

RISK FACTOR	75-84 AT EXAM PERSON EXAMS	# OF EVENTS	ANNUAL RATE	85-94 AT EXAM PERSON EXAMS	# OF EVENTS	ANNUAL RATE	35-64 COMBINED # OF EVENTS	CRUDE	AGE-ADJUSTED	65-94 COMBINED # OF EVENTS	CRUDE	AGE-ADJUSTED
NO	613	57	46	52	13	125	132	4	5	132	23	24
MILD	1005	87	43	57	12	105	91	6	5	209	26	26
DEFINITE	803	86	54	54	8	74	91	10	9	202	35	35
UNKNOWN	0	0	-	0	0	-	0	-	-	0	-	-
TOTAL	2419	230	48	163	33	101	314	6	6	543	28	28

LOGISTIC REGRESSIONS

AGE AT EXAM		COEFFICIENT OF RISK FACTOR	SE COEFFICIENT	P-VALUE	STANDARDIZED COEFFICIENT	STD SE
35-44		0.676	0.296	0.022	0.386	0.169
45-54		0.325	0.128	0.011	0.237	0.093
55-64		0.307	0.090	0.001	0.240	0.070
65-74		0.346	0.081	0.001	0.266	0.062
75-84		0.091	0.092	0.320	0.069	0.070
85-94		-0.319	0.245	0.194	-0.258	0.198
35-64	UNIVARIATE	0.446	0.070	0.001	0.332	0.052
65-94	UNIVARIATE	0.237	0.058	0.001	0.182	0.045
35-64	BIVARIATE	0.302	0.072	0.001	0.225	0.054
65-94	BIVARIATE	0.192	0.059	0.001	0.147	0.045
35-64	MULTIVARIATE	0.228	0.076	0.003	0.170	0.057
65-94	MULTIVARIATE	0.135	0.060	0.025	0.104	0.046

TABLE 23-4   DATE=12JUN85

AVERAGE ANNUAL INCIDENCE RATE FOR EVENT PER 1000 PERSONS AT RISK AT EXAM
BY AGE AND LEVEL OF RISK FACTOR AT EXAM: FRAMINGHAM STUDY 30-YEAR FOLLOWUP

SEX: MALE
EVENT: DEATH
RISK FACTOR: SERUM CHOLESTEROL
PERSONS AT RISK: ALL PERSONS ALIVE

RISK FACTOR	35-44 AT EXAM			45-54 AT EXAM			55-64 AT EXAM			65-74 AT EXAM		
	PERSON EXAMS	# OF EVENTS	ANNUAL RATE	PERSON EXAMS	# OF EVENTS	ANNUAL RATE	PERSON EXAMS	# OF EVENTS	ANNUAL RATE	PERSON EXAMS	# OF EVENTS	ANNUAL RATE
84-204	1253	6	2	1807	41	11	2484	74	15	1530	127	42
205-234	1258	3	1	2390	29	6	2695	93	17	1457	87	30
235-264	883	6	1	1843	29	8	2128	60	14	1059	64	30
265-294	488	1	1	1057	21	10	1075	43	20	525	51	30
295-1124	303	3	5	619	11	9	556	23	21	246	22	45
UNKNOWN	334	1	1	351	5	7	132	3	11	17	3	88
TOTAL	4519	20	2	8067	136	8	9070	295	16	4834	354	35

RISK FACTOR	75-84 AT EXAM			85-94 AT EXAM			35-64 COMBINED			65-94 COMBINED		
	PERSON EXAMS	# OF EVENTS	ANNUAL RATE	PERSON EXAMS	# OF EVENTS	ANNUAL RATE	# OF EVENTS	CRUDE	AGE-ADJUSTED	# OF EVENTS	CRUDE	AGE-ADJUSTED
84-204	537	95	88	23	10	217	121	11	11	232	56	53
205-234	422	69	82	14	4	143	125	10	10	160	42	42
235-264	232	32	69	7	3	214	95	10	9	99	38	40
265-294	130	16	62	6	0	0	65	12	12	47	36	37
295-1124	24	4	83	0	0	-	37	13	13	26	48	53
UNKNOWN	0	0	-	0	0	-	9	6	7	3	88	68
TOTAL	1345	216	80	50	17	170	452	10	10	567	46	46

LOGISTIC REGRESSIONS

AGE AT EXAM		COEFFICIENT OF RISK FACTOR	SE COEFFICIENT	P-VALUE	STANDARDIZED COEFFICIENT	STD SE
35-44		0.003	0.005	0.552	0.128	0.215
45-54		-0.001	0.002	0.636	-0.043	0.090
55-64		0.002	0.001	0.227	0.070	0.058
65-74		-0.002	0.001	0.091	-0.098	0.058
75-84		-0.004	0.002	0.026	-0.169	0.076
85-94		-0.015	0.008	0.055	-0.639	0.334
35-64	UNIVARIATE	0.001	0.001	0.477	0.034	0.047
65-94	UNIVARIATE	-0.005	0.001	0.001	-0.189	0.046
35-64	BIVARIATE	0.001	0.001	0.302	0.050	0.048
65-94	BIVARIATE	-0.003	0.001	0.008	-0.123	0.046
35-64	MULTIVARIATE	0.001	0.001	0.422	0.059	0.049
65-94	MULTIVARIATE	-0.003	0.001	0.014	-0.114	0.047

SEX: FEMALE
EVENT: DEATH
RISK FACTOR: SERUM CHOLESTEROL
PERSONS AT RISK: ALL PERSONS ALIVE

RISK FACTOR	35-44 AT EXAM			45-54 AT EXAM			55-64 AT EXAM			65-74 AT EXAM		
	PERSON EXAMS	# OF EVENTS	ANNUAL RATE	PERSON EXAMS	# OF EVENTS	ANNUAL RATE	PERSON EXAMS	# OF EVENTS	ANNUAL RATE	PERSON EXAMS	# OF EVENTS	ANNUAL RATE
84-204	2350	10	2	2120	21	5	1714	31	9	897	50	28
205-234	1419	6	2	2590	21	4	2741	41	7	1570	64	20
235-264	807	1	1	2268	28	6	3052	53	9	1912	50	13
265-294	372	0	0	1490	11	4	2364	31	7	1472	61	21
295-1124	171	2	6	1124	15	7	1999	38	10	1215	53	22
UNKNOWN	500	1	1	480	3	3	233	1	2	37	2	27
TOTAL	5619	20	2	10072	99	5	12103	195	8	7103	280	20

RISK FACTOR	75-84 AT EXAM			85-94 AT EXAM			35-64 COMBINED			65-94 COMBINED		
	PERSON EXAMS	# OF EVENTS	ANNUAL RATE	PERSON EXAMS	# OF EVENTS	ANNUAL RATE	# OF EVENTS	CRUDE	AGE-ADJUSTED	# OF EVENTS	CRUDE	AGE-ADJUSTED
84-204	463	59	64	60	17	142	62	5	6	126	44	39
205-234	588	59	50	39	4	51	68	5	5	127	29	28
235-264	660	51	39	42	7	83	82	6	6	108	21	21
265-294	419	42	50	12	4	167	42	5	4	107	23	31
295-1124	282	18	32	10	1	50	55	8	8	72	24	25
UNKNOWN	7	1	71	0	0	-	5	2	2	3	34	38
TOTAL	2419	230	48	163	33	101	314	6	6	543	28	28

LOGISTIC REGRESSIONS

AGE AT EXAM		COEFFICIENT OF RISK FACTOR	SE COEFFICIENT	P-VALUE	STANDARDIZED COEFFICIENT	STD SE
35-44		0.001	0.006	0.829	0.049	0.225
45-54		0.001	0.002	0.554	0.037	0.097
55-64		0.001	0.002	0.518	0.046	0.071
65-74		-0.001	0.001	0.477	-0.044	0.062
75-84		-0.005	0.002	0.001	-0.238	0.072
85-94		-0.011	0.005	0.028	-0.480	0.218
35-64	UNIVARIATE	0.003	0.001	0.003	0.154	0.051
65-94	UNIVARIATE	-0.005	0.001	0.001	-0.220	0.046
35-64	BIVARIATE	0.001	0.001	0.550	0.035	0.058
65-94	BIVARIATE	-0.003	0.001	0.004	-0.134	0.046
35-64	MULTIVARIATE	0.000	0.001	0.778	0.016	0.058
65-94	MULTIVARIATE	-0.003	0.001	0.011	-0.118	0.047

TABLE 23-5 DATE=20NOV85

AVERAGE ANNUAL INCIDENCE RATE FOR EVENT PER 1000 PERSONS AT RISK AT EXAM
BY AGE AND LEVEL OF RISK FACTOR AT EXAM: FRAMINGHAM STUDY 30-YEAR FOLLOWUP

SEX: MALE
EVENT: DEATH
RISK FACTOR: HEMATOCRIT
PERSONS AT RISK: ALL PERSONS ALIVE

RISK FACTOR	35-44 AT EXAM PERSON EXAMS	# OF EVENTS	ANNUAL RATE	45-54 AT EXAM PERSON EXAMS	# OF EVENTS	ANNUAL RATE	55-64 AT EXAM PERSON EXAMS	# OF EVENTS	ANNUAL RATE	65-74 AT EXAM PERSON EXAMS	# OF EVENTS	ANNUAL RATE
25-42	630	8	6	961	22	11	1121	53	24	558	50	45
43-44	551	2	2	996	17	9	1094	50	14	611	37	30
45-46	990	3	2	1942	33	8	2215	65	15	1194	81	34
47-48	786	1	1	1585	27	9	1825	59	16	1069	61	29
49-70	1192	5	2	2193	32	7	2656	85	16	1382	102	37
UNKNOWN	370	1	1	390	5	6	159	4	13	20	3	75
TOTAL	4519	20	2	8067	136	8	9070	296	16	4834	334	33

RISK FACTOR	75-84 AT EXAM PERSON EXAMS	# OF EVENTS	ANNUAL RATE	85-94 AT EXAM PERSON EXAMS	# OF EVENTS	ANNUAL RATE	35-64 COMBINED # OF EVENTS	CRUDE	AGE-ADJUSTED	65-94 COMBINED # OF EVENTS	CRUDE	AGE-ADJUSTED
25-42	283	55	97	19	7	184	83	15	15	112	65	57
43-44	169	30	89	8	2	125	49	9	9	69	44	44
45-46	313	51	81	7	2	143	101	10	10	134	44	45
47-48	266	31	58	9	3	167	87	10	10	95	35	36
49-70	314	49	78	7	3	214	122	10	10	154	45	47
UNKNOWN	0	0	-	0	0	-	10	5	8	3	75	58
TOTAL	1345	216	80	50	17	170	452	10	10	567	46	46

LOGISTIC REGRESSIONS

AGE AT EXAM		COEFFICIENT OF RISK FACTOR	SE COEFFICIENT	P-VALUE	STANDARDIZED COEFFICIENT	STD SE
35-44		-0.231	0.052	0.001	-0.848	0.193
45-54		-0.028	0.025	0.253	-0.101	0.088
55-64		-0.031	0.016	0.049	-0.116	0.059
65-74		-0.006	0.016	0.686	-0.023	0.057
75-84		-0.034	0.018	0.059	-0.140	0.074
85-94		-0.003	0.061	0.955	-0.017	0.301
35-64	UNIVARIATE	-0.039	0.013	0.002	-0.144	0.047
65-94	UNIVARIATE	-0.035	0.012	0.003	-0.131	0.044
35-64	BIVARIATE	-0.041	0.013	0.002	-0.149	0.047
65-94	BIVARIATE	-0.013	0.012	0.263	-0.049	0.044
35-64	MULTIVARIATE	-0.052	0.013	0.001	-0.191	0.048
65-94	MULTIVARIATE	-0.017	0.012	0.157	-0.064	0.045

SEX: FEMALE
EVENT: DEATH
RISK FACTOR: HEMATOCRIT
PERSONS AT RISK: ALL PERSONS ALIVE

RISK FACTOR	35-44 AT EXAM PERSON EXAMS	# OF EVENTS	ANNUAL RATE	45-54 AT EXAM PERSON EXAMS	# OF EVENTS	ANNUAL RATE	55-64 AT EXAM PERSON EXAMS	# OF EVENTS	ANNUAL RATE	65-74 AT EXAM PERSON EXAMS	# OF EVENTS	ANNUAL RATE
21-39	1698	6	2	1858	26	7	1609	26	8	751	47	31
40-41	967	3	2	1761	9	3	2144	30	7	1180	46	19
42-43	1043	4	2	1978	11	3	2371	33	7	1422	41	14
44-45	858	3	2	2222	21	5	3003	42	7	1835	60	16
46-65	487	3	3	1702	26	8	2700	63	12	1873	83	22
UNKNOWN	566	1	1	551	6	5	276	1	2	42	3	36
TOTAL	5619	20	2	10072	99	5	12103	195	8	7103	280	20

RISK FACTOR	75-84 AT EXAM PERSON EXAMS	# OF EVENTS	ANNUAL RATE	85-94 AT EXAM PERSON EXAMS	# OF EVENTS	ANNUAL RATE	35-64 COMBINED # OF EVENTS	CRUDE	AGE-ADJUSTED	65-94 COMBINED # OF EVENTS	CRUDE	AGE-ADJUSTED
21-39	375	51	68	33	10	152	58	4	6	108	47	42
40-41	449	39	43	19	9	115	42	4	4	94	28	27
42-43	437	34	39	26	4	77	48	4	4	79	21	22
44-45	607	47	39	34	3	44	66	5	5	110	22	22
46-65	545	58	53	31	7	113	92	9	8	148	30	31
UNKNOWN	6	1	83	0	0	-	8	3	3	4	42	47
TOTAL	2419	230	48	163	33	101	314	6	6	543	28	28

LOGISTIC REGRESSIONS

AGE AT EXAM		COEFFICIENT OF RISK FACTOR	SE COEFFICIENT	P-VALUE	STANDARDIZED COEFFICIENT	STD SE
35-44		0.014	0.064	0.831	0.050	0.232
45-54		0.031	0.030	0.289	0.112	0.106
55-64		0.062	0.021	0.003	0.215	0.073
65-74		-0.017	0.017	0.338	-0.058	0.061
75-84		-0.035	0.018	0.045	-0.137	0.068
85-94		-0.072	0.051	0.163	-0.280	0.201
35-64	UNIVARIATE	0.069	0.016	0.001	0.250	0.059
65-94	UNIVARIATE	-0.040	0.012	0.001	-0.147	0.043
35-64	BIVARIATE	0.044	0.017	0.008	0.159	0.060
65-94	BIVARIATE	-0.027	0.012	0.022	-0.100	0.043
35-64	MULTIVARIATE	0.020	0.017	0.219	0.073	0.060
65-94	MULTIVARIATE	-0.030	0.012	0.011	-0.109	0.043

TABLE 23-6   DATE=20NOV85
AVERAGE ANNUAL INCIDENCE RATE FOR EVENT PER 1000 PERSONS AT RISK AT EXAM
BY AGE AND LEVEL OF RISK FACTOR AT EXAM: FRAMINGHAM STUDY 30-YEAR FOLLOWUP

SEX: MALE
EVENT: DEATH
RISK FACTOR: BLOOD GLUCOSE (MG/DL)
PERSONS AT RISK: ALL PERSONS ALIVE

RISK FACTOR	35-44 AT EXAM PERSON EXAMS	# OF EVENTS	ANNUAL RATE	45-54 AT EXAM PERSON EXAMS	# OF EVENTS	ANNUAL RATE	55-64 AT EXAM PERSON EXAMS	# OF EVENTS	ANNUAL RATE	65-74 AT EXAM PERSON EXAMS	# OF EVENTS	ANNUAL RATE
29-69	1004	4	2	1562	28	9	1208	33	14	501	25	25
70-89	2531	12	2	4344	65	7	4526	144	16	2142	139	32
90-109	628	2	2	1402	28	10	2136	68	16	1296	88	34
110-129	107	0	0	331	7	11	608	26	21	428	34	40
130-524	49	1	10	204	7	17	491	23	23	450	45	50
UNKNOWN	200	1	3	224	1	2	101	2	10	17	3	88
TOTAL	4519	20	2	8067	136	8	9070	296	16	4834	334	35

RISK FACTOR	75-84 AT EXAM PERSON EXAMS	# OF EVENTS	ANNUAL RATE	85-94 AT EXAM PERSON EXAMS	# OF EVENTS	ANNUAL RATE	35-64 COMBINED # OF EVENTS	CRUDE	AGE-ADJUSTED	65-94 COMBINED # OF EVENTS	CRUDE	AGE-ADJUSTED
29-69	86	13	76	1	1	500	65	9	9	39	33	40
70-89	479	70	73	17	4	118	221	10	10	213	40	42
90-109	438	67	76	15	5	167	98	12	11	160	46	44
110-129	170	29	85	7	3	214	33	16	13	66	55	51
130-524	172	37	108	10	4	200	31	21	18	86	68	64
UNKNOWN	0	0	-	0	0	-	4	4	5	3	88	68
TOTAL	1345	216	80	50	17	170	452	10	10	567	46	46

LOGISTIC REGRESSIONS

AGE AT EXAM		COEFFICIENT OF RISK FACTOR	SE COEFFICIENT	P-VALUE	STANDARDIZED COEFFICIENT	STD SE
35-44		0.013	0.004	0.002	0.227	0.072
45-54		0.004	0.003	0.118	0.101	0.065
55-64		0.005	0.002	0.002	0.139	0.044
65-74		0.006	0.001	0.001	0.215	0.040
75-84		0.005	0.002	0.011	0.167	0.065
85-94		0.006	0.008	0.496	0.201	0.295
35-64	UNIVARIATE	0.007	0.001	0.001	0.176	0.030
65-94	UNIVARIATE	0.006	0.001	0.001	0.214	0.033
35-64	BIVARIATE	0.005	0.001	0.001	0.117	0.033
65-94	BIVARIATE	0.005	0.001	0.001	0.196	0.035
35-64	MULTIVARIATE	0.005	0.001	0.001	0.117	0.034
65-94	MULTIVARIATE	0.005	0.001	0.001	0.189	0.035

SEX: FEMALE
EVENT: DEATH
RISK FACTOR: BLOOD GLUCOSE (MG/DL)
PERSONS AT RISK: ALL PERSONS ALIVE

RISK FACTOR	35-44 AT EXAM PERSON EXAMS	# OF EVENTS	ANNUAL RATE	45-54 AT EXAM PERSON EXAMS	# OF EVENTS	ANNUAL RATE	55-64 AT EXAM PERSON EXAMS	# OF EVENTS	ANNUAL RATE	65-74 AT EXAM PERSON EXAMS	# OF EVENTS	ANNUAL RATE
29-69	1067	9	4	1806	14	4	1411	22	8	597	30	25
70-89	3414	10	1	5811	56	5	4371	88	7	3483	109	16
90-109	672	0	0	1646	13	4	2877	55	10	1982	73	18
110-129	96	0	0	303	2	3	777	19	12	582	24	21
130-524	39	0	0	167	10	30	496	10	10	427	42	49
UNKNOWN	331	1	2	339	4	6	171	1	3	32	2	31
TOTAL	5619	20	2	10072	99	5	12103	195	8	7103	280	20

RISK FACTOR	75-84 AT EXAM PERSON EXAMS	# OF EVENTS	ANNUAL RATE	85-94 AT EXAM PERSON EXAMS	# OF EVENTS	ANNUAL RATE	35-64 COMBINED # OF EVENTS	CRUDE	AGE-ADJUSTED	65-94 COMBINED # OF EVENTS	CRUDE	AGE-ADJUSTED
29-69	129	11	43	5	2	200	45	5	5	43	29	32
70-89	976	81	41	60	9	75	154	5	5	199	22	23
90-109	803	81	50	63	5	40	68	7	6	159	28	27
110-129	286	28	49	19	8	211	21	9	7	60	34	31
130-524	224	28	63	16	9	281	20	14	15	79	59	56
UNKNOWN	1	1	500	0	0	-	6	4	4	3	45	148
TOTAL	2419	230	48	163	33	101	314	6	6	543	28	28

LOGISTIC REGRESSIONS

AGE AT EXAM		COEFFICIENT OF RISK FACTOR	SE COEFFICIENT	P-VALUE	STANDARDIZED COEFFICIENT	STD SE
35-44		-0.064	0.024	0.007	-0.967	0.360
45-54		0.012	0.002	0.001	0.246	0.048
55-64		0.005	0.002	0.009	0.131	0.050
65-74		0.007	0.001	0.001	0.219	0.040
75-84		0.003	0.002	0.063	0.109	0.059
85-94		0.021	0.007	0.003	0.716	0.239
35-64	UNIVARIATE	0.008	0.001	0.001	0.183	0.032
65-94	UNIVARIATE	0.007	0.001	0.001	0.216	0.030
35-64	BIVARIATE	0.006	0.002	0.001	0.136	0.034
65-94	BIVARIATE	0.006	0.001	0.001	0.184	0.032
35-64	MULTIVARIATE	0.005	0.002	0.001	0.113	0.035
65-94	MULTIVARIATE	0.005	0.001	0.001	0.156	0.033

TABLE 23-7   DATE=20NOV85
AVERAGE ANNUAL INCIDENCE RATE FOR EVENT PER 1000 PERSONS AT RISK AT EXAM
BY AGE AND LEVEL OF RISK FACTOR AT EXAM: FRAMINGHAM STUDY 30-YEAR FOLLOWUP

SEX: MALE
EVENT: DEATH
RISK FACTOR: DIABETES MELLITUS
PERSONS AT RISK: ALL PERSONS ALIVE

RISK FACTOR	35-44 AT EXAM PERSON EXAMS	# OF EVENTS	ANNUAL RATE	45-54 AT EXAM PERSON EXAMS	# OF EVENTS	ANNUAL RATE	55-64 AT EXAM PERSON EXAMS	# OF EVENTS	ANNUAL RATE	65-74 AT EXAM PERSON EXAMS	# OF EVENTS	ANNUAL RATE
NO	4470	19	2	7812	128	8	8424	257	15	4269	278	35
YES	49	1	10	255	8	16	646	39	30	565	56	50
UNKNOWN	0	0	-	0	0	-	0	0	-	0	0	-
TOTAL	4519	20	2	8067	136	8	9070	296	16	4834	334	35

RISK FACTOR	75-84 AT EXAM PERSON EXAMS	# OF EVENTS	ANNUAL RATE	85-94 AT EXAM PERSON EXAMS	# OF EVENTS	ANNUAL RATE	35-64 COMBINED # OF EVENTS	CRUDE	AGE-ADJUSTED	65-94 COMBINED # OF EVENTS	CRUDE	AGE-ADJUSTED
NO	1170	177	76	44	16	182	404	10	10	471	43	43
YES	175	39	111	6	1	83	48	25	21	96	64	63
UNKNOWN	0	0	-	0	0	-	0	-	-	0	-	-
TOTAL	1345	216	80	50	17	170	452	10	10	567	46	46

LOGISTIC REGRESSIONS

AGE AT EXAM		COEFFICIENT OF RISK FACTOR	SE COEFFICIENT	P-VALUE	STANDARDIZED COEFFICIENT	STD SE
35-44		1.585	1.036	0.126	0.164	0.107
45-54		0.665	0.370	0.072	0.116	0.065
55-64		0.714	0.177	0.001	0.184	0.046
65-74		0.457	0.154	0.003	0.147	0.049
75-84		0.475	0.199	0.017	0.160	0.067
85-94		-1.050	1.139	0.357	-0.345	0.374
35-64	UNIVARIATE	0.984	0.156	0.001	0.201	0.032
65-94	UNIVARIATE	0.452	0.119	0.001	0.147	0.039
35-64	BIVARIATE	0.646	0.159	0.001	0.132	0.032
65-94	BIVARIATE	0.423	0.122	0.001	0.137	0.039
35-64	MULTIVARIATE	0.469	0.165	0.004	0.096	0.034
65-94	MULTIVARIATE	0.389	0.124	0.002	0.126	0.040

SEX: FEMALE
EVENT: DEATH
RISK FACTOR: DIABETES MELLITUS
PERSONS AT RISK: ALL PERSONS ALIVE

RISK FACTOR	35-44 AT EXAM PERSON EXAMS	# OF EVENTS	ANNUAL RATE	45-54 AT EXAM PERSON EXAMS	# OF EVENTS	ANNUAL RATE	55-64 AT EXAM PERSON EXAMS	# OF EVENTS	ANNUAL RATE	65-74 AT EXAM PERSON EXAMS	# OF EVENTS	ANNUAL RATE
NO	5574	19	2	9859	90	5	11497	174	8	6699	223	17
YES	43	1	12	213	9	21	606	21	17	604	57	47
UNKNOWN	0	0	-	0	0	-	0	0	-	0	0	-
TOTAL	5619	20	2	10072	99	5	12103	195	8	7103	280	20

RISK FACTOR	75-84 AT EXAM PERSON EXAMS	# OF EVENTS	ANNUAL RATE	85-94 AT EXAM PERSON EXAMS	# OF EVENTS	ANNUAL RATE	35-64 COMBINED # OF EVENTS	CRUDE	AGE-ADJUSTED	65-94 COMBINED # OF EVENTS	CRUDE	AGE-ADJUSTED
NO	2106	184	44	139	23	83	283	5	5	430	25	25
YES	313	46	73	24	10	208	31	18	18	113	60	56
UNKNOWN	0	0	-	0	0	-	0	-	-	0	-	-
TOTAL	2419	230	48	163	33	101	314	6	6	543	28	28

LOGISTIC REGRESSIONS

AGE AT EXAM		COEFFICIENT OF RISK FACTOR	SE COEFFICIENT	P-VALUE	STANDARDIZED COEFFICIENT	STD SE
35-44		1.941	1.038	0.061	0.169	0.090
45-54		1.566	0.357	0.001	0.225	0.051
55-64		0.848	0.235	0.001	0.185	0.051
65-74		1.076	0.155	0.001	0.300	0.043
75-84		0.588	0.177	0.001	0.197	0.060
85-94		1.282	0.473	0.007	0.456	0.168
35-64	UNIVARIATE	1.256	0.192	0.001	0.218	0.033
65-94	UNIVARIATE	0.970	0.112	0.001	0.287	0.033
35-64	BIVARIATE	0.985	0.195	0.001	0.171	0.034
65-94	BIVARIATE	0.849	0.114	0.001	0.252	0.034
35-64	MULTIVARIATE	0.830	0.199	0.001	0.144	0.034
65-94	MULTIVARIATE	0.707	0.117	0.001	0.209	0.035

TABLE 23-8   DATE=20NOV85
AVERAGE ANNUAL INCIDENCE RATE FOR EVENT PER 1000 PERSONS AT RISK AT EXAM
BY AGE AND LEVEL OF RISK FACTOR AT EXAM, FRAMINGHAM STUDY 30-YEAR FOLLOWUP

SEX: MALE
EVENT: DEATH
RISK FACTOR: GLUCOSE IN URINE, DEFINITE OR TRACE
PERSONS AT RISK: ALL PERSONS ALIVE

RISK FACTOR	35-44 AT EXAM			45-54 AT EXAM			55-64 AT EXAM			65-74 AT EXAM		
	PERSON EXAMS	# OF EVENTS	ANNUAL RATE	PERSON EXAMS	# OF EVENTS	ANNUAL RATE	PERSON EXAMS	# OF EVENTS	ANNUAL RATE	PERSON EXAMS	# OF EVENTS	ANNUAL RATE
NO	4426	17	2	7869	133	8	8786	275	16	4661	311	33
YES	60	2	17	177	3	8	260	19	37	167	23	69
UNKNOWN	33	1	15	21	0	0	24	2	42	6	0	0
TOTAL	4519	20	2	8067	136	8	9070	296	16	4834	334	35

RISK FACTOR	75-84 AT EXAM			85-94 AT EXAM			35-64 COMBINED			65-94 COMBINED		
	PERSON EXAMS	# OF EVENTS	ANNUAL RATE	PERSON EXAMS	# OF EVENTS	ANNUAL RATE	# OF EVENTS	CRUDE	AGE-ADJUSTED	# OF EVENTS	CRUDE	AGE-ADJUSTED
NO	1303	204	78	49	17	173	425	10	10	532	44	44
YES	39	12	154	1	0	0	24	24	22	35	85	87
UNKNOWN	3	0	0	0	0	-	3	19	21	0	0	0
TOTAL	1345	216	80	50	17	170	452	10	10	567	46	46

LOGISTIC REGRESSIONS

AGE AT EXAM		COEFFICIENT OF RISK FACTOR	SE COEFFICIENT	P-VALUE	STANDARDIZED COEFFICIENT	STD SE
35-44		2.191	0.759	0.004	0.252	0.087
45-54		0.003	0.589	0.996	0.000	0.086
55-64		0.892	0.246	0.001	0.149	0.041
65-74		0.804	0.232	0.001	0.147	0.042
75-84		0.873	0.355	0.014	0.147	0.060
85-94		-	-	-	-	-
35-64	UNIVARIATE	0.903	0.215	0.001	0.135	0.032
65-94	UNIVARIATE	0.740	0.191	0.001	0.133	0.034
35-64	BIVARIATE	0.745	0.217	0.001	0.112	0.032
65-94	BIVARIATE	0.827	0.195	0.001	0.148	0.035
35-64	MULTIVARIATE	0.669	0.225	0.003	0.100	0.034
65-94	MULTIVARIATE	0.804	0.196	0.001	0.144	0.035

SEX: FEMALE
EVENT: DEATH
RISK FACTOR: GLUCOSE IN URINE, DEFINITE OR TRACE
PERSONS AT RISK: ALL PERSONS ALIVE

RISK FACTOR	35-44 AT EXAM			45-54 AT EXAM			55-64 AT EXAM			65-74 AT EXAM		
	PERSON EXAMS	# OF EVENTS	ANNUAL RATE	PERSON EXAMS	# OF EVENTS	ANNUAL RATE	PERSON EXAMS	# OF EVENTS	ANNUAL RATE	PERSON EXAMS	# OF EVENTS	ANNUAL RATE
NO	5540	19	2	9947	91	5	11903	187	8	6957	266	19
YES	41	1	12	101	7	35	174	8	23	136	14	51
UNKNOWN	38	0	0	24	1	21	26	0	8	10	0	0
TOTAL	5619	20	2	10072	99	5	12103	195	8	7103	280	20

RISK FACTOR	75-84 AT EXAM			85-94 AT EXAM			35-64 COMBINED			65-94 COMBINED		
	PERSON EXAMS	# OF EVENTS	ANNUAL RATE	PERSON EXAMS	# OF EVENTS	ANNUAL RATE	# OF EVENTS	CRUDE	AGE-ADJUSTED	# OF EVENTS	CRUDE	AGE-ADJUSTED
NO	2375	224	47	160	31	97	297	5	5	521	27	27
YES	35	5	71	3	2	333	16	25	25	21	60	61
UNKNOWN	9	1	56	0	0	-	1	6	8	1	26	14
TOTAL	2419	230	48	163	33	101	314	6	6	543	28	28

LOGISTIC REGRESSIONS

AGE AT EXAM		COEFFICIENT OF RISK FACTOR	SE COEFFICIENT	P-VALUE	STANDARDIZED COEFFICIENT	STD SE
35-44		1.983	1.038	0.056	0.169	0.089
45-54		2.088	0.406	0.001	0.208	0.040
55-64		1.105	0.369	0.003	0.132	0.044
65-74		1.060	0.289	0.001	0.145	0.040
75-84		0.470	0.488	0.335	0.056	0.058
85-94		2.119	1.241	0.088	0.286	0.167
35-64	UNIVARIATE	1.582	0.263	0.001	0.168	0.028
65-94	UNIVARIATE	0.860	0.237	0.001	0.114	0.032
35-64	BIVARIATE	1.469	0.265	0.001	0.156	0.028
65-94	BIVARIATE	0.951	0.241	0.001	0.126	0.032
35-64	MULTIVARIATE	1.198	0.281	0.001	0.127	0.030
65-94	MULTIVARIATE	0.685	0.249	0.006	0.091	0.033

SEX: MALE
EVENT: DEATH
RISK FACTOR: GLUCOSE INTOLERANCE
PERSONS AT RISK: ALL PERSONS ALIVE

RISK FACTOR	35-44 AT EXAM			45-54 AT EXAM			55-64 AT EXAM			65-74 AT EXAM		
	PERSON EXAMS	# OF EVENTS	ANNUAL RATE	PERSON EXAMS	# OF EVENTS	ANNUAL RATE	PERSON EXAMS	# OF EVENTS	ANNUAL RATE	PERSON EXAMS	# OF EVENTS	ANNUAL RATE
NO	4381	18	2	7528	122	8	7973	242	15	3918	248	32
YES	138	2	7	539	14	13	1097	54	25	916	86	47
UNKNOWN	0	0	-	0	0	-	0	0	-	0	0	-
TOTAL	4519	20	2	8067	136	8	9070	296	16	4834	334	35

RISK FACTOR	75-84 AT EXAM			85-94 AT EXAM			35-64 COMBINED			65-94 COMBINED		
	PERSON EXAMS	# OF EVENTS	ANNUAL RATE	PERSON EXAMS	# OF EVENTS	ANNUAL RATE	# OF EVENTS	CRUDE	AGE-ADJUSTED	# OF EVENTS	CRUDE	AGE-ADJUSTED
NO	1035	156	75	36	10	139	382	10	10	414	41	42
YES	310	60	97	14	7	250	70	20	17	153	62	59
UNKNOWN	0	0	-	0	0	-	0	-	-	0	-	-
TOTAL	1345	216	80	50	17	170	452	10	10	567	46	46

LOGISTIC REGRESSIONS

AGE AT EXAM		COEFFICIENT OF RISK FACTOR	SE COEFFICIENT	P-VALUE	STANDARDIZED COEFFICIENT	STD SE
35-44		1.271	0.750	0.090	0.219	0.129
45-54		0.482	0.286	— 0.092	— 0.120	0.071
55-64		0.503	0.154	0.001	0.164	0.050
65-74		0.427	0.131	0.001	0.168	0.051
75-84		0.302	0.168	0.072	0.127	0.071
85-94		0.956	0.651	0.142	0.433	0.295
35-64	UNIVARIATE	0.741	0.132	0.001	0.203	0.036
65-94	UNIVARIATE	0.442	0.100	0.001	0.176	0.040
35-64	BIVARIATE	0.465	0.134	0.001	0.127	0.037
65-94	BIVARIATE	0.369	0.102	0.001	0.147	0.041
35-64	MULTIVARIATE	0.364	0.139	0.009	0.100	0.038
65-94	MULTIVARIATE	0.341	0.104	0.001	0.136	0.041

SEX: FEMALE
EVENT: DEATH
RISK FACTOR: GLUCOSE INTOLERANCE
PERSONS AT RISK: ALL PERSONS ALIVE

RISK FACTOR	35-44 AT EXAM			45-54 AT EXAM			55-64 AT EXAM			65-74 AT EXAM		
	PERSON EXAMS	# OF EVENTS	ANNUAL RATE	PERSON EXAMS	# OF EVENTS	ANNUAL RATE	PERSON EXAMS	# OF EVENTS	ANNUAL RATE	PERSON EXAMS	# OF EVENTS	ANNUAL RATE
NO	5525	19	2	9685	86	4	10988	165	8	6066	202	17
YES	94	1	5	387	13	17	1115	30	13	1037	78	38
UNKNOWN	0	0	-	0	0	-	0	0	-	0	0	-
TOTAL	5619	20	2	10072	99	5	12103	195	8	7103	280	20

RISK FACTOR	75-84 AT EXAM			85-94 AT EXAM			35-64 COMBINED			65-94 COMBINED		
	PERSON EXAMS	# OF EVENTS	ANNUAL RATE	PERSON EXAMS	# OF EVENTS	ANNUAL RATE	# OF EVENTS	CRUDE	AGE-ADJUSTED	# OF EVENTS	CRUDE	AGE-ADJUSTED
NO	1905	170	45	131	19	73	270	5	5	391	24	25
YES	514	60	58	32	14	219	44	14	13	152	48	46
UNKNOWN	0	0	-	0	0	-	0	-	-	0	-	-
TOTAL	2419	230	48	163	33	101	314	6	6	543	28	28

LOGISTIC REGRESSIONS

AGE AT EXAM		COEFFICIENT OF RISK FACTOR	SE COEFFICIENT	P-VALUE	STANDARDIZED COEFFICIENT	STD SE
35-44		1.137	1.031	0.270	0.146	0.132
45-54		1.356	0.302	0.001	0.261	0.058
55-64		0.595	0.201	0.003	0.172	0.058
65-74		0.859	0.138	0.001	0.303	0.049
75-84		0.299	0.159	0.060	0.122	0.065
85-94		1.523	0.434	0.001	0.607	0.173
35-64	UNIVARIATE	1.002	0.165	0.001	0.253	0.038
65-94	UNIVARIATE	0.739	0.100	0.001	0.273	0.037
35-64	BIVARIATE	0.735	0.167	0.001	0.171	0.039
65-94	BIVARIATE	0.622	0.101	0.001	0.230	0.037
35-64	MULTIVARIATE	0.623	0.171	0.001	0.145	0.040
65-94	MULTIVARIATE	0.517	0.104	0.001	0.191	0.038

TABLE 23-10   DATE=20NOV85

AVERAGE ANNUAL INCIDENCE RATE FOR EVENT PER 1000 PERSONS AT RISK AT EXAM
BY AGE AND LEVEL OF RISK FACTOR AT EXAM: FRAMINGHAM STUDY 30-YEAR FOLLOWUP

SEX: MALE
EVENT: DEATH
RISK FACTOR: METROPOLITAN RELATIVE WEIGHT (PERCENT)
PERSONS AT RISK: ALL PERSONS ALIVE

RISK FACTOR	35-44 AT EXAM			45-54 AT EXAM			55-64 AT EXAM			65-74 AT EXAM		
	PERSON EXAMS	# OF EVENTS	ANNUAL RATE	PERSON EXAMS	# OF EVENTS	ANNUAL RATE	PERSON EXAMS	# OF EVENTS	ANNUAL RATE	PERSON EXAMS	# OF EVENTS	ANNUAL RATE
54-104	723	7	5	1149	24	10	1272	62	24	831	86	52
105-114	979	3	2	1635	38	12	1858	63	17	1052	65	31
115-124	1153	4	2	2139	37	9	2467	72	15	1322	80	30
125-134	923	3	2	1691	14	4	1851	48	13	956	55	29
135-272	761	3	2	1452	23	8	1621	51	16	673	48	36
UNKNOWN	0	0	-	1	0	0	1	0	0	0	0	-
TOTAL	4519	20	2	8067	136	8	9070	296	16	4834	334	35

RISK FACTOR	75-84 AT EXAM			85-94 AT EXAM			35-64 COMBINED			65-94 COMBINED		
	PERSON EXAMS	# OF EVENTS	ANNUAL RATE	PERSON EXAMS	# OF EVENTS	ANNUAL RATE	# OF EVENTS	CRUDE	AGE-ADJUSTED	# OF EVENTS	CRUDE	AGE-ADJUSTED
54-104	277	53	96	19	6	158	93	15	15	145	64	62
105-114	357	58	81	15	5	167	104	12	10	138	59	59
115-124	399	51	64	14	5	179	113	19	7	85	37	40
125-134	196	29	74	2	1	250	65	10	10	73	46	51
135-272	116	25	108	0	0	-	77	10		73	46	
UNKNOWN	0	0	-	0	0	-	0	0	0	0	-	
TOTAL	1345	216	80	50	17	170	452	10	10	567	46	46

LOGISTIC REGRESSIONS

AGE AT EXAM		COEFFICIENT OF RISK FACTOR	SE COEFFICIENT	P-VALUE	STANDARDIZED COEFFICIENT	STD SE
35-44		-0.024	0.015	0.103	-0.393	0.241
45-54		-0.014	0.006	0.010	-0.232	0.090
55-64		-0.010	0.004	0.010	-0.156	0.061
65-74		-0.010	0.004	0.005	-0.161	0.058
75-84		-0.004	0.005	0.385	-0.065	0.075
85-94		-0.002	0.026	0.934	-0.025	0.301
35-64	UNIVARIATE	-0.011	0.003	0.001	-0.179	0.049
65-94	UNIVARIATE	-0.011	0.003	0.001	-0.175	0.045
35-64	BIVARIATE	-0.012	0.003	0.001	-0.186	0.050
65-94	BIVARIATE	-0.007	0.003	0.015	-0.112	0.046
35-64	MULTIVARIATE	-0.014	0.003	0.001	-0.226	0.052
65-94	MULTIVARIATE	-0.011	0.003	0.001	-0.167	0.048

SEX: FEMALE
EVENT: DEATH
RISK FACTOR: METROPOLITAN RELATIVE WEIGHT (PERCENT)
PERSONS AT RISK: ALL PERSONS ALIVE

RISK FACTOR	35-44 AT EXAM			45-54 AT EXAM			55-64 AT EXAM			65-74 AT EXAM		
	PERSON EXAMS	# OF EVENTS	ANNUAL RATE	PERSON EXAMS	# OF EVENTS	ANNUAL RATE	PERSON EXAMS	# OF EVENTS	ANNUAL RATE	PERSON EXAMS	# OF EVENTS	ANNUAL RATE
54-104	1694	10	3	2117	30	7	2140	47	11	1319	73	28
105-114	1431	2	1	2364	10	2	2541	39	8	1321	45	17
115-124	1059	2	1	2128	15	4	2556	36	7	1536	49	16
125-134	566	2	2	1383	16	6	1831	30	8	1140	37	16
135-272	861	2	1	2080	28	7	3035	43	7	1787	76	21
UNKNOWN	8	1	63	0	0	-	0	0	-	0	0	-
TOTAL	5619	20	2	10072	99	5	12103	195	8	7103	280	20

RISK FACTOR	75-84 AT EXAM			85-94 AT EXAM			35-64 COMBINED			65-94 COMBINED		
	PERSON EXAMS	# OF EVENTS	ANNUAL RATE	PERSON EXAMS	# OF EVENTS	ANNUAL RATE	# OF EVENTS	CRUDE	AGE-ADJUSTED	# OF EVENTS	CRUDE	AGE-ADJUSTED
54-104	632	78	62	51	14	137	87	7	7	165	41	38
105-114	457	38	42	33	6	91	51	4	4	89	25	24
115-124	487	29	30	31	6	97	54	5	5	84	20	21
125-134	342	32	47	26	3	58	48	6	6	72	24	25
135-272	501	53	53	22	4	91	75	6	6	133	29	30
UNKNOWN	0	0	-	0	0	-	63	13		0	-	
TOTAL	2419	230	48	163	33	101	314	6	6	543	28	28

LOGISTIC REGRESSIONS

AGE AT EXAM		COEFFICIENT OF RISK FACTOR	SE COEFFICIENT	P-VALUE	STANDARDIZED COEFFICIENT	STD SE
35-44		-0.016	0.014	0.240	-0.328	0.279
45-54		0.005	0.004	0.266	0.104	0.094
55-64		-0.006	0.003	0.069	-0.142	0.078
65-74		-0.004	0.003	0.194	-0.083	0.064
75-84		-0.002	0.003	0.446	-0.054	0.071
85-94		-0.018	0.011	0.087	-0.351	0.205
35-64	UNIVARIATE	-0.000	0.003	0.876	-0.009	0.057
65-94	UNIVARIATE	-0.006	0.002	0.002	-0.144	0.047
35-64	BIVARIATE	-0.003	0.003	0.210	-0.074	0.059
65-94	BIVARIATE	-0.003	0.002	0.109	-0.075	0.047
35-64	MULTIVARIATE	-0.005	0.003	0.082	-0.106	0.061
65-94	MULTIVARIATE	-0.003	0.002	0.124	-0.072	0.047

TABLE 23-11 DATE=20NOV85
AVERAGE ANNUAL INCIDENCE RATE FOR EVENT PER 1000 PERSONS AT RISK AT EXAM
BY AGE AND LEVEL OF RISK FACTOR AT EXAM: FRAMINGHAM STUDY 30-YEAR FOLLOWUP

SEX: MALE
EVENT: DEATH
RISK FACTOR: VITAL CAPACITY - HEIGHT INDEX (ML/INCH)
PERSONS AT RISK: ALL PERSONS ALIVE

RISK FACTOR	35-44 AT EXAM PERSON EXAMS	# OF EVENTS	ANNUAL RATE	45-54 AT EXAM PERSON EXAMS	# OF EVENTS	ANNUAL RATE	55-64 AT EXAM PERSON EXAMS	# OF EVENTS	ANNUAL RATE	65-74 AT EXAM PERSON EXAMS	# OF EVENTS	ANNUAL RATE
8-44	271	3	6	1108	34	15	2374	132	28	2179	196	45
45-49	416	2	2	1188	23	10	1692	48	14	974	58	30
50-54	751	2	1	1726	31	9	2061	47	11	857	40	23
55-59	1034	6	3	1823	24	7	1569	34	11	528	23	22
60-85	2034	7	2	2216	24	5	1356	34	13	284	16	28
UNKNOWN	13	0	0	6	0	0	18	1	28	12	1	42
TOTAL	4519	20	2	8067	136	8	9070	296	16	4834	334	35

RISK FACTOR	75-94 AT EXAM PERSON EXAMS	# OF EVENTS	ANNUAL RATE	85-94 AT EXAM PERSON EXAMS	# OF EVENTS	ANNUAL RATE	35-64 COMBINED # OF EVENTS	CRUDE	AGE-ADJUSTED	65-94 COMBINED # OF EVENTS	CRUDE	AGE-ADJUSTED
8-44	853	150	88	34	13	191	169	23	19	359	59	55
45-49	228	26	57	7	3	214	73	11	10	87	36	37
50-54	142	24	85	7	1	71	80	9	8	65	32	37
55-59	81	14	86	2	0	0	64	7	8	37	30	36
60-85	33	1	15	0	0	-	65	6	8	17	27	25
UNKNOWN	8	1	63	0	0	-	1	14	12	2	50	46
TOTAL	1345	216	80	50	17	170	452	10	10	567	46	46

LOGISTIC REGRESSIONS

AGE AT EXAM		COEFFICIENT OF RISK FACTOR	SE COEFFICIENT	P-VALUE	STANDARDIZED COEFFICIENT	STD SE
35-44		-0.053	0.023	0.022	-0.470	0.205
45-54		-0.045	0.009	0.001	-0.413	0.082
55-64		-0.040	0.006	0.001	-0.372	0.057
65-74		-0.043	0.006	0.001	-0.400	0.056
75-84		-0.028	0.008	0.001	-0.258	0.075
85-94		-0.074	0.043	0.087	-0.569	0.332
35-64	UNIVARIATE	-0.054	0.005	0.001	-0.525	0.045
65-94	UNIVARIATE	-0.045	0.005	0.001	-0.428	0.044
35-64	BIVARIATE	-0.039	0.005	0.001	-0.372	0.048
65-94	BIVARIATE	-0.035	0.005	0.001	-0.334	0.046
35-64	MULTIVARIATE	-0.031	0.005	0.001	-0.302	0.050
65-94	MULTIVARIATE	-0.032	0.005	0.001	-0.303	0.047

SEX: FEMALE
EVENT: DEATH
RISK FACTOR: VITAL CAPACITY - HEIGHT INDEX (ML/INCH)
PERSONS AT RISK: ALL PERSONS ALIVE

RISK FACTOR	35-44 AT EXAM PERSON EXAMS	# OF EVENTS	ANNUAL RATE	45-54 AT EXAM PERSON EXAMS	# OF EVENTS	ANNUAL RATE	55-64 AT EXAM PERSON EXAMS	# OF EVENTS	ANNUAL RATE	65-74 AT EXAM PERSON EXAMS	# OF EVENTS	ANNUAL RATE
4-29	122	0	0	546	13	12	1572	57	18	2068	130	31
30-34	291	1	2	1241	14	6	2547	46	9	1946	78	20
35-39	769	3	2	2103	21	5	3213	40	6	1728	44	13
40-44	1427	3	1	2838	33	6	2836	33	6	933	22	12
45-92	2989	12	2	3327	18	3	1914	19	5	422	5	6
UNKNOWN	21	1	24	17	0	0	21	0	0	6	1	83
TOTAL	5619	20	2	10072	99	5	12103	195	8	7103	280	20

RISK FACTOR	75-84 AT EXAM PERSON EXAMS	# OF EVENTS	ANNUAL RATE	85-94 AT EXAM PERSON EXAMS	# OF EVENTS	ANNUAL RATE	35-64 COMBINED # OF EVENTS	CRUDE	AGE-ADJUSTED	65-94 COMBINED # OF EVENTS	CRUDE	AGE-ADJUSTED
4-29	1212	138	57	109	24	110	70	16	12	292	43	39
30-34	686	55	40	37	7	95	61	7	6	140	26	26
35-39	370	28	38	16	2	63	64	5	5	74	18	20
40-44	123	5	20	1	0	0	49	5	5	27	13	14
45-92	26	3	58	0	0	-	49	3	4	8	9	19
UNKNOWN	2	1	250	0	0	-	1	8	5	2	125	124
TOTAL	2419	230	48	163	33	101	314	6	6	543	28	28

LOGISTIC REGRESSIONS

AGE AT EXAM		COEFFICIENT OF RISK FACTOR	SE COEFFICIENT	P-VALUE	STANDARDIZED COEFFICIENT	STD SE
35-44		0.005	0.032	0.883	0.034	0.230
45-54		-0.052	0.013	0.001	-0.387	0.097
55-64		-0.058	0.009	0.001	-0.433	0.069
65-74		-0.064	0.008	0.001	-0.470	0.060
75-84		-0.043	0.010	0.001	-0.304	0.069
85-94		-0.017	0.029	0.558	-0.113	0.193
35-64	UNIVARIATE	-0.065	0.007	0.001	-0.515	0.054
65-94	UNIVARIATE	-0.067	0.006	0.001	-0.505	0.044
35-64	BIVARIATE	-0.048	0.007	0.001	-0.383	0.059
65-94	BIVARIATE	-0.050	0.006	0.001	-0.379	0.047
35-64	MULTIVARIATE	-0.041	0.008	0.001	-0.327	0.062
65-94	MULTIVARIATE	-0.042	0.006	0.001	-0.315	0.048

TABLE 23-12   DATE=20NOV85
AVERAGE ANNUAL INCIDENCE RATE FOR EVENT PER 1000 PERSONS AT RISK AT EXAM
BY AGE AND LEVEL OF RISK FACTOR AT EXAM: FRAMINGHAM STUDY 30-YEAR FOLLOWUP

SEX: MALE
EVENT: DEATH
RISK FACTOR: HEART RATE (PER MINUTE)
PERSONS AT RISK: ALL PERSONS ALIVE

RISK FACTOR	35-44 AT EXAM			45-54 AT EXAM			55-64 AT EXAM			65-74 AT EXAM		
	PERSON EXAMS	# OF EVENTS	ANNUAL RATE	PERSON EXAMS	# OF EVENTS	ANNUAL RATE	PERSON EXAMS	# OF EVENTS	ANNUAL RATE	PERSON EXAMS	# OF EVENTS	ANNUAL RATE
30-67	1295	4	2	2016	17	4	2618	53	10	1425	77	27
68-75	1506	3	1	2594	36	7	2763	75	14	1447	87	30
76-83	728	2	1	1473	32	11	1552	47	15	802	58	36
84-91	586	7	6	1145	21	9	1211	49	20	607	59	49
92-220	404	4	5	839	30	18	926	72	39	553	53	48
UNKNOWN	0	0	-	0	0	-	0	0	-	0	0	-
TOTAL	4519	20	2	8067	136	8	9070	296	16	4834	334	35

RISK FACTOR	75-84 AT EXAM			85-94 AT EXAM			35-64 COMBINED			65-94 COMBINED		
	PERSON EXAMS	# OF EVENTS	ANNUAL RATE	PERSON EXAMS	# OF EVENTS	ANNUAL RATE	# OF EVENTS	CRUDE	AGE-ADJUSTED	# OF EVENTS	CRUDE	AGE-ADJUSTED
30-67	388	47	61	17	5	147	74	6	6	129	35	35
68-75	362	62	86	14	4	143	114	8	8	153	42	43
76-83	271	43	79	13	4	154	81	11	11	105	48	46
84-91	174	33	95	6	3	300	77	13	13	95	60	61
92-220	150	31	103	5	1	500	106	24	24	85	60	64
UNKNOWN	0	0	-	0	0	-	0	-		0	-	
TOTAL	1345	216	80	50	17	170	452	10	10	567	46	46

LOGISTIC REGRESSIONS

AGE AT EXAM		COEFFICIENT OF RISK FACTOR	SE COEFFICIENT	P-VALUE	STANDARDIZED COEFFICIENT	STD SE
35-44		0.038	0.015	0.013	0.468	0.188
45-54		0.033	0.006	0.001	0.410	0.078
55-64		0.030	0.004	0.001	0.387	0.053
65-74		0.016	0.004	0.001	0.208	0.053
75-84		0.011	0.005	0.035	0.152	0.072
85-94		0.041	0.029	0.162	0.449	0.321
35-64	UNIVARIATE	0.031	0.003	0.001	0.398	0.042
65-94	UNIVARIATE	0.014	0.003	0.001	0.184	0.042
35-64	BIVARIATE	0.031	0.003	0.001	0.395	0.042
65-94	BIVARIATE	0.014	0.003	0.001	0.184	0.043
35-64	MULTIVARIATE	0.025	0.003	0.001	0.314	0.044
65-94	MULTIVARIATE	0.012	0.003	0.001	0.156	0.044

SEX: FEMALE
EVENT: DEATH
RISK FACTOR: HEART RATE (PER MINUTE)
PERSONS AT RISK: ALL PERSONS ALIVE

RISK FACTOR	35-44 AT EXAM			45-54 AT EXAM			55-64 AT EXAM			65-74 AT EXAM		
	PERSON EXAMS	# OF EVENTS	ANNUAL RATE	PERSON EXAMS	# OF EVENTS	ANNUAL RATE	PERSON EXAMS	# OF EVENTS	ANNUAL RATE	PERSON EXAMS	# OF EVENTS	ANNUAL RATE
30-67	1092	2	1	1918	13	3	2450	22	4	1285	39	15
68-75	1806	5	1	3279	23	4	3641	48	7	1907	76	20
76-83	1051	3	1	1938	19	5	2442	40	8	1542	55	18
84-91	939	4	2	1620	21	6	1888	49	13	1196	54	23
92-220	731	6	4	1317	23	9	1682	36	11	1173	56	24
UNKNOWN	0	0	-	0	0	-	0	0	-	0	0	-
TOTAL	5619	20	2	10072	99	5	12103	195	8	7103	280	20

RISK FACTOR	75-84 AT EXAM			85-94 AT EXAM			35-64 COMBINED			65-94 COMBINED		
	PERSON EXAMS	# OF EVENTS	ANNUAL RATE	PERSON EXAMS	# OF EVENTS	ANNUAL RATE	# OF EVENTS	CRUDE	AGE-ADJUSTED	# OF EVENTS	CRUDE	AGE-ADJUSTED
30-67	459	34	37	36	9	125	37	3	3	82	23	22
68-75	586	56	48	41	8	98	76	4	4	140	28	28
76-83	532	45	42	30	4	67	62	6	6	104	25	25
84-91	410	40	49	29	6	103	74	8	8	100	31	30
92-220	432	55	64	27	6	111	65	9	9	117	36	35
UNKNOWN	0	0	-	0	0	-	0	-		0	-	
TOTAL	2419	230	48	163	33	101	314	6	6	543	28	28

LOGISTIC REGRESSIONS

AGE AT EXAM		COEFFICIENT OF RISK FACTOR	SE COEFFICIENT	P-VALUE	STANDARDIZED COEFFICIENT	STD SE
35-44		0.027	0.013	0.030	0.347	0.159
45-54		0.025	0.007	0.001	0.310	0.090
55-64		0.022	0.005	0.001	0.284	0.063
65-74		0.012	0.004	0.006	0.157	0.058
75-84		0.010	0.005	0.034	0.140	0.066
85-94		-0.006	0.014	0.682	-0.081	0.198
35-64	UNIVARIATE	0.023	0.004	0.001	0.296	0.049
65-94	UNIVARIATE	0.011	0.003	0.001	0.141	0.042
35-64	BIVARIATE	0.023	0.004	0.001	0.292	0.049
65-94	BIVARIATE	0.010	0.003	0.002	0.134	0.042
35-64	MULTIVARIATE	0.017	0.004	0.001	0.211	0.053
65-94	MULTIVARIATE	0.008	0.003	0.018	0.102	0.043

TABLE 23-13   DATE=12JUN85

AVERAGE ANNUAL INCIDENCE RATE FOR EVENT PER 1000 PERSONS AT RISK AT EXAM
BY AGE AND LEVEL OF RISK FACTOR AT EXAM: FRAMINGHAM STUDY 30-YEAR FOLLOWUP

SEX: MALE
EVENT: DEATH
RISK FACTOR: CIGARETTES SMOKED (NUMBER PER DAY)
PERSONS AT RISK: ALL PERSONS ALIVE

RISK FACTOR	35-44 AT EXAM			45-54 AT EXAM			55-64 AT EXAM			65-74 AT EXAM		
	PERSON EXAMS	# OF EVENTS	ANNUAL RATE	PERSON EXAMS	# OF EVENTS	ANNUAL RATE	PERSON EXAMS	# OF EVENTS	ANNUAL RATE	PERSON EXAMS	# OF EVENTS	ANNUAL RATE
NONE	1402	4	1	3220	33	5	4888	122	12	3235	203	31
1-10	437	1	1	746	13	9	856	32	19	463	35	38
11-20	1499	4	1	2136	46	11	1889	86	23	668	56	42
21-40	993	8	4	1672	36	11	1235	43	17	427	32	37
41-90	162	2	6	255	6	12	178	12	34	34	6	88
UNKNOWN	26	1	19	38	2	26	24	1	21	7	2	143
TOTAL	4519	20	2	8067	136	8	9070	296	16	4834	334	35

RISK FACTOR	75-84 AT EXAM			85-94 AT EXAM			35-64 COMBINED			65-94 COMBINED		
	PERSON EXAMS	# OF EVENTS	ANNUAL RATE	PERSON EXAMS	# OF EVENTS	ANNUAL RATE	# OF EVENTS	CRUDE	AGE- ADJUSTED	# OF EVENTS	CRUDE	AGE- ADJUSTED
NONE	1067	176	82	38	14	184	159	8	7	393	45	43
1-10	125	21	84	6	0	0	46	11	12	56	47	48
11-20	88	10	57	5	2	200	136	12	14	68	45	47
21-40	60	9	75	1	1	500	87	11	12	42	43	49
41-90	5	0	0	0	0	-	20	17	20	6	77	68
UNKNOWN	0	0	-	0	0	-	4	23	22	2	143	111
TOTAL	1345	216	80	50	17	170	452	10	10	567	46	46

LOGISTIC REGRESSIONS

AGE AT EXAM		COEFFICIENT OF RISK FACTOR	SE COEFFICIENT	P-VALUE	STANDARDIZED COEFFICIENT	STD SE
35-44		0.025	0.015	0.093	0.359	0.213
45-54		0.019	0.005	0.001	0.279	0.079
55-64		0.014	0.004	0.001	0.197	0.053
65-74		0.010	0.005	0.036	0.110	0.053
75-84		-0.009	0.009	0.338	-0.077	0.081
85-94		0.031	0.038	0.411	0.241	0.293
35-64	UNIVARIATE	0.010	0.003	0.001	0.149	0.045
65-94	UNIVARIATE	0.000	0.004	0.990	0.001	0.044
35-64	BIVARIATE	0.018	-0.003	0.001	0.263	0.044
65-94	BIVARIATE	0.008	0.004	0.044	0.090	0.045
35-64	MULTIVARIATE	0.020	0.003	0.001	0.286	0.045
65-94	MULTIVARIATE	0.011	0.004	0.006	0.124	0.045

SEX: FEMALE
EVENT: DEATH
RISK FACTOR: CIGARETTES SMOKED (NUMBER PER DAY)
PERSONS AT RISK: ALL PERSONS ALIVE

RISK FACTOR	35-44 AT EXAM			45-54 AT EXAM			55-64 AT EXAM			65-74 AT EXAM		
	PERSON EXAMS	# OF EVENTS	ANNUAL RATE	PERSON EXAMS	# OF EVENTS	ANNUAL RATE	PERSON EXAMS	# OF EVENTS	ANNUAL RATE	PERSON EXAMS	# OF EVENTS	ANNUAL RATE
NONE	2583	5	1	5599	43	4	8010	111	7	5562	195	18
1-10	1314	2	1	1732	17	5	1646	27	8	680	36	26
11-20	1373	8	3	2037	26	6	1767	39	11	650	36	28
21-40	298	3	5	629	9	7	624	18	14	191	12	31
41-90	22	1	23	38	2	26	18	0	0	1	0	0
UNKNOWN	29	1	17	37	2	27	38	0	0	19	1	26
TOTAL	5619	20	2	10072	99	5	12103	195	8	7103	230	20

RISK FACTOR	75-84 AT EXAM			85-94 AT EXAM			35-64 COMBINED			65-94 COMBINED		
	PERSON EXAMS	# OF EVENTS	ANNUAL RATE	PERSON EXAMS	# OF EVENTS	ANNUAL RATE	# OF EVENTS	CRUDE	AGE- ADJUSTED	# OF EVENTS	CRUDE	AGE- ADJUSTED
NONE	2139	202	47	154	31	101	159	5	5	428	27	27
1-10	161	15	47	7	2	143	46	5	5	53	31	33
11-20	99	12	61	2	0	0	73	7	8	48	32	36
21-40	16	1	31	0	0	-	30	10	10	13	31	30
41-90	0	0	-	0	0	-	3	19	14	0	0	0
UNKNOWN	4	0	0	0	0	-	3	14	13	1	22	19
TOTAL	2419	230	48	163	33	101	314	6	6	543	28	28

LOGISTIC REGRESSIONS

AGE AT EXAM		COEFFICIENT OF RISK FACTOR	SE COEFFICIENT	P-VALUE	STANDARDIZED COEFFICIENT	STD SE
35-44		0.066	0.017	0.001	0.644	0.167
45-54		0.024	0.008	0.004	0.247	0.037
55-64		0.022	0.006	0.001	0.207	0.061
65-74		0.023	0.007	0.001	0.172	0.050
75-84		0.004	0.014	0.797	0.017	0.067
85-94		-0.007	0.077	0.729	-0.018	0.202
35-64	UNIVARIATE	0.022	0.005	0.001	0.218	0.049
65-94	UNIVARIATE	0.008	0.006	0.168	0.057	0.041
35-64	BIVARIATE	0.027	0.005	0.001	0.270	0.048
65-94	BIVARIATE	0.021	0.006	0.001	0.150	0.042
35-64	MULTIVARIATE	0.028	0.005	0.001	0.279	0.048
65-94	MULTIVARIATE	0.023	0.006	0.001	0.161	0.042

TABLE 23-14   DATE*20NOV85
AVERAGE ANNUAL INCIDENCE RATE FOR EVENT PER 1000 PERSONS AT RISK AT EXAM
BY AGE AND LEVEL OF RISK FACTOR AT EXAM: FRAMINGHAM STUDY 30-YEAR FOLLOWUP

SEX: MALE
EVENT: DEATH
RISK FACTOR: ALBUMIN IN URINE, DEFINITE
PERSONS AT RISK: ALL PERSONS ALIVE

RISK FACTOR	35-44 AT EXAM PERSON EXAMS	# OF EVENTS	ANNUAL RATE	45-54 AT EXAM PERSON EXAMS	# OF EVENTS	ANNUAL RATE	55-64 AT EXAM PERSON EXAMS	# OF EVENTS	ANNUAL RATE	65-74 AT EXAM PERSON EXAMS	# OF EVENTS	ANNUAL RATE
NO	4411	17	2	7936	128	8	8828	271	15	4671	308	33
YES	71	2	14	105	8	38	215	23	53	157	26	83
UNKNOWN	37	1	14	26	0	0	27	2	37	6	0	0
TOTAL	4519	20	2	8067	136	8	9070	296	16	4834	334	35

RISK FACTOR	75-84 AT EXAM PERSON EXAMS	# OF EVENTS	ANNUAL RATE	85-94 AT EXAM PERSON EXAMS	# OF EVENTS	ANNUAL RATE	35-64 COMBINED # OF EVENTS	CRUDE	AGE-ADJUSTED	65-94 COMBINED # OF EVENTS	CRUDE	AGE-ADJUSTED
NO	1308	206	79	43	15	174	416	10	10	529	44	44
YES	34	10	147	7	2	143	33	42	40	38	96	97
UNKNOWN	3	0	0	0	0	-	3	17	18	0	0	0
TOTAL	1345	216	80	50	17	170	452	10	10	567	46	46

LOGISTIC REGRESSIONS

AGE AT EXAM		COEFFICIENT OF RISK FACTOR	SE COEFFICIENT	P-VALUE	STANDARDIZED COEFFICIENT	STD SE
35-44		2.014	0.757	0.008	0.251	0.095
45-54		1.616	0.378	0.001	0.183	0.043
55-64		1.330	0.229	0.001	0.203	0.035
65-74		1.034	0.223	0.001	0.183	0.039
75-84		0.802	0.384	0.037	0.126	0.060
85-94		-0.292	0.896	0.744	-0.102	0.314
35-64	UNIVARIATE	1.526	0.189	0.001	0.204	0.025
65-94	UNIVARIATE	0.903	0.186	0.001	0.158	0.033
35-64	BIVARIATE	1.375	0.191	0.001	0.183	0.026
65-94	BIVARIATE	0.898	0.192	0.001	0.158	0.034
35-64	MULTIVARIATE	1.101	0.203	0.001	0.147	0.027
65-94	MULTIVARIATE	0.715	0.197	0.001	0.126	0.035

SEX: FEMALE
EVENT: DEATH
RISK FACTOR: ALBUMIN IN URINE, DEFINITE
PERSONS AT RISK: ALL PERSONS ALIVE

RISK FACTOR	35-44 AT EXAM PERSON EXAMS	# OF EVENTS	ANNUAL RATE	45-54 AT EXAM PERSON EXAMS	# OF EVENTS	ANNUAL RATE	55-64 AT EXAM PERSON EXAMS	# OF EVENTS	ANNUAL RATE	65-74 AT EXAM PERSON EXAMS	# OF EVENTS	ANNUAL RATE
NO	5356	19	2	9766	95	5	11835	184	8	6955	268	19
YES	225	1	2	281	3	5	240	11	23	138	12	43
UNKNOWN	38	0	0	25	1	20	28	0	0	10	0	0
TOTAL	5619	20	2	10072	99	5	12103	195	8	7103	280	20

RISK FACTOR	75-84 AT EXAM PERSON EXAMS	# OF EVENTS	ANNUAL RATE	85-94 AT EXAM PERSON EXAMS	# OF EVENTS	ANNUAL RATE	35-64 COMBINED # OF EVENTS	CRUDE	AGE-ADJUSTED	65-94 COMBINED # OF EVENTS	CRUDE	AGE-ADJUSTED
NO	2360	221	47	156	30	96	298	6	6	519	27	27
YES	50	8	80	7	3	214	15	10	12	23	59	55
UNKNOWN	9	1	56	0	0	-	1	5	7	1	26	14
TOTAL	2419	230	48	163	33	101	314	6	6	543	28	28

LOGISTIC REGRESSIONS

AGE AT EXAM		COEFFICIENT OF RISK FACTOR	SE COEFFICIENT	P-VALUE	STANDARDIZED COEFFICIENT	STD SE
35-44		0.226	1.028	0.826	0.045	0.202
45-54		0.094	0.590	0.873	0.016	0.097
55-64		1.112	0.317	0.001	0.155	0.044
65-74		0.866	0.308	0.005	0.120	0.043
75-84		0.612	0.392	0.119	0.087	0.056
85-94		1.147	0.790	0.147	0.233	0.161
35-64	UNIVARIATE	0.607	0.267	0.023	0.098	0.043
85-94	UNIVARIATE	0.836	0.227	0.001	0.118	0.032
35-64	BIVARIATE	0.769	0.269	0.004	0.124	0.043
65-94	BIVARIATE	0.775	0.232	0.001	0.109	0.033
35-64	MULTIVARIATE	0.607	0.277	0.029	0.098	0.045
85-94	MULTIVARIATE	0.446	0.240	0.063	0.063	0.034

TABLE 23-15 ·   DATE=20NOV85
AVERAGE ANNUAL INCIDENCE RATE FOR EVENT PER 1000 PERSONS AT RISK AT EXAM
BY AGE AND LEVEL OF RISK FACTOR AT EXAM: FRAMINGHAM STUDY 30-YEAR FOLLOWUP

SEX: MALE
EVENT: DEATH
RISK FACTOR: HEART ENLARGEMENT BY X-RAY
PERSONS AT RISK: ALL PERSONS ALIVE

RISK FACTOR	35-44 AT EXAM			45-54 AT EXAM			55-64 AT EXAM			65-74 AT EXAM		
	PERSON EXAMS	# OF EVENTS	ANNUAL RATE	PERSON EXAMS	# OF EVENTS	ANNUAL RATE	PERSON EXAMS	# OF EVENTS	ANNUAL RATE	PERSON EXAMS	# OF EVENTS	ANNUAL RATE
NO	4101	14	2	6613	92	7	6543	180	14	3075	180	29
MILD	189	2	5	708	11	8	1126	35	16	705	43	31
DEFINITE	224	4	9	742	33	22	1400	81	29	1056	111	53
UNKNOWN	5	0	0	4	0	0	1	0	0	0	0	-
TOTAL	4519	20	2	8067	136	8	9070	296	16	4834	334	35

RISK FACTOR	75-84 AT EXAM			85-94 AT EXAM			35-64 COMBINED			65-94 COMBINED		
	PERSON EXAMS	# OF EVENTS	ANNUAL RATE	PERSON EXAMS	# OF EVENTS	ANNUAL RATE	# OF EVENTS	CRUDE	AGE-ADJUSTED	# OF EVENTS	CRUDE	AGE-ADJUSTED
NO	753	95	63	34	9	132	286	8	9	284	37	37
MILD	226	41	91	6	3	250	48	12	11	87	47	45
DEFINITE	366	80	109	10	5	250	118	25	22	196	68	66
UNKNOWN	0	0	-	0	0	-	0	0	0	0	-	
TOTAL	1345	216	80	50	17	170	452	10	10	567	46	46

LOGISTIC REGRESSIONS

AGE AT EXAM		COEFFICIENT OF RISK FACTOR	SE COEFFICIENT	P-VALUE	STANDARDIZED COEFFICIENT	STD SE
35-44		0.855	0.275	0.002	0.402	0.129
45-54		0.566	0.105	0.001	0.350	0.065
55-64		0.372	0.069	0.001	0.277	0.051
65-74		0.307	0.064	0.001	0.253	0.053
75-84		1.333	0.083	0.001	0.289	0.072
85-94		0.550	0.364	0.131	0.448	0.296
35-64	UNIVARIATE	0.556	0.056	0.001	0.366	0.037
65-94	UNIVARIATE	0.343	0.049	0.001	0.286	0.041
35-64	BIVARIATE	0.421	0.057	0.001	0.277	0.038
65-94	BIVARIATE	0.306	0.050	0.001	0.256	0.042
35-64	MULTIVARIATE	0.337	0.062	0.001	0.222	0.041
65-94	MULTIVARIATE	0.219	0.054	0.001	0.183	0.045

SEX: FEMALE
EVENT: DEATH
RISK FACTOR: HEART ENLARGEMENT BY X-RAY
PERSONS AT RISK: ALL PERSONS ALIVE

RISK FACTOR	35-44 AT EXAM			45-54 AT EXAM			55-64 AT EXAM			65-74 AT EXAM		
	PERSON EXAMS	# OF EVENTS	ANNUAL RATE	PERSON EXAMS	# OF EVENTS	ANNUAL RATE	PERSON EXAMS	# OF EVENTS	ANNUAL RATE	PERSON EXAMS	# OF EVENTS	ANNUAL RATE
NO	5166	17	2	8085	69	4	7836	99	6	3511	102	15
MILD	199	1	5	959	10	5	1879	32	9	1293	44	17
DEFINITE	246	2	4	1022	20	10	2384	64	13	2299	134	29
UNKNOWN	8	0	0	6	0	0	4	0	0	0	0	-
TOTAL	5619	20	2	10072	99	5	12103	195	8	7103	280	20

RISK FACTOR	75-84 AT EXAM			85-94 AT EXAM			35-64 COMBINED			65-94 COMBINED		
	PERSON EXAMS	# OF EVENTS	ANNUAL RATE	PERSON EXAMS	# OF EVENTS	ANNUAL RATE	# OF EVENTS	CRUDE	AGE-ADJUSTED	# OF EVENTS	CRUDE	AGE-ADJUSTED
NO	830	58	35	40	6	75	185	4	5	166	19	21
MILD	455	39	43	36	6	83	43	7	6	89	25	25
DEFINITE	1134	133	59	87	21	121	86	12	10	288	41	38
UNKNOWN	0	0	-	0	0	-	0	0	0	0	-	
TOTAL 1	2419	230	48	163	33	101	314	6	6	543	28	28

LOGISTIC REGRESSIONS

AGE AT EXAM		COEFFICIENT OF RISK FACTOR	SE COEFFICIENT	P-VALUE	STANDARDIZED COEFFICIENT	STD SE
35-44		0.452	0.361	0.211	0.200	0.160
45-54		0.402	0.127	0.002	0.258	0.082
55-64		0.381	0.081	0.001	0.305	0.065
65-74		0.368	0.068	0.001	0.327	0.061
75-84		0.290	0.082	0.001	0.259	0.073
85-94		0.317	0.249	0.203	0.266	0.209
35-64	UNIVARIATE	0.500	0.065	0.001	0.353	0.046
65-94	UNIVARIATE	0.414	0.050	0.001	0.372	0.045
35-64	BIVARIATE	0.353	0.068	0.001	0.249	0.048
65-94	BIVARIATE	0.315	0.052	0.001	0.283	0.046
35-64	MULTIVARIATE	0.306	0.073	0.001	0.216	0.052
65-94	MULTIVARIATE	0.223	0.054	0.001	0.201	0.049

TABLE 23-16 · DATE=12AUG85

AVERAGE ANNUAL INCIDENCE RATE FOR EVENT PER 1000 PERSONS AT RISK AT EXAM
BY AGE AND LEVEL OF RISK FACTOR AT EXAM: FRAMINGHAM STUDY 30-YEAR FOLLOWUP

SEX: MALE
EVENT: DEATH
RISK FACTOR: LEFT VENTRICULAR HYPERTROPHY BY ELECTROCARDIOGRAM
PERSONS AT RISK: ALL PERSONS ALIVE

RISK FACTOR	35-44 AT EXAM			45-54 AT EXAM			55-64 AT EXAM			65-74 AT EXAM		
	PERSON EXAMS	# OF EVENTS	ANNUAL RATE	PERSON EXAMS	# OF EVENTS	ANNUAL RATE	PERSON EXAMS	# OF EVENTS	ANNUAL RATE	PERSON EXAMS	# OF EVENTS	ANNUAL RATE
NO	4408	18	2	7812	112	7	8611	257	15	4527	279	31
BORDER	79	0	0	147	5	17	240	11	23	121	15	62
DEFINITE	32	2	31	108	19	88	219	28	64	186	40	108
UNKNOWN	0	0	-	0	0	-	0	0	-	0	0	-
TOTAL	4519	20	2	8067	136	8	9070	296	16	4834	334	35

RISK FACTOR	75-84 AT EXAM			85-94 AT EXAM			35-64 COMBINED			65-94 COMBINED		
	PERSON EXAMS	# OF EVENTS	ANNUAL RATE	PERSON EXAMS	# OF EVENTS	ANNUAL RATE	# OF EVENTS	CRUDE	AGE-ADJUSTED	# OF EVENTS	CRUDE	AGE-ADJUSTED
NO	1250	197	79	45	14	156	387	9	9	490	42	42
BORDER	44	6	68	2	1	250	16	17	16	22	66	65
DEFINITE	51	13	127	3	2	333	49	68	66	55	115	114
UNKNOWN	0	0	-	0	0	-	0	-	-	0	-	-
TOTAL	1345	216	80	50	17	170	452	10	10	567	46	46

LOGISTIC REGRESSIONS

AGE AT EXAM		COEFFICIENT OF RISK FACTOR	SE COEFFICIENT	P-VALUE	STANDARDIZED COEFFICIENT	STD SE
35-44		1.244	0.412	0.003	0.263	0.087
45-54		1.309	0.133	0.001	0.346	0.035
55-64		0.747	0.103	0.001	0.256	0.035
65-74		0.718	0.091	0.001	0.295	0.037
75-84		0.241	0.159	0.129	0.100	0.066
85-94		0.751	0.593	0.205	0.383	0.302
35-64	UNIVARIATE	1.023	0.080	0.001	0.298	0.023
65-94	UNIVARIATE	0.579	0.077	0.001	0.239	0.032
35-64	BIVARIATE	0.903	0.081	0.001	0.263	0.024
65-94	BIVARIATE	0.591	0.080	0.001	0.244	0.033
35-64	MULTIVARIATE	0.825	0.090	0.001	0.240	0.026
65-94	MULTIVARIATE	0.485	0.085	0.001	0.200	0.035

SEX: FEMALE
EVENT: DEATH
RISK FACTOR: LEFT VENTRICULAR HYPERTROPHY BY ELECTROCARDIOGRAM
PERSONS AT RISK: ALL PERSONS ALIVE

RISK FACTOR	35-44 AT EXAM			45-54 AT EXAM			55-64 AT EXAM			65-74 AT EXAM		
	PERSON EXAMS	# OF EVENTS	ANNUAL RATE	PERSON EXAMS	# OF EVENTS	ANNUAL RATE	PERSON EXAMS	# OF EVENTS	ANNUAL RATE	PERSON EXAMS	# OF EVENTS	ANNUAL RATE
NO	5566	18	2	9905	95	5	11679	172	7	6686	230	17
BORDER	38	0	0	111	1	5	250	9	18	182	14	38
DEFINITE	15	2	67	56	3	27	174	14	40	235	36	77
UNKNOWN	0	0	-	0	0	-	0	0	-	0	0	-
TOTAL	5619	20	2	10072	99	5	12103	195	8	7103	280	20

RISK FACTOR	75-84 AT EXAM			85-94 AT EXAM			35-64 COMBINED			65-94 COMBINED		
	PERSON EXAMS	# OF EVENTS	ANNUAL RATE	PERSON EXAMS	# OF EVENTS	ANNUAL RATE	# OF EVENTS	CRUDE	AGE-ADJUSTED	# OF EVENTS	CRUDE	AGE-ADJUSTED
NO	2217	195	44	142	23	81	285	5	5	448	25	25
BORDER	77	10	65	10	5	250	10	13	10	29	54	48
DEFINITE	125	25	100	11	5	227	19	39	41	66	89	85
UNKNOWN	0	0	-	0	0	-	0	-	-	0	-	-
TOTAL	2419	230	48	163	33	101	314	6	6	543	28	28

LOGISTIC REGRESSIONS

AGE AT EXAM		COEFFICIENT OF RISK FACTOR	SE COEFFICIENT	P-VALUE	STANDARDIZED COEFFICIENT	STD SE
35-44		1.825	0.413	0.001	0.240	0.054
45-54		0.764	0.304	0.012	0.138	0.055
55-64		0.888	0.156	0.001	0.244	0.037
65-74		0.815	0.093	0.001	0.315	0.036
75-84		0.473	0.114	0.001	0.222	0.053
85-94		0.848	0.305	0.004	0.471	0.165
35-64	UNIVARIATE	1.017	0.118	0.001	0.224	0.026
65-94	UNIVARIATE	0.723	0.069	0.001	0.298	0.029
35-64	BIVARIATE	0.879	0.119	0.001	0.194	0.026
65-94	BIVARIATE	0.664	0.071	0.001	0.274	0.029
35-64	MULTIVARIATE	0.738	0.154	0.001	0.163	0.029
65-94	MULTIVARIATE	0.551	0.075	0.001	0.227	0.031

TABLE 23-17 DATE=20NOV85
AVERAGE ANNUAL INCIDENCE RATE FOR EVENT PER 1000 PERSONS AT RISK AT EXAM
BY AGE AND LEVEL OF RISK FACTOR AT EXAM: FRAMINGHAM STUDY 30-YEAR FOLLOWUP

SEX: MALE
EVENT: DEATH
RISK FACTOR: INTRAVENTRICULAR CONDUCTION DEFECT (WITH QRS INTERVAL MORE THAN 0.11 SECOND)
PERSONS AT RISK: ALL PERSONS ALIVE

	35-44 AT EXAM			45-54 AT EXAM			55-64 AT EXAM			65-74 AT EXAM		
	PERSON EXAMS	# OF EVENTS	ANNUAL RATE	PERSON EXAMS	# OF EVENTS	ANNUAL RATE	PERSON EXAMS	# OF EVENTS	ANNUAL RATE	PERSON EXAMS	# OF EVENTS	ANNUAL RATE
	4201	20	2	7487	123	8	8247	261	16	4323	286	33
ETE	280	0	0	491	7	7	541	15	14	250	14	28
E	38	0	0	89	6	34	282	20	35	261	34	65
	0	0	-	0	0	-	0	0	-	0	0	-
	4519	20	2	8067	136	8	9070	296	16	4834	334	35

	75-84 AT EXAM			85-94 AT EXAM			35-64 COMBINED			65-94 COMBINED		
	PERSON EXAMS	# OF EVENTS	ANNUAL RATE	PERSON EXAMS	# OF EVENTS	ANNUAL RATE	# OF EVENTS	CRUDE	AGE-ADJUSTED	# OF EVENTS	CRUDE	AGE-ADJUSTED
	1180	183	78	41	13	159	404	10	10	482	43	44
.ETE	37	7	95	3	1	167	22	8	8	22	38	43
'E	128	26	102	6	3	250	26	32	27	63	80	74
4	0	0	-	0	0	-	0	-	-	0	-	-
	1345	216	80	50	17	170	452	10	10	567	46	46

LOGISTIC REGRESSIONS

	COEFFICIENT OF RISK FACTOR	SE COEFFICIENT	P-VALUE	STANDARDIZED COEFFICIENT	STD SE
	-	-	-	-	-
	0.444	0.211	0.035	0.139	0.066
	0.315	0.117	0.007	0.130	0.048
	0.312	0.096	0.001	0.154	0.047
	0.168	0.114	0.141	0.101	0.069
	0.357	3.428	0.403	0.242	0.290
UNIVARIATE	0.402	0.103	0.001	0.143	0.037
UNIVARIATE	0.299	0.072	0.001	0.155	0.037
BIVARIATE	0.288	0.102	0.005	0.102	0.036
BIVARIATE	0.231	0.073	0.002	0.120	0.038
MULTIVARIATE	0.267	0.106	0.012	0.095	0.038
MULTIVARIATE	0.218	0.074	0.003	0.113	0.038

SEX: FEMALE
EVENT: DEATH
RISK FACTOR: INTRAVENTRICULAR CONDUCTION DEFECT (WITH QRS INTERVAL MORE THAN 0.11 SECOND)
PERSONS AT RISK: ALL PERSONS ALIVE

	35-44 AT EXAM			45-54 AT EXAM			55-64 AT EXAM			65-74 AT EXAM		
	PERSON EXAMS	# OF EVENTS	ANNUAL RATE	PERSON EXAMS	# OF EVENTS	ANNUAL RATE	PERSON EXAMS	# OF EVENTS	ANNUAL RATE	PERSON EXAMS	# OF EVENTS	ANNUAL RATE
	5532	20	2	9873	98	5	11731	184	8	6794	264	19
LETE	75	0	0	151	0	0	195	3	8	134	2	7
TE	12	0	0	48	1	10	177	8	23	175	14	40
N	0	0	-	0	0	-	0	0	-	0	0	-
	5619	20	2	10072	99	5	12103	195	8	7103	280	20

	75-84 AT EXAM			85-94 AT EXAM			35-64 COMBINED			65-94 COMBINED		
	PERSON EXAMS	# OF EVENTS	ANNUAL RATE	PERSON EXAMS	# OF EVENTS	ANNUAL RATE	# OF EVENTS	CRUDE	AGE-ADJUSTED	# OF EVENTS	CRUDE	AGE-ADJUSTED
	2261	208	46	153	28	92	302	6	6	500	27	27
LETE	51	5	49	3	0	0	3	4	3	7	19	18
.TE	107	17	79	7	5	357	9	19	14	36	62	55
IN	0	0	-	0	0	-	0	-	-	0	-	-
	2419	230	48	163	33	101	314	6	6	543	28	28

LOGISTIC REGRESSIONS

	COEFFICIENT OF RISK FACTOR	SE COEFFICIENT	P-VALUE	STANDARDIZED COEFFICIENT	STD SE
	-	-	-	-	-
	-0.145	0.614	0.814	-0.027	0.112
	0.482	0.182	0.008	0.130	0.049
	0.287	0.146	0.050	0.096	0.049
	0.294	0.134	0.028	0.127	0.058
	1.033	0.390	0.008	0.439	0.166
UNIVARIATE	0.483	0.176	0.006	0.106	0.039
UNIVARIATE	0.394	0.092	0.001	0.144	0.033
BIVARIATE	0.352	0.175	0.045	0.077	0.038
BIVARIATE	0.324	0.093	0.001	0.118	0.034
MULTIVARIATE	0.384	0.178	0.031	0.084	0.039
MULTIVARIATE	0.299	0.095	0.002	0.109	0.035

TABLE 23-18   DATE=20NOV85
AVERAGE ANNUAL INCIDENCE RATE FOR EVENT PER 1000 PERSONS AT RISK AT EXAM
BY AGE AND LEVEL OF RISK FACTOR AT EXAM: FRAMINGHAM STUDY 30-YEAR FOLLOWUP

SEX: MALE
EVENT: DEATH
RISK FACTOR: DEFINITE NONSPECIFIC T-WAVE OR ST-SEGMENT ABNORMALITY BY ELECTROCARDIOGRAM
PERSONS AT RISK: ALL PERSONS ALIVE

RISK FACTOR	35-44 AT EXAM			45-54 AT EXAM			55-64 AT EXAM			65-74 AT EXAM		
	PERSON EXAMS	# OF EVENTS	ANNUAL RATE	PERSON EXAMS	# OF EVENTS	ANNUAL RATE	PERSON EXAMS	# OF EVENTS	ANNUAL RATE	PERSON EXAMS	# OF EVENTS	ANNUAL RATE
NO	4406	20	2	7641	122	8	8246	236	14	4170	268	32
YES	113	0	0	426	14	16	824	60	36	664	66	50
UNKNOWN	0	0	-	0	0	-	0	0	-	0	0	-
TOTAL	4519	20	2	8067	136	8	9070	296	16	4834	334	35

RISK FACTOR	75-84 AT EXAM			85-94 AT EXAM			35-64 COMBINED			65-94 COMBINED		
	PERSON EXAMS	# OF EVENTS	ANNUAL RATE	PERSON EXAMS	# OF EVENTS	ANNUAL RATE	# OF EVENTS	CRUDE	AGE-ADJUSTED	# OF EVENTS	CRUDE	AGE-ADJUSTED
NO	1063	156	73	31	8	129	378	9	9	432	41	42
YES	282	60	106	19	9	237	74	27	21	135	70	63
UNKNOWN	0	0	-	0	0	-	0	-	-	0	-	-
TOTAL	1345	216	80	50	17	170	452	10	10	567	46	46

LOGISTIC REGRESSIONS

AGE AT EXAM		COEFFICIENT OF RISK FACTOR	SE COEFFICIENT	P-VALUE	STANDARDIZED COEFFICIENT	STD SE
35-44		-	-	-	-	-
45-54		0.739	0.287	0.010	0.165	0.064
55-64		0.980	0.149	0.001	0.282	0.043
65-74		0.474	0.144	0.001	0.163	0.050
75-84		0.452	0.169	0.008	0.184	0.069
85-94		0.951	0.616	0.123	0.466	0.302
35-64	UNIVARIATE	1.107	0.130	0.001	0.269	0.032
65-94	UNIVARIATE	0.598	0.106	0.001	0.217	0.038
35-64	BIVARIATE	0.853	0.132	0.001	0.207	0.032
65-94	BIVARIATE	0.464	0.108	0.001	0.168	0.039
35-64	MULTIVARIATE	0.706	0.134	0.001	0.172	0.033
65-94	MULTIVARIATE	0.438	0.110	0.001	0.159	0.040

SEX: FEMALE
EVENT: DEATH
RISK FACTOR: DEFINITE NONSPECIFIC T-WAVE OR ST-SEGMENT ABNORMALITY BY ELECTROCARDIOGRAM
PERSONS AT RISK: ALL PERSONS ALIVE

RISK FACTOR	35-44 AT EXAM			45-54 AT EXAM			55-64 AT EXAM			65-74 AT EXAM		
	PERSON EXAMS	# OF EVENTS	ANNUAL RATE	PERSON EXAMS	# OF EVENTS	ANNUAL RATE	PERSON EXAMS	# OF EVENTS	ANNUAL RATE	PERSON EXAMS	# OF EVENTS	ANNUAL RATE
NO	5520	20	2	9642	88	5	11111	167	8	6163	218	18
YES	99	0	0	430	11	13	992	28	14	940	62	33
UNKNOWN	0	0	-	0	0	-	0	0	-	0	0	-
TOTAL	5619	20	2	10072	99	5	12103	195	8	7103	280	20

RISK FACTOR	75-84 AT EXAM			85-94 AT EXAM			35-64 COMBINED			65-94 COMBINED		
	PERSON EXAMS	# OF EVENTS	ANNUAL RATE	PERSON EXAMS	# OF EVENTS	ANNUAL RATE	# OF EVENTS	CRUDE	AGE-ADJUSTED	# OF EVENTS	CRUDE	AGE-ADJUSTED
NO	1945	168	43	124	26	105	275	5	5	412	25	26
YES	474	62	65	39	7	90	39	13	11	131	45	42
UNKNOWN	0	0	-	0	0	-	0	-	-	0	-	-
TOTAL	2419	230	48	163	33	101	314	6	6	543	28	28

LOGISTIC REGRESSIONS

AGE AT EXAM		COEFFICIENT OF RISK FACTOR	SE COEFFICIENT	P-VALUE	STANDARDIZED COEFFICIENT	STD SE
35-44		-	-	-	-	-
45-54		1.047	0.324	0.001	0.212	0.065
55-64		0.644	0.207	0.002	0.177	0.057
65-74		0.655	0.148	0.001	0.222	0.050
75-84		0.465	0.158	0.003	0.185	0.063
85-94		-0.193	0.472	0.683	-0.083	0.202
35-64	UNIVARIATE	0.911	0.173	0.001	0.207	0.039
65-94	UNIVARIATE	0.632	0.105	0.001	0.226	0.037
35-64	BIVARIATE	0.668	0.175	0.001	0.152	0.040
65-94	BIVARIATE	0.498	0.106	0.001	0.178	0.038
35-64	MULTIVARIATE	0.542	0.177	0.002	0.123	0.040
65-94	MULTIVARIATE	0.462	0.108	0.001	0.165	0.038